3 0116 00431 4983

Human
Centered
Technology

Key
to the
Future

PHILADELPHIA 1996

Volume 1

Proceedings

of the

Human Factors

and Ergonomics

Society

40th

Annual

Meeting

1996

Human Factors and Ergonomics Society
OFFICERS AND STAFF

President
Hal W. Hendrick

President-Elect
Arthur D. Fisk

Immediate Past President
F. Thomas Eggemeier

Secretary-Treasurer
Betty S. Goldsberry

Secretary-Treasurer-Elect
Mary Carol Day

Past Secretary-Treasurer
Jerry L. Duncan

Executive Council
Barry H. Beith
Kenneth R. Laughery, Sr.
Richard J. Hornick
David L. Post
Michelle M. Robertson
Mark S. Sanders

Executive Director
Lynn Strother

Publications Manager
Lois Smith

Editorial Assistance
Darcy L. Pettigrew
Steve Lussier

PREFACE

In publishing the *Proceedings of the Human Factors and Ergonomics Society 40th Annual Meeting,* the Human Factors and Ergonomics Society (formerly the Human Factors Society) seeks to facilitate an effective and timely dissemination of technical information within its area of concern. The papers contained in this document were printed directly from unedited reproducible copy submitted by the authors, who are solely responsible for their contents.

ISBN 0-945289-06-5

The papers contained in the 1996 proceedings are arranged alphabetically in order of technical specialty area categorized according to the Society's 20 technical groups. Volume 1 contains the areas of Aerospace Systems, Aging, Cognitive Engineering and Decision Making, Communication, Computer Systems, Consumer Products, Educators' Professional, Environmental Design, Forensics Professional, and Industrial Ergonomics. Also included is the category General Sessions, which contains papers on a variety of topics.

Volume 2 contains the areas of Medical Systems and Rehabilitation, Individual Differences in Performance, Organizational Design and Management, Safety, Surface Transportation, System Development, Test and Evaluation, Training, Virtual Environments, and Visual Performance. The category Special Sessions, containing descriptions of system and product demonstrations and alternative formats, is also included. The proceedings also contain half-page abstracts of poster presentations, which can be found under "POSTERS" in Volume 2.

Opinions expressed in proceedings articles are those of the authors and do not necessarily reflect those of the Human Factors and Ergonomics Society, nor should they be considered as expressions of official policy by the Society.

A cumulative index to the 17th–31st Annual Meeting proceedings (1972–1987) is available from the Human Factors and Ergonomics Society. To obtain copies of the *Proceedings of the 40th Annual Meeting* or the cumulative index, address inquiries to

Human Factors and Ergonomics Society
P.O. Box 1369
Santa Monica, CA 90406-1369 USA
310/394-1811, fax 310/394-2410
http://hfes.org

The HFES annual meeting proceedings are indexed or abstracted in the following publications or services: Applied Mechanics Reviews, Engineering Index Annual, EI Monthly, Cambridge Scientific Abstracts, EI Bioengineering Abstracts, EI Energy Abstracts, Ergonomics Abstracts, ISI Index to Scientific & Technical Proceedings, and International Aerospace Abstracts. This publication is also available on microfilm from University Microfilms International, 300 N. Zeeb Road, Department P.R., Ann Arbor, MI 48106; 18 Bedford Row, Department P.R., London WC1R 4EJ, England.

The paper used in this publication meets the minimum requirements of the American National Standard for Information Sciences—Permanence of Paper for Printed Library Materials, ANSI Z39.48-1984.

ISSN 1071-1813

DELAWARE VALLEY CHAPTER OFFICERS

President
Jennifer A. Smith

Secretary
Joan M. Ryder

Treasurer
David A. Rose

Board Members
Michael Andrew Szczepkowski
Lloyd Hitchcock
Janine A. Purcell

HOST COMMITTEE FOR THE 40th ANNUAL MEETING

General Cochair
Philip J. Federman

Technical Program Committee Liaison
Paul G. Stringer

Exhibits
Lloyd Hitchcock

Finance
Kevin J. Bracken

Technical Tours
Daniel D. Riley
Michael A. Szczepkowski

Public Relations
Floyd A. Glenn III

Information Systems
Preston Ginsburg
Nagi Kodali

Special Events
Robert J. Wherry, Jr.

Newsletter
Thomas E. Zurinskas

Volunteers
Parimal H. Kopardekar
Michael D. Snyder

TECHNICAL PROGRAM COMMITTEE

General Chair
John F. (Jeff) Kelley

Special Sessions
Clint A. Bowers

General Sessions
James R. McCracken

Workshops
Ronald G. Shapiro

Colloquia
Teresa L. Hood

Council of Technical Groups Liaison
Paula Sind-Prunier

Quality Control/Evaluation
Paul Green

Education
Anthony D. Andre

PROGRAM CHAIRS FOR THE 40th ANNUAL MEETING

Aerospace Systems
Anthony J. Aretz

Aging
Donald L. Lassiter

Cognitive Engineering and Decision Making
Kathleen Mosier

Communications
Cathleen Wharton

Computer Systems
Stephanie Guerlain

Consumer Products
Elizabeth B. N. Sanders

Educators' Professional
Susan G. Shapiro

Environmental Design
Marie A. Robinson

Forensics Professional
Michael S. Wogalter

Individual Differences in Performance
Clint A. Bowers

Industrial Ergonomics
Richard W. Marklin

Medical Systems and Rehabilitation
Laurie Wolf

Organizational Design and Management
Barrett S. Caldwell

Safety
Michael J. Kalsher

Surface Transportation
Renae Bowers-Carnahan

System Development
Laurel A. Allender

Test and Evaluation
Valerie J. Gawron

Training
Philip J. Smith

Virtual Environments
Kay M. Stanney

Visual Performance
Jennie J. Gallimore

PAPER REVIEWERS FOR THE 40th ANNUAL MEETING

The Human Factors and Ergonomics Society gratefully acknowledges the assistance of the following individuals in the technical review of papers submitted for the 40th Annual Meeting:

Edie M. Adams	Daryle J. Gardner-Bonneau	John Lee	Valerie J. B. Rice
Julie Adams	Daniel J. Garland	S. David Leonard	Edward J. Rinalducci
Jurine Adolf	Valerie Gawron	Michael Lewis	Jenifer W. Roberston
Thomas Albin	Jeffrey M. Gerth	Robert J. Logan	Michelle M. Robertson
Elizabeth Alicandri	Richard S. Gibson	Arnold M. Lund	Marie A. Robinson
Nancy S. Anderson	Richard Gilson	Kipp Lynch	Mark D. Rodgers
Anthony D. Andre	Richard Giorgi	Ed F. Madigan, Jr.	Wendy A. Rogers
Anthony J. Aretz	Barry P. Goettl	David E. Mandeville	Jannick Rolland
Alan E. Asper	Joseph H. Goldberg	Monica Marics	Emilie M. Roth
David N. Aurelio	Sallie Gordon	David W. Martin	Carrie Rudman
Larry Avery	John B. Gosbee	Roger Martin	Joan Ryder
M. M. Ayoub	Anand Gramopadhye	Pamela McCauley-Bell	Mark S. Sanders
Richard W. Backs	Chris Grant	Marshall McClintock	Mark W. Scerbo
Jack Barker	Joel S. Greenstein	James McCracken	Frank Schieber
Barry H. Beith	Douglas Griffith	Daniel E. McCrobie	Steven Selcon
Kevin B. Bennett	Scott S. Grisgby	Jennifer McGovern	Leon Segal
Dennis Beringer	Stephanie Guerlain	Dianne L. McMullin	Susan J. Shapiro
Jan Berkhout	Ellen C. Haas	Jon Meads	Joseph Sharit
Roman R. Beyer	M. Susan Hallbeck	David Meister	R. Jay Shively
Harry E. Blanchard	Sung H. Han	Laura G. Militello	Ned Silver
Paula C. Bohr	Peter A. Hancock	Christopher Miller	Barbara Silverstein
David B. Boles	Paul W. Hankey	James Miller	Michael Singer
Clint A. Bowers	Phillip Hash	Christine M. Mitchell	Stephen D. Small
Liwana S. Bringelson	Alan Hedge	Peter P. Mitchell	David B. D. Smith
Curt Braun	Susan T. Heers	Deborah A. Mitta	James L. Smith
R. Todd Brown	William Hefley	Patrick Monnier	Leighton L. Smith
Brian Buchholz	Lawrence Hettinger	William F. Moroney	Janan A. Smither
Jim Bushman	Douglas G. Hoecker	Jeffrey G. Morrison	Jeff Sokolov
John L. Campbell	Robert R. Hoffman	Daniel G. Morrow	Carolyn M. Sommerich
Mark W. Cannon	Richard J. Hornick	Kathleen Mosier	Noah A. Spivak
Janis Cannon-Bowers	Mary Hornsby	Ronald Mourant	Kay Stanney
John G. Casali	Avraham D. Horowitz	Michael J. Muller	Fredrick M. Streff
Sherry Perdue Casali	Richard Howell	Patricia Mullins	Carol A. M. Stuart-Buttle
Jerry Chubb	Beverly Huey	Randall J. Mumaw	Jane Fulton Suri
George A. Chuckrow	Richard Hughes	Steven A. Murray	Naomi Swanson
David J. Cochran	Susan Infield	Beverly Nicholas	David A. Thompson
Andrew M. Cohill	Philip A. Jacobs	Ron Noel	Jozsef A. Toth
H. Harvey Cohen	Florian Jentsch	Lorraine Normore	Pamela S. Tsang
Shari Converse	Jared B. Jobe	Elizabeth H. Nutter-Lewis	Marilyn L. Turner
Nancy Cooke	George Kaempf	Robert B. Ochsman	John J. Uhlarik
Andi Cowell	Harold S. Kaplan	Rick Omanson	Michael Venturino
Lesia L. Crumpton	Clare-Marie Karat	Michael O'Neill	Kim J. Vicente
Mary Czerwinski	Waldemar Karwowski	Judith Orasanu	Michael A. Vidulich
Mary Carol Day	Richard Kelly	David W. Osborne	Robert A. Virzi
David M. DeJoy	Robert Kennedy	Robert K. Osgood	Daniel F. Wallace
Sidney Dekker	Carter J. Kerk	Richard F. Pain	Joel S. Warm
Rebecca Denning	Gary Klatsky	Mark T. Palmer	Scott N. J. Watamaniuk
Rosalie Douglas	Gary A. Klein	Ginger Watson Papelis	Jennifer Watts
Bill Dowell	David Klinger	Zoe Pappas-Cipriano	Jayson Webb
Robert D. Dryden	R. Robert Knaff	Jay G. Pollack	Lisa Weinstein
David Dryer	Gina Kolasinski	David L. Post	Maxwell J. Wells
Joseph Dumas	Stephan Konz	Scott S. Potter	Christopher D. Wickens
Jerry R. Duncan	Jefferson Koonce	Michael L. Quinn	Steven Wiker
Carl Edlund	Philip T. Kortum	Mirielle Raby	Stephen B. Wilcox
Jenny Ehrlich	Sheila Krawczyk	Bernadette M. Racicot	Leonard D. Williams
Sandy Eisenhut	John G. Kreifeldt	Robert G. Radwin	Glenn Wilson
Mica Endsley	Richard F. Krenek	Josephine Randel	James Wilson
Eileen B. Entin	Karl H. E. Kroemer	David Ranson	Karen Wilson
Donald Farr	Mary LaLomia	Mark S. Redfern	John A. Wise
Fadi A. Fathallah	Don Lampton	James J. Reger	Michael S. Wogalter
Chris Forsythe	Jean-Francois Lapointe	Gary Reid	David Woods
J. Paul Frantz	Nancy Larson	John M. Reising	Yan Xiao
Marita Franzke	Irene Laudeman	Mark Resnick	Stephen L. Young
Andris Freivalds	Lila Laux	Jacqueline Reynolds-Morzall	
R. Marty Gage	Steven A. Lavender	Katie Ricci	
Sean Gallagher	Lee Leber	Laurie Rhodenizer	

CONTENTS

*Abstract only
**Manuscript not submitted for publication

AGING

COGNITIVE ENGINEERING AND DECISION MAKING

*Abstract only
**Manuscript not submitted for publication

*Abstract only
**Manuscript not submitted for publication

*Abstract only
**Manuscript not submitted for publication

CONSUMER PRODUCTS

EDUCATORS' PROFESSIONAL

*Abstract only
**Manuscript not submitted for publication

ENVIRONMENTAL DESIGN

FORENSICS PROFESSIONAL

GENERAL SESSIONS

INDIVIDUAL DIFFERENCES IN PERFORMANCE

INDUSTRIAL ERGONOMICS

*Abstract only
**Manuscript not submitted for publication

*Abstract only
**Manuscript not submitted for publication

Volume 2

MEDICAL SYSTEMS AND REHABILITATION

ORGANIZATIONAL DESIGN AND MANAGEMENT

*Abstract only
**Manuscript not submitted for publication

*Abstract only
**Manuscript not submitted for publication

SPECIAL SESSIONS

SURFACE TRANSPORTATION

*Abstract only
**Manuscript not submitted for publication

*Abstract only
**Manuscript not submitted for publication

*Abstract only
**Manuscript not submitted for publication

*Abstract only
**Manuscript not submitted for publication

*Abstract only
**Manuscript not submitted for publication

POSTERS

*Abstract only
**Manuscript not submitted for publication

*Abstract only
**Manuscript not submitted for publication

*Abstract only
**Manuscript not submitted for publication

*Abstract only
**Manuscript not submitted for publication

1996 WORKSHOPS

Human Factors Methods
Alphonse Chapanis and John B. Shafer

A Human Factors Approach to Accident Investigation
Douglas A. Wiegmann and Scott A. Shappell

Practical Usability Evaluation
Gary Perlman

After the Task Flow: Participatory Design of Data-Centered, Multiplatform, Graphical User Interfaces
Tom Dayton, Joseph Kramer, and Eugenie L. Bertus

Design Strategies and Methods in Interaction Design
Alp Tiritoglu and Richard Branham

Cognitive Task Analysis (CTA)
Gary A. Klein and Laura G. Militello

Integrated Cognitive/Behavioral Task Analysis in Real-Time Domains
Wayne W. Zachary and Joan M. Ryder

The MUSE Method for Usability Engineering
Kee Yong Lim

Converting Anthorpometric Statistics into Criteria for Design and Evaluation
John A. Roebuck, Jr.

The GOMS Family of Cognitive Task Analysis Techniques
Wayne D. Gray and Deborah A. Boehm-Davis

Using Computer Modeling and Simulation in Human Factors
Catherine Drury, Sue Dahl, and Rick Archer

Job Analysis in the Real World--What Works When?
Jerry Sue Thompson and Gary J. Lasswell

Preparing for Board of Certification in Professional Ergonomics Certification
Paula Sind-Prunier

Questionnaire Design and Use: A Primer for Practitioners
William F. Moroney, Joyce A. Cameron, and Leslie A. Whitaker

*Abstract only
**Manuscript not submitted for publication*

THE ERGONOMICS OF ECONOMICS IS THE ECONOMICS OF ERGONOMICS

Hal W. Hendrick
University of Southern California
Los Angeles, California

Several reasons for the failure of business and government decison-makers to more adequately recognize the benefits of human factors/ergonomics and strongly support and endorse its application are reviewed. In particular, the failure of members of our profession to more consisistently and adequately document and publicize the cost-benefits of their ernonomic applications is noted. To support the notion that *good ergonomics is good economics*, a variety of actual ergonomic applications, and the documented cost-benefits of each, are described.

INTRODUCTION

Human factors/ergonomics professionals have long recognized the tremendous potential of our discipline for improving the health, safety and comfort of persons, and both human and system productivity. For example, the tremendous improvements that have occurred in aviation safety since before World War II can, to a considerable extent, be attributed to the contributions of our profession to aircraft and related aviation systems design. Indeed, through the application of our unique scientific knowledge of human-system interface design, analysis, and evaluation, we have the *potential* to truly make a difference in the quality of life for virtually all persons on this globe. In fact, I know of no profession where so small a group of professionals has such a tremendous potential for truly *making a difference*.

In light of our potential, why is it, then, that more organizations, with their strong need to obtain employee commitment, reduce expenses, and increase productivity, are not banging down our doors for help, or creating human factors/ergonomics positions far beyond our capacity to fill them? Why is it that federal and state agencies are not pushing for legislation to ensure that human factors/ergonomics factors are systematically considered in the design of products for human use and work environments for employees? Why is it that both industry associations and members of congress sometimes view us as simply adding an additional expense burden and, thus, *increasing* the costs of production and *decreasing* competitiveness?

One possible contributing reason may be that some of these individuals and organizations have been exposed to *bad ergonomics* - or what, in a recent article on this topic, Ian Chong (1996) labels as "voodoo ergonomics" - either in the form of products or work environments which are *professed* to be ergonomically designed, but are not, or in which the so-called ergonomics was done by incompetent persons. This, indeed, is a concern - particularly when persons lacking professional training pass themselves off as ergonomists or human factors professionals; or tout their services as a panacea for almost anything. It is one of the major reasons that both establishing educational standards for professional education in human factors/ergonomics and professional certification have become top priority issues for the International Ergonomics Association and, indeed, for many national human factors/ergonomics societies and governmental groups - such as the European Union and, more recently, a major Japanese bureau. It also is one of the reasons we are starting to see a proliferation of ergonomics standards within the European Union and the International Standards Organization.

Another reason, well known to us, is that "everyone is an operator" (Mallett, 1995). Everyone "operates" systems on a daily basis, such as water faucets, an automobile, computer, television, VCR and telephone; and thus, it is very easy to naively assume from our operator experience that human factors is nothing more than "common sense". For example, I recently was prohibited from testifying as an expert witness because the judge naively concluded that the complex issues of attention, perception and reaction integral to the case could be assessed by the jurors on the basis of their "common sense". In fact, such "common sense" was likely to lead them to the scientifically wrong conclusion.

Most experienced ergonomists have their own personal list of "common sense" engineering design decisions that have resulted in serious accidents, fatalities, or just plain poor usability. In one very personal example, while serving in the US Air Force as a radar controller, I was directing an F-86 aircraft in for a landing at Suffolk County Air Force Base early one foggy morning. The pilot radioed me that he was passing through 10,000 feet, *and that was the last that I ever hear from him.* We found the crashed plane on a long island beach, which clearly indicated by its location that the pilot had actually been passing though 1,000 feet when he last radioed me. It turned out that he was flying with a new two-pointer altimeter with a direct readout for the 10,000 foot indicator. In the stress of landing with zero visibility, he apparently reverted back to a previously highly-learned occupational stereotype, interpreting his altitude reading as though he was looking at the usual three-pointer altimeter - a classic case of negative transfer of training; and of engineering "common sense" in lieu of sound ergonomics.

In addition to the above, I believe there are at least two other major interrelated reasons for this lack of appreciation of the potential benefits of ergonomics; and they are reasons over which we potentially have some control. First, I believe we sometimes expect organizational decision-makers to pro-actively support human factors/ergonomics simply because it is *the right thing to do*. Like God, mother and apple pie, it is hard to argue against doing anything that may better the human condition,

and so, that alone should be a compelling argument for actively supporting use of our discipline. In reality, managers have to be able to justify *any* investment in terms of its concrete benefits to the organization - to the organization's ability to be competitive and survive. That something "is the right thing to do" is, by itself, an excellent but decidedly *insufficient* reason for managers actually doing it.

Secondly, as a group, we have done a poor job of documenting and advertising the cost/benefits of good ergonomics - of getting the word out that most often, *good ergonomics is good economics.* In fact, that the ergonomics of economics *is* the economics of ergonomics.

In my 35 years of experience in this field, I can think of no application of our technology by myself, close colleagues, or in collaboration with my students that did *not* have a positive benefit for the organization that more than justified the cost.

As one attempt to rectify this situation, I want to share with you a broad spectrum of ergonomics applications that my predecessor as HFES President, Tom Eggemeier, and I have collected from within the U.S. and elsewhere where the costs and economic benefits were documented.

EXAMPLES FROM VARIOUS ERGONOMIC APPLICATIONS

Forestry Industry

A coordinated series of joint projects were undertaken by the Forest Engineering Technology Department of the University of Stellenbosch and Ergotech - the only true ergonomics consulting firm in South Africa - to improve safety and productivity in the South African forestry industry.

Leg protectors. In one project, an anthropometric survey was conducted of the very heterogeneous work force to provide the basic data for redesigning leg protectors for foresters. The original protector, obtained from Brazil, was modified to improve the types of fastening and anthropometric dimensions, as well as to incorporate improved materials. Included in the ergonomic design modification process was an extensive series of usability tests over a six month period. In a well designed field test, this ergonomically modified leg protector was introduced in a eucalyptus plantation for use by persons responsible for ax/hatchet debranching. Among the 300 laborers, an average of ten injuries per day was occurring with an average sick leave of five days per injury. During the one year period of the test, not one single ax/hatchet leg injury occurred, resulting not only in the considerable savings in human pain and suffering, but also in a direct net cost savings to the company of $250,000. Use of the leg protectors throughout the South African hardwood forestry industry is conservatively calculated to save four million dollars annually.

Vehicle improvements. A second study involved ergonomically improving the seating and visibility of 23 tractor-trailer forwarding units of a logging company with an investment of $300 per unit. This resulted in a better operating position for loading and improved vision and operator comfort. As a result, downtimes caused by accident damage to hydraulic hoses, fittings, etc. went down by $2,000 per annnum per unit; and daily hardwood extraction was increased by one load per day per vehicle. All toll, for a total investment of $6,900, a hard cost savings of $65,000 per year was achieved.

Other projects. Other innovations by this same collaborative effort between Stellenbosch University, Ergotec, and various forestry companies have included (a) the development of a unique light-weight, environmentally friendly pipe type of timber chute for more efficiently and safely transporting logs down slopes, (b) redesign of three-wheeled hydrostatic loaders to reduce both excessive whole-body vibration and noise, (c) classifying different terrain conditions - including ground slope, roughness, and other conditions - and determining the most effective tree harvesting system (method and equipment) for each, and (d) developing ergonomic check lists and work environment surveys tailored to the forest industry. All of these are expected to result in significant cost savings, as well as greater employee satisfaction and improved quality of work life.

I believe this is a good example of what ergonomics can potentially contribute to *any* given industry when there is a true collaborative effort and commitment.

Aircraft Design

Some 35 years ago, I joined the US Air Force's C-141 aircraft development system program office as the project engineer for both human factors and the alternate mission provisions. The C-141 was to be designed so that its cargo compartment, through the installation of alternate mission kits, could be reconfigured for cargo aerial delivery, carrying paratroopers and paratroop jumping, carrying passengers, or for medical evacuation. As initially configured, anything that did not absolutely have to be included in the aircraft for straight cargo carrying was placed in one of the alternate mission kits, making them heavy, complex, and requiring considerable time and effort to install. By meeting with the intended using organization, the Air Force Material Air Transport Command, and discussing their organizational design and management plan for actual utilization of the aircraft, I was able to identify numerous kit components that rarely ever would be removed from the airplane. Using these data, I worked with the Lockheed design engineers to reconfigure the kits to remove these components and, instead, install them permanently in the aircraft. As documented by the engineering change proposals, this effort greatly simplified the system and reduced actual *operational* aircraft weight - and thus, related operating costs for over 200 aircraft over the past 35 years - and installation time and labor and storage requirements for the kits. In addition, it saved over $2 million dollars in the initial cost of the aircraft fleet. I believe this is a good illustration of how macroergonomic considerations can result in highly cost effective micro-ergonomic design improvements to systems.

Application of ergonomic analysis methods also can sometimes result in considerable system design cost savings. For example, on this same C-141 development program, the design engineers had come up with a throttle system that both the human factors group and the FAA determined had too restricted an operating range, resulting in an obviously low control-response ratio. In short, the controls were too sensitive, so that a very small movement resulted in a large speed response by the aircraft - particularly when it was lightly loaded - and the pilot could not accurately control aircraft speed. The design engineers modified the controls so as to extend the range of movement without having to make a highly expensive *major* redesign of the throttle system. The redesign was questioned by both the Air Force and the FAA regarding its adequacy.   Working with the Lockheed human factors group, we were able to construct a full scale mockup of the controls, building in the expected friction components, and use it in a study to determine   the pilots' ability to control throttle movement. Using these data, along with the calculated aircraft responses to a given amount of control movement, we were able to determine that the revised range of control movement   had sufficiently increased the control-response ratio for safe and effective use. As a result of this very low cost effort, Lockheed did not have to incur a major, highly expensive redesign - one which, in addition to increasing weight and taking up more crew station space,   would have significantly increased aircraft price in the fleet production contract.

These, and numerous other cost-benefit human factors evaluations and improvements to the C-141's design, came at a total cost of less than $500,000 of professional human factors effort.

Materials Handling Systems Redesign

One group that does a somewhat better job of documenting the costs and benefits of their ergonomic interventions than many of us is the faculty of the Department of Human Work Sciences at Lulea University of Technology in Sweden. The following examples are from the Department's Division of Environment Technology's work with steel mills.   The basic approach to ergonomic analysis and redesign in these projects was to involve employee representatives with the Lulea faculty. For each project, the "pay off" period was calculated jointly with the company's management.

Steel pipes and rods handling & stock-keeping system.   A semi-automatic materials handling and stock keeping system for steel pipes and rods was ergonomically redesigned.   The redesign reduced the noise level in the area from 96 db to 78 db, increased production by 10%, dropped rejection from 2.5% to 1%, and paid back the redesign and development costs in approximately 18 months. After that, it was all profit.

Tube manufacturing handling & storage system. In a tube manufacturing facility, ergonomic redesign of the materials handling and stock keeping system reduced the noise level by 20 db and improved stock-keeping with a payback period of 15 months.

Forge shop manipulator. In a forge shop, the old manipulator was replaced with a new one, having an ergonomically designed cabin and over-all better work place design. In comparison with the old manipulator, whole body vibration was reduced, noise was reduced by 18 db, operator sick leave dropped from 8% to 2%, productivity improved, and maintenance costs dropped by 80%.

Product   Design or Redesign

The economic benefit of ergonomic design or redesign of a product can be assessed in several ways. For example, by its impact on (a)the value of the company's stock, (b) on sales, (c) on productivity or (d) on reductions in accidents. Four very different kinds of products are provided herein as illustrations of each of these beneficial economic impacts.

Replacement forklift truck design.   Allen Hedge and his colleagues at the Human Factors Laboratory at Cornell university participated with Pelican Design, a New York industrial design company, and the Raymond Corporation in the design and development of a new generation of forklift trucks to replace Raymond's two existing product lines. Human factors design principles were given prime considersation and   an "inside-out" human-centered approach was taken, with the form of the truck being built around the operator's needs. These included anthropometrics, task requirements, and comfort. The goal was to maximize operator comfort, minimize accident risks, and maximize productivity by optimizing task cycle times.

At the time the development project was begun, Raymond's market share had eroded from its former position of dominance in the market place of over 70% of sales, to about 30%, and shrinking. Both the new narrow isle and swing-reach truck lines were introduced in the U.S. in 1992, and the swing-reach in Europe in 1993. Order books at Raymond are full and once again the company is enjoying success. Raymond stock has risen from around $6 per share at the start of the project to around $21 today.

The swing-reach truck design was a finalist in the "product of the year" competition run by Modern Materials Handling magazine.

TV and VCR remote control. Thompson Consumer Electronics first developed their highly successful approach to user-centered design when they developed "System Link", an ergonomically oriented remote control that can operate various types of products made by different manufacturers. The original Thompson remote control design differed little from the competition's: A rectangular box with rows of small, identical buttons, most often black on black. Using their user-centered design approach, this initial design was replaced with the new ergonomic one which, among other things, was easier to grasp, used color-coded, soft-touch rubber buttons in distinctive sizes and shapes, and where the VCR and TV buttons were separated above and below the keypad.

When introduced in 1988, this new, ergonomically designed, "System Link" remote control gained the jump on the competition; and Thompson has since sold literally millions of them. As a result of this success, user-centered ergonomic design has become a key aspect of all new Thompson development projects.(March, 1994)

A more recent highly successful example is Thompson's RCA DSS satellite digital television system. In fact, after considering the competition, I purchased one myself.

CRT display redesign. The CRT display used by the Directory Assistants at Ameritech were ergonomically redesigned by Scott Lively, Richard Omanson, and Arnold Lund to meet the goal of reducing average call processing time. Included in the redesign was replacement of an all upper-case display with a mixed-case display and the addition of a highlighting feature for the listing selected by the Directory Assistant. Based on extensive before and after measurements of average call operating time

for 40 operators, results showed a 600ms reduction in average call operating time after introduction of the ergonomically redesigned CRT display. Although seemingly small, this reduction represents an annual savings of approximately $2.94 million across the five state region served by Ameritech.

In a related effort, done jointly with Northwestern's Institute for Learning Sciences, the traditional lecture and practice training program for new Directory Assistants was replaced by an ergonomically designed computer-based training program which incorporates a simulated work environment and error feedback. As a result, operator training time has been reduced from five days to one and a half days.

Centered high-mounted rear stop lamp. This is perhaps the best known ergonomic improvement to a widely used consumer product. In the late 1960's, the U.S. Department of Transportation funded six research projects that were the first steps of a program to improve vehicle rear lighting systems. The results of these studies led the newly founded National Highway Traffic Safety Administration (NHTSA) to sponsor two field research programs in the 1970's which demonstrated the potential of adding a center high-mounted stop lamp (CHML) to reduce response times of following drivers and, thus, avoid accidents. In the mid 70's, this ergonomic innovation and three other configurations were installed in 2,100 Washington D.C. area taxicabs. The CHML configuration resulted in a 50% reduction in both rear-end collisions, and collision severity. Following several additional field studies, Federal Motor Vehicle Safety Standard 108 was modified to require all new passenger cars built after 1985 to have CHMDLs.

Based on analyses of both actual production costs for the CHMLs and actual accident data for the 1986 and 1987 CHML equipped cars, NHTSA calculated that, when all cars are CHML equipped (1997), 126,000 reported crashes will be avoided annually at a property damage savings of $910 million per year. Addition of the savings in medical costs would, of course, considerably increase this figure. The total cost of the entire research program was $2 million and for the regulatory program, $3 million. (Transportation Research Board, National

Research Council, 1989). Not a bad ergonomics investment by the federal government!

Redesign of Food Service Stands via Participatory Ergonomics

Using a participatory ergonomics approach with food service personnel, my USC colleague, Andy Imada and George Stawowy, a visiting ergonomics doctoral student from the University of Aachen in Germany, redesigned two food service stands at Dodger Stadium in Los Angeles (Imada and Stawowy, 1996). The total cost was $40,000. Extensive before and after measures demonstrated a reduction in average customer transaction time of approximately 8 seconds. In terms of dollars, the increase in productivity for the two stands was approximately $1,200 per baseball game, resulting in a payback period of 33 games, or 40% of a single baseball season.

Modification of these two stands was relatively costly because, as the development prototypes, they consumed considerable time and effort. Modifying the other 50 stands in Dodger Stadium can now be done at a price of $12,000 per stand, resulting in a payback period of only 20 games. Potentially, the resulting productivity increases can be used to reduce customer waiting time, thereby also increasing customer satisfaction.

This modification effort is but one part of a macroergonomics intervention project to improve productivity. Imada anticipates that on-going work to improve the total system process, including packaging, storage and delivery of food products and supplies, and managerial processes, eventually will result in a much greater increase in productivity.

Reducing Work Related Musculoskeletal Disorders

Given the importence of this issue, and the rather considerable attention and debate which has resulted from the introduction of proposed workplace ergonomics regulations at both the Federal and State (California) levels, I have included four examples of documented, highly successful ergonomic intervention programs.

Reducing WMSD's at AT&T Global. AT&T Global Information Solutions in San Diego

employees 800 people and manufactures large mainframe computers. Following analyses of their OSHA 200 logs the company identified three types of frequent injuries: lifting, fastening, and keyboarding. The company next conducted extensive work site analyses to identify ergonomic deficiencies. As a result, the company made extensive ergonomic workstation improvements and provided proper lifting training for all employees. In the first year following the changes, worker's compensation losses dropped more than 75% , from $400,000 to $94,000. A second round of changes, which replaced conveyor systems with small, individual scissors-lift platforms, replaced heavy pneumatic drivers with lighter electric ones; this was followed by moving from an assembly line process to one where each worker builds an entire cabinet, with the ability to readily shift from standing to sitting. A further reduction in worker's compensation losses of more than 87% , to $12,000 resulted. All toll, these ergonomic changes have reduced worker's compensation costs at AT&T Global over the 1990-1994 period by $1.48 million. The added costs for these ergonomic improvements represent only a small fraction of these savings. In terms of lost workdays for each type of injury, in 1990, lifting injuries resulted in 166 lost days, fastening injuries, 94 days, and keyboarding, 38 days. For all three types of injuries, lost days dropped to a total of 34 for 1992, and to zero for both 1993 and 1994. (Center for Workplace health Information, 1995a).

Reducing WMSDs at Red Wing Shoes. Beginning in 1985 with (a) the initiation of a safety awareness program which includes basic machine setup and operation, safety principles and body mechanics, CTD's, and monthly safety meetings, (b) a stretching, exercise and conditioning program, (c) the hiring of an ergonomics advisor, (d) specialized training on ergonomics and workstation setup for machine maintenance workers and industrial engineers, the Red Wing Shoe Company of Red Wing, Minnesota made a commitment to reducing WMSD's via ergonomics. The company purchased adjustable ergonomic chairs for all seated operators and anti fatigue mats for all standing jobs, instituted Continuous Flow Manufacturing, which includes operators working in groups, cross training, and job rotation, ergonomically redesigned selected machines and workstations for flexibility, elimination of awkward postures, and greater ease of operation, and modified production processes to reduce cumulative trauma strain.

As a result of these various ergonomic interventions, workers compensation insurance premiums dropped by 70% from 1989 to 1995, for a savings of $3.1 million. During this same period, the number of OSHA reportable lost time injury days dropped from a ratio of 75 for 100 employees working a year, to 19.

The success of this program is attributed to upper management's support, employee education and training, and having everyone responsible for coordinating ergonomics. I also would note the total systems perspective of this effort. (Center for Workplace Health Information, 1995b).

Reducing WMSDs via ergonomics training and participatory ergonomics. In 1992, Bill Youngblood of Washington Ergonomics conducted a one day seminar for cross disciplinary teams of engineers, human resource management personnel, and safety/ergonomics committee members from seven manufacturing companies insured by Tokyo Marine and Fire Insurance Company, Ltd.. The seminar taught the basic principles of ergonomics and provided the materials to implement a participatory ergonomics process. The training focused on techniques for involving the workers in evaluating present workplace conditions and making cost effective improvements. The class materials provided the tools for establishing a baseline, setting improvement goals, and measuring results.

In six of the companies, the seminar data was used by the teams to implement a participatory ergonomics program with the workers; and received both funding from management and support from labor. The seventh company did *not* participate in the implementation of the training. Follow-up support was provided by a Senior Loss Control Consultant for Tokyo Marine.

For the six companies that *did* participate, annual reported strain type injuries dropped progressively from 131 in the six months prior to the training to 118, 65, and 42, respectively, in the next three six month periods. The cost of these injuries for the six months prior was $688,344. For the next three six month periods following the training, the injury costs were $327,980, $315,704, and $72,600, respectively, for a net savings over 18 months of $1,348,748, using the six months prior as the baseline. Worker involvement reportedly created enthusiasm and encouraged each individual to assume responsibility for the program's success. According to Bill Brough, the reduction of injuries resulted from a commitment to continuous improvement and was obtained by many small changes, not a major singular event

For the one company that did not participate in implementing the training, the number of reported strain injuries was 12 for the six months prior to training, and 10, 16, and 25 respectively for the next three six months periods.

Coupled with both management's' and labors' support, Tokyo Marine traces these reductions in strain type injuries for the six participating companies directly back to Bill Brough's participatory ergonomics training program and related materials. A good example of what can happen when you couple collaborative management-labor commitment with professional ergonomics.

Reducing WMSDs at Deere. One of the best known successful industrial safety ergonomics programs is that at Deere and Company, managed by Gary Lovested. Since 1979, when the world's largest manufacturer of farm equipment initiated an informal ergonomics program, Deere has recorded and 83% reduction in incidence of back injuries, and by 1984 had reduced workers' compensation costs by 32%. In 1979 Deere recognized that traditional interventions like employee lift training and conservative medical management were, by themselves, insufficient to reduce injuries. So the company began to use ergonomic principles to redesign and reduce physical stresses of the job. Eventually, ergonomics coordinators were appointed in all of Deere's U.S. and Canadian factories, foundries and distribution centers. These coordinators, chosen from the industrial engineering and safety departments, were trained in ergonomics.

Today, job evaluations and analyses are done in-house by both part-time ergonomics coordinators and wage-employee ergonomics teams/committees. The company has developed its own ergonomics check lists and surveys. The program involves extensive employee participation. According to Gary Lovestead, each year, literally hundreds to thousands of ergonomics improvements are implemented; and today, ergonomics is built into Deere's operating culture. (CTD News, 1995).

Human Factors Test and Evaluation

One of the 'baby bells", NYNEX had developed a new work station for its toll and assistance operators, whose job is to assist customers in completing their calls and to record the correct billing. The primary motivation behind developing the new workstation was to enable the operators to reduce their average time per customer by providing a more efficient workstation design. The current work station had been in use for several years and employed a 300-baud, character-oriented display and a keyboard on which functionally-related keys were color coded and spatially grouped. This functional grouping often separated common sequences of keys by a large distance on the keyboard. In contrast, the *proposed* workstation was ergonomically designed with sequential as well as functional considerations; it incorporated a graphic, high-resolution 1200-baud display, used icons and, in general, is a good example of a graphical user interface whose designers paid careful attention to human-computer interaction issues.

Under the name *Project Ernestine,* Wayne Gray and Michael Atwood of the NYNEX Science and Technology Center, and Bonnie John of Carnegie Mellon University (1993) designed and conducted a comparative field test, replacing 12 of the current work stations with 12 of the proposed ones. In addition they conducted a goals, operators, methods & selection rules (GOMS) analysis (Card, Moran, & Newell, 1980) in which both observation-based and specification-based GOMS models of the two workstations were developed and used. Contrary to expectations, the field test demonstrated that average operator time was 4% *slower* with the proposed workstation than with the currently used one. Further, the GOMS analysis accurately predicted this outcome, thus demonstrating the validity of the GOMS models for efficiently and economically evaluating telephone operator workstations. Had this test and evaluation *not* been conducted, and the proposed, presumably more efficient, workstation been adopted for all 100 operators, the performance decrement cost per year would have been $2.4 million.

Macroergonomics

Two macroergonomic organizational interventions, and their economic benefit in reducing lost time accidents and injuries will be described briefly.

Macroergonomic change in a petroleum distribution organization. Several years ago, my USC colleague, Andy Imada began a macroergonomic analysis and intervention program to improve safety and health in a company that manufactures and distributes petroleum products. The key components of this intervention included an organizational assessment that generated a strategic plan for improving safety, equipment changes to improve working conditions and enhance safety, and three macroergonomic classes of action items. These items included improving employee involvement, communication, and integrating safety into the broader organizational culture. The program utilized a participatory ergonomics approach involving all levels of the division's management and supervision, terminal and filling station personnel, and the truck drivers.

Over the course of several years, many aspects of the system's organizational design and management structure and processes were examined from a macroergonomics perspective and, in some cases, modified. Employee initiated ergonomic modifications were made to some of the equipment, new employee-designed safety training methods and structures were implemented, and Employees were given a greater role in selecting new tools and equipment related to their jobs.

Two years after initial installation of the program, industrial injuries had been reduced by 54%, motor vehicle accidents by 51%, off-the-job injuries by 84%, and lost work days by 94%. By four years later, further reductions occurred for all but off-the-job injuries, which climbed back 15%. (Nagamachi & Imada, 1992).

The company's Area Manager of Operations reports that he continues to save one-half of one percent of the annual petroleum delivery costs every year as a direct result of the macroergonomics intervention program. This amounts to a net savings of approximately $60,000 per year for the past three years, or $180,000, and is expected to continue.

Imada reports that perhaps the greatest reason for these *sustained* improvements has been the successful installation of safety as part of the organization's culture. From my first hand observation of this organization over the past several years, I would have to agree.

A macroergonomic approach for implementing TQM. Rooney, Morency, and Herrick (1993) have reported on the use of macroergonomics as an approach and methodology for introducing total quality management (TQM) at the L. L. Bean corporation. Using methods similar to those described above for Imada's intervention, but with TQM as the primary objective, over a 70% reduction in lost time accidents and injuries was achieved within a two year period in both the production and distribution divisions of the company. Other benefits, such as greater employee satisfaction and improvements in additional quality measures also were achieved.

CONCLUSION

The above are but a sample of the variety of ergonomic interventions which we, as a profession are capable of doing to not only improve the human condition, but the bottom line as well. From my 35 years of observation and experience, only rarely are truly good ergonomic interventions *not* beneficial in terms of the criteria that are used by managers in evaluating the allocation of their resources.

In terms of the theme of this year's annual meeting, I also would note from the above success stories that they share the common characteristic of being *human centered* in their approach and methodology.

As many of the above ergonomic interventions also illustrate, ergonomics offers a wonderful common ground for labor and management collaboration; for invariably, both can

benefit - managers, in terms of reduced costs and improved productivity, employees in terms of improved safety, health, comfort, usability of tools and equipment, including software, and improved quality of work life. Of course, both groups benefit from the increased competitiveness and related increased likelihood of long-term organizational survival that ultimately is afforded.

Clearly, to enable our profession to approach its tremendous potential for humankind, we, the professional human factors/ergonomics community, must better *document* the costs and benefits of our efforts, and *share* these data with our colleagues, business decision-makers and government policy makers. It is an integral part of *managing* our profession. Thus, it *is* up to *us* to document and spread the word that *good ergonomics* **is** *good economics!*

REFERENCES

Card, S. K., Moran, T. P., and Newell, A. (1980). Computer text editing: An information processing analysis of a routine cognitive skill. *Cognitive Psychology, 12,* 32-74.

Center for Workplace Health Information (1995a). An ergonomics honor roll: Case studies of results-oriented programs, AT&T Global. *CTD News Special Report: Best Ergonomic Practices,* pp. 4-6.

Center for Workplace Health Information (1995b). An ergonomics honor roll: Case studies of results-oriented programs, Red Wing Shoes. *CTD News Special Report: Best Ergonomic Practices,* pp. 2-3.

Center for Workplace Health Information (1995c). An ergo process that runs like a Deere. *CTD News 8,* August, pp. 6-10.

Chong, Ian (1996). The economics of ergonomics. *Workplace Ergonomics,* March/April, pp. 26-29.

Gray, W. D., John, B., and Atwood, M. (1993). Project Ernestine: validating a GOMS Analysis for Predicting and Explaining Real-World Task Performance. *Human-Computer Interaction, 8,* 237-309.

Imada, A. S.. and Stawowy, G. (1996). The effects of a participatory ergonomics redesign of food service stands on speed of service in a professional baseball stadium. In O. Brown, Jr. and H. W. Hendrick (Eds.), *Human factors in organizational design and management-V.* Amsterdam: North-Holland.

March, A. (1994). Usability: The new dimension. *Harvard Business Review,* September-October, pp. 144-152.

Mallett, R. (1995). Human factors: Why aren't they considered? *Professional Safety,* July, pp. 30-32.

Nagamachi, M., and Imada, A. S. (1992). A macroergonomic approach for improving safety and work design. *Proceedings of the 36th Annual Meeting of the Human Factors and Ergonomics Society.* Santa Monica, CA: Human Factors and Ergonomics Society, 859-861.

Rooney, E. F., Morency, R. R., and Herrick, D. R. (1993). Macroergonomics and total quality management at L. L. Bean: A case study. In N. R. Neilson and K. Jorgensen (Eds.), *Advances in industrial ergonomics and safety V* (pp. 493-498). London: Taylor & Francis,

Transportation Research Board, National Research Council (1989). Rear guard: Additional break lamps help prevent rear end crashes. *TR News,* November-December, pp. 12-13.

Warkotsch, W. (1994). Ergonomic research in South African forestry. *Suid-Afrikaanse Bosboutydskrif, 171,* 53-62.

THE POLICY GAME: ANYONE FOR ADVOCACY?

William C. Howell

American Psychological Association
Washington, D.C.

You're looking at a fellow who's spent the better part of a career feeding at the federal R&D trough. I thought I was pretty savvy about how the system worked--until I came to Washington. I found out pretty quickly that the real action takes place in something called the policy arena, and the game they play wasn't one that I recognized. And I realized that most of my former colleagues are probably just as naive as I was, and that what we don't know can hurt us.

For instance, one morning in 1994 I woke up to the news that the House of Representatives had voted to cut the Department of Defense budget for university-based research in half. Why? Simply because a couple of congressmen were having a spat over "earmarking" (that's the euphemism for "pork"), and one of them was in a position to use the DOD science appropriation cut to tweak the other. Fortunately, this tragedy was averted by the prompt mobilization of a number of advocacy groups who descended on the Senate defense appropriations committee (the next stop for this particular legislation) like bees on honey. Most of the money was restored.

Where would we have been without the watchful eye and prompt action of a science advocacy infrastructure? About $900 million poorer, that's where. Keep in mind that the DOD budget is such a massive and confusing document, with such mind-boggling numbers, that most senators would never have missed a piddling 900 mil unless someone had yelled.

And this is but one of many examples I could have used to make the point. I could just as easily have used last summer's attempt to kill the Social, Behavioral, and Economic Sciences Directorate at NSF, or a current bill that would virtually abolish important research on minors, or the ever-present threat to psychological testing, or the FAA's move to cut human factors research funding, or the present Congress's desire to gut safety regulations, or a move to abolish the Army Research Institute. Most of the time when we make noise in response to one of these pending disasters, other groups are yelling just as loudly in the opposite direction--so it's necessary to yell loudly and **effectively**.

And that's what I want to talk about today. It's called the policy game. There are literally thousands of teams, some of them--like the American Medical Association or the pharmaceutical industry-- brimming with cash and high priced professional talent; others-- like many of the smaller scientific societies--trying to get by on a shoestring with well-meaning but often inept amateurs. To make matters worse, these little guys have to play by different rules -- rules dictated by the IRS. The big guys can just buy a congressman or a vote, whereas the little guys have to be more creative. To be at all effective, they must band together, develop clever strategies, focus their efforts, and work harder. Most important of all, they must **recognize and capitalize on their strengths**, and **recruit dedicated, informed volunteers**.

Now let's talk some specifics. Where does the Human Factors community fit into this game? What are its main strengths and limitations? What are some reasonable goals? And what are the options that make sense for a society like HFES--five thousand or so professionals with a relatively small operating budget?

Before I try to answer these questions, let me put things in perspective by telling you a bit about my organization, the APA Science Directorate. I spend roughly $1 million annually on advocacy. I have five registered lobbyists who work on it full time plus four part-timers, two support staff and an occasional consultant. Each is assigned to work closely with a particular funding agency (NSF, DOD, etc.), but they also handle a lot of cross-agency issues (like peer review, or the Family Privacy Act).

My constituency (psychological scientists, academics, scientist- practitioners) numbers somewhere around 25,000. But when I talk or write to a policymaker, I don't say that. I tell them I represent all 148,000 members and affiliates of APA. Sometimes I add the phrase, "N of whom live and work in your district." Numbers matter--especially local numbers. What matters even more is if some of those local folks back up my pitch with well crafted letters of their own, or actual visits to the congressman's office. A genuine concern coming from a few voting constituents means even more than my theoretical 148,000 concerned members.

I never realized before I got here how personal and anecdotal the policy game can be. A good story beats good data anytime! Say you want to get support for increased mental health benefits. First you identify a member of congress who has someone close to them battling mental illness. Then you bring in the local volunteers--preferably with a few poignant anecdotes, and you work on the guy's staff and provide them with statistics and draft language they can use (staffers are young, overworked and underpaid and always grateful for help). If a few things break your way, you've got a good chance of winding up with a champion. This is, in fact, exactly what my counterparts in the mental health community did a while back, and some of the senators they've enlisted to their cause would surprise you were I to name them.

So in my shop we devote a lot of effort to recruiting volunteers, training them, organizing them, and keeping them informed and involved. We bring about two dozen scientists a year to Washington for intensive training and carefully orchestrated visits on Capitol Hill. We select them on the basis of their enthusiasm for the game

and how important their particular senator's or representative's committee assignments are for our agenda. Sen Burns of Montana, for instance, is critical--he chairs the Senate Science and Transportation Committee. So we make sure we have Montana well covered.

After our trainees go back home, we encourage them to follow up with local office visits, and we provide them with materials for doing so--and reminders. Our goal is for each of them to establish a personal relationship with that policymaker's office. These volunteers, along with 700 more (and growing) now participate in our electronic Science Advocacy Network. I have a young woman who does nothing but manage this network. We send out regular policy updates and occasional "action alerts" requesting letters or visits. Most of the time we do this selectively--based on the geography of the congresspersons we need to influence--but when appropriate, we pulse the whole network. We explain the issue and suggest language for our volunteers to use.

Incidentally, how to approach policymakers is an important aspect of advocacy, and one of the main things we stress in our training sessions. There's a right way and a wrong way to play the game, and the wrong way can do more harm than good. For example, congressional staffers hate orchestrated form letters and can spot them a mile away. And they hate anything critical about their boss or his or her voting record. So we encourage folks to write in a sincere, positive, personal way but following certain guidelines. Properly written, such letters will be read, and even a handful of good ones can make a big difference. Staffers pay close attention to their mail. On some issues they don't get very much, so even a couple of compelling letters will determine how they tell their boss to vote. A very good letter may even get to the boss him- or herself.

Speaking of staffers, it may come as a surprise to you, as it did me, that these kids--most of them still in their twenties--actually run the government. "Networking" with them is crucial, which is why we hire mostly young ex-staffers for our policy slots. We also like to get psychologists into staff positions in the congress, the White House and the agencies. This is easy to do if you can afford it. So every year we pay the salaries and expenses of a half-dozen young PhD's who do just that (they're called APA Policy Fellows). But they cost around $50,000 a pop.

One final thing we do is work closely with a lot of other organizations and coalitions, depending on what the issue is. Very few scientific societies are big enough to accomplish much on their own. Hence we belong to the Coalition of Social Science Associations, the Council of Science Society Presidents, the Federation, and a lot of less formal ones. All told, this costs me over $300,000.

So that's how one moderate-sized scientific association plays the policy game. What about HFES? Obviously, some of the things we do wouldn't make sense for your organization--just as the pharmaceutical industry's approach would be out of the question for us. There is, however, a lot one can do with a force of 5000 professionals if they're willing and used right.

In fact, your members, their geography and their technical expertise are your greatest assets. Another is their employers. Many work for large, influential organizations. Having a responsible position at Boeing, say, the largest employer in the state of Washington, isn't going to hurt my chances of being treated nicely by Sen Gorton's office. Nor does it hurt if I work for--or am--a small business. Politically speaking, small businesses, especially those related to technology, are "in" these days, whether you're talking to a Democrat or a Republican.

So, you have your members. You also have a good message--one that translates easily into compelling stories that even a harried congressman can grasp. Remember, the policy game is played with anecdotes, and if you get into somebody's office, you have maybe 15 minutes to make your pitch. "We make technology user-friendly;" "we make highways or nuclear power or air travel safer;" or "we give our downsized military a 'combat edge'"-- these are

the kinds of openings that will get their attention. Follow that up with a concrete success story or two, and you may have yourself a sale.

In contrast, this is where we in psychological science run into trouble. What we have to sell are mainly abstractions like "We help understand how people learn things;" or "We can tell you some of the factors that affect drug use and violence." The usual reaction is "So what? Can you fix public education or stop kids from taking drugs and killing one another?" We say "Well, not just yet--that will come later if you keep supporting our research". But "later" doesn't matter much to folks whose main priority is staying in office--**now**.

The third thing you have on your side is willing and highly compatible partners--outfits like APA and the Federation and probably a number of engineering and safety organizations--with whom you can leverage your assets. Speaking for behavioral science, we need your stories and your people; you need our policy resources and expertise. A lot of folks have the naive impression that you can't be a player in this game without a huge bankroll. Or the even more naive view that all it takes to play is a "hired gun" in Washington. You can play--and in my view, should--provided you seek out compatible partners and make good use of your resources. But as for the hired gun, you might as well try to hold off a buffalo stampede with a water pistol. And you'll pay plenty for it--think in the $100,000 range for openers.

In my view, the option that makes the most sense for HFES is to invest in the mobilization of your own volunteer force. This involves recruiting them, organizing them, educating them, keeping them informed, and using them to best advantage while working closely with other groups. I could see, for example, a large network of HFES members in a database catalogued by congressional district, employer, and area of expertise. I could see tutorials for these folks available on your Web site and presented at national and local meetings. I could see a close working relationship between the manager of this network and ours, the Federation's, and any others that you find appropriate. I could see this network including many federal employees, contractors and other "insiders" who are in a position to alert us all to pending crises, or targets of opportunity that we should address. These folks would constitute the "eyes and ears" for the Society's policy strategists.

In all of this, organization and an informed constituency is critical. You can't just get volunteers fired up about advocating and then turn them loose. They'll freelance all over the place in the name of Human Factors and create mayhem. I have a whole collection of horror stories I could relate about overzealous but under informed members who have nearly killed us by "doing it their way". Everyone, of course, has a citizen's right to advocate, but unless they can be convinced that the policy game is a team sport-- one that, like most, requires leadership, a solid game plan, and coordination--the HFES stands to lose along with the individual members, the discipline, and society at large.

If you succeed in marshalling your forces, however, you'll pretty quickly run up against another significant problem: deciding exactly what you want to advocate for. A key requirement for playing the game is knowing who you are and what your agenda is. When we go to bat for psychological science, pretty much everyone on the team knows what that is, and in general, supports what we're going after.

From what I can see, Human Factors hasn't quite reached that point. Some folks are primarily interested in establishing and promoting human factors as an **independent discipline** or **profession**--carving out and protecting "turf" through credentialing, or getting its stamp on various design standards and regulations. Others see it as more of an **applied science** that informs the broader design and policy communities. For them the primary objective is sustaining R&D funds regardless of disciplinary label. Still others see human factors as a **philosophy** that serves to draw a lot of scientific and design disciplines together in the interest of promoting better

systems.

We in psychology have another whole advocacy organization that deals with professional "guild issues"--it's considerably bigger than mine and supported by units in all 50 states. My policy people work much more closely with outside science organizations than they do with our own practice folks most of the time--not because we're at odds with the latter but because the issues (and policymakers who control them) are different. Were we forced to settle on a single agenda for all psychology, and a single strategy team to manage it, we'd be in big trouble.

You have to figure out a way to deal with this problem if you hope to get anywhere in the policy game. As I see it, that should be your first priority. What exactly is it you're going to try to sell? How does that map onto national policy issues? To whom is the Society going to delegate the job of writing the game plan and calling the shots? With whom are you going to team up in executing your strategy?

It's been said that if you don't know where you're going, any road will get you there. In the policy game, I'd phrase it a bit differently. If you don't know where you're going, you'll invest a lot time, energy, and money only to wind up exactly where you started. HFES now seems to be poised on the threshold of wanting to get somewhere, and once you decide where that is, and who you want as traveling companions, your path will become much clearer.

I believe you have a lot going for you, and human factors has some real contributions to make in the policy arena as well ultimately in the public interest. But don't expect the route to be easy or free of bumps and highjackers. You've got hard choices, and a lot of work ahead of you. Take it from someone who's been there and had to learn through on-the-job training!

A COMPARISON OF MILITARY ELECTRONIC APPROACH PLATE FORMATS

Kristen K. Liggett, John M. Reising, & Thomas J. Solz, Jr.

Advanced Cockpits Branch
Wright-Patterson AFB, OH 45433-7511

David C. Hartsock

Veda, Incorporated
Dayton, OH 45431

ABSTRACT

Electronic approach plate formats were compared to determine which facilitated the best pilot performance when flying precision and non-precision approaches. Four formats which varied in map orientation and color scheme were flown: monochrome north-up, monochrome track-up, color north-up, and color track-up. Although results revealed a statistically significant difference favoring the track-up orientation in the non-precision approaches, the differences were so small that they showed no <u>practical</u> impact on performance. However, when given their choice of which format to fly, pilot's overwhelmingly flew a color map format. In addition, half of the pilots flew a map orientation of north-up and half flew track-up.

INTRODUCTION

Instrument approach procedures (IAP) are designed to provide a descent from the enroute environment to a point where a safe landing, or if necessary, a missed approach, can be made. Currently, IAP information is presented in paper format and is published and distributed by both government and commercial cartographers as instrument approach charts. The Defense Mapping Agency (DMA) provides a complete set of military charts for the United States airspace which consists of 16 bound booklets, 5 inches x 8 inches, with each booklet containing approximately three hundred pages (U.S. Department of Transportation, 1993). The complete set of military charts is revised on an eight week cycle. The commercially distributed charts, primarily distributed by Jeppesen Sanderson, are revised on an as required basis (U.S. Department of Defense, 1993). Depending on flight plan requirements, pilots may be tasked with carrying many if not all of these books. Although pilots may only use one of these books during the approach, storage of multiple books is necessary in an already crowded cockpit.

The paper charts, which are black text on a white page, provide pilots with the aeronautical data required to execute instrument approaches to airports. Each procedure is designated for use with a specific electronic navigational aid, such as an instrument landing system (ILS), VHF omnidirectional radio range (VOR), global positioning system (GPS) or tactical air navigation (TACAN).

The size of the charts forces the symbology and text to be quite small in order to accommodate all the information. Fear of litigation and liability often precludes cartographers

from removing marginally useful information from the charts; as a result, current IAPs tend to be information dense.

Electronic approach plates (EAPs) offer a more flexible medium to present approach information to the aircrew (Mykityshyn, Kuchar, and Hansman, 1994). Databases could be used to provide the information to construct the format in real time in the cockpit. Instead of making revisions to a complete set of manuals, only the database would be affected by changes. Since the database is a computer software item, it could easily and inexpensively be updated.

Due to resolution limitations of current electronic displays, larger type fonts and symbol sizes than are currently used on the paper charts would be required to improve symbol legibility (Clay and Barlow, 1994). However, any increased size of the symbology results in increased clutter on information dense charts. Decluttering techniques or format modifications will be required for electronic charts. For instance, color and the use of a zooming capability could be added to the electronic displays to declutter the symbology.

Another feature of electronic displays is their ability to provide dynamic formats, while the information on the paper charts must remain static. The approaches on the paper charts are all shown in a north-up orientation, regardless of the actual track the pilot is flying. Electronic displays can rotate the symbology, thus providing both north-up and track-up formats.

The issues of color scheme and map orientation have been evaluated in the civilian aircraft industry using commercial (Jeppesen) approach plates (Mykityshyn and Hansman, 1991; Mykityshyn et al., 1994; Hofer, Palen,

Higman, Infield, and Possolo, 1992; and Hofer, Kimball, Pepitone, Higman, Infield, and Possolo, 1993). Because the military approach plates differ from Jeppesen plates, DMA wanted to evaluate some of the same issues, using their military approach plates, to facilitate the transition from paper approach plates to electronic formats in military aircraft. Therefore, DMA asked Wright Laboratory to evaluate color scheme and map orientation on military approach plates.

OBJECTIVE

The purpose of this study was to compare pilots' performance using four versions of EAP formats. The elements of interest were color scheme and map orientation. The four combinations tested were: monochrome north-up, monochrome track-up, color north-up, and color track-up. A secondary objective was to investigate the use of a zooming capability to control the size of the viewing area, thus managing the amount of overall clutter on the EAP display formats. A third objective was to evaluate these formats in both precision and non-precision approaches because of the different instrument procedures used for each type of approach.

METHODS

Apparatus

Dynamic Cockpit. This study was conducted in a fixed-based, single seat, generic fighter cockpit evaluation tool which contained a single throttle and the limited-displacement control stick (Figure 1). An F-16 aeromodel was employed.

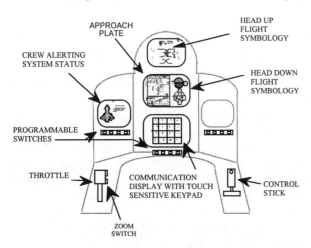

Figure 1. Cockpit.

Four of the five cathode ray tubes (CRTs) in the cockpit were utilized. The top CRT portrayed head-up flight symbology. The EAPs were positioned on the left 6"x8" portion of the center CRT. The center CRT also contained head down flight symbology on the right. The left CRT contained a crew alerting system status display with engine parameters. The bottom CRT employed a touch sensitive keypad used to change radio frequencies, and manipulate commanded altitudes and airspeed markers. Programmable switches were used to change navigation radios.

Display formats. Figure 2 shows an example of an EAP flown by the pilots during the study. The amount of

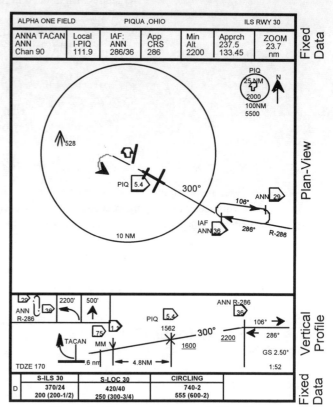

Figure 2. Electronic Approach Plate Format.

detail shown on the EAP was less than that on the paper charts, but all pertinent information was still present (Hofer et al., 1993). The plates were divided into three sections: two fixed data areas, a plan-view diagram, and a vertical profile diagram. The fixed data areas contained numerical data. The plan-view was a graphical look-down view of the entire approach. When the plan-view was viewed in the track-up mode, the text always remained upright as the map rotated around the fixed aircraft symbol in the center of the format. The vertical profile view was a side view of the approach and provided a graphical depiction primarily of altitude information.

In the monochrome version, all information was displayed in green on a black background. In the color version, colors recommended by the Federal Aviation Administration Advisory Circular 25-11 (1987) were used. To view a full color depiction of each plate, access the following World Wide Web site: http://www.wl.wpafb.af.mil/flight/fcd/figp/figp1/acd.htm.

A continuous zoom function was provided so that subjects could change the range of display coverage. This allowed them to vary the amount of detail displayed on both the plan view and the profile view during different segments of the approach. Pilots could change the display range of either the plan view or the profile view by simply touching the corresponding section of the display and then using the zoom switch to change the range. The point about which the zoom focused in both the plan-view and the profile-view was the aircraft symbol.

Experimental Design

This study employed a mixed experimental design. The two within-subjects variables were color scheme (color and monochrome) and map orientation (north-up and track-up). The between-subjects variable was type of approach (precision and non-precision). All variables were presented in a counterbalanced order.

Dependent Variables

Flight Performance. Although a number of objective flight performance measures were collected, the ones of primary interest were: root mean square (RMS) airspeed deviations, RMS altitude deviations, and RMS course deviations.

Subjective Measures. Three types of situational awareness (SA) probe questions were asked while the subjects were flying the approaches: world referenced, ego referenced, and focused attention questions (Hofer et al., 1993). World referenced questions asked location/position of two fixed objects, such as, "Is PARKK NDB north of SEATTLE TACAN?". Ego referenced questions asked location/position of a object with respect to the pilot's aircraft, for instance, "Is the airport to the left of your position?". Focused attention questions asked single pieces of information, such as, "What is the touchdown zone elevation (TDZE)?". Time to respond, as well as accuracy of response were recorded.

Subject-pilots also completed questionnaires after each profile was flown. After all profiles were flown, the pilots filled out a final questionnaire. This questionnaire contained a Subjective WORkload Dominance (SWORD) technique (Vidulich, 1989), a performance ranking of the display formats, and the solicitation of overall comments on the electronic approach formats. SWORD enables a subjective comparison of the difference in workload experienced when flying the profiles using the various formats.

Pilots' Choice. Previous research (Mykityshyn et al., 1994) comparing various electronic versions of commercial paper approach procedures (Jeppesen) has shown little primary flight performance differences. Realizing that primary flight performance might not be sensitive to the versions of the EAPs compared, a key feature of this study was the "Pilot's Choice" profile. This profile measured pilots' preference of color scheme and map orientation by allowing pilots to select monochrome, color, north-up, and track-up anytime during the flight of a fifth profile. Pilots' Choice data were gathered in two segments: from initial approach fix to final approach fix, and from final approach fix to missed approach fix.

Subjects

Sixteen Air Force pilots flew approaches using the different formats. All were required to have a minimum of 300 hours flying time. The subject pool consisted of 6 pilots with primarily fighter experience, 8 pilots with bomber/cargo experience, and 2 pilots with experience in both fighter and transport aircraft.

Procedure

Sessions lasted approximately 3 hours. The subjects were briefed first in a classroom environment. Then they were briefed in the cockpit while becoming familiar with the cockpit layout and the procedures for flying the profiles. Next, they flew an approach using a paper approach plate. This allowed subjects to become familiar with the aeromodel and the SA probe questioning procedures. Subjects could practice with this approach more than once if needed.

An approach plate with characteristics of the first condition (the combination of color scheme and map orientation) was introduced to the subject statically; the aeromodel was not running, but all graphics were present. This briefing allowed the subjects time to review the approach procedure and simulated what they would have done during mission planning or while flying enroute.

When training on a specific static format was completed, pilots flew the approach for data collection. A portion of the questionnaire was filled out pertaining to that specific format. The other formats were tested in the same fashion. After all four formats were flown, pilots flew the Pilot's Choice profile with their choice of color scheme and map orientation. The initial presentation of this format was in monochrome and north-up, but using the programmable switches, pilots could change the color scheme and orientation anytime during the approach. Upon completion of all flying tasks, pilots filled out the remainder of the questionnaire.

RESULTS

Flight Performance

Objective data from this experiment were analyzed using the Statistical Package for the Social Sciences (SPSS). Bartlett's test of sphericity was not significant ($F(2,14) = 31.42$, $p < 0.198$) indicating independence of the performance measures, therefore, results were analyzed using ANOVA. Results showed a significant interaction between type of approach (precision/non-precision) and map orientation (north-up/track-up), in terms of the RMS airspeed deviation ($F(1,14) = 5.48$, $p < 0.035$) (Figure 3). Analysis of simple effects revealed significantly less airspeed error when using the track-up orientation than when using the north-up orientation in the non-precision condition. No other significant results were found.

Zoom Data. Based on a frequency count during various phases of the procedure, pilots manipulated the zoom function primarily enroute to the initial approach fix (IAF).

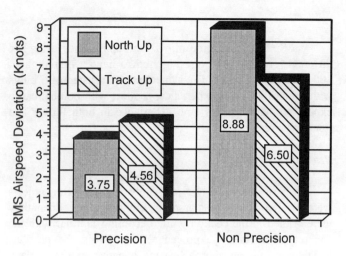

Figure 3. Interaction Between Type of Approach and Map Orientation.

On average, subjects manipulated the zoom function a total of 12 times during the "Start to IAF" portion of the flight, and 1.5 times for all other phases underlined combined. Figure 4 shows the average map range that the pilots settled on for each segment of the procedure.

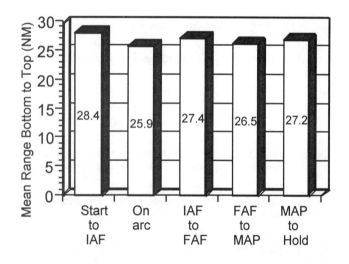

Figure 4. Map Range by Flight Segment

Subjective Measures

SA Probes. SA probe question data was scored by calculating the number of correct responses versus the number of questions asked. The focused attention questions were answered correctly 100% of the time. The world referenced questions were answered correctly 95.5% of the time. In all cases when a world referenced question was missed, the pilots were flying in the track-up mode (both monochrome and color versions). For the ego referenced questions, correct answers were given 98% of the time. The missed questions occurred when the pilots were flying in the monochrome north-up mode.

SWORD Data. The SWORD technique utilized a

series of pair-wise comparisons between the various system configurations to determine which configuration elicited the most workload. The SWORD data were analyzed using a 2X2X2 Analysis of Variance with the two within subjects factors as repeated measures. The ANOVA results showed a significant effect for color scheme ($F(1,14) = 112.95$, $p < 0.001$). Figure 5 shows that subjective workload was found to be significantly higher with the monochrome display than with the color display.

Higher Workload

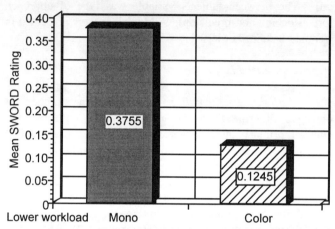

Figure 5. SWORD Main Effect for Color Scheme

Pilot's Choice

Results for the Pilots' Choice data revealed that half of the pilots flew north-up, half flew track-up, and all but one flew a color format.

DISCUSSION

Flight Performance

Although analysis of RMS airspeed error showed a statistically significant interaction between the map orientation variable and the type of approach variable, there was no underlined practical difference in airspeed. For non-precision approaches, pilots had 2.38 knots less airspeed error when flying a track-up EAP than when flying an EAP in the north-up mode. Since pilots are normally required to maintain their airspeed -5 to +10 knots during an approach for a qualified grade during an instrument evaluation (US Air Force, 1994), the airspeed deviations are well within limits.

Zoom Data. Pilots primarily used the zoom capability in the initial phase of the procedure. Because pilots are used to seeing the entire approach at one glance as on the paper charts, pilots zoomed the EAP until the entire procedure was in view, and then left the zoom at that range setting for the entire procedure. This is confirmed by Figure 4, which shows that the average map range for the different segments did not vary after the initial adjustment.

Subjective Measures

SA Probes. The SA probe data revealed that the pilots had more difficulty answering world referenced questions in the track-up mode. This is because in the track-up mode, the map rotates as pilots' fly each leg of the approach, thus, the orientation of objects also changed. When asked questions such as, "Is the parallel runway east of the landing runway?", the pilots had to mentally translate the map information to determine where north was located and then relate the two objects. For the ego referenced questions, pilots had a harder time while flying a north-up map. This may be due to the fact that in the north-up mode, the aircraft moved and at times it could be positioned upside down on the pilots' map, in which case, east was on the pilots' left and west was on the pilots' right. Therefore, questions pertaining to the location of an object with respect to the pilots' aircraft location were harder to answer.

SWORD Data. The SWORD data showed that pilots ranked the color formats as requiring less workload to accomplish the task than the monochrome formats. In the color condition, each different set of symbols was color coded, so pilots could recognize and identify navigation and geographical features easier. In the monochrome condition, there was much more cognitive workload involved in finding and identifying objects and information.

Pilots' Choice

Color Scheme Data. The monochrome format was chosen by only one pilot in the initial approach fix to final approach fix segment, and he changed to color in the final approach fix to the missed approach point segment. All other pilots chose color in both segments. Questionnaire data revealed that, even though one pilot did fly a monochrome format, all 16 preferred the color formats.

Map Orientation Data. The pilot's were evenly split in their choice of orientation during the segment from initial approach fix to final approach fix (8 chose north-up and 8 chose track-up). During the segment from final approach fix to the missed approach point, the results were virtually the same (7 chose north-up and 9 chose track-up). The reason that some of the pilots chose north-up could be due to a familiarization with the paper approach plates which are always north-up. Conversely, the reason for choosing the track-up orientation could relate to ease with which this orientation fits the pilots' mental model of where they are in space. The mental rotation involved in making the north-up orientation correspond to their current orientation is not needed in the track-up version. To clarify this idea, think of the <u>physical</u> rotation of a north-up map involved in turning a road map so that it corresponds to the direction of travel. Because the pilots cannot turn the electronic display, they must perform the rotation mentally.

IMPLICATIONS FOR DESIGNERS

Based on the results of this study, color versions of EAPs should be used. In addition, the designer should provide **pilot selectable** north-up and track-up versions of the EAP -- a track-up/north-up switch is an easy implementation. A zoom feature should be included with the center of focus of the zoom selectable by the pilot. In order to reduce the need for frequent adjustments of the zoom control, the initial presentation of the procedure should be at a scale sufficient to show the entire approach. Also, since the zoom capability provides the opportunity to see information at different scales, there must be upper and lower limits to the size of font and symbols used.

REFERENCES

Clay, M. C., and Barlow, T. (1994). <u>Handbook for the Design of Electronic Instrument Approach Procedures displays - phase II resource document for the design of electronic instrument approach procedure displays</u> (Contract No: 38125(4529)-2183). Cambridge, MA: Volpe National Transportation System Center.

Hofer, E. F., Palen, L. A., Higman, K. N., Infield, S. E., and Possolo, A. (1992). <u>Flight deck information management - phase II</u> (Final Report from Boeing Commercial Airplane Group to FAA on contract DTFA01-90-C-00055). Washington DC: Department of Transportation, Federal Aviation Administration, Flightcrew Systems Research Branch.

Hofer, E. F., Kimball, S. P., Pepitone, D. T., Higman, K. N., Infield, S. E., and Possolo, A. (1993). <u>Flight deck information management - phase III human factors guidelines for electronic approach information management and display</u> (Final Report from Boeing Commercial Airplane Group to FAA on contract DTFA01-90-C-00055). Washington DC: Department of Transportation, Federal Aviation Administration, Flightcrew Systems Research Branch.

Mykityshyn, M. G., and Hansman, R. J. (1991). <u>Development and evaluation of prototype designs for electronic display of instrument approach information</u> (Tech. Report DTRS-57-88-C-00078). Cambridge, MA: MIT.

Mykityshyn, M. G., Kuchar, J. K., and Hansman, R. J. (1994). Experimental study of electronically based instrument approach plates. <u>International Journal of Aviation Psychology</u>, 4(2), 141-146.

U.S. Air Force (1994). Air Force Instruction: Flight Management (AFI 11-401). Washington DC: Author.

U.S. Department of Defense, Defense Mapping Agency. (1993). <u>Flight information publication: general planning</u>. Washington DC: Author.

U.S. Department of Transportation, Federal Aviation Administration. (1987). <u>Transport category airplane electronic display systems</u> (AC No: 25-11). Washington DC: Author.

U.S. Department of Transportation, Federal Aviation Administration. (1993). <u>Airman's information manual</u>. Washington, DC: Author.

Vidulich, M. A. (1989). The use of judgment matrices in subjective workload assessment: the subjective workload dominance (SWORD) technique. In <u>Proceedings of the Human Factors Society 33rd Annual Meeting</u> (pp. 1406-1410). Santa Monica, CA: Human Factors Society.

NAVIGATIONAL CHECKING:
IMPLICATIONS FOR ELECTRONIC MAP DESIGN

Brian Schreiber, Christopher D. Wickens, Goetz Renner, and Jeff Alton
University of Illinois at Urbana-Champaign
Aviation Research Laboratory, Savoy, Illinois

In three experiments, subjects performed a navigational checking task in which the view presented on an electronic map was compared with the view of a simulated world to determine the congruence. Map viewpoints were varied in elevation and azimuth angle disparity relative to the world view to simulate the effects of map rotation and 3D map angle on navigational checking. On most trials the area depicted was the same, whereas on a small percentage, features of the map were altered, requiring the subject to judge "mismatch." In Experiment 1, using simple images, response time increased non-linearly with elevation angle distortion. In Experiment 3, using more complex images, similar effects were found and were augmented by effects of azimuth angle distortion. In Experiment 3, using dynamic realistic real world scenes, elevation angle and map scale effects were examined and revealed a complex pattern. The results are interpreted in terms of guidelines for 3D electronic map construction.

A critical facet of a pilot's flying task in visual meteorological conditions is *navigational checking*. This is the process of maintaining a fairly continuous comparison between geographical features in the forward field of view (FFOV), and some representation of those features on a map, in order to determine that the two are congruent (i.e., what I see, is what I should see; Aretz, 1991). The occurrence of breakdowns in this process is evident , for example, in the case of wrong airport landings (Antunano et al., 1989), many of which occur in conditions of good visibility. The physical nature of the map representation becomes particularly important on less standardized missions, such as military combat sorties (Kibbe, and Stiff, 1993), search and rescue operations, or EMS evacuations. In such cases time is critical, the consequences of navigational errors are severe, and automated navigational aides are not always available.

The navigational checking task of comparing two images to determine their congruence bears a parallel to various "same-different" judgment tasks examined in cognitive psychology (Posner, 1978; Cooper and Podgorny, 1976); however, in the former case the images are far more complex, and the concept of "sameness" is one that is explicitly defined by the rule: "is the **region** depicted on the map, the same as that depicted in the FFOV," independent of other physical differences between the two images?

As in the studies carried out in simpler laboratory paradigms, we assume that "same" judgments will be made most rapidly when there is actual physical congruence between the two images (Cooper and Shepard, 1978). That is, the map image is a virtual photographic rendering of the momentary and evolving FFOV. However, for an in-flight electronic map, this circumstance would require a high fidelity image generator, constantly updating the representation of the forward three-dimensional view, as the aircraft traverses the terrain or approaches a point on the ground. The expense of such a system is evident, and if data must be updated over limited bandwidth channels, it may not be feasible. Furthermore, a "map" of this sort would of

necessity exclude representation of a wider range of geographical information (i.e., that behind and beside the aircraft), necessary to support the awareness of hazards, such as weather, terrain, or other air traffic, that may not lie in the forward path (Wickens, 1996).

At issue in this paper is the extent to which one can reduce the degree of similarity of the viewpoint of the map and FFOV (away from the level of complete identity), in order to reduce the cost of dynamic image generation and updating, yet still preserve efficient navigational checking. We employ an information analysis framework called ARTS (Schreiber, Wickens, and Alton, 1995) to describe how reductions in this similarity can influence the Availability of geographical information, the sensory Resolution of that information, the cognitive Transformations necessary to compare the two images, and the Strategies used in that comparison.

Consider, for example, a map and FFOV shown in the two different viewpoints of Figure 1a. If the FFOV hides information that is present in the map (e.g., behind a building, or outside the FFOV), that information is not **available**. If critical features on the map are so small or compressed that they cannot easily be seen, then *resolution* becomes a problem. If the map is north-up and the FFOV faces south, then the mental rotation *transformation* is imposed. Finally, if the navigator thinks he might be lost, he will look for differences between the map and FFOV, a *strategy* that will make it more likely to find such differences if they exist.

In the current experiments our interest is on the cognitive transformations that are imposed by cost-induced simplifications of the presentation of maps, such that the physical rendering of the map and FFOV images no longer correspond, even when the same image is depicted. Our interest here is in three variables. 1) The **elevation angle** of map depiction: should it constantly match the slant angle of the FFOV, or can it be fixed, so that such disparity will occur? 2) The **azimuth angle** of the map; should it be fixed

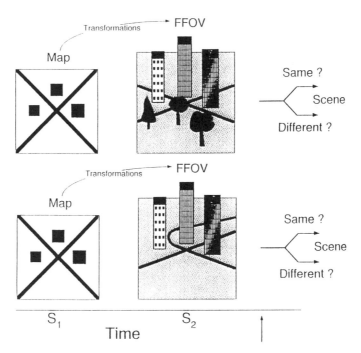

Figure 1: Schematic representation of navigational checking task: (a) (top row) same scene; (b) (bottom row) different scene. (Note curved road.)

(i.e., north up), or rotating in a track up fashion, thereby alleviating the need for mental rotation (Rossano and Warren, 1989; Wickens, Liang, Prevett, and Olmos, in press; Peruch et al., 1995)? 3) The **zoom level** or scale of the map. For reasons described below, we anticipate that the effects of both elevation angle and disparity may be non-linear.

General Method

In each of three experiments, subjects viewed two images shown schematically in Figure 1: a computer-generated map, and a computer-simulated world or the FFOV. On a majority of trials the two images "matched" in the region that was depicted (Figure 1a), but they differed in azimuth or elevation angle by varying degrees. Matching trials were more frequent and of more experimental interest than mismatch trials (Figure 1b) because (a) it was difficult to equate the degree of mismatches across all transformations and (b), most of the time in real flight a pilot would normally be "confirming" his/her location, and hence, would expect such a match to occur. Hence, mismatch trials were viewed more as "catch trials" to force the subjects to carefully scan all images, than as trials of intrinsic experimental interest. Subjects were to respond "same" or "different" as rapidly as possible via a keypress.

Detailed Method and Results

Experiment 1

Sixteen subjects (non-pilots) viewed very simple worlds consisting of two geometric objects (rectangular solids or cylinders) placed on a flat, textured surface. Our interest was in how the ability to judge sameness was affected by a variety of map elevation angles (ranging from a 90° "2D map" to a

30° "3D" map) and FFOV elevation angles (i.e., the "slant angle" of the pilot's view to the ground features); together these two manipulations created a set of elevation angle *disparities*.

The results indicated that the 30° map elevation angle supported the most rapid judgments (F=3.36; p<.01), although there was no effect on accuracy. There was also a significant cost to disparity between the map and FFOV (RT: F=5.18; p<.01. Accuracy: F=4.35; p<.01)). As shown in Figure 2, the effect of disparity was different at different map angles (F interaction = 3.47; p<01).

When elevation angle disparity is plotted on a linear scale, as in Figure 2, the RT effects are clearly non-linear, a function that is different from the linear function typically

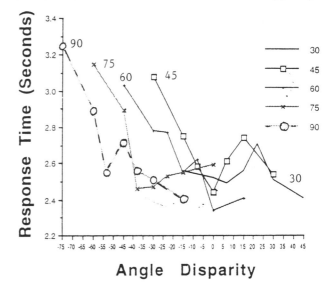

Figure 2: The interaction of map angle with the disparity between viewing angles.

obtained in mental rotation studies. However, in constructing the maps, a given difference in true space is represented by different amounts of display space, as a function of the elevation angle. This "gain" function, relating true space to displayed difference is non-linear, and follows a trigonometric function. We consider this influence on congruence judgments in the discussion.

Experiment 2

In Experiment 2, 20 subjects viewed worlds that were slightly more complex, consisting of an airport, with a particular orientation of crossing runways, and a small number of additional features (road, river, buildings, hills; Figure 3). Both the FFOV and the Map were generated on the same IRIS display. The map was rotated from the FFOV both vertically at different elevation angles, and laterally, at azimuth angles ranging from 0 to 90°.

Azimuth angle effects were observed on response time ($F(4,76) = 4.77$; $p<.02$) and accuracy ($F(4,76) = 3.36$; $p<.03$) replicating the frequently observed costs of mental rotation in these circumstances (Wickens et al., in press; Aretz and Wickens, 1992; Warren, Rossano, and Wear (1990). The effects were not linear, suggesting that small lateral disparities imposed negligible costs. Replicating the effects of Experiment 1, increasing levels of elevation angle disparity also imposed a cost on RT ($F(4,76)= 10.15$; $p<.01$), a cost which was greater at lower map angles. There was no systematic main effect of map angle.

Experiment 3

In Experiment 3, map elevation angle was varied at only two levels, 45 and 90°; and this manipulation was crossed with two levels of map scale; close and distant. Thirty subjects (15 pilots and 15 non-pilots) compared a rendering of a more realistic terrain area depicted on an Evans and Sutherland SPX 500 image generator, with a map rendered on an IRIS display system. Figure 4 represents the contour plot of the area depicted. The world consisted of mixtures of mountains, rivers, buildings, roads and other cultural features. Instead of the static renderings used in Experiments 1 and 2, in Experiment 3, the subject's viewpoint approached a point on the ground at a 12°slant angle.

The data revealed an interaction between map scale and map angle in both speed ($F(1,28) = 6.67$; $p<.02$), and accuracy ($F(1,28) = 7.35$; $p<.02$; see Figure 5). At the 45° (3D) angle, neither speed nor accuracy were affected by map scale. However, at the 90° (2D) map angle, performance was fastest, but accuracy was at near chance level with the small scale (close) map. Accuracy was the highest of all four map types at the large scale (distant) level for the 90° map, and performance was no slower than with either of the 45° maps.

Discussion and Conclusion

Experiments 1 and 2 exhibited a pattern of results supporting the use of dynamically updating maps that will minimize discrepancy between the elevation angle of the map and forward-field-of-view, as the latter changes via altitude change during flight. The results of Experiment 2 also yielded support for rotating maps, sensitive to the direction of travel.

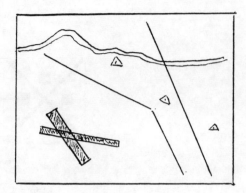

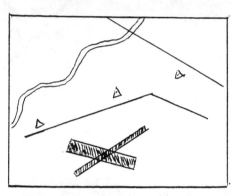

Figure 3: Example of stimuli used in Experiment 2. Azimuth rotation (30°) only.

However, both experiments suggested that such electronic map updating in either azimuth or elevation did not need to be continuous but could be intermittent and periodic. This is because small levels of discrepancy or misalignment in azimuth and elevation imposed no cost to performance relative to perfect alignment.

While larger disparity between map and FFOV, in both azimuth and elevation is detrimental, it appears that the quantitative effects of elevation angle disparity are complex and non-linear. they depend upon (1) the relative **importance** of lateral versus vertical variance in landmark identification (greater importance of lateral variance favors a more vertically-oriented--"2D"-- elevation angle), and (2) the degree of **compression** by which a given elevation or azimuth angle condenses variance in true space, when that variance is projected onto a planar display screen.

For example, a 90° look down (2D map) will provide the maximum possible ratio of display pixels per unit of real lateral space (minimum compression), and therefore the best possible resolution of lateral information. This viewpoint will provide a benefit, to the extent that it is variance in the lateral position of landmarks (i.e., triangulation) that is important for judgments of locations. Furthermore, at this elevation angle, a disparity of, say 15°, will produce little difference in the physical rendering of lateral position because the amount of that distortion will be predicted by the difference: $\sin 90°$ - $\sin 75°$ which is small (.033) compared to the distortion created by a similar elevation angle disparity from 45 - 30° ($\sin 45 - \sin 30 = .207$). Thus, if the map or FFOV is instead at, say 30°, this will produce a considerable compression of

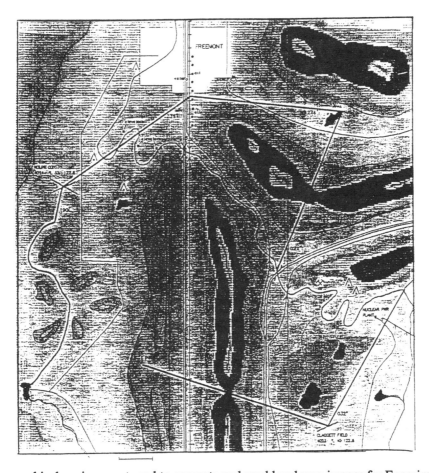

Figure 4: Topographical environment used to generate real world and map images for Experiment 3.

lateral information (Yeh and Silverstein, 1992), and a considerably greater change in displayed position, for each change in elevation angle. When we use this "sine-sine" function to characterize elevation angle disparity, rather than the linear function typically used to predict mental rotation transformations in Experiment 1 and shown in Figure 2, our ability to predict response time from elevation angle disparities increases from $r = .49$ to $r = .64$, an added 16% variance accounted for. We note also in Figure 2 that a given amount of disparity costs less for more vertically-oriented (2D) maps (e.g., 90°-75°) than for more forward looking maps (e.g., 60°-45°).

If static maps are to be used, then the optimum elevation angle will be dictated jointly by the relative importance of lateral versus vertical variance, and by the expected altitude (or viewing slant angle to the terrain) to be flown on the mission in question. Lower slant angles favor lower elevation angles.

Acknowledgments

This research was supported by contract N00014-93-1-0253 from the Naval Weapons Center at China Lake, California. Marion Kibbe and Robert Osgood were the technical monitors. The authors acknowledge the invaluable contributions of Jonathan Sivier and particularly Sharon Yeakel for the software development involved in this project.

References

Antunano, M.J., Mohler, S.R., and Gosbee, J.W. (1989). Geographic disorientation: Approaching and landing at the wrong airport. *Aviation, Space, and Environmental Medicine, 55*, 996-1004.

Aretz, A.J. (1991). The design of electronic map displays. *Human Factors, 33*(1), 85-101.

Aretz, A.J., and Wickens, C.D. (1992). The mental rotation of map displays. *Human Performance, 5*, 303-328.

Cooper, L.A., and Podgorny, P. (1976). Mental transformations and visual comparison processes: Effects of complexity and similarity. *Journal of Experimental Psychology: Human Perception and Performance, 2*, 503-514.

Cooper, L.A., and Shepard, R.N. (1978). Transformations on representations of objects in space. In E. C. Carterette and M. Friedman (Eds.), *Handbook of perception, Vol. VIII, Perceptual coding*. New York: Academic Press.

Kibbe, M.P., and Stiff, J.S. (1993). *Operator performance in pattern matching as a function of reference material structure* (Technical Report NAWCWPNS TP 8145). China Lake, CA: Naval Air Warfare Center Weapons Division.

Peruch, P., Vercher, J-L, and Gauthier, G.M. (1995). Acquisition of spatial knowledge through visual exploration of simulated environments. *Ecological Psychology,* 7(1), 1-20

Posner, M.I. (1978). *Chronometric explorations of the mind.* Hillsdale, NJ: Lawrence Erlbaum.

Rossano, M.J., and Warren, D.H. (1989). Misaligned maps lead to predictable errors. *Perception, 18*, 215-229.

Schreiber, B., Wickens, C.D., and Alton, J. (1995). *Navigational checking: The influence of 3D map rotation and scale.* University of Illinois Institute of Aviation Technical Report (ARL-95-9/NAWC-95-1). Savoy, IL: Aviation Res. Lab.

Warren, D.H., Rossano, M.J., and Wear, T.D. (1990). Perception of map-environment correspondence: The roles of features and alignment. *Ecological Psychology, 2*(2), 131-150.

Wickens, C.D. (1996). Situation awareness: Impact of automation and display technology. *NATO AGARD Proceedings 575: Situation Awareness: Limitations and Enhancement in the Aviation Environment* (pp. K2-1/K2-13). Neuilly-Sur-Seine, France: Advisory Group for Aerospace Research & Development.

Wickens, C.D., Liang, C-C, Prevett, T., and Olmos, O. (in press). Egocentric and exocentric displays for terminal area navigation. *International Journal of Aviation Psychology.*

Yeh, Y-Y, and Silverstein, L.D. (1992). Spatial judgments with monoscopic and stereoscopic presentation of perspective displays. *Human Factors, 34*, 583-600.

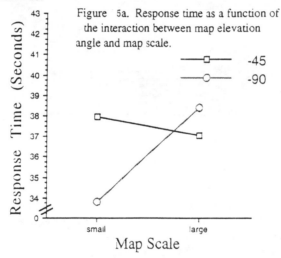

Figure 5a. Response time as a function of the interaction between map elevation angle and map scale.

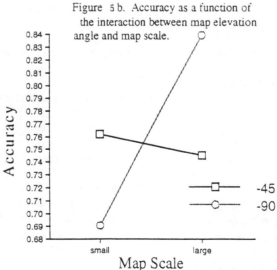

Figure 5 b. Accuracy as a function of the interaction between map elevation angle and map scale.

Figure 5: Performance effects as a function of the interaction between map elevation angle and map scale.

ANALOG TRACK ANGLE ERROR DISPLAYS
IMPROVE SIMULATED GPS APPROACH PERFORMANCE

Charles M. Oman[+] , M. Stephen Huntley , Jr.[*] ,
and Scott A. M. Rasmussen[+]

[+]Man Vehicle Laboratory, Department of Aeronautics and Astronautics
Massachusetts Institute of Technology, Cambridge, Massachusetts

[*] Cockpit Human Factors Program, DTS-45
Volpe National Transportation Systems Center, Cambridge, Massachusetts

Pilots flying non-precision instrument approaches traditionally rely on a course deviation indicator (CDI) analog display of cross track error (XTE) information. The new generation of GPS based area navigation (RNAV) receivers can also compute accurate track angle error (TAE). Does display of supplementary TAE information improve intercept and tracking performance ? Six pilots each flew 20 approaches in a light twin simulator to evaluate 3 different TAE/XTE display formats, in comparison to a conventional receiver CDI display and a more centrally located Horizontal Situation Indicator (HSI). Statistically significant performance improvements were seen in several phases of the approach when using the supplementary TAE information. Analog was preferred over numeric format. However, the advantage was offset by the need to widen the pilot's instrument scan to include the receiver display. Pilots found TAE helpful in establishing intercepts and the appropriate wind correction angle. Findings support the recent FAA TSO-C129 requirement that XTE be presented in the pilot's primary field of view, and the recommendation that avionics manufacturers include supplementary analog TAE display capability.

INTRODUCTION

Satellite based navigation systems and a new generation of microprocessor based cockpit avionics are revolutionizing air traffic control world wide. The FAA has established minimum performance and display airworthiness standards for stand alone GPS equipment (RTCA/DO-208 and TSO C-129), and has begun to publish a new class of GPS non-precision approaches. This initiative is particularly important for the general aviation (GA) community, since instrument approaches to thousands of new airports will eventually be possible. GPS RNAVs have flexible electronic displays, updatable databases, and many more operating modes than traditional VOR, DME, ILS, and ADF equipment. They will make instrument flying easier and safer, provided that the human factors aspects are properly considered at the design stage.

Most GA aircraft instruments are of the traditional "round dial" type, and panel space is limited. GPS receivers typically fit in a radio slot, often located outside of the pilot's primary scan. Due to severe space constraints, the GPS can have only a small front panel display and a limited set of control buttons and knobs. Since cross track error (XTE) information from the GPS functionally replaces that from VOR/ILS, XTE is typically converted to an analog signal, and displayed on a centrally located CDI or HSI, as shown in Fig. 1. Supplementary XTE information may also be shown on the receiver itself, as in Fig. 2.

Maintaining an aircraft on course centerline is a demanding triple integral manual control task, requiring a hierarchical, multiloop control strategy (Clement, et al, 1968). Pilots must monitor roll attitude and heading and partially base

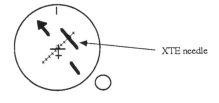

Fig. 1 (Top): Analog XTE on a HSI. (Bottom): Alphanumeric data on nearby GPS receiver display.

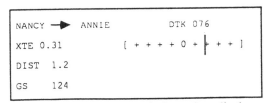

Fig. 2: Analog XTE CDI on GPS receiver display

their control on the rate of change of XTE to avoid large oscillations across the course centerline. The rate of movement of the XTE needle is often not easily perceived. On final approach, XTE needle sensitivity is deliberately increased to help the pilot see its movement and null small errors. However, the pilot must then more frequently scan all instruments, reducing the time available to perform other tasks. Inflight studies have demonstrated a direct relationship between CDI sensitivity and pilot workload and an inverse relationship with XTE during non-precision approaches (Huntley, et al, 1991).

GPS RNAV systems can compute "track angle error" (TAE), the difference between the desired track and the actual track (Fig. 3). TAE is mathematically proportional to the rate of change of XTE, the important manual control variable. In principle, if TAE is available, the pilot would not have to monitor the XTE needle as closely to judge its rate of movement. If the aircraft is significantly off course, TAE can be used to establish an appropriate intercept angle. The pilot reduces TAE to zero as the aircraft approaches centerline. Also, it is no longer necessary to use "cut-and-try" techniques to find the heading which eliminates needle "drift" due to crosswinds. By noting the heading at which TAE equals zero, the pilot can immediately establish the appropriate wind correction angle.

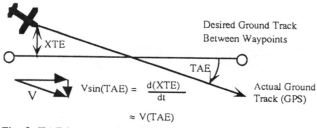

Fig. 3: TAE is proportional to the time derivative of XTE

The FAA's TSO C-129 standard for GPS RNAVs requires *at least* an alphanumeric display of TAE. The TSO suggests that the use of analog XTE and TAE data integrated "into one display may provide the optimum of situation and control information for the best overall tracking performance". However, the TSO does not specify the graphical format of these displays, in part because pilot performance with supplementary TAE information has never been formally investigated. The goal of this research was to begin this process via simulator experiments, evaluating formats appropriate for the small LCD or CRT displays typically found on the front of many GPS receivers and which conceivably could be implemented merely by a ROM software change. To what extent could such displays allow pilots to quantitatively improve their approach performance, or reduce their workload ? A complete report has been published elsewhere (Oman, et al, 1995).

METHODS

A total of five displays were investigated. The three supplementary TAE display formats evaluated were:

1&2) <u>Separate TAE and XTE sliding pointer displays</u> (2 versions). This format (Fig.4, top), originally suggested by FAA's G. Lyddane, added a TAE sliding pointer display beneath a conventional, "fly to" XTE CDI. The pointer was a triangle, located just beneath the XTE needle, and used the same "ten dot" scale. When the triangle was centered, TAE

was zero. Full scale pointer deflection was set at ± 90°, since this is the maximum useful course intercept angle. Which way should the pointer move ? If the triangle moves in the <u>same</u> direction in response as a stick roll command, it is easy to remember, and has the advantage that both the needle and the triangle appear on the <u>same</u> side of the display when converging with course. This display was therefore referred to as "Triangle/Same". However, it is a "fly from" display, whereas the XTE needle above it is "fly to", violating a well known human factors command-response consistency guideline. So a second version was also evaluated, with the opposite sign of the triangle movement. This version (Fig. 4, middle) was referred to as "Triangle/Opposite".

3) An <u>TAE/XTE integrated display</u>. In this format (Fig. 4, bottom), the sliding XTE needle was replaced with a sliding/rotating pointer, whose horizontal position was proportional to XTE as usual and whose tilt angle was equal to the TAE. The display resembled a "mail slot view" of a track-up moving map in an "inside-out", aircraft centered frame of reference, with the arrow corresponding to the desired track. As a result, the pointer tended to move horizontally in the direction it was tilted. This format was referred to as the "Track Vector" display.

Two <u>control conditions</u> were included in which no TAE information explicitly appeared in either numeric or analog format. These were an "HSI" format (Fig. 1) in which XTE was presented along with heading in the pilot's primary instrument scan, with only alphameric information appearing on the receiver, and an "XTE only" receiver display (Fig. 2) which presented analog XTE in the conventional 10 dot CDI format.

The HSI was 70 cm from the pilot's eye, and 9.5 cm beneath the attitude indicator. The GPS receiver display was created on a high resolution LCD display, located 35 cm (27 deg.) to the right of the HSI, in a 2 in. by 4 in. area subtending approximately 10° of horizontal visual angle. A consistent set of generic alphanumeric data was presented on all 5 displays: last and next waypoint, desired track (DTK), numeric XTE, groundspeed (GS), and distance (DIST) to waypoint. Numeric TAE was shown only on TAE displays. In all approaches, the pilot had to monitor DIST, and if a turn at the next waypoint was required, initiate a standard (3 deg/sec) turn at the appropriate point to intercept the next leg. Waypoints automatically sequenced when the aircraft crossed a line bisecting the angle between the inbound and outbound legs.

Experiments were performed in a fixed base Frasca 242 simulator, using a flight model resembling a Piper Aztec. Patchy, moderate-to-severe turbulence (Jansen, 1981) was added in the attitude axes, requiring the pilot to closely monitor the attitude indicator. A network of additional computers performed the GPS navigation calculations, created the displays, altitude dependent wind and collected data. Wind was always a 45° left or right head wind with respect to the final approach heading, but varied in strength using a power law atmospheric model. Pilots knew the wind direction varied, but were not told that only two relative wind directions were used. Since GPS system errors were not simulated, XTE was a direct measure of "Flight Technical Error" (FTE).

Six current, multiengine, instrument rated pilots (750-3387 hrs. experience), were recruited locally. Each pilot flew 20 approaches, four with each display format over two ten-approach sessions. Eight approach charts were used, each with a different final approach heading and altitudes. Half the approaches were based on a GPS "T" approach geometry. The pilot was required to intercept the initial approach leg, and fly five miles to an intermediate approach waypoint (IF) at the

Fig. 4: (Top): Triangle/Same; (Middle): Triangle/Opposite; (Bottom): Track Vector.

center of the "T", maintaining 3100 ft. above ground level (AGL). At the IF, the pilot turned 90° right or left, and flew five miles to the final approach fix (FAF). XTE display scale sensitivity changed from ±1 nm to ±0.3 nm at a waypoint 2 miles before the FAF. At the FAF, the pilot began descending to the 750 ft. minimum descent altitude (MDA), and flew five miles to the missed approach point (MAP) at the bottom of the T. At the MAP, the pilot then executed the missed approach procedure, and flew to the Missed Approach Holding Fix (MAHF). The remaining approaches used a more difficult "Crooked T" geometry in which the pilot was required to make a 45° turn at the FAF, and then fly a descending, two mile dogleg before turning back to the runway heading. Pilots rated overall workload on a modified Bedford workload scale (Roscoe and Ellis, 1990; Huntley, et al 1993) after each approach. At the end of the second session, pilots ranked the 5 displays on three preference scales: ease of interpretation (EOI), effect on flight path control accuracy (FPA), and overall preference (OP). Also, they indicated the strength of their preference in "head to head" (HTH) comparisons of pairs of displays on a ± 7 point analog scale. The scores from the 10 pairs were summed and ranked using a tournament method, yielding a second measure of overall preference.

Ground tracks from each approach were normalized by rotation to a common final approach heading, left/right reversal where appropriate, and then compared by subject and

display. The combined track records were used to retrospectively separate the approach into a series of 13 segments of varying lengths, chosen to isolate the various intercept, tracking, turning, descending and climbing phases of the approach. The mean, standard deviation, and RMS values of XTE, TAE, altitude, pitch and roll attitude along each segment were computed, and analyzed using Systat v.5.2. In addition, XTE was sampled at half mile intervals and averaged by display format. Mean and 95% limits of the XTE distribution were estimated based on the sample variance at each slice, using a method originally proposed by Huntley (1993). Differences in tracking performance, as measured by the variance of XTE for different pairs of displays were assessed based on their F ratio.

RESULTS

In debriefing evaluations, pilots said that, as we anticipated, they were able to choose an intercept TAE, and then reduce it in several steps as they approached centerline. While tracking along a leg, an offset of the triangle or a tilt of the vector allowed pilots to detect and anticipate the magnitude and direction of slow changes in XTE. It was possible to immediately determine the cross wind correction angle without a cut-and-try approach. They learned to distinguish the "diverging" and "converging" XTE/TAE pointer configurations at a glance and react appropriately. If the XTE indicator was off scale, an appropriate indication on the TAE pointer reassured the pilot that XTE would soon be on scale again. The track vector display could be visualized as a track up moving map, but some pilots occasionally had difficulty maintaining the map interpretation, particularly immediately after waypoint changeover following a 90° turn. There is reason to think performance with this format could be improved by making it appear more map-like and adding a vertical reference mark to permit vernier judgments of tilt.

As shown in Table 1, questionnaire results

Table 1. Pilot Display Preference Ranks by display, using 4 different scales (see text). Rank = 1 is best.

Display Format	Display Preference Scales			
	OP	HTH	FPA	EOI
Δ/Same	2	2	1	2
Δ/Opposite	3	4	2	5
Vector	4	3	4	3
HSI	1	1	3	1
XTE only	5	5	5	4

underscored the relative importance of having XTE in the primary instrument scan area, rather than alone on the GPS receiver. By a small numerical margin, pilots preferred the HSI display over the "triangle/same" TAE display on the OP and HTH scales. They preferred the triangle/same display over either of the other two TAE formats. All pilots always ranked one or more of the TAE display formats above the XTE only format on both scales, so the consensus was that TAE information was subjectively useful. (Four pilots preferred the "triangle/same" in direct comparison to the "triangle/opposite" format, and only one preferred the latter.) In terms of FPA, pilots preferred the triangle/same display, though the three

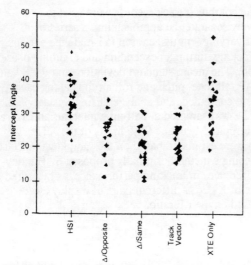

Fig. 5: Intercept Angle (degrees) vs. Display Type

subjects who actually tracked most consistently ranked the HSI first. Four of the five pilots said the HSI was the easiest of the displays to interpret, though three of the four cited long-standing training and experience with the HSI format as a reason for this. A significant effect of display was found for the HTH, FPA and EOI scale rank scores from the individual subjects (Friedman rank ANOVAs, $p < 0.04$). For the OP scale results, the display effect was at the $p < 0.06$ level. Four of the six pilots said they never referred to the numeric TAE information at all, since it was not easy to remember the meaning of the L/R TAE indication, and because the redundant analog TAE pointer was available. The L and R symbols were intended to indicate the direction of TAE.

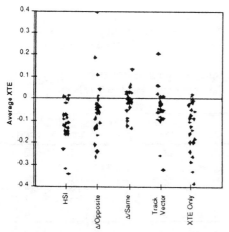

Fig. 6: Average XTE (nm) for miles 2 - 4 of Initial Approach Leg, vs. Display Type.

ANOVA of modified Bedford Workload scores revealed significant differences between subjects ($F(5,108)=29$; $p<0.0001$), and T and Crooked T approach types ($F(1,108)=28.4$; $p<0.0001$). However, adding display to the ANOVA did not reveal a significant effect. No trends were found by sequential approach or session number, suggesting training had asymptoted practice effects. Ranking workload scores within subjects did not reveal a display dependent effect. It was concluded that although the workload metric was demonstrably sensitive to approach geometry,

display effects, if they exist, were small compared to geometry and inter subject effects.

When using the TAE displays, pilots intercepted the initial leg of the approach and flew along it differently. We defined two intercept measures: 1) "Intercept Angle", the absolute value of TAE when XTE equaled 0.3 nm and 2) "Intercept Distance", the distance where the aircraft first crossed within 0.05 nm of the initial approach centerline. Intercept Angle (Fig.5) and Distance were consistently steeper and shorter, respectively, for approaches made with the non-TAE displays. The average increase in intercept angle with non-TAE formats approximated the drift angle to be expected (12 deg.) for pilots unaware of the crosswind. ANOVA of intercept angle data showed significant effects by display ($F(4,85) = 24.1$, $p < 0.0001$), subject ($F(5,85) = 4.6$, $p < 0.001$), and subject by display ($F(20,85) = 1.7$, $p < 0.04$). The significant subject by display interaction effects indicate that some subjects used the TAE information differently than others. Contrasts comparing TAE and non-TAE intercept angle and distance results by display were both significant ($p < 0.0001$).

Once established on the initial leg, pilots flying with the triangle/same and vector displays had smaller downwind biases, as shown in Fig. 6. The difference in average tracking bias was significant by display ($F(4,90) = 7.8$, $p < 0.001$) and subject. ($F(5,90) = 6.4$; $p < 0.001$). The subject by display interaction was not significant.

In terms of tracking performance during the initial, intermediate, and second missed approach, the HSI or Triangle/same displays generally ranked best, followed by the track vector display. Average display effects were smaller than intersubject effects. Inter subject differences in XTE and TAE tracking performance correlated with inner loop attitude control for the corresponding segments.

The 95% XTE envelopes were calculated for all approaches and compared by display. An example is shown in Fig. 7. TAE display envelope widths were significantly different during portions of the approach. On the initial approach leg, F ratio tests showed the triangle/same display XTE envelope was significantly narrower than for all other displays, and the centers of the initial approach envelopes for the HSI and XTE only formats were clearly displaced downwind, as compared to all three TAE based displays. During and immediately after the 90° IF turn, pilots using the HSI had significantly narrower XTE envelopes than with the other displays, by F ratio test. The track vector display envelope remained wide between the IF and the sensitivity changeover waypoint. Thereafter, XTE rapidly converged.

On T geometry final approaches, the track vector display envelope was narrower than for any of the other displays. The differences disappeared at the MAP itself, perhaps because the pilots had shifted their attention to missed approach activities. Average 95% XTE envelope width during the last three miles of final approach provides a useful metric for T approach performance comparisons. Average envelope widths were: HSI: 0.17 nm, Triangle/opposite: 0.22 nm, Triangle/same: 0.21 nm, Track vector: 0.15 nm, and XTE only: 0.23 nm. Comparing the track vector result with that of XTE only, deletion of TAE vector information from the receiver display resulted in a 53% increase in the average envelope width.

On the Crooked T approaches, the envelopes consistently widened after the 45° dogleg turns. However, re-intercept performance and short final tracking was better with the track vector display. The track vector was the only one of the five displays for which the CDI is predicted to remain on scale through both turns in 95% of approaches flown. Next

best performance was with the triangle/same display. The relatively poor performance with XTE only and HSI displays after 45° turns using high CDI sensitivity suggests clear advantages for track vector and triangle/same displays when circumstances compel pilots to maneuver during the critical final stages of an instrument approach.

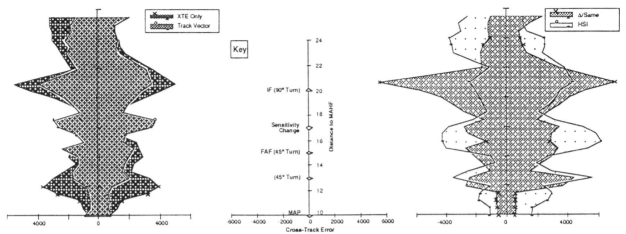

Fig. 7: Estimated 95% limits of XTE distribution for all approaches combined. (Left): XTE only vs. Track Vector. (Right): Triangle/Same vs. HSI. (Middle): Key. Ordinates are distance from MAHF along the desired track. Circles and crosses denote slices were an F ratio test indicated a significant difference.

CONCLUSIONS

Results showed that under turbulence conditions requiring diligent attitude instrument scan, addition of analog TAE information to the receiver XTE display significantly improved approach intercept and tracking performance, probably by allowing the pilot to predict XTE changes and create outer loop control lead. Determination of wind correction angle was simplified. Pilots elected to use analog rather than numeric TAE. The "triangle/same" TAE display produced the largest initial leg intercept and tracking improvement, and was preferred overall for flight path control. The "track vector" display produced the greatest reduction (35%) in XTE envelope width during the last three miles of T approaches, as compared to the XTE only display. Control tests with a HSI showed the improvement due to addition of TAE to the receiver display is offset by the need to widen the pilot's instrument scan. It is likely that better performance be obtained by simultaneously displaying XTE on the HSI and XTE/TAE information on the GPS receiver. Although Bedford workload scores were sensitive to approach geometry, no consistent effect of display format on workload was found. Pilots may have chosen to keep workload constant, and allowed performance to vary instead. Overall, our findings support the FAA TSO-C129 recommendation that manufacturers provide analog TAE display capability and the requirement that analog XTE appear in the primary field of view. An experiment with 12 subjects comparing numeric TAE, analog TAE and XTE predictor displays is underway.

ACKNOWLEDGMENTS

We thank G. Lyddane, R. Disario, D. Hannon, T. Carpenter-Smith, J. Bastow, S. Robinson, F. Sheelen, J. Giurleo, and J. Turner, and also our subjects. The human use protocol was approved by committees at MIT and DOT/Volpe. Supported by MIT CTS Contract DTRS-57-92-C-0054 TTD#27A and C.

REFERENCES

Clement, W.F., Jex, H.R., Graham, D. (1968). A Manual Control-Display Theory Applied to Instrument Landings of a Jet Transport. IEEE Trans. Man Machine Sys. 9:4 93-110

FAA (1994) Technical Standard Order C129: Airborne Supplemental Navigation Equipment Using the GPS. FAA, Washington, DC.

Etkin, B. (1980) The turbulent wind and its effect on flight. AIAA Aircraft Systems Meeting, Anaheim, CA, AIAA-80-1836.

Huntley, M.S. Jr. (1993) Flight Technical Error for Category B Non-Precision Approaches and Missed Approaches Using Non-Differential GPS for Course Guidance. Report DOT-VNTSC-FAA-93-17

Jansen, C.J. (1981) Non-Gaussian Atmospheric Turbulence Model for Flight Simulator Research. AIAA J. Aircraft 19 (1981): 374-379.

Oman, C.M., Huntley, M.S., Rasmussen, S.A.M., and Robinson S.K. (1995). The use of analog track angle error displays for improving simulated GPS approach performance. DOT/FAA report AR-95/104 US Department of Transportation, Federal Aviation Administration.

Roscoe, A.H. and Ellis, G.A., (1990) A subjective rating scale for assessing pilot workload in flight. Tech. Report TR90019, RAE Farnborough, UK.

RTCA Special Committee 15. (1991) Minimum Operational Performance Standards For Airborne Supplemental Navigation Equipment Using Global Positioning System, DO-208 July, 1991, 1140 Connecticut Avenue, N.W., Washington, DC.

EFFECTS OF VIRTUALLY-AUGMENTED FIGHTER COCKPIT DISPLAYS
ON PILOT PERFORMANCE, WORKLOAD, AND SITUATION AWARENESS

Lawrence J. Hettinger, Bart J. Brickman, Merry M. Roe, W. Todd Nelson
Logicon Technical Services, Inc.
Dayton, Ohio

Michael W. Haas
Armstrong Laboratory
Wright-Patterson Air Force Base, Ohio

Virtually-augmented display concepts are being developed at the US Air Force Armstrong Laboratory's Synthesized Immersion Research Environment (SIRE) Facility at Wright-Patterson Air Force Base, Ohio, for use in future USAF crew stations. These displays incorporate aspects of virtual environment technology to provide users with intuitive, multisensory representations of operationally relevant information. This paper describes an evaluation that was recently conducted to contrast the effects of conventional, F-15 types of cockpit displays and virtually-augmented, multisensory cockpit displays on pilot-aircraft system performance, workload, and situation awareness in a simulated air combat task. Eighteen military pilots from the United States, France, and Great Britain served as test pilots. The results indicate a statistically significant advantage for the virtually-augmented cockpit configuration across all three classes of measures investigated. The results are discussed in terms of their relevance for the continuing evolution of advanced crew station design.

INTRODUCTION

The introduction of virtual environment technology into the cockpit of future fighter aircraft and other airborne crew stations offers a number of potentially significant performance benefits (Furness, 1986; Haas & Hettinger, 1993; Hettinger, Nelson, Brickman, Haas, & Roumes, 1995). This technology is predicated on the presentation of information in intuitive, readily comprehensible formats using multisensory displays. Our intent is to develop interfaces that take full advantage of the normal human capability to extract meaningful information from the environment using multiple sensory modalities. In so doing, it offers the potential to significantly enhance an individual's awareness of the nature and status of events and processes that impact on task performance.

In the US Air Force Armstrong Laboratory's Synthesized Immersion Research Environment (SIRE) Facility at Wright-Patterson Air Force Base, Ohio, we are engaged in the development of "virtually-augmented" displays for use in future USAF crewstations. These displays use aspects of virtual environment technology, such as localized, 3-D auditory displays and direct pictorial visual displays, to supplement information obtained directly from the environment. Parallel programs of engineering development and human performance evaluation are being

pursued to converge on designs that optimize the impact of virtual environment technology on the fighter pilot's mission.

As part of this development effort, a human performance evaluation was recently conducted to contrast pilot performance in a simulated air combat task using two alternative cockpit displays. One cockpit configuration employed an array of displays characteristic of those found in a standard F-15, while the other employed a variety of virtually-augmented displays designed to enhance pilot performance and situation awareness, while maintaining workload at acceptable levels. Military pilots from the United States, France, and Great Britain participated as test pilots in the evaluation. Results from the experiment will be used to help guide the design of a combined US-French Super Cockpit as part of an ongoing Memorandum of Understanding between the two nations.

METHOD

Subjects

Eighteen male military pilots (8 from the United States, 6 from France, and 4 from Great Britain) served as subjects. Their ages ranged from 30 to 48 years with mean age of 36.8 years. Eleven pilots had fighter aircraft experience including F-4, F-15, Jaguar, Mirage, Rafale, and Tornado. Fighter aircraft hours ranged from 500 to 2,000 with a mean of 1353.2 hours. Overall flight hours for the entire group ranged from 1720 to 4808 with a mean of 2980.4 hours. All subjects had normal or corrected-to-normal vision, and were not paid for their participation.

Apparatus

The experiment was conducted using the Fusion Interfaces for Tactical Environments (FITE) flight simulator located within the SIRE Facility. The cockpit used in the present experiment was a fixed-base simulator that consisted of an F-16 fiberglass body, head-down displays, and F-16C throttle, and a sidestick controller. The cockpit was housed in a cubic projection room that measured 8' x 8' x 8'. All displays and controls were driven by a network of 23, 486-33 MHz microcomputers. As described above the conventional cockpit consisted of an array of standard F-15 instruments, while the modified cockpit consisted of an array of virtually-augmented, multisensory displays (visual, localized audio, and non-localized audio). All displays were presented on an array of six rectangular liquid crystals displays (LCD's). Each LCD measured 12 cm x 9 cm and had a 1280 x 1024 pixel resolution. Non-localized and localized audio signals were presented using helmet-mounted speakers.

The out-the-window display, which included building, clouds, ground patterns, and other aircraft, was projected onto the surface of the cubic projection room by six, black and white limelight projectors. This arrangement provided subjects with a 240 degree (horizontal) by 120 degree (vertical) field-of-view.

Procedure

Each of the 18 pilots received a minimum of eight hours training with the use of the conventional and virtually-augmented cockpits prior to data collection. Each of the eight US pilots participated in 18 simulated air intercept scenarios, while each of the ten French and British pilots participated in 12 simulated scenarios. The task required pilots to fly an air-to-air combat scenario through hostile air space. The pilot's task was to locate and destroy four enemy bombers (simulated models flying a variety of preprogrammed routes) and return to safe air space. The mission also included two hostile fighters, operated in real time by two of the other pilots in a particular group, whose task was to eliminate the experimental pilot and protect the bombers. During the experiment, each pilot flew either 18 or 12 simulated trials using the primary cockpit. Half of the trials used the conventional F-15 types of displays, and the other half used the virtually-augmented displays. The flight scenario also included a friendly F-15 fighter that flew a pre-determined

route, but did not deploy weapons at either the bomber or the hostile fighters.

The mission began with the pilot flying a prescribed Combat Air Patrol (CAP) phase and continued through the intercept, weapons employment, and egress phases. The mission was terminated when the primary pilot was either shot down or out of fuel, or when the primary pilot shot down all the bombers and returned to safe air space. A mission "win" was scored when the primary cockpit had shot down all the bombers and returned to safe airspace. Any other mission outcome was considered to be a "loss."

Flight parameter data from all three active cockpits in the experiment, including altitude, airspeed, angle of attack, and weapons status and deployment were recorded in real-time and stored on a separate computer. All aircraft were equipped with medium-range missiles, short-range missiles, and guns.

At the conclusion of the experiment, each pilot completed a Subjective Workload Dominance (SWORD) scale (Vidulich, Ward, & Scheren, 1991) and a 13 Dimension Cognitive Compatibility Situation Awareness Rating Techniques (13D CC-SART) scale (Taylor 1995). SWORD is a subjective scale that was used to elicit judgments about the workload experienced by cockpit type and mission phase. The 13D CC-SART is a multidimensional, rating scale that was used to compare the levels of situation awareness afforded by the two cockpits and within each phase.

Design

Two cockpit interface designs (conventional and virtually-augmented) were combined factorially with three threat aircraft initial altitudes (low, medium, and high) to provide six unique experimental conditions. Each subject participated in all experimental conditions over a five day period. The order of presentation of experimental conditions was counterbalanced to guard against order effects.

RESULTS

Three classes of dependent measures were collected: Air combat performance measures, the workload associated with performance of the task, and situation awareness during performance of the task. The following discussion presents illustrative results from all three classes of measures.

Air-to-Air Combat Performance Metrics

Mission Outcome. A mission was defined as a win if the primary cockpit intercepted and destroyed all four of the enemy bombers and returned to safe airspace. A mission was defined as a loss if the primary cockpit was shot down by either of the two threat fighters or ran out of fuel before reaching safe airspace.   Mission outcome data are presented in Table 1. A Chi-Square test for independence was performed on the data. This analysis revealed a statistically significant difference between the virtually-augmented and conventional cockpit configurations in mission outcome, X^2 (1, n=263) = 5.42, p < .05. An examination of Table 1 indicates that the virtually-augmented cockpit configuration resulted in Wins on 28 out of 132 trials (21.2%), while the performance in the conventional cockpit was significantly worse, with only 14 Wins out of 131 missions, or 10.7%.

Table 1
Mission Outcome as a Function of Cockpit Configuration

Cockpit Configuration	Outcome N (%)	
	Win	Loss
Virtually-Augmented	28 (21.2)	104 (78.8)
Conventional	14 (10.7)	117 (89.3)

Ground Strikes. The frequency with which the primary cockpit suffered a catastrophic ground strike is presented in Table 2. An examination of the Table reveals that only 2 out of 132 missions (1.5 %) flown with the virtually-augmented cockpit ended with a crash. Comparatively, 13 out of 132 missions (9.85 %) flown with the conventional cockpit resulted in ground collisions.

Table 2
Number of Ground Collisions as a Function
of Cockpit Configuration

Cockpit Configuration	Frequency of Crashes N (%)
Virtually-Augmented	2 (1.5)
Conventional	13 (9.85)

Workload Metrics

Subjective Workload Dominance Technique (SWORD). The SWORD rating procedure was conducted at the conclusion of each subject's final experimental session. A 2 (cockpit configuration) x 4 (mission phase) repeated measures analysis of variance procedure was performed on these data. The analysis yielded a statistically significant main effect for Cockpit Configuration, F (1,17) = 9.138, p < .008. An examination of the means revealed that workload was rated significantly higher for the conventional cockpit (.1512) than the virtually-augmented cockpit (.0988). The analysis of variance procedure also yielded a statistically

significant main effect for Mission Phase, F (3,51) = 30.014, p < .001. The mean workload ratings for each mission phase may be found in Table 3. An examination of the means in Table 3 reveals that mean workload ratings increased over the course of a mission, from the lowest rating of .0606 for the CAP phase, to .1228 for the Intercept phase, to the highest mean rating of .2381 for the Weapons Employment phase. Finally, the workload ratings fell during the Egress phase, with a mean workload rating of .0785.

Table 3
SWORD Workload Ratings x Mission
Phase

Mission Phase	Mean SWORD Rating
CAP	.0606
Intercept	.1228
Weapons Employment	2381
Egress	.0785

Situation Awareness Metrics

13-Dimension Cognitive Compatibility-Situation Awareness Rating Technique. The 13-D CC-SART rating procedure was also completed at the conclusion of each subject's final experimental session. This technique allowed the determination of a subjective rating of situation awareness for each cockpit configuration, within each mission phase, on 13 dimensions that are thought to impact upon a pilot's situation awareness (e.g., for example, intuitiveness of the interface). The data were subjected to a 2 (cockpit configuration) x 4 (mission phase) x 13 (rating dimension) repeated measures analysis of variance procedure. The analysis revealed a statistically significant main effect for Cockpit Configuration, F (1,17) = 7.514, p < .014. An examination of the means revealed that pilots rated SA as significantly higher for the virtually-augmented cockpit with a mean rating of 5.235 than for the conventional cockpit, with a mean rating of 4.727. The analysis of variance also yielded a statistically significant main effect for Mission Phase, F (3, 51) = 13.177, p < .001. The mean SA ratings for each mission phase may be found in Table 4. An examination of the table reveals that the pilots rated their situation awareness as 5.25 for the CAP phase, 4.74 for the intercept phase, 4.52 for the weapons employment phase, and 5.40 for the egress phase. This result is intuitive in that the subjects rated their SA higher for the CAP and the Egress phases, when there was somewhat less activity for them to consider, and lower for the combat phases (Intercept and Weapons employment) when a multitude of simultaneous events demand attention. A post-hoc analysis revealed that the SA ratings for the CAP and Egress phase were both statistically significantly different than the intercept phase and weapons employment phase, p <

.01, but were not significantly different from each other. Furthermore, the post-hoc test revealed that the Intercept and the Weapons phases were not significantly different from each other.

Table 4
13 D CC-SART Situation Awareness Ratings by Mission Phase

Mission Phase	Mean SA Rating
CAP	5.25
Intercept	4.74
Weapons Employment	4.52
Egress	5.40

DISCUSSION

The results of this experimental evaluation indicate significant advantages across several classes of dependent variables for the virtually-augmented crewstation over the conventional crewstation. Specifically, these advantages were found for system performance data, situation awareness measures, workload measures, and the debriefing questionnaire. The virtually-augmented crewstation yielded a greater number of Wins, fewer groundstrikes, and longer life expectancies during the missions than the conventional crewstation. In addition, the 13-Dimension Cognitive Compatibility Situation Awareness Rating Technique revealed significantly higher SA ratings for the virtually-augmented crewstation over the conventional cockpit. Furthermore, while the SA and performance measures were higher, the subjective workload associated with the mission, as measured by the Subjective Workload Dominance Technique (SWORD), was found to be significantly lower for the Virtually-augmented crewstation. Finally, the debrief questionnaire revealed highly favorable subjective impressions for the virtually-augmented crewstation. Specifically, the questionnaire yielded 19 statistically significant and 18 non-significant advantages for the virtually-augmented crewstation and the qualitative questionnaire data showed highly favorable subjective impressions of the virtually-augmented crewstation.

It is important to note that each of these statistically significant advantages for the virtually-augmented crewstation were obtained despite the absence of previous flying experience. Comparatively, the pilots had thousands of hours of flight experience using the conventional crewstation. Among the most favored aspects of the virtually-augmented crewstation was the presence of the Ground-Collision Avoidance System (G-CAS). In addition, the pictorial format display and the color coding of friendly and enemy aircraft received highly favorable ratings. Finally, the use of three dimensional audio cueing for the radar warning receiver was also highly accepted.

Future work in the SIRE Facility will continue to focus on the development of new virtually-augmented, multisensory display concepts as well as refinement of existing prototype displays. In addition, the emphasis on multisensory displays will be combined with adaptive interface logic and alternative control technologies (such as brain-actuated control) to develop the crewstation of the future.

References

Furness, T.A. (1986). The super cockpit and its human factors challenges. In *Proceedings of the Human Factors Society 30th Annual Meeting* (pp. 48-52). Santa Monica, CA: Human Factors Society.

Haas, M.W. & Hettinger, L.J. (1993). Applying virtual reality technology to cockpits of future fighter aircraft. *Virtual Reality Systems*, 1(2), 18-26.

Hettinger, L.J., Nelson, W.T., Brickman, B.J., Haas, M.W., & Roumes, C. (1995). Assessing human performance as a design aid for airborne applications of virtual environment technology. In *Proceedings of the Eighth International Symposium on Aviation Psychology* (pp. 170-175). Columbus, OH: The Ohio State University.

Taylor, R.M. (1995). CC-SART: The development of an experimental measure of cognitive compatibility in system design. *Report to TTCP UTP-7 Human Factors in Aircraft Environments, Annual Meeting, DCIEM*, Toronto, Canada, 12-16 June.

Vidulich, M. A., Ward, F. G., Scheren, J. (1991). Using the subjective workload dominance (SWORD) technique for projective workload assessment. *Human Factors*, 33(6), 677-691.

PERSONAL MINIMUMS FOR AVIATOR RISK MANAGEMENT

Richard S. Jensen and James Guilkey
The Ohio State University
Columbus, Ohio

David R. Hunter
Federal Aviation Administration
Washington, DC

This paper discusses the results of the development of a personal minimums pilot training program to help general aviation pilots manage risks in their decisions about takeoff. Result of an initial field testing exercise in various locations around the USA are also presented. The key safety point to the training is the commitment of pilots to their self-developed minimums for flight which are expected to be more conservative than those established by the FAA. The field tests revealed high acceptance of the concept and the engaged learning method used in the presentation.

INTRODUCTION

Risk management is an important part of the over all task of flying airplanes. All pilots recognize that there are risks in their flying activities and that some activities are riskier than others. Many pilots will not perform certain kinds of flying activities under certain circumstances but may do those same flying activities under other circumstances. For example, some pilots will fly a single-engine aircraft at night over mountains with a student in training who needs such experience but may be reluctant to do so with their own family (unless they have an important business meeting the next day at their destination). Although most pilots have not established guidelines for these types of decisions, these are, in fact, risk management decisions.

All of us can be pressured into taking risks by our circumstances. What factors cause you to take risks that you might not otherwise take? When you are in a long line of cars at a busy intersection, how long does the gap in traffic need to be before you will risk turning left across and into the traffic? Does the risk gap get shorter as you wait longer? If the driver behind you is honking? If you are in a hurry to catch a plane? If you have impatient passengers? We take risks in flight for many very legitimate reasons (e.g., transportation, convenience, economics). Sometimes, we make risky choices based on less legitimate reasons as well. After a close call, many people rationalize, "I had no choice, I had to do it." Safety in our aviation

system depends, to a great extent, upon the amount of control we exercise over our choices to take risks.

In well established flying organizations such as airlines and corporations some of the risk management is governed by the establishment of clearly defined limits or minimums (Standard Operating Procedures or SOPs) for flying activity. On the other hand, most general aviation pilots do not have a large organization that sets down the limits of their flying choices. Risk management for these pilots is left almost entirely to their discretion under the scrutiny of the FAA. For example, most Category I ILS approaches have posted weather minima of 200 feet cloud height and 1/2 mile visibility for execution. Nevertheless, in setting these limits, the FAA still places most of the responsibility for risk management in the hands of the individual pilot. The FAA limits are based on the capabilities of ideal pilots and equipment because, in addition to regulating for safety, they are tasked to promote aviation. Before each flight the individual pilot must determine whether his or her currency and aircraft meets this ideal level of skill and capability. If ideal levels are not met, pilot judgment is expected to produce more conservative operational minimums for that situation.

The importance of decisions made prior to flight is underscored by a study by McElhatton & Drew (1993) showing that many airline pilot decisions (or lack of decisions) made during the preflight phase lead directly to incidents much later in flight. Of the

125 Aviation Safety Reporting System (ASRS) incident reports reviewed in their study, ninety percent of all time-related human errors occurred in the preflight or taxi-out phase of operation (e.g., the KLM crew who felt pressure to takeoff before their "duty time" expired and crashed into another aircraft on a runway in Tenerife in 1977).

Incident and accident data such as these are not surprising. A flight is a sequence of events that are related and influenced by prior events. Once the commitment to flight is made, a strong psychological force exists in the pilot's mind to continue to the intended destination. Also, social or peer pressure may make admitting an error and turning around difficult or seemingly impossible. Furthermore, once airborne, certain options disappear, including the option to add more fuel. Pilots need to recognize the extent of the forces that will be applied to them once they make the decision to fly and factor them into their preflight go/no-go decision.

Preflight Personal Minimums Training

A training program to teach pilots to develop and use personal minimums was developed consisting of four major parts: 1) an introduction, 2) identification of risk factors, 3) establishment of personal minimums, and 4) a commitment to use the guidelines. The program is designed to be completed within the 90-minute time frame of most FAA seminars.

Because of the nature of the training, the program is best conducted by a facilitator who interacts with the students in an engaged learning format. Participants grapple with case studies in small groups and with the instructor. In the case studies, students are encouraged to place themselves in the situation being studied as if they were the pilot involved as they identify risk factors that they would face. To develop a commitment using cognitive dissonance (Festinger, 1957), students are encouraged to say to others what their decision would be when faced with the given risk factors. Strong emphasis is placed on social, economic, and organizational pressures as risk factors. They then develop their own personal minimums to counter these risk factors and write them in a checklist form. The form is signed as a commitment to the minimums identified.

Introduction to preflight decision making. In the first part of the session students are given the basic

concepts and motivated to learn and establish their own personal minimums through an examination of why preflight decisions are so important to flight safety. Preflight decisions are defined as: "Any and all decisions made prior to taxiing the airplane onto a runway with the intent to take off." Under this broad definition, preflight starts with the first idea to make a flight until the pilot initiates the first action to take the aircraft into the air or until that possibility no longer exists. This definition includes normal preflight decisions, but emphasizes the pilot's complete situation, including all of the assessments and commitments pilots make.

Personal minimums are defined as an individualized set of decision criteria (standards) to which the pilot is committed to use in preflight decisions. Because each pilot manifests a unique set of skills and personality, he or she is the best judge of the risks involved in the contemplated flight. Psychological theory suggests that, when they know they are not being watched, people are much more likely to follow standards that they have made themselves than those imposed upon them by someone else. Therefore, the emphasis in this program is placed on "personal" minimums.

In contrast with imposed minimums (organizationally or FAA established limits or SOPs) personal minimums are self-generated. We believe that this difference establishes the key to the commitment of each pilot to follow the guidelines. If the pilot has determined and established his or her own minimums, there is a strong likelihood that they will be followed when faced with pressures to the contrary. It is understood that, in most cases, personal minimums are (and will be) more conservative than the FARs. The FAA, aircraft passengers, the legal system, and the general public expect pilots to use more stringent minimums than those mandated by the FARs.

Flying is not unique as an enterprise of trust. In automobile driving, the actions of the driver are not observed by the police at every corner. Instead, the safety of the driver mostly depends upon voluntary compliance with the rules of the road and setting some personal standards depending on the situation. While some pilot operations take place under the direct supervision of regulators, management, supervisors, or others in authority, most flights in general aviation, remain unsupervised and are seldom observed.

In business aviation, commercial flight operations, flying clubs, and other situations, operational limitations or sets of rules define operational minimums. In these operations, pilots with less experience must operate with higher ceilings, higher approach minimums, lighter winds, and lower crosswind components. In airline and military operations, an extensive set of SOPs provides tools to assist pilots in making critical preflight go/no-go decisions. Flights may be canceled by dispatchers, chief pilots, or supervisory personnel, even before the pilot has an opportunity to address the go/no-go decision. Such supervision has proven to be effective in improving safety in these organizations. At least in part, this may be the reason for the very good safety record of these organizations.

How can we use the experience of these highly supervised operations to improve safety in general aviation? Application of personal minimums offers an approach to improving risk assessment and management by self-supervision. Personal assignment of minimums requires an awareness of the risk factors involved before one can attempt to manage them. Furthermore, because some pilots may think that their responsibility ends with understanding their own safety risk, personal minimums training teaches them to acknowledge and consider the level of risk that is acceptable to passengers, the company, or others potentially affected by any flight consequences.

Risk factors. The second part of the training program is an individual or group activity in which the students are required to generate and classify a list of preflight risk factors into an organized taxonomy. Research has shown that when people make decisions, they usually consider only a very small number of factors. The primary purpose of this activity is to expand the pilot's knowledge base of risk factors that go into preflight decision making and to classify them into categories that are easily recalled when they are needed. Risk awareness is an effective tool to increase safety, but requires experience and knowledge.

To stimulate thinking about risk factors and how they apply to preflight decisions the instructor can offer one or more case studies in the form of "trigger tapes" of situations developed in the FAA's "Back to Basics" program or stories found elsewhere. One excellent example was published in *Flying Magazine's,* "I learned about Flying From That" (McCutcheon, 1991). In this case a pilot feels pressure to make a medical evacuation flight in a Cessna 210 with inoperative radios, in questionable weather with night approaching. After arriving for the pick up, he finds that there is both child and a mental patient to transport plus families and other cargo causing the plane to be overloaded as well. Despite the child's condition, because of the numerous risk factors, he decides not to takeoff.

As students read this story, they are asked to identify the risk factors facing the pilot including the subtle psychological factors such as the condition of the child. When all have listed these factors individually, the class is opened for discussion and all are invited to share their discoveries. The same could be done with any aviation preflight scenario including those from individuals in the class either in front of the whole class or in small groups.

After listing risk factors, the students are to organize them into a list with six suggested categories: pilot, aircraft, environment, operation/mission, organization/social, and miscellaneous. The first three categories are those normally used to represent the pilot's world. The operation/mission category is added as a place to put factors regarding personal pressures to complete a mission. The organization/social category is included as a place to put risk factors that organizations can add to the pilot's decision process including subtle pressures to complete flights on schedule. Finally, the miscellaneous category is included to underscore the emphasis on personal freedom in the construction of this tool in anticipation that some may not wish to identify a particular risk factor with one of the given categories. The structure presented is offered as a starting point, not a required set of categories. What is placed under any category will be a function of each pilot's mental model of how these factors are classified.

Personal minimums. The final section of the training is the establishment of personal minimums to fit many different circumstances in which the individual pilot may be found and writing these on a checklist form provided. These could take the form of simple acronyms, rules of thumb, sayings, if-then statements, forms, or memory aids representing what the pilot intends to do about the risk factors developed in the previous exercise. Some examples are provided to assist in getting started. The checklist has a place to sign indicating that, on this date, the pilot committed to the minimums shown.

Evaluation. The final exercise consists of a review and evaluation of items included in the personal minimums checklist. Pilots are encouraged to have peers review and evaluate completed guidelines and to sign a statement of commitment to use them. This exercise can be highly rewarding because it offers opportunities for verbal expression and public commitment. However, the instructor/facilitator must be very sensitive to students who may not wish to share their ideas; some might consider them too personal to share. The activity includes presentation of suggested evaluation criteria and evaluation of a sample personal minimums checklist.

In closing, the instructor reminds each pilot that the personal minimums checklist is unique to her or him. Other pilots' minimums will be different. The set of risk factors and checklist provided is never complete. They must be reviewed continuously and changed as required to accommodate the changing pilot activities and capabilities. To provide each pilot with further sources of information, they are referred to advisory circulars, magazines, books, newsletter (IFR Refresher, FAA Safety Review, etc.), and standard pilot references (Airman's Information Manual, FAR's, etc.).

FIELD TESTING

Testing was first done by presenting the live program as described above to OSU students and to pilot groups in the Columbus local area including the Columbus FSDO, who organized an audience of Part 135 operators, the Wright-Patterson AFB flying club in Dayton, a chapter meeting of a local Civil Air Patrol, a local chapter meeting of Glassair builders and pilots, and a local flying club (Central Ohio Flyer's Association). The program was then taken on the road to the EAA Convention at Oshkosh and to FAA FSDOs in Baltimore-Washington, Anchorage, Long Beach, and Chicago. Finally, it was presented at the FAA's Hanger 6 safety meeting for their own pilots. In each case changes were made to the program, the student material, and the visual material. The final program reflects the ideas and responses from participants in these seminars.

Data Collection and Analysis

At each seminar presentation, participants were asked to fill out an evaluation sheet, shown in Appendix C. The questions were designed to gather data on specific questions regarding how well the participants understood the concepts as well as how they felt about the program itself. We wanted to know if participants thought that the program would be well received by the general aviation community and whether they would invite their friends to the program at a future date.

Although attempts were made to ease the burden of filling out the evaluation form, it still appeared to be a tedious task for some participants following each seminar. Some participants did not do the task. However, 187 responses were received from the various field programs which provided useful data on the program.

The background of the participants covered the spectrum of the general aviation community. Of the 187 participants, 165 indicated that they had logged flight time. The average flight time among the participants who had logged flight time was 2,237 hours, 20 had between 5,000 and 10,000 hours, 5 had between 10,000 and 20,000 hours, and two had over 20,000 hours. Table 1 presents the number of seminar participants who indicated that they hold various pilot certificates and ratings.

The analysis of the data consisted of tabulating the responses to each of the questions that had a rating form and visual inspection of all of the responses that were provided in written form. These data are presented in Appendices D and E respectively.

RESULTS

The evaluation form used measured the participant's beliefs about what they learned, their perceptions, and their attitudes following the presentation. We cannot tell how much they actually learned and understood or how much their behavior may have changed as a result of participating in the program. However, from the evaluation forms it is clear that the Personal Minimums program is interesting, thought provoking, and valuable to audiences of pilots. Many comments were received indicating that it should be encouraged as a safety program by the FAA.

Question 1 was, "To what degree will this program be helpful to you? As shown in Figure 1, nearly 90 percent said that it was helpful or extremely helpful.

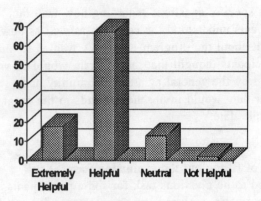

Figure 1. Responses to question 1, "To what degree will this program be helpful to you?"

Question 2 was, "Will you use Personal Minimums to assist in your pre-takeoff decisions in the future?" Nearly 90 percent of the participants indicated that they will use personal minimums. Written responses of some participants indicated that they already use personal minimums, perhaps casting some doubt on the validity of these response.

Question 3 was, "Would you recommend this seminar to other aviators?" Figure 2 shows participant responses to this question. One can see that over 90 percent of the participants would recommend the program to other pilots.

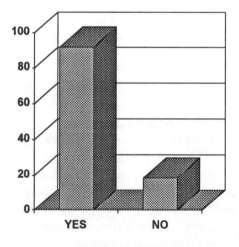

Figure 2. Participant responses to question, "Would you recommend this seminar to other aviators?"

DISCUSSION

Closing the gap between the criteria provided by FAA regulations and what society expects is crucial to safety in all of aviation. The FAA expects this gap to be closed through pilot judgment. In the expertise literature, this is the difference between the competent and the expert (Dreyfus and Dreyfus, 1986). The airlines and most corporate aviation programs have closed that gap by strict standard operating procedures and training programs. General aviation pilots must develop their own standards of judgment to close this gap. Although only a small part of the results are shown, it was clear to the investigators that this new, well accepted program has the potential to make general aviation safer. It remains to be seen whether the training will actually change pilot behavior in a positive direction.

Judgment cannot be taught by conventional methods. Students must be coaxed into teaching each other through sharing situations that they have encountered. The bottom line is that behavior must be changed. Crucial to changing behavior is to get students to commit to safe decisions before they face them. Telling someone in a seminar environment what they will do in those situations will later create a "cognitive dissonance" is in the pilot when faced with the temptation to depart from that position (Festinger, 1957). It is only by using such a powerful psychological force that one can counter the very strong pressures in the pilot's mind such as "get-there-itis".

ACKNOWLEDGEMENT

The research reported here is supported by the FAA Office of Aviation Medicine, Dr. David Hunter is the Contract Technical Monitor.

REFERENCES

Dreyfus, H.L. and Dreyfus, S.E. (1986). *Mind over machine: The power of human intuition and expertise in the era of the computer*. New York: The Free Press.

Festinger, L. (1957). *A Theory of Cognitive dissonance*. Evanstan, IL: Row, Peterson.

McCutcheon, D. (1991). I Learned About Flying From That: Better Late Than Never. *Flying*, May 1991, pages 118-119.

McElhatton, J. & Drew, C. (1993). Time pressure as a causal factor in aviation safety incidents the "hurry-up" syndrome, In R. S. Jensen & D. Neumeister (eds.), *Proceedings of the Seventh International Symposium on Aviation Psychology* (pp. 269-274). Columbus, OH: Ohio State University.

EMPIRICAL VALIDATION OF THE PREDICTION OF OPERATOR PERFORMANCE (POP) MODEL

C.S. Jordan*, E.W. Farmer*, A.J. Belyavin**, S.J. Selcon*
A.J. Bunting**, C.R. Shanks*, and P. Newman*,
Defence Evaluation and Research Agency (DERA)
Farnborough
United Kingdom
GU14 6TD

* Systems Integration Department, Defence Research Agency
**Applied Physiology Department, Centre for Human Sciences, Protection and Life Sciences Division

This paper describes an experiment conducted to validate the Prediction of Operator Performance (POP) model in a flight simulation context. The POP model uses subjective ratings of the demand imposed by single tasks to predict both the demand and performance associated with concurrent tasks. Previous experiments on the POP model have investigated a wide range of experimental tasks including tracking and verbal reasoning. In this experiment eight subjects performed flight control, threat assessment and threat identification tasks singly and in combination. Performance measures and POP scores were collected at the completion of each task condition. The results demonstrated performance decrements in the dual task conditions that were consistent with the predictions. The implications for the POP model are discussed in terms of workload modelling and human performance modelling within the context of the Integrated Performance Modelling Environment (IPME) currently being developed within the Defence Evaluation and Research Agency

INTRODUCTION

The Prediction of Operator Performance (POP) model has been developed as a result of a three-year research programme to produce an empirically derived predictive workload technique. The POP model was based on the premise that an algorithm could be developed to predict both perceived workload and performance under multi-task conditions from ratings of the demands imposed by single tasks. Three aspects of workload are addressed in the model: Input Demand (how much demand was imposed by the acquisition of information from external sources (e.g., from a visual display or auditory signals)?); Central Demand (how much demand was imposed by the mental operations (e.g., memorisation, calculation, decision making) required by the task?) and Output Demand (how much demand was imposed by the responses (e.g., keypad entries, control adjustments, vocal utterances) required by the task?). Subjects rate each demand on a 0-100+ scale. Ratings exceeding 100 represent excessive workload.

It was shown (Farmer et al., 1995) that subjects could make meaningful estimates of different aspects of the workload imposed by a single task. For dual tasks (tracking and verbal reasoning), ratings of the demands imposed by the single tasks, supplemented by information concerning the degree of conflict between the tasks, could be used to predict demand ratings and performance decrements when these tasks were performed concurrently (Farmer et al., 1995). This paper describes an experiment to validate the POP model in terms of its applicability to flight and threat assessment tasks.

Previous workload prediction models such as the VACP approach (see Schuck, 1991) have been based on subject-matter experts' ratings of task demands and the conflict between different types of task. The VACP model attempts to characterise the demand of any given task by assigning a value on four seven-point scales corresponding to the visual, auditory, cognitive and psychomotor domains. The conflict (interference) between concurrent tasks is then estimated using conflict matrices, again derived from subject-matter experts' ratings. Conflict is classified as 'acceptable', 'marginal' or 'unacceptable', based on the extent to which the subject-matter experts agreed that the particular combination of task categories could be performed together. In contrast to this technique and other workload modelling approaches, the main advantages offered by the POP model are:

1. The basis of the algorithm governing workload prediction within the POP model has been derived on the basis of empirical evidence and subjected to a validation exercise
2. When using the POP model, subjects assign the demand value that they think most appropriate for a specific task, and can therefore discriminate between easy and more difficult variants of the task
3. A rating of 100 corresponds to the maximum demand acceptable for completion of the task, but subjects are free to provide ratings greater than 100 if they assess the demand as being excessive for completion of the task
4. Estimates of combined demand are derived using validated rules based on empirical data; the workload

prediction depends on the competition for the same mental resources

5. The conflict between tasks is based on empirical data and psychological theory

6. An estimate of the degree of performance degradation is derived, thus enabling designers to evaluate the effect of task combinations on performance

METHOD

Subjects

Eight male subjects, aged between 21 and 41 years, participated in the experiment. All were employees of the Defence Research Agency (DRA), Farnborough, UK.

Apparatus

The experiment was conducted using the Fused Imagery Simulation Testbed (FIST) developed at DRA Farnborough (Smith et al., 1995). FIST consists of a cockpit mock-up, fitted with both head-up and head-down displays, with a Hands-on-Throttle-and-Stick (HOTAS) system and a 40° field of view projection system (see Figure 1).

Figure 1: Diagram of the Fused Imagery Simulation Test-bed (FIST)

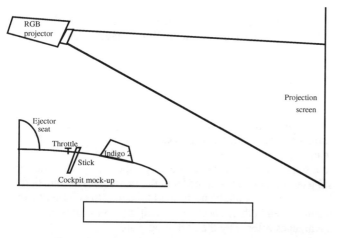

The aircraft software for this experiment was generated by a Silicon Graphics 310 VGX(T) computer and simulated a fast-jet flight model. The head-down display was presented on a Silicon Graphics Indigo2.

Tasks

Three tasks were used in this experiment: flight control; threat assessment; and threat identification. For the flight control task (primary task) a head-up display was projected onto a screen approximately 2 metres from the subject. Features of the terrain were generated on the display. The head-up display provided information on altitude relative to the terrain, heading, speed, distance from waypoint and waypoint number, and the aircraft's attitude. Subjects were required to fly the aircraft along the waypoint track

maintaining an altitude of 1000ft (+/- 200ft) and a heading+/- 10 degrees from the waypoint track. The flight control task had two modes: flight with variable throttle control, where the subjects were required to maintain a speed of 450 knots using a throttle; flight with maximum throttle (throttle fixed at maximum), where the subject had no control of speed and there was no requirement to maintain a constant speed. A joystick mounted in front of the subject, and a throttle positioned to the left of the subject, were used to control the aircraft.

The threat assessment task was adapted from experiments by Selcon et al., (1995) and was based on the pictorial representation of threat information. The subject was presented with representations of three hostile aircraft and his own aircraft on a 'head-down display' (see Figure 2). Each threat was displayed with a launch success zone (an irregular representation of its missile range) and the subject was required to identify which threat envelope his aircraft was currently within, and hence the most dangerous aircraft. Subjects were presented with a verbal 'warning' signal prior to the presentation of the threats, and responded by pressing one of the buttons on the three-button mouse positioned to the left of the throttle. Stimuli were presented for 500ms every 30s during a 5-minute trial block.

Figure 2: Scenario example from the threat assessment task

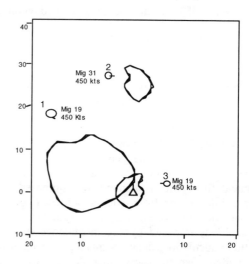

The threat identification task was based on the threat assessment task, except that, in this task, subjects were required to identify whether the hostile aircraft were all Mig 19s, all Mig 31s, or a combination.

Design

A within-subjects design was used. The independent variables were: number of tasks (two levels) and type of task (four levels). The four types of task were: flight with variable throttle; flight with maximum throttle; threat assesment task; and threat identification task. The dependent variables were: deviation from waypoint, deviation from altitude, deviation from velocity (flight simulation task); and number of correct responses and response time (threat assessment and threat identification tasks). Subjects completed the following conditions:

1. flight simulation (with variable throttle control)
2. flight simulation (with maximum throttle)
3. threat assessment
4. threat identification
5. flight simulation (with variable throttle control) and threat assessment
6. flight simulation (with maximum throttle) and threat assessment
7. flight simulation (with throttle control) and threat identification
8. flight simulation (with maximum throttle) and threat identification

The eight conditions were randomly assigned to a letter and presented in an order based on an 8x8 Williams square - a special form of Latin square.

Procedure.

Following a description of the single and dual tasks and the aircraft controls, eight practice sessions, each lasting 2.5 minutes, were presented in random order providing the subject with practice on both single and dual tasks. The subjects then completed the eight conditions of the experiment in an assigned order, providing POP ratings after each condition. Each experimental condition was of 5 minutes duration.

RESULTS

Objective measures

Height error. Analysis of Variance (ANOVA) was performed on log-transformed absolute height error data meaned over each transit between waypoints, with factors of waypoint, flight task, type of secondary task, and subjects. There was a reliable difference between waypoints (F = 33.68, df = 4, 28, p < 0.001).

ANOVA conducted on the mean absolute height error included the factors of '5 seconds before/5 seconds after' onset of the secondary task, flight task, secondary task, occurrence of secondary task, and subjects. There were reliable differences between presentations of the secondary task (F = 10.03, df = 7, 49, p < 0.001), and this factor interacted with before/after (F = 5.92, df = 7, 49, p < 0.001) and with flight task (F = 2.60, df = 7, 49, p < 0.05). Flight task also interacted with secondary task (F = 12.56, df = 1, 7, p < 0.01). Error for flight with throttle control/threat assessment was greater than that for flight with maximum throttle/threat assessment (p < 0.05) and that for flight with throttle control/threat identification (p < 0.05). Table 1 shows the effect of secondary task on flight task.

		Occurrence of Secondary Task (1-8)							
		1	2	3	4	5	6	7	8
Flight with throttle control	Before	236	293	181	199	321	712	454	375
	After	517	233	225	226	648	715	377	344
Flight with max. throttle	Before	279	197	200	281	852	419	303	366
	After	551	200	355	310	796	385	204	287

Table 1: Mean height deviation (feet) for each secondary task presentation

Speed error. ANOVA was performed on the log-transformed absolute speed error meaned over each transit between waypoints (up to waypoint 5), again using only flight with throttle control and factors of waypoint, secondary task and subjects. There was a reliable difference between waypoints (F = 8.89, df = 4, 28, p < 0.001).

Analysis was performed on the mean square-root-transformed absolute speed error before and after the onset of the secondary task; this analysis was comparable to that reported above for height data. There were reliable effects of before/after secondary task (F = 6.18, df = 1, 7, p < 0.05) and secondary task presentations (F = 3.60, df = 7, 49, p < 0.01); the latter factor interacted with the type of secondary task (F = 2.22, df = 7, 49, p < 0.05). Table 2 shows the effect of secondary task and secondary task occurrence.

		Occurrence of Secondary Task (1-8)							
		1	2	3	4	5	6	7	8
Flight with threat assessment	Before	14.2	21.1	17.7	14.4	37.4	32.0	16.3	15.3
	After	20.4	21.0	20.0	13.4	39.2	39.0	15.7	16.0
Flight with threat identification	Before	19.9	20.7	10.6	13.8	30.2	21.3	32.6	21.7
	After	23.8	16.8	12.3	12.9	34.8	19.1	31.8	22.7

Table 2: Mean speed deviation (knots) for each secondary task presentation

Course error. The proportion of time spent (log-transformed) more than $10°$ away from the target course was investigated using ANOVA with factors of flight task, secondary task and subjects. The interaction of flight task and secondary task was found to be significant (F = 4.46, df = 2, 14, p < 0.05). Using Newman-Keuls range test and Bonferroni t-tests, it was shown that, for the flight task with maximum throttle, the threat identification task produced greater course deviation than either the threat assessment task or flight only (p < 0.05). Means for the secondary task and flight task are shown in Table 3.

	Flight with throttle control	Flight with maximum throttle
Flight only	22.5	21.9
With threat assessment	24.1	20.9
With threat identification	22.6	29.8

Table 3: Mean percentage of time outside course bounds

Subjective measures

POP ratings Ratings of Input, Central and Output Demand were analysed separately by ANOVA with factors of conditions and subjects. A reliable difference between conditions was apparent for Input Demand ($F = 14.69$, df = 7, 49, $p < 0.001$), Central Demand ($F = 9.30$, df = 7, 49, $p < 0.001$) and Output Demand ($F = 4.68$, df = 7, 49, $p < 0.001$).

The Newman-Keuls range test showed a single vs dual task difference for Input Demand and Central Demand with single tasks rated lower than dual tasks. For Output Demand, the rating for the threat identification task only was lower than those for the flight task with throttle control (performed either singly or with a secondary task) and for the flight task with maximum throttle when it was performed with the threat identification task ($p < 0.05$). For the Output demand scale The threat assessment task was rated lower than flight with throttle control/threat identification task ($p < 0.01$) and flight with maximum throttle was rated lower under single-task conditions than when performed with the threat identification task ($p < 0.05$).

Condition	Input	Central	Output
Flight with throttle control	59.8	47.4	46.3
Flight with maximum throttle	56.6	42.9	35.9
Threat assessment	34.1	33.8	26.1
Threat identification	26.5	24.8	21.6
Flight with throttle control /threat assessment	87.8	77.9	50.5
Flight with maximum throttle/ threat assessment	69.0	65.2	43.1
Flight with throttle control/ threat identification	74.4	64.8	47.1
Flight with maximum throttle/ threat identification	81.2	75.0	61.2

Table 4: POP ratings in each condition

Figures 3 and 4 provide an indication of the POP ratings allocated to the single and dual task conditions. Figure

3 presents the results for the threat assessment task with and without the flight task. For POP Input and POP Central demand, the sum of the single task ratings of flight control and threat assessment/threat identification are nearly equivalent to the dual task ratings. The difference in rating score between the single flight task and the dual task combination suggests an appreciable increase in demand with the introduction of the secondary task.

Figure 3: Single and Dual task ratings for threat assessment and flight

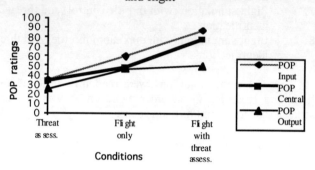

In Figure 4, the introduction of the threat identification task produced ratings of Input demand that are almost the sum of the flight and threat assessment single task ratings. The Central demand ratings were less than additive and the POP Output demand ratings were only slightly affected by the the introduction of the secondary task.

Figure 4: Single and Dual task ratings for threat identification and flight

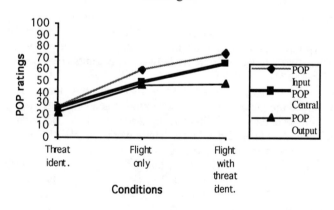

DISCUSSION

The flying performance measures revealed evidence of interference between tasks. Absolute speed error, course error, and performance on the secondary tasks were all affected by the requirement to perform tasks concurrently. The POP scores were consistent with the finding of performance decrements, since dual-task scores were reliably higher than their single-task equivalents. Indeed, Central Demand ratings for the dual tasks were little short of the sum of the ratings for the single tasks. This near-additivity suggests that subjects were unable to share resources between the primary and secondary task, and is probably attributable to the use of both head-up and head-down displays. If the two tasks were presented on a single display, the performance degradation and dual-task demand ratings would be predicted to be lower.

Comparison of the patterns of results for particular tasks suggests that subjects were able to provide meaningful estimates of demand. For example, the flying tasks were rated as relatively high on Output Demand, a finding consistent with previous studies of laboratory-based tracking tasks, whereas the secondary tasks, with relatively minor response requirements, were assigned low Output Demand ratings. When the tasks were performed concurrently, the Output Demand rating was in general only a little higher than that for the flying task alone; this finding contrasts with the near-additivity of Central Demand ratings.

The basic analysis of measures reported above provides qualitative support for the validity of the POP model. In the final stage of development of the model, meta-analysis of selected experimental data collected during the programme was used to extend the mathematical modelling. The outcome of this stage was an impressive linear relationship between predicted and observed values for Central Demand. The following data sets were examined:

(a)   Accuracy data for the threat assessment task

(b)   Height error for the flight task

The threat assessment task was treated as a vigilance task with similar characteristics to the Bakan vigilance task (Bakan, 1959) used extensively in the development of the POP model. For the threat assessment task, the normal deviates for the fitted Bakan task values of POP Central Demand and the equivalent normal deviates for the observed accuracy scores are shown in Table 5. There is a reasonable agreement between the observed and predicted values.

Task	Fitted values (Bakan task)	Observed (Threat assess.)
Threat assessment	-1.67	-1.51
With flight task + throttle control	-0.75	-0.47
With flight task with max. throttle	-0.93	-0.81

Table 5: Normal deviates for the vigilance task

For the flight task, height error was taken as the measure of accuracy.

The 5-second periods before and after onset of the secondary task were compared. Table 6 shows the dual-task decrements obtained.

Task	Log difference *100
Flight with threat assessment	10.3
Flight with threat identification	13.6

Table 6: Dual-task decrements in height data

If it is assumed that all tasks are dominated by a spatial resource requirement (an assumption consistent with additivity of the ratings and previous experiments in the research programme), the interference effect on the flight task (treated as a spatial tracking task) should be 0.5 of the rating on the other task (Farmer et al., 1995). The predicted decrements of 16.9 and 12.4 are not inconsistent with the observed values.

With the ability to predict dual task performance now demonstrated and the availability of task networks to model human performance, the POP model is currently being implemented within the Integrated Performance Modelling Environment, a task network simulation architecture which will model a range of performance elements including stressors, workload and cognitive models.

REFERENCES

Bakan, P. (1959) Extroversion-introversion and improvement in an auditory vigilance task. *British Journal of Psychology*, 50, 325-332.

Farmer, E.W., Belyavin, A.J., Jordan, C.S., Bunting, A.J., Tattersall, and Jones, D.M. (1995) *Predictive Workload Assessment: Final Report.* DRA Report DRA/AS/MMI/CR95100/1.

Schuck, M.M. (1991) *A New Method For The development of task conflict matrices. (Unpublished manuscript).* North York, Ontario: Defence and Civil Institute of Environmental Medicine.

Selcon, S.J., Coxell, A.W., Bunting, A., Smith, F.J., Lal, R., Dudfield, H.J. and Shorter, N. (1995). *The development of an explanatory tool for image-fused displays.* DRA Report DRA/AS/MMI/CR95059/1

Smith, F.J., Shanks, C.R., Selcon, S.J., Dudfield, H.J., and Bygrave, A.E. (1995) *A description of the Fused Imagery Simulation Testbed (FIST) Facility.* DRA Report DRA/AS/MMI/CR95060/1.

EVALUATION OF AIRCRAFT PILOT TEAM PERFORMANCE

Robert W. Holt and Edward Meiman
George Mason University
Fairfax, Virginia

Thomas L. Seamster
Cognitive and Human Factors
Sante Fe, New Mexico

Accurate assessment of team performance in complex, dynamic systems is difficult, particularly teamwork such as Crew Resource Management (CRM) in aircraft. Seventy pilots from two fleets were evaluated as two-person crews by a Maneuver Validation (MV), which focused on proficiency on separate maneuvers, and by a Line Operational Evaluation (LOE), which focused on the crew flying a simulated line flight. Instructor/Evaluator (I/E) pilots helped design LOE content and a structured evaluation worksheet. I/E reliability training resulted in high evaluator agreement (average r_{wg} = .80) and acceptable inter-rater correlations (average r = .54). Path analysis supported the assessment flow from Observable Behaviors to Technical and CRM performance to Captain (PIC), First Officer (SIC), and Crew evaluations for each event set. Fleet evaluations were different on the LOE assessment, but equivalent on the MV assessment. Detailed analysis of assessments also indicated a different role of the SIC across fleets. One fleet assessed SIC more on CRM performance and weighted SIC performance more in evaluating Crew performance. The other fleet assessed SIC on technical performance and weighted SIC performance less in evaluating Crew performance.

Complex, dynamic systems such as nuclear power plants or commercial aircraft require skilled team performance for optimal results. Evaluation of team performance in these work environments is difficult (see Taggart (1995) and Anderson and Henley, (1995) for different approaches). It is particularly difficult to accurately assess the teamwork aspects of performance such as Crew Resource Management (CRM) which are most critical for performance. This study reports on the reliability and validity of systematic evaluation of commercial aircraft crews by trained instructors in a Line Operational Evaluation (LOE) simulation.

Accurate evaluation of such teams depends on several critical steps. First, accurate evaluation requires carefully developed, relevant, and realistic assessment materials which are consistently administered by assessors. The initial content of the LOE in this study was based on a task analysis and was refined by input from the evaluator pilots about typical problems encountered in previous LOFT and line checks. For details on the LOE development process, see Seamster, Edens, McDougall, and Hamman (1994).

Second, evaluation materials must be easily and accurately used by evaluators under high workload. Since the performance of teams is complex and multi-faceted, the focal point of evaluation must be clearly established and trained to avoid stereotypic or heuristic-based responses. Seamster,

Edens, and Holt (1995) found that a five-point scale with a focus on separate event sets of the LOE apparently produced less halo rating error. For this study, a four-point rating scale and a structured evaluation worksheet were iteratively developed with evaluator pilot input. Each scale point was labeled such that instructors could agree on a unique, clearly distinguishable meaning. The worksheet structured and focused the evaluations for each event set in a flow from observable behavior to technical and CRM elements to Captain, First Officer, and Crew evaluations.

Third, evaluators must be trained to ensure acceptable inter-rater reliability. Reliability implies both agreement of evaluators on a given team performance (Law and Sherman, 1995) and positive correlation of evaluations across teams with differing performance. In this study, evaluator training focused on both covariation of ratings across different performances and agreement of ratings for a given performance.

We hypothesized that properly-trained evaluators using appropriate materials in a carefully-developed assessment situation would show high levels of inter-rater agreement and inter-rater correlation. Further, we hypothesized that a path analysis of relationships among the evaluations would support the trained evaluation sequence. Finally we hypothesized that the LOE evaluation would be distinct from traditional maneuver proficiency and possibly show differences among assessed fleets.

METHOD

Sample and Materials

Seventy pilots in 35 two-person crews (Pilot-In-Command and Second-In-Command) were evaluated in a 2-day assessment at a commercial airline as part of that carrier's Advanced Qualification Program (AQP) process. About 55% of the crews represented one of the carrier's fleets (Fleet 1) while 45% represented a second fleet (Fleet 2). Day 1 was a maneuver-based proficiency check in the aircraft simulator. Day 2 was an LOE simulation of a normal flight. The LOE script for this flight was divided into 11 segments called "event sets". Each event set was the focal point of an evaluation. The start of each event set was defined by a triggering event which presented a flight problem and the end was defined by meeting or not meeting defined "success criteria" for that problem within a specified time. Missing records reduced the usable event sets for data analysis from 385 to 369.

Development of evaluation materials. Initially, evaluators had problems categorizing the targeted behaviors for each event set as either "observed" or "not observed". We added a "partially observed" category for observable behaviors and carefully redefined the "fully observed" and "not observed" categories.

The scale for technical, CRM, PIC, SIC, and Crew evaluations was iteratively developed into a 4-point scale of unsatisfactory (failed FAA requirements), satisfactory (met FAA requirements), standard (met carrier requirements), and above standard (exceeded carrier requirements). These scale point labels resemble the labels for the new Line/LOS Checklist (Taggart, 1995). To enhance reliability, the scale points for "3" and "1" ratings were explicitly and behaviorally anchored as in a Behaviorally-Anchored Rating Scale (BARS). The "2" rating was defined as "FAA minimally acceptable performance". The "4" rating was defined as substantially better than "3" performance on at least one important criterion.

Evaluator calibration training

The instructors for each fleet were trained on the LOE worksheet using videotaped segments from LOEs with volunteer pilots. After clarification of the form and its use, the first set of taped segments were individually rated, followed by focus-group discussions of any differences. Evaluators were taught to avoid stereotypic patterns of assessment such as "all 3s" which would increase inter-rater agreement but lower inter-rater correlations. Evaluators were also taught to avoid different internal standards for evaluation (e.g. harsh vs. lenient grading) which would not affect inter-rater correlations but would decrease inter-rater agreement. After training, the second set of taped segments were individually rated (without group discussion) and used for the reliability estimates.

Evaluation Procedure

Day 1 maneuver proficiency was rated by trained evaluator pilots using the four-point scale. For each maneuver, each pilot was assessed once as pilot flying (PF) and once as pilot not flying (PNF). Each pilot's initial performance (before repeats) as PF and PNF was averaged across all maneuvers to obtain a PF score and a PNF score.

Day 2 LOE performance was evaluated by the same corps of instructor pilots using an LOE script and the structured evaluation worksheet for each event set. During the event set, instructors checked off specific observable behaviors as fully, partially, or not observed for that crew and wrote comments. As soon as possible after the end of the event set, the instructor used the 4-point scale to rate the crew for that event set on two specific technical aspects of performance, one CRM aspect of performance, the PIC and SIC individually, and the Crew as a team.

RESULTS

LOE Assessment Reliability

Average inter-rater correlations (Pearson r) among pairs of raters across segments and average agreement indices (r_{wg}) among the set of raters for each segment are given in Table 1. Confirming our first hypothesis, agreement is quite high for these groups of raters, averaging .80 across all scales. The inter-rater correlations are acceptable although not as high as the agreement indices, averaging .54. This data seems to indicate that the rater training was quite effective in producing evaluator agreement on each segment, and reasonably effective in producing consistent shifts among evaluators across segments.

LOE Assessment Validity

Internal evidence. The evaluations of the instructors were analyzed to see if the hypothesized evaluation sequence

Table 1. Consistency and Agreement Indices of Inter-rater Reliability

	Observable Behaviors	Technical performance	CRM performance	Pilot-In-Command	Second-In Command	Crew Performance
Consistency **r**	.47	.53	.54	.62	.55	.55
Agreement **r_{wg}**	.63	.82	.74	.87	.83	.89

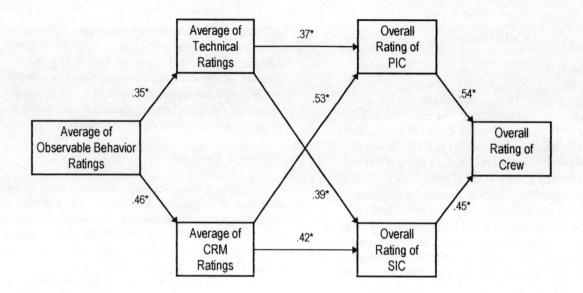

Figure 1. Path analysis of LOE evaluation sequence (Standardized path coefficients).
(* p < .05)

of the LOE worksheet from observable behaviors to technical and CRM ratings to PIC, SIC, and then Crew ratings was plausible. Confirming our second hypothesis, all hypothesized path coefficients were positive (Figure 1) and statistically significant (p < .05). These results support the evaluators' use of observable behaviors as a basis for technical and CRM evaluations, the use of technical and CRM ratings for PIC and SIC evaluations, and the use of PIC and SIC ratings for Crew evaluations.

External evidence. The average LOE score was regressed on PF and PNF maneuver proficiency averages. As expected, maneuver proficiency was significantly related to overall LOE performance (Table 2). However, only the PF maneuver performance predicts LOE performance, not the PNF performance. The ability of maneuver proficiency to predict LOE performance on each event set is much lower, although still significant (Table 2). This confirms that capability to perform maneuvers can predict some of the average performance on the LOE, but less of the performance

on each LOE event set.

When the variance predicted by maneuver proficiency (PF and PNF) is removed from the PIC and SIC ratings for each event set on the LOE, the residual ratings still contain systematic variance (Figure 2). These residual PIC and SIC ratings are strongly predicted by technical and CRM performance on each event set, and residual PIC and SIC ratings still strongly predict the overall crew rating. The strength of the relationships with the residual ratings shows that LOE event sets assess aspects of crew performance largely unrelated to maneuver proficiency.

Fleet Differences

For maneuver proficiency, average performance was not significantly different across the fleets. Comparing the fleets on each of 14 maneuvers, we found only three significant differences for pilot flying (PF) performance, but they were not all in the same direction. We found similar inconsistency in the

Table 2. Prediction of LOE performance from maneuver proficiency.

	LOE performance averaged across Event Sets: (N=35)		LOE performance for each Event Set: (N=369)	
	PIC	**SIC**	**PIC**	**SIC**
B for PF maneuver average	.408	.676*	.244*	.227*
B for PNF maneuver average	.117	-.118	-.010	-.011
Proportion of criterion variance predicted by maneuver scores= R^2	.264*	.343*	.055*	.116*

*p<.05

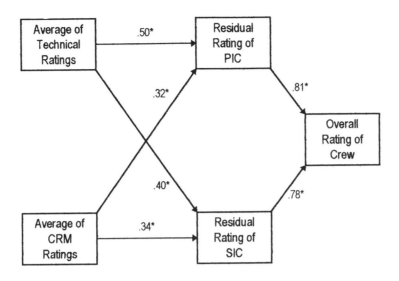

Figure 2. Systematic relationships of residual PIC, and SIC LOE ratings.
(* p < .05.)

three significant differences for pilot not flying (PNF) performance. Overall, these mixed results suggest no strong, consistent differences between the fleets.

However, LOE evaluations show strong, consistent differences between the fleets. For the LOE, Fleet 1 was rated significantly lower on observable behaviors, technical performance, CRM performance, and overall PIC, SIC, and Crew ratings compared to Fleet 2 (Table 3). These results clearly indicate that Fleet 1 LOE performance was rated inferior to Fleet 2.

Fleets 1 and 2 also have significant differences in the relationships among the LOE variables as indicated by the separate path analyses for each fleet (Figure 3). In Crew evaluations, Fleet 1 results emphasized PIC performance significantly less than Fleet 2 ($z = -3.41$, $p < .05$) but SIC performance significantly more ($z = 2.72$, $p < .05$). That is, Captain performance played a more important role in crew evaluations in Fleet 2, but First Officer performance played a more important role in crew evaluations in Fleet 1, a surprising finding.

The relative contribution of CRM to evaluation of the SIC was significantly lower in Fleet 2 compared to Fleet 1 ($z =$

-2.52, $p < .05$). Further, within Fleet 2, the contribution of CRM to SIC evaluation was significantly lower than technical performance ($t(197) = -2.67$, $p < .05$). Both of these results imply that in Fleet 2 technical performance was emphasized more than CRM performance for the SIC.

These empirical results suggest qualitative differences in the evaluation of the fleets. Specifically, Fleet 1 evaluations emphasize SIC performance in the crew and also hold the SIC accountable for CRM performance. Fleet 2 evaluations emphasize PIC performance in the crew and hold the SIC responsible mainly for technical performance.

DISCUSSION

Reliable and valid evaluations can answer important questions relevant to team performance in complex, dynamic systems. In this study, careful evaluator training and development of materials produced LOE evaluations with high levels of agreement and acceptable levels of consistency. Careful, ongoing rater training is critical for establishing and preserving this level of reliability so that the data will clearly show important aspects of team process.

Table 3. Fleet differences in LOE evaluations

	Fleet 1		Fleet 2
Average of Observable Behavior Rating	2.51	<	2.69*
Average of Technical Rating	2.84	<	2.94*
Average of CRM Rating	2.63	<	2.88*
PIC rating for the Event Set	2.69	<	2.91*
SIC rating for the Event Set	2.72	<	2.95*
Crew rating for the Event Set	2.73	<	2.91*

* p < .05.

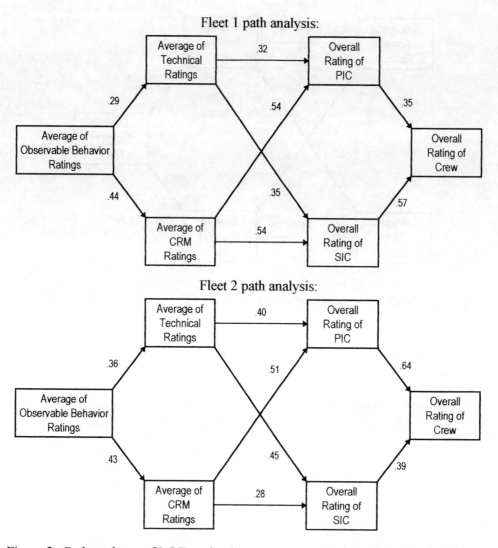

Figure 3. Path analyses of LOE evaluation process across fleets. (all coefficients, $p < .05$)

The pattern of internal and external relationships supported the validity of the LOE evaluations. Evaluations of fleets differed in average level and in details such as the relative importance of CRM to evaluations of the SIC, and in the relative importance of PIC vs. SIC performance to Crew performance. When shown these results, experienced training pilots felt the indicated different roles for SIC in the two fleets were real. There are several possible causes for these differences in crew process between the fleets which will be the focus for future research.

These differences may be due to systematic differences in the way the I/Es use the rating scales to evaluate pilots. If so, having I/Es evaluate the same videotaped segments of LOEs should confirm these differences.

Alternatively, these differences may be due to differences in pilot composition in the fleets, different fleet aircraft and procedures, or different fleet cultures (Sherman and Helmreich, 1995) The important point of having reliable teamwork evaluations is that these different possible explanations can be statistically evaluated in future research, and the effect of training interventions can be objectively established.

REFERENCES

Anderson, P. and Henley, I. (1995). Assessing the development of team skills using a time-limited exercise. In Proceedings of the Eighth International Symposium on Aviation Psychology. Columbus, Ohio: Ohio State University

Law, J.R., and Sherman, P.J. (1995). Do raters agree? Assessing inter-rater agreement in the evaluation of air Crew Resource Management skills. In Proceedings of the Eighth International Symposium on Aviation Psychology. Columbus, Ohio: Ohio State University

Seamster, T.L., Edens, E.S., and Holt, R.W. (1995). Scenario event sets and the reliability of CRM assessment. In Proceedings of the Eighth International Symposium on Aviation Psychology. Columbus, Ohio: Ohio State University

Seamster, T.L., Edens, E.S., McDougall, W.A., and Hamman, W.R. (1994). Observable crew behaviors in the development and assessment of Line Operational Evaluations (LOEs). Washington, D.C.: The Federal Aviation Administration.

Sherman, P.J., and Helmreich, R.L. (1995). Attitudes toward automation: The effect of national culture. In Proceedings of the Eighth International Symposium on Aviation Psychology. Columbus, Ohio: Ohio State University.

Taggart, W.R. (1995). The NASA/UT/FAA Line-LOS-Checklist: Assessing system safety and crew performance. In Proceedings of the Eighth International Symposium on Aviation Psychology. Columbus, Ohio: Ohio State University.

ACKNOWLEDGMENTS

This research was supported by FAA grant 94-G-034.

AUTOMATION AND CREW PERFORMANCE:
THE IMPORTANCE OF WHO AND WHAT

Florian Jentsch and Clint Bowers, Ph.D.
University of Central Florida
Orlando, Florida

In the past, "automation" in the cockpit has been associated with both an <u>increase and a reduction</u> in workload, performance, and crew communications. A reason for these apparently contradictory results could be that different types of automation were studied, for example, automation that required little vs. considerable interaction with a machine. The current study investigated whether automation of different task domains for different crewmembers has differential effects. Sixteen pilots formed two-pilot crews. The crews each completed four experimental flights in a PC-based simulator, encountering different levels of automation. Systematically, an automated system for the pilot (a simple autopilot) and one for the copilot (a navigation computer) were made available or were unavailable, alone or in combination. The results showed that the automation manipulations improved task performance only when they were combined. Additionally, the use of the navigation computer reduced communications within the cockpit, while the autopilot was associated with larger contributions by the pilot. Further, mental workload and frustration were lower when the navigation computer was available. Also, perceived effort was lower with the autopilot on. The results of the current study suggest a more detailed picture of automation effects than previously known. Additionally, the results suggest that different types of automation can interact to create unexpected effects that may not be uncovered if one studies automated systems individually.

Wiener and others were the first to raise the concern that cockpit automation may not always and uniformly improve performance, lower workload, and improve crew coordination (Wiener, 1985; 1988; 1989; Wiener, Chidester, Kanki, Palmer, Curry, and Gregorich, 1991). For example, because many automated systems require extensive interaction with a machine, crews in automated systems were often observed to reduce the time that at least one of the pilots was visually checking for traffic. Other potentially negative issues associated with automated systems were (a) that it is difficult for one pilot to see what the other pilot is doing, and that, as a result, (b) explicit crew coordination and communication may gain more importance in automated aircraft.

Furthermore, automation has been associated with increased as well as reduced workload, with improved or degraded crew coordination, and with more or less communication among crew members. For example, Costley and his colleagues suggested that crews in automated aircraft displayed fewer communications than those in non-automated aircraft (Costley, Johnson, and Lawson, 1989). Yet, at the same time, Bowers and his group, as well as Veinott and Irwin (1993), presented experimental results that suggested more communication among crew members in automated aircraft (Bowers, Deaton, Oser, Prince, and Kolb, 1993; Kanki and Palmer, 1993; Thornton, Braun, Bowers, and Morgan, 1992).

A possible explanation for these apparently contradictory findings is that "automation" is not a uniform and dichotomous construct. Instead, its effect may depend on the function of a particular system and interact with the characteristics of the task environment. For example, while Thornton et al. (1992) and Bowers et al. (1993) compared performance across the automation conditions provided by a simple autopilot, Costley et al. (1991) based their findings on observations with advanced flight management systems. Thus, the different results from these studies may have reflected <u>what</u> was automated: In one case, a psychomotor flight control task was automated through the use of an autopilot; in the other, a more cognitive task was automated by using a flight management system. The different demand characteristics inherent in these two task domains may have moderated the effectiveness and efficiency of the automated system.

Notice also that not only <u>what</u> was automated was different among the various investigations, but also <u>who</u> in the crew was primarily affected by the systems' function. The autopilot system in the Bowers et al. (1993) and Thornton et al. (1992) studies directly affected only the pilot-flying, conceivably freeing resources that the pilot could devote towards other task aspects, for example, coordinating with the pilot-not-flying. In the Costley et al. study, on the other hand, the use of a flight management and engine control computer most directly affected the pilots-not-flying, requiring them to interact through a keyboard/VDU human-machine interface with a complex system and thereby removing them from the coordination loop. Such a pattern of interactions would explain why crew communication and coordination increased in the Thornton et al. and Bowers et al. studies, but decreased in the investigations by Costley et al. (1993).

Purpose of the Present Study

The purpose of the present study was to investigate whether automating different functions for different crewmembers could lead to the differences in performance, workload, and communications observed in previous investigations. Two different task domains, one psychomotor, one more cognitive, were automated, and two-person crews flew under all four possible combinations of the automation conditions.

We expected that the manipulation of automation would lead to a pattern of results similar to those observed in previous studies. In particular, we expected that:

1. The automation of the pilot's psychomotor task would free resources that the pilot could devote towards increasing coordination with the copilot.

2. The automation of the copilot's cognitive task, while reducing mental demands, would reduce crew interactions by partially removing the copilot from the crew interactions because of the increased need for interactions with the machine.

METHOD

Participants

Sixteen pilots from an airline-affiliated flight academy in the U.S. volunteered to participate in the current study for payment. Participants were placed in two-person teams and then randomly assigned to either the pilot or copilot position. Although the participants may have been acquainted, none of the crews had previously flown together.

Design

The study followed a 2 Conditions x 2 Positions x (2 Copilot Automation x 2 Pilot Automation) mixed design. Between-subjects factors were Performance Condition with two levels (as manipulated by additional financial incentives for outstanding task performance) and Position with two levels ("Pilot" and "Copilot"). Within-subjects factors were manipulation of the copilot automation and manipulation of the pilot automation, each with two levels (automation "Off" or "On").

The crews flew five flights: one practice and four experimental flights. Each team completed four experimental flights in a 2 Copilot Automation x 2 Pilot Automation fully factorial within-subjects design: During each flight, the copilot automation (a navigation computer) and the pilot automation (an altitude-hold autopilot) were either "on" or "off." The sequence of experimental flights was randomized.

Apparatus

A low-fidelity flight simulation, Flight Simulator 4.0 (Artwick, 1989), was used. The software was configured to simulate the characteristics of an early model Cessna 210 aircraft.

Simulated Flight Scenarios

The simulated flights were conducted in a generic environment that simulated a grassy area in the Northeast of the United States. Buildings were placed in the area and varied according to location, orientation, and shape. The crews' task was to find buildings, to identify their location using the radio navigation equipment, and to translate the rho/theta-coordinates into a generic x-, y- rectangular grid system. Once they had entered the search sector, the crews had to remain in a specified range of altitudes (as close as possible to 500 ft AGL) and were allowed 30 minutes to complete their task.

Procedure

Each crew completed the practice and all four experimental flights in one day. Upon completion of each flight, the team members completed the NASA Task Load Index (TLX).

Dependent Measures

Three types of dependent measures were taken: (a) A primary task performance measure (i.e., the number of buildings identified); (b) perceived workload as measured by the NASA TLX task-load index scale; and (c) two types of communication measures: The first set used the number of words spoken by each crew member during three five-minute segments of each flight as the dependent variable. The five-minute segments were minutes 0 to 5, 10 to 15, and 20 to 25 after the crews entered the search sector. For the second set of data, the proportion of words contributed by the pilot to the crew communications was calculated.

RESULTS

Overview

To remove any variance that an order effect of the scenarios could have accounted for, order was used as a covariate. The analyses showed the following statistically significant effects:

1. *Pilot automation.* The crews felt that during the flights with the autopilot engaged, significantly less effort was required than during the flights without the autopilot. Additionally, the contribution the pilots made to the conversation was significantly greater during flights with the autopilot on than when it was off.

2. *Copilot automation.* Mental workload and frustration as perceived by the crews were significantly lower during the flights with the copilot automation (i.e., the navigation computer) available. Also, the overall number of statements made by the pilots and copilots was lower with the copilot automation than without it.

3. *Pilot automation x Copilot automation.* Primary task performance and subjective perceptions of performance (i.e., NASA TLX subscale "performance") showed significant interaction effects: The best performance and the best performance perceptions were obtained when both types of automation were available, followed by the flights during which neither of the automated systems was available. The worst performance (and associated performance perceptions) occurred during those flights when only one of the two automated systems was available.

4. *Performance condition.* No significant main effects for performance condition were observed. However, among the subjective workload data, several Condition x Position x Pilot Automation x Copilot Automation four-way interactions were observed.

5. *Crewmember position.* No significant main effects for crewmember position (Pilot vs. Copilot) were observed. However, the subjective workload data showed several significant three-way and four-way interactions involving crewmember position.

In the following sections, we focus only on effects involving the automation conditions.

Primary Task Performance

A significant Pilot Automation x Copilot Automation interaction was observed, $F(1, 5) = 8.11$, $p = .036$. The best average performance was observed when both the pilot and copilot automations were "on." In this condition, an average of 16.75 buildings were identified. The worst performance resulted with only the copilot automation available, $M = 15.125$ (Figure 1).

Subjective Workload

Mental workload. The analyses of the sixteen pilots scores on the mental subscale of the NASA TLX showed a significant main effect for the copilot automation, $F(1, 11) = 5.10$, $p = .045$, with mental workload higher during the flights without the copilot automation system, $M = 57.97$, than when the copilot automation was available, $M = 50.78$.

Frustration. One significant effect was found: During the flights with the copilot automation available, the teams felt significantly less frustrated ($M = 28.59$) than when the copilot automation was not available ($M = 42.03$), $F(1, 11) = 8.38$, $p = .015$.

Effort. With respect to the effort subscale of the NASA TLX, there was a significant effect of the pilot automation. The teams perceived that the flights without the autopilot required more effort ($M = 67.5$) than those which were

conducted with the autopilot engaged ($M = 60.6$), $F(1, 11) = 4.89$, $p = .049$.

Performance. The analyses of the performance perceptions for the eight teams showed a significant Copilot Automation x Pilot Automation two-way interaction, $F(1, 11) = 9.25$, $p = .011$. The crews believed to have performed better when no automation or both types of automation were available ($M = 34.1$ for no automation and $M = 29.1$ for both types of automation on, respectively). The crews thought they had performed worse when only one of the two types of automation was available ($M = 43.1$ when only the autopilot was on and $M = 41.6$ when only the navigation computer was on). This pattern of means closely matched that of the primary task performance (cf. Figure 1).

Communication Analyses

Speech rates. The analysis regarding the number of words spoken by each crewmember during the flights showed a significant main effect for the copilot automation, $F(1, 11) = 5.93$, $p = .033$. On average, the crew members uttered more words when the copilot automation was unavailable ($M = 36$ words / minute / crewmember) than when the navigation computer was available ($M = 32$ words / minute / crewmember).

Pilot contribution. Because we expected that the pilot's spare capacity and thus contribution to the communications would be higher when the pilot had the autopilot on than when he/she had to manually control the aircraft, we analyzed the relative contributions made by the pilots to the crew communications during each flight. As expected, the pilot's contribution was higher when the autopilot was engaged ($M = 43$ percent) than when the autopilot was off ($M = 39$ percent), $F(1, 5) = 5.75$, $p = .031$ (one-tailed).

DISCUSSION

The goal of the present study was to assess the degree to which the equivocal results of existing studies of automation and crew performance might be attributed to differences in the automation technology employed. The results provide interesting information in this regard.

The first finding of interest is related to the actual performance effects attributable to automation. Past results have demonstrated little advantage of automation in terms of task performance (i.e., Thornton et al., 1992). The present results replicate and amplify this finding. Individual automation manipulations again demonstrated no significant advantage over the "no automation" condition. However, it appears that the combination of automation technologies did allow a mild, albeit not statistically significant, performance advantage. This suggests that there might be a "resource threshold" that must be achieved by the automation technology before the systems can be used to an advantage by the crew. Put another way, it might be that automation must make a substantial difference to the crew or it just gets in the way. There is a need to study greater combinations of automation technologies and to assess the cost-benefits ratio of these systems more directly.

The key function that most automatic technologies are thought to serve is the reduction of crew workload. Presumably, the automation of specific functions frees processing resources for application to other aspects of the flight. The results reported by Thornton and her colleagues (1992), among others, provided little support for this notion, however. The present results might make this effect somewhat more clear. Relative to other conditions, copilot automation was associated with less perceived mental demand and frustration. Pilot automation was associated with less effort, but the predicted effect for physical demand was not obtained. Additionally, both automation manipulations together resulted in perceptions of better task performance.

Overall, the results indicate that automation is somewhat effective in impacting the crew members' perceptions of workload, but these perceived workload savings are not manifested in better task performance. The question, then, is "Why don't crews utilize automation to greater advantage?" One possible explanation is suggested in the results of Kleinman and Serfaty (1989). These researchers found that teams were more likely to use "explicit" coordination during low workload, but switched to an intuited "implicit" coordination when workload became high. One might suggest that a similar pattern might exist in air crews. That is, the workload savings afforded by automation might be used in switching to more effortful, but safer, explicit communication strategies. In this case, one would expect greater levels of communication in the automated conditions. In general, however, this hypothesis was not supported. In fact, copilot automation was associated with significantly less crew communication, most likely because the interactions with the navigation computer removed the copilot from the coordination loop.

Directions for Further Study

The results of this experiment indicate several issues requiring additional study in the area of automation. First, it is imperative to specify the type of automatic system being employed. The present results indicate that different automatic systems are likely to have different effects on crew performance and processes. It is only by greater specificity that the effects of automation can be isolated. Toward this end, Bowers, Jentsch, and Salas (1995) have proposed the use of a taxonomy of automation to identify the specific aspects of flight being targeted. It seems clear that such a system is required in the scientific literature.

A second critical issue is related to the interaction of automated systems. The present data indicate that for automation, the whole might be greater than the sum of its parts. The combination of psychomotor and cognitive automation created effects unlike those observed in either system individually. Yet, existing studies have studied only one system or large combinations of systems studied as a whole. Neither of these approaches is likely to be useful in

specifying automation effects. Rather, a series of studies is required to test the interactions of automation technologies in a logical, controlled manner.

Finally, there is a need to investigate the degree to which automation imposes an additional demand for training. Thus far, the data seem to indicate that crews in automated systems do not necessarily perform more effectively. Additionally, workload savings provided by automation do not seem to translate into more effective crew processes. It might be that crew members require additional training to apply partial workload savings more efficiently. This represents an interesting area for future research.

ACKNOWLEDGEMENTS

The authors would like to thank the students, flight instructors, and staff at Comair Aviation Academy for their support of this study, in particular Thomas Hall and Wayne Ceynowa. Special thanks also go to Heather Bergen, Paul Caballero, Julie Henning, Barbara Holmes, Debbie Margulies, and Ben Redshaw for their support.

This research was funded through Contract No. N61339-93-C-0101 by the Naval Air Warfare Center Training Systems Division in Orlando, Florida, COTR Randall Oser. The opinions herein are those of the authors and do not necessarily reflect the opinion of the University of Central Florida or the U.S. Navy.

Correspondence regarding this paper should be sent to Clint Bowers, Ph.D., Department of Psychology, University of Central Florida, P.O. Box 161390, Orlando, Florida 32816-1390.

REFERENCES

Artwick, B. (1989). Microsoft Flight Simulator, V4.0 [computer software]. Bellingham, Washington: Microsoft.

Bowers, C.A., Deaton, J., Oser, R., Prince, C., and Kolb, M. (1993). The impact of automation on crew communication and performance. Proceedings of Seventh International Symposium on Aviation Psychology (pp. 573-577). Columbus: Ohio State University.

Bowers, C.A., Jentsch, F., and Salas, E. (1995) "Studying automation in the lab - can you? should you?". In N. McDonald, N. Johnston, and R. Fuller (Eds.), Applications of Psychology to the Aviation System. Hampshire, England: Avebury Aviation.

Costley, J., Johnson, D., and Lawson, D. (1989). A comparison of cockpit communication B737-B757. Proceedings of Fifth International Symposium on Aviation Psychology (pp. 413-418). Columbus: Ohio State University.

Kanki, B.G. and Palmer, M.T. (1993). Communication and crew resource management. In E.L. Wiener, B.G. Kanki, and R.L. Helmreich (Eds.), Cockpit Resource

Management (pp. 99-136). San Diego: Academic Press.

Kleinman, D.L. and Serfaty, D. (1989). Team performance assessment in distributed decision making. Proceedings of Interactive Networked Simulation for Training (pp. 22-27). Orlando: University of Central Florida.

Thornton, C., Braun, C., Bowers, C., and Morgan, B. (1992). Automation effects in the cockpit: A low-fidelity investigation. Proceedings of the 36th Annual Meeting of the Human Factors Society, Vol. 1 (pp. 30-34). Santa Monica, CA: The Human Factors Society.

Veinott, E.S. and Irwin, C.M. (1993). Analysis of communication in the standard versus automated aircraft. Proceedings of Seventh International Symposium on Aviation Psychology (pp. 584-588). Columbus: Ohio State University.

Wiener, E.L. (1985). Human factors in cockpit automation: A field study of flight crew transition. (NASA

Contractor Report No. 177333). Moffett Field, CA: NASA-Ames Research Center.

Wiener, E.L. (1988). Cockpit automation. In E.L. Wiener and D.C. Nagel (Eds.), Human factors in aviation (pp. 433-461). San Diego: Academic Press.

Wiener, E.L. (1989). Human factors of advanced technology ("glass cockpit") transport aircraft (NASA Contractor Report No. 177528). Moffett Field, CA: NASA-Ames Research Center.

Wiener, E.L., Chidester, T.R., Kanki, B.G., Palmer, E.A., Curry, R.E., and Gregorich, S.A. (1991). The impact of cockpit automation on crew coordination and communication I. Overview, LOFT Evaluations, error severity, and questionnaire data (NASA Contractor Report No. 177587). Moffett Field, CA: NASA-Ames Research Center.

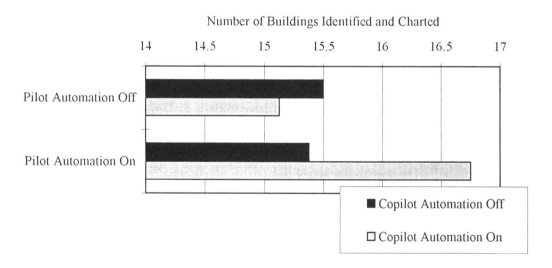

Number of Buildings Identified and Charted

Figure 1.   Average primary task performance (number of buildings identified) by automation condition.

PITFALLS, PRATFALLS, AND PROSPECTS FOR PROGRESS in PILOT HIRING, SCREENING, AND SELECTION

A roundtable debate on current airline practices

Dr. Robert O. Besco (Capt. AAL, Ret.)
President, PPI
Dallas, TX

Abstract

Recent airline accidents have been attributed to pilot errors of junior pilots whose professional flying experience, at the time of hiring, did not measure up to the standards of the hiring airline. How these pilots slipped through the screening process of conscientious and well intentioned personnel departments is a major question. Several tested and proven methods of applicant screening are suggested to be used in the hiring process for new pilots.

INTRODUCTION

In several recent accident investigations, low standards of prior professional performance histories have been detected in the accident pilots by the NTSB accident investigation process. Many of the pilots had pre-employment track records of unsatisfactory low flight deck proficiency and/or in-training fulfillment. The pre-employment screening process had not revealed these negative professional performance histories. **The operating principles, philosophies, and practices of pre-employment screening have not been successful in eliminating applicants with poor professional performance histories and with documented track records of failed cockpit performance.** This panel discussion calls for a reemphasis on the principle that the burden of proof is on the applicant to demonstrate high quality prior professional performance. Several proven, simple, low cost improvements to pilot selection and screening programs will be described.

THE PROBLEM

The NTSB has recently uncovered cases where the accident pilots did not measure up to the hiring standards of the individual flight department (NTSB, 1967, 1985, 1986, 1988, 1991, 1994a, 1994b and 1995a and 1995b). These accidents all happened at organizations where competent, well intentioned screening staffs did not discover the negative aspects of the applicant pilots professional careers. Several rationales have been suggested for these breakdowns in the pilot screening process. Some of these current counterproductive assumptions of screening (in alphabetical order) are:

- All pilot applicants are already screened and qualified by the FAA licensing and certification procedures.

- Conducting background checks on pilot applicants is very expensive and time consuming.

- Fighter pilots are macho and are dysfunctional in airline CRM programs.

- It is difficult to reliably measure professional quality of pilots during flight.

- No industry agreement exists on what constitutes acceptable pilot performance.

- No validated pilot aptitude measures exist for screening out pilots with the "wrong stuff."

- Our flight department managers and pilot hiring interviewers can discover the "bad apples" in the applicant pool by using insightful and intuitive interviewing techniques based on the practical wisdom from decades of being in aviation operations.

- Pilot applicants subscribe to services from hiring consultants that help them mask deficiencies in their professional and personal histories.   Only their best foot is forward.

- Pilots who have either thousands of hours of high performance flying time, or high level education with advanced degrees will not be good "team players." They may have ambitions to take over the management of the flight department.

- Previous employers are reluctant to release negative information about any individual's substandard professional performance.

- Quality pilot applicants should want the job badly enough to make themselves available on two weeks or less notification of report dates.

- The personnel department's function is more recruitment than selection.

- There are plenty of qualified and licensed applicants who will work for probationary wages for a year and for "B" scale wages for several years.

- We should not hire pilots with extensive flying or supervisory backgrounds in high performance and complex aircraft. "Overqualified" applicants will be malcontents when assigned for long periods to subordinate cockpit roles.   They will not be content to spend years in subordinate flying assignments.

- Young applicants who meet minimum standards make better corporate employees than older pilots with more experience.

No one airline has all of these misconceptions. However, practically all aviation operations follow at least one of these false assumptions. Every one of these negative practices or policies has been observed in multiple organizations.   All of these detract from effective flight operations. Nothing on this list can be considered a benefit to the passengers, stock holders, or employees of an airline.

All of the dysfunctional and counter productive hiring, screening and training factors can be ameliorated by changing philosophies and practices in the pilot selection process. The critical nature of flight crew performance requires that the selection process be as rigorous and reliable as is possible. The pilot selection process deserves as much corporate emphasis as the acquisition of other critical corporate assets.

SIMPLE, INEXPENSIVE, TRIED AND TRUE IMPROVEMENTS

There are time honored and repeatedly proven philosophies, principles and practices of selection. These tried and true procedures have been used successfully for generations by organizations that must select only exceptional, high performing professionals (Chaney, 1991, Hunter and Burke, 1995, and Owens, 1976).   Some of these recommendations may seem obvious, but every single one of them has been ignored or discounted at one time or another by flight departments as revealed during the NTSB post-accident investigations. Many of them require an ego investment that is sometimes more difficult to make than an economic one.

Screening selection process have been successfully employed in both industry and academic settings. Techniques which have been proven to be effective exist. Some of those pertinent to this discussion are below.

- Establish the philosophy and the practice that the burden-of-proof is on the applicant to demonstrate a background of high performance that would meet your standards.

- Require verification of written referrals and recommendations from all schools and previous employers. These should cover all elements of essential performance in your operations.

- Require that applicants submit a complete file of letters of recommendation as well as employment and education histories with the application blank.

- Enforce the practice that without a complete and verifiable file of recommendation letters, as well as training and certification documents, there will be no further review of the application.

- With those applicants who meet all other criteria, follow up on these letters of referral and recommendation by making personal, manager to manager, contact with the individuals from the key schools and employers.

- Build your pilot screening program around the idea that the best single predictor of future performance of a pilot is their past performance as a pilot.

- Openly announce that only the cream-of-the-crop pilots are eligible to be hired by your organization.

- Follow up on this policy with practices that insure only superlative pilots with impeccable professional and personal track records are actually hired.

- Implement the practice that your organization will not compromise its flight department quality standards to meet growth objectives.

- Treat the inventory of quality pilot skills as a resource that is vital and critical to accomplishing the corporate goals.

- Put recently hired pilots on a frequent review schedule of their professional performance. Use simulators to test performance under high workload stress.

- Assume applicants will have their best foot forward. The selection process must examine both feet and review all aspects of an applicant's professional background.

- Give the pilot selection process as much emphasis, priority, planning, and corporate support that you give to the aircraft selection process. Assume the pilot hiring process is as important as the screening of other key executives.

- Plan months ahead for pilot staffing needs of the flight department.

- Don't wait to start a pilot hiring and screening program until you need pilots bad, because by then, for sure, **you will get them bad!**

The above improvements can be very readily implemented with a modest economic investment. The economic benefits in improved pilot performance are arguably worth several times the financial investment.

CONCLUSIONS

The application of proven screening and selection policies and practices can be a major factor in keeping marginally qualified pilots out of airline flight decks.

The resolution of these pilot screening issues can make a huge step forward in reducing the risks of future pilot performance breakdowns. The fewer and fewer military trained pilot applicants available every year make the issues and problems of pilot screening ever more important.

REFERENCES

Chaney, F. B. (1991). <u>Selecting Leaders.</u> Address to The Executive Committee #17, San Diego, CA.

Hunter, D. R. and Burke, E. F. (1995). <u>Handbook of Pilot Selection.</u> Brookfield, VT: Avebury Publishing..

National Transportation Safety Board. (1967). <u>Aviation Accident Report, Paul Kelly Flying Service, Inc., Lear Jet 23, N243F, Palm Springs, CA, November 14, 1965.</u> (NTSB/AAR 06-67). Washington, DC: Author.

National Transportation Safety Board. (1985). <u>Aircraft Accident Report: Air Illinois, Inc., Hawker Siddley HS748-2A, N748LL, near Pinckneyville, IL, October 11, 1983.</u> (NTSB/AAR-85-03). Washington, DC: Author.

National Transportation Safety Board. (1986). <u>Aircraft Accident Report: Henson Airlines Flight 1517, N339HA, Beech B99, Grottoes, VA, September 23, 1985.</u> (NTSB/AAR-86-07). Washington, DC: Author.

National Transportation Safety Board. (1988). <u>Aircraft Accident Report, Continental Airlines, Inc., Flight 1713, McDonnell-Douglas DC-9-14, N626TX, Stapleton International Airport, Denver, CO, November 15, 1987.</u> (NTSB/AAR-88-09). Washington, DC: Author.

National Transportation Safety Board. (1991). <u>Aircraft Accident Report, Ryan International Airlines DC-9-15, N565PC, Loss of Control on Takeoff, Cleveland-Hopkins International Airport, Cleveland, OH, February 17, 1991.</u> (NTSB/AAR-91/09). Washington, DC: Author.

National Transportation Safety Board. (1994a). <u>Aircraft Accident Report, Uncontrolled Collision With Terrain, American International Airways, Flight 808, Douglas DC-8061, N814CK, U.S. Naval Air Station, Guantanamo Bay, Cuba, August 18, 1993.</u> (NTSB/AAR-94/04). Washington, DC: Author.

National Transportation Safety Board. (1994b). <u>Aircraft Accident /Incident Summary Report, In-Flight Loss of Control Leading to Forced Landing and Runway Overrun, Continental Express, Inc. N24706, Embraer EMB-120RT, Pine Bluff, AR, April 29, 1993.</u> (NTSB/AAR-94/02/SUM). Washington, DC: Author.

National Transportation Safety Board. (1995). <u>Aircraft Accident/Incident Report, Uncontrolled Collision with Terrain, Flagship Airlines, Inc., dba American Eagle Flight 3379, Bae Jetstream 3201, N918AE, Morrisville, North Carolina, December 13, 1994.</u> (NTSB/AAR-94/07). Washington, DC: Author.

Owens, W. A. (1976). "Background Data." In M.D. Dunnette, (Ed.), <u>Handbook of Industrial and organizational psychology.</u> Chicago, Rand-McNally.

CLUTTER AND DISPLAY CONFORMALITY:
CHANGES IN COGNITIVE CAPTURE

Brittisha N. Boston and Curt C. Braun
Department of Psychology
University of Idaho, Moscow

HUD technology has provided pilots with a variety of performance advantages over traditional head-down cockpit displays. Reduced visual switching times, increased situational awareness, and a variety of flight performance improvements have all been associated with HUDs. Despite these advantages, research has identified some possible HUD shortcomings. Objective and subjective data suggest HUDs increase the difficulty associated with switching one's attention among stimuli within the same visual space, causing pilots to inappropriately fixate on the display, a state termed cognitive capture. Possible factors that might affect cognitive capture include display conformality, clutter, and perceptual separation. The present research used reaction time data to assess the effects of these factors on cognitive capture. Sixty participants completed a simulated ship navigation task using either a conformal or non-conformal HUD. Each of four trials was factorial for display clutter (low, high) and obstacle probability (low, high). Reaction time data showed a performance advantage for conformal displays particularly in the high clutter condition. Subjective data revealed that the conformal display was less distracting, appeared to depict fewer instruments, and required less effort to attend to the environment than the non-conformal display. Findings concerning perceptual separation were inconclusive.

INTRODUCTION

Head-up display (HUD) technology has afforded pilots with a variety of performance advantages. By design, the superimposition and collimation of information reduces the time normally required to switch visual focus from traditional cockpit displays to the environment (Abrams, Meyer, and Kornblum, 1989; Haines, Fischer, and Price, 1980; Larry and Elworth, 1972). Moreover, conformal symbology (representations of real-world features that conform to environmental contours) has been shown to enhance situational awareness (Weintraub and Ensing, 1992). Finally, the integration of flight path information on the HUD reduces the requirement that pilots perform this task (McCann and Foyle, 1994).

In addition to improving switching time and situational awareness, HUDs have been shown to enhance flight performance. Boucek, Pfaff, and Smith (1983), for example, reported that the HUD produced significantly better flight path tracking than traditional displays. Similar performance differences have also been reported for measures of airspeed and altitude maintenance, touchdown location variability, number of missed approaches, and time to detect expected events (Boucek, Pfaff, and Smith, 1983; Brickner, 1989; Desmond, 1983; Johnson, 1990; Larish and Wickens, 1991; Wickens, Martin-Emerson, and Larish, 1993).

This improved performance is echoed in subjective data from pilots who report an overwhelming preference for HUD systems over traditional displays. When contrasting display formats, pilots find flying easier when using a HUD; believe the HUD increases their approach precision while reducing their workload, and find the HUD particularly helpful in low visibility departures because it reduces head-down time (Boucek, Pfaff, and Smith, 1983; Fischer, Haines, and Price, 1980; Foyle and McCann, 1994; Haines, Fischer, and Price, 1980; Long and Wickens, 1994).

HUD Disadvantages

The HUD is not without its shortcomings. Misaccommodation, disorientation leading to poor recovery from unusual attitudes, and cognitive capture have all been reported (Becklen and Cervone, 1984; DeBack, 1987; Fischer, Haines, and Price, 1980; Hull, Gill, and Roscoe, 1982; Iavecchia, Iavecchia, and Roscoe, 1988; Long and Wickens, 1994; Norman and Ehrlich, 1986; Roscoe, 1987a, 1987b; Weintraub and Ensing, 1992; Weintraub, Haines, and Randle, 1985). The latter phenomenon, cognitive capture, is the focus of this research.

Cognitive capture. Difficulties associated with cognitive capture center around the ability to effectively switch one's attention from the HUD to other elements in the same visual space. Unlike switching attention from the head-down instrument panel to the environment, this cognitive switch of attention does not involve a change in fixation, visual accommodation, or convergence but involves only the mental act of shifting one's attention among stimuli within the same visual space (Haines, Fischer, and Price, 1980; Weintraub, Haines, and Randle, 1984). Cognitive capture occurs when a pilot fails to switch his attention, thus inappropriately fixating on HUD symbology at the expense of other HUD or external scene information.

Difficulties switching attention have been reported by pilots flying HUD-equipped aircraft. Fischer, Haines, and Price (1980), for example, noted that several pilots reported having to consciously force the attentional switch from the HUD to the external scene. Similar reports were also noted by Lauber, Bray, Harrison, and Hemingway (1982).

Assessing cognitive capture, Becklen and Cervone (1984) found that only 30% of the participants viewing superimposed images noticed unexpected events. In an aircraft simulation, Fischer, Haines, and Price (1980) found that cognitive capture occurred with the HUD but not with conventional head-down displays. Specifically, all four of the subject pilots flying with a conventional display detected an unexpected runway obstacle within 1 to 3 seconds. Of the four pilots flying with HUDs, two detected the obstacle within 5 seconds, the other two never saw the obstacle. Although no statistical analysis was performed to determine the significance of these results, they are consistent with those obtained by Haines, Fischer, and Price (1980) and Long and Wickens (1994). Wickens, Martin-Emerson, and Larish (1993) found a 6-second disadvantage for the HUD as compared to a traditional head-down display in detecting runway obstacles although the difference did not reach conventional significance levels.

Although most of the HUD research clearly points to cognitive capture as a problem, no specific cause has been given. Subjective reports have indicated that cognitive capture might be caused by compelling symbology (Boucek, Pfaff, and Smith, 1983; Fischer, Haines, and Price, 1980; Weintraub, 1994), excessive clutter on the HUD (Fischer, Haines, and Price, 1980; Johnson, 1990; Weintraub, Haines, and Randle, 1985), or the amount of perceptual separation between the HUD symbology and the environment scene (McCann and Foyle, 1994; Wickens, Martin-Emerson, and Larish, 1993).

Purpose Statement

The aim of the present research was to empirically evaluate the effect of display symbology (conformal vs. non-conformal), clutter, and to a lesser degree, perceptual separation, on cognitive capture. To assess the effect of these variables, participants completed a ship navigation task where conformal and non-conformal displays were used. Each display type was presented with two levels of clutter (high and low). The probability of an obstacle appearing in the path of the ship was also varied.

METHOD

Participants

Participants included 23 male (mean age = 21.74) and 37 female (mean age = 21.48) volunteers. All participants were screened to ensure normal or corrected-to-normal vision. Participants were given course credit for participating.

Apparatus

The experimental task consisted of a computerized simulation of a ship maneuvering through a marked shipping channel. The use of a shipping rather than flying scenario allowed for the presentation of multiple obstacles per trial. Moreover, it reduced the level of skill needed to perform the task.

Displays. The simulation was created using a graphically-based programming language. The HUD imagery consisted of a course guidance instrument, speed and ocean depth indicators, and clutter elements (see Figure 1). In the conformal symbology condition, the course guidance instrument depicted two lines indicating the course and a circle identifying the ship's position. Course guidance in the non-conformal condition was depicted by a needle and a range of dots representing course deviations. The non-conformal display mimicked the traditional course deviation indicator used in aviation. A forcing function was used to introduce a course perturbation every 20 seconds.

Located to each side of both displays were speed and ocean depth instruments. Indicators on both instruments fluctuated and moved out of a predetermined range once every 20 seconds. Finally, clutter was manipulated by presenting meaningful, static, alphanumeric character sets at the top and bottom of the display. The four display combinations are shown in Figure 1.

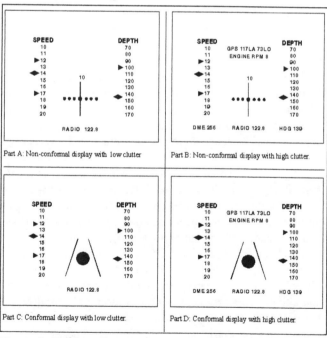

Figure 1. Conformal and non-conformal displays shown with low and high levels of clutter.

Environment. The environment scene depicted an open water shipping channel marked along the right side with red buoys. Three different obstacles appeared in the environment at random intervals. These obstacles included a log, a speed boat, and a freighter. The size of these obstacles were scaled to be equivalent to the visual angle subtended by a Cessna 182, Boeing 737, and a Boeing 747 viewed at a height of 30.48 meters on a 3 degree glideslope. Obstacles were presented using two different probabilities: .03 and .09. When presented, a obstacle would remain on the screen for up to 10 seconds or until it was cleared by the subject. The environment and the obstacles are shown in Figure 2.

Figure 2. Environment scene with buoy and three obstacles.

Hardware. The displays were presented using an IBM Model 80 PC and the environment was presented on a 486 66 MHz PC. Both systems used high resolution SVGA monitors. The monitor depicting the displays was mounted perpendicular to the environment monitor and a beam splitter was used to superimpose the images. All display images were mirrored so that they appeared correctly when reflected from the beam splitter. Participants were seated 80cm from the center of the beam splitter and instructed to adjust the height of the chair until their eyes were perpendicular to the center of the beam splitter.

Design and Procedure

The factorial combination of display type, clutter, probability, and obstacle type created a 2 (display) x 2 (clutter) x 2 (probability) x 3 (obstacle) mixed model design. Display served as the between-groups factor. All remaining independent variables were presented as within-groups factors.

Participants were randomly assigned to either the conformal or non-conformal conditions. Within each condition, participants completed one training trial and four data collection trials. During the training trial, participants received instruction concerning the displays and the environment. Concerning the displays, participants were instructed to monitor the ship's course, depth, and speed. Subjects were also instructed to maintain the ship's course by pressing the left and right cursor keys and to note depth and speed deviations by pressing the up cursor key. Although subjects

noted deviations in course, depth, and speed, these responses were not recorded. For the environment, participants were instructed to watch for possible obstacles in the ship's path. Examples of each obstacle were shown and participants were instructed to remove the obstacles by depressing the appropriately marked key on the keyboard. The time required to clear the obstacle was recorded. Each trial lasted five minutes.

Following each trial, participants completed a performance questionnaire. Items on the performance questionnaire assessed the level of distraction caused by the HUD; perceptions concerning number of instruments displayed; the extent to which participants fixated on the HUD; the extent to which the displays appeared separated from the environment; the level of environmental occlusion caused by the HUD; the level of difficulty associated with attending to the environment, and the ease with which obstacles could be detected. All responses were made using 9-point Likert scales.

RESULTS

A preliminary examination of the data revealed large differences in the variability among data from the three obstacles. The data variability for the log obstacle was seven times as large as those for the boat or freighter. This variability is due the large difference in the number of instances each obstacle was not detected at all (i.e., RT = 10 sec). For this reason, data for the log obstacle were dropped from further analyses.

Reaction Time

The obstacle latency data was analyzed in a 2 (display) x 2 (clutter) x 2 (probability) x 2 (obstacle) mixed-model Analysis of Variance (ANOVA). Display was a between-groups factor and the remaining were within-groups factors. The analysis revealed a significant main effect for Display $F(1, 57) = 5.19$, $p < .05$ and a significant Display x Clutter interaction, $F(1, 57) = 4.51$, $p < .05$. This interaction is shown in Figure 3.

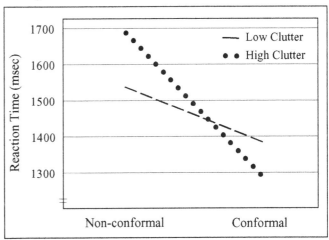

Figure 3. Mean response times to obstacles in low and high clutter conditions by display type.

A test of simple effects indicated that the difference between display types in the high clutter condition was significant, $p <$ .05. Means and standard deviations are shown in Table 1.

Display	Clutter	
	Low	High
Non-conformal	1535 (927)	1687 (1158)
Conformal	1384 (735)	1282 (636)

Table 1. Means and (Standard Deviations) for Reaction Times to Obstacles by Display and Clutter (milliseconds).

Participant Performance

The items on the performance questionnaire were correlated to identify potential redundancies. Items concerning HUD-caused distractions and environmental occlusion were all highly correlated and therefore averaged into one item called HUD distractions. All performance items were evaluated using 2 (display) x 2 (clutter) x 2 (probability) mixed-model ANOVAs.

HUD distractions. A significant main effect for Display was noted, $F(1, 58) = 4.72$, $p < .05$. Participants reported significantly higher levels of distraction associated with the non-conformal (M = 3.34, SD = 1.72) than the conformal display (M = 2.51, SD = 1.57), $F(1, 57) = 4.71$, $p < .05$. No other main effects or interactions were significant.

Excessive HUD instruments. This item asked participants to report the extent to which they felt the instrumentation on the HUD was excessive. Despite the fact that the two displays contained the same number of instruments, significant differences between displays were observed, $F(1, 57) = 4.71$, $p < .05$. Subjects rated the non-conformal HUD instrumentation as more excessive (M = 3.87, SD = 2.22) than the conformal HUD instrumentation (M = 2.88, SD = 1.98). No other main effects or interactions were significant.

Separation of display and environment. Data concerning the extent to which the display appeared separate from the environment revealed a significant Clutter x Probability interaction, $F(1, 58) = 5.79$. $p < .05$. As shown in Figure 4, ratings of display or perceptual separability increased in the low clutter condition as the probability went from low to high. The opposite, however, was observed in the high clutter condition where ratings of separability decreased from the low probability to high probability condition. No other main effects or interactions were observed.

Difficulty attending to environment. Significant differences were noted between the two displays, $F(1, 58) = 7.53$, $p < .05$. Participants reported greater difficulty attending to the environment in the non-conformal (M =4.19, SD = 2.32) than the conformal condition (M = 2.97, SD = 1.85). No other main effects or interactions were observed.

Other items. Data from questionnaire items concerning the extent to which participants fixated on the HUD and the ease with which obstacles could be detected did not yield any significant main effects or interactions.

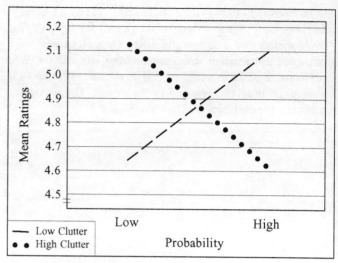

Figure 4. Mean ratings of display/environment separation by clutter and probability. (1 = strongly disagree, 9 = strongly agree).

DISCUSSION

The current findings provide additional evidence supporting the utility of conformal display formats. Overall, obstacle response times were lower in the conformal than the non-conformal condition. Moreover, the greatest advantage of the conformal over the non-conformal display was seen in the high clutter condition. In addition to superior performance, subjective ratings indicate that the conformal display was less distracting, appeared to depict fewer instruments, and required less effort to attend to the environment than the non-conformal display.

Findings concerning the effect of perceptual separation on obstacle detection were less clear (see Figure 4). Although no simple explanation can be given for this interaction, further analyses failed to reveal a relationship between ratings of perceptual separation and obstacle detection for any of the probability or clutter conditions.

Concerning cognitive switching and capture, the current findings do not support earlier reports that conformal HUD displays increase the time needed to detect obstacles (Fischer, Haines, and Price, 1980; Haines, Fischer, and Price, 1980; Long and Wickens, 1994). The current findings do, however, reflect those reported by Wickens, Martin-Emerson and Larish (1993) who noted shorter obstacle detection times for a HUD than a traditional head-down display.

Care must be exercised when comparing the present findings with those presented by other HUD researchers. Because both displays were presented in a head-up manner, these findings are unlikely to generalize to studies contrasting traditional head-down cockpit displays and HUDs. Furthermore, the lack of performance data concerning the ship's path and the accuracy of monitoring might warrant additional caution when interpreting the findings.

With these limitations in mind, the current findings have shown: 1) superior performance of conformal displays as measured by the time to detect obstacles; 2) that high levels of display clutter are more likely to adversely affect non-conformal rather than conformal displays, and 3) that subjective ratings of perceptual separation do not appear to be related to objective measures of performance.

Future research should evaluate the effects of symbology type and level of clutter on cognitive capture by replicating this study using an aviation scenario.

REFERENCES

Abrams, R. A., Meyer, D. E., and Kornblum, S. (1989). Speed and accuracy of saccadic eye movements: Characteristics of impulse variability in the oculomotor system. Journal of Experimental Psychology: Human Perception and Performance, 15, 529-543.

Becklen, R., and Cervone, D. (1984). Selective looking and the noticing of unexpected events. Memory and Cognition, 11, 601-608.

Boucek, G. P., Pfaff, T. A., and Smith, W. D. (1983, October 3-6). The use of holographic head-up display of flight path symbology in varying weather conditions. SAE Technical Paper Series, 831445, 103-109.

Brickner, M. S. (1989). Apparent limitations of head-up-displays and thermal imaging systems. In R. S. Jensen (Ed.), Proceedings of the Fifth International Symposium on Aviation Psychology (pp. 703-707). Columbus, OH: Ohio State University, Department of Aviation.

DeBack, N. J. (1987, October 26-29). Heads-up display (HUD) human factors issues. Paper presented at the Flight Safety Foundation 40th International Air Safety Seminar, Tokyo, Japan.

Desmond, J. P. (1983, October 3-6). A holographic head-up display for low visibility landing operations. SAE Technical Paper Series, 831451.

Fischer, E., Haines, R. F., and Price, T. A. (1980). Cognitive issues in head-up displays. NASA Technical Paper, 1711. NASA Ames Research Center, Moffett Field, CA.

Foyle, D. C., and McCann, R. S. (1994, February 1-3). Attentional effects with superimposed symbology. Paper presented at the Flight-Worthy Helmet-Mounted Display and Symbology for Helicopters Workshop, NASA Ames Research Center, Moffett Field, CA.

Haines, R. F., Fischer, E., and Price, T. A. (1980). Head-up transition behavior of pilots with and without head-up displays in simulated low-visibility approaches. NASA Technical Paper, 1720. NASA Ames Research Center, Moffett Field, CA.

Hull, J. C., Gill, R. T., and Roscoe, S. N. (1982). Locus of the stimulus to visual accommodation: Where in the world, or where in the eye? Human Factors, 24, 311-319.

Iavecchia, J. H., Iavecchia, H. P., and Roscoe, S. N. (1988). Eye accommodation to head-up virtual images. Human Factors, 30(6), 689-702.

Johnson, T. (1990, October 1-4). Alaska Airlines Experience with the HGS-1000 head up guidance system. SAE Technical Paper Series, 901828.

Larish, I. A., and Wickens, C. D. (1991). Divided attention with superimposed and separated imagery: Implications for head-up displays. Aviation Research Laboratory Technical Report. University of Illinois, Urbana, IL: ARL Institute of Aviation.

Larry, C., and Elworth, C. L. (1972). The effects of pilot age, lighting, and head-down time on visual accommodation (Report No. D162-10378-1 TN (REV LTR)). Seattle, WA: The Boeing Company.

Lauber, J. K., Bray, R. S., Harrison, R. L., and Hemingway, J. C. (1982). An operational evaluation of head-up displays for civil transport operations. NASA Technical Paper, 1815. NASA Ames Research Center, Moffett Field, CA.

Long, J. and Wickens, C. D. (1994). Implications of object vs. space-based theories of attention in the design of the aircraft head-up display. Aviation Research Laboratory Technical Report. University of Illinois, Urbana, IL: ARL Institute of Aviation.

McCann, R. S., and Foyle, D. C. (1994, October 3-6). Superimposed symbology: Attentional problems and design solutions. SAE Technical Paper Series, 942111, 43-50.

Norman, J., and Ehrlich, S. (1986). Visual accommodation and virtual image displays: Target detection and recognition. Human Factors, 28, 135-151.

Roscoe, S. N. (1987a). The trouble with HUDs and HMDs. Human Factors Society Bulletin, 30, 1-3.

Roscoe, S. (1987b). The trouble with virtual images revisited. Human Factors Society Bulletin, 30, 3-5.

Weintraub, D. J. (1994, February 1-3). HMD symbology design from the ground up. Paper presented at the Flight-Worthy Helmet-Mounted Display and Symbology for Helicopters Workshop, NASA Ames Research Center, Moffett Field, CA.

Weintraub, D. J., and Ensing, M. (1992). Human factors issues in head-up display design: The book of HUD (SOAR 92-2). Wright-Patterson Air Force Base, Dayton, OH: CSERIAC.

Weintraub, D. J., Haines, R. F., and Randle, R. J. (1984). The utility of head-up displays: Eye-focus vs. decision times. Proceedings of the 28th Annual Meeting of the Human Factors Society (pp. 529-533). Santa Monica, CA: Human Factors Society.

Weintraub, D. J., Haines, R. F., and Randle, R. J. (1985). Head-up display (HUD) utility II: Runway to HUD transitions monitoring eye focus and decision times. Proceedings of the Human Factors Society 29th Annual Meeting (pp. 615-619). Santa Monica, CA: Human Factors Society.

Wickens, C. D., Martin-Emerson, R., and Larish, I. A. (1993). Attentional tunneling and the head-up display. In R. S. Jensen (Ed.), Proceedings of the Seventh International Symposium on Aviation Psychology (pp. 865-870). Columbus, OH: Ohio State University.

THE EFFECT OF CLUTTER AND LOWLIGHTING SYMBOLOGY ON PILOT PERFORMANCE WITH HEAD-UP DISPLAYS

Patricia May Ververs and Christopher D. Wickens
University of Illinois at Urbana-Champaign
Aviation Research Laboratory, Savoy, Illinois

The problems of focusing attention while using head-up displays has been noted in the literature. To examine the issue, twenty-four instrument-rated pilots from the University of Illinois flew a high-fidelity simulator during cruise flight using head-up and head-down displays. The pilots were presented with either a symbology set with the bare essentials to fly the simulation or an enhanced set of information. The intensity of the display symbology was manipulated, including a condition lowlighting non-essential flight task information. Flying with a HUD in good weather conditions provided a clear advantage to tracking performance and event detection over the head-down conditions. This result was found presumably through the extraction of attitude information from the two domains, the near symbology and the far horizon. Clutter was found to slow the detection of changes on the symbology and detection of targets in the environment. However, lowlighting the non-essential information begins to ameliorate this problem. The appropriate combination of location, intensity, and contrast modulated visual attention between the symbology and the environment and produced a win-win situation for HUDs.

The use of head-up displays (HUDs) has a long history in the military dating back to the 1960s (Weinstein, Ercoline, McKenzie, Bitton, & Gillingham, 1993). However, only recently has their application been introduced to general aviation. Much of general aviation (GA) operations involves flight in visual meterological conditions (VMC), where one of the tasks of the pilot is to "see and avoid" other traffic, a task which is impossible when the eyes are down in the cockpit. There has been only one other laboratory study, of which we are aware, in which mid-air detection has been systematically examined in HUD use (May & Wickens, 1995). Therefore, we have created an environment in which to study pilot performance during typical GA tasks.

How a pilot processes the information on a HUD is a direct result of how effectively focused attention is modulated. The pilot can focus attention toward the near domain symbology information while filtering the far domain environment or the pilot can focus attention on far domain objects such as the horizon, landmarks, or other mid-air targets while ignoring the head-up instrumentation. Inappropriate focused attention can lead to attentional tunneling or attentional capture and has been reported as the culprit in the loss of performance in detecting anomalies in a number of HUD studies (Fischer, Haines, & Price, 1980; Larish & Wickens, 1991; Wickens & Long, 1995). It is the goal of this study to investigate how focused attention is modulated between the HUD and the environment while three display parameters, discussed below, are manipulated.

Information and Clutter. There is a danger that displays may become cluttered with information which is less relevant

to proper flight control, an unfortunate occurrence despite research revealing the potential problems of added information both with basic attentional tasks (B.A. Eriksen and C.W. Eriksen, 1974) and real-world tasks such as reading an instrument panel for driving (Kurokawa & Wierwille, 1991) or flying (Wickens & Andre, 1990).

Location. The location to display information continues to be a source of debate in the research of head-up displays. In the head-up location, the symbology is scanned without the pilot having to move his or her eyes inside the cockpit and provides continual information regarding some far domain objects such as the horizon, regardless of the weather conditions. However, this added information also overlaps the far domain scene making it potentially harder to detect objects in the environment such as other aircraft.

Intensity and Contrast. The contrast of the symbology with its background has important implications for the ability to process the information. Weintraub and Ensing (1992) recommend at least a 1.5/1 luminance-contrast ratio in moderate ambient illumination conditions. Since the background (i.e., the sky) of the head-up display is continually changing it is important to provide adequate contrast ratios for the HUD. However, a very bright display of a large block of information can be distracting to a pilot and may mask the detection of potential aircraft targets in the far domain. One solution may be to highlight essential information for flight information and lowlight all other information (Martens & Wickens, 1995). This allows all information to be visible yet makes less crucial information perceptually distinct and therefore easier to filter.

THE EXPERIMENT

This experiment replicates many aspects of the paradigm and manipulation used by May and Wickens (1995), in which the joint effects of location and image contrast were examined as these were modulated by weather conditions. However, their simulation was low fidelity and a higher fidelity simulation is used here. In the current experiment, we also explicitly vary the amount of clutter along with the different weather conditions. Furthermore, we introduce a lowlighting intensity manipulation.

Clutter. Two levels of information were presented to the pilots. The low clutter condition provided the minimal amount of information needed to fly the simulation. The high clutter condition displayed the same symbology information as the low condition, but also included less relevant task information such as bank angle, vertical velocity, and navigational waypoints. It was expected that the added information on the HUD would disrupt scanning patterns and the general processing of information leading to slower detections of far domain targets.

Location. The symbology sets were located in either the head-up or head-down locations. The head-up location was expected to provide a less effortful means of extracting information from both domains by reducing scanning requirements. Previous research suggests that sufficient contrast between the outside environment and the symbology is required for the HUD performance to equal that of the HDD (Fischer, 1979; May & Wickens, 1995; Wickens & Long, 1995). Therefore, the HUD was expected to result in at least equal flightpath tracking performance and response times to the detection of near domain symbology events to those obtained when pilots used the head-down display. Additionally, with appropriate contrasts between the symbology and the environment, good weather conditions would be expected to provide the best tracking performance since attitude information would be available from the far domain horizon as well as the symbology.

Intensity and Contrast. Two levels of lighting intensity were used in three different formats: low intensity, high intensity, high intensity for essential information with the non-essential information lowlighted. Of course the actual contrast ratios varied depending on the particular weather conditions. The same contrast ratios were achieved in both display locations by manipulating the luminance of the background against which the head-down symbology was presented so that the ratio was equal to that of the head-up. Therefore any difference in display location would be due to the location of the symbology and not the contrast ratio. The contrast ratios used surpass the minimum standards as recommended by Weintraub and Ensing (1992) for typical viewing conditions and averaged about a 2:1 ratio. Of course the actual contrast ratios vary depending on the particular weather conditions, though the differences were slight. In the clear conditions, the low intensity symbology had a contrast ratio of 1.9/1 when the ratios of the different areas of

the background were averaged. The high intensity symbology had a weighted average of 2.2/1. In the cloudy conditions, much of the ground was not decipherable through the clouds so there were only two contrast ratios, low intensity provided a ratio of 2.0/1 and high intensity was 2.1/1. It was assumed that the high intensity conditions would provide the easiest extraction of information by the pilots. However, in the high clutter condition this may become distracting and therefore disrupt detection of far domain events. Lowlighting was expected to provide two perceptually distinct information sets. The brighter intensity (and most essential) information would be perceived as inherently more important information, while the lowlighted symbology would have a lower perceived importance but would still be available for processing. Therefore, the lowlighting condition was expected to provide the means for processing all the information with minimal disruption of both far and near domain event detection. This would be evident if the high intensity in the low clutter condition does not differ from the lowlighting condition in the high clutter condition.

METHOD

Design. A repeated-measures 2 x 2 x 3 x 2 factorial design was employed. There were two locations of the symbology, two levels of instrumentation information, three lighting conditions, and two weather conditions, which dictated the background luminance.

Subjects. Twenty-four paid volunteers participated in the study. All were licensed pilots with between 150 to 15,000 flight time hours, the average being 1,430 hours.

Apparatus/Symbology. The outside environment of the high-fidelity simulation was generated by an Evans and Sutherland SPX500 graphics display generator and presented on two 3.0 x 2.2 m screens at a distance of 3.2 meters. The symbology was created by a Silicon Graphics IRIS workstation. In the head-up location the symbology set was overlaid on the outside environment covering a visual angle of 25 and 16 degrees across in the high-level and low-level information, respectively and 19 and 13 degrees vertically. In the head-down location the symbology was presented on a 16-inch monitor located 64 cm from the pilot with the high and low information sets with visual angles of 25 and 18 degrees across, respectively and subtended 11 and 8 degrees vertically. The visual angles of the high information sets in both of the display locations were equated to the Frasca simulator used by all students of the Institute of Aviation. Figure 1 shows the cluttered version of the instrumentation set used by the pilots. (See figure on the last page of paper.)

Procedure. The pilots flew a windy segmented path as commanded by discrete changes in three parameters: heading, airspeed, and altitude, where only one parameter changed at a time. A two-axis joystick with a throttle button was used to implement standard flight dynamics. New command values appearing in boxes near the individual display instruments defined the near domain events. The pilot responded to the event by pressing a button as soon as the command value was detected on the heading, airspeed, or altitude indicator. The far domain event was the appearance of a helicopter at a distance within the pilot's forward field of view. In the head-up location the helicopter always initially appeared behind the instrumentation. The pilot detected this event by squeezing the trigger of the joystick.

The pilots completed a practice block of trials where they became familiar with the flight dynamics and event types. Halfway through each experimental block the pilot flew into a cloudy region where the horizon and ground were no longer visible and then the second half of the events would occur. The helicopter targets were not obscured by the clouds. There were ten blocks of trials with four iterations of each of the heading, altitude, airspeed and helicopter events in each block. A post-experiment questionnaire surveying personal preferences was completed at the end of the session. The entire session was completed in 2.5 hours.

RESULTS

Tracking Error. Mean Absolute Error (MAE) was recorded for the tracking of the three axes of control. As depicted in figure 2, tracking performance in clear weather conditions was consistently better than in cloudy conditions for all three control axes (heading: $F(1,23) = 246.7$, $p < 0.01$; altitude: $F(1,23) = 24.7$, $p < 0.01$; airspeed: $F(1,23) = 26.6$, $p < 0.01$).

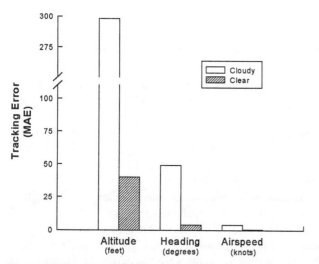

Figure 2. Tracking performance on the three axes of control in cloudy and clear conditions.

In cloudy conditions the far domain horizon is obscured from the pilot and information about the relative position of the aircraft is only available from the symbology. However, in good weather conditions, the pilots are able to extract information about the pitch and roll attitude of the aircraft from either the far domain or the pitch ladder on the symbology. The superior performance in clear conditions

provides evidence that the pilots are visually attending to the far domain horizon to stabilize the aircraft, an information channel that provides higher resolution change (particularly in the roll axis) then does the artificial horizon.

Response Time to Near Domain Symbology Events. Repeated Measures ANOVAs revealed a significant main effect of clutter ($F(1,23) = 15.08$, $p < 0.01$) indicating a 1/2 second faster detection time for the near domain events (command path changes) when the symbology set was limited to only essential information. A main effect of location ($F(1,23) = 38.77$, $p < 0.01$) indicated a 700 millisecond advantage in detecting near domain events for the head-up location. Response times to symbology events were slower in the clear conditions ($F(1,23) = 12.63$, $p < 0.01$). This would be expected since attention was apparently focused on the far domain in the clear weather conditions, as was shown in tracking data, and would need to be drawn inward to the symbology in order to detect changes. There was also a significant interaction between the weather conditions and the location of the symbology ($F(1,23)$ 6.20, $p < 0.05$). As shown in figure 3, while clear weather had a small effect on response time to symbology events in the head-up condition, it had a particularly degrading effect in the head-down condition.

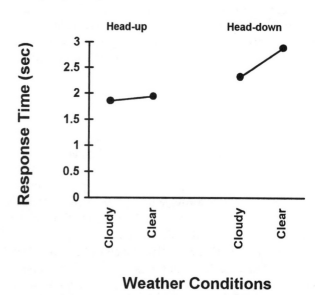

Figure 3. Effects of weather on detection of near domain symbology events in head-up and head-down conditions

The effects of the different intensity levels and clutter conditions were first analyzed without the lowlighting condition. Clutter was found to slow detection of events on the symbology ($F(1,23)$ 15.08, $p < 0.01$), while there was a marginally significant effect of the lower intensity slowing the detection of the symbology events ($F(1,23)$ 3.13, $p = 0.09$). Three planned comparisons were made to analyze the effect of lowlighting. The significance level was adjusted using the Bonferroni procedure. Paired t-tests were evaluated using a familywise rate of alpha = 0.05 netting a significance level of .017 for each of the comparisons. As would be expected, response times to the events on the brighter and less cluttered (high intensity, low clutter) information set were faster than to events on either of the cluttered displays (high intensity, high clutter: $t(24) = 3.23$, $p < 0.01$; low intensity, high clutter: $t(24) = 3.32$, $p < 0.01$). But the responses using this same bright, small information set were only marginally faster than the lowlighting condition ($t(24) = 2.21$, $p = 0.04$). Therefore, highlighting the main components of the symbology set, while lowlighting the less essential information may have made it easier for the pilots to concentrate attention of the primary instruments.

Response Time to Far Domain Event. There was a significant decrement to detection of the helicopters when the added clutter information was presented to the pilots ($F(1,23) = 14.20$, $p < 0.01$). The effect was evident and of equal magnitude in both the head-up and head-down conditions. The weather (clear and cloudy) conditions did not influence the detection of far domain events ($F(1,23) = 0.51$, $p = 0.48$). Overall, the pilots were significantly slower in detecting the mid-air targets when the symbology was presented in the head-down location as compared to the head-up presentation ($F(1,23) = 54.53$, $p < 0.01$). This effect of location mirrors that found with near domain events. See Figure 4.

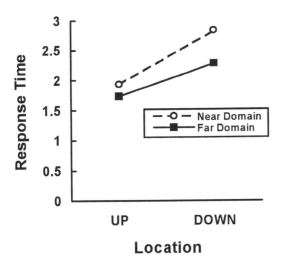

Figure 4. Effect of display location on the detection of near and far domain events.

The combined and individual effects of intensity and clutter were first analyzed without the data from the lowlighting condition. While clutter was shown above to significantly slow the detection of far domain targets, there was no main effect of intensity of the symbology ($F(1,23) = 1.72$, $p = 0.20$). There was also no main effect of the interaction between clutter and intensity ($F(1,23) = 1.58$, $p = 0.22$). Paired t-tests were used to investigate the effects of the lowlighting condition on detection of helicopters. The significance level was again adjusted to yield a total familywise comparison level equal to .05 and .017 for individual comparisons. Significant faster response times were found for the high intensity, low clutter condition as

compared to the high intensity, high clutter condition ($t(24) = 3.23$, $p < 0.01$), but there was not a significant difference between high intensity, low clutter display set and the lowlighting condition ($t(24) = 1.46$, $p = 0.16$). Therefore, the lowlighted additional information could be provided without disrupting response time to the helicopters in the far domain.

DISCUSSION AND PRACTICAL IMPLICATIONS

This experiment investigated the potential problems encountered when using head-up displays in general aviation aircraft and identified steps to be taken to ameliorate these problems. A potentially damaging situation is the presentation of too much information in a cluttered display. This experiment revealed that added clutter adversely affected the pilots' performance not only in processing information beyond the symbology, but also in processing the symbology itself. Lowlighting offered the ability to continuously present all the information with less apparent distraction to the pilot. The presence of lowlighted material had no influence on detecting the far domain targets, and a modest slowing effect on detection events on the symbology itself. Evidence from the attention literature confirms that a person can easily select information for processing according to color or intensity (Christ, 1975; Wickens & Andre, 1990). Lowlighting creates perceptually distinctive subsets of information with an intuitive weighting of the more important information (Martens & Wickens, 1995).

The head-up location of the symbology also improved pilots' performance, as the overlapping display presumably reduced the amount scanning. This result came without imposing a cost to response time to either the near or far domain targets. It appears that the pilots could readily switch attention between the two superimposed fields, particularly when lowlighting or decluttering were implemented.

Manipulation of weather conditions provided ample evidence that the pilots' attention was located in the far domain environment in the clear weather conditions. Here, the true horizon provided higher resolution information for control of attitude resulting in better tracking performance and faster detection to events on the symbology when the weather was clear and the display was head-up. There are two potential explanations of this effect. First, the data suggest that attention needed to be redirected when switching between the far horizon and the near symbology set. Noteworthy is the fact that the response time to near domain events was the slowest when the symbology was presented in the head-down location and the weather was clear. Under these conditions we assume that the pilots' attention was focused on the far domain and would need to redirected not only inward but also downward to respond to the events on the symbology set. An alternative explanation deals with reaccommodation. Since the display in the head-up location was overlaid on the far screen 5 meters away and the head-down was located less than one meter away, the pilots' eyes would need to reaccommodate between the near and far locations. This might account for the difference in response time when the pilots need to extract information from both domain locations. More detailed analyses need to be performed in order to resolve these issues.

These experimental results offer a number of suggestions when HUDs are implemented into the general aviation aircraft. First, well-designed HUDs can be advantageous during cruise flight. Second, the amount of information displayed in the head-up location should be limited. A cluttered display makes it difficult to process the information. Third, if it is determined that more information should be available, then the designers should lowlight instrumentation that is non-essential to the pilot's primary task. One of the most important responsibilities of general aviation pilots is to "see and avoid" other traffic. We see here that the head-up location of the symbology better allows the pilot to continue to scan the environment and monitor the aircraft's instruments without having to go back inside the cockpit. Finally, further investigation of the modulation of attention needs to be performed utilizing actual HUDs in the real world in order for the combined effects of all variables to be realized.

ACKNOWLEDGMENTS

The authors wish to acknowledge the Federal Aviation Administration for research funding and in particular Dennis Beringer as the contract monitor of the FAA grant DTFA 95-G-049, Sharon Yeakel for her programming support, and Steve Owen for assistance with data collection.

REFERENCES

Christ, R.E. (1975). Review and analysis of color coding search for visual displays. Human Factors, 17, 542-550.

Eriksen, B.A. & Eriksen, C.W. (1974). Effects of noise letters upon the identification of a target letter in a nonsearch task. Perception & Psychophysics, 16, 143-149.

Fischer, E. (1979). The role of cognitive switching in head-up displays (NASA Report 3137). Moffett Field, CA: NASA Ames.

Fischer, E., Haines, R.F., & Price, T.A. (1980). Cognitive issues in head-up displays (NASA Technical Paper 1711). Moffett Field, CA: NASA Ames Research Center.

Kurokawa, K. & Wierwille, W.W. (1991). Effects of instrument panel clutter and control labeling on visual demand and task performance. Proc of the Society for Info Display, 22, 99-102.

Larish, I. & Wickens, C.D. (1991). Divided attention with superimposed and separated imagery: Implications for head-up displays. Techical Report 91-4, Savoy, IL: ARL.

Martens, M.H., & Wickens, C.D. (1995). Lowlighting solutions to display clutter. University of Illinois Institute of Aviation Technical Report (ARL-95-10/NASA-95-4). Savoy, IL: ARL.

May, P.A. & Wickens, C.D. (1995). The role of visual attention in HUDs: Design implication for varying symbology intensity. Proceedings of the 39th Annual Meeting of the Human Factors and Ergonomics Society, Santa Monica, CA.

Weinstein, L.F., Ercoline, W.R., McKenzie, I., & Gillingham, K.K. (1993). Standardization of aircraft control and performance symbology on the USAF head-up display. Armstrong Laboratory Techical Report (AL/CF-TR-1993-0088). Brooks AFB, TX.

Weintraub, D.J. & Ensing, M. (1992). Human Factors Issues in Head-Up Display Design: The Book of HUD. State-of-the-Art Report, Crew System Ergonomics Information Analysis Center, Wright-Patterson AFB, Dayton,OH.

Wickens, C.D. & Andre, A. (1990). Proximity compatibility and information display: Effects of color, space, and objectness of information integration. Human Factors, 19, 61-77.

Wickens, C.D. & Long, J. (1995). Object- vs. space-based models of visual attention: Implications for the use of head-up displays. Journal of Experimental Psychology: Applied, 1, 179-194.

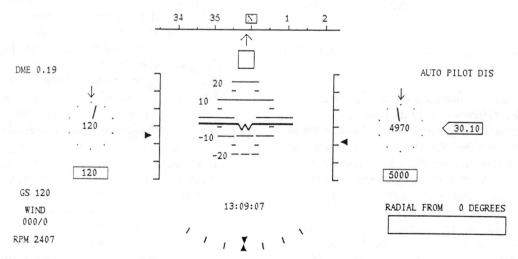

Figure 1. Cluttered instrumentation display.

THE EFFECT OF ACTIVE NOISE REDUCTION TECHNOLOGY ON NOISE INDUCED PILOT FATIGUE AND ASSOCIATED COGNITIVE PERFORMANCE DECREMENTS

John Jordan, Wayne C. Harris, Phillip N. Goernert, and John Roberts
Engineering Psychology and Human Performance Laboratory
Mankato State University
Mankato MN 56001

Fatigue in the general aviation cockpit is caused by a variety of factors including cockpit noise. Active Noise Reduction (ANR) headsets have been proposed as a method to decrease cockpit noise and thereby decrease fatigue and the cognitive performance decrements associated with fatigue. The present study contrasted the subjective fatigue and cognitive performance of Certified Flight Instructors (CFI) during a day of in-air flight instruction with ANR headsets versus conventional headsets. CFIs exhibited increased subjective fatigue and a pattern of more rapid but less accurate cognitive performance at the end of an instructional day. However, subjective fatigue and cognitive performance changes did not differ as a function of type of headset worn. The results suggest that subjective fatigue increases and cognitive performance changes occur during a day of flying but they provide no evidence that the use of ANR headsets modify these changes.

Despite steady technological advances in aviation, general aviation, commuter flights, and major airline mishaps averaged over 1000 fatalities annually between 1980 to 1988 (Baker, O'Neill, Ginzburg, and Li, 1992). As aircraft systems, avionics and powerplants have become more sophisticated and reliable, the human has become the weak link in the aviation safety chain. After reviewing 35,000 incident reports collected by the Federal Aviation Administration's Aviation Safety Reporting System, Billings and Reynard (1984) concluded that the majority of airline incidents are the result of human error. Although errors may be committed by a variety of personnel in the aviation system, the brunt of the responsibility for human errors accidents is borne by pilots. The general aviation average of 1285 deaths per year between 1980 and 1988 was responsible for 81% of aviation fatalities, and the rate of death per 100,000 hours in general aviation was 1.68 compared to a rate of 0.04 in scheduled airline service (Baker, et al., 1992).

While deficient piloting skills cause accidents, fatal accidents are more likely to be the result of decision making errors. O'Hare, Wiggins, Batt, and Morrision (1994), reported that decision making errors comprised only 35% of the aviation accidents they investigated, but they accounted for 62.5% of the fatalities. Jensen and Benel (1977) found that over half of all fatal accidents reviewed in an FAA study of general aviation accident data were related to poor pilot

judgment. Cognitive performance of pilots is therefore an important safety issue. Cognitive effects are of increasing concern in aviation because while the modern cockpit decreases physical demands, it increases cognitive demands on pilots (Jensen, 1989). Greater aircraft performance capabilities increase the complexity of pilots' planning and decision making. Moreover, increased aircraft speed give them less time to process information. As a result, pilot performance is limited by memory and attention capacity (Jensen, 1989).

While selection and training can eliminate individuals who consistently exhibit poor cognitive performance, transient periods of cognitive impairment create periods of "short term accident liability" (Allnutt, 1982). Transient cognitive impairment can be created by a variety of factors. Fatigue is an important factor that should be included in any discussion of pilot cognitive performance (Stokes and Kite, 1996). A recent review of Aviation Safety Reporting System records indicated that roughly 21% of reported incidents were fatigue related (Gander, Nguyen, Rosekind, & Connell, 1993). Fatigue has a variety of cognitive effects of importance to pilots: channelized thought processes, increased distractibility, fixation, decreased visual perception, short-term memory loss, and impaired judgment and decision making (Reinhart, 1993).

One of the stressors pilots experience which contribute to fatigue is cockpit noise (FAA, 1989; Hawkins, 1993). In

small reciprocating engine airplanes, cockpit noise can reach levels of 90 to 115 dB (Reinhart, 1993). This problem has typically been dealt with through the use of headsets and intercoms. In noisier aircraft, such as turboprops, used by many regional airlines, some form of "muff" type headset fitting around the entire ear is worn.

A fairly recent development is the introduction of Active Noise Reduction technology (ANR). ANR headsets have tiny microphones in their ear-cups which sense the noise coming into the ear. Electronic circuitry produces an equal but opposite sound wave which is directed into the ear-cup and "cancels" the original noise (Gower & Casali, 1994). ANR headsets can reduce low frequency noise 15 dB or more below what is possible with passive muff-type designs (Day, 1994) and are most effective below 300 Hz (Gower & Casali, 1994).

The adverse effects of continuous noise are particularly noticeable in complex, multi-component tasks (Hockey, 1986), so the possibility that ANR aviation headsets provide greater noise protection than conventional passive headsets has significant implications for general aviation safety. The present study tested the hypothesis that ANR headsets, by protecting pilots from cockpit noise, prevent noise-induced pilot fatigue and the associated drop in cognitive performance better than passive headsets.

METHOD

Subjects

Eleven male and one female Certified Flight Instructors (CFI) associated with the Mankato State University and Saint Cloud State University aviation programs participated in the study. Ages ranged from 21 to 28 years, with a mean of 24.58. Average total flight time at the beginning of the experiment was 918 hours, with a range of 400 hours to 3020 hours. All subjects held at least commercial and certified flight instructor certificates with associated single engine and instrument ratings. All but one of the subjects also possessed multi-engine pilot and instructor ratings. One subject held an Airline Transport Pilot certificate.

Apparatus

A Telex ANR 4000 active noise reduction headset was used to administer the experimental treatment. Cockpit noise exposure was varied by activating or deactivating the headset's ANR feature. With the ANR feature deactivated, the headset performed like a typical passive unit.

Cognitive Assessment

The effect of noise exposure on cognitive performance was measured using 10 tests from the Automated Neuropsychological Assessment Metric, Version 3 (ANAM) cognitive assessment battery, developed by the Office of Military Performance Assessment Technology (OMPAT)

(Reeves, Winter, LaCour, Raynsfor, Vogal, & Grissett, 1991). Computer-based cognitive assessment has been shown to be sensitive to sleep deprivation and circadian rhythmicity (Elsmore, Naitoh, & Linnville, 1992), and to be a more accurate predictor of flying performance than paper and pencil tests (Stokes, Banich, & Elledge, 1991).

The ANAM battery is designed as a measure of subjective fatigue and cognitive processing efficiency. For this experiment, tasks recommended for use in assessing fitness for duty during sustained operations were selected (Reeves et al., 1991). These tasks assess different aspects of information processing related to pilot performance (including perceptual, memory, and symbolic, and spatial information integration/manipulation). The Procedural Reaction Time and Two Choice Reaction Time tasks assess subjects' ability to detect and categorize stimuli. The Sternberg Memory Recall and Running Memory tasks measure short term memory. The Spatial and Mathematical Processing tasks assess subjects' ability to integrate and manipulate spatial and symbolic information. The ANAM also assesses subjective fatigue. This is important because "feeling tired does not necessarily correlate with physiological impairment, nor with reduced efficiency in work output or other kinds of human performance" (Hockey, 1986). The Stanford Sleep Scale measures subjective fatigue, and the Mood 2 sub-test assesses subjective state on the following scales: activity, fatigue, happiness, depression, anger, and fear.

Procedure

Subjects received 3 practice administrations of the ANAM test battery in order to minimize the effect of learning during repeated assessment. Subjects were randomly assigned to one of two experimental conditions. Members of the first group conducted a routine day of flight instruction wearing the Telex ANR 4000 headset with its noise canceling feature activated, and a second day of instruction with the ANR function deactivated. Subjects in the second group flew with the ANR feature deactivated during the first flight and had the feature activated on during their second flight. Each subject received a cognitive performance assessment using the ANAM at the beginning and end of both days.

RESULTS

The study examined two issues. First, do pilots report greater subjective fatigue and exhibit decreased cognitive performance at the end of a workday which includes at least 4 hours of flight? And second, if differences in fatigue and cognitive performance do occur over the course of the day, are these differences less pronounced when ANR headsets?

Average pre-flight to post-flight fatigue and cognitive performance changes during the 2 days of testing were assessed for each pilot, independent of headset type. Difference scores were calculated for each ANAM task by subtracting post-flight scores from pre-flight scores.

Averaging corresponding scores from "active" and "passive" days yielded mean pre-post difference scores on each task. Single-sample T-tests were performed to determine if mean pre-post difference scores varied significantly from zero. Pilots reported an increase in post-flight sleepiness on the Stanford Sleep Scale and an increase in fatigue on the fatigue item of the Mood Scale (p< .006, .003). They also reported feeling less active on the Mood Scale (p< .006). There were no significant post-flight changes in happiness, depression, anger, or fear.

Post-flight cognitive performance measures showed a consistent pattern of faster response times on all tasks. Memory tasks exhibited the most change with significantly faster responses on both the 2 and 6 item Sternberg tasks (P<.03 and .003 respectively). The speed increase approached significance on Running Memory (P<.07). Post-flight cognitive performance accuracy decreased for all memory and information integration/manipulation tasks. The accuracy decrease approached significance for Spatial Processing (P<.06). Post flight accuracy was higher on both perceptual tasks. The effect was significant for Procedural Reaction Time (P<.05) (see Tables 1, 2, and 3).

With the exception of the Sternberg 6 reaction time difference (P< .02), no significant differences in fatigue and cognitive performance were observed as a function of headset type.

Table 1. Flight Instructor cognitive performance changes during an instructional day.

Task	RT (Sec)	P <	Accuracy (%)	P <
Perceptual tasks				
Procedural	5.36 faster	.30	2.5 higher	.05
2 Choice	5.27 faster	.60	0.46 higher	.60
Memory tasks				
Sternberg (2)	16.21 faster	.03	1.22 lower	.23
Sternberg(6)	31.41 faster	.003	0.37 lower	.82
Running memory	13.48 faster	.07	1.34 lower	.14
Continuous memory	9.99 faster	.12	0.58 lower	.45
Information integration/manipulation tasks				
Math Processing	90.49 faster	.09	0.21 lower	.91
Spatial	80.66 faster	.10	5.21 lower	.06

Table 2. Flight Instructor cognitive performance response time changes in seconds with ANR headsets activated and ANR headset deactivated

Task	Active	Passive	P <
Perceptual tasks			
Procedural RT	3.47 faster	7.24 faster	.71
2-Choice RT	12.19 faster	1.66 slower	.55
Memory tasks			
Sternberg(2)	20.99 faster	11.44 faster	.56
Sternberg(6)	50.49 faster	12.32 faster	.02
Running Memory	26.16 faster	0.79 faster	.10
Continuous memory	15.71 faster	4.26 faster	.38
Information integration/manipulation tasks			
Math processing	92.26 faster	88.72 faster	.97
Spatial Processing	15.55 faster	45.77 faster	.39

Table 3. Percent accuracy changes during days with ANR activated and ANR deactivated

Task	Active	Passive	P <
Perceptual tasks			
Procedural RT	4.17 more	0.83 more	.21
2-Choice RT	0.19 more	0.74 more	.57
Memory tasks			
Sternberg(2)	2.06 less	0.37 less	.23
Sternberg(6)	0.56 more	1.29 less	.68
Running Memory	1.84 less	0.85 less	.57
Continuous memory	0.52 less	0.63 less	.91
Information integration/manipulation tasks			
Math processing	0.83 less	0.42 more	.67
Spatial Processing	4.17 less	6.25 less	.59

DISCUSSION

Predicting the cognitive performance changes that are likely to occur during a day of flying is complicated by the simultaneous presence of a number of factors that affect performance during that period. High accident rates have been noted the first hour after awakening (Ribak, Ashkenazi, Klepfish, Avgar, Tall, Kallner, and Noyman, 1983), performance declines as the number of hours awake increase (Allnutt, Haslam, Rejman, and Green, 1990), the time of day affects performance with low performance in the early morning but times of best performance occurring at different times during the day for different tasks (Hockey, 1986), and performance is sensitive to physical (Bonnet, 1980) and mental (Mascord and Heath, 1992) workload and to environmental conditions including noise (Hartley, 1973).

It is possible to comfortably predict that pilots perform poorly when all these factors are present in extreme levels. Indeed, cognitive decrements have been reported after lengthy periods in the cockpit (Neri, Shappell, and DeJohn, 1992). However, it is not clear whether cognitive performance changes occur during a day of general aviation flying. The current study examined changes in a variety of cognitive tasks during a day of general aviation flying. Additionally, it assessed whether the noise reduction produced by ANR headsets attenuates fatigue and cognitive changes produced during the day.

Pilot fatigue and cognitive performance changes were noted between 0800 and 1600. Pilots exhibited significant post-flight increases in subjective fatigue on both the Stanford Sleep Scale and the Mood scales. However, this change cannot be attributed to the unique demands of flying since university students completing the battery at 0800 and 1600 also indicated significantly higher fatigue at 1600 (Harris, Goernert, Nelson, Netsch, and Scheel, 1996). In contrast to the absence of cognitive performance changes in university students between morning and afternoon assessments (Harris

et al., 1996), pilot cognitive performance changes were noted during this period. Although changes did not reach statistical significance in all cases, a pattern of faster post-flight responses on cognitive performance tasks was evident. This trend was most pronounced for memory tasks, with significant speed increases on both Sternberg tasks and speed increases approaching significance on Running Memory. A less pronounced post-flight speed increase was exhibited on information integration/manipulation tasks.

Accuracy changes were smaller and less consistent. Accuracy increased on the 2 perceptual tasks with a significant improvement occurring on Procedural Reaction Time. In contrast, post-flight accuracy scores decreased for all memory and information integration/manipulation tasks. This effect approached significance on the Spatial Processing task (P< 0.06).

Faster but possibly less accurate responding does not indicate diminished cognitive performance but it does suggest a change in strategy that may affect performance for a variety of tasks. Post-flight speed increases were generally more pronounced than were changes in accuracy, thus the overall performance change was faster responding with only minimal accuracy cost. While faster responding might improve performance on non-critical tasks, any decrease in accuracy is a serious concern in the cockpit This explanation is consistent with fatigue studies which have reported increased acceptance of risk and avoidance of effort (Holding, 1983; Neri, Shappell & DeJohn, 1992). A similar faster and less accurate grammatical reasoning was noted on workdays by Rosa and Colligan (1988). Observation of the largest accuracy decrement in Spatial Processing is consistent with the high spatial demands present in the aviation environment and for the same reason decrements of pilot spatial processing would be a source of concern. Indeed, Neri, Shappell, and DeJohn (1992) observed spatial test deterioration in pilots completing cognitive tests during a simulated sustained mission.

The results of the present research suggest that fatigue and cognitive performance changes occur in general aviation pilots during a relatively brief period of flying. Increased subjective fatigue may not be more pronounced than fatigue generate during a non-flying day but cognitive performance changes are apt to take the form of faster but less accurate responding on tasks involving short term memory or information integration/manipulation. Cognitive operations during different decisions vary, and the effect of a cognitive deficit such as reduced memory capacity will therefore not have identical effects on all tasks (Babkoff, Mikulincer, Caspy, and Kempinsky, 1988). It is not clear whether the fatigue and cognitive performance changes observed in this experiment were sufficient to degrade flying performance, although computer-based cognitive performance has been shown to be related to performance in a flight simulator (Hyland, Kay, & Deimler, 1994).

The importance of general aviation safety is not as obvious as major carrier safety. While a single, dramatic crash can kill or injury hundreds of people, general aviation is more dangerous than commercial aviation on a per mile flown basis and in the overall number of deaths per year (Baker et al., 1992). While lack of aircraft control skill contributes to

general aviation accidents, decision making errors, poor cognitive performance, are over-represented in fatal accidents (O'Hare et al., 1994). Bad decisions can be the result of a knowledge deficit, but they are frequently associated with periods when some combination of lack of sleep and stress have produced a brief, but potentially fatal, period of cognitive performance deterioration.

The data offers no indication that ANR headsets are more effective than passive headsets at protecting pilots from noise-induced fatigue and associated cognitive performance decrements. However, even though the study did not detect a significant reduction in noise-induced fatigue or cognitive performance decrements in pilots wearing ANR headsets, the researchers believe that examination of the effects of ANR headsets warrants further investigation. During longer periods of exposure or in aircraft that generate more noise, these headsets may have beneficial effects. In addition, noise has other deleterious effects besides fatigue. It can lead to headaches, nausea, disorientation, irritability (Reinhart, 1993), hearing loss, and degradation of communication. ANR technology may therefore offer benefits to pilots aside from any ability to reduce noise-induced fatigue or cognitive performance deterioration.

REFERENCES

Allnutt, M. (1982). Human factors: Basic principles. In R. Hurst and L. R. Hurst. (Eds.) Pilot error: The human factors. (pp. 1-22), New York: Jason Aronson.

Allnutt, M. F., Haslam, D. R., Rejman, M. H. and Green, S. (1990). Sustained performance and some effects on the design and operation of complex systems. In Broadbent, D. E., Reason, J. and Bandeley, A. (Eds.) Human Factors in hazardous situations. Proceedings of The Royal Society of London, 327, 529-541.

Babkoff, H. Mikulincer, M. Caspy, T., Kempinski, D., and Sing, H. (1988). The topology of performance curves during 72 hours of sleep loss: A memory and search task. Quarterly journal of Experimental Psychology, 40-A, 737-756.

Baker, S. P., O'Neill, B. Ginzburg, M. .J. and Li, G (1992). The injury fact book. New York: Oxford Press.

Billings, C. E., Reynard, W. D. (1984). Human factors in aircraft incidents: results of a seven year study. Aviation, Space, and Environmental Medicine, 55, 960-965.

Bonnet, M. H. (1980). Sleep, performance and mood after the energy expenditure equivalent of 40 hours of sleep deprivation. Psychophysiology, 17, 56-63

Elsmore, T., Naitoh, P. & Linnville (1992) Performance assessment in sustained operations. Naval Health Research Center Report No. 92-30

Federal Aviation Administration, (1989). Advisory circular; cockpit noise and speech interference between crew members. FAA A.C. No. 20-133.

Gander, P. H., Nguyen, D., Rosekind, M. R., Connell, L. J. (1993). Age, circadian rhythms, and sleep loss in flight crews. Aviation, Space and Environmental Medicine, 64(3), 189-195.

Gower, D. W., Casali, J. G. (1994). Speech intelligibility and protective effectiveness of selected active noise reduction and conventional communications headsets. Human Factors, 36(2), 350-367.

Harris, W. C., Goernert, P. N., Nelson, K., Netsch, S. and Scheel, M. (1996). Daytime cognitive performance and fatigue change. Proceedings of the Human Factors and Ergonomics Society 40th annual meeting, Philadelphia, PA, September, 1996.

Hartley, L. R. (1973) Effects of noise or prior performance on serial reaction. Journal of experimental physiology, 101, 255-261.

Hawkins, F. H. (1993). Human factors in flight. Hants, England: Ashgate.

Hockey, G. R. J. (1986) Cognitive Processes and performance, In K. R. Boff, L. Kaufman and J. P Thomas (Eds.) Handbook of perception and human performance, volume 2, NY: Wiley.

Holding, D. H. (1983) Fatigue. In G. R.. Hockey (Ed.) Stress and fatigue in human performance. 145-167. Chichester, Wiley.

Hyland, D.T., Kay, E.J., and Deimler, J.M. (1994). Age 60 study, Part IV: Experimental evaluation of pilot performance. Report no. 8025-4b. Federal Aviation Administration, Washington D.C.

Jensen, R. S. (Ed.). (1989). Aviation Psychology. Aldershot: Gower Technical.

Jensen, R. S., & Benel, R. A. (1977). Judgment evaluation and instruction in civil pilot training (Final Report FAA-RD-78-24). Springfield, VA: National Technical Information Service.

Mascord, D. J and Heath, R. A. (1992) Behavioral and physiological indices of fatigue in a visual tracking task. Journal of Safety Research, 23, 19-25.

O'Hare, D., Wiggins, M., Batt, R. and Morrision, D. (1994).Cognitive failure analysis for aircraft accident investigation. Ergonomics, 37, 1855-1869.

Neri, D. F., Shappell, S. A. and DeJohn, C.A. (1992). Simulated sustained flight operations and performance. Part 1: Effect of fatigue. Military Psychology, 4, 137-155.

Reeves, D. L., Winter, K. P., LaCour, S. J., Raynsfor, .M., Vogal, K., Grissett,. J. D. (1991). The UTC- PAB/AGARD stress battery: user's manual and system documentation. NAMRL Special Report 91-3.

Reinhart, R. O. (1993) Fit to fly: a pilot's guide to health and safety. Blue Ridge Summit, PA. Tab Books.

Ribak, J., Ashkenazi, I.E., Klepfish, A. Avgar, D., Tall, J., Kallner, B. and Noyman, Y. (1983). Diurnal rhythicity and air force flight accidents, Aviation, Space and Environmental Medicine, 54, 1096-1099.

Rosa, R. R. and Colligan M. J. (1988) Long workdays Vs restdays; Assessing fatigue and alertness with a portable performance battery. Human Factors, 30, 305-317.

Stokes, A. F. and Kite, K. (1994). Flight stress: Stress, fatigue and performance in aviation.. Hants, England: Ashgate

Stokes, A. F., Banich, M. T. and Elledge, V. C. (1991). Testing the tests - An empirical evaluation of screening tests for the detection of cognitive impairment in aviators. Aviation, Space and Environmental Medicine, 62, 783-788.

Evaluation of Hand-Dominance
On Manual Control of Aircraft

Valerie J. Gawron and James E. Priest
Calspan SRL Corporation, Buffalo, New York 14225

In the transport aircraft community the non-dominant hand control of aircraft is the norm. This historical precedence may be biasing the cockpit designs of the newer fly-by-wire aircraft which utilize a small sidestick controller rather than a wheel-column. Very little data are available to determine what effect non-dominant hand control using small throw controllers has on the pilot operator. To provide such data, a part-task simulation study was undertaken. Three different compensatory tracking tasks were performed with both left and right hand-controllers. Six right-hand dominant and three left-hand dominant subjects performed all three tasks, with both controllers. The results indicate that performance degraded and workload increased when the pilots were forced to use their non-dominant hand.

TEST FACILITY

The test facility was a fixed based, part task aircraft simulator (Figure 1). The simulation was produced by a desktop 486PC driving two VGA monitors (one for primary flight display and the other as a data display).

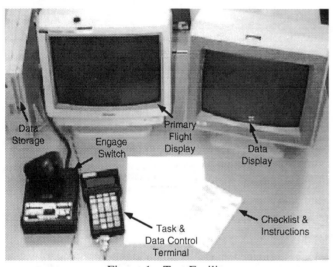

Figure 1 - Test Facility

The subjects were seated in a chair with the sidestick controller placed on either side of the chair depending on the task being flown (Figure 2). No arm rest was used. The grip was canted forward 25°. The center-of-rotation was nominally 4.5 inches below the subject's hand.

Figure 2 - Controller and Seat

EXPERIMENT DESIGN

The experiment was a 3 x 2 factorial design in which three different compensatory tracking tasks were performed with both left and right hand-controllers. For each hand-controller combination, two Sum-of-Sines (SOS) tasks and one Discrete-Tracking-Task (DTT) were performed yielding six distinct hand - task combinations (Table 1).

Table 1 - Task Identifiers and Descriptions

Task Label	Description
R-SOS1	Right Hand Control - Lower Frequency SOS Tracking Task
R-SOS2	Right Hand Control - Higher Frequency SOS Tracking Task
R-DTT	Right Hand Control - Discrete Tracking Task
L-SOS1	Left Hand Control - Lower Frequency SOS Tracking Task
L-SOS2	Left Hand Control - Higher Frequency SOS Tracking Task
L-DTT	Left Hand Control - Discrete Tracking Task

The SOS tasks were distinguished by the attenuation of the high frequency portion of the summed sinewave forcing function. The SOS2 task had more high frequency content than the SOS1 task. The discrete tracking task consisted of a combination of step-and ramp-type commands in both pitch and roll. The order of the SOS and DTT task was counterbalanced to control fatigue and warm up effects. For the SOS task, a different phasing of the summed sinewaves was used for each task. The frequency content of the SOS task was not changed by the sinewave phase randomization. For the DTT task, the initial pitch and roll direction was counterbalanced.

The primary flight display had a single line horizon, an aircraft control symbol, target symbol, and a terrain grid (Figure 3). The control symbol was a fixed reference on the display. The horizon line and terrain grid pitched and rolled about control symbol. The "fly to" target symbol was driven in pitch and roll. As the target symbol moved away from the control symbol, pitch and roll errors were generated. The subject's task was to keep the control symbol on top of the target symbol and thereby keep the errors to a minimum.

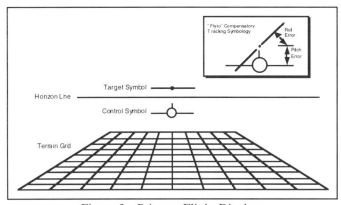

Figure 3 - Primary Flight Display

The ordering of tracking task versus the hand-controller combination was counter balanced across subjects using a Latin Square design. The test matrix was sufficiently symmetric to allow an equal number of subjects to begin with their dominant hand as began with their non-dominant hand. All subjects completed the three tasks with a given hand-controller combination prior to commencing trials with the other hand-controller combination.

A dominance test was given to each subject prior to the experiment (Reference 1). All of the subjects were strongly hand-dominant with one-third of the subjects left-hand dominant. The percentage of left-hand dominant subjects in this study was higher than the general population (roughly 10% left-hand dominant). This breakdown between left and right-hand dominance was used to provide a balanced test matrix.

All of the experiment subjects were experienced pilots familiar with the test facility whose experience ranged from a low of 3,500 hours to a high of 9,000 hours of flight time (Table 2). Pilots with both fighter and transport aircraft experience were used.

Table 2 - Subject Experience and Hand-Dominance

Subject	Hand Dominance	Total Hours	Military Branch[1]	Primary Aircraft
1	Right	8,000	Air Force	Fighter
2	Right	4,500	Air Force	Transport
3	Right	3,700	Navy	Transport
4	Right	4,400	Air Force	Fighter
5	Right	5,100	Air Force	Fighter
6	Left	4,500	Navy	Fighter
7	Right	7,600	Air Force	Fighter
8	Left	3,500	Air Force	Transport
9	Left	9,000	Navy	Fighter

Note 1 - All subjects were United States Military except subject 7

The subjects were instructed to optimize their tracking performance to attain the lowest possible Normalized Root Mean Square Error (NRMSE) scores by keeping the control symbol on top of the target symbol (zero error occurred when the wings on the control symbol overlaid the target symbol). They were told to use a consistent technique across task and hand-controller combination, and to try not to use a less aggressive strategy to compensate for a lack of dexterity in their non-dominate hand. They were instructed to minimize both the pitch and roll error since these were weighted equally.

Each subject was trained to asymptotic performance in each task using the given hand-controller combination. Strict bounds for what constituted "asymptotic" performance were difficult to establish since scatter between subjects was not consistent. As a general guide, they were told use NRMSE scores over three trials within 10% to define an approximate envelope. Both pitch and roll error scores had to fall within this envelope. If in doubt, they were told to use more trials. On subsequent tasks, with the same hand-controller combination, they generally reached asymptotic performance in fewer trials.

Once they reached asymptotic performance in a given task, they completed five more trials. These five trials were used for data analysis.

RESULTS

The variations in subject experience (primary aircraft type, total flight time, etc.) and handedness did not have a significant effect on overall tracking performance. When collapsed across task and hand, the NRMSE scores were very similar (Figure 4). All subjects had slightly better performance in the roll axis than in the pitch axis.

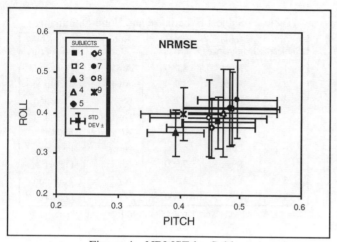

Figure 4 - NRMSE by Subject

The subjects showed similar trends in pitch and roll NRMSE for dominant and non-dominant hand control (Figure 5). All subjects were able to track as well or better with their dominant hand than with their non-dominant hand in the pitch axis. The data did not reveal the dominance trend in the roll axis where for the most part, the subjects had equal performance across hands.

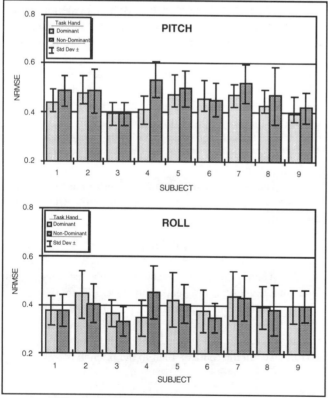

Figure 5 - NRMSE by Hand Dominance

Subjective comments were collected from an exit questionnaire. The subject's comments regarding "handedness effects" were intriguing. Most subjects perceived a difference in their ability to perform the tasks with either hand. They felt more closely coupled to the airplane (in a closed-loop

sense) and better able to cope with any sub-optimum aspects of the controller when flying with their dominant hand. Subject 1 (right-hand dominant) summarizes these perceptions well in his comments:

> ... My right hand "filtered out" anomalies [in the controller] better.
>
> ....I felt I used larger, open loop, type inputs with my left hand.
>
> ...With my right hand, I varied inputs in a more closed loop manner. Everything else being equal, I would prefer a right-hand sidestick controller.
>
> ... Left sidestick would take a little more initial practice to become comfortable.

Subject 4 (right-hand dominant), who was the only subject to have a dominance effect in both axes, even noticed a subconscious urge to switch hands.

> ...Once I noticed that my right hand wanted to cross over and help my left hand.

In contrast to comments such as these, Subject 8 (left-hand dominant) was not able to discern any difference between hands.

> ...Felt I was able to perform equally well with either hand.

The workload measure used was Subjective Workload Dominance (SWORD) technique (Reference 2). The workload for dominant or non-dominant hand control showed a very strong trend (Figure 6). Not a single subject perceived less workload using the non-dominant hand. The workload varied from equal between hands for Subject 9 (left hand dominant) to a slight increase in workload for non-dominant hand control with Subjects 2 and 3, to nearly a 5 times increase in workload for Subject 4 (right hand dominant). Subjects 1 and 4 commented the strongest on their workload increase when using their non-dominant hand for control.

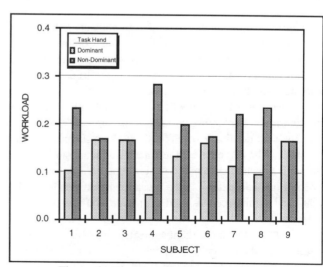

Figure 6 - Workload by Hand-Dominance

Two types of data were collected for analysis. First, RMSE data were collected for each task performed using each hand for each subject for each of five replications thereby yielding 3 x 2 x 9 x 5 = 270 data points. Given the addition of event in RMSE data, two separate analyses were performed. Second, SWORD data were collected for each task performed using each hand for each subject thereby yielding 3 x 2 x 9 = 54 data points.

A repeated-measures multivariate analysis of variance (MANOVA) was calculated using the pitch NRMSE and roll NRMSE as the two dependent measures (Table 3). There were both between and within-subject effects. The between-subject effect was dominance. The within-subject effects were: hand used for control, replication, and task.

Table 3 - MANOVA Source Table for NRMSE

Effect	Hotellings T	Approx. F	Deg. of Freedom	p	power
Dominance (D)	0.07122	7.44288	2, 209	0.001	0.94
Hand (H)	0.00734	0.76676	2, 209	0.466	0.18
Replication (R)	0.01249	0.32465	8, 416	0.957	0.16
Task (T)	2.10022	109.21141	4, 416	0.000	1.00
D x H	0.10798	11.28359	2, 209	0.000	0.99
D x R	0.00812	0.21109	8, 416	0.989	0.12
D x T	0.02919	1.51799	4, 416	0.196	0.47
H x R	0.01253	0.32583	8, 416	0.956	0.16
H x T	0.02597	1.35055	4, 16	0.250	0.42
R x T	0.03490	0.45368	16, 416	0.967	0.31
D x H x R	0.01607	041780	8, 416	0.910	0.20
D x H x T	0.01702	0.88487	4, 416	0.473	0.28
D x R x T	0.01289	0.16759	16, 416	1.000	0.12
H x R x T	0.03909	0.50821	16, 416	0.943	0.35
D x H x R x T	0.02580	0.33535	16, 416	0.993	0.22

There were two significant main effects: dominance and task. Univariate F values indicate significant dominance effects on both pitch NRMSE (F = 14.658, p = 0.000, power = 0.97) and roll NRMSE (F = 4.390, p = 0.037, power = 0.546). Univariate F values indicate significant task effects also on both pitch NRMSE (F = 98.741, p = 0.000, power = 1.000) and roll NRMSE (F = 122.133, p = 0.000, power = 1.000). The task effect is illustrated in Figure 7.

There was also significant dominance by hand-controller combination interaction effect. Examination of the univariate F values for each of the two dependent variables reveals a significant interaction for pitch NRMSE (F = 16.626, p = 0.000, power = 0.982) but not for roll NRMSE (F = 0.283, p = 0.595, power = 0.039). These effects are illustrated in Figure 8.

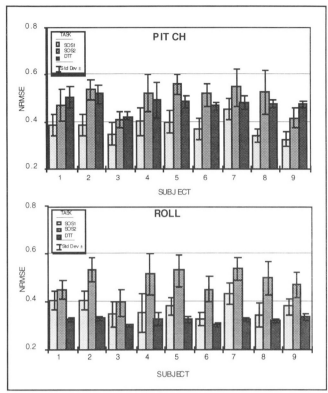

Figure 7 - Task NRMSE By Subject

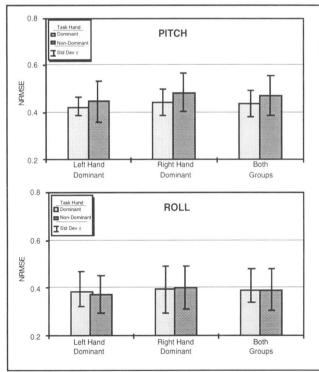

Figure 8 - NRMSE by Dominance

A repeated-measures analysis of variance (ANOVA) was calculated using the SWORD data. There were both between-subject and within-subject effects. The between-subject effect was dominance. The within-subject effects were hand used for control and task. The dependent variable was the SWORD score. The results are presented in Table 4.

Table 4 - ANOVA Source Table for SWORD Scores

Effect	Sum of Squares	Deg. of Freedom	Mean Square	F	p	power
Dominance(D)	0.000	1	0.000	3.090	0.122	0.331
Hand (H)	0.000	1	0.000	0.410	0.541	0.092
Task (T)	0.300	2	0.150	9.650	0.002	0.949
D x H	0.060	1	0.060	5.880	0.046	0.551
D x T	0.010	2	0.000	0.190	0.830	0.075
H x T	0.010	2	0.000	0.090	0.430	0.174
D x H x T	0.010	2	0.010	2.360	0.131	0.397

As can be seen from Table 4, there was one significant main effect, task. The mean SWORD scores for each task were: SOS1 = 0.0742, DTT = 0.1729, and SOS2 = 0.2531. A Scheffe post hoc test was performed on these means. The results indicate that each of these means is significantly different from each other mean, $p < 0.050$. Both the left and right hand dominant groups showed this trend. There was also a significant interaction effect between dominance and hand (Figure 9). On average, the subjects rated the tasks lower in workload when using their dominant hand.

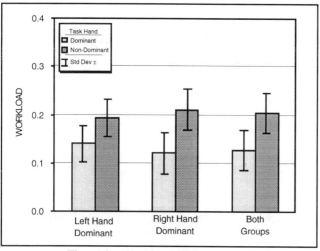

Figure 9 - Workload by Dominance

CONCLUSIONS

The objective of this study was to evaluate what effect, if any, handedness has on pilot performance and workload when using a small-throw, sidestick controller. The data support the following conclusions:

1. The subject pilots' performance degraded when they were forced to use their non-dominant hand for control.

2. The subject pilots' workload was higher when they were forced to use their non-dominant hand for control.

3. After training to an asymptotic condition, differences in performance and workload between dominant and non-dominant hand control were still present.

One could argue that the pilots would have eventually adapted to using non-dominant hand control. In other words, they would have "trained out" the differences. The data do not support either conclusion. However, perhaps a more compelling question is with the advent of highly augmented fly-by-wire aircraft, what design constraint is mandating this condition? If the two handed controller (i.e., a wheel-column) is no longer required, then why force the pilot into this sub-optimum condition?

Further analysis on these data may help quantify other issues such as pilot compensation techniques and transfer-of-training that were not fully addressed in this analysis. The pilots eluded to variations in their internal loop-gains and differences in their technique when using their non-dominant hand for control.

Modeling of the open-loop pilot-vehicle system to determine gain and phase margins would help quantify what differences, if any, existed in pilot compensation techniques between hand-controller combinations. MANOVAs performed on the objective performance data (e.g., rms stick activity, open-loop performance measures) using the configuration as the independent variable would help determine the differences in pilot technique between dominant and non-dominant hand control. The transfer-of-training issue could be evaluated by accomplishing additional MANOVAs on the number of trials to reach asymptotic performance and the change in performance with condition transfer using the three independent variables: 1) starting condition; 2) subject "handedness;" and 3) the interaction of 1 and 2. The transfer-of-training and performance analyses would provide quantitative data on the influence of the controller design and cockpit layout to further support future design efforts.

REFERENCES

1. Coren, Stanley, "The Left-Hander Syndrome - Causes and Consequences of Left-Handedness", First Vintage Books Edition, July 1993.

2. Vidulich, M.A. (1989). The use of Judgment Matrices in Subjective Workload Assessment: The Subjective Workload Dominance (SWORD) Technique. In the Proceedings of the Human Factors Society 33rd Annual Meeting (pg. 1406-1410). Santa Monica, CA: Human Factors Society.

INFORMATION EXTRACTION DURING INSTRUMENT FLIGHT: AN EVALUATION OF THE VALIDITY OF THE EYE-MIND HYPOTHESIS

Julianne Fox, David Merwin, Roger Marsh, George McConkie & Arthur Kramer

Beckman Institute & Institute of Aviation
University of Illinois at Urbana-Champaign

A study was performed to determine the extent to which flight-relevant information on instruments peripheral to fixation is extracted and used during fixed-wing instrument flight. Twenty student and twenty instructor pilots flew a series of missions in a fixed-wing flight simulator which was interfaced with an eye-tracker. In one mission flight-relevant information was removed from instruments peripheral to fixation while in the other mission peripheral information was intact. Pilots' performance was degraded and eye scan strategies were modified when peripheral information was removed. Furthermore, in several situations instructor pilots' performance was more adversely influenced by the removal of peripheral information than was student pilots' performance. The data are discussed in terms of attentional strategies during flight.

A large body of literature on visual scanning of cockpit instrumentation has accumulated over the past 50 years (Fitts et al., 1950; Jones et al., 1949; Senders, 1964; Carbonnell et al.,1968; Allen et al., 1969; Senders, 1970; DeMaio et al., 1978; Harris & Christhilf, 1980; Tole et al., 1982; Hameluck, 1990; Kramer et al., 1994; Wickens et al., 1995). The results of this research have been used successfully both in the design of cockpit instrument panels as well as to characterize information extraction strategies in a variety of flight situations with novice and experienced fixed wing and helicopter pilots.

An important assumption that underlies this research is referred to as the eye-mind hypothesis. The eye-mind hypothesis states that information is extracted from a relatively small area of space centered on fixation. That is, the use of eye tracking technology to characterize information extraction strategies in flight carries with it the (usually implicit) assumption that information is extracted from the fixated instrument.

While the eye-mind hypothesis may be a reasonable simplifying assumption under many conditions there are also a number of situations in which it is likely incorrect. For example, there is now a relatively voluminous literature which suggests that even in tasks which require high acuity, such as reading, information is extracted from the near periphery as well as from the fovea (McConkie, 1983; Rayner, 1983). In a similar vein studies of visual information extraction strategies in athletics and automobile driving have likewise found that information is extracted from both the fovea as well as from the visual periphery (Abernathy, 1988; Cole & Hughes, 1990). Laboratory research on visual attention has also clearly shown that attention can be decoupled from fixation on the basis of both physical and symbolic cues. For example, Posner (1980) has shown both benefits and costs in response times when the potential location of subsequent peripheral targets are validly or invalidly pre-cued.

In fact, there has been speculation that pilots use peripheral information to some extent during both visually referenced and instrument flight (Carbonell et al., 1968; Fitts et al.,1950, Allen et al., 1969; Jones et al., 1949). Consistent with this speculation a few studies have reported that pilots can accurately extract information on the position of an instrument pointer well into the visual periphery (Senders et al., 1955; DeMaio et al., 1978). However, although these studies suggest that pilots can extract flight relevant information from instruments peripheral to fixation they are silent with respect to whether pilots actually do use such information extraction strategies during flight.

In the present study we addressed the question of whether information is extracted from instruments other than those that are directly fixated. This was accomplished by using an eye movement contingent control technique (McConkie et al., 1984) to track, in real-time, pilots' eye movements as they flew a variety of flight profiles that required changes in heading, airspeed and altitude. Given the high spatial and temporal resolution of the eye tracker, we were able to make changes in specific instruments as pilots made saccades between instruments. In essence, we were able to capitalize on saccadic suppression (i.e. the suppression of vision during the time course of a saccadic eye movement) to ensure that changes in instruments peripheral to the instrument to be fixated would not be detected. Thus, in this way we could remove flight relevant information (i.e. by removing the pointers from the instruments) from peripheral instruments as pilots naturally scanned the instrument panel of a desk-top fixed wing flight simulator.

Two conditions were contrasted in the study. In the experimental condition we removed flight relevant peripheral information while in the control condition pilots flew the flight profiles with all of the instruments intact (i.e. both the fixated and peripheral instruments). If pilots use peripheral information to guide their scan strategy then we would expect poorer flight performance, and perhaps a modified scan pattern, in the experimental as compared to the control condition. On the other hand, if the eye-mind hypothesis is correct then it would be expected that the removal of peripheral information, in the experimental condition, would not have any influence on either flight performance or scan strategies. That is, flight performance and scan strategies should be equivalent in the control and experimental flights.

One additional issue was examined in the present study. There is now a substantial body of data which suggests that the frequency of automobile accidents can be predicted by the breadth of a driver's attentional field, with fewer accidents being associated with larger attentional fields (Ball & Owsley, 1991; Owsley et al., 1991). Furthermore, it appears that the size of the attentional field can be expanded with practice (Ball et al., 1988). Given these findings as well as other research (Delaney, 1992; Gopher, 1993) which suggests a relationship between piloting proficiency and attention we examined the hypothesis that the attentional field expands with flight experience. Within the present context this would be realized by an increased probability of information extraction from the periphery with increasing flight experience. Assuming that experienced pilots use information from peripheral instruments to guide their scan strategy this would predict either a greater disruption of flight performance and/or a more extensive modification of scan strategy for the instructor than for the student pilots when information is degraded on peripheral instruments.

METHOD

Subjects.

Twenty flight instructors and twenty beginning instrument students enrolled in Aviation 130 at the Institute of Aviation at the University of Illinois participated in the study. The flight instructors had a total flight time of between 225-3000 hours with a mean time of 959 hours and total instrument time between 55-2250 with a mean instrument time of 205 hours. Their ages ranged between 21-31 years old with a mean age of 24. The beginning instrument students had a total flight time of between 73-164 hours with a mean flight time of 110 hours and a total instrument time of between 7-24 hours with a mean instrument time of 15 hours. Their ages ranged between 19-31 years old with a mean age of 24. All subjects voluntarily participated in the study and were paid $7.50 per hour.

Stimuli and Apparatus.

Flight Simulation. A Zeos 100 MHz Pentium with a Matrox MGA Impression Plus graphics card (120 Hz refresh rate) was used to generate the flight simulation scenarios. A 20 inch monitor was used to display the instrument panel, instruction box and an overview of the simulation to be flown. The instrument panel is illustrated in Figure 1. The display subtended 20 degrees of visual angle horizontally and 15 degrees of visual angle vertically at a viewing distance of 42 inches. The instruments were 4 degrees in diameter with a minimum separation between instruments of .7 degrees. A joystick, mounted directly in front of the pilot, was used to fly the simulation. An analog toggle switch on the top of the joystick was used to make power inputs, a button next to the toggle switch was available for indicating anything abnormal observed by the subject during the simulation. The trim was located at the base of the joystick. Joystick inputs were sampled at 120 Hz.

Eye tracker. An SRI Purkinje Gen5 eyetracker with bite bar was used to record eye-movement data. The sampling and output rate of the eye tracker was 1000 Hz. The eyetracker was capable of measuring eye movements with an accuracy of 10 min. of arc. Eye position in x and y instrument panel referenced coordinates was calculated at 1000 Hz. Changes to the aircraft instrument panel were made within 8.3 msec. of the eyes crossing the instrument boundaries, which were well within the peak velocity of the saccade. That is, our eye-tracking/flight simulation algorithm was able to detect the landing site of the eye and remove the pointers of the peripheral instruments during saccades, thus capitalizing on saccadic suppression.

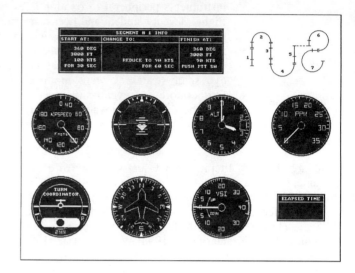

Figure 1. A graphic illustration of the instrument panel used in the flight simulator.

Flight Task.

Both instructor and student pilots were required to fly the same eight segment flight in both the control and experimental condition. The flight consisted of heading, airspeed and altitude changes and combinations thereof in two 15 minute simulated instrument flights (see Table 1). Prior to each flight segment, there was a 30 second straight and level lead-in leg allowing the pilot to both read the instruction box displayed directly above the instrument panel and to prepare for the required changes to be made. The pilots were instructed to remain within +/- 50 feet of assigned altitude, +/-5 degrees of assigned heading and +/- 5 knots of assigned airspeed. In

Segment	Parameters Changing	Abbreviations used in Figures
1	None	None
2	Airspeed	S
3	Heading	H
4	Altitude	A
5	Heading & Airspeed	H-S
6	Altitude & Airspeed	A-S
7	Heading & Altitude	H-A
8	Heading, Altitude & Airspeed	H-A-S

Table 1. Parameters changed during each flight segment.

addition, all climbs and descents were to be made at 500 feet per minute and all turns at standard rate (3 degrees per second). Each pilot received 20 minutes of practice immediately prior to flying the two eight segment flights. During the control condition, all instruments remained unaltered. During the experimental condition, all instruments other than those fixated by the pilot were presented without needles and/or numbers (see figure 2). The airspeed indicator, altimeter and vertical speed indicator were all displayed without their needles, the airplane on the turn coordinator was missing, the miniature airplane and lubber line on the attitude indicator was missing and the directional gyro was without the heading information. The order of presentation of the control and experimental flights was counterbalanced across pilots.

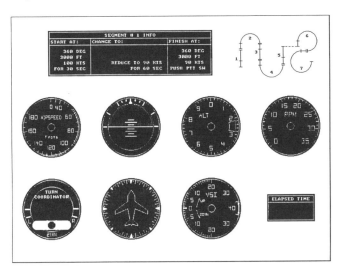

Figure 2. A graphic illustration of the instrument panel used in the flight simulator with peripheral information degraded.

Questionnaires. Both a pre-flight and post-flight questionnaire were given to each pilot. The preflight questionnaire obtained information regarding age, flight experience and education background. The post-flight questionnaire was directed at finding out whether or not the pilot noticed the missing information in their periphery during the experimental flight and secondly, whether or not they felt that they routinely were able to extract information from instruments in their periphery.

RESULTS AND DISCUSSION

We collected a number of flight performance measures during the eight segments in the control and experimental flights. These measures included: RMS heading, altitude, airspeed, turn rate and climb rate error as well as measures of percent time in criterion for heading, altitude and airspeed. Eye scan measures which were collected included: frequency and duration of fixations on the airspeed indicator, attitude indicator, altimeter, tachometer, turn coordinator, directional gyro, vertical speed indicator, clock and direction box. We also computed first and second order Markov coefficients. Markov coefficients provide an index of the degree of homogeneity of the pilot's scan pattern.

These measures were analyzed in three-way mixed mode ANOVAs. The factors were group (novices and instructors),

condition (control and experimental flights), and flight segment (straight & level, airspeed change, heading change, altitude change, heading & airspeed change, altitude & air speed change, heading & altitude change, heading, airspeed and altitude change). Given the large amount of data obtained in

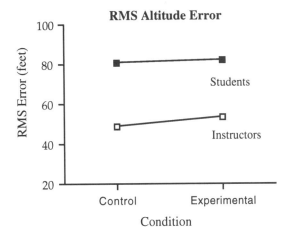

Figure 3. RMS Altitude error for students and instructors.

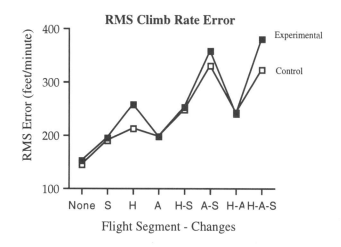

Figure 4. RMS Climb Rate error for the control and experimental condition.

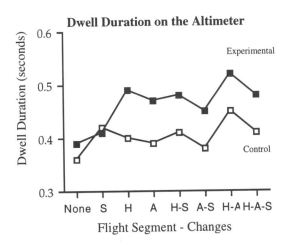

Figure 5. Dwell Duration on the altimeter for both groups.

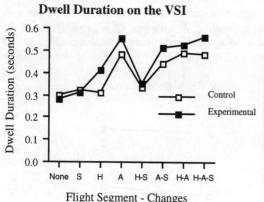

Figure 6. Dwell Duration on the VSI for both the instructors and students.

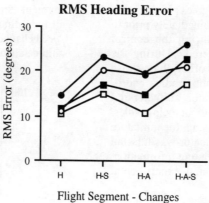

Figure 7. RMS Heading error for both groups during both conditions.

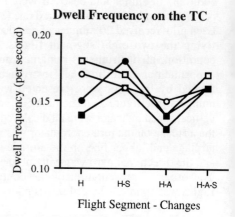

Figure 8. Dwell Frequency on the Turn Coordinator for both groups during both conditions.

the study our analyses will focus on the significant main effects and interactions that are directly relevant to our hypotheses. That is, we will focus on the effects of the condition variable (i.e. whether the peripheral instruments were degraded). To save space we will not include the F statistics. However, any effect that we discuss is significant at $p < .05$.

Vertical Axis control.

Several interesting effects of degraded peripheral information were associated with the control of the vertical axis. First, as displayed in Figure 3 the pilots showed significant increases in RMS altitude error when information on the peripheral instruments was removed. Furthermore, as indicated in the figure the removal of the information from the peripheral instruments had a larger impact on instructor than on student pilot performance. Performance effects for the condition variable can also be seen in RMS climb rate error (see Figure 4). Both the students and the instructors showed poorer performance in the absence of peripheral information on a number of the flight segments. Condition effects were also obtained for the scan measures on instruments relevant to the control of altitude. Figure 5 illustrates that degraded peripheral information increased the duration of fixations on the altimeter for both the students and instructors for all flight segments except the straight and level and air speed change segments. As illustrated in Figure 6, a similar pattern of effects was obtained for fixation duration on the vertical speed indicator. Fixation durations increased with the removal of peripheral information. Interestingly, the increase in the duration of fixations was accompanied by a decrease in fixation frequency in the absence of peripheral information.

Thus, with regard to the control of altitude both the student and instructor pilots showed a degradation of flight performance on a subset of flight segments when information was removed from instruments peripheral to fixation. Furthermore, there was a trend for larger performance costs for the instructor than for the student pilots. Scan strategies were also modified in response to the removal of peripheral information with the general trend being an increase in the duration and in some cases a decrease in the frequency of fixations on the instruments associated with altitude control.

Lateral axis control.

As illustrated in Figure 7, the removal of information from peripheral instruments resulted in an increased error in heading control, for both student and instructor pilots, on those flight segments in which heading was to be changed. Interestingly, the amount of disruption of heading control increased at a faster rate from single to multiple flight parameter change segments for instructor than for student pilots.

The flight performance effects were accompanied by changes in the frequency of fixations on the turn coordinator. As illustrated in Figure 8 fixation frequencies decreased on the heading change segments when information was removed from the peripheral instruments. The magnitude of these changes was larger, on two flight segments (heading change, heading/altitude/airspeed change), for the instructors than for the student pilots. The decreased fixation frequency on the turn coordinator with the removal of peripheral information was accompanied by increases in fixation duration for the student and instructor pilots. Thus, the trend is similar for lateral as for vertical control. In both cases, performance degraded and scan strategies changed when information was removed from peripheral instruments during saccades.

Longitudinal axis control.

Overall, the pilots were quite accurate in the control of airspeed. The removal of peripheral information did not influence performance. However, both the fixation duration and fixation frequency increased on the airspeed indicator when information was removed from the peripheral instruments.

Finally, with regard to scan pattern, the Markov coefficients increased in magnitude with the removal of information from the peripheral instruments. These data suggest that scan patterns become more homogeneous or stereotypical in the absence of task-relevant information on peripheral instruments.

CONCLUSIONS

The data are relatively clear in suggesting that both student and instructor pilots employ information from the visual periphery during instrument flight. In the present study, performance was degraded and scanning strategies were modified when information was removed from instruments peripheral to fixation. Although such a finding would not be particularly surprising in visually referenced flight given previous reports that velocity estimation entails the use of information in the visual periphery, the general assumption which underlies the use of eye movement analysis in instrument flight is that information is only extracted from the fixated instrument. Our data suggest that this is not always the case.

Of course, an important question concerns the nature of the information that is extracted from the peripheral instruments during instrument flight. Although we are far from being able to provide a complete answer to this question we can make some educated guesses both on the basis of our data as well as well as our knowledge of the visual system. Given acuity limitations it seems reasonable to suppose that information regarding orientation and movement can be discerned from instruments peripheral to fixation while the extraction of precise numerical information cannot. That is, the breadth of attention is constrained by the limits of retinal acuity.

The modifications of the scan patterns that were observed in our study are consistent with these limitations. The instruments that showed the most reliable modifications of scan strategies in the absence of peripheral information were those instruments from which orientation (i.e. turn coordinator, altimeter, airspeed indicator, vertical speed indicator) and relatively high frequency movement (i.e. vertical speed indicator, turn coordinator) information could serve as an indication of flight control status. On the other hand, scan strategies did not change, with the removal of peripheral information, for those instruments such as the directional gyro and attitude indicator in which higher acuity information is needed. Future experiments, in which a subset of instruments are degraded, will be necessary to provide a more fine-grained analysis of the types of information that are extracted from peripheral instruments (and the situations in which such information is extracted).

A second question addressed in our study was whether the breadth of the attentional field differed for instructor and student pilots. Although the data have not yet been exhaustively examined, instructors showed larger performance costs on both the lateral and vertical axes, when flight relevant information was removed from the instruments peripheral to fixation. Such a finding is consistent with the hypothesis that the breadth of attention expands with increasing expertise.

There are, however, several interesting questions that we have yet to address with regard to the interaction of expertise and attentional breadth. For example, during debriefing a subset of both the student and instructor pilots reported that they had noticed that the pointers were missing from the peripheral instruments at different times. Thus, an important question concerns whether conscious awareness of the missing peripheral information was related to whether or not performance and scan strategies changed with our condition manipulation (i.e. whether peripheral information was removed or not). Finally, we are currently in the process of administering a simple laboratory test of the functional attentional field (modeled on the Ball et al. (1988) test used with automobile drivers) to the student and instructor pilots who served in our eye-movement contingent control study. It will be interesting to examine the relationship between performance in this simple laboratory test and the susceptibility of flight performance and eye scan strategies to degraded peripheral information during instrument flight.

ACKNOWLEDGMENTS

This work was supported by Grant #N00014-93-1-0253 from the Office of Naval Research to Arthur Kramer.

REFERENCES

Abernathy, B. (1988). Visual search in sport and ergonomics: Its relationship to visual attention and performer expertise. Human Performance, 1, 205-235

Allen R. W., Clement, W. F., & Jex, H. R. (1970) Research on scanning, sampling, and reconstruction using separate main and secondary tracking tasks (NASA CR-1569). Washington, DC: NASA.

Ball, K., Beard, B., Roenker, D., Miller, R., Griggs, D. (1988). Age and visual search: Expanding the useful field of view. Journal of the Optical Society of America, 5, 2210-2219.

Ball, K. & Owsley, C. (1991). Identifying correlates of accident involvement for the older driver. Human Factors, 33, 583-595.

Carbonnell, J. R., Ward, J. L., & Senders, J. W. (1968). A queueing model of visual sampling: Experimental validation. IEEE Transactions on Man-Machine Systems, **MMS-9**, 82-87.

Cole, B. & Hughes, P. (1990). Drivers don't search: They just notice. In D. Brogan (Ed.), Visual search. Bristol, PA: Taylor & Francis.

Delaney, H. (1992). Dichotic listening and psychomotor task performance as predictors of naval primary flight training criteria. The International Journal of Aviation Psychology, **2**, 107-120.

DeMaio, J., Parkinson, S. & Crosby, J. (1978). A reaction time analysis of instrument scanning. Human Factors, 20, 467-461.

Fitts, P.M., Jones, R.E., & Milton, J.L. (1950). Eye movements of aircraft pilots during instrument landing approaches. Aeronautical Engineering Review, **9**, 24-29.

Gopher, D. (1993). The skill of attention control: Acquisition and execution of attention strategies. In D. Meyer & S. Kornblun (Eds.), Attention and Performance XIV. Hillsdale, N.J.: Erlbaum.

Hameluck, D. E. (1990). Mental models, mental workload, and instrument scanning in flight. Dissertation, York University, Ontario.

Harris R. L. & Christhilf, D. M. (1980). What do pilots see in displays? Proceedings of the Human Factors Society 24th Annual Meeting, (pp. 22-26). Santa Monica, CA: Human Factors and Ergonomics Society.

Jones, R. E., Milton, J. L. & Fitts, P. M. (1949). Eye fixations of aircraft pilots, I: A review of prior eye-movement studies and a description of a technique for recording the frequency, duration, and sequences of eye fixations during instrument flight (USAF Technical Report No. 5837). Wright Patterson AFB, Ohio.

Kramer, A.F., Tham, M.P., Konrad, C., Wickens, C.D., Lintern, J., Marsh, R. & Fox, J. (1994). Instrument scan and pilot expertise. Proceedings of the Human Factors Society, 38th Annual Meeting. Nashville, TN.

McConkie, G. (1983). Eye movements and perception during reading. In K. Rayner (Ed.), Eye movements in reading: Perceptual and language processes. N.Y.: Academic Press.

McConkie, G., Wolverton, G. & Zola, D. (1984). Instrument considerations in research involving eye-movement control. In A. Gale & F. Johnson (Eds.), Theoretical and applied aspects of eye movement research. Amsterdam: Elsevier.

Owsley, C., Ball, K., Sloane, M., Roenker, D. & Bruni, J. (1991). Visual perception/cognitive correlates of vehicle accidents in older drivers. Psychology and Aging, **6**, 403-415.

Posner, M. (1980). Orienting of attention. Quarterly Journal of Experimental Psychology, 32, 3-25.

Rayner, K. (1983). The perceptual span and eye movement control during reading. In K. Rayner (Ed.), Eye movements in reading: Perceptual and language processes. N.Y.: Academic Press.

Senders, J. W. (1970). Visual Scanning Behavior. In: Visual Search. Proceedings of the National Research Council Committee on Visual Symposium, National Academy of Sciences, Washington, DC.

Senders, J., Webb, I. & Baker, C. (1955). The peripheral viewing of dials. The Journal of Applied Psychology, **39**, 433-436.

Tole, J. R., Stephens, A. T., Harris, R. L., & Ephrath, A. R. (1982). Visual scanning behavior and mental workload in aircraft pilots. Aviation, Space, and Environmental Medicine, **53**(1), 54-61.

Wickens, C. D., Bellenkes, A. H., & Kramer, A. F. (1995). Visual scanning of expert pilots: The role of attentional flexibility and mental model development. (Tech. Rep. No. UIUC-BI-HPP-95-05). Urbana: University of Illinois, Beckman Institute, Human Perception and Performance.

ATTENTION DISTRIBUTION AND SITUATION AWARENESS
IN AIR TRAFFIC CONTROL

Mica R. Endsley
Texas Tech University
Lubbock, TX

Mark D. Rodgers
Federal Aviation Administration
Washington, D.C.

A study was conducted to investigate the way in which controllers deploy their attention in processing information in en route air traffic control scenarios. Actual air traffic control scenarios in which operational errors occurred were re-created using SATORI and displayed to twenty active air traffic control specialists. SAGAT was used to measure the subjects' ongoing understanding of the scenarios along pertinent situation awareness requirements. The data revealed an interesting pattern of attention distribution in processing the displays that can be linked to prior findings regarding operational errors in air traffic control.

INTRODUCTION

Air traffic control (ATC) provides a demanding environment in which controller must continuously monitor their displays to obtain an accurate and up-to-date picture of the situation. To more thoroughly understand how controllers employ their attention in developing this picture to maintain situation awareness (SA), the present study was conducted. Actual ATC data was used to re-create scenarios using the Situation Assessment Through the Re-creation of Incidents (SATORI) system (Rodgers & Duke, 1993) to gain more insight into the nature of these processes. SATORI graphically recreates a visual display of the radar data recorded during actual air traffic control (based on computer tapes normally recorded at each air traffic control facility) synchronized with the recorded audio tapes of communications between the controller and pilots.

In this study SATORI was combined with a modification of the Situation Awareness Global Assessment Technique (SAGAT) (Endsley, 1988). SAGAT is a technique used during simulations in which the simulation is frozen at random, unexpected intervals with all display screens blanked and the operator of the simulation is queried about the state of the current situation. The operator's perceptions are then compared to the actual state of the environment to provide an objective assessment of the operator's situation awareness. The use of SAGAT to measure situation awareness in aircraft simulations has been extensively validated (Endsley, 1990a; 1990b).

In this study, SAGAT was modified to include queries that pertain to major factors associated with SA in en route ATC, based on an analysis by Endsley and Rodgers (1994). As a second modification, the technique was employed in conjunction with SATORI which involves the passive viewing of a situation as opposed to an interactive simulation in which the subject is involved. While it is not clear how the SA of a passive observer differs from that of an active participant, this measure should still provide an indication of the way in which controllers distribute their attention to various factors involved in the scenarios. As this study involves currently certified controllers viewing re-creations of real ATC scenarios, the combined use of SATORI and SAGAT may provide insight into controller attention deployment and SA in operational settings.

In addition, it may be considered that SA under passive viewing conditions may be analogous to that expected if the ATC system ever becomes highly automated. Under conditions of high automation the role of the controller would become one of monitor of an air traffic situation that is controlled by an automated system. The scenarios recreated for subjects using SATORI similarly involve the passive monitoring of a situation that is actually controlled by another. The study may therefore provide some insights into SA with a hypothetical automated system. This is of concern as there is some indication that SA may be compromised under highly automated systems (Endsley & Kiris, 1995).

METHOD

Subjects

Twenty subjects participated in the study on a voluntary basis. All subjects were experienced, full performance level (FPL) status Air Traffic Control Specialists at Atlanta ARTCC. The twenty subjects included four subjects viewing scenarios in each of five areas of specialization in the facility. All subjects were certified in the area of specialization for the re-created errors that they observed during the study. Subjects were relieved from the air traffic control room floor to participate in the study. Once subjects completed their participation, they returned to their assigned duties in the control room.

Procedure

Fifteen scenarios involving operational errors (OE) that occurred in the Atlanta Air Route Traffic Control Center (ARTCC) in 1993 and 1994 were recreated using SATORI. SATORI recreations were presented on a DEC 3000-300 Alpha computer system using dual Sony 19-inch high-resolution (1280 x 1024) color monitors. Three errors in each of five areas of specialization of the center were selected. (One error was eliminated during testing due to problems with the data tapes.) Subjects were provided with a set of instructions and signed a voluntary subject consent form. They were then shown scenarios involving three errors from sectors in the area of specialization on which they were certified. Each scenario consisted of a recreation of the ten minutes immediately prior to the occurrence of the OE. Twice during each scenario, the recreation was halted and the screen blanked. The first freeze occurred two minutes prior to the occurrence of the error and the second freeze occurred at the time of the OE in each scenario. Although subjects were informed that freezes would occur, they were not informed as to the timing of the freezes or the occurrence of the error.

During each freeze subjects were provided with a map of the sector. Sector boundaries, navigation aids, airways and intersection markings were shown on the map, however, no aircraft were included. Subjects were asked to indicate the location of all known aircraft on the map, and, for each aircraft, to indicate or make a judgment of:

(1) if the aircraft was:
    (a) in the displayed sector's control,
    (b) other aircraft in the sector not under sector control, or
    (c) would be in the sector's control in the next two minutes,
(2) aircraft call sign,
(3) aircraft altitude,
(4) aircraft groundspeed,
(5) aircraft heading,
(6) the next sector the aircraft would transition to,
(7) whether the aircraft was climbing, descending or level,
(8) whether the aircraft was in a right turn, left turn or straight,
(9) which pairs of aircraft had lost or would lose separation if they stayed on their current (assigned) courses,
(10) which aircraft would be leaving the sector in the next two minutes,
(11) which aircraft had received clearances that had not been completed, and, for those, whether the aircraft received its clearance correctly and

whether the aircraft was conforming to its clearance, and
(12) which aircraft were currently being impacted by weather or would be impacted in the next five minutes.

RESULTS AND DISCUSSION

Subjects' responses to each question were scored for accuracy based on computer data for each aircraft at the time of each freeze. Responses were scored as either correct or incorrect based on operationally determined tolerance intervals. Missing responses were scored as incorrect.

Means and standard deviations for subject response accuracy are shown in Table 1. On average, 12.8 aircraft were actually present at the time of the freezes (range 4 to 23) across scenarios. Of these, subjects on average reported 8.0 aircraft or 67.1% of the aircraft present. Mean distance error was 9.6 miles (.68 inches) from the aircrafts' reported location to their actual location. This error most likely reflects inattention to aircraft not currently under consideration (aircraft movement occurring during lapses in the visual scan of that aircraft), or may also be an artifact of the passive viewing procedure used in this study.

For the aircraft reported, the correctness of subject responses on the remaining questions was calculated. Subjects correctly identified the control level of the aircraft (in sector control, other aircraft in sector, will be in sector control in the next 2 minutes) for 73.8% of the aircraft reported. Aircraft callsigns were often incomplete. The initial alphabetical part of the callsign (indicating airline company, military or civil aircraft designation) was reported correctly 73.8% of the time. The numerical part of the callsign (the aircraft identification number) was reported correctly for only 38.4% of the aircraft. It should be noted that other studies have found that in general 4% of OEs involve readback errors associated with aircraft identification (Rodgers & Nye, 1993). The low level of accuracy in recall knowledge of aircraft callsigns is probably highly indicative of these readback errors, as it indicates that controllers may not attend to or retain much information on aircraft callsign in working memory, particularly the identification number.

Aircraft altitude was correctly reported (+/- 300 ft) for 59.7% of the aircraft (mean error of 655 ft). The aircraft were correctly identified as ascending, descending or level 66.4% of the time. Correct groundspeed (+/- 10 knots) was reported for only 28.0% of the aircraft (mean error 21.8 knots). Correct aircraft heading (+/- 15 degrees) was reported for 48.4% of the aircraft (mean error 15.6 degrees). Only 35.1% of the aircraft were correctly identified as being in a left turn, right turn or proceeding straight ahead. These results indicate that subjects were fairly poor at keeping up with the dynamics of the aircraft in the scenario, at least for many of the aircraft.

An argument can be made that perhaps subjects simply did not retain this type of detailed information about

Table 1. Awareness of Situation Across all Subjects and Scenarios

Variable	Mean	Std. Dev.
Actual aircraft present (number)	12.857	5.591
Aircraft reported (%)	67.1	18.0
Distance error (miles)	9.649	4.479
Control level (% correct)	73.8	17.3
Call sign: alphabetic (% correct)	79.9	23.4
Call sign: numeric (% correct)	38.4	32.0
Altitude (% correct)	59.7	22.1
Change in altitude (% correct)	66.4	25.6
Speed (% correct)	28.0	25.6
Heading (% correct)	48.4	30.6
Turn (% correct)	35.1	40.2
Separation problems (% correct)	86.2	32.3
Transition to next sector (% correct)	63.5	45.1
Assigned clearances complete (% correct)	23.2	22.9
Assigned clearance correct (% correct)	74.4	43.9
Assigned clearance conformance (% correct)	82.9	37.9
Weather impact (% correct)	60.7	49.1

each aircraft (Level 1 SA), instead maintaining awareness of higher level situation comprehension and projection issues (e.g. aircraft separation and future projections of actions). Previous research, however, indicates that people do maintain task relevant information about Level 1 SA elements that can be reliably recalled under this type of testing (Endsley, 1990a). There is also evidence that this measurement technique is reflective of subject attention allocation across sources of information (Fracker, 1990). It is more likely, therefore, that these measures do provide some indication of the ways in which subjects in this study were deploying their attention across displayed information, as least on a relative basis.

The subjects' higher level of understanding of the scenarios was also evaluated. The aircraft pairs that the subjects identified as having "lost or will lose separation if they stay on their current (assigned) courses" was compared to those aircraft that actually had lost or would lose separation (in the following 2 minutes) at the time of the freeze. Subjects correctly identified 86.2% of these aircraft pairs. (Aircraft pairs that the subject identified as having potential separation problems, but did not, were not scored.) Subjects correctly identified 63.5% of aircraft that would be leaving the sector in the next two minutes. Thus, they did not appear to be fully aware of upcoming sector transitions.

Subjects correctly identified only 23.2% of aircraft that had not yet completed control assignments. Of those that they identified as not having completed an assignment, subjects were correct in 74.4% of the cases in their

identification of whether the aircraft had correctly received its assignment, and in 82.9% of the cases in their identification of whether the aircraft was conforming to its assignment. Overall, subjects did not attend well to an aircraft after a clearance was given in terms of monitoring for compliance or progress in completing the control action.

Subjects were incorrect in identifying weather as a current impact (or impact in the next 5 minutes) in 39.3% of the scenarios. This is perplexing in that even though light and heavy weather symbols were displayed in some scenarios, poor weather did not impact traffic in any of the scenarios presented. This finding most likely indicates a fairly poor ability by controllers to estimate the impact of weather on air traffic, an issue which has been raised by controllers as problematic.

CONCLUSIONS

The frequency of correct responses on each variable provides some insight into the tradeoffs that controllers make in allocating their limited attention across multiple aircraft and pieces of information that compete for that attention. This analysis is not meant to be critical regarding the information controllers did not attend to or retain in working memory. Attention allocation strategies, such as those indicated here, are needed and are effective the vast majority of the time in dealing with the demands of controlling air traffic as can be demonstrated by the effective daily performance of controllers and relatively low error rates nationwide. A point that can be made, however, is that these strategies may lead to a lack of situation awareness that occasionally (due to a probabilistic link between SA and performance (Endsley, 1995) results in errors. This point is reinforced in that the patterns of attention demonstrated here can be correlated with certain systematic characteristics of OEs. It should also be noted that a fairly high degree of variability was present on many of the variables, across aircraft, subjects, freezes and scenarios. Possible sources of these variations need to be examined more closely.

One limitation of this study is that it involved subjects who, although current, experienced controllers, passively viewed the ATC scenario re-creations. While this procedure provided insight into factors affecting actual OEs (which can be difficult to produce under simulated conditions), one can speculate as to whether their situation recall accuracy is the same as it would be for controllers actively working the same scenarios. While it is probably difficult to stipulate that the absolute levels of SA and workload reported would be the same, the general patterns presented here are probably valid as reflections of subjects' attention distribution.

It is somewhat likely, however, that levels of SA may be lower under passive viewing conditions rather than active decision making, such as has been demonstrated under higher levels of automation in recent research (Endsley & Kiris, 1995). In this light, difficulties in accurately identifying aircraft separation problems shown by subjects in this study may be at least partially reflective of the difficulties associated with passive monitoring. This possibility needs to be seriously investigated with regard to systems being developed for automating future air traffic control.

REFERENCES

Endsley, M. R. (1988). Design and evaluation for situation awareness enhancement. In Proceedings of the Human Factors Society 32nd Annual Meeting (pp. 97-101). Santa Monica, CA: Human Factors Society.

Endsley, M. R. (1990a). A methodology for the objective measurement of situation awareness. In Situational Awareness in Aerospace Operations (AGARD-CP-478) (pp. 1/1 - 1/9). Neuilly Sur Seine, France: NATO - AGARD.

Endsley, M. R. (1990b). Predictive utility of an objective measure of situation awareness. In Proceedings of the Human Factors Society 34th Annual Meeting (pp. 41-45). Santa Monica, CA: Human Factors Society.

Endsley, M. R. (1995). Toward a theory of situation awareness. Human Factors, 37(1), 32-64.

Endsley, M. R., and Kiris, E. O. (1995). The out-of-the-loop performance problem and level of control in automation. Human Factors, 37(2), 381-394.

Endsley, M. R., & Rodgers, M. D. (1994). Situation awareness information requirements for en route air traffic control (DOT/FAA/AM-94/27). Washington, D.C.: Federal Aviation Administration Office of Aviation Medicine.

Fracker, M. L. (1990). Attention gradients in situation awareness. In Situational Awareness in Aerospace Operations (AGARD-CP-478) (Conference Proceedings #478) (pp. 6/1-6/10). Neuilly Sur Seine, France: NATO - AGARD.

Rodgers, M. D., & Duke, D. A. (1993). SATORI: Situation assessment through the re-creation of incidents. Journal of Air Traffic Control, 35(4), 10-14.

Rodgers, M. D., & Nye, L. G. (1993). Factors associated with the severity of operational errors at Air Route Traffic Control Centers. In M. D. Rodgers (Ed.) An analysis of the operational error database for Air Route Traffic Control Centers (DOT/FAA/AM-93/22). Oklahoma City, OK: Human Factors Research Laboratory, Civil Aeromedical Institute, Federal Aviation Administration.

AUTOMATION IN GENERAL AVIATION:
RESPONSES OF PILOTS TO AUTOPILOT AND PITCH TRIM MALFUNCTIONS

Dennis B. Beringer
Human Factors Research Laboratory
FAA Civil Aeromedical Institute, Oklahoma City

Interactions between the pilot and onboard-automated systems, designed to reduce pilot workload and to decrease variability of aircraft performance, have been implicated in a number of accidents and incidents. An examination of autopilot use by 29 general-aviation pilots having complex aircraft experience was performed in the Civil Aeromedical Institute's Advanced General Aviation Research Simulator, configured as a Piper Malibu, for four simulated autopilot/pitch-trim failures. Response times were longer for subtle (sensor) failures than for overt (commanded roll) ones, and two distinct response strategies, immediate disconnect and manual override, were revealed. It was also found that there was an appreciable delay between first response to runaway pitch trim and final resolution. Implications for certification are discussed.

INTRODUCTION

The most visible and recollected aircraft accidents are those resulting in the loss of large commercial aircraft, such as China Airlines' Flight 140, April 26, 1994, on approach to Nagoya/Komaki airport, Nagoya, Japan (Katz, 1995). The data indicated that the aircraft, an Airbus A-300-600R, ultimately stalled and crashed after attaining a pitch-up attitude of approximately 52 degrees at 78 knots. The problem appeared to be the pilot's continued attempts to fly the airplane manually with the autopilot engaged in go-around mode. The captain, who had apparently inherited the approach from the first officer after an autothrottle, but not autopilot, disengagement, ultimately lost the struggle with the aircraft as the autopilot trimmed the aircraft nose up after the captain's continued attempts to force the nose down. Problems with automated systems are not restricted to commercial carriers, however. Similar incidents involving pitch trim malfunctions and other autopilot difficulties have been reported for general aviation (GA) aircraft (Wilson, 1995; Katz, 1995).

Present certification standards require that an autopilot system, in a hard-over failure where the control surface servo is driven at its maximum rate, cannot place the aircraft in greater than a 60-degree bank nor place undue loads (0-2 Gs limits) on the airframe "within a reasonable period of time" (F.A.R. 23.1329). This time has been defined, via Advisory Circular 23.1329-2 (DOT/FAA, 1991), as the three seconds following initial detection of the uncommanded bank. Similarly, this restriction applies to pitch and pitch trim tests to the degree that the aircraft cannot stall, exceed limit speeds, or require excessive control force during recovery at the end of the three-second period. This interval supposedly provides time during which the pilot can diagnose the problem and take corrective action (assumes autopilot disconnect). A de-

lay of one second was adopted for malfunctions during coupled approach on the theory that the pilot is likely to be attending to the instruments more closely on approach than during cruise. Cooling and Herbers (1983) noted, in their discussion of human factors, that "...there are no studies available to support the FAA certification standard of a three second delay (enroute) or a one second delay (on approach) before initiation of recovery by the pilot from an autopilot malfunction." However, it has been suggested that the data were actually derived from a study of airline pilots' responses collected during a study performed at Wright-Patterson AFB in the 1960's (ACE-100, 1996). The focus of our research, in support of Aircraft Certification, was the responses of pilots to overt and subtle autopilot malfunctions, and the factors influencing the speed and the selection of those pilot responses.

METHOD

Design/Subjects

The experimental approach, a single-factor within-subject design using autopilot malfunction type (4) as the independent variable, was selected because high between-subject variability in response times to malfunctions was expected. The malfunction types were: "command over" roll (rate = 6 deg/sec), soft roll (sensor) (rate = 1 deg/sec), soft pitch (sensor) (rate = 0.2 deg/sec), and runaway pitch trim up. The last was selected for practical reasons to increase the likelihood of completing data collection. Runaway pitch-trim down can, if not attended to, create significant pitch-down attitudes and possible over-speed conditions, increasing the potential for a prematurely terminated or interrupted data run. Dependent variables recorded included flight performance indices and states of critical switches; autopilot disconnect, engage, pitch-trim and circuit breaker. Pilots were obtained from the local area who were instrument rated and had experience with complex aircraft and autopilot systems. Age ranged from 24 to 72 years (median = 42) and the sample contained 27 men and 2 women. No subject had less than 300 hours of flight experience.

Equipment/Procedures/Tasks

Data-collection sessions were conducted in the Advanced General Aviation Research Simulator (AGARS) in the Human Factors Research Laboratory, Civil Aeromedical Institute. The simulator was configured as a Piper Malibu with Bendix/King avionics (KFC-150 autopilot); software approximated behavior of both but exact flight equations were not available. High-fidelity primary flight displays were presented in the cockpit on three masked CRTs that replicated the Malibu panel layout and gave the appearance of hard, dedicated instrumentation. The out-the-window depiction spanned 150 degrees of horizontal visual arc and was a high-resolution textured representation of the Oklahoma City area.

Pilots participated in one 2- to 2.5-hour session. They were told that the study was to examine use of autopilots in routine flying and to gather opinion data on useful features. The first hour consisted of experiment-related paperwork and familiarization-training activities, including: reading excerpts from the autopilot (AP) manual, cockpit familiarization, and a half-hour familiarization flight using all AP modes. The second half of the session was used to collect performance data for the malfunction conditions. A simple round-robin instrument clearance was flown from Will Rogers World Airport to two local very-high-frequency ominrange (VOR) stations and back in instrument-flight-rules (IFR) conditions between textured cloud layers (distinct visual horizon but no ground detail). Pilots were required to interact with air traffic control, fly vectors, track inbound to two VOR stations, and fly a fully-coupled instrument-landing-system (ILS) approach, and were instructed to fly as much of the course as possible with the AP engaged.

Malfunctions were spaced such that sufficient time elapsed between failures (13-15 minutes) to prevent interference between episodes. Command roll and soft pitch were encountered in level flight, soft roll during descent, and half pitch trim during the ILS approach and half during ascent from 6000' to 7000'. Only the pitch-trim malfunction produced both auditory and visual warnings, consisting of a steady TRIM light and steady pure tone of 3.1 Khz at approximately 77 dB. The simulated system did not immediately disconnect during the runaway, representing a worst-case scenario (the KFC-150 AP does automatically disconnect, although some others do not), allowing the pitch servo to compensate for (and mask) the initial trim deflection. Data collection flights averaged 1.2 hours, followed by an AP-experience questionnaire and interview to determine each pilot's knowledge of AP and autotrim malfunction consequences and to gather task difficulty ratings.

RESULTS

Response Times

Command roll (roll servo). Of all the failures, commanded-roll and pitch-trim failures were rated as easiest to diagnose (by 11 of 26 pilots). The commanded-roll failure emulated an AP-commanded roll that failed to stop at the target bank angle. Analyses for both roll malfunctions and the soft-pitch malfunction are based upon time from initial failure to disconnect of the AP by any means (yoke-mounted disconnect, panel disengage, circuit breaker). Times ranged from 1.8 seconds to 107.1 (means, medians, and ranges are summarized in Table 1). However, 69 % of the pilots disconnected within 13 seconds of the initial failure and half within 8 seconds. These "immediate" disconnects by 18 of the 29 pilots were defined by sequences where no other significant actions occurred between failure onset and AP disconnect. The distribution of these times is shown in Figure 1. Using a response time of 8.7 seconds or less as a cutoff value, 93.7% of the sample of "immediate" responders was included.

Table 1. Response time mean, median, and range by failure and response category types.

Failure Type	Response Category	n	Response Time		Range	
			Mean	Med	Low	High
Command Roll	All (Disc)	29	16.5	8.5	1.8	107.1
	Immediate	18	5.9	5.9	1.7	11.8
	Manual Override	10	26.3	23.0	8.9	53.8
Soft Roll	Immediate	16	11.7	11.5	4.5	21.2
	Manual Override	13	37.5	26.0	13.2	85.1
Soft Pitch	Immediate	12	17.7	17.4	6.5	31.5
	Manual Override - 1	16	46.2	50.0	15.2	76.2
Pitch Trim Up	All (Disc)	25	10.5	6.9	0.2	39.2
	All (CB pull)	25	35.4	23.5	4.9	109.7
	All (CB lag)	25	25.0	15.7	0	102.3
	- extremes	23	22.7	15.7	5.1	71.3

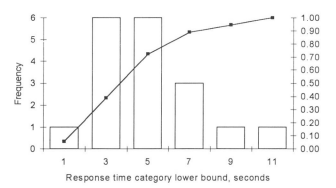

Figure 1. Commanded-roll response-time distribution and cumulative frequency plot, immediate disconnects.

Ten pilots chose to manually override the AP, whether by using the control-wheel steering option or by overpowering the roll servo without disconnecting the AP. 90% had response times of 48.3 seconds or less (Figure 2). A post-hoc comparison of the log-transformed disconnect times of the two groups, with the highest and lowest extreme times removed, indicated a significant difference ($F[1,24] = 53.27$, $p<0.0001$) between the immediate disconnects (untransformed mean = 5.93 seconds) and the manual overrides (untransformed mean = 28.26 seconds).

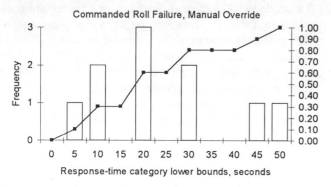

Figure 2. Commanded-roll response-time distribution and cumulative frequency plot, manual override.

Soft roll (roll sensor). The soft-roll failure was rated as third in difficulty to diagnose, but was rated easiest to correct (by 13 of 26 pilots). Following removal of one outlier (194 seconds), pilot performance was again categorized as immediate disconnect (16) or manual override (12). Those categorized as immediate disconnect responses averaged 11.72 seconds (range: 4.52 to 16.69) (Figure 3) while those categorized as manual overrides averaged 37.45 seconds (range 13.16 to 85.14) (Figure 4). Approximately 88% of all immediate disconnects occurred in less than 17 seconds, with 75% occurring in less than 14 seconds. Post-hoc comparison indicated the mean difference to be significant for both raw and log transformed scores (log scores: $F[1,26] = 27.07$, $p<.00005$).

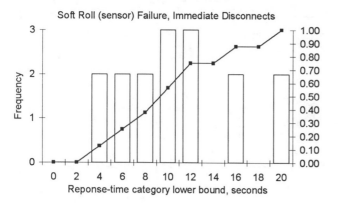

Figure 3. Soft-roll (sensor) response-time distribution and cumulative frequency plot, immediate disconnects.

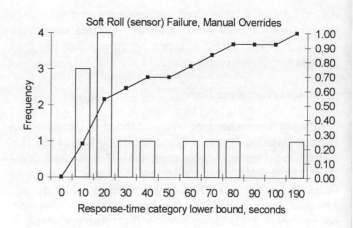

Figure 4. Soft-roll (sensor) response-time distribution and cumulative frequency plot, manual overrides.

Soft pitch (pitch sensor). The soft-pitch failure was rated as most difficult to diagnose (by 12 of 26 pilots), and was rated third easiest to correct, missing a tie for second by one tally. Performances were again categorized as either immediate disconnect (12) or manual override (17), and the distributions are shown in Figures 5 and 6. Three pilots never diagnosed the failures, manually flying the airplane without disconnecting the autopilot, and their scores and one other outlier were removed, leaving 13. Immediate disconnects averaged 17.38 seconds (range: 6.5 to 31.5) and manual overrides averaged 46.19 (range: 15.2 to 76.2). Approximately 50% of immediate disconnects occurred in less than 16 seconds, with approximately 85% occurring in less than 24 seconds (Figure 5). Post-hoc comparison of the log-transformed data showed the distributions of the two types of responses to be significantly different ($F[1,22] = 20.69$, $p<.0005$).

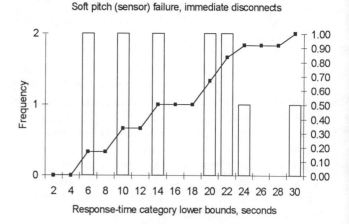

Figure 5. Soft-pitch (sensor) response-time distribution and cumulative frequency plot, immediate disconnects.

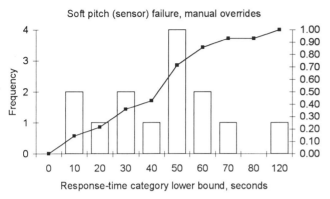

Figure 6. Soft-pitch (sensor) response-time distribution and cumulative frequency plot, manual overrides.

Runaway pitch trim. This failure was different from the others in that only pulling the pitch-trim circuit breaker would correct the problem. The interim solution was the AP disconnect/trim interrupt switch. Only three pilots chose the optimal response, depressing and holding the disconnect, then pulling the circuit breaker. Four others depressed and held the disconnect at various times during the recovery. The vast majority of initial responses were yoke AP disconnect (15), followed in frequency by panel-mounted AP-engage switch (5), mode manipulation (2), manual override (2), and pitch trim circuit breaker (1). Overall, 21 of the 25 pilots considered were classified as "immediate" responders, two were classified as manual overriders, and two as mode changers. It should also be noted that two pilots never heard the warning tone due to high-frequency hearing loss, responding only to aircraft performance changes.

Two stages of response were of interest; first, the time required to detect a malfunction and initiate some action (AP disconnect, control-wheel steering, AP engage or circuit breaker) and second, the time lag between the initial action and the pulling of the pitch-trim circuit breaker. Average time to initial action for the usable 25 pilots was 10.46 seconds, with all except one response over 3 seconds One can see from Figure 7 that 50% of the responses occurred in less than 7 seconds, with 65% of the cases in less than 9 seconds.

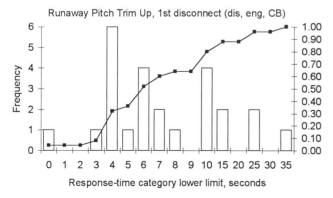

Figure 7. Runaway pitch-trim response-time distribution and cumulative frequency plot, first disconnect.

Time to pull the pitch-trim circuit breaker averaged 35.4 seconds (range: 4.91 to 109.69) (Figure 8), with an average lag of 22.69 seconds (high and low scores removed) between the initial response to the runaway pitch trim (disconnect or control movement) and the required remedy.

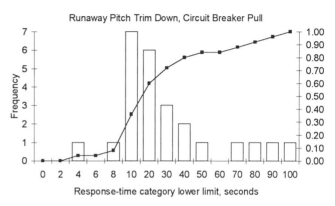

Figure 8. Runaway pitch-trim response-time distribution and cumulative frequency plot, circuit breaker pull.

Initial examination of the *questionnaire and interview data* indicated that all pilots understood they could manually overpower the autopilot servos, and 22 were aware of the potential interaction between a runaway pitch-trim motor and autopilot pitch-attitude (elevator servo) inputs. Four pilots had not considered the potential interaction previously, but grasped the concept immediately during the interview. When asked what their strategy for dealing with autopilot malfunctions was, the group voiced two anchor strategies and a combination of the two as a third. The *immediate-disconnect* strategy was endorsed by nine individuals, while two others expressed a *procedural* approach that was closely related to the immediate disconnect strategy. Another five individuals suggested that they would *fly the aircraft through the malfunction* while attempting to diagnose the problem. A third group took a middle-of-the-road stance, saying that the strategy was malfunction dependent. These seven expressed their strategies as, "Fly through mild failures; disconnect for severe failures," or "diagnose while the unit is still engaged, then disconnect."

Correlations.
Mode-of-flight effects. The mode of flight during which the failure is encountered is also of particular interest. Recall that the delay used during certification is to be *1* second during a coupled approach and that the experimental procedure was set up to examine the pitch-trim failure during both cruise climb and a coupled ILS approach. The aircraft, in either of these conditions, would be more likely to reach slow airspeeds than if the failure was encountered in level cruise or cruise descent, and the approach mode was specifically covered by the advisory circular.

Independent-samples *t* tests indicated no significant mean difference between response times for these two flight modes.

Levene's Test of variability, however, indicated a significant difference for the circuit-breaker lag ($F = 3.406$, $p<0.1$). The group experiencing the failure on climbout (SE of Mean = 7.37) was more variable in their responses than was the group receiving it on approach (SE of Mean = 5.07). When these scores were log transformed, as is usually advisable for response times, no significant effects of mean or variance differences were found. Although the first analysis could lend some credibility to the assumption that pilots were somewhat more attentive on approach, the lack of an effect for the log-transformed scores would tend to downplay this explanation. That the difference might represent an inherent difference between the post-hoc groups (climb, approach) was examined by performing comparable analyses of all other RT variables (commanded roll, soft pitch/sensor, soft roll/sensor). No significant mean or variance differences were found for either the raw or transformed scores, suggesting that these two groups of pilots were not significantly different in their performance on the experimental tasks.

Flight Performance Data

The Advisory Circular 23.1329-2 specifies that attitude and performance specification limits shall not be exceeded during recovery from excursions induced by an autopilot malfunction. Examination of pitch, bank, altitude, and indicated airspeed for each recovery indicated that only one individual exceeded 60 degrees of bank during one recovery, and for all other cases and all other malfunctions the aircraft was in a flyable condition and did not exceed attitude or airspeed performance limitations. Thus, one can say that recoveries were timely enough to prevent the aircraft from assuming extreme attitudes or airspeeds (overspeed or stall).

DISCUSSION/CONCLUSIONS

Present certification practice assumes that a malfunction will be either severe enough to produce supra-threshold cues or that an alert will warn the pilot, starting the three-second "recognition" period. Flight test personnel (FAA Aircraft Certification Service, 1996) have reported test malfunctions that have gone undetected until the test administrator or safety pilot pointed them out, sometimes after reaching criterion limits. These autopilots failed to obtain certification. Our data indicate pilots require an average of 5.9 seconds to a clearly supra-threshold event, some requiring as long as 11.8 seconds. General certification practice for "obvious" malfunctions allows one second for detection. Combined with the 3-second waiting period, this produces a four-second interval within which the pilot must detect and respond to the malfunction; less than the mean sample response. For the commanded-roll failure, one could accommodate 90% of this pilot sample using 9 seconds as the interval upper bound. Using even 7 seconds as the criterion, 70% of the sample would be accommodated. One should note that at the usual 5 deg/sec commanded roll rate, a 60-degree bank would not be ex-

ceeded for 12 seconds. A roll-servo hard failure at 15 deg/sec for this aircraft type, however, does so in four seconds.

It was not surprising that significantly longer intervals were required for pilot response to the more subtle failures. However, because the attitude indicator (ADI) continued to depict actual attitude during these malfunctions (in a true sensor failure the ADI would not), detection times were probably shorter than would otherwise be expected. Given this ADI anomaly, the potential consequences of the pitch-trim down runaway, and the "moderate" roll rate in the commanded-roll failure, additional data collection is planned for runaway pitch trim down as well as true attitude sensor failure and hard-over roll-servo failure (12-15 deg/sec roll rate). The following are recommended based upon the presently available data:

- Increase the waiting period for "command-over" and "sensor-loss" failures to accommodate at least 75% of the general pilot population using cumulative frequency curves on response time distributions.

- Consider eliminating separate treatment of approach and other flight modes given no detectable pilot response differences.

- Pursue additional failure annunciation or "fail-safe" modes from manufacturers.

- Continue use of attitude and performance limitations as ultimate criteria for acceptance.

Additional specific recommendations may be available upon conclusion of the second planned phase of experimentation.

ACKNOWLEDGMENTS

The author thanks Mr. Barry Runnels for engineering support of the AGARS during the course of the study and Mr. Howard Harris for his efforts during data collection, reduction and analysis. FAA Aircraft Certification Service (AIR-3) sponsored the research, and technical coordination was provided by the Small Airplane Directorate (ACE-100).

REFERENCES

ACE-110 (Small Airplane Directorate, FAA) (1996). Personal communication.

Cooling, J. E. & Herbers, P. V. (1983). Considerations in autopilot litigation. *Journal of Air Law and Commerce, 48,* 693-723.

DOT/FAA. Automatic Pilot System Installation in Part 23 Airplanes. Advisory Circular 23.1329-2, March 3, 1991.

FAA Aircraft Certification Service (1996). Personal communication with flight test personnel.

Katz, P. (1995). NTSB Debriefer: The dark side of "Otto pilot". In *Plane & Pilot,* (February), *31(2),* 18-19.

Wilson, B. G. (1995). I learned About flying from that (#660): Unacquainted with the autopilot. In *Flying,* (June), *122(6),* 122.

AN EMPIRICAL VALIDATION OF SUBJECTIVE WORKLOAD RATINGS

Anthony J. Aretz
Chris Johannsen
Keith Ober
United States Air Force Academy
Colorado Springs, CO

A correlational design was used to regress NASA TLX subjective workload ratings onto several potential independent variables (i.e., the number of concurrent tasks, task combination, task resource demands, and flight experience) to determine task characteristics that influence pilot subjective workload ratings. A part task simulator was used to present up to six concurrent tasks, in different combinations, to 27 cadets at the USAF Academy. The results indicated the number of concurrent tasks had the largest impact on subjective workload ratings. In terms of multiple resource theory, spatial, verbal, and visual demands (in that order) contributed the most variance. The implication for theoreticians and designers is that the number of concurrent tasks, mental resource demands, and time constraints seem to be key contributors to subjective workload ratings.

INTRODUCTION

Subjective workload ratings have become an integral part of the aerospace system design process, mainly due to their sensitivity, ease of use, and face validity. Since subjective ratings are so important to the design process, the human factors community has been involved in an extensive effort over the past 20 years to understand the validity and reliability issues associated with subjective workload measurement (O'Donnell and Eggemeier, 1986). An initial effort was completed last year to build on existing dual-task paradigm research (Aretz, Shacklett, Acquaro, and Miller, 1995). This initial effort supported Yeh and Wickens' (1988) predictions that the number of concurrent tasks and effort are the most important variables that influence subjective ratings. However, in contrast to Yeh and Wickens (1988), Aretz et al. found subjective ratings to be sensitive to increases in workload of up to five concurrent tasks used in the study.

A drawback of this initial study was it did not examine subjective ratings from a multiple resource perspective (Wickens, 1992). Multiple resource theory suggests that concurrent tasks with fewer resource conflicts should have better performance, but that subjective ratings may not be sensitive to the differences in the conflicts (Yeh and Wickens, 1988). Since the previous study contained primarily visual-spatial-manual tasks, this hypothesis could not be explored. The current study examines this possibility by including an additional task demanding auditory-verbal-vocal resources. The question of interest was how the number of concurrent tasks and resource demands combine to influence subjective workload ratings.

Another hypothesis examined in this study is the possibility that hemispheric lateralization influences flight performance. The literature suggests that Right Hemisphere Dominant (RHD) subjects have more laterally organized brains (Hellige, 1993) and that the two hemispheres act as independent resource pools (Friedman and Polson, 1981), with the RH containing spatial resources and the LH containing verbal resources. If this is true, RHD subjects might perform better and have lower subjective ratings because of the efficiency of their spatial resources; the opposite would be true for Left Hemisphere Dominant (LHD) subjects.

METHOD

Participants

Twenty-seven USAF Academy cadets (24 males and 3 females, 19 to 22 years old) voluntarily participated in the study. Eighteen subjects had no prior flight experience and nine had from 2 to 106 hours (M=15.0). The 27 subjects were selected from 40 volunteers who had completed a split visual field Sternberg task. There were nine LHD subjects, nine Equally Dominant (ED) subjects, and nine RHD subjects. The Hemispheric Dominance (HD) of each subject was determined by the distribution of visual field response time decrements (the difference between left visual field and right visual field response times). The LHD and RHD subjects were selected based on their location in the tails of the distribution and the ED were selected based on their central location.

Apparatus

An Israeli Aircraft Industry's (IAI) Pilot Evaluation System (PES) was used to present eight different flight scenarios, including up to five concurrent tasks (see Figure 1). The PES consists of a 486 computer, an F-16 type control stick and throttle and a generic fighter radar display containing altitude, velocity, heading, and radar information (see Figure 2).

Figure 1. *PES Part Task Simulator.*

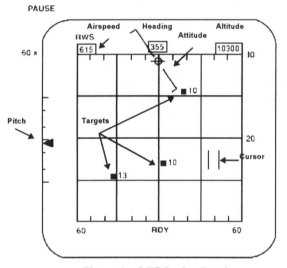

Figure 2. *PES Radar Display.*

The PES was designed to test a pilot candidate's attentional capacity to assess their fighter pilot potential (IAI, 1991). This assessment assumes individual differences in attentional capacity and that a candidate with greater attentional capacity would perform better on the PES and become a better pilot. The PES has received preliminary validation as a flight screener by Garvin, Acosta, and Murphy (1995). This research showed the PES distinguished between current USAF pilots and USAF non-pilots (but not between fighter and non-fighter pilots) and predicted the success of USAF Academy cadets in the Academy's new strenuous T-3 flight indoctrination program (but not the previous less strenuous T-41 program).

PES Flight Tasks

Target intercept. Intercepting and launching a missile at a target was the primary task. The subject manipulated the cursor over a target using a switch on the throttle. When the cursor was over the target, the subject locked on the target by pressing a switch on the stick. When the subject locked on a target, a more detailed view of the target was activated, including information about its altitude, airspeed, and distance from the aircraft. The subject tracked the target and when the target was within an 18-mile range, the subject disengaged and re-engaged the target using the two switches. When the target was within the firing envelope, the subject armed and fired a missile using two additional switches on the control stick. When the missile was launched, the target disappeared. Six latency measures and one accuracy measure were recorded: 1) the time from target onset to initial cursor movement; 2) the time from cursor initiation to lock on; 3) the time from entering the 18-mile range to disengage; 4) the time from disengagement to relock; 5) the time from entering the launch envelope to arming a missile; and 6) the time from onset of the ready signal to pressing the fire switch. The accuracy measure reflected the number of false alarms (i.e., initiating an action before the appropriate time).

Match target altitude. The subject manipulated the stick so the aircraft matched and maintained the target's altitude during the course of the engagement. Target altitude was displayed next to a target when the target appeared on the display. Three performance measures were recorded: 1) the time from target onset to first reaching the target's altitude (+/- 50 ft); 2) the sum of the squared deviations from target altitude from the point of onset to task completion; and 3) the standard deviation of the distribution of deviations.

Match target velocity. The subject manipulated the throttle to match and maintain the target's velocity. (When the subject was tasked with maintaining both a target's altitude and velocity, the stick and throttle inputs were coupled to affect altitude and velocity as in a generic fighter aircraft.) Three performance measures were recorded: 1) the time from target onset to first reaching the target's velocity (+/- 50 mph); 2) the sum of the squared deviations from the target's velocity from task onset to completion; and 3) the standard deviation of the distribution of deviations.

Point at target. The subject used the stick to maneuver the aircraft to match the target's heading so the target remained in the middle of the display. Three performance measures were recorded: 1) the time from target onset to first matching the targets heading (+/- 5°); 2) the sum of the squared deviations from the target's heading from task onset to completion; and 3) the standard deviation of the distribution of deviations.

Tone response. The subject responded to random tones (a high or low tone) presented through a headset at approximately 15-30 second intervals. The subject pressed the right button on the throttle for a high tone and the left button for a low tone. Two performance measures were recorded: 1) the time from the onset of the tone to a response; and 2) the proportion of errors in executing the response.

Overall performance. Overall performance on each flight scenario was computed by the PES using a figure of merit technique based on ideal performance (derived from Israeli fighter pilot baselines; personal communication, IAI, April, 1995). Two principles were behind this computation: 1) each dependent variable had a desired performance window and subjects received more points for staying in the middle of this performance window and lost points for deviations, 2) side tasks were weighted twice that of the primary intercept task, resulting in a poorer score if the side tasks were not performed well (indicating less attentional capacity). The result was a single percent score based on the subject's score divided by the total possible.

Auditory Digit Task

In addition to the PES flight tasks, subjects performed an auditory digit subtraction task. A tape player presented single digits at a rate of one digit every four seconds. Subjects were required to respond vocally with the absolute value of the difference between the digit just presented and the previous digit. Consequently, the task required subjects to keep a single digit in working memory. Digit performance was scored using the percentage of correct responses during a scenario flight.

Scenarios

Table 1 shows the hypothesized resource demands of each task according to Wickens' (1992) multiple resource theory. Table 2 shows the nine scenarios used in the study and the total resource demands (in terms of the number of tasks demanding that resource); however, no attempt was made to control for task difficulty and it cannot be assumed that different combinations of tasks have equal difficulty. These scenarios were created by combining eight PES scenarios (selected from 22 possible PES scenarios) with the digit task for a total of nine scenarios containing from one to six concurrent tasks in different combinations.

Table 1. *Task Resource Demands.*

Task	Input	Central Processing	Response
Digits	Auditory	Verbal	Vocal
Tones	Auditory	Spatial	Manual
Target Int.	Visual	Spatial	Manual
Match Alt.	Visual	Spatial	Manual
Match Vel.	Visual	Spatial	Manual
Center Targ.	Visual	Spatial	Manual

Subjective Workload Ratings

A computer version of the NASA-TLX (Hart and Staveland, 1988) was used to collect subjective workload ratings. The NASA-TLX was selected because it is easy to administer and has been well validated (e.g., Hill, Iavecchia, Bittner, Zakland, and Christ, 1992; Vidulich and Tsang,

1985). Using the TLX also allowed a direct comparison to the Aretz et al. (1995) study.

Table 2. Scenario Descriptions.

Scen	Tasks	Total Tasks	Resource Demands* (a/v;v/s;v/m)
1	digits	1	1/0;1/0;1/0
2	intercept targ.	1	0/1;0/1;0/1
3	digits/match vel.	2	1/1;1/1;1/1
4	match alt/vel /center target	3	0/3;0/3;0/3
5	digits/tones/intercept /match vel	4	2/2;1/3;1/3
6	tones/intercept /match alt/vel	4	1/3;1/3;1/3
7	digits/tones/intercept /match alt/vel	5	2/3;1/4;1/4
8	tones/intercept /match alt/vel/center targ	5	1/4;0/5;0/5
9	digits/tones/intercept /match alt/vel/center targ	6	2/4;1/5;1/5

* a/v=auditory/visual; v/s=verbal/spatial; v/m=vocal/manual

Procedures

The subjects were first presented with a computer slide show explaining the basic operation of the PES. Subjects then flew four practice scenarios (different from the nine data collection scenarios) in the same sequence to familiarize them with the five PES tasks. After each practice scenario, the subjects completed practice TLX ratings to provide a relative individual baseline for data collection. (There was no practice for the digits task.) Next, the subjects flew the nine data collection scenarios. The order of the data collection scenarios was counterbalanced across subjects using a modified Latin square.

NASA-TLX ratings were collected at the completion of each scenario. At the conclusion of all scenarios, subjects completed a paired comparison of the six subscales of the NASA-TLX, generating a weight for each dimension used to compute an overall workload rating. Each subject took approximately one hour to complete the study.

RESULTS

Three stepwise regressions were performed. In the first, PES performance was regressed onto the number of tasks, task combination (scenario), flight experience, and resource demands. The number of tasks demanding visual resources was the only significant variable to contribute to the total variance, $R^2=.26$. Table 3 shows the results of the TLX ratings regressed onto the same variables. The number of concurrent tasks accounted for the most variance, followed by verbal resource demands, and task combination (scenario). Table 3 also shows the results of a stepwise regression with the number of tasks and scenario removed. This regression revealed spatial, verbal, and visual resource demands contribute the most to TLX variance. A negative correlation was also found between simulator performance and subjective workload ($r = -.314$, $p < .05$) and Figure 3 shows a plot of both PES performance and TLX ratings as a function of the number of tasks.

Table 3. *NASA TLX Regression Analysis.*

Step #	Variable	Cum R^2	Final Beta
1	# Tasks	0.42	0.40
2	Verb Demands	0.46	0.29
3	Scenario	0.49	0.30
With Tasks and Scenario Removed			
1	Spat Demands	0.32	0.73
2	Verb Demands	0.46	-0.07
3	Vis Demands	0.48	0.45

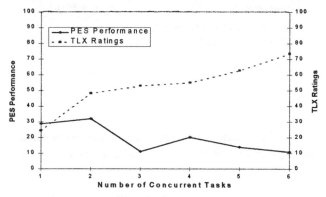

Figure 3. *PES Performance and TLX Ratings.*

Table 4 presents the results of the overall weighted TLX workload ratings regressed onto the TLX subscales. The time demands subscale accounted for most of the variance, followed by performance, effort, and the other subscales.

Table 4. *NASA-TLX Subscale Regression Analysis.*

Step #	TLX Subscale	Cum R^2	Final Beta
1	time	0.870	0.191
2	performance	0.929	0.226
3	effort	0.968	0.195
4	frustration	0.978	0.174
5	physical	0.988	0.157
6	mental	0.995	0.202

Another analysis of the data examined the scenarios with the same number of concurrent tasks, but with less resource competition. This situation occurred for two pairs of scenarios in the study--5/6 and 7/8. Figure 4 shows the mean TLX ratings for each scenario and that the TLX ratings were higher for the two scenarios with less resource conflict but the same number of tasks (scenarios 5 and 7), $F(1,25)=63.11$, $p<.001$ and $F(1,25)=44.78$, $p<.001$, respectively. Further, performance on the PES decreased

from scenario 5 to 6 and remained relatively constant from 7 to 8, indicating a possible dissociation between subjective ratings and performance. It is also interesting to note that the TLX ratings for scenario 3 (the concurrent performance of scenarios 1 and 2; $M=48.0$) was approximately the sum of the TLX ratings for scenarios 1 ($M=29.7$) and 2 ($M=19.3$).

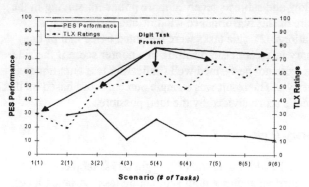

Figure 4. *PES Performance and TLX Ratings by Scenario.*

Figure 5 shows performance for the digit task and indicates performance decreased as the number of concurrent tasks increased, $F(4,100)=50.12$, $p<.001$. (One subjects' data were lost.) Finally, an analysis of hemispheric dominance revealed no significant differences in the TLX ratings, $p>.05$, but did find a significant difference in PES performance, $F(2,24)=3.62$, $p=.042$. A post hoc Tukey test revealed RHD subjects had higher PES scores ($M=21.3$) than LHD subjects ($M=15.5$; $M=19.7$ for ED subjects).

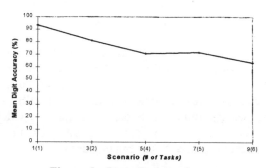

Figure 5. *Digit Task Performance.*

DISCUSSION

If a designer's goal is to assess the workload of a design (and eventually predict performance), this study showed that the number of concurrent tasks is a major variable accounting for 40% of subjective workload variability (see Table 3). This R^2 is higher than those found by Vidulich and Tsang (1994) for their similar immediate ratings condition (28-30%), suggesting that practice with the workload ratings minimizes some of the disadvantages of immediate (vs. delayed relative) ratings. Although this study did not use the Subjective Workload Dominance (SWORD) technique (Vidulich, Ward, and Schueren, 1991), an even higher R^2 might be expected based on Vidulich and Tsang's findings.

These results are consistent with Yeh and Wickens' (1988) predictions and Aretz et al.'s (1995) findings that the number of tasks is a prime contributor to subjective workload ratings. The finding that the time demands subscale of the TLX accounted for the most variance in the ratings was not consistent with the previous study (which found the effort subscale to be the most sensitive), but this difference could be due to the time paced digit task.

The TLX data were also consistent with Aretz et al.'s findings that subjective ratings are sensitive to workload increases in task saturated conditions (see Figure 3). The moderate negative correlation between PES performance and TLX ratings also indicates an association between task performance and subjective workload ratings. Consequently, these data do not support Yeh and Wickens' (1988) suggestion that subjective measures may not be as sensitive in high workload conditions.

The finding that the same number of tasks with fewer resource conflicts led to higher subjective ratings supports Yeh and Wickens' (1988) suggestion that subjective workload ratings are less sensitive than performance to changes in resource competition. Subjective ratings decreased for scenarios 6 and 8 (relative to scenarios 5 and 7), but PES and digit task performance decreased or remained relatively stable. This finding provides support for Yeh and Wickens suggestion that dissociations between performance and workload can result when tasks differ in the nature of their resource demands. As predicted by multiple resource theory, PES performance was better in scenario 5 as compared to scenario 6 (i.e., when there was less resource competition), but the workload ratings were higher.

This dissociation may be due to Yeh and Wickens' hypothesis that subjective ratings are most sensitive to working memory demands. The digit task seemed to be a major factor that contributed to high workload ratings (see Figure 4, scenarios 1, 3, 5, 7, and 9). In contrast, TLX ratings were relatively lower when the digit task was not required (scenarios 2, 4, 6, and 8). The digit task was the only task that placed relatively high demands on working memory (the information required for the PES tasks was continuously available on the radar display). Another possibility for the large influence of the digit task, and one the TLX subscale stepwise regression supports, was the forced paced nature of the digit task. The TLX time subscale contributed the most variance to the weighted workload ratings, while mental demands contributed the least. The constant pressure to keep up with the digit task seems to have been the primary contributor to the relatively high workload ratings when the digit task was present.

Finally, the performance difference found among the different hemispheric lateralities indicates RHD people may have a relative advantage when performance is heavily weighted toward spatial tasks. However, the subjective ratings were not sensitive to these performance differences, again supporting the hypothesis that subjective ratings and performance can dissociate based on resource demands.

More importantly, this finding suggests the possibility that HD might be a valid predictor of pilot performance.

CONCLUSIONS

The implication of these results for designers is that subjective workload and performance can be predicted in a relative manner during the design process using the number of concurrent tasks to be performed. However, it is important to remember that the workload predictions will be moderated by the mental resource demands of the tasks. Subjective ratings seem to have varying sensitivity to changes in the nature of the resource competition among the tasks and can dissociate from performance. Concerning the TLX ratings themselves, the data also showed the TLX subscales can be used in a diagnostic fashion to determine task characteristics that contribute most to workload ratings. In summary, the time honored method of task analysis that projects tasks required onto the time available is likely to be a good approximation for workload and performance estimation during the design process.

REFERENCES

Aretz, A. J., Shacklett, S., Acquaro, P., and Miller, D. (1995). The prediction of subjective pilot workload. In *Proceedings of the 38th Annual Meeting of the Human Factors Society*, pp. 94-97. Santa Monica, CA: Human Factors Society.

Friedman, A. and Polson, M.C. (1981). Hemispheres as independent resource systems: Limited capacity processing and cerebral specialization. *Journal of Experimental Psychology: Human Perception and Performance*, 7, 1031-1058.

Garvin, J.D., Acosta, S.C., and Murphy, T.E. (1995). Flight training selection using simulators - A validity assessment. Presented at the Eighth International Symposium on Aviation Psychology. Columbus, OH: Ohio State University.

Hart, S.G., and Staveland, L.E. (1988). Development of NASA-TLX (Task Load Index): Results of empirical and theoretical research. In P.A. Hancock and N. Meshkati (Eds.), *Human Mental Workload*, pp. 139-183. Amsterdam: Elsevier.

Hellige, J. B. (1993). *Hemispheric asymmetry: What's right and what's left*. London: Harvard University Press.

Hill, S.G., Iavecchia, H.P., Byers, J.C., Bittner, A.C. Jr., Zaklad, A.L., and Christ, R.E. (1992). Comparison of four subjective workload rating scales. *Human Factors*, 34, 429-439.

Israeli Aircraft Industries. (1991). *Pilot Evaluation System Validation and Design Principles*. [Brochure]. Lod, Israel: Author.

O'Donnell, R.D. and Eggemeier, F.T. (1986). Workload assessment methodology. In *Handbook of Perception and Human Performance*, K.R. Boff, L. Kaufmann, and J.P. Thomas (Eds.) New York, NY: John Wiley and Sons.

Vidulich, M.A., and Tsang, P.S. (1986). Techniques of subjective workload assessment: A comparison of SWAT and the NASA-bipolar methods. *Ergonomics*, 29, 1385-1398.

Visulich, M.A., and Tsang, P.S. (1994). The roles of immediacy and redundancy in relative subjective workload assessment. *Human Factors*, 36, 503-513.

Vidulich, M.A., Ward, G.F., and Schueren, J. (1991). Using the Subjective Workload Dominance (SWORD) technique for projective workload assessment. *Human Factors*, 33, 677-691.

Wickens, C.D. (1992). *Engineering Psychology and Human Performance*, 2nd Ed. New York, NY: Harper Collins Publishers.

Yeh, Y., and Wickens, C.D. (1988). Dissociation of performance and subjective measures of workload. *Human Factors*, 30, 111-120.

SYMPOSIUM:
DECISION MAKING IN FREE FLIGHT

Kip Smith, Peter Hancock, & Stephen Scallen
Human Factors Research Laboratory
University of Minnesota

In its web-page introduction to free flight, the Federal Aviation Administration (FAA) describes free flight as "an innovative concept to improve the efficiency of the National Airspace System [in which] pilots operating under instrument flight rules will be able to select the aircraft's course, speed, and altitude in real time" (FAA, 1996). This symposium presents the results of experimental and theoretic investigations on (1) the ability of pilots and air traffic managers to master the demands of free flight and (2) the design of technology and decision aids tailored to need those demands.

INTRODUCTION

This symposium presents the vanguard in experimental and theoretic investigations of the decision making of commercial pilots, professional traffic managers, and designers of technology for the free flight environment. The broad scope of the papers reflects the nascent character of free flight. Dr. Denning discusses case studies of pre-flight planning in field tests of free flight. Dr. Braune presents a rigorous methodology for the design and development of glass cockpit technology. Mr. Knecht introduces a mathematic measure of aircraft separation that he applies to assess pilot performance in empirical studies and offers as a candidate for the conflict probe. Mr. Scallen reviews a series of experiments on the ability of commercial pilots to maintain separation while operating in a simulated free flight environment.

The breadth of topics in the symposium underscores the central role of human factors research in the evolution of free flight. All four topics address the FAA's goal of ensuring that the transition to free flight will not compromise safety.

TARGET AUDIENCES

The FAA indicates that "free flight is designed to provide the user community with the flexibility to better manage its operations and the capability to benefit from advanced avionics" (FAA< 1996). The symposium addresses the diverse interests of the aviation community.

THE HUMAN FACTORS RESEARCH COMMUNITY

The symposium has at least two practical benefits for researchers. First, the papers document alternative methodologies for conducting research in dynamic environments. Second, the presenters will identify research niches that need further investigation for the evolution of free flight.

DESIGNERS

The papers discuss how free flight will change demands on (1) users and service providers and on (2) decision technology. Designers who aim to enhance the fit of technology to users will come to appreciate the need for collaborative planning in all stages of the design process.

USERS AND SERVICE PROVIDERS

The symposium will present an informal forum for users and service providers to interact with development professionals at a pivotal stage in the evolution of free flight. Such interaction should help define key issues and ensure that research responds to the interests of end users.

ADMINISTRATORS

Representatives of administrative agencies will see the implications of several of the empirical findings presented during the symposium. Among these findings are constraints on the safe implementation of free flight and recommendations for procedures for effecting the transition to free flight.

All who attend the symposium will participate in a cooperative exploration and development of free flight concepts.

REFERENCES

FAA, (1996). Federal Aviation Administration Free Flight Main Page. http://asd.orlab.faa.gov/files/ffmain.htm (June).

INITIAL EXPERIENCES WITH THE EXPANDED NATIONAL ROUTE PROGRAM

Rebecca Denning *
Philip J. Smith *
Elaine McCoy **
Judith Orasanu ***
Charles Billings *
Amy Van Horn **
Michelle Rodvold ***

* Cognitive Systems Engineering Laboratory, The Ohio State University, Columbus, Ohio 43210

** Department of Aviation, Ohio University, Athens, Ohio 45710

*** NASA Ames Research Center, Moffett Field, California 94035

One of the goals of the free flight concept is to give the airlines (and other operators) more flexibility in selecting the routes to be flown. This applies to both preflight planning and to flight amendments made while enroute. Exploration of the former, flexibility in preflight planning, has already begun with the implementation of the expended National Route Program (NRP). This paper documents the experiences associated with the introduction of the expanded NRP, and points to areas that must be addressed to help make the program successful from a traffic management perspective.

BACKGROUND

Several months after implementation of the expanded NRP, we conducted a study of the impact of this flexibility in preflight planning (which was being implemented in a step-by-step process at progressively lower altitudes) on air traffic patterns. Specifically, under the expanded NRP flights over 500 miles, that met certain altitude requirements were given the ability to file flight plans without seeking permission from the air traffic management system, subject to certain constraints:

1. Flights had to be planned over published SIDs for the first 200 miles and STARs for the last 200 miles (published airways could be used for the first 200 miles and last 200 miles if appropriate SIDs or STARs were not available;

2. The route had to contain at least one waypoint or NAVAID (navigational aid) per ARTCC within 200 miles of the previous ARTCC;

3. Routes had to avoid active restricted areas by at least three miles unless permission was obtained and the relevant ARTCC was advised;

4. The flight plan had to indicate that it was an NRP flight.

Two enroute air traffic control centers (Chicago and Cleveland) were included in this study. The observations presented below were the result of structured interviews with traffic managers, an airline air traffic control coordinator (also a dispatcher), and pilots.

OBSERVATIONS

Regarding changes in traffic patterns, there were major shifts in the flow of traffic through these Centers. Several are summarized below.

Observation 1

The source of the major concerns was not the expanded NRP per se; it is the spinoffs of the expanded NRP, such as the increase in direct routes, the decrease in the use of preferential routes, and the cancellation of restrictions and procedures contained in advisory circular 90-91. The impact was that flights are now going direct through sectors where they previously were not direct. The problem was that the complexity in sectors increased with NRP and Non-NRP aircraft going direct. The major change in air traffic patterns wasn't due to flights filed under the new expanded NRP. It was due to the spinoff of the expanded NRP.

The "old" NRP (advisory circular 90-91) had several pages of restrictions, restricting flights to certain preferential routes, at specific peak times of the day. With the expansion of the new NRP, these route restrictions were canceled and the use of the preferential routes had decreased. Mile-in-trail restrictions were also removed. Cleveland, for example, always had a 20 mile-in-trail restrictions at specific times, for flights landing in Newark. All flights to Newark were sequenced by Chicago Center over Carlton. This restriction was now canceled. The flights to Newark now came through Chicago on various routes, often through different sectors on different days reference the winds, regardless of whether they were NRP or Non-NRP. Some pilots request direct routes without any approval from the next sector. This was interpreted by some controllers as the "Free Flight" concept. This was being accomplished by amending flights to file direct after the were airborne. As a result, maintaining the big picture in terms of sequencing and impact was much more difficult.

The net result is that 70 to 80% of the flights over many centers were now on direct routes. For some centers, such as Kansas City and west, this wasn't a major problem. For others like Chicago and east, there was significant impact. Some aircraft had to be held at the Chicago Center, Cleveland Center, and Indianapolis Center boundaries because sequencing multiple flows became nearly impossible.

Observation 2

Several airline dispatchers and ATC coordinators that we talked to about Observation 1 were unaware that this was happening (even with flights form their own airlines). This concerned them, as the best route in terms of fuel consumption, time, sequencing, etc. may not be the direct route. Furthermore, most pilots have neither the data nor the computer support software to determine the best route while airborne:

ATC Coordinator: "If a direct route had been better, I would have filed it through the NRP. For example, there was a flight that flew from DFW direct to Parker. I had planned it over Albuquerque because of a favorable southerly jetstream. The plane was six minutes late."

Observation 3

A number of concerns highlighted internal airline communication problems, as well as issues regarding AOC-TMU or AOC-ATCSCC interactions:

ATC Coordinator: "A higher level of AOC/Pilot

communication is needed to ensure that pilots understand why the dispatcher has planned a flight in a particular way, and that the AOC has feedback on what is happening to flights while enroute."

ATC Coordinator: "The problem with the expanded NRP is that there's no feedback to the AOCs. Nobody's getting smarter. ... When we went to free flight on Jan. 9, we cut off the feedback loop for those flights filed under the expanded NRP. ... How do we get this local knowledge that the TMUs and controller have out there for the dispatcher and pilots? ... There are problems in the ATC system that I don't know about. I need a mechanism to get feedback. ... How do we give the airlines more timely information? Depending on where they're going on which day, how do we get the information to everybody? How do we all get the same picture?"

Observation 4

At some Centers, the expanded NRP itself was raising some issues. As an example, one challenge arising form the expanded NRP was that airlines sometimes wanted to cross their high altitude flights over departure and arrival routes. For instance, for certain flights over the top of O'Hare, the Center had always preferred that the traffic be routed over Badger in order to avoid having enroute traffic cross the departure lanes. One airline, however, preferred (and was now been filing) these flights over Iowa City-Waterloo under the expanded NRP. Such flights criss-crossed through the departure lanes, creating a "very tricky, complex operation" for ATC.

This scenario raises a difficult tradeoff: Do you let three or four planes cross at the cost of slowing departures by about 20%? This tradeoff was particularly interesting given such flights would most often be slowing departures from Chicago of flights by two other airlines.

A second concern had to do with potential inefficiencies due to a lack of information:

ATC Coordinator: "A global perspective is important in revising a flight plan. There may be a perception that a restriction is unnecessary while a plane is early in the flight, but you then hit a wall at a later point. You're better off with a 10 degree turn over Joliet instead of 40 degrees over Niagara Falls and then another 40 degrees to get sequenced at Slate Run."

ATC Coordinator: "At some point, paying for

flexibility isn't economical. If flexibility increases capacity, then there's a benefit."

ATC Coordinator: "If a flight hits a bottle neck because it was filed using the expanded NRP, it may cost us more money, not less. I'd rather be slowed at Chicago than holding at Slate Run. ... We may be trying to optimize the enroute portion, but at the same time de-optimizing the arrivals."

TMO: "As another example, we ended up moving three NRPs up to the northwest arrival fix to land. It would have been cheaper for them to file to the northwest fix to begin with."

ATC Coordinator: "Someone has to be responsible for identifying and communicating constraints and bottlenecks, so that the air traffic management system and the AOCs can respond to them effectively."

A third issue that was arising as a direct result of the expanded NRP at some airpoints, and also as a result of the increased numbers of direct flights, had to do with balancing of loads at cornerposts:

For example, "if we get a jetstream right out of the southwest part of the country, everyone rides it [into O'Hare]. 75% of these airplanes are all coming in at the southwest cornerpost, creating a major volume saturation point. The old solution was to create a delay program to avoid launching too many flights into traffic, for example creating 20 minute delays at an airport, and to [increase capacity by moving] half a dozen flights to the northwest cornerpost. [Under the expanded NRP] we're not allowed to do this because of free flight. If they [the airlines] create the bottleneck, then they have to live with it."

Another situation which had specifically arisen due to the expanded NRP occurred when one particular airline "has five flights which originate in the LA Basin, PHX and LAS. When they all file to the Southwest cornerpost at DTW during certain arrival banks, the result is an overload at that fix. We respond by moving a couple of those flights, or other flights originating in Florida, to another fix."

A fourth example of an issue associated with the actual expanded NRP concerns what was happening when there were arrival rate restrictions (due to weather, etc.). For instance, in one case Kennedy had set a reduced arrival rate of 50% at 2 p.m. because of the weather forecast. To deal with this, Chicago Center began limiting flights bound for

Kennedy that were flying the standard pref routes. In addition, however, there were flights filed under the expanded NRP that were not limited. The net result was that the capacity for Kennedy was exceeded, with many planes

"winding up in high altitude airborne holding, and that's a major problem."

Observation 5

The expanded NRP is sometimes causing unpredictable changes in controller workload:

"One day a controller is inundated, the next day he's twiddling his thumbs."

IMPLICATIONS FOR A FREE FLIGHT ENVIRONMENT

As indicated above, the initial implementation of the expanded NRP posed some significant challenges, some resulting from the order applied as intended, while others the result of unanticipated behaviors. Since the time of this study, a number of solutions have been developed to deal with these challenges, such as the creation of new superhigh sectors for the enroute traffic, so that flights "climbing to altitude won't be dodging these guys going East-West." and the design of a training videotape for controllers. (We are currently completing a study of system performance "one year later.")

Several lessons can be learned form these experiences that could be applied to further efforts to move toward a free flight environment:

1. In shifting control, it is critical to shift access to information as well. An effective free flight environment is likely to require more, rather than less, communication between the traffic management/monitoring system, dispatchers, pilots, and controllers (traffic monitors);

2. Major changes in traffic patterns need to be anticipated, and appropriate solutions prepared, prior to implementation so that they are available in a more timely fashion;

3. Significant changes require major investments in training and supervision so that dispatchers, controllers, pilots and traffic managers understand their intended new roles.

ACKNOWLEDGEMENTS

This work has been supported by the FAA

Office of the Chief Scientist and Technical Advisor for Human Factors (AAR-100) and NASA Ames Research Center. We would like to express appreciation to Eleana Edens, Larry Cole and Tom McCloy of the FAA; to the participating air traffic managers, pilots, dispatchers and airlines; and to the Airline Dispatchers Federation for their assistance in conducting this work.

Human Factors Systems Engineering as a Requirement for the Safe and Efficient Transition to Free Flight

Rolf J. Braune
Braune & Associates, Inc
Redmond, WA 98052

Dieter W. Jahns
SynerTech Associates
Bellingham, WA 98227

Alvah C. Bittner, Jr.
Battelle AS & E
Seattle, WA 98105

This paper discusses a top-down analysis methodology for the design, development and evaluation of advanced technological systems like those being considered for General Aviation Free Flight and also the Advanced General Aviation Technology Experiment (AGATE). A current project sponsored by the FAA's Civil Aeromedical Institute is being introduced as an example.

BACKGROUND

The aviation community is pursuing a number of new technology initiatives which promise to significantly enhance operational safety and efficiency. Included are the Advanced General Aviation Transport Experiments (AGATE), Situational Awareness for Safety (SAS), and Free Flight.

The AGATE program focuses on single-pilot, light, all-weather transportation aircraft. The program is aimed at providing the foundation for United States industry leadership in technologies for improved utility, safety, affordability, performance, and environmental compatibility in General Aviation aircraft (The AGATE Flier, August 1994). Applicable cockpit technologies will respond to the need for more intuitive displays and controls to reduce the cost and time for obtaining and maintaining all-weather flying skills. These might include multi-function displays, touch-panel interface, voice interface, computerized decision aids, and a single-lever power control.

Situational Awareness for Safety (SAS) is a new Federal Aviation Administration initiative which focuses on increasing pilot situational awareness of position, terrain, weather, and other information through next-generation avionics; establishing enabling standards, specifications, and technologies for Free Flight; and facilitating means and opportunities for affordable avionics (Chang, Livack, and McDaniel, 1996).

Free flight (FF) is a system concept which addresses the Federal Aviation Administration's (FAA) strategic goal for System Capacity and Air Traffic Services, by providing improved accessibility, flexibility and predictability for the aviation community, while maintaining or improving operational safety. It is a concept that embraces the entire range of aviation operations addressing the interests of all system users (e.g., air carriers, general aviation, and military). The transition to mature free flight is guided by benefits-driven and time-phased introduction of new technologies and procedures, including ground movement and preflight planning activities.

Free flight is defined as "...a safe and efficient flight operating capability under instrument flight rules (IFR) in which the operators have the freedom to select their paths and speed in real time. Air traffic restrictions are only imposed to ensure separation, to preclude exceeding airport capacity, to prevent unauthorized flight through Special Use Airspace (SUA), and to ensure safety of flight. Restrictions are limited in extent and duration to correct the identified problem. Any activity which removes restrictions represents a move toward free flight." (RTCA Task Force 3, 1995). The free flight concept outlines the move from today's rather rigid and largely procedural, analog, and ground-based Air Traffic Control system comprising HF/VHF-voice communications, terrestrial-based navigation systems, radar surveillance, and limited decision support to that of the future flexible collaborative system by applying mostly space-based technologies and a new form of Air Traffic Control and Flow Management that will be increasingly seamless and truly world-wide, know as Air Traffic Management (ATM).

There are a number of benefits that properly equipped general aviation aircraft stand to gain in a benefits driven evolution to free flight. Those include: (1) increased use of low altitude direct routing in terminal and en route areas, (2) access through special use airspace; (3) increased access to general aviation airports in poor weather conditions; (4) improved weather information and displays in cockpits, and (5) reduced separation and improved aircraft monitoring (Aircraft Owners and Pilots Association, 1995). Even more critical than gaining new benefits is the requirement that the general aviation community will not lose any of the access or flexibility it has now.

In order to ensure a safe and efficient transition to free flight, and to facilitate the introduction of new technologies like AGATE and SAS, it is understood that the

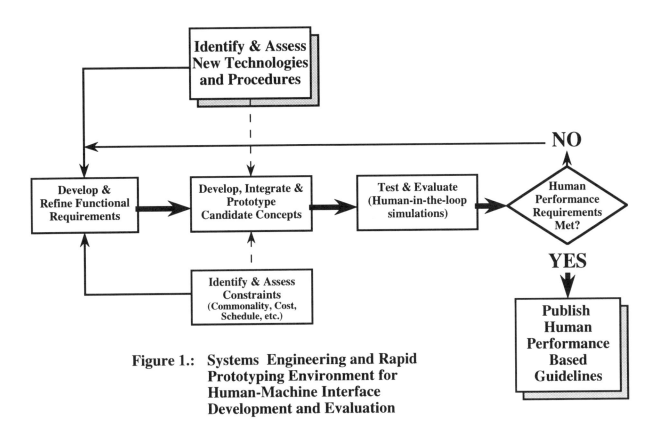

Figure 1.: Systems Engineering and Rapid Prototyping Environment for Human-Machine Interface Development and Evaluation

human factors concerns must be raised and addressed from the beginning and throughout the research and development of new technologies, policies and procedures. As system improvements are developed, human factors issues must be included during the analysis and design phases (RTCA Task Force 3, 1995). Past history indicates that if human factors concerns were addressed during the development cycle, it was often too late in the process to facilitate required improvements, contributing significantly to the well known "automation problem" in aviation (e.g., Braune, 1987; Braune and Fadden, 1987; Norman and Orlady, 1988; Wise et al, 1992; Wise et al., 1993; Funk, Lyall, and Riley, 1995). The overwhelming consensus in those publications is that poorly designed and integrated, i.e., non-human factored systems, negatively impact the safety and efficiency of the operation.

In response to the potential for negative side effects advanced technology design guidelines have been developed. Braune and Fadden (1987) grouped those guidelines into four major categories: (1) global recommendations, (2) design analysis methods, (3) human performance centered guidelines, and (4) design philosophies. For the purpose of this symposium the focus will be on design analysis methods.

THE PROCESS

The technical approach to the successful introduction of new technologies and procedures should be guided by sound systems engineering methods which establish functional requirements and operating conditions first as a prerequisite to

developing and evaluating candidate system designs. Such a "front-end" analysis helps identify and resolve any critical human factors issues that could lead to a less than optimal design. The approach emphasizes the fundamental need for a top-down analysis and design methodology with particular focus on clear operational performance requirements (e.g., Booher, 1990; Braune, Graeber, and Fadden, 1991; Meister 1985; Van Cott and Kinkade, 1972). In that it follows previous efforts to develop mission and functional requirements in aerospace systems (e.g., Taylor and Poole, 1984; Alter and Regal, 1992; Swink and Goins, 1992). Figure 1 provides a top level outline of the systems engineering process with particular emphasis on the human-machine interface. For the purpose of this symposium the discussion has been focused on the ***Develop & Refine Functional Requirements*** step.

The first step in this analysis process is to create a list of aircraft missions that identify the high-level purposes for the airplane's very existence. For example, a military airplane may be created to perform a variety of combat roles which have been specified by the Department of Defense. In the case of a commercial transport airplane, the purpose is to fly a payload (people and/or cargo) from one airport to another safely, efficiently, and comfortably. In the case of general aviation aircraft the purpose could be similar to that of commercial transport airplanes, but it could also include those of providing initial and advanced flight training, or aerobatic flight. In the context of this discussion the aviation mission to be considered will be one of providing safe, efficient, and comfortable transportation for people and/or cargo between two airports.

All aircraft could be said to have two sets of "missions": the "normal" flight which contains all standard phases of an error-free flight (e.g., takeoff, climb out, cruise, hold, approach, etc.); and the "non-normal" conditions, which include unusual circumstances and malfunctions to which the pilot needs to respond (e.g., weather, equipment failures).

To facilitate organization, the missions are broken down into a number of subsets. For normal flight, the mission is broken down into flight phases, such as taxi, takeoff, approach, etc. For non-normal flight, a list of non-normal situations is generated. This list includes conditions resulting from human limitations as well as system abnormalities, and is meant to be as comprehensive as possible.

Each mission phase is then separated into all the functions which need to be accomplished by the pilot or the on-board automation to complete the phase. A good strategy is to separate the larger functions into sub-functions. For example, the phase "Descent" would include the function "Control flight path", which could be further broken down into "Follow lateral flight path", "Maintain appropriate longitudinal profile", "Meet airspeed/altitude restrictions", and "Modify route for weather/traffic". Each of these lower level functions would, in turn be divided into its next lower level functions.

With a list of functions performed by the pilot and on-board systems, an examination of the information required by these functions can be made. For example, the high level function "Identify collision hazards on or near flight path" suggests a need for some display which allows the pilot to determine if the plane will pass too close to an obstacle in the future, and implies that enough advance warning be available to avoid an incident or accident. The low level function "Monitor current fuel status" requires information on the current fuel level in each tank.

Once the information requirements are available, a first cut can be made at allocating functions between the pilot, the air traffic controller, and automation. Alternative system concepts can be explored including the preliminary development of cockpit controls and displays requirements.

OBJECTIVES

Experience has shown that unless a systematic human engineering and design methodology is followed, promised gains in efficiency and safety may be jeopardized resulting in serious human-machine interface and automation problems. In order to reduce the likelihood of such problems, the FAA's Civil Aeromedical Institute in Oklahoma City is sponsoring a user-centered analysis approach in the General Aviation context. A brief outline of this effort follows below.

The overall, long-range objective, to which the project will contribute, is to answer the question: "Can the separation assurance function (SAF) in a Free Flight

environment be shared or shifted in real time between the pilot and air traffic controller (each supported by a particular level of automation/technology) within the general aviation mission structure?" If this can be accomplished, what are the hazards, risks and benefits which must be considered in further developments of Free Flight? The specific objectives of the project are to:

1. Develop a functional analysis framework which can be generalized to several free flight missions.
2. Organize the functions by mission segment and identify specific lower level functions which must be performed within a given set of operating conditions and assumptions.
3. Further analyze the functions down to the individual task level.
4. Organize the functions and tasks into a mission segment time line and identify the major tasks involving communication, navigation, and surveillance (CNS).
5. Identify all interactions with other aircraft and ground-based components along with information transfer and decision making required of the pilot and the associated response times.
6. Formulate a set of functional and information requirements based on which candidate design concepts can be developed.

Using the above approach a functional analysis framework is being built around the following mission criteria:
(1) select a route between two destinations that represents a challenging IMC environment where area navigation capability (RNAV) would provide additional benefits;
(2) plan and analyze the route using a high performance single engine general aviation aircraft;
(3) contact pilots and controllers who are familiar with the route for review and request operational inputs into the analysis (e.g. weather, no radar coverage);
(4) contact safety organizations (e.g., Flight Safety Foundation, NTSB, ASRS) to obtain data in regard to accidents and incidents along the route.

CONTRIBUTION

This project establishes a functional analysis framework and methodology which can be generalized to several free flight missions. It will be a definitive map of the operation of a free flight system concept for a given set of conditions. Based on the requirements derived from this analysis candidate general aviation free flight system concepts can be formulated based on realistic, alternative uses of resources and operational conditions. The primary focus will be on possible conceptual designs reflecting alternative functional assignments between ATC, the pilot, and cockpit automation.

ACKNOWLEDGEMENT

This research is being supported by the FAA's Civil Aeromedical Institute under Order No. DTFA-02-96F90156. Dr. Robert E. Blanchard, Director of Human Factors, is the technical monitor.

REFERENCES

Aircraft Owners and Pilots Association (AOPA) (1995). Free Flight General Aviation Benefits, In Final Report of RTCA Task Force 3 Free Flight Implementation, RTCA, Inc., Washington, D.C., October 26.

Alter, K.W. and Regal, D.M. (1992). Definition of the 2005 Flight Deck Environment.NASA Contractor Report 4479, NASA Langley Research Center, Hampton, VA.

Booher, H.R. (1990). MANPRINT: An Approach to Systems Integration. New York, Van Nostrand Reinhold.

Braune, R.J. and Fadden, D.M. (1987). Flight Deck Automation Today: Where Do We Go From Here? SAE Aerotech, Long Beach, CA, October 5-8.

Braune, R.J. (1987). Summary of the Workshop on Cockpit Automation in Commercial Airplanes. Proceedings of the Fourth International Symposium on Aviation Psychology, Columbus, OH, April 27-29.

Braune, R.J., Graeber, R.C., and Fadden, D.M. (1991). Human Factors Engineering - An Integral Part of the Flight Deck Design Process. AIAA Aircraft Design and Systems Meeting, Baltimore, Maryland, Sept. 23-26.

Chang, G.C., Livack, G.S., and McDaniel, J.I. (1996). Emerging Cockpit Technologies for Free Flight: Situational Awareness for Safety, Automatic Dependent Surveillance-Broadcast, Air-to-Air Data Link, and Weatherlink. 2nd Situational Awareness for Safety System Requirements Team Meeting, Annapolis, MD.

Funk, K., Lyall, E., and Riley, V. (1995). Perceived Human Factors Problems of Flightdeck Automation. A Comparative Analysis of Flightdecks with Varying Levels of Automation, Federal Aviation Administration Grant 93-G-039, Phase 1 Final Report.

Meister, D. (1985). Behavioral Analysis and Measurement Methods. John Wiley & Sons, New York.

Norman, S.D. and Orlady, H.W. (1988). Flight Deck Automation: Promises and Realities. Final Report of a NASA/FAA/Industry Workshop, Carmel Valley, CA August 1-4.

RTCA Task Force 3 (1995). Final Report, Free Flight Implementation. RTCA, Inc., Washington, DC, October 26.

Swink, J.R. and Goins, R.T. (1992). Identification of High-Level Functional/System Requirements for Future Civil Transports. NASA Contractor Report 189561, NASA Langley Research Center, Hampton, VA.

Taylor, R.R. and Poole, E.R. (1984). A mission oriented approach to cockpit design as applied to observation and attack helicopters. 40th Annual Forum of the American Helicopter Society, Arlington, VA., May 16-18.

Van Cott, H.P. and Kinkade, R.G. (1972). Human Engineering Guide to Equipment Design. American Institutes for Research, Washington, DC.

Wise, J.A., Guide, P.C., Abbott, D.W., and Ryan L. (1992). Human Factors in Aviation Safety: The Effects of Automation on Corporate Pilots. Volume 1 - Interim Technical Report, CAAR-15406-92-1, Embry-Riddle Aeronautical University, Daytona Beach, Florida.

Wise, J.A., Abbott, D.W., Tilden, D., Dyck, J.L., Guide, P.C., and Ryan, L. (1993). Automation in Corporate Aviation: Human Factors Issues. Final Report, CAAR-15405-93-1, Embry-Riddle Aeronautical University, Daytona Beach, FL.

A DYNAMIC CONFLICT PROBE AND INDEX OF COLLISION RISK

William Knecht, Kip Smith, & P.A. Hancock
University of Minnesota
Minneapolis, Minnesota

We present a mathematical index of colllision risk and propose it as a candidate conflict probe for use in free flight. The index can be used to aid and measure ATC and pilot performance during simulated or real flight. The instantaneous value of the index is a relative (ordinal) measure of global situational risk. When the value of the index is graphed over time, conflicts can be easily identified and their severity estimated. Peak values indicate approximate times and degree of risk associated with the minimum separation of aircraft.

INTRODUCTION

Planned modifications of the National Airspace System by the Federal Aviation Administration (FAA) call for evaluations of a system of flight control called *free flight*. Free flight essentially involves shifting the burden of high-altitude, en-route traffic control from air traffic controllers (ATC) to flight crews, except in the case of emergencies.

Part of the evaluation of the free flight concept involves attempting to assess its risks before moving to such a system, rather than afterward. To this end, the Human Factors Research Laboratory (HFRL) of the University of Minnesota is engaged in FAA-funded research concerning theoretical and practical aspects of airspace risk. Part of such risk assessment concerns the development of conflict probes and mathematical indices of airspace risk.

The FAA defines a *conflict probe* as any measure which gives indication of the time when an airspace conflict occurs, along with some measure of the severity of the conflict. Typically, severity is defined as a function of how close two aircraft come together.

Indices of collision risk are more comprehensive. In addition to being conflict probes, they mathematically express the instantaneous degree of danger inherent in traffic scenarios which may be complex and which change rapidly over time.

The index of collision risk is a useful tool in assessing the performance of pilots and ATC in simulator-based air traffic scenarios. By the use of such computer simulations, we can easily and safely acquire information about traffic situations which pose high threat of collision. We also gain insight into the types of tasks involved in managing aircraft and the perceptual and cognitive resources necessary to safely solve air traffic problems.

THE RISK INDEX

The index of collision risk in the overall airspace is:

$$\sum_{i=1}^{N-1} \sum_{j=i+1}^{N} \frac{1}{\left(\dfrac{d_{ij}}{c}\right)^a} \tag{1}$$

where N is the number of aircraft at a given altitude layer, d_{ij} is the distance from the *ith* aircraft to the *jth* aircraft, a is the *exponent of suppression* and c is the *constant of criticality*. The details of a and c are discussed in more detail later. The index is a sum of $N(N-1)/2$ components, each component representing a contribution associated with one aircraft pair. By considering all possible pairings of aircraft at a given altitude layer, the equation yields a global index of risk at that altitude.

The Single Aircraft Pair Condition

The simplest conflict involves a single pair of aircraft. In this case, Equation 1 reduces to:

$$\frac{1}{\left(\dfrac{x}{c}\right)^a} \tag{2}$$

where $x = d_{12}$. Equation 2 embodies the characteristic that conflict is a function of separation. The constant of criticality c tunes Equation 2 to "explode" at a specified aircraft separation. As long as $(x/c) > 1$, Equation 2 remains relatively small. But as $(x/c) \rightarrow 0$ the value of Equation 2 increases rapidly. Typically we set c to 5 nautical miles (nm.), the minimum allowable separation according to FAA regulations.

The purpose of the exponent of suppression is to increase the relative contribution of "close" aircraft (e.g. when $x < c$), while suppressing the relative contribution of more distant aircraft.

Parameters a and c cause Equation 2 to respond to a decrease in aircraft separation with an increased, factor-weighted indicator of airspace risk. The constant of criticality defines the separation at which risk becomes unacceptable. The exponent of suppression tunes the equation to reject "noise" due to large separation distances.

If the separation distance x is allowed to approach zero, the value of Equation 2 will approach infinity. There is no such thing as "infinite risk". Therefore, we exclude $x = 0$ from the set of meaningful values that separation distances may take.

Figure 1 illustrates the use of the index in an enroute flight scenario in which a single pair of aircraft approach nearly head-on, missing each other by 5 nm. (9.265 km.).

Figure 2 illustrates the behavior of Equation 2 with $a = 3$ and $c = 5$ for the scenario in Figure 1. The two horizontal axes represent the value of the index over time and distance, respectively.

Observe that a positive slope (segment A) indicates that relatively distant aircraft are moving closer together, while a negative slope (segment B) indicates aircraft are moving farther apart. In this way the index discriminates between approaching and diverging aircraft.

The peak in Figure 2 represents minimum separation. The time at which the index reaches a local maximum is not affected by the value of either c or a; Equation 2 always reaches a local maximum at the time when x is minimum.

An Air Traffic Controller's Use Of The Index

ATC typically considers every aircraft in the control area. From this global viewpoint the distance between every pair of aircraft must be considered. Equation 1 is appropriate for this condition.

A Pilot's Use Of The Index

The index can be adjusted to reflect the "local" viewpoint of pilots. From this viewpoint only the interactions between the pilot's ownship and other aircraft need be considered. The index becomes:

$$\sum_{j=2}^{N} \frac{1}{\left(\dfrac{d_{1j}}{c}\right)^{a}} \qquad (3)$$

where the subscript 1 reflects the ownship and the js reflect other aircraft.

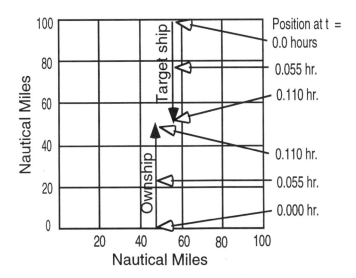

Figure 1. Map view plot of the progress of two approaching aircraft, each travelling at 450 knots and passing within 5 nm. at time = 0.11 hours. The black-headed arrows indicate the velocity vectors of the two planes.

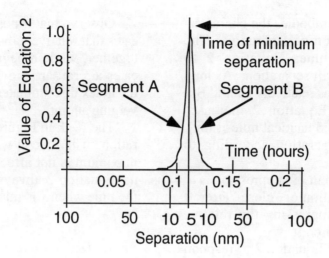

<u>Figure 2.</u> Plot of the value of Equation 2 (y axis) vs. time and separation distance (x axis) for the two aircraft in Figure 1. Note that the maximum value of the index occurs at t = 0.11, the time of minimum separation in Figure 1. Slope > 0 as aircraft converge, and < 0 as they diverge.

To illustrate, consider Figure 3, where four pairwise comparisions are to be made, one for each of four target aircraft as they approach and sequentially cross the path of the ownship. In this scenario the speeds of all aircraft are identical.

Figure 4 shows the time plot of Equation 3 in the scenario of Figure 3 ($c = 5$). The effect of varying the exponent of suppression, a, is illustrated. With lower values of a, information about the various aircraft is blurred together. With higher values of a, the effect of relatively distant aircraft is more suppressed and the peak due to each sequential crossing stands out prominently at its time of occurrence.

Experimenters' use of the index

Current simulator studies, using the HFRL glass cockpit display flight simulator, confirm the usefulness of the index. Even in relatively crowded simulated airspaces, the index peaks nicely at times of minimum separation and gives an estimate of the relative global collision risk at any given time. This is highly useful in assessing the efficacy of pilots' maneuvers during simulated air traffic conflict situations. By analyzing and comparing data from solved and unsolved conflicts, we gain insight into the types of traffic situations we anticipate will be problematic during free flight.

SUMMARY

We offer an index to probe conflict in the airspace and to estimate collision risk. It can be used

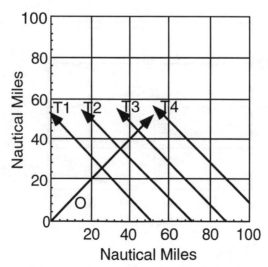

<u>Figure 3.</u> Map of the interaction between an ownship (O) and four targets (T1-T4).

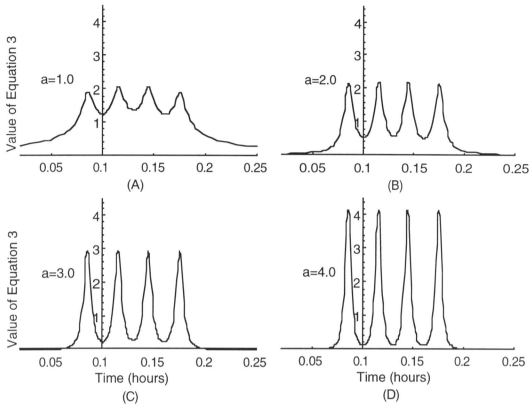

Figure 4. The values of Equation 3 as each of four target ships, in turn, crosses an ownship. Note the effect of varying the exponent of suppression, *a*

to aid and measure ATC and pilot performance during simulated or real flight. The instantaneous value of the index is a relative (ordinal) measure of global situational risk. When the value of the index is graphed over time, conflicts can be easily identified and their severity estimated. Peak values indicate approximate times and degree of risk associated with the minimum separation of aircraft.

ACKNOWLEDGEMENTS

This work was funded through FAA grant 93-G-048, Dr. Eleana Edens, Technical Monitor.

REFERENCES

Aviation Systems Research Corporation (1995). Reinventing ATC: The economic impact. RMB Associates, Golden, CO.

Buckley, E.P., DeBaryshe, B.D., Hitchner, N., and Kohn, P. (1983). Methods and measurements in real-time air traffic control system simulation. U.S. Department of Transportation - Federal Aviation Administration technical note (DOT/FAA/CT-83/26), Atlantic City, NJ.

Dailey, J.T. (1984). Characteristics of the air traffic controller. In S.B. Sells, J.T. Dailey, and E.W. Pickerel (Eds.), Selection of air traffic controllers, (pp. 128-141). Washington, D.C.: FAA.

Danaher, J.W. (1980). Human error in ATC system operations. Human Factors, 22(5), 535-545.

Department of Transportation (1995). Air traffic control handbook (Order number 7110.65 and supplements). Federal Aviation Administration, Washington, DC.

Federal Aviation Administration (1995). National plan for civil aviation human factors: An initiative for research and application. Federal Aviation Administration, Washington, DC.

Hopkin, D.V. (1980). The measurement of the air traffic controller. Human Factors, 22(5), 545-560.

Hopkin, D.V. (1988). Air traffic control. In E.L. Wiener and D.C. Nagel (Eds.), Human factors in aviation (pp. 639-663). New York: Academic Press.

Hopkin, D.V. (1991a). Automation and systems issues in air traffic control: Blue sky concepts. In

J.A. Wise, V.D. Hopkin, and M.L. Smith (Eds.) <u>Automation and systems issues in air traffic control</u> (pp. 541-545). Berlin: Springer-Verlag.

Hopkin, D.V. (1991b). Closing remarks. In J.A. Wise, V.D. Hopkin, and M.L. Smith (Eds.) <u>Automation and systems issues in air traffic control</u> (pp. 553-559). Berlin: Springer-Verlag.

March, J.G. and Shapira, Z. (1987). Managerial perspectives on risk and risk taking. <u>Management Science, 33(11)</u>, 1404-1418.

RTCA (January, 1995). <u>Report of the RTCA board of directors' select committee on free flight.</u> RTCA, Incorporated, Washington, DC.

Small, D.W., Carlson, L.S., and Kerns, K. (1995). <u>Human factors issues in free flight (MP 95W0000184)</u>. Mitre Corporation, McLean, VA.

Vortac, O.U., Edwards, M.B., Fuller, D.K., and Manning, C.A. (1993). Automation and cognition in air traffic control: An empirical investigation. <u>Applied Cognitive Psychology, 7, 631-651</u>.

PILOT ACTIONS DURING TRAFFIC SITUATIONS IN A
FREE-FLIGHT AIRSPACE STRUCTURE

S. F. Scallen   Kip Smith   P. A. Hancock
Human Factors Research Laboratory
School of Kinesiology
University of Minnesota
Minneapolis, MN

One facet of the proposed restructuring of the National Airspace System currently generating much interest is called 'Free Flight'. At the heart of the Free Flight system is an increased flexibility in pilot decision making and responsibility for the definition and maintenance of separation, of preferred routes and speeds, and the conduct of maneuvers in response to potential conflicts and other emergencies in the airspace. Here, we describe a simulation experiment where fifteen commercial pilots were presented with traffic conflict situations in the en route environment. Within the scenarios we manipulated density, type of conflict, and relative bearing of conflict aircraft. Pilots were required to navigate a simulated 757 aircraft to destination airport, avoiding all possible traffic conflicts. Their ability to maintain separation was the principle dependent measure. Results indicated that density and bearing did not appear to have any substantive effect on pilot response. However, overtaking conflicts produced a higher frequency of operational errors than crossing or converging conflicts. Further analysis of individual pilot responses revealed different strategies. Analysis of individual overtaking scenarios revealed patterns of pilot action associated with efficient and often creative conflict resolutions. An example of an inefficient conflict resolution was also identified.

INTRODUCTION

Increases in traffic volume, projections of higher demand in the future and the continued deterioration of air traffic control technology will continue to pressure operational capabilities in the National Airspace System (NAS). The aviation community has responded by emphasizing the development and implementation of new technologies such as global positioning systems, traffic avoidance systems, and flight management systems. Together, these technologies have been successful in distributing traffic and control information to the flightdeck. The redistribution of information has stimulated the reconceptualization of the NAS structure in which alternative control structures, not just alternative technologies, are being proposed.

The proposed restructuring concept currently generating the most talk is called 'Free Flight'. At the heart of the Free Flight system is an increased flexibility in pilot decision making and responsibility for the definition and maintenance of separation, of preferred routes and speeds, and the conduct of maneuvers designed to avert uncertainties and emergencies in the airspace (RTCA, 1995). At the present time, Free Flight is still in its conceptual and developmental infancy. The FAA has indicated its commitment to continuing collaborative efforts with the aviation community, through the auspices of a consensus building body, the RTCA. According to the FAA, the move towards Free Flight is already underway via, for example, the expanded National Route Program (NRP), the Central Pacific oceanic program, and the development of communications, navigation, surveillance, and air traffic management technologies (FAA, 1996). The NRP program, for example, is designed to ultimately

permit aircraft flying above 29,000 feet to select their own routes as alternatives to published preferred IFR routes, thereby removing the restrictions and constraints currently imposed on these users. However, the FAA has also identified a need for simulation and experimental research in order to ensure that FAA activities and programs produce the safest and most efficient Free Flight capabilities, a view which is becoming more prominent in the aviation community (e.g., Small, Carlson, & Kerns, 1995).

The Human Factors Research Laboratory (HFRL) at the University of Minnesota has developed a research program to evaluate possible constraints on the safe and efficient implementation of Free Flight. At the heart of the program is the pilot decision-making process in a Free Flight structure, evaluated with respect to safety (separation maintenance) and pilot workload. This paper describes and evaluates one aspect of the research program: how pilot actions are related to separation maintenance when control for routing (path and altitude) and speed are solely the responsibility of the pilot.

METHOD

Apparatus - Glass Cockpit Instrumentation

HFRL has developed and instrumented a platform for investigating pilot performance. The platform has two major components, software written in the C language and hardware featuring a glass cockpit. The software enables pilots to navigate a simulated Boeing 757 aircraft and to maintain or deviate from the preprogrammed flight plan using IFR rules and procedures. Glass cockpit instrumentation includes a primary flight display (PFD), a navigation display, and a flight management system (FMS) display with a multi-control display unit (MCDU) keyboard input device. The navigation display is presented on the center CRT (see Figure 1) and has two views, plan view and forward view, with available radii of 15, 30, 50, 100, 150, and 300 miles. The FMS system combines a Honeywell MCDU keyboard input device, a flight management information screen (right hand CRT), and three information toggle buttons (vertical navigation-VNAV, lateral navigation-LNAV, and Status).

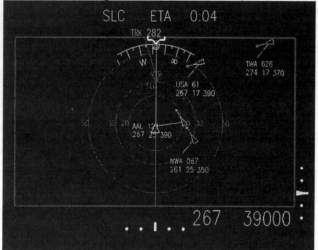

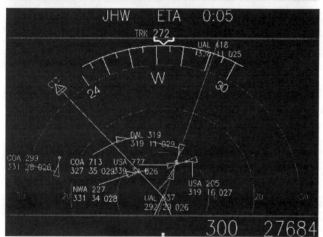

Figure 1. *The navigation display in both map and forward display views. In these examples the map view has a range of 50 miles while the forward view has a range of 30 miles. The displays show the pilot's aircraft (large triangle) and all other aircraft in the airspace. Data tags indicate callsign, airspeed, heading, and altitude.*

Free Flight Situation

The goal for each pilot is to navigate his/her aircraft in a manner consistent with a preprogrammed flightplan and arrive at a destination airport. Pilots are instructed that the relevant flightplan information, including all waypoints, has been previously entered into the FMS. Each scenario begins with the pilot's aircraft having already taken off, climbed to the en-route altitude, and progressed part way through the flightplan. Pilots are given a high altitude map which indicates the programmed flightplan, the starting position, and other relevant information including waypoints and no fly zones (military airspaces that must be

avoided). Pilots are instructed that they have full authority to make and execute decisions for routing (course, altitude, and speed) with none of the current FAA mandated routing and altitude restrictions in effect. The only requirement is that the pilot must maintain a minimum separation of 5 miles horizontally and 2000 ft. vertically. The pilot is also instructed that no ATC service exists.

Scenarios

While en-route, pilots encounter traffic scenarios which portray realistic conflict situations. Traffic is composed of drones (aircraft that pose no potential conflict for the ownship) and one or more 'targets' (aircraft that will eventually conflict with the ownship). Twelve selected scenarios encompassed three factors of airspace complexity. One factor was traffic situation which represented the course of the conflict aircraft (crossing, converging, and overtaking). Another factor was traffic density defined as low (≤ 8 aircraft) and high (≥ 11 aircraft). The third factor was the relative bearing of traffic (90°, 45°, and 0°).

Experimental Participants

The subjects for this experiment were professional commercial airline pilots. We present the data for five pilots who were male, with a mean age of 42 years and mean experience of 9117 logged flight hours. At the time of testing, all subjects were in reported good health.

Procedure

Pilots received instruction and practice on each component of the glass cockpit system. Pilots then navigated their aircraft in a practice scenario. At the start of each scenario pilots also received a map with a flight plan and the approximate position at the starting time for each scenario. Pilot were instructed to navigate toward the scenario destination by making autonomous path (course and altitude) and speed changes.

Action Measures

Each pilot interaction with cockpit controls (button/switch pushes) were recorded, time stamped, and written to an output. Number of maneuvers executed and number of input errors was also recorded. Together, these data constituted 'pilot action'.

Performance Measures

Pilot performance was assessed via two outcome measures. A dichotomous variable classified the ownship aircraft as either 'separated' or 'not separated' based on 5 nautical miles (nm.), the minimum separation according to FAA regulations (DOT, 1995). As well, a mathematic index was used to provide a continuous measure of aircraft separation. The index is the sum of the inverse distances of aircraft within an altitude strata. As separation decreases, the value of the index increases. Our objective is to track the changing configuration of aircraft in a specified space. We label this continuous measure of separation 'risk'. Thus, the index of risk is operationally defined as loss or opportunity for loss of separation. For a complete discussion see Knecht, Scallen, Smith, and Hancock (1996).

RESULTS

Figure 2 presents graphs of three measures for the four types of conflicts in the experimental scenarios (control, overtaking, crossing, converging). The graphs reveal that scenarios involving an overtaking conflict were associated with a larger number of separation violations, larger number of maneuvers, and more data input errors. Similar analyses indicate that neither density nor bearing appeared to b related to number of separation violations, maneuvers, or data input errors.

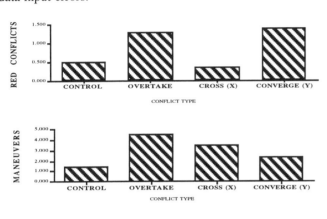

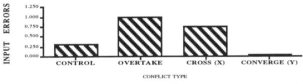

Figure 2. *Number of red conflicts (loss of minimum separation), maneuvers executed, and FMS data input errors by conflict type. Note how overtaking and converging to path conflicts result in more operational errors,. Overtaking errors are largely errors of commission.*

Risk index values for each pilot were plotted for two overtaking scenarios (scenarios 317 and 237, see the upper portions of Figures 3, 4, 5, 6, and 7). Time is plotted along the x-axis with the risk values plotted on the y-axis. Increasing values of the risk index can be interpreted as increased risk of the loss of separation. The index is constructed so that any value over 1 indicates a loss of separation. The baseline condition in each plot which indicates the pattern of separation built into the scenario (what would occur if the pilot did nothing). Pilot plots reflect the change in the separation pattern as a result of pilot maneuvers. For example, pilot 7 (Figure 3) was able to 'solve' the programmed conflict (maintain a risk value <1). In contrast, pilot 3 (Figure 4) lost separation.

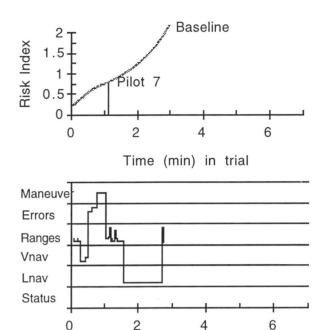

Figure 3. *Risk values and action plot for pilot 7 in scenario 317.*

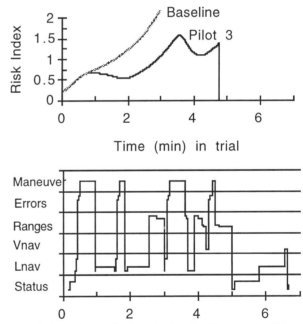

Figure 4. *Risk values and action plot for pilot 3 in scenario 317.*

DISCUSSION

The upper graph in Figure 2 indicates that overtaking and converging scenarios presented challenging and sometimes unresolvable problems for pilots. Inspection of the middle graph in Figure 2 indicates that overtaking scenarios involved significantly more executed maneuvers than converging scenarios. Thus, converging scenarios may reflect errors of omission (failure to detect or predict the programmed conflict), whereas overtaking scenarios reflect errors of commission where pilot actions to avoid the programmed conflicts were simply ineffective. This distinction suggests that pilots find overtaking scenarios particularly troublesome. These data indicate that airspace conflict type (specifically, overtaking conflicts) may be an important factor in determining or predicting pilots' ability to maintain safe flight in free-flight structures.

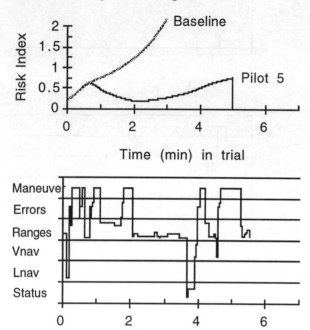

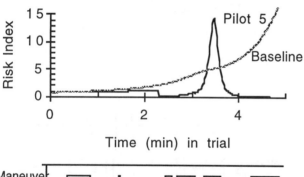

Figure 5. *Risk values and action plot for pilot 5 in scenario 317.*

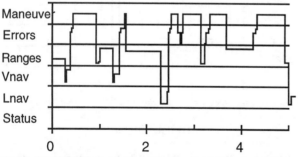

Figure 6. *Risk values and action plot for pilot 5 in scenario 237.*

The outcome data do not provide a complete picture of pilot action because they do not identify specific actions or describe how action occurred over the course of the scenario. Furthermore, outcome measures such as 'separated' or 'not separated' do not indicate the how separation changed over time and, specifically, how separation was lost over time. To gain this type of information we must look more closely at process data from individual scenarios.

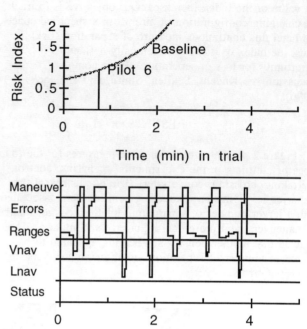

Figure 7. *Risk values and action plot for pilot 6 in scenario 237.*

We represent the process of pilot action in 'action plots' (see the lower portion of Figures 3, 4, 5, 6, and 7). In each plot we group similar behavior together and then represent the transition between behavior groups over the course of the scenario. For example, all behaviors associated with manipulating the Lateral Navigation (LNAV) portion of the Flight Management System are grouped together, all VNAV are grouped together, and all behaviors related to changing the range function are grouped together. Figures 3, 4, and 5 show action plots for three pilots in scenario 317: pilots 7 and 5 who remained separated and pilot 3 who lost separation. Figures 6 and 7 show action plots for two pilots in scenario 237: pilot 6 who remained separated and pilot 5 who lost separation.

Taken together, risk index values and pilot action plots demonstrate relationships between pilot action and outcome. For example, the upper plot of Figure 3 indicates that pilot 7 avoided loss of separation, minimizing the value of the index, in essence, 'solving' the conflict. The lower plot of Figure 3 indicates this pilot explored different view ranges of the scenario, selected vertical navigation information, executed a vertical maneuver, and again explored different view ranges. Thus, pilot 7 efficiently and effectively resolved a potential conflict with a single vertical maneuver. Index values for pilot 5 (Figure 5) indicate that he too initially avoided loss of separation, but only for a short time, where by the third minute he was steadily losing separation. Shortly after minute five he managed to 'solve' the conflict. Actions for pilot 5 indicate that he chose an alternate solution to the initial conflict, a lateral maneuver, that was at first effective but soon put him in conflict with another aircraft. Eventually pilot 5 explored alternate range of views and finally executed an effective vertical maneuver. Pilot 3 (Figure 4), like pilot 5, initially executed a lateral maneuver and again, like pilot 3, momentarily gained separation but soon began loosing separation. Unlike pilot 5, however, pilot 3 executed another lateral maneuver, and yet another, even after he had already lost separation. Finally, pilot 3 executed an efficient vertical maneuver, though a full two minutes after separation had been lost. It appears as though pilot 3 'bought in' to a lateral resolution and refused to abandon his belief in it, even when presented with a clear indication that it was not working.

As a whole, it appears that the overtaking conflict in scenario 317 could effectively be avoided with a simple vertical maneuver, a maneuver three pilots efficiently executed. Another pilot, 5, though first executing a lateral maneuver, eventually resolved the conflict vertically before loss of separation, demonstrating that pilots can creatively resolve conflicts. However, one of the five pilots demonstrated an unsafe pattern of action, not because of the initial choice of lateral maneuver, but because of the protracted effort to use a lateral resolution, to the exclusion of other options. This pilot's perseverance strategy appeared to be very ineffective.

Analysis of data from the second overtaking scenario (237) also yields a number of interesting observations. In this scenario most pilots lost separation. The scenario was extremely difficult. Almost all pilots initially executed a vertical maneuver. It was momentarily effective but they quickly lost separation again, almost at the time of the programmed conflict. Figures 6 and 7 present risk values and action plots for two pilots: pilot 6 who was the only pilot to maintain separation and pilot 5 who is a representative example of the remaining pilots. The lower portions of this figures depict the action process for these two pilots At first glance, the action plots for these pilots seem remarkably similar, and in many respects they are. Both are marked by multiple maneuvers and errors inputting data. Together, these data indicate the scenario was, in general, difficult to resolve, requiring overt action throughout. However, the two action plots indicate one interesting difference in relation to the type of view control (map and forward) and the range of view controls (15 to 300 miles). Figure 8 shows Pilot 6 was characterized by dedicated use of the map display with range at 30 and 50 miles. Pilot 5 divided his time between the two types of view with range at 15 and 30 miles.

It appears that the combination of map display and ranges toward the mid to high settings were useful to pilot 6. This may not fully account for the findings for this particular scenario, but they may, in some part, contribute to the success of individual pilots. One thing is certain. This scenario was extremely challenging for pilots, producing best case and worst case scenarios, a pilot that acted effectively, and pilots that acted but were largely ineffective.

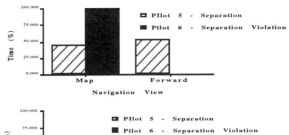

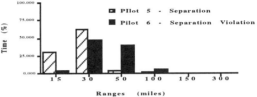

Figure 8. *Percentage of time pilots 5 and 6 spent in the map and forward view and the different range radii for overtaking scenario 237. Data indicate marked difference in strategy between the two pilots.*

CONCLUSIONS

With respect to pilot effectiveness in free-flight:
- type of conflict (specifically overtaking conflicts) appears to be an important complexity factor.
- the effective resolution of traffic conflicts by pilots may be characterized by creative solutions.
- when resolving traffic conflicts, individual pilots can display a pattern of decision making characterized by protracted attempts at a single resolution option, to the exclusion of other options. Perseverance can be an ineffective strategy.
- the manipulation of traffic information displays can be highly related to effective conflict resolution.

With respect to Free Flight and distributed control:
- distribution of control should be reflective of complexity factors (overtaking conflicts may justify centralized intervention).
- creativity should be valued.
- individual actions and patterns of action can be related to effective and ineffective separation maintenance.

ACKNOWLEDGMENTS

This work is funded by the Federal Aviation Administration through grant 93-G-048 to the third author. Dr. Eleana Edens, Larry Cole, and Tom McCloy are technical monitors. The views expressed are those of the authors and do not necessarily represent those of the FAA.

REFERENCES

Department of Transportation (1995). <u>Air traffic control handbook (Order number 7110.65 and supplements)</u>. Federal Aviation Administration, Washington, DC.

FAA. (1996). Federal Aviation Administration's Free Flight Main Page. http://asd.orlab.faa.gov/files/ffmain.htm, May.

Knecht, W., Scallen, S.F., Smith, K., & Hancock, P.A. (1996) <u>Metrics of airspace risk</u>. Unpublished manuscript, Human Factors Research Laboratory, University of Minnesota, Minneapolis, MN.

RTCA (January, 1995). <u>Report of the RTCA board of directors' select committee on free flight.</u> RTCA, Incorporated, Washington, DC.

Small, D.W., Carlson, L.S., and Kerns, K. (1995). <u>Human factors issues in free flight (MP 95W0000184)</u>. Mitre Corporation, McLean, VA.

Aging and Expertise

Timothy A. Salthouse
Georgia Institute of Technology
Atlanta, Georgia 30332

The positive benefits of experience are so well accepted that they have been reflected in various cliches such as "USE IT OR LOSE IT" and "THOSE WHO LIVE BY THEIR WITS DIE WITH THEIR WITS." However, surprisingly little definitive information exists regarding the role of experience with respect to the prevention or elimination of age-related declines in perceptual, motoric, and cognitive activities. This talk has two major goals. The first is to describe different methodological approaches for examining the influence of experience on age-related differences in performance. The second goal is to summarize empirical research concerned with the effects of experience on the relations between age and perceptual, motoric, and cognitive performance.

Aging and pilot performance: The role of expertise and environmental support

Daniel Morrow
University of New Hampshire

Piloting is a complex task that often places heavy demands on pilot cognitive resources such as working memory capacity. Aging may limit pilot performance because of age-related decrements in cognitive resources. Yet studies with pilots tend to find fewer age effects than might be expected from aging studies in the general population, suggesting that knowledge and experience mitigate age decrements in resources. Expertise-based mitigation may depend on external or contextual factors. The expertise literature shows that highly domain-relevant tasks (with materials and procedures that are compatible with domain goals) enable experts to outperform novices, but these benefits decline when experts are confronted with less organized materials or unfamiliar procedures. Domain-relevant tasks may mitigate age differences in piloting by providing environmental support that reduces the resources needed to use domain knowledge, or by supporting expert strategies that compensate for age decrements in resources.

Morrow, Leirer, Altieri, & Fitzsimmons (1994) examined whether expertise reduces age differences for tasks that are similar to piloting but not for less relevant tasks. Older and younger pilots and nonpilots performed a laboratory task similar to routine Air Traffic Control (ATC) communication. They either read or listened to typical ATC messages and then read back (repeated) the message. They also performed tasks that were less related to piloting. Expertise reduced age differences only for the more domain-relevant tasks. This finding provides evidence that knowledge mitigates age decrements in cognitive resources such as working memory capacity when supported by domain-relevant tasks, perhaps because older pilots use compensatory strategies. Additional evidence for expertise-based mitigation is provided by Lassiter, Morrow, Hinson, & Miller (1996), who examined if expertise (high vs. low hour pilots) reduced age differences in the ability to handle increasing workload in a simulated flying task. Expertise tended to mitigate the effects of age-related declines in cognitive resources on performance, particularly as task demands increased. These studies suggest that flight displays, procedures and other aspects of the flying environment that are designed to take advantage of pilot expertise will particularly benefit older pilots.

COGNITIVE AGING: GENERAL VERSUS PROCESS-SPECIFIC SLOWING IN A VISUAL SEARCH TASK

Michael F. Gorman and Donald L. Fisher
University of Massachusetts Amherst

The fact that response times increase as one ages has long been established. Previously, a model of general slowing in the nonlexical domains has done a really good job of explaining the differences between older and younger adults. However, an alternative process-specific model has not been conclusively ruled out. This experiment tested general and process-specific models of slowing in the nonlexical domain using older and younger adults performing a visual search task. The task manipulated the presence of the target, the number of search items, and the structure of the display of the search items. It was found that a process-specific model explained significantly more of the variability than a general model of slowing. It was also discovered that the process most greatly affected was that of deciding to terminate a search when no target was present in the display.

It is well known that as humans age their performance decreases in a wide variety of tasks (Davies, Taylor & Dorn, 1992; Salthouse, 1991; Cerella, Poon & Williams, 1980). For example, performance decrements have been observed in tests of memory (Howard & Wiggs, 1993), intelligence (Hertzog, 1989; Schaie, 1989) and attention (Giambra, 1993; Madden & Plude, 1993).

In general there are two theories which attempt to explain the decline in performance of older adults. More specifically, we are talking about molar linear models of general slowing (and the related latent linear models of common slowing) and latent linear models of process-specific slowing (Fisher and Glaser, 1996). First, consider the molar linear model of general slowing. Defining O_i as the time on average for an older adult to respond, Y_i as the time on average for a younger adult to respond, and i as representing the task, the molar linear model of general slowing can be expressed by the equation:

$$O_i = \beta Y_i. \tag{1}$$

The model is a molar one because only the overall response times appear in the equation. The molar model is a general one because the same slowing factor, β, is used from one task to the next.

Now, consider the related latent linear model of common slowing. In order to define the latent model, we need to know the structure of the underlying cognitive processes in each task. Suppose that in one task the processes x_1, x_2 and x_3 with durations, respectively, of X_1, X_2 and X_3, were arranged in series. Then for younger adults, the latent model can be expressed as:

$$Y = X_1 + X_2 + X_3. \tag{2}$$

If the latent model is one of common slowing, then for older adults we can write:

$$O = \beta X_1 + \beta X_2 + \beta X_3. \tag{3}$$

The molar linear model of general slowing and latent linear model of common slowing are clearly conceptually identical to one another in this simple case. Specifically if all processes are arranged in series, then the slowing of each of the individual processes by a common factor β is identical to the slowing of the overall response time by a common factor β. However, the molar and latent models are no longer conceptually identical if each of the processes is slowed by a different factor. In this case (the latent linear model of process-specific slowing) we can write:

$$O = \beta_1 X_1 + \beta_2 X_2 + \beta_3 X_3. \tag{4}$$

The molar model of general slowing with its simplicity and elegance does a good job of predicting the differences between older and younger adults (Cerella et al. 1980). Support for this model has also been established by Salthouse and Somberg (1982) who showed a molar model of general slowing was able to explain 98.2% of the variability between older and younger adults. More recently, investigators have argued persuasively that slowing may not be general across all domains, but instead may be general only to broad domains such as the lexical (Lima, Hale and Myerson, 1991) and nonlexical (Lima, Hale and Myerson 1991; Hale, Myerson, Faust, and Fristoe, 1995) domains. We will focus here just on the nonlexical domain. Still, however well the molar models of general slowing explain the effect of age on performance in this domain, it is possible that a latent model of process-specific slowing may do a better job if more complete tests of the latter model are used (Fisk, Fisher & Rogers, 1992; Fisk & Fisher, 1994; Fisher, Fisk & Duffy, 1995). These tests we propose to undertake here. Specifically, our experiment attempts to determine whether a process-specific model will do a better job of predicting the effect of age on performance than a general slowing model by manipulating various factors of a visual search task. Simply, subjects were presented a target, and then asked to search a grid of letters and indicate whether or not the target was present.

<u>Figure 1</u>.   Predicted and observed response times of younger adults as a function of grid size. (Predictions come from Table 2, where YAS = observed response time when the target is <u>a</u>bsent and the grid is <u>s</u>ystematic, ypr = predicted response time when the target is <u>p</u>resent and the grid is random.)

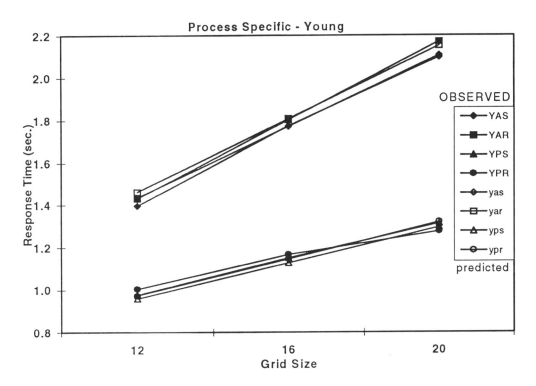

<u>Figure 2</u>.   Predicted and observed response times of older adults as a function of grid size. (Predictions come from Table 2, where OAS = observed response time when the target is <u>a</u>bsent and the grid is <u>s</u>ystematic, opr = predicted response time when the target is <u>p</u>resent and the grid is random)

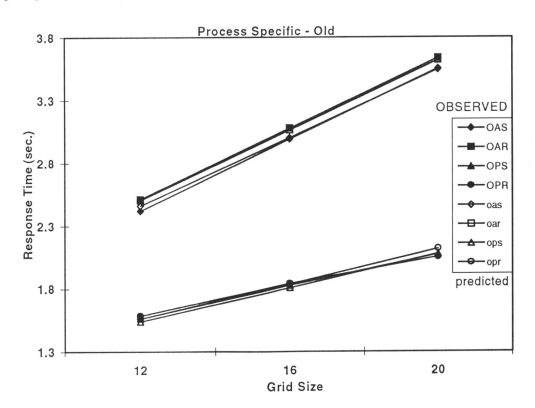

DISCUSSION

As the results indicate, the model of general (or common) slowing does a really good job of explaining the variability between the age groups, accounting for 99.73% of the variability. However, the process-specific model not only explains significantly more of the variability, it also provides us with additional insight as to what is occurring in this particular task. Looking at the coefficients of the models in Table 2, it is apparent that across models, the values are almost equal, except for β_n. That is, while $\beta = 1.670$, in the general model, and $\beta_a = 1.583$ and $\beta_c = 1.616$ in the process-specific model are almost equal, it is $\beta_n = 12.695$ that stands out and provides us with the important information. The fact that the slowing parameter for β_n, the time to terminate a search, is so much larger than the other slowing parameters indicates that this is the subprocess of the task that is giving older adults problems.

The above finding has a practical as well as theoretical import. For example, older adults are typically involved in many more accidents at signalized left turn intersections than younger adults (Staplin and Fisk, 1991). Perhaps it takes older adults so long to come to a decision that an oncoming car is not present after scanning the roadway, that other fast approaching cars that were not a threat become a threat during the long decision time. While, tests are now being proposed to screen older drivers, these tests are meaningful only if they require older adults to exercise the cognitive operations that are critical to safe driving. The critical cognitive operations can be <u>known</u> only if latent models like the ones we proposed are used to identify the problematic processes.

One finding we expected but did not observe was what we termed the "getting lost factor". We expected that in a random search grid, the older adults would have more difficulty remembering where they had been, and as a result would have to recheck some percentage of the search grid items. This however turned out not to be the case.

In conclusion, a model of general slowing often does a very good job of explaining the differences between older and younger adults. Although it is tempting to stop there and say good enough, in doing so we are missing out on the added insight and understanding that a process-specific model provides.

ACKNOWLEDGEMENTS

This research was supported in part by a grant from the National Institute of Aging (AG12461) to Donald L. Fisher.

REFERENCES

Cerella, J., Poon, L. W. & Williams, D. M. (1980). Age and the complexity hypothesis. In L. W. Poon (Ed.), *Aging in the 1980s: Psychological issues* (pp. 332-340). Washington DC: American Psychological Association.

Davies, D. R., Taylor, A. & Dorn, L. (1992). Aging and human performance. In A. P. Smith & D. M. Jones (Eds.), *Handbook of Human Performance: Volume 3 State and Trait* (pp. 25-61). San Diego, CA: Academic Press Limited.

Fisher, D. L., Fisk, A. D. & Duffy, S. A. (r1995). Why latent models are needed to test hypotheses about the slowing of word and language processes in older adults. In Ph. Allen & Th. R. Bashore (Eds.), *Age differences in word and language processing* (pp. 1-29). Elsevier Science B. V.

Fisher, D. L. & Glaser, R. (1996). Cognitive aging: Models of general, task-specific and process-specific slowing. *Psychonomic Bulletin and Review.* Accepted with revisions.

Fisk, A. D. & Fisher, D. L. (1994). Brinley plots and theories of aging: The explicit, muddled and implicit debates. *Journal of Gerontology: Psychological Sciences, 49*, P81-P89.

Fisk, A. D., Fisher, D. L. & Rogers, W. A. (1992). General slowing alone cannot explain age-related search effects: A reply to Cerella (1991). *Journal of Experimental Psychology: General, 121*, 73-78.

Giambra, L. M. (1993). Sustained attention in older adults: performance and processes (pp. 259-272). In J. Cerella, J. Rybash, W. Hoyer and M. L. Commons (Eds.), *Adult information processing: Limits on loss.* San Diego: Academic Press.

Hale, S., Myerson, J., Faust, M., and Fristoe, N. (1995). Converging evidence for domain-specific slowing form multiple nonlexical tasks and multiple analytic methods. *Journal of Gerontology: Psychological Sciences, 50B*, P202-P211.

Hertzog, C. (1989) Influences of cognitive slowing on age differences in intelligence. *Developmental Psychology, 5*, 636-651.

Howard, D. V. & Wiggs, C. L. (1993). Aging and learning: insights from implicit and explicit tests (pp. 512-528). In J. Cerella, J. Rybash, W. Hoyer and M. L. Commons (Eds.), *Adult information processing: Limits on loss.* San Diego: Academic Press.

Lima, S. D., Hale, S. & Myerson, J. (1991). How general is slowing? Evidence from the lexical domain. *Psychology and Aging, 6*, 416-425.

Madden, D. J. & Plude, D. J. (1993). Selective preservation of selective attention (pp. 273-302). In J. Cerella, J. Rybash, W. Hoyer and M. L. Commons (Eds.), *Adult information processing: Limits on loss.* San Diego: Academic Press.

Neter, J. and Wasserman, W. (1974). *Applied linear statistical models.* Homewood, Illinois: Richard D. Irwin.

Salthouse, T. A. (1991). *Theoretical perspectives on cognitive aging.* Hillsdale, NJ: Lawrence Erlbaum.

Salthouse, T. A. & Somberg, B. L. (1982). Isolating the age deficit in speeded performance. *Journal of Gerontology, 37*, 59-63.

Schaie, K. W. (1989). Perceptual speed in adulthood: Cross-sectional and longitudinal studies. *Psychology and Aging, 4*, 443-453.

Staplin, L. & Fisk, A. D. (1991). A cognitive engineering approach to improving signalized left turn intersections. *Human Factors, 33(5)*, 559-571.

Wechsler, D. (1955). *Manual for the Weschler adult intelligence scale.* New York: The Psychological Corporation.

PAIRED ASSOCIATE LEARNING: AGE DIFFERENCES

Konstantinos V. Katsikopoulos, Donald L. Fisher, Michael T. Pullen
University of Massachusetts Amherst
Amherst, MA

The issue of age related differences in performance during the acqustion phase of a paired associate learning task is discussed within the framework of a precise mathematical tool. A two-stage, four-state Markov model is employed to analyze the data sets from two age groups consisting of 24 subjects each. The relative efficiencies of the acqustion processes of the younger and the older groups of adults are reflected in the different values of parameters. (These values were obtained by optimizing the fit of the model to the two data sets). The two major findings are: (i) the younger adults form associations (even temporary ones more easily) and (ii) these associations tend to decay less quickly, again in the younger adults. The results speak against the general decrement hypothesis, allthough further investigation is needed.

INTRODUCTION

This study was designed to generate evidence that could be used to understand better the differences in paired associate learning between younger (up to 30 years old) and older (above 60 years old) adults. Our goal was not only to investigate if such differences exist but also their nature as well. Specifically, the differences in performance can occur as a result of changes with age in any one of a number of different learning processes. These processes include the initial forming of an association between the stimulus and response in short term memory and the subsequent coding of an association first in what we call a critical state (an intermediate state between the short and long term memories) and then in long term memory. If the older adults are performing less efficiently, then for both theoretical and practical purposes it is important to know if all processes are degraded equally and, if not, which processes are degraded the most.

At this point it is important to understand *why* we study the paired associate task. We are motivated not only by the desire to make theoretical contributions, but because we hope to add to the practical store of benefits too. If the points where the elderly are especially weak are known, we can design more suitable training methods. We are indeed implying that the paired associate task is in essence a laboratory simulation of many real world tasks which the elderly often find difficult to perform, such as voice mail and the operation of ATMs (Fisher, in press). Learning the vocabulary of a foreign language is another example. Generally, we believe that the results can generalize to many everyday activities in a fairly straightforward fashion.

There has been considerable work on the matter. The first study to investigate age differences in paired associate learning was done by Ruch (1934). He concluded that age differences do exist. Most of the studies since have replicated

this result (Korchin and Basowitz, 1957; Salthouse, Kausler and Saults, 1988). These differences though are not always of the same magnitude as demonstrated by Canestrari (1963) and Monge and Hultsch (1971) but depend on factors such as the time available to produce a response and the difficulty of the items. These findings actually imply that the psychological processes that underlie the task are *not* equally degraded. On the other hand, Fisher (in press) analyzed the results of a study by Kausler and Puckett (1980) and found the opposite pattern of results, that is all the processes were affected equally by the aging factor. Kausler (1992) calls this hypothesis the *general decrement hypothesis*. In this study, we want exactly to examine the validity of this hypothesis.

Arenberg and Tchabo (1994) have demonstrated that age differences in learning would probably be explained more adequately by integrating theoretical developments from many areas not currently considered. We believe that one of those areas that could prove useful is mathematical learning theory. During the 1950's and 1960's many models were developed that attempted to capture the nature of paired associate learning (see for example Estes, 1950; Bush and Mosteller, 1951; and Norman, 1963). But up to date we are not aware of any attempts to view age related differences in the context of such mathematical tools. In earlier studies, the emphasis was given to the quality of fits of the models to data generated mainly by using *younger* adults as subjects. Furthermore, the values of the parameters were not viewed as characterizing the core psychological processes of the task, and therefore would not have been seen as being able to provide a basis for comparing the efficiencies of the processes in the two age groups. Using a standard experimental paired associate paradigm (Pullen, 1995) but treating the data within the context of a Markov model proposed by Fisher (in press), we will view the problem in just this way. That is, we will argue that the parameters of the Markov model reflect the operation of the underlying processes and that, therefore, the effect of age on the efficiency of the core psychological processes can be determined by analyzing the values of the parameters for the younger and older adults.

METHOD

Twenty four younger and twenty four older adults were recruited from the University of Massachusetts and the local area. Each subject attended five consecutive hour long sessions approximately 24 hours apart.

The stimuli were presented on a monitor with a black background and amber or white characters using an IBM compatible computer. Responses were made on a standard computer keyboard.

Each stimulus in the paired associate was a CVC (consonant-vowel-consonant grouping) and it was always meaningless. Each response was one of four letters. The CVCs were taken from the Richardson and Erlebacher (1957) study. The four response sets were chosen so that the subject's index and middle fingers would be identically spaced for each session.

During each of four sessions, subjects were asked to learn a list of 24 pairs. Half of the stimuli in each list were classified as easy and half as difficult in terms of how familiar they were. Each paired associate was presented 16 times. Each list was divided in blocks with possible sizes of 24,12,8 or 4 pairs. Each block was repeated a fixed number of times before a new block was introduced and all blocks were randomized each time. The repetition level took the values of 16, 8 or 4. The repetition level was varied between subjects, the block size within subjects. Each subject was assigned to a repetition level and on each of four days was trained on a different list with 1 (of size 24), 2 (of size 12), 3 (of size 8) or 6 (of size 4) different blocks respectively. Eight subjects per age group were assigned to each repetition level. A given block size and a given repetition level defined a daily session (for example 4B4R defined a block size of 4 and a repetition level of 4). A unique set of CVCs was used in each training session. In all equivalent training sessions (fixed block size and repetition level) the presentation schedule remained the same. The dependent variable was accuracy.

Each subject was seated in front of the computer. At the beginning of each trial a '+' was displayed on the screen. Then the CVC would appear in the same position and the subject was prompted for a response. The subjects were given as much time as they wanted to provide the response. After responding, they would be informed of the correct response and would be given 2 seconds to study the pair.

RESULTS

Before presenting the data, we should talk about the Lag Sensitive Model that we shall use to analyze the results (Fisher, in press). Although more complex than many models of paired associate learning, this complexity is needed to explain results such as those we observed in the condition 4B4R (also see Young, 1971). It is a Markov model which assumes that the training process for each pair consists of two stages (the learning stage which occurs on a trial where the pair is trained and the forgetting stage which occurs on a trial where the the pair is not trained) and four states in each of the two stages. These states represent the *level of proficiency* of the subject in relation to a given pair or alternatively the *different memories* where the association the subject has formed resides. These different memories we use are much influenced by the short term memory theories that were developed in the 1960's; at the same time the mathematical learning models were developed too. So, if the association is learned (or equivalently it resides in the long term memory), the paired associate is assumed to be in the *Learned state L*. If the correct association between the stimulus and the response member of the pair is not learned, the paired associate is said to be in the *Unconditioned state U*. If the association that has been formed resides in short term memory (but has not yet been coded for storage in the long term memory), the paired associate is said to be in the *Short term state S*. For the cases where the association has been coded for storage in long term memory but is not yet completely learned, we introduce the *Critical state C*.

We need not feel uncomfortable if cognitive psychology theories of the past do not provide consistent support for the postulation of these states. When Estes (1960) proposed his famous two state model, he postulated the existence of only two states: *Unconditioned* and *Learned*. Of course there exist more possibilities for the status of an association and Estes was aware of that, but in an attempt to model a task mathematically one has to approximate the psychological theories which are usually much more descriptive. In our model it is mainly the existence of the *Critical state C* which can be questioned by theorists. Perhaps doubts could be overcome more easily if we focus first more on its mathematical than its psychological interpratation.

The transitions between the states are governed by two 4x4 (one for each stage) transition matrices. Each entry m_{ij} is the probability that after a trial (training or nontraining trial, depending on the stage the matrix represents) the pair will move from state i to j, where i,j take the values L, C, S and U. Thus, for example $p_T(C,L)$ would be the entry in the row C and the column L for the training matrix M_T and it would denote the probability that after a training trial a paired associate that resided in the *critical state C* would enter the *learned state L*. All these entries are probabilities, so they range from 0 to 1. A value of 0 means that the corresponding transition is impossible, while a value of 1 means that the corresponding transition is certain. Furthermore, all the entries of a given row

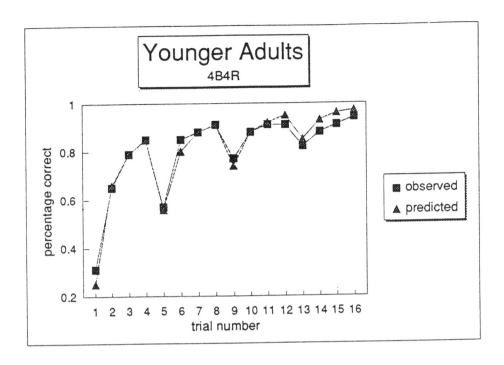

Figure 1. Predicted and Observed Learning Curve for Younger Adults

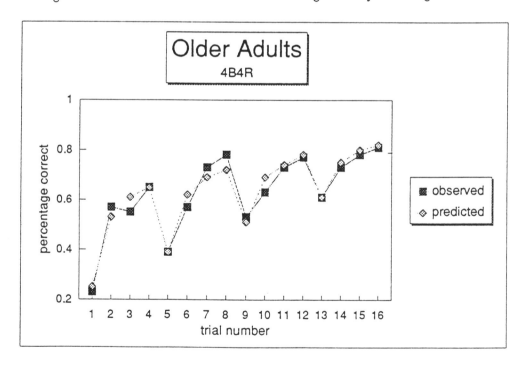

Figure 2. Predicted and Observed Learning Curve for Older Adults

should sum to 1, since it is impossible that none of the transitions occurs (including remaining in the same state). These constraints are actually very useful in our attempt at reducing the number of free parameters. Note that it seems we need $4 \times 4 = 16$ parameters for each matrix, so 32 for the model. But the fact that every four parameters have to sum to 1 and the fact that we have a consistent set of learning and forgetting axioms enable us to describe the process fully with six free parameters [which are $p_T(C,L)$, $p_T(C,C)$, $p_T(S,C)$, $p_T(U,S)$, $p_N(C,C)$ and $p_N(C,S)$].

The set of the axioms are: state L is *absorbing* in both stages, meaning that since it is entered it can not be left. We also assume that a pair which is not learned, can not be learned during a nontraining trial and that even during training a paired associate can not advance more than one state at a time. Finally, during nontraining a permanent association can not be formed and during training an association which already exists can not be lost (even if it is not permanent). We also assume that the subject guesses the correct response with probability g when in state U (*guessing axiom*).

The model is called the Lag Sensitive Model because it can exactly predict the lag effects on performance (see also Young, 1971). To do that we actually modify the chain of the forgetting stage to a non-markov model by letting $p_N(S_i, S_{i+1}) = \exp(-NII/8)$ and $p_N(S_i, C_{i+1}) = p_N(S_i, U_{i+1})$ where NII is the Number of Intervening Items.

Now, if we know the above 6 parameters and the training schedule we can predict the probability of a correct response on each of the 16 training trials in the 12 different conditions (4 block sizes $\times$ 3 repetition levels). In our analysis we will focus on the results for the 4B4R condition since other one stage models can not predict them (see Figures 1 and 2). This data has the form of 16 probabilities per age group and each probability is based on 24 (pairs) $\times$ 8 (subjects) = 192 observations. We can see a common (for both age groups) pattern in the data: accuracy increases during the first four trials, then it decreases in the fifth trial, then it increases again until the eighth trial, it drops in the ninth trial and so on. The most profound and logical explanation is the fact that the repetition level is 4, so each paired associate is trained 4 times with an average lag of 4 (block size), but then it will be trained again (for the fifth time) after *all* other paired associates that belong to other blocks are presented 4 times and similarly for the other circles until the last trial. Except for this common pattern, we can also see that the performance of the younger group is consistently superior.

Using an iterative search we determined the values of the parameters which yield the minimum chi-square (best fit). Both fits (for younger and older adults) were quite satisfactory and thus the Lag Sensitive Model can not be rejected out of hand (see Figures 1 and 2). Additionally, the observations did not differ form the predictions by more than an amount that one would expect by chance alone (the chi-square values were

5.50 and 5.79 for the younger and the older group respectively).

The values of the parameters were:

	Younger	Older
$p_T(C,L)$	0.6	0.9
$p_T(C,C)$	0.3	0.1
$p_T(S,C)$	1.0	1.0
$p_T(U,S)$	0.8	0.6
$p_N(C,C)$	0.7	0.1
$p_N(C,S)$	0.3	0.8

DISCUSSION

Now that we have a uniform framework to work within (our model explains the performance of both the younger and the older adults), we can examine the *general decrement hypothesis*. As stated before, this hypothesis argues that all the processes involved in the task are degraded equally in the case of the elderly. We will argue that this hypothesis must be rejected (since one of the processes is degraded in the case of the older adults and a second process operates more efficiently in older adults).

It is very important that we clarify some points: by *process* we understand all the transitions from one of the *memories* we described above, to one of the other *memories*. These transitions represent the rules that govern the processing of information in each state. And second we assume that a given transition probability reflects the succefullness of the corresponding transition or equivalently the percentage of material that is *on the average* transmitted between states. These two assumptions will lead us to our conclusions.

First suppose that a paired associate is not being trained. Thus we consider the matrix M_N. We have $p_N(C,C) = .7$ and $p_N(C,U) = 1 - p_N(C,C) - p_N(C,S) = 0$ for the younger group. So, the younger adults retain 70% of the material that already existed in the critical state memory after a nontraining trial and it is impossible for them to lose completely an association. But for the older adults we have $p_N(C,C) = .1$ and $p_N(C,U) = .1$, so we can conclude that the processing of information in the critical state memory is degraded in the case of the older adults.

If we consider a paired associate that is trained, for the hypothesis to be valid we would want the transition from the critical state to the learned state to be more frequent for the younger group. But clearly that is not true: for the younger group $p_T(C,L) = .6$ and for the older group $p_T(C,L) = .9$

Much remains to be done. Some of the future directions could be: we should analyze in a similar fashion the results from the remaining conditions and ensure that our conclusions are not a product of chance factors. Regardless of the outcome of this more complete analysis, we believe that we can benefit considerably by viewing behavior in general and learning more specifically as a probabilistic process. The fact that uncertainty underlies the paired associate learning task need not make us uncomfortable; in fact this random element is a central feature of our approach and we believe it can help us explain age related differences.

ACKNOWLEDGEMENTS

Portions of this research were supported by a grant from the National Institute of Aging (R01-AG12461).

REFERENCES

Arenberg D. and Robertson-Tchabo E.A (1994). *Learning and Aging: Behavioral Pocesses. 421-432.*

Bush R.R and Mosteller F.A (1951). A mathematical model for simple learning. *Psychological Review, 58,* 313-323.

Canestrari R.E (1963). Paced and Self-Paced Learning in Young and Elderly Adults. *Journal of Gerontology, 18,* 165-168.

Estes W.K (1950). Toward a Statistical Theory of Learning. *Psychological Review, 57,* 94-107.

Estes W.K (1960). Learning theory and the new mental chemistry. *Psychological Review, 67,* 207-223.

Fisher D.L (in press). State models of skill acquisition: Optimizing the training of older adults. In W.A Rogers, A.D Fisk and N.Walker (Eds.), *Aging and skilled performance: Advances in theory and applications.*

Kausler D.H (1992).Comments on aging memory and its everyday operations. In L.W Poon, D.C Rubin and B.W Wilson (Eds.), *Everyday cognition in* adulthood and late life. Cambridge University Press, pp. 483-495.

Kausler D.H and Puckett J.M (1980). Frequency judgements and correlated cognitive abilities in younger and elderly adults. *Journal of Gerontology, 35,* 376-382.

Korchin S.J and Basowitz H. (1957). Age differences in verbal learning. *Journal of Abnormal and Social Psychology, 54,* 64-69.

Monge R.H and Hultsch D.F (1971). Paired associate learning as a function of adult age and the length of the anticipation and inspection intervals. *Journal of Gerontology, 26,* 157-162.

Norman M.F (1963). Incremental learning on random trials. *Journal of Mathematical Psychology, 1,* 336-350.

Pullen M.T (1995). *An optimal training theory: A paired associate learning task.* Unpublished master's thesis. Amherst: University of Massachusetts.

Richardson J. and Erlebacher A. (1957). Associate connection between paired verbal items. *Journal of Experimental Psychology, 56,* 62-69.

Ruch F.L (1934). The differential effects of age upon human learning. *Journal of General Psychology, 11,* 261-286.

Salthouse T.A, Kausler D.H and Saults J.S (1988). Utilization of path analytic procedures to investigate the role of processing resources in cognitive aging. *Psychology and Aging, 3,* 158-166.

Young J.L (1971). Reinforcement test intervals in paired associate learning. *Journal of Mathematical Psychology, 8,* 58-81.

RETENTION OF MULTIPLE-TASK PERFORMANCE: AGE-RELATED DIFFERENCES

Richard A. Sit and Arthur D. Fisk
Georgia Institute of Technology
Atlanta, GA

This study examined the relationship between retention of both multiple-task performance and the micro-components of a complex task. Young and older adults trained on a synthetic work task (Elsmore, 1994) with both groups acquiring skill in performing the complex task. After a five month retention period, older adults' initial performance on the multiple-task declined significantly more than younger adults. Both groups of adults regained their final trained level of performance after only four 5-minute trials. However, throughout the retention trials older adults only emphasized a single component of the complex task. Young adults successfully allocated attention to all task components. These and other aspects of the data suggest that a major locus of age-related decline in complex task performance is due to differential loss in strategic allocation of attention to component tasks. The data also show how measuring multiple-task performance may underestimate lack of component processing efficiency.

Over the last few years, the major objective of much research in cognition and aging has been to understand the moderating influences of skill acquisition (e.g., consistency, task structure, learning requirements, amount of practice) as they relate to models developed to describe and predict the influences of aging on performance improvement and learning underlying the acquisition of skill. This research has been successful in delineating age-related performance improvement and learning underlying acquired skills for several task domains (e.g., see Fisk & Rogers, 1991; Rogers, Fisk & Hertzog, 1994). However, a major gap in understanding concerns long-term retention of acquired knowledge and acquired skills. As a consequence, we know relatively little about how maintenance of knowledge and skills is affected by conditions of original learning, training strategies, intervening activities, the differential decay or survival of component processes of skill and so on.

Why is the Study of Retention Important?

Understanding retention of knowledge and skill is clearly important from both a pragmatic and a theoretical perspective. Pragmatically the importance is clear. In most activities of daily living, the opportunity and the need for performance after training are either delayed or infrequent. This is often the case when individuals learn how to use new technology. Another major area, emergency procedures have a low probability of occurrence; but, when they do occur, accurate and fluent performance is crucial. The goals of designing instructional and training procedures for skills and knowledge that must be maintained during periods of disuse might be quite different than goals for simply ensuring the attainment of mastery during training. From this perspective, it is crucial to have an accurate age-related account of how skills and knowledge decay during periods of disuse.

From a theoretical perspective, the importance of understanding age-related issues of retention are equally clear.

Although the number of studies providing information on retention of skill after various intervals is small, the study of retention of learned materials has been important for theory development and refinement. For example, through the assessment of retention we have learned about the importance of massed and distributed practice and the differences in the learning that occurs in these situations.

What Do We Know About Age-Related Retention Capability?

Salthouse (1991) remarked that no firm conclusion could be reached concerning the presence or absence of age differences in mechanisms related to retention of information. It is not surprising that he would reach such a conclusion. Consider that of the 23 studies he reviewed, 12 showed similar rates of loss for young and old adults and 11 demonstrated greater loss for older adults. Currently, it is difficult to integrate age-related studies that measure the retention of knowledge and skills. The number of studies using long-term retention intervals is relatively small, in part due to the difficulty of performing such studies.

Yet, the picture is not as bleak as it might seem. Although the results of previous studies that have examined age-related maintenance of knowledge or skills are mixed, there seems to be the potential for an integrative framework. One potential explanation for the inconsistent findings may be due to the assessment of different original learning (either in degree or kind) between young and old adults on the target task (Fisk, Hertzog, Lee, Rogers, & Anderson-Garlach, 1994). Alternatively, old adults may truly be less able to retain trained performance for some types of learning but not others (Anderson-Garlach & Fisk, 1994; Rogers, Gilbert, & Fisk, 1994). Finally, activities subsequent to original learning of the target task may affect older adults more than younger adults (Fisk, Cooper, Hertzog, & Anderson-Garlach, 1995).

Our preliminary long-term retention studies examining maintenance of acquired search-detection skill have yielded interesting, yet theoretically challenging, results. Our studies suggest: (a) in some situations old and young adults retain an impressive amount of skill even after 16 months without exposure to the task; (b) retention performance declines within a three month period and that decline remains stable between three and six months for both age groups; (c) old and young adults equally retain general, task-relevant skills, at least when single task performance is evaluated; (d) old adults' performance declines more than young adults' for both extensively trained and moderately trained stimuli; (e) when an interfering processing activity is inserted prior to the retention interval, old adults' performance declines disproportionately more than young adults' performance especially when compared with a task not subjected to such interference; and (f) depending on the type of search task, for both age groups, the initial retention deficit is largely attenuated by the end of the retention retraining periods we have used (see Fisk, Hertzog, et al., 1994; Fisk, Copper, et al., 1994 for a review). This pattern of results suggests that a simple model of retention may not best describe age-related long-term retention. A difference in the qualitative nature of learning is sufficient to produce age-related differences in retention capability (Fisk, Hertzog et al., 1994). However, our data suggest that learning differences may not be a *necessary* condition for age-related retention effects.

The previous research involving age-related patterns of retention of skill has involved single task components of more complex tasks. The present research attempts to broaden our understanding of age-related retention capability by directly examining retention of performance in multiple-task situations. In today's society a growing number of occupational and daily living activities require the performance of multiple, simultaneous actions. In many work situations operators are required to monitor, control, and manipulate information via complex systems. The important questions for the present investigation concerned the relationship between retention of multiple-task performance and the micro-components of the task: Is it sufficient to know the retention of a single task component to predict multiple-task performance (is there an average, general decline)? Does multiple-task performance after some period of disuse give lawful insight into performance of the underlying micro-components? Finally, it is important to understand whether the same retention function, whatever it is, can be applied across age groups with equal accuracy.

Synthetic Work Environments

Synthetic work tasks have been found to be effective in simulating real-world multiple-task environments (e.g., Alluisi, 1967; Chiles, Alluisi, & Adams, 1968; Elsmore, 1994; Morgan & Alluisi, 1979). The goal of such simulations is to abstract the critical aspects of a wide range of activities (e.g., memory and classification of items, performance of a self-paced task, arithmetic problems; and monitoring and reaction to both visual and auditory information) while bringing the study under controlled laboratory conditions.

Purpose of the Study

The purpose of this study was to investigate the performance of young and older adults trained on a synthetic work task, after a 5-month retention interval. Specifically, this study examined the effects of Age, Task Type (memory, arithmetic, visual monitoring, and auditory monitoring), and Trial (performance before versus after a 5-month retention interval), on performance (multiple-task score and the four individual component task scores).

METHOD

Participants

Twelve young adults (M=32.0 years of age, SD=4.6) and 19 older adults (M=68.8, SD=5.8) participated in this retention study. This group was recruited from a sampling frame of 20 young and 26 older adults. These individuals were trained on the task in a previous study (Salthouse, Hambrick, Lukas, and Dell, in press). Subjects were paid $25.00 for their participation in retention testing, had at least the equivalent of 20/40 correct vision, and reported their health as good to excellent. Participants were administered several general ability tests which revealed age-related patterns consistent with the general population.

Apparatus

Synwork1. This study used Synwork1, a computer-based synthetic work task (Elsmore, 1994), and operated on 486 PC-compatible computers equipped with color monitors. All interaction with the testing program was done via a standard mouse. During a test trial, the computer screen is divided into four quadrants or "windows", each assigned to a different task. A small window in the center of the screen is used for displaying a composite "score" for performance on all of the tasks within the synthetic work environment. In the upper left of the screen is a memory task. The initial display in this task consists of a set of 5 letters. The set is then removed and followed by periodic displays of a probe letter which is to be classified as YES, a member of the set, or NO, not a member of the set. The upper right of the display contains an arithmetic task. In this task, two three-digit numbers are to be added by adjusting plus and minus buttons to produce the correct sum, in the row below the addends. This task is completely self-paced. The lower left quadrant of the display contains a visual monitoring task. In this task, the participant monitors the position of a pointer moving continuously along a horizontal scale, and attempts to reset it before it reaches the end of the scale. The lower right quadrant of the display contains an auditory monitoring task. High and low tones are presented periodically throughout the trial, and the task is to respond whenever a high tone occurs. On any given trial, all participants were exposed to the same sequence of events.

Procedure

Initial Training. The training portion of the study is described in Salthouse et al. (in press). To summarize, all participants came to the laboratory on three separate days within a 10-day period. Testing was conducted in groups of one to five individuals, with each participant seated at a microcomputer. On the first day participants performed a variety of abilities tests, performed eight trials maneuvering a pointer through a W-shaped maze, and read specific instructions on how to perform Synwork1. Each task was then presented in isolation for two 1-minute trials, followed by the four tasks presented together for one minute. The remainder of the first day was spent training on the four tasks together for five 5-minute trials. Over the next two days, participants trained on the multiple-task for 10 5-minute trials per day.

Retention Testing. After a five month retention interval, participants returned to the laboratory and the same procedure used for training was followed. Testing was conducted in groups of one to four individuals, with each participant seated at a computer workstation. Participants completed a series of abilities tests and were given verbal and written instructions on the study and Synwork1. Participants performed eight mouse maneuvering trials, the component tasks in isolation for two 1-minute trials each, and the four tasks presented together for one 1-minute trial. Next, all four tasks were performed together for four 5-minute trials.   Participants performed four additional 5-minute trials following retention testing; however, those trials were part of a different investigation and will not be discussed.

The scoring parameters were as follows: in the memory task, probe stimuli occurred every 10 seconds, 10 points were awarded for correct responses, and 10 points were subtracted for incorrect responses and memory list retrievals after the initial display at the beginning of a trial. In the arithmetic task, 10 points were awarded for correct responses and five points were subtracted for incorrect responses. In the visual monitoring task, the participant was awarded one point for every 10 pixels the pointer (moved 6.7 pixels per second) was away from the middle of the scale at the time of reset and 10 points were deducted for every second that the pointer was at the end of the 100.5 pixel scale (15 seconds to reach the end of the scale). The auditory monitoring task required the participant to discriminate between a low (523 Hz) non-target tone and a high (2092 Hz) target (.2 probability) tone every five seconds. Ten points were awarded for a hit and 10 points were subtracted for a miss or false alarm.

RESULTS

Initial Training

Retention participants' performance during initial training was examined to determine if this subset of people had developed performance indicative of skill in the multiple-task environment. Presented in the first column of Table 1 are both

adult group's percent improvement attained during initial training. These percentages represent the change in scores from initial training performance to final trained performance relative to initial training performance.

During initial training, at most, a participant could earn 1650 points on the multiple-task. Possible points for each of the component tasks were: memory (1200), arithmetic (130), visual monitoring (200), and auditory monitoring (120). During training, both groups of adults acquired skill on all four component tasks. After 20 training trials, young and old adults' mean multiple-task scores improved to 1365 and 951, and their mean component scores to: 1110 and 996 (memory), 38 and 9 (arithmetic), 119 and 7 (visual monitoring, and 98 and -62 (auditory monitoring), respectively.

An ANOVA test examining multiple-task performance across the initial training trials also indicated that both young and older adults significantly improved their performance on the complex task ($F(1, 29) = 106.19$, $p <.001$). This analysis also found a significant main effect of Age ($F(1, 29) = 15.13$, $p < .001$). Throughout training, young adults' multiple-task and component-task performance was higher than older adults' performance. These data are consistent with the entire sample tested by Salthouse et al. (in press).

Retention Performance

Initial- Multiple-Task Retention Performance. Multiple-task performance on the last trial prior to the retention interval and on the first trial subsequent to the five month retention interval (see Table 1) were examined. This analysis found main effects of Age ($F(1, 28) = 16.99$, $p < .001$), Task Type ($F(3, 84) = 43.66$, $p < .001$), and Trial ($F(1, 28) = 4.93$, $p < .05$). Also, there were significant Age x Trial ($F(1, 28) = 4.43$, $p < .05$) and Age x Task Type ($F(3, 84) = 3.18$, $p < .05$) interactions. Younger adults had better retention of previously acquired complex skills over a five month period than did older adults. Older adults' multiple-task performance declined 68.5%, whereas, younger adults' performance did not statistically change (actually improved 1.8%).

To better understand these significant main effects and interactions, it is necessary to examine the corresponding performance on individual task components as they were performed during multiple-task performance. An analysis of individual component scores (see Table 1) revealed strong performance trends. Older adults emphasized the memory search component. In fact, by the end of training they performed this task as well as younger adults (component score of 251 versus 238, respectively). After the retention interval, older adults obtained their observed multiple-task score by continuing to emphasize the memory search component. Although the older adults' memory search score declined 33% (from 251 to 168), this component score was 158% of the mean multiple-task score (the other component scores were negative). It is clear that older adults' decline on the other task components was quite dramatic. However, young adults successfully allocated attention to all four task components.

Table 1. Mean scores for multiple-task and individual component performance on the synthetic work task

	Percent Improvement During Initial Training	Final Trained Performance (SD)	5-Min. Trials After Retention Interval (SD)			
			1	2	3	4
Young Adults (n = 12)						
Multiple-Task Score	453%	621 (182)	632 (162)	631 (149)	546 (220)	644 (167)
Component Score: Memory	111%	238 (55)	254 (70)	264 (48)	225 (77)	277 (19)
Math	538%	131 (91)	93 (84)	115 (83)	98 (83)	130 (86)
Visual	146%	149 (44)	176 (24)	160 (47)	127 (67)	135 (91)
Auditory	567%	101 (27)	91 (44)	92 (34)	98 (29)	100 (20)
Older Adults (n = 19)						
Multiple-Task Score	240%	336 (279)	106 (394)	194 (313)	209 (200)	318 (239)
Component Score: Memory	254%	251 (85)	168 (121)	182 (131)	168 (85)	231 (83)
Math	150%	31 (55)	- 5 (46)	17 (38)	19 (38)	27 (48)
Visual	101%	73 (151)	- 9 (245)	59 (132)	51 (132)	113 (117)
Auditory	30%	- 19 (99)	- 52 (113)	- 61 (107)	- 30 (97)	- 53 (109)

Single Task Retention. Performance on isolated individual components before training and before retention testing were examined. When tasks was performed in isolation, there were only significant main effects of Age on the memory ($F(1, 29) = 7.86$, $p < .05$) and arithmetic tasks($F(1, 29) = 12.68$, $p < .001$). Younger adults consistently performed higher than older adults on all components performed in isolation. The most interesting finding from these analyses was that after the retention interval neither age group performed worse on the isolated components relative to the same performance assessed during initial training.

Relearning During Retention Testing. Throughout retention testing young adults maintained their multiple-task score and their component scores at or near the level of performance they attained at the end of training. Although the older adults demonstrated dramatic decline in performance when initially tested after the retention interval, by the fourth 5-minute trial their mean performance level was at or near their level of final trained multiple-task performance. However, older adults did exhibit a higher degree of inter-subject performance variability during retention testing than during the final 5-minute trial of training. The return to trained performance level for the older adult participants shows the relatively transient nature of the initial decline.

CONCLUSIONS

The main findings from this experiment examining age-related retention capability were: (a) On multiple-task performance the older adults' performance declined across the retention interval more than the younger adults. (b) The initial multiple-task performance attained by the older adults was possible only because the older adults seemed to focus mainly on one component. (c) Single task performance of each component did not suffer for either age group due to nonperformance during the retention interval. (d) Performance quickly returned to the final trained level for the older adults.

These data answer the initial questions that motivated the study and add to our growing understanding of age-related skills retention capability. First, the data suggest that, for complex task performance requiring multi-tasking, assessing retention of the combined task performance data (as well as retention of components performed in isolation) may underrepresent older adults performance on microcomponents of the multiple-task. Such data are important for suggesting the critical nature of refresher training for older adults learning a new task requiring coordination of attention allocation during task performance (cf. Korteling, 1994; Kramer & Larish, in press).

Second, the data point clearly to the locus of loss in these types of task (where each major task component is variably mapped and cannot benefit from automatic processing) as being related to the loss in strategies for allocation of attention and task coordination (Schneider & Detweiler, 1988). This conclusion is straightforward when performance of each single component performed in isolation is considered (no loss due to retention interval) and the quick reinstatement of performance to that seen during the final training trial. Given a loss of a strategy to perform the overall multiple-task, it is not surprising that older adults would maintain performance on the memory search task. During initial training (the first 20 5-minute trials) the memory search component was emphasized by giving that component more points for successful performance and increasing its demand frequency relative to the other components (see Salthouse et al., in press). Hence, what appears to have decayed across the retention interval was the older adults knowledge of how to maximize multiple-task score. Third, for this class of tasks, multi-component tasks requiring acquisition of a performance strategy that remains heavily dependent on control, attention demanding processes, it is clear that a generalized component decline function will not emerge.

ACKNOWLEDGMENTS

This research was supported in part by NIH Grant No. P50 AG11715 under the auspices of the Center for Applied Cognitive Aging Research on Aging (one of the Edward R. Roybal Centers for Research on Applied Gerontology) and in part by NIH Grant No. R01 AG 07654. We would like to thank T. Salthouse for making available the list of participants from the original study and for providing the original training data.

REFERENCES

Alluisi, E. A. (1967). Methodology in the use of synthetic tasks to assess complex performance. *Human Factors, 9*, 375-384.

Anderson-Garlach, M. M., and Fisk, A. D. (1994, April). *Age-related retention of skilled performance: Within-subject examination of visual search, memory search, and lexical decision.* Presented at the Fifth Cognitive Aging Conference, Atlanta.

Chiles, W. D., Alluisi, E. A., and Adams, O. S. (1968). Work schedules and performance during confinement. *Human Factors, 10*, 143-196.

Elsmore, T. F. (1994). SYNWORK1: A PC-based tool for assessment of performance in a simulated work environment. *Behavior Research Methods, Instrumentation, and Computers, 26*, 421-426.

Fisk, A. D., and Rogers, W. A. (1991). Toward an understanding of age-related memory and visual search effects. *Journal of Experimental Psychology: General, 120*, 131-149.

Fisk, A. D., Cooper, B. P., Hertzog, C., and Anderson-Garlach, M. M. (1995). Age-related retention of skilled memory search: Examination of associative learning, interference, and task-specific skills. *Journal of Gerontology: Psychological Sciences, 50B*, P150-P161.

Fisk, A. D., Hertzog, C., Lee, M. D., Rogers, W. A., and Anderson-Garlach, M. M. (1994). Long-term retention of skilled visual search: Do young adults retain more than old adults? *Psychology and Aging, 9*, 206-215.

Korteling, J. (1994). Effects of aging, skill modification, and demand alternation on multi-task performance. Human Factors, 36, 27-43.

Kramer, A. F., and Larish, J. L. (in press). Aging and dual-task performance. In W. R. Rogers, A. D. Fisk, and N. Walker (Eds.), *Aging and skilled performance: Advances in theory and application*. Hillsdale, NJ: Erlbaum.

Morgan, B. B., and Alluisi, E. A. (1979). Synthetic work: A methodology for assessment of human performance. *Perceptual and Motor Skills, 35*, 835-845.

Rogers, W. A., Fisk, A. D., and Hertzog, C. (1994). Do ability-performance relationships differentiate age and practice effects in visual search? *Journal of Experimental Psychology: Learning Memory and Cognition, 20*, 710-738.

Rogers, W. A., Gilbert, D. K., and Fisk, A. D. (1994, April). *Long-term retention of general skill and stimulus-specific abilities in associative learning: Age-related differences.* Presented at the Fifth Cognitive Aging Conference, Atlanta.

Salthouse, T. A. (1991). *Theoretical perspectives on cognitive aging.* Hillsdale, NJ: Erlbaum.

Salthouse, T. A., Hambrick, D. Z., Lukas, K. E., Dell, T. C. (in press). Determinants of adult age differences on synthetic work performance. *Journal of Experimental Psychology: Applied.*

Schneider, W, & Detweiler, M. (1988). The role of practice in dual-task performance: Toward workload modeling in a connectionist/control architecture. *Human Factors, 30*(5), 539-566.

W. Poon, D. C. Rubin, and B. A. Wilson (Eds.), *Everyday cognition in adulthood and late life* (pp. 545-569). Cambridge, England: Cambridge University Press.

EXPERTISE AND AGE EFFECTS ON PILOT MENTAL WORKLOAD
IN A SIMULATED AVIATION TASK

Donald L. Lassiter
Methodist College
Fayetteville, NC

Daniel G. Morrow
University of New Hampshire
Durham, NH

Gary E. Hinson
Methodist College
Fayetteville, NC

Michael Miller
Catholic University of America
Washington, DC

David Z. Hambrick
Georgia Institute of Technology
Atlanta, GA

ABSTRACT

This study investigated the effects of expertise and age on cognitive resources relevant to mental workload of pilots engaged in simulated aviation tasks. A secondary task workload assessment methodology was used, with a PC-based flying task as the primary task, and a Sternberg choice reaction time task as the secondary task. A mixed design using repeated measures was employed, with age and expertise as between-subjects factors and workload as the within-subjects factor. Pilots ranging in age from 21 to 79 years and 28 to 11,817 hours of flight time served as subjects. Of interest was whether expertise would mitigate the adverse effect of aging on pilots' mental workload handling ability as defined by two measures of secondary task performance: choice reaction time and accuracy. Results indicated that expertise did mitigate the effects of age regarding secondary task accuracy. Implications of results are discussed, and directions for future research are presented.

INTRODUCTION

The task of flying involves the time-sharing of several tasks, i.e., it is a multiple-task situation, placing a great deal of mental workload demand on the pilot. The concept of mental workload implies that limitations exist in the human information processing framework (Gopher & Donchin, 1986). It has often been reported in the aging literature that physical abilities, perceptual processes, and memory processes decrease with age (e.g., see Wickens, Braune, & Stokes, 1987; Tsang, 1992). To some degree, laboratory research with non-pilots has indicated that aging detrimentally effects the ability of the human information processing system to handle appreciable amounts of mental workload.

A frequent finding in the expertise literature is that expertise improves performance of domain-relevant tasks by reducing workload demands on short-term memory capacity (e.g., see Yekovich, Walker, Ogle, & Thompson, 1990). Domain-relevant tasks are tasks that tap the specific encapsulated knowledge within an expert's particular area of expertise, or domain (Rybash, Hoyer, & Roodin, 1986), whereas domain-general tasks do not tap a specific area of knowledge, but rather draw on generalized background knowledge and ability. To access the knowledge to perform a domain-relevant task might require less effort for an expert than a novice, who may not possess the required knowledge in

as useful a form to perform the same task. This is probably because the knowledge possessed by the novice is far less structured and automatized than that of the expert (Rybash et al., 1986).

Flying can be considered a well-defined task domain, so it may be that expertise in flying could counteract the detrimental effects of aging resulting from increases in mental workload of domain-relevant aviation tasks. The number of studies looking at the relationships among expertise, aging, and workload involving pilots as subjects has been small, but is now increasing (see Morrow, Leirer, & Altieri, 1992; Morrow, Leirer, Altieri, & Fitzsimmons, 1994; Tsang & Shaner, 1994; Tsang, 1995). However, there is still a need for systematic research programs involving pilots to investigate the relationships among these variables (Tsang, 1992).

Two specific aims of the current research effort were to determine: 1) if aging adversely affects the ability to handle increases in mental workload in a simulated aviation task; and 2) if expertise in piloting (as defined by the number of hours of flight time) can mitigate the adverse effect of aging on the ability to handle increases in mental workload. Such mitigation would be demonstrated by a significant interaction between expertise and age regarding the ability to handle mental workload.

METHOD

This study utilized the secondary task workload methodology (i.e., the subsidiary task technique) using a Sternberg (1969) choice reaction time task as the secondary task, and flying courses of two different levels of difficulty on Microsoft's Flight Simulator 4.0 (Microsoft, 1990) as the primary task. The Sternberg task has been used extensively as a secondary task in mental workload research and has shown sufficient sensitivity and diagnosticity (e.g., Wickens et al., 1987). The rationale for picking this task was that it may tap cognitive resources related to working memory and monitoring and compete with the primary task for these resources. The major independent variables in this study were age (in years), expertise (in hours of total flight time), and workload. Workload consisted of six levels based on combinations of levels of difficulty of the primary task (3 levels) and secondary task (2 levels). The three levels of primary task were zero (i.e., it was not performed with the secondary task; the secondary task was performed alone); easy course, and difficult course. The two levels of secondary task were 2 letters (low load) and 4 letters (high load) in the memory set. Therefore, the six levels of workload were: 1) secondary task alone (low load); 2) secondary task alone (high load); 3) primary task (easy course) with secondary task (low load); 4) primary task (easy course) with secondary task (high load); 5) primary task (difficult course) with secondary task (low load); and 6) primary task (difficult course) with secondary task (high load). The two major dependent variables in the study reflected secondary task performance: choice reaction time (in msecs) and accuracy (percent correct responses).

Subjects

A total of 42 paid volunteers served as subjects in this study. All had general aviation experience. The age range was 21 to 79 years, while the range of expertise was 28 to 11,817 hours of flight time. Subjects were recruited from the local general aviation community, primarily from a large local flying club. Other subjects came from the sizable active/retired military aviation community in the area. All subjects had general aviation experience, and all but three subjects had such experience within the twelve months prior to participation in this study. None of the subjects reported having prior laboratory testing, although those with military experience reported having simulator time. None of the subjects reported having any experience with the PC-based flying simulation used in this study.

Apparatus

Equipment for this study consisted of two computers for presenting the primary and secondary tasks, as well as collecting and analyzing data. The lab was partitioned into subject and experimenter stations. The subject station (see Figure 1) had two monitors, one each for the primary and secondary tasks, as well as a flight yoke and pedals for performing the primary task. Two large telegraph keys were mounted next to the yoke for responding to the secondary task. The experimenter station had two monitors which allowed the experimenter to see what the subject viewed, as well as mice/keyboards for controlling the simulation.

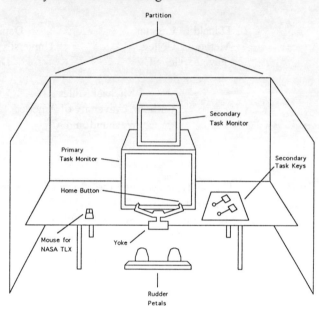

Figure 1. The subject station. The mouse was used to collect data not reported here in another phase of the study.

Procedure

Subjects participated in a total of seven two-hour experimental sessions: five practice sessions and two data collection sessions. Subjects were run one at a time, one session per subject per day. The first practice session consisted of a standardized briefing (as the subject read along), updating the subject's biographical information form, and the subject signing a consent form. Following instructions read aloud by the experimenter, the subject became familiar with the simulator apparatus and primary task software. Then the subject practiced the primary task, flying eight ten-minute courses once (four easy and four difficult courses). Before practicing each course, the subject was given a map of the course and the experimenter read aloud detailed instructions describing that course. During this first session, the software program monitoring the subject's performance did not interrupt the session if performance was not up to criterion. Feedback concerning performance was given verbally to the subject. After this first session, the subject was given maps of the eight courses and instructed to memorize them before the second session. An example of a course map is shown in Figure 2. The purpose of memorizing these courses was to have subjects use their working memory to call up and maintain a representation of the course as they flew it, therefore competing with working memory demand of the secondary task (described below). The subject's memory was tested before flying each course in subsequent sessions by having the

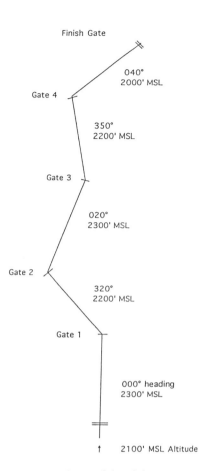

Figure 2. A map of one of the eight courses used in the study.

subject sketch the course map from memory without error. As sessions proceeded, subjects had no difficulty reproducing the maps. An easy course consisted of five two-minute splits, where at each of the four middle splits, the subject had to execute a change in heading of 60 degrees. Gates located at the positions of the course changes provided guidance, but were not visible to the subject until they were only a short distance away to insure that the subject relied on instruments and memory to fly the course. For these easy courses, all gates were at the same altitude. A difficult course was the same length as an easy course, but the subject had to make different changes in heading at each gate, as well as change altitude. Four variants of the courses within each level of difficulty were constructed by using mirror and inverted rotations. The second and third practice sessions each gave the subject ten courses to fly. From the ninth course flown in the second session onward, the software program monitoring the subject's performance interrupted and terminated the flight if the subject did not fly to criterion, and the subject was required to start that course over. The criterion was defined as flying each 20-mile course with no more than a 2000 feet deviation from ideal course, a 20 degree deviation from course (to allow for turns), a 200 feet deviation from assigned altitude, and a 10kts/hr deviation from assigned airspeed. The fourth practice session contained four additional ten-minute runs of primary task practice. Thus, over the first three and a half sessions of the

experiment, 32 total practice runs of the primary task were performed. The purpose of the extensive practice was to insure that subjects flew the primary task at criterion and could maintain that performance during the dual-task sessions of primary and secondary task performance that came later. (The secondary task must not intrude on primary task performance; otherwise, any differences in secondary task performance between single- and dual-task conditions would not be interpretable). Then, after a standardized set of instructions read aloud, the subject was given four 10-minute blocks of practice on the secondary task by itself (two blocks each of memory sets of two and four letters). The subject was given the letters to memorize right before each run. On a trial, the subject was presented a brief flash of a single letter on the secondary task monitor. While keeping the right thumb on the red "home" button located on the right handle of the flight yoke, the subject decided if the letter presented was a member of the memory set. The subject then released the home button and pressed either the "yes" or "no" telegraph key located next to the flight yoke. Subjects were instructed to make their decisions and key presses as quickly as possible. Choice reaction time (CRT) was operationally defined as the time between onset of the letter flash and release of the home button. Movement time was defined as the time between release of the home button and pressing the "yes" or "no" telegraph key. Of interest in this study was CRT. Also, accuracy of the subject's responses was recorded. An accurate response was defined as either responding "yes" when a memory set letter was presented (a "hit") or "no" when it was not (a "correct rejection"). All other response types (including infrequent non-responses) were considered incorrect responses. The percentage of correct responses for each run of secondary task performance in the study was calculated. The fifth practice session consisted of practicing the secondary task alone for two runs, then practicing the primary and secondary tasks together for four runs in a dual-task situation. These dual-task runs consisted of combining both levels of primary task (easy and difficult course) and both levels of secondary task (memory set of two and four letters). The sixth and seventh sessions were each comprised of six data collection runs like those in session five. All seven sessions utilized counterbalancing of primary task courses and secondary task stimuli across the runs to minimize possible order effects. After the last session, the subject was debriefed and compensated.

RESULTS

The data were analyzed by a mixed design ANOVA with two between subjects variables (Age and Expertise) and one repeated measures (within subjects) variable (Workload) with six levels. Age and Expertise were allowed to freely vary in the subject sample because a regression analysis had been originally planned. To perform the ANOVA, however, Age and Expertise were partitioned into two levels each using the median split technique to form four groups of subjects (young, less expertise; young, more expertise; old, less expertise; and old, more expertise). Initial analysis of the data from the entire sample revealed a virtual absence of significant findings involving expertise, perhaps because a median split was used to partition expertise into two levels - a weak manipulation. Therefore, another analysis was performed on a subset of the data consisting of those subjects with extreme amounts of expertise (i.e., least and most hours of flight time) within each age group. Within each age group, ten subjects were selected: the five with the most and the five with the least flight time. For this subset of subjects, the age range was 21 to 75 years, while the expertise range was 28 to 11,817 hours of flight time. For the groups, the mean ages and hours of flight time were: 1) young, less expertise: 27.2 years and 60.02 hours; 2) young, more expertise: 36.6 years and 5277.88 hours; 3) old, less expertise: 58 years and 217.6 hours; and 4) old, more expertise: 52.6 years and 7323.4 hours.

First, analyses were conducted to test if the secondary task intruded on primary task performance in the dual-task conditions (i.e., workload levels 3 - 6). These analyses were done because a requirement of the secondary task methodology is that the secondary task does not intrude on primary task performance. Separate ANOVAs were conducted comparing primary task performance measures in the dual-task conditions with those same measures collected during the last practice runs of performing the primary task alone. Primary task performance measures analyzed were deviations from criterion values (i.e., root-mean-square-errors, or RMSEs) for heading, altitude, airspeed, and distance off course. None of these analyses found a significant main effect for task condition (i.e., primary task alone vs. dual-task) on any of these measures, thus indicating that the addition of the secondary task did not intrude on these measures of primary task performance. Analyses were then performed to see if primary task performance varied across the different levels of workload in the dual-task conditions. Results indicated that as workload increased, distance off course (a general summary measure of primary task performance) and heading were unaffected, but altitude (F[3, 48] = 3.31, p < .03) and airspeed (F[3, 48] = 25.62, p < .001) deviations increased with workload.

Next, mixed design ANOVAs were conducted separately on CRT and percent correct responses on the secondary task for those conditions where subjects performed the secondary task alone during the last two sessions. Here, since the primary task was not performed in these conditions, Workload was equivalent to memory load of the secondary task (2 or 4 letters), and thus had two levels. For CRT, the results indicated significant effects for Age (F[1, 16] = 8.30, p < .011) and Workload (F[1, 16] = 21.12, p < .001). No other effects were observed. The presence of an age effect was expected, as the literature has repeatedly shown that age slows CRT. For percent correct responses, the results indicated only a significant effect for Workload (F[1, 16] = 10.35, p < .005). The absence of an expertise effect and interactions involving age and expertise in these analyses demonstrated that the Sternberg task was domain-general, an important finding indicating that the expert pilots in this study were not simply more capable people in terms of secondary task performance.

Next, mixed design ANOVAs with Workload as a six-level within-subjects factor were conducted separately on CRT and percent correct responses on the secondary task. For CRT, the following results were observed (see Figure 3a). Age (F[1, 16] = 11.53, p < .004), Workload (F[5, 80] = 99.14, p < .001), and Expertise (F[1, 16] = 6.84, p < .019) all had significant main effects. Also, there was a significant Expertise x Workload interaction (F[5, 80] = 4.90, p < .001), indicating that the CRT of pilots with more expertise was less affected by increases in workload. Regarding CRT, however, expertise did not significantly mitigate the effects of age on the ability to handle increases in workload, as indicated by the absence of a significant three-way interaction of Age, Expertise, and Workload (although it approached significance). For percent correct (see Figure 3b), Age (F[1, 16] = 13.32, p < .002), Expertise (F[1, 16] = 11.06, p < .004), and Workload (F[5, 80] = 16.39, p < .001) were all significant. Also, these interactions

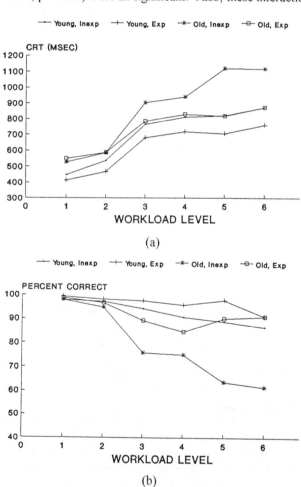

Figure 3. Secondary task (a) choice reaction time performance and (b) percent correct performance of subjects as a function of workload level. See text for subject group and workload level definitions.

were significant: Age x Workload (F[5, 80] = 4.58, p < .001); Expertise by Workload (F[5, 80] = 5.22, p < .001); and a three-way interaction of Age x Expertise x Workload (F[5, 80]

= 2.52, p < .036). This three-way interaction is discussed below. Also, repeated measures, mixed design ANOVAs with Workload as a four-level within-subjects factor (i.e., restricted to dual-task conditions) were conducted. For CRT, Age (F[1, 16] = 10.30, p < .005), Expertise (F[1, 16] = 9.09, p < .008), and Workload (F[3, 48] = 8.65, p < .001) were all significant. For percent correct, Age (F[1, 16] = 14.24, p < .002) and Expertise (F[1, 16] = 11.62, p < .004) were significant.

DISCUSSION

The goal of this study was to find evidence that expertise mitigates the affects of aging on pilot performance; specifically, a three-way interaction among age, expertise, and workload. Evidence of this interaction was provided by the analysis of subjects with extreme levels of flight expertise, with six levels of workload, for percent correct (accuracy) but not choice reaction time (although it approached significance; see Figures 3b and 3a, respectively). Figure 3b shows that: 1) the accuracy of younger subjects with more expertise was least affected by increases in workload; 2) the accuracy of older subjects with less expertise was most affected as workload increased; and 3) younger subjects with less expertise and older subjects with more expertise were moderately affected by the increase in workload. So it would seem that this subset of the data indicates that in going from a single-task situation performing a choice reaction time task to a dual-task situation performing a simulated aviation task and choice reaction time task, expertise (i.e., hours of flight time) mitigated the detrimental effects of age to some degree regarding cognitive resources involved in working memory and monitoring. This important finding coincides with those reported by Tsang and Shaner (1994) and Morrow et al. (1994). Although this three-way interaction was not found in the strictly dual-task workload manipulations, strengthening these manipulations in future studies may uncover it.

To improve the research, the following refinements will be incorporated: 1) an extreme groups design to better manipulate age and expertise; 2) stronger primary task workload manipulations; 3) more primary task practice for subjects to stabilize specific measures of primary task performance across levels of workload; 4) embedding the secondary task display into the primary task display; and 5) higher fidelity simulation to increase domain relevancy and support for domain-specific strategies used by expert pilots.

The results of this preliminary study provide a partial indication that expertise mitigates the adverse effects of aging regarding the ability to handle mental workload (regarding working memory and monitoring) in a simulation task. These findings will help "fine tune" the methodology and direct further research investigating additional cognitive resources involved in flying that may be affected by age, expertise, and workload. Ultimately this research may provide support for development of: 1) performance criteria for aging pilots to augment the age criteria currently in place in several branches of aviation; and 2) training regimens for older pilots to help them maintain their skills and certification.

ACKNOWLEDGEMENTS

This research was supported by National Institute of Aging Grant AG12388. We would like to thank John Demos for his help in data collection.

REFERENCES

Gopher, D., & Donchin, E. (1986). Workload: An examination of the concept. In K. R. Boff, L. Kaufmen, & J. P. Thomas (Eds.), Handbook of Perception and Human Performance (pp. 41-1, 41-44). New York: John Wiley & Sons.

Microsoft (1990). Flight Simulator 4.0. Redmond, WA: Microsoft Corporation.

Morrow, D. G., Leirer, V. O., & Altieri, P. A. (1992). Aging, expertise, and narrative processing. Psychology and Aging, 7, 376-388.

Morrow, D. G., Leirer, V. O., Altieri, P. A., & Fitzsimmons, C. (1994). When expertise reduces age differences in performance. Psychology and Aging, 9, 134-148.

Rybash, J. M., Hoyer, W. J., & Roodin, P. A. (1986). Adult cognition and aging. New York: Pergammon Press.

Sternberg, S. (1969). On the discovery of processing stages: Some extensions of Donders' method. Acta Psychologica, 30, 276-315.

Tsang, P. S. (1992). A reappraisal of aging and pilot performance. The International Journal of Aviation Psychology, 2(3), 193-212.

Tsang, P. S. (1995, April). Boundaries of cognitive performance as a function of age and piloting expertise. Paper given at the Eighth International Symposium on Aviation Psychology. Columbus, OH.

Tsang, P. S., & Shaner, T. L. (1994, March). Age and expertise in time-sharing performance. Paper given at the Biennial Cognitive Aging Conference. Atlanta, GA.

Wickens, C. D., Braune, R., & Stokes, A. (1987). Age differences in the speed and capacity of information processing: 1. A dual-task approach. Psychology and Aging, 2, 70-78.

Yekovich, F., Walker, C., Ogle, L. & Thompson, M. (1990). The influence of domain knowledge on inferencing in low aptitude individuals. In A. Graesser & G. Bower (Eds.), The psychology of learning and motivation: Advances in research and theory (Vol. 24, pp. 259-278). San Diego, CA: Academic Press.

AGING AND TECHNOLOGY: A DEVELOPMENTAL VIEW

J.L. Fozard

NIA Gerontology Research Center, Baltimore, MD 21224

Applications of technology to improve the living and working environment and medical care of aging and aged people define a newly developing discipline called gerontechnology. Both this field and the human factors applications to aging that are embedded in it require a developmental view of the relationship between a person and her/his environment. From a developmental viewpoint, technology can affect aging through prevention of chronic problems that limit mobility; enhancement of social activities, work, education and recreation, and compensation for impaired functioning. Integration of technology into the lives of aging persons reacquires a developmental approach to the design of products and environments, consumer involvement in design and significant changes in the infrastructure for technology development and dispersal.

AGING AND ERGONOMIC THEORY

Smith (1990) identified two broad models that have guided conceptual developments relative to aging in human factors and ergonomics theory. The first is a decrimental model (Faletti, 1984; Lawton and Nahemow, 1973) that emphasizes age associated decline in human sensory, cognitive and mobility function. The role of ergonomics is to improve the match between the declining human capacities and environmental demands; task analysis is the main tool used in determining how to achieve the match (Faletti, 1984; Clark, Czaja and Weber, 1990).

The second model emphasizes the total developmental process of aging which includes but is not limited to decline (Fozard, 1981; Fozard and Popkin, 1978). According to Smith (1990) the central ideas of Fozard's version of the developmental model would "...accommodate aspects of constancy and growth as well as decline; be sensitive to social and psychological needs as well as performance needs; and because both people and environments change, acknowledge the temporary nature of generalizations about aging." (p. 511). Smith points out that implementing this approach requires needs assessments, training and counseling as well as task analysis.

Both models are based on the systems theory that is central to human factors--the optimal utilization of human and machine capabilities to achieve best system performance. The implication of this view is that aging cannot be defined independently of the environment in which it occurs--age grading of human abilities only has meaning in reference to specified environmental challenges and supports. Accordingly the 1990 report of the National Research Council on Human Factors for an Aging population identified as the highest priority, the need for ergonomics data base on "...problems, tasks and abilities... so that task analyses can be performed where the benefit is likely to be the greatest..."

Adding Age to Ergonomic Analyses

Human factors analyses emphasize the reciprocal processes of a person receiving information from a machine and using that information to control the machine. The primary focus is on the interface between the person and the machine. Aging requires additional consideration of age differences in both the internal and external aspects of the environment that influence the interface. Important variations of the external environment include climate, lighting and acoustic factors-- all of which are age-sensitive. Important variations in the internal environment include age-related differences in organs and physiological systems that affect a person's performance, e.g., neural, cardiopulmonary, muscular-skeletal, hormonal, etc.

As indicated by Smith (1990), the developmental view adds a further dynamic to the person/environment analysis. Both technology and the environment of which it is a part are changing over time at the same time a person is aging; hence, personal aging and the epoch in time during which a person ages are interdependent. Thus, technology introduced at the present may be adapted to very differently by young and old people who have had different experiences, and, the technology in turn, may alter the course of aging itself for both the young and old.

Applications of the human factors analytic model just described sensitize the user to the heterogeneity of aging--a collection of universal processes that are certainly not uniform. Aging diminishes the similarity among coevals--people who are the same age by the calendar including identical twins. Differential exposure to disease, environmental pollutants, lifestyle choices, experiences with the built man-made environment--all combine with genetic differences to make each person's aging a very individual experience.

AGING AND GERONTECHNOLOGY

Gerontechnology is the multi disciplinary study of aging and technology for the benefit of a preferred living and working environment and adapted medical care of aging and aged people, and for care givers of elderly persons who cannot function without assistance. The term, gerontechnology, is a composite of gerontology, the scientific study of aging, and technology, the d evelopment and application of products, environments and services. The term was coined by Graafmans and Brouwers (1989), and has described in numerous publications including a book, e.g., Bouma and Graafmans (1992); Vercruyssen, et al (in press).

Many of the concepts and analytical tools of gerontechnology are the same as those used in human factors, particularly the developmental view of ergonomics described above and the application of the analysis of the person-environment interface. At the core of gerontechnology theory is the recognition of the heterogeneity of aging. As a discipline, gerontechnology is broader than ergonomics and aging inasmuch as it considers how technology oriented toward aging and aged people should be developed, dispersed and distributed (Bouma, 1992; Fozard, 1994) . It also addresses how education and knowledge transfer in gerontechnology can best be accomplished (Vercruyssen et al, in press).

Research goals in gerontechnology

Research in gerontechnology is application driven and addresses the interaction of aging and aged persons with products and their technical or built environments. Consonant with the developmental view of aging described above, gerontechnology considers both the challenges and opportunities of normal and pathological human aging. "Challenges" refers to age-related declines in physical, physiological, perceptual, cognitive, and motor processes that may limit functioning during aging. "Opportunities" refer to such positive outcomes as increased time to pursue new activities in self-discovery, e.g., artistic activities, post-retirement or second career work activities; and relationships with grandchildren and others outside of the family.

A complementary goal is to understand how age affects the extent and pattern of use of technological devices, especially new ones. Older people may evaluate and adapt to new technology on the basis of previous experience with similar devices relatively more than younger people who have not had the same experiences.

The five goals of research in gerontechnology are to provide technology to: (1) improve the way in which aging is studied; (2) prevent the effects of declines in strength, flexibility endurance, perceptual and cognitively abilities that are associated commonly with aging; (3) enhance the performance of new roles (the opportunities) provided by aging; (4) compensate for declining capabilities (the challenges) of aging and (5) assist care givers.

1. *Improve aging research.* Examples of contributions of technology to improved research include computer-based imaging of organs and tissues, signal processing of neurological events, monitoring of blood flow, noninvasive acquisition of biochemical measures.

2. *Prevention.* Technology plays a role in the primary prevention of many 'problems of the elderly' that are modifiable through long range, nonmedical interventions involving

nutrition, physical activity, strength training, behavior modification and life style remodeling which avoids exposure to chronically dangerous environmental conditions such as auditory noise, excessive alcohol and tobacco consumption, etc. The preventive role of technology includes the design of equipment to facilitate the interventions and the design of monitoring equipment that provides feedback about compliance with interventions and their effectiveness. Whether prevention is primary or secondary depends in part on the timing of the use of technology. For example, use of movement activated light switches near dangerous stairs or passages will prevent many falls from occurring. Where age-related declines have already occurred, monitoring devices may prevent additional problems.

3. *Enhancement.* Technology can enhance the performance of new roles (opportunities) provided by aging, including changes in work, leisure, living and social situations. Examples include adaptable housing that meets the differing needs of people during the life cycle of the family, user-friendly communication technology that remotely connect older persons with family and friends, technology that allows work at home, enhances the potential for artistic activities in music and art, etc.

4. *Compensation.* Examples of technology that compensate for declining abilities (challenges) of aging include products and techniques that offset the consequences of sensory losses, task redesign that speeds response time, and devices which can be operated with reduced strength and motor skill. This is the most developed area of gerontechnology.

5. *Care giver assistance.* While there a number of mechanical devices available for lifting and transporting persons who cannot move them selves, few are suitable for home use.Work related injuries particularly to the lower back occur very frequently in persons who lift and assist in the transport of less able elderly persons even when the helper has been trained in the proper techniques of lifting. The development of equipment suitable for home use deserves high priority.

Another significant development in home-based medical care is the use of complicated medical equipment for monitoring and administration of drugs by family and other nonprofessional care givers. Improvements in the ease and safety of use of existing equipment is a high priority for gerontechnological research and application.

Implementing gerontechnology

Turning gerontechnology's concepts into action require considerations of advocacy and the infrastructure for development, dispersal and distribution of technology.

Who are the advocates for gerontechnology--applications of technology directed toward aging and aged people? The heterogeneity of aging and the widespread ignoring or denying of the consequences of aging by most persons create a unique problem in advocacy.

Gerontechnology is differs from biomedical and environmental technology in the way people's interests are identified and used to shape the development of technology. In biomedical technology, physicians and scientists identify the needs of patient groups. People have a choice to accept or refuse treatments based on biomedical technology either through ethics committees or directly. In environmental technology, the needs of people are identified partly through public health , science, and public concerns for conservation of natural and human resources. Groups of people may benefit from noise reducing barriers between highways and residential areas, or reduction of air pollution. Working through elected public officials, people limit explication of natural resources and increase protection from environmental pollutants.

In gerontechnology, the needs of people are identified and articulated in two ways in the stream of events involved in technology development and dispersal. The first is feedback, the evaluations by people of available technology. Feedback requires an evaluation by users that goes beyond the success of sales, which is the most common criteria. To the extent that gerontechnology is successfully based on good scientific information about peoples needs, abilities and aspirations, feedback of such information to designers and manufacturers will also influence the development of technological application..

The second is feedforward, the identification of needs based on involvement by people in the developmental process itself--needs and preferences are identified by serving on focus groups or evaluators of prototypes. Examples of how these concepts are applied are given by Coleman and Pullinger (1993) and Pirkl (1994).

Changes in the infrastructure of the dispersal and distribution of technologically based products and services are required. In particular, client oriented approaches to the marketing of products are needed. Using a salon approach to marketing of lighting devices, vision enhancing devices and products such as television sets and computers are examples. The salon approach would be an alternative to the blend of warehouse and showroom that is emphasized in marketing today.

ACKNOWLEDGMENTS

Most of the concepts presented in this paper are shared intellectual property with many colleagues in Europe and the United States. I have tried to identify the major developers of theses ideas in the references, but the citation of references is a pale reflection of the discussions and arguments that are occurring in the development of the major concepts of gerontechnology.

REFERENCES

Bouma, H. (1992) Gerontechnology: Making technology relevant for the elderly. In H. Bouma, & J.A.M. Graafmans, (Eds) Gerontechnology. Amsterdam: IOS Press, pp. 2-5.

Bouma, H., & Graafmans, J.A.M. (Eds) (1992) Gerontechnology. Amsterdam: IOS Press.

Coleman, R., & Pullinger, D.J. (Special issue eds) (1993) Designing for our future selves. Applied Ergonomics, 24, 1-62.

Clark, M.C., Czaja, S.J., & Weber, R.A. (1990) Older adults and daily living task performance. Human Factors, 32, 537-549.

Faletti, M.V. (1984) Human factors research and functional environments for the aged. In A. Altman, M.P. Lawton, & J.F. Wohlwill (Eds), Elderly people and the environment. New York: Plenum, pp.191-237.

Fozard, J.L. (1981) Person-environment relationships in adulthood: Implications for human factors engineering. Human Factors, 23, 3-27.

Fozard, J.L. (1994). Future perspectives in gerontechnology. Report EUT/BMGT/94.689, Institute for Gerontechnology, Eindhoven University of Technology, PO Box 513, 5600MB Eindhoven, NL.

Fozard, J.L. & Popkin, S.J. (1978) Optimizing adult development: Ends and means of an applied psychology of aging. American Psychologist, 33, 975-989.

Graafmans, J.A.M., & Brouwers, A. (1989) Gerontechnology: The modeling of normal aging. Proceedings of the 33rd Annual Meeting of the Human Factors Society, Denver, CO.

Lawton, M.P. & Nahemow, L. (1973) Ecology and the aging process. In C. Eisdorfer & M.P. Lawton (Eds), The psychology of adult development and aging. Washington, D.: American Psychological Association, 1973.

National Research Council . (1990) Human Factors Research Needs for an Aging Population. Washington, D.C.: National Academy Press.

Pirkl, J.J. (1994) Transgenerational design: Products for an aging population. New York: Van Nostrand Reinhold.

Smith, D.B.D. (1990) Human factors and aging: An overview of research needs and application opportunities. Human Factors, 32, 509-526.

Vercruyssen, M., Graafmans, J.A.M., Fozard, J.L., Bouma, H., & Rietsema, J. (In press) Gerontechnology. In J.E. Birren (Ed), Encyclopedia of Gerontology. San Diego, CA: Academic Press.

The Ordering of Over-the-Counter Pharmaceutical Label Components

William J. Vigilante Jr. and Michael S. Wogalter

Department of Psychology
North Carolina State University
Raleigh, NC 27695-7801

ABSTRACT

Recently there has been increasing interest in enabling consumers to more easily acquire information from over-the-counter (OTC) nonprescription pharmaceutical labels. Standardization of the format of labels is being considered by industry, government, and health-related professional organizations as a way to facilitate their usability. Potentially standardization could assist consumers in quickly locating information that they need to use the medication safely. The purpose of the present research is to determine whether consumers have a consistent preference for the ordering of information (component headings) on OTC drug labels. If so then this could serve as a partial basis for standardization. Results showed relatively consistent orders across four drugs and three participant groups (adults attending a flea market, senior citizens, and undergraduates). In general, the data indicate that people prefer that labels first provide what the drug is used for (indications); second provide information on associated hazards (warnings, cautions, drug interaction precautions) and use (directions); and third provide information on active ingredients. The remaining components were preferred in the following order: whether the package is safety sealed, inactive ingredients, storage instructions, manufacturer information, and then finally the bar code. Given the reasonable consistent orders generated by participants it seems plausible that if standardization were implemented that the ordering would roughly reflect this basic ordering.

INTRODUCTION

In recent years, there has been a trend for consumers to take on more responsibility for their health and medical care. In accord with this, there has been increased interest in better enabling consumers to more easily acquire information from over-the-counter (OTC) nonprescription pharmaceutical labels (FDA Public Hearing, 1995). One set of proposals being considered by industry, government, and health-related professional organization is OTC label standardization. This interest derives in part from the highly successful nutrition label in the U.S. that was mandated in 1990 through passage of the Nutrition Labeling and Education Act (NLEA). The NLEA requires most food products to have "Nutrition Facts" labels (Federal Register, 1991) with a standardized content and format, e.g., placement of information, wording, serving sizes, etc. (Wogalter, Kalsher, and Litynski, 1996).

Widespread belief that standardized labels is beneficial is also apparent in the American National Standards Institute's (ANSI, 1991) guidelines for consumer product warnings, ANSI Z535.4. This standard specifies particular formats, styles, colors, and words for warning labels, based on the idea that having a consistent look will aid consumers.

What are the potential benefits of standardized labels, and in particular OTC labels? One possible advantage of a uniform format is that consumers will be able to quickly and efficiently locate the information on the label (Wogalter and Kalsher, 1994). This may be important when comparing OTC products in the store, or critical when determining in an

emergency medical situation whether a particular medication is appropriate. Consistency in format has been shown to be beneficial in other domains. For example, search speed and accuracy is facilitated by preserving information groupings across computer display panels (Tullis, 1984) and by consistent placement of commands menus and other categorized lists (Somberg, 1987). Also, standardization may help certain groups of individuals (e.g., the lay public, but most particularly, the elderly) to become more familiar with the expected location of relevant information on drug labels.

In recent testimony given to the U.S. Food and Drug Administration on OTC Drug Labeling, the American Pharmaceutical Association (APhA) (1995) focused on four categories of information for possible standardization: (1) primary use of the product; (2) dosage; (3) cautions and major side effects associated with the product's proper use; and (4) active ingredients. However, the APhA offered no recommendations on the order or format of such information on OTC drug labels. In addition, the Nonprescription Drug Manufacturers Association has proposed a standard format for OTC labels, but has offered no empirical performance data to support its utility with lay consumers.

Unlike posted warning signs which generally describe a single hazard with only a few words, OTC drug labeling usually contains substantial amounts of information. Important information can be buried in other less important information. The question addressed in the present research is how to best sequence this information so that consumers will be able to find what they need when they need it. Possibly some sort of prioritization scheme can be found based on pre-existing consumer expectations that will facilitate information search.

Recent research on the ordering of warnings in product operators' manuals offers some guidance for OTC label prioritization. Product manuals, like many OTC labels, contain substantial amounts of information. Research suggests that how safety warnings are sequenced in a list can determine the extent to which product manual warnings are read. Using focus groups, Showers, Celuch, and Lust (1992) noted that presenting obvious (already well known) warnings first in a list might deter the reading of subsequent (lesser known) warnings in the list. However they were unable to verify this finding in a subsequent experiment (Lust, Celuch, and Showers, 1995). In other recent research, Vigilante and Wogalter (1996) used an empirical procedure to determine a preferred ordering of safety warnings for various power-tool manuals based on perceived importance by lay consumers.

The purpose of the present study is determine whether a consistent ordering of components based on consumer expectations can be found for OTC drug labels. The procedure employs a technique similar to that used by Vigilante and Wogalter (1996) for the prioritization of product manual warnings Similarly, Morrow, Leirer, Altieri, and Tanke (1991) found that elders tended to group and order prescription drug information into three categories: (a) general information: with doctor's name first, medication name second, and purpose third; (b) how to take: dosage ordered fourth, schedule fifth, duration sixth, and warnings seventh; and (c) possible outcomes: mild side-effects eighth, severe side-effects ninth, and emergency information last.

It is possible that peoples' judgments of the importance of OTC label components depend on the particular drug. For some medications, the warnings and cautions may be viewed as the most important, whereas for other medications, the indications (what the drug is used for) may be viewed as the most important. Moreover, consumers' judgments may depend on demographic membership, or the situation in which the drug is taken. If so, then it might not be possible to find a consistent ordering that could be used for all OTC medications to benefit consumers under most circumstances. However, if a consistent ordering of label information is found, then the issue becomes: what is its form?

In the present investigation, information from four actual OTC drug labels was used, and three populations of consumers (adults attending a flea market, senior citizens, and undergraduate students) were sampled. They ordered label components according to four specific label-use scenarios and one general (overall) best order.

METHOD

Participants

A total of 140 individuals participated. They were composed of three subgroups. One consisted of 50 adults solicited at a flea-market in Raleigh, NC (42% females); they had a mean age of 38 (SD = 10.57) ranging from 23 to 60. They reported their highest attained educational level as follows: 6% did not complete high school, 8% completed high school, 28% had some college or trade school, 44% had a bachelors degree, 2% had some post-graduate study, 8% had a masters degree, and 4% had a doctoral degree.

A second subgroup consisted of 40 senior citizens recruited from a retirement community in Chapel Hill, NC (60% females); they had a mean age of 78 (SD = 7.35) with ages ranging from 61 to 91. They reported their highest attained educational levels as follows: 5% completed high school, 7.5% had some college or trade school, 32.5% had a bachelors degree, 15% had some post-graduate study, 17.5% had a masters degree, and 22.5% had a doctoral degree.

A third subgroup consisted of 50 undergraduate students from North Carolina State University, who received credit in their introductory psychology course (60% females); they had a mean age of 19 (SD = 1.76) ranging from 17 to 25.

Stimulus Materials

The material used as the stimuli came directly from the text of four actual (store-bought) OTC pharmaceutical products: (a) Marezine® (for motion sickness), Himmel Pharmaceuticals Inc., Hypoluxo, FL; (b) Tavist-D® (antihistamine/nasal decongestant), Sandoz Consumer Pharmaceuticals Div., East Hanover, NJ; (c) Nytol® (sleep aid), Block Drug Company Inc.; Jersey City, NJ; and (d) New-Skin® (liquid bandage), Medtech Laboratories Inc., Jackson, WY. The four drugs represent a sample of available OTC products that consumers might purchase and administer without the advice of a trained professional health-care provider. The drugs Tavist-D® and Nytol® are frequently-advertised products and are probably familiar (in name and its potential use) to most U.S. citizens, whereas the drugs Marezine® and New-Skin® are lesser known products. Informal interviews during debriefing confirmed the differences in familiarity between the two pairs of products.

Table 1 shows the headings in the order that they originally appeared on the labels. Tavist-D® and Nytol® each contained ten components while Marezine® and New Skin® only contained nine. Headings and associated textual material were re-printed in 46-point bold and 12-point regular Times font, respectively. The print size was enlarged (and held constant) from the actual drug labels so as not to introduce another confounding variable, print size. Issues associated with print size on drug labels has been investigated in other research (e.g., Wogalter and Dietrich, 1995; Wogalter, Magurno, Scott, and Dietrich, 1996). Each heading was accompanied by its associated text and printed on separate 10.2 x 15.2 cm (4 x 6 inch) cards.

Procedure

Participants first completed a consent form and then a questionnaire requesting demographic information such as gender, age, and highest educational level. Participants were told that they would be ordering a set of label components from four actual nonprescription medications. They were told to arrange these headings considering five scenarios in which they might need to consult the label.

Participants were asked to sort the heading-text cards (label components), in the best possible order, given each of the following five situations:

(1) *Purchasing*: When you are deciding whether to buy the drug;
(2) *Taking the medication*: When you are about to take the drug;
(3) *Administering to another individual*: When you are deciding whether to give the drug to another person;
(4) *Emergency*: When you are involved in a medical-crisis situation (e.g., an overdose or allergic reaction);
(5) *For all situations*: Given there will be only one order on the label and considering all possible situations that the label would be consulted.

The participants were first given one of the first four scenarios (in a randomized order for each participant) and asked to sort the cards for each of the four drugs (randomized before every scenario). After completing the ordering of components for the four drugs for one scenario, the sequence was repeated for another scenario, and this procedure continued until all drug labels were sorted with respect to the first four scenarios. The first four scenarios set the stage for the fifth judgment, always presented last, that asked for the best possible ordering for each of the drugs. After the participants sorted each drug for every scenario, they were debriefed, thanked, and released.

RESULTS

Only the data from the fifth scenario (all situations in which the label would be consulted) are presented in this

Table 1

Order of Component Headings on Actual Labels of the Four Drugs.

Marezine®	Tavist-D®	Nytol®	New Skin®
Indications	Indications	Safety Sealed	Indications
Directions	Directions	Active Ingredients	Caution
Warnings	Warnings	Inactive Ingredients	Directions
Active Ingredients	Drug Interaction Precaution	Indications	Warnings
Inactive Ingredients	Active Ingredients	Directions	Storage
Storage	Inactive Ingredients	Warnings	Active Ingredients
Manufacturer	Storage	Storage	Manufacturer
Bar Code	Bar Code	Caution	Bar Code
Safety Sealed	Safety Sealed	Manufacturer	Safety Sealed
	Manufacturer	Bar Code	

article. An enlarged description of this study with analyses for all scenarios will be available in a future report.

The orders were converted to rank scores with low numbers representing positions closer to the top of label. Table 2 shows the mean rankings for each component for each participant group separately as well as composite mean rank for all participants (using an unweighted means computation).

The component orders for each drug and participant group were first analyzed using the nonparametric multi-condition within-subjects Friedman test. All were significant, $ps < .0001$. These analyses were followed by paired comparisons among label components using the Wilcoxon Matched-Pair Signed-Rank test. Because there were as many as 36 pairwise comparisons among components for each drug, experiment-wise alpha error rate was controlled by using the Bonferroni correction technique which indicated the use of a .001 probability level for establishing significance.

Results of the Wilcoxon test for the four drugs can be found in Table 2. The headings in this table are ordered by mean rank for all participant groups combined. The subscripts following each of the components in the table indicate which components are significantly different, $p < .001$, from other components within each drug/participant grouping. Components with the same letter subscript are not significantly different. As can be seen in the table, across the three population groups and four drugs, the ordering of components is reasonably consistent. Generally, the components are arranged in the following order: (1) the Indications component was always ranked first; (2) the next set of components was personal hazard information (including Warnings, Caution, and Drug Interaction Precautions) and Directions, (3) the third grouping tended to consist of Active Ingredient, Safety Sealed, and Inactive Ingredients, and (4) lastly by separate groupings of Storage, Manufacturer, and Bar Code, in this order.

While the relative order among components did not vary much, the clusters of statistically significant differences among the components varied depending on drug and group examined. The senior citizens showed the fewest distinct groupings among the label components (indicating that they were somewhat more variable in their orderings). The flea market adults and students were less variable and most similar in terms of order and number of distinct groupings of components. The student population's orderings most closely resembled the overall (all) headings' orderings for the drugs Marezine®, Nytol®, and New-Skin®. The flea-market group most closely resembled the overall (all) heading's ordering for the drug Tavist-D®.

DISCUSSION

This study provides evidence for the existence of a preferred order of drug label components that is reasonably consistent across drugs and participant groups. If people did not have an ordering preferences, then the components would

Table 2

Label Components Ordered by Mean Rank for Each Drug and Population Group.

All (N = 140)		Students (N = 50)		Flea Market (N = 50)		Seniors (N = 40)	

Marezine®

1.63^a	Indications	1.52^a	Indications	1.70^a	Indications	1.68^a	Indications
2.63^b	Warnings	2.40^b	Warnings	2.40^{ab}	Directions	2.55^{ab}	Directions
2.81^b	Directions	3.42^c	Directions	2.68^b	Warnings	2.85^b	Warnings
4.41^c	Active Ingredient	4.40^c	Active Ingredient	4.30^c	Active Ingredient	4.53^c	Safety Sealed
4.76^c	Safety Sealed	4.58^{cd}	Safety Sealed	5.12^{cd}	Safety Sealed	4.58^c	Active Ingredient
5.59^d	Inactive Ingredients	5.66^{de}	Inactive Ingredients	5.30^d	Inactive Ingredients	5.85^{cd}	Inactive Ingredients
6.47^e	Storage	6.26^e	Storage	6.66^e	Storage	6.50^d	Storage
7.76^f	Manufacturer	7.78^f	Manufacturer	7.88^f	Manufacturer	7.58^e	Manufacturer
8.96^g	Bar Code	9.00^g	Bar Code	8.96^g	Bar Code	8.90^f	Bar Code

Tavist-D®

1.87^a	Indications	1.80^a	Indications	1.74^a	Indications	2.13^a	Indications
2.95^b	Warnings	2.58^{ab}	Drug Interaction Precaut.	2.34^{ab}	Warnings	2.58^{ab}	Directions
3.16^b	Directions	3.12^b	Warnings	2.52^b	Directions	3.25^{ab}	Drug Interaction Precaut.
3.19^b	Drug Interaction Precaut.	4.28^c	Directions	3.76^c	Drug Interaction Precaut.	3.53^b	Warnings
5.41^c	Active Ingredient	5.28^c	Active Ingredient	5.36^d	Active Ingredient	5.63^c	Active Ingredient
5.79^c	Safety Sealed	5.34^{cd}	Safety Sealed	6.30^{de}	Safety Sealed	5.73^{cd}	Safety Sealed
6.80^d	Inactive Ingredients	6.60^{de}	Inactive Ingredients	7.10^{ef}	Inactive Ingredients	6.68^d	Inactive Ingredients
7.34^d	Storage	7.26^e	Storage	7.46^f	Storage	7.30^d	Storage
8.55^e	Manufacturer	8.74^f	Manufacturer	8.42^g	Manufacturer	8.48^e	Manufacturer
9.96^f	Bar Code	10.00^g	Bar Code	10.00^h	Bar Code	9.85^f	Bar Code

Maximum Strength Nytol®

1.70^a	Indications	1.70^a	Indications	1.58^a	Indications	1.85^a	Indications
3.06^b	Warnings	2.88^b	Warnings	2.34^b	Directions	3.08^b	Warnings
3.27^b	Directions	3.20^b	Caution	3.22^c	Warnings	3.23^{bc}	Directions
3.42^b	Caution	4.24^c	Directions	3.40^c	Caution	3.73^{cd}	Caution
5.24^c	Active Ingredient	5.18^c	Active Ingredient	5.40^d	Active Ingredient	5.13^e	Active Ingredient
5.79^{cd}	Safety Sealed	5.42^{cd}	Safety Sealed	6.32^{de}	Safety Sealed	5.58^{def}	Safety Sealed
6.61^d	Inactive Ingredients	6.54^{de}	Inactive Ingredients	6.70^e	Inactive Ingredients	6.58^{fg}	Inactive Ingredients
7.49^e	Storage	7.16^e	Storage	7.86^f	Storage	7.43^g	Storage
8.36^f	Manufacturer	8.68^f	Manufacturer	8.20^f	Manufacturer	8.18^h	Manufacturer
9.99^g	Bar Code	10.00^g	Bar Code	9.98^g	Bar Code	9.98^i	Bar Code

New Skin®

1.93^a	Indications	1.70^a	Indications	1.96^a	Indications	2.18^a	Indications
2.94^b	Caution	2.82^b	Caution	2.24^{ab}	Directions	2.88^{abc}	Directions
3.06^{bc}	Directions	3.20^{bc}	Warnings	2.98^b	Caution	3.03^{ab}	Caution
3.59^c	Warnings	4.04^{cd}	Directions	3.80^c	Warnings	3.83^{bd}	Warnings
5.14^d	Safety Sealed	4.74^{de}	Safety Sealed	5.20^d	Active Ingredient	4.78^{cde}	Safety Sealed
5.20^d	Active Ingredient	5.22^e	Active Ingredient	5.84^{de}	Safety Sealed	5.18^e	Active Ingredient
6.61^e	Storage	6.52^f	Storage	6.60^e	Storage	6.75^f	Storage
7.59^f	Manufacturer	7.76^g	Manufacturer	7.44^f	Manufacturer	7.58^f	Manufacturer
8.94^g	Bar Code	9.00^h	Bar Code	8.94^g	Bar Code	8.88^g	Bar Code

Note. Pairwise comparisons between mean ranks for components of each drug and population group. Components with different letter superscripts are significantly different at the .001 level.

be ordered randomly and there would be no (or a few as a result of chance) statistically significant differences between the components.

The orderings found in this study indicate that people expect/desire labels first to tell what the drug is used for (Indications); second, to tell about the hazards associated with the drug (Warnings, Cautions, Drug Interaction Precautions) and how to use the drug (Directions); third about the chemicals involved (Active and Inactive Ingredients) and whether the container is Safety Sealed; followed by information on Storage, the Manufacturer, and lastly by the Bar Code. These results are similar to those reported by Morrow et al. (1991) which found that elderly people prefer prescription drug information to be ordered according to (1) what the product is and/or used for, (2) how the drug should be taken, and (3) warnings and hazards associated with the drug along with emergency information. As with the results found by Morrow et al. (1991) the orderings are likely to be based on peoples' mental models of how to take the drugs. Making use of consumers pre-existing cognitions when designing OTC labels is likely to benefit proper use.

Three of the drugs tested in present study (Marezine®, Tavist-D®, and New Skin®) yielded component orderings that are similar to the original orderings of the components on the drugs' original labels. The component ordering Nytol®, however, varied greatly from the label.

Given the reasonable consistent orders generated by participants it seems reasonable that if standardization were implemented that the ordering would roughly reflect this basic ordering. There are, however, other factors that may be important for OTC label standardization-related decisions that still need to be addressed in research. These include:

(1) a consideration of the size of sections relative to the label configuration and size (e.g., some sections maybe too long for a single column of text on some containers and may not fit or look right in some label arrangements);
(2) whether pictorials/icons, if any, should be included;
(3) the possible need for flexibility when a drug has critical lesser-known risks that need to be communicated to consumers;
(4) whether to use bullet-type marks to highlight main points; and
(5) how to make the tradeoff between print size and white space for label design.

Future research should be conducted to determine whether a standard preferred ordering of information does in fact facilitate information search and acquisition. For this determination, performance (e.g., reaction time and accuracy) in information search tasks could be measured.

Investigation is also needed on potential negative effects of label standardization. A potential downside of standardizing OTC drug labels is that consumers may habituate to them and not notice important safety information. Problems can also arise if a product is changed in some fashion (e.g., a revision in ingredients or dosage); consumers may become so accustomed to a particular format that they

may not notice subtle differences. Trying to communicate new information might (or might not) be more difficult with standardized labels than without.

Other potential problems with standardizing drug labels include: deciding which headings should be contained on all labels (and which might be optional), what the names of the headings should be, and where information should be placed. Of the four drugs used in this study none had exactly the same set of headings. Also, information found under one heading for a particular drug was sometimes listed under a different heading for another drug.

The problems of standardizing OTC-drug labels can be addressed through research. Label designs and formats should not be based only on information from focus groups or expert judgments but should also include evaluations from consumers using performance measures to empirically determine whether the labels are usable. Research conducted for the purpose of finding the best ways to present information is likely to benefit consumers by facilitating knowledge acquisition and prevent potential negative outcomes from inappropriate medication use.

REFERENCES

Federal Register (1991) Vol. 56, no 229, Nov. 27, Dept. of Agriculture, Food Safety and Inspection Service.

Engle, J. P. (1995). Testimony of the American Pharmaceutical Association: FDA Public Hearing on Over the Counter Drug Labeling, Sept. 29, 1995.

Lust, J. A., Celuch, K. G. and Showers, L. (1995). An investigation of the effects of placement of obvious warnings and safety warning format in product manuals. *Proceedings of the Marketing and Public Policy Conference* (Georgia State University, Atlanta), 11-20.

Morrow, D., Leirer, V., Altieri, P. and Tanke, E. (1991). Elders' schema for taking medication: implications for instruction design. *Journal of Gerontology, 46(6),* 378-385.

Showers, L. S., Celuch, K. G. and Lust, J. A. (1992). Consumers' use of product owner manuals. *Advancing the Consumer's Interest, 4(1),* 22-28.

Somberg, B. L. (1987). A comparison of rule-based and positionally constant arrangement of computer menu items. *Chi 1987 Proceedings,* 150-153.

Tullis, T. S. (1984). Predicting the usability of alphanumeric displays. Ph.D. dissertation, Rice University.

U.S. Food and Drug Administration (1995). Public Hearing on Over-the-Counter Drug Labeling, Washington, DC, Sept. 29.

Vigilante, W. J. Jr., and Wogalter, M. S. (1996). The ordering of safety warnings in product manuals. In A. Mital, H. Krueger, S. Kumar, M. Menozzie, and J. E. Fernandez (Eds.), *Advances in Occupational Ergonomics and Safety.* Amsterdam: IOS Press.

Wogalter, M. S., and Dietrich, D. A. (1995). Enhancing label readability for over-the-counter pharmaceuticals by elderly consumers. *Proceedings of the Human Factors and Ergonomics Society, 39,* 143-147.

Wogalter, M. S., and Kalsher, M. J. (1994). Product label list format of item arrangement and completeness on comparison time and accuracy. *Proceedings of the Human Factors and Ergonomics Society, 38,* 389-393.

Wogalter, M.S., Kalsher, M. J., and Litynski, D. M. (1996). Influence of food label quantifier terms on connoted amount and purchase intention. *Proceedings of the Human Factors and Ergonomics Society, 40,* in press.

Wogalter, M. S., Magurno, A., Scott, K., and Dietrich, D. A. (1996). Facilitating information acquisition for over-the-counter drugs using supplemental labels. *Proceedings of the Human Factors and Ergonomics Society, 40,* in press.

ONLINE LIBRARY CATALOGS: AGE-RELATED DIFFERENCES IN QUERY CONSTRUCTION AND ERROR RECOVERY

Sherry E. Mead[1], Brian A. Jamieson[2], Gabriel K. Rousseau[2], Richard A. Sit[1], and Wendy A. Rogers[2]

[1]Georgia Institute of Technology   [2]University of Georgia
Atlanta, GA   Athens, GA

Online library catalogs have become pervasive in today's library. Unfortunately, these systems have been developed by computer programmers or librarians with little analysis of user behavior on the system. The present study compared the search performance of younger and older adults with general computer experience who were novice online catalog users on a set of ten search tasks of varying difficulty. This study examined types of errors made by novice users in database query construction and subsequent error recovery. Younger adults achieved a higher overall success rate than did older adults and were more efficient in performing these searches. Older adults made more query construction errors and recovered from them less efficiently than did younger adults. These data have important implications for identifying the specific needs, limitations, and capabilities of online library catalog users and the design of online library catalog systems for adults of differing ages.

There is a lack of human factors studies that have focused on older adults and their use of technology, including online library catalogs. Online library catalogs provide fast access to library information from any location, using many access points, and powerful search commands. The few studies of older adults and their use of computer technology that do exist mostly have explored applications such as word processing or spreadsheets (Charness and Bosman, 1990; Zandri and Charness, 1994). A prominent finding among existing research studies is that older adults have problems with and do not use new forms of automated technology (e.g., Adams and Thieben, 1991; Dyck and Smither, 1992; Smither and Braun, 1994). Consequently, there is a need to explore and assemble basic data specifically on the needs, limitations, and capabilities of older adults with an increasingly common type of computer system in today's society, the online library catalog.

Seymour (1991) reviewed 16 studies of online library catalog users and concluded that such studies frequently employ inadequate methodology. She describes surveys used as poorly designed, samples as excessively small, and sampling methods as questionable. User observation studies employed few, if any, experimental controls. Lack of information about user goals make their results difficult to interpret.

The quantitative analysis of transaction logs and the introduction of rigorous experimental methods has been suggested by Lewis (1987) as a remedy for the limited usefulness and generalizability of earlier studies. The present study included measures of computer and database experience, and controlled for exposure to training materials, task difficulty, user goals, and system functionality. Additionally, transaction logs, which included all screens viewed and characters typed during searches, were recorded and analyzed. The selected online library catalog possesses many features common to most online library catalogs and computerized databases in general. For example, the present system allows character string matching (keyword search), specification of the field to be searched, and the use of Boolean AND and OR.

The issues that will be discussed here are database query construction errors and error recovery rates and strategies for older and younger adults. Other researchers have reported that character string matching is a difficult concept for new users to acquire (Elkerton and Williges, 1984); Boolean AND and OR can cause confusion since they are inconsistent with everyday usage (Avrahami and Kareev, 1993); and that searchers frequently employ sub-optimal cross-referencing and keyword selection strategies (Markey, 1984). Search algorithms (e.g., string matching vs. synonym matching) and the contents of specific database fields may contradict user expectations as evidenced by keywords selected and fields searched. Nearly all of these earlier studies have focused on younger adults. Sit (1994) studied 54 older adults and concluded that the categories of problems experienced by older adults were similar to those reported for younger adults. The present study will allow a direct comparison of the performance of older and younger adults.

Of interest also is whether the complex command syntax often required by mainframe computer systems poses problems for younger and older adult novice users. We will gain a better understanding of the needs, expectations, and capabilities of online library catalog users of differing ages and to identify aspects of this particular online library system in need of improved human factors design.

METHOD

Participants

Ten young adults aged 18 to 33 years ($M = 22$, $SD = 5.14$) and ten older adults aged 63 to 76 years ($M = 68.7$, $SD = $

5.08) participated in this study. Six of the younger adults and six of the older adults were women. Average educational attainment by older participants was college graduation (range: high school diploma to Master's degree). All younger adults reported having some college. Health ratings (range: 1 = poor to 4 = excellent) reported by older adults (M = 3.4, SD = .52) were similar to those reported by young adults (M = 3.6, SD = .52).

All participants passed a Snellen acuity test at a level of 20/40. Participants were able to read 10 point type on a reduced Snellen chart at a distance of approximately 20 cm.

Materials

Ability tests. The Digit-Symbol Substitution (Wechsler, 1981), the Extended Range Vocabulary Test (Ekstrom, French, Harman, and Derman, 1976) the Nelson-Denney Reading Comprehension Test (Brown, Bennet, and Hanna, 1981), and the Alphabet Span Test (Craik, 1986) were used to assess differences in perceptual speed, vocabulary, reading comprehension, and working memory capacity among the groups. Scores for younger and older adults are reported in Table 1.

Table 1. Mean ability test scores by age group.

	Older adults		Younger adults		p <
Vocabulary	30.8	(9.15)	18.2	(6.84)	.01
Digit-Symbol	48.5	(8.50)	72.5	(8.05)	.01
Reading Comprehension	24.4	(9.01)	32.6	(4.62)	.02
Reading Rate	271.3	(80.45)	252.4	(51.47)	.55

Online library system. The University of Georgia Academic Libraries Information Network (GALIN) has a command line interface and requires the user to specify a search method (e.g.,, keyword search, browse an alphabetized list, guided search) and the field to be searched. Boolean AND and OR are available. Online help consists of a menu-based help system and on-screen examples and suggestions.

A search command must take the form: Find (keyword search) or Browse (browse an alphabetized list) followed by a two-letter field code (e.g., AU = author) followed by a search term. Boolean AND or OR may be followed by another search term. A second field code must follow Boolean AND and may follow OR but is not required.

In general, search commands take the form:

Find *fieldCode searchTerm Boolean fieldCode searchTerm*

Search terms may include more than one keyword (e.g., "insomnia therapy" is a search term). Multi-keyword search terms will match a record if the keywords appear adjacent and in order in the specified field. That is, "insomnia therapy", "therapy insomnia", and "therapy for insomnia" will match different records. Spaces and punctuation (commas, hyphens) may be used interchangeably.

Computers. The system was accessed via an IBM-compatible PC with a 33 MHz 486DX CPU located in a cubicle in the laboratory and having a direct hardware connection to the IBM mainframe computer on which the system resides. Participants viewed 12 point green-on-black text on a 12 inch Super VGA monitor. Experimenters viewed the participant's display on a secondary monitor in an adjacent cubicle.

Training program. A training demonstration consistent with demonstrations given by University of Georgia reference librarians was developed. Each participant was given step-by-step instructions for performing each of four sample searches. Keyword searching of the Title, Author, Subject, and Notes fields; the use of Boolean operators; navigation commands; and accessing the online help system were demonstrated. Screen layout was explained and the location of useful on-screen information was pointed out. Database record organization and the type of information in the to-be-searched fields were described.

The paper documentation currently available in the UGA library was edited to remove references to functionality that participants were instructed not to use, specifically the browse an alphabetic list and guided search methods, changing databases, and quitting the system. The edited documentation was printed in black 12 point type on white paper, compiled into a booklet, and made available throughout the experiment. Experimenters recorded the number of references to paper documentation made by participants for each search task.

Search tasks. Ten search tasks were constructed. Each could be successfully completed via a single search command. There were three simple searches: Title search, Author search, and Subject search. A fourth search required participants to use the ALL field delimiter which searches the Notes field in addition to the Title, Author, and Subject fields. Three conjunctive searches required participants to locate records containing two target search terms in the same field (e.g., both "Frank Herbert" and "Bill Ransom" in the Author field). Three disjunctive searches required participants to locate records having one of two target search terms in the specified field (e.g., either "Susan Faludi" or "Naomi Wolf" in the Author field). The task descriptions given to participants and their optimal search commands are listed in Table 2.

Transaction log coding. In order to assess search efficiency and to categorize errors, participant command histories (transaction logs) were coded by two raters (not the experimenters). Transaction logs included each screen viewed by the participant, all command line entries, and all system messages.

Each component of a search command was coded separately. That is, if a participant entered "find au grisham, john", the search method, "find", the field code, "au", and the search term, "grisham, john", were coded individually.

Table 2. Search task descriptions and optimal search commands by search type.

Order	Task description	Optimal search command
Simple searches		
1	Look for 2 books by **John Grisham**.	f au grisham john
2	Look for a book called **Internet Navigator**.	f ti internet navigator
3	Look for 2 books about **insomnia therapy**.	f su insomnia therapy
ALL search		
4	Look for a book involving **Mount Olympus**.	f all mount olympus
Conjunctive searches		
5	Look for 3 books by **Frank Herbert** and **Bill Ransom**.	f au herbert frank and au ransom bill
7	Look for a book with the word **jelly** and the word **beer** in the title.	f ti jelly and ti beer
9	Look for a book about **abdomen muscles** and **back muscles**.	f su abdomen muscles and su back muscles
Disjunctive searches[1]		
6	Look for 3 books by **Susan Faludi** or **Naomi Wolf**.	f au faludi susan or au wolf naomi
8	Look for 4 books with the word **U-Haul** or the word **Bekins** in the title.	f ti u-haul or ti bekins
10	Look for 2 books about **Hershey chocolate** or **chocolate candy**.	f ti hershey chocolate or ti chocolate candy

[1]*Note*. The field code following OR is not required.

Each component was assigned a command type (e.g., field code, search term) and a descriptor (e.g., inefficient, target). Inter-rater agreement was 82% for descriptors and 93% for command types. All discrepancies were discussed until 100% agreement was reached.

Procedure

All participants were tested individually during a single three-hour session. The session began with vision and ability tests followed by demographic and technology use surveys.

Participants read all paper documentation. An experimenter conducted the training demonstration. Participants read over the task instructions and were informed that they had unlimited access to paper documentation and online help, but would receive no verbal assistance from the experimenter. Participants were told that all ten search tasks could be completed successfully and that they had an unlimited amount of time to do so. Breaks were encouraged.

RESULTS AND DISCUSSION

Self-reported relevant experience (from the technology use survey) is presented in Table 3. Although both age groups had computer experience, younger adults reported more relevant experience than did older adults. Therefore, performance differences between the age groups cannot be attributed exclusively to aging, per se, but may be due in part to differences in experience levels.

Scoring

Search tasks were scored as correct if the participant viewed the required number of target records (records that met the criteria given in the task descriptions) and recorded their call numbers on their answer sheets. Search commands were scored as optimal if they allowed the retrieval of target records via the minimum number of command line entries.

Table 3.
Computer, library, and other technology usage by age group.

	Older adults	Younger adults
Freq. of computer use	weekly	more than weekly
# applications used	1.8 (1.23)	5.5 (2.95)
(e.g., word processor, spreadsheet)		
# who use databases	n = 1	n = 4
Freq. of library use	monthly	more than monthly
# technologies used	2.5 (3.89)	9.6 (4.79)
(e.g., photocopier, answering machine)		

Search Success

Younger adults successfully completed more searches (M = 9.2, SD = 1.55) than did older adults (M = 7.7, SD = 1.64, $t(18)$ = 2.1, $p < .05$). Specifically, younger adults completed more conjunctive searches and were more likely to complete the search requiring use of the ALL field delimiter. Older adults completed as many disjunctive and simple searches as did younger adults (See Table 4). Difficulty of conjunctive searches was increased by the required field delimiter following Boolean AND. The ALL search task was initially interpreted as a subject search by all participants in both age groups.

Table 4.
Percent correct search tasks by age group and search type.

Search type	Older adults	Younger adults
Simple	97%	100%
ALL	10%	70%
Conjunctive	70%	93%
Disjunctive	87%	90%

Search Efficiency

Optimal search commands retrieve a list containing all the target records for a given search task and no non-target records. Thus, each search task could be successfully completed by entering a single search command.

Older adults entered more search commands per task ($M = 3.0$, $SD = .63$) than did younger adults ($M = 1.69$, $SD = .48$, $t(18) = 5.21$, $p < .0001$). Younger adults entered at least one optimal search command on a higher proportion of search tasks ($M = .85$, $SD = .21$) than older did adults ($M = .60$, $SD = .24$; $t(18) = 2.47$, $p < .01$).

Rate of search success via sub-optimal commands was similar for younger and older adults. Since young adults successfully completed 92% of tasks and entered optimal search commands on 85%, they succeeded via sub-optimal commands on 13% of search tasks. Likewise, since older adults successfully completed 77% of tasks and entered optimal search commands on 60%, they succeeded via sub-optimal commands on 17% of search tasks. Thus, older adults were only slightly more likely to successfully employ sub-optimal search commands than were younger adults.

Boolean operators were required to optimally perform 6 of the 10 search tasks. Younger adults used Boolean operators to successfully complete more search tasks ($M = 5.0$, $SD = 1.89$) than did older adults ($M = 2.3$, $SD = 2.26$, $t(18) = 2.90$, $p < .01$). Older adults used Boolean operators unsuccessfully on more search tasks ($M = 1.6$, $SD = 0.84$) than did younger adults ($M = 0.2$, $SD = 0.42$, $t(18) = 4.70$, $p < .001$). This result is consistent with previous findings that Boolean AND and OR are problematic for database searchers. Further, they are more problematic for older searchers than for younger searchers.

Search Command Errors

Older adults entered more search commands that contained errors ($M = 18.8$, $SD = 6.99$) than did younger adults ($M = 6.3$, $SD = 5.10$, $t(18) = 4.57$, $p < .001$). Older adults were also more likely to search non-target fields ($M = 9.0$, $SD = 5.33$) than were younger adults ($M = 3.3$, $SD = 3.23$; $t(18) = 2.89$, $p < .01$). This result suggests that database structure, or the type of information contained in specific database fields, is inconsistent with older user's expectations.

Older adults were more likely to enter non-target search terms ($M = 4.6$, $SD = 3.03$) than were younger adults ($M = .60$, $SD = 1.35$; $t(18) = 3.82$, $p < .005$). Search commands entered by older adults were more likely to contain syntax or typographical errors ($M = 5.4$, $SD = 4.70$) than were commands entered by younger adults ($M = 1.4$, $SD = 1.07$; $t(18) = 2.63$, $p < .02$). Character string matching algorithms may be less intuitive for older users than for younger users. As mentioned earlier, differential experience with databases and with computers in general may exaggerate these effects.

Older adults were also more likely to leave out required command components, specifically, search methods (e.g., f or find) and field delimiters, ($M = 14.0$, $SD = 4.44$) than were younger adults ($M = 1.0$, $SD = 2.31$; $t(18) = 2.33$, $p < .04$). This result suggests that complex command structure is especially undesirable for older users.

Error Recovery

Older adults successfully completed 61% ($SD = 0.21$) of search tasks that included at least one search command error. Error recovery rate for younger adults was 86% ($SD = 0.26$, $t(18) = 2.32$, $p < .04$). Older adults entered more search commands when successfully recovering from an error ($M = 2.69$, $SD = .95$) than did young adults ($M = 1.98$, $SD = .94$), but this difference was not significant ($p = .11$).

Interestingly, if their first search command failed to retrieve target records, both younger and older adults were most likely to try searching a different field. Relative to younger adults, older adults were more likely to try different search terms and Boolean operators, and to reverse the order of two-word search terms. Table 5 lists specific changes made to search commands during error recovery attempts.

Although older adults tried a wider variety of error recovery strategies, younger adults were more likely to select the correct strategy. Of course, this result is due in part to the fact that older adults made more different types of search command errors than did younger adults.

Table 5. Mean number of error recovery attempts that included specific types of changes to search commands by age group.

	Older adults		Younger adults		$p <$
Changed field	4.6	(1.71)	3.0	(2.05)	.0
Changed search method	3.8	(4.49)	1.1	(2.13)	.1
Changed search term	4.0	(1.89)	1.2	(2.30)	.0
Changed Boolean	3.0	(1.41)	1.1	(1.29)	.0
Added missing command	4.1	(4.46)	0.8	(0.79)	.0
Reversed keyword order	1.1	(0.88)	0.3	(0.48)	.0
Corrected typo	0.6	(0.97)	1.0	(1.15)	.4

CONCLUSIONS

Searching the present online library system required the entry of complex, multi-term search commands which must be typed in a specific order and must exactly match database records in order to return any useful results. The results of this study indicate that this task is more difficult for older adults than for younger adults, even when both groups of participants had some previous computer experience. Errors made by novice users suggest several system characteristics that are especially problematic: 1) Requiring an exact match between character strings entered and character strings in records. More flexible string-matching algorithms and the inclusion of dictionary and thesaurus functions may be helpful as long as they do not create new tasks for the user; 2) Requiring entry of a list of disparate command elements in a specific order. Simplicity and flexibility, in combination with informative error messages are necessary; and, 3) Requiring the user to specify a field to search. System defaults may be especially helpful for older adults.

A tendency to search non-target fields and to search for non-target terms indicates that the structure of computerized databases and the nature of computerized search algorithms may not have been obvious to the older adults who participated in this study. A combination of free-text search capabilities and improved training and documentation may make keyword search of existing databases more productive and efficient, especially for older adults. In conclusion, this study provided a better understanding of the capabilities and limitations of online library catalog users of differing ages. Further, this study made several specific design recommendations based on user performance that promise to help improve both online library systems and other online information retrieval systems.

ACKNOWLEDGMENTS

This research was funded by NIH/NIA grant #P50AG11715 to the Center for Applied Cognitive Research on Aging, one of the Edward R. Roybal Centers for Applied Gerontology.

REFERENCES

Adams, A. S., & Thieben, K. A. (1991). Automatic teller machines and the older population. *Applied Ergonomics, 22*, 85-90.

Avrahami, J. & Kareev, Y. (1993). What do you expect to get when you ask for a cup of coffee and a muffin or a croissant - on the interpretation of sentences containing multiple connectives. *International Journal of Man-Machine Studies, 38*, 429-434.

Brown, J. I., Fishco, V. V., & Hanna, G. (1993). *Nelson-Denney Reading Test*. Chicago: Riverside.

Charness, N. & Bosman, E. A. (1990). Human factors design for older adults. In J. E. Birren & K. W. Schaie (Eds.), *Handbook of the Psychology of Aging* (3rd ed., pp. 446-463). New York: Van Nostrand Reinhold.

Dyck, J. L. & Smither, J. A. (1992). Computer anxiety and the older adult: Relationships with computer experience, gender, education, and age. *Proceedings of the Human Factors Society 36th Annual Meeting*, 185-189.

Ekstrom, R. B., French, J. W., Harman, H. H., & Derman, D. (1976). *Kit of factor-referenced cognitive tests*. Princeton, NJ: Educational Testing Service.

Elkerton, J. & Williges, R. C. (1984). Information retrieval strategies in a file search environment. *Human Factors, 26*, 171-184.

Lewis, D. W. (1987). Research on the use of online catalogs and its implications for library practice. *The Journal of Academic Librarianship, 13*, 152-157.

Markey, K. (1984). *Subject searching in library catalogs: Before and after the introduction of online catalogs*. Dublin, OH: OCLC Online Computer Library Center, Inc.

Seymour, S. (1991). Online public access user studies: A review of research methodologies, March 1986 - November 1989. *Library and Information Science Research, 13*, 89-102.

Sit, R. A. (1994). *Relationships among Education, Experience/Expertise, User Performance, and User Satisfaction in Older Adult Online Public Access Catalog Users*. Unpublished master's thesis, University of Southern California, Los Angeles, California.

Smither, J. A. & Braun, C. C. (1994). Technology and older adults: Factors affecting the adoption of automatic teller machines. *Journal of General Psychology, 121*, 381-389.

Wechsler, D. (1981). *Manual for the Wechsler Adult Intelligence Scale-Revised*. New York: Psychological Corporation.

Zandri, E., & Charness, N. (1989). Training older and younger adults to use software. Special Issue: Cognitive aging: Issues in research and application. *Educational Gerontology, 15*, 615-631.

MOUSE ACCELERATIONS AND PERFORMANCE OF OLDER COMPUTER USERS

Neff Walker, Jeff Millians & Aileen Worden
Georgia Institute of Technology
Atlanta, Georgia

ABSTRACT

In general, as people age, their movement control performance gets worse. Older adults take longer than younger adults to make similar movements. In this study we compared older and younger experienced computer users on their ability to use a mouse to position a cursor. The distance of the movements and the size of the targets were varied to represent a broad range of cursor control tasks that would be used on a computer. We also investigated the effects that dynamic gain adjustment had on performance for both age groups. Our results showed that older adults are both slower and less accurate when using the mouse. There was evidence that the age-related difference in performance was greater when the target size was smaller. Some of the difference in age-related performance could be ameliorated by using a specific dynamic gain function. The results are used to discuss possible age-related computer interface design guidelines

INTRODUCTION

There is large body of literature that shows that as people age, their movement control performance gets worse. Generally, older adults take longer than younger adults to make the same movements. While there are arguments about the specific factors and their relative importance as the source of this age-related decline in performance (see Jagacinski, Liao & Fayyad, 1995 for a discussion) almost all researchers believe that a major factor in the decline is due to an increase in motor noise with age (e.g., Welford, 1981).

Motor noise is defined as random, unintentional error that occurs during the transmission of the signal to the muscles that control the movement (e.g., Fitts, 1954). Motor noise is assumed to increase with increased force of movement (e.g., Schmidt, Zelasnik, Hawkins, Frank, & Quinn, 1978). The result is that degree of accuracy of the movement endpoint decreases as the force of the movement increases. This is the noise-to-force ratio. Older adults seem to have a larger noise-to-force ratio. In order to maintain the same level of accuracy in a movement task, they must slow down their movements.

It may be that older adults have more difficulty using a computer mouse due to their increased motor noise. In our previous work (Walker, Philbin & Fisk, unpublished) we found that older adults were much slower than younger adults when positioning a cursor with a mouse. There was also some evidence that there was a critical target size, below which the older adults had great difficulty in placing the cursor on the target. This may have been caused by the underlying motor noise. While the findings of our previous work were suggestive, the study did not use a wide range of target sizes nor did it have experienced mouse users. The primary

purpose of this study was to determine if these findings would be obtained with experienced older mouse users.

A second purpose of the study was to investigate acceleration functions as a possible solution to this problem. One could implement software changes in the gain and acceleration profiles that translate mouse movement into cursor movement. In this study we evaluated the effects of different dynamic gain adjustments on older and younger computer users for a variety of cursor positioning tasks.

This study will seek to answer three questions. First, is there a critical minimum target size below which older adults cannot effectively use a mouse to position the cursor. Second, do different dynamic gain functions affect cursor positioning performance of older adults. Finally, do older adults have a larger noise-to-force ratio than younger adults and is the relationship between noise and force linear.

METHOD

Subjects

Twelve younger and twelve older adults participated in this experiment. The younger group consisted of undergraduates recruited from a Georgia Institute of Technology psychology course. Their ages ranged between 18 and 21 years (mean = 18.9). Five points of extra credit were given toward their course grade for participating in this experiment. The older adults were recruited from an organization that hosts enrichment classes for senior citizens. Their ages ranged between 63 and 79 (mean = 70.3). This group of subjects was paid $40 each for their participation. All subjects had a minimum of one year computer mouse experience. All subjects were informed that there was a $50 reward for the "best performer" in each group.

A drug questionnaire administered to both groups before the experiment revealed that none of the subjects were taking more than one drug rated to cause minimal effects on attention. All of the participants were required to have at least 20/40 natural or corrected vision.

Design

This study utilized a four factor, mixed design. The three within-subject variables were distance of movement to the target (50 pixels, 100 pixels, 200 pixels, 300 pixels, or 400 pixels), target width (3 pixels, 6 pixels 12 pixels or 24 pixels) and acceleration function (unaccelerated, low slope, medium slope and high slope). The corresponding slopes of these functions were 0.0, 0.1, 0.2, and 0.5 pixels to mouse movement ratio are shown in Figure 1. The between-subject variable was age (young and old). The two dependent measures were cursor positioning time and percentage of correct movement trials (hit rate).

Figure 1

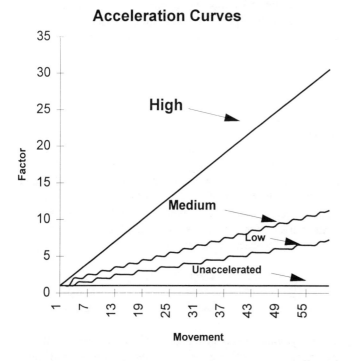

Apparatus

This study was run on PC's with 14 inch/color monitors. The input device was a Microsoft bus mouse that recorded mouse movements at 5 ms intervals. The program was run in graphics mode using 480 X 640 pixel resolution.

Procedure

The experiment was run over a maximum of 5 days. Each session lasted between 20 and 60 minutes. Some subjects participated in two session per day, but always had a 30 minute rest period between sessions. The first session was a training day. On this day, all of the subjects performed 3 correct trials with each of the distance x width x acceleration functions combinations for a total of 240 correct practice trials. For a correct trial, the subject clicked on the starting box which read "go" and moved the cursor across the screen to the target. When the target was reached a reverse video 10 X 10 pixel frame appeared and the participant released the mouse button. For an incorrect trial a beep signaled that the target was missed and the subjects repeated the trial.

Feedback for performance was given in terms of points after each trial and total points earned were given at the end of each session. Forty points were deducted for every incorrect trial. Points were also deducted or earned contingent upon the amount time it took to reach the target. The time-contingent points were based on established mean times for reaching targets. This established mean was based on previous studies that determined average times for reaching targets for young and old computer users. (Walker, et al., unpublished). Subjects that reached the target within 2.5% of the established mean did not receive or loose points. Deviations greater than 5% resulted in gained or lost points. One point was earned for every 5% deviation below the mean and one point was deducted for every 5% deviation above the mean. The point system was designed to reward speed and accuracy equally. The subjects were competing for a $50 reward that was given to the participant with the most points at the end of the experiment.

Each subject was tested on one acceleration function per session. They were presented with a different acceleration every session until all four accelerations were presented to each subject. Accelerations were assigned based on a Latin Square design to control for order effects.

Each session consisted of sixty blocks. There were five types of blocks corresponding to each target distance. There were four conditions within each block corresponding to each target width. Each type of block was repeated twelve times in random order resulting in 240 correct trials per day.

At the end of the last session participants performed 160 trials of a different task. They were instructed to aim for a target line and then to move the cursor toward the target as fast as possible and release the cursor as close as possible to the 1 pixel target without making any corrective movements. The emphasis was on the speed of the movement, with the only constraint being the aiming for the target. Subjects made 32 movements for each of the five movement distances. The purpose of this task was to generate a measure of the noise-to-force ratio. This was done by comparing mean maximum velocity of the movement (a measure of force) to the absolute value of the distance of the cursor from the target (a measure of error). The measures were based on the first submovement only.

RESULTS

The primary dependent measures that were analyzed were hit rate and mean movement time for correct trials. The mean movement time and hit rates were calculated for all trials that had the same acceleration function, movement distance, and target width for each subject. These resulting means were used in two age by distance by width by acceleration function analysis of variance.

The analysis of variance on mean movement time revealed eight significant effects. As expected, there were significant main effects of age (F $(1,22)$ = 36.88, p < .001), movement distance (F $(4,88)$ = 210.34, p < .001), and target width (F $(3,66)$ = 318.08, p < .001). In addition, there was a significant main effect of acceleration function (F $(3,66)$ = 2.76, p < .05. The interpretation of the main effects should be made in light of the significant interactions. There was a significant interaction of movement distance and age (F $(4,88)$ = 16.00, p < .001). Follow-up analyses revealed that while younger adults always had shorter movement times than older adults, the size of this difference increased with movement distance.

There was a significant interaction of target width and age (F $(3,66)$ = 12.05, p < .001). Follow-up analyses revealed that while younger adults always had shorter movement times than older adults, the difference due to age increased as target width decreased. There was also a significant interaction of movement distance and target width (F $(12,264)$ = 2.89, p < .001).

Finally, there was a significant interaction of age by target width by acceleration function (F $(9,198)$ = 2.10, p < .05). Follow-up analyses revealed that the locus of this interaction was in the effect that acceleration function had on movement time for the older adults when target width was smallest. For the older adults, movement time to the smallest targets was less when using acceleration function 2 (mean = 1773) than when using the other three acceleration functions (means for acceleration functions 1, 3, and 4 were 1882, 1870 and 1998, respectively). In general, there was a trend towards acceleration function 2 yielding shorter movement times for the older adults, especially when target width was less, while there was less of a difference in acceleration function for the younger adults.

The analysis of variance on hit rate revealed three significant main effects and a significant interaction. There was a significant main effect of acceleration function (F $(3,66)$ = 4.60, p < .01). Pair-wise comparisons revealed that hit rate was significantly lower for acceleration function 3 than for acceleration function 4.

There was a significant main effect of age (F $(1,22)$ = 7.07, p < .05). Younger adults had higher hit rates (mean = 97.2%) than did older adults (90.4%). In addition, there was a

main effect of target width (F $(3,66)$ = 28.16, p < .001). In general, hit rates went up as target width increased. However, the key to interpreting both of these main effects can be found in the interaction of age and target width (F $(3,66)$ = 9.67, p < .001). Follow-up analyses revealed (and as is shown in Figure 2) hit rates for older adults was much lower when target width was smallest. Other than this condition, age and target width had little effect on hit rate.

Figure 2

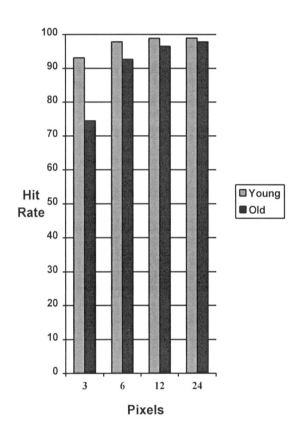

Finally, we ran regression analyses for both age groups using mean peak velocity as the predictor variable and mean standard deviation of the distance from the target as the outcome variable to test the underlying noise-to-force ratios of the two groups. Here only the data from the last 160 trials when there were no accuracy constraints were used. Regressions for both groups revealed significant equations (R^2 values of .66 and 0.62 for younger and older groups respectively). The key value was the slope functions as this gives the measure of how noise (standard deviation from the aimed for target) is affected by force (peak acceleration during movement). The slopes for the older adults were twice those of younger adults. This data supports the idea that age-related differences in the noise-to-force ratio does affect movement control performance.

DISCUSSION

A major finding of this study is that there is a critical minimum target size below which older adults cannot effectively use a mouse to position the cursor. In addition, the deleterious effects of small targets on accuracy increases significantly as distance to the target increases. Although hit rates for both age groups decreased as target size decreased a minimum critical target width and a distance-to-the-target / width-of-target interaction was not found for younger adults. These results suggest that even experienced older computer users have problems using a mouse to position a cursor.

The study also showed that different acceleration functions can be used to improve performance for older adults. When moving to smaller targets, older adults performed better with the second acceleration function. While the performance of the older adults was still worse than for the younger adults, the results suggest that adjustments of gain and acceleration may be a fruitful way to increase the usability of computer systems for older adults.

Finally, the study also showed that even experienced older computer users show a larger noise-to-force ratio when using a mouse to position a cursor. The overall results of this experiment can be used to evaluate possible solutions (e.g., use of area cursor, dynamic gain adjustment, sticky icons) to decrease the difficulty older computer users have in positioning a cursor and to begin to develop age-related interface design guidelines.

ACKNOWLEDGMENTS

This research was funded by a grant from the National Institute on Aging to the Southeastern Center for Applied Cognitive Aging Research. The center is one of the Edward R. Roybal Centers for Research on Applied Gerontology.

REFERENCES

Fitts, P. M. (1954). The information capacity of the human motor system in controlling the amplitude of movement. Journal of Experimental Psychology, 47, 381-391.

Jagacinski, R. J., Liao, M. J., & Fayyad, E. A. (1995). Generalized slowing in sinusoidal tracking by older adults. Psychology and Aging, 10, 8-19.

Schmidt, R. A., Zelasnik, H., Hawkins, B., Frank, J.S., & Quinn, J. T. (1979). Motor output variability: A theory for the accuracy of rapid motor acts. Psychological Review, 86, 415-451.

Walker, N., Philbin, D. A.., & Fisk, A. D. (unpublished manuscript). Age-related differences in movement control: Adjusting submovement structure to optimize performance.

Welford, A. T. (1981). Signal, noise, performance, and age. Human Factors, 23, 97-109.

MATCHING AERONAUTICAL DECISION MAKING (ADM) TO THE NATURALISTIC DECISION MAKING (NDM) PERSPECTIVE: HOW CLOSE IS THE FIT?

Panel Chair: Renée J. Stout
Naval Air Warfare Center Training Systems Division
Orlando, Florida

Panel Co-Chair: Clint A. Bowers
University of Central Florida
Orlando, Florida

Panelists: Clint A. Bowers, Gary Klein, Marvin S. Cohen, Judith Orasanu and Carolyn Prince

OVERVIEW

The importance of effective decision making in the aircraft has been widely recognized, with examples of disastrous consequences of ineffective decision making among crew members ever present (Hartel, Smith, & Prince, 1991). This accepted significance of decision making in aviation has led to a proliferation of training which focuses on enhancing ADM, which is evidenced by the fact that ADM training modules can be found in all of the major areas of aviation (Kaempf & Klein, 1994). While existing training on ADM may vary in specific methods and content, the core of this training still stems from a Federal Aviation Administration (FAA) advisory circular (FAA, 1991). The focus of this training is on changing pilot motivation and hazardous attitudes, as well as determining whether pilots choose correct outcomes when making decisions, which in turn is based on models from the classical decision making literature (Kaempf & Klein, 1994). In addition, while current ADM training acknowledges the importance of cognitive factors in decision making, insufficient attention is given to understanding the knowledge, skills, and cognitive processes that underlie effective decision making. Indeed, it has only been in very recent years that research has devoted attention to studying cognitive factors related to decision making in aviation (Orasanu & Fischer, in press). As such, there is a resultant dearth of specific training strategies for enhancing ADM that consider the underlying cognitive mechanisms of decision making in aviation, and much work remains to be done.

In recent years, there has been a significant advancement in the field of decision making that has broadened the scope of decision making research to include a delineation of cognitive processes that facilitate the implementation of appropriate actions. This perspective has become known as Naturalistic Decision Making (NDM), and its current favor is immediately apparent given that a new HFES technical group is devoted to this subject. Briefly, the position of NDM is that research must focus on what effective decision makers actually do, and that the available empirical evidence that has been found, when taking this approach, strongly suggests that classical decision making models have no direct relevance to real world decisions (Klein, 1989). That is, classical decision making models support that decision makers generate a list of alternative actions and optimize decision outcomes by comparing the value of each option against a set of criteria. However, data collected in a wide variety of operational settings, such as decisions required by fireground commanders, tank platoon leaders, Naval command-and-control personnel, Army officers, critical care nurses, design engineers, business employees, and jurors (see Kaempf & Klein, 1994) reveals that options are seldom compared by experienced decision makers. Instead, the importance of situation assessment has been recognized in that once a situation has been accurately assessed, the solution to the problem is evident to the experienced decision maker. This is characterized by one prevailing model of decision making that has been generated under the NDM approach, known as Recognition Primed Decision Making (RPD; Klein, 1989). Thus, the value that the NDM perspective can add to existing ADM training lies in the identification of the knowledge, skills, and cognitive processes inherent in effective expert decision making -- or what expert decision makers actually *do*. That is, once these processes are identified, training strategies can be developed to improve these processes in decision makers in aviation.

While some progress has been made in the ADM area through taking an NDM perspective (e.g., Orasanu & Fischer, in press), a question that has not been fully addressed is, what in NDM actually specifically applies or transfers to the aviation environment. On the one hand, models developed under NDM intuitively transfer to aviation, because NDM has been shown to be relevant to a variety of task domains that are characterized by factors representative of aviation as well. These features include time pressure, uncertainty, cue ambiguity, dynamic or rapidly changing conditions, stressful conditions, and potentially severe consequences of errors. On the other hand, without specific data to support the accuracy of NDM models in describing decisional processes in aviation, it is not clear whether these models fully apply. In addition, it is unclear whether NDM holds in aviation when conditions are routine and many of the above characteristics are no longer present.

Given the discrepancy between what common sense dictates regarding the appropriateness of NDM to ADM and what the available empirical data provides, advancement in this field requires a concentrated effort to address several important questions. These include: 1) what is the theoretical fit between ADM and NDM?; what are the similarities, and what are the differences?; 2) what are the implications of NDM for measurement of aviation decision making and performance?; and 3) does NDM drive the development of training strategies for enhancing ADM, or is it that training research can address issues posed by NDM and that NDM simply provides a focus for training research?

The purpose of this panel session is to begin to explore these issues by providing multiple perspectives from experts in the area of NDM, as well as ADM. Both theoretical positions and empirical evidence will be discussed to address the questions posed above. It is hoped that these questions will stimulate critical thinking and discussion among panelists, as well as audience members interested in ADM and in NDM. General discussions that will be provided by each of the presenters to begin to address these questions are further described below.

References

FAA (1991). Advisory circular: Aeronautical decision making (AC Number 60-22). Washington, DC: Federal Aviation Administration.

Hartel, C., Smith, K. A., & Prince, C. (1991, April). Defining aircrew coordination: Searching mishaps for meaning. Paper presented at the 6th International Symposium on Aviation Psychology. Columbus, OH.

Kaempf, G. L, & Klein, G. (1994). Aeronautical decision making: The next generation. In N. Johnston, N. McDonald, & R. Fuller (Eds.), Aviation psychology in practice (pp. 223-254). Brookfield, VT: Ashgate Publishing.

Klein, G. A. (1989). Recognition-primed decisions. In W. B. Rouse (Ed.), Advances in man-machine system research (Vol. 5, pp. 47-92). Greenwich, CT: JAI Press.

Orasanu, J. (in press). Finding decisions in naturalistic environments: The view from the cockpit. To appear in C. Zsambok & G. A. Klein (Eds.), Naturalistic Decision Making. Hillsdale, NJ: Lawrence Erlbaum Associates.

PANEL PRESENTATIONS

Paper Title: How Relevant is Decision Making Theory to ADM?
Author: Clint A. Bowers.
University of Central Florida.

Although improving the judgment of pilots has been an area of concern for some time, it is only recently that Aviation Decision Making (ADM) has emerged as a science unto itself. For the most part, efforts in this area have been dedicated to the development of small, intuitively derived ADM training packages. Although there are some data to indicate that these types of programs can be helpful, there have been criticisms of this state-of-the-art and it seems clear that there is a need for a program of research dedicated to the improvement of ADM. Similarly, there is a need for a theoretical framework to direct this effort. However, with a plethora of theoretical models of decision making available, it is difficult to isolate one to guide ADM research. The proposed discussion will adopt an approach to summarizing decision making models (cf. Jentsch, Vincenzi, & Bowers, 1996), such as Probabilistic theories, Naturalistic Decision Making, etc., and apply it to the problem of ADM. It is hoped that this will serve to organize and guide empirical research in this area.

Paper Title: The Intersection Between ADM and NDM.
Author: Gary A. Klein.
Klein Associates, Inc.

The decision making of pilots is not unique. Similar strategies are found in other domains. For example, Orasanu and Fischer (in press) describe three types of pilot decisions: recognitional decisions, comparisons between options, and derivation of new options. Klein, Calderwood, and Clinton-Cirocco (1985) had specified these same three categories in research with firefighters. What is unique about Aeronautical Decision Making (ADM) is the opportunity to synthesize different lines of research. No other population has been as well studied as pilots. Now there is a growing interest in decision processes to improve training and design. Early ADM research was limited to the classical viewpoint wherein pilots compared alternative options. Kaempf and Klein (1994) have suggested that the Naturalistic Decision Making (NDM) framework can help ADM investigators examine how pilots actually make decisions. Pilots are becoming the *Drosophila* of NDM research.

Paper Title: Distinctive Aspects of Aeronautical Decision Making.
Author: Marvin S. Cohen.
Cognitive Technologies, Inc.

NDM has emphasized the key role, in situation assessment and decision making, of recognizing familiar patterns and retrieving appropriate responses. In the aeronautical domain, however, simple pattern matching may not be the whole story. For example, perhaps the most distinctive aspect of that domain is the existence of multiple widely separated sources of authority and information. Pilots share decision making responsibility in different ways and to varying degrees with airline operations centers / dispatch, ground controllers, tower, TRACON, regional centers, and national flow control. Within this complex system and with the equipment currently at their disposal, pilots have difficulty constructing a complete and consistent picture of the airspace environment as it affects them, and resolving competing demands. Other aspects of aeronautical decision making, though less unique, compound this problem: Pilots must grapple with qualitatively different types of uncertainty (e.g., regarding weather, traffic in both local and large-scale portions of the airspace, clearing runways, and so on); they must balance competing goals of safety and efficiency, as well as lesser goals such as passenger convenience; and they must handle different degrees of time stress ranging from moderate (e.g., diversion due to weather) to extreme (e.g., aborting a take-off or landing). Future changes in the roles of pilots, ATC, and airline operations, such as the introduction of "free flight", may make a pattern recognition approach to pilot situation understanding and decision making even more difficult.

Paper Title: Aviation Decision Making: A Special Case of Naturalistic Decision Making or a Generic Process Model?
Author: Judith Orasanu.
NASA-Ames Research Center.

The NDM approach seeks to understand how experts make decisions in their domains of expertise. To develop a theory of Aviation Decision Making, we have analyzed 40 professional flight crews "flying" challenging scenarios in full-mission simulators. These observations were supplemented by analyses of cockpit voice recordings and accident reports by the National Transportation Safety Board, and several hundred incident reports submitted to the Aviation Safety Reporting System. The resulting decision model involves two components: situation assessment (What's the problem? How much time do I have to make a decision? How much risk is involved?) and choice of a course of action, which depends on how well specified the options are in the particular context. While rule-based decisions predominate in aviation (which is highly proceduralized), more difficult decisions involve choice from among options that differ in the degree to which they satisfy constraints or competing goals. Ideal strategies for dealing with different classes of problems provide a basis for both training and assessing ADM. Similarities and distinctions between our model and other naturalistic models will be discussed.

Paper Title: Documenting and Evaluating Aeronautical Decision Making: Does NDM Provide a Valuable Approach?
Author: Carolyn Prince.
Naval Air Warfare Center Training Systems Division.

There is no human-controlled flight that does not require decision making. In fact, it has been estimated that for a simple flight, with perfect weather and no unexpected problems, at least ten decisions must be made (Smith, 1992). Because of the context, decisions made in the aircraft can have more significant consequences than decisions made in many other situations. There is evidence, however, that decisions made in aviation cannot be solved with a single, optimal strategy. Thus, aviation decisions are prevalent and important, but varied. These three characteristics make evaluation of the aviation team's decision-making process both crucial and problematic. Crucial, because of their critical nature; problematic, because there is no single strategy to be applied. An approach to decision process documentation for feedback and evaluation is being built upon research evidence that has accrued in the naturalistic decision making literature as well as research that has been conducted in the aviation environment. This approach is described and discussed.

Distributed Team Decision Making:
Understanding the Whole as well as the Parts

Judith Orasanu, Organizer and Chair
NASA-Ames Research Center
Mail Stop 262-4
Moffett Field, CA 94035-1000
jorasanu@eos.arc.nasa.gov

Daniel Serfaty, Co-Chair
Aptima, Inc.
25 Mall Rd., Suite 300
Burlington, MA 01803
Aptima@Tiac.net

This panel will address issues relating to the nature of distributed team decision making in complex dynamic environments. Panelists will address several issues: what is meant by distributed decision making, how it differs from individual decision making, methods used to study distributed decision making, problems observed in developing shared models of the problem, methods for integrating diverse perspectives, goals, and knowledge sources, and techniques to enhance decision making by distributed teams.

Examination of public decisions that in hindsight have been poor ones reveals that practically all of those decisions were made by, or at least influenced by, a team rather than a single individual. (Consider, for example, the USS Vincennes incident or NASA's Challenger disaster.) Both of these cases illustrate central features of distributed team decision making. In both cases the situations were so complex that no single individual could manage the entire problem alone. Despite the fact that one person bore responsibility for the final decision, many specialists contributed specific information, knowledge or expertise. Competing goals associated with different stakeholder positions tugged at the decision maker. In both cases the decision maker had to rely on data that were perceived and interpreted by someone else. Interpretations had to be communicated, merged and evaluated, especially in light of differing stakes in the outcome.

At first blush one might assume that decisions made by a team would necessarily be better than decisions made by a single individual, given the additional cognitive resources. This is not always the case. Differing views can paralyze a team, working against rather than toward a solution. Teams may suffer from a collective error. Even when all team members are pulling together to solve a problem, communication of information among team members may be faulty or not timely, resulting in poor solutions. Status differences between team members may inhibit sharing of information or

appreciation of information that is offered. In more extreme cases, cultural differences between team members may interfere with accurate understanding of each other's perceptions, intentions or actions.

This panel will address the nature of distributed decision making and how it differs from individual decision making in dynamic high-risk environments. Teams offer redundancy and enhanced resources that can prevent or mitigate errors. But harnessing the potential of the augmented resources requires communication and coordination (Orasanu & Salas, 1993). Panelists will address whether there are general principles that describe successful distributed decision making across diverse domains or whether the principles are domain specific. If specific, what constrains and differentiates them? Factors that contribute to the complexity of distributed decision making relative to individual decision making will be discussed.

For this panel distributed decision making is defined as a process involving multiple participants who differ in their specific roles within an organizational structure and in the knowledge or resources available to them (see Sundstrom, DeMeuse, & Futrell, 1990). Participants are interdependent; none has sufficient knowledge or span of control to manage the problem alone. By "distributed" we mean that resources, knowledge and skills are distributed among team members who may be either physically collocated or dispersed. The problems addressed by the team may be either routine ones for which team members have received training, such as radar operators on the Vincennes, or unexpected problems for which no specific solution or training was provided (e.g., the case of the DC-10 aircraft that lost all flight controls at 39,000 ft.).

Of particular concern are problems that are encountered in complex engineered systems, where the consequences of failure can be disastrous for the decision makers themselves, as well as others. Problems faced by the decision team are dynamic, the state of the situation changing over time, with or without actions on the part of the participants. The true state of affairs may be known only vaguely through cues that are unreliable or ambiguous. Outcomes are uncertain. In some domains an adversarial relationship exists, while in others cooperation and negotiation prevail.

Concern with complex real-world problems poses severe challenges to the researcher:
• How to assess the specific problem-relevant knowledge of team members
• How to observe and characterize the decision processes involved in distributed decision making.

In addition to questions of methodology, panelists will address issues reflecting their particular domain expertise:
• How does the structure of the team contribute to effective decision making? Who has responsibility for making decisions? Is this function fixed or does it vary across types of problems within the domain?

• How does the architecture of the team influence the integration of information? Team architecture determines who talks to whom, what kind of information is passed in each direction, and the respective roles of various team members. Are certain kinds of decisions best served by particular communication structures? This issue is central to how team members develop a shared model of the problem they face.

• Are there optimal strategies for sharing information to build common problem models? How are these strategies affected by differing perspectives, expertise and goals? Social factors such as status, trust, and persuasion may influence the success of the communication process. Extent of knowledge about other team members (their capabilities and limitations) may also influence communication strategies and the ease with which coordination is achieved.

• How do distributed teams manage differences in their perception of risks or tolerance of ambiguity? How can these different positions be integrated to achieve a successful decision?

The panel will open with a framework for analyzing distributed decision making by Daniel Serfaty. He will show how it applies in the areas of Naval command and control and Army planning. Rhona Flin will discuss the problems of developing a common situation model as teams cope with emergencies on North Sea oil platforms when experts are distributed on land and on the platforms. Yan Xiao and Colin Mackenzie and will address the difficulties of diagnosing medical trauma when the consultants are remote, observing the patient and local caregivers by videotape with varying levels of commentary and restricted interaction capability. Phil Smith will address strategies for negotiation of flight plans by dispatchers and traffic managers who have different goals, constraints, and costs. Marvin Cohen will address the problem of integrating elements of an air wing who are trained separately but thrust together and must function in a coordinated fashion.

Members of the audience will be invited to share their views on the following issues:

-- methods for characterizing distributed decision making
-- methods for analyzing features that contribute to or interfere with effective decision making
-- strategies for building shared situation models
-- techniques for information integration
-- methods for coping with competing goals and perspectives
-- methods for identifying points of vulnerability in distributed decision making.

REFERENCES

Orasanu, J. & Salas, E. (1993). Team decision making in complex environments. In G. A. Klein, J. Orasanu, R. Calderwood, & C. E. Zsambok (Eds.), Decision making in action: Models and methods. Norwood, NJ: Ablex Publishers. (327-345).

Sundstrom, E., DeMeuse, K. P., & Futrell, D. (1990). Work teams: Applications and effectiveness. American Psychologist, 45(2), 120-133.

What is a Distributed Team?

Daniel Serfaty
APTIMA, Inc.
Burlington, MA 01803

The term "distributed" is widely used by both the research and operational communities. A clarification is needed to explain what is meant by the word as in distributed team decision-making. While the "distributed" property is often associated with the absence of "co-location" among decision-makers, we contend that this definition is incomplete. The concept of distribution includes: distribution of location, information, authority, resources, function, responsibility, expertise, knowledge, communication, and goals. The dimension of geographical distribution can range from complete co-location (e.g., team members facing each other around a conference table), to extreme distribution (e.g., a team made of astronauts in a space vehicle with their counterparts flight controllers on Earth). This generalization of the concept of distribution has implications beyond its theoretical merit. To enable team members to perform as a team, one must design overlaps or train the team along several of these dimensions. For example, the distribution of information is often engineered to provide optimal overlap, thereby supporting a shared mental model, through the use of common displays. Applications to team performance assessment, team design, and team training will be discussed.

Distributed Decision Making for Offshore Oil Platform Emergencies

Rhona Flin
The Robert Gordon University
Faculty of Management
Viewfield Road
Aberdeen AB9 2PW, UK

An emergency on an offshore oil installation is controlled by a distributed team, which comprises three core groups. The main team is based in the emergency control center on the platform; there is an on-scene commander and response teams out on the installation; and thirdly, there are personnel off the platform in the shore emergency response center (more than 100 miles away), as well as commanders on adjacent vessels and installations. Our efforts to study such teams in simulated emergencies have demonstrated the difficulties of analyzing distributed decision making carried out by phone and radios in this type of environment. The build up of situation awareness has a mosaic quality in such teams and the mental model of the evolving situation may only be maintained by a few key individuals. Reliance on standard operating procedures may be particularly critical for teams where inter- and intra-group communications are so highly constrained. Our current research on decision making in incident commanders continues to demonstrate the need to better understand how distributed teams can establish and maintain a sense of situation awareness during an escalating incident.

Flin, R., Slaven, G., & Stewart, K. (in press). Emergency decision making in the offshore oil and gas industry. Human Factors.

Distributed Cooperative Problem-Solving in the Interactions of Airline Dispatchers within the Aviation System

Philip J. Smith
Cognitive Systems Engineering Laboratory
The Ohio State University
210 Baker, 1971 Neil Ave.
Columbus OH 43210
Phil+@osu.edu

Elaine McCoy
Department of Aviation
Ohio University
Athens OH

Airline dispatchers are responsible for working cooperatively with flight crews, with other airline operations center staff (crew scheduling, weight and balance, etc.) and with traffic managers at the FAA facilities for traffic management (the Systems Command Center and the regional traffic management units). Most of these individuals are geographically separated, have different sources of information and have different decision support tools. In addition, their goals and priorities do not completely overlap. Data from a series of observational studies, surveys, focus groups and structured interviews will be used to develop an abstract description of these interactions, and to develop and support hypotheses about the factors contributing to successful and unsuccessful interactions of dispatchers with these other groups.

Remote Diagnosis in Trauma Patient Resuscitation

Y. Xiao and C.F. Mackenzie
University of Maryland School of Medicine
Room 5-34 MSTF
Anesthesiology Research Labs
10 So. Pine St.
Baltimore, MD 21201

Distributed decision-making requires some level of shared understanding of the situation and tasks at hand from each of the decision makers. To investigate potential obstacles to a shared understanding in dynamic tasks, we tested the ability of domain experts to make diagnoses when viewing trauma patient resuscitation through video and data links. After viewing segments of videos captured during real-life patient management, subjects in the experiment were requested to answer questions about their understanding of the patient status and team activities shown on video, and to provide intervention instructions. Results indicate a number of difficulties for the subjects (all familiar with the domain as well as with the particular work environment) to understand patient management progress and team activities. Some of the difficulties were related to the limitations in technology used in the experiment; others seemed to be generic. The latter included the cognitive overload in understanding and tracking multiple activity threads. Although distributed decision-making promises many advantages, it remains a challenge for a decision-maker to understand remote events and activities, particularly when events evolve rapidly and decisions have to be made when information is not complete.

Xiao, Y., & Mackenzie, C.F. (1966). Remote diagnosis in dynamic task environments. Proceedings of the 40th Annual Meeting of the Human Factors and Ergonomics Society. Santa Monica, CA: Human Factors and Ergonomics Society.

Interactive Decision Making among Subteams in a Distributed Environment [1]

Marvin S. Cohen
Cognitive Technologies, Inc.
4200 Lorcom Lane
Arlington, VA 22207

We are investigating cognitively based techniques for training geographically separated subteams to improve the coordinated decision making skills they will need in a full team environment. An important dimension governing the design of such training concerns the degree of interaction among the subteams for particular tasks. At one extreme (output dependence), the effect of a subteam's performance depends on other subteams' performance (e.g., in an air attack, strike depends on prior suppression of air defense). At the other extreme (input dependence), a subteam's performance depends on inputs from another subteam (e.g., fighters depend on continuous intelligence from command and control elements). This dimension helps predict the required granularity and detail of knowledge of the other subteams that must be acquired, appropriate strategies for building situation awareness, whether tasks should be initially trained in a subteam or full team context, whether demonstration/instruction is sufficient or realistic practice is required, and the sequence of difficulty that practice should introduce. Training of mental models and decision making strategies must be tailored to the character of a particular distributed environment.

(1) This work is supported by Contract No. N61339-95-C-0102 with the Naval Air Warfare Center Training Systems Division.

NEW METHODS FOR MODELING HUMAN-MACHINE INTERACTION

Raymond S. Nickerson
Tufts University
Medford, Massachusetts

Use of the term <u>mental model</u> has proliferated in the discussion of human-machine interaction. Although it seems clear that humans must depend on mental models when doing problem solving in the domain of complex systems, the literature on the topic presents a confusing variety of perspectives, and there is little empirical evidence of the structure of the models people use or of how they influence human performance. The objectives of this symposium are to (a) provide a taxonomy for mental models and suggest a theory that is intended to unify what appear now to be disparate views, (b) outline an information-theoretic method for determining the structure of complex systems, and (c) describe an application of the theory and method to a process-control simulation. In the first presentation, Moray makes the case for the need for modeling methods that can deal effectively with systems of unusual complexity. In the second, Conant describes such a method. Jamieson, in the third, reports the results of an experiment in which this method was applied.

A taxonomy and theory of mental models

Moray argues that mental models must exist -- that they are logically essential to successful performance of system control -- but that the literature regarding them is confusing. In particular, the term has been used to connote constructs varying in complexity from very simple representations with few interconnections to extremely complex representations with many elements interacting dynamically in numerous ways. Models vary in complexity, in large part, because the systems modeled do so.

Moray classifies the various uses that have been made of the mental-model concept in the literature in terms of a taxonomy that highlights type and temporal dynamics of task, human-to-system and system-to-human effects, and degree of environmental disturbance. He notes that the difference between a mental model and the reality it models must increase with the complexity of the system modeled, and that models of very complex systems must be fairly high-level abstractions in order to be manageable cognitively. Lattice theory and mapping theory are proposed as the appropriate formalisms for describing models and their relations to the realities they represent.

An information-theoretic method for revealing system structure

Noting that in order to be effective, a mental model of a complex system must be simple enough to be useable but also sufficiently representative of the modeled system to be useful, Conant describes an approach to model construction that is applicable to systems that can be assumed to be <u>nearly decomposable</u>. A nearly-decomposable system consists of subsystems between which interactions are sufficiently weak that they can be safely ignored; it is assumed that many, if not most, complex systems are of this type.

Conant's method -- Extended Dependency Analysis (EDA) -- is based on a generalization of Shannon's information theory. It begins with a conceptualization of a system as a set of N probabilistically-related "target" variables, and, for each of these variables, it algorithmically searches for the smallest of the maximally-informative set of related variables; this set (the D-set) is considered the best -- most efficient -- predictor set for the target variable of interest. The collection of D-sets, one for each of the N target variables, constitutes a decomposition of the modeled system into sets of lower-order, and presumably more readily observable, dependencies.

Using the Conant method to discover and model H-M structures

Jamieson reports an experiment in which he applied both the lattice-theory formalism, as proposed by Moray, and the EDA method developed by Conant. The problem context was provided by a process-control simulation of a thermo-hydraulic system for processing orange juice (Pasteuriser). Successful performance required reacting appropriately to the development of a leak in one of the pipes of the system.

The results, while non-quantitative, generally supported the idea that both the lattice formalism and the EDA decomposition can be useful analytical tools for studying and representing complex human-machine interaction. Analyses using these tools captured both individual differences in the models developed by the participants and the subset of variables that, across participants, constituted relatively adequate system representations.

Conclusion

The presentations support the assumption that the mental-model concept can facilitate the understanding of complex human-machine systems and that useful characterizations of such systems are within reach. From a practical perspective, a particularly interesting aspect of the tools considered is the potential they appear to provide for determining how people develop mental models that represent the dependencies that must be represented to make the controller's task manageable, without being encumbered with those that do not serve this purpose.

A TAXONOMY AND THEORY OF MENTAL MODELS

Neville Moray
LAMIH-PERCOTEC,Université de Valenciennes
Le Mont Houy, 59304 Valenciennes Cedex, France

Abstract

The work of systems researchers such as Conant and Ashby (1970) and Wonham (1976) has shown that good controllers of dynamic systems logically require internal models of the processes controlled. Since human operators are good controllers it follows that they must use mental models and hence the remaining problem is to elucidate their nature, but the literature on mental models is confused because mental models are used in situations other than manual control. A taxonomy is developed to describe the relation between different types of mental models, from those proposed by Johnson-Laird (1983) for logical problems to those proposed by Bainbridge (1991) and others for complex industrial processes. Such a taxonomy provides a coherent description of mental models which can clear up the conceptual difficulties noted by Wilson and Rutherford (1989). A brief outline is given of the application of Mapping Theory and Lattice Theory as formalisms for mental models, especially in relation to the use of Conant's Method to identify systems and models (Conant, 1976).

A TAXONOMY OF MENTAL MODELS

The notion of "mental model" has become more and more common in recent years, particularly in the field of the ergonomics of complex human-machine systems, but the literature is extremely confusing. The conceptual confusion has been clearly demonstrated in two reviews by Rutherford and Wilson (1991) and Wilson and Rutherford (1989). Mental models must exist in manual control as shown by the work of Conant and Ashby (1970) and Wonham (1976), and the success of Optimal Control Theory, which all imply the *logical* necessity of an internal model in a successful controller. Since humans are successful controllers, they needs must be using internal, that is mental, models. However, the problem is that the notion of mental models, or mental representations, has been extended far beyond closed loop manual control. One finds the concept used in human-computer interaction, in perceptual-motor skills, in supervisory control and attention allocation, in fault detection and diagnosis, and even in modelling military, economic and social intervention systems. The purpose of this paper is to provide a unifying conceptual framework, and to indicate a new approach to identifiying mental models.

Consider, for example, the difference between the use of the term "mental model" in Johnson-Laird (1983), Gentner and Stevens (1983), Bainbridge (1991), and Sheridan and Stassen (1979). (See Figure 1 and Table 1.)

In Figure 1 three aspects of a work situation or context are distinguished. The Task is defined as the particular set of operations required by the work context and containing all the specifically task-relevant information. Thus in Johnson-Laird's work the task is the reasoning problem which is set to the subjects in the laboratory, and may consist of patterns on cards, verbal descriptions written on a sheet of paper, etc..

For Johnson-Laird the information about a logical reasoning task becomes a mental model because it is turned into an internal representation, and that representation is manipulated by the person to arrive at the answer. Nothing the person does influences the information provided, (the task,) nor does the environment act either on the state of the task or the state of the person. Gentner and Stevens (1983) edited a series of studies of how people understand the causal processes in simple physical devices. The predominant flow of information and control is from the device to the person. The information thus received becomes part of the "mental model", and the contents of the mental model are then examined and manipulated to

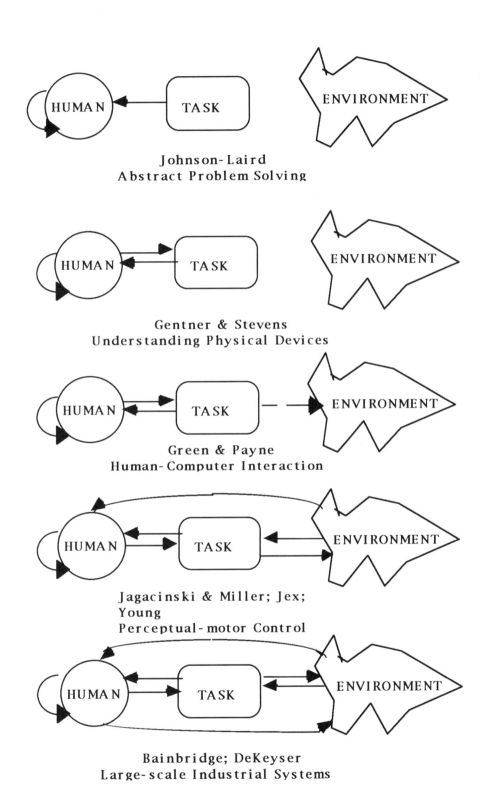

Figure 1. Coupling between mental model, task and environment in various contexts.

construct a causal account of the functioning of the device. In HCI (Schiele and Green, in press) the mental model is a generator of action, and thus in Figure 1 the main emphasis is on action generation, an arrow from the operator to the task. In the field of human factors operators of power plants, of chemical factories, of petroleum distillation plants, of economies, etc., are all dealing with immensely complex systems. There may be very many degrees of freedom (nuclear power plants have of the order of 45 degrees of freedom), hundreds or even thousands of displays and controls, and computer driven automation with hundreds of pages of displays. Plants may run for months or even years at a time without shutting down, in the face of major disturbance inputs from the environment. Such systems are far too large for an operator to be able to track the exact values of all state variables from moment to moment; they may have enormous time lags and phase lags in response to inputs; and they may have many levels of automation from manual control to autonomous robots. A disturbance which is caused by the environment may result in the loss of a major component in the plant. A disturbance may propagate through the plant generating other transients in its wake. There is tight coupling between the environment and the plant, and even tight coupling between the environment and the operator (e.g. through stress). Clearly even an experienced operator cannot be fully up to date in his or her knowledge of plant state and the desirable control actions in such large and complex systems. How then can operators manage to act efficiently? The answer, acceptable alike to engineers and human factors practitioners, is that they have a "mental model" of the plant, and this model allows them sufficient understanding of the plant for them to control it.

There seem to be five main uses of "mental models", which can be characterised by the type of task, the degree to which the human interacts with the task and the environment, the temporal dynamics of the task, and the extent to which the environment interacts with the system, (the latter comprising the human and the task or plant with which he or she interacts). These relations are summarised in Table 1.

Table 1. Relations between human, task and environment in different mental model studies

Researcher	Type of Task	Temporal dynamics	Human to system	System to human	Environment effects
Johnson -Laird	Logical Reasoning	Negligible	Negligible	Negligible	Negligible
Gentner & Stevens	Causal analysis	Negligible	Slight	Slight	Negligible
Schiele, F., Green	HCI text/ drawing	Negligible	Strong	Slight	Negligible
Jagacinski & Miller; Jex	Perceptual- motor skill	Strong but short term	Strong	Strong	Slight
Young; Stassen	Complex dynamic skills	Strong	Strong	Strong	Considerable
Bainbridge, De Keyser	Industrial H-M Systems	Very strong & long	Very Strong	Very Strong	Very strong & long

A FORMALISM FOR MENTAL MODELS

Models Are Mappings And Lattices

As we move from the logical reasoning task to industrial systems it is less and less likely that the internal representation is a complete representation of the task and its environment. Indeed as context is allowed to influence the task and we move from the laboratory to massive industrial plants or even political economies, the range of required knowledge comes increasingly to approach the domain of the natural world. A theory of models must describe the loss of information, as a mental model is formed from experience of the world, and *the canonical form of a mental model is a homomorphic mapping from one domain to another*, resulting in an "imperfect" representation of the thing modelled. The appropriate formalisms for discussing both models and relations are lattice theory and mapping theory (Ashby, 1953; Moray, 1987, 1989,1996b). An appropriate mapping onto a lattice (rather than a tree) can account for the fact that two pieces of knowledge which are quite disparate can interact. The model can be used as a base for further mappings, thus making models of models with differing amounts of detail (or abstraction) which represent parts and wholes, etc.. This can be formally represented in lattice and mapping formalisms, and corresponds in some ways to the development of Abstraction Hierarchies and Part-Whole Hierarchies (Rasmussen, 1986; Rasmussen et al., 1994; Moray, 1987, 1989).

The model can include, by such mappings, both models of knowledge (hypotheses, facts and expectancies about the task and the world), and also plans for action - models for what to do. Such mappings can be related to other theories of cognitive performance and to long and short term memory in applied psychology. The long term models waiting to be activated by task demands and intentions are sometimes called "schemas" (more correctly "schemata") by other writers. The long-term models which pop up automatically when activated by a Task or an Environment are called "frames" by Shank and Abelson (1977), and the long-term action orientated models which pop up automatically are called "scripts" by some writers (Minsky, 1975). Operators in very complex dynamic industrial systems such as nuclear or fossil power plants, oil and chemical refineries, etc. normally operate at a high level of abstraction (Rasmussen, 1986) because only thus can they handle the workload of very high degree of freedom systems with many components. They therefore use a model of their model of a model of . . . . the properties of the process.

Models can be models of other models. Furthermore, while the relation between the models and sub-models is hierarchical, the overall knowledge structure, at least of causality, is neither a tree nor a heterarchical network, but a lattice (Ashby, 1953; Moray, 1987, 1989, 1996).

IDENTIFYING MENTAL MODELS

To identify the lattice structure of an operator's model requires that we identify how the operator maps relations, particularly causal relations, from the observed process into the mind. Traditionally the most common approach has been to rely on verbal protocols (Bainbridge, 1974), or on a formal analysis of what the context must contain and which is therefore potentially the content of the model (Johnson-Laird, 1983; Schiele, and Green, in press). We wish to introduce a new method, based on Conant's method of system identification (Conant, 1976), and which is described by Conant in another paper in this symposium. An example of such an application is provided Jamieson's paper in this symposium, in which Conant's method is applied to the analysis of an operator's mental model of the PASTEURISER simulated process control plant (Lee and Moray, 1992). The proposed method does not directly identify the mental model pure and simple. Rather it identifies the nature of the human-machine *system* in which the two are coupled together as a unit. When operators interact with a complex system, they choose certain strategies of behaviour, and certain system subunits, on which to lavish attention and through which to interact with the state of the system. These strategies vary greatly between operators as was shown by Lee (1991). If we assume that the choice of strategies is based on the particular mental model which the operator has acquired, then if we can map the couplings within the human-machine system we have an indirect representation of the operator's model. Furthermore, the analysis can be performed purely on the basis of the observed interactions, behaviours, and system state variable dynamics, without the need to add extraneous verbal tasks such as verbal protocol generation.

In essence, we can first analyse the causal coupling within the system when it is running automatically using Conant's method. We then analyse the coupled human-machine system in a variety of situations. The changes we observe represent the quantitative causal addition of theoperator's mental model, from which the structure of the model itself can be deduced.

REFERENCES

Ashby, W.R. (1956). *Introduction to Cybernetics*,. London: Chapman and Hall.

Bainbridge, L. 1991. *Mental models in cognitive skill*. In A.Rutherford and Y.Rogers (eds), *Models in the Mind*, NewYork: Academic Press.

Bainbridge, L. 1974. *Analysis of protocols from a process control task*. In E.Edwards and F.Lees (eds), *The Human Operator in Process Control*, London: Taylor and Francis. 146-158.

Conant, R.C. 1976. Laws of Information That Govern Systems, *IEEE Transactions on Systems, Science and Cybernetics*, SMC-6, 240-255.

Conant, R.C. and Ashby, W.R., 1970. Every good regulator of a system must be a model of that system. *International Journal of System Science*, 1, 89-97.

Gentner, D. and Stevens, A.L. (1983). *Mental Models*. Hillsdale, N.J.: Lawrence Erlbaum..

Johnson-Laird, P.N., 1983. *Mental Models*. Cambridge, Mass.: Harvard University Press.

Lee, J.D. 1991. *Trust, self-confidence and operators' adaptation to automation*. Unpublished Ph.D. Thesis, University of Illinois at Urbana-Champaign. Engineering Psychology Research Laboratory Technical Report EPRL-92-01.

Lee, J.D. and Moray, N. 1992. Trust, control strategies and allocation of function in human-machine systems. *Ergonomics.35*, 1243-1270.

Minsky, M. (1975). *A framework for representing knowledge*. In: P.Winston, (ed) *The Psychology of Computer Vision*. New York: McGraw-Hill.

Moray, N., (1987). Intelligent aids, mental models, and the theory of machines, *International Journal of Man-machine Studies*, 27, 619-629.

Moray, N. (1989). A lattice theory approach to the structure of mental models. *Philosophical Transactions of the Royal Society of London, series B, 327*, 447-593.

Moray, N. (1996a) Mental models in theory and practice. *Proceedings of Attention and Performance*, July, Tel-Aviv, Israel. In press.

Moray, N. (1996b) *Models of models of .....mental models*. In T.B.Sheridan (editor) Liber Amicorum in Honour of Henk Stassen. In Press. Cambridge, Mass: MIT Press.

Rasmussen, J. (1986). *Information processing and human-machine interaction: an approach to cognitive engineering*. Amsterdam: North-Holland.

Rasmussen, J., Pejtersen, A.M., and Goodstein, L. (1994). *Cognitive Systems Engineering*. New York: Wiley.

Rutherford, A. & and Wilson J.R. (1991). *Models of mental models: and ergonomist-psychologist dialogue*. In , M.J. Tauber & D. Ackermann, (eds) (1991). *Mental Models and Human-Computer Interaction 2*. Amsterdam: Elsevier Science Publishers B.V. (North Holland).

Schiele, F., and Green, T.R.G. *Using task-action grammars to analyze "the Macintosh style"*. To appear in H.Thimbleby (ed), *Formal Methods in HCI*. Cambridge: Cambridge University Press.

Shank, R. & Abelson, R. (1977). *Scripts, Plans, Goals and Understanding*. Hillsdale, NJ: Lawrence Erlbaum Associates.

Sheridan, T.B. and Stassen, H.G. (1979). *Definitions, models and measures*. In N.Moray, (ed) *Mental Workload*. London: Plenum Press. pp. 219-234.

Wilson, J.R. and Rutherford A. 1989. Mental models: theory and application in human factors. *Human Factors, 31*, 617-634.

Wonham, M. 1976. Towards an abstract internal model principle. *IEEE Transactions on Systems, Man and Cybernatics*, SMC-6, 241-250.

AN INFORMATION-THEORETIC METHOD
FOR REVEALING SYSTEM STRUCTURE

Roger C. Conant
University of Illinois at Chicago
Chicago, Illinois

A generalized version of Shannon's information theory has proven useful in deducing the structure of general systems having many variables. In a system consisting of a human operator interacting with a physical apparatus, the detection of structure will presumably be useful in understanding and modeling the operator's performance. An overview of an information-theoretic technique for detecting system structure is provided.

If successful interaction of a human operator with a system requires the operator to form a mental model of the system, as we believe (Conant and Ashby, 1970), then clearly the mental model must be simple enough to be manageable by the human. If the system appears to be complex, some means must be found to construct a model which is both sufficiently simple to be useable, and sufficiently representative of the pertinent aspects of the system to be valid. When a human is faced with the task of understanding a complex system, i.e. making a mental model of it, one useful strategy is to take advantage of Simon's observation (1962) that many, perhaps most, complex systems are nearly decomposable; that is, complex systems commonly consist of subsystems, within which interactions are relatively intense and between which interactions are relatively weak. For such "nearly-decomposable" systems the high-order, complex, and holistic system relation over all variables can be approximately reduced to a number of lower-order relations, each of much smaller order and therefore easier for the human to discover and understand. Our work presumes this near-decomposability and tries to find the lower-order relations by means of a generalized form of information theory. We believe that through exposing the subrelations we will be able to gain insight into operator behavior and modeling.

For purposes of this paper, our definition of "system" is that a system consists of a set of N variables, together with a probabilistic relation between them. The data used by our method is the probability distribution inferred from observing the N variables of the system over many sequential points in time. For example, in the experiment reported (Jamieson, 1996), 11 variables were observed over about 300 sequential points in time; the matrix of observational data implicitly represents the system relation.

Since initial work along this line (Ashby, 1964; Conant, 1972) much progress has been made in the development of algorithms to detect subrelations embedded within a complex relation. The current tool, labeled Extended Dependency Analysis (EDA) (Conant, 1988) and generously referred to by my colleagues as the Conant method, is an algorithm which uses information theory techniques to detect subrelations of a complex system, such that the collection of these low-order subrelations equals or approximates the holistic relation of the system, when this is possible (Ashby, 1964). There are some relations which cannot be decomposed into lower order subrelations, such as the parity relation used in computer error coding, but such non-decomposable relations seem to be rare in nature. In terms of probabilities, EDA is a technique for reducing a high-order probability distribution over all N variables of a system into a set of N lower-order distributions which in the aggregate are equivalent to it, within acceptable tolerance. It may be thought of as a means to find N low-order probability distributions which when suitably combined would yield an order-N distribution having the same statistics as that of the original system, and therefore preserving information about its dynamics. The decomposition of the system relation represented by the results of EDA may be viewed as a form of enlightened reductionism - enlightened because the reduction to subrelations is entirely quantitative and objective.

Although this is not the place to present the details of EDA, a summary is in order here. Information theory (Shannon, 1951) was motivated by a desire to make two variables, a sent message and a received message, be as closely correlated as possible - to minimize their probabilistic independence, in other words. It was soon realized by McGill (1954) and others that information theory could be generalized easily to quantify the strength of relationship not only between two variables, as in the original version of information theory, but more generally between N variables, and therefore that N-dimensional (or "generalized") information theory could be used as a tool for investigating relationships in complex systems containing many variables (Ashby, 1965; Conant, 1972). In EDA, information theory is employed to answer the question, for each of the N system variables, "Which set of variables carries the most information about this 'target' variable?" sin8ce the variables in that set are those most closely linked to the target variable and most likely to be in a subrelation with it, in the decomposition of the system into parts. In other language, EDA asks: which set of variables is the best predictor of the target variable?

More technically, for each of the N variables, EDA finds a set of related variables, called the Dependency set or D-set for that target variable, having three properties: (1) the D-set is a significantly better predictor of the target than any

smaller set; (2) no larger set of variables is a significantly better predictor than the D-set; (3) no other set of the same size is a better predictor than the D-set. In other words, the D-set is the smallest set of variables which is maximally informative about the target variable. Any smaller set is worse, and no larger set or other set of the same size is any better. Although it is tempting to call the D-set the "best" set, we can't quite do that because sometimes there are several sets equally well qualified. The use of the word "significance" above is significance in the chi-squared sense. It was shown by Miller (1955) that the chi-squared statistic is intimately related to the information theory quantity variously called transmission, transinformation, or mutual information, which is the key quantity calculated internally by the EDA algorithm. It represents the quantity of information conveyed about the target variable by the D-set, and therefore is a measure of the strength of the relation.

The D-set for a target variable is constructed by finding the single variable which maximizes the transmission between the target and one other variable, then finding the pair of variables which maximizes the transmission between the target and any two variables, then finding the best triple of variables, and so on. These successively larger sets do not necessarily include all the variables in their one-order-lower predecessors, although they usually do. The chi-squared test is employed to discover whether these sets of increasing size provide significantly more information than their predecessors, and the D-set is selected at the point of diminishing returns using this criterion. For small systems the D-sets can be discovered by the exhaustive search of all pairs, triples, etc., but for systems with more than a small number of variables, exhaustive search is computationally infeasible, because the number of pairs, triples, etc rises exponentially with the number of variables in the system. In this case various heuristics must be employed to extend the analysis (putting the E in EDA). A full description of the algorithm and heuristics is available elsewhere (Conant, 1988).

Given N variables, EDA derives a collection of D-sets, one for each target variable, and therefore [approximately] decomposes a high-order system relation into a collection of relations each of lower order. For example in the present instance (Jamieson, 1996), the 22-variable dynamic system relation (involving 11 state variables measured at time t, and the same 11 variables measured some time-lag later) was decomposed into 11 subrelations each involving one time-lagged target variable and its associated D-set of variables at time t, with D-sets typically containing two or three variables. Taken collectively these 11 low-order relations are suggestive of the causal relations in the system. Another analysis by EDA, with no time lag between the target and D-set variables, broke down the 11-variable static relation into 11 lower-order relations.

The EDA algorithm, which runs on a microcomputer, has been tested, with satisfactory results, on systems with up to 200 variables. It has always performed well in tests where the dynamics of the system were known a priori. Nevertheless when EDA is applied it is prudent to compare the EDA results with known relations in the system. In the current investigation EDA reveals subrelations which "make sense" and are plausible according to what is known of the system, for example showing that the inflow rate (a random input variable, in the experiment) is not causally influenced by any other variable, while the fluid volume in the vat at time t is a strong predictor of the volume at time t+1. The fact that known relations are revealed by the EDA analysis gives us confidence that the other relations exposed are also valid.

It is inevitable that when measurements are made on a system which is being controlled by a human operator, the operator has an influence on the dynamics of the system; indeed, the human is an integral part of the system. In the current application in which the system variables were state variables such as flow rates and vat volumes, the human operator is not explicitly visible; rather, the presence of the operator is implicit in the relations which reside in the system when the

operator is running it. Although the physical laws of behavior of the Pasteurizer machinery being controlled are fixed, different operators with supposedly different mental models employ different strategies to control it, so that for each operator there is a different order-N system relation and consequently a different decomposition. The EDA analysis therefore makes it possible to investigate hypotheses about operator behavior, for example a hypothesis that good operators succeed by reducing the system behavior to simpler form (fewer important relations, and of lower order) than poor operators. Although the investigation reported illustrates system decomposition in a system of 11 variables, another investigation in progress on the same collected data involves 27 variables, many of which are "request for information by operator" variables. The 27-variable investigation may make the role of the operator more visible and lead to more insights about the interaction of the operator with the physical apparatus being controlled. We believe that the preliminary 11-variable investigation reported has been valuable in confirming the applicability of EDA to the study of mental models.

References

Ashby, W.R. (1964). Constraint analysis of many dimensional relations, General Systems Yearbook, Vol. 9, 99-105.

Ashby, W. R. (1965). Measuring the internal information exchange in a system. Cybernetica, Vol. 8, No. 1, 5-22.

Conant, R., and W. R. Ashby (1970). Every good regulator of a system must be a model of that system. Int. Journal of Systems Science, Vol. 1, No. 2, 89-97.

Conant, R. (1972). Detecting subsystems of a complex system. IEEE Trans. Systems, Man and Cybernetics, Vol. SMC-2, No. 4, 550-553.

Conant, R. (1976). Laws of information which govern systems. IEEE Trans. on Systems, Man, and Cybernetics, Vol. SMC-6, No. 4, 240-255.

Conant, R. C. (1988). Extended Dependency

Analysis of large systems, Part I: dynamic systems; Part II: static systems. Int. J. General Systems, 14, 97-141.

Garner, W. (1962). Uncertainty and Structure as Psychological Concepts, Wiley, New York.

Jamieson, G. (1996). Using the Conant method to discover and model human- machine structures. Proc. Human Factors and Ergonomics Society, Philadelphia, Sept. 2-6, 1996, submitted.

Klir, G.J. (1986). Reconstructability analysis: an offspring of Ashby's constraint theory. (Third W. Ross Ashby Memorial Lecture.) Systems Research, Vol. 3, No. 4, 267-271.

Klir, G. J. (ed.) (1981). Special Issue on Reconstructability Analysis, Int. J. of General Systems, Vol. 7, No. 1.

McGill, W. (1954). Multivariate information transmission. Psychometrika, Vol. 19, 97-116.

Miller, G. (1955). Note on the bias of information estimates. In Information Theory in Psychology, ed. H. Quastler, The Free Press, Glencoe, Ill.

Moray, N. (1989). A lattice theory approach to the structure of mental models. Philosophical Transactions of the Royal Society of London, Vol. B327, 577-583.

Shannon, C.E. and W. Weaver (1951). The mathematical theory of communication. Urbana, IL: Univ. of Illinois.

Simon, H. (1962). The architecture of complexity. Proc. Amer. Phil. Soc., Vol. 106, 467-482.

USING THE CONANT METHOD TO DISCOVER AND MODEL HUMAN-MACHINE STRUCTURES

Greg A. Jamieson

Department of Mechanical and Industrial Engineering
University of Illinois at Urbana-Champaign

ABSTRACT

The term Mental Model has been used to account for human performance in a variety of domains. Cross disciplinary interpretations of the term have appeared incompatible with each other, threatening to render it vacuous. Moray (1988, 1990, 1991, in press) proposes lattice theory as a comprehensive formalism of the structure of Mental Models, assimilating the apparently disparate interpretations. To date, there has been little objective, quantitative support for mental models of complex human-machine systems. The Conant (1972, 1976, 1988) Method of Extended Dependency Analysis (EDA) suggests a viable quantitative means of discovering human-machine structures, a crucial step toward an empirical validation of the Mental Models concept. Results of a process control experiment suggest that a complex system can be reduced to a manageable number of key variables. Further, successful operators identify these variables and demonstrate patterns of information flow which exploit crucial system relationships.

INTRODUCTION

The term 'mental model' has accumulated a number of distinctly different interpretations. Take, for example, the influential work of Johnson-Laird (1983). In this interpretation, information necessary to carry out a mental task is referred to as a mental model because it is turned into some internal representation that can be manipulated to explore alternative solutions. Compare this with the collection of papers edited by Genter and Stevens (1983), in which mental models are incorporated in developing causal accounts of the functioning of simple devices. These perspectives contrast the use of the term in Edwards and Lees (1974) and the field studies of Bainbridge (1992). In these cases we see mental models applied to more complex systems, specifically industrial process control (our present concern). Often, an individual's connotation of the term is a function of discipline. Wilson and Rutherford (1989) elucidate differing perspectives of engineers and psychologists.

More recently; however, Moray (in press) suggests that the various usages of the term 'mental model' are more coherent than disparate and that their intentions can be unified in the context of the lattice theory formalism (Moray, 1988, 1990, 1991). Drawing from Rasmussen's (1986) proposal that humans aggregate detailed properties of a complex system into less detailed representations, Moray developed a model based on Ashby's (1956) premise that such representations are homomorphs of the real system.

If operators of complex systems indeed use mental models, the lattice theory formalism suggests that they should be able to apply past experience to new situations by drawing on learned causal relationships between system components. Extrapolating these relationships should allow prediction of, and planning for, future states of the system. In cases where those causal relations no longer apply (e.g. component failures), the structure of the lattice should manifest itself in impaired performance of a localized nature. More specifically, it is expected that operators will identify a few variables that give appropriate (if partial) knowledge with which they can monitor the system instead of sampling all of the state variables. Thus, under conditions in which the learned causal relations hold, operators can reduce their workload without sacrificing safety or efficiency (Moray, 1991).

These predictions follow from the theory. However, there has been no analytical means of assessing how the less detailed representations of system properties are truly organized. Further, no clear method for measuring the operator's behavior in conjunction with such systems has been provided. These shortcomings are further complicated by the fact that operator control strategies are inextricably linked with the systems they control. Conant's (1988) Method of Extended Dependency Analysis (EDA) tackles both problems simultaneously.

Conant (1972, 1976, 1988) demonstrates that it is possible to identify the decomposition of a system into subsystems by calculating the Shannon information transmission (Shannon and Weaver, 1951) between elements (T) and normalizing those values by measuring the entropy of the variables (H). Variable interaction within the subsystem tends to be strong whereas interactions between subsystems is weak.

Careful grouping of variables into subsystems affords the observer an opportunity to reduce the complexity of such a system. Moray (1991) hypothesizes that, given sufficient experience with the system, the operator will discover these decompositions and incorporate them as a mental model.

METHODS

Subjects: Thirty-two (4 female, 28 male) undergraduate engineering majors participated in this study. Each operator was paid a flat sum of $15.00.

Experimental Environment: The Pasteuriser process control simulation (see Figure 1) produces a laboratory sampling of the complexities that challenge operators of complex systems (cf. Woods, 1988). The global characteristics of the simulation correspond to Brehmer's (1992) classification of a 'microworld'.

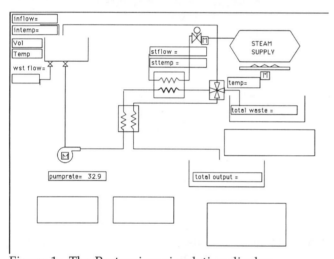

Figure 1: The Pasteuriser simulation display.

The Pasteuriser is a standard thermo-hydraulic system. Operators must establish and maintain a dynamic balance between the interconnected mass (orange juice) and energy (steam) cycles in order to produce mass at a given energy level. The process is governed by realistic heat transfer equations and thermodynamic properties. The simulation was created using QuickBasic code on a Macintosh LCIII.

A simulated pipe leak in the mass flow system was introduced at the start of post training trials. Under fault conditions, 20% of the fluid displaced by the feedstock pump is lost from the system. The fault is located such that assimilation of multiple state variables is necessary to detect the problem.

Procedure: Subjects were given a description of the physical plant and detailed instruction on how to enter commands to change the settings of a pump,

valve, and heater. As operators, they were charged with three goals: 1) Maximize the amount of juice pasteurized. 2) Keep the input vat from running dry. 3) Minimize waste. Each subject participated in three hours of operation of the simulation; one hour on each of three consecutive days. Day 1 consisted of six eight minute trials under normal conditions following instructions. Day 2 consisted of three eighteen minute trials, again under normal conditions. On Day 3 subjects completed three eighteen minute trials under the fault condition.

RESULTS

The results of a Conant Method analysis for two representative subjects are given (Figures 2-9). Both subjects were presented with a mass fault on Day 3 of the experiment. One (Subject A) performed well under both no-fault and fault conditions. The other, (Subject B) performed well under normal conditions, but faltered when presented with the fault. The intention is to demonstrate the usefulness of the Conant Method in describing human-machine systems.

The results of the analysis show that, for any given element in the system, the Conant Method will identify the subset of elements which carry the most information about that variable over the course of the trial. The Conant Method includes tests of the significance of information transmission based on the Chi-square statistic at $p<0.01$ level of confidence. In addition, the method successfully identifies the set of variables at time t which contain the most information about a target variable at time t+τ ($p<0.01$). τ values of 1, 2, 4, 8, 16 and 32 iterations (1 iteration=1.8 seconds) have been utilized.

The Conant Analysis itself is a quantitative information-theoretic approach. However, a quantitative means of representing the results has not yet been fully developed. A graphical approach has facilitated a number of preliminary qualitative observations.

Static Decompositions: Figures 2 through 5 are static decomposition diagrams for the representative subjects. These diagrams show, over the course of an 18 minute trial, which variables provide the most information about the target variable in question. Thus, for subject A in the normal condition (Figure 2), VATVOL and PUMPRATE contain the most information about the state of OVERFLOW. This is indicated by arrows leading from VATVOL and PUMPRATE to OVERFLOW. This relationship accords well with the physics implied by Pasteuriser; one would anticipate that the rate of overflow from a vessel of fixed volume would depend upon the current level of fluid in the vessel and the rate at which fluid is being removed. A connecting line with an arrow on each end indicates that each connected variable was a predominant source of information about the other.

A comparison of Figures 2 & 3, and 4 & 5, reveals differences in static decomposition within subjects between conditions (normal and fault). A given target variable may have one set of multiple determining variables under normal conditions but a different set under fault conditions. This is to be expected as the system couplings change when a fault is introduced. However, the decompositions also differ between subjects within condition (cf. figures 2 & 4; figures 3 & 5). Recall the physical plant that each operator is working with under both conditions is identical.

Causal Decompositions: The time lagged decomposition indicates which state variables contain the most information (at present) about the target variable at τ iterations in the future. Thus, for Subject A in the normal condition (Figure 6) at a time lag of 1 iteration, PUMPRATE and 3WAY contain the most information about PASTEUR. Once again this corroborates the physics of the system; one would expect that the volume of successfully pasteurized juice depends upon the rate at which product is being cycled and the temperature of that product when it reaches the valve that directs its output.

Note that this diagram is essentially 6 'static type' diagrams combined. The decomposition at any iteration is independent of the decomposition at all other iterations. To conserve space dots were used instead of repeating the name of each state variable at each interval. To read the causal diagram, look for any line drawn left to right from one variable to another (including itself). This indicates causality with respect to time t=0. Thus, in Figure 6 there is a line running downwards from VATVOL at t=8 to PASTEUR at t=16. This means that the value of VATVOL at t=0 affects the value of PASTEUR 16 seconds later (<u>NOT</u> that the value of VATVOL at t=8 affects PASTEUR at t=16). By contrast, in the same figure, the setting of the 3 way valve affects its own setting until t=2, does not affect it at t=4, but does affect it again at t=8.

As with the static decompositions, the same differences within subjects between fault conditions and between subjects within fault conditions are evident. The causal decompositions are different for operators working on identical system. It is also very interesting to note that, at short time intervals, the state variables often include themselves as the variable or one of the variables containing the most information about its state. In other words, if an operator wants to know what the state of VATVOL

will be next iteration, the Conant Analysis indicates she should look at VATVOL now.

At longer time intervals there appears to be a resolution of the diagram. It becomes apparent as the time lags increase that there are 4 or 5 variables which contain information about nearly all of the other variables. Despite the differences noted between subjects, this trend is generally consistent. The decomposition appears to be identifying a subset of system variables which contain a significant amount of information about the rest of the system.

DISCUSSION

It appears that the Conant Method of Subsystem Decomposition effectively transfers to human-machine systems, identifying the static and time lagged decompositions for each system element. The lagged aspect of the analysis is an indication of the causal structure of the system. If an operator is aware of a reduced set of variables which carry the most information about the future value of any target element, that operator need only attend to those crucial variables in order to have a fairly reliable understanding of the state of the system as a whole. Such an organization bears a strong resemblance to Moray's lattice theory formalism of a mental model. But our perspective is clouded.

The Conant Decompositions presented here consider only values of the state variables over time. Thus, we might anticipate that, if the method divulges the subsystem decomposition, the result would be identical for identical plants. This does not appear to be the case. It seems quite evident that the behavior of the operator affects the decomposition of the system; there is no other agent to initiate a change. No operator control action can affect the algorithms which govern the relationships between the state variables (i.e., the physical plant). However, the control actions of a subject could affect which variables drive the target variable, in some cases decoupling certain system elements. As suggested by Vicente (personal communication, December 7, 1995), in terms of patterns of information flow, the Conant method provides analytical support to the claim that human operators are an inherent part of the system. Each operator inadvertently rearranges some of the couplings of the elements. Thus, there are as many systems, call them 'utilized structures' as there are combinations of plants, operators, and strategies. More explicitly:
Utilized Structure $\geq$ Plant Structure + Mental Model + Strategy.
To describe the system without considering each of these actors is not to describe the true system.

It is further evident that many of these combinations result in successful control, and many do not. This prospect is a daunting challenge to the researcher who intends to relate the structure of the human-machine system to task performance. For this reason a quantitative method of evaluating the product of the Conant Analysis is crucial.

The causal decompositions suggest that, despite individual differences, there appears to be a subset of state variables that contain information about most of the system variables at sufficient time lags. This raises a number of questions. First, as noted above, the information provided by this reduced set of variables may be sufficient to facilitate successful operation of the plant. Thus, in lattice theory terms, one homomorph of an operator's mental model may contain only these state variables, arranged in such a manner as to culminate in successfully processed product. This homomorph would be simpler than a complete system model, allowing the operator to reduce her workload without sacrificing productivity. Second, since the resolution of the causal decomposition diagrams becomes more acute at extended time delays, we must consider the manner in which operators process information about the state variables. If a critical decomposition becomes evident at a bandwidth which exceeds human information processing limits, recognition of crucial relationships will be inhibited.

Recall also Moray's first general phenomenon expected of operators using mental models; that they should be able to apply past experiences to new situations by drawing on established causal relationships between system components. If learning can be described as focusing on causal relationships that are most informative, EDA provides a rigorous, quantitative method for testing this hypothesis. Given a quantitative means of comparing individual decompositions, one could examine the relationship between the decompositions and changes in performance resulting from the introduction of the fault. Lattice theory suggests that if a fault is introduced in a component associated with a crucial state variable, it will be detected and accommodated more quickly and effectively than a fault located outside the realm of the core decomposition.

Continued evaluation of the Conant Method necessitates a quantitative representation of the results. Further, the incorporation of each operator's unique monitoring and control data into the EDA may help isolate the effects of individual strategies.

ACKNOWLEDGMENTS

The author is presently affiliated with the Cognitive Engineering Laboratory, Department of Mechanical and Industrial Engineering, University of Toronto.

REFERENCES

Ashby, W.R. (1956). Introduction to cybernetics. London: Chapman and Hall.

Bainbridge, L. (1992). Mental models in cognitive skill. In A. Rutherford, A. Rogers, & P.A. Bibby, (eds)(1992). Models in the Mind. London: Academic Press.

Brehmer, B. (1992). Dynamic decision making: Human control of complex systems. Acta Psychologica, 81, 206-223.

Conant, R.C. (1972). Detecting subsystems of a complex system. IEEE Transactions on Systems, Man, and Cybernetics, SMC-2(7), 550-553.

Conant, R.C. (1976). Laws of information which govern systems. IEEE Transactions on Systems, Man, and Cybernetics, SMC-6(4), 240-255.

Conant, R.C. (1988). Extended Dependency Analysis of large systems, Part I: dynamic systems; Part II: static systems. International Journal of General Systems, 14, 97-141.

Edwards, E. & Lees, F.P. (Eds.). (1974). The Human Operator in Process Control. London: Taylor and Francis.

Gentner, D. & Stevens, A.L. (1983). Mental models. Hillsdale, N.J.: Lawrence Earlbaum.

Johnson-Laird, P.N.,(1983) Mental models. Cambridge, MA: Harvard University Press.

Moray, N. (1988). A lattice theory of mental models of complex systems. Technical Report EPRL-88-08. Engineering Psychology Research Laboratory, University of Illinois, Urbana, Illinois.

Moray, N. (1990). A lattice theory approach to the structure of mental models. Philosophical Transactions of the Royal Society of London, B327, 577-583.

Moray, N. (1991). Mental models of complex dynamic systems. Paper presented at meeting on Mental Models, Cambridge University.

Moray, N. (in press). Models of models of....mental models. In T.B. Sheridan (ed.) Liber Amicorum in Honour of Henk Stassen. Cambridge, MA: MIT Press.

Rasmussen, J. (1986). Information processing and human-machine interactions: An approach to cognitive engineering. New York: North-Holland.

Shannon, C.E. & Weaver W.,(1951). The Mathematical Theory of Communication. Urbana, IL: Univ. of Illinois.

Wilson, J.R. & Rutherford, A. (1989). Mental models: Theory and application in human factors. Human Factors , 31, 617-633.

Woods, D.D. (1988). Coping with complexity: The psychology of human behavior in complex systems. In L.P. Goodstein, H.B. Anderson, and S.E. Olsen (Eds.), Tasks, errors, and mental models. (pp.128-148). London: Taylor and Francis.

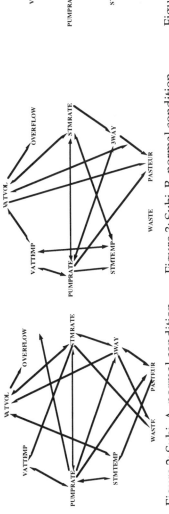

Figure 2: Subj. A, normal condition.

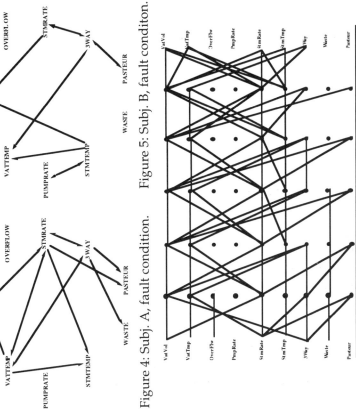

Figure 3: Subj. B, normal condition.

Figure 4: Subj. A, fault condition.

Figure 5: Subj. B, fault condition.

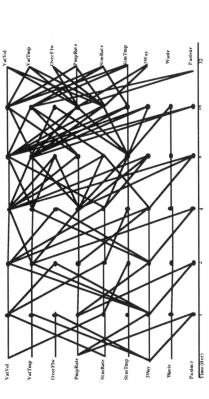

Figure 6: Causal Analysis; Subject A, normal condition.

Figure 8: Causal Analysis; Subject B, normal condition.

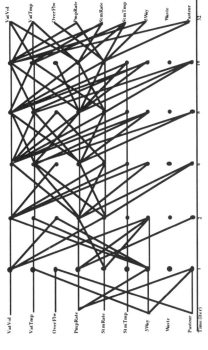

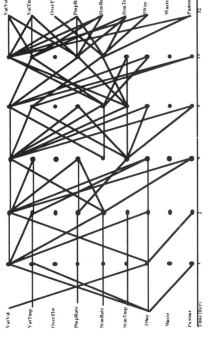

Figure 7: Causal Analysis; Subject A, fault condition.

Figure 9: Causal Analysis; Subject B, fault condition.

THE NATURE OF UNCERTAINTY IN NATURALISTIC DECISION MAKING

SYMPOSIUM OVERVIEW

Gary Klein, Ph.D.
Klein Associates Inc.
Fairborn, Ohio

One of the primary complications in naturalistic settings is the degree of uncertainty about different aspects of the decision task. Somehow, skilled decision makers have to wrestle with these uncertainties. One approach is to gather more information, in order to reduce uncertainty; but that carries its own penalty, which is to delay a decision, which means that certain options will be foreclosed. This is particularly problematic in time-pressured settings, where delay may not be possible. Decision makers have to be able to act with the information given.

In order to gain a better understanding of the way decision makers handle limitations in the data and information, we need to clarify what is meant by the term "uncertainty." Different researchers have defined it in different ways. As long as we leave it as a vague concept, we will have difficulty in studying it effectively. The purpose of this symposium is to try to sort out different meanings of the term. We want to reduce the uncertainty about the term "uncertainty."

The papers in this session all present viewpoints about how to define uncertainty, and they provide data collected for tasks conducted in field settings. Hopefully, we will be able to achieve some convergence that will result in a strong basis for research in the future.

THINKING NATURALLY ABOUT UNCERTAINTY

Marvin S. Cohen and Jared T. Freeman
Cognitive Technologies, Inc.
Arlington, VA

Methods for handling uncertainty should be evaluated in terms of their cognitive compatibility with real-world decision makers. Bayesian models of uncertainty demand precise up-front assessments of all problem elements and discourage the dynamic evolution of problem understanding. They handle missing or conflicting data by mathematical aggregation, while real-world decision makers regard gaps in knowledge and conflicting evidence as problems to be solved. Finally, they produce as output a statistical average rather than a coherent picture of the situation. Another approach to decision making, based on pattern-matching, does not address the ways in which situation pictures are evaluated and modified. A third approach, however, called the Recognition / Metacognition model, treats decision making under uncertainty as a problem-solving process that starts with the results of recognition, verifies them, and improves them where necessary. Critiquing strategies identify problems of incompleteness, conflict, and unreliability in situation models, and lead to correcting steps that retrieve or collect additional information or adopt assumptions. Training methods based on this model have been developed and tested with active-duty Naval officers.

INTRODUCTION

Is probabilistic reasoning the only way to handle uncertainty? It may seem to some that the answer must be *yes* by definition; uncertainty is defined as probabilities other than one or zero. To others, the hoary tradition of probability theory, stretching from Pascal and LaPlace to its elaboration in modern statistics and decision theory, provides convincing support for this answer. Yet few researchers are still prepared to argue, as many did three decades ago, that real-world decision makers correctly use probabilistic reasoning in real-world tasks. Naturalistic models, based on observation of experienced decision makers rather than laboratory studies of artificial tasks, have focused on mechanisms for situation assessment like pattern recognition or schema instantiation that match cues to stored knowledge. In doing so, they have turned attention away from uncertainty as such. Yet no one even vaguely familiar with domains such as combat, business, or medicine is prepared to argue that uncertainty is an unimportant or incidental feature of real-world decision making environments.

How do real world decision makers handle uncertainty, and how can they be trained or aided to handle it better? Must they chose between inappropriate "intuitive" approaches based on pattern matching, on the one hand, and technically difficult and "unnatural"

analytical methods, on the other? We contend that the study of naturalistic decision making is now mature enough to face uncertainty head on. Is there a *naturalistic* way to handle uncertainty? If so, it will be distinct from formal probabilistic reasoning, and it will go beyond pattern matching while nonetheless building on it. We shall first review two of the most popular current approaches, before describing a new model, training based on the model, and an empirical test of the training.

Probabilities, Patterns, or Problem-Solving?

The probabilistic approach starts with an abstract view of how uncertainty *ought* to be dealt with and provides an elaborate formal apparatus to do the job. By means of this apparatus, the unknown probability of a hypothesis can be expressed in terms of other probabilities that are more easily estimated or assessed. The decision maker generates a set of exhaustive and mutually exclusive hypotheses, makes numerical assessments of the probabilistic relationships between the hypotheses and evidence, and combines the numerical assessments to get probabilities for the hypotheses.

When compared with such normative models, real-world decision makers consistently fall short, revealing systematic errors, or "biases." But the models themselves may also fall short in various ways as *intellectual tools*. The idea of an effective intellectual tool has to do with *cognitive compatibility* between the tool and the decision maker: (1) Does the tool require inputs about which the decision maker has confident and precise intuitions? (2) Are the operations applied by the tool to the inputs understandable and plausible? (3) Does the tool produce as outputs judgments or choices that the decision maker cannot confidently make, but which he needs? In short, a good intellectual tool matches the pattern of a decision maker's knowledge and ignorance: it uses what he knows to generate what he needs to know, using reasoning processes that he trusts.

Decision analytic methods do not fair well when evaluated in terms of cognitive compatibility. (1) Can decision makers confidently and easily provide the inputs required by formal uncertainty models? Often not. The most obvious problem is sheer effort and time: Very large numbers of exact numerical probability judgments are demanded even by the simplest real-world problems. A more subtle problem, however, is that these methods require that the problem be completely modeled up front. They foreclose (or at least discourage) the dynamic process by which decision makers discover new hypotheses, new kinds of evidence, new options, and even new goals as a problem unfolds in time.

(2) Do the reasoning processes applied to the inputs make sense? Again, there are serious problems. For example, laws of probability treat conflicting data the same way they treat congruent data, by quantifying the force of the evidence and then mathematically aggregating. From the point of view of a real-world decision maker, this resort to mathematical aggregation may often seem overhasty. It is not dealing with uncertainty so much as sweeping it under the rug. If two sources who are each regarded as highly reliable contradict one another, the decision maker is not likely to take an average. He usually wants to understand why the sources disagree, and may well learn (or assume) than one or both of them is not as reliable as he supposed. In other words, conflict is a problem that needs to be solved. It can be a symptom of erroneous assumptions about the evidence or its meaning. Finding and correcting mistaken beliefs can lead to a more plausible judgment in the present case and enduring lessons for future situations.

(3) Finally, are the outputs of formal analyses useful for decision makers? Again, the verdict is not wholly positive. The output is not a coherent picture of the situation but an unrealizable abstraction: an assignment of probabilities across propositions (e.g., 30% chance the enemy will attack; 70% chance the

enemy will not attack). Such an abstraction cannot be visualized, anticipated, or planned for in the real world. In exchange for this loss of concreteness, we should at least get a good depiction of uncertainty itself. Paradoxically, we do not. Bayesian models do not make allowances for varying degrees of knowledge underlying the conclusions. The probability of an event might be judged .5 because of complete ignorance regarding two possible hypotheses *or* because there is a mass of conflicting evidence pointing in both directions. The decision maker is expected to treat both cases the same way.

A second approach looks in a different direction for the nature of decision making skill, toward a set of virtually automatic responses to recognized patterns. This view has been popular in research on differences between experts and novices, beginning with Chase and Simon's work on chess. Experts are thought to be distinguished from novices by the accumulation of a large repertoire of patterns or schemas over the course of their experience in a specific domain. This point of view — with its emphasis on rapid, automatic, and domain-specific processes rather than slow, deliberative, and general-purpose ones — seems diametrically opposed to the decision theoretic approach. Unfortunately and ironically, normative methods and pattern recognition have some parallel shortcomings. They both give short shrift to the problem-solving processes that are often triggered in the real world by gaps and conflicts in data. What do real-world decision makers do when the elements of a pattern are incomplete, or when the data do not perfectly match any single pattern, but partially fit conflicting patterns instead? How do decision makers create a picture of the situation, evaluate it, then change their minds? How do they balance the need to act with the need to think more about a problem? Although recognition may be at the heart of proficient decision making, other processes are often crucial for success.

What is needed is a descriptively based account of the problem-solving strategies that experienced decision makers bring to bear on uncertainty. Such an account may in turn provide the foundation for cognitively compatible tools for training and decision aiding.

A Naturalistic Model of Critical Thinking

Our observations of decision making performance, in Naval anti-air warfare as well as other domains, suggest that recognition is supplemented by processes that verify and improve its results. Because of their function, we call these processes *meta-recognitional*. Meta-recognitional skills probe for flaws in recognized assessments and plans, try to patch up any weaknesses that are found, and evaluate the results. They are analogous to the *meta-comprehension* skills that proficient readers use when they try construct a mental model based on the information in a text. For example, according to researchers in this area, skilled readers continually test and evaluate the current state of their comprehension, and they adopt a variety of strategies for correcting problems that are found, such as inconsistencies or gaps in their understanding.

To reflect the complementary roles of recognition and metacognition in decision making, we have called this framework the Recognition / Metacognition (R/M) model. In the R/M model, the basic level of cognition is recognitional, including processes that activate assessments in response to internal and external cues. Assessments may in turn be associated with structures that organize actual and potential information into a situation model or plan. According to the R/M model, the integration of observations into situation models and plans often occurs under the influence of metacognitive control. Meta-recognitional processes include:

1. Identification of evidence-conclusion relationships (or *arguments*) within the evolving situation model and plan. This is simply an implicit or explicit awareness that cue A was *observed* on this occasion, while some conclusion (e.g., intent to attack) along with expectations of observing cue B were *inferred*. On some other occasion cue B might be observed and cue A inferred.

2. Processes of *critiquing* that identify problems in the arguments that support the situation model or plan. Critiquing can result in the discovery of three kinds of problems: *incompleteness, unreliability,* or *conflict*. A model or plan is incomplete if expected arguments are missing; that is, some expected information has not been obtained that would ordinarily confirm or disconfirm the assessment. An argument is unreliable if it depends on implausible assumptions. An argument in support of an assessment is conflictual if there are other arguments that support contradictory assessments.

3. Processes of *correcting* that respond to these problems. Correcting may instigate additional observation, additional information retrieval, revision of assumptions, and/or the generation of new hypotheses, options, or goals. These processes fill gaps in models or plans, resolve conflict among arguments, and substitute more reliable assumptions for less reliable ones.

4. A control process called the *quick test*, which regulates critiquing and correcting. The quick test considers the available time, the costs of an error, and the degree of uncertainty or novelty in the situation. If conditions are appropriate, the quick test inhibits recognitional responding and interposes a process of critical thinking. If conditions are inappropriate, it allows immediate action based on the current recognitional response.

The R / M model explains how experienced decision makers are able to exploit their experience in a domain and at the same time handle uncertainty and novelty. They construct and manipulate concrete, visualizable models of the situation, not abstract aggregations (such as 70% chance of hostile intent, 30% chance non-hostile). Uncertainty is represented explicitly at the metacognitive level, by "annotating" the situation model or plan to highlight points of incompleteness, conflict, and unreliability. In response to specific problems of this kind, metacognitive strategies try to improve the current situation model and plan or find better ones. Such strategies are highly dynamic and iterative: For example, gaps may be filled by collecting further data or by making assumptions (e.g., the worst case). The resulting arguments may then be found to conflict with other arguments. Such conflict may be resolved by evaluating the reliability of the assumptions in the conflicting arguments. This process stops when the quick test indicates that immediate action on the basis of the current best model is called for. The output is a coherent, consistent model of the situation together with an understanding of its strengths and weaknesses.

Our hypothesis is that decision aiding and training based on the R/M model will be cognitively compatible with the way real-world decision makers think about uncertainty. In terms of inputs, such tools begin with a recognitional response rather than a complete decomposition of the problem; they require judgments only in response to specifically identified problems in the recognitional response and only as long as time and stakes permit; and they encourage a dynamic evolution of understanding. In terms of processes, they rely on problem solving rather than mathematical aggregation, stimulating collection and retrieval of new information, adjustment of assumptions, or generation of new hypotheses, options, or goals. Finally, their outputs combine the concreteness of a coherent situation picture with an explicit recognition of qualitatively different types of uncertainty.

Training to Handle Uncertainty

Each definition of decision making skill has its own implications for training. From the formal, normative point of view, training should convey a set of general purpose methods for structuring and quantifying evidence, hypotheses, options, and outcomes. Examples of applying the techniques are useful only incidentally, for practice and to demonstrate the generality of the methods. By contrast, examples are central from the recognitional point of view, in which decision training focuses intensely on a particular application area. Training can accelerate the accumulation of experiences with the characteristic patterns of the relevant domain by using realistic simulations and outcome feedback. From the point of view of the R/M model, the focus of training is neither a small set of general-purpose methods nor a vast quantity of specialized patterns. Our focus is a moderately sized set of strategies for critical thinking, which are general but at the same time *build on the recognitional skills of a particular domain.*

We have developed and tested training methods based on the R/M framework in Army battlefield situation assessment (Cohen & Freeman, 1995) and in Navy anti-air warfare (Cohen, Freeman, Wolf, & Militello, 1995). The Navy training focuses on the situation in which an air or surface platform with unknown intent is approaching own ship. The most recent version of the Navy training begins with an overview of the process of building and improving situation models, and then, in three additional units, focuses on particular aspects of that process.

We call the overview of the critical thinking process *STEP*. STEP consists of a cycle of four strategies: (1) Building a Story based on an assessment and filling the gaps in that story. If an officer has assessed the intent of an approaching platform, in order to take that assessment seriously, he must consider the events in the past and future that it implies. Such a story consists of events that could have motivated the intent and other events that would effectively achieve the intent. (2) Testing the story to identify data or knowledge that conflict with it, and attempting to revise the story to explain the conflicting information. Decision makers do not simply drop an assessment in the face of conflict; if they did, they would be paralyzed in situations where no familiar pattern fits all the data. Rather, they explore assumptions that would be sufficient to explain the conflicting evidence. (3) Evaluating the story and the assumptions on which it is based. Simply being able to construct a story around an assessment doesn't mean the assessment is true. Decision makers now step back and ask if the story makes sense. Can the assumptions be confirmed by other data, or accepted as the most plausible possibilities? If not, the decision maker may consider

another assessment, and attempt to construct, test, and evaluate a new story. (4) Formulating contingency Plans to protect against unreliable assumptions in the current story. At any given time, the decision maker has a best available assessment along with an appreciation of its weaknesses. Contingency plans guard against the possibility that assumptions in the story turn out to be false.

After receiving an overview and practice with STEP, in the second unit of training officers study and apply a particularly important kind of story, based on hostile intent. Training helps them identify issues that experienced decision makers typically consider in assessing whether intent is hostile. A complete hostile intent story includes information or assumptions about why the country involved would want to attack, why they chose the relevant platform as the attack vehicle, why they chose a particular asset as the target of attack, how the platform was able to localize the target, how the platform is managing to protect itself while arriving at a position for engagement, and what the platform must do in order to execute an engagement.

The third unit focuses on explaining conflict and generating alternative hypotheses. It presents a variant of the devil's advocate technique that forces officers to generate alternative interpretations of the evidence.

The fourth and final unit of training provides guidance regarding when critical thinking is appropriate (i.e., the quick test). It provides strategies for assessing the available time, the stakes, and the uncertainty in a situation in order to decide when it is necessary to commit to an action.

In all four units, training consists of presentation and discussion of the R/M concepts followed by practice and feedback with a simulated anti-air warfare scenario.

EMPIRICAL TESTS

Method

The training methods were tested in two experiments at United States Navy training facilities, one involving 60 officers (study 1) and the other involving 35 officers (study 2). In study 1, we were able to utilize a control group (which was exposed to a general discussion of knowledge representation and problem solving-strategies with practice examples from the participants' skill areas), as well as a pretest-posttest comparison. In that study, however, only 90 minutes were available for training, and practice utilized paper-and-pencil examples rather than automated simulations. In study 2, four hours were available for training and more realistic practice was possible due to the availability of automated simulations. In this study, however, there was no control group, and only the pretest-posttest comparison was available. Both studies counterbalanced two scenario sequences for the pretest and posttest.

The evaluation examined the effects of training on both performance (i.e., assessments and actions) and on critical thinking processes in realistically simulated anti-air warfare scenarios. In both studies, each pretest and posttest scenario had three breaks, at which participants responded to questions. These questions probed for the likely intent of a designated track, degree of confidence in that assessment, and reasons for it, other possible intents, reasons in favor of one of the non-accepted intents, data that conflict with a designated intent, possible explanations of the conflicting data, and actions the participant would take or plan for. Dependent measures included the number of arguments generated for a hypothesis, the amount of conflicting evidence identified, success in explaining conflicting evidence, the number of alternative hypotheses generated, the accuracy of the accepted assessments, appropriateness of actions, and participants' own evaluations of the training (in a subsequent debriefing).

Summary of Results

Trained participants identified more evidence in favor of a hypothesis than untrained participants ($p=.001$ for the pretest-posttest comparison in study 2; $p=.092$ for the treatment-control group comparison in study 1). Perhaps more interestingly, trained participants identified significantly more evidence that conflicted with a designated hypothesis than untrained participants ($p<.001$ for study 2, $p=.015$ for study 1). Training also improved participants' ability to patch up stories by constructing explanations for the conflicting evidence ($p<.001$ for study 1, ns for study 2). Trained participants also generated more alternative assessments of possible intent than untrained participants ($p<.001$ for study 2, ns for study 1).

In both studies, training had a significant effect on assessments ($p=001$ in study 1, $p=.002$ in study 2). In two of the four test scenarios (one of the two used in each study), training significantly increased agreement with the subject matter expert who had designed the scenarios ($p=.034$ in study 2, $p=.015$ in study 1). Actions taken by participants significantly varied in a way that reflected these assessments. Training did not reduce participants' confidence in the assessments that they accepted. It did, however, increase contingency planning in case assumptions proved wrong. Trained participants in study 1 (who were more likely to regard the designated contact as non-hostile) were more likely to make contingency plans for engagement of the contact than untrained participants ($p=.005$ in study 1, not yet analyzed in study 2).

Finally, participants generally evaluated the training favorably both in numerical evaluations and in qualitative comments.

Conclusion

Trained participants were significantly better than untrained participants in all four aspects of the STEP method: building stories, testing and patching up stories, evaluating stories, and making contingency plans in case stories were wrong. This study provides a preliminary demonstration that meta-recognitional skills can be taught effectively, that decision makers will use them in relatively realistic tactical situations, and that use of such skills will improve outcomes. This is a promising avenue for the development of cognitively compatible tools for handling uncertainty based on naturalistic models of decision making.

ACKNOWLEDGMENTS

This work was supported by contracts N61339-92-C-0092 and N61339-95-C-0107 with the Naval Air Warfare Center / Training Systems Division. Klein Associates, Inc., served as a subcontractor in these efforts. Sonalysts, Inc., provided support in setting up the experimental simulations.

REFERENCES

Cohen, Marvin S. and Freeman, Jared T. (1995). *Methods for Training Cognitive Skills in Battlefield Situation Assessment.* Arlington, VA: Cognitive Technologies, Inc.

Cohen, Marvin S., Freeman, Jared T., Wolf, Steve, and Militello, L. (1995). *Training Metacognitive Skills in Naval Combat Decision Making.* Arlington, VA: Cognitive Technologies, Inc.

SIMPLE STRATEGIES OR SIMPLE TASKS?
DYNAMIC DECISION MAKING IN "COMPLEX" WORLDS

Alex Kirlik, Ling Rothrock, Neff Walker, and Arthur D. Fisk
Center for Human-Machine Systems Research
School of Industrial & Systems Engineering
School of Psychology
Georgia Institute of Technology
Atlanta, GA 30332-0205

ABSTRACT

Decision makers in operational environments perform in a world of dynamism, time pressure, and uncertainty. Perhaps the most stable empirical finding to emerge from naturalistic studies in these domains is that, despite apparent task complexity, performers only rarely report the use of complex, enumerative decision strategies. If we accept that decision making in these domains is often effective, we are presented with a dilemma: either decision strategies are (covertly) more complex than these performers claim, or these tasks are (subtlety) more simple than they might appear. We present a set of empirical findings and modeling results which suggest the latter explanation: that the simplicity of decision making is not merely apparent but largely real, and that tasks of high apparent complexity may yet admit to rather simple types of decision strategies. We also discuss empirical evidence that sheds light on the error forms resulting from the tendency of performers to seek and employ heuristic solutions to dynamic, uncertain decision problems.

INTRODUCTION

People are generally delighted when shown a simple solution to a complex problem. For those so inclined, one great appeal of mathematics is the feeling of surprise when finding a simple answer to a complex problem, as if the simplicity of the answer reveals that the problem was not really complex after all. Scientists routinely characterize problems as either simple or complex, not on the basis of problem features themselves, but rather, on the complexity of the methods needed for their solutions. While the solution of a mathematician or physicist may be difficult for you or I to understand, we still believe that scientists such as these do *discover* an underlying simplicity to the world that was not initially apparent. To solve a math problem, you or I might have to labor through complex calculations, where a mathematician might exploit a symmetry that allows for a simple solution. The ability to see such simplicities owes to considerable experience, training and talent, in the end allowing the mathematician to dispense with many of the calculations that you or I, lacking such abilities, must perform.

In recent years, an increasing amount of cognitive engineering research has been devoted to studying a somewhat different product of experience, training and talent: the practitioner in naturalistic decision making environments. Like the mathematician, these practitioners have a large array of knowledge about their domains of expertise. And also like the mathematician, we find that these practitioners also characterize problems or decisions as "simple" if their solutions are simple, and as "complex" if their solutions are not readily apparent but must instead be contemplated. Perhaps the single most stable finding (Klein, 1993) of naturalistic decision making research is how frequently practitioners report simple solutions to problems that appear extremely complex to people without the requisite training or experience (say, a cognitive researcher or a knowledge engineer).

Although we typically delight in being shown simple answers to complex problems, many researchers react with considerable skepticism to practitioner's claims of simple strategies for naturalistic decision making. While we are willing to realize that various math problems are not complex after all (after being shown a simple solution), we are much less willing to believe that the decision problems we study are not complex after all, even after a practitioner demonstrates and relates simple solutions. When presented with a decision problem in an operational environment, an analyst might have to work through a highly enumerative, analytical decision strategy in order to obtain a (defensible) solution, and thus tends to believe that such a process is in some sense necessary. As a result, the naturalistic researcher is continually cautioned about the possibility that practitioners are actually using covert analytical strategies despite the simplicity of verbally reported strategies (e.g., Doherty, 1993).

THE DILEMMA OF DECISION MAKING IN "COMPLEX" WORLDS

If we accept that dynamic decision making in complex, operational environments is often effective, there are two ways to interpret practitioner's verbal claims about using simple decision strategies. Either strategies are covertly more complex than they appear on the surface, or these decision tasks are subtly less complex than they might seem. The lack of conscious access to underlying cognitive activities licenses nearly unbridled speculation relating to the first interpretation, since the theorist is free to posit covert mechanisms of tremendous complexity. Entertaining the second interpretation, on the other hand, is more constraining because it relates not to the nature of unobservable phenomena but instead to features of the observable task environment. Explaining how the mathematician can *dispense* with the complex calculations that were necessary for one not equipped to exploit the subtle simplicities of nature requires one to actually discover these simplicities: one cannot use the previous tactic of postulating theoretical entities that are unobservable by definition. While we do recognize the possibility that performers may indeed use covert strategies that are more complex than they verbally report, speculation on the nature of these strategies should be as constrained as much as possible by empirical analysis of the task environment so that the complexity of covert processes is not overestimated.

The empirical studies and modeling results reported below justify, to some extent, the claim that decision makers in complex, dynamic decision tasks indeed do use relatively simple decision strategies. We show that even quite complex dynamic decision tasks are sometimes amenable to quite simple heuristic strategies, and show that performers exploit these strategies when available.

In addition, we provide some evidence that performers opt for simple strategies even when they are only very crude approximations to normative behavior.

EMPIRICAL STUDIES

A number of theorists have suggested that dynamic decision making tasks, while apparently more complex than static tasks from a formal point of view, may actually favor action-oriented, closed-loop decision strategies that are less cognitively-intensive than those required for single-shot, static decisions (e.g., Hogarth, 1981; Connolly, 1988). Speculation of this type received support from our previous research modeling human performance in a real-time, micro-world dynamic decision task (Kirlik, Miller, and Jagacinski, 1993). Given that the model presented there was validated mainly through a sufficiency demonstration (computer models of various subjects were successfully capable of mimicking behavior), the psychological validity of that model is open to debate. However, in the context of the preceding discussion, it should not be overlooked that this modeling demonstrated that a dynamic task far too complex to be treated by available formal methods could be proficiently performed by a computer model using quite simple decision heuristics. Contributing reasons were that: a) uncertainties in this task did not reward extensive prediction; b) the real-time nature of the task cascaded and amplified the negative effects of delayed action thus favoring "good but fast" rather than "optimal but slow" decisions; and c) rich perceptual information was available to specify many if not most constraints on productive decision-making. These are also features of many decision tasks in operational environments. A demonstration that such features may actually favor simple, heuristic strategies suggests that even if performers are capable of complex analytical reasoning below the level of conscious awareness, it may not pay them to do so.

Recently, we have conducted two additional experiments on dynamic decision making to shed light on this issue. The first experiment (Krosnick, 1994; Kirlik, Walker, Fisk, and Nagel, 1996), was conducted using the simulation EJSTARS, which was modeled after a U.S. Army situation assessment and decision making task. Participants were required to monitor and identify initially unknown vehicles shown moving on a computer-version of a U.S. Geological Survey map. After identifying vehicles, participants had to determine if any posed threats to a variety of friendly assets also shown on the map display. The criteria relevant to this decision included:

1. Locomotion: A vehicle is threatening only if it can locomote to a friendly asset. The EJSTARS map display indicates areas of level plains, forests, mountains, rivers, and ice. The type (weight) of an enemy vehicle determines where it can and cannot locomote within the terrain, and thus, whether it poses an impending threat to a particular friendly asset.

2. Weapons Range: A vehicle is threatening only if it can locomote to a point within its weapons range of a friendly asset. The type of weapons carried by an enemy vehicle determines its weapons range, and thus, whether the vehicle poses an impending threat to a particular friendly asset.

3. Penetrability: A vehicle is threatening only if its weapons are capable of penetrating the defensive armor of a particular friendly asset. Armor thickness varies among friendly assets, and weapons penetration depth varies among enemy vehicle types; the relationship between these two properties determining whether the vehicle poses an impending threat to a particular friendly asset.

4. Priority: Friendly assets differ in terms of their priority (in EJSTARS, the number of points lost should the friendly asset be lost to enemy attack). In the case of multiple impending threats, attacks should be directed toward the enemy vehicle threatening the most highly valued friendly asset.

The information pertaining to these criteria (e.g., vehicle weight, weapons range, etc.) was available from call-up text windows overlaid on the map display.

We intended to create a reasonably realistic task of sufficient complexity to challenge experimental participants and to allow us to investigate concepts for display-based decision aiding (for details see Kirlik et al., 1996). Experimental results indicated that display-based aiding improved performance only in the initial two sessions, as by the third session the unaided, or baseline display participants performed at the level of the aided participants. To explain this result, we conducted a detailed analysis of participant behavior in this task to understand why aiding was not more successful, and how the unaided participants managed to integrate the large array of graphical and textual information to determine, in real time, whether or not to take action on a potentially threatening enemy vehicle (Kirlik, Krosnick, and Fisk, 1995).

With respect to three of the four decision criteria discussed above (1, 2, and 4), we found that our participants appeared to have an extremely efficient method for handling them: ignoring them all together. This tendency was revealed by an analysis which compared how subjects treated vehicles that they had previously correctly identified (and thus for which correct call-up information was provided) and vehicles for which they had previously misidentified (and thus for which incorrect call-up information was provided). This analysis revealed that accuracy with respect to range, locomotion, and priority decision criteria was the same whether or not participants were provided with correct information regarding these criteria. As discussed in Kirlik et. al (1995), this and additional observations all pointed to the conclusion that these subjects were using a simple "wait and see" strategy to determine whether an enemy vehicle would attempt to attack a friendly resource. The limited value of the decision aiding seemed to result from the fact that the aiding approach presumed performers would have to integrate the four decision criteria to perform acceptably, when instead performers dispensed with such analysis by discovering a simple, empirically-based, approximation strategy.

Finally, we have also conducted experiments to investigate tactical decision making in a simulated U.S. Navy Combat Information Center (CIC) context (Hodge, Rothrock, Kirlik, Walker, Fisk, Phipps, and Gay, 1995). In this task, a participant performed the task of an Anti-Air Warfare coordinator, responsible for viewing a radar display and other information in order to identify aircraft in a military context. In theory, to reach acceptable levels of identification performance, the participant had to consider the following types of information: a) Identification Friend or Foe (IFF) status; b) Electronic Sensor Emissions; c) Visual Identification; d) Altitude; e) Speed; f) Course; g) Bearing; h) Range; i) Knowledge of Civilian Air Corridors; and j) Knowledge of Hostile and Friendly Countries.

We intended to create a reasonably realistic task of sufficient complexity to challenge experimental participants, to mimic CIC task properties, and to allow us to investigate a variety of training concepts (for details see Hodge et al., 1995). At the end of the experiment, we asked participants what strategies they used to identify aircraft. An example response is as follows:

1. The [Sensor] is always right. ARINC564 [a sensor emission] always means assume friendly/ NMIL [non-military].

2. If IFF corresponds to an actual flight # and it is in a reasonable place for that flight to be, assume friendly/NMIL [non-military].

3. If it has no IFF and its [Sensor] is off, it is still assume[d] friendly/NMIL if it is flying high and its speed is < 400 [kts]. I usually try to VID [visually identify] these or keep checking EWS [Sensor] if they're going to come close.

4. If all else fails, VID.

This participant also related some details on how he allocated his attention to the system display in order to put the above strategy into practice.

We did not train participants to use any particular identification strategies such as the one given above. Initially we were not even sure what strategies would be both efficient and effective given the large number of decision criteria that were potentially relevant to the hundreds of identification tasks experienced over the course of the experiment. We did, however, perform detailed analyses to determine if some of the strategies offered by various participants were plausible (i.e., consistent with their own behavior).

For this purpose we used an inferential model based on genetic algorithms to infer a simple yet empirically sound set of rules that described participant's identification judgments (for details see Rothrock, 1995). For the participant whose strategy is given above, we selected a representative experimental session and found that this participant identified 20 aircraft without error. We applied the model to the empirical data from this session and inferred a set of 7 simple non-compensatory (if-and-or-then) rules that were consistent with this participant's identification judgments. While the inferred set of rules are not unique, the 7 rules identified by the model were not inconsistent with the strategy verbally described by the participant himself. In addition, applying the verbal strategy to the actual information available concerning the 20 aircraft identified by this participant resulted in reproducing the judgments actually made during task performance. In short, we could find no empirical evidence to suggest that this performer used a judgment strategy any more sophisticated than the heuristic strategy he verbally offered.

We emphasize, however, that while all participants did verbally provide relatively simple heuristic strategies, many erroneous identification judgments were made during this experiment. Some of the heuristics apparently used by participants were based on false assumptions about either the properties of the task environment or about the reliability of judgment cues. While we are exploring the factors contributing to these errors, the fact remains that we have yet found no empirical evidence to suggest that participants were using strategies that were in any way more complex than the heuristic strategies they verbally provided.

CONCLUSION

Experienced decision makers in complex, dynamic tasks typically report the use of relatively simple decision strategies. The tendency to treat these reports skeptically may result from the inability of a less experienced analyst or researcher to conceive how problems that are complex when viewed abstractly, might yield to simpler solutions when viewed in their concrete context. Our modeling results and empirical findings provide evidence that "complex" dynamic decision tasks may actually be subtly more simple than they first appear. Detailed task analysis is needed to avoid falsely attributing overly-sophisticated covert reasoning to skilled performers.

ACKNOWLEDGMENTS

The authors are grateful for the assistance of Burton W. Krosnick and Kevin Hodge in conducting this research. Support for portions of this work was provided by Contract N61339-0225 from the Naval Air Warfare Center, Training Systems Division (Jan Cannon-Bowers, technical monitor).

REFERENCES

Connolly, T. (1988). Hedge-clipping, tree-felling, and the management of ambiguity. In, *Decision Making: An Interdisciplinary Inquiry*, M.B. McCaskey, L.R. Pondy, and H. Thomas, (Eds.). New York: Wiley.

Doherty, M.E. (1983). A laboratory scientist's view of naturalistic decision making. In, *Decision Making in Action: Models and Methods*, G.A. Klein, J. Orasanu, R. Calderwood, and C.E. Zsambok (Eds.). (pp. 362-388). Norwood, NJ: Ablex.

Hodge, K.A., Rothrock, L., Kirlik, A., Walker, N., Fisk, A.D., Phipps, D.A., and Gay, P.E. (1995). Training for tactical decision-making under stress: Towards automatization of component skills. Human Attention and Performance Laboratory Technical Report HAPL-9501. School of Psychology, Georgia Institute of Technology, Atlanta, GA.

Kirlik, A., Krosnick, B.W., and Fisk, A.D. (1995). Positive and negative effects of perceptually augmented displays on dynamic decision making performance. Manuscript submitted for publication.

Kirlik, A., Miller, R.A., and Jagacinski, R.J. (1993). Supervisory control in a dynamic and uncertain environment: A process model of skilled human-environment interaction. *IEEE Transactions on Systems, Man, and Cybernetics, Vol. 24*, No. 4.

Kirlik, A., Walker, N., Fisk, A.D., and Nagel, K. (1996). Supporting perception in the service of dynamic decision making. *Human Factors*, 38(2).

Klein, G.A. (1993). A recognition-primed decision (RPD) model of rapid decision making. In, *Decision Making in Action: Models and Methods*, G.A. Klein, J. Orasanu, R. Calderwood, and C.E. Zsambok (Eds.). (pp. 138-147). Norwood, NJ: Ablex.

Krosnick, B.W. (1994). *Perceptual augmentation to support skill acquisition and robust decision making and control skills.* Unpublished masters thesis, School of Industrial & Systems Engineering, Georgia Institute of Technology, Atlanta, GA.

Rothrock, Ling (1995). *Performance measures and outcome analyses of dynamic decision making in real-time supervisory control.* Unpublished doctoral dissertation. School of Industrial & Systems Engineering, Georgia Institute of Technology, Atlanta, GA.

HOW DECISION-MAKERS COPE WITH UNCERTAINTY

Raanan Lipshitz, University of Haifa, Haifa, Israel
and
Orna Strauss, Israel Police

Abstract

We analyzed 112 self reports of decision-making under uncertainty to find how decision makers conceptualize uncertainty and cope with it in the real world. The results show that decision makers distinguish between three types of uncertainty, inadequate understanding, incomplete information and undifferentiated alternatives, to which they apply five strategies of coping, reducing uncertainty, assumption-based reasoning, weighing pros and cons of competing alternatives, suppressing uncertainty, and forestalling. The relationships between these types of uncertainty and tactics of coping suggest a *R.A.W.F.S.* (Reduction, Assumption based reasoning, Weighing pros and cons, Forestalling and Suppression) heuristic of contingent coping with uncertainty in naturalistic settings.

INTRODUCTION

Uncertainty is ubiquitous in realistic setting, where it constitutes a major obstacle to effective decision-making (Brunsson, 1985; McCaskey, 1986; Thompson, 1967). Despite -- or perhaps because of -- this widespread interest in uncertainty, there is no consensus across disciplines as to what is uncertainty. Our study examined three questions: (1) How do decision-makers (e.g., managers and military officers) conceptualize the uncertainty which they encounter in their work? (2) How do decision-makers cope with this uncertainty? (3) Are there systematic relationships between different conceptualizations of uncertainty and different methods of coping?

Conceptualizing Uncertainty

Defining uncertainty in the context of action as *a sense of doubt that blocks or delays action,* we posited that different types of uncertainty could be classified according to their issue (i.e., what the decision maker is uncertain about) and source (i.e., what causes this uncertainty.) Specifically, we identified three basic issues, outcomes, situation and alternatives, and three basic sources, incomplete information, inadequate understanding and undifferentiated alternatives.

Coping with Uncertainty

Since uncertainty blocks or delays action, coping with uncertainty is a major task of decision makers in the real world. Several classifications of methods of coping with uncertainty can be found in the literature (e.g., Allaire & Firsirotu, 1989; Thompson, 1967). The standard procedure for coping with uncertainty in formal and behavioral decision theories can be labeled the *R.Q.P* heuristic: *Reduce* uncertainty by a thorough information search, *Quantify* the residue that cannot be reduced and *Plug* the result into some formal scheme that incorporates uncertainty as a factor in the selection of a preferred course of action (Cohen, Schum, Freeling & Innis, 1985; Hogarth, 1987; Raiffa, 1968; Smithson, 1989.)

Notwithstanding its formal elegance and flexibility, the *R.Q.P.* heuristic has several drawbacks as a guide for describing and prescribing for decision-making in naturalistic settings: Reducing uncertainty is often unfeasible because information is either unavailable, ambiguous or misleading. Quantifying and plugging are even more problematic, basically because "there are many areas of both practical and theoretical inference in which nobody knows how to calculate a numerical probability value" (Meehl, 1978, p. 831.) Thus, managers are often reluctant to use quantified measures of uncertainty, thereby handicapping the application of decision support systems that rely on quantification (Isenberg, 1985; March & Shapira, 1987).

Assuming that decision makers do first try to reduce uncertainty by collecting additional information, the question, then, is what do they do with information that cannot be reduced this way, granted that they do not resort to quantification? Based on a review of the literature we distinguished between three basic strategies that decision makers use to cope with uncertainty, *reducing* uncertainty; *acknowledging* uncertainty and *suppressing* uncertainty. Each strategy consists of more specific tactics of coping with uncertainty.

Reducing Uncertainty: Tactics for reducing uncertainty include collecting additional information before making a decision or deferring decisions until additional information becomes available. When additional information is not available, it is possible to reduce uncertainty by extrapolating from available information One tactic of extrapolation is to use statistical methods to predict future events from information on present or past events (Allaire & Firsirotu, 1989.) Another tactic of extrapolation is assumption-based reasoning, filling gaps in firm knowledge by making assumptions, that (1) go beyond (while being constrained by) what is more firmly known, and (2) are subject to retraction when and if they conflict with new evidence or with lines of reasoning supported by other assumptions (Cohen, 1989). Using assumption-based reasoning experienced decision makers can act quickly and efficiently within their domain of expertise with very little information. A tactic of reducing uncertainty that combines prediction and assumption-based reasoning is mental simulation or scenario building, imagining possible future developments in a script-like fashion (Klein, 1993). Finally, uncertainty can be reduced by improving predictability through shortening time-horizons (preferring short term to long term goals, and short term feedback to long range planning (Cyert & March, 1963), and by selecting one of the many possible interpretations of equivocal information (Weick, 1979). The tactics listed so far rely, one way or another, on information processing. An entirely different approach to reducing uncertainty is to control the sources of variability that reduces predictability by instituting standard operating procedures, incorporating critical elements into the organization (i.e., acquisition) or by negotiating long term contractual arrangements (Thompson, 1967).

Acknowledging Uncertainty: Decision makers can acknowledge uncertainty in two ways. (1) Take it into account in selecting a course of action. (2) Prepare to avoid or confront potential risks. The Rational Choice model presents a sophisticated tactic of accounting for uncertainty by including it as a factor in concurrent option evaluation (Raiffa, 1968). Less sophisticated tactics of incorporating uncertainty as a factor in concurrent option evaluation are the minimax regret and maxmin strategies and avoiding ambiguity by preferring options with clear outcome probabilities (Curley, Yates & Abrams, 1986).

Thompson (1967) and Allaire and Firsirotu (1989) proposed several tactics of acknowledging uncertainty by preparing to avoid or confront potential risks. These include buffering (e.g., building stockpiles to shield production from unstable supply of required input), rationing (rearranging priorities following unanticipated contingencies), planning very carefully for all reasonable contingencies, and adopting a flexible strategy that allows for easy and inexpensive change.

Suppressing Uncertainty: This strategy includes tactics of denial (ignoring or distorting undesirable information) and tactics of rationalization (coping with uncertainty symbolically by going through the motions of reducing uncertainty or acknowledging it.) The quintessential suppression tactic of ignoring undesirable information was described as the Pollyana effect, the acquisition of an (often false) sense of security through the belief that "this [unfortunate outcome] cannot happen to me" (Matlin & Stang, 1978). Janis & Mann (1977) described various suppression tactics that decision makers use to align their preferences and beliefs with their decisions.

Method

Subjects

112 students in a course in Decision-Making at the Israel Defense Forces Command & General Staff College participated in the study. Most of them were male officers from all branches of the military in the ranks of Captain - Lt. Colonel.

Procedure

Data: Subjects wrote a case of decision-making under uncertainty based on their personal experience prior to the beginning of the course. Writing instructions encouraged students to write fully and frankly, without defining either what is decision-making, or what is uncertainty.

Case analysis: Conceptualizations of uncertainty and tactics of coping with uncertainty in the cases were identified with an instrument consisting of 16 conceptualizations and 12 tactics of coping developed on the basis of the conceptual schemes presented in the Introduction (see Lipshitz & Strauss, 1996 for details). Preliminary analysis showed interjudge agreement between five judges of $.89 \leq \kappa \leq 1.00$ for conceptualizations of uncertainty and $.87 \leq \kappa \leq 1.00$ for tactics of coping.

RESULTS

Data included 122 pairs of uncertainty and coping tactics. Interjudge agreement between the second author and an independent judge who analyzed a randomly selected sample of 40 cases, was $\kappa = .83$ for the conceptualizations of uncertainty and $\kappa = .93$ for the tactics of uncertainty.

The two most frequent conceptualizations were inadequate understanding of the situation owing to equivocal information (24.6%), and conflict among undifferentiated alternatives (24.6%). Tactics of reduction were reported most frequently (46.8%), followed by tactics of acknowledgment (41.8%) and tactics of suppression (11.5%). The latter result probably reflects the low social desirability of such tactics. Three tactics were reported most frequently, assumption-based reasoning (22.1%), preempting (21.3%) and weighing pros and cons (18.9%).

The relationships between conceptualizations and tactics were explored by cross-tabulated conceptualizations of uncertainty grouped according to their sources (incomplete information, inadequate understanding, and conflict among undifferentiated alternatives) and tactics of coping grouped into five categories (reduction, suppression and three spin-offs of the original strategy of acknowledgment assumption-based reasoning, forestalling and weighing pros and cons.) A systematic relationship emerged ($\chi^2_{(8)} = 32.3$, $p < .001$). Forestalling and suppression were not related differentially to the three categories of uncertainty. Each of the remaining categories was associated primarily with a different category of uncertainty. The conditional probability of reduction given lack of understanding was $p = .37$ compared to its marginal probability of $p = 25$; the conditional probability of weighing pros and cons given conflict was $p = .47$ compared to its marginal probability of $p = .19$ and the conditional probability of assumption-based reasoning given lack of information was $p = .33$ compared to its marginal probability of $p = .22$. In addition, these conditional probabilities were considerably larger than the two remaining conditional probabilities in each category of uncertainty.

We also analyzed the decision rules used by decision makers in the 30 instances in which they weighed the pros and cons of the outcomes of conflicting alternatives. To this end we counted the number of positive and negative attributes associated with every alternative, counting uncertainty as a negative attribute if the decision maker referred to uncertainty as such explicitly. In general, the chosen alternative was associated either with more positive or with less negative attributes (or both) than the rejected alternatives.

DISCUSSION

Analysis of case reports of decision-making under uncertainty revealed a variety of tactics that decision makers use to cope with uncertainty in naturalistic settings. These tactics are associated differentially with different types of uncertainty in a pattern that suggests an alternative to the R.Q.P heuristic. This alternative, which we label the R.A.W.F.S. (Reduction, Assumption-based reasoning, Weighing pros & cons, Forestalling, Suppression) heuristic is presented in Figure 1.

The R.A.W.F.S. heuristic hypothesis: Extrapolated from our findings and previous naturalistic decision-making research, the R.A.W.F.S. heuristic attempts to capture the essentially dynamic nature of decision-making in naturalistic settings. The heuristic assumes that decision makers first try to make sense of their situation [1] (Klein, 1993; Weick, 1979), proceed to use serial option evaluation in search for an appropriately matching option evaluating, [2] and critique tentatively selected alternatives by mental simulation of their implementation [3] (Klein, 1993). If decision makers cannot make sense of the situation they experience confusion or lack of understanding (Weick, 1979) which they try to reduce by collecting additional information, delaying action, searching for advice, or falling back on familiar habits or standard operating procedures [4a & 4b]. This aspect of the R.A.W.F.S. heuristic is consistent with our findings as well as with the R.Q.P heuristic and numerous writers who suggest that decision makers first try to cope with uncertainty by searching for additional information (e.g., Smithson, 1989). If reduction is unfeasible as is often the case because reliable information is scarce in the real world, decision makers experience uncertainty as lack of information and resort to assumption-based reasoning [5]. This aspect of the R.A.W.F.S. heuristic is consistent with the pattern of

contingent coping in our findings, as well as with Cohen et al. (1993). If reduction or assumption-based reasoning produce two or more admissible and insufficiently differentiated alternatives [6], decision makers experience uncertainty as conflict among alternatives which they attempt to resolve by weighing the pros and cons of each alternative [7]. This aspect of the R.A.W.F.S. heuristic is consistent with our findings as well as with Image Theory (Beach, 1990), which posits a similar pattern of sequential followed by concurrent alternative evaluation similar to the sequence [2] → [6] → [7] in the heuristic. The positions of *reduction, assumption-based reasoning* and *weighing of pros and cons* in Figure 1 reflect their differentiated associations with lack of understanding, lacking information, and conflict among alternatives. The positions of *suppression* [8] and *forestalling* [5], [8] reflect their undifferentiated associations with a particular form of uncertainty. In addition, the links leading to suppression are based on Janis & Mann (1977). The sequence depicted in Figure 1 is not obligatory. Thus, for example, if a decision maker frames his or her uncertainty as undifferentiated alternatives to begin with, he or she will "enter" the process at node [6].

The R.A.W.F.S. heuristic suggests several directions for future research. First, the heuristic clearly needs to be tested with decision-making *in vivo*. Alternatively, it is possible to use (preferably high fidelity) simulators such as those that are used for training. The R.A.W.F.S. heuristic can also be tested and elaborated by way of applications. Cohen, Adelman, Tolcott, Bresnick, & Marvin (1993) have already outlined how assumption-based reasoning can be used to support situation-assessment based decision-making. This work can be extended to include other strategies and meta-strategies which specify conditions of optimal use for each strategy. Finally, since the tactics (and certainly the strategies) are fairly abstract, studying how practitioners in different domains operationalize them concretely can be used to design training programs.

REFERENCES

Allaire, Y., & Firsirotu, M.E. (1989). Coping with strategic uncertainty. *Sloan Management Journal*, (3), 7-16.

Beach, L.R., (1990). *Image theory: Decision-making in personal and organizational contexts*. London: Wiley.

Brunsson, N. (1985). *The Irrational organization*. Chichester: Wiley & Sons.

Cohen, M.S., Adelman, L., Tolcott, M.A., Bresnick, T.A., & Marvin, F.M. (1993). *A cognitive framework for battlefield commanders' situation-assessment*. TR 93-1, United States Army Research Institute, Fort Leavenworth Field Unit.

Cohen, M.S., Schum, D.A., Freeling, A.N.S., & Chinnis, J.O. (1985). *On the art and science of hedging a conclusion: Alternative theories of uncertainty in intelligence analysis*. Falls Church, VA: Decision Science Consortium.

Curley, S.P., Yates, F., & Abrams, R.A. (1986). Psychological sources of ambiguity avoidance. *Organizational Behavior and* Hirst, E., & Schweitzer, M.

Cyert, R., & March, J. (1963). *A behavioral theory of the firm.* Englewood Cliffs, NJ: Prentice-Hall.

Hogarth, R.M. (1987). *The psychology of judgment and choice.* San Francisco, Ca: Jossey Bass.

Isenberg, D. (1985). Some hows and whats of managerial thinking: Implications for future army leaders. In J.G. Hunt and J. Blair (Eds.), *Military leadership In the future battlefield* NY: Pergammon Press.

Janis, I.L., & Mann, L. (1977). *Decision-making: A psychological analysis of conflict, choice and commitment.* New York: Free Press.

Klein, G.A. (1993). A recognition-primed decision (RPD) model of rapid decision-making. In Klein, G. A , J. Orasanu, R. Calderwood, & C. Zsambok (eds.), *Decision-making in action: Models and methods,* (138-147). Norwood, CT: Ablex.

Lipshitz, R., & Strauss, O. (1996). *Coping with uncertainty: A naturalistic decision-making analysis.* University of Haifa, Haifa, Israel.

March, J.G. & Shapira, Z. (1987). Managerial perspectives on risk and risk taking. *Management Science, 33,* 1404-1418.

Matlin, M.W., & Stang, D.J. (1978*). The Pollyana principle: Selectivity in language, memory and thought.* Cambridge, MA: Schenkman.

McCaskey, M.B. (1986). *The executive challenge: Managing change and ambiguity.* Harper Business.

Meehl, P.E. (1978). Theoretical risks and tabular asterisks: Sir Karl, Sir Ronald, and the slow progress of soft psychology. *Journal of Counseling and Clinical Psychology, 46,* 806-834.

Raiffa, H. (1968*). Decision analysis: Introductory lectures on choices under uncertainty.* Reading, MA: Addison Wesley.

Smithson, M. (1989). *Ignorance and uncertainty: Emerging paradigms.* New York: Springer Verlag.

Thompson, J. (1967). *Organizations in action.* New York: McGraw Hill.

Weick, K.E. (1979). *The Social psychology of organizing.* Reading, MA: Addison Wesley.

Figure 1
Coping with Uncertainty: The
R.A.W.F.S. Heuristic Hypothesis.

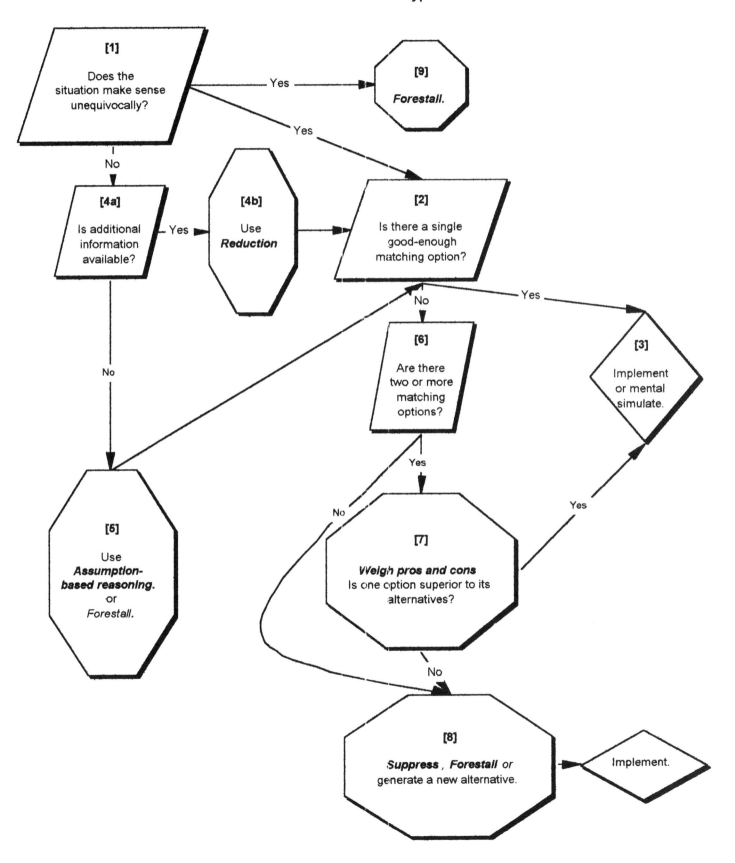

AN ANALYSIS OF UNCERTAINTY IN THE MARINE CORPS

Michael J. McCloskey
Klein Associates Inc.
Fairborn, Ohio

In an effort to better understand the nature of uncertainty faced by Marine Corps warfighters, Klein Associates conducted an exploratory study of uncertainty. We developed a method for classifying levels and sources of uncertainty and we studied Marines in action to determine the specific types of uncertainty that occurs on the battlefield. Uncertainty can exist at the levels of raw data, inferences drawn from data, or projections/diagnoses from multiple inferences and/or data. The sources of uncertainty include missing, unreliable, ambiguous/conflicting, or complex information.

We observed decision makers at regimental and battalion command posts during a military exercise and recorded instances where they faced uncertainty. We coded these instances for both the level and source of the uncertainty. The uncertainty events we observed clustered around missing, unreliable and ambiguous data, ambiguous inferences, and ambiguous projections. These areas have implications for the design of supports for operating in the midst of uncertainty.

INTRODUCTION

Marine Corps officers must constantly make critical battlefield decisions based on missing or otherwise degraded information. An inexperienced officer may be hesitant to act until more information can be gathered and completely informed decisions can be made. Experienced officers, however, realize the importance of making quick, effective decisions in the midst of uncertainty. As General George Patton observed, "A good plan violently executed now is better than a perfect plan next week."

In order to help decision makers operate effectively under uncertainty we must first understand *what makes an event uncertain*. The objectives of this effort, therefore, were to identify primary sources and levels of uncertainty and to determine the main types of uncertainty facing Marine Corps officers. To do this, we conducted an exploratory study of uncertainty in the Marine Corps regiment and battalion. We developed a means for classifying instances of uncertainty and applied this classification scheme to the uncertainty faced by Marine decision makers.

METHOD

Uncertainty Coding Scheme

In the course of our investigation, we developed two dimensions for classifying uncertainty. One dimension described four potential *sources* of uncertainty. The other dimension described the *levels* at which uncertainty can exist. We developed these particular dimensions because we believe that they provide a descriptive, meaningful representation of the nature of uncertainty.

In the development of the *sources* of uncertainty, we reviewed narratives in which historical military commanders were faced with uncertain situations on the battlefield. Combining these findings with our observations of a Marines exercise, we identified four primary sources of uncertainty for battlefield decision makers. These sources of uncertainty are consistent with the framework presented by Cohen, Freeman and Wolf (in press):

Missing information - Information is missing when the decision maker does not possess it, or does possess it, but cannot find it when it is needed. A commander waiting for an observer to phone in and tell him whether a target has been destroyed is uncertain due to missing information.

Unreliable information - Information is unreliable when the credibility of the source is in question. The information source is either low, or perceived to be low even if the information is highly accurate. An intelligence officer receiving a report from an inexperienced observer in the field hesitantly reporting, "Yeah, it..um..looks like the enemy advance guard passing through the channel" would be uncertain about that force due to unreliability of the reporter.

Ambiguous/conflicting information - Information is ambiguous and/or conflicting if there is more than one reasonable way to interpret it. A commanding officer who receives a report of a large dust trail along one potential enemy avenue of approach, but is told by his intelligence officer that the terrain along that avenue is impassable by enemy troops, would likely suffer uncertainty due to ambiguity or conflict.

Complexity - Information is complex when it is difficult to integrate the different facets of the data. The information often becomes complex when there are several different possible outcomes and many separate pieces of data must be integrated in order to develop a clear picture of the situation. An operations officer trying to predict the outcome of a battle with a vast number of unpredictable, intangible factors, such as weather, friendly and enemy morale levels, enemy intentions, and friendly and enemy reactions, will face uncertainty due to the complexity of the situation.

The second dimension describes the *levels* of uncertainty. This dimension parallels Endsley's (1995) components of situation awareness. The categories of this dimension are:

Data - This refers to uncertainty regarding elements of raw data. These might include grid coordinates of troop locations, enemy force sizes and strengths, and amount of available supplies. This level frequently involves quantitative data. This category

is consistent with Endsley's level one, the perception of elements of the environment.

Inference - A body of data becomes knowledge when inferences are drawn from it. For example, if a commander hears a report of four unknown enemy vehicles in a lightly wooded area, he may infer that, because tanks are the only type of vehicle that can traverse that terrain, the enemy vehicles must be tanks. This category is consistent with Endsley's level two, comprehending the situation.

Projection - At this level, inferences are synthesized into projections of the future, into diagnoses and explanations of events, or into identification of critical battle focal points. A commander will take several inferences regarding enemy and friendly assets, strengths, and capabilities, as well as a multitude of environmental and other factors, and develop an understanding of what should happen on the battlefield. This is consistent with Endsley's level three, the projection of future events.

Combining these two axes, *source* and *level*, we developed a matrix for codifying the instances of uncertainty.

Data Collection

We conducted observations at a combined arms exercise (CAX) at the exercise range at Twentynine Palms, California in September, 1995. A team of trained observers spent five days, working at both the regimental and the battalion command posts. Among the five observers, we observed and recorded nearly 100 instances in which command post staff members were faced with uncertainty.

Three of the observers then reviewed each uncertainty narrative and made judgments as to the presence of uncertainty. The goal was to select only the instances where uncertainty was definitely present. Between the three observers, 59 narratives were agreed upon as containing instances of uncertainty. These narratives were then rewritten by a fourth observer to remove any subjective assessments regarding the nature of the uncertainty.

Table 1. Matrix of Observed Instances of Uncertainty During Exercise

Sources of Uncertainty Levels of Uncertainty	Missing	Unreliable	Ambiguous /conflicting	Complex
Data	19	7	7	0
Inference	1	1	8	0
Projection	2	1	13	0

Data Coding

Two researchers independently coded the content of uncertainty. The narratives were coded independently along the two dimensions (the three levels of uncertainty and the four sources of uncertainty). Along both dimensions, agreement was statistically acceptable (source agreement 73%, $\kappa=0.536$; level agreement 67%, $\kappa=0.589$) (Fleiss, 1981).

RESULTS

Table 1 presents the uncertainty matrix with the frequencies of type of uncertainty. 56% of the uncertainty incidents for the Marines were about the data. Of these uncertainty events relating to data, 58% were a result of missing data. This is a common example of missing data:

> *During the exercise, members of the regimental CP delayed further operations because they did not know if a handoff of control had occurred on the battlefield. They were waiting for a message over the radio stating that the handoff had occurred, and until they got that message, they would not proceed with their plans.*

Of the data-related incidents of uncertainty, 21% were caused by unreliable data. This sometimes occurred when a commander or other staff member would receive data from a relatively new piece of equipment in the command post. This equipment had a history of occasionally providing inaccurate data. Rather than trust the data, the commander would request verification from some other source, often from an observer on the battlefield who could provide "eyes-on" confirmation.

The remaining 21% of instances of data-related uncertainty were the result of ambiguous or conflicting data. In these cases, there would typically be multiple pieces of data obscuring the true situation. The fire support coordinator, the staff member responsible for directing artillery to targets and coordinating these fires with the air officer, tended to fall victim to this type of uncertainty:

> *In one case, the fire support coordinator had to suddenly cancel a mission where aircraft were supposed to deliver weapons on a particular target. The artillery unit was supposed to cease firing on the target for a brief time, creating a time window for the aircraft to deliver their weapons without the threat of friendly fire. The data received by the coordinator, though, were unclear. They did not convince him that the artillery had indeed stopped firing and that the aircraft could proceed safely to the target.*

At the inference level, the dominant source of uncertainty was ambiguous or conflicting inferences. These cases comprise 80% of the inference-related instances of uncertainty. A characteristic example of this type of uncertainty follows:

A battalion commander believed he had control of some air assets and began to make plans accordingly. In reality, control had not been passed to him from the regimental commander. The battalion commander inferred from what he knew about the available assets and his own battlefield situation that some aircraft would be his to direct, and so he began to act and plan tentatively as if they were his. He assumed, but was uncertain, that a higher-level commander had already passed the control.

At the level of projection, 81% of the instances of uncertainty were caused by ambiguous or conflicting projections. These projections were either about the progression of future events or the diagnosing of the current state of battlefield situations. The intelligence officer in the command post is frequently called upon to project about future events. Since the events have not occurred yet, there is an inherent degree of uncertainty in much of his work. For example:

The intelligence officer must often predict the routes that enemy units will be taking. In one case, a less experienced intelligence officer relied on simple heuristics to predict an enemy route. He guessed that the enemy would take the easiest, most direct route of advance. When he eventually received intelligence data back from the battlefield, the intelligence officer discovered that the adversary had selected a route that was never even considered in the command post.

DISCUSSION

Table 1 showed the five cells within our matrix where the uncertainty was most frequent. At the levels of inference and projection, the trouble centers around ambiguous/conflicting inferences and projections. At the data level, a majority of the

uncertainty is the result of missing data (although unreliable and ambiguous data also contributed significantly to uncertainty). None of the observed instances of uncertainty resulted from complex information. This was likely a result of the simple nature of the exercise, and also because regimental responsibilities are more simple than those of division or higher echelons. In actual military operations, some uncertainty will likely be caused by complexity. Therefore, we suspect that this sixth cell may be important to consider — uncertainty in understanding due to complexity.

At the level of inferences, uncertainty frequently resulted from ambiguous or conflicting inferences. Future support development could focus on facilitating decision making in this area. We recommend the use of decision scenario training augmented with expert modeling to provide explanations for the occurrences of events and help decision makers clarify potentially confusing inferences.

There is another aspect to be considered — the efforts to manage uncertainty. Cohen, Freeman and Wolf (in press) have directed attention to metacognitive processes. Although we did not code for these, we did find a high rate of activity devoted to metacognitive issues such as verifying reports, questioning assumptions, and developing alternative means to gather information. These metacognitive activities need to be considered, particularly in deriving training programs.

The fact that a great deal of uncertainty was caused primarily by missing data, but also by unreliable or ambiguous data, has implications for the design of computer supports. The battlefield will never be completely certain. As new systems are implemented, questions about missing information will not go away. They will simply be replaced with increasingly more detailed questions. In addition, the tempo of decision making will increase as information technologies mature; commanders will be pressed to make faster decisions at the same level of uncertainty as they have previously encountered. Most commanders do not need more information. They need practice at making decisions with incomplete or imperfect information. Some of the Marine decision makers we observed would not accept any degree of uncertainty, and thus became paralyzed when faced with uncertainty on the battlefield. Throughout history, exceptional commanders shared the characteristic of being able to operate comfortably in the midst of uncertainty. We have developed a script for reacting to uncertainty (Klein, Schmitt, McCloskey, Heaton, & Wolf, 1996), in which decision makers can respond to uncertainty with actions that prepare for, represent, reduce, or manage it. (See Lipshitz & Strauss (1996), for a related framework.) As we learn more about the nature of uncertainty, we will be in a better position to develop useful guidance and support.

ACKNOWLEDGMENTS

This research was sponsored by Contract N66001-95-C-7024 from the Naval Command, Control and Ocean Surveillance Center, San Diego, CA.

REFERENCES

Cohen, M.S., Freeman, J.T., & Wolf, S.P. (in press). Meta-recognition in time-stressed decision making: recognizing, critiquing, and correcting. Human Factors Special Issue. Santa Monica, CA: HFES.

Endsley, M.R. (1995). Toward a theory of situation awareness in dynamic systems. Human Factors Special Issue, 37 (1), 85-104.

Fleiss, J.L. (1981). Statistical methods for rates and proportions. New York, NY: John Wiley & Sons.

Klein, G., Schmitt, J., McCloskey, M., Heaton, J., & Wolf, S. (1996). Fighting in the Fog: A Study of Uncertainty in the U.S. Marine Corps. NCCOSC under Contract #N66001-95-C-7024.

Lipshitz, R. & Strauss, O. (1996). Coping with uncertainty: A naturalistic decision-making analysis. Human Factors & Ergonomics Society Proceedings.

IMPACT OF NATURALISTIC DECISION SUPPORT ON TACTICAL SITUATION AWARENESS

Jeffrey G. Morrison
NCCOSC RDT&E Division (NRaD)
San Diego, CA

Richard T. Kelly
Pacific Science & Engineering Group, Inc.
San Diego, CA

Susan G. Hutchins
Naval Postgraduate School
Monterey, CA

A prototype decision support system (DSS) was developed to enhance Navy tactical decision making based on naturalistic decision processes. Displays were developed to support critical decision making tasks through recognition-primed and explanation-based reasoning processes and cognitive analysis of the decision making problems faced by Navy tactical officers in a shipboard Combat Information Center. Baseline testing in high intensity, ambiguous scenarios indicated that experienced decision makers were not well served by current systems, and their performance revealed periodic loss of situation awareness. A study is described with eight, expert Navy tactical decision making teams that used either their current system alone or in conjunction with the prototype DSS. When the teams had the prototype DSS available, we observed significantly fewer communications to clarify the tactical situation, significantly more critical contacts identified early in the scenario, and a significantly greater number of defensive actions taken against imminent threats. These findings suggest that the prototype DSS enhanced the commanders' awareness of the tactical situation, which in turn contributed to greater confidence, lower workload, and more effective performance.

INTRODUCTION

Efforts to develop a prototype decision support system (DSS) were initiated as one thrust of the Navy's Tactical Decision Making Under Stress (TADMUS) project. The objective of this effort is to evaluate and demonstrate display concepts derived from current cognitive theory with expert decision makers in an appropriate test environment. The focus of the DSS was on enhancing the performance of tactical decision makers (viz., the Commanding Officer (CO) and Tactical Action Officer (TAO) working as a team) for single ship, air defense missions in high density, ambiguous littoral warfare situations. The approach taken in designing the DSS was to analyze the cognitive tasks performed by the decision makers in a shipboard Combat Information Center (CIC) and then to develop a set of displays to support these tasks based on the underlying decision making processes naturally used by the CO/TAO team.

Cognitive task analyses identified two higher order tasks performed by the CO/TAO team: situation assessment and selection of alternative courses of action (Kaempf, Wolf, & Miller, 1993). The analyses indicated that 87% of the information transactions associated with situation assessment involved feature matching strategies (trying to match the observed events in the scenario to those previously experienced), while 12% of their actions were related to story generation strategies (developing a novel hypothetical explanation to explain the observed events). With regard to selecting courses of action, command level decision makers relied almost exclusively on recognition of applicable tactics based on rules of engagement (94%), while much more rarely developed a general selection strategy extrapolated from previous experience (6% of actions selected).

Baseline tests in representative littoral scenarios corroborated these analyses (Hutchins & Kowalski, 1993; Hutchins, Morrison, & Kelly, 1996). The communications analysis indicated a predominance of feature matching strategies in assessing the situation typically followed by the selection among preplanned response sets (tactics) that were considered to fit the situation. These tests also suggested that experienced decision makers were not particularly well served by current systems in demanding missions. Teams exhibited periodic losses of situation awareness, often linked with limitations in human memory and shared attention capacity. Environmental stressors such as time compression and highly ambiguous information increased decision biases, e.g. confirmation bias, hypervigilance, task fixation, etc. Problems associated with short term memory limitations included: (a) mixing up track numbers (track being recalled as 7003 vs. 7033) and forgetting track numbers; (b) mixing up track kinematic data (track recalled as descending vs. ascending in altitude, closing vs. opening in range, etc.) and forgetting track kinematic data; and (c) associating past track related events/actions with the wrong track and associating completed own-ship actions with the wrong track. Problems related to decision biases included: (a) carrying initial threat assessment throughout the scenario regardless of new information (framing error) and (b) assessing a track based on information other than that associated with the track, e.g., old intelligence data, assessments of similar tracks, outcomes of unrelated events, past decision maker experiences, etc. (e.g. confirmation bias).

DECISION SUPPORT SYSTEM DESIGN

Based on these analyses, a prototype DSS was developed with the objectives of: (1) minimizing the mismatches between cognitive processes and the information available in the

CIC to facilitate decision making; (2) mitigating the short-comings of current CIC displays in imposing high information processing demands and exceeding the limitations of human memory; and (3) transferring the data in the current CIC from numeric to graphical representations wherever appropriate to facilitate the interpretation of spatial data. It was determined that the DSS should not filter or extensively process data; i.e., it should support rather than aid (automate) decision making and leave as much decision making with the decision makers as possible. The design goal of the DSS was to take the data in the system and present it as meaningful information relative to the decision making tasks being performed based on a theoretical understanding of human decision making.

The current generation DSS was designed expressly for the evaluation of display elements to support feature matching, story generation (viz. Explanation-Based Reasoning (EBR)), and Recognition-Primed Decision making (RPD) with the goal of reducing errors, reducing workload, and improving adherence to rules of engagement. The design was significantly influenced by inputs from subject matter experts to ensure its validity and usefulness for the operational community. It is implemented on a Macintosh computer which may operate independent of, synchronized with, or linked to a scenario driver simulation.

Figure 1 shows the first DSS prototype display. The DSS is a composite of several display modules, which are arranged in a tiled format so that no significant data are obscured by overlapping windows. The DSS was conceived as a supplementary display to complement the existing geo-plot and text displays in current CICs. DSS modules have been discussed and demonstrated in detail elsewhere (cf. Moore, Quinn, & Morrison, 1996). Nevertheless, three modules will be discussed here as an illustration of how the information requirements of tactical decision making tasks were mapped with cognitive processes described in naturalistic decision making theory to generate the DSS.

Track Profile

The track profile module consists of two graphical displays in the upper portion of the DSS that show the current position of a selected track in both horizontal and plan-form displays. Information requirements addressed by this module included the need to: (1) see where the target track is relative to own-ship, (2) see what the track has been doing over time, (3) recognize whether the target can shoot you, and (4) recognize whether you could shoot the target. An important aspect of this display is that it shows a historical plot of what the target has done in space and time since it was first acquired by the system (the history is replayed each time the target is selected). This greatly offloads the short term memory requirements on the CO and TAO in interpreting the significance of

the selected target. This historical dimension of the display allows the decision maker to see what the track has done and primes his recognition of a likely mission for that track which would account for its actions. In addition, the profiles show own-ship weapon and target threat envelopes displayed in terms of range and altitude so that the decision maker can visualize and compare mental models (templates) as he considers possible track intentions and own ship options.

Response Manager

The response manager is located immediately below the track profile and is tied to it via a line indicating the target's current distance from own ship. It represents a Gantt chart type display showing a template of pre-planned actions and the optimal windows in which to perform them. The display serves as a graphical embodiment of battle orders and doctrine, and shows which actions have been taken with regard to the selected track. The display is intended to support RPD and serves the need to: (1) recall the relevant tactics and strategies for the type of target being assessed, (2) recognize which actions need to be taken with the target and when they should be taken, and (3) remember which actions have been taken and have yet to be taken for the selected target.

Basis for Assessment

This module is located in the lower left area of the DSS and is intended to support EBR (story generation). The basis for assessment module presents the underlying data used to generate the DSS's threat assessment for the displayed track. The display shows three categories of assessment decision makers focus on: potential threat, non-threat, or unknown. The decision maker selects the hypothesis he wishes to explore and data are presented in a tabular format within three categories: supporting evidence, counter evidence, and assumptions. These categories were found to be at the core of all story generation in which commanders engage while deciding whether a target with the potential to be a threat is, in fact, a real threat. This EBR related to threat assessment is also typically one of the decision making tasks performed when deciding whether to fire on a target or not. The display was designed to present the relevant data necessary for a commander to consider and evaluate all likely explanations for what a target may be, and what it may be doing (i.e., "intents") through the generation of alternative stories to explain the available and missing data regarding the target in question. The display is also intended to highlight data discrepant with a given hypothesis to minimize confirmation and framing biases. Assumptions listed are those assumptions necessary to "buy into" the selected assessment. As a result, the basis for assessment module is expected to be particularly effective in helping sort out and avoid "Blue-on-Blue" and "Blue-on-White" engagements.

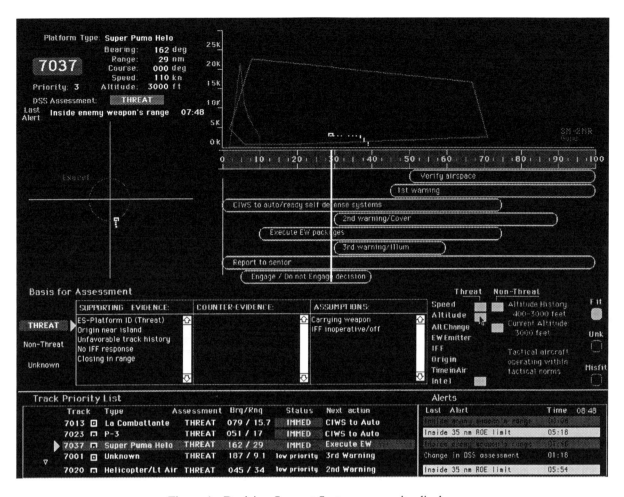

Figure 1. Decision Support System composite display.

DSS EVALUATION EXPERIMENT

The ultimate goal of any display design is to positively impact the performance of the person-machine system of which it is a part. Therefore, a study was performed to examine how the DSS impacted the decision making of COs and TAOs relative to performance in a traditional CIC in a medium-fidelity simulation. Although the contributions of individual display modules could not be assessed objectively due to resource limitations, overall effects of the DSS on decision performance were examined in terms of a variety of performance criteria.

Method

Eight expert Navy tactical decision making teams (with emphasis on the CO and TAO) used either their current display systems alone or in conjunction with the prototype DSS at NRaD's *Decision Evaluation Facility for Tactical Teams* (DEFTT) CIC simulator. A within-subject factorial design was employed across four test scenarios such that each team performed two scenarios with the DSS and two scenarios without it. Scenarios were constructed to simulate peace keeping missions with a very high number of targets to be dealt with in a short period of time (i.e. were time compressed), and with a significant number of highly ambiguous tracks regard-

ing assessment and intent. Subjects were given appropriate geo-political and intelligence briefings prior to each test run. The order of the scenarios and DSS conditions was counterbalanced using a Latin Square. Criterion-referenced training with the baseline DEFTT display system and with the DSS was provided, and two practice scenarios were run prior to beginning the test session. In addition to collecting objective data on tactical actions, display usage, control inputs, and voice communications, subjective assessments (via questionnaires and a structured interview) were solicited from each CO and TAO at the conclusion of the test session.

Results

Results indicated no evidence of a practice effect over the four-scenario test session and no consistent differences between the scenarios themselves. Substantial differences were observed, however, between teams – notably in their subjective workload assessments and in their communications.

The results of primary interest concerned the extent to which the DSS promoted greater awareness of the tactical situation by the CO and TAO. Awareness of the tactical situation was examined via several performance measures. Specifically, it was predicted that if the CO/TAO team was more aware of the tactical situation, they would:

- identify the critical contacts earlier and more accurately;
- take more of the tactical actions required by the rules of engagement in a timely manner (i.e., later); and
- ask fewer questions to clarify previously reported track data and the relative locations of tracks.

Critical contacts. During the scenario runs, the experimenter probed the CO/TAO team at prespecified times to identify the tracks that were considered to be of greatest tactical interest at that time. Their responses were contrasted with those of an independent group of five subject matter experts. As shown in Figure 2, significantly more of the critical contacts were identified when the DSS was available. Significant differences ($p < .05$) were noted at both the early and mid-scenario probes; performance was comparable at the late probe, however. Late in the scenario the critical tracks may become more obvious even without the DSS. Nevertheless, earlier recognition of critical tracks earlier in the scenario affords decision makers a broader array of response options and permits more effective coordination of response actions.

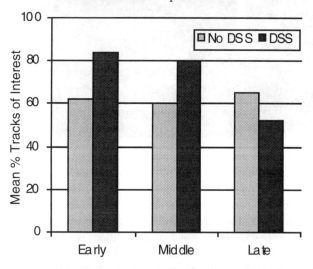

Figure 2. Percent of critical contacts reported as tracks of interest.

Tactical actions. Using the rules of engagement as a benchmark for decision performance in the scenarios, a group of subject matter experts assessed whether the CO/TAO teams warned and/or illuminated threat tracks at specified times and took appropriate defensive actions. A modified form of the AAW Team Performance Index (Dwyer, 1992) was used for scoring tactical performance, and these data are summarized in Figure 3. In scenarios when the DSS was available, CO/TAO teams were significantly more likely to take defensive actions in a timely manner against imminent threats ($p < .05$). This indicates that the DSS promoted an earlier recognition of the emerging risks of the tactical situation. By contrast, no difference was observed in the number of tracks that were warned or illuminated when the DSS was available. However, several subject matter experts contended that these may not be diagnostic performance indices since they represent provocative tactical actions that commanders may consider to be inappro-

priate against certain tracks in a littoral situation. Not taking provocative actions would be appropriate and expected if commanders had assessed that the track was not an imminent threat, and felt comfortable with prolonging those actions because they had a good tactical picture - as would be expected if the DSS was being effective in meeting its design objectives.

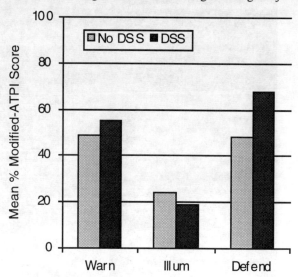

Figure 3. Team performance of tactical actions required by the rules of engagement.

Clarifying communications. The voice communications during each of the scenario runs were coded by their message content (exchanging tactical data or track status, correlating or assessing tracks and issuing orders, and clarifying the tactical situation). Overall, about 20% of the communications were for clarification purposes, reflecting uncertainty about track location, kinematics, identification, status, or priority. When the teams had the DSS available, fewer communications were aimed at clarifying the tactical situation, particularly track kinematics, identification, and priority – each of which are directly aided by the DSS. On the other hand, with the DSS, decision makers tended to spend more time clarifying ambiguous communications and checking on the status of actions. While this result may seem counterintuitive, it reveals a greater situation awareness where ambiguous, incomplete, or erroneous communications are more likely to be caught and corrected when the DSS was available.

User responses. Feedback from the expert CO/TAO teams who participated in this experiment also indicated that the DSS provided them an excellent summary of the overall tactical situation as well as of key data for individual tracks. In particular, COs and TAOs considered that both the Track Profile and the Basis for Assessment modules provided important information not readily available in present day systems (see Figure 4). Since the Track Profile module supported feature matching, which is the most commonly used decision strategy, its high rating was anticipated. Yet, when the track data are conflicting or ambiguous and when the decision maker has time available, the Basis for Assessment module was rated as helping substantially. Note that by encouraging decision makers to consider the full range of available evidence along with various explanations for it, this module reduces the likelihood of mistakenly engaging friendly or neutral tracks, and was rated highly with regard to avoiding Blue-on-Blue and Blue-on-White engagements.

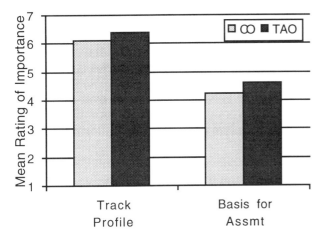

Figure 4. Mean CO and TAO ratings of the importance of information in Track Profile and Basis for Assessment.

CONCLUSIONS

Operational decision making predominantly relies on feature matching strategies. To a lesser extent when faced with conflicting or ambiguous data, decision makers employ story generation or explanation based reasoning strategies. Displays that are consistent with these naturalistic decision making strategies provide the most useful support to commanders, facilitating the rapid development of an accurate assessment of the situation. Displays that support both feature matching and explanation based reasoning are recommended for complex decision making tasks. While the feature matching displays will likely be used far more often, the explanation based reasoning display is of substantial value under certain circumstances, particularly with less experienced decision makers.

The DSS was developed for application to Navy tactical decision making on a single ship in support of AAW in dense, fast-paced littoral settings. With some adaptation, it could support other military decision situations, including concurrent decisions involving other warfare areas, higher-level, supervisory decisions involving multi-ship battle groups, and even collaboration among tactical decision makers in joint service or multi-national operations. Several new research projects are underway to explore these applications. In addition to these direct applications to support military decision making, the decision support and display principles identified through this effort are relevant to other complex decision making settings, such as nuclear power control, flight control, process control, and disaster relief planning. Further, additional work is looking at developing derivative displays reflect emerging theories of decision making, extension of the DSS concepts to other

workstations within the CIC, as well as better integration of DSS modules with shipboard data processing systems.

ACKNOWLEDGMENTS

This effort was performed as part of the Tactical Decision Making Under Stress (TADMUS) project, sponsored by the Office of Naval Research, Cognitive and Neural Science Technology Division with Gerald S. Malecki as program manager. The authors gratefully acknowledge the substantial contributions of Jeffrey Grossman, Steve Francis, Brent Hardy, Pat Kelly, C. C. Johnson, Ron Moore, Connie O'Leary, Mike Quinn, Will Rogers, and Dan Westra to various phases of this project, as well as our colleagues at NAWC-TSD. We acknowledge the support of the TADMUS Technical Advisory Board, and dedicate this paper to the memory of Dr. Martin A. Tolcott, its Chair from 1988 through 1996.

REFERENCES

Dwyer, D. J. (1992). An index for measuring naval team performance. *Proceedings of the Human Factors Society 36th Annual Meeting*, 1356-1360. Santa Monica, CA: Human Factors Society.

Hutchins, S. G. and Kowalski, J. T. (1993). Tactical decision making under stress: Preliminary results and lessons learned. *Proceedings of the 10th Annual Conference on Command and Control Decision Aids*. Washington, DC: National Defense University.

Hutchins, S. G., Morrison, J. G., and Kelly, R. T. (1996). Principles for aiding complex military decision making. *Proceedings of the Second International Command and Control Research and Technology Symposium*, Monterey, CA: National Defense University.

Kaempf, G. L., Wolf, S., and Miller, T. E. (1993). Decision making in the Aegis combat information center. *Proceedings of the Human Factors and Ergonomics Society 37th Annual Meeting*, 1107-1111. Santa Monica, CA: Human Factors and Ergonomics Society.

Moore, R. A., Quinn, M. L., and Morrison, J. G. (1996). A tactical decision support system based on naturalistic cognitive processes. *Proceedings of the Human Factors and Ergonomics Society 40th Annual Meeting*, this volume. Santa Monica, CA: Human Factors and Ergonomics Society.

Automation Bias, Accountability, and Verification Behaviors

Kathleen L. Mosier
San Jose State University
Foundation
at NASA Ames Research Ctr.

Linda J. Skitka
Mark D. Burdick
University of Illinois at
Chicago

Susan T. Heers
Western Aerospace Labs
at NASA Ames

Automated procedural and decision aids may in some cases have the paradoxical effect of increasing errors rather than eliminating them. Results of recent research investigating the use of automated systems have indicated the presence *automation bias*, a term describing errors made when decision makers rely on automated cues as a heuristic replacement for vigilant information seeking and processing (Mosier & Skitka, in press). *Automation commission errors*, i.e., errors made when decision makers take inappropriate action because they over-attend to automated information or directives, and *automation omission errors*, i.e., errors made when decision makers do not take appropriate action because they are not informed of an imminent problem or situation by automated aids, can result from this tendency.

A wide body of social psychological research has found that many cognitive biases and resultant errors can be ameliorated by imposing pre-decisional accountability, which sensitizes decision makers to the need to construct compelling justifications for their choices and how they make them. To what extent these effects generalize to performance situations has yet to be empirically established. The two studies presented represent concurrent efforts, with student and "glass cockpit" pilot samples, to determine the effects of accountability pressures on automation bias and on verification of the accurate functioning of automated aids. Students (Experiment 1) and commercial pilots (Experiment 2) performed simulated flight tasks using automated aids. In both studies, participants who perceived themselves "accountable" for their strategies of interaction with the automation were significantly more likely to verify its correct functioning, and committed significantly fewer automation-related errors than those who did not report this perception.

Automated decision aids have been introduced into many work environments with the explicit goal of reducing human error. Flight management systems, for example, are taking increasing control of flight operations, such as calculating fuel efficient paths, navigation, detecting system malfunctions and abnormalities, in addition to flying the plane. Nuclear power plants and even medical diagnostics are similarly becoming more and more automated. One of the purportedly beneficial facets of automation and automated decision aids is that they replace or supersede traditional displays of information with new, typically more salient cues. A potential danger in this is that the information search process of human operators may get short-circuited, i.e., operators may stop short at the automated display, and not double-check the operation of the automated system. Because of this, automated procedural and decision aids may in some cases have the paradoxical effect of increasing errors rather than eliminating them. Results of recent research investigating the use of automated systems have indicated the presence *automation bias*, a term describing errors made when decision makers rely on automated cues as a heuristic replacement for vigilant information seeking and processing (Mosier & Skitka, in press). Potential negative effects of automation bias can be broken down into *automation commission errors*, i.e., errors made when decision

makers take inappropriate action because they over-attend to automated information or directives and do not attend to other environmental cues, and *automation omission errors*, i.e., errors made when decision makers do not take appropriate action because they are not informed of an imminent problem or situation by automated aids. Evidence of the tendency to make both omission and commission errors has been found in commercial air-crew self-reports (Mosier, Skitka, & Korte, 1994), as well as in non-flight decision making contexts with novices (Skitka & Mosier, 1994; Mosier, Skitka, & Heers, 1995).

A wide body of social psychological research has found that many cognitive biases and resultant errors can be ameliorated by imposing pre-decisional accountability, which sensitizes decision makers to the need to construct compelling justifications for their choices and how they make them. Accountability demands cause decision makers to employ more multidimensional, self-critical, and vigilant information seeking, and more complex data processing, and have been shown to reduce cognitive "freezing" or premature closure on judgmental problems (Kruglanski & Freund, 1983), and to lead decision makers to employ more consistent patterns of cue utilization (Hagafors & Brehmer, 1983). To what extent these effects generalize to performance situations has yet to be empirically established.

The two studies described below represent concurrent efforts, with student and with "glass cockpit" pilot samples, to determine the effects of accountability pressures on automation bias and on verification of the accurate functioning of automated aids.

EXPERIMENT 1: STUDENT STUDY METHOD

Participants were 121 university students. The primary experimental task was presented on a 13" color display monitor with four quadrants (see Figure 1), using the *Window/PANES: Workload/ PerformANcE Simulation* (NASA-Ames Research Center, 1989). The software presents a screen with four quadrants. Participants were required to track a target circle, presented in the upper right quadrant (all participants were right-handed), that moved as a function of the disturbance imposed by a sum of sines algorithm using a two-axis joystick.

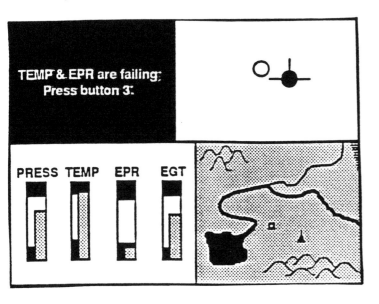

Figure 1. Window/PANES display.

Fluctuating bar graphs representing four gauges were presented in the bottom left quadrant. Subjects were trained to push a reset button on the control board when three of the four gauges went out of range. On the bottom right quadrant, subjects monitored their progress on a map, and were to press a button once they passed "waypoints" on the display. The upper left quadrant contained the "Automated Monitoring Aid" (AMA), which presented directives prompting gauge resets and waypoint button presses. The AMA also informed operators of impending gauge failures, and prompted a choice button response. Subjects were told, however, that the AMA was not 100% reliable, and that they could verify the accuracy of its choice

button response directives on another screen. The verification display, which used the MouseLab decision laboratory software (Johnson, Payne, Schkade, & Bettman, 1991) was run on a second microcomputer and displayed on a second monitor. Participants were instructed to open as many cells as required by the directive (1-3) in order from left to right to completely verify the directive's accuracy (i.e., whether or not the gauges were really failing and whether the response button indicated by the AMA was correct).

Design and Procedures

Participants were randomly assigned to Accountability conditions. Four groups of Accountable participants were instructed that their task was to maximize (1) their Overall Performance, or (2) Accuracy of responses, or (3) Quick Response to events, or (4) Tracking. All four of these groups were told that they were being constantly monitored by the experimenter, and that they would be asked to explain and justify their performance to the experimenters in an interview following the task. Participants in the Non-Accountable group were told that their performance could not be monitored by the experimenter, that their data were not being saved past the end of each trial, and no mention of an interview was made.

The number of boxes that had to be opened to completely verify an automated directive was the verification level for that event. The choice-response gauge task provided a way to examine commission errors (i.e., pressing an incorrect button) while assessing verification behaviors. Opportunities for omission errors occurred when subjects were not prompted to reset when necessary, or to indicate when they passed a waypoint. All participants were informed that the verification display data were 100% accurate.

Participants completed a total of three training and five experimental trials of ten minutes each. Across experimental trials were 50 total "events," i.e., gauge events and waypoint checks. Omission error opportunities occurred when three gauges went into the red zone with no automated prompt to reset, or when subjects passed waypoints and received no prompt to report in. Each type occurred twice. Six commission error opportunities were presented--all of which consisted of automated directives for choice response tasks that prompted the wrong button press (subjects who verified would see the correct button press indicated on the MouseLab display). At the end of each trial, participants were given visual feedback on their mean response time, percentage correct, and root mean squared tracking error.

After the experimental trials, participants were given a questionnaire that included manipulation checks, and tapped perceptions of their experience during the experimental task and their attitudes toward automation. No "justification" interviews were actually conducted.

RESULTS

Commission Errors and Verification Behavior. A mixed design analysis of variance (ANOVA) revealed main effects for Accountability condition on the tendency to make commission errors, $[F(4, 111)=3.26, p<.05]$. Subjects in the Overall Performance and Accuracy conditions made significantly fewer commission errors than subjects in the remaining three conditions, $[F(1, 111)=12.18, p<.001]$.

A similar pattern emerged with respect to which subjects were most likely to completely verify automated directives. Accountability had a significant effect on number of complete verifications, $[F(4, 111)=4.53, p<.01]$. Tukey tests indicated that subjects in the Overall Performance and Accuracy conditions were significantly more likely to completely verify than subjects in the Quick Response group. A significant Verification Level x Accountability interaction $[F(2, 222)=2.87, p< .01]$ was also revealed. As shown in Figure 2, subjects in the Overall Performance accountability condition were dramatically more likely to completely verify in the three level verification events than in the two or one level

events, and also more likely to verify (especially at higher levels) than other groups, regardless of level.

Omission Errors. Did accountability similarly affect subjects' tendency to make omission errors? A one-way analysis of variance investigating the effect of accountability condition on the number omission errors revealed that participants in the Overall Performance and Accuracy conditions made significantly fewer errors $[F(4,116)=2.49, p<.05;$ see Table 1] than those in other groups.

Table 1.
Number of Omission Errors as a Function of Accountability Condition.

Accountable for:	
Overall Performance	1.18
Accuracy	1.08
Quick Response	1.45
Tracking	2.12
Not Accountable	1.78

In sum, imposing accountability for Overall Performance or for Accuracy had the effect of making participants more vigilant and more likely to verify the accuracy of automated information, and resulted in fewer errors. Correlational analyses of post-experiment questionnaires indicated that highly confident and comfortable subjects were less likely to make commission errors $[r (111) = -.20, p<.05]$. Significant correlations were found between omission errors and items relating to accountability, such as comfort in justifying strategies, and perceptions that performance was being monitored, $[r (111) = -.15, p<.05]$.

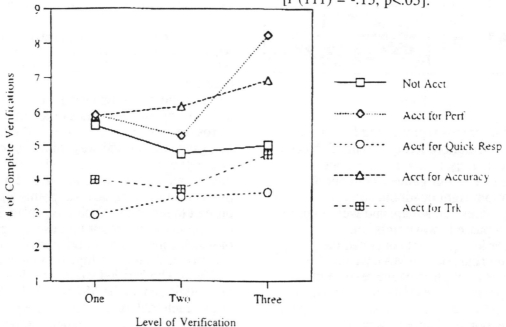

Figure 2. Number of complete verifications by accountability condition and verification level.

EXPERIMENT 2: COMMERCIAL PILOTS
METHOD

Participants in this study were 21 commercial glass cockpit pilots (i.e., pilots of automated aircraft). The part-task flight simulation facility used in this experiment is modeled after the Advanced Concepts Flight Simulator at Ames Research Center, and employs two Silicon Graphics color monitors to present glass displays of primary and secondary flight displays, navigation and communication information, and electronic checklists, as well as Engine Indicating and Crew Alerting System (EICAS) and Fight Management System (FMS) instruments. Subjects interacted with the controls and displays of the aircraft through a touchscreen overlaying the instruments needed to accomplish the flight.

In addition to the aircraft displays, a secondary tracking task was presented on one of the monitors to provide a means of increasing the workload of the pilots, and was incorporated into the automation events described below. This task involved using a joystick to keep displayed cross-hairs inside the boundaries of a blue circle. When the cross-hairs crossed the boundaries of the circle, the circle turned red. Feedback on how much time the subject was able to stay within the target circle was accumulated and displayed to the subject.

Design and Procedures

Pilots were assigned to either one Accountable condition or to a Non-Accountable group. Accountable subjects were told that their performance would be monitored, and that they would be interviewed and asked to explain and justify their strategies and performance in the use of automated systems at the end of the experiment. Additionally, a video camera was placed in a prominent position in the booth with Accountable participants. Pilots in the Non-Accountable group were told that their performance data could not be collected (due to a computer malfunction) or linked to them personally in any way, and no mention of an interview was made. Pilots were trained individually on how each of the components of the experimental task functioned and were given time to practice. Following training, subjects flew two legs (order was counterbalanced): Los Angeles (LAX) to San Francisco (SFO), and SFO to Sacramento (SMF). The flight route was pre-loaded into the FMS prior to beginning the trial. Subjects were instructed to communicate with Air Traffic Control (ATC) through textual datalink messages sent and received on the CDU screen of their FMS. Clearances from ATC (e.g., a change in altitude, speed, heading, or

frequency) could be auto-loaded into the appropriate flight system, and correct loading could be verified by checking the the Mode Control Panel (MCP) or navigation display. Pilots manually performed the secondary tracking task from SFO-SMF whenever they were above 5,000'. The secondary task was automated on the LAX-SFO leg.

Four automation failures during these legs offered the possibility for pilots to make omission errors if they did not verify proper automation functioning: 1) an altitude clearance misloaded into the flight control systems, and was reflected by incorrect numbers on the MCP; 2) the flight system incorrectly executed a commanded heading change, and the improper execution was reflected on the navigational display; 3) a frequency change misloaded into the flight control systems, reflected by incorrect numbers on the MCP; and 4) the tracking task automation failed at 7,000' during the LAX-SFO flight, which was signalled by the boundary circle turning red. In all cases, verification information was available on the appropriate display, as it would be in the aircraft. One opportunity for a commission error, a false "Engine Fire" message, yielded no variance in responses (every pilot committed the error), and is discussed in detail elsewhere (Mosier & Skitka, 1996). Debriefing forms included questions on flight experience, as well as questions probing perceptions of accountability and attitudes toward automation. No "justification" interviews were conducted.

RESULTS

Numbers of errors did not vary significantly as a function of manipulated accountability. They were, however correlated with total flight hours [$r(20)=.49$, $p<.05$] and with years of flight experience [$r(20)=.46$, $p<.05$], indicating that increased experience *decreased* the likelihood of catching the automation failures. Descriptive analyses of the entire sample revealed that the altitude load failure and the heading capture failure, the two events most critical to aircraft operation safety, remained undetected by 44% and 48% of the participants respectively. The frequency misload was undetected by 71% of pilot participants. The tracking task automation failure, completely irrelevant to flight functioning, was detected by all of the participants.

In order to ascertain the underlying factors that discriminate participants who were more likely to verify automated tasks (and thus catch errors) from those less likely to do so, pilots were classified according to the number of omission errors they committed. Those who missed two or three out of three flight-related events were categorized as "high-bias" participants, and those who missed none or only one automation failure were placed into the "low automation-bias" group. ANOVAs were conducted using bias group as the independent variable and responses on the debriefing questionnaire as dependent variables.

Bias groups were statistically equivalent on items such as comfort with the experiment, confidence in their strategies, and confidence in computers. However, low-bias participants reported more nervousness [F(1,19)=7.08; p<.015], a higher sense of being evaluated on their performance [F(1,19)=2.21; p<.00] and strategies in use of the automation [F(1,19)=9.63; p<.006], and a stronger need to justify their interaction with the automation [F(1,19)=6.24; p<.02]. In other words, pilots that reported a higher internalized sense of accountability for their interactions with automation verified correct automation functioning more often and committed fewer errors.

DISCUSSION

Results of these studies demonstrate that the perception that one is "accountable" for particular aspects of performance affects one's strategies in performing the task. In the student study, experimentally manipulated accountability decreased the tendency to make errors under specific conditions, i.e., when subjects were accountable for their Overall Performance or Accuracy. Analysis of the verification data indicates that accountability effects occur largely due to the expected increase in cognitive vigilance--subjects in these conditions were more likely to completely verify than Non-Accountable subjects or subjects accountable for a Quick Response or for Tracking. This effect was especially pronounced as verification required opening more boxes, making it more difficult or costly to accomplish. Conversely, imposing accountability for Tracking led to an increase of automation-related errors, possibly because it encouraged participants to concentrate attention on the tracking task rather than on checking for proper automation functioning.

Although experimentally-manipulated accountability did not impact the automation bias of the pilot subjects, there is evidence that internalized accountability led to an increase in verification of automated functioning and fewer resultant errors. Apparently, the sense that one is accountable for one's interaction with automation does encourage vigilant, proactive strategies and stimulates verification of the functioning of automated systems. The fact that, for the pilot sample, this perception was not correspondent with our external accountability manipulation indicates the need to

establish whether accountability is a variable that can be significantly influenced in professional decision makers (i.e., pilots, who are already at a high level of personal responsibility for their conduct), or if it is part of some innate personality construct. It is clear, however, that the perception of accountability for one's interaction with automation does encourage vigilant, proactive strategies and stimulates verification of the functioning of automated systems.

ACKNOWLEDGEMENTS

This research was supported by NASA grants NCC2-798, NCC2-837, and NAS2-832. Special thanks to reviewers Mary Connors, Ute Fischer, and Irene Laudeman. Susan Heers is currently affiliated with Monterey Technologies.

REFERENCES

Hagafors. R., & Brehmer, B. (1983). Does having to justify one's decisions change the nature of the decision process? *Organizational Behavior and Human Performance*, *31*, 223-232.

Johnson, E. J., Payne, J. W., Schkade, D. A., & Bettman, J. R. (1991). *Monitoring information processing and decisions: The MouseLab system*. Philadelphia: University of Pennsylvania, The Wharton School.

Kruglanski, A. W., & Freund, T. (1983). The freezing and unfreezing of lay inferences: Effects on impressional primacy, ethnic stereotyping, and numerical anchoring. *Journal of Experimental Social Psychology*, *14*, 448-468.

Mosier, K. L., & Skitka, L. J. (in press). Human Decision Makers and Automated Decision Aids: Made for Each Other? In R. Parasuraman & M. Mouloua (Eds.), *Automation and Human Performance: Theory and Applications*. NJ: Lawrence Erlbaum Associates, Inc.

Mosier, K. L., Skitka, L. J., & Heers, S. T. (1995). Automation and accountability for performance. In R. S. Jensen & L. A. Rakovan (Ed.), *Proceedings of the Eighth International Symposium on Aviation Psychology* (pp. 221-226). Columbus, Ohio.

Mosier, K. L., Skitka, L. J., & Korte, K. J. (1994). Cognitive and social psychological issues in flight crew/automation interaction. *Proceedings of the Automation Technology and Human Performance Conference*, Sage.

NASA Ames Research Center (1989). Window/PANES: Workload PerformANcE Simulation. Moffett Field, CA: NASA Ames Research Center, Rotorcraft Human Factors Research Branch.

Skitka, L. J., & Mosier, K. L. (1994). *Automation bias: When, where, why?* Presented at the Annual Conference of the Society for Judgment and Decision Making.

FIXIT: An Architecture to Support Recognition-Primed Decision Making in Complex System Fault Management Activities

Andrew J. Weiner, David A. Thurman, and Christine M. Mitchell
Center for Human-Machine Systems Research
School of Industrial and Systems Engineering
Georgia Institute of Technology
[weiner, dave, cm]@chmsr.gatech.edu

Abstract

This research explores the use of the theory of recognition primed decision making and the technology of case-based reasoning to store fault management experience and make it available to operators confronting similar anomalous situations. Specifically, this project uses case-based reasoning technology to construct a knowledge base of actual fault management experience. This knowledge base is organized so as to enable operators to rapidly recognize a fault and retrieve information about previous or similar fault occurrences. The Fault Information Extraction and Investigation Tool (FIXIT), a case-based architecture for computationally encoding fault management experience, is described. FIXIT was implemented in proof-of-concept form for satellite ground controllers. An empirical study with NASA controllers showed that fault management using FIXIT was consistently superior to fault detection and response using existing fault management resources of NASA satellite ground control centers.

Introduction

The need to better understand and model the process of fault management is motivated by three prevalent trends in the control of complex dynamic systems. The first results from the increasing levels of automation found in most modern control systems. Most modern automation handles non-nominal situations poorly. Human operators are frequently left to deal with anomalous and unexpected system conditions (Sheridan, 1992). This trend points to the importance of developing automated systems which can detect, diagnose, and respond to anomalies *and* support the human operator when the limits of automation are reached.

A second trend relates to the changing role of the operator in human-machine systems. Reduced budgets and a corresponding desire to reduce the necessary number of operations personnel mean that operators are increasingly controlling multiple, independent systems concurrently (see Thurman & Mitchell, 1995). The result has been a marked reduction in the detailed knowledge operators have with which to respond to unexpected situations or to manage faults.

The third trend results from high levels of turnover among operations personnel. In general, when personnel changes occur, relatively expert personnel are replaced by less experienced people. In order to judge whether or not a controlled process is in error, operators must have knowledge of the process being performed or expectations about its correct outcome (Hutchins, 1995). Operators' ability to successfully perform fault management is directly related to their level of experience with the controlled system. As the amount of time and experience that operators have with a specific system decrease, the speed and efficiency of fault management also decreases.

One of the results of giving operators responsibility to control multiple independent systems, coupled with high operator turnover, may be that operators do not have the relevant knowledge to successfully manage faults. Given the context in which fault management is conducted (i.e., high workload and high likelihood of distractions and interruptions) even if the operator possesses the relevant knowledge it is easy for a memory lapse to occur (Woods, Johannesen, Cook & Sarter, 1994). The brittle nature of modern automation also means that operators may be called upon to perform without the aid of automation at the very time they are in most need of assistance (Woods et al., 1994).

The Fault Information Extraction and Investigation Tool (FIXIT) is proposed as one strategy to help mitigate these problems by encoding previous fault management activities and making that information available to operations personnel. FIXIT stores and makes available to the operator past fault management experiences in such a way as to facilitate recognition of the current fault as an instance or variant of previous experience. Based on both the theory of recognition-primed decision making and the technology of case-based reasoning, FIXIT is intended to embody the properties of human-centered automation--that is, automation that reflects how humans solve problems and make decisions. As such, it both aids operators who lack widespread knowledge of previous faults and forms a bridge between automated fault management systems and the humans that must assume control of them when the limits of automation are reached.

Conceptual Architecture

Figure 1 shows the theoretical underpinnings of the FIXIT architecture. They combine the cognitive science assumptions on *recognition* of previous experience as a major component of problem solving (Klein, 1989; Kolodner, 1993) and the *representation* of complex systems on a continuum of decompositions of both aggregation and abstraction (Rasmussen, 1986b).

Problem Solving by Real-World Experts

Research on expert problem solving shows that a significant aspect of what specialists do when performing in actual work domains is to use their experience and expertise to assess a situation, identify if a problem exists, and, if so, determine whether and how to act upon it. Klein (1993) suggests that rather than optimizing or satisficing, expert operators frequently recall previous experiences and adapt them to solve problems. Klein (1989, 1993) proposes recognition-primed decision making (RPD) as an alternative theory for how decision makers solve problems in real-world settings. Klein (1993) suggests that decision making in real-world setting is the result of three processes: situation assessment, formulation of a plausible solution, and mental simulation. Situation assessment is used to generate a plausible course of action while mental simulation is used to evaluate that course of action.

This research focuses upon assisting problem solvers or automated systems in the task of situation assessment in unfamiliar circumstances in order to formulate a potential solution based on previous experience, and in simulating the proposed solution to test if it meets the needs of the current problem.

Complementing the situation assessment stage of recognition-primed decision making, case-based reasoning combines a *cognitive model* describing how people use and reason from past experience and a *technology* for finding and presenting such experiences (Domeshek & Kolodner, 1992; Kolodner, 1993). Information about previous experiences is organized as a library of cases or, more specifically, as a 'case base'. Cases represent specific operational knowledge tied to specific situations (Kolodner, 1993). Cases make explicit how a task was carried out, how a piece of knowledge was applied, and what strategies were used to accomplish a task. Every case is accompanied by a set of indexes, which are combinations of descriptors which uniquely distinguish a case from other cases.

FIXIT records previous fault management experience as cases. The cases contain information to support diagnosis and identification of a particular problem, as well as information describing corrective actions to be taken (i.e., operational knowledge). An important issue in the organization of fault management information is the level of description needed so an anomaly can be recognized and managed in a recognition-primed way. FIXIT uses two levels of abstraction to organize fault management information. At the lowest level, anomaly information is stored as *reports*, which are detailed descriptions of a single anomaly incident. At a higher level, anomaly reports which share common characteristics are abstracted to form *anomaly categories*.

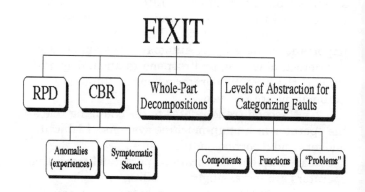

Figure 1. FIXIT conceptual components

Representations of Complex Dynamic Systems

Symptomatic search (Rasmussen, 1986) is an efficient strategy to diagnose current anomalies and a method for retrieving cases from an anomaly case base. In symptomatic search, a set of symptoms characterizing an observed fault is used as a template to search through a set of symptom sets to find a match or near match (Sheridan, 1992). FIXIT extends this notion of symptomatic search by linking anomalies and anomaly categories by membership in two complementary whole-part decompositions and an abstraction hierarchy.

The *component decomposition* is structured according to the whole-part component relationships in the controlled system. The *functional decomposition* describes bottom-up how components and functions are used, and describes top-down how purposes are implemented as functions and components. FIXIT uses these two decompositions to situate the anomaly in the context of the controlled system.

Levels of abstraction have been proposed as a useful way to describe how expert operators organize system complexity (e.g. Mitchell & Miller, 1986; Rasmussen, 1986a). FIXIT uses three levels of abstraction to categorize anomalies. 'Component' and 'function' characteristics of the anomalies are linked to the two whole-part decompositions discussed earlier. A 'problem' category is used to describe anomalies which do not fit into the categories based on the two decompositions. These include anomalies linked to the operating environment, or to multiple components, and/or multiple functions. For example, electromagnetic interference in a nuclear power plant, or the interference from the South Atlantic Anomaly in satellite system, would both be categorized as 'problem' anomalies. These categories are not mutually exclusive, and most anomalies are categorized using multiple levels of abstraction.

Computational Architecture

The FIXIT computational architecture is shown in Figure 2. FIXIT diagnoses anomalies using two categories of information obtained from the control system. anomaly symptoms and state information. Anomaly symptoms are error indicators and event messages detected by the control system (e.g., a sensor showing that the battery current is too high), as well as information received from other automated systems or expert systems (e.g., a flag from an expert system used to monitor fluid levels in a process control plant). FIXIT also uses information about the state of the system (e.g., current altitude or position) to aid in diagnosis.

The *system decomposition* is a combination of two whole-part decompositions. The physical

decomposition divides the controlled system into subsystems, components, and sensors. The functional decomposition divides the controlled system into functional subsystems, component functions, and sensor functions. The system decomposition is used to situate the anomaly symptoms in the context of the controlled system. FIXIT uses the system decomposition bottom-up to determine which system components and functions are related to each of the anomaly symptoms.

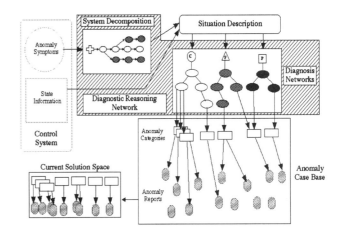

Figure 2. FIXIT computational architecture

The product of FIXIT's situation assessment process is the *situation description*. The situation description is used as the template for the symptomatic search of the diagnosis networks. The *diagnosis networks* are FIXIT's repository of fault management *expertise*. The organization of the networks reflects diagnostic procedures used by system operators. The three diagnosis networks are based on the three levels of abstraction used to categorize the anomalies. Each network is used to retrieve a particular type of anomaly from the anomaly case base. The diagnosis networks serve as indexing schemes for the anomaly case base. Symptomatic search is performed by pruning the diagnosis networks of any branch that does not relate to a symptom contained in the situation description. The resulting tree structures provide a means for the operator to observe FIXIT's reasoning strategy.

The *anomaly case base* is FIXIT's repository of fault management *experience*. The experiences in the anomaly case base provide lessons and advice on how to solve various problems, and also how *not* to solve problems.

The anomaly case base stores two different types of anomaly information: anomaly reports and categories. An anomaly report is a detailed record of a single occurrence of an anomaly. It contains a detailed description of the incident, a description of the impacts of the anomaly on the system, a list of corrective actions that were performed, and a description the final resolution of the anomaly. Whereas an anomaly report is a single incident, an anomaly category is an abstraction of a set of anomaly reports that share common characteristics. Anomaly categories contain two types of information. The first is primarily *descriptive*: a description of the anomaly, information about the anomaly's previous impacts on a system, and a description of how the anomaly was resolved. An anomaly category also contains *prescriptive* information: a system-independent description of the anomaly, the anomaly's potential impacts on the system, as well as a list of corrective actions for the operator to perform. In a future automated system, corrective actions could be executed automatically by the control system.

The Current Solution Space is the set of anomaly categories and reports retrieved from the Anomaly Case Base. The retrieved solutions contain information for identifying, diagnosing and responding to individual anomalies. The solutions can be adapted and used by operators for fault management or, potentially, by automated systems for use in control.

Evaluation

FIXIT has been implemented in proof-of-concept form as a decision-aid to support controllers of a NASA satellite system. The anomaly case base

contains information from actual NASA spacecraft anomaly logs currently used to diagnose and remediate faults. To evaluate the effectiveness of the FIXIT architecture in supporting real-time fault management, an experiment was conducted using satellite ground control operators at NASA Goddard Space Flight Center. Eight subject teams (a pair of operators, one a certified NASA satellite ground controller and the other an experimental confederate) controlled a simulated satellite ground control system during eight satellite control scenarios. Scenarios were matched by fault type and each team saw four fault categories under each of the two conditions: FIXIT and conventional control room fault management.

Six performance measures related to subjects' ability to detect, identify, and formulate an appropriate solution to the presented anomalies were collected. As shown in Figures 3 and 4, the use of FIXIT resulted in significantly improved subject performance for all six measures. Subjects detected anomalies ($p<0.03$), identified them as replicas or variants of previous anomalies ($p<0.001$), and determined the correct response ($p<0.003$) more quickly with FIXIT than when using their conventional fault management tools. Subjects' anomaly identifications and responses were also evaluated for correctness. Figure 4 shows that subjects accuracy of diagnosis and response was significantly higher with FIXIT than in the conventional condition.

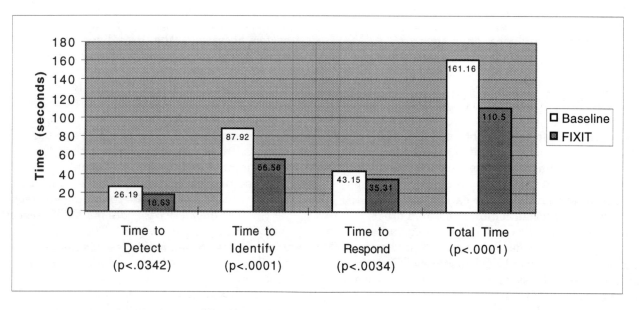

Figure 3. Comparison of fault detection, identification, and response formulation times.

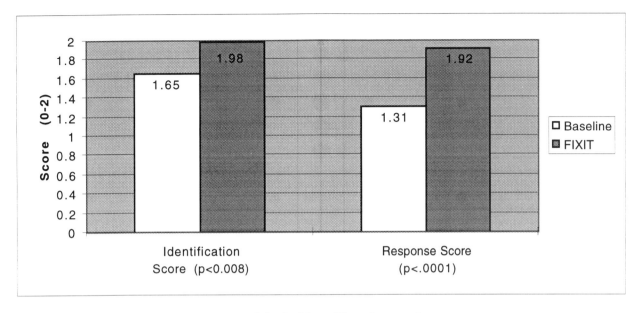

Figure 4. Comparison of fault identification and response accuracy.

Discussion

FIXIT is a domain-general computational architecture for encoding fault management experience. That is, the components which make up the architecture are appropriate for a *class* of complex domains. The semantics of a specific set of fault management experiences and whole-part system decompositions, however, do not generalize, except in concept, outside the application for which they were developed, e.g., process control, satellite ground control, manufacturing. Thus, though a particular instance of FIXIT is domain-specific, the computational architecture is applicable to a variety of domains. In addition, the FIXIT knowledge-base is extensible. As new anomalies are experienced, they can be added to FIXIT's case base, thus providing an evolving institutional memory of fault management experience.

References

Domeshek, E., & Kolodner, J. (1992). *Towards a case-based aid for conceptual design* : College of Computing, Georgia Institute of Technology, Atlanta, Georgia.

Hutchins, E. (1995). *Cognition in the Wild.* Cambridge, MA: The MIT Press.

Klein, G. A. (1989). Recognition-primed decisions. In W. B. Rouse (Ed.), *Advances in man-machine system research,* (Vol. 5, pp. 47-92). Greenwich, CT: JAI Press.

Klein, G. A., Orasanu, J., Calderwood, R., & Zsambok, C. E. (Eds.). (1993). *Decision Making In Action: Models and Methods.* Norwood, New Jersey: Ablex.

Kolodner, J. (1993). *Case-Based Reasoning.* San Mateo, CA: Morgan Kaufmann.

Mitchell, C. M., & Miller, R. A. (1986). A discrete model of operator function: A methodology for information display design. *IEEE Transactions on Systems, Man, and Cybernetics, SMC-16*(3), 343-357.

Rasmussen, J. (1986a). A framework for cognitive task analysis. In E. Hollnagel, G. Mancini, & D. D. Woods (Eds.), *Intelligent Decision Support in Process Environments,* (pp. 175-196). Berlin: Springer-Verlag.

Rasmussen, J. (1986b). *Information Processing and Human-Machine Interaction: An Approach to Cognitive Engineering.* New York: North Holland.

Sheridan, T. B. (1992). *Telerobotics, Automation, and Human Supervisory Control.* Cambridge, Massachusetts: The MIT Press.

Thurman, D. A., & Mitchell, C. M. (1995). Multi-system management: The next step after supervisory control? In *Proceedings of the 1995 IEEE International Conference on Systems, Man, and Cybernetics,* Vancouver, BC, to appear.

Woods, D. D., Johannesen, L. J., Cook, R. I., & Sarter, N. B. (1994). *Behind Human Error: Cognitive Systems, Computers, and Hindsight.* Wright-Patterson AFB, OH: Crew Systems Ergonomics Information Analysis Center.

USING RELIABILITY INFORMATION IN DYNAMIC DECISION SUPPORT: A COGNITIVE ENGINEERING APPROACH

Robert P. Mahan, University of Georgia, Athens, Georgia
Susan S. Kirschenbaum Naval Underwater Systems Center, Newport, Rhode Island
Jeff M. Jilig[1], Texas A & M University, College Station, Texas
Christopher J. Marino, University of Georgia, Athens, Georgia

The absence of feedback on the quality of decision making complicates the efficacy of methods designed to represent probabilistic information, especially in dynamic decision environments. The present study investigated the effects of changes in task information reliability on the performance of multi-cue judgment across conditions where the reliability of stimulus information was presented as a feedforward information source to enhance performance. Significant decrements in judgment performance were found across discrete changes in the reliability of source information. A Graphic format for presentation of reliability information produced high task performance under high and medium information reliability conditions, whereas an Animated presentation of reliability information produced high task performance under high and low reliability conditions. The results are evaluated in the context of work settings that call for dynamic decision making skills where the absence of immediate feedback is a constraint on performance.

INTRODUCTION

There is little doubt that providing information on the quality of decision making has powerful effects on improving and/or maintaining levels of performance (Kahneman and Tversky, 1973; Tversky and Kahneman, 1974; Brehmer, 1971, 1974, 1978; Balzer, Doherty and O'Conner, 1989). Explicitly understanding the consequences or outcomes of decision making performance helps one maintain control over the manner in which information is and should be used in the decision making process (Balzer et. al., 1989; Hammond and Wascoe, 1980; Hammond, McClelland and Mumpower, 1981). For example, the complex and uncertain nature of multi-dimensional judgment tasks often induces implicit control over the organizing principles used to produce criterion judgments (Hammond and Summers, 1972; Brehmer, 1971, 1974, 1978; Brehmer and Joyce, 1988). Feedback helps keep the features of an otherwise implicit organizing principle, explicit (Brehmer, 1976; Hammond et al 1981).

However, in many real jobs, the absence of feedback often accompanies operational decision making. The inertial qualities of complex physical and social systems often demand many decisions per unit time before a system response is actually observed, or an outcome is collected (Mahan, 1994). The lack of immediate feedback seems to aggravate what is termed the knowledge control problem, or the ability of operators to consistently integrate system information into a judgment about system criteria on the basis of a learned information organizing principle (Hammond and Summers, 1972; Balzer et al, 1989; Brehmer and Joyce, 1988; Mahan 1992, 1994).

Reliability Feedforward Information

It is useful to consider that in many real tasks uncertainty is associated with the fidelity of the information system used to deliver information to the user. The information system can refer to the instruments, observations, algorithms, as well as the display systems that are used to produce system information. Reliability identifies the fidelity of system indicator information and generally addresses the quality of the measurement processes used to acquire and represent system data. Disturbances in the transmission of system data, representational preferences in software programs that must choose a coordinate to display information from correlated data points, temporal delays in data transfer and instrumentation errors are a few examples in which observed system data can be corrupted (see Stewart and Lusk, 1994 for review).

In many practical situations information exists concerning the reliability of data used to make operational judgments in delayed feedback systems, although it may not be available to the operator. One possible decision support technique for delayed feedback systems may be to provide operators information on the statistical properties of system information immediately preceding its diagnostic use.

METHODS

Subjects

Twelve male student participants were paid volunteers for this study. The availability of female participants was limited due to several concurrent all female studies. Participants ranged in age from 22 years to 26 years with a mean age of 24.3 years. None had knowledge of the experiment prior to the briefing they received from the experimenter. The participants were paid $50.00 for their participation. *Judgment task.* The judgment task required that participants integrate information from four graphically displayed information sources (Cues) in estimating the time required to navigate an ocean vessel to a docking resupply platform at sea (discussed below). Participants formed their judgments under three reliability conditions and across four reliability presentation formats. Cues for the judgment task were randomly generated input values from a hypothetical navigation task. Randomly generating the cues produced low cue intercorrelations which was viewed as a developmental effort aimed at understanding the manipulations in this study. The selection of cues was based on their representing rather distinct variables effecting surface navigation.

Graphical Cue Information

Several distinct 2D geometric forms were selected for cue representation that provided a reasonable discrimination among cues (Baily, 1989). An area metric served to communicate the magnitude of the fixed ecological linkage between the cues and the criterion (i.e., the larger the object area the stronger the linkage between a cue and the criterion). The graphic cues were all scaled from 1 to 10 and displayed as follows: Weather status (1= very bad; 10 = very good), displayed as a solid black square; Traffic status (1= very few vessels in waterway; 10= many vessels), displayed as a solid black triangle; Visual contact (1= very poor visual contact of resupply platform; 10= very clear view of platform), displayed as solid black ellipse with a horizontal major axis; Auditory contact, (1= sounding of platform Beacon is very low; 10= beacon is very high) displayed as a solid black circle.

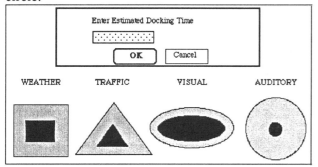

Figure 1. Graphic display showing 1 case of the 40 cases for which participants made criterion judgments. The interface shows cue reliability (outer graphic image) superimposed on corresponding graphically displayed cue value (center graphic image).

Representing Reliability Feedforward Information

The reliability of the cue information was represented and displayed in four different ways: (a) Numeric: a numeric value positioned below each graphic cue indicated the percentage amount of noise in the cue source- the larger the number, the less reliable the cue; (b) Graphic (Figure 1): the cues were superimposed over a gray background picture of corresponding cue geometry. The larger the absolute difference between the background and inner cue images, the less reliable the cue; (c) Animated- the cues in this format pulsed at a frequency of 3 Hz. The larger the magnitude of the pulse, the less reliable the cue; and (d) No Reliability Information- cues were presented in graphic format without any information given about their reliability. The participants were all <u>trained</u> to use the cue and reliability information in forming judgments of navigation time (see below).

Environmental Criterion

Task environment. A simulation, taken from real open-sea navigation exercises, was constructed that produced a <u>true value</u> (Y) which was the time in minutes it would take to reach a resupply point at sea. The mathematical components of the task were defined by the following equation:

$$Y = 100 + (.4(r_1))X_1 + (.15(r_2))X_2 + (-.35(r_3))X_3 + (-.1(r_4))X_4$$

where r is the cue reliability coefficient, and X_1 is the Weather cue value; X_2 is Traffic; X_3 is Visual; and X_4 is the Auditory cue value. In computing the true values for the criterion variable, cue weights were fixed constants with the above values.

Cue diagnosticity was a function of the product of reliability and fixed ecological weights. Cue reliabilities were changed by <u>randomly</u> selecting an r value in the interval {0, 1} from a uniform probability distribution. Altering the distribution interval changed the range of the reliabilities, and thus, the diagnostic value of a given cue source. For example, randomly selecting r values from the interval 0.7 to 1.0 would produce cues with higher average validities than selecting r values from, the interval 0.3 to 1.0. The reliability <u>factor</u> in this experiment reflected three reliability configurations that were generated by changing the distribution interval for the reliability (r) parameter.

Training. Training for the judgment task was conducted by providing immediate feedback on the accuracy (outcome) of each judgment of navigation time. <u>Outcome feedback</u> consisted of the <u>true</u> navigation time value generated from the navigation rule. Participants were considered trained when the r^2 between their judgments and true criterion (navigation) scores for a given reliability condition/presentation format was equal to 90% of task predictability (i.e., regression of criterion values on cues).

Experiment

In the experiment, cue magnitude and cue reliability were the only information sources available to the participants for making predictions. In order to simulate a delayed feedback decision task where the quality of a judgment is not immediately known, outcome feedback was not used.

During an experimental session, a single trained subject would perform a block of 40 judgments under each cue reliability and reliability presentation condition combination. All 12 conditions were performed by each subject during each experimental session. After six of the 12 conditions were completed, each subject was given a 15 minute break. The remaining six conditions were then performed. The experiment took an average of approximately 2 hrs. 20 minutes to complete by the trained subjects. The participants received the conditions in a counterbalanced format in order to control for linear order effects. Statistical tests for main effects due to the position of experimental conditions were not significant.

RESULTS

A Lens Model analysis of operator judgment was conducted because it offered an approach to capture accuracy within the context of the Social Judgment Theory (SJT) perspective. SJT is a compelling ecological theory configured around the probabilistic structure of the task environment (see Brehmer and Joyce, 1988 for review). The correlation between navigation judgments and true navigation scores (r_a) represented the level of achievement participants obtained in their predictions of the criterion. These scores were referred to as achievement scores. Achievement scores underwent Fisher Z transformation and then were backtransformed to pearson correlations. Mean achievement scores for each of the reliability conditions across reliability presentation formats were evaluated for between group differences. A 3 X 4 repeated measures factorial analysis of variance (ANOVA) with three levels of cue reliability crossed with four levels of reliability presentation was used to analyze overall effects. Correlational achievement, r_a, was the dependent variable. Results indicate there were significant differences in achievement for the main effect reliability $F (2, 22) = 52.98$, $p < 0.001$). These differences indicated that as information reliability became lower, judgment performance became poorer as well. In addition, significant differences in achievement for the reliability presentation format was seen $F (3, 33) = 17.01$, $p < 0.001$). Finally, a significant interaction effect between reliability and presentation format was seen $F (6, 66) = 17.27$, $p < 0.001$).

Simple main effect analyses (i.e., separate one-way ANOVA's) adjusted for familywise error on the three reliability conditions were found to be statistically significant (p<.05). Bonferonni adjusted single degree of freedom contrasts (p<.05) within each simple main effect was evaluated for reliability presentation format effects. Significant differences for achievement across reliability presentation formats were found for the high reliability (High-R) condition. Graphic, Animated and Numeric reliability presentation formats were associated with the highest achievement scores. Differences were also found among the medium reliability (Med-R) conditions, with the Graphic reliability presentation format associated with the highest achievement scores. Finally, low reliability (Low-R) differences were found. In this case, the Animated reliability presentation format was associated with the highest achievement scores. Finally, the No feedforward format produced the poorest achievement performance of all the displays across all reliability conditions.

DISCUSSION

Reliability feedforward information was successfully used by participants to maintain navigation expertise that could be accurately executed in the absence of outcome feedback. The Graphic reliability presentation format appeared to generate high achievement scores during judgment performance under both High-R and Med-R conditions, whereas the Animated display format produced high achievement scores during performance under High-R and Low-R conditions. Finally, the Numeric format produced high achievement during the High-R condition alone. One possible interpretation of these data is that when the task was very predictable (High-R), any feedforward format could be used effectively. In this case, very little reliability information needed to be encoded by the operator (i.e., there wasn't much information on the screen to process). When the task became less predictable, such as in the Med-R condition, participants may have found the Graphic display more informative and possibly easier to use. While during Low-R performance, participants needed the animated quality of the Animated display in order to effectively use the information or to fully understand the diagnostic nature of the cue array at hand.

The findings may be related to a combination task properties including cue salience and task complexity. During Med-R, the time dependent nature of the Animated display may have made it more difficult to utilize because participants had to encode the meaning of the animation. In contrast, the Graphic display presented all relevant diagnostic information (reliability and magnitude) to the subject at once. Simultaneous presentation of diagnostic information has been found to facilitate to the subject at once. Simultaneous presentation of diagnostic information has been found to facilitate information use, and is an important characteristic of the integrated display concept (Moray, 1981; Rasmussen, 1986, 1988).

During performance under the Low-R cue reliability condition, the participants tended to utilize the Animated reliability presentation format more effectively then other display formats. The animated characteristics of the dynamic display may have served to enhance the salience of cue magnitude and reliability. In this case, the benefit of cue salience may have outweighed the cost in effort associated with using the dynamic display. That is, when the task became most difficult due to large changes in reliability, the dynamic display may have provided the participants with a more robust image of cue validity leading to better control over the execution of their learned judgment protocols.

Future Research

A next step might be to examine the feedforward idea in the context of recent developments in integrated display research. Representing information reliability as a configural property of a multicue object display may be a step in evaluating a more efficient decision support device for the feedforward approach described here.

ACKNOWLEDGMENTS

[1] Now at IBM, RISC Systems 6000 Division, Austin, TX.

REFERENCES

Baily, R. W. (1989). Human Performance Engineering (2nd ed.). New Jersey: Prentice Hall.

Balzer, W. K., Doherty, M., and O'Conner, R. (1989). Effects of Cognitive feedback on performance. Psychological Bulletin, 106, 410-433.

Brehmer, B. (1971). Subjects' ability to use functional rules. Psychonomic Science, 24, 259-260.

Brehmer, B. (1974). Hypotheses about relations between scaled variables in the learning of probabilistic learning tasks. Organizational Behavior and Human Performance, 22(3) 445-464.

Brehmer, B. (1976). Subjects' ability to find the parameters of functional rules in probabilistic inference tasks. Organizational Behavior and Human Performance, 17, 388-397.

Brehmer, B. (1978). Task information and performance in probabilistic inference tasks. Organizational Behavior and Human Performance, 11, 1-27.

Brehmer, B., and Joyce, C. R. B. (1988). Human Judgment: The SJT View. New York: North Holland.

Hammond, K. R., McClelland, G. H., and Mumpower, J. (1981). Human Judgment and Decision Making: Theories, Methods and Procedures. New York: Praeger.

Hammond, K. R. and Wascoe, N. E. (1980). Realizations of Brunswik's Representative Design. San Francisco: Jossey-Bass Inc., Publishers

Hammond, K. R., and Summers, D. A. (1972). Cognitive control. Psychological Review, 79, 58-67.

Kahneman, D., and Tversky, A., (1973). On the psychology of prediction. Psychological Review, 80(4), 237-251.

Mahan, R. P. (1994) Stress induced strategy changes toward intuitive cognition: A cognitive continuum framework approach. Human Performance, Vol. 7, 3. 85-118.

Mahan, R. P. (1992). Effects of task uncertainty and continuous performance on knowledge execution in complex decision making. International Journal Computer Integrated Manufacturing, 5, 58-67.

Moray, N. (1981). The role of attention in the detection of errors and the diagnosis of errors in man-machine systems. In J. Rasmussen and W. Rouse (Eds.), Human Detection and Diagnosis of System Failures. New York: Plenum Press.

Rasmussen, J. (1986). Information Processing and Human Machine Interaction. New York: North Holland.

Rasmussen, J. (1988). A cognitive engineering approach to the modeling of decision making and its organization in process control, emergency management, CAD/CAM, office systems, and library systems. In: W. B. Rouse (Ed.), Advances in Man-Machine Systems Research. 4, Greenwich, CT.: JAI Press.

Stewart, T. R., and Lusk, C. M. (1994). Seven components of judgmental forecasting skill: Implications for research and the improvement of forecasts. Journal of Forecasting, 13, 579-599.

Tversky, A., and Kahneman, D. (1974). Judgment under uncertainty: Heuristics and Biases. Science, 185, 1124-1131.

REMOTE DIAGNOSIS IN DYNAMIC TASK ENVIRONMENTS

Y. Xiao[1] and C. F. Mackenzie[2]

University of Maryland School of Medicine, Baltimore, Maryland
[1]yxiao@umabnet.ab.umd.edu
[2]colin@anesthlab.ab.umd.edu

Increasingly telecommunication systems have become an integral part of many professions. However, little empirical data and guidelines exist for designing telecommunication systems to facilitate decision makers in cooperative efforts in dynamic task environments. A preliminary experiment was conducted in which the subjects (all experienced in the domain concerned) were presented with video-tapes of previously recorded real-life trauma patient resuscitation. The experiment examined the subjects' ability to understand the status of the patient and resuscitation efforts shown in the video. The experiment was to simulate remote diagnosis tasks in which experts provide consultation through video linkage. The subjects were found to have a number of difficulties in achieving a full understanding. Hypotheses about the reasons that could explain these difficulties are proposed and they include (1) background noise, viewing range restriction, and insecure viewing access to remote sites (2) visual information overload due to the multiple action threads at remote sites (3) lack of adequate dynamic mental models of remote events and activities (4) lack of context information.

Telecommunication advances enable remotely located individuals to collaborate on problem-solving with expertise unavailable locally. Increasingly telecommunication systems have become an integral part of many professions. Interesting and challenging research issues arise in the use of telecommunication systems in decision making and problem solving, many of which have been discussed in the context of distributed decision making and computer supported cooperative work (Kiesler *et al.*, 1984; National Research Council, 1990; Rasmussen *et al.*, 1991; U.S. Congress, Office of Technology Assessment, 1995).

For a decision maker to effectively participate in a decision making process, a prerequisite is to be able to assess the situation and problems at hand. In a distributed decision making context, this requirement means that the decision maker has to rely on telecommunication links (e.g. computer, telephone, and video networks) to achieve situation assessment and to understand problems to be tackled. This requirement may be fulfilled relatively easily when events evolve slowly, but it can be difficult to satisfy when situations change rapidly (similar argument is put forward by Allely, 1995).

Little empirical data have been reported on how people can assess dynamically changing situations and problems through telecommunication links. Therefore little empirical basis exists to guide the design of telecommunication systems in support of distributed decision-making in this regard.

As a first step to address these issues, a project was initiated to examine the ability of trauma experts to remotely manage trauma patients through telecommunication links, and identify how telecommunication systems should be designed to facilitate such tasks. Noted features of the domain of trauma patient resuscitation are that the patient condition changes rapidly and is often uncertain, and that the resuscitation effort is carried out by a multi-disciplinary team. Apart from being used as a research vehicle, trauma patient resuscitation could benefit from telecommunication because in many situations injured patients are spatially remote from expert care providers. In this paper, we will report the methodology used in and the results from a preliminary experiment.

METHOD AND MATERIALS

The basic methodology adopted in the experiment was to present to subjects videos of real-life trauma patient resuscitation and then measure the subject's ability to assess the status of the patient and the progress of the resuscitation effort. The video presentation was to simulate remote diagnosis through telecommunication which provides experts with live video images. During the course of the presentation of the stimulus materials, stop points were inserted, at which the subjects filled in questionnaires specially designed to capture their understanding of patient status and resuscitation activities contained in the stimulus materials. The questionnaire contained open questions and were generic (i.e. same across all stop points and not case-specific; see Table 1 for a list of questions in the questionnaire).

From a video-library (over 100 cases) of real-life trauma patient resuscitation in a Level-I shock trauma center, case segments were selected as stimulus materials. To prepare stimulus materials, these case segments were extensively analyzed based not only on the video-audio recordings, but also on patient admission records, discharge summaries, and the transcripts from the interviews with the case participants while they reviewed the videotaped cases. The case analysis yielded the causes and rationale for patient status changes and resuscitation efforts. The stimulus materials used in the experiment contained (1) audio-video recordings captured in real life (see Figure 1 for a screen dump from the video recordings), (2) continuous measurement of patient vital signs during the course of resuscitation, and (3) descriptions of the patient history upon admission to the shock trauma center. Patient history was given to the subjects at the beginning of each experimental session.

Stop points were chosen in each case segments based on the stages in the resuscitation effort. For each stop point, 1–3 items of descriptions were generated based on the analysis results to represent the ideal understanding of the status of the patient and of the resuscitation activities, and these items were used to score the questionnaires filled by the subjects. Thus even though questionnaires were generic, the scoring items were dependent on the specific stop point (see Table 2).

Four case segments (5–8 minutes each) were used in the experiment, with 3–4 stop points in each case segment. Table 2 (left four columns) describes the scoring items for all the stop points. These case segments were selected to represent a wide range of trauma patient resuscitation scenarios, and they were relatively complex.

Experimental subjects were recruited from clinicians who were well experienced in resuscitating trauma patients. While answering questionnaires, the subjects were encouraged to think aloud and audio-recordings were made for later interpretation of experimental results. The subjects were given one practice run to familiarize themselves with viewing videotaped resuscitation and with answering questionnaires at stop points. One case segment was shown in each experimental session. At the end of the session, the subject was debriefed on the case and on the final outcome of the patient. The presenting order of case segments was randomized across subjects.

Figure 1. A screen dump of the videotapes used in the experiment. The patient's head is located at the bottom of the screen. Patient vital signs data were overlayed at the top of the screen and were updated every 5 seconds.

1	I would describe the current patient status as (list up to 5 most important descriptors, in the order of decreasing importance)
	The following is unclear to me (list up to 3 most important, specific areas, in the order of decreasing importance)
2	I would describe the current team activities as (list up to 3 most important descriptors, in the order of decreasing importance)
	The following is unclear to me (list up to 3 most important, specific areas, in the order of decreasing importance)
3	I would describe the decisions just made by the team as (list up to 3 most important decisions, in the order of decreasing importance)
	The following is unclear to me (list up to 3 most important, specific areas, in the order of decreasing importance)
4	The team at the moment should consider the following differential diagnoses (list up to 5 most important differential diagnoses, in the order of decreasing importance)
	The following is unclear to me (list up to 3 most important, specific areas, in the order of decreasing importance)
5	I am anticipating the following immediate patient problems (list up to 3 most important, specific problems, in the order of decreasing importance)
6	List, in priority order, three most important objectives of the team and the instructions you would give to achieve the objectives.
7	List, in priority order, three decisions that the team could be making next.
8	List, in priority order, three most important pieces of information you would like to obtain, and the reasons why you need them.
9	Please rate your responses to the following statements on the five-point scale: 1. I am comfortable to giving instructions to the team 2. Given the opportunity, I would obtain more information 3. I know the tasks being carried out by the team

Table 1. Questions in the questionnaire used in the experiment to measure the subjects' understanding of remote events and activities.

RESULTS AND OBSERVATIONS

Three subjects went through a total of 12 experiment sessions (4 case segments each subject). Two subjects had one year and one subject 10 years of experience in the same shock trauma center where the stimulus materials were videotaped. All were anesthesiologists. The subjects spent between 10 to 20 minutes at each stop point to fill in questionnaires. The filled questionnaires were scored against the left four columns in Table 2 and the results are in the right three columns of Table 2.

Cases	Stop Points	Time	Scoring Items	Scores		
				S1	S2	S3
Case 1	SP 1	1'13"	Detected the acute hemorrhage	+	+	+
			Anticipated "MAST^a off" event	-	-	-
			Detected the slow progress of the surgeons	+	-	-
	SP 2	3'13"	Detected "MAST off" event	-	-	-
			Detected the urgent need for rapid infusion	-	+	+
	SP 3	5'03"	Detected ACP^b's effort in establishing IV accesses	-	-	-
Case 2	SP 1	1'18"	Detected the pressure on ACP to intubate	+	-	-
			Detected the lack of IV^c access and obstacles to intubation	+	+	+
	SP 2	2'10"	Detected nasal intubation and IM^d injection in the tongue	-	-	-
			Anticipated possible patient vomiting	-	-	-
	SP 3	3'26"	Recognized IV established	+	+	+
	SP 4	5'21"	Detected the delay in achieving patient muscle relaxation	-	-	-
			Put forward differential diagnoses for the delay	-	-	-
Case 3	SP 1	3'16"	Identified cues for missed intubation	+	+	+
			Identified cues for confirming correct ETT^e position	+	-	-
	SP 2	4'32"	Detected the lack of positive ETT position confirmation	-	+	+
			Put forward differential diagnoses for the lack of positive ETT position confirmation	+	+	-
	SP 3	6'26"	Detected the need to remove ETT	-	+	+
Case 4	SP 1	0'38"	Detected the need for IV bolus	+	+	-
	SP 2	4'28"	Detected the increasing, very high BP^f	+	+	-
			Detected the need for intervention	-	+	-
	SP 3	8'05"	Detected the decreasing, very low BP	+	+	-
			Detected the need for intervention	-	-	-

Table 2. Items used for scoring questionnaires at stop points (SP 1–4) for the four case segments (case 1–4). The results from three subjects (S1–3) are in the three right columns. + and - indicated whether the subject scored the item or not, respectively. a: MAST=military anti-shock trousers, b: ACP=anesthesia care providers, c: IV=intravenous, d: IM=intramuscular, e: ETT=endo-tracheal tube, f: BP=blood pressure.

We will first concentrate on those items in Table 2 that none of the subjects scored.

- Case 1, SP1 and SP2. The "MAST-off" (MAST: military anti-shock trousers) event represents a major disturbance to the patient physiology and the care providers have to coordinate the timing of the event to prepare and to compensate for the disturbance. In this case, the event occurred outside the field of view of the video camera, but there were cues in the team activities indicating that the event was impending and occurred. None of the subjects anticipated the event and none of the subjects detected the event, either.

- Case 1, SP3. Establishing intravenous (IV) access was by tradition the responsibility of the surgeons. The slow progress by the surgeons was identified by one of the subjects, and none of the subjects detected that the anesthesia care providers (ACP) expanded their roles to compensate for the lack of efficient progress by the surgeons.

- Case 2, SP2 and SP3. Intubation is a process to protect the patient's breathing airway from obstruction and prevent regurgitation of stomach contents into the lungs. In this case segment, the patient needed to be paralyzed for intubation, and the obstacle was that the team had not established IV access at the moment and therefore could not delivery paralyzing drugs via an IV port. All subjects at SP1 detected the obstacle to intubation. However, none of them detected the two parallel, circumventing plans being carried out (i.e. nasal intubation and intramuscular drug injection into the tongue), and none of them identified the danger of intubating an unparalyzed patient (e.g. the patient may vomit and aspirate into the lungs).

- Case 2, SP4. A patient is usually paralyzed within 60–90 seconds after the injection of paralyzing drugs. At SP4, 90 seconds after the injection of paralyzing drugs, the team was still testing whether or not the patient was paralyzed. None of the subject detected such a delay nor proposed explanations for it.

- Case 4, SP3. Two of the subjects detected the very low and still decreasing blood pressure (BP) reading, but they chose not to intervene. In comparison, the anesthesiologist in the actual case intervened immediately after the stop point.

Among the items that one or more subjects scored, we will describe those in Case 3, partly because the case segment contained an error that could lead to fatal outcome. (This case was reported in Xiao, Mackenzie, & LOTAS, 1995 as an illustrative case to describe the findings on fixation errors.) All the subjects identified cues indicating that the endo-tracheal tube (ETT) was misplaced (at SP1). The actual case progressed from this point without correcting this error until after the last stop point (SP3). One of the subjects (S1) also identified cues indicating that the ETT may be placed correctly. At SP3, this subject, unlike the other two, did not propose to correct the error and to remove the ETT. It appears that the subject who noticed the false confirmatory cues made a similar type of error in judgment as in the real case (see the analysis in Xiao et al., 1995).

The think-aloud data revealed that the subjects used correlating information to compensate for lack of complete patient data. For example, when tachycardia (high heart rhythm) was observed (from the heart rate data, one of the most readily available patient monitoring data), the subjects inferred that the patient must be hypovolemic (low blood volume) given the type of injury of the patient. The subjects were also found to utilize secondary cues reflected in facial expressions of the resuscitation team. In Case 3 at SP1, one of the cues that the subjects mentioned was the hesitating and not-so-confident look in the face of the person who performed the intubation.

In debriefing, one of the subjects commented that the reason she missed several obvious cues was that she was concentrating on one line of activities on the video screen and did not notice other concurrent activities.

DISCUSSION

The results reported here were from a small sample. Nonetheless, these results lead to several interesting observations that can be used as hypotheses in guiding future efforts.

Among the possible reasons why the subjects did not score all the items, many important cues seemed to be missed by the subjects. Possible explanations for this are proposed here. Firstly, the verbalization and verbal communications in the stimulus materials were degraded by the background noise of a typical patient resuscitation setting. Secondly, the viewing range for the subject was restricted. Thirdly, the visual access was not secure because the people in the video often moved into the line of sight and obstructed the crucial viewing focus. Fourthly, typical video displays used in the experiment contained the activities of 3–5 people working on different aspects of resuscitation (e.g. cannulation, ventilating the patient, preparing syringes, etc.). The multiple action threads in the video screen may have overwhelmed the subjects and caused visual information overload.

It should be emphasized that all these factors are likely to be present in the circumstances under which distributed decision makers have to work and they are not simply a matter of implementation technology.

Another reason why the subjects did not score some of the items may be because of the out-of-control-loop phenomenon (e.g. Endsley & Kiris, 1995). In Case 2 at SP4, none of the subjects detected the delay in achieving muscle relaxation, whereas in the actual case, immediately after SP4, the team discussed the possible reasons why the patient was not paralyzed. It seems that the subjects did not have the anticipation of patient status changes due to an action (i.e. injection of paralyzing drugs) that they did not perform, and consequently did not detect the delay. In generalized terms, this hypothesis can be reworded as that the subjects did not have an adequate dynamic mental model of the patient status to guide their information searching process.

A third reason could be that the subjects did not have complete context information as the on-site care givers. In Case 4, SP3, the cue of the very low, decreasing BP did not trigger the intervention as it did in the real case. One subject offered the explanation that she did not intervene because she would like to know what other team members were doing (e.g. whether the surgeons were ready to start operation, which would stimulate the patient and could cause a rise in BP). This piece of context information was not readily available in the stimulus materials. The lack of complete context information may also explain the inability of the subject in anticipating potential significant events in Case 1, SP1–2 and Case 2, SP2–3.

Several limitations in the current experiment should be pointed out here. First and foremost, the subjects did not have the opportunity to interact with the remote team and there were no two-way communications between the subjects and the remote team. In contrast with real remote diagnosis tasks, the subjects could not intervene the resuscitation activities, and the remote team could not volunteer information. Second, the subjects had worked in the work environment where the stimulus materials were recorded and thus they knew the work environment very well. In real remote diagnosis tasks the ability to understand remote events and activities is likely to be less than what was observed here. Third, the stop points used in the experiment introduced pauses in the subjects' mental efforts. Currently the questionnaire took 10–20 minutes to fill. The effect of such long delays was difficult to measure and assess.

In summary, the results from the preliminary experiment put forward the following possible reasons to explain the difficulties for a decision maker to assess the situation and tasks at remote sites:

- Background noise, viewing range restriction, and insecure access to remote sites
- Visual information overload due to the multiple action threads at remote sites
- Lack of adequate dynamic mental models of remote events and activities
- Lack of context information about remote sites

The findings from the current experiment have implications

for the design of telecommunication systems in support of distributed decision making and for future empirical studies on remote diagnosis in dynamic task environments. Although the information carried in video is rich, our findings indicate that there are still needs to provide non-video supporting information to help users in comprehending video images and in compensating for the lack of complete data about remote events and activities. Video information alone will have serious limitations on an expert's ability to make situation assessment. In addition, when multiple, concurrent threads of activities are involved, a remote expert could be overloaded and miss important cues. A *team* of remote experts may be needed in such circumstances.

Future efforts are needed to obtain empirical data to fine-tune the hypotheses generated in this study. Apart from continuing the experimental effort to include more subjects with various experience backgrounds, recording the visual scanning pattern of the subjects through the use of, for example, a gaze analyzer, may provide more definite answers to issues such as visual information overload and information seeking strategies.

ACKNOWLEDGMENT

The research reported here was supported by NASA Grant NCC-921 and ONR Grant #N00014-91-J-1540. The authors thank the contribution made by Drs. Fouché, Freeland, and Jaberi.

REFERENCES

Allely, E. B. (1995). Synchronous and asynchronous telemedicine. *Journal of Medical Systems*, **19**, 207–212.

Endsley, M. R., and Kiris, E. O. (1995). The out-of-the-loop performance problem and level of control. *Human Factors*, **37**, 381–394.

Kiesler, S., Siegel, J., and McGuire, T. (1984). Social psychological aspects of computer-mediated communication. *American Psychologist*, **39**, 1123–1134.

National Research Council. (1990). *Distributed Decision Making*. Washington, DC: National Academy Press.

Rasmussen, J., Brehmer, B., and Leplat, J. (Eds.). (1991). *Distributed Decision Making: Cognitive Models for Cooperative Work*. Chichester: Wiley.

U.S. Congress, Office of Technology Assessment. (1995). *Bringing Health Care Online: The Role of Information Technologies (OTA-ITC-624)*. Washington, DC: U.S. Government Printing Office.

Xiao, Y., Mackenzie, C. F., and the LOTAS Group. (1995). Decision making in dynamic environments: Fixation errors and their causes. In: *Proceedings of Human Factors and Ergonomics Society 39th Annual Meeting*, 469–473. Santa Monica, CA: Human Factors and Ergonomics Society.

ASSESSING A PERCEPTUAL MODEL
OF RISKY REAL-TIME DECISION MAKING

Joshua B. Hurwitz

Armstrong Laboratory

Brooks AFB, TX

A new model of real-time risky decision making is introduced that predicts tradeoffs between processing and risk taking during driving. This model, called Decision-Making under Risk in a Vehicle, or DRIVE, was fitted to data from a task in which subjects decided when to cross an intersection as a car approached from the cross street. Results showed that subjects attempted to cross less often before the oncoming car when it started closer to the intersection, even though objective risk was the same regardless of starting distance. Also, when the car started closer, subjects who reported having more real-life automobile accidents were less likely to take advantage of a longer opportunity to cross first. These results, along with results from fitting DRIVE to the data, suggest that risk-taking effects can be accounted for by a model of risk perception, and not by a model of risk acceptance.

Two questions in research on risky real-time decision making are how a decision maker (1) integrates real-time information and (2) judges risk levels to decide on the nature and timing of risky real-time responses. For example, drivers are often faced with making maneuvers that have a small probability of producing an accident. However, there can be incentives to take such risks because of time pressures. The goal of the research presented here is to identify and model processes underlying such risk taking.

Up until now, major theories have treated this issue as a problem of risk acceptance, but have lacked detailed mechanisms for assessing effects and individual differences that are due to risk perception. For example, models such as Risk Homeostasis Theory (RHT; Aschenbrenner & Biehl, 1993; Jannssen, 1988; Streff & Geller, 1988; Wilde, 1976; Wilde, Claxton-Oldfield & Platenius, 1985) do not explain how, under specific conditions, information is processed and integrated to time risky responses. Lacking well-specified processing mechanisms, these theories cannot separate errors due to poor distance- and velocity-estimation abilities from those due to risk taking.

Consider the case in which a driver decides when to cross an intersection. She may take some time to process information about oncoming cars, but if she takes too long, she might not arrive on time at her destination. If she crosses before sufficiently processing information about these cars, she might risk colliding with one of them. She may risk crossing quickly because they are far away, but there is a potential for error because of the tendency to underestimate distances for objects that are in motion relative to an observer (Harte, 1975) and are far from the observer (Foley, 1980). Given such errors, the driver may underestimate a car's speed and mistakenly assume that she can cross safely and quickly.

The Model. The study presented here introduces a model of real-time decision making that incorporates these distortions, and accounts for how drivers trade off processing time and risk taking. This model, Decision-Making under Risk in a Vehicle, or DRIVE, simulates how drivers make choices in real time when faced with incentives to take risks.

According to DRIVE, a driver decides when to cross by comparing an oncoming car's perceived and projected distances. The more similar these two distances are, the more cautious the driver will be. The perceived distance is the oncoming car's actual distance, coupled with error (Foley, 1980). The projected distance is the driver's estimate of how far the oncoming car would go in the driver's travel time, which is the time it takes the driver's car to reach and cross over the path of the oncoming car. Formally, the projected distance is the driver's travel time multiplied by the driver's estimate of the oncoming car's velocity.

The estimate of a car's velocity, v_ρ, at time t is derived from moment-to-moment changes, ΔD_ρ, in the car's perceived distance and from the driver's preexisting bias, k, about how fast that car should

be traveling. Formally,

$$v_t = v_{t-1} + [(\Delta D_t - v_{t-1})f_t + (k - v_{t-1})g_t]h_t \qquad (1)$$

where f_t, g_t, and h_t are sampling rates. The changes in perceived distance can be interpreted as affecting risk perception because they rely on information from the environment. The bias, on the other hand, can be interpreted as influencing risk acceptance, because it is independent of perceptual input.

While the bias is constant over time, changes in perceived distance increase as the car approaches because of distortions in computing distance in 3-dimensional space (Foley, 1980). The distorted distance, D_t, at time t is d_t^γ (Stevens & Galanter, 1957), where d_t is the true distance and $0 < \gamma \le 1$. Given these distortions, a distant car traveling at a given velocity appears to be traveling more slowly than a closer car traveling at the same velocity, and an oncoming car traveling at a fixed velocity appears to be accelerating.

Effects of Bias on DRIVE. In estimating the velocity of an oncoming car, a driver adjusts the mixture of bias and perceptual input over time. It is more difficult to process the movements of a distant car, so the driver relies more on bias and less on perceptual input. However, as the car gets closer, the driver relies more on perceptual input and less on bias.

The bias can influence DRIVE's prediction of risk taking. If the model is biased to underestimate the velocity, then it predicts that the driver will more likely respond prior to obtaining sufficient information from perceptual input. However, if the bias overestimates the velocity, then DRIVE predicts that the driver will be more cautious, and will wait until the car is closer before deciding when to cross. In this case, DRIVE simulates the vacillation process that is often part of time-limited decision making (Busemeyer & Townsend, 1993; Janis & Mann, 1977; Swenson, 1992).

One final component of DRIVE is its assumption that, at each moment in time, drivers learn more about faster cars, so that $h_t = v_{t-1} + \beta$. This assumption about velocity-based learning makes sense given that a faster oncoming car poses more of a danger to a driver should he decide to cross first.

METHOD

Subjects

Subjects were 145 Air Force recruits, 89 males and 56 females, ranging in age from 17 to 35 years, with a mean age of 20.8 years. All except 3 subjects knew how to drive. The average age at which subjects reported having started driving was 15.3 years, with a range of 8 to 25 years. The average age at which they reported receiving their driver's license was 16.5 years, with a range of 14 to 25 years.

Procedure

On each trial, subjects were presented with a 3-dimensional scene depicting a "car" on a road approaching an intersection from the side (Figure 1). The "virtual eye", the direction of view on the computer screen, displayed this scene from the point of view of a driver waiting to cross the intersection from the cross street. The subject could cross at any time during this scene by pressing the mouse button. Each trial ended with feedback indicating whether the subject had crossed safely, collided or not responded.

Figure 1. A scene from the crossing task in which a car is approaching the intersection. The number at the bottom shows the points earned if a successful crossing is initiated at this time during the trial.

On trials in which subjects received points, a timer was presented below the 3-dimensional scene. The numbers on the timer started at some value p_0 and ended either at 0 if the subject had not responded quickly enough, or at some value p_t. At the end of the trial, subjects were told how many points they gained or lost, and how many they had accumulated thus far. They gained points for fast, successful crossings and lost points for collisions. Their cumulative points never fell below 0.

Trial Sequences. The sequences of trials were set up so that subjects had an initial 64 trials to practice without points, followed by 64 trials with points feedback. The practice and points trials were divided into 4 16-trial blocks. On half the blocks, the subject's car took 4 secs. to cross, and on the remaining blocks, it took 8 secs.. Subjects were told which car they would have at the beginning of each block.

On half the trials in each block, the oncoming car started 250 feet away from the intersection, and on the other half, it started 500 feet away. The trials were further subdivided according to the window of opportunity available for crossing prior to the oncoming car: 0, 400, 900 and 2100 msecs. During a block of 16 trials, each window was presented on 2 trials in each distance condition, and all window X distance conditions were presented in a random order.

RESULTS AND DISCUSSION

The results and model fits presented below will show that a model of risk perception, such as DRIVE, provides detailed and accurate predictions of data that cannot be accounted for by risk-acceptance models such as Risk Homeostasis Theory. The reason why RHT cannot account for these data is that differences in risk taking were observed across conditions that were equally dangerous.

One outcome was that subjects attempted to cross first more often when the oncoming car started from farther away ($F(1,144) = 280$, MSE = 0.10, $p < .0001$, $R^2 = 0.66$). Another result showed that subjects were sensitive to the window of opportunity, making more attempts to cross first as the window got longer ($F(3,432) = 90$, MSE = 0.06, $p < .0001$, $R^2 = 0.38$). However, this sensitivity was greater when the oncoming car started closer to the intersection ($F(3,432) = 16$, MSE = 0.04, $p < .0001$, $R^2 = 0.10$).

Attempts to cross first produced collisions more often with a relatively small window of opportunity ($F(3,432) = 439$, MSE = 0.09, $p < .0001$, $R^2 = 0.75$). Collisions were also more likely when the oncoming car started from farther away ($F(1,144) = 140$, MSE = 0.04, $p < .0001$, $R^2 = 0.49$).

These results suggest that risk-taking arose mainly from differences in risk perception. The fact that the oncoming car started from farther away could not have produced risk compensation because objective risk levels, as represented by the windows of opportunity, were equal across those conditions. So there was no change in objective risk to compensate for.

Individual differences and model fits. Further analyses were performed using an index of accident rates derived from a self-reported driving history. Only 48 subjects were used for this analysis, because of suspected lack of reliability in the self reports. Results showed that drivers who reported having more accidents in their real-life driving were less likely to attempt to cross first when the oncoming car started from 250 feet away and when there was a 2100-msec. window of

opportunity ($r = 0.42$, $t(46) = 3.14$, $p < 0.003$). An attempt to cross first is defined as a response that leads either to a successful first crossing or to an accident. Given that almost all attempts were successful in the 2100-msec. condition, it appears that subjects with fewer accidents were more sensitive to the opportunity to cross first under this condition.

Using the same accident-rate index derived from the self-report measure, subjects were divided into high-accident (N=12) and low-accident (N=36) groups for the purposes of the model fits. For each group, DRIVE was fitted to cumulative response probability curves for all 16 cells in the design (2 car speeds X 2 distances X 4 windows of opportunity). Each cumulative probability was an average for a given condition over all subjects in a group, and was computed for every 33-msec. time slice as the oncoming car approached the intersection.

In the 250-foot condition, DRIVE predicted the tendency for low-accident subjects to attempt more first crossings than high-accident subjects when a window of opportunity was available (Figure 2). Also, analyses of DRIVE's parameter estimates revealed that the fit to the high-accident group data produced a greater tendency to use the bias, when compared to the fit to low-accident group data. Finally, in the high-accident fit, DRIVE was less able to discriminate when it was safe versus when it was unsafe to cross. These results suggest that group differences may have been due to differences in the ability to process the information necessary to accurately assess the risk.

Model comparisons. Aside from showing DRIVE's prediction of group differences, Figure 2 also shows that DRIVE was better than three other models at predicting attempts to cross first. The results show that eliminating some of the perceptual mechanisms in DRIVE reduced its ability to predict key attributes of the results.

In particular, DRIVE_2D, a model that did not distort perceived distances ($\gamma = 1$), underestimated the probability of attempting to cross first. Furthermore, this model did not predict the increase in attempts with longer windows of opportunity. This occurred mainly because not distorting perceived distances brought them closer to the projected distances, making the model more cautious than DRIVE.

Unlike DRIVE_2D, DRIVE_PS overestimated the probability of attempting to cross first. This second version of DRIVE excluded bias from velocity estimates, so that $g_t = 0$ in Equation 1. The bias in DRIVE was set fairly high, so eliminating it made the model more risky.

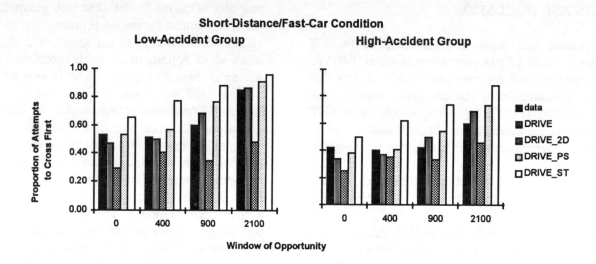

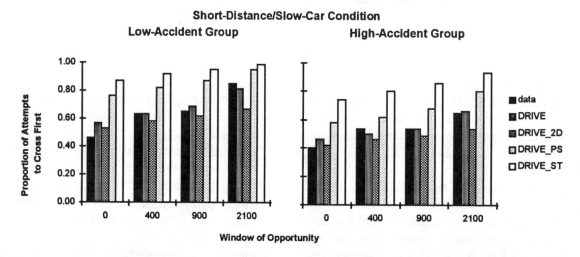

Figure 2. Data and model predictions for the short-distance (250-foot) condition.

Finally, DRIVE_ST, a model without velocity-based learning ($h_t = \beta$), also overestimated attempts to cross first. DRIVE's velocity estimate increased as the oncoming car approached the intersection, causing DRIVE to learn faster. The estimates for DRIVE_ST, however, remained low, making it riskier than DRIVE.

Future efforts. These model-fitting results were obtained by using DRIVE's parameter estimates, and then constraining some of the parameters to produce the restricted model. For example, to produce DRIVE_PS, the parameter that regulates DRIVE's sampling rate for the bias, g_ρ was set to 0. A stronger test would be to fit each restricted model to the data, and then assess how well it predicts an independent set of data. For example, in a follow-up study, we are asking subjects for continuous velocity estimates of oncoming cars, as well as for intersection-crossing responses. Each model can be compared for its accuracy in predicting velocity estimates using parameters estimated from the crossing task data, with only a slight change in that model's response generation mechanism.

One better test of the model's accuracy at predicting individual differences would be to fit it to individual subject data, rather than to mean data as was presented here. Parameter estimates could then be correlated with criterion variables to assess the model's predictive validity. Also, the estimates could be factor analyzed along with other laboratory variables to assess how they relate to general cognitive and psychomotor abilities.

Finally, we hope to improve our criterion measures by obtaining objective driving history data to supplement the self-report data used in this study. We will also select subjects on the basis of their history of moving violations. Hopefully, selecting from a population that is more diverse than our restricted sample of young Air Force recruits will allow for a more rigorous test of DRIVE's validity.

REFERENCES

Aschenbrenner, K.M. & Biehl, B. (1993). *Challenges to Accident Prevention: The issue of risk compensation behaviour.* Groningen, The Netherlands: STYX Publications.

Busemeyer, J.R. & Townsend, J.T. (1993). Decision field theory: A dynamic-cognitive approach to decision making in an uncertain environment. *Psychological Review, 100,* 432-459.

Foley, J.M. (1980). Binocular distance perception. *Psychological Review, 87,* 411-434.

Harte, D.B. (1975). Underestimation of length by subjects in motion. *Perceptual and Motor Skills, 41,* 519-522.

Janis, I.L. & Mann, L. (1977). *Decision making: A psychological analysis of conflict, choice, and commitment.* New York: Free Press.

Jannssen, W. (1988). *Risk compensation and the effect of an incentive: A laboratory study.* TNO Reportnr. IZF 1988 C-26. Leidschendam, Nl.

Stevens, S.S. & Galanter, E.H. (1957). Ratio scales and category scales for a dozen perceptual continua. *Journal of Experimental Psychology, 54,* 377-411.

Streff, F.M. & Geller, E.S. (1988). An experimental test of risk compensation: between-subject versus within-subject analyses. Accident Analysis and Prevention, 20, 277-287.

Swenson, O. (1992). Differentiation and consolidation theory of human decision making: A frame of reference for the study of pre- and post-decision processes. *Acta Psychologica, 80,* 143-168.

Wilde, G.J.S., Claxton-Oldfield, S.P. & Platenius, P.H. (1985). Risk homeostasis in an experimental context. In L. Evans and R.C. Schwing (Eds.), Human Behavior and Traffic Safety. New York: Plenum.

Wilde, G.S. (1976). Social interaction patterns in driving behavior: An introductory review. Human Factors, 18, 477-492.

A TEST-BATTERY APPROACH TO COGNITIVE ENGINEERING: TO META-MEASURE OR NOT TO META-MEASURE, THAT IS THE QUESTION!

Stephen J. Selcon, Thomas D. Hardiman, Darryl G. Croft
Systems Integration Department
Defence Research Agency
Farnborough, Hants. UK.

Mica R. Endsley
Department of Industrial Engineering
Texas Tech University
Lubbock, TX. USA.

This paper describes a task battery approach to system evaluation and cognitive engineering. The limitations of current techniques are discussed. An alternative paradigm using both task-specific measures and design driver measures is suggested. Such high level design driver measures have been referred to as meta-measures (Hardiman et al, 1995). In an experiment conducted to evaluate the utility of using meta-measures within the context of cockpit system assessment, an explanatory display for air combat decision support was tested within a realistic flight simulation. Multiple measures of performance were taken, together with the Situational Awareness Rating Technique (SART) as a meta-measure of Situational Awareness. Results showed that although task-specific measures clearly indicated one candidate design as producing better performance, they allowed little explanation of the underlying mechanism. Concurrent analysis of the meta-measure, however, enabled clarification of the cause of the performance differences, in this case through improved understanding, more information being available, and the information being more task relevant. The potential advantages of this approach for cognitive engineering are discussed.

INTRODUCTION

Bittner (1990) described the two main cultures of human factors (HF) measurement. The first aims to understand how work-related tasks are accomplished and systematically influenced by behavioural and non-behavioural variables. These measures focus upon narrowly-defined and well-controlled problems. However, this form of measurement does not lead to new design-guidance principles applicable to broad operational systems.

The second approach, referred to as applications measurement, applies HF principles to the design, development and testing of broad operational systems. This form of measurement is conducted under loosely-controlled conditions where the primary interest of the research is concerned with the success of the system rather than the successful measurement of performance. It could be argued that the aim of HF measurement is to accomplish both these goals, making it a multi-faceted discipline. However, Bittner lists the problems related to the multi-faceted approach to measurement, including the difficulty of selecting appropriate criteria and measures (Vreuls and Obermeyer, 1985), the integration of subjective and performance data (Meister, 1989), and the difficulty of integrating or trading-off measures within and between facets (Kantowitz and Weldon, 1985). Further, due to the divergent goals of HF practitioners and researchers, human factors research is limited through little validation of understanding-orientated research results obtained in operational settings, failures to exploit operationally derived results (i.e. understanding of performance differences), and the proliferation of under-evaluated measures and assessment measures. Moreover, Card, Moran and Newell (1983) refer to the pursuit of 'binarism' in cognitive psychology, where two current theories are tested within a well-established paradigm. This has resulted in a large base of quantitative data over many diverse phenomena, with many local theories. As a backlash from this, cognitive psychology has been concentrated in more recent years upon broader models of performance such as workload and situational awareness (SA).

One potential solution to the difficulties described above is to use performance measures supported by the concept of design drivers and their associated meta-measurement (Hardiman, Dudfield, Selcon, and Smith, 1995). Design drivers are broader models of performance that have been identified as generic rules or guidelines that should be followed in display design. The major design drivers in aviation display design have traditionally been workload and more recently SA. In other words, design drivers predict that if certain rules or guidelines are followed in design then performance benefits will result. This concept is not novel. Taylor and Selcon (1990) describe SA as a "meta-goal" which is not part of the mission goals but is necessary to allow mission goals to be attained.

The usefulness of the measurement of design drivers, meta-measurement, is the primary interest of the current paper. Task-specific performance measures quantify key aspects of human performance that are associated with successful task completion (e.g. speed, accuracy, RMS error etc). In general, these measure the visual, cognitive, and psycho-motor performance of the subject when using systems. Although essential, these measures reflect only such performance aspects and cannot directly convey information on other important task-dimensions such as SA. Meta-measurement offers an alternative yet complementary approach to task-specific performance measures. These measures are 'higher level' or 'meta-' measures that can be applied across all evaluation tasks. They can potentially provide greater sensitivity and diagnosticity to differences between candidate systems. To this end, the current paper will examine, using an experimental approach, how task-specific and meta-measures can be used together to provide a more complete assessment of human system synergy.

METHOD

Twelve experienced RAF pilots completed a dynamic, fixed-based simulation. The simulator consisted of a cockpit mock-up with a head-down display, a head-out screen, control stick and throttle. The head-out screen displayed both an outside world and overlaid head-up display (HUD) symbology. The pilots' flight task consisted of following a course of waypoints in a low level ingress to target at 450 knots at 2000 feet. A threat assessment task was presented on a pseudo-radar track-up display which presented ownship location, height and speed and the waypoint course. One to three threat aircraft were presented for 30 seconds at between 20 and 40 second intervals to prevent the subject from predicting when the threat assessment task would occur. When the threats were presented, the pilot was required to avoid the threats whilst remaining as close as possible to the original waypoint course. The hostile aircraft displayed on the head-down display were either Mig 31 or Mig 19 and their speed was either 450 knots or 650 knots. The pseudo-radar display threat information consisted of the aircraft's type, location, relative aspect (i.e.

heading of ownship and threat ship), and speed. All display components were presented dynamically.

The independent variable, displayed threat information, contained two levels. The first presented the location and aspect of the threat aircraft in graphical form and aircraft type and speed in text form. The second condition presented the Launch Success Zone (LSZ) of each threat aircraft in the form of graphical LSZ envelopes as developed by Selcon (1995a, 1995b). The shape of each LSZ was drawn based upon algorithms that took into account the threat aircraft's type, speed, location, and relative aspect.

A within-subjects experimental design was used. The flight task performance measures taken were RMS Error for maintaining the waypoint course at 2000 feet and 450 knots. The threat assessment task performance measure was threat exposure time (i.e. time spent within a threat's LSZ). The meta-measure taken was 14-D SART (Taylor, 1989; Selcon and Taylor, 1989) which was completed at pre-set intervals during the flight task. Every subject completed SART eight times during each condition.

RESULTS

All performance data were analysed using balanced Analysis of Variance (ANOVA). The flight-task performance measure did not show significant performance differences. However, the pilots were exposed less to threats ($F_{1,11}=19.962$, $p<0.001$) when using the LSZ display (mean=6.5 seconds) than when using the no LSZ display (mean=11.75 seconds).

The SART data were analysed using balanced ANOVAs. All significant effects were tested post-hoc using t-tests adjusted using the Bonneferroni inequality to produce an experimentwise error of less than 5%. An overall measure of SA was calculated from the SART score using the formula

$$SA(c) = \_U/n_U - (\_D/n_D - \_S/n_S)$$

or

$$\text{Overall SA= Understanding Mean -(Demand Mean - Supply Mean)}$$

where $\_U/n_U$ is the mean of the scores on the Understanding related SART dimensions, $\_D/n_D$ is the mean of the scores on the Demand related SART dimensions, and $\_S/n_S$ is the mean of the scores on the Supply related SART dimensions. The SART Dimensions are provided in Figure 1. Significant differences between the conditions were found. These are shown in Figure 1 in bold text.

Figure 1.   The SART Dimensions

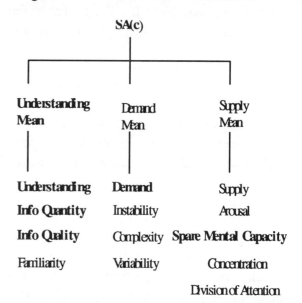

SA(c) (i.e. overall SA) was found to be significantly greater ($F_{1,11}=12.066$, $p<0.01$) when the pilots were using the LSZ display than when using the no LSZs display. The significant overall SA advantage gained in the use of the LSZ display can be attributable to the significant advantage in the SART understanding dimensions. There was a significant difference in $\_U/n_U$ ($F_{1,11}=8.76$, $p<0.05$) with better understanding for the LSZ display than the no LSZ display. Within the individual understanding dimensions, significant increases for the LSZ display over the no LSZ display were found for rated understanding (P<0.05), information quantity, (P<0.05), and information quality (P<0.005). While there was no significant difference in Demand Mean score, the no LSZ display was found to require greater demand upon attentional resources ($F_{1,11}=12.924$, $p<0.005$) than the LSZ display. Further, while there was no significant difference in Supply Mean score, pilots using the no LSZ display has less spare mental capacity ($F_{1,11}=12.924$, $p<0.05$) than when using the LSZ display.

DISCUSSION

The task-specific measure, threat exposure, clearly demonstrated that the LSZ condition provided a performance advantage over the no LSZ display. These data may be sufficient to allow the human factors practitioner to recommend the LSZ display over the other. This may satisfy those working under the first culture described by Bittner (1990). It can provide support for a decision regarding the choice of one candidate system against another. In other words, it allows a 'one-time' or 'one-off' decision. However, if these data are considered alone, they fail to provide an explanation of why performance differences were gained. Such an understanding would greatly increase the applicability of the findings to other systems. For example, optimal design aspects of one successful system (such as information quality) can be applied in the design of other similar systems.

Similarly, if considered alone, the SA data do not allow a decision of the superiority of either one of the candidate displays, as increased SA is only meaningful when it improves performance. However, when the SA measure is considered concurrently with the performance data, it can be used to support those data. In the above experiment, the results indicate that the LSZ provided a greater understanding of the situation. In specific, understanding was improved through the LSZ display providing more information which was of greater quality. Further, the no LSZ display placed a greater demand upon the attentional resources of the

pilot and (possibly as a consequence) the pilot had less spare mental capacity.

Thus the two results in conjunction allow a complete assessment of the display to be achieved. By knowing that not only does the information manipulation work in the specific experimental setting, but also that it improves understanding and awareness, more weight can be given to the findings. The task-battery approach helps to provide confidence to the researcher that the results will not only be accurate but, perhaps more importantly, that they are robust and likely to be replicable in other tasks involving that display. Since experimental evaluation of all mission relevant uses of a display is unlikely to be possible, a means of providing generalisable design recommendations is required by the cognitive engineer. The results of this study imply that the combined use of performance and meta-measures may provide such a means at little extra cost to the researcher.

REFERENCES

Bittner, A.C. (1990). Human factors measurement: Nature, problems, and strengthening. Proceedings of the Human Factors Society 34th Annual Meeting, 1253-1257.

Card, S.K., Moran, T.P., & Newell, A. (1983). The Psychology of Human-Computer Interaction, Hillsdale N.J.: Erlbaum.

Hardiman, T.D., Dudfield, H.J., Selcon, S.J., & Smith, F.J. (1995). Designing novel head-up displays to promote situational awareness. AGARD 79th Aerospace Medical Panel Symposium on "Situational Awareness: Limitations and Enhancements in the Aviation Environment, 15.1-15.7.

Kantowitz, B.H. & Weldon, M. (1985). On scaling performance operator characteristics. Human Factors, 27, 531-547.

Meister, D. (1989). Conceptual Aspects of Human Factors, Baltimore: John Hopkins University Press.

Selcon, S.J., Smith, F., Bunting, A., Irving, M., and Coxell, A. (1995) An explanatory tool or decision support on data-fused panoramic displays. In Proceedings of the SPIE International Symposium on Aerospace/Defense Sensing and Control and Dual-use Photonics, Orlando, FL, 17-21 April 1995.

Selcon, S.J., Bunting, A., Coxell, A., Lal, R., and Dudfield, H. (1995) Explaining decision support: an experimental evaluation of an explanatory tool for data-fused displays. In Proceedings of the 8thInternational Symposium on Aviation Psychology, Vol 1, 92-97. Columbus, OH, 24-27 April 1995.

Selcon, S.J. and Taylor, R.M. (1990). Evaluation of the Situational Awareness Rating Technique (SART) as a tool for aircrew systems design. AGARD Conference Proceedings, CP478, Copenhagen, DK.

Taylor, R.M. (1990). Development of the Situational Awareness Rating Technique (SART) as a tool for aircrew systems design. AGARD Conference Proceedings, CP478, Copenhagen, DK.

Taylor, R.M. & Selcon, S.J. (1990). Understanding Situational Awareness. Contemporary Ergonomics, 105-110.

Vreuls, D. & Obermayer, R.W. (1985). Human-system performance measurement in training simulators. Human Factors, 27. 241-250.

Optimizing Aided Target-Recognition Performance

Eileen B. Entin
Elliot E. Entin
ALPHATECH, Inc.
Burlington, MA 01803-4562

Daniel Serfaty*
APTIMA, Inc.
Burlington, MA 01803

Combining the capability of an automated target recognition (ATR) system with the expertise of human operators has the potential to improve combined human-machine target-recognition performance. The objective of this work was to develop empirically validated guidelines for the display of information provided by an ATR system. Motivated by the premise that operators will have higher trust in an ATR system that justifies its decisions, this work explored alternative ATR displays that provided different types and amounts of information in support of the system's target decisions. Operators' usage of and preference for the displays were contrasted for low- and high-accuracy simulated ATR systems representative, respectively, of current and future technologies. Performance results suggested that the amount of supporting information that is displayed must be congruent with the ATR system's accuracy. Operators expressed a clear preference for an ATR system that provides a confidence rating for its target decisions. Based on the results design guidelines are suggested for displaying ATR information to help operators balance confidence in their own judgments with trust in the ATR's decisions.

INTRODUCTION

In domains such as medical diagnosis and military target-identification systems, an individual may be required to detect the presence of certain objects based on a combination of direct image information and judgments provided by an automated system about the image. In machine-aided target recognition, human operators work with an automated target recognition (ATR) system to locate targets in cluttered and degraded imagery. Operators combine the automated information with their own judgment in order to reach a final decision. An understanding of factors affecting the operator's utilization of automated information is important in determining how ATR information can be presented so as to optimize overall system performance.

Lee and Moray (1991) found that operators' use of an automated control system depends on the balance between their trust in the automated system and their confidence in their ability to handle the system manually. Research on operators' use of ATR systems indicates that operators undervalue the information provided by a highly accurate ATR (Kibbe and Weisgerber, 1991). This leads to significantly suboptimal performance, especially when the image quality is poor (MacMillan, Entin, and Serfaty; 1994). If operators underrate the

accuracy of an ATR system and fail to rely on it when its performance is superior to unaided human vision, then overall target-recognition performance will be less than the optimal level that could be achieved.

Which design factors can increase an operator's trust in and effective use of an ATR system? Entin, Entin, and Serfaty (1995) advanced the theoretical premise that the relationship between factors such as the quality of visual or automated information and integrated human-machine performance is mediated by a set of intervening variables, including operators' confidence in their own and in the automated system's abilities. They hypothesized that operators will be more willing to trust an ATR system that explains or supports its target decisions. A pilot experiment conducted to test this hypothesis showed that aided performance was improved when an ATR system supported its target decisions. On the question of how much supporting information should be provided, especially in time-constrained situations, and how that information should be displayed, Entin, Serfaty, and MacMillan (1995) found that the most effective method for presenting the automated information depends upon the amount of decision time available. In particular they found that operators perform better under increased time pressure when the information provided by the ATR system is less detailed.

* formerly at ALPHATECH

In an attempt to follow this body of research with design guidelines, this paper develops a set of recommendations for the design and display of information that an ATR system presents in support of its target decisions. To provide an empirical foundation for these guidelines, we investigated operators' usage of and preferences for alternative ATR systems that varied in both the amount and the type of supporting information they provided. We explored their usage in a low-quality simulated ATR system representative of currently available technology and a high-quality simulated ATR system representative of potential technology.

METHOD

Experiment Testbed

The experiment was conducted using ALPHATECH's ATR-Interface Testbed (Entin & MacMillan, 1993). The Testbed presents "scenes" composed of objects randomly selected from sets of target and non-target objects. The objects in a scene are black silhouettes presented against a white background. The quality of the object images in the scenes can be distorted in order to simulate the uncertainty and noise associated with the sensor systems used for aided target recognition. The distortion is accomplished by randomly changing a specific percent of the black pixels in the scene to white and the white pixels to black. At a distortion rate of .37 the recognition task based on the visual information alone is relatively easy, and subjects make few recognition errors. At a distortion rate of .43, the rate used in this experiment, the recognition task is much more difficult. The Testbed also provides the capability of presenting information supplied by a (simulated) automated system.

Procedure

In the experiment we presented scenes comprised of seven to 10 objects with, on average, a 50:50 mix of targets and non-targets. The quality of the visual information was highly degraded. The subjects' task in the experiment was to select the objects in the scene that were targets and to rate their confidence in their decisions. Subjects made both unaided and aided target identifications.

Figure 1 shows a sample scene. For those trials in which an ATR system was available, the ATR's target decisions were displayed automatically. The objects selected by the automated system as targets are marked by black squares. The subject could click on the 'Remove ATR Judgment' button

to remove this information, and on the 'Show ATR Judgment' button to redisplay it. For those trials in which no automated-system information was available, the 'Show' and 'Remove' buttons were grayed out.

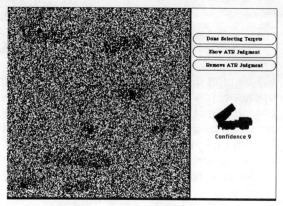

Figure 1. Example of Interface Used in Experiment

For each object the ATR identified as a target, subjects could bring up supporting information for that decision. In the example shown in Fig. 1 the subject has requested the supporting information for the object in the upper left hand corner of the scene that was identified by the automated system as a target. To obtain that supporting information, the subject double-clicked on the object. As shown in the example, the supporting information is displayed in the bottom right hand corner of the display. In this example the automated system displays the closest matching target and its confidence in its decision.

To make their target selections, the subjects clicked on those objects they decided were targets, and the objects were marked with a blue bull's eye. When finished selecting targets, the subject clicked on the 'Done Selecting Targets' button, and the next scene appeared.

Design

There were two levels of ATR accuracy: *high* (hit rate = .90; false-alarm rate = .10) and *low* (hit rate = .90; false-alarm rate = .40). We designed four types of ATRs that varied in terms of how much and what type of supporting information they presented about each object they identified as a target. The first three were used in the experiment. The *Alpha* ATR presented the closest matching target. The *Beta* ATR provided the closest match and its confidence in the match. The *Delta* ATR provided the three closest matches, rank ordered in closeness of fit. Examples of the information supplied by the three ATR system are shown in Fig. 2. Within each accuracy level, the ATR's hit and false-alarm rates were the same for the three types of ATRs. We designed, but did not use in the experiment, a hypothetical *Gamma* ATR that would have provided the three closest matches and confidence ratings for each match.

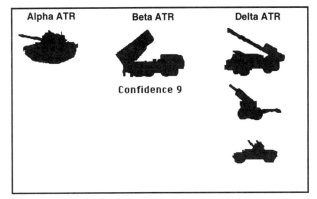

Figure 2. Supporting Information Supplied by the Three ATR Systems.

Performance was analyzed using Receiver Operating Characteristic (ROC) curves based on signal detection theory (Green & Swets, 1974). Two performance measures were used: the hit rate (the proportion of times in which a target was correctly selected as a target) and the false-alarm rate (the proportion of times that a non-target shape was mistakenly selected as a target). For a given signal quality, there is a tradeoff between the hit rate and the false-alarm rate. An individual may prefer an operating point that minimizes false alarms at the expense of missed detections or an operating point that maximizes detections at the expense of a higher false-alarm rate. The factors affecting this decision tradeoff include the perceived frequency of targets (vs. non-targets) and the relative costs of making a type I (vs. type II) error. In this experiment subjects were not given any specific instructions about the relative costs of the two types of errors. Higher-quality target-recognition information can put an individual on a new ROC curve.

Subjects' preferences and opinions about the different types of ATRs were captured in questionnaires that were administered after each type of ATR was presented and at the end of the experiment.

Fifteen active-duty military officers served as subjects in the experiment. ATR accuracy was a between-subjects variable. ATR availability and type were within-subjects variables. Each subject completed six unaided trials and six trials using each of the three types of ATRs that were implemented.

RESULTS

Performance

Figure 3, abstracted from an ROC curve, shows the operating points for the high- and low-accuracy ATR systems, and the subjects' unaided and aided performance. Clearly, aided performance is superior to unaided performance. Averaging over the experiment conditions, there was a significant difference between the aided and unaided performance both in terms of hit rate and false-alarm rate (p < .009 and p<.03, respectively).

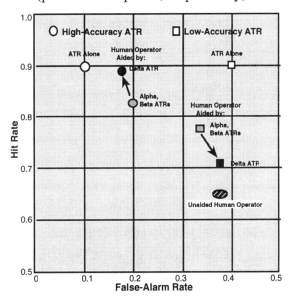

Figure 3. Operating Points as a Function of ATR Availability, Accuracy, and Type

In designing the three types of ATR system displays, we hypothesized that subjects would put more trust in the Beta (best match plus confidence) or Delta (three closest matches) ATRs than in the Alpha ATR (best match) because the former two provided more support for their decisions. As shown in Fig. 3, the effectiveness of the type of ATR on performance was mediated by the accuracy of the ATR. With the high-accuracy ATR, the highest level of aided performance was achieved with the Delta ATR, whereas with the low-accuracy ATR the lowest level of performance was attained with the Delta ATR. The interaction between ATR accuracy and type was significant for the hit rate (p = .05).

These results suggest that presentation of three alternative matches rather than the single best match increased performance for a high-quality ATR that had few false-alarms, but decreased performance for a low-quality ATR that had a large number of false alarms.

Although the subjects aided performance was significantly better than their unaided performance, nonetheless the subjects' aided performance fell below the performance of the ATR systems. With the high-accuracy ATR, subjects would have achieved higher hit rates and lower false-alarm rates had they relied solely on the ATR. This pattern of results is similar to those from previous experiments in which poor-quality visual and high-quality automated information was available (Entin et al, 1995; MacMillan et al, 1994). But in this experiment the difference between the ATR's performance and the subjects' aided performance was considerably smaller than in a previous experiment using an ATR of similar accuracy (MacMillan et al., 1994), suggesting that the availability of supporting information bolstered the subjects' reliance on the ATR system's judgments.

Usage of ATR System Information

The degree of integration of the ATR's judgments into the operators' decisions is an indicator of the level of trust operators put in the ATR system (Lee and Morey, 1991). Subjects should agree with the high-accuracy ATR system's target judgments since it has few false alarms. As would be expected, for the ATRs' target decisions there was much stronger agreement with the high-accuracy (90%) than with the low-accuracy (70%) ATR. On the other hand, both the low- and high-accuracy ATR systems rarely missed targets, so it is equally safe to agree with their non-target decisions. But subjects were less likely to agree with the low-accuracy (75%) than with the high-accuracy (85%) ATR's non-target decisions. The subjects' identifying as targets 25 percent of the objects that the low-accuracy ATR system called non-targets is reflected in the finding that the subjects' aided false-alarm rate with the low-accuracy ATR was not much below their unaided rate.

Although the ATR's target decisions were automatically reported, subjects had to request the supporting information for each ATR target decision. A relatively high query rate would be an indicator of the usefulness of the supporting information. On average supporting information was requested for 73 percent of the objects identified by the ATR as targets, and there was no difference in query rate attributable to ATR accuracy. Across ATR types, the highest percentage of queries (80%) was made for the Beta ATR, the one that presented confidence ratings for its target decisions.

If the supporting information lends credibility to the ATR system's decisions, then we would expect that subjects would agree with the ATR's decision more frequently when they queried the ATR than when they did not. In all comparable ATR type and accuracy conditions, subjects showed a higher percent of agreement with the ATR system when they queried it than when they did not. This finding that agreement rates were higher for objects queried than for those not queried leads us to conclude that providing supporting evidence increases subjects' trust in the ATR's target decisions. Paralleling the performance findings, the highest percentage of agreement was for the high-accuracy Delta ATR.

Operator Preferences

After using each of the three ATRs, subjects completed a questionnaire that queried their opinions about the usefulness of the ATR system and their degree of trust in and reliance on it. For all three types of ATRs subjects indicated they placed significantly more trust in the high- than in the low-accuracy ATR. On a 7-point scale the mean rating was 5.7 for the high-accuracy and 4.1 for the low-accuracy ATR systems. For the high-accuracy ATR systems, subjects placed higher trust in the matches provided by the Beta and Delta ATRs than by the Alpha ATR, but the difference was not statistically significant. Subjects found the high-accuracy ATR systems significantly more useful (p<.05) and relied on them to a somewhat greater extent (p<.10) than the low-accuracy ATR systems. Their ratings for the usefulness of the three types of ATR systems were similar.

In order to obtain as much information as possible about the perceived utility of different types of ATR displays, the hypothetical Gamma ATR system that would have provided both the three closest matches and confidence ratings for each match was also included in the subjective evaluation process. We asked subjects which ATR system (if any) they would prefer to use under constrained and unconstrained decision times. Most subjects preferred the Beta ATR that provided the single best match and confidence under high time pressure, and the Gamma ATR that provided the three closest matches and associated confidences under low time pressure. Clearly some form of ATR confidence rating is perceived as useful regardless of whether decision time is ample or limited.

DISCUSSION

This research explored how operators' usage of and trust in an ATR system are affected by the quality, amount, and type of supporting information that the ATR system provides. The finding that concurrence rates were higher for objects queried than for those not queried leads us to conclude that supporting evidence increases operators' trust in the ATR's target decisions.

We found that the ability of an ATR system to support its target decisions narrowed the gap between the performance of a high-quality ATR system and aided human performance. However, the finding that the operator's aided performance fell below that of a highly accurate ATR system suggests that in some cases operators are overvaluing their own abilities to recognize targets when the visual imagery is highly degraded and undervaluing that of the ATR system, leading to suboptimal performance. In this experiment we did not provide feedback to the operators on their decisions, so they had no way of comparing the accuracy of their target decisions with those of the ATR system. The results suggest the need for a training mechanism that would allow operators to calibrate their own and the ATR system's abilities.

The performance findings suggest that to optimize aided performance, the type of supporting information should be congruent with the quality of the automated information. The amount of decision time available seems to be a factor in how much detail (implemented here as the number of alternative matches provided) operators want about the ATR's decisions. Preferences expressed by the subjects for the best match with limited decision time available, and the three closest matches with liberal decision time available, support the conclusion that the amount of information provided should be tailored to the amount of decision time available.

The operators' preference for an ATR system that provides confidence ratings indicates the need for a straightforward, easily understood metric that serves as an index to the amount of weight they should place on the ATR's decision. This preference, combined with the need to support operators in better calibrating their trust in their own decisions and those of the ATR system, points to continued and enhanced use of confidence ratings in displaying ATR information. For example a display in which an ATR system highlights those objects for which its own confidence is low would focus the human operator's attention on objects requiring human judgment. An ATR system that monitored the operator's decisions in real time and called attention to those objects for which the operator's decision conflicted with a decision in which the ATR

system is highly confident would help operators in weighing their own judgments against those of the ATR system. Such adaptive designs for ATR displays hold the promise of augmenting the synergy between human and automated judgments, leading to high overall system performance.

ACKNOWLEDGMENT

This research was sponsored by the Human Engineering Directorate, Army Research Laboratory, Aberdeen Proving Ground, MD 21005-5001, contract number DAA15-92-C-0078, under the direction of Dr. James Walrath.

REFERENCES

Entin, E. B., Entin, E. E., and Serfaty, D. (1995). *Human-Computer Interfaces for Machine-Aided Target Acquisition*. TR-697. Burlington, MA: ALPHATECH, Inc.

Entin, E. B. and MacMillan, J. (1993). *Human image processing in unaided target detection*. TR-600. Burlington, MA: ALPHATECH, Inc.

Entin, E. B., Serfaty, D., and MacMillan, J. (1995). The effect of time pressure on visual information utilization in machine-aided target recognition. In *Proceedings of the Human Factors and Ergonomics Society Annual Meeting*. San Diego, CA.

Green, D. and Swets, J. A. (1974). *Signal detection theory and psychophysics*. Huntington, NY: E. Krieger Publishing Co.

Kibbe, M. P. and Weisgerber, S. A. (1990). *Targeting decisions using multiple imaging sensors: Operator performance and calibration*. Naval Weapons Center Technical Publication 7054, China Lake, CA.

Lee, J. and Morey, N. (1991). Trust, self-confidence and supervisory control in a process control simulation. In *Proceedings of the 1991 IEEE International Conference on Systems, Man, and Cybernetics*, Charlottesville, VA.

MacMillan, J., Entin, E. B., and Serfaty, D. (1994). Operator reliance on automated support for target recognition. In *Proceedings of the Human Factors and Ergonomics Society Annual Meeting*, San Diego, CA.

Understanding Flight Deck Task Management:
A Prelude to Human-Centered Design

William Rogers
BBN Corp.
Cambridge, Massachusetts
Session Organizer & Co-chair

Kathleen Mosier
NASA Ames Research Center
Moffett Field, California
Session Chair

Session Abstract

Several related concepts, including task management (TM), agenda management, and workload management, have recently been developed to describe a major function performed by pilots on commercial flight decks. This function is related to the fact that real-time operation of complex systems often involves multiple tasks which must be performed concurrently or in rapid succession, and therefore, must be managed. It is not that this function is new. It is more that we are better able, with recent advances in cognitive psychology and associated tools and methods, to identify and quantify it. While TM appears to compose an increasingly large part of the flight crew's role on the flight deck, our understanding of it is in its infancy, and it has been largely overlooked in flight deck function, task, and information requirements analyses. Thus it is not systematically considered in the design process. This symposium explicitly addresses task management as a high level flight deck function. The basic premise is that we must thoroughly understand what TM is, how pilots currently perform it on the flight deck, and what problems occur in pilot-performed TM, before we can address TM issues with a human-centered design approach.

This symposium consists of four papers which seek to describe TM processes and document TM practices and problems in real flight deck environments. The Rogers paper provides normative and operational descriptions of task management that provide useful insights into the set of processes that compose TM and the way that pilots view TM in a real-time setting. His distinction of strategic and tactical TM, based on the time-constraints associated with a particular flight situation, may prove useful in determining the types of design aids that are appropriate under different flight conditions. Pre-planning and contingency planning aids may benefit strategic TM while allocation, prioritization, and memory aids may benefit tactical TM.

The Schutte and Trujillo paper and the Latorella paper provide data from full-mission simulation studies that quantify TM activities and problems in non-normal and normal flight, respectively. The Schutte and Trujillo paper suggests that subjects spend considerable time performing TM during non-normal situations, and those subjects that perform best in fault

and mission management tasks seem to use a common rule of thumb (aviate, navigate, communicate, manage systems) to determine the order in which initial monitoring and situation assessment tasks occur, and a situation/event-dependent strategy to determine the order that particular discrete tasks should be performed. Application of inappropriate strategies in ordering tasks appears to contribute to operational errors.

The Latorella paper investigates an aspect of TM, interruption management, by quantifying deleterious effects of interruptions on procedural tasks during descent. She found over 50% more errors (omissions, misorderings, or redundantly performed activities) occur when procedures are interrupted than when not interrupted. She also found that subjects spent a greater proportion of time engaging in unnecessary flight path management during interrupted procedure conditions than uninterrupted conditions. Finally, she reported that rather than slowing procedure performance, interruptions tended to hasten performance on remaining procedural tasks. This could be related to the observation reported in the Rogers paper that pilots seem to use a TM practice of hurrying the pace of task performance during tactical TM, when there are pressing time constraints.

The Funk and McCoy paper describes an on-going effort to: (1) document agenda management errors in aviation accidents and incidents; (2) develop a formal normative model of agenda management; and (3) facilitate agenda management. Funk and McCoy take a broader perspective of task management, re-naming it "agenda management" to include activities performed by humans (i.e., tasks) and functions performed by any goal-directed "actor," human or automated. Thus agenda management includes management of goals, functions, actor assignments and resource allocations. This is the most formal, complete analysis of task or agenda management activities to date. The level of detail provided here is necessary before human-centered TM aids can be designed.

In combination, these studies begin to provide a basis for understanding the heretofore neglected flight deck function of task management. This understanding, at both a conceptual and an operational level, is required before TM can be systematically addressed within the flight deck design process.

FLIGHT DECK TASK MANAGEMENT: A COGNITIVE ENGINEERING ANALYSIS

William H. Rogers
BBN Corp.
Cambridge, Massachusetts

Flight deck task management (TM) is a mostly cognitive function that is not well understood. There is increasing evidence of difficulties in unassisted pilot performance of TM, and it is anticipated that the complexity and scope of TM on future flight decks will increase. This all points to the need for a formal analysis of TM as part of the flight deck design process. To this end, cognitive engineering techniques were used to analyze the mental processes involved in flight deck TM. Normative and operational descriptions of TM were developed based, respectively, on previous analyses and pilot interviews. The two descriptions present interesting contrasts, but they are complementary, and in combination, provide a useful framework for beginning to address TM more formally within the flight deck design process.

INTRODUCTION

Piloting commercial aircraft, like operation of many complex systems, involves performance of multiple tasks in a real-time environment. Since humans are basically serial processors, this environment dictates that pilot attention and cognitive resources be allocated to what is referred to as task management or TM (Funk, 1991; Abbott, 1993). Funk (1991) defined TM to include: creating a TM agenda, assessing the situation, activating tasks, assessing progress and status of active tasks, terminating tasks, assessing task resource requirements, prioritizing tasks, allocating resources, and updating the TM agenda. Abbott (1993) defined TM to include "managing tasks and associated resources involved in conducting the mission ... involving monitoring, scheduling, and allocating the tasks and task resources between and for each major function." Others have defined comparable functions to include managing mission objectives, and managing normal and non-normal procedures and operations. A related concept, workload management (Raby & Wickens, 1994), has also received recent attention, especially in the context of pilot training.

TM has always occurred on the flight deck. However, it has not been analyzed as a major flight deck function in the design process like other high level functions such as flight control, flight guidance, communications, and systems management. This may be because it is primarily cognitive in nature, and is not well understood. Hence flight deck functional requirements documents do not address it, and few automated systems are explicitly designed to perform it or assist pilots in performing it. While it has been the exclusive purview of the flight crew, human-centered design principles (e.g., Palmer, Rogers, Press, Latorella & Abbott, 1995) and function allocation guidelines (e.g., McGuire et al., 1991) suggest that it may be a prime candidate for automated aids.

THESIS

Understanding TM and addressing it formally within the flight deck design process are important for several reasons. First, a recent analysis (e.g., Madhaven & Funk, 1993) suggests TM errors contribute to a significant number of aviation incidents. It is expected as systems become more intelligent and complex, and more tasks and automated resources must be managed, flight crew TM load will increase. Adams, Tenney & Pew (1994), stated that "To the extent that people can properly attend to such aspects of the system only one at a time, the resulting situation must be characterized by frequent interruptions, unplanned shifts of attention, and system-driven changes in purpose and modes of thought. In effect, as the scope and autonomy of machine intelligence increase, the scheduling and initiative for information and activity management is increasingly usurped from the human operator. But the ultimate responsibility for their safe and sound operation is not." Thus systematically analyzing TM in the flight deck design process for future aircraft takes on additional import.

Second, a major tenet of a human-centered design philosophy is that pilots should be responsible for, involved in, and informed about functions and tasks that are critical to mission safety (Billings, 1991; Palmer et al., 1995). The requirements for pilots to be involved and informed in critical functions such as flight control and flight guidance, and to perform all the supporting functions on the flight deck such as TM, can become overwhelming during busy and non-normal situations. While pilots may perform individual tasks well, in combination, the performance requirements for multiple tasks can be beyond human abilities. To resolve this potential crunch on human attention and activity, human-centered design and function allocation principles suggest that TM is a prime candidate for automation or automation aids: It is less important for the flight crew to be involved in supporting functions such as TM than those that are directly critical to mission safety, such as flight control. Since TM can indirectly affect flight safety, automation aids are a better solution than fully autonomous systems, allowing pilots to retain final authority.

Finally, human-centered technologies have advanced to the point that many of the cognitive activities required to perform functions such as TM can be successfully aided by automated systems, so that shared performance of TM may be superior to human-alone or automation-alone performance. Such aids, albeit simple, are already on the flight deck: Electronic checklists, advanced caution and warning systems, and advanced flight management systems include features that aid pilots in remembering to perform tasks, prioritizing activities, and supplying information relevant to TM decisions.

The recent acknowledgment of TM as a major function on the flight deck, the increasing evidence of difficulties in unassisted pilot performance of TM, the anticipated increases in complexity and scope of TM on future aircraft, the human-centered design and function allocation rationale for considering automated solutions for TM, and the advances in intelligent automation and decision aids that could enable such solutions, all point to the need for a formal analysis of TM as part of the design process. This study analyzed the cognitive processes that compose flight deck TM, both from a normative and an operational perspective. The analyses described are a necessary precursor to systematic application of human-centered technology to design of TM aids.

SOURCES OF INFORMATION

This exploratory study used a two-pronged approach to analyze TM. First, previous conceptual analyses of TM were reviewed and used as a basis from which to develop a normative description of flight deck TM. The analyses by Abbott (1993) and Funk (1991) were the primary sources used. Recent flight deck function and information analyses (e.g., McGuire et al., 1991; Swink & Goins, 1992; Alter & Regal, 1992), which include functions similar to TM, were reviewed as well. Normative models and descriptions identify idealized processes performed as part of a particular function, but they do not necessarily match how the function is performed in real-time operational environments.

The objective of the second part of this study, therefore, was to begin to identify how TM is actually thought about and performed by pilots in the real-time flight deck environment. The main method used for this part of the study was structured pilot interviews. Three retired pilots were interviewed. One was a Concorde pilot and the other two flew a variety of modern commercial aircraft including the B757/B767 and B747-400. Since the entire flight domain was too broad to address in this effort, these interviews were performed in the context of two specific flight scenarios for a supersonic commercial aircraft. The scenarios were chosen to explore situations in which the TM load and overall pilot workload would be high, so that there would be a maximum number and diversity of TM activities and decisions.

The two scenarios selected were a supersonic phase and an approach and landing phase of a flight from San Francisco to Tokyo (the scenarios were independent, i.e., the events that occurred in the supersonic cruise scenario were not germane to the approach and landing scenario). Two "snapshots" were described for each scenario. The total of four snapshots were used to assess the degree of generality of TM processes. For the supersonic cruise phase, the first snapshot was characterized as normal flight, and the point in the flight was five minutes from the next waypoint and ten minutes from the point of equal time (PET), at which the closest alternate airport changed from Anchorage to Shemya (there are a variety of tasks associated with crossing both waypoints and PET's). The second snapshot was defined as twenty minutes later, with two of the four engines flamed out for unknown reasons, resulting in the inability of the aircraft to maintain supersonic speed and altitude. For the approach and landing scenario, the first snapshot was defined as five minutes prior

to top of descent into Tokyo. There were reports of thunderstorms in the terminal area and the airport was near minimum visibility for landing. For the second snapshot the airport had gone below minimums while the aircraft was on final approach, the runway could not be sighted, and a missed approach was just initiated.

The interviews consisted of having the pilots "think out loud" while performing experimental tasks with index cards, and then answering specific questions related to TM. The interviews were videotaped for later analysis. First, pilots were given a set of 40 index cards, each describing a potential TM cognitive process (Table 1). They were asked to sort the cards into piles based on the similarity of the items on each card. Subjects were asked to describe the similarity or commonality among cards in each pile and provide a title describing what each pile represented. Subjects were also asked to choose a small number of cards (the number was left unspecified) that captured the essence of TM. This exercise was aimed at understanding what TM means to pilots.

Monitor tasks	Schedule tasks
Allocate tasks	Initiate tasks
Prioritize tasks	Terminate tasks
Assess tasks	Coordinate tasks
Integrate tasks	Anticipate tasks
Modify tasks	Revise tasks
Plan tasks	Organize tasks
Review tasks	Categorize tasks
Compare tasks	Evaluate tasks
Predict tasks	Rehearse tasks
Analyze tasks	Assign tasks
Perform tasks	Infer tasks
Select tasks	Update tasks
Delay tasks	Preplan tasks
Drop tasks	Interrupt/restart tasks
Order tasks	Identify tasks
Direct tasks	Assess task resources
Assign task resources	Assess situation
Assess goals	Prioritize goals
Manage interruptions	Assess workload

Table 1. Forty potential TM processes presented to interviewees on index cards.

The rest of the interview items described below were repeated four times, once for each snapshot for each scenario. For these activities, the subjects were instructed to assume that they were the Captain, pilot-not-flying. First, the pilots were given a deck of 55 index cards, each labeled with a task that the author had determined might be performed on the flight deck within the context of the two scenarios. Inclusion of tasks was based on previous task analyses (e.g., Alter & Regal, 1992), requirements documents, review of standard operating procedures, and discussions with pilots. These tasks were described in fairly general terms to keep the size of the set manageable and to allow description of the tasks as independently as possible of specific aircraft systems and flight deck designs.

Table 2 shows a random sample of the task descriptions. Pilots were first asked to separate the cards into two piles-- those that they would think about or perform in the next ten minutes, from those that they would not. They were asked to think out loud as they performed this task to provide insights as to why each task was or was not "active" at this point in the flight. After they completed this task, they were asked to describe the decisions, mental processes, actions, and information that would be required to manage the active tasks in the specific context of the scenario and point in the flight.

Control aircraft through mode control panel
Review approach procedures
Review company operational specs
Review airport charts
Plan/review descent
Track/analyze/predict fuel (e.g., fuel burn, estimated fuel at destination, etc.)
Passively monitor systems
Compensate/isolate system failures
Determine effects of system failures on aircraft performance
Obtain/consult reference information
Tune radios
Monitor sonic boom footprint
Request air traffic control (ATC) clearance
Analyze/predict progress (e.g. position and time) along flight path
Determine weather at alternate or emergency field
Monitor/manage fuel and center of gravity
Plan path for alternate or emergency field
Perform emergency & non-normal procedures & checklists

Table 2. Sample of 55 tasks presented to interviewees on index cards.

Next, the interviewees were asked to take the index cards representing active tasks and sort them into piles based on similarity, that is, they were asked to put the cards that they thought about in a similar manner in the same pile and those that they thought were dissimilar in different piles. They were asked to think out loud as they performed this activity, and then they were asked to describe what each resulting pile of cards represented. Subjects were then asked to sequence the cards according to the order in which they would do or think about the tasks over the next 10 minutes of the flight. They were asked to think out loud as they performed this task, and to provide a general summary as to why they ordered them the way they did. Finally, they were asked how they would decide whether to perform the tasks themselves or assign them to the co-pilot or automated systems, and how they would deal with interruptions that might occur during the period that they were managing these tasks. Each interview took one to two days.

RESULTS

Normative description

A preliminary set of flight deck TM processes/activities was defined based on the review of previous conceptual TM analyses and related functions. These processes are hypothesized to be conducted in the order they are described,

but obviously much cycling occurs among the various processes in dynamic environments.

Assess situation. The operational context and goals must be assessed in order to identify the set of tasks that needs to be performed. The flight situation is dynamic and must continually be reassessed as the flight proceeds. In order to know what tasks need to be performed, one must be aware of the phase of flight, aircraft position, aircraft attitude and speed, aircraft and systems states, environmental conditions, unusual and other significant events, and short- and long-term goals. Further, one must be able to project this situation knowledge into the future, making predictions about the course of events.

Identify tasks. The tasks required to achieve the goals under the current and anticipated circumstances of the particular flight must be identified. Also, as a precursor to scheduling tasks, task characteristics such as time required to complete, deadline, difficulty or complexity, resources required, and interdependencies (i.e., one task requires that another be completed before it can be started) should be identified.

Prioritize tasks. Tasks must be prioritized in terms of urgency (how quickly they must be done). This means the estimates of task deadlines must be compared to one another to determine which ones are more urgent. Tasks must also be prioritized in terms of criticality, that is, how necessary are they to mission safety or success. These relative prioritizations, along with the task characteristics identified above, help in making scheduling decisions.

Assess resources. The availability and capability of task resources, including automation, the other crew member, company resources, and oneself, must be determined so that assignments of tasks to those resources can be made appropriately. This involves determining the status and predicted status of each task resource and the ability of the resource to perform part or all of the various tasks that have to be performed. It also includes predicting one's own workload to ensure that tasks assigned to oneself can be performed.

Allocate resources. Human and machine resources must be assigned to perform tasks. This may be a straightforward assignment, as to an autopilot or autothrottle, or it may require negotiation and discussion, as in deciding the division of duties between the crew members if not specified by procedures or regulations. The allocation process is ongoing; resources may need to be reassigned as conditions and goals change. Back-up allocations should be considered, and if resources assigned to tasks are not completely reliable, they must be monitored.

Schedule tasks. This element involves setting an order in which the identified tasks should be performed. It also includes the determination of when tasks need to be started, delayed, temporarily stopped, or resumed. It depends not only on completion times, deadlines, priorities, resources, interdependencies, and the overall context, but also, in the case of performance by humans (pilots), momentum and continuity. That is, other things being equal, humans perform better if they can continue working on a task once they have started rather than switching back and forth between tasks, causing discontinuity of thought.

Perform tasks. This is not considered part of task management. It is included to explicitly distinguish between TM and task performance. For example, if a call to ATC must be performed, identifying that as a task, prioritizing it, scheduling it, allocating it, etc., constitute task management, but the actual call to ATC is a communication task, not a task management activity.

Monitor tasks. Task performance should be monitored relative to the schedule to assure that tasks are started on time, completed on time, and are progressing as expected. Bottlenecks and resource limitations should be identified. If the tasks are not progressing toward completion as expected, then one should consider ways to hurry, delay, or modify task performance in order to meet the schedule. The schedule, the overall situation, or the tasks to be performed may need to be re-evaluated.

Manage interruptions. In real-time environments such as a commercial flight, interruptions will occur as tasks are being performed. If an interruption is processed, then attention to the ongoing task may be, at least momentarily, suspended. Often the interruption, such as a call from ATC, can signal the requirement to perform a new task, such as reporting position, or changing heading. These interruptions must be managed-- one must determine whether to stop the current task to process the interruption, and whether to immediately go back to the current task or to perform a new task if one is associated with the interruption. Further, one must remember at what stage the first task was stopped so that it can be efficiently resumed, and if the interruption changes overall goals or tasks, the need to continue, terminate or modify ongoing tasks must be assessed.

Operational description

TM processes. All three pilots sorted the 40 "TM process" index cards into six groups. While their descriptions of the groups varied, there was a common "time-based" theme that included a dichotomy between planning processes and real-time processes. One pilot described the underlying organizing strategy as "steps in dealing with a situation." Another pilot in responding to the request to choose a small number of cards to describe the "essence" of TM, chose one card that best described each of the six groups: (1) assess goals, which he described as high level planning; (2) assess situation, which included assessing workload and task resources; (3) analyze tasks, which included identifying, categorizing, and ordering tasks; (4) assign tasks, which included scheduling and allocating tasks and assigning task resources; (5) rehearsing tasks, which included monitoring, revising and modifying tasks; and (6) perform tasks, which included prioritizing tasks, managing interruptions, and delaying tasks. He described the first four groups as "planning ahead," and the last two as "real-time," when in the midst of a busy situation. This same dichotomy emerged in the other segments of the interview, and these two basic aspects of TM will be referred to as strategic and tactical TM.

Scenario-specific TM. The three pilots identified an average of about 31 of the possible 55 tasks per snapshot as "active." There were more active tasks for the second snapshot of each scenario than the first. The normal cruise snapshot resulted in the fewest active tasks (avg.= 27) and the two engine flame out during cruise resulted in the most active tasks (avg.= 38). For the approach and landing scenario, there was an average of 30 active tasks for the top of descent snapshot and 31 active tasks for the missed approach snapshot.

Based on review of the "thinking out loud" comments subjects made as they performed the card tasks, their description of the decisions, mental processes, actions, and information that were required to manage the active tasks, the results of the sorting tasks in which they separated the task cards into piles based on similarity, and the results of the ordering tasks, the following commonalties in TM emerged:

1) There is a very pronounced dichotomy in TM, referred to here as strategic and tactical. Pre-planning, building a mental model, monitoring, contingency planning, filling gaps with continuous and pre-planning items, and performing tasks early to avoid real and potential workload bottlenecks later, were used to describe TM activities when the flight is proceeding normally and there is little or no time pressure. Splitting duties (between the crew members), using a well-learned, well-rehearsed mental list of discrete items to be performed, doing time-critical, high priority items, operating in "real time," hurrying the pace of tasks, and deferring or dropping tasks, were used to describe TM activities when there was an emergency or time-pressured situation.

2) TM is driven by time. The overriding explicit TM process is scheduling or ordering tasks. The ordering is primarily urgency- and immediate event-driven for tactical TM, and workload management-driven for strategic TM.

3) Tasks are divided into discrete real time tasks, discrete pre-planning tasks, and continuous or repetitive tasks. Discrete tasks are naturally ordered along a priority or time dimension and continuous tasks are interleaved with discrete tasks but not explicitly ordered.

The four snapshots provided interesting combinations of these common TM processes. The first snapshot in each scenario, that is, the normal supersonic flight and the pre-top of descent, were characterized primarily by strategic TM. For the normal supersonic flight, the upcoming waypoint and PET were events which required specific ordered tasks to be conducted, but the tasks were routine, well-practiced, and planned in advance. More pre-planning tasks were included in the active set than for the time-pressured snapshots, primarily related to the fact that if there was an unexpected problem, the closest airport would change when the PET was crossed. There were a number of planning tasks done in regard to the contingency to divert to the nearest airport to reduce workload later if a problem actually occurred. For the pre-top of descent, pre-planning was even more pronounced. All the activities performed around the top of descent were done in order to be prepared for the descent, when things started getting busy. This scenario provided the most obvious example of strategic TM that was driven by workload management. Pilots knew that they would be busy later, so they wanted to get as many tasks done early as possible.

The second snapshot in each scenario, that is, the two engine flame out and the missed approach, were characterized by tactical TM. There were immediate, time-critical items that had to be performed. For the two engines out, there appeared to be a quick allocation and performance of time-critical items in the four major flight deck functions (aviate, navigate, communicate, manage systems), followed by a phase where items that required more time were addressed. This phase started with systems management, since analysis of the engine problem drove decisions in navigation and communication (e.g., if the engines could be re-started, the diversion decision might be different than if they could not). Interestingly, pilots consistently referred to a point in this scenario where things "settled down," or "got back to normal," which is interpreted as a mental transition point between tactical and strategic TM, where the time urgency is gone and one can begin to think and plan ahead again. The missed approach epitomized tactical TM. Since the alternate airport was close, there was never enough time for things to "get back to normal." Again, time-critical items, including flying the missed approach, contacting ATC, and assessing fuel, were accomplished first. But since time remained critical, items such as the approach review for the alternate airport were shortened or hurried, and other items dropped completely. There was a conscious strategy to disregard interruptions during these real-time, tactical TM phases.

DISCUSSION

This study represents on-going work and the current results, especially given the small number of subjects and the informal analyses associated with the interviews, must be considered preliminary.

The normative description of TM provided here modifies and expands previous work. Each process is given equal weighting and it is assumed that there is a logical sequence, that is, each process depends on completion of the processes preceding it. The operational description of TM derived from pilot interviews highlights the overriding time-based nature of TM on the flight deck. All of the TM processes from the normative description were referred to by pilots during the interviews, but the pervading process was scheduling or ordering of tasks. This process appears to be conducted differently depending on time-constraints. With no or little time pressure TM was performed strategically, and scheduling of tasks was based primarily on assessment of resources, that is, the active task list included many items that were being thought about or performed early in order to avoid workload bottlenecks that could occur later because of naturally busy periods (e.g., descent) or because of unanticipated problems. With time pressure, TM appeared to be performed tactically, and scheduling of tasks was based primarily on time-based priorities which were often pre-determined from a rehearsed "mental list." Pilots described performing and managing tasks during busy periods as "having to operate in real time," and tasks were managed by explicit task allocation and by hurrying, deferring, or eliminating tasks. Pilots reported that they consciously raised their resistance to interruptions during these periods. It is likely that many of the in-depth cognitive processes defined in the normative description of TM are abbreviated or eliminated during tactical TM. There is not enough time to devote cognitive resources to TM in light of the high workload related to performance of immediate tasks. This is likely one reason TM errors occur and automated TM aids would be useful.

Once validated, the set of TM processes defined by the normative description should assist in development of TM functional requirements as part of the flight deck design process. The operational depiction of two distinct TM modes (strategic and tactical) which are driven by different goals (workload balancing and urgency, respectively) and involve different managing mechanisms (e.g., pre-planning and early performance of tasks for strategic TM and hurrying,

deferring, and dropping tasks for tactical TM), should be useful in expanding and validating normative TM models, as well as in identifying and categorizing current TM problems as a prelude to human-centered design.

ACKNOWLEDGMENTS

This work was performed under NASA contract NAS1-20653. The author wishes to thank Paul Schutte, the NASA technical monitor.

REFERENCES

Abbott, T. (1993). Functional categories for future flight deck designs (NASA Technical Memorandum TM-109005). Hampton, VA: NASA Langley Research Center.

Adams, M. J., Tenney, Y. J., & Pew, R. W. (1994). State-of-the-art report: Strategic workload and the cognitive management of advanced multi-task systems (CSERIAC-SOAR 91-6). Wright Patterson Air Force Base, OH: Crew System Ergonomics Information Analysis Center Publication Series.

Alter, K. W., & Regal, D. M. (1992). Definition of the 2005 flight deck environment (NASA Contractor Report 4479). Seattle, WA: Boeing Flight Deck Research.

Billings, C. E. (1991). Human-Centered Aircraft Automation: A Concept and Guidelines (NASA Technical Memorandum TM-103885). Moffett Field, CA: NASA Ames Research Center.

Funk, K. (1991). Cockpit TM: preliminary definitions, normative theory, error taxonomy, and design recommendations. The International Journal of Aviation Psychology, 1(4), 271-285.

Madhaven, D., & Funk, K. (1993). Cockpit task management errors in critical in-flight incidents. In Proceedings of the Seventh International Symposium on Aviation Psychology (pp. 970-974). Columbus, Ohio.

McGuire, J. C., Zich, J. A., Goins, R. T., Erickson, J. B., Dwyer, J. P., Cody, W. J., & Rouse, W. B. (1991). An Exploration of Function Analysis and Function Allocation in the Commercial Flight Domain (NASA CR-4374). Long Beach, CA: McDonnell Douglas Aircraft Company.

Palmer, M. T., Rogers, W. H., Press, H. N., Latorella, K. A., & Abbott, T. S. (1995). A crew-centered flight deck design philosophy for high-speed civil transport (HSCT) aircraft (NASA Technical Memorandum TM-109171). Hampton, VA: NASA Langley Research Center.

Raby, M. & Wickens, C. D. (1994). Strategic workload management and decision biases in aviation. The International Journal of Aviation Psychology, 4(3), 211-240.

Swink, J. R., & Goins, R. T. (1992). Identification of high-level functional/system requirements for future civil transports (NASA Contractor Report No. 189561). Long Beach, CA: Douglas Aircraft Company.

FLIGHT CREW TASK MANAGEMENT IN NON-NORMAL SITUATIONS

Paul C. Schutte
Anna C. Trujillo
Langley Research Center
National Aeronautics and Space Administration
Hampton, Virginia

Task management (TM) is always performed on the flight deck, although not always explicitly, consistently, or rigorously. Nowhere is TM as important as it is in dealing with non-normal situations. The objective of this study was to analyze pilot TM behavior for non-normal situations. Specifically, the study observed pilots' performance in a full workload environment in order to discern their TM strategies. This study identified four different TM prioritization and allocation strategies: 'Aviate-Navigate-Communicate-Manage Systems;' 'Perceived Severity;' 'Procedure Based;' and 'Event/Interrupt Driven.' Subjects used these strategies to manage their personal workload and to schedule monitoring and assessment of the situation. The 'Perceived Severity' strategy for personal workload management combined with the 'Aviate-Navigate-Communicate- Manage Systems' strategy for monitoring and assessing appeared to be the most effective (fewest errors and fastest response times) in responding to the novel system failure used in this study.

INTRODUCTION

There are four primary functions performed by the flight crew in civil transport aircraft to accomplish its mission. These functions are flight management, communications management, systems management, and task management (Abbott 1993). Flight management involves what is generally known as aviation (primary control of the aircraft) and navigation. Communications management is concerned with all communications occurring on the flight deck, including communications with air traffic control and with other crew members. Systems management occurs when the system that the flight crew is interacting with must be monitored or managed in performing a function.

"Task management is the first-level function of managing tasks and associated resources involved in conducting the mission. This is both a supervisory and a supporting function to the other three major flight deck functions. This function involves monitoring, scheduling, and allocating the tasks and task resources between and for each major function" (Abbott & Rogers 1993). Of the four functions, task management (TM) is perhaps one of the most underrated in that it has received the least amount of attention in flight deck design and airline training. TM is always performed on the flight deck, although not always explicitly, consistently, or rigorously. While some aids — such as procedures, checklists, and levels of alerts — do assist the crew in managing specific tasks and workload, the aids do not address the problem in a more general sense. For example, a procedure for shutting down an auxiliary power unit may not account for incoming air traffic control (ATC) calls or approach procedures. Likewise a standard approach procedure may not have a contingency for dealing with an oil system failure. It is the job of the flight crew to merge these procedures together in a safe and efficient manner. However, little training is given on how to do so. Pilots often use rules of thumb — such as "first aviate, then navigate, then communicate, then administrate" — to prioritize tasks. These rules of thumb are helpful in many cases but are sometimes inappropriate for specific contexts, and are too general to help in others. Raby and Wickens (1994) concluded that "different pilots clearly differed in their scheduling" and that pilots prioritize tasks when faced with high workload situations.

Nowhere is TM as important as it is in dealing with non-normal situations. Here, the workload of the flight crew members is generally increased and they are required to respond to abnormalities that may affect several functions. For example, a systems failure may also affect the planned destination and aircraft handling (flight management) and may precipitate the need for communications. The crew often has to reschedule and reallocate a number of tasks and resources based on their assessment of the criticality of the problem. Non-normal situations require the flight crew to devote more time and attention to the function related to the cause of the non-normal condition (for example, systems management if an engine has failed, or flight management if a storm is encountered). However, it is rare that this function is the only one affected (Rogers, Schutte, Latorella, in press). TM has a large role to play in how the crew's attention is dispersed among the different functions.

OBJECTIVE

The objective of this study was to analyze pilot TM behavior in response to systems failures for which there was no specified procedure. Specifically, the study observed pilots' performance in a full workload environment to discern their TM strategies (if any existed).

METHOD

Subjects

Sixteen airline pilots from 4 major US airlines served as test subjects. Each of the subjects was rated in Extended Twin-engine Operations (ETOPS). The subjects were tasked with acting as Captain/Pilot-Not-Flying on an ETOPS oceanic flight. The over-water route was chosen because it forced the subject to stay aloft and make decisions as opposed to simply landing the aircraft at the first suitable airport. Also, ETOPS operations have specific rules which state that in the event of having only one operational primary system (e.g., engine or hydraulic) the aircraft must proceed immediately to the closest suitable alternate. A confederate first officer acted as Pilot-Flying. The confederate was fully capable in normal operations and was obedient to the Captain; however, he offered no help in decision making or situation assessment.

Apparatus

This study was conducted in a transport-class flight-deck simulator which was designed to provide full mission and full workload capability in a two-crew layout. The simulator's primary flight displays, navigation map displays, autopilot/flight director, and flight management system were designed to be analogous to current state-of-the-art transport aircraft such as the Boeing 747-400, the McDonnell Douglas MD11, and the Airbus A320. Radio communications with a confederate air traffic controller and confederate company dispatcher were provided.

Scenario

The scenario for this study was a large fuel leak near the engine. As in most modern aircraft, there is no appropriate checklist for a fuel leak. In this scenario, a large leak developed in the fuel line downstream of the fuel flow sensor but prior to the engine combustor. The location is important because it causes the fuel flow reading on the flight deck to appear normal for the throttle setting, while the thrust developed by the engine was abnormally low for the throttle setting. The fault indications — as seen by the subject — were an advance of the autothrottle (in order to maintain speed at reduced thrust), followed by a degradation in the engine thrust parameters (engine pressure ratio, engine rotational speeds, and exhaust gas temperature), followed by the inability of the aircraft to maintain altitude and speed. If nothing were done, the flight management system would alert the crew that there was a difference in the calculated fuel (based on fuel flow burn) and measured fuel in the tanks. The most appropriate response for this failure was to shut down the engine in order to conserve fuel. With the engine shut down, the subject should command a descent, declare an emergency, and proceed to the nearest suitable alternate in accordance with ETOPS rules.

Measures

The subject's performance measures included factors such as response time, fault identification accuracy, and compensation accuracy. The subject's performance in non-systems tasks was also measured and included factors such as ETOPS compliance, route management, and fuel used.

The subjects were asked at regular intervals (approximately every 5 minutes) to state the tasks on which they were focusing. The subjects had three choices of response: systems (e.g., operations and status); spatial (e.g., altitude, aircraft position); or mission awareness (e.g., flight planning, fuel, weather). This measure was designed to provide some insight into their TM strategies. The subjects were instructed that they could postpone their response to the probes if they desired.

During the scenario, several "natural" conversational probes (e.g., "What happened back there?" "Could you give me any additional information that might help maintenance figure this out?") regarding the non-normal situation were made by the air traffic controller, the dispatcher, or the first officer. Audio-video recordings of the scenarios were made and transcribed to capture all of the utterances of the subjects.

The subject's performance was scored based on three criteria: correct diagnosis of the failure; correct system response to the failure; and correct mission response to the failure. For this scenario, the criteria were 'fuel leak,' 'engine shutdown,' and 'divert to alternate airport,' respectively.

The video tapes and transcripts were used to determine the management function that the subject was performing, and the TM strategy employed by the subject. The subject's utterances on the transcripts were scored according to a modified version of flight, communications, systems, and task management categories. TM strategies were attributed to each of the subjects subjectively by the authors based on review of the video tapes and the scoring data. A TM strategy was assigned based on the order and amount of time spent in each functional category and the subjects' TM statements about intent and priority. The TM strategies were compared with the subject's performance to see if there were any correspondences.

RESULTS

As mentioned earlier, there were no cautions or warnings or appropriate checklists for the fuel leak. Twelve of the subjects correctly identified the problem as a fuel leak. Nine of the sixteen subjects correctly shut down the engine, averting a low fuel situation. Of those nine, two did not divert to the closest alternate, in violation of ETOPS rules. Seven subjects did not shut down the engine which led to a low fuel situation (especially for the two subjects who did not divert). Table 1 shows a summary of the subjects' performance ordered by correctness. Subjects who did not shut down the engine received the lowest ratings since that left them in a fuel critical situation. These data were used to determine the effectiveness of the TM strategies derived below.

Two sets of data are the main emphasis of this paper. The first was their periodic assessment of their current focus based on the probes. The second was the transcripts and video tape of their performance. They were used to describe the subject's focus of attention during the scenarios.

The self assessment data showed that subjects were focusing their attention on systems an average of 46% of the time. This is not surprising, since the major events in the scenario were systems failures and the subjects were acting as pilot not flying, meaning they were charged with performing systems management duties. They stated that they were focusing on the mission an average of 36% of the time. Finally, only 18% of the their responses indicated that they were focusing on spatial awareness.

Correct System Response	Correct Mission Response	Correct Diagnosis	Number of Subjects
Yes	Yes	Yes	6
Yes	Yes	No	1
Yes	No	Yes	2
No	Yes	Yes	3
No	Yes	No	2
No	No	Yes	1
No	No	No	1

Table 1 - Summary of Subjects' Performance

For the second set of data, each of the subject's utterances were categorized into one of six functional categories. The categories were derived from Abbott (1993) and are described below. The flight management category was decomposed into aviate and navigate in accordance to Abbott's structure. After reviewing a few of the video tapes it became clear that an additional category was necessary for addressing the subject's questions about the simulation. These utterances were scored in a new category called, Flight Deck Systems.

Aviate - statements describing the Captain's manipulation or consideration of the motion of the aircraft; e.g., ascend, descend, turn left, turn right, pitch up, pitch down, roll.

Navigate - statements describing the Captain's manipulation or consideration of the aircraft route, change in destination, change in waypoints, point of equal time, etc. Also discussion of capabilities to complete the route (e.g., do we have enough fuel to make it to Honolulu?)

Communicate - since all utterances are a form of communication, this category had to be restricted to statements describing the initiation or termination of communications outside of the flight deck. The statements in-between the initiation and termination (i.e., the content of the communication) were scored using the appropriate management function.

System - any statements pertaining to the operation of airframe systems (i.e., engines, fuel, hydraulic, electrical, environmental).

Flight Deck Systems - any statements pertaining to the operation of flight deck instrumentation and automation that were peculiar to the simulator. This includes the FMS operation, display symbology, etc. Generally, these statements are in the form of questions about operation of the simulation and differences between the simulator and the aircraft that the subjects were used to flying.

Task - any statements pertaining to assigning duties, stating an intent or order of functions in the future, evaluation of resources, mission planning, etc.

Scoring the utterances in this manner provides some insight into how often such statements are made. Figure 1

shows the percentage of utterances for each category over all of the subjects. The standard deviations across subjects were no greater than 6.5%. The large percentage of Flight Deck Systems utterances was not surprising given that the subjects were unfamiliar with certain aspects of the aircraft simulator.

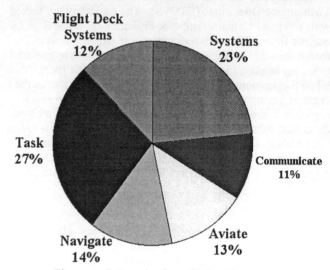

Figure 1. Categorization of Pilot Utterances

A review of time plots of the utterance categories showed the diversity of management functions that are addressed, acknowledged, or attended to over time. Figure 2 shows a sample plot of the utterance categories over time.

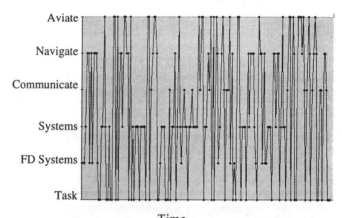

Time

Figure 2 - Attention Transition Diagram

While it is tempting to use these plots to ascertain the subject's TM behavior, the plots alone do not provide sufficient data to do so. A single aviate utterance could be a deflection of an aviate task, a delegation of an aviate task, or a single step completion of an aviate task. An utterance could have been made to acknowledge a comment by the first officer or ATC, but not to attend to that comment.

In order to discern the TM strategy, careful reviews of the video tapes were required in conjunction with the scores. It is important to note that the strategies attributed to each of the subjects were assigned subjectively by the authors. A more rigorous approach in which independent raters score the utterances and categorize the strategies would be useful, but was beyond the scope of this project. Still, the results from this study bring out a number of interesting points. The authors recognized four different TM strategies that were employed by the subjects when faced with a non-normal system failure. The strategies represented how the subjects allocated resources and prioritized tasks. The first two strategies appeared to prioritize and allocate activities based on a goal-driven paradigm while the second two strategies appeared to be stimulus-driven. The four TM strategies follow.

Aviate-Navigate-Communicate-Manage Systems (ANCS) : subjects prioritize their attention with Aviate as the highest priority and Systems as the lowest priority. This prioritization occurs with little regard for the context of the situation.

Perceived Severity : subjects place highest priority on what they perceive to be the most threatening problem.

Procedure Based : subjects migrate towards tasks for which there are well defined procedures. These range from systems procedures to Federal Aviation Regulations (FARs).

Event/Interrupt (E/I) Driven : subjects' attention is given to a particular task based on an event or an interruption. They will typically continue pursuing that task until either the task is completed, the subject can do no more on the task, or until another event or interruption disrupts the task.

Subjects appeared to use the strategies for two distinct TM activities: monitoring and assessment of the situation; and personal workload management (i.e., where the subject's attention was focused). The monitoring activity comprises the 'assess current situation' and 'assess progress and status of active tasks' of Funk's (1991) Cockpit TM procedure. For example, one subject prioritized his personal workload based on Perceived Severity and monitored his situation based on ANCS strategy. When faced with the fuel leak, he first monitored to see that there were no Aviate, Navigate, or Communicate problems and that the First Officer had the airplane. Once the assessment was complete, he focused his attention on the most severe problem; i.e., the systems. Another subject, whose personal workload management was based on an ANCS strategy and monitoring was based on a Perceived Severity strategy, quickly assessed the systems problem to see if it was critical (e.g., engine fire). Since it was not, he allocated most of his time to the aviate task, with the navigate and communicate tasks taking the lesser amounts of his time, respectively. When the other functional tasks were complete, he addressed the systems problem. Table 2 is a summary of the different TM strategies employed by the sixteen subjects ordered by their performance.

None of the subjects who had correct performance in all three measures (i.e., Correct Diagnosis / Correct System Response / Correct Mission Response) used the ANCS strategy or the Procedure Based strategy for their personal workload management. The seven subjects who did not shut down the engine (thus, placing them in a fuel critical situation) used the ANCS strategy for their personal workload management. ANCS was used for monitoring by five of the nine subjects who shut down the engine. There was no apparent correlation between the subject's airline and the strategies used.

	Personal Workload Management	Monitoring	Performance
S U B #	ANCS, Perceived Severity, Procedure Based, E/I Driven	ANCS, Perceived Severity, Procedure Based, E/I Driven	System Response/ Mission Response/ Diagnosis
1	E/I Driven	Procedure Based	Yes/Yes/Yes
2	E/I Driven	Perceived Severity	Yes/Yes/Yes
3	E/I Driven	ANCS	Yes/Yes/Yes
4	Perceived Severity	ANCS	Yes/Yes/Yes
5	Perceived Severity	ANCS	Yes/Yes/Yes
6	Perceived Severity	ANCS	Yes/Yes/Yes
7	ANCS	Procedure Based	Yes/Yes/No
8	Perceived Severity	ANCS	Yes/No/Yes
9	Procedure Based	Perceived Severity	Yes/No/Yes
10	ANCS	Procedure Based	No/Yes/Yes
11	ANCS	Perceived Severity	No/Yes/Yes
12	ANCS	E/I Driven	No/Yes/Yes
13	ANCS	Procedure Based	No/Yes/No
14	ANCS	Procedure Based	No/Yes/No
15	ANCS	Procedure Based	No/No/Yes
16	ANCS	Procedure Based	No/No/No

Table 2 - Summary of TM Strategy and Performance for the 16 Subjects

DISCUSSION

Perhaps the most important result is that TM appears to play a significant role in how the flight crew deals with an abnormality. The scoring of the utterances show that a large percentage of the Captain/Pilot-Not-Flying's time in this experiment was spent in TM. However, the amount of training and procedures for TM is significantly less than for the other management categories. Thus, TM is largely dependent on individual differences between flight crews and personal style. This may be the only difference between an incident and an accident.

The strategy that appeared to be most useful in dealing with the systems failure presented in this experiment was to prioritize and allocate personal workload based on either Perceived Severity or E/I Driven strategy and to monitor based on an ANCS strategy. The subjects whose personal workload management was E/I Driven took longer to shutdown the engine (average 5320 seconds — standard deviation 510) than those based on Perceived Severity (average 3416 seconds — standard deviation 1297). Thus, the goal-driven combination of Perceived Severity/ANCS would appear to be the most useful for the role of Captain/Pilot-Not-Flying.

The success of the Perceived Severity/ANCS strategy combination is likely due to an appropriate emphasis being placed on the problems at hand and sufficient coverage from the strategy for monitoring (ANCS) to assure that all other flight deck tasks are assessed and performed. Subjects using the ANCS strategy as their monitoring seemed to cycle through the four categories to spot check for problems and to insure that all else was well.

Another reason that this strategy appears to work well is that the most severe problem (in this study, the fuel leak) is kept in the subjects' minds as they cycle through ANCS. Often, there is additional helpful information to be found by doing so. In the fuel leak scenario, navigation tasks such as position plots helped subjects identify excessive fuel usage prior to receiving the fuel disagree message from the flight management system. The aviate task showed that rudder was being deployed to compensate for the asymmetric thrust from the engines. Some subjects used this information to confirm that there was a real problem and not a sensor failure. Finally, the subjects used the communicate task with dispatch and with the flight attendants (e.g., "Could you go look out the window and see if there is any fluid streaming out of the engine?") to obtain additional information for dealing with the problem. Thus, the diversity of information provided by ANCS was beneficial in solving a problem in the systems.

Based on these results, it appears that using the ANCS for personal workload management was not helpful in dealing with the systems failure in this experiment. This is not surprising since systems is given the lowest priority in the ANCS strategy. It is possible that an aviate problem, such as traffic or weather, could be sufficiently handled by an ANCS strategy for personal workload management. However, it could be that an ANCS (personal workload) and Perceived Severity (monitor) strategy would lead the pilot to fixate on the aviate problem and miss a subsequent navigate or systems problem, whereas the reverse strategy could appropriately draw attention away momentarily to check on the status of those other areas. Controlled testing of these hypotheses would be required to definitively answer this question.

After reviewing the video tapes it became clear that interruptions played a significant part in the TM of the subjects — not only for those using E/I Driven strategies but for everyone. Some subjects had to work hard to stay true to their strategies in light of some very compelling interruptions. Not surprisingly, ATC was an ample supplier of interruptions. (The timing of ATC calls was not controlled.) However, there was another supplier from a less likely source. The confederate first officer was instructed to perform all the duties of first officer, flying the aircraft. He was well qualified for this and performed flawlessly according to the latest in training — especially in the aspect of Crew Resource Management (CRM). CRM dictates that a crew member clearly communicates his intent and actions to the rest of the crew. Since the first officer's task was an aviate task, these communications acted as distractions to the task that the subject was performing. Often, while scoring the utterances, it was seen that the subject would make an aviate comment, only in response to the first officer. In some cases, these comments actually disrupted the task that the subject was performing and led him to tend to the aviate task. Thus, CRM can potentially be misapplied if communications are not prioritized and managed.

An important methodological note is that providing a full-mission scenario where the subject (in this case, the pilot not flying) has access to other relevant personnel for realism (co-pilot, air traffic controller, dispatch) can produce natural conversations and "thinking out loud" that are a rich source of information for analysis of cognitive processes and decision making. Since the "think out loud" protocol is not explicitly instructed, it has the advantage of occurring unobtrusively. "Naturalistic" probes that require pilots to describe their thought processes and decisions can be successfully inserted into the scenarios as part of ATC or dispatch conversation without disruption to the flow of the experiment. The danger of providing realistic resources, such as dispatchers, is that some pilots may use them as resources and delegate responsibilities to them which, as in this study, were the very tasks that the subject was to perform.

In summary, task management is a significant and important part of the flight crew's duties. Improvements in the task management strategies used by pilots holds significant potential for the reduction of the negative consequences of pilot actions when responding to a system failure. More research needs to be performed to determine the most effective task management strategies. Those strategies should be considered for explicit incorporation into flight crew training.

ACKNOWLEDGMENTS

The authors are indebted to Dr. William Rogers for co-developing and co-running the experiment and to John Barry for providing programming and data analysis.

REFERENCES

Abbott, T. (1993). Functional categories for future flight deck designs (NASA Technical Memorandum TM-109005). Hampton, VA: NASA Langley Research Center.

Abbott, T. S., & Rogers, W. H. (1993). Functional categories for human-centered flight deck design. Proceedings of the 12th Digital Avionics Systems Conference. New York: AIAA/IEEE.

Funk, K. (1991). Cockpit task management: preliminary definitions, normative theory, error taxonomy, and design recommendations. The International Journal of Aviation Psychology, 1(4), 271-285.

Raby, M. & Wickens, C. D. (1994). Strategic workload management and decision biases in aviation. The International Journal of Aviation Psychology, 4(3), 211-240.

Rogers, W. H., Schutte, P. C., & Latorella, K. A. (in press). Fault management in aviation systems. In M. Mouloua & R. Parasuraman (Eds.), Automation and Human Performance: Theory and Applications. Hillsdale, NJ: Lawrence Erlbaum Associates.

INVESTIGATING INTERRUPTIONS: AN EXAMPLE FROM THE FLIGHTDECK

K. A. Latorella
Industrial Engineering Department
State University of New York at Buffalo
Buffalo, New York 14260

This study investigates an aspect of multiple-task management, interruption management, in an operational context. Fourteen commercial airline pilots each performed 16 approaches in a simulated commercial flightdeck. Air traffic control (ATC) clearances interrupted subjects as they performed three procedures during these approaches. Common ATC interruptions were found to be significantly disruptive to ongoing procedure performance on the flightdeck by producing significantly more procedure performance errors and increased flightpath management activity. These results corroborate, for the flightdeck, that which is true in laboratory experiments, and which is evidenced in aviation accident/incident reports.

INTRODUCTION

Human operators increasingly act as managers of multiple tasks in complex and dynamic environments. One aspect of multiple-task management (MTM) is the handling of interruptions, or interruption management. Research in attention management (*e.g.*, Broadbent, 1958; Schneider & Detweiler, 1988) and human error (*e.g.*, Reason, 1992) indicate that humans do not handle interruptions easily, or, often, very well. Previous research investigating interruption management takes four approaches: (1) development of a theoretical framework for MTM, including interruption management (2) laboratory studies aimed at understanding mechanisms of interruptions (3) human/machine interface evaluations using interruption-recovery as an evaluation metric and (4) identification of interruptions as a causal factor in accident/incident analyses and field investigations.

Adams, Tenney, & Pew (1995) describe problems associated with developing situation awareness in MTM environments. One problem associated with situation awareness is to accurately develop and retain a task queue in memory. Consistent with theories of limited working memory capacity, these authors suggest that interruptions may cause other tasks' representations to be deleted from the queue, and the tasks not performed. An interruption is not merely an additional task competing for a limited resource, it also redefines that which is resident in active memory. Knowledge structures associated with the interrupting task impose on those already resident at the time of interruption. Based on these two facets of the MTM problem, Adams, Tenney, & Pew (1995) develop a framework for the management of multiple tasks based on Neisser's (1976) expanded model of the perceptual cycle. Neisser's model includes an *explicit focus* memory bin and an *implicit focus* memory bin. Explicit focus describes a limited-capacity storage, and corresponds to working memory. Implicit focus

relates to the knowledge structures that are related to those tasks represented in explicit focus. This framework provides the foundation for speculation on when interruptions might be handled most easily. Specifically, that interruptions related to the items in explicit focus should be easily integrated with ongoing activities, and, interruptions tangent to the immediate explicit focus should be relatively easily assimilated due to common representations in implicit focus.

While Adams, Tenney, & Pew (1995) postulate mechanisms affecting interruption management, laboratory experiments demonstrate deleterious effects of interruptions and empirically identify significant causes of these effects for simple tasks. Most research indicates that interruptions increase post-interruption performance times (Detweiler, Hess, & Phelps, 1994; Gillie & Broadbent, 1989; Field, 1987; and Kreifeldt & McCarthey, 1981) and error rates due to interruptions (Detweiler, Hess, & Phelps, 1994; Cellier & Eyrolle, 1992; Gillie & Broadbent, 1989; Field, 1987; and Kreifeldt & McCarthey, 1981). However, Cellier & Eyrolle (1992) found increased performance speed following an interruption. They attribute this to activation of previously untapped resources after interruption. Significant factors influencing the effects of interruptions include; task complexity (Cellier & Eyrolle, 1992; Gillie & Broadbent, 1989), similarity of interrupted and ongoing tasks (Detweiler, Hess, & Phelps, 1994; Cellier & Eyrolle, 1992; Gillie & Broadbent, 1989), memory load at interruption (Detweiler, Hess, & Phelps, 1994), ability to rehearse departure conditions (Detweiler, Hess, & Phelps, 1994; Gillie & Broadbent, 1989), and time constraints (Cellier & Eyrolle, 1992). Laboratory investigations identify factors that mediate interruption management performance in artificial settings.

In addition to research that explicitly studies mechanisms of interruption performance, studies using interruption recovery to evaluate human/machine interface investigations provide useful insights. In a comparison of reverse-polish notation and algebraic notation calculators (Kreifeldt & McCarthey, 1981), and in a comparison of different search strategies in a menu system (Field, 1987), users' performance was generally worse following the introduction of an interruption. Field studies and accident and incident reports indicate the significance and magnitude of deleterious effects of interruptions in more complex operational contexts.

Interruptions are cited as a contributing cause of power plant incidents (*e.g.*, Bainbridge, 1984). In addition, the frequency, types (Monan, 1979), and deleterious effects (*e.g.*, Madhaven & Funk, 1993) of interruptions on the flightdeck are well documented. A recent search of the Aviation Safety and Reporting System (ASRS) revealed at least 315 reported incidents due to interruptions on the flightdeck since 1986. Worse than this, interruptions are implicated in disastrous accidents which result in loss of life. An air traffic control (ATC) interruption during checklist procedures appears to have caused the aircrew of an airliner departing from the Detroit Metropolitan Airport forgot to lower the flaps before takeoff (National Transportation Safety Board, 1988). System failures can also be considered interruptions to ongoing flightdeck tasks. On an Eastern L-1011 aircraft, the crew became so distracted by an alert and performance of an irregular procedure that flightpath management duties went unattended (National Transportation Safety Board, 1973). Recognizing the effects of interruptions on the flightdeck, *task interruption* was included in a taxonomy of cockpit task management (CTM) errors (Funk, 1991; Chou & Funk, 1990). The is relatively little experimental research investigating interruptions on the flightdeck. Existing studies use frequency-of-interruption as a dependent measure to evaluate datalink and checklist usage (Williams, 1995; Linde & Goguen, 1987, respectively). Williams (1995) reports significantly greater resumption times for datalink than voice ATC interruptions, however does not quantitatively report the degree of disruption imposed by these interruptions.

The study of interruptions is motivated by the significant effects of interruptions, both in terms of their ubiquity and their consequences in operational environments, and the implications for extending basic cognitive theories to an understanding of MTM. Several laboratory studies cite the deleterious effects of interruptions, and field and incident/accident investigations evidence the consequences of interruptions in operational environments, the effects of interruptions on the flightdeck have not been experimentally quantified. This research investigates the effects of interruptions on the commercial flightdeck. The fundamental question answered by this research is, to what extent do interruptions disrupt pilot activity on the flightdeck? To address this issue, an experiment was conducted in which commercial airline pilots performed approach and descent procedures in a commercial flight simulation environment. ATC clearances were systematically inserted into the scenario to interrupt procedure performance.

METHODS

Subjects

Fourteen male commercial airline pilots from various carriers served as subjects. Subjects currently flying advanced Boeing aircraft with minimally one year glass-cockpit, flightpath management system / control display unit (FMS/CDU) experience, and 5,000 flying hours.

Apparatus

The experiment was conducted in NASA Langley's Transport Systems Research Vehicle (TSRV), a fixed-base simulator similar to a B-737. A remote confederate interacted with subjects for real-time ATC and airline operations personnel (company) contacts. Thirty-second continuous loop tapes provided Automatic Terminal Information Service (ATIS) information. A menu system on a touchscreen contained approach and final descent checklists. Another touchscreen display contained a simple datalink system for introducing visual interruptions. Datalink messages were introduced by the mechanized voice utterance, "incoming message". Subjects were not provided with an out-the-window view.

Scenario

The scenarios included three components: flightpath management (FPM), procedure performance, and interruption management. Flightpath profiles required subjects to manually fly a complex, step-down approach with multiple turns, and hard crossing restrictions at each waypoint in single crew member operations. Subjects flew in the attitude control wheel steering (ACWS) mode of the autopilot with no other autopilot functions or autothrottles. ACWS is a rate-controlled flight mode which retains an established attitude or lateral deviation. Flightpath profiles provided three procedural intervals, which required minimal flightpath management (FPM) activity. Natural procedural interval deadlines were imposed by creating difficult FPM intervals between the procedural intervals and extremely difficult FPM regions surrounding waypoints.

Three procedures, the top of descent (TOD) procedure, 18,000' (18K) procedure, and final approach fix (FAF) procedure were designed to be performed in the procedural intervals (Table 1). The TOD procedure was performed at 19,000', 290 knots calibrated, indicated airspeed (KIAS), level flight. The 18K procedure was performed in a stable descent from 18,000' to 12,000' at 240 KIAS. The FAF procedure was performed in a stable descent from 8,000' to 4000' at 150 KIAS. Field elevation was 3,500'. Tasks identified for inclusion in these procedures were obtained through extensive interviews with two retired United Airlines pilots and reference to several airlines' approach and descent checklists. Some tasks included in these procedures were somewhat artificially placed in order to define experimental conditions or satisfy experimental control, *e.g.*, that the 18K and FAF procedures be isomorphic.

Interruptions were interjected into this scenario in each of the three procedures. In any one run, subjects could experience as few as zero and as many as three interruptions, one *per* procedure. Interruptions were ATC-initiated clearances associated with one of five FMS/CDU interrupting tasks (IT); enter an initial runway, side-step to the parallel runway, program a standard holding pattern, alter a crossing speed restriction, or alter a crossing altitude restriction. While requests to enter initial runways could be either through the auditory or visual modality (via the datalink system), and only occurred in the first procedure, all other interruptions were voice-conveyed and occurred in both the 18K and FAF procedures. These interruptions were interjected into the procedures at specific intervention positions (IP), triggered by subject performance of pre-defined events constituent to procedural tasks.

Table 1. Procedural Tasks

TOD Procedure
Pre-tune Company Frequency
Pre-tune ATIS Frequency
Listen to ATIS
Pre-tune Tower Frequency
Obtain Status Information from FMS/CDU

18,000' Procedure
Set Altimeters in FMS/CDU
Contact Company
Obtain ETA-Zulu Time from FMS/CDU
Calculate ETA-local Time.
Turn on Seatbelt sign
Announce to Cabin (Seatbelt sign, gate, ETA)
Turn on Landing Lights
Turn on Anti-skid
Select appropriate Autobrakes
Perform Approach Checklist

FAF Procedure
Select EPR for Go-Around from FMS/CDU
Contact Tower
Obtain Vref30. from FMS/CDU
Calculate Adjusted Target Speed.
Turn on No-Smoking sign
Announce to Cabin (No-Smoking, landing)
Lower Gear
Arm Speedbrake
Select Flaps 25
Perform Final Descent Checklist

Pairing ITs and IPs defined different experimental interruption conditions. These conditions operationalized several task factors hypothesized to affect flightdeck interruption management, *e.g.*, modality and semantic similarity of an interruption and ongoing task, embeddedness of an interruption in a procedure, and the relationship between

procedural tasks severed by an interruption. Null IP and IT specifications provided uninterrupted procedure control conditions. All subjects received all experimental conditions. Since this paper addresses only the first stage of analysis for this investigation, *i.e.*, quantifiying the general effects of interruptions on the flightdeck, the above factors will not be elaborated on further.

Procedure

Each subject participated in the experiment for two days. On the first day, subjects received explicit descriptions of scenario performance goals and FPM, procedure performance, and interruption simulation training. To enable the introduction of interruptions at specific points in these procedures, and to provide a standard performance goal by which to evaluate performance errors, subjects were asked to perform procedural tasks in exactly the order and method trained. Subjects were told that the scenario would also include *incidental tasks* to increase scenario realism, and that these tasks were termed *incidental* only because they would occur non-deterministically throughout the scenario. Subjects were instructed that they must perform incidental tasks and must confirm any incoming ATC request prior to actually accomplishing the incidental task. The second day included three refresher runs and sixteen data runs. Each run was approximately seventeen minutes in length and was preceded by a three-minute reset period. Subjects received a break after the refresher runs and after each set of four data runs.

Dependent Measures

This experiment measured the deleterious effects of interruptions on the ensemble task; that is, the integration of procedure and interrupting task requirements. Three measures defined these deleterious effects: (1) procedure performance errors, (2) procedure performance time, and (3) ensemble FPM activity. Procedure performance errors were identified as procedural task omissions, misorderings, or redundant performance of procedural tasks. To ascertain the effect of interruptions on procedure performance time, ensemble performance times, for which interruptions occurred within a procedure, were compared to constructed *composite times*. Composite times were constructed by adding the average of uninterrupted procedure times and interruption performance times for those occurring before procedures. Composite times were defined for all possible (subject, procedure, interrupting task) triplets, to eliminate effects of these variables. The ensemble FPM activity measure counted the number of active attitude and lateral control inputs made during the ensemble interval, *i.e.*, from the first activity of the procedure or interruption to the last activity of either the procedure or interruption, and divided by the elapsed seconds in that interval. Since the scenario afforded hands-free FPM after deviations were nullified, and procedure intervals were time-constrained, reversion to active FPM during procedures was considered non-optimizing performance.

RESULTS

Procedure performance errors were analyzed using analysis of variance rather than a non-parametric statistic because error data was too sparse to calculate expected cell frequencies. Significant effects using this parametric assessment are conservative estimates of effects expected from an non-parametric analysis. Overall error rates were very low, less than one major error per procedure. However, interrupted procedures contained significantly more, on average 53% more, procedure performance errors than uninterrupted procedures, $F(1,13) = 25.809$, $p = 0.0002$ (Table 2). Generally, one would expect one error in every three uninterrupted procedures, but in one of every two interrupted procedures. Some task omissions were more consistently evident than others and seemed associated with the previous occurrence of an interruption (Table 3).

Table 2. Procedure performance errors.

condition	n	mean	std.dev.
interrupted	504	0.518	0.860
uninterrupted	168	0.339	0.716

Table 3. Frequently omitted tasks.

		% of omissions		
task omitted	% omitted	no IT	after IT	before IT
tune tower	9.56	4.17	66.67	29.17
obtain vref	3.10	0	71.43	28.57
descent check	4.93	9.10	90.90	0

The mean ensemble time for each subject's performance on each procedure and interrupting-task type was compared to its corresponding composite time using a paired t-test, matching on subject, interrupting task-type, and procedure. Marginally significant results indicated that composite times were slightly longer, 1.63 seconds on average, than ensemble times, $t(242) = -1.672$, $p = 0.0958$ (Table 4). One explanation for this result might be that interrupted conditions' average performance times were less than uninterrupted average performance times due to more omissions. To eliminate the possibility of this bias, a paired t-test was performed on the data after all conditions having either a procedural or interruption performance error were removed. On error-free data, the same trend exists, $t(132) = -1.665$, $p = 0.0984$ (Table 5).

Table 4. Procedure Time Comparison.

measure	n	mean	std.dev.
ensemble time	250	111.47	19.14
composite time	243	112.64	16.97

Table 5. Error-free Procedure Time Comparison.

measure	n	mean	std.dev.
ensemble time	136	114.09	18.27
composite time	133	115.87	13.21

On average, subjects made 17.25 active FPM control inputs during interrupted conditions and only 13.10 inputs during uninterrupted conditions. While the absolute difference between interrupted and uninterrupted conditions' average ensemble FPM *per* second is small, it represents a significant increase, $X^2(1) = 14$, $p < .005$, on average 10%, in the proportion of the ensemble interval devoted to FPM.

Table 6. Ensemble FPM Activity *per* second.

condition	n	mean	std.dev.
interrupted	467	0.160	0.142
uninterrupted	161	0.146	0.122

DISCUSSION

This research empirically quantifies the disruptive influence of interruptions on the commercial flightdeck. The effects of interruptions found in laboratory experiments and exemplified in field studies and accident/incident investigations were supported by this experiment's results. As is reported for more generic tasks, ATC interruptions significantly increased performance errors in flightdeck procedures. Several examples illustrate the attribution of task omissions to previous occurrence of an interruption. These omissions have operational consequences, and may have serious operational consequences if combined with other irregular occurrences. For example, a mis-tuned tower frequency minimally causes confusion and increased radio traffic, and maximally, if left uncorrected, may result in pilots' inability to receive life-saving instructions.

Interruptions did not degrade subjects' speed in performing ongoing procedural tasks, even when considering only error-free data. In fact, a marginally significant result indicated that the presence of an interruption actually slightly sped procedure performance time. Although contrary to findings of most previous research, this result was consistent with that reported by Cellier & Eyrolle (1992), and may have indicated that subjects' adopted a compensatory strategy after interruption. Consistent with results indicating a significant effect of time constraints (Cellier & Eyrolle, 1992), and strategic workload theory (*e.g.*, Raby & Wickens, 1991) subjects may have recognized that additional demands of an interruption might interfere with performance on impending high-FPM workload regions, and actively compensated by performing remaining procedural tasks faster. Given this, increased error rates in interrupted conditions could be, derivative of a speed/accuracy trade-off effect rather than directly due to an interrupt's imposition on working memory.

The ensemble FPM activity measure is not directly related to any dependent measure previously associated with studying the effects of interruptions. However it does implicate the disruptive effects of interruptions on the flightdeck. Given the scenario conditions, active FPM during procedures was unnecessary and therefore considered sub-optimizing behavior.

Having quantified the effects of ATC interruptions and datalink presentations degrade performance on the flightdeck, and in light of accidents and incidents attributable to flightdeck interruptions, it is evident that further research is required to identify specific task, environment, and operator performance-shaping characteristics that modulate interruption management behavior. Subsequent analyses of results from this investigation more sensitively study task factors hypothesized to affect interruption management on the flightdeck.

Understanding both the significance of and the factors modulating interruption management improves our understanding of how humans manage multiple tasks, and provides a means for more sensitively introducing and integrating interrupting tasks in MTM contexts. For the aviation domain, this information contributes to an understanding of human CTM (*e.g.*, Abbott, 1993), models of CTM (*e.g.*, Funk, 1996), and informs design of CTM aids (*e.g.*, Funk and Lind, 1992), however the the ubiquity of MTM contexts and interruptions suggests a wide range of applications.

ACKNOWLEDGMENTS

This work was conducted at the NASA Langley Research Center's Crew/Vehicle Integration Branch and funded by a NASA Graduate Student Researcher Fellowship, grant number NGT-50992. This work benefited significantly from the auspices of Mr. Paul Schutte and Dr. Kathy Abbott, the expertise of Captain Dave Simmon (ret.), the programming wizardry of Mr. John Barry, useful comments from Dr. William Rogers, and the assistance of TSRV hardware and software support personnel.

REFERENCES

Adams, M. J., Tenney, Y. J., and Pew, R. W. (1995). Situation awareness and the cognitive management of complex systems. *Human Factors, 37*(1), 85-104.

Abbott, T. (1993). *Functional categories for future flightdeck designs.* NASA-TM-109005. Washington, DC: The National Aviation and Space Administration.

Bainbridge, L. (1984). Analysis of verbal protocols from a process control task. In E. Edwards and F.P. Lees (Eds.), *The Human Operator in Process Control* (pp. 146-158). London: Taylor and Francis.

Broadbent, D. E. (1958). *Perception and communications.* New York: Pergamon Press.

Cellier, J. and Eyrolle, H. (1992). Interference between switched tasks. *Ergonomics, 35*(1), 25-36.

Chou, C. and Funk, K. (1990). Management of multiple tasks: Cockpit task management errors. In *Proceedings of the IEEE International Conference on Systems, Man, and Cybernetics.* Piscataway, NJ: Institute of Electrical and Electronics Engineers.

Detweiler, M. C., Hess, S. M., and Phelps, M. P. (1994). *Interruptions and working memory.* Unpublished technical report. University Park, PA: The Pennsylvania State University.

Field, G. (1987). Experimentus Interruptus. *SIGCHI Bulletin, 19*(2), 42-46.

Funk K. H. (1996). A functional model of flightdeck agenda management. In *Proceedings of the Human Factors and Ergonomics Society 40th Annual Meeting.* Santa Monica, CA: Human Factors and Ergonomics Society.

Funk, K. H. (1991) Cockpit task management: Preliminary definitions, normative theory, error taxonomy, and design recommendations. *The International Journal of Aviation Psychology, 1*(4), 271-285.

Funk, K. H. and Lind, J. H. (1992). Agent-based pilot-vehicle interfaces: Concept and prototype. *IEEE Transactions on Systems, Man, and Cybernetics, 22*(6), 1309-1322.

Funk, K. (1989). Development of a task-oriented pilot-vehicle interface. In *Proceedings of the IEEE International Conference on Systems, Man, and Cybernetics.* Piscataway, NJ: Institute of Electrical and Electronics Engineers.

Gillie, T. and Broadbent, D. (1989). What makes interruptions disruptive? A study of length, similarity, and complexity. *Psychological Research, 50,* 243-250.

Linde, C. and Goguen, J. (1987). *Checklist Interruption and Resumption: A Linguistic Study.* NASA-CR-177460. Washington, DC: The National Aviation and Space Administration.

Kreifeldt, J. G. and McCarthey, M. E. (1981). Interruption as a test of the user-computer interface. In *Proceedings of the 17th Annual Conference on Manual Control.* JPL Publication 81-95. Jet Propulsion Laboratory, California Institute of Technology.

Madhaven, D. and Funk, K. (1993). Cockpit task management errors in critical in-flight incidents. *Seventh International Symposium on Aviation Psychology.* Columbus, Ohio.

Monan, W.P. (1979). *Distraction - A Human Factor in Air Carrier Hazard Events.* NASA-TM-78608. Washington, DC: The National Aviation and Space Administration.

Neisser (1976). *Cognition and reality: Principles and implications of cognitive psychology.* San Francisco: Freeman Press.

National Transportation Safety Board. (1988). *Aircraft accident report - Northwest Airlines, McDonnell Douglas DC-9-82, N312RC, Detroit Metropolitan Wayne Co., Airport, Romulus MI, August 16, 1987.* Washington, DC: National Transportation Safety Board.

National Transportation Safety Board. (1973). *Aircraft Accident report: Eastern Airlines Inc. Lockheed L-1011, N310EA, Miami International Airport, Miami, Florida, December 29, 1972.* Washington, DC: National Transportation Safety Board

Raby, M. and Wickens, C. D. (1991). Strategic behavior in flight workload management. *Sixth International Symposium on Aviation Psychology.* Columbus, Ohio.

Reason J. (1992). *Human error.* New York: Cambridge University Press.

Schneider, W. and Detweiler, M. (1988). The role of practice in dual-task performance: Toward workload modeling in a connectionist/control architecture. *Human Factors, 30,* 539-566.

Williams, C. (1995). Potential conflicts of time sharing the flight management system control display unit with data link communications. *Eigth International Symposium on Aviation Psychology.* Columbus, Ohio.

A FUNCTIONAL MODEL OF FLIGHTDECK AGENDA MANAGEMENT

Ken Funk and Bill McCoy
Oregon State University
Corvallis, Oregon USA

Our research represents an effort to understand and facilitate the management of flightdeck activities by pilots. We developed a preliminary, normative theory of Cockpit Task Management (CTM) and from it defined an error taxonomy. Based on analyses using this error taxonomy we found CTM errors in 76 (23 per cent) of 324 aircraft accident reports and 231 (49 per cent) of 470 aircraft incident reports. Concluding that CTM is a significant factor in flight safety and recognizing the need to broaden as well as refine the concept, we developed a model of Agenda Management, which includes management not only of tasks, but goals, functions, actor assignments, and resource allocations as well. Major components of the functional model include maintaining situation awareness, managing goals (recognizing, inferring, and prioritizing), managing functions (activating, assessing status, and prioritizing), assigning actors (pilots and flightdeck automation) to functions, and allocating resources (such as displays and controls) to functions.

INTRODUCTION

Pilots of modern aircraft must not only perform multiple, concurrent tasks, they must also manage those tasks as well as other functions being performed by non-human actors on the flightdeck. This paper describes our efforts to understand and facilitate the management of flightdeck activities, a process we call Agenda Management.

Our studies parallel and to a certain extent follow a line of research established by Johannsen and Rouse (1979), Hart (1989), Moray and his colleagues (Moray, Dessouky, Kijowski, & Adapathya, 1991), and Wickens and his colleagues (Raby and Wickens, 1994). The concept common to all of these is that the human operator of a complex system must perform multiple, concurrent tasks to control the system and that, as human perceptual and cognitive resources are limited, must therefore manage those tasks.

Our own efforts to formalize this notion in the context of aviation resulted in a preliminary, normative theory of Cockpit Task Management (Funk, 1991). We defined Cockpit Task Management (CTM) in terms of the following activities:

- task initiation: recognizing that a particular goal must be accomplished and therefore that a task must be performed to achieve it.
- task monitoring: assessing progress towards achieving each goal and the level of performance in executing the task.
- task prioritization: assessing relative task priority in terms of overall mission and safety importance, urgency, and momentum or continuity.

- resource allocation: allocating human and machine resources to the completion of tasks based on task priority.
- task termination: recognizing that a goal is achieved, unachievable, or no longer relevant, and ceasing action on the task.

According to our preliminary theory, these activities comprise a high-level cognitive process which serves to determine which low-level activities (i.e., tasks) are being done at any given time.

To validate the theory, we analyzed aircraft accidents and incidents (Chou, Madhavan, & Funk, in press). First, we developed a CTM error taxonomy consisting of the following error categories:

- task initiation errors: early, late, lacking
- task prioritization errors: incorrect
- task termination errors: early, late, incorrect

We applied this taxonomy to 324 US National Transportation Safety Board (NTSB) aircraft accident reports. These were, to our knowledge, all NTSB reports on aircraft accidents occurring between 1960 and 1989. First we reviewed abstracts of the reports and eliminated those that did not have some clear indication of task management errors. Of the remainder we examined either the abstracts or the complete reports themselves in detail, reinterpreting the NTSB's conclusions in light of CTM theory and the error taxonomy. We found CTM errors in 76 (23 per cent) of the reports.

Next we applied the taxonomy to 470 Aviation Safety Reporting System (ASRS) aircraft incident reports. These reports were obtained from ASRS in three separate search requests: in-flight engine emergencies, controlled flight toward terrain, and incidents in the terminal phases of flight (descent, approach, and landing). We reviewed the narrative sections of these reports, where the reporter describes the incident in his/her own words. We looked for explicit references to neglected tasks, misprioritizations, and delays, again interpreting the conclusions of the reporter in terms of CTM theory. We found CTM errors in 231 (49 per cent) of the reports.

From these studies we concluded that CTM is a significant factor in flight safety and thereby warrants both further study and efforts to facilitate it to reduce the likelihood of error.

Events subsequent to these studies, in particular, several accidents involving highly automated aircraft, have led us to change our perspective, definitions, and terminology somewhat.

In particular, if we define an actor as an entity capable of goal-directed activity, it is very clear that human pilots are not the only actors in the cockpit or on the flightdeck. Monitoring and control of the aircraft and its subsystems are performed by machine actors as well, such as autopilots, flight management systems, and automated warning and alerting systems. A common definition of task is a function performed by a human, where a function is a process performed to achieve a goal. Therefore, we must acknowledge that flightcrews in automated aircraft manage functions, not just tasks.

Furthermore, it is also clear, especially from some recent accidents, that actors frequently have conflicting goals, and that these conflicts may lead to conflicting actions, resulting in unsafe conditions. Goals must be managed too.

From these insights, we have changed our terminology and now refer to Agenda Management. An agenda is a set of goals, functions, actor assignments, and resource allocations. Managing this agenda is an important process performed by the flightcrew.

OBJECTIVES

The objectives of our research are to

1. develop and validate a formal model of Agenda Management.
2. investigate means of facilitating Agenda Management.

METHOD

Since Agenda Management is an activity or a function itself, we decided that a functional modeling approach would be appropriate. We performed a functional decomposition of the process using IDEF0, a graphical modeling tool. An IDEF0 model consists of block diagrams representing activities or functions that transform entities, and the entities those functions act on or are constrained by. The functions are denoted by verb phrases, the entities are denoted by noun phrases. IDEF0 provides a framework that helps the modeller identify key transformations that take place, the objects of the transformations, factors which limit or guide the transformations, and the mechanisms that perform the transformations.

Starting at the most general level of flightdeck activities, we used knowledge derived from the studies described above to decompose higher level activities to lower level activities, continuing to a level we felt was necessary for validation and adequate to help guide the development of Agenda Management aids.

RESULTS

A portion of our functional model of Agenda Management is presented below. In particular, major functions in the process are identified and defined. Each function is denoted by its IDEF0 identifier, which consists of the letter 'A' (for Activity) and a sequence of digits showing hierarchical relationships between functions (A11 is the first subfunction of A1, A112 is the second subfunction of A11, etc.).

A0 perform flightdeck activities -- Perform the activities of operating a commercial transport aircraft from its flightdeck. These activities are performed by human actors (flightcrew) and machine actors (flightdeck automation) using flightdeck resources (displays, sensors, controls, actuators, radios, and other non-'intelligent' devices). The actors may be viewed as a single, integrated cognitive system.

- **A1 manage agendas** -- Manage the agendas of all actors.

  - **A11 manage individual agendas** -- Manage the agenda of each individual actor. Each actor manages his/her/its own agenda and these agendas may or may not be consistent.

    The following subfunction descriptions (A111 through A1144) reflect the activities performed by a single actor in the management of his/her/its own agenda.

    - **A111 manage goals** -- Recognize, infer, activate, and terminate goals. Prioritize active goals. This must be coordinated with the goal management of other actors through shared agenda information.

- **A1111 infer goals** -- Infer the other actors' goals from actor and other system state information in the situation models: "What are the other actors' goals that they have not explicitly declared?"

- **A1112 assess goals** -- Determine what goals should be pursued. Initially, this is just the mission goal, which is decomposed into subgoals. But at any given time, this activity involves adding goals inferred from other actors and this actor's newly derived goals to the set of current (pre-existing) goals, then assessing each to determine if it is pending, active, or terminated: "What should we be getting ready to do (pending goals)? What should we be doing now (active goals)? What can we forget about (terminated goals)?"

- **A1113 prioritize goals** -- Rank the goals based on the importance and urgency of each goal. A goal has high importance if its achievement is a necessary condition for achieving the mission goal. It has high urgency if it must be achieved soon. "What is most important? What is most urgent? What is most worthy of our attention right now?"

- **A1114 identify goal faults** -- Identify any goal problems, such as erroneous or conflicting goals: "Are our goals appropriate and are we in agreement about them?"

- **A112 manage functions** -- Initiate, assess, prioritize, and terminate functions to achieve goals. This must be coordinated with the function management of other actors through shared agenda information.

- **A1121 activate/deactivate functions** -- Based on the active goals, determine what functions should be performed now: "Are we actually doing what we should be doing?"

- **A1122 assess function status** -- Determine how well each function is being performed, with respect to achieving the goal, based on accuracy, speed, and other factors. As well as considering the current state of affairs, look ahead. In addition to using global information, use specific status information derived in the process of performing each function. "How well are we doing now? Are things likely to get better, worse, or stay the

same? Is it likely that we will achieve the goals?"

- **A1123 prioritize functions** -- For each function, determine its priority, based on its goal's priority, its status, and its momentum (i.e., functions nearly completed have a greater momentum than do functions just begun). "What should we be doing right now?"

- **A1124 identify function faults** -- Identify any problems with the current functions, such as inappropriate functions, misprioritized functions, or discrepancies about functions: "Are we in agreement about what we should be doing right now and how well we're doing?"

- **A113 assign actors to functions** -- Decide which actors are to perform each function. This must be coordinated with the actor assignments of other actors through shared agenda information.

- **A1131 identify feasible assignments** -- Identify different ways that actors could be feasibly assigned to perform functions: "How could we assign actors to functions?"

- **A1132 evaluate feasible assignments** -- Evaluate the different ways actors could be assigned to functions: "What are the advantages and disadvantages of particular actor assignments?

- **A1133 select assignments** -- Select the best actor assignments: "What are the best assignments?"

- **A1134 identify assignment faults** -- Identify problems with the assignments, such as inappropriate assignments and inconsistencies between actors: "Do we agree on the correct actor assignments?"

- **A114 allocate resources to functions** -- Decide what resources are to be used to perform each function. This must be coordinated with the resource allocations of other actors through shared agenda information.

- **A1141 identify feasible allocations** -- Identify the feasible ways in which resources could be assigned to functions: "How could we allocate resources to functions?"

- **A1142 evaluate feasible allocations** -- Rate the different feasible allocations: "What are the advantages and disadvantages of different resource allocations?"

- **A1143 select allocations** -- Select the best resource allocation: "What are the best resource allocations?"

- **A1144 identify allocation faults** -- Identify any problems with the resource allocations, such as inappropriate allocations or inconsistencies between actors: "Do we agree on the best resource allocations?"

- **A12 share agenda information** -- Communicate information (overtly and covertly) about agendas among the actors. It is only through sharing agenda information that the individual agendas can approach consistency.

- **A2 perform other functions** -- Perform specific functions (other than managing agendas) to achieve the mission goal and its subgoals. These can include monitoring the state of aircraft subsystems, changing the state of the aircraft and its subsystems by manipulating controls, making decisions, solving problems, and planning. The last function yields additional (derived) goals to accomplish. Performing such functions involves maintaining situation models.

The following descriptions (A21 through A25) pertain to the performance of a single function to achieve a single goal, possibly by multiple actors.

- **A21 coordinate actors** -- Coordinate the activities of the actors assigned to perform the function. Decide what roles and responsibilities each actor will have in performing the function.

- **A22 assess function** -- Assess the status of this function: how well it is being performed, what the future prospects look like, and the likelihood that the goal will be achieved.

- **A23 maintain situation models** -- Update and exercise each actor's situation model. Each actor has an internal representation of the current state of the world and at least human actors can project their models into the future. Maintenance of these models is driven by the need for performing this function. By extension, similar situation model maintenance activities are conducted in parallel for other functions.

- **A231 determine information requirements** -- Determine what information is needed to perform this function.

- **A232 acquire situation information** -- Obtain information from the environment, the aircraft, and other actors.

- **A233 integrate situation information** -- Integrate new situation information and shared information from other actors' situation models into the current situation models.

  - **A2331 update existing situation information** -- Use new information about the various systems to update their state representations in the situation models.

  - **A2332 add new situation information** -- Add other, new information (not just updates of old information) to the situation models.

  - **A2333 project situation models** -- Use possible courses of action to project the current situation into the future, yielding one or more possible scenarios.

  - **A2334 identify situation model faults** -- Identify problems with the situation models, such as inaccuracies, omissions, internal inconsistencies, and lack of agreement between the models of different actors.

- **A234 share situation information** -- Communicate about the actors' situation models.

- **A24 decide/plan** -- Decide on what actions to perform immediately to achieve the goal, or plan what to do in the future. Planning may yield subgoals derived from this function's goal. These will be added to the actors' agendas.

- **A25 act** -- Perform the actions necessary to achieve this function's goal. These may include control manipulations, utterances, etc.

DISCUSSION

The model emphasizes that the flightcrew must manage goals and functions, assign actors to functions, and allocate resources to functions. It also underlines the importance of maintaining situational awareness and communicating information about individual agendas to identify and resolve conflicts.

The elements of the full IDEF0 model provide further details to be used in analyzing accident and incident reports to identify where Agenda Management may have broken down. Therefore, it is a potentially useful tool in developing means of facilitating the process of Agenda Management through procedures, training, and computational aids.

MODEL VALIDATION

The full IDEF0 model reflects our understanding of Agenda Management, an understanding built largely from analyzing accident and incident reports and observing subject behavior in our laboratory. It seems to comport well with normal flightdeck operations. However, it must be viewed as a hypothesis, subject to validation.

Our initial approach to validation is a continuation of our incident report studies. We have prepared a list of keywords designed to elicit incident reports in which the reporters describe goals that were not met because a function was not completed or was interfered with due to misprioritizations or other Agenda Management errors.

For each such report we find, we are attempting to determine what goals the flightcrew was pursuing and what functions were not performed satisfactorily as a result of failures in Agenda Management. We will attempt to do a rough quantification of goals and functions in order to ascertain the limits to human Agenda Management performance. We also hope to use the reporters' descriptions to determine if the structure of our model is consistent with flightdeck practice. From this analysis we hope to refine the current model and move towards a model that may be ultimately validated.

We recognize the limitations inherent in incident report studies. Therefore, we anticipate that further validation efforts will involve pilot surveys and simulator experiments.

AIDS TO FACILITATE AGENDA MANAGEMENT

In parallel with our recent modeling efforts, we have been using knowledge gained in our accident and incident studies to develop experimental, computational aids to facilitate Agenda Management (Funk & Kim, 1995). We have learned that if an aid can accurately ascertain flightcrew goals and monitor functions being performed to achieve those goals, it can help improve Agenda Management performance by bringing to the flightcrew's attention goal conflicts, unsatisfactory function performance, and other Agenda Management problems. Our current

efforts center on developing methods for overt and covert goal communication and mechanisms for assessing function performance.

ACKNOWLEDGMENT

This work is supported by the NASA Ames Research Center under grant NAG 2-875. Kevin Corker is our technical monitor.

REFERENCES

Chou, C.D., Madhavan, D., & Funk, K. (in press). Studies of Cockpit Task Management Errors. *The International Journal of Aviation Psychology*.

Funk, K. (1991). Cockpit task management: preliminary definitions, normative theory, error taxonomy, and design recommendations. *The International Journal of Aviation Psychology, 1(4)*, 271-285.

Funk, K. & Kim, J.N. (1995). Agent-based aids to facilitate cockpit task management. *Proceedings of the 1995 IEEE International Conference on Systems, Man and Cybernetics*, Piscataway, NJ: The Institute of Electrical and Electronics Engineers, pp. 1521-1526.

Hart, S.G. (1989). Crew workload management strategies: a critical factor in system performance. In R.S. Jensen (Ed.), *Proceedings of the Fifth International Symposium on Aviation Psychology* (pp. 22-27). Columbus, OH: The Ohio State University, Department of Aviation.

Johannsen, G. & Rouse, W.B. (1979). Mathematical concepts for modeling human behavior in complex man-machine systems. *Human Factors*, 21(6), 733-747.

Moray, N., Dessouky, M., Kijowski, B., & Adapathya, R. (1991). Strategic behavior, workload, and performance in task scheduling. *Human Factors*. 33(6), 607-629.

Raby, M., & Wickens, C.D. (1994). Strategic workload management and decision biases in aviation. *The International Journal of Aviation Psychology, 4(3)*, 211-240.

TEACHING COGNITIVE SYSTEMS ENGINEERING

David D. Woods
Jennifer C. Watts
John M. Graham
Daniel L. Kidwell
Philip J. Smith

Cognitive Systems Engineering Laboratory
The Ohio State University
Columbus, Ohio

Our motivation for this paper is to stimulate discussions within the human factors community about teaching Cognitive Engineering at the undergraduate level. For the last fourteen years, the Cognitive Systems Engineering Laboratory at the Ohio State University has offered an undergraduate course in Cognitive Engineering (multiple offerings per year to Industrial Engineering, Industrial Design, Computer Science and Psychology students). In this paper, we will draw from our teaching experiences and describe our framework for teaching Cognitive Engineering.

COGNITIVE SYSTEMS ENGINEERING: BASIC TENETS

Since its origins in the early 1980's, the field of Cognitive Engineering has greatly matured (Norman, 1981; Hollnagel & Woods, 1983, originally 1982). Signs of this growth include books for the specialist with the title of "Cognitive Systems Engineering" (Rasmussen, Pejtersen, and Goodstein, 1994), books which convey Cognitive Engineering principles to the general public (Norman, 1988; 1993), and even job announcements looking for Cognitive Engineers. Recently, HFES has introduced the Cognitive Engineering and Decision Making Technical Group which has quickly become a significant part of the society's annual meeting program. Finally, classes teaching Cognitive Engineering principles are fundamental aspects of curricula at universities like The Ohio State University, Georgia Institute of Technology, University of Toronto, and University of Illinois, Urbana-Champaign.

As the field continues to mature and becomes a more frequent part of curricula, it becomes appropriate to explore the challenges of teaching Cognitive Engineering, particularly at the undergraduate level. We hope to begin a reflective dialogue about these issues based on our own experiences. Since 1982, the Cognitive Systems Engineering Laboratory at The Ohio State University has regularly offered undergraduate courses in Cognitive Engineering. These courses serve students from Industrial Engineering (a requirement since 1984), Industrial Design (required since 1990), Computer Science (elective) and Psychology (elective). These classes focus exclusively on Cognitive Engineering problems, concepts, and techniques and are not a hybrid or survey of Human Factors in general.

A driving force behind Cognitive Engineering, first noted in the early 80's, was the explosion of technological powers which penetrated many fields of practice, particularly

- the computer's power for creating more autonomous machines,

- techniques in Artificial Intelligence (AI) as a force for "thinking" machines, i.e., machines that were cognitive agents forming assessments independently,

- the computer as a medium for interaction,

- the exploding visualization power brought on through the shift from hardwired to computer-based display of data.

A second driver was the widespread experience that these powers were often used clumsily, creating new problems, errors and paths to failure for beleaguered practitioners. The rapid step change introduced by computerization culminated in a succession of real world failures of complex systems and the persistent observation that neither simply "human error" nor over-automation (machine error) adequately accounted for these failures.

Unfortunately, since the field's original papers were introduced, there have been two alternative views of Cognitive Engineering as a response to these forces. When Norman coined the term in 1981, he wrote that it was, in part, a response to a need for an applied cognitive science. Unfortunately, many have seized on this comment as the base definition of the Cognitive Engineering label. Hollnagel and Woods (1983), writing about the same forces and needs when Norman's technical report was introduced, saw the introduction of Cognitive Engineering as the introduction of a new theoretical area as well as a new challenge for affecting real systems consisting of distributed cognitive agents, both human and machine. For Hollnagel and Woods (1983; see also Woods, 1994, as well as Hutchins, 1995, Zhang and Norman, 1994; and Norman, 1993), the Cognitive Systems synthesis was not simply an application of cognitive psychology and cognitive science. Instead, it was, as the subtitle played on a cliché, "new wine in new bottles."

Woods (1994) summarizes the new conceptualization succinctly:

• Cognitive Systems Engineering is an approach to the interaction and cooperation of people and technology.

• It is concerned with COGNITIVE WORK (cognition in context):
  1. the use of knowledge in pursuit of goals,
  2. groups of people cooperating to solve problems with tools.
  Studies of cognitive work discover how the behavior and strategies of practitioners are adapted to the various purposes and constraints of the field of activity.

• Its goal is better design by AIDING COGNITION at work.

• It is governed by Kubrick's bone, which is the idea that the construct of a COGNITIVE SYSTEM can function as a new unit of analysis:
  1. people are defined, in part, by intelligently creating and using technology,
  2. the opposition of people versus automation is fundamentally wrong,
  3. the new unit of analysis is a system of human and machine agents that perform cognitive work -- joint and distributed cognitive systems.

The results of this new conceptualization are four basic principles of a new Cognitive Systems Synthesis:

1. The Cognitive System principle:
   Operational systems (people, technology, policies and procedures) can be described, studied and designed in terms of concepts about cognitive processing distributed over multiple agents and artifacts,

2. The Cooperative Cognition principle:
   Cognitive systems are distributed over multiple agents, both people and machines. The canonical case for human cognition is not a single person alone, rapt in thought (or rapt in thought in front of a CRT), but rather the externalized cognitive activity across interacting agents and the influence of external artifacts on cognition.

3. The Representation principle:
   Properties of artifacts influence the cognitive strategies of practitioners -- the REPRESENTATION EFFECT. This principle is concerned with how artifacts function as cognitive tools and how to develop a research base about cognitive tools, their uses and effects. In effect, one can see a movement over the last 50 years of experimental psychology from--what stimulus attributes control behavior, to--how cognition guides behavior, to--how properties of artifacts shape cognition which guides behavior

4. Coping with Complexity:
   Cognitive systems adapt to the demands of the field of practice, and practitioners re-shape artifacts to function as tools.

This conceptual base shows how the Cognitive Systems synthesis arises at the intersection of what are normally boundaries or dividing lines between more traditional areas of inquiry. Cognitive System problems and concepts overlap and connect:

1. between technological and behavioral sciences,

2. between individual and social perspectives,

3. between the laboratory and the field,

4. between design activity and empirical investigation

5. between theory and application.

As Woods (1994) has noted, Cognitive Engineers are "inter-cultural brokers."

These intersections define the fundamental challenge for teaching Cognitive Engineering: students normally only work within one of these perspectives. However, our goal is to demonstrate the power of the new synthesis across these perspectives for studying and designing the impact of technology change.

TEACHING COGNITIVE ENGINEERING

The inter-cultural role of Cognitive Engineering has led us to actively promote classes as mixing grounds for students from different backgrounds, despite the difficulties of handling the different starting points and goals of these diverse audiences. We take advantage of this multi-disciplinary mix by building the class around collaborative exercises which are scaled versions of real Cognitive Engineering problems.

These exercises first function as a recursive use of cognitive walkthroughs. We take students on explorations of cognition in the workplace, where teams of competent practitioners handle evolving problems in situations such as aircraft cockpits, space mission control centers, and surgical operating rooms. We examine how computer power in many forms (AI, automation, computer graphics, HCI design) influences cognitive work.

Second, the exercises reveal how real computer systems often fail to support real human practitioners--the clumsy use of technology and automation surprises. Borrowing from Norman (1988), the technique shows students how to do it wrong, that is how to design cognitive artifacts that will degrade human performance and create human error. The goal is to help students hone their critical eye, to learn to see how the properties of artifacts shape cognition and therefore error and expertise in a field of practice.

These exercises encourage students to step outside of their perspectives and explore how other disciplines provide relevant concepts and ideas for the design of cognitive systems. Ideally, each exercise is a guided exploration allowing students to gain hands-on experience with the subtasks involved in developing cognitive artifacts. Ultimately, the exercises help the students directly confront the issues involved in developing effective cognitive tools that enhance human and system performance.

The exercises are carried out in a cooperative problem solving style and are organized as a series of phases:

1. individuals or small groups explore a case to discover important issues,

2. students publicly share insights and perspectives across all class participants and groups, and

3. individuals then synthesize all of the points and insights to move closer to a proposed resolution for the problem motivating the case.

LEARNING AT THE INTERSECTIONS

The inter-cultural role of Cognitive Engineering leads us find ways to inspire students to confront the possibilities afforded by the Cognitive System Synthesis for various goals, and therefore learn at the intersections mentioned above.

The Intersection of Technological And Behavioral Sciences

Our focus on the intersection between technological and behavioral sciences teaches students how to make use of technical capabilities to develop systems which are sensitive to basic principles about human behavior and performance. We teach students that fundamentally, the Cognitive Systems synthesis invokes a new unit of analysis--that all of the devices that conduct or shape cognitive work should be grouped as part of a single joint human-machine cognitive system. They learn that incidents, automation surprises, and accidents are not due to either human error or to over-automation. Rather they exhibit characteristics of a human-machine coordination breakdown--a kind of weakness in a distributed cognitive system.

We also teach students that the Cognitive Systems view is concerned with how properties of artifacts shape cognition and therefore behavior. They learn that the possibilities of technology afford designers many degrees of freedom. However, with this freedom comes the responsibility to use the technology skillfully to support practitioners in their contexts. The course work shows students how to view people and technology, not as separate and independent topics, but rather as interconnected parts of a larger and more useful system boundary -- a system of machine and human elements that together perform cognitive work.

The Intersection of Individual And Social Perspectives

Our focus on the intersection between individual and social perspectives teaches students how to study and design systems to support cognition in context. Students usually think that the canonical case of cognition is an individual rapt in thought (or a single person alone, rapt in thought in front of a CRT). Since they also recognize that an individual's activities occur with some relation to other people, they layer on top of individuals the perspective of a group composed of interacting individuals. Above these two layers, they point to the role of organizational factors which affect different groups of individuals. However, this method of parsing human-machine systems may be an artifact of how scientists have primarily studied cognition—by studying individuals alone in tasks removed from any larger context.

However, if we look at cognition in context, (for example, commercial flightdecks, or surgical operating rooms), we do not see cognitive activity isolated in a single individual. Rather cognitive activity is distributed across multiple agents (Hutchins, 1995). Also, we do not see cognitive activity existing solely in a thoughtful individual, but rather as an integrated part of a stream of activity (Klein, et al, 1993).

Similarly, students' initial image of advanced machine capabilities is usually that of a machine alone, rapt in thought or action. However, the reality of the workplace is that automated subtasks exist in a larger context of interconnected tasks and multiple actors. Introducing automated and intelligent agents into a larger system changes the composition of the distributed system of monitors and managers and shifts the human's role within that cooperative ensemble (Billings, 1996). It seems paradoxical, but studies of the impact of new automation and 'intelligent' machines reveal that design of automated systems is really the design of a new human-machine cooperative system-- a joint cognitive system that distributes cognitive work across multiple agents

When we look at cognition in context, we also see these sets of active agents embedded in a larger group, professional, organizational, or institutional context which constrains their activities, sets up rewards and punishments, and defines goals which are not always consistent (e.g., Hutchins, 1990; Klein and Thordsen, 1989). Even the moments of individual cognition are set up and conditioned by the larger system and communities of practice in which that individual is embedded.

Therefore, in our classes, we teach students to shift from the canonical case of seeing cognition as internal and isolated, to the Cognitive Engineering perspective, which sees it as public and shared, distributed across agents, and distributed between external artifacts and internal strategies. They learn that understanding cognition depends as much on studying the context in which cognition is embedded and the larger distributed system of artifacts and multiple agents, as on studying what goes on between the ears.

The Intersection of the Laboratory and the Field

The phenomena of interest for Cognitive Engineering lie in the field where practitioners use tools to do cognitive work in the face of significant complex demands (uncertainty, conflicting goals, limited resources, and interleaved tasks). Therefore, we teach our students new research strategies that combine experimental values with techniques for methods like work analysis, ethnography, and retrospective analysis (Woods and Sarter, 1993). Students learn to use these methods as converging operations to discover how the behavior and strategies of practitioners are adapted to the various purposes and constraints of the field of activity.

In learning how to do this type of empirical investigation students must confront two critical constraints. First, the problem with studying new technology is that it transforms the nature of the practice, changing what is canonical and exceptional. Cognitive systems adapt over time developing new routines and tailoring systems and their activities. Second, the goal of system design is in part to transform the nature of practice, aiding cognition and enhancing performance. This design agenda coupled with the goal of understanding makes Cognitive Engineering relatively unique.

The Intersection of Design Activity And Empirical Investigation

Cognitive Engineering needs to teach students a synthesis of the open-ended innovation of fundamental design and the technical basis for design grounded in the growing Cognitive Systems research base and in the criterion for success--aiding performance.

We teach students that when they design cognitive systems, they should adopt the attitude of an experimenter whose goal is to understand and model the dynamics of these systems. This is because new technology typically embodies an hypothesis or prediction about how technology affects cognition and performance. Students learn that they can use the designs they create as probes to study the impact of new technology on a field of practice. It is important for students to learn how to combine open-ended design techniques with techniques for studying the effects of technology in operational settings. This combination of skills helps students develop useful and usable systems in the face of realistic constraints (see Woods and Sarter, 1993; Woods et al., 1996 for a larger discussion of these issues).

The Intersection Of Theory And Application

Many people hold the typical view that "basic" work flows down a "pipeline" into application. The Cognitive Systems synthesis challenges this conventional wisdom. "It is, ..., the fundamental principle of cognition that the universal can be perceived only in the particular, while the particular can be thought of only in reference to the universal" (Cassirer, 1953, p. 86). As Hutchins (personal communication, 1992) has stated, "There are powerful regularities to be described at a level of analysis that transcends the details of the specific domain. It is not possible to discover these regularities without understanding the details of the domain, but the regularities are not about the domain specific details, they are about the nature of human cognition in human activity." The converse applies as well, "if we are to enhance the performance of operational systems, we need conceptual looking glasses that enable us to see past the unending variety of technology and particular domains" (Woods and Sarter, 1993).

Thus, we teach Cognitive Engineering students to be context bound yet theoretical. Being context bound in the study of cognitive systems is not simply to do "applied" studies in particular domains. Rather each investigation has a parallel status: 1) as a study of an aspect of the Cognitive Systems synthesis in a "natural" laboratory and, 2) as means to innovate and constructively influence short term change and immediate problems in the host field of practice.

Students learn that Cognitive Systems synthesis is intensely committed to building a theoretical base through investigations and design interventions into ongoing fields of practice, as it innovates new designs of artifacts, representations and cooperative structures.

This cycle of empirical investigation, modeling and design, in general, deals with four mutually interdependent themes, which are fundamental topics taught in our classes:

- Theme 1: How to make automated and intelligent systems team players: theories of cooperative human-machine problem solving.

- Theme 2: Cognitive factors behind the label human error: how the clumsy use of technology creates human error.

- Theme 3: Visualizing function: how to design computer-based aids to human performance.

- Theme 4: How to study cognitive systems in the context of real world settings to support both research and design goals.

CONCLUSION: REFLECTIVE GROWTH AT THE INTERSECTIONS

Ultimately, students represent one aspect of the following intersecting concerns: what to build as a technologist, how to design information and interactions, and how to understand aspects of cognition. Ultimately, students in Cognitive Engineering need a thorough grounding in technological areas, in cognitive psychology, cognitive science, and social science areas, and in design skills. Yet they need to learn to keep from getting lost in only one aspect of the intersecting cultures. There is a need to learn how to avoid:

1) getting lost in the details and short term horizon of particular fields of practice,

2) getting lost in the technology itself, blinded to larger views by the effort required to actually create new systems,

3) getting lost in their personal visions of what they imagine to be the impact of technology on human performance,

4) studying the current work culture as a static entity when that field of practice is actually evolving, and

5) allowing resource pressures to squelch innovation and the search for what would be useful for practitioners.

The intersections at the heart of the Cognitive Systems synthesis lead to a base set of activities that proficient Cognitive Engineers should be able to carry out:

- They should understand, question, and add to the conceptual and empirical base by modeling 1) the role of artifacts, their use and effects, on human-human and human-machine cooperative activity; and 2) the cognitive character of various activities (e.g., diagnosis as a process of disturbance management).

- They should demonstrate proficiency in the techniques for studying cognitive systems in the wild--using concepts to guide empirical investigations oriented to learning about the interaction of demands, agents and tools in differing fields of practice.

- They should demonstrate craftmanship in action, engaging in open ended innovation to develop new concepts and systems that aid cognition in the target field of practice and add to the conceptual and empirical base.

The purpose of this paper is to initiate discussions within the Cognitive Engineering community about education at the undergraduate level, the training of professionals, and the teaching of other kinds of professionals so that they can collaborate and integrate Cognitive Engineering effectively into research and design activities. These discussions should help the Cognitive engineering community develop new ways to teach effectively as our field continues to mature.

ACKNOWLEDGMENTS

This work was partially supported by a National Science Foundation Graduate Fellowship. Any opinions, findings, conclusions or recommendations expressed in this publication are those of the authors and do not necessarily reflect the views of th National Science Foundation.

REFERENCES

Billings, CE. (1996). Aviation Automation: The search for a human-centered approach, Hillsdale, NJ: Lawrence Erlbaum Associates, Inc.

Hollnagel, E., & Woods, D. D. (1983). Cognitive systems engineering: New wine in new bottles. *International Journal of Man-Machine Studies*, 18, 583-600 (originally Riso National Laboratory report, R-42, Roskilde, Denmark, 1982).

Hutchins, E. (1995). Cognition in the Wild. Cambridge, MA: The MIT Press.

Klein, G., & Thordsen, M. (1989). Cognitive processes of the team mind No. PO # A72145C). NASA Ames Research Center.

Klein, G., Orasanu, J., Calderwood, R., & Zsambok, C. (1993). Decision Making in Action: Models and Methods. Norwood, NJ: Ablex Publishing Company.

Norman, D. A. (1981) Steps Towards a Cognitive Engineering: Technical Report, Program in Cognitive Science, University of California, San Diego, 1981).

Norman, D. A. (1988). The Psychology of Everyday Things. New York: Basic Books, Inc.

Norman, D. A. (1993). Things That Make Us Smart. Reading, MA: Addison-Wesley.

Rasmussen, J., Pejtersen, A., & Goodstein, L. (1994). Cognitive Systems Engineering. New York: John Wiley & Sons, Inc.

Woods, D.D. (1994). Observations from Studying Cognitive Systems in Context. In *Proceedings of the Sixteenth Annual Conference of the Cognitive Science Society*. [Keynote Address]

Woods, D.D., Patterson, E.S., Corban, J.M., Watts, J.C. (1996). Bridging the gap between user-centered intentions and actual design practice, In *Proceedings of the Human Factors and Ergonomics Society 40th annual Meeting*, Philadelphia, PA

Woods, D. D., & Sarter, N. B. (1993). Evaluating the Impact of new technology on human-machine cooperation. In J. Wise, V. D. Hopkin, & P. Stager (Eds.), Verification and Validation of Complex and Integrated Human-Machine Systems Berlin: Springer-Verlag.

Zhang, J., & Norman, D. (1994). Representations in distributed cognitive tasks, *Cognitive Science*, 18, 87-122.

USE OF VISUALIZATION AND CONTEXTUALIZATION IN TRAINING OPERATORS OF COMPLEX SYSTEMS

Alan R. Chappell and Christine M. Mitchell
Center for Human-Machine Systems Research
Industrial and Systems Engineering
Georgia Institute of Technology
Atlanta, GA 30332-0205

The complexity of control systems in complex dynamic systems can lead to increased cognitive demands on the operator. Lacking accurate and complete system knowledge and/or interfaces that clearly present the system state and constraints, the operator may misunderstand the automation and its behaviors. This paper presents a research effort that uses visualization and contextualization in training as tools for addressing these problems. A problematic feature of modern glass cockpit aircraft's automated systems, vertical path navigation (VNAV), is used as a testbed. The VNAV Tutor uses contextualization and visualization to improve the pilot's understanding of VNAV control and the interaction of various levels of automation during VNAV usage. Furthermore, it attempts to help pilots build a robust conceptual model of vertical navigation operation. An evaluation showed that the VNAV Tutor enhanced both the conceptual understanding and use of the vertical navigation function by pilots.

INTRODUCTION

In the effort to enhance safety by increasing the level of automation in many complex systems, designers often increase the importance and the demands of the human operator's role (Bainbridge, 1987). The operator is needed to act as the connector between automated subtasks, often performing the functions too difficult or costly to automate, and to handle unexpected or anomalous system behaviors. Additionally, the automation's interface to the operator often provides insufficient support for communicating necessary information or control inputs (Roth & Woods, 1988; Woods, Johannesen, Cook & Sarter, 1994). Such systems increase the cognitive demands on the operator (Billings, 1991).

The modern glass cockpit aircraft typifies the challenges to the operator of many complex dynamic systems. Various research efforts have addressed the potential problems associated with the proliferation of technology and automation on the flight deck. Among these efforts, extensive pilot opinions have been solicited including, for example, ratings of glass cockpit pilots' attitudes and opinions concerning current cockpit automation (Wiener, 1989). Wiener reports that even experienced pilots are surprised by the automation. His famous characterization of pilot interaction with the auto flight system highlights potential automation problems: Pilots ask, "What is it doing *now*? *Why* is it doing that? What will it do *next*?" (Hughes & Dornheim, 1995). The confusion displayed by these questions is indicative of a mismatch between the operator's knowledge/expectations of the system and the system's behavior/capabilities.

Several strategies are available for addressing the difficulties presented by current levels of automation in complex dynamic systems, including design, aiding, and training. This research focuses on training.

VERTICAL NAVIGATION IN GLASS COCKPIT AIRCRAFT

The most frequent problem that pilots have with the flight management system is with the vertical navigation mode (Wiener, 1989; Sarter & Woods, 1992). In a recent study Sarter and Woods (1992) found that out of 159 mode problems reported by airline pilots, 63 (40%) were related to VNAV operation. In a review of mode awareness incidents in the NASA Aviation Safety Reporting System database, 74% involved vertical navigation (Vakil, Hansman, Midkiff & Vaneck, 1995). The flight management system in general, and the vertical navigation mode in particular, supports smooth, fuel-efficient flight. Thus, effective use of the flight management system, particularly the VNAV mode, is very desirable. The vertical navigation mode, however, is not well supported by current pilot interfaces (Sarter & Woods, 1992; Vakil, et al., 1995). Furthermore, pilots often receive little explicit training in its use.

The VNAV Tutor is a training system that uses visualization, animation, and other inexpensive computer-based technologies and a range of standard and non-standard flight scenarios to teach pilots about (1) the vertical navigation mode and associated submodes, and (2) the vertical profile programmed into the flight management system and its dynamic relation with other cockpit controls and displays.

THE VNAV TUTOR

The VNAV Tutor was implemented using a part-task flight simulator running on a UNIX workstation with three monitors. The simulator replicates to a high level of fidelity the electronic flight instruments, and the auto flight systems of the Boeing 757/767 aircraft. The underlying simulation is

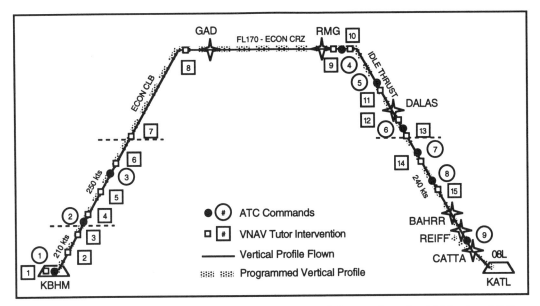

Figure 1. LOFT scenario used for the first VNAV Tutor training session. See Tables 1 and 2 for descriptions of the numbered tutor interventions and ATC commands.

based on a three degrees of freedom, point mass model of the B757. This model provides sufficient fidelity in the aircraft dynamics since no hand flying is required by the tutor tasks.

The VNAV Tutor was designed specifically to remediate two major pilot concerns identified by Sarter and Woods (1992): (a) inability to understand/visualize the vertical profile; and (b) inability to properly conceptualize the interaction of the mode control panel constraints with VNAV and LNAV operations. Toward this aim, the VNAV Tutor elucidates the control processes normally hidden by the flight management system interface, i.e., makes the abstract concepts visible and understandable (Norman, 1988; Hollan, Hutchins & Weitzman, 1984). The following sections demonstrate the use of contextualization and visualization as tools for achieving these goals.

Line-Oriented Flight Training: Contextualization

The VNAV Tutor employs complete flight scenarios—a type of training called line-oriented flight training (LOFT) by the airline industry—in order to appropriately contextualize information as it is presented. VNAV scenarios help the student/pilot (1) explore the content of the vertical profile, (2) use the VNAV mode to execute that profile, (3) study the interaction between the FMS and other vertical navigation modes, and (4) use the FMS vertical navigation to complete various in-flight maneuvers.

LOFT scenarios include entire flights: take-off, climb, cruise, descent, and landing. LOFT-based design makes the scenarios similar to the pilot's operating environment, increasing the realism of the task being performed. The four LOFT scenarios used by the VNAV Tutor are based on those developed by a major airline to teach pilots transitioning to aircraft with sophisticated automation the concepts and operations of the flight management system. During the scenario, air traffic control messages and tutorial instructions

occur at specific altitudes or distances from the destination airport in order to contextualize and sequence the information provided by the tutor and the actions expected of the student/pilot.

For example, Figure 1 depicts the first LOFT scenario used by the VNAV Tutor. The flight begins on the runway at the left of the figure (KBHM: Birmingham, AL), flies the vertical profile shown, and lands on the runway at the right of the figure (KATL: Atlanta, GA). ATC commands and major tutor interventions are marked and numbered on the vertical profiles. These numbers are identified in the associated tables. Table 1 lists the topics of the tutor interventions while Table 2 describes the ATC commands given. This scenario is specifically intended to be relatively simple (lower work load), to allow the student to learn to use the simulation environment and basic VNAV concepts.

Table 1. VNAV Tutor intervention topics for the first scenario

Intervention	Intervention Topic
1	Introduction to vertical profile and associated CDU pages
2	Engagement of VNAV
3	Speed restriction
4	Handle ATC altitude clearance with MCP altitude window
5	Transition speed restriction
6	Raise MCP altitude limit
7	VNAV cruise operation
8	Descent upcoming, VNAV limited by MCP altitude limit
9	Lower MCP altitude limit
10	VNAV descent operation
11	MCP altitude limit interaction with VNAV descent
12	Transition speed restriction
13	Lower MCP altitude limit
14	Arm the approach mode
15	Control of speed in approach mode

Table 2. ATC commands from the first scenario.

Command	ATC Command
1	Cleared for take-off; Climb to 10,000 ft
2	Maintain 10,000 ft
3	Climb to 17,000 ft
4	Descend to 13,000 ft
5	Descend to 8,000 ft
6	Descend to 5000 ft
7	Cleared for ILS 08L approach
8	Slow to 180 kts
9	Cleared to land

The Vertical Profile Display: Visualization

The heart of the VNAV Tutor is a vertical profile display—a display added to the conventional cockpit displays as part of the training environment. The vertical profile display (Figure 2) provides an otherwise unavailable visual representation of the flight management system vertical profile and its interaction with other vertical navigation modes of the aircraft.

Through the vertical profile display the VNAV Tutor provides a graphical depiction of the current and predicted aircraft behavior in the vertical plane. This display, by providing a visualization of the vertical path as programmed into the flight management system and constrained by other cockpit automation, more adequately supports development and maintenance of a good conceptual representation than do the alphanumeric displays found in the cockpit.

The vertical profile display depicts, in real-time, the vertical path programmed into the flight management system, including waypoints and restrictions. The vertical profile display depicts a standard VNAV climb, cruise, and descent sequence in a format familiar to pilots from other cockpit displays. Aspects such as transition altitude, crossing restrictions, step climbs and descents, and VNAV points of inflection (e.g., top of climb-T/C) are illustrated. Figure 2 shows the vertical profile display during a VNAV-controlled climb.

The vertical profile display integrates VNAV relevant information distributed throughout the cockpit. It uses color coding and symbology that is analogous to that used on the horizontal situation indicator. A magenta line with associated waypoints illustrates the programmed path normally found as alphanumeric text on four control and display unit pages. Near the upper left corner, the vertical profile display shows the currently active vertical control mode, e.g., VNAV SPD in Figure 2. This indication summarizes information provided on the attitude direction indicator. Additionally, a green line drawn horizontally across the display shows the value currently entered into the mode control panel altitude limit: 10,000 feet altitude as depicted in Figure 2. Instead of requiring the pilot to search the cockpit for information from multiple displays, the vertical profile display integrates all the major pieces of information needed by the pilot to monitor VNAV operation. The vertical profile display integrates this information into a pictorial format that allows the pilot to visualize, and thus, more directly understand, the situation.

EVALUATION

An evaluation of the VNAV Tutor was conducted on-site at a major commercial airline's training facility, with subjects drawn from Boeing 757/767 ground school. Subjects were taken at the point in training where all information fundamental to VNAV use had been taught, but no VNAV specific topics had been presented. (At the time of this study, no significant effort was made to present VNAV information anywhere in training, although some basics were available through simulator flying. Transitioning pilots were told they

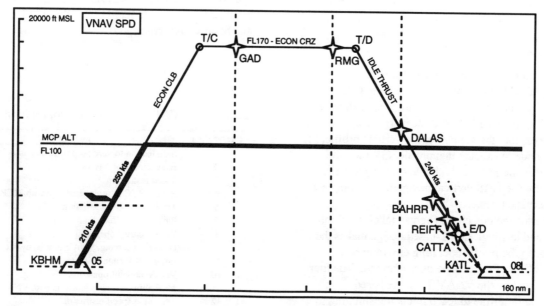

Figure 2. Vertical Profile Display during a VNAV climb. FMS VNAV Speed is the current vertical control mode. Although a cruise altitude of 17000 ft is programmed into the FMS, altitude is constrained to 10,000 ft by the mode control panel altitude limit.

would "learn VNAV on the line", i.e. through on-the-job learning.) Subjects were 757/767 students with little or no previous experience in flight management system equipped aircraft, i.e., pilots transitioning to the glass cockpit. All participants were volunteers.

The evaluation consisted of six one hour sessions per participant. The initial meeting assessed the subject's knowledge regarding the FMS and VNAV using a written questionnaire. Four training sessions with the VNAV Tutor followed. Each of these sessions presented one of the four training LOFT scenarios that comprise the curriculum. In the sixth session, the pilot flew an evaluation LOFT scenario that did not incorporate the tutor or the vertical profile display. Next, the pilots completed a questionnaire, similar in content to the one used during to the first session to gauge the degree to which pilots felt they understood VNAV operations. The evaluation for a particular subject concluded by soliciting pilot reactions and opinions about the VNAV Tutor.

The primary measures of tutor performance were observations of pilot performance and responses to questions. The evaluation session was similar to the four training sessions except that no tutor dialog was provided and the vertical profile display was not available.

During the evaluation session air traffic control commands prompt the pilot to use key features/capabilities of the VNAV system. These features include: engaging VNAV, altitude limiting, changing cruise altitude, speed intervening, descending early, and selecting a new arrival/runway. During the evaluation session, the researcher recorded the pilot's accuracy in performing these tasks. Additionally, at prespecified points in the flight, the researcher paused the scenario and asked the pilot questions about the state of the flight management system and other auto flight equipment. These questions involved impending or recently occurring mode transitions in order to gauge mode awareness. The researcher recorded the answers and resumed the scenario.

Subject Profile

The evaluation consisted of five subjects, three captains

and two first officers. All subjects had over 10,000 hours of flight time except one first officer who listed 2500 hours. Previous commercial experience included Boeing 727, 737, 747 (no FMS); DC-9, 10; and L10-11. All of these are traditional ('non-glass') aircraft, and hence, the subjects had no experience with the VNAV system. Results of the questionnaires to determine the initial level of VNAV knowledge are summarized in Figure 3. This graph shows that the subjects claimed only limited knowledge of VNAV control, its features, and its interaction with other vertical navigation functions. Although some subjects knew selected topics, none felt comfortable with their level of VNAV knowledge.

Since the subjects did not know how to use VNAV, performance measures of prior knowledge could not be taken. Hence, Figure 3 is taken as indicative of the baseline VNAV knowledge. Additionally, since the pilots receive no further explicit VNAV training in ground school, this is indicative of the level of VNAV knowledge of pilots transition from traditional to glass cockpit aircraft upon entering revenue flight, i.e., when beginning to carry passengers.

Results

Data from the evaluation session are summarized according to the accuracy in performing specified VNAV tasks and the accuracy or completeness of responses to the questions posed. These data are presented in Figures 4 and 5.

Figure 4 shows subject performance on a group of eight VNAV tasks. These tasks include VNAV mode activation, speed intervention, descend now activation, and approach change operation. These tasks provide a performance-based evaluation of how well the subjects (student/pilots) learned VNAV.

Figure 5 shows the percent of student/pilots giving correct responses to mode awareness questions in nine different situations. The answers indicate how well the student/pilots understood the VNAV modes and the interaction of the mode

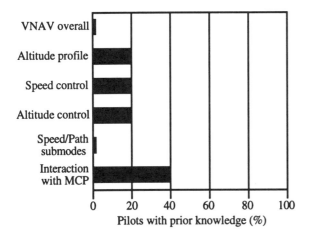

Figure 3. Pilots indicating prior knowledge of VNAV topics in a pre-training questionnaire.

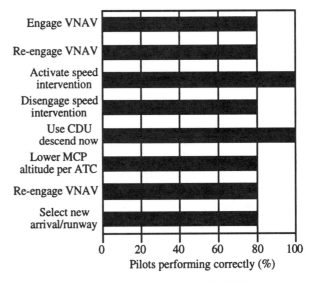

Figure 4. Pilots correctly performing VNAV tasks after training.

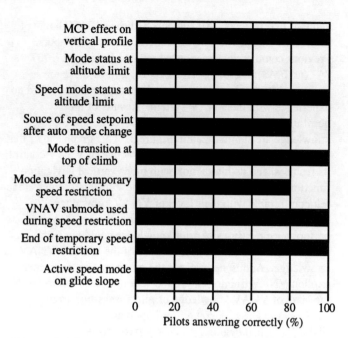

Figure 5. Pilots correctly answering question about the MCP state or mode transitions after training.

control panel with VNAV. Since these questions were asked during an ongoing scenario which the pilot was flying, the answers represent the degree to which the student/pilots were able to apply the material presented to an operational task.

Discussion

Figure 4 shows that 80% or greater of the student/pilots successfully completed each of the VNAV tasks tested. The data provide strong evidence to suggest that the VNAV Tutor effectively teaches a wide range of VNAV operations. Recall that all subjects were transitioning from older aircraft into the glass cockpit and as Figure 3 shows, had little or no prior VNAV knowledge or experience. At the conclusion of VNAV Tutor training, subjects both demonstrated and indicated that they understood the fundamentals of VNAV operation.

Figure 5 shows that in five of the nine situations tested, all the student/pilots were aware of the situation and its implication to VNAV control, while 80% or greater were aware in all but two situations. The only question on which fewer than half the student/pilots gave correct responses (Figure 5, last question) involved speed control after the auto pilot began tracking the glide slope for landing. This topic was presented, but not emphasized in the VNAV Tutor.

Pilots also completed a questionnaire to solicit their reactions to the VNAV Tutor. The response was unanimously positive. The VNAV Tutor added an important component to normal airline training. They particularly liked the vertical profile display. Many suggested its incorporation as a standard cockpit display.

Comparing Figures 4 and 5 to Figure 3 indicates that, using the VNAV Tutor, student/pilots learned to use and understand VNAV for most tasks. In addition, the VNAV

Tutor helped student/pilots acquire a good level of mode awareness. Finally, one of the goals of the VNAV Tutor was to help pilots develop a robust conceptual model of vertical navigation modes, operations, and mode interactions. Performance in the evaluation scenario, conducted without the vertical profile display, suggests that the tutor accomplished this goal: students effectively managed vertical navigation even without the visualization support provided during training.

Recall that Figure 3 is taken to represent the knowledge level typical of pilots at the completion of ground school who are transitioning from traditional to glass cockpit aircraft. Although the pilots completing the VNAV Tutor were not perfect in their VNAV knowledge (Figures 4 and 5), the performance demonstrated on realistic tasks in a realistic task environment indicates a substantial improvement over similar pilots entering actual flight service with only standard ground school training.

These results suggest that the VNAV Tutor helps pilots overcome the opaque interface to the auto flight system and more reliably utilize its capabilities. More generally, the results indicate that computer-based training utilizing appropriate visualization and contextualization can be effectively used to increase operator performance in complex domains.

ACKNOWLEDGMENTS

This paper reports and extends work performed by Edward G. Crowther while at Georgia Institute of Technology. This research was supported by the NASA Ames Research Center grant NCC2-824 (Everett Palmer, Technical Monitor). The authors would also like to thank the pilots and ground training personnel at the anonymous US airline who provided extensive training and documentation on the Boeing 757/767.

REFERENCES

Bainbridge, L. (1987). Ironies of automation. In J. Rasmussen & J. Leplat (Eds.), *New technology and human error.* New York: Wiley.

Billings, C. E. (1991). *Human centered aircraft automation: A concept and guidelines* (NASA TM 103885). Moffett Field, CA: NASA Ames Research Center.

Hollan, H. D., Hutchins, E. L. & Weitzman, L. (1984). STEAMER: An interactive inspectable simulation-based training system. *AI Magazine, 5*(2), 15-27.

Hughes, D. & Dornheim, M. A. (1995). Accidents direct focus on cockpit automation. *Aviation Week and Space Technology.* January 30, 52-54.

Norman, D. A. (1988). *The Psychology of Everyday Things.* New York, NY: Basic Books.

Roth, E. M. & Woods, D. D. (1988). Aiding human performance I: Cognitive analysis. *Le Travail Humain, 51,* 39-64.

Sarter, N. B. & Woods, D. D. (1992). Pilot interaction with cockpit automation: Operational experiences with the flight management system. *International Journal of Aviation Psychology, 2*(4), 303-321.

Vakil, S. S., Hansman, R. J., Midkiff, A. H., & Vaneck, T. (1995). Mode awareness in advanced autoflight systems. *Proceedings of the 1995 IFAC/IFIP/IFOR/IEA Symposium on Analysis, Design, and Evaluation of Man-Machine Systems,* to appear.

Wiener, E. L. (1989). *The human factors of advanced technology ("glass cockpit") transport aircraft* (NASA Contractor Report 177528). Moffett Field, CA: NASA Ames Research Center.

Woods, D. D., Johannesen, L. J., Cook, R. I. & Sarter, N. B. (1994). Behind human error: Cognitive systems, computers, and hindsight. Wright Patterson AFB, OH: CSERIAC Program Office.

COGNITIVE TASK ANALYSIS OF AIR TRAFFIC CONTROL INSTRUCTION TO IDENTIFY RULE-BASED MEASURES OF STUDENT SIMULATOR PERFORMANCE

Richard E. Redding
Department of Psychology, University of Virginia
Charlottesville, Virginia
rer6e@virginia.edu

Thomas L. Seamster
Cognitive and Human Factors
Santa Fe, New Mexico

This study is the first cognitive task analysis of the air traffic controller student evaluation process. The goal was to identify cognitive and behavioral rule-based measures for assessing student simulator performance. A knowledge structure analysis of experienced instructor's sorting of assessment concepts was followed by a protocol analysis of instructor's simulator training sessions with students. The data were translated into IF-THEN cognitive and/or behavioral assessment rules for use in an automated evaluation or intelligent tutoring system. Traditional summary measures of performance activate the rules, providing real-time diagnosis of student performance. This event-driven approach determines not only that some action is incorrect but also what is missing from the student's knowledge or skill base.

As air traffic controller training transitions from a behavioral to a cognitive approach, student evaluation must also shift from relying on behavioral measures to include measures of cognitive processes (Redding & Seamster, 1994, 1995; Seamster, Redding, Cannon, Ryder, & Purcell, 1993). Behavioral measures specify *what* the student has done. But cognitive measures describe *why* the student performs as he does, necessary for providing diagnostic feedback and instructional remedies. Both behavioral and cognitive measures are required for a complete picture of the "what, how and why" of student performance (Ryder & Redding, 1993).

This pilot study is the first cognitive task analysis of the student evaluation process. Cognitive task analysis (see Redding, 1995; Redding & Seamster, 1994; and Ryder & Redding, 1993) identifies cognitive structures and processes underlying expertise, and determines optimal structures and processes through a systematic comparison of expert and novice performers. The goal of this study was to identify cognitive and behavioral diagnostic rule-based measures for assessing student performance during a Federal Aviation Administration (FAA) terminal radar air traffic control simulation. The rule-based measures were derived from the strategies used by experienced instructors to assess student performance. We systematically captured air traffic control instructor's expertise in a collection of rules, illustrating how the tools of cognitive task analysis may be applied to difficult, hard to define activities.

METHOD

Participants included five instructors, with a mean of 15.3 years controller experience and 2.5 years instructor experience.

Participants completed a sorting task of 40 assessment concepts, obtained from the FAA guide to student assessment. The sorting data were submitted to hierarchical cluster analysis.

Additionally, both retrospective and think-aloud protocols were obtained from instructor over-the-shoulder sessions with students. Immediately after each session, a retrospective protocol was collected to gather the instructor's memory for the overall session. Then, the instructor was asked to think-aloud while viewing a replay of the simulator session. Protocols were transcribed in their entirety, and protocol analysis was performed used standard techniques (see Ericsson & Simon, 1993). Verbalizations were segmented with each segment corresponding to a statement or clause (Ericsson & Simon, 1993). Segments were coded into various categories of instructor strategies, with 91% inter-coder agreement. The independent measure was the level of student performance (superior vs. poor students). The dependent measures were the frequencies of the various instructor strategies.

RESULTS

Hierarchical Cluster Analysis

The knowledge structure analysis was conducted first to provide insights into the organizational structure of how experienced instructors assess student performance and to focus the analysis of strategies conducted through the protocol analysis. Knowledge structure analysis reveals the psychological organization of key job concepts (see Redding & Seamster, 1994; Ryder & Redding, 1993).

The cluster analysis of the sorting data identified two clusters that characterize how experienced controllers organize assessment concepts for purposes of evaluating students: a Primary Control cluster, and a Secondary Control and Planning cluster. The Primary Control cluster includes 11 student performance elements that determine whether the student can maintain control of the airspace. As shown in *Figure 1*, the grouping of these elements differs substantially from the organization of the same elements in the FAA's current form for evaluating students. The FAA form organizes assessment by grouping together common control functions (separation, control judgment, methods & procedures, communication & coordination). In contrast, the experienced instructor evaluates performance *across* these job functions, focusing on the common cognitive-behavioral tasks involved in air traffic control (see Seamster, Redding, Cannon, Ryder, & Purcell, 1993). Experienced instructors evaluate performance according to the performance elements (primary control, secondary control and planning, additional services, communication & record-keeping) cutting across the different job functions.

The Secondary Control and Planning cluster includes 7 items that are considered less important for assessment. This cluster has a strong cognitive orientation that includes decision-making, alertness, planning, and problem prioritization. The secondary importance of this cluster reflects the strong behavioral orientation of the current training program.

These results are supported by the instructors' ratings, on a 3-point scale, of the importance of the 40 assessment concepts. Nine of the 20 concepts rated as most important are included in the Primary cluster.

Protocol Analysis

The retrospective and think-aloud protocols were reviewed, with segments containing three or more adjoining instructional strategies identified as strategy segments. The resulting 406 strategy segments formed the basis for the derived rule-sets. They were coded into categories according to the type of strategies they contained, and their frequencies were tabulated. Descriptive strategies (5%) are the least complex and are limited to explanations of the airspace or aircraft. Diagnostic strategies (18%) specify, either cognitively or behaviorally, why the student performs as he/she does. Prescriptive strategies (16%) are instructional strategies prescribing the behaviors or knowledge

Figure 1. Comparison between organization of FAA form with experienced instructors' organization.

Organization of Evaluation Form 3120-25		
SEPARTION	Separation	A
	Safety advisories	
CONTROL JUDGMENT	Awareness	B
	Good control judgment	
	Control actions	
	Positive control	
METHODS & PROCEDURES	Prompt action to correct errors	C
	Effective traffic flow	
	Aircraft identity	
	Strip posting	
	Clearance delivery	
	LOA's/Directives	
	General control information	
	Equipment failure recovery	
	Visual scanning	
	Effective working speed	
	Traffic advisories	
EQUIPMENT	Equipment status information	D
	Computer entries	
	Equipment capabilities	
COMMUNICATION & COORDINATION	Required coordinations	E
	Cooperative, professional manner	
	Communication	
	Prescribed phraseology	
	Makes only necessary transmissions	
	Communications methods	
	Relief briefings	

Organization of Experienced Instructors		
A	Ensuring separation	PRIMARY CONTROL
B	Speed control	
	Vectoring techniques	
	Altitude assignment	
	Command of control situations	
	Traffic sequencing	
C	Response to errors	
	Traffic flow orderliness	
	Traffic advisory delivery	
E	Handoff performance	
	Pointout performance	
B	Decision making	SECONDARY CONTROL & PLANNING
	Alertness for possible problems	
	Control instruction planning	
	Issuance of instructions	
	Prioritization	
C	Maintaining aircraft identity	
D	Determining equipment status	
A	Informing about unsafe situations	ADDITIONAL SERVICES
B	Recognition of overload	
	Control procedure orderliness	
C	Weather assistance	
	Expedite traffic	
B	Strip verification	COMMUNICATION & RECORD-KEEPING
C	Strip posting	
E	Phraseology and word usage	
	Speech rate	
	Transmission efficiency	
	Use of radio/interphone	

and skills the student needs to improve. Meta-Assessment strategies were the most common (60%), and deal with how the instructor makes the assessment (what he looks for, what to do to further assess, the student actions triggering an assessment technique).

The strategy segments relating to situation awareness provides the best overall picture of the differences between superior and poor students. Segments directed at poor performing students indicate their unawareness of a situation developing or of aircraft locations, and include comments about student's radar scanning activity and not paying attention to the proper events. This is consistent with Seamster, Redding, Cannon, Purcell, and Ryder's (1993) finding that maintaining adequate situation awareness is one of the most central controller tasks, and that poor performers use inefficient scanning techniques. Segments dealing with aircraft separation and airspace integrity generally also were directed at poorer, as opposed to better students.

Translation of Protocol Segments into Assessment Rules

About half the strategy segments were translatable into sets of diagnostic rules, resulting in 200 rule-sets (most of which are meta-assessment strategies). These diagnostic rules fall into two categories: behavior rules and cognitive rules. Behavior rules provide explanations based on the student's actions under specific air traffic conditions, and are more complex than the computer-based measures that only report the state of the airspace (e.g., an arrival conflict). As shown in *Figure 2*, the behavior measure indicates the student's actions and the outcome. Cognitive rules explain student behavior based on the knowledge and skill required to control the airspace. The cognitive rule indicates that a student lacks certain knowledge or skills.

Figure 2. Sample rule-based measures.

Sample Behavioral Measure
IF: There is a departing aircraft
AND: There is a potential conflict at X feet
AND: Student has assigned departing aircraft
         an altitude of X feet
AND: Student stops departing aircraft at
         X minus Y feet
AND: Student is above average
AND: Scenario presents a light workload
THEN:Student ensures separation
AND: Student shows good response to errors

Sample Cognitive Measure
IF: There is an overflight
AND: That aircraft enters the sector between

X and Y feet and at or above Z speed
AND: That aircraft is not a potential conflict
         with other aircraft
AND: Student descends that aircraft to a
         point that causes loss of speed
AND: Student performs a late handoff of the
         overflight
AND: Student is above average
AND: Scenario presents a light workload
THEN:Student is not in command of control
         situation
AND: Student shows poor handoff
         performance
AND: Student may lack knowledge of policy to
         keep overflights high and at speed
AND: Student may lack knowledge of the effect of
         altitude on aircraft speed

DISCUSSION

We used techniques of cognitive task analysis (see Redding, 1995; Redding & Seamster, 1994; Ryder & Redding, 1993) including a hierarchical cluster analysis of sorting data and a protocol analysis of instructor assessment of students, to identify rule-based behavioral and cognitive measures of student simulator performance. The measures are based on strategies used by experienced controllers to assess student performance and were developed within the framework of the knowledge structures guiding that assessment. The rule-based measures include up to five levels of complexity: computer-based data elements; computer-based summary measures; traditional instructor over-the-shoulder measures; diagnostic behavior measures, and diagnostic cognitive measures. A standard computer interim or summary measure signifying an aspect of student performance, triggers a set of rules that provide a diagnostic behavioral or cognitive assessment.

Though there are many computer summary measures currently available, instructors usually cannot use them directly to diagnose student performance. A benefit of the rule-based system's event-driven approach, however, is that meaningfulness is assured. Instructors are only presented with certain measures if specified conditions occur. Since no two training sessions are the same, conditional measures also help ensure that the right combination of measures is used under the appropriate conditions. This assessment method is consistent with intelligent tutoring approaches where the system functions as an interactive coach. As the instructor's role expands to that of coach, real-time diagnosis becomes more important. The instructor must determine what is missing from the student's knowledge or skill base. The rule-based measures, which are interactive and provide real-time feedback about the *why* of student performance, are a significant departure from current summary measures (see Boone & Steen, 1981; Vreuls & Obermayer, 1985) that provide only a static picture of the *what* of student performance.

These rules could be developed into an assessment system that advises and prompts in a consistent manner based on strategies used by experienced instructors, providing explanations of why students perform in certain ways, and prescribing instructional remedies (Redding & Seamster, 1995). However, further research is needed to test a greater number of subjects, identify other rule-based measures, better quantify the rules, and validate the findings from this pilot study.

The results also represent a preliminary cognitive task analysis of the student assessment process, with the knowledge structure analysis providing a useful organizational structure for grouping the rule-based measures. It reveals that assessment is not performed in reference to air traffic control job function groupings, though that is how FAA forms organize assessment. This finding is consistent with those of Eastman (1984) who found that the organization and sequencing of a U.S. Air Force airmen's repair manual did not match experienced airmen's actual cognitive organization of the repair job. Our findings and those of Eastman, further demonstrate the importance and utility of a cognitive task analysis for determining the proper sequencing and organization of instructional manuals and forms.

REFERENCES

Boone, J.O., & Steen, J.A. (1981). A comparison between over-the-shoulder and computer-derived measurement procedures in assessing student performance in radar air traffic control. *Aviation, Space, and Environmental Medicine, 52*, 583-589.

Eastman, R.W. (1984). *Investigation of airmen's conceptual understanding of a complex task.* Paper presented at the annual meeting of the American Educational Research Association, New Orleans, LA.

Ericsson, K.A., & Simon, H.A. (1993). *Protocol analysis: Verbal reports as data.* (2d ed.). Cambridge, MA: MIT Press.

Redding, R.E. (1995). Cognitive task analysis for instructional design: Applications in distance education. *Distance Education, 16(1)*, 88-106.

Redding, R.E., & Seamster, T.L. (1995). Cognitive task analysis for human resource management in aviation: Personnel selection, training and evaluation. In N. Johnston, N. McDonald, & R. Fuller (Eds.), *Aviation psychology: Training and selection*, (pp. 170-175). Aldershot, UK: Avebury Technical.

Redding, R.E., & Seamster, T.L. (1994). Cognitive task analysis in air traffic controller and aviation crew training. In N. Johnston, N. McDonald, & R. Fuller (Eds.), *Aviation psychology in practice*, (pp. 190-222). Aldershot, UK: Avebury Technical.

Ryder, J.M., & Redding, R.E. (1993). Integrating cognitive task analysis into Instructional Systems Development. *Educational Training Research & Development, 41(2)*, 75-96.

Seamster, T.L., Redding, R.E., Cannon, J.R., Ryder, J.A., & Purcell, J.A. (1993). Cognitive task analysis of expertise in air traffic control. *Intern'l. J. of Aviation Psychology, 3(4)*, 257-283.

Vreuls, D., & Obermayer, R.W. (1985). Human-system performance measurement in training simulators. *Human Factors, 27*, 241-250.

CORPORATE MEMORY: LESSONS LEARNED FROM
INNOVATIVE HUMAN FACTORS RESEARCHERS

Robert J. B. Hutton and Gary Klein
Klein Associates Inc.
Fairborn, Ohio

Lessons learned from experienced researchers can provide an invaluable resource for any organization. The purpose of this project was to interview successful researchers from the Armstrong Laboratory (AL/HEA) at Wright-Patterson AFB to learn from some of their successful projects. The ultimate goals of the project were threefold: to provide AL/HEA with an alternative way to capture and describe the successes of their researchers; to identify themes that emerged from these projects regarding researchers' problem-solving, project leadership, and project management skills; and finally, to provide recommendations to the organization which would promote and support ways to increase opportunities for successful projects.

Eleven interviews were conducted. Each of the researchers was interviewed about a project that had provided some concrete benefit to the Air Force. We used a form of the Critical Decision method to elicit 15 accounts. Several themes were identified that characterized the research projects, and recommendations were made to encourage the initiative of laboratory personnel and increase opportunities for successes.

INTRODUCTION

This project was prompted by a concern of the Human Engineering Division Armstrong Laboratory (AL/HEA at Wright-Patterson AFB) that professional communications were not sufficient to provide an insight into the experience and lessons learned by their successful scientists and researchers, and that future generations of researchers would not be able to benefit from their predecessors' hard-earned experiences.

We looked at the innovators and high-impact researchers at the Armstrong Laboratory. The goal of the project was to highlight the common themes from several projects that had provided a concrete benefit to the Air Force. It was hoped that these themes would answer questions such as: What made these projects successful? How were opportunities detected? What characterized researchers' approaches to selecting questions or problems to follow up? What was the time course of the project and of the subsequent benefits? How were the researchers carrying out the strategic goals

of the Division? How did the researchers work through bureaucratic barriers? What were the team dynamics of the effort?

One approach to trying to preserve this corporate memory has been to use case studies which highlight the knowledge and skills of organizational members that provide some insight into how to succeed in the organizational environments (Klein, 1992). A knowledge engineering approach was taken in order to mine the experiences and knowledge of key personnel.

Our approach relied on an in-depth interview with individuals who have been identified as being successful according to some criterion that was important to the organization. An adapted Critical Decision method (Klein, Calderwood, & McGregor, 1989) was performed during which the interviewers elicited an instance of a "success" and then followed up with questions about that specific incident regarding the problem definition, problem-solving, and solution implementation skills involved in the success.

METHOD

Eleven researchers were recommended by Division management as people who had been influential in projects that had provided some concrete benefit to the Air Force. The interviewees' involvement with AL/HEA spanned it's fifty year existence.

A form of the Critical Decision method interview was used in order to highlight the important aspects of problem definition, problem solving, project leadership and management. The same two-person interview team conducted each interview. First, the interviewee was asked to provide some examples where s/he had been instrumental in the initiation of a project which had provided some concrete benefit to the Air Force. On this first sweep through the stories, we were looking for examples where the interviewee had been a key player in the project. We were also looking for examples where the project had not merely ended in a technical report sitting on a shelf gathering dust, but had found its way into a useful product, whether it be general human factors design guidelines or design specifications for a specific system.

After this first sweep through each of the stories, one story was chosen and the interviewee was asked to recall the project in more detail. This allowed us to put the key decision points on a timeline in order to solidify the story and flesh out the key decisions, turning points, or changes in strategy to approaching the problem.

Having identified these key points in the project, we asked specifically about each one in order to understand what information or events had precipitated the change or decision, and what information was used in order to guide the decision and subsequent course of action. Key areas of interest included how the researchers had identified the problems, how solution methods had been generated, how opportunities were detected, how barriers (such as funding or bureaucracy) had been overcome, what individual initiatives were taken, what the time course of different phases of the project and the benefits were, and what the dynamics of the team were.

A final sweep was taken through the story in order to identify where the interviewee's experience at AL/HEA had helped them successfully accomplish the project, or where a new researcher, fresh from a Ph.D. for example, would have stumbled and possibly run into problems.

On completion of all the interviews, 15 cases had been accumulated (some researchers were interviewed about more than one project). Once all the interviews had been conducted, each story was written up with the key points highlighted.

The purpose of the story write-ups was to provide an alternate view of the laboratory's activities, as opposed to the sometimes dry technical reports and professional publications. A major goal of our stories was to provide lessons learned to researchers who might learn through vicarious experience about some of the "tricks of the trade" and about some of the endeavors of many successful researchers at the lab. A further goal was to provide the Human Engineering Division with recommendations to promote further successes. As the 15 incidents and key lessons learned were being recorded, certain themes of success emerged. Based on these emergent themes, recommendations to support and promote research and project management practices were identified.

SUMMARY OF PROJECTS

The accounts that we generated reflected a broad range of projects with many different implications for human factors research. They included: the discovery of the means to provide a high resolution, miniature color display which can be used with technologies such as helmet-mounted display applications; the initiation of the B-52, B-1, B-1(b), and B-2 simulation, test, and evaluation facilities at AL/HEA; the design of a new B-2 synthetic aperture radar cursor to improve target designation accuracy; the implementation of new training, and personal equipment, for the high-G cockpit, and generation of plans and specifications for an articulating seat intended to improve the pilot's capability to withstand high-G forces; new HUD symbol configuration for cuing air-to-air missile launches; the recognition and study of the problems of arctic navigation leading to improved arctic and space navigation instruments using gyroscope technology; the implementation of human factors design standards (MIL-STDs); an improved understanding of mental workload and the development of fitness-for-duty measures; the development of anthropometric databases including moments of inertia and weight distribution data; a task network modeling capability called SAINT (Systems Analysis of Integrated Networks of Tasks); a cockpit windscreen refraction and displacement measurement device to improve the accuracy of HUD corrections; the development of an austere runway glide-slope indicator that was accurate, robust, and easy to set up by ground troops; the incorporation of display symbology in night vision goggles; and Warrick's principle of control/display movement compatibility. The incident accounts will be published in a forthcoming technical report (Klein & Hutton, in press).

THEMES

As we thought about each project, identified the key decision points or researcher intervention points, and identified how projects were shepherded through the system, several themes emerged from these stories about how opportunities were taken, how research programs were initiated, and how the results of these programs were turned into benefits for the Air Force.

User-Centered versus Generic Projects

The first theme was that the projects we sampled provided a contrast between those projects undertaken with a specific user in mind versus projects intended to have a broader application. This distinction is not necessarily the same as the "applied versus basic" research distinction, but can better be characterized as "firefighting versus fire prevention." All the projects culminated in some application; however, in some cases the projects were inspired by specific user problems and redesign issues, while other projects were anticipatory in nature.

Skilled Problem Finding

A second theme was the skill of the researchers at problem finding. Instead of wasting their time pursuing lines of investigation that had little chance of success, these researchers seemed to choose projects with the knowledge that there was a workable solution. The solution may have been achieved via a new technology or a collaboration with another lab that had access to the required equipment or skills, but in each case the researcher had a vision of how the project might turn out. They appeared to avoid projects that they did not judge solvable.

Organizational Support and Vision

A third theme that emerged was the type of organizational support needed by these projects. AL/HEA has always taken the view that its researchers should be allowed to pursue creative ideas and their own initiatives, with the availability of some internal funding if external funding was not forthcoming. The lab has realized that its primary resource is its researchers and their intellectual capital, and therefore these researchers have been encouraged to identify opportunities, create programs, and forge alliances with like-minded communities.

Forging Alliances with Colleagues and Communities

The final theme was the importance of colleagues and communities, particularly partnerships with other laboratories, and also with user communities. Several of the projects emerged through discussions with colleagues at different labs, through alliances with user groups, sometimes through military personnel who had passed through the lab and into these other research or user communities, and also through professional meetings and conferences.

RECOMMENDATIONS

From these stories and their themes, several recommendations were made:

- expand the toolkit of the researcher and the laboratory by taking opportunities to learn and use new techniques from other colleagues or laboratories,

- encourage researchers to expand their horizons and understanding of issues by interacting with operational communities,

- encourage management to interact with the leadership of operational communities to foster goodwill and support the rapid transition of the research into concrete solutions or benefits to that community,

- reduce contract management burdens by protecting researchers from undue or excessive administrative burdens,

- increase dissemination and the rate of happy accidents (coincidental meetings, being in "the right place at the right time") by describing research and projects to user communities,

- make good use of emissaries, for example by encouraging and maintaining contact with personnel who move on to different labs or into user communities,

- manage internal project funds to support innovative thinking and problem solving while not prohibiting the need to look for external support for research with specific user communities, and

- take advantage of individual personnel's strengths and capabilities within the strategic vision of the organization.

CONCLUSIONS

This project resulted in a report which provided three valuable resources to the Human Engineering Division. The first was a potted history of some of the key research personnel and projects in the laboratory. More important, it provided a lessons learned document which could provide up-

and-coming researchers with some vicarious project problem-solving, leadership, and management experiences, and an insight into how opportunities were identified and carried through. Probably of the most use to the organization were the recommendations that could improve the chances of the organization repeating its successes. This project provided a set of organizational recommendations based on past successes, which indicated how to maintain and foster an environment which encourages its champions to seek and run with opportunities, to take advantage of creative problem-solving opportunities, and to interact with other research and user communities in their areas of expertise.

ACKNOWLEDGMENTS

We would very much like to thank all those researchers at AL/HEA who gave their time to talk to us, and were generous with their memories. We would also like to extend our gratitude to Ken Boff, Lew Hahn, and Tanya Ellifritt at AL/HEA for their help. This project was supported by funding from the University of Dayton Research Institute, UDRI-82252X, and Logicon Technical Services, Inc.

REFERENCES

Klein, G. A. (1992). Using knowledge engineering to preserve corporate memory. In R. R. Hoffman (Ed.), *The psychology of expertise: Cognitive research and empirical AI.* New York: Springer-Verlag.

Klein, G. A., & Hutton, R. J. B. (in press). The innovators: High-impact researchers at the Armstrong Laboratory Human Engineering Division.(AAMRL Technical Report). Wright-Patterson AFB, OH: Armstrong Aerospace Medical Research Laboratory.

Klein, G. A., Calderwood, R., and MacGregor, D. (1989). Critical decision method for eliciting knowledge. *IEEE Transactions on Systems, Man, and Cybernetics,* Special Issue, 462-472.

Cognitive Engineering Of A New Telephone Operator Workstation Using COGNET

Joan M. Ryder, Monica Z. Weiland,
Michael A. Szczepkowski and Wayne W. Zachary
CHI Systems, Inc.
Lower Gwynedd, PA

Many cognitive engineering methodologies for user-centered design involve modeling procedural knowledge; others deal with domain semantics or conceptual models. COGNET (COGnitive NEwork of Tasks) is a framework for modeling human cognition and decision-making which provides an integrated representation of the knowledge, behavioral actions, strategies and problem solving skills used in a domain or task situation, yielding a powerful cognitive engineering tool. A case study of the design of the user interface for a new telephone operator workstation is presented to illustrate the derivation of the design from the components of the COGNET model. The model does not directly convey any specific feature of the interface design, but rather a formal representation of the what the user must do with the resulting interface. This information is then evolved through a set of transformations which systematically move toward design features, in a fully traceable manner.

INTRODUCTION

The benefits of user-centered design for the development of interactive systems is well recognized within the human factors community. Understanding of the user and the users' tasks are a critical component of user-centered design. Many cognitive engineering methodologies for user-centered design involve modeling procedural knowledge (e.g., Card, Moran and Newell, 1983; Kieras, 1988); others deal with domain semantics or conceptual models (e.g., Chechile, Eggleston, Fleischman and Sasseville, 1989; McDonald, Dearholt, Paap and Schvaneveldt, 1986). COGNET (COGnitive NEtwork of Tasks) is a framework for modeling human cognition and decision-making (Zachary, Ryder, Ross and Weiland, 1992) which provides an integrated representation of the knowledge (mental model), procedures (behavioral actions), strategies and problem solving skills (cognitive operations) used in a domain or task situation, yielding a powerful cognitive engineering tool for the design of user interfaces.

A case study of the design of the user interface (UI) for a new telephone operator workstation is presented to illustrate the process by which the COGNET model is used to derive a detailed UI design. The workstation, being developed by Pacific Bell, is a multipurpose workstation integrating Operator Assistance (OA), Directory Assistance (DA), and other phone services requiring operator intervention. Currently, these services are performed separately using dedicated single function terminals. Thus, part of the challenge of this design effort was the fact that the new workstation integrated previously separate jobs into one position and needed to be flexible enough to support the addition of new services in an efficient yet consistent fashion.

COGNET-BASED COGNITIVE ENGINEERING METHODOLOGY

COGNET is both a modeling framework and a cognitive task analysis methodology. It is a methodology for acquiring knowledge from users and system planners and a format for representing the knowledge as analyzed. A complete description of the COGNET methodology and representational formalism can be found in Zachary et al.. (1992). The framework has proven useful in generating validated models of cognitive processes and human-computer interaction in a number of complex domains (e.g., air traffic control: Seamster, Redding, Cannon, Ryder and Purcell, 1993; air AntiSubmarine Warfare: Zachary, Ryder and Zubritzky, 1989). The main products of a COGNET analysis in a particular domain are: 1) a "mental model" that is a representation of the declarative knowledge required for task performance, 2) a set of task models that represent the procedural knowledge involved in a task domain hierarchically from high level goals down to low-level workstation actions and cognitive operations. Each task model also includes the set of triggers, specified as conditions in the mental model, that indicate when it should be performed.

The design effort described below shows how the COGNET formalism provides the basis for a systematic cognitive engineering methodology for UI design. Figure 1 illustrates the specific steps in the process of deriving a UI design from the COGNET representation. The left side of the diagram shows the phases (in bold) of the process. The right side of the diagram shows the specific products of each phase and how these products feed into the products of the next phase. The application of this process to UI design for the operator workstation is discussed below.

APPLICATION OF METHODOLOGY TO TELEPHONE OPERATOR WORKSTATION

Data Collection

Model development was based on observation and video/audio recording of live operator performance, followed by debrief sessions deriving structured verbal protocols from the operators observed. Experienced operators performing DA and OA were selected to serve as participants in the data collection. Twelve hours of performance (>1200 call instances) were recorded and analyzed, representing the range of component

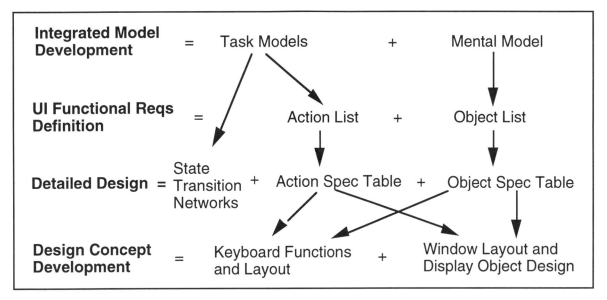

Figure 1. UI Design Process

jobs, various workstations then in service, and different times of the day. In addition, interviews were conducted with system planners to obtain a vision of potential future services the workstation should be able to support.

Design Process

In the first phase of the UI design effort, Integrated Model Development, separate COGNET models were built for OA and DA based on the data collected, as described above. These models were integrated by combining the mental model contents and eliminating the information that overlapped. Then, the task models were analyzed to find those tasks that were performed by more than one type of operator and redundant tasks were eliminated. The analysis resulted in a *mental model* with six panels representing six categories of knowledge used in handling all types of calls from both predecessor jobs (Call Status, Information Request Status, Call Completion Status, Billing Data, Position/Center Data, and Reference Data), with each panel divided into levels indicating the factors considered in performing all operator tasks, and each level specifying what type of information could be posted there. One panel is shown as part of Figure 2. Analysis of procedural knowledge resulted in 66 *task models* in eight categories. Figure 3 provides an excerpt of one task model. Two model validation meetings were held to review and the refine the model, prior to the subsequent design activities.

Figure 2 shows an example of the evolution of the UI design components from the COGNET model components. One part of the model, a Mental Model Panel and the hypotheses populating each level (details not shown) provide the data for the object list. As the design evolves, the object list is transformed into an object specification table, which then informs the display design (one window of which is shown). The task models evolve similarly into the action list,

then the action specification table, then the keyboard layout and UI design concept. The evolution of these components is described below.

The next phase, UI Functional Requirements Definition, consisted of defining all of the display objects (*object list*) that must appear on the screen at any time, and defining all of the operator actions (*action list*) that the system must support. The list of display objects was derived directly from the mental model contents, in that all of the pieces of information appearing in the mental model were labeled as coming either from the customer (verbally) or from the system. Those not coming verbally from the customer were then put on the display object list and were designated as output objects. In addition, the action list was analyzed to determine what objects were required for operator input (e.g., Enter Locality requires a Locality fill-in field) and what output objects were needed for feedback of operator actions (e.g., displaying an icon to show that a particular type of search had been initiated). These were termed input and output objects, respectively. Similarly, all actions represented in the task models were listed, and redundancies eliminated, to generate a complete list of actions that the system needed to support from the keyboard.

In the following phase, Detailed Design, all possible interactions between the display objects and control actions were enumerated and defined. Because display object behavior and what actions were available to the operator often depended on the state of the system, the states of the system were first defined. It was determined that the state of the system corresponded very closely to the type of task the operator was performing, as defined by the task models. Some examples of states included a general billing state for entering the type of charge (e.g., person-to-person or collect) and a coin billing state, which required different functions to be available (such as coin collect and coin return) as wll as a display showing the amount of money paid and owed. Starting with these high

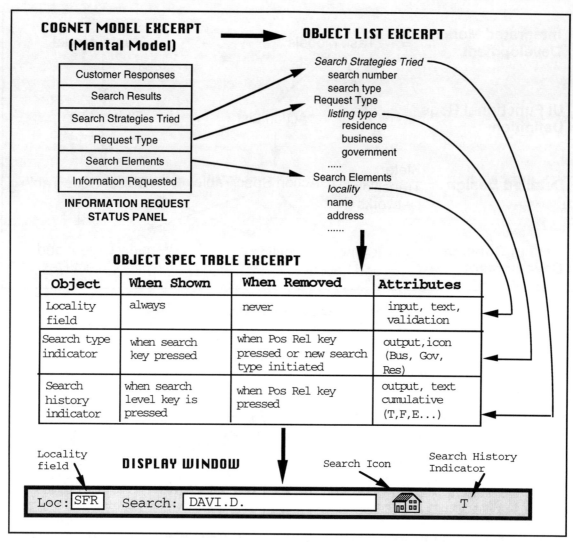

Figure 2. Example of UI Design Evolution

level states, *state transition networks (STNs)* were defined. The STNs specified: what displays needed to be designed (i.e., one for each display 'state'), and what workstation input-operations needed to be designed (i.e., to provide the operator inputs associated with each state transition), but did not define the format of the displays or operator inputs involved in any of these transactions. The definition of the STNs served to define the different contexts that affect object/action behavior.

Another step in the detailed UI design involved deriving the complete object/action space. The object and action lists derived from the COGNET model in the functional requirements phase provided the starting point for an object-action analysis in which the objects and actions were combined to determine:

1) what actions were permissible on what object-classes in what UI states (as defined by the STNs), and

2) which of these permissible object-actions involved human-computer interactions.

Each object-action pair that fell into the second category above then represented a kind of transaction that must be included in the UI. The object/action analysis resulted in a set of tables. One specified all objects, their attributes and when they would be displayed (*object specification table*). Another specified

which actions could be applied to which objects, and what each action on a given object 'meant' in terms of the operator's understanding of the job and the UI state (*action specification table*). Figure 2 shows an excerpt from the object specification table. The action specification table included (example in parentheses) an action (press business search key), the event the action triggered (initiate business search), the UI state the action was possible in (search entry/edit), the UI response (display business search icon to right of search entry field), the system response (query database for business listing matching search input), the system response (retrieve listings), and the secondary UI response (display and format listings in data window).

The final phase, <u>Design Concept Development</u>, appears in the diagram to be a separable phase that took place at the end of the design process. In reality this phase took placethroughout the process. This is because the seeds of ideas for the ultimate look and feel of the interface began as the COGNET model evolved. These rough ideas took shape as the requirements got refined further and the inevitable tradeoff decisions were made. The result was definition and layout of display windows, design and placement of display objects, and

GOAL: Process Inquiry for Number Given Name

<u>**Triggers:**</u>
        <NUMBER> on INFORMATION REQUEST STATUS: *Information Requested* AND
            <Name> on INFORMATION REQUEST STATUS: *Search Elements*

<u>GOMS:</u>

GOAL: Determine Request Type (Business, Residence, Government)

    <u>Determine</u>: Request Type from Search Information Given
    <u>Question</u>: Customer for type ... IF (Type is ambiguous from info requested)
    <u>Post</u>: <REQUEST TYPE> on INFORMATION REQUEST STATUS: *Request Type*

GOAL: Gather Additional Search Parameters ... IF (<INFORMATION REQUEST STATUS :
    Search Elements> Does not include Locality and FF Word)

    <u>Question</u>: customer for needed search element(s)
    <u>Post</u>: <ADDITIONAL SEARCH ELEMENTS> on INFORMATION REQUEST STATUS:
    Request Information ... If any

GOAL: Search for Appropriate Listing

    Use Search Method Based on Request Type

    Method A: Business Search Method

    GOAL: Enter Initial Search Parameters for Basic Business Search

        Enter: Locality in Locality Field

        Enter: Transition to Next Field

        Enter: Business Name in FF Word Field

        Enter: Transition to Next Field

        Determine: if FF Word is Unique Enough for initial Search

        Enter: SF Word in SF Word Field.. IF (SF Word) and (FF Word not unique)

    GOAL: Perform Initial Business Search

        Initiate: Business Search

        Post: <Initial Search> on Information Request Status: Search Strategies Tried

        Give: Progress Report of Search... IF (DA Search is prolonged)

                 •
                 •
                 •

note: FF=First Finding Word; SF=Second Finding Word

**Figure 3. Example of procedural knowledge from the task
"Process Inquiry for Number Given Name"**

definition of interface control structure including keyboard layout. Each of these products are discussed below.

    Part of the concept was the identification of *windows*, the basic objects they would contain, and the functions they performed. The mental model had identified the main functional categories of operator information across existing and anticipated services—billing, call handling, information search, and position/center data. Our UI design provided a window associated with each category. Although the window organization was different from either of the main predecessor services, usability testing indicated that it was favorably received by most operators. We believe that was due to the fit of the organization of the windows with the operators' conceptual models of their domain and the efficient manner in which it integrated different services.

    The STNs defined the set of displays that needed to be designed (one for each state). Following the initial display concept and the complete object specification (from the Object/Action Space), the actual *display objects* were designed and arranged in the display windows. As the window layouts and object hierarchy were derived directly from the COGNET mental model, the displays reflected the organization of domain objects according to operators' conceptual framework. The objects included in each display were those that specifically represented the information relevant to conduct of each call flow.

Another aspect of interface design that was driven by the COGNET operator model was the definition and layout of actions on the *keyboard*. Requirements for task speed dictated a keyboard as the only input device and justified a custom keyboard design. In addition to alphabetic and numeric keys, all actions in the Object/Action Space were allocated to dedicated and soft function keys based on expected action frequency and co-occurrence patterns (as determined from the Task Models, UI states and transitions between states, and call-type frequency data provided by the client). The keys were arranged to minimize the time required to complete the expected tasks.

DISCUSSION

In summary, a COGNET model of a task domain provides the basis for a user interface design methodology. The model does not directly convey any specific feature of the interface design, but rather a formal representation of the what the user must do with the resulting interface. This information is then *evolved* through a set of transformations which systematically move toward design features, in a fully traceable manner.

Usability testing sessions (walkthroughs with a preliminary procedures manual and paper designs) were held to evaluate the detailed UI design and procedures and suggest improvements. Subjective review of the overall UI was very favorable, with specific changes in UI details suggested. Additional modeling of keystoke-level performance for the completed design (similar to Gray, John and Atwood, 1993) or empirical testing of working prototypes is needed to verify that the design does not degrade performance times relative to the existing single function workstations.

ACKNOWLEDGEMENT

The work reported here was performed under contract P.O. #6370 from Pacific Bell Operator Services Inc. The ideas presented in this paper are those of the authors alone and are not the official position of Pacific Bell Operator Services or its affiliates.

REFERENCES

Card, S.K., Moran, T.P., & Newell, A. (1983). *The Psychology of Human-Computer Interaction*. Hillsdale, NJ: Erlbaum.

Chechile, R.A., Eggleston, R.G., Fleischman, R.N., & Sasseville, A.M. (1989). Modeling the cognitive content of displays. *Human Factors, 31(1)*, 31-43.

Gray, W.D., John, B.E., & Atwood, M.E. (1993). Project Ernestine: Validating a GOMS analysis for predicting and explaining real-world task performance. *Human-Computer Interaction, 8*, 237-309.

Kieras, D.E. (1988). Towards a practical GOMS model methodology for user interface design. In M. Helander (Ed.), *Handbook of Human-Computer Interaction*. North-Hollan

McDonald, J.E., Dearholt, D.W., Papp, K.R., & Schvaneveldt, R.W. (1986). A formal interface design methodology based on user knowledge. In *Human Factors in Computer Systems, CHI '86. Proceedings* (pp. 285-290). New York: ACM.

Seamster, T.L., Redding, R. E., Cannon, J.R., Ryder, J.M., & Purcell, J.A. (1993). Cognitive task analysis of expertise in air traffic control. *International Journal of Aviation Psychology, 3(4)*, 257-283.

Zachary, W., Ryder, J., & Zubritzky, M. (1989). *Information processing models of human-computer interaction in naval air anti-submarine warfare* (Technical Report 891215.8704). Spring House, PA: CHI Systems, Inc.

Zachary, W.W., Ryder, J.M., Ross, L., & Weiland, M.Z. (1992). Intelligent computer-human interaction in real-time multi-tasking process control and monitoring systems. In M. Helander and M. Nagamachi (Eds.), *Human Factors in Design for Manufacturability*. New York: Taylor and Francis.

USER-INITIATED NOTIFICATION: A CONCEPT FOR AIDING THE MONITORING ACTIVITIES OF PROCESS CONTROL OPERATORS

Stephanie Guerlain and Peter Bullemer
Honeywell Technology Center
Minneapolis, MN

Monitoring activities in a process control environment are quite unique depending on the current situation and the operator's current understanding of that situation. Furthermore, the operator may be required to monitor multiple simultaneous events over potentially long periods of time. Currently, operators must periodically scan displays to gather such information, or manipulate the alarm or control system in ways not originally intended in order to gather that information as appropriate. Furthermore, if the monitoring activities span multiple operating shifts, then there is the potential for operators to forget to communicate these requirements at shift change. Despite the uniqueness of the situations that will require process events to be monitored, it is hypothesized that there is a limited set of conditions that can be pre-defined in a tool that will allow operators to set up their own monitoring "agents" according to their current diagnostic needs. Such a tool is predicted to decrease the working memory load of operators, and reduce the time it takes them to detect important process changes (or lack of them). Furthermore, it is proposed that this concept is extensible to other plant personnel and to other domains that have similar monitoring requirements. Although some potential pitfalls can be predicted with the introduction of this tool, the number of predicted benefits warrant the further exploration of this concept. This will be the next step in our design process.

INTRODUCTION

In process control or other environments requiring supervisory control of automation, operators need to monitor both significant events in the process as well as what the automation is doing to control the process (Woods, 1994). Monitoring activities take place in both normal and abnormal situations, particularly when an operator makes a set of control moves to change the state of the plant (e.g., either in response to normal plant operations requirements, such as changing the levels and kinds of products being manufactured, or in response to an anomaly situation to move the plant back to a safer state). In both types of situations, the operator needs to get feedback as to whether the intended effect of the control move has indeed taken place. This is primarily done by periodically checking overview, process and trend displays, talking to field operators, and by relying on the alarm system to alert the operator that the plant has reached an unsafe state. In several field studies of normal and abnormal operations, we have identified shortcomings of today's technology in supporting operations teams (Soken, Bullemer, Ramanathan and Reinhart, 1994).

The purpose of this paper is to describe an interface design concept that would allow operators in a process control plant to initiate monitoring events based on their current diagnostic needs, thereby having a means to use the computer as a monitoring tool and notify them that the event has occurred instead of having to do the monitoring themselves. We call this concept *user-initiated notification*.

PLANT OPERATIONS

Normally, a process control plant is functionally divided into a number of units, such that an operations team is responsible for each unit of the plant. Each operations team consists of two to five field operators and one or two board operators (depending on the size of the unit). The field operators work on the unit itself, manually changing valves, watching for leaks, etc. and the board operator(s) monitor and control the unit through the use of a distributed control system (DCS) (Zwaga and Hoonhout, 1994), which is a computerized control system that has a number of operating displays, such as schematics, alarm summaries, and trend graphs. With this system, each automated controller or sensor is given a tag name, such as FI-

201 (Flow Indicator 201). For each of these tags, the control system contains information about the sensed state of the process variable (PV), the setpoint (SP) and the output (OP) of the controller. Although operators monitor the plant through schematic, group, trend, and alarm summary displays, it is generally plant engineers who are responsible for designing and implementing the control system, as well as all of the operating displays used and seen by each shift team.

The majority of communications that occur are between the field operators and board operators of a particular unit, but cross-unit communication is sometimes necessary, since products generally flow from one unit to another. Another important communication flow is during shift change, when the previous shift team must communicate important events to the next shift team. This information is passed verbally from one operator to the next, and the head operator out on the unit will write notes in a log book located on the unit and the board operator will write notes in a log book located in the control room. Finally, in some plants, operators will rotate their positions, such that one day an operator will be out in the field and the next day that same person may be operating the board, or responsible for a different position in the field. This cross-training has benefit for mutual understanding of roles and responsibilities and how each part of the job affects the other.

MONITORING ACTIVITIES

Field operators will conduct periodic rounds of their part of the plant, and check for any leaks through observation and check that pumps are operating correctly by sound and touch (field operators will touch a pump to feel for heat and vibration that is abnormal). Field operators also rely on the board operator to monitor process states that need to be checked in the field, such as a regulator or a valve being stuck. Sometimes, a field operator will call the board operator to tell him about an abnormal process state (e.g., "The wet gas compressor sounds like it's cavitating!"). Sometimes, a DCS display is available in "the shed" on the unit, from which field operators can monitor, but not control the unit. Sometimes, a field operator will see an alarm on this display that is relevant to his area that he will respond to even before the board operator has a chance to contact him.

Board operators will monitor the plant from the control room primarily through the displays available to him on the DCS. There can be hundreds of such displays, and 6-8 screens can be seen at a

time (depending on the number of monitors he has). Generally, operators will leave 3 or 4 overview displays up at all times, and change the views seen on the other screens as necessary.

Board operators will rely heavily on the alarm system (and to a less extent field operators) to notify them of abnormal states, and respond to those as necessary. Board operators may also need to monitor particular plant states more carefully, based on their current tasks (such as following a procedure, changing the operating conditions of the plant, etc.).

In our own field studies of petrochemical operations and in a field study of nuclear power plant operators conducted by Vicente & Burns (1995), a number of interesting observations were made about the ways that operators of a plant will manipulate their environment to support their monitoring activities, thus "finishing the design" of the tools available to them. For example, Vicente & Burns (1995) observed operators leaving the doors open on strip charts that needed closer monitoring to remind both themselves and operators on later shifts that these particular process variables deserved closer attention. Another observation was that operators would sometimes manipulate alarm limit setpoints during operations, not because the previous alarm limits were inappropriate, but because a situation had occurred which required them to monitor a particular process variable for certain characteristics that were not necessarily an alarm state.

In our own observations of plant operators, such manipulation of alarm limits is a rare event (since a shift supervisor usually needs to physically turn a key to allow for this capability). What should be noted, however, is that since the alarm system is designed for orienting of abnormal states, operators will sometimes manipulate the alarm system to support orienting of other kinds of states that the operator feels are informative. That is because appropriate, dynamic mechanisms are not currently available to support this functional requirement of plant operations. The following situations were identified where such a strategy might be useful:

- increasing the setpoint after an initial alarm to get a "second chance" (since the process value may continue to increase after having "passed" the alarm limit and one would like to monitor for that).
- changing the alarm setpoint to a value at which time an action needs to be taken (as in following a procedure).
- manipulating the setpoint on a variable that is correlated with one that needs to be monitored but is not alarmed.

Once the event of interest has happened, the intent is to set the alarm limits back to where they were. However, operators may forget to take this step, thus risking the lack of alarms when appropriate, a situation which can become worse at shift change, since the next operator will have no idea that the previous operator has forgotten to reset some of the alarm limits.

USER-INITIATED NOTIFICATION

Since much of the information necessary to monitor the plant is centrally located in the DCS, it becomes possible to use this information to the operator's advantage. The idea behind user-initiated notification is to design a tool that will give the operator of a distributed control system a means to easily generate information about the process related to their current diagnostic needs. Thus, it gives operators more flexibility and control over what types of information they would like to hear about, rather than having to rely on the pre-defined alarm limits set up ahead of time by plant engineers. This is useful since alarm limits are set up based on "steady state" operations (and are thus not as helpful in other situations, such as startup). Moreover, there are significant, situation-specific needs for notification events that do not qualify as alarm events, (i.e., a vessel filling to the halfway mark is informative and important to know in the current task context but not an alarm condition).

The initial concept was based on the idea that there are a set of alarm-initiated activities, taken from (Stanton, 1994): 1) Observe (hear) alarm 2) Accept alarm 3) Analyze alarm message 4) Investigate cause of alarm 5) Take corrective action 6) Monitor that the corrective action has had the desired effect. In other words, operators will often make a control move (such as opening a valve) that they think will have the desired effect (e.g., increasing flow to another area of the plant). However, in order to ensure that the desired affect has indeed taken place, the operator must actively monitor those values. People are not very good at vigilance and monitoring tasks and there is a cost to sampling — attention will have to be diverted away from some other activity (Stanton, 1994).

The intent, then, for the user-initiated notification tool, is to provide the operator with an easy means to ask the computer to monitor various process values for certain conditions (e.g., reaching a particular value, increasing, decreasing, high rate of change, etc.) and be notified of that condition when it occurs. The potential benefits of such a display are the following:

- Avoids the "inappropriate" use of the alarm system to generate this type of information.
- Offloads operator's memory by providing a tool to do specific types of monitoring.
- Allows for "temporary" monitoring activities that are only in effect until the event is triggered or the operator deletes that monitor.
- Allows operators to have control and flexibility over what type of information is displayed.
- Could be used as a "handoff" tool at shift change (one operator could discuss the monitors that are currently in place and why, giving the next operator a context of what kinds of important things are happening in the process).
- Gives operators a way to be more "proactive" in their monitoring activities without lots of effort.
- Can be used as a knowledge acquisition tool - of what kinds of information operators feel they'd like to see.

When this user interface concept was described to plant engineers and operators at a couple of petrochemical companies, they instantly saw the value of the display and could think of many situations in which this kind of a tool would be useful:
- To monitor that a control move has had the desired effect.
- To monitor for certain constraints in the process (e.g., the operator often increases the rate of feed input until certain constraining limits are reached. The operator could have the computer monitor for these conditions).
- To aid in transferring the operating task at shift change (since the next operator could see what monitors had been initiated by the previous operator).
- To aid in following procedures (e.g., since procedures often are of the form, "Bring the level to 20%, then start pump X - the operator could have the computer monitor for this condition, so the operator would know when it was time to move to the next step).

In fact, plant personnel suggested uses for the tool that far exceeded the original intent of the concept:
- To facilitate coordination between board operators, field operators, instrumentation men, and plant engineers (e.g., the board operator could tell the computer, "If this value gets to X, tell the field operator to take a sample.")
- Plant engineers could use it to "look over the shoulder" of board operators to see if they are employing good standards of practice.

- Plant engineers could set up generic monitors that are specific to a particular situation, such as startup, and save and call up a set of these reminders as appropriate.
- Operators and/or engineers could use it to set up periodic reminders (e.g., certain pieces of equipment need to be tested every two weeks).
- Plant supervisors could use it to see if plant goals were being reached.

What was interesting was that the original intent of the concept was quite simple, but plant personnel could immediately envision multiple, diverse extensions to the concept. The common theme was that a tool like this could be designed for each type of person in the plant (field operator, board operator, supervisor, engineer, etc.) to monitor for their information requirements. In fact, this concept appears to be generic enough that it could be applied not only in the process control domain, but in many domains that have similar monitoring requirements.

OPENING UP PANDORA'S BOX?

In the brief period of time that we have explored this interface concept, it could already be envisioned that such a monitoring system could actually have a dark side. What if operators set up a number of monitors that the next operator does not want? Perhaps there should be a time-out for these monitors, or they are personalized to each operator. While the initial concept was for operators to initiate a small set of temporary monitors, plant engineers could envision thousands of these monitors being set up to monitor for various plant characteristics. If so, then perhaps the alarm flood problem will only be added to with a "notification flood". What was originally designed as a concept to aid the monitoring activities of operators could potentially increase the problem of information overload. Perhaps there should be a limit to the number and kind of monitors that can be invoked at one time. Finally, if engineers could use such a system to "watch over the shoulders" of operators, it could be used as a method to assign blame to operators for poor performance. Perhaps these kinds of options should not be designed into the monitoring system, so such capabilities are not possible. On the other hand, displays that support monitoring of group activity are needed. The very intriguing aspect of this concept is that it enables others to see more of the formerly hidden monitoring activities of plant personnel. Despite some of the foreseeable problems with this interface concept, the number of foreseeable benefits certainly warrants exploring this concept further,

while considering design features that will limit the potential negative consequences that this tool could introduce.

CURRENT DEVELOPMENT

Based on the initial positive response we had to the concept of user-initiated notification, we are currently developing a test prototype of this kind of a tool in a petrochemical plant. This is a tool in addition to the current alarm system, that will eventually allow operators to set up to 20 monitors at a time. These are either event-based logical, event-based analog, or time-based. Event-based analog monitors can check for whether the setpoint, output, or process variable of any flow, level, pressure or temperature is equal to, greater than, or less than either a particular value (such as 100°) or another process parameter (i.e., "Tell me if this flow equals another flow"). Logical combinations allow for two values to be combined with AND or OR. Logical flags for each process parameter can also be checked, such as, "Tell me if this value goes out of alarm". This is particularly useful when bringing a piece of equipment back on line. When the level in a tank reaches the low alarm limit, for example, you know that it is filling up as intended.

Currently, the system is working for most of these cases, for a particular board operator. Further planned functionality includes having time-based monitors which can either be one-time events, such as, "Remind me at 1:00 to take a sample" or periodic, such as, "Remind the field operator every Monday morning to take these samples." The next step in our design process will be to design a front-end interface to allow operators to specify these monitors in an easy-to-use format. As a first step, operators will be able to specify a query by choosing from a list of options, and have the ability to turn monitors on and off or erase them completely. They can also specify how they would like to be notified, whether as a sound, or in a message list on the alarm summary. This simple version will be tested for user acceptance and utility before extending the concept to be more integrated with their current operating displays. For example, it may be possible to set up monitors directly from trend graphs and other schematics and utilize different means for notifying operators, such as marking flags directly on schematics. These concepts will be explored further as we learn more about the actual use of this tool by operations.

CONCLUSION

Monitoring activities taken by operators in a process control environment are quite unique depending on the current situation and operator's current understanding of that situation. Furthermore, the operator may be required to monitor multiple simultaneous events over potentially long periods of time. Currently, the operator must periodically scan displays to gather such information, or manipulate the alarm or control system in ways not originally intended in order to gather that information as appropriate. Furthermore, if the monitoring activities span multiple operating shifts, then there is the potential for operators to forget to communicate these requirements at shift change. Despite the uniqueness of the situations that will require process events to be monitored, it is hypothesized that there is a limited set of circumstances that can be pre-defined in a tool that will allow operators to set up their own monitoring "agents" according to their current diagnostic needs. Such a tool is predicted to decrease the working memory load of operators, and reduce the time it takes them to detect important process changes (or lack of them). Furthermore, it is proposed that this concept is extensible to other plant personnel and to other domains that have similar monitoring requirements. Although some potential pitfalls can be predicted with the introduction of this tool, the number of predicted benefits warrant the further exploration of this concept. This will be the next step in our design process.

REFERENCES

Cox, J. and Easter, J. (1989). Influence of operator training in the design of advanced control rooms and I&C at Westinghouse. Eighth Symposium on the Training of Nuclear Facility Personnel, Gatlinburg, TN

Sassen, A., Buiël, E., and Hoegee, J. (1994). "A laboratory evaluation of a human operator support system." International Journal of Human-Computer Studies 40: 895-931.

Soken, N., Bullemer, P., Ramanathan, P. and Reinhart, W. (1994). Human-computer interaction requirements for managing abnormal situations in chemical process industries. Petroleum Division Symposium on Computers and Engineering, American Society for Mechanical Engineers, Houston, TX, pp. 120-128 .

Vicente, K. and Burns, C. (1995). A field study of operator cognitive monitoring at Pickering Nuclear Generating Station - B. Toronto, Ontario, Canada, Cognitive Engineering Laboratory, University of Toronto.

Woods, D. (1994). Cognitive demands and activities in dynamic fault management: abductive reasoning and disturbance management. Human Factors in Alarm Design. N. Stanton. London, Taylor & Francis: 63-92.

Zwaga, H. and Hoonhout, H. (1994). Supervisory control behavior and the implementation of alarms in process control. Human Factors in Alarm Design. N. Stanton. London, Taylor & Francis: 119-134.

IN AND OUT OF THE LOOP: STRATEGIC SKILL AND WINDOWS OF OPPORTUNITY IN DISCRETE PROCESS CONTROL

Penelope Sanderson
Engineering Psychology Research Laboratory
University of Illinois at Urbana-Champaign

A long-standing question in cognitive engineering is the effect of computer aids on work skills. We examined the performance of human operators in a discrete process control environment. During an extended training period, humans controlled production under one of four different conditions (manual, hybrid, supervisory, and automatic) that varied the degree of in-the-loop control. After a training period all subjects transferred to a supervisory condition with two scheduling logic errors to compensate for. Training results showed that production was best overall with supervisory control, intermediate with hybrid control, and worst with manual control. Supervisory performance did not differ across the two scheduling aids. When scheduling logic errors were introduced, subjects who had trained in the automatic condition (simply observing) took longer to respond and let production drop most. Overall, performance depends on an interplay between (1) ability to identify goal-relevant cues and (2) ability to connect cues to action. Moreover, whether out-of-the-loop unfamiliarity is seen depends on how tight the "windows of opportunity" for action are in the domain.

INTRODUCTION

An important question for cognitive engineering is whether computer aids erode human operators' work skills (Wickens, 1992). In the field of aviation, survey results suggest pilots' concern at the impact of flight deck automation on basic flight skills (Wiener, 1988; Hughes, 1989). Experimental results with tracking tasks show that operators are faster to detect failures under automatic control conditions if they have had prior manual experience rather than prior experience only under automatic control conditions (Kessell & Wickens, 1982). In an experimental study of process control, Shiff (1983) found that operators who had been working under supervisory conditions were less effective in compensating for failures than subjects who had worked in manual mode. These results were attributed to the fact that manual controllers experience the dynamics directly so that visual cues become associated with the need for motor actions and are thus better differentiated even when motor actions are unnecesary. In contrast, however, Curry and Ephrath (1977) found that subjects who monitored an automated system later did better in a flight task than subjects who interacted manually with the system. These results were attributed to the fact that the workload of manual control may have narrowed subjects' attentional focus to a point where other important information was missed.

This question has also been addressed in the context of discrete process control, such as flexible manufacturing systems (see Sanderson, 1989 for a review). In these cases, the focus has been on finding the best combination of human and computer for normal operations rather than examining transfer to manual control after a failure. For example, Dunkler, Mitchell, & Govindaraj (1988) compared computer-based scheduling with human-computer interactive scheduling and found that the latter condition led to reliably superior results. Further promising studies in this area have produced suggestive further results that, unfortunately, have not since been clarified (Hwang & Salvendy, 1988).

Finally, Endsley and Kiris (1995) have examined performance across a series of discrete navigational problems under levels of control ranging from unaided to fully automatic (expert system based). When all subjects are subsequently transferred to unaided problem-solving performance is briefly worse in the more automated conditions before settling down to similar levels across conditions. However because subjects solved six unrelated problems in a single session and acquired no specific skill while doing so, it is unclear how much Endley and Kiris' (1995) results would generalize to continuous or discrete control tasks:.

There are clearly discrepancies and uncertainties in the published data. In the work presented here, we addressed three general questions:

1. In a discrete control task, what factors account for any performance differences between unaided humans, humans working with computer aids, and computers alone?

2. To what extent do the above results depend upon the quality of the computer aid?

3. If trained subjects are all transferred to the computer-aided condition and must compensate for scheduling logic errors, are there differences in groups according to their prior training?

THE CIM SIMULATION ENVIRONMENT

The environment used for this experiment ("CIM") was a real-time interactive simulation of a full-scale physical computer-integrated manufacturing system housed in our university department that produces toy robots. All system parameters and all time constants except one were the same in CIM as in the full-scale system.

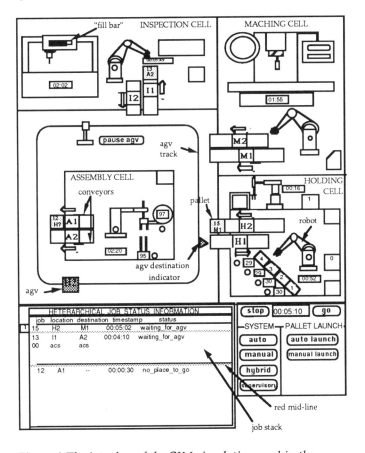

Figure 1 The interface of the CIM simulation used in the experiment. Identifiers with arrows did not appear in actual interface.

CIM contains four principal cells (see Figure 1). Pallets (metal plates) containing toy robot parts are transported by the Automated Guided Vehicle (AGV) along a track from cell to cell in the order given below.

Holding Cell. Here the pallet is fixtured so it can hold the unassembled toy robot parts securely. The parts are then loaded onto it.

Machining Cell. The parts of the toy robot are placed successively on a Computer Numerically

Controlled machining center and are machined to specifications.

Inspection Cell. Here the parts are inspected successively to ensure they have been machined correctly.

Assembly Cell. The parts on the pallet are moved onto a special assembly fixture where they are assembled into a toy robot, which is then returned to a finished products store within the Holding Cell.

The AGV travels in a counter-clockwise direction only and delivers pallets to either of the two conveyors at each cell. Information available to the operator can be seen in Figure 1. The job status display shows the stack of jobs the AGV is working through.

Depending upon experimental condition some or all of the following actions are available to operators. Most actions involve direct manipulation (Shneiderman, 1988) or clicks on command buttons in the interface.

Launch pallets. New pallets with toy robot parts can be launched into the system. (Low-level function)

Conveyor control. Subjects can send pallets in and out of cells by clicking on the arrows alongside the conveyors. (Low-level)

Pause and reactivate AGV. By hitting the "Pause AGV" button or clicking on the AGV track, the subjects can pause or reactivate the AGV. (Low-level)

Pallet assignments. Subjects can assign the cell and conveyor to which a pallet should be directed by clicking the pallet and then the desired conveyor. (High-level)

AGV routing. Subjects can change AGV routing by reordering the pallets in the job status display or by clicking the track and then the pallet to which the AGV should be directed. (High-level)

Finally, UIUC-CIM could run under either of the following two scheduling algorithms:

First Come First Served (FCFS). Under the FCFS algorithm, the AGV would often bypass jobs that could have been transported to their destinations along the route towards the first-completed job.

No passing waiting jobs (NOPASS). In the NOPASS conditions, the scheduling algorithm used is one that does not allow the AGV to pass waiting jobs.

EXPERIMENTAL DESIGN AND HYPOTHESES

This was a transfer experiment. Subjects were allocated to one of five conditions where they controlled high-level functions and/or low-level functions (see above). and were then transferred to conditions where they had to deal with scheduling errors. The five training conditions were:

Manual: all functions performed manually.

Hybrid: low-level functions performed automatically, high-level by human.

Supervisory: both low- and high-level functions performed automatically, but human can intervene in high-level functions.

There were two automatic conditions. The Auto-ool condition was added when the NOPASS conditions were run (after all the FCFS conditions).

Auto-il. In the Automatic in-the-loop training sessions, subjects watched the automatic scheduler and deliberately imagined how they would improve performance. They therefore lacked physical in-the-loop control, but had directed their attention to high-level relational aspects of CIM in operation.

Auto-ool. In the Automatic out-of-the-loop training session, subjects' attention was diverted to a low-level bookkeeping task. They therefore lacked both physical in-the-loop control and had not directed their attention to high-level relational aspects

Subjects ran in 10 25-minute sessions. In Sessions 1-7 they performed under their assigned training condition. In Session 8 they transferred to the Supervisory condition, if not already in it, and in Sessions 9 and 10 they exercised Supervisory control while handling scheduling errors. For approximately half the subjects the Supervisory control algorithm was FCFS whereas for the other half it was NOPASS.

In preliminary simulation runs we established that the FCFS algorithm by itself consistently produces six products per session and the NOPASS algorithm seven products. Any deviation from these "throughputs" reflects human intervention. Pilot studies with extremely experienced human operators showed that under Supervisory control, throughput was always eight products if no scheduling errors were made and no opportunities for improvement were missed. We therefore knew that *in principle* throughput could be better than either of the scheduling algorithms by itself, but we wished to know whether *in normal practice* it would be, and when. We hypothesized the following.

H1. Human supervisory control will lead to higher throughput than manual or hybrid control alone.

H2. Human supervisory control with the FCFS or NOPASS algorithm will lead to better throughput than with the algorithm alone.

H3. Throughput will be best overall under supervisory control with the already more efficient NOPASS algorithm than with the FCFS algorithm.

A second concern was to determine the effect of training condition on operators' ability to intervene and maintain acceptable system performance when the scheduling algorithm develops an logical error. Supervisory mode was chosen at transfer because it is the most similar to the conditions seen in industry—normally the system will be left to run automatically, but intervention is possible if needed.

We transferred all subjects to supervisory control in Session 8 so they would become familiar with the manual aspects of working the controls. Therefore, any differences between training groups in Sessions 9 and 10 would more probably be due to lack of higher-level strategic skill than to lack of manual skill. In Session 9,

the AGV regularly picked up the farthest rather than nearest pallet. This is a subtle error to notice and to predict because it is only seen when more than one pallet is waiting. If uncorrected, system throughput drops to only four products per session. In Session 10, every third pallet was transported beyond its destination cell. This error was much easier to see and quickly had the effect of choking the system, leading to no more throughput until fixed. Therefore our fourth hypothesis was the following.

H4. Throughput will be preserved better in conditions where subjects have habitually focused attention on high-level relational aspects of CIM's operation (Supervisory, Hybrid, and Automatic-il) and will be most compromised in conditions where attention has been diverted to low-level or irrelevant tasks (Manual and Automatic-ool).

RESULTS

Training conditions. Data from the automatic groups were not included in the training analyses because they were deterministically generated and thus had no variance. Between-within ANOVAs were performed on throughput, using Algorithm (FCFS vs NOPASS), Condition (Manual, Hybrid, Supervisory), and Sessions (1-7 or 3-7) as factors.

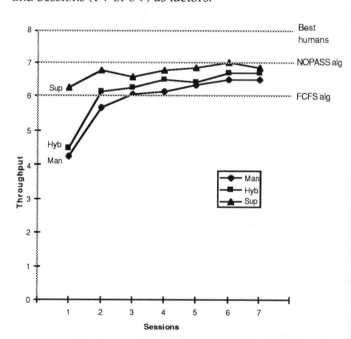

Figure 2 General throughput during training for the Manual, Hybrid, and Supervisory conditions. NOPASS alg and FCFS alg refer to performance by the scheduling rules by themselves.

As Figure 2 shows, throughput improves across sessions, whether analyzed from Sessions 1 to 7, $F(6,180)=27.121$, MSe=0.462, $p<0.001$, or Sessions 3 to 7, $F(4,120)=2.472$, MSe=0.397, $p<0.05$, once the initial effect of practice has worn off. Throughput is highest for the Supervisory condition, less for the Hybrid condition, and least for the Manual condition, both over Sessions 1 to 7, $F(2,30)=22.139$, MSe=0.616, $p<0.001$ and Sessions 3 to 7, $F(2,30)=4.247$, MSe=0.841, $p<0.05$. Some initial differences between the Hybrid and Manual conditions in Session 1 account for a significant Condition by Session, $F(6,180)=28.51$, MSe=0.462, $p<0.001$ and Condition by Algorithm interactions, $F(2,30)=3.91$, MSe=0.616, $p<0.05$, but these do not persist after Day 1. Overall, these results support hypothesis H1 that human-computer scheduling in general is better than human alone.

A separate analysis was performed across the FCFS and NOPASS Supervisory conditions from Session 1-8 and 3-8. Unexpectedly, there was no difference in throughput with the FCFS and NOPASS algorithms, despite the fact that the effectiveness of the two algorithms is substantially different. Therefore, the answer to whether human-computer scheduling results in higher throughput than the scheduling algorithms by themselves is mixed, as predicted. After Day 1 the average human-computer interactive throughput for the FCFS algorithm is equal to or greater than throughput with the FCFS algorithm alone. However, for the NOPASS algorithm, the average human-computer interactive throughput is equal to or *less* than the throughput of the NOPASS algorithm alone. These results were borne out at the level of individual subjects, as well. Significantly more FCFS than NOPASS subjects improved upon the algorithm's performance, and this effect is still present in the final training session.

Overall, these results support hypothesis H2 that human-computer interactive scheduling is better than an algorithm alone for the FCFS condition but fail to support it for the NOPASS condition. The results fail to support hypothesis H3 that human intervention will improve the performance of an already very good algorithm, even when there is room for improvement.

Transfer conditions. Surprisingly, when all subjects were transferred to Supervisory control in Session 8 (for the Supervisory conditions this was just a continuation of their prior mode of interaction) there were no significant differences between conditions. However, in Sessions 9 and 10 some differences emerged, even through they were not strong (note that we had set up a conservative test of transfer effects). One ANOVA was performed across Sessions 8, 9, and 10 with the factors Algorithm (FCFS vs NOPASS), Condition (Manual, Hybrid, Supervisory, and Auto-il) and Sessions (8-10). A further ANOVA was performed within the NOPASS conditions alone, with factors Condition (Manual, Hybrid, Supervisory, Auto-il, and Auto-ool) and Sessions (8-10).

Figure 3 shows throughput across Sessions 8 to 10. Differences are small but, as predicted, Hybrid and

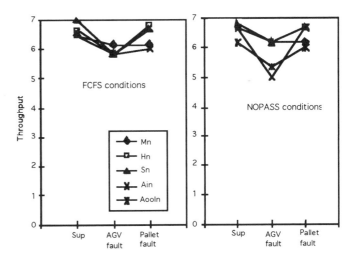

Figure 3 Throughput in Sessions 8, 9, and 10 for FCFS conditions (left) and NOPASS conditions (right).

Supervisory subjects appear to perform best overall, whereas Manual and Automatic–il subjects perform worse in the Pallet destination error under FCFS conditions and Automatic–il subjects perform worse with the AGV direction error under NOPASS conditions. The above pattern is supported in a significant Condition by Session by Algorithm interaction, $F(6,80)=2.273$, MSe=0.306, $p<0.05$. However, there was only a marginal effect of Condition, $F(3.40)=2.754$, MSe=0.460, $p=0.055$, with the Automatic–il subjects tending to do worse overall, and a marginal interaction of Condition and Session, $F(6,80)=2.030$, MSe=0.306, $p=0.071$, with the Manual subjects tending to do disproportionately worse with the Pallet error. An analysis of subjects' latency to respond showed a significant effect of Condition, with Hybrid subjects responding fastest, followed by Supervisory, Manual, and Automatic–il, as predicted, $F(3,40)=4.018$, MSe=18.307, $p<0.05$. For the pallet error the means fell in this same order, but Condition was not significant.

In the NOPASS analysis there is a main effect of Condition, $F(4,25)=3.015$, MSe=0.529, $p<0.05$, in which Automatic–ool subjects condition are always least productive or next to least productive. A marginal interaction of Condition by Session, $F(8,50)=1.907$, MSe=0.342, $p=0.08$ reflects the especially poor relative throughput of the Automatic–il subjects on the AGV error in Session 9. No latency effects were seen.

Overall, these results only partially support hypothesis H4 that subjects will handle scheduling logic errors better after having focused attention on high-level relational aspects of CIM's operation. The fact that subjects in the Auto-il condition did relatively poorly suggests a residual influence of not having had manual in-the-loop experience. However, the fact that the Auto-ool subjects did especially poorly in the NOPASS conditions suggests that the experience of focusing attention on critical cues also plays an important role.

DISCUSSION

The above results indicate that human-computer interactive (Supervisory) control leads to better throughput than unaided (Manual and Hybrid) control under normal conditions, as Dunkler et al (1988) suggested. However, whether human-computer interactive performance is better than computer alone under normal conditions depends on the quality of the computer's performance. Humans could improve the performance of a simple algorithm (FCFS) but found it considerably more difficult to improve the performance of a more powerful algorithm (NOPASS). Therefore, the results are only partly in accordance with the findings of Dunkler et al (1988) and other studies that claim overall benefits for human-computer interactive scheduling (Sanderson, 1989).

Critically, the results also show that it is possible for human-computer interactive performance to be *worse* than computer alone—even with quite experienced subjects—if improvement requires highly skilled, error-free intervention by humans. Further analyses indicate that successful FCFS subjects tend to intervene more often and more accurately over sessions, whereas successful NOPASS subjects learn to intervene less with the more powerful NOPASS algorithm, reserving intervention for the few situations where it is genuinely needed. Therefore, the NOPASS subjects not only have to learn to identify goal-relevant cues and connect those cues to action (Brunswik, 1943; Dawes, 1975; Kirlik, Miller, & Jagacinski, 1993), but they also have to distinguish between cues that should lead to action *and cues that should lead to inaction*. These differential adaptations to a controller with which one is coupled are loosely analogous to findings on the flexibility of the human transfer function in classical control theory (Sheridan & Ferrell, 1974).

When looking at the effects of scheduling logic errors, no direct comparisons can be made with the Shiff (1983) and Endsley and Kiris (1995) studies because of the different domains used and the fact that their studies transferred subjects to fully human-in-the-loop conditions rather than supervisory conditions where the impact of occult scheduling errors were examined. However, those studies and the present one share the finding that out-of-the-loop experience can lead to problems at transfer. Overall, there is a tendency for Automatic conditions subjects to compensate least effectively for software failures, suggesting some residual inability to connect cues to action. However, the fact that these results are not strong in the present study could be due to the "equalizing" effect of the initial transfer to Supervisory control before the faults were introduced, during which Manual subjects learned to identify goal-relevant cues and Automatic condition subjects learned at least partly to connect cues to action.

However, an important further possible reason for the lack of differences between conditions at transfer is that discrete control systems do not necessarily require corrective action at a precise moment in time, but instead within a "window of opportunity," during which a later action is just as effective as an earlier action, leaving time for less effective operators to work out what to do. In general, the ability to identify goal-relevant cues and/or connect cues to action appears to be "buffered" by the windows of opportunity that discrete processes provide, suggesting that qualitatively different figures of merit may be needed when examining human performance in discrete systems.

REFERENCES

Brunswik, E. (1943). Organismic achievement and environmental probability. *Psychological Review, 50*, 255-272.

Curry, R., & Ephrath, A. (1977). Monitoring and control of unreliable systems. In T.B. Sheridan & G. Johannsen (Eds.), *Monitoring behavior and supervisory control*. New York: Plenum.

Dawes, R.M. (1975). The mind, the model, and the task. In F. Restle, R.M. Shiffrin, N.J. Castellan, H.R. Lindman, & D.P. Pisoni (Eds.), *Cognitive Theory, 1*. Hillsdale, NJ: LEA.

Dunkler, O., Mitchell, C., Govindaraj, T., & Ammons, J. (1988). The effectiveness of supervisory control strategies in scheduling flexible manufacturing systems. *IEEE Transactions on Systems, Man, and Cybernetics, SMC-18*, 223-237.

Endsley, M., & Kiris, E. (1995). The out-of-the-loop performance problem and level of control in automation. *Human Factors, 37*, 381-394.

Hughes, D. (1989) Glass cockpit study reveals human factors problems. *Aviation Week and Space Technology, August 7*, 32-36.

Hwang, S.-L., & Salvendy, G. (1988). Operator performance and subjective response in control of flexible manufacturing systems. *Work and Stress, 2*, 27-39.

Kessell, C. & Wickens, C.D. (1982). The transfer of failure detection skills between monitoring and controlling dynamic systems. *Human Factors, 24*, 49-60.

Kirlik, A., Miller, R.A., & Jagacinski, R.J. (1993). Supervisory control in a dynamic and uncertain environment: A process model of skilled human-environment interaction. *IEEE Transactions on Systems, Man, & CyberneticsI, I231, 929*

Moray, N. (1986). Monitoring behavior and supervisory control In K. Boff (Ed.), *Handbook of perception and human performance*. New York: Wiley.

Sanderson, P. M. (1989). The human planning and scheduling role in advanced manufacturing systems: An emerging human factors domain. *Human Factors, 31*, 635-664.

Sharit, J. (1985). Supervisory control of a flexible manufacturimg system. *Human Factors, 27*, 47-59.

Sheridan, T. & Ferrell, R. (1974). *Man-machine systems* Cambridge, MA: MIT Press.

Sheridan, T.B. (1987). Supervisory control. In. G. Salvendy (Ed.), *Handbook of Human Factors*. New York: Wiley.

Shiff, B. (1983). *An experimental study of the human-computer interface in process control*. Unpublished thesis, University of Toronto, Department of Industrial Engineering, Toronto, Canada.

Shneiderman, B. (1988). *Designing the user interface*. New York: Addison Wesley.

Wickens, C.D. (1992). *Engineering psychology and human performance*. New York: Harper-Collins.

Wiener, E. (1988). Cockpit automation. In E.L. Eiener & D.C. Nagel (Eds.), *Human factors in aviation*. San Diego: Academic Press.

WORK DOMAIN ANALYSIS OF A PASTEURIZATION PLANT: USING ABSTRACTION HIERARCHIES TO ANALYZE SENSOR NEEDS

Dal Vernon C. Reising and Penelope M. Sanderson
Department of Mechanical and Industrial Engineering
University of Illinois at Urbana-Champaign

This paper discusses the use of Rasmussen's abstraction hierarchy (AH) in performing an analysis of the work domain of a Pasteurization plant. Our goal is to examine the strengths and weakness of ecological interface design (EID) for systems in which critical variables are unreliable, faulty, or not measurable. In this paper we report our use of AHs to analyze the functioning of a pasteurization plant and the impact of unreliable or faulty sensors on the intelligibility of information available to the human operator. Although there is considerable current interest in EID, building thorough AHs for complex systems is a large task and there are currently very few detailed published examples to help human factors professionals wishing to take this approach. In this paper we present details of how we built the Pasteurization plant AH and show we are using the AH in our research. We argue that because AHs can indicate information that should be displayed for a process to be intelligible, techniques like EID that use AHs are in a position to bring about a profoundly user-centered approach to system design—an approach in which the ultimate information needs of human controllers will drive the engineering agenda of sensor and instrumentation design in a feedforward manner.

INTRODUCTION

The field of cognitive engineering has been strongly influenced by the systems-oriented work of Rasmussen (Rasmussen, Pejtersen, and Goodstein, 1994) who has emphasized the primacy of analyzing a work domain before proceeding to information system design, display design, allocation of function, training, etc. A critical tool proposed by Rasmussen (1986) and Rasmussen et al (1994) for work domain analysis is the abstraction hierarchy (AH). An AH is a layered description of the functional relations in a system, moving from the overall functional purpose of the system down through layers describing how the purpose has been achieved (abstract, generalized, and physical function levels) to the lowest level, at which elements of the physical form of the system are identified.

The AH's centrality in display design is evident in, for example, ecological interface design (EID: Dinadis and Vicente, 1996; Bisantz and Vicente, 1994; Vicente and Rasmussen, 1990). However, an important practical problem is that at present there are very few published examples of AHs that are sufficiently detailed to really drive the process of design. Most of Rasmussen's published examples show the five levels of the hierarchy (functional purpose, abstract function, generalized function, physical function, and physical form) in which

the items at each level are usually unconnected either within or between levels. This means that topology within a level, and means-ends relations between levels, are not directly shown. However, Vicente and Rasmussen (1990) and Bisantz and Vicente (1994) have offered detailed and informative AH-based work domain analyses of the DURESS system and there is a community of people trying to use the approach as a strong basis for information system design. Therefore, in this paper we wish to do the following.

1. Argue that if AHs are to be a viable foundation for work domain analysis, cognitive engineers have to have a good idea of how AHs speak to information system design. Cognitive engineers will also benefit from a corpus of examples. The work described herein is intended to contribute to that corpus.

2. Present the development of an AH based on a real physical pasteurization system, now instantiated as a microworld model of pasteurization, rather than on a microworld with no prior physical equivalent.

3. Present different forms of the Pasteurizer AH to highlight different properties of the Pasteurizer and explore the way the AH can support different information needs of the analyst.

4. Explore implications of information accessibility and sensor reliability (Vicente and Rasmussen, 1992) on display design. (Some information such as energy and entropy is inherently unmeasurable and must be derived

from variables such as temperature, pressure, etc. Other information is in principle measurable, but current sensor technology does not allow us to measure it directly or completely.)

5. Argue that operators' information needs should drive sensor and instrumentation engineering research.

PASTEURIZER ABSTRACTION HIERARCHY

The Pasteurization plant that served as a stimulus for this research was initially developed as a microworld by Moray for studies in trust (Lee and Moray, 1992; Muir and Moray, 1987). We have consulted with colleagues who work with a physical pasteurization plant to gain real-world detail in our analytic work. See Figure 1 for a mimic diagram of the Pasteurizer.

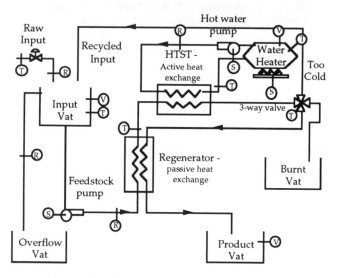

Figure 2. A mimic diagram of the Pasteurizer II system. Main components are labeled. Sensor type and location are indicated by the circled letters: T is a temperature sensor, V is a volume sensor, R is a volume flow rate sensor, S is a setting indicator. In this diagram, the "minimal" configuration of sensors is shown.

Building our Pasteurizer AH has been an iterative process, during which several preliminary versions were produced before the version to be presented here. Moreover, thermodynamics and heat transfer references had to be intensively consulted (Schmidt, Henderson, and Wolgemuth 1993). Overall, our approach is not a prescription for building AHs, but instead represents what we found useful given our research question and the nature of the Pasteurizer itself (for other representational formats, see Dinadis and Vicente, 1996; Bisantz and Vicente, 1994; Rasmussen, Pejtersen and Goodstein, 1994; Vicente and Rasmussen, 1990). We wished to be able to specify, and to visualize within the one representation, means-ends relations *between* different levels of abstraction and topological relations *within* levels of abstraction. We also wanted to be able to see the effect of state changes and sensor failures, and to infer implications for display design. To achieve this, we

produced a series of graphs, each emphasizing different properties of the same AH.

The Pasteurizer can be either successfully producing product, rejecting output that is too hot, or recycling output that is too cold (see Figure 1). In the first two cases, because fluid leaves the system there is a clear mass and energy sink, whereas in the third case the too-cold output is recycled into the inflow, preserving its mass and energy for continued use. The AHs presented here represent the Pasteurizer when successfully producing pasteurized product with no wastage; thus all nodes relating to overflowing, burned, waste, or recycled liquid at the abstract and generalized function levels have been dimmed.

Figures 2 shows a detailed AH for the Pasteurizer, in which means-ends relations ("why-how" relations between levels) only are shown. Levels of abstraction are indicated by the initials at the left of the figure. Mass and energy properties are shown at the abstract function level, material and flows at the generalized function level, and the functional properties of physical elements such as vats, pumps, heaters, and conduits at the physical function level. Properties related to each other are connected between AH levels. In addition, we have also developed AH representations of the Pasteurizer that show (1) topological relations *within* levels and (2) the aggregation of properties within levels, but because of space limitations these variants cannot be reproduced here. However, all these AHs formed the foundation for analytic work on sensors, to be described in the next section.

SENSOR FAILURES AND SENSOR UNAVAILABILITY

Some important uses of the AH that have not yet been widely advocated are as follows (although the underlying issues have been pointed out—see Moray, Lee, Vicente, Jones, and Rasmussen, 1994; Vicente and Rasmussen, 1992).

1. *Sensor technology limitations.* Two kinds of sensor limitations are sensor accessibility and sensor reliability (Vicente and Rasmussen, 1992). To date, research on EID has assumed that the variables and constraints that should be displayed will be measurable.

2. *Derivations based on inaccessible or unreliable information.* EID advocates displaying higher-order variables and relationships for a system, such as mass and mass balance, and energy and energy balance. However, these variables and relationships have to be derived from lower-order variables such as volume, temperature, density, and specific heat, and so are only as good as the data available.

3. *Impact on interface content and form when sensors used for supporting the interface fail.* Little has been written about designing robust displays in the face of sensor failures other than to highlight the significance of evaluating for robustness and being aware of the

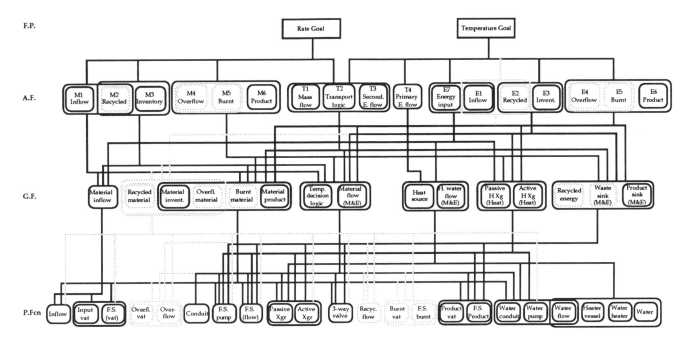

Figure 2. *The means-ends or between level AH representation of the Pasteurizer. The links between nodes of different abstraction levels indicate either the "why" relation, the "how" relation or both for the corresponding (aggregated) nodes.*

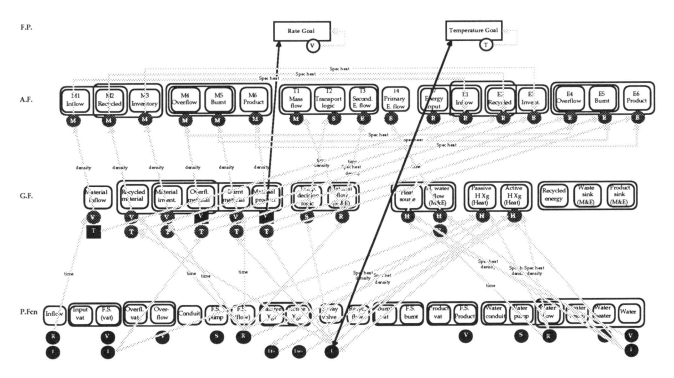

Figure 3: *The abstraction hierarchy representation that shows the relationship between a minimal set of sensors and the higher-order variables. Labels on these arrows indicate the relevant physical constants needed to derive the particular higher-order variable:* (1) *Volume(V) = time * Rate(R); (2) Heat transfer(H) = specific heat * density * Rate(R) * ΔTemperature(T); (3) Mass(M) = density * Volume(V); (4) Energy(E) = specific heat * Mass(M) * Temperature(T) or Energy(E) = specific heat * density * Rate(R) * Temperature(T); (5) S indicates a setting for a control element.*

potential failure modes of configural displays (Moray, Lee, Vicente, Jones, and Rasmussen, 1994).

The above three issues are the subject of the first author's ongoing Ph.D. dissertation (Reising, 1996) in which the following specific questions are being posed.

1. Will the configural displays typical of EID be compromised if the sensors they are based on fail or are unreliable?

2. Will representing the higher level information suggested by the EID framework aid in sensor failure detection?

3. Will derived higher order variables be more or less effective in supporting control and diagnosis than the same variables directly sensed?

4. Will alphanumeric or graphical formats lead to more accurate diagnosis of sensor abnormalities?

Working AH representation. As work proceeds towards the above goals, we have used AH representations of the Pasteurizer to help us describe the effects of sensor unavailability, unreliability, or failure on the information available to the operator and to guide the design of ecological displays (the latter is not the focus of the present paper).

As an example of our use of the AH to analyze sensor issues, Figure 3 shows an AH that was used to determine the effects of sensor failures or sensor unreliability on the derivation of higher-order variables. Note that properties related to each other at each level of abstraction have been aggregated, certain within-level relations have been introduced, and the full set of possible within- and between-level connections has not been displayed. Sensors and/or derived information are indicated by black and gray nodes. Figure 3 shows the impact of the *minimal* sensor set version of the Pasteurizer that was shown in Figure 1. The higher levels of abstraction in Figure 3 mostly contain variables that are derived mathematically (indicated by grayed sensor nodes) from low level variables that are directly sensed (indicated by solid black sensor nodes).

The AH representation in Figure 3 allows us to trace the effect of a compromised sensor on all variables dependent on it. For example, the temperature at the three-way valve (black circle "T" at bottom center of figure) is clearly used to derive a great deal of heat, mass, and energy information at higher levels of abstraction. However, such a sensor configuration will not lead to an interface that readily supports diagnosis and control during unanticipated abnormal system states. For example, if there is energy loss between the passive and active heat exchangers, with the minimal sensor set there will be no way to determine the actual heat exchange occurring in the active heat exchanger (see also Figure 1). Figure 3 can be contrasted with an AH developed under the assumption that a maximal set of sensors is available. There is insufficient space here to reproduce the latter, but it can be found in Reising (1996). Specifically, having a maximal sensor set removes

the need to derive some higher-level variables, such as energy loss between passive and active heaters, so limiting the propagation of wrong information. This should greatly increase human operators' ability to control and interpret the system.

AHs of this kind show where sensors need to be placed, they indicate what the sensors must be measuring, and they help us think about how higher-level variables should be derived from lower-level sensed information. Such analyses also open up software engineering issues concerning how derived information should be sent to graphical displays in such a way that sensor compromise does not lead to "visual pathogens"—in other words, confusing disturbances to graphical arrangements that reflect inconsistent underlying data. Finally, although this has not been the focus of the present paper, the AHs are a crucial part of the generation of ecological displays.

ENGINEERING INSTRUMENTATION AGENDA

As the above examples suggest, an AH representation can point to the information needs that should be supported in interface design. EID, which uses AHs, is therefore in a position to bring about a profoundly user-centered approach to system design in which the ultimate information needs of human controllers drive the engineering agenda of sensor and instrumentation design in a feedforward manner.

Two real-world examples of the instrumentation problem are as follows. First, in nuclear power plant control a useful variable to display would be the departure from nucleate boiling ratio (DNBR) in the reactor core. However, because of the harsh conditions of the core, no current sensors exists that could give accurate, sustained measures of this variable. Second, in nuclear power plant control the neutron flux of the reactor core must be monitored. Although there are sensors located around the core, there are no sensors in the center of the core to provide continuous, accurate indications of the neutron flux throughout the various quadrants of the reactor core. Instead, sensors that measure the flux are dropped down into the core at regular intervals, from which neutron flux profiles are generated and updated. Because of the importance of DNBR and neutron flux in achieving higher-level functions, engineering more direct means of instrumentation them should be a high priority.

GENERATING ABSTRACTION HIERARCHIES

In conclusion, we make some observations about building the Pasteurizer AH that should generalize to other such efforts.

1. Building an AH is a highly iterative process— one that can probably benefit from considerable *general*

expertise in thinking in AH terms and performing such analyses.

2. To build the AH for a thermal system, sound basic knowledge in thermodynamics and heat transfer is needed. Clearly, seeking the aid of domain experts is crucial whatever the domain.

3. The thermodynamic "system" represented by the Pasteurizer has three discrete configurations because of the three-way valve, having no (significant) energy sink when fluid is being recycled, but having a sink when fluid is being sent to external vats. Such non-linearities are difficult to represent easily.

4. The AH must to be molded to a form that will clearly lead to implications for system design. During our analyses we continually asked whether our AHs would let us make inferences about what operators' information needs will be and the (in)adequacy of sensor information.

5. In order to build the AH confidently we found it necessary to be considerably more specific about the physical plant itself than was available in the initial microworld, so that we effectively designed a physical plant. In the case of a preexisting physical plant or system, most information of this kind would already be available.

6. Keeping track of links in the AH quickly became confusing. Many links overlapped visually, making it hard to trace relations and keep mass- and energy-related connections separate. To help in the generation of AHs we used MacFlow, but could not easily perform quick rearrangements and queries of the AH structure. It has been argued that the AH would be more widely used and understood if an AH-building software environment were developed in which hierarchies under development could be fluently manipulated both visually and logically (Sanderson, 1994). We welcome emerging efforts to produce such tools.

REFERENCES

Bisantz, A. M. & Vicente, K. J. (1994). Making the abstraction hierarchy concrete. *International Journal of Human-Computer Studies, 40(1)*, 83-117.

Dinadis, N., & Vicente, K. J. (1996). Ecological interface design for a power plant feedwater system. *IEEE Transactions on Nuclear Science. 43(1)*, 266-277.

Lee, J. D., & Moray, N. (1992). Trust, control strategies, and allocation of function in human-machine systems. *Ergonomics, 35*, 1243-1270.

Moray, N., Lee, J., Vicente, K. J., Jones, B. G., & Rasmussen, J. (1994). A direct perception interface for nuclear power plants. In *Proceedings of the Human Factors and Ergonomics Society 38th Annual Meeting* (pp. 481-485). Santa Monica, CA: Human Factors and Ergonomics Society.

Muir, B., & Moray, N. (1987). Operators' trust in relation to system faults. *Proceedings of the IEEE International Conference on Systems, Man, and Cybernetics*. Alexandra, VA, October.

Rasmussen, J. (1986). *Information processing and human-machine interaction: An approach to cognitive engineering*. New York: North-Holland.

Rasmussen, J., Pejtersen, A. M., & Goodstein, L. P. (1994). *Cognitive Systems Engineering*. New York: John Wiley & Sons.

Reising, D. C. (1996). *The effect of sensor availability, reliability, and failure on ecological interfaces designed for supervisory control of continuous process control systems*. Ph.D. Preliminary Examination proposal, Dept. of M.& I.E., University of Illinois at Urbana-Champaign. (Also available as Engineering Psychology Research Laboratory technical report EPRL-96-0).

Sanderson, P.M. (1994). *Requirements for ICE2*. Unpublished document, Engineering Psychology Research Laboratory, University of Illinois at Urbana-Champaign, April, 1994.

Schmidt, F. W., Henderson, R. E., & Wolgemuth, C. H. (1993). *Introduction to Thermal Sciences: Thermodynamics, Fluid Dynamics, Heat Transfer* (2nd Ed.). New York: John Wiley & Sons.

Vicente, K. J. & Rasmussen, J. (1990). The ecology of human-machine systems II: Mediating "direct perception" in complex work domains. *Ecological Psychology, 2(3)*, 207-249.

Vicente, K. J. & Rasmussen, J. (1992). Ecological interface design: Theoretical foundations. *IEEE Transactions on Systems, Man, and Cybernetics, 22(4)*, 589-606.

COMPUTERS AND TELECOMMUNICATIONS IN THE YEAR 2000 - VIRTUAL ENVIRONMENTS AND INTELLIGENT TRANSPORTATION SYSTEMS

Kay M. Stanney
University of Central Florida
Orlando, FL

Robert S. Kennedy
Essex Corporation
Orland
William Yurcik
University of Pittsburgh
Pittsburgh, PA

In this, the third of three sets of position papers for the CTG-CSTG co-sponsored symposium on Computers and Telecommunications in the Year 2000, the authors provide us with a description of the future in two relatively new areas - virtual reality and intelligent transportation systems. In the first of the two papers, Kay M. Stanney and Robert S. Kennedy describe the performance efficiency, health and safety, and social issues that virtual technology will have to resolve, if virtual environments are to become commercially viable. In the second paper, William Yurcik describes the planned components of smart cars and smart highways , as well as the barriers we currently face in deploying associated intelligent transportation systems.

Human Factors Issues Associated with Virtual Environments Technology

Kay M. Stanney
University of Central Florida
stanney@fems.engr.ucf.edu
and
Robert S. Kennedy
Essex Corporation
rkennedy@msis.dmso.mil

Virtual Reality (VR) technology will be used to advance many fields, including medicine, education, design, training, and entertainment. The reality is, however, a considerable amount of systematic research must be done before VR technology receives widespread use. If VR systems are to be effective and well received by their users, researchers need to focus significant efforts on addressing a number of human factors issues. This position paper will explore many of these human factors issues, including: 1) human performance efficiency in virtual worlds, which is likely influenced by task characteristics, user characteristics, human sensory and motor physiology, multi-modal interaction, and the potential need for new design metaphors; 2) health and safety issues; and 3) the social impact of the technology.

Human Performance Efficiency in Virtual Worlds

A fundamental advance still required for Virtual Environments (VEs) to be effective is to determine how to maximize the efficiency of human task performance in virtual worlds. While it is difficult to gauge the importance of the various human factors issues requiring attention, it is clear that if humans cannot perform efficiently in virtual environments, then further pursuit of this technology may be fruitless. Focusing on understanding how humans can perform most effectively in VEs is thus of primary importance in advancing this technology. Human performance in VEs will likely be influenced by several factors, including: task characteristics; user characteristics; design constraints imposed by human sensory and motor physiology; integration issues with multi-modal interaction; and the potential need for new visual, auditory and haptic design metaphors uniquely suited to virtual environments.

Task Characteristics

Some tasks may be uniquely suited to virtual representation, while others may not be effectively performed in such environments. It is important to determine the types of tasks for which VEs will be appropriate. In order to obtain this understanding the relationship between task characteristics and the corresponding virtual environment characteristics which effectively support their performance (e.g., stereoscopic 3D visualization, real-time interactivity, immersion, etc.) must be attained.

User Characteristics

Significant individual performance differences have already been noted in early studies. User characteristics that significantly influence VR experiences need to be identified in order to design VR systems that accommodate the unique needs of users. In order to determine which user characteristics are influential in VEs, one can start by examining studies in human-computer interaction and from there perform the necessary additional empirical work.

Design Constraints Imposed by Human Sensory and Motor Physiology

In order for designers to be able to maximize human efficiency in VEs it is essential to obtain an understanding of design constraints imposed by human sensory and motor physiology. Without a foundation of knowledge in these areas there is a chance that the multi-modal interactions provided by VE systems will not be compatible with their users. Such design incompatibilities could place artificial limits on human VE performance. The physiological and perceptual issues which need to be explored for their implication to the design of multi-modal VEs, include: visual perception, auditory perception, and haptic perception.

Integration Issues with Multi-Modal Interaction

While developers are focusing on synthesizing effective visual, auditory, and haptic representations in virtual worlds, it is also important to determine how to effectively integrate this multi-modal interaction. One of the aspects that makes VEs unique from other interactive technologies is its ability to present the user with multiple inputs and outputs. This multi-modal interaction may be a primary factor that leads to enhanced human performance for certain tasks presented in virtual worlds. There is currently, however, a limited understanding on how to effectively provide such sensorial parallelism.

Virtual Environment Design Metaphors

It is known that well-designed metaphors can assist novice users in effectively performing tasks in human-computer interaction. Thus, designing effective VE metaphors could similarly enhance human performance in virtual worlds. Unfortunately, at the present time many human-VE interface designers are using old metaphors (e.g., windows, toolbars), that may be inappropriate for human-virtual environment interaction.

HEALTH AND SAFETY ISSUES IN VIRTUAL ENVIRONMENTS

Maximizing human performance in VEs is essential to the success of this technology. Of equal importance is ensuring the health and welfare of users who interact with these environments. If the human element in these systems is ignored or minimized it could result in discomfort, harm, or even injury.

It is essential that VE developers ensure that advances in VE technology do not come at the cost of human well being. Two critical unresolved health and safety issues are 1) potential "cybersickness", a form of motion sickness experienced in virtual worlds, and 2) transfer of maladaptive cognitive and/or psychomotor compensations from VR to real world environments (e.g., ataxia, degraded hand-eye coordination, reduced visual functioning) with, as yet, unknown adverse legal, economic, individual, and social consequences. Users of VE systems generally experience various levels of sickness ranging from headaches to severe nausea. Early studies indicate that 85% to 95% of the general public may experience some level of these symptoms. Currently, however, VE system designers cannot avoid this issue because the exact causes of cybersickness are not known. In fact, while there are many suggestions about the causes of motion sickness, to date there are no definitive predictive theories of cybersickness. In addition, the lingering deleterious aftereffects could potentially compromise the safety of users at post exposure. At present, however, the time-course of these aftereffects is not known. Thus, in order for VE systems to receive wide-spread use, urgent and appropriate multisensory human factors research clarification of the potential central nervous system interactive sources of cybersickness and transfer of maladaptive cognitive and psychomotor compensations from VR to real world environments is required.

THE SOCIAL IMPACT OF VIRTUAL TECHNOLOGY

While researchers are often concerned about human performance and health and safety issues when developing a new technology, an often times neglected effect of new technologies is their potential social impact. Virtual reality is a technology, which like its ancestors (e.g., television, computers, video games) has the potential for negative social implications through misuse and abuse. Its higher level of user interaction may even pose a greater threat than past technologies. Currently, however, the potential negative social influences resulting from VE exposure are not well understood. Unfortunately, early studies indicate that there is reason to be concerned. Interaction with violent VEs has been shown to increase aggressive thought content, while catharsis was not found. These issues need to be proactively explored in order to circumvent negative social consequences from human-virtual environment interaction.

CONCLUSIONS

This symposium presents many of the human factors issues which must be addressed in order for VR technology to reach its full potential without inflicting harm along the way. VR technology promises to permeate both professional and personal aspects of our lives. If this influx is to be a positive influence rather than a forceful intrusion, it is essential that each of these human factors issues receive significant systematic research.

Position Paper on the Future Applications of

Telecommunications in

Intelligent Transportation Systems

William Yurcik
University of Pittsburgh
{yurcik@tele.pitt.edu}

The answer to future transportation problems lies in better management of the highway capacity that currently exists. Congestion and safety can be addressed through the application of telecommunications technology integrated with vehicle and highway sensors and computers.

Smart Cars

"Smart cars" of the future will contain one or all of the following information systems which focus on increasing the efficiency and safety:

(1) longitudinal collision avoidance systems address accidents in which vehicles are moving in parallel paths prior to collision or in which one struck vehicle is stationary. The systems will use radar and sonar to warn drivers or take action in cases of insufficient distances between vehicles in front (following too close) or in back (approaching vehicle). Adaptive cruise control systems are being developed to allow vehicles to *lock* onto a lead vehicle and form a cohesive *platoon* of vehicles all traveling in the same path.

(2) lateral collision avoidance systems augment a driver's ability to avoid angular collisions including warnings to the driver or emergency steering/throttle control of the vehicle. Sensors for blind spots or imminent collisions will be signaled to the driver. Advanced versions may include vehicle-to-vehicle communications notifying of lane change events.

(3) intersection collision avoidance systems provide warnings of imminent collisions with crossing traffic, as well as warnings of control signals (either a stop sign or a traffic signal) in the intersection ahead.

(4) vision enhancement systems provide a heads-up windshield display when driving visibility is low (such as nighttime and foggy conditions) by using infrared and computer vision technology to decisively define the lane of travel and significant edges.

(5) safety readiness systems monitor the condition of the driver, vehicle, and road surface. These systems look for psychophysiological features of the driver for impairment indications which have been identified as features of unsafe driving (i.e. drift-and-jerk steering, alcohol level, etc...). Safety readiness systems also extend monitoring internal to the vehicle from standard features (i.e. oil level, oil pressure) to additional features (i.e. tire pressure, brake temperature) and external to the vehicle to monitor road surface conditions and tire traction.

(6) pre-collision restraint deployment systems receive information from collision avoidance systems to configure and activate passenger safety systems prior to actual impact. For example, airbag(s) may be inflated differently depending on the number of passengers and where the passengers are located within the car.

Smart Highways

Highways of the future will have the following characteristics:

(1) the ability to accept smart cars on smart highway segments and any vehicles on general purpose highway segments

(2) smart highway design characteristics consistent with general purpose highways so that smart highways can also be used when automation is off

Drivers will enter the smart highway segment through a check-in area where officials will only accept certified smart cars. The driver will transmit the desired destination to the smart highway by voice (if driving) or touchpad (if stopped). For commuters the destination may be predetermined and the driver need only to confirm. The smart highway then assumes control of the approved vehicle merging with other smart highway traffic. The smart highway assumes fully automated control of the entire vehicle as part of an aggregate traffic flow. When the destination is reached, the system moves the vehicle to an off-ramp where control is returned to the driver after the driver's ability to resume control has been demonstrated (wake up the driver!). In extreme cases of malfunction, the default state will be to bring all vehicles to an immediate stop where manual control can be resumed.

Traffic Management

Traffic management to avoid traffic congestion includes monitoring the flow, identifying and interpreting disruptions, and attempting to change traffic patterns. To keep track of traffic, electronic sensors embedded in highways and cameras allowing visual monitoring will be employed. Future systems will detect changes in traffic flow and automatically alter signs and traffic control signals at highway entry ramps to divert or smooth the stream of vehicles.

Enroute Traveler Information Systems

Enroute traveler information systems will consist of a vehicle computer containing detailed maps and programmed information linked to a Global Positioning System (GPS). The driver will state (via keyboard or voice) a desired destination and receive map information highlighting vehicle location and locations of desired landmarks (rest stops, gas stations, etc...).

Enroute traveler information systems will also include driver advisory and sign notifications. Driver advisories will be provided as audible messages. Highway sign information will be presented to drivers as voice output and/or a heads-up visual display. Such use of audio/visual presentation to overcome hearing and visual impairments will make highways more accessible. Candidate technologies include variations of Highway Advisory Radio (HAR), FM subcarrier, microwave beacon, spread spectrum radio, cellular telephone, and transponder-based smart car <-> smart highway systems.

Route Navigation Systems

Route navigation is closely related to enroute traveler information since the two systems will rely on the same information, however route navigation systems process information into real-time directions for the traveler. Thus a map display is considered enroute traveler information while a route navigation system will use map information (supplemented by indications of traffic congestion) to suggest routes and give instructions.

There are a number of different procedures that can be used to determine the traveler's routing. Mobile-based systems will utilize programs based on the best information available to provide routing instructions to the traveler based upon customizable parameters provided by the traveler (time, avoid highways, etc...).

Interfaces

Information will be conveyed through heads-up displays where the instructions are projected onto the windshield and proprioceptive feedback through the steering wheel/accelerator pedal/brake pedal so the driver does not need to look away from the highway. Voice instruction is another means of providing information that does not require the traveler to divert his/her gaze. Similarly, speech recognition technology can permit the traveler to enter information into systems without requiring any diversion of sight or additional movements by the traveler. These systems will either be continuously active or only activated when a driver action is necessary.

Unintended Consequences

One unintended consequence of ITS will be that drivers may defer a disproportionate amount of safety concerns to automation. Drivers will initially seek a given level of vehicle safety but then compensate for safety features by driving carelessly. This reaction would greatly diminish the net benefits from ITS.

Another unintended consequence is false alarms. Methods for distinguishing true pre-collision incidents from non-collision incidents are important to maintain human confidence in the systems and to prevent systems meant to augment human limitations from confusing or distracting the driver or even causing an accident.

Lastly, driver education will need to be revolutionized because the improper use of technology will have different effects than those desired. For instance, pumping brakes under icy conditions is a standard drivers education technique to maintain vehicle control but vehicles with anti-lock brakes are already designed to automatically pulsate and actually have decreased traction on ice when brakes are pumped. I would expect that drivers of the future will learn appropriate driving techniques via computer simulations similar to current video arcade games.

Barriers to ITS Deployment

There are industry, societal, and economic barriers to the implementation of ITS that may prove to be the toughest hurdles.

Though the transportation industry is conservative, if it is possible to significantly improve transportation efficiency and safety to the public via ITS then it will only be a matter of time before these barriers will be successfully addressed.

The major industry barrier is conflicting standards. An ITS standards process must be initiated coupling industry with governments and universities so that timely standards can be set which balance the need for interoperability versus incentives for innovation.

The major societal barrier is liability. While ITS systems will be designed to be fault tolerant, even a remote possibility of litigation from catastrophic highway injuries has a chilling effect on organizations that need to be involved. Jurisdiction of litigation will be problematic in that the overwhelming majority of highways, even interstate highways, are owned by state and local governments. In the long term, this issue will be settled by cost/benefit analysis, whether the benefits of ITS outweigh the costs of potential liability litigation.

Lastly, the major economic barrier to the deployment of ITS is uncertain consumer demand. Vehicle information systems are likely to be optional items that add to the price of a vehicle unless they are mandated and/or subsidized by the government. Consumer willingness to buy "smart car" devices is tough to gauge, but external benefits gained from integration with highway and traveler information systems will make the price of "smart car" devices significantly below their functional value and thus very attractive.

Other Issues

Other closely related applications of telecommunications to ITS that are acknowledged but not covered include electronic toll collection, automatic truck weigh stations, and travel demand management kiosk systems.

SIMPLIFIED ENGLISH FOR AIRCRAFT WORKCARDS

Steven Chervak, Colin G. Drury and James P. Ouellette
State University of New York at Buffalo
Department of Industrial Engineering
Buffalo, NY 14260

For technical communications in international civil aviation maintenance, most manufacturers have adopted a restricted language: Simplified English (SE). This uses a standard vocabulary and syntax rules with the aim of improving understanding, particularly for people with restricted abilities in English. This paper describes the first test of the efficacy of Simplified English for comprehension of documentation used at the worksite by Aircraft Maintenance Technicians (AMTs). Sixteen workcards, representing two levels of difficulty (Easy and Difficult), two levels of language (SE and Non-SE) and two levels of document layout (standard and revised) were tested on 175 practicing AMTs in a between subject design using a comprehension test. Comprehension was significantly improved with Simplified English, particularly for the Difficult workcards and for non-native English speakers. No effects of layout were found.

INTRODUCTION

The importance of good document design practices to the writing of aircraft work control cards (workcards) has already been documented (Bohr, 1978; Patel, Drury and Lofgren, 1994). Experimental documents produced in the latter study had a better choice of case and font, a more consistent paragraph structure, and better integration of text with graphics: when evaluated by users, they were universally preferred. There are, however, issues in document design which go beyond layout and typography, and include the structure of the language itself.

Most major transport aircraft manufacturers now use Simplified English (SE) in their documentation. However, the impact of this restricted language on Aircraft Maintenance Technicians (AMTs) has not been directly measured. The current study provides such an evaluation to determine whether SE enhances (or degrades) comprehension of workcards by AMTs.

Simplified English. There have been various attempts to produce artificial languages to allow people of different countries to intercommunicate. For general use, the early twentieth century saw Esperanto and later Basic English (Ogden, 1932). More recently, restricted technical languages have appeared, such as Caterpillar Fundamental English (CFE) for documentation of agricultural vehicles, and Simplified English (SE) for documentation of procedures on commercial aircraft.

Issues in Evaluation. While restricted languages such as SE make logical sense, there is still a need to evaluate their effectiveness. Despite potentially-reduced ambiguity, there are still feelings among some technical writers that SE prevents them from expressing instructions in the most obvious manner. Restricted languages can appear restrictive to some. As documentation is designed for the user, the effect of SE on the AMT is the ultimate criterion. Hence direct evaluation of SE using actual workcards and practicing AMTs is the obvious step.

For SE, a major evaluation study by Shubert, Spyridakis, Holmback and Coney (1995), used a timed comprehension test. SE and Non-SE versions of two maintenance manual procedures were tested on 127 engineering students. The two procedures differed in a number of measures of writing complexity, while having comparable overall lengths. The comprehension test was timed and its performance was measured both by whether each question had the correct answer and whether the information used for the answer could be located correctly within the procedure.

Analyses were performed separately for native English speakers and non-native English speakers and for the two documents which differed in complexity. Measures of both comprehension and content location showed a significant effect of Simplified English and a significant Simplified English × Procedure interaction. The native English speakers scored higher than their non-native English speaking counterparts. Simplified English gave higher comprehension and location scores than non-Simplified English for the more complex procedure only. Performance time was not a significant factor, except that non-native English speakers were slower overall.

From these studies it was concluded that to evaluate a restricted language we must control both user native language and document complexity. In addition the evaluation should focus on accuracy of comprehension using a comprehension test based upon the documents themselves. "Accuracy of comprehension" should measure correctness of both comprehension questions and location questions.

METHODOLOGY

The basis of the methodology was to extend the comprehension test technique to the use of workcards by practicing AMTs. Differences from the Shubert, et al (1995) study were in the choice of subjects (AMTs versus students), levels of document complexity (four workcards versus two procedures) and the addition of two levels of workcard layout to provide a test performance of the Patel, et al (1994) results.

Choice of Workcards. Following discussions with computational linguists at Boeing Inc. and with Aerospace Industries Association of America (AIAA) Simplified English Committee members, it was decided to use actual examples of existing workcards in the evaluation. For two aircraft types, Boeing had produced workcards in

pre-SE maintenance manual language and had later modified these to Simplified English standards. Thus the workcards were realistic to AMTs and represented actual writing practice by those who write maintenance manual procedures. In this way, difficulties of translating Simplified English workcards back into artificial non-Simplified English versions was avoided.

Seventeen potential workcards were analyzed for possible inclusion in the study. The Boeing computational linguists and University of Washington technical communications researchers analyzed the non-SE versions of each in terms of total words, mean words per sentence, percentage passive voice, and Flesch-Kinkaid reading score. A task difficulty rating of each workcard by an experienced engineer was also used for guidance. Each of these variables was split at the median to be able to match workcards at high or low level of each variable. From this analysis, four workcards were chosen, two "easy" on all the measures and two "difficult." Within each pair the lengths were different, which would presumably mainly affect performance times, although document length could also affect comprehension through additional cognitive load.

Choice of AMTs. One hundred seventy-five licensed AMTs, from eight major air carrier maintenance sites were tested. Mean age was 37.7 years, and mean experience 13.2 years. The data from our sample is comparable with demographic data on aircraft mechanics (in all branches of aviation) compiled for 1988 by the Bureau of Labor Statistics (BLS). Wilcoxon tests of the median age in our sample shows that it was not significantly different from the BLS data (t = 7879, p > 0.50). For the experience distribution, the sample median was significantly greater than the BLS data (t = 10,142, p < 0.001), showing that our sample was more experienced than the AMTs of the earlier data. In particular, there were far fewer AMTs with three years or less experience, a finding probably representing reduced hiring patterns in major airlines during the 1990s.

The Accuracy Level Test was given to each of the 175 mechanics in order to determine their individual reading level. Scores are equivalent to reading grade levels. Carver (1987) provides norming data for this test for two appropriate comparison groups: freshmen undergraduate and beginning graduate students. The mean score of our sample (13.35) was significantly higher than for college freshmen (12.5) with t (174) = 6.95, p < 0.001. However, it was significantly lower than for graduate students (14.3) with t (174) = -7.85, p < 0.001. Thus the reading level of our AMT sample was typical of an educated adult group, i.e. above college freshmen but below graduate students.

Evaluation Procedure. All testing took place at airline maintenance facilities, in whatever room was made available. AMTs were tested individually or in groups depending upon their arrival times. Each AMT was given written instructions for completing:

1. A one-page demographic questionnaire
2. A test of reading comprehension, the Accuracy Level Test (Carver, 1987). This was a ten-minute timed vocabulary test which measured reading level as an equivalent grade level. This test has high reliability (0.91) measured on college students (Carver, 1987) and has a high validity (0.77 to 0.84) when compared to the Nelson-Denny Reading Test.
3. The workcard comprehension test. Completion time was measured with a stopwatch, with accuracy (comprehension plus location) scored later.

Each AMT was given one of the sixteen possible workcards. Workcards were distributed in order, with a different starting point at each carrier. For the comprehension test, each AMT was given the workcard and a set of questions (20 each for three workcards, 19 for the other). Generally, a question concerning a specific technical point was followed by a question asking where in the workcard this information was located. Questions demanded either a short answer, a "fill in the blank" or a multiple choice. Because the four workcards represented different procedures, there was no way to match the individual questions across workcards.

Experimental Design. This was a three factor factorial experimental design with AMTs nested under all three factors.

RESULTS

The first analyses presented here assess the effects of the three independent variables on the two performance measures, using selected performance predictors as covariates. Subsidiary analyses explore the role of native language. All analyses used analysis of variance or covariance procedures. The four possible individual variables which may affect performance, and therefore could be useful covariates, were: AMT experience, inspection experience, age, and reading ability score. An intercorrelation matrix of these and the two performance variables (time, accuracy) showed that inspection experience was uncorrelated with other variables and that AMT experience was highly correlated with age. The other two variables, age and reading ability were moderately correlated with time and accuracy. Correlation coefficients were calculated as 0.217 between age and task completion time and -0.158 between age and task completion accuracy. Age and reading ability were tested, singly and together, as covariates, and gave almost identical results. Only the analyses using age as a covariate are presented here for simplicity.

Table 1 shows a summary of the significant effects for task completion time and task completion accuracy. The covariate was significant in each case showing that times increased and accuracy decreased with increasing age. Both performance measures showed a significant workcard effect and a significant interaction between Simplified English/non-Simplified English and workcard. For times, each workcard had a somewhat different effect of Simplified English. Workcards Easy 1 and Difficult 2 gave slower performance, and the others faster performance. However, for accuracy (Figure 1) the effects were much clearer. For the two Easy workcards, there was no significant change in accuracy between Simplified English and non-Simplified English versions, but for the two Difficult workcards, Simplified English gave clearly superior accuracy.

In the Shubert, et al study it had been found that SE was of greatest benefit to non-English speakers, so that a similar test was appropriate in our study. Of the 175 AMTs tested, 157 were native English speakers and only 18 non-native English speakers. Because there was an even distribution of the 16 workcards to AMTs, nine non-native English speaking AMTs were given SE workcards and nine non-SE workcards. The number of non-native English speakers was too small for this characteristic to be used within the main ANCOVA, either as a covariate, or as a fourth factor. Hence, a separate ANOVA was performed with only two variables, each at two levels:

Language of workcard: SE or non-SE
Native language: English or non-English

Figure 1. Effects of Workcard and Simplifed English on Task Completion Accuracy

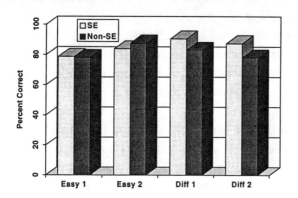

Only the AMT's native language affected task completion time significantly. Native English speakers took an average of 20.5 min while non-native English speakers took longer, an average of 24.7 min, to complete the comprehension test. Accuracy was different between the two types of English, between native and non-native English speakers, and for the interaction of these two factors. There was a clear superiority for Simplified English, with accuracy increasing from 76% to 86% overall. Equally important is the finding that the effect was most marked for non-native English speakers, where the improvement in accuracy was from 69% to 87%. Indeed, Simplified English allowed non-native English speakers to achieve about the same level of performance as native English speakers. Performing multiple comparisons among the four means shows that only the differences between the lowest mean (non-SE/non-native English speakers) and the other three were significant at $p = 0.05$.

Table 1. Significance levels for all factors and interactions in GLM ANOVA

Factor	Performance Measure	
	Task Completion Time	Task Completion Accuracy
Age (covariate)	p = .001	p = 0.006
Workcard (W)	p = .001	p = 0.004
SE/Non-SE (S)	not significant	not significant (p = 0.073)
Layout (L)	not significant	not significant
W × S	p = 0.011	p = 0.024
W × L	not significant	not significant
S × L	not significant	not significant
W × S × L	not significant	not significant

DISCUSSION

This large and realistic study measured the effects of SE across a range of AMT backgrounds, types of workcard and workcard layouts. The aim was to determine whether SE helps (or hinders) comprehension of workcard information, and whether it does so uniformly or in mainly particular circumstances. In doing so, it was intended to confirm and extend existing comprehension studies, and to make sound recommendations on the use of SE by the aviation maintenance community.

The major result was that SE was indeed useful, having a positive effect on comprehension accuracy without any consistent negative effect on speed of performance. On a representative sample of 175 practicing AMTs from sites across the USA, it was accuracy which was impacted by SE, showing that performance changes with SE should be in the direction of error reduction. In this aspect, the current work supports and extends that of Shubert, et al (1995), where comprehension correctness and content location were also the affected outcome measures.

In terms of which factors interacted with the SE factor, again previous research was confirmed and extended. Both the native language of the AMT and the complexity of the workcard interacted with the SE/Non-SE factor. The effect of SE was to improve the accuracy by about 2% for native English speakers, but by about 18% for non-native English speakers. If we consider error rates, the inverse of accuracy rates, the results look even more dramatic, as shown in Table 2.

Table 2. Error rates across native language

Language Spoken	Non-Simplified English	Simplified English
Native English (157)	17%	15%
Non-Native English (18)	31%	13%
Whole Sample (175)	18%	14%

The conclusion is simple and direct: Simplified English workcards allowed non-native English speakers to achieve the same level of performance as native speakers: the non-Simplified English versions of the workcards did not.

An analogous effect was seen for the interaction between workcard and Simplified English. The two Difficult workcards were the only ones where Simplified English made a significant difference. Again, in terms of error rates, we have the data in Table 3.

Table 3. Error rates for workcards

Type of Workcard	Non-Simplified English	Simplified English
Easy Workcards	17%	19%
Difficult Workcards	18%	11%

Here, for the easy workcards there was no difference between Simplified English and non-Simplified English, but for the difficult workcard the errors were reduced by a third.

Overall, Simplified English proved to have the most effect where the most effect was needed, i.e. for those whose native language was not English and where the material was more difficult. Because of our large and varied sample, this result is generalizable across a range of age and experience levels, and appears independent of the particular make of aircraft with which the AMT is familiar.

CONCLUSIONS

1. Aircraft manufacturers, and technical operations departments in airlines, can use SE for workcards, and be confident that it will improve comprehension accuracy.

2. The effectiveness of SE is greatest where it is most needed: for non-native English speakers and for difficult workcards. Under more favorable conditions, i.e. with native English

speakers and easier workcards, SE will not adversely affect performance.

REFERENCES

Bohr, E. (1978). Application of instructional design principles to nuclear power plant operating procedures manuals. In Easterby, R. and Zwaga, H. (Eds), *Information Design*, New York: John Wiley & Sons Ltd, 517-577.

Carver, R. P. (1987). *Technical Manual for the Accuracy Level Test*. REVTAC Publications, Inc.

Ogden, C. K. (1932). *Basic English, A General Introduction with Rules and Grammar*. London: Paul Treber & Co., Ltd.

Patel, S., Drury, C. G. and Lofgren, J. (1994). Design of workcards for aircraft inspection. *Applied Ergonomics 1994*, **25(5)**, 283-293.

Shubert, S. K., Spyridakis, J. H., Holmback, H. K. and Coney, M. B. (1995). The comprehensibility of Simplified English in Procedures. *Journal of Technical Communications*, in press.

EFFECTS OF SPEECH INTELLIGIBILITY, MORPHOLOGICAL CONFUSIONS, AND REDUNDANCY ON TASK PERFORMANCE

Leslie A. Whitaker
University of Dayton

Mike McCloskey
Klein Associates, Inc.

Leslie J. Peters
Tripler AMC

Speech is a vital means of communication for completing many tasks. The speech intelligibility needed for successful communication may be degraded by ambient noise levels, poor communication equipment, or hearing impairments. The present research tested the impact of speech message content on task performance under conditions of degraded speech intelligibility. Sixteen subjects participated in a laboratory experiment using on a board game. Message redundancy, morphological confusions, and speech intelligibility were varied. Task performance and subjective workload were measured. Morphological confusions adversely affected performance (both speed and accuracy) and subjective workload (SWAT) to an increasingly greater extent as speech intelligibility decreased. High redundancy improved the accuracy of performance and subjective workload when speech intelligibility decreased; however, high redundant messages were longer and required more time to process than low redundant messages. These results extend earlier work which measured the impact of morphological confusions and redundancy on speech intelligibility itself to the measurement of their impact on performance. Implications for the development of message content guidelines and their impact on performance are discussed.

Speech communication is important to the successful completion of many tasks when information must be transmitted between team members. Unfortunately, there are many reasons that information may not be clearly transmitted by speech. For example, ambient noise in the environment can interfere; communication systems like radio or telephone can add their own noise or distortion to speech; and a listener's hearing impairment may interfere with accurate reception of speech information. This distortion or masking by noise can be measured as loss in speech intelligibility. Speech intelligibility is defined as the count of the percentage of words that the listener hears correctly (Kling & Riggs, 1972).

The theoretical construct the present research is a model of auditory workload which proposes that performance will deteriorate when auditory workload increases above an acceptable level (Peters, 1991). In this model there are groups of factors which are potential drivers of auditory workload. A group called *transmission factors* includes aspects of the communication itself and the nature of the task. For example, speech intelligibility, number of people communicating, complexity of task, direction of information flow. These factors were tested in a series of field experiments reported in previous HFES meetings

and summarized in Whitaker (1995). The second group is called *linguistic factors* and includes aspects of the message content. For example, semantic familiarity, expectancy, morphological confusions, and redundancy. The present study was the first in a new series which will examine the impact of these linguistic factors on task performance when speech intelligibility is suboptimal.

The linguistic factors included as possible drivers of auditory workload are the syntactic and semantic features of the message. Although there has been considerable study of the impact of semantics and syntactical variables on reading comprehension and speed, there has been little published experimental research which establishes the relationships among these variables and levels of speech intelligibility (Rose, 1992).

Two variables which have been examined are redundancy and morphological confusions. Redundancy refers to either repeating the same words more than once or to providing elaboration of the same information but using different words. Morphological confusion is defined as the level of confusions which occur between the stimuli in the possible response set (words, numbers, or letters). In this case, *auditory* morphological confusion levels are of interest. Miller,

Heise, and Lichten (1951) studied the effect of redundancy on the intelligibility of spoken words in varying levels of noise. They varied redundancy by allowing the subjects to have the target word repeated either one or twice. The repetition of the same word did not substantially improve the recognition of the target words. However, Schmidt-Nielsen and Everett (1982) reported an analysis of radio communication under conditions of suboptimal speech intelligibility. They described that more experienced communicators (Naval Research Laboratory amateur radio club members) compensated for poor speech intelligibility differently than did inexperienced communicators (college students). The experienced communicators spontaneously used redundancy consistently, while the inexperienced group seldom did. The experienced group, for example, would use such phrase repetitions as "My shot, my shot, delta one, delta one." The experienced communicators also avoided words with a high degree of morphological confusion. That is, words that might be easily misunderstood were replaced with dissimilar sounding words. For example, the words "hit" and "miss" might be replaced with "affirmative" and "negative." It is important to note that the experienced communicators utilized redundancy and morphological features of words spontaneously. This led us to consider the degree to which these manipulations enhance the effectiveness of auditory communication. To answer this question, these variables were tested in the present experiment.

Being mindful of the results found by Miller et al. (1951) which showed that repetition was not an effective means of employing redundancy to increase speech intelligibility, we used an alternative means of varying redundancy; Rose (1992) proposed that redundancy was extra information that further identifies the goal--this elaboration form of redundancy was used in the present experiment.

METHODS

Sixteen university students (10 males and 6 females) participated as subjects in the experiment. They were paid $15 for their participation. As a motivational reinforcer, the two subjects with the best performances received additional awards of $30 and $20. The experimental session lasted approximately three hours, including practice, intelligibility testing (Modified Rhymes Test), experimental tasks, and debriefing.

The experimental task was to move a marker to a specified location on a board game (see Figure 1) by following tape recorded message instructions. Examples of the message for trials at the two morphological confusion and the two redundancy levels are illustrated in Table 1. The levels of morphological confusion were determined from published results for numbers (Morgan, Chambers & Morton, 1973) and for letters (Shaw, 1975).

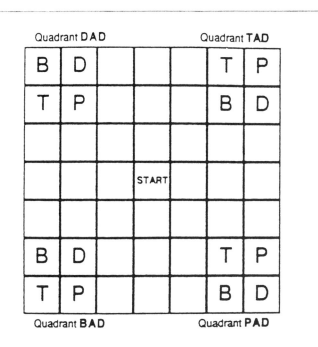

Figure 1. Morphologically Confusing Game Board

Table 1. Examples of the Four Message Sets.

Low Redundancy: Low Morphologic Confusion
"Ready, move to quadrant tan, one square, then three squares."

Low Redundancy: High Morphologic Confusion
"Ready, move to quadrant dad, two squares, then three squares."

High Redundancy: Low Morphologic Confusion
"Ready, move to quadrant tan, one square, then three squares, to the U."

High Redundancy: High Morphologic Confusion
"Ready, move to quadrant dad, two squares, then three squares, to the T."

NOTES: Subjects were told that the first distance was always up or down and the second was left or right.
Low Morph Confusion Quadrants: Tan, Rot, Led, Sip
High Morph Confusion Quadrants: Dad, Tad, Bad, Pad
Low Morphologic Confusion Letters: P, Y, U, X
High Morphologic Confusion Letters: B, D, T, P
Low Morphologic Confusion Distances: 1, 3
High Morphologic Confusion Distances: 2, 3

Subjects listened to the recorded messages for each trial over standard military headphones (David Clark Model # H10-66) set at low impedance. All trial messages were recorded at 100% intelligibility. For presentation, the recorded speech was passed through a potentiometer which varied the duty cycle (on-time) of the speech transmission within each 1/60 of a second. The effect of this circuit was to distort the spoken message and to decrease speech intelligibility. For each subject, the correct potentiometer settings needed to produce four specific levels of speech intelligibility (25%, 50%, 75%, and 100%) were established in the following way: Each subject was tested separately on a standardized set of 50-word lists known as the Modified Rhymes Test (House, Williams, & Kryter, 1963). The potentiometer was set at approximately the correct duty cycle to produce the desired level of speech intelligibility. The subject listened to the MRT list and completed the standard choice recognition task to establish his/her actual level of speech intelligibility at that duty-cycle. If the speech intelligibility was correct, the setting was recorded. If it was not correct, the potentiometer was reset and the test rerun with the new MRT list. This procedure was completed for each subject for all three (25, 50, and 75) levels of speech intelligibility. These potentiometer settings were then used for each subject during their experimental trials. The trial messages were passed through the potentiometer and the distorted speech was heard by the subject.

In a 2 x 2 x 4 within-subjects design, two levels of morphological confusion, two of redundancy, and four levels of speech intelligibility were tested. Each subject completed eight trials at each level of the design for a total of 128 trials per subject. Each subject was tested individually. A block of eight training trials was presented first, and then the 128 test trials. Rest breaks were given after each Intelligibility level (i.e., after each 32 trials.) The order of intelligibility, morphological confusion, and redundancy levels was counterbalanced across subjects and the specific test trials were randomized within these conditions. Three response measures were obtained:(1) time to respond and (2) response accuracy were recorded for each trial. A (3) subjective workload rating using the Subjective Workload Assessment Technique (SWAT) was obtained after each block of eight trials. All trials within a block had a fixed level of the three independent variables (speech intelligibility, morphological confusion, and redundancy).

RESULTS

The effects of speech intelligibility, morphological confusion level, and redundancy on response speed and accuracy were analyzed using a three-way, within-subjects, multivariate analysis of variance (MANOVA). Multivariate analyses of simple effects were conducted for all significant interactions. Univariate tests of significance were conducted for all significant multivariate effects to assess the individual effects of accuracy and response time separately. A separate analysis of variance (ANOVA) was conducted

for the workload measure (SWAT). Analyses of simple effects were conducted for all significant interactions.

Significant main effects were found for all three independent variables: Intelligibility ($F(6,90) = 20.24$, $p<.001$), Morphological confusion ($F(2,14) = 39.32$, $p<.001$), and Redundancy ($F(2, 14) = 36.67$, $<.001$). As Intelligibility decreased, subjects took significantly longer to respond ($F(3,45) = 33.64$, $p<.001$) and were less accurate ($F(3,45) = 93.39$, $p<.001$). Increases in Morphological confusion level resulted in responses which were significantly slower ($F(1,15) = 46.29$, $p<.001$) and less accurate ($F(1,15) = 43.91$, $p<.001$). Higher Redundancy in the instruction message resulted in responses being more accurate ($F(1,15) = 42.71$, $p<.001$), but slower ($F(1,15) = 32.93$, $p<.001$). The increase in response time was less than the total increase in average message length (1.4 seconds) between the low and high redundancy conditions.

The Intelligibility X Morphological Confusion interaction was found to be significant in the MANOVA and follow-on univariate tests confirmed this for both response time ($F(3,45) = 3.79$, $p<.05$) and accuracy ($F(3,45) = 21.31$, $p<.001$). These results are shown in Figures 2 and 3.

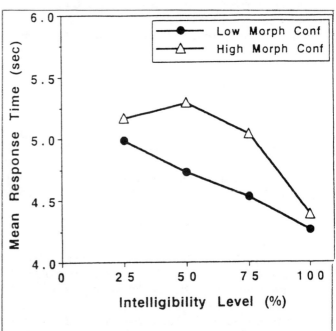

Figure 2. Mean response times for correct trials as a function of intelligibility and morphological confusion.

Notice that the response times for high morphologic confusions at 25% intelligibility decreased and did not differ significantly from those of low morphologic confusion trials at this intelligibility level. We assume that this occurred because subject's accuracy levels had dropped so low (to approximately 40%) that many responses in this condition were guessing and subjects did not need as much time to process the guess (see Luce, 1986, for analysis of response time strategies including fast guessing.)

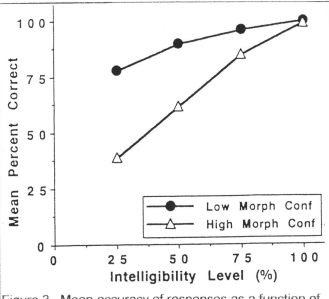

Figure 3. Mean accuracy of responses as a function of intelligibility and morphological confusion level.

The Intelligibility X Redundancy interaction was significant in the MANOVA and follow-on univariate tests also confirmed this for both response time (F(3,45) = 5.37, p<.01) and accuracy (F(3,45) = 3.74, p<.05). Subjects were always slower for the high redundancy condition (longer message) than the low redundancy condition (see Figure 4).

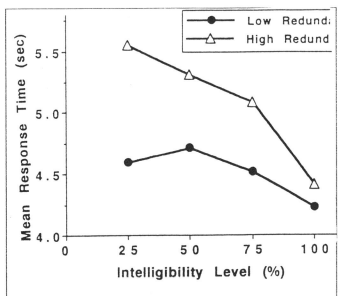

Figure 4. Mean response times for correct trials as a function of intelligibility and redundancy.

However, they were more accurate at all intelligibility levels below 100% when redundancy level of the message was high than when it was low (see Figure 5).

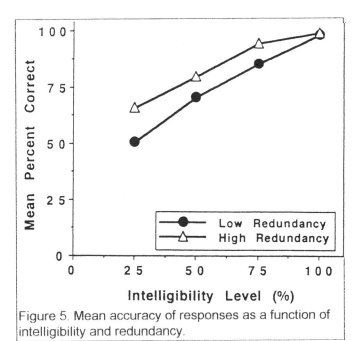

Figure 5. Mean accuracy of responses as a function of intelligibility and redundancy.

One workload rating was obtained for each block of trials at each level of the three independent variables. The workload metric known as SWAT originally required an initial card sorting procedure to establish the basis for analysis of this metric. However, research reported by Biers & McInerney (1988) has shown that SWAT can be an effective workload measure without the performance of the card sort. Furthermore, research by Whitaker, Peters, & Garinther (1990) found that a simple sum of the three subscale values in the rating exhibited equal sensitivity to the metric based on the card sort (a conjoint measurement analysis). Therefore, the simplified procedure of eliminating the card sort and using the ratings sum (minimum = 3 and maximum = 9) was used in the present analysis.

The SWAT scores were analyzed in an ANOVA to determine the effect of the three independent variables. SWAT scores increased as intelligibility decreased (F(3,45) = 62.55, p<.001), increased when morphological confusion level increased (F(1,15) = 29.35,p<.001), and increased when redundancy decreased (F(1,15)=8.59, p<.05). All these results are in the predicted direction. The significant interaction of Intelligibility X Morphological Confusion (F(3,45) = 11.16, p<.001) is shown in Figure 6. The mean SWAT score for low redundancy was 5.14 and for high redundancy was 4.83.

DISCUSSION

As the intelligibility level of the recorded message was decreased, subjects took longer to complete the tasks and they made more errors. Furthermore, subjective workload increased with decreasing intelligibility levels. Thus the task obviously became more difficult when the speech instructions became less clear. The effect of the message content variables, morphological confusions and redundancy, also affected task performance and subjective workload.

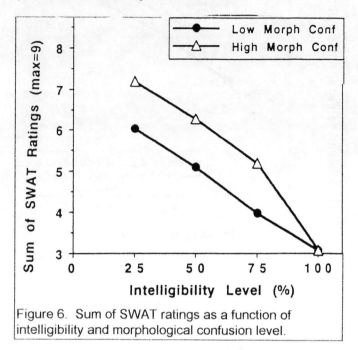

Figure 6. Sum of SWAT ratings as a function of intelligibility and morphological confusion level.

When the message and the accompanying task contained phonetically similar words, subjects made significantly more errors than when the words were phonetically distinct. The increased number of errors resulted from the subjects' misinterpreting a letter, number or quadrant name for another similar-sounding one. The increased confusion induced by the high level of morphological confusions also resulted in longer response times for correct trials. Even if the subjects were able to make the correct movement, more time was required to discriminate the phonetically confusing messages and perceived workload was increased. Until intelligibility dropped to 25%, subjects continued make use of lower morphological confusions in their performance. At that intelligibility level, it appears that subjects began to guess and to respond quickly in the hope of improving at least their speed scores, even if their accuracy measure was a matter of chance.

Redundancy in a message resulted in more accurate task performance whenever speech intelligibility was less than 100%. When faced with uncertainty in where the proper goal location was, subjects used redundancy when it was available. Because the redundant message contained more words, it did take longer to present and response times for redundant messages was longer than for non-redundant messages. However, redundant messages resulted in more accurate performance and less reported workload. These results show an important difference between the mere repeating of the information in the same words (like that used by Miller et al ,1951, and reported as a spontaneous communication strategy by experienced communicators in Schmidt-Nielsen and Everett, 1985),

which had little effect on word recognition, vs. the redundancy used in the present study which *elaborated* the information in the redundant form of the message.

In conclusion, wherever speech is a vital component of successful task performance, special attention should be paid to both the physical quality and the content of spoken messages. Whenever possible, clarity of speech should be maintained because losses in speech intelligibility result in decreased performance and increased workload. However, in environments where degraded auditory intelligibility cannot be prevented, certain measures can help to compensate for the decreased clarity. By avoiding messages with similar-sounding key words, listeners will perform more quickly and accurately and with less effort. Also, by introducing redundancy into the messages, the listener is in essence given additional chances to catch the crucial information, resulting in more accurate performance. However, this is accomplished at the cost of increased response time. Therefore, the use of redundancy should depend on the intended task outcome. If it is more important to finish the task quickly, and errors can be tolerated, then redundancy should not be used. However, for tasks where errors are costly, and time is available, then message redundancy (in the form of elaboration) can be a valuable tool to enhance task performance and decreased operator workload.

REFERENCES

Biers, D. W., & McInerney, P. (1988) An alternative to measuring subjective workload: Use of SWAT without the card sort. Proceedings of the 32nd Annual Meeting of the Human Factors Society, 1136-1139.

House, A. S., Williams, M. H. L., & Kryter, K. D.(1963) Psychoacoustic speech tests: A modified rhyme test. Report ESD-TDR-63-403. Cambridge, MA: Bolt, Beranek, and Newman.

Luce, R. D. (1986) Response Times: Their Role in Inferring Elementary Mental Organization. Oxford University Press: New York.

Miller, G. A., Heise, G. A.,, & Lichten,W. (1951) The intelligibility of speech as a function of the context of the test materials. Journal of Experimental Psychology, 41, 329-335.

Morgan, B. J. T., Chambers, S. M., & Morton, J. (1973) Acoustic confusion of digits in memory and recognition. Perception & Psychophysics, 14, 375-383.

Peters, L. J. (1991) Auditory performance: A model to predict task performance as a function of auditory workload. Proceedings of the 35th Annual Human Factors Society Meeting, 609-613.

Rose, A. (1992) Developing a message complexity index. American Institutes of Research Report.

Schmidt-Nielsen, A., and Everett, (1982) Problems in evaluating the real-world usability of digital voice communication systems. Behavior Research Methods, Instruments, & Computers, 17, 226-234.

Shaw, D. J. (1975) A phonological interpretation of two acoustic confusion matrices. Perception & Psychophysics, 17, 537-542.

Whitaker, L. A., Peters, L. J., and Garinther, G. A. (1990) Effects of speech intelligibility among Bradley fighting vehicle crew members: SIMNET performance and subjective workload. Proceedings of the 34th Annual Meeting of the Human Factors Society, 186-188.

Whitaker, L. A. (1995) Effect of Speech Intelligibility on Task Performance. Final Report: Klein Associates, August, 1995.

CAN A USABLE PRODUCT FLASH *12:00*?
PERCEIVED USABILITY IS A FUNCTION OF USEFULNESS

Amy L. Schwartz and Lynne Thomson
Ameritech
Hoffman Estates,IL

Colleen M. Seifert and Michael G. Shafto
Cognitive Science Associates
Ann Arbor, MI

Most usability quality measures are performance-based and center on detection of interface problems. This study focused on the customer's subjective assessment of ease of use for Ameritech telephony products relative to other consumer products that they use every day. This assessment was accomplished through an empirical study that measured customer attitudes and opinions about their experiences using 12 Ameritech telephony products and 16 other consumer products. The results demonstrate where these telephony products sit relative to other consumer products in customers' perceptions of usability. In addition, a Multidimensional Scaling Analysis of the data revealed three separate dimensions that underlie people's perceptions of usability: automaticity of use, value to the customer, and self-evident operation.

INTRODUCTION

Product usability is a central issue in human factors. Most of the scientific literature on usability has focused on defining criteria for successful designs (e.g., Schneiderman, 1988), integrating evaluation into the design cycle (e.g., Grudin, 1991), and developing methods for usability testing (e.g., Gould, Boies, and Lewis, 1991; Virzi, 1992). We tend to define usability almost exclusively in terms of performance-based measures like time to accomplish the task, error rates, and retention of procedures. These measures have led to general principles of good design such as "functionality, consistency, naturalness, minimal memorization, feedback, user help, and user control" (Holcomb and Tharp, 1991). However, very little attention has been paid to users' *subjective* impressions of product use (Norman, 1988).

Design guidelines and performance-based measures may lead to products that score very highly on user performance measures, and which receive good ratings in the laboratory with new users. However, for evaluating products used by real people in everyday settings, these criteria may not be sufficient to predict what consumers mean by "ease of use."

The *psychological perception* of usability in the user is a critical factor for determining ease of use. In order to understand how people view a product's use, we need to sample consumers' real-world experiences with specific products. These subjective judgments about usability must also be placed in a context of other products that comprise their normal comparable experiences with use. For example, when a consumer retrieves her Voice Mail messages after programming her VCR, she is likely to compare the two products in her internal metric for ease of use.

The economic approach in marketing focuses on measuring customer satisfaction in a variety of ways to determine impressions about products (Johnson and Fornell, 1991). In this area, studies have attempted to identify the factors that determine customer retention (Rust and Zahorik, 1993), related by implication to satisfaction and usability. In a study of bank customers, items such as "convenience to work" and "listens to my needs" were rated, and a factor analysis of satisfaction measures revealed "convenience" and "warmth" as dimensions underlying customer's attitudes towards their bank. However, more general findings, such as psychological factors underlying assessments of more than one product, have proved elusive. As Peterson and Wilson (1991) note, "Despite the plethora of variables, perspectives, and methodologies investigated, it is not possible to succinctly identify specific causes of the striking characteristic of customer satisfaction ratings."

Prior research on usability and marketing has not provided an adequate framework for answering these questions of subjective usability.

How, then, can products be accurately assessed for subjective ease of use? And how can areas of difficulty for consumers be identified and improved? The study reported here had three foci:
• To examine people's everyday experience with familiar products
• To set people's evaluations within the context of both telephony and non-telephony products
• To investigate the underlying psychological dimensions of usability.

METHOD

Subjects

A total of 379 Ameritech customers were recruited to serve as paid subjects. Each of the 12 target telephony products tested was subscribed to by at least 25 of the subjects.

Materials and Procedure

The task was completely questionnaire-based. Twelve Ameritech consumer telephony products and 16 non-telephony products were selected for the questionnaire. Each product was presented with a task description that is a typical or default use for the product. The task descriptions were provided to focus users in their judgments. For example, when judging a VCR, we asked them to think about programming the VCR to record a show broadcast the next day (rather than on playing a tape in the machine).

The questionnaire contained three parts:
• Part 1: <u>Product Use Estimates</u> - This instrument was designed to measure the frequency of use of each of the products in the study. Subjects indicated frequency of use of each product on a five point scale.
• Part 2: <u>Comparison Judgments</u> - This instrument measured the comparison of a target product (one of the 12 Ameritech services) against all other Ameritech and consumer products (27 comparisons in total). Each subject completed only one set of comparison ratings, based on the target Ameritech service group they were assigned to. For example the booklet for Automatic Callback included the following instructions:

Think about a time recently when you used Automatic Callback (*69) to return the last incoming call. Remember what it was like to use it, and what happened when you used it. Now, for each product listed below, consider how much harder or easier that product is to use compared to using Automatic Callback to return the

last incoming call. For your answer, circle a number on the scale between
1 - *Much Easier* and *7 - Much Harder*.

A sample item followed to demonstrate use of the rating scale:

Compared to changing the station on a Car Radio, using Automatic Callback to return the last incoming call is:
```
1........2........3........4........5........6........7
Much                                    Much
Easier                                  Harder
```

• Part 3: <u>Scale Item Ratings</u> - These scale items were statements like "I use it successfully every time" or "Using it is effortless." Subjects were asked to rate each telephony and non-telephony product on 20 scale items regarding ease of use. The six point scale ranged from strongly disagree to strongly agree.

For all three parts, subjects only rated products that they had used. They completed the questionnaires at a market research facility where the completion time averaged about 35 minutes.

RESULTS AND DISCUSSION

Multidimensional Scaling Analysis

Multidimensional Scaling (MDS) provides an analysis that allows comparison of all of the products and rating scale items. In the scaling analysis, all 12 telephony services, all 16 non-telephony products, and all 20 scale items were included, for a total of 48 objects in the analysis. The results provide a measure of the "distances" or "dissimilarities" among these 48 items as measured by subjects' responses in the survey. Including all rating scale items and product comparisons in a single analysis is a variant of Coombs' (1974) "unfolding" technique, which determines the metric relations inherent in data.

By combining the product comparison data with the product rating scale data, a larger, more varied dataset is produced. This method also allows direct determination of the relationship between specific scale items and the underlying dimensions that arise in terms of the scores for each item on each dimension. This circumvents the need to correlate the comparison results with the rating scale results, and the scale items assist in interpreting the emergent dimensions.

Scaling solution. The data were entered into a Linear MDS procedure that minimizes Kruskal Stress, resulting in a three dimensional solution.

Figure One
Three Dimensional Solution

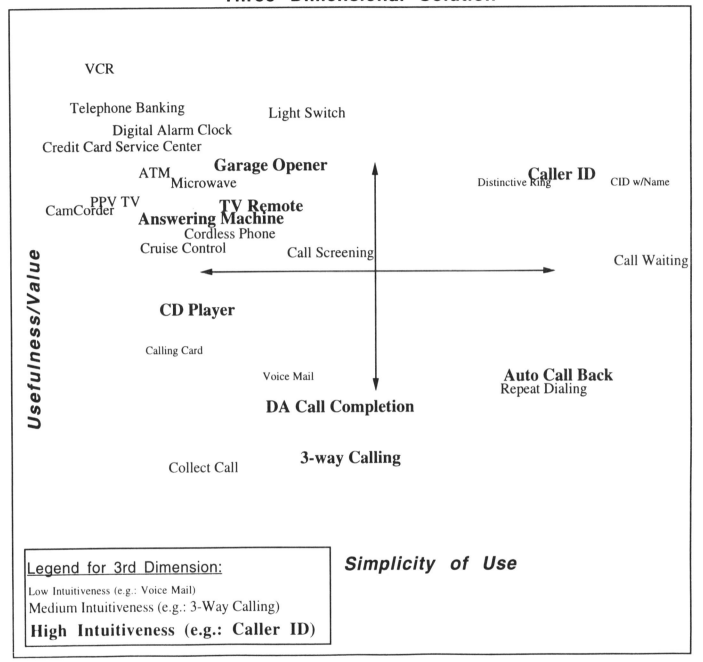

The stress of the final configuration was .166 after 23 iterations, and the proportion of the variance accounted for (RSQ) was .82. Thus, the scaling procedure resulted in an unambiguous fit that covers the data well. The sum of the squared loadings on the three dimensions indicates their relative contribution to the fit: 26.27, 12.02, and 9.66 respectively. As in any MDS solution, scale values may vary as a function of the subject population, the comparison products, and the rating scales used in the survey instrument.

Figure 1 shows the three dimensional solution. The first dimension, shown on the horizontal axis, is labelled "simplicity of use." There is a low demand for attention or skill for products that show up on the positive end of this dimension. The scale items that loaded heavily on this dimension were:

I don't have to pay attention when using it.
It usually requires just one quick step.
Using it is effortless.
It's easy to use without written instructions.

The telephony products do well on this dimension, especially products like Caller ID or Call Waiting that demand little effort on the user's part to operate.

Dimension 2, shown on the vertical axis, appears to reflect a sense of the product's utility or value. We have labelled this dimension "usefulness." Heavily loading scale items for this dimension were:

> I need to have it.
> I get my money's worth from it.
> It gives me more control in my life.
> I use it whenever I can.

The figure shows that a VCR was rated as the most useful product while telephony products like Call Screening and Collect Call were rated the least useful. It is important to note that these results come from a diverse group of Ameritech customers. Some products are extremely useful to a specific customer group. For example, we would expect a Collect Call to be extremely useful to a prisoner.

The third dimension, which we label "Intuitiveness," is coded on the figure by type size and bolding. For ease of illustration, we have coded products as low, medium, or high on this dimension. Specific scale items related to this dimension are:

> I remember how to even if I haven't used it lately.
> It is fun to use.
> I use it successfully every time.
> I understand the technology it involves.

Products like Caller ID, Directory Assistance Call Completion, and a CD player did well on this dimension, while telephony products like Distinctive Ringing and Calling Card were rated the least intuitive.

CONCLUSIONS

This study was successful at suggesting meaningful psychological dimensions underlying usability. The implications of this work are that a customer will judge a product as highly usable when that product requires little attention or effort, is useful, and is intuitive with low memory requirements. The multidimensional nature of this solution suggests that a high value on one dimension can help offset a low value on the other dimensions. For example, a product that was extremely useful to a user could require a large amount of attention and

memory to operate and still be judged overall as usable. The finding that users perform this type of cognitive cost-benefit analysis is perhaps not surprising, but it is surprising that the result of the analysis is subjectively viewed in terms of overall usability.

The finding that these telephony products are perceived as less useful than many other consumer non-telephony products may be a function of the marginal utility that the telephony products provide over and above dial-tone. We suspect that basic telephone service would have been extremely high on the usefulness dimension. Subjects may have been considering the increment in usefulness that the other telephony products provide over dial-tone in their judgments.

The usefulness finding has several implications for new product development. First, it reinforces the move away from a new product development model based on the concept of "Use new technology to make *something* and customers will come." No one in telecommunications can afford any longer to make products that are not useful to customers. Second, human factors engineers are often in the position where we need to recommend getting rid of certain functions in a product in order to simplify the interface and increase usability. Our findings suggest that we need to strongly consider the perceived usefulness that additional functionality can bring to a product, since the cognitive cost incurred by including some functionality may be offset by the increased utility of the product.

Our future directions with this work center on trying to extend this methodology to laboratory evaluations of new and existing products. The study reported here involved customers who had used these products for at least several months in their daily lives. In addition, we believe we need to integrate performance-based variables into an overall model of ease of use.

REFERENCES

Coombs, C. H. (1974). Psychological scaling without a unit of measurement. In G. M. Maranell (Ed.), Scaling: A sourcebook for behavioral scientists, 281-299. Chicago, IL: Aldine.

Gould, J. D., Boies, S. J., & Lewis, C. (1991). Making usable, useful productivity-enhancing computer applications. *Communications of the ACM, 34(1),* 74-85.

Grudin, J. (1991). Interactive systems: Bridging the gaps between developers and users. *Computer,* IEEE, 59-69.

Holcomb, R. & Tharp, A. (1991). Users, a software usability model and product evaluation. *Interacting with Computers, 3(2),* 155-166.

Johnson, M. D. & Fornell, C. (1991). A framework for comparing customer satisfaction across individuals and product categories. *Journal of Economic Psychology, 12(2),* 267-286.

Norman, D. A. (1988). *The psychology of everyday things.* New York: Basic Books.

Peterson, R. A. & Wilson, W. R. (1991). Measuring customer satisfaction: Fact and artifact. *Journal of the Academy of Marketing Science, 20(1),* 61-71.

Rust, R. T. & Zahorik, A. J. (1993). Customer satisfaction, customer retention, and market share. *Journal of Retailing, 69(2),* 193-215.

Schneiderman, B. (1988). We can design better user interfaces: A review of human-computer interaction styles. *Ergonomics, 31(5),* 699-710.

Virzi, R. A. (1992). Refining the test phase of usability evaluation: How many subjects is enough? *Human Factors, 34 (4),* 457-468.

BENEFITING DESIGN EVEN LATE IN THE DEVELOPMENT CYCLE:
CONTRIBUTIONS BY HUMAN FACTORS ENGINEERS

Merissa Walkenstein & Ronda Eisenberg
AT&T Lucent Technologies
Middletown, NJ

This paper describes an experimental study that compares a graphical user interface for a computer-telephony product designed without the involvement of a human factors engineer to a redesign of that interface designed with a human factors engineer late in the development cycle. Both interfaces were usability tested with target customers. Results from a number of measures, both subjective and objective, indicate that the interface designed with the human factors engineer was easier to use than the interface designed without the human factors engineer. The results of this study show the benefits of involving human factors engineers in the design of graphical user interfaces even towards the end of a development cycle. However, this involvement is most effective when human factors engineers are included as an integral part of the design and development process even at this late stage in the process.

INTRODUCTION

When designing user interfaces, human factors engineers rely on their training in user interface design principals and guidelines in order to ensure ease of use of the product being designed. Human factors engineers also utilize user testing techniques and undergo an iterative design process that involves user testing and redesigns until the usability goals are achieved. While the value of this training and process is realized by many in a development organization, this is not always the case with the organization as a whole. Those with human factors training are not always included in the design of a user interface, and often when they are included, are peripheral to the actual design and development process. Those who recognize the value of including the human factors engineer as an intrinsic part of the design and development process, often find themselves having to demonstrate this value to the rest of the organization. Demonstrating the value of human factors engineering can be done most effectively when experimental data is provided that clearly indicates the effect of including human factors engineering in the process, as well as showing the extent to which it needs to be involved.

In an experimental study comparing design capabilities of designers both trained and untrained in human factors techniques, Bailey (1993), conducted usability testing to compare user performance on alternative designs made by both developers and human factors engineers. Bailey found that the interfaces designed by the developers were less usable (e.g., higher failure rates, larger number of errors) than those designed by the human factors engineers. Bailey also found that developers were unsuccessful at improving upon their own design when given feedback from user testing and allowed to iterate on their design, whereas human factors

engineers were successful at this process. These results indicate that designs originally created by human factors engineers result in greater usability than those designed by developers, and that human factors engineers can iterate on their own designs and improve upon them.

Miller & Stimart (1994), also ran an experiment that compared three different designs of a product that varied in level of human factors involvement. One of the designs was the original design that had no human factors engineering involvement. Another design was a redesign of the original design that had minimal human factors engineering involvement late in the development cycle, with the development team being independent from the human factors engineers. Another design was a complete redesign of the original design and had human factors engineering fully integrated into the development cycle from the start of the design process. As measured by speed, accuracy and subjective responses, results indicated that minimal human factors involvement produced minimal improvements in the product, and that complete human factors involvement produced substantial improvements in the product. These results show that not only are designs that involved human factors engineers more usable, but that these designs become more usable the more involved the human factors engineer is in the design and development process. The results also demonstrate that human factors engineers can improve upon designs made by others, even when their involvement is minimal.

Although involvement by a human factors engineer on a design is optimal at the beginning of the design phase, it is often the case for them to be called in to fix problems with an interface after the design has been completed and development on the design is well on its way. At this point the fixes usually need to be done in a hurry and the human factors engineer may not have time to perform usability testing, nor get the opportunity to iterate on the design. Miller & Stimart demonstrated that this type of involvement may only provide minimal increases in usability. However, if the human factors engineer plays a direct role in the design and development process even at this late stage in the cycle, usability improvements could be more than minimal. The experiment outlined in this paper explores this issue further, and compares a user interface designed without human factors engineering involvement to that same interface modified by a human factors engineer involved in-depth late in the development process.

Background

A working relationship was formed between the AT&T Corporation and an independent software vendor (ISV) in order to develop a computer-telephony application that linked the PC with the phone. AT&T and the ISV company agreed to build upon an existing application developed by the ISV company. The purpose of the application was to provide customers with additional functionality over the phone as well as to enhance operations of existing phone features. The application offered standard business phone features such as multiple lines (up to ten), dial, hang up, hold, conference, and transfer. In addition it provided users with a directory of contact names and numbers and the ability to dial these directly from the PC. It also offered a call log of all calls received and made, as well as caller ID displays for incoming calls.

Usability inspection of the existing user interface (UI) by an AT&T human factors engineer, revealed potential usability problems with the UI (Nielsen, 1995). Through discussions with the ISV team, it was clear that a human factors engineer had not been involved in the design of their UI. Although UI changes were allowed to be made to the existing product, changes to the existing code base needed to be minimal, and new UI specifications needed to be completed immediately due to scheduling constraints. Based on results of the usability inspection, the human factors engineer proposed changes to the UI. Though it was late in the development cycle, the human factors engineer worked closely with the development team and wrote detailed specifications and developed a detailed prototype that demonstrated the changes. The human factors engineer was also able to lay out the actual screens for the developers, using their tools. Because of the strong involvement the human factors engineer had with the development team, many of the specified changes were implemented and these changes were implemented as specified. Some of the changes however were considered too complex and were rejected at the last minute due to time constraints.

Several sources explain the reasoning for the changes made to the New design. These include: user

interface design principals found in human factors texts such as Mayhew's <u>Principles and Guidelines in Software User Interface Design</u>; standard graphical user interface guidelines such as Apple's <u>Macintosh Human Interface Guidelines</u> and Microsoft's <u>The Windows Interface - An Application Design Guide</u>; in-house AT&T human factors guidelines on graphical telephony interfaces; the designer's own experience with prior designs for telephony applications and user testing of these applications. The New design is shown in Figure 1, and the Original design is shown in Figure 2. Listed below are just some of the changes made to the New design.

Sufficient use of perceptual cues.
- Added disabling and graying out of telephone controls that were not functioning in the given context, such as the hang up button when a call was on hold.
- Made grouping of related items more obvious through spatial proximity, such as items related to an individual call appearance.
- Used color as an indication of state, such as a white background for an active call, and a gray background for a call on hold.

Error handling.
- Added prompting for users when an operation could be destructive, such as accidentally hanging up a call.
- Provided a way for users to cancel out of operations, such as canceling in the middle of a multi-step operation like conference.

Consistency with environment.
- Removed check boxes that implied multiple selections but were being used for mutually exclusive selection.
- Labeled buttons with functional words instead of function key codes.
- Used pop-up message boxes instead of re-painting the screen with a message.

Guidance.
- Presented instructional text in dialog boxes about progress and ensuing steps such as for completing a conference call or a transfer.
- Utilized the status bar in the main window for instructional information.

Graphics.
- Used various gray shades for organization of elements and to add depth.
- Added icons to telephone controls to aid in comprehension and to add color.
- Removed graphics that appeared as operational icons but were not.

Figure 1. New Design : Main screen with two calls.

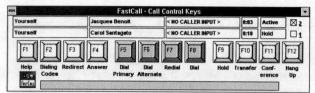

Figure 2. Original Design: Main screen with two calls.

The study that is reported in this paper was able to be conducted after the UI specifications had been made for the New design. As per the agreement between AT&T and the ISV company, the New design was to be implemented into the product immediately in order to release the product to market as soon as possible. The study was run in anticipation of the following release of the product, and results from the study were to feed into the design for the following release of the product.

METHOD

Twenty-three target customers of the product were used as participants in the study. Participants signed up for the study voluntarily in response to a poster at a telecommunications convention. Participants were required to be telecommunication managers (responsible for buying the communication systems at their companies), regular users of Microsoft Windows™, and regular users of AT&T's large business DEFINITY® phones. Participants were tested individually on PC demos of the New design (programmed in Visual Basic) or the Original design (programmed in C). Both demos appeared and operated almost exactly as the actual interface would, although there was no connection to the phone. Seventeen males and six females were randomly assigned to perform tasks with either the Original or New design, resulting in eleven participants in the Original condition, and twelve in the New condition. Participants in both conditions performed the same series of thirteen tasks in the same order. Tasks involved using the following fifteen sub-tasks, or features: answer, hold, hang up, answer multiple calls, place call, answer and place multiple calls, redial, enter touchtones, dial from the directory, transfer, supervised transfer, conference, redirect a call to the operator, learning call handling for multiple calls, and caller identification. At the end of each task, participants were told whether they had been successful or unsuccessful. After a series of tasks, participants rated the ease of use of the individual features they had used. At the end of all the tasks, participants responded to questions about the application overall.

RESULTS

Measures included participant ratings, participant comments, and participant performance (task success rate). All rating scales ranged from 1 to 7, with 7 being most positive.

Mean individual feature ratings were primarily high for the New design, with most rated as 6 or higher. For nine of the fifteen features rated, participants in the New condition gave higher ease of use ratings than participants in the Original condition (all $ps < .05$), and an additional four of the features rated showed trends in the same direction. One of the features, touchtones, though given low ratings by participants in both conditions ($M = 1.92$, New versus $M = 3.70$, Original), was rated lower by New design participants than Original design participants ($p < .05$). Caller ID ratings showed a trend towards lower ratings by New design participants than Original design participants, though ratings by both types of participants were high ($M = 5.92$, New versus $M = 6.45$, Original).

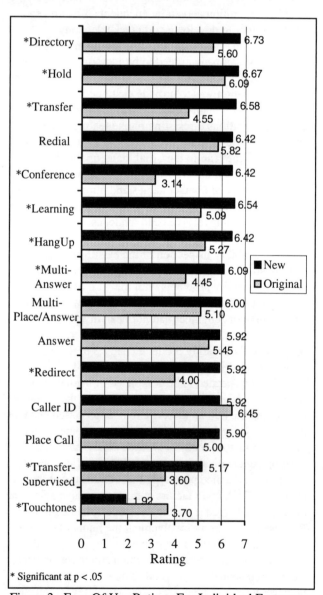

Figure 3. Ease Of Use Ratings For Individual Features.

Results for overall ease of use of the two designs showed higher mean ratings for the New design ($M = $

6.17) than the Original design ($M = 4.71$), $p < .05$. A pre-set usability objective of having at least 80% of participants using the New design rate overall ease of use with a 6 or a 7, was met. Results also showed that participants in the New condition rated the design as easier to use than the phone ($M = 5.03$), $p < .05$ (where 1 = "Much harder on PC", 4 = "Same", and 7 = "Much easier on PC"). Participants in the Original condition rated the design as harder to use than the phone ($M = 3.14$), $p < .001$. Ratings of overall appearance for the two designs, though in the predicted direction, were not significantly different from each other ($M = 5.25$, New versus $M = 4.73$, Original).

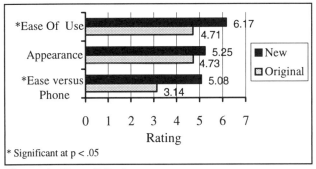

Figure 4. Overall Ratings.

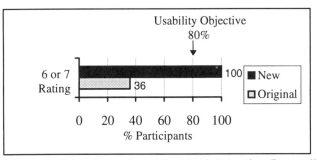

Figure 5. Percentage of participants giving a 6 or 7 overall ease of use rating.

Results showed that participants in the New condition successfully completed more of the 13 tasks ($M = 11.42$) than participants in the Original condition ($M = 8.36$), $p < .001$.

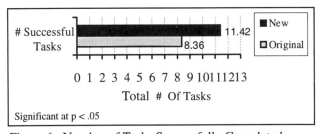

Figure 6. Number of Tasks Successfully Completed.

Responses for the open ended questions were organized into categories. For what they "Liked Most", participants in the New condition (n = 6) more frequently mentioned "user interface design" than those in the Original condition (n = 1), $X^2 (1, N = 23) = 4.54$, $p < .05$. For what they "Liked Least", participants in the Original condition (n = 5) more frequently mentioned "not intuitive/doesn't act like phone" than those in the New condition (n = 0), $X^2 (1, N = 23) = 6.97$, $p < .01$. For what they "Wanted Changed", participants in the New condition (n = 6) more frequently mentioned "nothing" than those in the Original condition (n = 1), $X^2 (1, N = 23) = 4.54$, $p < .05$.

DISCUSSION

The results of the study indicate that the New design showed improvement in usability over the Original design. Participants not only rated the New design as easier to use than the Original design, but their performance on the New design was better. Pre-set usability objectives for the New design were not only met but exceeded, as compared with the findings for the Original design which were at an unacceptable level. Task succes rate was also significantly higher in the New condition than the Original condition. Most of the individual features were rated as very easy to use in the New design, and were significantly easier to use than those in the Original design. However one of the features, touchtones, though it was given low ratings for both designs, was considered significantly harder to use in the New design than the Original design. Design changes to this feature had been proposed but had been rejected by the development team due to time constraints. The human factors engineer anticipated that users would be unable to locate this feature and that they could accidentally drop a call when trying to operate the feature. While general design changes had been rejected and the location and operation of the feature remained the same, an error handling message box was allowed to be incorporated to warn users about dropping a call. Performance data was analyzed and showed that although slightly more participants in the Original condition than in the New condition were able to operate the feature, almost all of them accidentally dropped a call when doing so. Only one participant in the New condition dropped the call. Though one of the anticipated problems had been improved, the problem of locating the feature had not been improved. Overall differences in the two designs seemed to impact participants differently when they operated the touchtone feature, though the operation of the feature did not differ between the two designs. The Original design was more keyboard oriented and the New design was more mouse oriented. The touchtone feature being used was also keyboard oriented since its redesign changes had been rejected. One explanation for the differences in performance is that participants in the New condition had not been primed with keyboard use throughout the tasks as had been the participants in the Original condition.

If human factors engineering had been involved earlier in the design process, and there had been time to incorporate the results of this user testing and iterate on the New design, the problem with the touchtone issue could have been eliminated. In spite of this problem, however, improvements in usability for the New design were substantial. Though it was late in the development cycle and changes to the interface had to be minimal, the direct involvement of the human factors engineer with the design of the New interface and the in-depth involvement with the development team encouraged more changes to be implemented and ensured that those changes were implemented as specified. Though the data reported in this study and others indicate that it is optimal to involve human factors engineers at the beginning of the development cycle, results of this study also suggest that human factors engineers who are fully integrated into the design process even late in the development cycle can substantially impact a design. Experimental results such as these help demonstrate the value and need of not only involving human factors engineering in the design and development process, but of making it an integral part of that process.

REFERENCES

Apple Computer, Inc. (1992). Macintosh Human Interface Guidelines. Addison-Wesley Publishing Company, Massachusetts.

AT&T Internal Document (1995). Call Handling User Interface Look. *Graphical User Interface Look, Listen, And Feel, Standards Document.*

Bailey, G. (1993). Iterative methodology and designer training in human-computer interface design. *Proceedings of INTERCHI'93*, pp. 198-205.

Mayhew, D.J. (1992). Principles and Guidelines in Software User Interface Design. Prentice Hall, New Jersey.

Microsoft Press (1992). The Windows Interface-An Application Design Guide. Microsoft Press, Washington.

Miller, M.A. and Stimart, R.P. (1994). The user interface design process: The good, the bad, & we did what we could in two weeks. *Proceedings of the Human Factors and Ergonomics Society 38th Annual Meeting*, pp. 305-309.

Nielsen, J. (1995). Usability inspection methods. *Proceedings of CHI'95*, pp. 377-378.

Software Ergonomics Comes of Age: The ANSI/HFES-200 Standard

Chair: Paul Reed, Lucent Technologies: Bell Laboratories
Co-Chair: Patricia Billingsley, The Merritt Group

Panelists:
Evelyn Williams, Hewlett-Packard
Arnold Lund, Ameritech Services
Eric Bergman, Sunsoft, Inc.
Daryle Gardner-Bonneau, Michigan State University

Introduction

In 1994, the Human Factors and Ergonomics Society (HFES) proposed a software ergonomics standards project to the American National Standard Institute (ANSI), which was approved and designated as ANSI/HFES-200. The purpose of this panel session is to review the background, status, contents, and issues relating to the HFES/ANSI-200 proposed standard on software ergonomics. Recent HFES administrations have identified the development of ergonomics standards as a strategic objective of the society. In addition, the considerable success of the ANSI/HFES-100 Standard for the Human Factors Engineering of VDT Workstations and the ongoing software standards ergonomics activity in the International Standards Organization (ISO) have contributed momentum for the development of an ANSI/HFES software user interface standard.

The stakeholders in a software user interface standard include computer system (both hardware and software) manufacturers, end-users of computing systems, employers with personnel using software-based systems, and others. The purpose of ANSI operating rules and procedures is to insure that the interests of directly- and materially-affected stakeholders are balanced and fairly represented; and it is the duty of the standards-development organization to facilitate a fair representation of these interests. For an HFES-initiated software user-interface standardization effort, the HFES is also a stakeholder to the extent that the relative success of such an undertaking will have a significant effect on the perceived professional integrity and prestige of the society. The users of the software ergonomics standard are expected to include people in a wide range of functional roles such as: user interface designers, usability professionals, software procurement personnel, system testers, quality team personnel, operational efficiency experts, and others. The philosophy of the standard is to specify software user interface objectives and design goals that can be used as criteria for designing or evaluating user interfaces. As in the ANSI/HFES-100 VDT Standard, the software ergonomics standard contains two types of specifications: requirements and recommendations. Requirements are identified by including the word "shall", and recommendations are identified by the word "should". Conformance to the standard is achieved by meeting all of the requirements, as identified by "shall" statements. The vast majority of specifications in the software ergonomics standard are "should" statements. Our committee has adopted a conservative policy such that any "shall" statement must be rigorously justified by compelling empirical and practical support.

Finally, because of the wide range of system development processes employed today, the standard does not specify any particular development methodology and is non-prescriptive in general.

Evolution of ANSI/HFES Software Ergonomics Standard

The 12-year saga of the ANSI/HFES-200 standard began in late 1984, when the German national standards organization (DIN) released a draft standard for public review. The draft standard, DIN 66 234 Part 8, contained general guidelines for the design of software user interfaces. The content of the guidelines was not particularly controversial, but Germany's expressed intent to treat the guidelines as requirements caused great consternation among major computer vendors in the U.S.

In February 1985, several computer companies sent members of their human factors staffs to a meeting at CBEMA headquarters to prepare a response. It was decided that each company should write its own critique of the DIN draft, focusing on why it should not be treated as a standard. As it turned out, every company had the same primary objection - the draft contained no explicit criteria against which software could be tested to determine whether or not it complied with the standard, making the compliance decision a very arbitrary one.

This group continued to meet on a regular basis, and in 1986 organized as the HFES Human-Computer Interaction Standards Committee. The committee originally had no intention of contributing to the development of new standards. However, the members began to realize that several bodies, both in the US and abroad, were already in the process of developing HCI standards, some with little or no input from the human factors community. The most important of these efforts was being undertaken by an ISO committee (ISO TC159/SC4/WG5) tasked with developing

software-related material for inclusion in ISO standard 9241, Ergonomic requirements for office work with visual display terminals.

The HFES/HCI committee decided that the most effective way to influence this process was to first define robust criteria for standards versus guidelines, then use those criteria to generate sample content for the software sections of ISO 9241. Between 1986 and 1993, the committee developed standards and guidelines covering most of the software topics planned for inclusion in ISO 9241. Their material, which was submitted as technical contributions to the ISO committee, was highly influential. In many cases, it provided the primary content for the software parts of the ISO 9241 standard.

In 1993, the HFES Executive Council asked the HFES/HCI committee to turn its attention to developing a US-national software ergonomics standard, as a companion document to the ANSI/HFES-100 standard. Since then, the HFES/HCI committee has been engaged in developing a national standard that merges the software parts of ISO 9241 with new material in related technical areas.

Relationship of ANSI/HFES-200 to ISO 9241

A critical element in the strategy employed by the HFES-200 software ergonomics standards committee is the harmonization and re-use of ISO 9241 standards within the HFES/ANSI-200 proposed standard. This approach was taken because for two reasons. First, in the early days of the HFES 200 committee members developed working documents that reviewed by the committee and then were contributed to ISO. This material was accepted and used extensively in their initial committee drafts. While the material has changed, the changes have typically being in the direction of improvements to the guidelines. The second reason for this approach is the fact that many US companies developing equipment that would be affected by HFES-200 are marketing and delivering this equipment on a world wide basis. It would create difficulties for these companies if the HFES-200 standard were substantially different from ISO 9241 in the areas where these standards overlap.

Taking international companies into account has not only affected the approach being taken to HFES-200 but it has affected the format to be used for the standard. All specifications that are taken from ISO 9241 will be cross referenced so that it will be easy for companies who are trying to comply with both standards. While harmonization is a goal, committee policy is to incorporate the substantive specifications from ISO Parts unless the committee identifies a compelling technical objection to particular specifications. Differences in the two standards are to be explicitly identified to make it easy for users to know when compliance to HFES-200 may not assure compliance with the software parts of ISO 9241.

There are approximately 7 separate software parts in ISO-9241 (Parts 10-17, excluding 11) which are planned to be incorporated in the proposed draft standard for HFES-200 (see the listing below). Taking a harmonized approach to standards development with this many separate parts has led to a number of issues and concerns.

Parts of ISO-9241 to be incorporated into ANSI/HFES-200:
 - Part 10: Dialog Principles (International Standard)
 - Part 12: Presentation of Information (Committee Draft)
 - Part 13: User Guidance
 - Part 14: Menu Dialogs
 - Part 15: Command Languages
 - Part 16: Direct Manipulation Dialogs
 - Part 17: Form Filling Dialogs

The largest issue is the fact that committee is trying to track a moving target. Only two of these parts (Part 10 and Part 14) are essentially finalized. The other parts are at various places in standards development process. Parts 12 and 16 are ones that have

been fluctuating the most. The continuously changing nature of these documents makes it difficult for the committee to cross reference the various guidelines and to determine whether or not there are compelling technical objections to particular specifications.

A second issue is the content of ISO 9241. Since the document has been under development for a long period of time, material is somewhat dated. In order to address the types of user interfaces and software that is currently being developed, the various parts of 9241 need to be updated by adding new specifications, new clauses to existing specifications and new examples and notes for existing specifications. The committee has been of the general agreement that these changes need to take place for HFES to be of use and interest to software developers.

Other issues include concerns about the phrasing of the ISO specifications. The majority of the participants in ISO committees do not speak English as their native language. As a result some of the wording used in the specification is awkward, unclear, or uses improper grammar. The committee has debated whether it improves the usability of the standard to change the wording of the ISO specifications or if it reduces usability.

Use of Color in ANSI/HFES-200

It is a rare application today that is offered without color. While there is debate about whether color enhances performance, there is no doubt that color is demanded by the majority of users. ISO 9241, parts 8 and 12, contains guidelines about the use of color, but the scope of the guidance provided by ISO is narrow (focusing on the high degree of specificity required for effective conformance testing) and their practical value to the user interface designer is limited. The effort undertaken by the ANSI-200 committee is an attempt to determine whether a broader range of guidelines can be identified that can be justified as being part of a standard, and whether these guidelines can be structured in a way that makes them usable to designers.

There are several formidable challenges that need to be addressed when writing guidelines for color. The colors in an application design can appear differently to a user depending on the characteristics of the monitor, the environment in which the monitor is being viewed, and other applications that my be simultaneously presented to the user. A significant number of users may be less sensitive to specific hues, and users differ in the color combinations they like and their ability to choose colors (when they have a choice) that are optimal for performance. In the multi-media world, pictures and video designed to represent the real world contain a rich palette of colors, and are typically presented in visually complex windowed environment. While hundreds of articles been written and many studies conducted on the use of color, they have most frequently been based on the limited display technologies that are increasingly rare. Further, it has been noted that it is not uncommon for them to contradict one another.

Nevertheless, it does seem to be possible to identify guidelines that are consistent with good practice, and the industry in general has largely converged on a core set of these guidelines that companies regularly include in corporate style guides. The guidelines that are forming the basis of the ANSI/HFES-200 standard are driven by the characteristics of the perceptual and cognitive systems. Additional guidelines logically flow from considerations of the problems designers will face when implementing many of these fundamental guidelines. Research has generally confirmed the performance impact of violating guidance based on human physiology, and has additionally identified common automatic responses and preferences associated with the use of color that need to be reflected in design. Given a basic set of principles that in general should guide design, the final challenge is how to translate the

principles into guidelines that are structured in a way that reflects the thought processes designers actually use when designing an interface.

Accessibility in ANSI/HFES-200

A significant percentage of people will experience disability during their lives. Such disabilities may be permanent or temporary; short-lived or chronic. Just as any other heterogeneous population, the characteristics and capabilities of people with any given disability vary significantly. In spite of their numbers, the needs of people with disabilities have only recently begun to receive attention in the design of software user interfaces.

Users with disabilities require access to information technologies just as they require access to buildings and services. Just as legislation has influenced building architecture, so too does a growing volume of legislation promise to influence development of increasingly accessible computer systems and software in both the U.S. (e.g., the American's with Disabilities Act (ADA), Section 508 of the Federal Rehabilitation Act, the Tech Act) and Europe (e.g., The European Commission's TIDE Initiative).

Given this push towards accessible information technologies, an obvious question becomes: what constitutes accessible computer software? The answer is not simple. The variety of interaction styles, needs, and capabilities of users with disabilities means that a broad net must be cast to capture a significant portion of those user requirements. The ANSI 200 committee is drawing from a broad range of existing practice, guidelines, and research to address this wide range of needs.

In an ideal world, every software environment would provide the support necessary for accessibility. That is, any unmodified software would be usable by the broadest possible user population without users having to provide additional assistive hardware or software. In cases where such assistive software or hardware were used, systems would integrate as effectively as possible with those assistive technologies.

As this notion of an "ideal world" for computer accessibility suggests, accessibility means maximizing the number of people who can use computer systems by taking into account the varying physical and sensory capabilities of users. By this definition, accessibility is simply a category of usability. Inaccessible software is by definition not usable! Given that the primary goal of HFES/ANSI-200 is to increase system usability, the inclusion of access for users with disabilities within the standard was critical to insuring truly comprehensive coverage of users and user issues.

Voice Input/Output and Telephony

One of the shortcomings of ISO 9241 is that it is limited, almost exclusively, to design guidance for graphical user interfaces. It contains only a handful of guidelines for interfaces involving voice and non-speech audio, and does not address telephony at all. Many voice technology developers and users have expressed the need for human factors guidelines for the design of voice interactive systems and applications, and the section on Voice Input/Output and Telephony in the ANSI/HFES 200 standard is an attempt to respond to this need. An additional "driver" for incorporating voice interface guidance in this standard is the sheer number of abominable interactive voice response (IVR) applications we've all encountered that, sadly, reflect the need for design guidance on the part of both systems and application developers alike.

Nevertheless, writing design guidance in this area is difficult for at least two reasons. First, the technology itself is in a state of flux, so there was initially some question as to whether useful design guidance could be written at this point in time. The Committee has, as a result, taken great care to avoid writing technology-

driven or technology-bound guidelines. It has attempted to write guidelines that are robust enough to survive changes in technology. Second, because ISO is currently working on a multi-media standard, we have steered clear of as many issues as possible related to combining media, although this is difficult to do in some instances.

The Voice Input/Output and Telephony section of the standard will be divided into four subsections. In the first, automated speech recognition is addressed, with the vast majority of guidance devoted to command and control voice interfaces. Speech dictation is minimally addressed. The second subsection addresses speech output, and the third non-speech audio. Both of these sections are relatively short and limited in scope. Finally, the largest subsection concerns interactive voice response systems.

In addition to addressing auditory interfaces in the ANSI/HFES 200 standard, we had, as a secondary goal, that this section of the standard be "liftable" for use as a standalone document by people developing voice interactive applications. Because of this secondary goal, the Voice I/O and Telephony section duplicates some material in other parts of the standard, mostly high level design guidance, that can be applied equally to voice and graphical user interfaces.

Designing Mimic Diagrams: Moving From Art to Science

Philip Moore
Colin Corbridge
DERA Centre for Human Sciences, UK
pmoore@dra.hmg.gb

This paper describes work to develop a graphical user interface for an experimental ship control system which utilises mimics extensively in its displays. Lack of a systematic method for mimic design resulted in mimics being designed using the collective expertise of a human factors development team. A methodology for mimic design recently proposed by Javaux et al (1996) is evaluated retrospectively with reference to the experience gained designing mimics for the Ship Control Centre (SCC). The seven stages of this novel methodology are discussed in terms of their applicability to the SCC development work and limitations within this methodology are identified. Suggestions are made concerning the future development of a principled method for the design of mimics.

INTRODUCTION

Mimic diagrams are used in many contexts for presenting information to users of technological systems. As part of the development of a graphical user interface for the control of a ship machinery control simulator a number of mimic diagrams were designed. Guidelines to assist in the development of mimics, and the identification of applications to which they are particularly suited, were found to be few and far between. The design of mimics tends to rely on the graphic design skills of the developers rather than any explicitly stated human factors principles *per se*. In this paper, experience gained in designing mimics for the control of ship's machinery is used to provide retrospective comments on a seven stage method for mimic design recently proposed by Javaux, Colard and Vanderdonckt (1996). The aim is to contribute to the development of a systematic method for mimic design, containing explicit technical guidelines.

The dictionary definition of mimic is "to imitate, copy minutely" or "(of a thing) resemble closely" (C.O.D., 1995). It is more difficult to derive an unambiguous definition of a mimic in terms of a display format. While it can be said that a mimic will resemble closely or take on the appearance of a system it is aiming to represent, other factors need to be considered. For example, mimics will be composed of individual elements many of which will be domain specific symbols.

The schematic map of the London Underground (Tube) is perhaps one of the most widely known mimics and is a classic example of this type of representation. This highly schematic mimic supports the user if their task is to find how to get from one Underground station to another. However, it is of no real use in aiding the user if their task is to travel, for example, to the station which is nearest to Buckingham Palace. This is because much geographical information has been omitted from the map and only important information concerning tube lines and stations has been abstracted, simplified and presented on the mimic. What is important in a mimic diagram is that the information which has been abstracted and presented on the display is appropriate to the users' task demands.

In an industrial context a network of pipes with controls for operating valves in the system might be represented as a mimic. The information displayed could be based on a similar process of abstraction in which geographical information relating to absolute location is omitted. Such a display would be sufficient for monitoring and controlling flow under normal conditions. However, if this mimic is to be used as a basis for directing emergency response to the outbreak of a fire then this representation will be sub-optimal. Geographical information required during fire management is not incorporated. For a task such as fire control a representation that gives *actual* locations of objects is likely to be required. (It is important to note that abstraction is not limited to spatial information but may involve functional or causal attributes of a system.)

A principled approach to the design of mimics requires detailed consideration of the following: the task the operator must perform; the types of knowledge required to perform the task and the understanding of the system the operator must maintain.

Our interest in mimic diagrams arose from the requirement to design new displays for a Ship Control Centre (SCC). Current Royal Navy ships use conventional, "hard-wired", panel mounted displays and controls including some simple mimic diagrams. The panel displays are large and cover several square metres. The aim was to replace the existing hard wired displays with four VDUs for the operator and one for the supervisor. The lack of display "real estate" in the new system required a radical review of the existing displays and resulted in the requirement for new mimic diagrams to be designed.

METHODOLOGY AND CRITIQUE

Traditionally the design of mimic diagrams follows an approach based on simplification and abstraction of the salient system attributes, usually viewed from an engineering perspective. However, a recent paper by Javaux et al (1996) proposes a systematic, "operator centred" method by which mimic diagrams may be produced. This method is not yet sufficiently comprehensive to cover the full range of mimic diagrams found in modern technological systems. Nevertheless it provides a valuable framework by which the design process can be discussed and is therefore described in detail below.

The method developed by Javaux et al was originally divided into seven stages and devised specifically to support the design of mimic diagrams. Figure 1 shows the seven stages with an additional sub-stage (2A) added (see below). Although this methodology was not followed during our own design process, each stage is discussed in relation to our experience in designing mimics for the SCC.

Stage 1. Contextual task analysis.

In Stage 1 the tasks that operators perform are described and analysed. Javaux et al (1996) argue that tasks should be characterised on three major dimensions: task type (monitoring, diagnostic etc.), frequency of execution and complexity. A contextual task analysis would typically be undertaken as part of the process of designing the

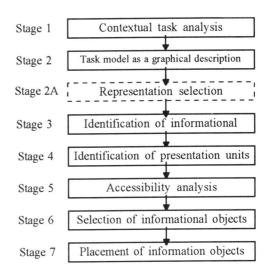

Stage 1	Contextual task analysis
Stage 2	Task model as a graphical description
Stage 2A	Representation selection
Stage 3	Identification of informational
Stage 4	Identification of presentation units
Stage 5	Accessibility analysis
Stage 6	Selection of informational objects
Stage 7	Placement of information objects

Figure 1: Seven stage method for mimic design proposed by Javaux et al (1996).

displays for any new system.

Developing displays for a new SCC involved video recording operators performing a representative range of tasks in a full scale SCC simulator. These recordings were then reviewed, and discussed with subject matter experts to provide detailed analyses of the tasks. In the case of new systems the acquisition of such prospective data, particularly those involving significant cognitive components, may be a significant problem (Hollnagel, 1996).

Stage 2. Expression of the task model as a Plan Based Description (PBD).

The tasks elicited from Stage 1 are formalised in a graphical description to illustrate temporal and hierarchical relationships between the task elements. The method focuses on the information which is to be displayed on the mimic diagram.

In the case of displays for the new SCC it was envisaged that Direct Manipulation Interfaces (DMI) would be employed (Shneiderman, 1982) to provide integrated display and control functionality on the mimic diagrams. Combining these control/display elements also serves to reduce the load on the operator who does not need to associate separate control/display elements. It is likely that many future applications in which mimics are used will incorporate controls into the displays, as was done in the SCC. This requirement will necessitate modifications to the method as currently proposed.

Display and control information required to perform particular tasks within the SCC was defined without reference to the format in which the information was to be presented. This information was provided in tabular format rather than in the more formal notation advocated by Javaux et al (1996). The information requirements were determined empirically from the task analysis. However, this was sufficient for the purpose of specifying the contents of the SCC displays.

A possible theoretical approach to the problem of defining what information is required on the mimic is provided by Sturrock and Kirwan (1996) who categorised the knowledge required by nuclear power plant operators to deal with complex scenarios. They defined six levels of knowledge used by operators ranging from "textbook theory" (level 1) to "tactical knowledge; problem solving and specific fault knowledge" (level 6). It was argued that categorising knowledge used in this manner would assist in defining the information required on the interface at a given time. The authors were also able to identify deficiencies in existing interfaces using this taxonomy. Such theoretical approaches are an essential prerequisite to the future development of a comprehensive method for mimic diagram design. They provide potential guidance on the information requirements faced by operators of systems which are currently under design.

Stage 2A. Selection of appropriate representation.

This stage was not included in the method proposed by Javaux et al (1996). It is an inherent assumption in the method they propose that the display being designed is a mimic diagram.

Our experience gained in designing the SCC displays suggest that this decision may be delayed until after Stages 1 and 2 have been completed. At this point it is appropriate to question whether a mimic is a suitable form of representation for displaying information about a particular system. There are at least two principle factors which need to be considered when selecting the most suitable representation. Firstly there is the requirement to minimise the short term memory load placed on the operator of the system (Shneiderman, 1992). If a user needs to hold significant information about the state of a system when performing tasks, and this can be supported by a spatial representation that mimics the system, then a mimic should be used to minimise operator loading. Various system "states" may fall into this category, including the position of an object(s) within a system or various system configurations.

A typical example from the SCC is the fuel storage and distribution system. In this system a number of different valves can be opened or closed to re-configure the system to supply fuel to the gas turbines and diesel generators. Fuel can be routed by a number of alternative paths to specific plant items. It is unrealistic to expect users to memorise the current state in which the system is configured hence the decision was taken to employ a mimic diagram for this system.

A second reason to use mimics is to support the users' mental model of the system. This is particularly evident when the user needs information about the locations of parts of the system that they are controlling and, for example, needs to set these into a particular configuration. Continued exposure to the mimic, which presents operators with the salient information about plant status, may be used to direct operators' attention to important process variables and develop their underlying knowledge of the plant layout. Mimics are also suited to representing linkages or relationships both within a particular system and between systems. An example of this would be using a mimic for diagnostic purposes. Here the user needs to know how the components of the system connect together and how these systems interact. Mimics may therefore be useful in supporting operators' knowledge about the underlying causal processes operating within the system.

Stage 3. Identification of informational needs and their representation in a graphical formalism.

This stage involves the initial layout of the mimic diagram(s) in a form which supports the operator performing specific, individual tasks. At this point no firm decisions are made concerning use of colour, symbology etc. These initial layouts are part of the scoping exercise designed to provide insights into the magnitude of the information which must be contained within the system.

In designing displays for the SCC it was possible to develop individual mimic diagrams for a range of systems including plant overview, fuel storage and generation, electrical generation, chilled water distribution etc.

Stage 4. Identification of presentation units and the subsequent allocation of tasks.

The purpose of this stage of the method is to determine what information must be contained on a single mimic diagram. The aim is to provide a mimic which supports a range of operator tasks. This will minimise the number of mimics which the operators must view to perform all of their tasks.

In practice Javaux et al (1996) suggest this "involves finding tasks that can be grouped together and supported by the same image".

For each system in the SCC a single mimic was employed to support monitoring and fault detection. The mimics were not designed to support in-depth fault diagnosis. It was not therefore possible to comment in detail on this aspect of the method.

Stage 5. Accessibility analysis.

This stage is concerned with navigation between mimic displays in the system. This is usually represented in terms of a hierarchy with vertical movements denoting moves to different levels of the system and horizontal moves to displays on the same level of the hierarchy.

In the SCC application an overview mimic, providing information about overall plant status, forms the apex of the hierarchy of screens. This allows access to mimics controlling electrical generation, fuel storage and distribution, chilled water distribution etc. The hierarchical structure was designed to be "broad but shallow" to minimise the number of levels the operators had to move through to obtain relevant information. The hierarchical structure is composed of screens providing information in a variety of formats including: mimics, control panels and text displays. It is likely that in any complex system the operators will require access to displays in a range of formats including mimics. Navigation between screens of different formats will require the provision of appropriate, consistent controls.

Stage 6. The selection of informational objects.

This stage is concerned with the selection of objects such as icons, formats and colour codes, quantities (e.g. digital or analogue values) and the functional grouping of relevant objects.

Existing user interface design guidelines (e.g. Smith and Mosier, 1986) although not specifically targeted to support the design of mimics were utilised in the development of the SCC mimics. Guidelines relating to consistency; the use of colour coding; text labelling etc. were particularly useful.

Much of the symbology used in the mimics for the SCC application is domain specific and this is likely to be the case in many process control applications. Symbology utilised should, where possible, be consistent with operators' past experience. Where control and display functionality is to be integrated on a novel mimic display animation of existing symbology to provide

feedback concerning control actions may be required.

An important consideration here is how control functions should be depicted to the user in order to indicate to the operators that these elements of the mimic can be manipulated to effect control actions. A theoretical perspective on this problem is provided by work in Ecological Interface Design (e.g. Flach, Hancock, Caird & Vicente 1995). Gibson (1979) introduced the theory of direct perception, which proposes that humans have evolved to perceive their environment directly from all the cues available within it. This bottom up theory of perception contrasts with the top down theoretical view which suggests that perception is hypothesis driven. Gibson also proposed the notion of "affordances", suggesting that all objects afford certain properties, and that these properties are also perceived directly. The idea of direct perception is particularly applicable to the design of mimics since the aim is for the user to be able to pick up the necessary information directly from the interface, thus minimising the need for high-level cognitive processing. The ultimate aim is to design system interfaces where the interface "affords" the properties of the system state directly to the user.

Stage 7. Placement of informational objects.

This stage is concerned with the spatial organisation of informational objects on the screen. Here, factors which influence screen layout such as consistency, clarity and simplicity are of primary concern.

Developing a consistent orientation for the mimics, with reference to the geographical layout of the ship, provided an important first constraint on the layout of information. Other more detailed aspects of consistency had then to be addressed e.g. text positioning on the mimics. In developing mimic diagrams for the SCC application particular problems arose in designing to minimise clutter on the display. Clutter may occur as a consequence of a number of factors including too much information being presented to the operator on a single screen and information being inappropriately presented etc. In designing mimics to reduce display clutter the following guidelines, often employed in designing other types of displays and derived from a range of sources, were particularly useful:

- Simplicity: information should be presented in the simplest way possible.
- Spacing and grouping: can be utilised to increase legibility of mimic displays and assist the

operator in assimilating the information presented on the mimic.
- Alignment: display elements should be positioned in lines or in matrices according to functional groups.

Both this stage and the previous stage are likely to be influenced not only by human factors considerations but also the aesthetics of the final design. Principles of graphic design are likely to be important in this context and any proposed method for designing mimics should make appropriate reference to this source of expertise.

CONCLUSIONS

The methodology for developing mimics, as originally described by Javaux et al (1996), represents a welcome step in the search for a principled design method. Based on our experience in developing mimics for a Ship Control Centre application a number of limitations in the method were identified by retrospective analysis. Firstly, it is assumed at the beginning of the process that a design decision has been made to employ mimic diagrams in a given context. On the basis of experience gained during the design of displays for a SCC it is argued that this decision should be made after the initial stages of the method, when contextual task analysis and task description have been completed. At this point an appropriate representation would then be selected which might be a mimic diagram, control panel or some other display format. Secondly, the methodology does not include explicit provision for controls to be integrated with displays on the mimic diagram. With the development of Direct Manipulation Interfaces this is an important concern. Thirdly, the activities within each stage of the model are specified at a relatively high level. For industrial practitioners and those involved in the production of mimic diagrams a more detailed set of guidelines is required to ensure that the design of mimic diagrams is optimised. Although Javaux et al provide some examples, and others are contained in this paper, there is a need for a comprehensive list of guidelines pertinent to the design of mimic diagrams to be compiled. Despite these limitations the method encourages a systematic approach to mimic design, and thus represents a significant first step towards the goal of providing operators with mimics which are optimised to support the tasks they are required to undertake.

ACKNOWLEDGEMENTS

Any views expressed in this paper are those of the authors and do not necessarily represent those of DERA.

REFERENCES

C.O.D. (1995) Concise Oxford Dictionary. Oxford: Oxford University Press.

Flach, J., Hancock, P.A., Caird, J.K., & Vicente K. (Eds) (1995) Global perspectives on the ecology of human-machine systems. New Jersey: Lawrence Erlbaum.

Gibson J.J.(1979) The ecological approach to visual perception. Boston: Houghton-Mifflin.

Hollnagel, E. (1996) Decision support and task nets. In Robertson, S. A. (Ed.) Contemporary Ergonomics 1996. Proceedings of the Annual Conference of the Ergonomics Society, University of Leicester 10-12 April 1996. London: Taylor and Francis

Javaux, D., Colard, M. I. and Vanderdonckt, J. (1996) Visual display design: a comparison of two methodologies. In Proceedings of the 1st International Conference on Applied Ergonomics (ICAE '96) Human Aspects at Work. Istanbul, May 21-24, 1996.

Shneiderman, B (1982) The future of interactive systems and the emergence of direct manipulation. Behaviour and. Information Techology, 1, 227-236.

Shneiderman, B (1992) Designing the User Interface. Strategies for Effective Human-Computer Interaction. Addison-Wesley: New York.

Smith, S. L. and Mosier, J. N. (1986) Guidelines for designing user interface software. Report number ESD-TR-86-278. MITRE Corporation: Bedford, MA.

Sturrock, F and Kirwan, B (1996) Mapping knowledge utilisation by nuclear power plant operators in complex scenarios. In Robertson, S. A. (Ed.) Contemporary Ergonomics 1996. Proceedings of the Annual Conference of the Ergonomics Society, University of Leicester 10-12 April 1996. London: Taylor and Francis.

A New Mechanism for Attention Direction in Information Management

Bruce G. Coury and Ralph D. Semmel
The Johns Hopkins University
Applied Physics Laboratory
Laurel, MD 20723-6099

ABSTRACT

Information systems can present an enormous amount of information to a user. One of the important issues in the design of the user interface is directing the user's attention to the critical information in a situation, especially when a user's request for information produces an uncertain result. The purpose of this paper is to describe how a unique approach provides a solution to the problem of attention direction while simultaneously representing the database designer's intent to the user to facilitate user/system communication.

Introduction

Information technologies have the capacity to generate an enormous amount of information from a wide variety of sources. The amount of information can be overwhelming, especially when a user must quickly assess a situation and select the appropriate course of action while operating under adverse conditions and severe time constraints. As a result, the user interface for such technologies must be able to effectively select the most relevant information and present it in a comprehensible form (Coury, 1996; Coury and Semmel, in press).

The enormity of the problem has been acknowledged by the information technology community. Information filtering has been a highly active area of research and development in information retrieval research area (Belkin and Croft, 1992), and a variety of methods for visualizing and navigating information spaces have been developed and reported in the literature (Lohse et al., 1994). How best to represent information spaces and provide access to them has been singled out as a critical area of future research in a recent National Research Council report (see discussion by Wickens and Seidler, 1995) on emerging needs for human factors research.

A fundamental component of such a design will be the mechanisms that direct the user's attention to critical information, especially when a request spans multiple information sources and produces an ambiguous or uncertain result. The purpose of this paper is to show how two areas of research -- cognitive research in attention and advanced query formulation -- can be combined to provide new ways of representing information at the user interface. In this paper we describe how such a mechanism can be developed by using conceptual schemas derived from an Extended Entity-Relationship (EER) Model (Elmasri and Navathe, 1994) to represent the knowledge in a database and generate the context for directing the user's attention. The paper will also show how this approach can communicate the database designer's intent (i.e., the designer's mental model of the database) and enhance user/information system communication.

The paper will begin with a discussion of the rationale for our approach, and then follow with a brief review of the background to the approach and how it will be used as the basis for attention direction in the user interface.

The Motivation for the Research

The research is motivated by two general trends in the development of information technologies. First, the types of information technologies that are being developed for large scale systems place many layers of data processing and information integration between the user and the actual sources of data and information. As a result, the user has potentially less direct knowledge of a situation and must rely on processed information stored in databases to assess a situation.

Second, and probably most important, is that users are becoming less knowledgeable about the specifics of a system and the way in which information is processed, represented and stored. This is especially troublesome when information from multiple sources can contribute to a situation in a number of different ways. For the user to adequately understand the value and contribution of information and make effective use of an information source, he or she needs to know how knowledge and information is organized, structured and used in an information system. Such knowledge is contained within the database design, but is rarely available in a useful form, especially for naive users. Current approaches to information systems design attack this problem with predefined views of database information. Unfortunately, the dynamic and volatile nature of the kinds of situations faced by users diminishes the utility of predefined views.

From our perspective, knowledge of the database design is critical to effective communication between a user and a system. This may seem to contradict current notions of user interface design that advocate insulating the user from any knowledge of the underlying characteristics of a database. We oppose widespread adoption of such a view for the following reasons:

1. Many of the information systems of concern to us (e.g., military command and control, and ship system supervision) are an integral part of a much larger information processing and decision making system. The information stored in databases represent the knowledge used by both users and decision aids to assess situations and plan responses to events. Consequently, understanding the source of information, how it was produced and then used by the various processing and decision systems is important to the user and the user's trust in the information system and decision aid.

2. Most of the users are highly trained and experienced decision makers who understand, at least at a global level, the overall information processing space and the sources of information that may be of significance in a particular situation. Thus, the target population of users for our research possess reasonable knowledge and understanding of the information space and are focused on assessing the credibility and utility of the information being provided to them.

3. If one assumes that a basic requirement of effective communication is mutual understanding of the knowledge possessed by the communicants, then some mechanism and representation must be found for the information system to communicate that knowledge to the user. In our work, the solution to that problem is contained within the

design knowledge of the database and the designer's *intent* in organizing information in a particular way. In this sense, our meaning of design intent is consistent with Suchman's (1987) notion of the designer's intentions regarding the use of a computer-based artifact.

As a result, our approach initially focuses on the ability to convey to the user the database designer's intent in organizing and structuring information in a particular way. Contained within that intent is the database designer's preconceived notion of the importance of information and how information elements are related. It is that understanding of the structure of information which provides the basis for our development of attention direction capabilities.

The Basis for Our Approach

In cognitive science, the structure and organization of information and knowledge in semantic memory is contained within the description of objects, object attributes, and the relationships among objects. Being able to describe the attributes of objects and the way in which those objects are related have been fundamental to such cognitive processes as categorization (Rosch and Lloyd, 1978) and concept formation (Estes, 1994; Smith and Medin, 1981). The notion of a *schema* or *prototype* also has a long history in cognitive research, and has been the basis for a number of models of human cognition (Anderson, 1993; Holland et al., 1986; Rasmussen, 1986).

The representations of concepts, categories and objects found in the cognitive literature can take many forms, including attribute lists (Estes, 1994), rule-based mental models (Holland et al., 1986), and working-memory elements (Anderson, 1993). All the representations provide a fundamental way to capture the knowledge contained in objects, the relationships among objects, and the structural form (such as the levels of abstraction of categorical information) inherent in the organization of a knowledge representation.

Similar schemes for representing the organization and structure of databases are found in the computer science literature (Elmasri and Navathe, 1994). Much like the cognitive representations of human memory, the EER model uses a *conceptual schema* as a knowledge-rich representation of the design of a database that characterizes the world being modeled in terms of entities (objects) and relationships. In Figure 1, an example of an EER conceptual schema is shown for a portion of a banking database; entity types are indi cated by rectangles, and relationships among entity types are represented by diamonds (current development is actually being done for military command and control information systems).

The conceptual schema plays an important role in the design of a database. The conceptual schema, as part of the EER model, is used in the database design process to capture the data and information requirements of users and provide a concise description of data types, relationships, and constraints. As part of the database requirements analysis, the conceptual schema is frequently used to communicate with users and refine the requirements for the database and information system. Once the conceptual design has been finalized, the logical and physical design and implementation of the information system can proceed.

It is important to point out that as a conceptual level model, the conceptual schema expresses the knowledge and structure of a database in a way that is closely related to the user's cognitive model of semantically-related conceptual objects. This similarity of representation provides the bridge between the user's cognitive model and the system's model of an information space, and provides the appropriate *cognitive fit* between problem and representation (Vessey, 1991). In our approach, the ability to exploit the system's model and the database de

signer's intent with a representation that is consistent with human cognitive models of semantic memory provides the basis for communicating system knowledge (and the designer's intent) to an end user. In addition, conceptual schemas provide the underlying structure for developing attention direction capabilities for the user interface.

Context in Attention Direction

In previous discussions of intelligent user interfaces (Coury, 1996; Coury and Semmel, in press), we have been particularly concerned with the role of attention in human and decision making. As a result, we have emphasized the development of mechanisms that will direct the user's attention to the information that is critical to a situation and the decision making process.

In its final form, such a mechanism (we call it the *Attention Director*) will assess the level and source of uncertainty and ambiguity in a situation, communicate that understanding to the user, and work cooperatively with the user to manage that uncertainty and reduce ambiguity. The emphasis on managing uncertainty is based on cognitive research that has demonstrated the importance of uncertainty in human information processing. Much of that research has emphasized the link between the cognitive demands of a task

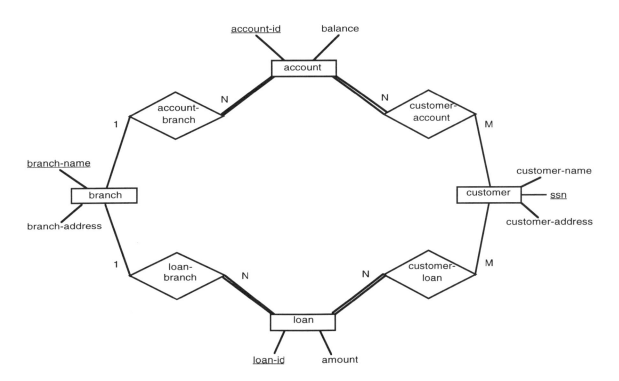

Figure 1: an example of an EER conceptual schema for a portion of a banking database; entity types are indicated by rectangles, and relationships among entity types are represented by diamonds.

and the correlational structure of that task, where the correlational structure of a task is defined by the mapping of task information to response categories (Garner, 1974). A person's ability to exploit that correlational structure and effectively allocate attentional resources can be significantly effected by the way in which information is represented (Coury and Boulette, 1992; Rasmussen, 1986).

A promising approach for an attention direction mechanism is contained within a system known as QUICK developed by Semmel (1993, 1994). QUICK enables a user to submit high-level requests that contain only the attributes of interest and the constraints that must be satisfied. By exploiting conceptual-level design knowledge contained in the EER model conceptual schema, QUICK formulates a semantically reasonable query that can be submitted to the database for subsequent execution. From an attention direction standpoint, QUICK can use knowledge about sets of semantically related conceptual-level objects to perform higher level reasoning tasks. One such set of objects is called a *context*, a meta-level construct that represents a set of strongly associated EER objects. Contexts can be used to determine how distinct EER objects are related.

We illustrate the use of context with the previous banking example. Suppose a user posts a request to identify all customer and branch associations. First, the contexts containing the specified objects are found. In this case, the two contexts indicated by Figure 2 are identified, showing that the relationships between customers and branches can be expressed either in terms of loans or accounts. Then the found context is pruned of extraneous objects, and the resulting subgraphs consisting of the relationship type loan-branch and customer-branch and its participating entity types are returned. The system then maps the pruned subgraphs to the underlying database, formulates a query, and pro

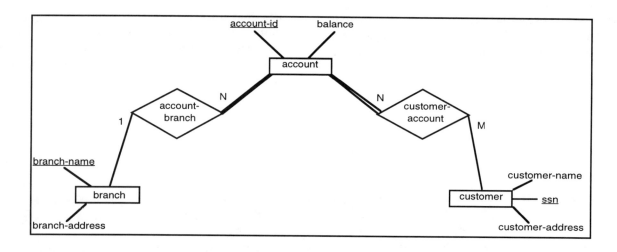

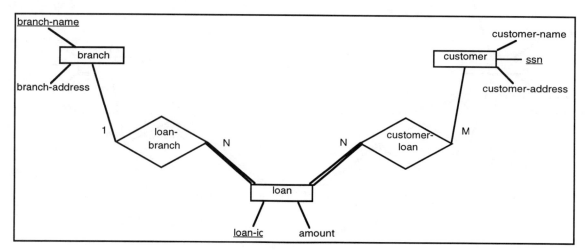

Figure 2: the two EER conceptual schema subgraphs showing that the relationships between customers and branches can be expressed either in terms of accounts (top) or loans (bottom).

vides the user interface with augmented subgraphs to indicate the reasoning used for information retrieval. The two subgraphs represent the sources of ambiguity in the request posed by the user. By presenting the two subgraphs to a user, we can illustrate the two ways in which the request can be answered, and express the ambiguity in terms of object relationships. This approach allows the system to convey to the user its knowledge of the possible ways in which the objects of interest are related. A simple method for resolving this ambiguity is to allow the user to select the appropriate context. In future work, the user will be able to not only specify which context is appropriate, but to modify a context or create a new context by specifying new relationships in the database.

Summary and Conclusions

We believe there are a number of distinct advantages of such an approach to attention direction. First, a conceptual schema provides an intuitive understanding of a database that can be used to explain the structure and organization of knowledge in a situation, and convey to a user the database designer's intent. Second, the conceptual graphs provide users with a way to select areas of interest or to dynamically regenerate contexts. Third, multiple contexts may indicate ambiguity in a situation, and can be used to identify the type, source and location of uncertainty in a situation. Thus, multiple contexts provide an important basis for directing the user's attention to the areas of uncertainty and ambiguity in a situation. In addition, the conceptual graphs can be used both as a representation of the problem and as a means for interacting directly with knowledge in an information system.

Acknowledgments

The authors would like to thank both the DARPA Ship Systems Automation Program managed by Capt. Robert Lowell, and the Army Research Laboratory Intelligent Databases Program managed by Robert Winkler for providing the resources to conduct this research. Without their aid and support, and the facilities provided by the Milton S. Eisenhower Research Center at The Johns Hopkins University Applied Physics Laboratory, this research would not be possible.

References

Anderson, J.R. (1993). *Rules of the Mind* (Hillsdale, NJ: Lawrence Erlbaum).

Belkin, N., and Croft, B. (1992). Information filtering and information retrieval: two sides of the same coin? *Communications of the ACM*, 35(12), 29-38.

Coury, B.G., and Boulette, M.D. (1992). Time stress and the processing of visual displays. *Human Factors*, 34(6), 707-725.

Coury, B.G., and Semmel, R.D. (in press). New directions in the design of intelligent user interfaces for supervisory control. To appear in M. Mouloua and R. Parasuraman (Eds.), *Automation and Human Performance: Theory and Applications* (Hillsdale, NJ: Lawrence Erlbaum).

Elmasri, R., and Navathe, S. (1994). *Fundamentals of Database Systems* (Redwood City, CA: Benjamin/Cummings).

Estes, W.K. (1994). *Classification and Cognition* (New York: Oxford Univ. Press).

Holland, J.H., Holyoak, K.J., Nisbett, R.E., and Thagard, P.R. (1986). *Induction: Process of Inference, Learning, and Discovery* (Cambridge, MA: MIT Press).

Lohse, G. L., Biolsi, K., Walker, N., and Rueter, H. H. (1994). A classification of visual representations. *Communications of the ACM*, 37(12), 36-49.

Rasmussen, J. (1986) *Information Processing and Human-Machine Interaction*. (New York: North-Holland 1986).

Rosch, E., and Lloyd, B.B. (1978). *Cognition and Categorization* (Hillsdale, NJ: Lawrence Erlbaum).

Semmel, R.D. (1994). Discovering context in an entity-relationship conceptual schema. *Journal of Computer and Software Engineering*, 2(1), 47-63.

Semmel, R.D., and Silberberg, D.P. (1993). An extended entity-relationship model for automatic query generation. *Telematics and Informatics*, 10(3), 301-317.

Suchman, L.A. (1987). *Plans and Situation Actions* (New York: Cambridge University Press).

Smith, E.E., and Medin, D.L. (1981). *Categories and Concepts* (Cambridge, MA: Harvard University Press).

Vessey, I. (1991). Cognitive fit: a theory-based analysis of the graphs versus tables literature. *Decision Sciences*, 22, 219-240.

Wickens, C.D., and Seidler, K.S. (1995). Information access and usability. In R.S. Nickerson (Ed.), *Emerging Needs and Opportunities for Human Factors Research* (Washington, D.C., National Academy Press), 200-219.

SCENARIO-BASED DESIGN FOR HUMAN-COMPUTER INTERFACE DEVELOPMENT

Dennis C. Neale and Jonathan K. Kies
Human-Computer Interaction Laboratory
Department of Industrial and Systems Engineering
Virginia Polytechnic Institute and State University
Blacksburg, Virginia

ABSTRACT

Scenario-based techniques have been receiving increased attention in the design of human-computer interaction. A cohesive methodology or framework, however, has yet to materialize, and scenario methods have not been well defined. Claims are being made about the ability of scenarios to play a role throughout the development life cycle. The objective of this paper is to examine the ability of scenarios to serve as the primary design representations early in the system design life cycle for envisioning the system, requirements specification, user-designer communication, and design rationale. These findings represent a case study in the design of a world-wide web site for the Human Factors Engineering Center at Virginia Tech. Example-based narratives were elicited using a "micro-scenario" generating task that involved prospective end-users brainstorming user-system interactions. Conclusions are drawn about the effectiveness of the technique for system development, and guidelines are provided for using scenarios to specify behavioral requirements.

INTRODUCTION

User-centered system design places the emphasis on the people who will eventually use the system, rather than on the technology surrounding a particular artifact or system (Norman and Draper, 1986). More recently, user interface designers are trying to expand the involvement of users in the design process by prioritizing their central role throughout system design. Participatory and Cooperative design are two examples of such an effort (Kyng, 1991; Kyng, 1995; Muller, Tudor, Wildman, White, Root, Dayton, Carr, Diekmann, and Dystra-Erickson, 1995). Scenarios can be an effective tool for involving users early on in system design. Many questions, however, regarding extensive user involvement in system design have been left unanswered. For example, how do different user classes with a variety of backgrounds and knowledge perceive a design domain and arrive at descriptive use-cases (scenarios), and how can these design representations then be exploited to make design decisions and claims? Scenarios can provide the initial requirements gathering for a system (Holbrook, 1990; Carroll and Rosson, 1991), indicating what needs to be designed to meet user's needs. This paper focuses on a "micro-scenario" generating methodology used in the initial stages of the development life cycle for: 1) *envisioning* what the system will contain and determining *system requirements* for how it will perform, 2) creating the *design specifications,* 3) providing user-designer *communication*, and 4) establishing *design rationale*.

Scenarios and Their Uses

A scenario is a representative instance of an interaction between a system and one of its intended users. One of the central premises of a scenario is its inherent "concretizing" of a design representation for purposes of analysis, synthesis, and communication (Nardi, 1992; Young and Barnard, 1992). Scenarios have a temporal component: A scenario is an episode or sequence of events that characterize a particular situation (Wright, 1992; Wexelblat, 1987). Carroll (1995) defines a scenario for a computer system as "...a narrative description of what people do and experience as they try to make use of computer systems and applications." The narrative component of a scenario is considered to be a defining characteristic. Example-based narratives (scenarios) allow users to express ideas from their point of view. Scenario-based techniques can facilitate user involvement, providing design representations for a variety of system design needs (see Carroll (1995) for a comprehensive list of the roles scenarios can play throughout the system development life cycle).

Generating Scenarios

Tognazinni (1992) and subsequently Nielsen (1995) discuss a "discount" scenario construction method which addresses the concerns for many users of a potential system. These designers used a scenario-building exercise that had members of the design team, with various backgrounds, brainstorm to generate lists of items under categories, such as *users of the system* and *information to be accessed by the user*.

Each list item, in essence, was a micro-scenario. Micro scenarios are less elaborate narratives which lack the richer contextual detail found in full-blown scenarios. Micro and elaborate scenarios can be developed with either empirical or analytical approaches. Carroll and Rosson (1992) point out that empirical and analytical approaches to scenario generation are complementary for developing a scenario set of adequate scope. Nevertheless, an empirical task analysis approach is imperative for developing the initial set of scenarios for system development. Users should be involved at the very onset of the system development to insure that the designers are "on track." Our initial approach to determining the scope of system requirements involved a use-scenario task analysis methodology. An important distinction between Tognazinni's and Nielsen's methods and the method used in this project is that they had design team members (analytic) develop scenarios, whereas actual prospective end-users (empirical) were used for this project. Our cost-effective (discount) approach involved users brainstorming micro-scenarios. The level at which task scenarios are generated is at the "...level people construe their work to themselves: the level at which tasks become meaningful to the people who engage in them." (Carroll and Rosson, 1992). This level of task description and analysis lends itself well to an empirical approach that directly involves users. This initial effort provides the set of basic-level task scenarios for system development.

METHOD

Many user classes are expected to access the world-wide web (WWW) site for the Virginia Tech Human Factors Engineering Center (HFEC). Therefore, a variety of user groups were asked to spend approximately one hour generating scenarios by employing a semi-structured brainstorming method, an approach that focused on involving different groups of people cooperatively from a variety of backgrounds. Table 1 outlines the basic differences in these user classes. By no

were rough guesses made by the experimenters and were not based on formal evaluations.

Because this technique involved only a blackboard and some imagination on the part of participants, it was a very low-cost approach to developing scenario-based design representations. In addition, it has much in common with Muller's et al. (1995) notion that low-tech representations keep the language in the user's domain, allowing them to easily express their viewpoint. The user-designer meetings began with information being provided on the background of the WWW, HFEC, and the micro-scenario brainstorming procedure. The following categories were used to develop lists.

- Users
- User's backgrounds
- User's goals / needs
- Information to be included
- Your goals

Each user class from table 1 was asked to cooperatively fill in the lists from their perspectives as it related to the content domain. Finally, with the experimenters facilitating the process, users were guided in gradually constructing scenarios by combining list items and elaborating on each item to provide further detail and context. The advantage of this method is that it encompasses a systematic approach to the development of scenarios, which may not be tangible or immediately obvious to all the users of a potential system. All sessions were videotaped so that scenario details from list items could be captured.

FINDINGS

Table 2 shows some sample brainstorming output from user classes. Figure 1 provides one example of a detailed scenario generated from the lists in Table 2. The relationship of participants with respect to the HFEC website produced interesting biases in the content of list items and scenarios

Table 1: Characteristics of User Classes

User Class	Knowledge of Content Domain	Knowledge of WWW	Knowledge of Scenario-Based Design	Number in Group
Human Factors Faculty	High	Medium	Low	3
University Department Head and Lab Manager (Alumni)	Medium	High	High / Medium	2
Records and Advising Personnel	High	None	None	2
Human Factors Graduate Students / Student Chapter Officers	High	High	Low	2
Prospective Students	High	High/Medium	Low	5

means do the groups listed in Table 1 represent all user classes who may potentially access the system. These groups were selected because they represent the major user classes and were easily accessible. The knowledge levels of users in Table 1

generated. Often, participants would think in terms which corresponded most closely to their specific role, needs, and experiences. This finding is not surprising, especially considering that participants were explicitly instructed to "wear the hat" of their position when brainstorming. This result stresses the need to have a full range of representative users for a potential system during scenario generation.

In general, users with knowledge of system design removed themselves from their positions to a greater extent, forcing more abstract items to be generated. This abstractness resulted in the absence of detail and specific narrative content. These users produced the fewest scenarios and often generated list items with considerable brevity. List items generated were not embedded in rich descriptions or narratives. The narratives that were generated were based on previous experience the individuals had with dissemination of the kind of information to be included on the website. The participants who had the least knowledge and experience with design projects were able to effectively produce scenario descriptions in great detail. Not surprisingly, the group with

methodology is not an essential prerequisite for creating rich scenarios due to the fact that the individuals with no knowledge of design methods were effective at providing detailed descriptions. The utility and robustness of the method was demonstrated by the fact that useful scenarios were generated by users in spite of a large range in knowledge for the content domain and system design expertise.

Group size appeared to have little effect on the brief exercises. A minimum of two individuals is helpful, but more does not add significantly to the brainstorming output. In fact, with the largest group (5), some individuals were able to "blend in" and avoid contributing equally. In the smaller groups of 2 individuals, if one person made many comments, the other seemed to feel compelled to contribute. An optimal group size is perhaps somewhere between 2 and 5 participants.

Table 2: Examples from brainstorming lists generated by different user classes

Users	User Backgrounds	User Goals/Needs	Information to be included	Your Goals
• Prospective students • Other HCI people • Other faculty • Visitors • Sponsors • Other researchers • Publishers • Parents • High school counselor • Alumni • Employers • Campus visitors • Former students • Users looking for human factors consulting • Lost users • International students • Transfer students	• Computer experience • High school students • English reading level	• Learn • Navigation ease • Program comparison • Obtain software and research data • Repository • Link to related information • Concise explanation • Smaller home page vs. glitzy • Program vitality vs. need for funding • Search capability • Send for more information - form filling • Understanding admissions process	• Papers • Curriculum • Professors vitas • Calendar of events • Map • Directions to Virginia Tech • Course catalog • Course descriptions • Faculty profiles • Tour of labs and campus - video • Current sponsors descriptions • Frequently asked questions • Student personal information • Available funding • Previous research • Organizational structure • Application procedure • Admission requirements	• Utilize others information • Rich content • Search for collaboration - research • Advertising the HFEC • Software techniques • Learn from others • How to contact people in the HFEC • Video conference • Get a job

Alumni could be a low level or mid level manager in a high tech company in northern Virginia. Fifty or seventy people and do a lot of work with the government. That person may be more interested in recruiting, in government relations, and what areas we are working in. Problems of organizations like ours is one of just keeping in touch. We rely on alumni for many things, not just money. Hiring our graduates. We get input from them. They help us stay in touch where the field is, where the industry is going. Because in the technical disciplines, if we don't stay up to date, we are dead. I think we could rely on them for many things. It's a great potential medium for facilitating interaction. And the alumni want to be involved. They would like to know current activities. They want to see if Tech is keeping up with current technology.

Figure 1: Alumni micro-scenario

the most overall experience and knowledge of the scenario-based design approach was able to capture much of the essence of the scenario by describing considerable detail and context in each list item. Many of the comments were described in highly concrete language, relating prior experiences and situations. However, intimate knowledge with the

DISCUSSION

After working with the different groups in the brainstorming session, several observations can be made concerning the method's effectiveness. In general, regardless of differences in user class, the lists generated were equal in length and richness; however, content did vary somewhat depending on the background of users. Furthermore, as outlined above, how the lists were generated varied considerably between the different user classes. Some groups generated lists by providing list items abstractly, while others provided all list items embedded in micro-scenarios. Clearly, if the intent is to generate micro-scenarios, the procedure needs to be modified to facilitate more context-rich information. When the information was given in micro-scenarios, lists were generated rapidly and simultaneously. An interesting byproduct of list items being embedded in micro-scenarios was that the interviewer had to try and determine what components of the micro-scenario went into which list categories. This further facilitated discussion between the designers and the users. If the information was provided context free, the lists were generated in a sequential manner. The most important lists, and the easiest for users to generate, were 1) *Users* and 2) *Information to be included*. Regardless of whether list items were couched in micro-scenarios, users continually returned to the *Users* list to further develop the other lists or other micro-scenarios. When trying to develop the list items, information naturally overlapped between lists.

Particularly interesting is that regardless of user backgrounds, much of the information in lists across groups was similar (although scenarios tended to be more distinct to a user class). This reassured the designers by confirming that the most important information had been gathered during the initial stages of design. Nielsen (1995) referred to the list items as micro-scenarios, but until the designers of this project constructed lists of their own, they did not fully appreciate why the lists contained micro-scenarios. This leads to a very important point: Lists, in and of themselves, do not convey the rich context contained in micro-scenarios. Context-based narratives are needed to supplement the more abstract list items. Videotaping or audiotaping the sessions is essential for preserving the descriptive element lost when compiling list items in real time.

Functional specification and envisioning the system

One use for task analysis traditionally has been to specify system requirements by subdividing tasks into their basic components for purposes of analysis and design. Wexelblat (1987) makes a distinction between functional and behavioral specifications. Functional specifications are the traditional list-like functions and features of a system; they are system oriented. Behavioral specifications describe how the system behaves from the user's perspective. Wexelblat conceptualizes system behavior using a scenario-based specification procedure early in the life cycle, establishing a dialog between users and designers to provide not only the functional specification, but the behavioral requirements that are usually absent from standard requirements analysis and specification. List items and the narrative scenarios, along with the software and hardware capabilities and limitations, were quite adequate for developing the functional and behavioral design specifications. In fact, the "meat" of the

requirements specification document resulted from analysis and synthesis of the scenario material. Furthermore, the narrative aspects of the scenarios excelled at providing the behavioral specifications—the video material being especially well suited. Salomon (1993), introducing a novel design concept, has demonstrated the usefulness of relying primarily on visual materials for communicating design specifications. This form of requirements analysis provides an explicit picture for envisioning what the system will look like and do.

User-designer communication

Erickson (1995) outlines the social process of design in which communication is central among users and designers, within an organization, and between the designers themselves. In particular, he identifies the story-like nature of scenarios as valuable for the initial exploration of a system with these various groups. Scenario stories (real or imaginative) are especially suited for user-designer communication because users are comfortable with this form of design expression. Scenarios as design objects, shared between the user and designer, allowed the designers in this project to maximize user input early in the development life cycle before physical presentation of the system to the user was possible. As a result, scenarios were a valuable communication tool for potential users and designers alike.

Design rationale

The design of large systems often implies shifting personnel, dynamic motivations, and changing technology, which create a difficult environment for designing usable systems. By codifying design decisions in an explicit format, many of these problems can be overcome. For example, new design team members can quickly understand why certain decisions were made without wasting the time of other team members. This principle is particularly relevant to the design of this WWW interface because the project is dynamic and team members change on regular basis. Documents, video, and audio transcripts resulting from this study preserve the scenarios of use which are supported by the current system and are available for future designers to review.

CONCLUSIONS

Based on our experience using the list-building technique, the following changes to the method outlined in this paper are recommended:

1. Users naturally understand how to describe their experience in a scenario-based format. Avoid the technical language when describing the procedure to users. Inform the users that they are being asked to brainstorm ideas and ways of using a particular system: encourage a storytelling format.

2. For the micro-scenario list-building method (one hour) using targeted end-users, it is suggested that only the

Users and *Information to be included* list categories be used. Because the scenarios often lacked significant detail, categories overlapped, and users found it very difficult to generate list items for particular categories, for example, *User backgrounds*. However, it should be noted that the more experience users had as system designers, the better they generated lists for all the categories. Therefore, the categories for scenario generation should be modified according to the background of the users in the scenario exercise, designers goals, and the system of interest. Insuring that appropriate categories for list items are chosen is a critical step to safeguard that representative scenarios get developed and expanded upon. Pretest list categories.

3. List items should be developed simultaneously during storytelling (scenario) generation, not sequentially (given as simple list items under categories). Even when users were instructed to develop the lists sequentially, items were often given simultaneously in micro-scenarios. Rather than suggesting that the lists be constructed, and then returned to for scenario development, the micro-scenarios themselves should provide the material for list items.

4. It is important that the list-building sessions be captured on at least audiotape and preferably videotape. This material contains much of the context for list items and is lost if not recorded.

In terms of a scenario-based approach for task analysis, scenarios excelled in developing the initial set of functional and behavioral requirements. Scenarios also allowed the users and designers to envision how the system would look and function before other more tangible representations were available. In the process of working with scenarios for the design stages outlined in this paper, decisions for design rationale were captured in a format that was understandable and accessible to a variety of audiences concerned with the design.

By focusing on scenarios as design representations and on the concrete elements of the situations from the users' point of view early in the development cycle, the designers for this system feel that scenario-based design has been and will continue to be a useful tool. Guindon (1990) claims that top-down decomposition is ill-suited for the unstructured nature of design in the early stages. Clearly, this project has benefited from the more opportunistic decomposition design afforded by scenarios as the primary design object. The micro-scenarios generated in this project have been very useful in evoking reflection on the tasks, concerns, experiences, and needs of the prospective users, rather than simply on system functions.

Note: URL for the HFEC at Virginia Tech:
http://hci.ise.vt.edu/hfec/

ACKNOWLEDGMENTS

The authors wish to acknowledge the support of the National Science Foundation while conducting this research. Dr. John M. Carroll is thanked for conceptual input and Dr. Robert C. Williges is thanked for encouraging the development of the Human Factors Engineering Center's WWW site.

REFERENCES

Carroll, J.M. (1995). The scenario perspective on system development. In J.M. Carroll (Ed.), Scenario-based design: Envisioning work and technology in system development. New York: John Wiley.

Carroll, J.M. and Rosson, M.B. (1991). Deliberated evolution: Stalking the view matcher in design space. Human-Computer Interaction. 6, 281-318.

Carroll, J.M. and Rosson, M.B. (1992). Getting around the task-artifact cycle: How to make claims and design by scenario. ACM Transactions on Information Systems, 10, 181-212.

Erickson, T. (1995). Notes on design practice: Stories and prototypes as catalysts for communication. In J.M. Carroll (Ed.), Scenario-based design: Envisioning work and technology in system development. New York: John Wiley.

Guindon, R. (1990). Designing the design process: Exploiting opportunistic thoughts. Human-Computer Interaction, 5, 305-344.

Holbrook, H. (1990). A scenario-based methodology for conducting requirements elicitation. ACM SIGSOFT Software Engineering Notes, 15(1), 95-103.

Kirwan, B. and Ainsworth, L. K. (1992) A Guide to Task Analysis. London: Taylor and Francis.

Kyng, M. (1991). Designing for cooperation - cooperating in design. Communications of the ACM, 34(12), pp. 64-73.

Kyng, M. (1995). Creating contexts for design. In J.M. Carroll (Ed.), Scenario-based design: Envisioning work and technology in system development. New York: John Wiley.

Muller, M.J., Tudor, L.G., Wildman, D.M., White, E.A., Root, R.A., Dayton, T., Carr, R., Diekmann, B., and Dystra-Erickson, E. 1995. In J.M. Carroll (Ed.), Scenario-based design: Envisioning work and technology in system development. New York: John Wiley.

Nardi, B.A. (1992). The use of scenarios in design. SIGCHI Bulletin, 24(4), 13-14.

Nielsen, J. (1995) Scenarios in Discount Usability Engineering. In Carroll, J. M. (Ed.) Scenario-Based Design: Envisioning Work and Technology in Systems Development. John Wiley and Sons: New York, NY.

Norman, D.A. and Draper, S.W. (1986). User centered system design. Hillsdale, NJ: Lawrence Erlbaum Associates.

Salomon, G. B. (1995). Iterative design of an information kiosk. In R.M. Baecker, J. Grudin, W.A.S. Buxton, and S. Greenberg (Eds.), Readings in human-computer interaction: toward the year 2000 (2nd ed.), (pp. 25-34). San Francisco, CA: Morgan Kaufman Publishers, Inc.

Tognazzini, B. (1992) Tog on Interface. New York, NY: Addison-Wesley.

Wexelblat, A. (1987). Report on scenario technology. Technical Report STP-139-87. Austin, TX: MCC.

Wright, P. (1992). What's in a scenario. SIGCHI Bulletin, 24(4), 11.

Young, R.M. and Barnard, P.J. (1992). Multiple uses of scenarios: A reply to Campbell. SIGCHI Bulletin, 24(4), 10.

EXTENDING USER-CENTERED METHODS BEYOND INTERFACE DESIGN TO FUNCTIONAL DEFINITION

John F. ("Jeff") Kelley, Susan L. Spraragen, Lauretta Jones, Sharon L. Greene, Stephen Boies
IBM T.J. Watson Research Center
Yorktown Heights, New York

The contributions of human factors or usability practitioners to application development often begin with a functional specification handed down from an external source. User-centered design methods are commonly applied to *how* function is delivered but not *what* functions will be delivered. We in the Interactive Transaction Systems (ITS) group at the T.J. Watson Research Division of IBM have succeeded, during several application development efforts, in expanding the scope of our user-centered, iterative design approaches to include functional as well as interface definition for both software and hardware (kiosk/workstation) design. By learning our customer's business and owning the entire development process, we can better design our solutions to solve their problems (and their client's problems in the case of service industry solutions). We achieve this by including the functional definition in the first of four phases we have defined for all of our development projects.

A significant facilitation for this in the arena of software development has been the CADT (Customer Access Development Toolset) development platform we use to build our applications. This set of tools for iterative application design and development gives us the flexibility to quickly and effectively address emerging functional requirements.

INTRODUCTION

Students in human factors design courses have long been taught that a successful designer must begin by stepping back and taking a systems approach to defining the problem and subtask suite to which a technological solution will be applied (c.f. Chapanis, 1970). This perspective has a traditional importance, predominantly in supporting allocation of functions or tasks to humans or to automation (McCormack, 1976). All too often, however, a human factors design group is brought in after the overall functional specification for an application has been determined, sometimes by a marketing division, sometimes by the customer. This paper describes a four-phase process we have evolved to bring human factors considerations and continuing user participation (Holtzblatt & Beyer, 1995) to all aspects of functional and interface design.

PHASES OF APPLICATION DEVELOPMENT

We have evolved a four-phase, iterative process for developing and delivering software applications, based on the four basic principles of usable system design (Gould, 1987): early and continual focus on users, integrated design, early and continual user testing, and iterative design. The four phases we use are:

1. Discovery: Learn customer's business; Establish user-constituencies; Teach customer about capabilities of our technology; Establish initial functional specification; Establish value (if any) of proceeding for each constituency.

2. Proof of concept: Quickly bring customer a proto-application stub to bootstrap the application.

3. Pilot iteration: Build initial, coherent application in concert with continuing user testing.

4. Application rollout: Enterprise-wide delivery of fully functional system; continue iterations of function and interface after delivery.

Each of these phases brings value to our customers and it is not uncommon for us to establish a remunerative contract for each phase. However, it is the sum total of all of these phases that make for a successful application.

This paper describes the expansion of the scope of Human Factors contributions into the area of functional specification. Six example (four kiosk-based and two desktop workstation) applications are used to illustrate how we have brought this idea into practice:

Expo '92 (Guest Services System) -services and information kiosks for visitors to the 1992 World's Fair (Kelley, et al, 1993)

Touch Illinois -Customer access kiosks for the Department of Employment Services

NYPL -New York Public Library resource reservation kiosks

AMNH -American Museum of Natural History lobby way-finding kiosks

EMS -Electronic Marketplace desktop system for online corporate purchasing

Auto Finance -Networked dealership desktop workstations for arranging auto loans

DISCOVERY PHASE

Human factors practitioners are integral members of our development team from start to finish; they join in a thorough task analysis at project inception. It is crucial to go to the customer and really understand the business environment from the perspective of an "operational" or domain expert, and not just as a "technical" expert (Wallace, et. al, 1995). Application development can only be truly successful if it addresses a real problem. Ideal customers recognize that they have a problem and that they want to solve it; their desire is great enough that they are open to new and exciting solutions.

Often however, customers will not fully understand the problems the design team has been approached to solve or won't appreciate the extent to which technology can address their problems. The developer's job here is to make effective "value propositions" to each of the affected user groups or constituencies, to show how the proposed technology will make their job faster, more effective (and more fun).

Constituency Definition

We begin the discovery phase by understanding the constituencies on whose behalf the development will proceed. Each brings a new perspective to the functional definition. In our engagements with customers on the EMS, Touch Illinois, NYPL, and AMNH projects, we have developed the following generic set of constituencies.

Customer -Corporate entity that hires us.

Champion -High-level executive at Customer who believes in us and in the mission we undertake.

Executive -Customers who make go/no-go decisions about project; in the best of worlds, the Executive is also the Champion.

Supervisor -Manages work of staff; often has come up the ranks from staff position; often no longer works directly with clients, but has domain expertise based on the old way of doing things.

Staff -Customers who currently do the work the old way with clients.

Client -End users of application ("customers of our customer"). In client-user applications, there are often interfaces used by staff with support/management functions (and styles) that differ from the ones used by clients.

It is critical to gain executive support for the developer's requirement to have full access to all constituencies; we cannot derive a valid idea of client needs and perspectives by proxy.

In the Auto Finance project, we used a proprietary consulting methodology based, in part, on the work of Winograd & Flores (1987). We derived a more complex, three-tiered taxonomy, specific to that customer. The "bank" tier comprised executive, legal, marketing, finance, and information services roles; the "dealer" tier consisted of owner/general manager, finance, sales manager, sales, and office staff; the third tier was "consumer".

As we penetrated each layer of our constituency, new functional requirements came to light. Executives at our first banking customer for Auto Finance first became interested in our work after seeing a demonstration of Touch Illinois. Database-assisted input and other interface techniques opened up the possibility of more accurate end-user data entry, moving this task out into the field and decreasing loan approval turn-around time. The Information Services department saw an opportunity to have loan application transactions formatted in the field eliminating bottlenecks in workflow. Dealer general managers wanted same-day funding to improve cash flow. Dealer Finance folks saw 5-minute loan commitments streamlining back-office functions, but insisted on an open system that would connect to an arbitrary number of lending sources. Sales staff envisioned a more effective tool for closing sales.

Some sales managers wanted the system to be usable by the consuming public (who saw it as a less-threatening way to deal with the personal matters in loan applications), and other sales managers wanted to "own" the loan interface in order to increase "face time" with customers.

The point of this example is that the developers must persist in efforts to gain full access to all the layers of constituency all the way up to the level supervisor/staff if all useful potential function is to be uncovered. To each constituency, the value added by the proposed application is specified and each constituency is included as partners in development; this has the nice side-effect of building a broad base of support for continued development phases.

During the first phase of the AMNH project, the Executive introduced us to heads of all the departments she thought would have an interest in the visitor way-finding kiosk application, not just the department currently responsible for this function. We approached each interview in an open way to learn as much as we could about each department's tasks. We asked questions about what, of the existing work, was repeated, mundane, or difficult. Implicit was the assumption that there were technological solutions possible in the proposed application. In the process, we developed a long laundry list of potential functions and established "advanced publicity" in different quarters for the application still-to-come. We also learned things that had direct impact on the core function from unexpected places. The Events Manager informed us that certain exhibits and hallways are blocked off from time-to-time (a definite impact on way-finding!) which would require moving the kiosks (wheels suddenly a must) and notifying visitors (dynamic updating of pathfinding algorithms).

Application Champion: One key to a successful customer engagement is getting a sufficiently high level executive on board as early as possible. This application champion must be enthused about the project and must have a sufficient ego stake in the outcome (and confidence in the developers) to sustain the project throughout all the phases. In the EMS project, our champion was a high-level executive in the Corporate Procurement division who was able to open doors for us (sometimes literally during site maintenance visits!). Customer champions are not always corporate executives; in the Touch Illinois and Expo '92 projects our champions were political appointees.

Staff Support: In client-user applications, support of staff roles is particularly important. If they feel threatened or out of the loop, they can block access to clients or otherwise make life difficult for developers. We were fortunate in the Touch Illinois project to have staff members on board early on (they were key contributors to the functional design and had a significant interface component of their own in the final application). In AMNH, the museum staff (guards, information desk volunteers) helped us gain access to the ultimate end users -- the visitors. They also helped us learn about the questions most frequently asked by visitors. Again, these were critical inputs for the functional design and content of the application.

Information Services Support: In network-centric, transactional applications such as the ones described here, it is also vital to gain the support of the Information Services or Data Processing department. We have to be sensitive not only to the impact our application will have in front of the screen, but also to the impact this technology will have on the folks maintaining legacy (existing/previous generation computing) systems behind the scenes. Often, active support from Information Services groups is required to get the application working, sometimes in ways that show no obvious direct benefit to the existing Information Services functions. We have been fortunate to get lots of proactive and patient Information Services support in Touch Illinois and EMS projects including advanced notice of upcoming changes in legacy systems to which we were connected, after-hours help in setting up special userids and network gateways, special insights into pitfalls and quirks in existing database update procedures, etc..

In addition to understanding the taxonomy of user constituencies, it is important to develop an understanding of the political ecology in which these roles exist; turf wars and other interactions among roles can have an impact on application design.

Methods

The primary information-gathering techniques we use in the discovery phase are one-on-one (or few-on-few) interviews and informal critical incidents. Principles of contextual inquiry (Baecker, et. al., 1995) and apprenticing with the customer (Beyer & Holtzblatt, 1995) guide our efforts to fully understand our customer's needs apart from whatever

technological impediments may currently be in place. As an example, one of the tasks undertaken by developers in the Touch Illinois project was to learn how to fill out the paper-based unemployment application forms.

For public-access applications, the site-walk is crucial; this is not a guided tour with executives or staff, this is an independent, in-depth walk-through where developers observe client-users in the target environment.

The AMNH site-walk taught us many things about the ecology (graphical and social) in which our application would exist that we could not have learned effectively in back-office interviews. As an example, our team heard seven different languages spoken by visitors at the museum, we made note of color schemes used in existing signs and banners, and we learned about the different physical constraints our kiosk and application would have to operate under (lighting, noise, traffic flow patterns, etc.).

The Functional Specification Document

We don't use them. More accurately, the functional specifications for the applications we develop are embodied in the applications themselves. Rather than write the emerging functional requirements down on massive amounts of paper, we code them right into the emerging application. This saves time in authoring, review and approval of documents (which have no value in themselves); this also shortens our iteration and delivery time tremendously. It also provides a much clearer portrayal of the proposed functions since the specification is in the "language" of the ultimate implementation of those functions. This, of course, only works if the development platform supports it.

The primary outcome of the discovery phase is, instead, an informal list of possible functions; a small number of which will be encoded in the bootstrap proto-application delivered in the next phase. This list will be added to, prioritized, and implemented in the pilot iteration phase.

Go/No-go Decision

One non-trivial outcome of the discovery phase is the decision on the part of the designers on whether or not to proceed with the engagement. An example might be a customer who, for political or other reasons, has a rigid requirement for a glitzy-looking lobby kiosk and is not interested in uncovering and meeting real needs of any particular group of users. Another example might be that, after considerable dialog in the first phase, a customer still underestimates the capabilities of his or her clients to accomplish meaningful work on a well-human-factored application and resists commitment to a truly useful product. We have learned that we must be prepared to walk away.

PROOF OF CONCEPT

In this phase, we bring the customer a functioning proto-application to demonstrate the key concepts.

This coinage of "proto-application" is a studied avoidance of the term "prototype"; the latter implies a function-impoverished model of an application which will be discarded once the "real programming" begins. We believe that a substantial improvement in effectiveness is gained through the use of a development platform that supports iterations on the deliverable code base.

This quick demonstration helps the customer application champion to "sell" the idea broadly within the organization. It is important that the development platform enables delivery of a proto-application early (i.e., within hours or days, not weeks or months) in order to guide the customer's expectations. In Expo '92, we quickly created the proto-application that was demonstrated to high-level executives in late 1989 -- 2.5 years before opening day. This served to excite the customer and gain political support for the project throughout the various constituencies involved.

As soon as the project is proposed and the functions begin to be uncovered, the customer is going to have a heightened awareness of related technologies, some of which may not be appropriate for solving their immediate needs. Seeing our technology transposed into their domain gives them a framework for thinking about how technology can be applied to their problems and can generate ideas that go beyond the boundaries of the previously existing task.

After working with the proto-application, the customer is equipped to begin prioritizing the ideas generated in the brainstorming portion of the first phase.

An early application stub is also a vehicle for establishing contact with end users (clients). Preliminary, isolated explorations of interaction techniques can begin here, in advance of iterative development (where techniques are employed and studied in the context of the actual emerging application). We have been successful using traditional human factors techniques (c.f. Shneiderman, 1980) in this context. An example of an independent orthogonal study which spun off from an earlier online timecard project can be found in Kelley & Ukelson (1992). An example at NYPL was an early test of some soft keyboard designs for library patrons, that population comprising more sophisticated keyboard users than users of previous ITS kiosk soft keyboards. We found a requirement for keyboard layout and function that more closely matched what is found on mechanical, QWERTY keyboards. (An interesting anecdote of this small study was that, during preliminary tests of a sub-function that collected people's opinions, we found our first requirement for an exclamation point on our soft keyboard!)

PILOT ITERATION

By iterating the design on an installed application, we maintain a vehicle for continuing discussion of the emerging functionality (and interface design) with actual users or clients in the actual environment of intended use. In the Expo '92 project, there were many challenges we had to face concerning the design of the kiosk hardware for use in the hot, Mediterranean sun and in the design of usable, useful functions for a multi-national user population. We had two proto-applications installed at the Expo site fully two years in advance of the opening of the Fair. In fact, when we installed our first model kiosk, there was nothing on the 2 square-mile site except dirt and a visitor's center. This didn't stop busloads of tourists from coming by to see the construction however. We stationed a developer/experimenter at these sites and the Expo staff would shepherd people over to use and comment on our kiosk information system.

Our experiences with client-user applications show that the staff often learns new things about clients (and about the job they are doing with clients) at the same time we do. This is where the customer can develop confidence in the client's ability to accomplish meaningful transactions with little guidance; this often comes as a pleasant surprise to customers who, we have found, often underestimate the capabilities of their clients.

Customers who may initially underestimate the capabilities of the developers also begin, in this phase, to learn the real value-potential of the application. At NYPL, we began our iterations on a terminal reservation function to replace a burdensome, paper-based process that was a bottleneck for librarians. We argued that this sub-function immediately replace, not supplement the old method. Staff soon learned that this worked and saved them and their patrons time; pressure soon followed to propagate this application throughout the library.

Iterating on function does not only imply modifying or adding function. In the NYPL pilot, in addition to adding a fast reservation path to meet observed client needs, we ended up *dropping* a database display-ordering interaction. Although it seemed like a good idea when it came up in the brainstorming part of the discovery phase, we observed it to be an impediment to users with domain expertise.

It is important to get proto-applications in front of users and begin doing design work in the field as soon as possible. ITS developers at this juncture can frequently be heard reminding each other of our prevailing philosophy: "Make it Work, Make it Good, Make it Great". Aside from bringing focus to discussions about nuances or future functionality, this also expresses the confidence that our methodology and CADT development tools will be there to support refinements and the ultimate pursuit of quality.

APPLICATION ROLLOUT

The usefulness of an application relies on the currency of the content. It is crucial to design (and iterate) the content management functions so that these can be effectively owned by the customer and easily updated. It is also crucial that the application be fully integrated into the customer's workflow; if the new solution is perceived as being difficult to update or as requiring extra work to incorporate, there will be pressure to retire the application once active participation on the part of the developers winds down.

It is important to point out that functional and interface development does not stop when the full application is shipped. There will usually be ecological implications to the introduction of the new technology that cannot be anticipated. In addition, new customer problems may arise that can be solved readily by adding function. In Expo '92, after the fair began, the Expo administration requested the addition of a function to display a daily lottery. This was a case of the customer requesting a functional change after learning about the flexible nature of our development platform.

It is for this reason, among others, that we don't speak in terms of "releases" of applications since that implies that teams of programmers are tucked away in labs busily working on improvements and fixes. Rather we think of the application in the field as a living organism, continually susceptible to growth and evolution. Supported by the right development tools, this can either be driven by the original developers or the customer's Information Services staff can be empowered to own this phase in the application's life.

PROPER DEVELOPMENT TOOLS

Maintaining an openness to new and evolving functionality requires a development environment and toolset that supports flexibility without sacrificing reliability.

CADT, the IBM Customer Access Development Toolset for OS/2 and Windows is a set of tools (Wiecha, et al, 1990) developed in the ITS group which enable cooperative development of transaction-based, highly interactive applications. The fundamental concept behind the CADT tools is the division of the task of application development into components of style and content. This division facilitates the addition, subtraction, and modification of function independent of specific stylistic aspects of the interface. The division in the tools between style and content specification also facilitates cooperative development of applications with contributions from a multidisciplinary team (with varying skills) as well as generous code re-use and the ability to iterate development under a user-centered model of Research in the Marketplace.

CONCLUSION

By following the principles outlined in this paper, human factors practitioners in the ITS group have been successful in meeting not only the usability needs of their customers, but also in meeting their functional needs, both those that were known, a-priori, and those that needed to be uncovered. The key principles are:

--Owning the entire development process, including functional definition

--Determining and making contact with all levels of user-constituency

--Becoming "experts" in task domain; not being afraid to bring our own prior experience and insights to the table

--Using iterative, user-centered design methodologies from project inception through the entire development and product life-cycle

--Remaining open to emerging function; not setting arbitrary limits on what the system can do and what the clients will do with the system

--Using development tools that support flexible design and cooperative work

By applying this approach over the course of multiple projects, our methodology and body of functional and interface techniques has grown, making subsequent work easier and more effective.

REFERENCES

Baecker, R.M., Grudin, J., Buxton, W.A.S., & Greenberg, S. (1995). *Readings in Human-Computer Interaction: Toward the Year 2000.* San Francisco: Morgan Kaufmann.

Beyer, H. R., & Holtzblatt, K. (1995). Apprenticing with the customer. *Communications of the ACM 39:5*, 45-52.

Bødker, K. & Pedersen, J. S. (1991). Workplace cultures: Looking at artifacts, symbols and practices. In Greenbaum, J. & Kyng, M. (eds) *Design at Work.* Hillsdale, NJ: Lawrence Erlbaum.

Chapanis, A. (1970). Human Factors in Systems Engineering. In DeGreen, K. B. (ed) *Systems Psychology* New York: McGraw-Hill.

Gould, J. D. (1987). How to Design Usable Systems. *Human-Computer Interaction - INTERACT '87.* Bullinger, H.J. & Shackel, B. (Eds).. Elsevier Science Publishers B.V. (North-Holland).

Holtzblatt, K., Beyer, H.R. (1995). Requirements gathering: the human factor. *Communications of the ACM, 38:5*, 33-44.

Kelley, J.F., Ukelson, J. (1992). COAS: Combined Object-Action Selection: A Human Factors Experiment. *Proceedings of the Human Factors Society 36th Annual Meeting.* 316-320.

Kelley, J.F., Bennett, W., Boies, S., Cesar, C., Gould, J., Greene, S., Jones, L., Kesselman, J., Mushlin, R., Spraragen, S., Ukelson, J., Wiecha, C.. (1993). IBM EXPO "92 Guest Services System. Video/Demonstration at *Human Factors and Ergonomics Society Annual Meeting.* (videotape available from author).

McCormack, E. J. (1976). *Human Factors in Engineering and Design* (fourth edition). New York: McGraw-Hill.

Shneiderman, B. (1980). *Software Psychology* Cambridge: Winthrop.

Wallace, D.F., Dawson, J.A., & Blaylock, C.J. (1995). *Proceedings of the Human Factors and Ergonomics Society 39th Annual Meeting*, 1185-1189.

Wiecha, C., Bennett, W., Boies, S., Gould, J., & Greene, S. (1990). ITS: A Tool for Rapidly Developing Interactive Applications. *ACM Transactions on Office Information Systems; 8*, 204-236.

Winorgrad, T. & Flores, F. (1987). *Understanding Computers and Cognition.* Reading: Addison-Wesley.

COMPUTERS AND TELECOMMUNICATIONS IN THE YEAR 2000 - VIEWING THE FUTURE, EMPOWERING PEOPLE, AND INTERNATIONALIZING USER INTERFACES

Arnold Lund
Ameritech
Hoffman Estates, IL

Lila Laux
US WEST Technologies
Denver, CO

Nuray Aykin
AT&T Labs
Holmdel, NJ

In this, the first of three sets of position papers for the CTG-CSTG co-sponsored symposium on Computers and Telecommunications in the Year 2000, Arnold Lund describes a view of the future, based on technologies available today which, however, have yet to be integrated. Following his presentation, Lila Laux provides a paper reminding us that computers and telecommunications in the future will need to accommodate a wider variety of users, if we are to ensure that the elderly, individuals with disabilities, and other technology users with special needs are not to become an information-poor underclass in our world. Finally, Nuray Aykin discusses the problems and challenges we face in internationalizing and localizing user interfaces so that software developed in the United States can be successfully marketed abroad.

One View of the Future
Arnold M. Lund
Ameritech
arnold.lund@ameritech.com

An approach to identifying the human factors issues that will predominate in the future is to begin to picture what life in the future might be like, and then back the issues out of that picture. What follows is a story that is possible today, but probably won't exist for a few years. All the technologies are known, but the infrastructure for supporting the applications cost-effectively is just being deployed now.

A Story

Maria Pulaski has just finished hugging her son and has sent him off with his day care class for a trip to the zoo. She smiles as she thinks about the Power Rangers safety watch she strapped onto his wrist before she sent him off. It contains a smart chip with his allergies and medical records; and if he gets lost, a touch of a button will send a signal with his exact position to the teacher to help him locate her son quickly. Her son can even spontaneously send short messages to her during the day about the fun he is having. As she drives home, her thoughts are interrupted by the chime of her wallet communicator. She slips it from her purse and flips it open.

As she talks with the client about the design problem they have been trying to solve, she notices on the small screen of the device that she has two voice messages and three electronic messages waiting. When she finishes with her call, she uses a voice command to hear the list of the most important messages. One is from her spouse who thinks a visit to the zoo is a great idea. He wants to know if they can get a babysitter on Saturday for their son, and just spend the day together at the zoo. After listening to the latest traffic and weather reports that she has had sent automatically to her messaging service, she asks her electronic agent to coordinate calendars and make reservations at the smorgasbord restaurant nearest to the zoo.

Since her home automation system knows her position, as she nears her house the lights go on, a message that she is arriving home is sent to her security monitoring service, and there is an exchange of messages between her house and the local utilities making sure that the temperature in the rooms she is likely to be using, the hot water heater, and so on are all adjusted based on her past patterns of behavior. She enters the house as music from the Byrds plays in the background. She has been tweaking the playlist of her customized audio channel and she is happy with the result. She has been amazed at how adding just a couple of additional songs she specifically wanted to hear has resulted in a variety of long forgotten favorites appearing on the channel. On her television, an ever-changing set of digital photographs appears, complementing the music. She stops off in the kitchen and looks at the screen of the communications center. She notices that her local grocery is advertising a special on eggs, and sets a reminder for later. Using the device, she quickly double checks her bank balance and moves an extra $500 from savings to checking. When she moved money from her account to her son's cash card this morning just before leaving for the day care center, she had noticed that her checking account was getting low. She picks up the message from the electronic agent confirming that both their calendars and the babysitter's are clear, and that the arrangements have now been made; and that a tentative reservation at a smorgasbord has also been made. In addition, she plays a message from a new Finnish restaurant near the zoo extolling the virtues of its cuisine. Her electronic agent thought she might be interested in the restaurant as an alternative to the smorgasbord. Using a voice command, she asks to talk to her spouse. The call is placed for her and she reaches her husband. She agrees that t he zoo would be fun and says that the babysitting has already been arranged. They agree to try the Finnish restaurant if there are low-fat dishes on the menu.

After she hangs up the phone, she turns on the TV to see if she can learn more about the restaurant. She notices that several new movies have been added to the video-on-demand listings, and sees a confirmation that the complete series of P. G. Wodehouse programs is now available as part of the television programming listings. Her agent notifies her that an infommercial for that new Nordic Walker she had been looking at off and on in the electronic catalog is available for watching. It also reminds her that the community council is going to be meeting that evening and that she can attend electronically if she wants. She goes immediately to the electronic yellow pages, however, and watches the infommercial about the restaurant. Looking at the menu, she notices that if she makes an early reservation a special version of lutefisk can be prepared that takes a couple of days to get ready. She has her agent complete the arrangements for the date with her husband.

Grabbing her smart badge, she goes into her home office. As the computer comes on, since her smart badge is present, the display automatically configures itself to duplicate the environment she usually works with at her company's main office, and business calls are automatically forwarded to the home office. Calls from friends and family will be automatically routed to her voice messaging system, calls from day care and from her son will ring the phone, and all other calls are simply given a "not available" message. She confirms her identity with a voice print. She notices that she has a new message from the client she had been talking with in the car. She listens to it, and then asks to be connected. The video conferencing window opens up between her and the client, she pulls up the design so they both can work on it together, and her work day begins.

Issues

All of the applications referred to in this story exist today in some form. The drop in computing cost and improvements in miniaturization, and the construction of high bandwidth "on-ramps" to the information infrastructure, will enable the emergence of applications like these. The technology for personal, wireless communications devices will get smaller and cheaper. This will allow a variety of new kinds of devices to evolve, and will allow the functionality currently available in cellular telephones, pagers, wireless modems in PDAs and geopositioning systems to be combined in a variety of ways. More and more homes will have screen phones that integrate computer capabilities into the phone. We will have made available to 5 million homes in our region interactive television, plus a link to a broadband network supporting 500Mb data rates to the home. In many systems, it will be possible to video conference out of the home. Now that the standards are defined and development of sophisticated home automation systems is underway, new construction will link the intelligent home to service providers. We are actively exploring applications of intelligent agents and other forms of user support now. Telecommuting will continue to grow, and a variety of products and services will be available, providing a seamless connection between the corporate office and the home office. Services will continue to become available that integrate all these environments, allowing functions to be allocated to the delivery environments best suited for those functions and to environments best suited to user's needs. The particular consumer devices that emerge, and the design issues that human factors professionals will have to face, will be driven by where the technology helps people satisfy fundamental needs. We will continue to need to conduct research to understand the design principles that apply to traditional computer-human interfaces, but we will also need to understand how those principles vary by media (e.g., how designing for a bit-mapped display is different from designing for NTSC). Traditional iterative usability methodologies will continue to apply to the design of individual applications, but we will also need to design across media domains. A given messaging application, for example, might have workstation, telephone, digital assistant, and television interfaces, that are both optimized for media and the environments in which they are used, but that also need to be usable together and possess product family branding. In many cases, we will need to continue to design interfaces that "disappear" and that are only experienced as tools enabling a user to achieve goals, but in other cases we will need to create interfaces that provide an entertainment experience in themselves. Finally, to the extent that we are able to move beyond responding to design requests for isolated products, and to understand the interaction of an individual user with a product in the context of their use of other products and the rich social and cultural environment in which they live, we will be able to not only make a product usable, we will be able to make life a little better for people.

Human Beings, Human Factors, Computers, and Telecommunications in the Year 2000

Lila Laux
U.S. West Technologies
lilalaux@aol.com

The year 2000 is just over three years away. Yet, during that three year period we will no doubt see substantial changes in the availability and use of computers and telecommunications systems. The non-technical reasons that will drive these changes include the Telecommunications Act enacted on February 8, 1996, the aging of the world population, especially the populous cohort known in America as the baby boomers, and ongoing activity related to enforcement of the Americans With Disabilities Act (ADA). From a technical perspective, increases in the functionality, power, and portability of computers and telecommunications devices will increase the ability to utilize these technologies in more diverse and innovative ways.

But, it is not easy to predict the future in such a rapidly expanding and changing environment. If we look at what was predicted to happen in terms of telecommunications just three years ago, we see that the predictions missed the mark in some areas and failed to anticipate major changes in others. We expected to see video on demand and other services coming into homes via cable, a technically feasible but difficult to implement service which has not yet turned out to be financially viable. Interactive television is a technology which is expensive to implement and for which there is not yet a robust market. On the other hand, the numbers of users who access the World Wide Web (WWW) has increased dramatically and is now doubling every four months, a rate so rapid that some computer gurus predict that the system will collapse this year! So although we can predict with certainty that there will be advances in terms of what CAN be offered, we can't be sure which offerings will find markets that will make them economically feasible.

One source of uncertainty is how to attract and accommodate the large number of new users that will be required to support changes in technology if it is to be financially viable. For example, low cost telephone devices to allow people to send and receive email and other one-function small computer "appliances" are already being developed to meet the current needs of users who are not computer literate and/or have no desire to invest the time, energy, and money to reach the degree of computer literacy which would allow them to use traditional computer-based email systems.

Older people may not typically want to access the broad range of services which most computers provide today, so for them, more may NOT be better. On the other hand, people with disabilities want to be able to do everything a person with no disabilities can accomplish, and new computer technology

offers the possibility that this can be achieved. Determining what level of functionality users want and need and making it available to them is the challenge to human factors practitioners for the year 2000.

There are several serious barriers to making the computer and telecommunications services such as interactive TV and cell phones as ubiquitous as the telephone is in America today. These barriers are principally human, as opposed to technology, issues: the number of people in the US, and in the world, who are not technically sophisticated is large.
Fifty percent of Americans are functionally illiterate, and as many as 40% of students who begin school in the US do not graduate from high school. For these populations, usability of both software and hardware poses a serious problem to designers. Many potential users, both older and younger, have physical, sensory, or cognitive disabilities which also create serious usability problems, as our ability to create technology leads us to create ever more complex products. And the numbers of users with disabilities can be expected to increase as the baby boomers age and as legislation aimed at accessibility for people with disabilities is enforced.

But another problem is not addressed by the ADA. Many users, but particularly those with disabilities, do not have the financial resources to take advantage of the assistive and accommodative technology which would enable them to use computer and telecommunications devices to overcome the limitations imposed on them by their disability. Only 20% of people with disabilities in the US are employed at a full-time job, and the median income for this group is significantly lower than the median income for their age cohorts who have no disabilities. Education and training costs, as well as hardware and software costs, are prohibitive for many of the potential users of computers and telecommunications technology who could benefit the most from it. An act as simple as accessing the WWW becomes an insurmountable task if the equipment is unavailable, the person is untrained or apprehensive, or there is a disability such as blindness or paralysis which prevents information transfer. For many people, computers and telecommunications devices are simply not accessible, either because of life circumstances or because of a disability.

How will this picture develop by the year 2000? The technology is certainly there to make many services widely available. Will computer and telecommunication technology become almost universal, like the telephone today? (Note, however, that in inner cities, telephone penetration is less than 80% and in rural areas a significant percentage of the population does not have access to telephones). One reason the telephone is widely available today is that universal service is mandated by law and local personal phone service is underwritten by commercial and long distance services. Will we have such a system for computer and telecommunications devices, especially in light of the new telecommunications act which essentially opens up the telephone market to competition (although maintaining the concept of universal basic telephone service)? How can the people who provide adaptive and accommodative technology keep up with rapid changes in technology, especially since they ARE always having to adapt and accommodate because there is, as yet, no plan for designing computers and telecommunications devices for universal access.

Most developers of computer and new telecommunications devices are young, white, educated males. Most current users of these technologies are young, white, affluent males. Will we be able to meet the needs of elders, the poor, people with disabilities, the illiterate, the technically unsophisticated? And, will we choose to? And if we choose to do it, how shall we proceed? If universal access to computer and telecommunications technology is not made a national priority, there is a real danger of creating a very large underclass who are unable to access or use the technology which will be critical to functioning in the next century. The potential costs of this disparity to society are already being experienced and will surely worsen if the problem is not addressed before the year 2000.

Challenges in Internationalization of User Interfaces and Software

Nuray Aykin
AT&T Labs
nuray.aykin@att.com

Many of the US software companies today have more than half of their product sales outside the US. In the past, it was OK to sell a product in English, with the US data conventions and cultural influences. Because most users were software professionals and engineers, and their usage of software was limited to programming, data analysis, etc. However, today, computer users vary from a geophysicist to an elementary school student to a musician, and each uses computers for different purposes. A student may prepare a project paper gathering information from the internet, or play games on the computer. The users now requesting not too many features on a product, but easy-to-use products. Easy-to-use product may mean in many locales that it is in that locale's language, and it has that locale's data conventions as well as cultural preferences.

As in the US market, to do business effectively, we need to know the needs of our customers in the global market. Customer requirements should drive user interface and software requirements. Designing products for international markets requires understanding of the internationalization and localization concepts. The core software should be designed with the internationalization in mind, and it has to be ready for localization. For example, the address formats change from

country to country. The software must be flexible enough to accommodate two to eight lines of addresses without changing the core of the software.

It is often costly and time-consuming to customize a product that was meant to be marketed only in the US as compared to a product that was designed with internationalization in mind. Today, many products/services are designed on a per-product, per-country basis. This piece-meal approach produces little potential for reuse. The result is delays in product deployment, increased cost, reduced customer satisfaction, and complicated software management and maintenance resulting from managing several versions of the same product. This approach also makes it harder to get into new markets due to the high cost of reengineering.

Having a well defined software architecture for internationalization, following the internationalization processes, understanding different cultural needs, and "how to do business in a country" will decrease the cost of localization, shorten time to market, simplify software management and maintenance of localized versions, make the software easier to extend to different locales, improve quality and ease of use, and provide a competitive edge in the global market. It will also help the company to enter into new markets with confidence that the products can be localized with a predefined cost.

Design Considerations

The following list provides the internationalization/localization considerations that should be taken into account while designing products for the international market.

Language Translation - The user information must be easily understood by non-native as well as native speakers of the language, and must be easily translated into a different language. Some work is being done towards simplifying and standardizing English to help translation and comprehension of English as a foreign language.

Character Sets - The software should be able to support multi-byte character sets. US uses 7-bit ASCII; Cryllic, Latin, Arabic, Greek and Hebrew use 8-bit; Japanese, Korean, and simplified Chinese require 16-bit; and traditional Chinese requires a 32-bit character set.

Writing Directions - Some languages are written/read from left to right (e.g., English), some languages from right to left (e.g., Arabic), and some from top to bottom (e.g., Japanese). The software should anticipate different writing conventions.

Keyboard Conventions - Strategies for designing for international products include using function names rather than cap names in user manuals, selecting a default-keyboard setting, and placing keyboard information on-line for easy customization. Designs could include keyboard control such as key-sequence-to-function mapping and character-set mapping.

Symbols and icons - The designers should avoid culture- or language-specific icons. For example, a rural mailbox icon can be meaningful choice for the US market, but not for the European and Japanese markets where such mailboxes do not exist. An icon showing a tree log to represent 'log' in the US market could be a cute choice, but it loses its meaning when it gets translated into another language.

Collating Sequences - The collating sequence defines the value and position of each character relative to other characters. Characters to be sorted include letters, numbers, punctuation marks, and special symbols such as #, &, *, and %.

Local data Conventions - The design should be flexible enough to accommodate the various data conventions. Local data conventions may include but are not limited to:

Thousands and decimal separators
Grouping separators
Positive and negative value representation
Currency format
Marking format
Metric representation
Date Time
Telephone numbers
Addresses

The user interface software architecture is made up of modular software that can be partitioned into core modules (free of locale-specific elements), and localization modules to fit the needs of a country/locale. The core modules include the user interface elements that are free of country-specific requirements, such as the contents of the forms and menus, messages, internationalized icons and symbols, on-line help, and documentation. When there is a need to localize the product, the only change in these modules is the language. The base components such as executable objects, internal data files, etc. are also in the core modules. These modules stay the same when the product gets localized.

The localization modules contain country-specific requirements such as date notation, time formats, numeric formats, etc. For example, date in the US can be shown as 5/12/95, but in Europe it can be shown as 12.5.95. The week in the US starts on Sunday, but in many European countries, it starts on Monday. There should also be support elements for the architecture. In order to be successful in a local market and understand the customer needs, we need to know more than their date and time formats. We need to understand their culture, standards, quotes and tariffs, etc. We need to leverage the skills of in-country people to gather this information, and also help us in translation, customer support, and maintenance.

It is important to have a sophisticated knowledge of user needs to separate what is core and what is localizable. The first visible effect of internationalization is on the user interface. This includes the translation of screens and help messages, and choosing colors, icons and symbols for a locale. This architecture emphasizes the user interface modules, but it also includes the modules necessary for the remainder of the underlying software to be localized. Each module must use user-centered design to ensure that the user needs are met. Each module must be standardized and tested.

Existing software platforms by companies such as Microsoft, Macintosh, Hewlett Packard, and Sun offer tools and ways to internationalize and localize software products. Understanding and using these tools provides the basis for the right architecture for the software. However, it will not guarantee high-quality, compelling, easy to learn and use localized products. The responsibility is on developers, user interface designers, systems engineers, writers, translators, project managers, and locale experts.

With the growth in using computers everywhere, and information reachable everywhere, we also see an explosion of localized products. On the internet, you can now download different fonts and character sets to support the language you are using, you can translate your messages into other languages, and you can put up your home page in your own language. Some internet sites lets you choose the language before you view their page. Microsoft simultaneously releases its products in more than 20 languages. Hewlett Packard lets you choose your language at the login screen.

It is not easy to localize a product. It is a very challenging job. But, understanding the needs of the users, the culture they live in, and their expectations, and being able to translate these into easy-to-use products is a very rewarding experience. We hope to continue to share our challenges and experiences as user interface professionals designing for the international market.

COMPUTERS AND TELECOMMUNICATIONS IN THE YEAR 2000-MULTI-MODAL INTERFACES, MINIATURISATION, AND PORTABILITY

Leah S. Kaufman and Jim Stewart
Microsoft Corporation
Redmond, WA

Bruce Thomas
Philips Corporate Design
Eindhoven, The Netherlands

Gerhard Deffner
Texas Instruments, Inc.
Dallas, TX

In this, the second of three sets of position papers for the CTG-CSTG co-sponsored symposium on Computers and Telecommunications in the Year 2000, we begin with a paper by Leah Kaufman and Jim Stewart on the human factors challenges involved in creating an effective multi-modal communications environment. Bruce Thomas continues with a position paper outlining the advantages and disadvantages of technology miniaturisation, and how these advantages and disadvantages impact our approaches to user interface design. In the final paper in this set, Gerhard Deffner describes the portability-functionality dilemma, in which designers are confronted with two distinct user goals that are difficult to meet simultaneously.

Meeting Human Needs in a Multi-modal Communications Environment

Leah S. Kaufman and Jim Stewart
Microsoft Corporation

Our lives, both at work and at home, depend on the effective communication of information - information about people, places, ideas, and time. Communications and computing technologies enable the delivery of more information of greater complexity at faster rates to our home, our office, and at all points between. How well do the user interfaces (UIs) to the computer, phone, and pager keep pace with this barrage? How easily can we sort through these different forms and determine which information is most important and which can safely be reviewed later?

As the power of communication technologies increases, the challenge of designing user interfaces that can successfully take advantage of this improved capability also increases. We are constantly called on to design, develop, and test that magical new UI that attempts to deliver the promise of these immense computing resources (i.e., today's expectation for people to work at ever higher levels of productivity).

From the human factors perspective, this boon of communications can also be our bane. Now that the computer has become the hub for receiving information from email,

faxes, phones, and other computers, the user needs powerful, personalized ways to cope with this flood. In order for computers to meet users' telecommunications needs we need innovation on at least two elements of the UI: user-defined rules for managing the information, and multi-modal means for reviewing the information and responding. While these UI features are not new, it is our contention that their implementation needs to be centered on the human information processing system and the user's needs.

Rules and Filters for Managing Incoming Information

The human information processing system excels at pattern recognition. As we scan a visual array, selective attention allows us to move through the display until we recognize some target bit of information. When applied to work at the computer, this means we can visually scan lists with intent to locate a piece of information. The intent may be as specific as 'the important message from Tom' or as general as 'new items'. Whenever the area to be searched is reduced, we increase the rate at which we can locate information. A related finding is that the more distinguishable target information is from its background, the quicker the search rate. This means that it is easier to locate a single piece of information among categorized items than when the items are combined in an unorganized display - that is, by categorizing the information into distinct groups, we have effectively reduced the size of the list. This basic principle of cognition has strong implications for how computers should mediate the increased information flow provided by telecommunications. Any item received by the computer has characteristics that can be used for grouping and organizing it with similar items; the date and time sent, sender, type of message (e.g., email, voicemail, fax), header information, even keywords within the content ofthe item. By automatically using rules for grouping and allowing the user to modify these rules or create new ones, we improve the user's ability to search and review the lists of items coming through the computer. Regardless of whether the user is reviewing these lists via the monitor or over the phone, rule-based categorization of the messages is an invaluable mechanism for managing large arrays of information.

Given the range of users, tasks, and work environments, the computer UI should capture the different ways the user might organize and prioritize information and allow the user to describe new categories or criteria for deciding what happens to the incoming information. Additionally, the rules should be used in other ways: to group or order the information for review, for integration of different kinds of information, and by automating steps the user routinely carries out. For example, messages from a particular phone or fax number could be sent a particular, pre-recorded or written response, email with a specific phrase in it might automatically receive a certain fax in reply. By devising easy, usable UI implementations that help users to group their incoming messages and automate their responses, we will have addressed one of the main issues inherent in telecommunications and computing systems.

Appropriate use of Audio, Visual, and Physical Interactions

The contemporary computer-based telecommunications system will be able to provide voice output of text and messages, speech recognition of verbal commands, probably a visual display, and at least one method for physical input such as a keyboard, keypad, mouse, stylus, or pen. Information sent in verbal form may be displayed and reviewed as text, while text-based information can be heard instead of visually scanned. The assumption behind this flexibility is that the information, the environment in which it's reviewed, and the concurrent user tasks may all influence which mode is optimal for receiving and responding to the message.

In order to decide on a mode for receiving and responding to information, we need explicit descriptions of the users' telecommunications-based tasks and activities. By matching these descriptions with research on the types of information that are best comprehended and attended to in each mode, we can learn if certain activities should be restricted to a particular mode or will work well in more than one. It is equally important to review the combinations of tasks that users do and determine, for multi-tasking situations, how to support these in a cross-modal manner.

Conclusion

The increased power in communication and computing technologies means that computer users can receive more information in a wide array of formats. In order to manage this influx, users need tools that let them easily manage, organize, review, and respond to messages. These tools need to accommodate for the perceptual and cognitive resources brought to bear when humans use the new technology. User interfaces that allow people to use rule-based systems for categorizing and integrating incoming information, and a multi-modal means for reviewing and responding will increase usability and usefulness, and perhaps deliver the promise behind the technology wave.

Telecommunications 2000
SMALL IS BEAUTIFUL, BUT IS IT USABLE?

Bruce Thomas
Philips Corporate Design
e-mail: c862820@nlccmail.snads.philips.nl

Introduction

Mobile telephony is perhaps the fastest growing sector of the telecommunications market. By the year 2000, ownership of mobile phones will become commonplace. Ownership will not be restricted to the professional user; it will be available to all members of the community.

The primary benefits of owning and using a mobile phone are efficient and portable communication. Driving forces in the design and manufacture of a mobile phone are, therefore, the size and weight of the product. For this reason, manufacturers of mobile phones are striving towards smaller and lighter products, while at the same time emphasising these features when advewrtising products.

It is to be expected that this trend will continue into the early years of the next century. However, the very size of the mobile phones of the future raises some serious human factors issues. This paper considers how these issues may be resolved without losing the benefit of portability the small size offers.

Advantages of Miniaturisation

Miniaturisation enables products to fit unobtrusively into people's lives. This is particularly the case for products which the user carries on a regular basis, such as a walkman or a wristwatch. The prime consideration for such products is that they can be carried comfortably, but they should also be carried elegantly. This means that, if worn covertly, they should not irritate the user through pressing into the body, nor should they cause unsightly bulges in clothing. A small, light product will do neither.

A further advantage of miniaturisation for portable products is that they can become an item of apparel, like a wristwatch. This is a highly personal product which the user always carries and which has a strong personal association. Miniaturisation allows a mobile phone to have the same virtues as a watch of being useful as well as decorative.

Disadvanteges of Minisaturisation

A small product has a reduced space for user interface elements. All control keys and the display must fit in the space available. In the case of a mobile phone, this means accommodating not only a 12 digit keypad, but also a number of other function keys which enable the user to gain access to the functionality within the phone, such as a directory, as well as an ever increasing diversity of services which can be accessed via telecommunications networks.

A further disadvantage of a small mobile phone is that the earpiece to mouthpiece ratio can be disturbed. When the earpiece is held on the ear, unless the phone is very long and thin, the mouthpiece is not in the vicinity of the mouth. Even when the microphone is sufficiently sensitive to cope with this, users sometimes feel disturbed about "talking into thin air".

The Conventional Approach

The solutions to the problems posed by miniaturisation currently adopted by many manufacturers are to reduce the size and number of keys. Each of these approaches is problematic from the human factors point of view. Reducing the size of the keys makes them less accessible, particularly if this is associated with a reduction in the pitch. Reducing the number of keys creates a challenge for interface designers, in that more functions have to be accessed with the few keys remaining.

The problems noted above are exacerbated with the increasing functionality of mobile phones. Current GSM telephones, for example, already typically provide 80 or more functions, the vast majority of which are mandatory requirements of the GSM specification itself. In addition to the functionality within the phone, network operators are providing ever more services to the user, all of which must also be accessed by means of the phone.

Alternative Solutions

Mobile phones in the year 2000 are likely to be extremely small, while at the same time supporting an extensive range of functions and services. If current trends continue, it is clear that the conventional approach to interface design will not be adequate to provide usable products of this kind. We are therefore, likely to see the development of a number of alternative interaction mechanisms. Noyes (1996) describes two technologies which would seem to offer potential solutions: pen and speech.

Pen interaction is already being used in other areas, for example, on Personal Digital Assistants (PDAs) and organisers. The use of a pen has some appeal, in that interactions can make use of both selection and command based dialogues. A pen has a finer point than a finger, and therefore needs less space for keys. The use of a larger screen, occupying the area that would otherwise be taken up with a screen plus keypad, also provides the possibility to offer more selections to the user. Nevertheless, the presentation of a multitude of functions on what could still be a very small screen and the operation of the pen itself with the degree of accuracy demanded will still require human factors issues to be addressed to obtain the optimum from such solutions.

Speech has an intuitive appeal in the world of mobile phones, in that the products already contain elements required for spoken communication - microphone and speaker. Currently, speech interfaces are still fairly primitive, and mature solutions which would allow the keypad to be abandoned are not yet available. Further, a number of human factors issues are still to be resolved: the systems are largely speaker-dependent rather than speaker independent, vocabulary is limited, power drain can be high (requiring the user frequently to recharge batteries), there is a limited acceptance or the idea of "talking to machines", and inevitably a command dialogue must be used (requiring the user to remember the instructions to be given to the machine). Nevertheless, the technology is maturing to such an extent that it is probable that by the year 2000 it would be a feasible option for miniaturised products.

Conclusion

Miniaturisation enables users to gain a great deal more comfort and convenience when carrying a portable product such as a mobile phone. However, the trend towards ever smaller products is also accompanied by a rapid expansion in functions and services offered to users. It is unlikely that the conventional approach to interface design for these products will continue to be adequate to provide usable technology into the next century.

Of the alternatives available, only speech recognition offers the possibility to eliminate totally the need for space for user interface elements. Although a number of human factors issues remain to be resolved with the use of this technology, it is conceivable that a small speech-operated product, carried as an item of apparel, might be available within the next few years. If this is the case, then we may already see the development of a Star Trek Next Generation communicator by the year 2000.

Reference

Noyes J. (1996) Communicating with computers: Speech and pen-based input. In S.A. Robertson (Ed.)., Contemporary Ergonomics 1996. (pp. 518-523) London: Taylor & Francis.

The Portability / Functionality Dilemma - User Interface Bandwidth

Gerhard Deffner

Texas Instruments, Inc.

GPHD%mimi@magic.itg.ti.com

To fully understand the scope of user interface problems to come, assume that the worlds of computing, telecommunications, and portable personal electronic widgets will have merged by the year 2000, and battery technology is no longer a barrier. A major portion of society will carry small, powerful devices combining the functionality of laptop computers, cellular phones, Personal Digital Assistants (PDAs), and modems. These devices will, in turn, be supported by a backbone of networked computers of maximum capacity in all respects-processing power, bandwidth, fault tolerance, storage, interconnectivity, etc.

Portability, however, requires smaller size, severely limiting the ease and efficiency of user-system interaction. The two main problems found with smaller equipment are

1) Fewer and smaller buttons or keys, as well as smaller displays.

2) Lack of equally portable external support materials, manuals, etc. to provide user guidance and training.

This situation is tolerable in the case of a wrist watch, for instance, where functionality and user-system interactions are limited and little training and even less recurring user guidance are required. But the next generation of portable electronic computing/communications devices will high light the disconnect between a) functionality and complexity of user-system interaction, and b) availability of user interface and information display bandwidth.

Devices will provide much more power and functionality than even today's desktop computers, but wont even have keyboards! And given the lack of large displays and supporting materials, the user must depend on memory and mental models to guide navigation.

This mismatch between functional power and user interface bandwidth calls for a novel approach to user-system interaction. We must go beyond the traditional limits of keyboards-plus-screens and DTMF keys-plus-microphones-plus-earpieces.

All available user interface technologies (e.g., high resolution displays, touch input, voice recognition, speech synthesis) will have to be combined. This will allow for increased user interface bandwidth. Usability will depend on successful integration into human behavioral stereotypes and mental models. Intelligent agents will not provide the high degree of flexibility and creative use expected from new information

tools. Only human, everyday metaphors can help users benefit from the wealth of information, communication, storage, and processing available through the new generation of tools. It is time for the next desktop metaphor that includes more than just sitting at a desk, shuffling files, and throwing documents into the trash. It will be modeled after how humans interact with each other, not just how they interact with objects.

DESIGN ISSUES in the DEVELOPMENT of COOPERATIVE PROBLEM-SOLVING SYSTEMS

Panel Chair: Rebecca Denning
Center for Cognitive Science
The Ohio State University
Columbus Oh 43210

Panelists

Norman D. Geddes
Applied Systems Intelligence, Inc.
Roswell, GA

Patricia M. Jones
University of Illinois at Urbana-Champaign
Dept of Mechanical & Industrial Engineering
Urbana, IL

Christine Mitchell
Center for Human-Machine Systems Research
School of Industrial & Systems Engineering
Georgia Institute of Technology
Atlanta, GA

Philip Smith
Cognitive Systems Engineering Laboratory
The Ohio State University
Columbus, OH

Overview

The history of the design of decision support systems encompasses the invention of a variety of different technologies. Some have been based on optimization techniques, others on the use of heuristic methods, and still others on algorithms for learning based on a set of training cases.

This history also records frequent failures to use these technologies in many of the applications for which they were designed. One historical reaction to such failures has been to blame the underlying technology. As an example, the field of expert systems was in part a response to such failures in the use the optimization techniques explored in the 1960s:

"Several recent incidents have involved complex computer control systems. The suggestion is that reliance on the escalating power of brute force may be heading toward danger. However effective and reliable such systems may be in normal conditions, use of brute force may not be worth the price paid during the rare episodes when a computer-controlled power station or military installation or air-traffic control system malfunctions. On these occasions a new factor becomes paramount: The human operator or supervisor needs to follow what the computing system 'thinks it is doing'," (Michie, 1986).

Based on this conclusion, Michie suggested that the use of artificial intelligence techniques to develop consulting systems should be considered to help deal with this concern.

The theme of this panel will be that it is not enough to look at the underlying technologies alone. These failures have often been directly due to a failure to consider human factors issues - issues concerning the designer as well as the user.

One of the primary weaknesses has been a failure to recognize the limitations of the designer as expressed in that particular technology. Systems based on optimization techniques generally used oversimplified models of the relevant "world", either intentionally (due to resource or technology limitations) or unintentionally (due to designer error). Similarly, an expert system may have an incomplete knowlege base, and a connectionist system may be trained on an inadequate set. In addition, each of these technologies imposes strong constraints on the form of the underlying reasoning or computation which, interacting with the associated explicit or implicit knowledge base, leads to brittle performance.

In response to these limitations, there has been an increasingly strong trend to develop systems that integrate both human and computer agents as a problem-solving team. Such systems are specifically designed to support cooperative problem-solving. Such systems can be contrasted with traditional consulting systems in that they interact with the user at intermediate steps in the problem-solving process, rather than simple providing an answer (and perhaps an explanation justifying this answer).

To design effective cooperative problem-solving systems, however, many challenging human factors issues must be considered. These include questions about the "cognitive compatibility" of the underlying problem-solving technologies, about the roles of the human and computer agents and the nature and timing of their interactions, about the design of the interfaces between the person (or people), the computer (or computers) and the "world", and about the need to support some level of mutual understanding and coordination between the cooperating agents.

All four of the panelists have been actively engaged in the design, implementation and evaluation of cooperative systems. They will be asked to set the stage for discussion by explicitly identifying the design principles, design concepts and design methodologies that they have found to be most effective in guiding the development of such systems, and to provide illustrative examples.

Associate Systems: An Architecture for Human-Machine Cooperative Problem Solving

Norman D. Geddes

Over the past 10 years, an architecture for a form of comprehensive joint human-computer problem solving has emerged, the associate system. The central concept behind the associate system is that the human remains in charge of the system, while the computer follows the human's lead and supports the human as requested or as needed. The first such system, the ARPA-sponsored Pilot's Associate, provided an initial framework for integrating computer-based automatic functions with a range of manual and partially automated functions in a seamless architecture. Since the conclusion of this project, a number of other associate systems have been initiated that have shown the value of this architecture for both individuals and teams of operators. The behavior of an associate system is controlled by a suite of dynamic cognitive models that represent the intentions of the human operators and the cognitive and physical task demands on the operators. The intent model allows the associate to direct its activities to support the actual on-going operator tasks, while the demand models suggest what tasks require support and the form that the support might take. Experience with associate systems in complex real-time domains indicate that the acceptance of this type of aiding by both novice and expert users is high, and that the associate architecture provides robust behavior even when confronted with problems not thought to be within its capabilities.

Cooperative Distributed Planning Between Humans and Machines

Patricia Jones

In the context of cognitive engineering, "cooperative problem solving" refers to cooperative interaction between humans and machines in the process of jointly solving problems. Here, "machine" refers to an intelligent software system that supports human interaction. One general class of problems to be solved are planning problems -- i.e., the coordination of activity towards some common goal (which may have distinct subgoals). This panelist will describe one approach to cooperative planning and activity management distributed between human and machine, and will raise as issues for discussion:

* Machine as agent vs machine as tool.
Does this distinction matter? If so, how?
(e.g.,locus of control, malleability/adaptivity)

* Metaphors of conversational interaction (machine as partner in dialogue) versus world (machine as providing space of representations for human use)

* What is "good" cooperative problem solving unfolds in the context of practice and cannot be guaranteed by design. Yet, one tries to design systems to promote "good" joint problem solving. What are metrics for goodness?

* Design of representations versus design of interaction with those representations.

* How to design malleable representations that are extensible by end users. How to be "radically tailorable" in the context of cooperative problem solving.

* According to Schmidt and Bannon, two key issues for CSCW are "sharing an information space" and "articulation work". How can insights from CSCW, and in particular efforts in these two areas, inform the design of human-computer cooperative problem solving systems?

The Interface between Human and Machine Agents: The Foundation of Human-Centered Automation?

Christine M. Mitchell

Automation in complex systems is proliferating. As with most former instances of control systems, the interface with the human operator responsible for overall system control is often unfriendly, creating a range of automation induced problems. Remediation of this situation can take at least one of two forms: a drastic change in the design of the automation itself or design of the interface between the automation and the user. History demonstrates that the former strategy, redesigning the system design process to ensure that the 'human factor' is considered throughout the process, typically fails to produce 'human-centered automation.' This presentation describes an approach to the latter: design of interfaces to automatic systems, or machine agents, that promotes cooperative problem solving. We propose that an interface which promotes effective cooperation between human and machine agents must allow the machine agent to be inspectable (What is it doing?), capable of self-explanation (Why is it doing that?), predictable (What will it do next?), repairable (Don't do this, do that.), and able-to-learn (Do not make the same mistake next time).

This panelist will raise these issues in the context of a case-study of satellite ground controllers using an 'intelligent' ground control system. The case study highlights the difficulties and brittle interfaces between machine and human agents. She will outline a methodology to design the interface between human and machine agents and describe its use to redesign the interface for an expert system for satellite ground control. The methodology uses the operator function model (OFM) to structure the human interface to the expert system to promote the features identified above, e.g., inspectable, predictable, repairable, and able-to-learn.

Alternative Roles for Cooperative Problem-Solving Systems

Philip J. Smith

Based on examples from two domains (aviation and medicine), alternative roles for cooperative systems will be briefly discussed. These examples will then be used to highlight certain key design questions and design principles that need to be addressed in deciding how to integrate human and computer agents, and will provide concrete examples to help ground discussions among the panelists and the audience.

As an example from aviation, the design of a flight planning system will be discussed. An underlying architecture will be outlined that includes both optimization and expert systems technologies, and is explicitly designed to actively engage the user (an airline dispatcher) in an assessment of the situation and in the generation and evaluation of alternative flight plans.

As an example from medicine, an expert critiquing system will be described, with an emphasis on design solutions to ensure that the user has an accurate mental model of the computer system, and on the design of an interface that unobtrusively encourages the user to provide data to allow the computer to infer and critique that user's problem-solving processes.

References

Michie, D. (1986). On Machine Intelligence (2nd Edition). New York, NY: John Wiley and Sons.

CRITICAL SUCCESS FACTORS AND LESSONS LEARNED FROM CREATING A LARGE WINDOWS HELP SYSTEM

Kenneth R. Ohnemus
CSC
Cincinnati, Ohio

The sheer volume of information, 15 books and over 1,500 graphics, contained in CSC's corporate system development methodology, Catalyst,[sm] has created a need to have this information on-line. On-line access, it was felt, would dramatically impact productivity. To help support users and facilitate the use of Catalyst, the Microsoft (MS) help system, in conjunction with RoboHELP[TM], was used to put Catalyst on-line in a hypertext format. Incorporating feedback and usability concerns was key in determining how its more than 20,000 users could best utilize this product. The design evolved over several months and in its final form also extended the limited functionality of the MS help system, providing a more robust product. This paper discusses the challenges surrounding the design of a large windows help system, approximately 53 MB in size (33 MB when fully compressed). The lessons learned can form the basis for creating an effective help system development process for meeting users' needs.

INTRODUCTION

Many of the issues surrounding the use of hypertext, such as navigation, getting lost and excessive cognitive demands, as well as benefits including better information access, reduced costs and improved maintenance, have been discussed elsewhere and will not be presented here (Brockmann, 1990; Conklin, 1987; Nielsen, 1989; Ohnemus and Mallin, 1994; Shneiderman and Kearsley, 1989).

CSC Catalyst is CSC's approach to initiating, motivating, designing, implementing, managing and coordinating change in large organizations. The methodology is a non-prescriptive, process driven framework for implementing business change.

Catalyst was a good application for hypertext because of the large size of the methodology, 15 books. By having this information on-line CSC could dramatically increase usability and the ability to search and quickly find information. The large number of inter- and intra-book jumps was another reason. Within a given book are many references to other chapters and sections. In addition, many references existed between books. Having Catalyst on-line was a strategic investment aimed at encouraging and facilitating the use of Catalyst. The project utilized many of the techniques provided by the Catalyst methodology.

PROJECT START-UP

The project began with a kick-off meeting and included the executive sponsor, the developer and the developer who would integrate HyperCatalyst with the CSC Toolkit. Along with responsibilities, the team discussed project goals, objectives and expectations. The primary developer was the project manager and lead architect. Prior to starting development, a statement of work was formulated, describing in detail both the work to be undertaken and the deliverables. A project plan using Project Workbench[TM] was developed and updated with actual times to manage and track the progress of the project.

With the executive sponsor in San Francisco and the project manager in Cincinnati, it was important to establish ground rules and expectations prior to starting the project. Communication with the executive sponsor centered around electronic methods due to the distance factor, the three hour time difference, and lack of face-to-face contact. Telephone and voice mail were primary communications vehicles in managing expectations, keeping the executive sponsor up-to-date and discussing various issues. Lotus Notes[TM] was used to electronically send work products back and forth between the developer, the executive sponsor and end users.

PACKAGE SELECTION/ENVIRONMENT

The team performed a help authoring package evaluation and selection to find a tool best suited to our needs. Tools considered had to run on MS Windows, be capable of compiling a large help file without crashing, deal effectively with a large number of graphics and have a short compile duration. The development tools considered were Doc-To-Help 1.5 ™ and RoboHELP 2.0 ™. These help tools automate the process of developing Windows help files through the use of embedded macros within MS Word for Windows. ™ RoboHELP was selected because it best met our criteria and was a powerful yet easy to learn and use tool that could handle a large project containing approximately 1,500 graphics. Doc-To-Help was not chosen primarily because of significantly longer compile times, tool crashes with large files and the repeated occurrence of Word out of memory errors. Following a RoboHELP tutorial, the team had an understanding of what the tool could do for us. As this tool works from within Word for Windows, a good understanding of that product was also essential.

PROJECT IMPLEMENTATION

Data Gathering/Prototyping/Design

The team wanted to provide the product a corporate look and feel that would reflect CSC's way of doing business based on how the tool would be used. From the outset, the developers worked with users to gather input and develop the initial design. To gain a better understanding of how Catalyst was being used, several data gathering sessions were held with users and the Catalyst methodology development team before design was undertaken. This initial information was utilized to prototype the first version of the product. Initially, the team developed a limited function prototype which evolved over the course of one and a half months. Early on, the team prototyped alternative ways of presenting information to determine the best design metaphor to meet the stated objectives. Prototype iterations were timeboxed to a day or less.

Early on discussions focused on formatting issues and the help file structure. Formatting issues such as heading sizes, button placement and usage, and navigation/location information were refined through user involvement. The file structure was based on how users' would use the Catalyst books in the field. Catalyst is structured so that users can go to a particular topic

providing the 'what' to do. Once at a given topic, users can access associated diagrams and introductory and work product information. Another very powerful, but little known feature utilized is the mid-topic context ID, which is a topic within a topic. This feature is extremely useful when many steps or sections of a topic exist. Instead of making each step or section its own topic or forcing the user to scroll through a long section of text, one can allow users to jump to a given step or section. Using a secondary window, the team created various steps or sections buttons which when "clicked" brought the user to a specific location. The only drawback to this feature is there is no automated mechanism for assigning mid-topic context id numbers, which are used by applications to jump to a specific location within a help file. The benefits of this feature, however, far outweighed the time required to manually input and assign mid-topic id numbers.

The basic structure consists of a contents window, navigation buttons, location information and hypertext links. The design metaphor is a modified pictorial representation of the Catalyst framework and book structure which provides a corporate look and feel. From the contents window, users gain access to the various portions of the Catalyst methodology. Users can make selections and view additional information in either main, secondary or pop-up windows depending on the nature of the information. The design is graphically oriented and leveraged the large number of graphics available by including hotspots for navigating through the help system.

User disorientation and getting lost, major problems in using hypertext systems, were handled primarily by providing location information on the top left of windows. This information is especially important since the user can access Catalyst in many different ways. Examples include jumping from one book to another as well as accessing Catalyst from a supporting software tool.

Using Lotus Notes, the team obtained input from users across various locations within CSC. The developers were able to maintain a user-centered perspective and to manage the users' expectations by having this continuous user involvement. At every possible opportunity the team demonstrated and tested the evolving product to as many people as possible. Using both the telephone and Lotus Notes, design alternatives and feedback were provided. This team approach and the subsequent usability testing refined the design to its current structure.

Graphics

Converting the graphics into a form the MS help system would recognize proved to be a greater challenge than anticipated. The original graphics were developed using Aldus Freehand™ for the Macintosh.™ The MS help system prefers either bitmaps (BMPs) or windows metafiles (WMF), meaning the graphics had to be translated into one of these formats. Though they looked great in printed form, when viewed on-line the graphics were, at best, barely legible. A multi-step process unfolded for dealing with the problem. Once a graphic was a BMP, it could then be opened in an enhanced hotspot editor and saved as a segmented hypergraphic (SHG) format. This strips the graphic of excess BMP baggage not required by the MS help compiler. Hotspots are also applied using this editor. Finally, given that the help compiler does not compile tables embedded within a file, graphics of the tables had to be developed and subjected to the same process outlined above. Be aware that color graphics can significantly increase the size of a help system, so developers should use color judiciously. The only color utilized was for titles or major words and was used consistently across all graphics.

Usability and System Testing

The team performed usability evaluations repeatedly throughout development. In addition to end users, the developers utilized upper management to generate enthusiasm toward the product. Usability testing began with gaining user feedback on competing designs. As the design progressed, users provided with little design information were asked to perform specific tasks. In addition to local users, the evolving design was sent to remote locations for usability testing via Lotus Notes. As the size of the project grew, diskettes replaced Lotus Notes as the most effective means of delivery to other locations. Prototype iterations were timeboxed to less than a day. After the initial design and usability problems were eliminated and the design became more stable, users at various locations were recruited to perform system testing.

Table 1 presents the major usability concerns addressed, how they were identified and the resolution. With large graphics, for example, the user in many cases is able to go to a specific activity by clicking within one of the boxes. Often, however, the entire graphic is not visible without scrolling. If users scroll down to and click on activity 8 (not visible on the window) they are brought to the appropriate information. But if they return

to the graphic, users would be placed at the top of the graphic and have to scroll back down to the next highest numbered activity. By sectioning the graphics into multiple pieces, users could return to the graphic so that the previous and next activities were visible. Without this usability enhancing feature, each return to the graphic would have required scrolling down to a specific activity prior to making a selection.

A test plan and procedure was utilized for reporting problems and suggestions. As problems were encountered, they were logged, corrected and re-tested to ensure proper functioning. Most problems related to incorrect jump locations, either via text or graphic hotspots and buttons not functioning properly. System testing focused on intra-book testing. Jumps between books were tested to make sure they went to the proper location and then the tester returned to the primary book. The hard copy of books was essential so that testers could ensure the text and graphics were accurately represented. Prior to the end product's release to the field, beta testing was conducted. This identified some minor issues, but more importantly, generated new ideas for future enhancements such as the ability to launch a tool from within HyperCatalyst.

User Feedback and Satisfaction

To date, informal user feedback has been extremely favorable. Users appreciate the design and the corporate identity provided by HyperCatalyst and feel the product is straightforward and easy to use. Additionally, formal feedback was gathered from interviews with 160 people representing over 60 different projects across the globe. The results indicate hearty endorsement for the usability of HyperCatalyst and point to near unanimous use of the product where it is available. Comments included, "we make heavy use of HyperCatalyst" and "we prefer HyperCatalyst over the books." Others noted they used HyperCatalyst in conjunction with the books. Complaints commonly associated with the product are its unavailability to some, its being too large for certain computers and that the graphics are not reusable; none of which is related to quality or usability. With the MS help system, only text can be copied for use elsewhere.

Lastly, a database on Lotus Notes was developed to encourage electronic feedback. No design or usability problems have been recorded, though a few incorrect hypertext jumps have been found. There was also a suggestion to map the on-line graphics to a graphics repository to facilitate their re-use for projects and proposals.

Usability Concern/Problem	Issue	Identified By Testing	Resolution
Large graphics usability (user scrolling required to see in entirety.)	Returning to graphic: placed at the top of graphic and have to scroll down to the bottom to activate additional hotspots.	Usability.	Sectioning the graphic into multiple pieces (the graphic still appeared as a whole to the user) from a system perspective.
Navigation information.	Prevent getting lost and user disorientation.	Usability.	Provide location information in a non-scrolling region (top left corner of window).
Button locations.	Optimal button locations.	Usability.	Locate in non-scrolling region.
Searching (identical entries).	Unable to distinguish between entries such as Work Products.	System and beta.	Add text to make multiple entries unique such as Work Products - Integration.
Incorrect jumps.	Wrong information.	Usability, system and beta.	Correct jumps.
Graphics legibility.	User unable to read.	Usability, system and beta.	Correct graphics.
Text.	Misspelled words.	System and beta.	Correct spelling.

Table 1. Major usability concerns and problems addressed.

Initially, there were plans to distribute a questionnaire, but it was decided to take advantage of a worldwide initiative to obtain feedback on the use of the Catalyst methodology. This provided a unique opportunity to get access to a much wider audience. After an upcoming release, there are plans to distribute a questionnaire as was originally anticipated.

CRITICAL SUCCESS FACTORS AND RECOMMENDATIONS

To summarize, critical success factors can loosely be grouped into two categories, general and development. General activities include communication, project planning, and understanding the MS help system. Development activities encompass project start-up, package selection and implementation activities, especially prototyping, testing and user feedback.

General

One of the most critical aspects of delivering a successful hypertext project is proper planning. A good model is required to design a tool which increases productivity. An effective model supports users' work habits, is task oriented and provides appropriate options for the users. By understanding the users' environment, the developers were able to structure a help file readily accepted by users. As top-level design decisions are

made, borrowing ideas from other help systems is extremely useful. Begin by looking at numerous help systems, paying particular attention to innovative and effective ones. The importance of viewing the entire system, including keyword search terms and how topics are written, can be quite valuable, especially to those new to developing help systems.

Determining how many topics to construct must also be considered (i.e., Will fewer but longer topics be defined or will the topics be of shorter duration which leads to many more topics?). Determining how to break down the information into topics can dramatically affect the usability of the system. Utilizing mid-topic context IDs allowed us to keep large amounts of related information grouped together, while still providing a mechanism for easy access to the various steps or sections. If each step or section was its own topic, the user would have been forced to view information linearly, one piece at a time, as opposed to having access to the complete picture.

Developers must also plan and decide upon the screen layout and graphic structure (including text styles, sizes, indentations, and heading levels), navigation routes, and possible hypertext links prior to beginning the development process. For this situation, most small graphics were embedded within the text, however, for larger graphics the team chose to place them into secondary windows. Decisions about which features and capabilities of the help system to be utilized must also be addressed early on. For example, will topic titles appear in non-scrolling regions and how will jumps and secondary/pop-up windows be used? Having a solid vision of how the final product will look is also essential to producing a high quality, usable product. This initial planning provides a blueprint for moving forward. If solid planning is not completed prior to development, a tremendous amount of time will be lost. There is no other road map to aid the way.

Understanding the limitations and capabilities of the MS help system prior to development is also essential. Weeks into development the developers were still learning about the strengths and weaknesses of the MS help system. Features of the MS help system dictate to a large degree what can be done. One very powerful, but little known feature is the mid-topic context id (a topic within a topic). This feature is extremely useful when many steps or sections of a topic exist. This allowed the developers to keep the information integrated while allowing an easy, highly usable way to get to a specific location within a given topic. As the team learned more about the system, they were able to extend the limited functionality of secondary windows through the use of buttons and macros. Developers can also modify/add buttons to the toolbar using macros, as well as adding menus and menu items to the menu bar. For example, help menu items were added for a copyright notice as well as simple help on using the product.

The constant communication and weekly formal status reports helped us keep a handle on progress and quickly resolve issues as they arose. In this instance, with the project sponsor remotely located, keeping him up-to-date on events and managing his expectations, helped make the effort more successful. The electronic communication mechanism, Lotus Notes was without a doubt one of the most critical aspects to the success of this development effort. The impact of the ability to send a product back and forth several thousand miles in near real-time cannot be underestimated. The team could see changes and competing ideas throughout the day, as often as necessary.

Development

The kick-off meeting was critical to jump start the project, help to make it a success, and solidify and personalize our team relationship. The kick-off meeting allowed us to define our scope, and identify risks and constraints that might affect the project. Identifying tools that meet project criteria early and quickly also influence the success of a project, especially projects under a short timeframe. Determining the quality of technical support

and obtaining vendor training may be necessary in some cases. Prototyping and usability testing are also key success factors.

System testing turned out to be more difficult and time consuming than originally anticipated. Testing the hypertext links and checking spelling, along with testing the ease of use is an important issue if the help system is to be accepted by the users. In general, the help compiler generates help files relatively quickly which allows for fast turn around of problems. Prior to releasing the product, it is advisable to compile using full compression to reduce the size of the help file.

Typically the help compiler generates a file in a matter of minutes. In uncompressed mode this project took approximately 7 hours to compile resulting in a help file 53 MB in size. Compiling with full compression took approximately 72 hours and resulted in a 33 MB help file. The bottom line is to test early and often and to allow plenty of time for testing prior to releasing the product. The importance of releasing accurate, high quality material cannot be overstated. Releasing a sub-optimal product will only serve to discourage use and reduce acceptance. Users will use the product only if it assists them in performing their roles.

The main challenge remaining with this product is to port the application to the Windows 95 platform. Prior to beginning that effort, however, a significant amount of time will be spent in planning in order to help ensure a smooth migration and also to take advantage of new enhancements to the MS Help System.

REFERENCES

Brockmann, R.J. (1990). *Writing Better Computer User Documentation: From Paper to Hypertext.* John Wiley & Sons, Inc.: New York, NY.

Computer Sciences Corporation. (1994). *Catalyst Release 3.0.* Books 1-16. Computer Sciences Corporation: El Segundo, CA.

Conklin, J. (1987). Hypertext: An introduction and survey. *IEEE Computer.* 20 (9), 17-40.

Nielsen, J. (1989). Hypertext II. *SIGCHI Bulletin,* 21(2), 41-47.

Ohnemus, K.R. and Mallin, D.F. (1994). Benefits of Implementing On-line Methods and Procedures. In *SIGDOC '94: 12th Annual Conference on Systems Documentation,* 49-55.

Shneiderman, B. and Kearsley, G. (1989). *Hypertext Hands-On!: An introduction to a New Way of Organizing and Accessing Information.* Addison-Wesley: Reading, MA.

TRADITIONAL VS. WEB STYLE GUIDES: HOW DO THEY DIFFER?

Julie A. Ratner
Eric Grose
Chris Forsythe
Statistics and Human Factors Department
Sandia National Laboratories
Albuquerque, NM 87185-0829

This paper describes a study in which Web style guides were characterized, compared to traditional human-computer interface (HCI) style guides, and evaluated against findings from HCI reviews of web pages and applications. Findings showed little consistency among the 21 Web style guides assessed, with 75% of recommendations appearing in only one style guide. While there was some overlap, only 20% of Web-relevant recommendations from traditional style guides were found in Web style guides. Web style guides emphasized common look and feel, information display, and navigation issues, with little mention of many issues prominent in traditional style guides such as help, message boxes, and data entry. This difference is reinforced by other results showing that Web style guides address Web information-only pages with much greater success than web-based control enabling features, like buttons and entry fields. It is concluded that while the WWW represents a unique graphical user interface (GUI) environment, development of Web style guides has been less rigorous, with issues associated with web-based control enabling features neglected.

INTRODUCTION

The prolific expansion of the World Wide Web (WWW) has fueled an unprecedented growth in the number of developers creating computer-based materials for public access (December, 1994). In the brief existence of the WWW, there has not been an opportunity for style guides and conventions comparable to other HCI domains to be developed and gain acceptance from the development community, concerns with the effectiveness of style guides aside (Lowgren, 1993). This predicament is worsened by Web developers with limited knowledge of traditional graphical interface concerns and unfamiliarity with problems raised by cross-platform and browser compatibility requirements of Web-based interfaces. To fill these voids, numerous HCI style guides have appeared for Web development.

This paper examines Web style guides in three ways: 1) It compares the development environment of Web style guides with their traditional GUI counterparts. 2) It compares the content of Web and traditional style guides to see where they differ. In particular, it examines the extent to which Web style recommendations include the most essential of the traditional recommendations, as judged by a group of human factors practitioners. 3) It examines how useful Web style recommendations are in actual practice by comparing them against human factors reviews of Web pages generated for use in a large R&D company.

METHOD

Phase 1. A Comparison of Style Guide Development

The comparison of HTML and established HCI style guides began with a consideration of the development process. Ten authors of style guides, five Web and five traditional, responded to a survey asking them about their educational and professional backgrounds. Authors were asked to describe why they created their style guide, who would use it, and who sponsored it. Also, questions were asked about the time it took to create the style guide, the resources used, and how often it was updated.

Phase 2. A Comparison of Style Guide Content

This phase of the research sought to compare the content and emphases of Web and traditional style guides.

Style guides were sampled in the following manner. First, twenty-one WWW sites offering Web interface recommendations were identified in the fall of 1995. From these sources, a set of 357 unique recommendations were obtained. Second, recommendations from traditional style guides were collected from five printed sources. Style guides not applicable to the WWW, or which were specific to a particular platform or operating system, were discarded. This process resulted in a collection of 270 unique interface recommendations. For each domain, Web and traditional, style guides were combined to form a comprehensive set of interface recommendations. Web and traditional style recommendations were compared qualitatively to examine their similarities and differences.

To assess the representation of the most essential traditional style recommendations within the Web style recommendations, it was first necessary to determine which recommendations were most important. Eleven practicing human factors professionals responded to a survey asking them to review the 270 recommendations obtained from the traditional style guides. For each recommendation, participants gave their opinion as to how essential they thought the recommendation was to the usability of a system. Answers were indicated by assigning an integer to each item using a scale from 0 (non-essential) to 4 (absolutely essential). Descriptive statistics were used to identify those items which the participants generally agreed were essential to usability. These "essential" recommendations from traditional sources were compared to the Web recommendations to determine how well they were accounted for in the Web domain.

Phase 3. A Comparison of Web Recommendations and Actual Human Factors Evaluations

As a further assessment of their practical effectiveness, web style recommendations were compared to actual human factors reviews of the web-based, corporate information infrastructure at Sandia National Laboratories. (The reviewers were knowledgeable of the traditional GUI style guides, but all reviews occurred prior to exposure to any of the Web style guides.)

There are generally two different kinds of Web pages. First are those which derive information for the user by performing routines, such as accessing databases; this paper refers to these pages as applications. Examples of such applications are an electronic phone book, or a conference room scheduler. The second kind of page contains information only, examples of which are a corporate newsletter, engineering procedures, or departmental home pages.

Findings from these two types of pages were compared to Web style guides to assess how well the guides would have accounted for deficiencies identified by the human factors experts.

RESULTS

Phase 1. A Comparison of Style Guide Development

Survey results from Web and traditional style guide authors revealed distinct differences. All Web style guide authors were from educational environments and described the development process as informal with an average of 3.26 weeks spent writing the style guide, with revisions prompted by feedback from users or impending conference submissions. In contrast, the traditional style guide authors were from military, government or corporate organizations. Three of five specified that a formal process was employed to develop the style guide, with an average of 54.2 weeks for the initial version, with subsequent versions following annual reviews and structured working groups.

Web style guide authors explained that resources to write the guide drew from the fields of human factors, marketing, graphic arts, and public relations. Traditional guidelines were primarily based on human factors, cognitive psychology, and military sources. The audience for traditional style guides is programmers, interface designers, and technical experts. The audience for Web style guides seems less clear: in only one instance did an author identify a specific audience for the document, and that was the staff at his university.

Phase 2. A Comparison of Style Guide Content

Style recommendations from both Web and traditional style guides were assigned to one of 20 categories and compared. Figure 1 shows the distribution of recommendations across categories. Four categories accounted for 63% of the Web recommendations: Common Look and Feel (CL&F), Information Display, Navigation, and Labels. Table 1 lists the most frequently cited recommendations from the Web style guides.

Recommendation	#Style Guides
1. Signature should be included at bottom of Web page	12
2. "Here" or similar words should not be used to designate a link	12
3. Pages should be timestamped/dated	11
4. Every page should have a title	11
5. Design for all platforms and browsers	10
6. Menu bar or buttons should be provided for navigation	8
7. Text used for links should be descriptive and meaningful	7
8. Pages should have common look and feel	7

Table 1. Eight Highest Ranked Recommendations from Web Style Guides

As shown in Figure 1, the categorization of recommendations reveals some overlap, but also considerable differences between Web and traditional style guides. A Chi Square test of independence showed a significant difference in the distributions of recommendations across categories (X^2=245.5, p<0.05; df=19). The biggest differences arise from the greater emphasis on information display and consideration of help, data entry, and message boxes within traditional style guides.

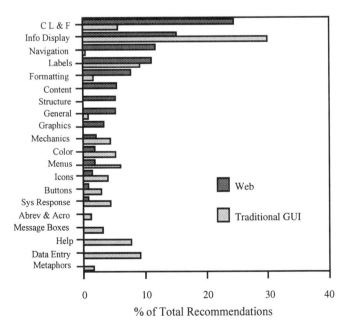

Figure 1. Distribution of Style Guide Recommendations Across Categories

The next comparison assessed the degree to which traditional style recommendations were addressed within Web style guides. Of the 270 Web-relevant recommendations found in traditional style guides, only 53 (20%) also appeared within Web style guides, with another 20 partially addressed within the Web style guides. The greatest overlap occurred for recommendations categorized as common look and feel (15 of 16), menu design (11 of 17) and labels (15 of 25).

With regard to the surveys used to identify the most essential guidelines from traditional style guides, a plot of this data revealed 26 guidelines whose mean score was greater than three and standard deviation less than 0.75. These items received consistently high ratings from the experts (see Figure 2).

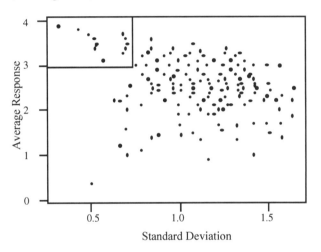

Figure 2. A plot of the survey data by mean and standard deviation. (The boxed area shows the most essential guidelines)

Table 2 shows the eight highest ranked recommendations from traditional style guides.

Recommendation	mean	s.d.
1. If a user can't log-on, a prompt should be provided explaining why	3.9	0.32
2. Text should be readable from normal viewing distances	3.8	0.42
3. Where precise readings are required, precise values should be displayed	3.7	0.67
4. Displayed information should be readable to the degree of accuracy required by the task	3.7	0.67
5. A positive indication of function actuation should be provided	3.7	0.67
6. The system should provide positive feedback regarding the acceptance or rejection of input	3.7	0.48
7. Data entry formats should match source document formats	3.6	0.52
8. An easy mechanism should be provided for correcting erroneous entries	3.6	0.70

Phase 3. A Comparison of Web Recommendations and Actual Human Factors Evaluations

In the review of corporate web-based applications, only 57 of 141 findings (40%) corresponded to recommendations presented in the Web style guides. In contrast, for web pages with primarily information content, 38 of 59 findings (64%) were addressed within Web style guides (see Figure 3). These proportions are statistically different at the 0.001 level of significance.

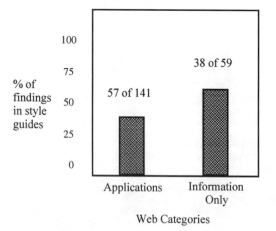

Figure 3. Percentage of findings accounted for in Web style guides

CONCLUSION

These findings reveal distinct differences in the development and content of Web and traditional GUI style guides. There are several explanations for this result.

First, it may be argued that while there are overlaps, the WWW introduces unique human-computer interface concerns. This explanation is supported by the emphasis placed on common look and feel within the Web style guides. It is reasonable to suppose that in the hypertext environment of the Web, where users may readily move between unrelated sites, common look and feel is of greater importance than with stand-alone software applications where differences are less apparent.

The second explanation for the apparent differences between Web and traditional style guides would assert that Web style guide development has occurred with less rigor and little reference to traditional style guides and principles. That there was poor agreement between the Web style recommendations and the actual findings of human factors reviews of control-enabling applications supports this explanation. This was not true, however, for those Web pages that contained information only. For these pages the Web style guides accounted for 64% of findings, which is as good as might be expected (Lowgren, 1993; Thovtrup, 1991). One explanation for this result might be due to the greater representation of information only pages on the Web than application pages. Inexperience with Web applications may explain the scarcity of application-specific recommendations in Web style guides.

It does appears that Web authors tended not to reference traditional style guides when creating their own. The explosive growth of the Web should leave no surprise that some authors, from diverse backgrounds, would not think to look beyond the Web for answers to its interface problems. But the appearance of these Web style guides would not be possible without the Web itself; the ability to publish one's own document and disseminate it widely is a new phenomenon (December, 1994).

To be fair, many of the style guides were written for specific audiences (to assist authors within a particular company, for example.) Contrast this with the traditional style guides which were formally sponsored to address specific, but more comprehensive, concerns. There is also the fact that the Web is still new. Its use to conduct work and business is still being explored. As it matures it may undergo the same pressures to standardize as did the early graphical environments.

ACKNOWLEDGEMENTS

This work was supported by the United States Department of Energy under Contract DE-AC04-94AL85000. Also, the authors would like to thank Stephen Crowder, of Sandia National Laboratories, for his help with statistical analyses.

Special thanks to Stephen Crowder of Sandia's Statistics and Human Factors Department for help with the statistical analyses.

REFERENCES

Bodart, F. & Vanderdonckt, J. M. Expressing Guidelines into an Ergonomic Styleguide for Highly Interactive Applications, in Proc. ACM INTERCHI'93 Human Factors in Computing Systems. (1993), 35-36.

December, J & Randall, N. The World Wide Web Unleashed. Sams, Indiana, 1994.

Lowgren, J. & Lauren, Ul. Supporting the Use of Guidelines and Style Guides in Professional User Interface Design. Interacting with Computers. 5,4 (Dec. 1993), 385-396.

Tetzlaff, L. & Schwartz, D. R. The Use of Guidelines in Interface Design. in Proc. CHI'90 Human Factors in Computing Systems, (1990), 329-333.

Thovtrup, H. & Nielsen, J. Assessing the Usability of a User Interface Standard. in Proc. ACM CHI'91 Human Factors in Computing Systems, (1991), 335-341.

DATA ENTRY INTERACTION TECHNIQUES
FOR GRAPHICAL USER INTERFACES

Donna L. Cuomo, Eliot Jablonka, Jane N. Mosier
The MITRE Corporation
Bedford, MA   01730

Despite the widespread use of graphical forms-based interfaces, there are no widely accepted and proven interaction methods for supporting even the most common data entry tasks. We present three common data entry tasks (task flow, domain-related data entry, hierarchical list selection) and a variety of commonly-used designs to support them, and some pros and cons of each. We recommend that good designs to commonly performed data entry tasks be identified and catalogued

Introduction

Many systems with graphical user interfaces involve entry of data into forms. Examples include tax forms, purchase requisitions, maintenance data collection, messaging systems, travel orders and vouchers, work time recording, and scheduling/ meeting applications. Despite the widespread use of graphical forms-based user interfaces, there are no widely accepted and proven interaction methods for supporting even the most common data entry tasks, and there has been little research comparing different interaction methods. It is even difficult to find descriptions of possible data entry interaction methods and guidance on when one method might be more suitable than another method for a particular task. Most comparative literature has been in the input device arena (e.g., Epps, 1986; Ewing et al., 1986; Goodwin, 1975; Karat et al., 1986; Card et al., 1978). A study comparing interaction methods was done by Gould et al. (1989). They compared experienced and inexperienced user's performance on seven different interaction methods for specifying event dates. A study comparing interaction techniques for reordering fields in a table was performed by Tullis and Kodimer (1992). But there has been no attempt to classify and describe common data entry tasks, to identify interaction methods currently in use to support these tasks, to research or catalog the techniques, to provide guidance on which are best for particular tasks, or to develop higher-level graphical user interface software objects for the various interaction techniques. We feel that such an effort would be very helpful to designers and this paper attempts to illustrate the concept and further explore three data entry tasks.

Common Data Entry Tasks

We have designed, evaluated, and used a large variety of military and civilian/commercial systems which have forms data entry components. In the systems we have seen, the design of data entry system components are treated in widely varying ways.

These components include:
- data entry task flow,
- entry of domain-related data such as dates, and
- entering or selecting data items from hierarchical lists.

Each of these common data entry tasks for forms-based systems is described in more detail below. For each component, we include examples of different design approaches, along with potential pros and cons of each approach. Some considerations in choosing between design alternatives will be discussed including screen real estate, and entry efficiency for novice vs. experienced users. These techniques were not empirically compared.

Task Flow

Small dialog boxes are sufficient for supporting simple data entry task flows. A user selects a command from a menu, enters appropriate data or command information, presses the "OK" button, and the system performs the requested action or enters the information. However, when that action requires completion of dozens or hundreds of fields, and when there is field dependence, simple dialog boxes are insufficient.

Imagine your task is to complete your taxes for this year. Which forms do you complete? How do you know when entries on one form now require you to complete another? On paper forms, instructions such as "If your answer is 'Yes' then append form ABC" are provided. You then complete form ABC and staple it to your cover form. How do you do that online? The approach selected must require minimum actions, but also allow users to change decisions previously made without starting over.

There are two major issues in the task flow area:

- How users view and access many related data items on a single screen. Simply permitting scrolling through screens of fields is one approach. Another is an outline approach which allow users to show and hide groups of fields on demand (similar to file management in Macintosh's System 7, where clicking on an icon shows or hides the next level of detail). Tabbed user interfaces (such as used in Windows 95 dialog boxes) is another approach. A fourth is a wizard-like approach where users are forced through a fixed sequence of data entry dialog boxes.

- How to show which fields are relevant, given the answers already provided. Appropriate techniques will be somewhat dependent on which of the above approaches is used.

Providing all the fields on a display with scrolling is one approach that works better when there are fewer fields. It is simple and straightforward, and allows for a direct simple mapping of the screen version to a printed copy of a form. It also allows for simple methods of repeating fields. Problems arise when there are too many fields. Users can lose their context in a long form if they try to jump around. Also, users may need to scroll through a lot of irrelevant fields to find the ones they are looking for.

The outline approach (figure 1) can help with the scrolling problem while retaining the benefits of the display all fields approach. The outline approach allows users more flexibility in which sets of fields to view on screen. Field management could become awkward as the number of sets of fields increases.

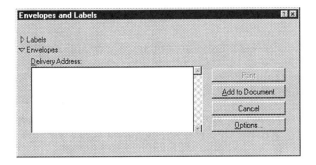

Figure 1. Outline Approach.

Tabbed user interfaces (figure 2) provide an intuitive way of organizing fields that allow users to easily flip between sets of fields. Tabbed windows work especially well when fields can be grouped into a few groups, and fields do not need to be repeated. Problems arise when field validation on one tab relies on field entries on other tabs. Only one tab can be seen at a time, so users may need to flip back and forth a lot between tabs to see what they have completed.

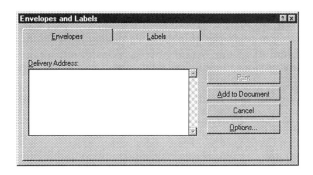

Figure 2. Tabbed Approach.

The wizard interface (figure 3) is a good choice when the field sequence is very important and one field is dependent on data in another. Another feature of wizards is that they can give the user a preview of the finished product. Wizards are the most structured approach to filling in fields which is good for novice users, but could become cumbersome for expert users. Wizards may also require users to backup to edit previously completed fields.

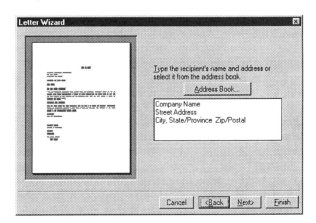

Figure 3. Wizard-Like Approach.

Domain-Related Data Entry

A number of different data entry approaches are provided in commercially-published style guides – radio/option buttons, text entry fields, drop down lists. However, none of those address the fact that data often belong to a domain that allows us to interpret and constrain. The value 07/04/96, for example, isn't just an alphanumeric combination; it's the Fourth of July. Domains in common use on forms include dates, times and prices. Because there exists no standard way to enter any of these, each designer is left to design something new.

Gould et al's excellent article on entry of dates was a good start in this area. However, that study did not include what would now be considered modern graphical user interface (GUI) objects – those that combine the ability to select and type. The objects they evaluated required a great deal of screen real estate, something not practical in most forms applications.

Three approaches are common in date and time entry. The first is exemplified by most spreadsheet applications: users enter a value into an edit field, and the application may understand several formats. Because the application can interpret a variety of different formats, no format prompts are displayed. This approach only works if the application truly does understand and accept the large variety of formats any user might enter. The advantages of this approach are that it uses little space and does not require a default date or time, since defaults are sometimes inappropriate.

A second approach is similar in that an edit field is used, but the field includes spin controls and is divided into separately manipulable subcomponents of date or time (e.g., month, day, year). Users can either use the spin controls or enter a value. This approach is used in the Windows 95 date/time control panel, the Macintosh System 7 date/time control panel (figure 4) and in ON Technology's Meeting Maker scheduling application, among others. Tabbing moves the cursor from one subcomponent to the next. Hence, to tab out of the field, one must tab several times. That is one disadvantage in a rapid data entry environment, though this approach is more efficient than the calendar approach described below. Another disadvantage is that a default date or time must always be displayed, since users actually enter data by changing the current value. When no logical default exists, the current date or time is commonly used. One important advantage is that this object uses space efficiently. In addition, it is tailorable when different date or time formats are required. And if accurate date entry is important, adding the day of the week as separate subcomponent is a simple matter that has little impact on the required screen real estate.

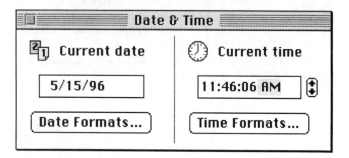

Figure 4. Set/Date Time with Spin Controls Approach.

A third approach that applies to date only is direct manipulation of a graphical calendar (figure 5). This approach requires a means of specifying the month, and then allows the user to select a day displayed in a Sunday through Saturday table. This approach makes date selection easy for novice users, unless dates in past or future months will be selected often. In addition, the calendar visualization is easy to understand and should increase the likelihood that a correct date is selected. However, this approach requires a great deal of space. Early versions of ON Technology's Meeting Maker displayed this as part of an optional popup window. Intuit's Quicken displays this in place of a drop down menu as part of their standard date widget. In both cases, this approach is a secondary means of date entry, with the primary means being direct entry of a date into an edit box.

May				96		
Sun	Mon	Tue	Wed	Thu	Fri	Sat
			1	2	3	4
5	6	7	8	9	10	11
12	13	14	**15**	16	17	18
19	20	21	22	23	24	25
26	27	28	29	30	31	

Figure 5. Set/Date Time with Direct Manipulation.

The discussion here focused on time and date domains, but these are not the only domains commonly used. When entering prices, the designer must decide whether and how to encourage or discourage entry of units ($), and how to ensure that decimal points and commas are or are not entered, among other things. Given the frequency of use of these common domains, including standard means of data entry and display in the basic styles would make forms easier to design, implement and use.

Data Entry via Hierarchical Lists

Users often have to enter values which are based on hierarchical coding schemes. For example, to enter a work unit code in an aircraft maintenance data system, users need to enter a 5 character code. The code is actually a composite value where the first two characters encode a system, the next character encodes the subsystem, the next character the sub-sub-system, and the last character is a component (e.g., 74FAA). Another example is a project number code where the first two digits might represent the center that owns the project, the next two numbers the year the project was started, the next four numbers a project specific number and the last two numbers or letters are specific tasks on the project (e.g., 02957789AA). Valid values for these fields can be numerous, numbering in the hundreds or thousands. Although very experienced users may have some subset of these numbers memorized and can directly type them in, new users, infrequent users, and users using values unfamiliar to them perform better when a selection mechanism is provided. In general, a combination method of direct entry and hierarchical list selection should be provided. We have seen this type of hierarchical selection implemented or designed in the following ways.

The first method (figure 6) is a single list dialog box accessible from the form which supports "drill down," similar to the file selection dialog box on the Macintosh. The user double clicks on list items in the lower list to bring up the next lower level list associated with that item. The option menu above shows the value of the list item clicked on to obtain the current list. The path used to reach the current level is available as options on the menu and allows users to move back up the path hierarchy. The text field near the top shows the entire set of values selected along the path. The data entry field at the top is the same as the data entry field on the form and supports direct value type in, filling out the other fields and lists below to match the typed value. This allows expert users to have direct type in and still see the text results to confirm the entered value. Upon pressing OK, the 5 character code is inserted into the field on the form.

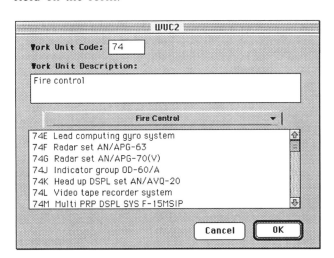

Figure 6. "Drill Down" List Selection.

A second way to complete this data entry field is via a dialog box containing four drop-down lists whose values are dependent on the selection of the previous list (figure 7). Upon opening the dialog box, the only active drop down list would be the System list. Once a system is selected, the Sub-system list would become active, containing list values associated with the System selected. Each value selected is displayed in full so a text box displaying what was selected previously like that

above is not necessary. One potential problem with this method is that the user can change the System value and this will wipe out the values below it which might be an unexpected result. The method above makes the hierarchical relationship between lists more obvious.

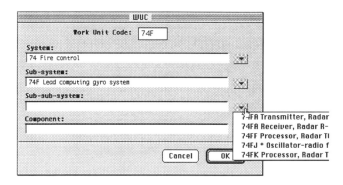

Figure 7. Dependent Drop-Down Lists.

A third way we have seen hierarchical list selection implemented is via cascading menus (figure 8). This is illustrated below. In this case, four levels of hierarchy are required and because of the length of the text items, the menus start to overlay one another (e.g., the last menu is overlaying the second menu). The problems with this technique are that fine motor control is needed to control the menus and most guideline documents recommend against cascading menus that are four levels deep. Going backward to select a different menu item on a previous menu can also be awkward. This method does not support displaying the corresponding text when the coded characters are directly typed into the field as did method 1 and possibly method 2. The advantage is selection is done directly on the form containing the input field - traversing to a separate dialog window is not necessary.

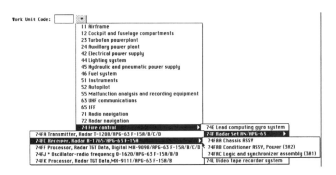

Figure 8. Cascading Menus.

A fourth method to accommodate this type of selection is a single combo-box with drill-down menus (figure 9). In this method, the list is dropped-down and posted, and the user double clicks on the item of interest, opening the next level list in the same space this list occupies. This is the method we like the least (and we have seen it proposed) because there is no path to show where you have come from, it is not possible to back up a level (you must start over from the beginning), and there is no cancel mechanism. The lack of a scroll bar also makes scrolling through the choices awkward on long lists. The advantage, which does not outweigh the disadvantages, is it uses little screen real estate.

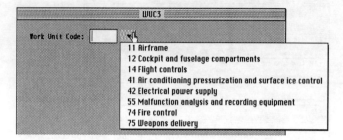

Figure 9. Single Combo Box with Drill Down Menus.

Discussion

There are many common data entry tasks which occur across systems yet the human factors/CHI community has done little to provide examples of and guidance on concrete interaction techniques to support these tasks. This results in design work for each data-entry system designer/developer, some poor interaction methods, and inconsistencies in interaction techniques across systems for users for basic data entry tasks. While complete standardization may not be desirable or achievable, various commonly used or innovative techniques could be documented, critiqued, and guidance provided for selecting among the techniques. In addition to reducing design time and poor designs, once some efficient interaction techniques have been identified, higher level graphical user interface objects to support these techniques could be created.

References

Bowen, C.(1995) *Theater Battle Management Human Computer Interface Specification, Version 1.1* (MP95B0000035). C. Bowen. The MITRE Corporation, 202 Burlington Road, Bedford MA 01730. February, 1995.

Card, S. K., Moran, T. P., and Newell, A. (1983). *The Psychology of Human Computer Interaction*. Hillsdale, NJ: Erlbaum.

Carroll, J. M., (Ed.) *Designing Interaction*. Cambridge University Press, 1991.

Epps, B. W. (1986) Comparison of six cursor control devices based on Fitts' law models. Proceedings of the Human Factors *Society 30th Annual Meeting,* pp. 327-331.

Ewing, J., Mehrabanzad, S., Sheck, S., Ostroff, D., and Shneiderman, B. (1986). An experimental comparison of a mouse and arrow-jump keys for an interactive encyclopedia. *International Journal of Man-Machine Studies*, 24, 29-45.

Goodwin, N. C. (1975). Cursor positioning on an electronic display using lightpen, lightgun, or keyboard for three basic tasks. *Human Factors,* 17, 289-295.

Gould, J., Boies, S., Meluson, A., Rasamny, M., and Vosburgh, A. (1989). Entry and Selection Methods for Specifying Dates. *Human Factors,* 31(2), 199-214.

Helander, M. (Ed.) *Handbook of Human Computer Interaction*. Elsevier Science Publishers BV, 1988.

Karat, J., McDonald, J. E., and Anderson, M. (1986) A comparison of menu selection techniques: Touch panel, mouse, and keyboard. *International Journal of Man-Machine Studies*, 25, 73-88.

Smith, S. L., and Mosier, J. N. (1986) *Guidelines for Designing User Interface Software* (ESD-TR-86-278). The MITRE Corporation, 202 Burlington Road, Bedford MA 01730.

Tullis , T., and Kodimer , M. L. (1992). A Comparison of Direct-Manipulation, Selection, and Data-entry Techniques for Reordering Fields in a Table. *Proceedings of the Human Factors Society 36th Annual Meeting - 1992*, pp. 298-302.

A CASE STUDY OF TRANSPARENT USER INTERFACES IN A COMMERCIAL 3-D MODELING AND PAINT APPLICATION

Beverly L. Harrison [1, 2]

Kim J. Vicente [2]

[1] Alias\Wavefront

110 Richmond Street East
Toronto, Ontario,
Canada
M5C 1P1

[2] Cognitive Engineering Laboratory

Dept. of Mechanical & Industrial Engineering,
5 King's College Road,
University of Toronto,
Toronto, Ontario, Canada
M5S 3G8

This paper describes a case study of transparent user interface tools in a commercial 3-D modeling and paint application. It represents another step in an ongoing research program evaluating transparent human-computer interfaces. Results from previous controlled experiments were used to inform our design choices in the working product. We collected data from 11 users one month after they received the application, using a semi-structured interview. Working sessions were also video taped and analyzed. Based on these results, modifications were made and a follow-up interview was conducted three weeks later. A number of transparency–related issues were identified. This case study illustrates the value and challenges that one encounters in transitioning from basic research to commercial applications.

INTRODUCTION

This paper describes a case study of transparent user interface tools in a commercial 3-D modeling and paint application. It represents another step in an ongoing research program evaluating transparent human-computer interfaces.

The interaction between limited screen real estate, graphical user interfaces, and human attention limitations results in a visual attention problem, towards which this research is aimed. Many HCI applications are designed with a work space or data area which is the primary focus of attention. The limited display size means that the tools to manipulate the data appear in windows, menus, dialogue boxes, and palettes over top of the work area, thereby diverting or blocking users' attention from the underlying data area. This area often provides feedback about the actions users apply (e.g., in drawing systems there are traditional tools to change paint brushes and colors), meaning users frequently have to move interface objects in and out of the way to complete domain tasks. As a result, users spend a great deal of time performing interface management tasks rather than productive work.

Transparency offers a potential remedy to this problem. If interface objects were semi-transparent, users might be able to allocate their attention either to the top layer of information (interface objects and tools) or to the bottom layer (data area) at will, since the latter could be visible under the former. As a result, the fluency of work should be improved since there is less need for interface management tasks.

There are a few examples of interfaces which exemplify this seamless task-tool integration through the application of transparency (Bier, Stone, Pier, Buxton, & DeRose, 1993; Bier, Stone, Fishkin, Buxton, & Baudel, 1994; Ishii & Kobayashi, 1991; Ishii, 1990; Knowlton, 1977; Zhai, Buxton, & Milgram, 1994; Schmandt, 1983). However, most of these have not been the subject of empirical evaluation, and none have found their way into a commercially available product. The research program represented here addresses this gap in the literature by conducting a systematic series of experiments evaluating the impact of transparency on human performance, culminating in a case study of transparency in a commercial 3-D modeling and paint application that is the focus of the present paper.

PROGRESSION OF RESEARCH

The implementation of transparency in the commercial application observed here was preceded by several controlled experiments. The first set of experiments (Harrison, Ishii, Vicente, & Buxton, 1995) investigated the impact of transparency on interference between visual layers using a Stroop paradigm (Figure 1). Although this is not a very representative task, the study was well controlled and provided an initial conservative estimate of the level of transparency that is required to minimize interference between foreground and background layers. The subsequent set of studies (Harrison, Kurtenbach, & Vicente, 1995; Harrison & Vicente, 1996) sacrificed some experimental control but used a more representative set of stimuli to investigate the impact of transparency under more realistic conditions (Figures 2 and 3). The next step in the research program, described here, was to conduct a case study of transparency in a commercial application under even less controlled but more representative conditions.

This case study was conducted jointly with a company using their application in 3-D modeling and painting, "MagicPaint". (Product name is disguised for confidentiality.) An evaluation of existing "MagicPaint" users was conducted using this modified software.

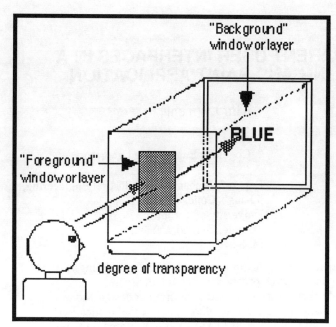

Figure 1. Model of Stroop Experiment Configuration. Same approach was used for all experiments, substituting image content.

Figure 2. Sample Image from Icon Palette Selection Experiment. 50% transparent, wire frame background. Indep. variables: transparency level, icon type, background type. Dependent variables: selection response time and error rate.

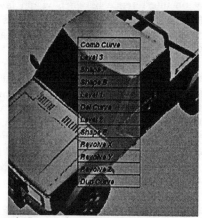

Figure 3. Sample Image from Text Menu Selection Experiment. 50% transparent, regular Motif style menu, solid background. User task - select a target icon from the palette.

RELATED WORK

Many applications are designed with a large work space or data area which is the primary focus of attention, while the tools to manipulate the data appear in windows and palettes over top of the work area. To simplify our conceptual model, we have classified the work space or data area as the *background layer* (a single large window typically) and the UI tools (menus, dialogs, palettes) as the *foreground layer* (multiple objects or windows). This reflects the visual representation of many task contexts, where the "work" is displayed in a full-screen window, thereby requiring visually overlapping UI tools. These tools divert or block our attention from our work, which is often providing feedback about the actions we apply, for example, painting or drawing systems and the traditional opaque UI tools to change paint brushes and colors. However, there are several examples of highly-advanced systems which exemplify more seamless task-tool integration through the application of transparent user interfaces.

In Heads Up Display (HUD) design, aircraft instrumentation (a graphical computer interface) is superimposed on the external real world scene, using specially engineered windshields (e.g., Larish and Wickens, 1991; Wickens, Martin-Emerson, and Larish, 1993). In the ClearBoard work (Ishii and Kobayashi, 1991), a large drawing surface, resembling an electronic drafting table, is overlayed on a life sized video image of the user's collaborative partner. The TeamWorkStation system (Ishii, 1990), predecessor to ClearBoard, created semi-transparent computer work space windows superimposed with video image windows (e.g., a person, an object being discussed). The ToolGlass and MagicLens work (Bier at al., 1993; Bier et al., 1994) reflects a tight coupling between tool function, target object specification, and transparency. These projects used clear or see-through palettes which could be aligned with underlying objects. Tools were invoked by clicking "through" the desired function, using alignment to specific the target object for the function. Other designs include such things as video overlays like those used in presenting sports scores in broadcast television.

Some designs combine transparency and 3-D projected views of the user interface. Several examples are: the work on "3-D silk (volume) cursors" (Zhai, Buxton and Milgram, 1994), the work by Knowlton (Knowlton, 1977), which used graphical overlays projected down onto half-silvered mirrors over blank keyboard keys to dynamically re-label buttons and functions keys (e.g., for telephone operators), and the work by Schmandt (1983), who built a system to allow users to manually manipulate and interact with objects in a 3-D computer space using a 3-D wand. Again a half–silvered mirror was used to project the computer space over the user's hand(s) and the input device. Disney has also developed a product called the "ImaginEasel" for animators and artists. ImaginEasel keeps the user's hand and input device in the workspace (using mirrors). In every case transparency seems to provide a more seamless integration between the data or work and the UI tools.

METHOD

The case study was conducted in two stages. Approximately one month after the first release of the transparency version of MagicPaint, we interviewed 11 users. Several modifications were made based on these comments. We then waited an additional three weeks and again interviewed these same users.

A semi-structured interview was used. All users were asked the same questions about their usage of MagicPaint to obtain a "user profile". They were also all asked to list the three aspects of the interface they liked the most and disliked the most. Finally, they were asked for their opinions about transparency in particular and any additional comments or observations. All responses were recorded. Those aspects of the interview relating to transparency are summarized below.

We intermittently video taped work sessions from randomly selected users at random times before and after the transparent user interfaces were introduced. These video tapes were analyzed to augment our interview data on work patterns or changes in work patterns. The results of the video tapes analyses are integrated into the results reported below.

User Profiles

The case study was conducted with product users who worked for the company that developed MagicPaint. We collected user profile information to categorize both levels of expertise and how extensively users had worked with the new transparent interface. Users were grouped into three categories based on these profiles (Table 1).

The first category was comprised of four users who worked primarily as programmers or product testers. Their use of MagicPaint was characterized by short "bursty" paint sessions used to produce scribble style drawings. A large combination of features and painting tools were used, with each being used only a few times. Their task goal was to ensure that the system worked at a basic level and that there were no apparent software bugs.

The second user group consisted of three "demo gurus". These users were primarily marketing and sales staff who were expert MagicPaint users. Their task goal is to produce fast, interesting, and flashy demonstrations to potential customers and at trade shows. These users typically import almost completed, high quality artwork and demonstrate key features through fast, dynamic touch ups to the image.

The last user category is the group comprised of four hired artists who use MagicPaint to produce images. These highly skilled users most accurately reflect the task goals and

usage patterns of MagicPaint customers. Their task goal is to produce high quality images. They accomplish this through lengthy sessions which span days, weeks and sometimes months, depending upon the image complexity. They heavily customize the system and the defaults to suit their particular task and work style.

RESULTS

Both prior to and after the introduction of transparency the users maintained similar screen layouts. This consisted of a drawing area which was a full-screen window, with menu items and icon palettes minimized in size and pushed to either the top or the side of the drawing window (Figure 4). Interactive windows, overlapping the drawing area, were used to change tool attributes (Figure 5). According to the users, and in particular the artists who were extensive expert users, "no screen size is big enough" and "anything which minimized the number and size of overlapping windows is good".

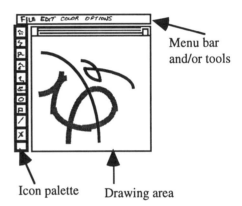

Figure 4. Diagram of typical screen configuration and layout of windows. Reflects task during drawing or painting, typical of a right handed user.

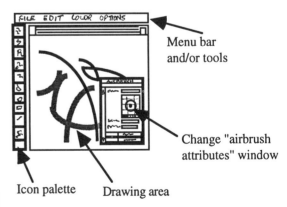

Figure 5. Diagram of typical screen configuration and layout with overlapping tool modifier window.

Based in part on the results of our previous research (Harrison et al., 1995, and Harrison and Vicente, 1996), we decided to create UI windows which were a hybrid of opaque and transparent components. The guideline we applied was that any selectable item within the window would be opaque, while non-selectable areas of the window

Type of User	# of Users	# who used transparency at beginning of trial	reason for NOT using transparency
programmers, testers	4	4	
product "demo" experts	3	1	2 users did not know how to turn feature on
professional artists	4	2	2 users turned feature off after one week of usage

Table 1. User profiles for subjects in case study.

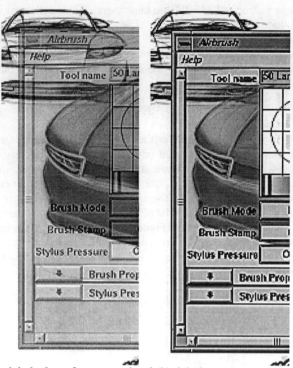

6 (a). labels & surface around labels are both transparent

6 (b). labels are opaque, surface around labels is transparent

Figure 6. Actual Images from MagicPaint showing design alternatives for transparency.

would be 50% transparent (see Figure 6). The window size and location were determined by the user (as in any normal windowing system). Users generally maintained the same window sizes throughout the trial. All UI windows were semi-transparent, with varying numbers and types of opaque components, depending upon the specific window properties. These opaque components were typically buttons, sliders, color wheels, and radius/opacity selectors. Whenever the user selected a particular tool item from the menu or from the icon palette the appropriate UI window would appear and remain until the user elected to explicitly close it.

In the first interview phase, we discovered that about half of the users interviewed had not been using transparency (see summary in Table 1). The people who did use it were: all four programmer/testers, only one "demo guru", and two artists. Two users had not yet received information about how to turn the feature on! Two users had turned transparency off after about one week of use. Specific comments and reasons are described below.

Issues Related to Technical Limitations

There were complaints that the font within transparency windows was difficult to read, particularly whenever it was over dark images. This was attributed to using black fonts primarily combined with a dark gray or black stippled window color. The result was distortion to the font edges. This was later fixed by changing the default window color to light gray, thereby avoiding conflicts with any of the font colors.

Another common complaint was about "flashing windows". Here users were referring to an unfortunate artifact

produced by drawing windows in the overlay plane. In particular, any window in the overlay plane can be assigned it's own color map (roughly, a list of colors that window uses). Whenever the cursor moves over a particular window, that window and its color map become active. If the color maps are different between windows, different color maps are swapped in and out. This generates a flashing of the window borders and title bar. Although it may sound subtle, to artists this was so annoying that one turned off transparency altogether. This problem can be partially addressed by assigning a common color map. However, some slight but noticeable highlighting changes will still occur which are meant to indicate the currently active window. The complete solution to this problem is unclear at this time.

Users requested the ability to set their own level of transparency by using something like a slider. Unfortunately, due to technical limitations this is not feasible at this time. Some users also requested the ability to set their own window color. We have not decided whether to provide this capability as yet, since the choice of window color may interact with transparency to negatively affect performance.

Users commented that some windows have very little transparent area and many opaque objects, limiting the utility of transparency. In the current design of the product, there are a number of objects within a window which we left opaque. These were selectable objects (buttons, sliders) which could be entered from the keyboard (text entry fields), and selectable areas, such as the color wheel. In particular, there are a number of embedded diagrams, such as the color wheel or the radius diagram for adjusting brush size, where precise selection and accurate visual perception are critical. In these cases, the objects were left opaque (see Figure 6b). The area around these objects was transparent. Windows which contain many such objects are primarily opaque as a result. We did implement a version of the system where both the button and object borders were transparent (see Figure 6a). While this made more of the window transparent, it was not obvious that the "buttons" and text fields were selectable to users. This was partly because users often equate "grayed-out" or partially invisible items with "un-selectable at this time".

Issues Related to Work flow

In general, transparency was felt to be most useful when windows were poorly placed or there were too many open. The number of windows open at any one time and their locations are completely determined by the user. A typical session has anywhere from one to four windows open simultaneously (or a minimum of 10% of the screen obscured). One artist commented that with experience more than one open window should not happen often. Ideally, almost all the artists commented that *any* window was in their way. They wanted a system where no windows were required. Transparency was seen to be one step in that general direction.

Most users commented that transparency works best in situations where they wanted visual feedback about some operation, as opposed to situations where they were actually trying to draw or paint directly behind the window.

Several users wanted transparency in the "support" windows. This would include the Help System windows and any message dialogs presented to the user. The message dialogs are generally system related issues. Serious errors would likely need to remain opaque to ensure that they were clearly seen and also because often with such errors it is important to interrupt the user's work. We are currently working on making the Help system transparent.

EXTENSIONS AND IMPLICATIONS

The exercise of implementing transparency in a product revealed many more technical challenges than we had anticipated. We made a series of compromises in the design parameters to adjust for this. A number of operating system level bugs were discovered and various temporary work-arounds had to be created. Finally, we discovered that our implementation approach is greatly impacted by the hardware configurations of specific machines. In order to successfully run transparent interfaces, we found it necessary to include procedures that determine the type of machine running the program and then adjust for it's capabilities (or lack thereof). However, we believe that this reasonably reflects the complexity of creating new technologies and integrating these into commercial products for trials. Many useful insights can be attained and many creative solutions can emerge in such a situation.

A number of the users' comments resulted directly from the technical issues. It is clear that even seemingly minor issues to a programmer can be highly irritating to someone with an artist's eye. These can influence the user to the point of shutting off features! It is interesting to note that with transparency, it is sometimes difficult to "hide" the technical and implementation aspects from the user (as good UI practice might suggest). Again, this provided us with a number of valuable insights into our user group and we were able to address many of the apparently aesthetic yet critical issues.

The case study confirmed our earlier experimental findings that default settings have an enormous impact of the usability of transparency. Window background color, font color and type style, and transparency level itself interact and make the difference between a system that works and one that does not. Relatively minor adjustments result in major perceptual differences. We determined that 50% transparent windows in lighter colors seemed to work best.

We were not able to assess whether transparency has a significant impact on improving the user's overall productivity, as intended. Since we implemented transparency into a new product release, it is very difficult to separate out which aspects of work are influenced by the introduction of transparency and which are influenced by some other changes to the system or its features. The main changes introduced (unrelated to transparency) were: new icon palettes and a complete reorganization of menu item ordering, hierarchic menus, user definable and save–able "toolkits" (collections of tools in a sequence), and user constructable and save–able pop-up menus.

In summary, this case study illustrates the challenges that one encounters in transitioning from basic research to commercial applications. We believe that transparent interfaces hold great promise, but it is equally clear that, for now, technical implementation details stand in the way of realizing those benefits. Therefore, further technical innovation is required before we can put the insights gained from our previous experiments (Harrison, Ishii, et al., 1995;

Harrison, Kurtenbach, and Vicente, 1995; Harrison and Vicente, 1996) into practice. It is essential that these obstacles be overcome if human factors research is to have an impact on actual design practice.

ACKNOWLEDGMENTS

This research was sponsored by research grants from Alias|Wavefront and the Natural Sciences and Engineering Research Council of Canada. We would also like to thank Bill Buxton, Hiroshi Ishii, Gordon Kurtenbach, and members of the Cognitive Engineering Lab for their comments.

REFERENCES

Bier, E. A., Stone, M. C., Pier, K., Buxton, W., & DeRose, T. D. (1993). Toolglass and magic lenses: The see-through interface. In Proceedings of SIGGRAPH'93. (pp. 73-80). New York: ACM.

Bier, E. A., Stone, M. C., Fishkin, K., Buxton, W., & Baudel, T. (1994). A taxonomy of see-through tools. In Proceedings of CHI'94. (pp. 358-364). New York: ACM.

Harrison, B. L., Ishii, H., Vicente, K. J., & Buxton, B. (1995). Transparent layered user interfaces: An evaluation of a display design space to enhance focused and divided attention. In Proceedings of CHI'95. (pp. 317-324). New York ACM.

Harrison, B. L., Kurtenbach, G., & Vicente, K. J. (1995). An experimental evaluation of transparent user interface tools and information content. In Proceedings of User Interface Software and Technologies - UIST'95 (pp. 81-90). New York: ACM.

Harrison, B. L., & Vicente, K. J. (1996). An experimental evaluation of transparency menu usage. In Proceedings of CHI'96. (pp. 391-396). New York: ACM.

Ishii, H., & Kobayashi, M. (1991). Clearboard: A seamless medium for shared drawing and conversation with eye contact. In Proceedings of CHI'91. (pp. 525-532). New York: ACM.

Ishii, H. (1990). TeamWorkStation: Toward a seamless shared workspace. In Proceedings of Computer Supported Collaborative Work - CSCW'90. New York: ACM.

Knowlton, K. C. (1977). Computer displays optically superimposed on input devices. Bell System Technical Journal, 56, (pp. 367-383).

Larish, I. and Wickens, C. D. (1991). Divided Attention with Superimposed and Separated Imagery: Implications for Head-Up Displays. University of Illinois Institute of Aviation Technical Report (ARL-91-4/NASA HUD-91-1).

Schmandt, C. (1983). Spatial input/display correspondance in a stereoscopic computer graphic workstation. Computer Graphics, 17, (pp. 253-259).

Wickens, C. D., Martin-Emerson, R., and Larish, I. (1993). Attentional tunneling and the Head-up Display. In Proceedings of the 7th Annual Symposium on Aviation Psychology, Ohio State University, Ohio, (pp. 865-870).

Zhai, S., Buxton, W., & Milgram, P. (1994). The "silk cursor": Investigating transparency for 3D target acquisition. In Proceedings of CHI'94. (pp. 459-464). New York: ACM.

IMPACT OF NEW INPUT TECHNOLOGY ON DESIGN OF CHAIR ARMRESTS: INVESTIGATION ON KEYBOARD AND MOUSE

Rajendra Paul
Haworth
Holland, Michigan

Rani Lueder
Humanics ErgoSystems
Encino, California

Allen Selner and Jayant Limaye
Medstar BioMedical, Inc.
Studio City, California

This study investigated the effects of addition of a mouse input device on the design of chair armrests. Eleven subjects performed a VDT task for four hours sitting in a chair under three armrest conditions: 1) no armrests, 2) height adjustable armrests, and 3) height and rotation adjustable armrests. The VDT task consisted of 90 minutes of graphics work using a mouse and 60 minutes of keying. The three experimental conditions were performed on three separate days in a random sequence. Muscle fatigue in the forearm (flexor and extensor) and neck-shoulder(trapezius) muscles was measured using surface electromyography. Subjects' working posture was measured using a SVHS camera and postural analysis was conducted using the Ariel Digitizing System. Discomfort and subjective preferences were recorded using the Corlett Scale and other questionnaires. The results suggest that height-adjustable armrests did not provide effective forearm support during mouse use; while the height and rotation adjustable armrests provided superior arm support. For keyboard work, both armrests reduced neck and shoulder fatigue measured in terms of the frequency shift. These results highlight the need for proactive research to tailor workplace design to match the demands imposed by new office technologies.

Key words: Armrests; Office Seating; VDT work; Mouse; Input devices.

INTRODUCTION

The ANSI/HFS 100-1988 standard provides recommendations for chair armrests predominantly based on anthropometric data(Human Factors Society, 1988). Among the few studies conducted on arm supports, the study by Occhipinti et al. (1985) demonstrated that arm supports reduced the compressive stresses in the spinal discs by as much as 40 percent. In a related study on wrist supports, Weber et al. (1984) reported that resting the hands on a wrist rest reduced loading on the shoulder muscles. The more the hands rested on the wrist rest (as measured by pressure on the wrist rest), the lower the EMG activity in trapezius muscles. Similar use of chair armrests in place of wrist rests to support the arms will likely reduce shoulder loading. In another study on wrist rests, Paul and Menon (1994) cautioned that some designs of wrist rests could create very high levels of contact pressure on the sensitive wrist area and thereby increase the risk of injury.

A critical feature of chair armrests is height adjustability. The armrests should be height-adjustable to cover the range of elbow rest heights of the intended user population. Non-height adjustable armrests also pose another practical problem because the height requirements for the work surface and chair armrests are similar. Consequently, fixed height armrests touch against the work surface and preclude many users from pulling the chair under the desk during reading and writing tasks when users intend to rest their hands on the work surface.

Casual observations suggest that even the height adjustable armrests might not provide good support during computer work, particularly when using a mouse. See Figure 1. When a mouse is placed on the right side of the keyboard, the right hand must traverse laterally at least 12 inches from its normal typing position to grasp the mouse; ten inches of which is over the keyboard surface. For left-handed individuals, if the mouse is placed on the left side of the keyboard, the corresponding lateral travel of the left hand is at least eight inches. This considerable lateral hand movement of hand is difficult to achieve solely by rotating the forearm at the elbow joint. For this reason, the majority of office employees extend their arms at the shoulder joint and use the mouse without armrest support. In absence of a forearm support, the arm is supported only at the shoulder joint, which can substantially increase the stress on the shoulder muscles and also increase the spinal loading. The increasing use of a mouse for office VDT tasks makes it important to study such observed changes in the role of chair armrests and their effectiveness when working with new input devices like graphics tablets and joysticks.

This study investigated the role of armrests for office employees using both a keyboard and a mouse. Stresses on the forearm (flexor and extensor) and shoulder (trapezius) muscles, postures assumed during VDT work, and subjects' responses and preferences were documented. These measurements were recorded under three experimental conditions: 1) no armrests, 2) height adjustable armrests, and 3) height and rotation adjustable armrests. The height adjustable armrests included a height adjustment range of two inches, viz. 8.0" to 10.0" above the seat pan. The height

and rotation adjustable armrests included a height adjustment range of five inches, viz. 7.2" to 12.2". In addition, its arm support surface could also be rotated 360 degrees around a vertical axis. See Figure 2.

EXPERIMENTAL PROCEDURE

Subjects

Eleven healthy subjects, six male and five female, screened to represent the office-going population, participated in this study. They were also screened for any symptoms or history of cumulative trauma disorders (CTDs). Their average age (years), height (inches) and weight (lbs) were 34.3 (standard deviation, SD = 11.7), 66.2 (SD = 4.9) and 157.0 (SD = 39.3) respectively. Each individual had at least one year of computer experience.

Equipment

Surface electromyography (EMG) equipment was used to measure the activity of forearm flexor and extensor muscles and trapezius muscles on both the left and right sides. The sensors for trapezius muscle were placed approximately two inches lateral to the spine at the base of the neck. The shift in frequency of EMG signals was used as the measure muscle fatigue. With fatigue, muscle fibers activate slowly, i.e. at lower frequency, with a corresponding decrease in frequency of EMG signals. Interested readers are referred to Grandjean (1987) for further information.

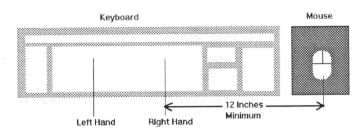

Figure 1. Lateral displacement of hand during mouse use (not to scale).

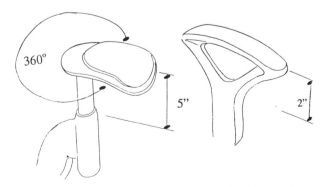

Figure 2. Armrest designs evaluated in this study (not to scale).

The working postures were recorded using a SVHS camera and joint angles were derived from the video using the Ariel Digitizing System. Corlett comfort questionnaire (Corlett and Bishop, 1976) and preference surveys were administered at the beginning and end of each keyboard and mouse trial.

Procedure

The experiment was conducted in a quiet laboratory setting. A height adjustable workstation and a highly adjustable chair, i.e. Haworth Accolade, with seat and back rest height and tilt adjustments were adjusted according to commonly accepted ergonomic principles to fit the subjects' morphology. The subjects evaluated each armrest for four hours, during which the EMG, postural and survey data were collected. See Figure 3. The four-hour period included an initial briefing and training period of about 45-60 minutes. Thereafter, the subjects performed a self-paced graphics task using

only the mouse for 90 minutes. Subjects were then given a 30 minute rest break to recover from the muscular fatigue developed during the mouse task. Finally subjects performed a text-entry task for 60 minutes. The subjects sat in the same chair and performed the same task under three experimental conditions on three separate days. The sequence in which the three conditions were tested was randomized.

RESULTS

The shifts in the frequency of muscle activity during the keyboard and mouse tasks are shown in Figures 4 and 5 respectively. The changes in perceived discomfort after 90 minutes of mouse use are shown in Figure 6. Results indicating subjective preferences are shown in Figure 7. The changes in perceived discomfort with and without armrests after 60 minutes of keying were not significantly different. The significance levels indicated in these

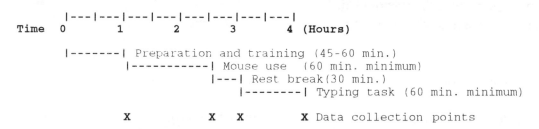

Figure 3. The time sequence followed during the experiment.

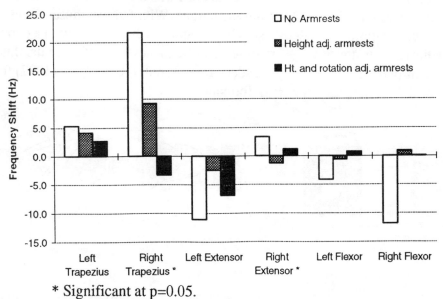

* Significant at p=0.05.

Figure 4. Frequency shift after 60 minutes of keying.

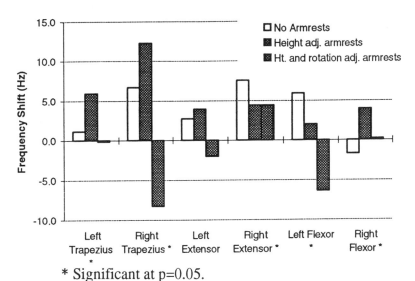

* Significant at p=0.05.

Figure 5. Frequency shift after 90 minutes of mouse use.

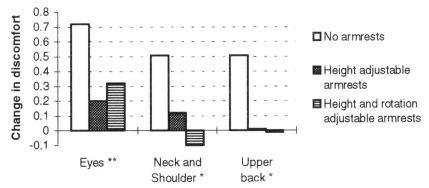

* Significant at p<0.05. ** Significant at p<0.10

Figure 6. Change in perceived discomfort after 90 minutes of mouse use. The results for upper arm, elbows, hands and wrist and lower back were not significant.

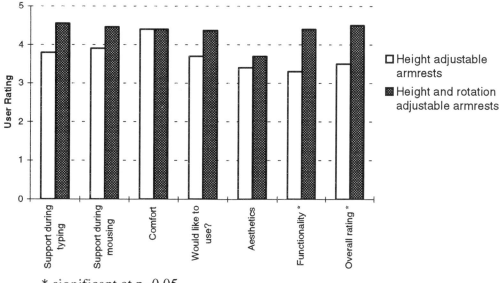

* significant at p=0.05

Figure 7. Subjective user ratings [1 = Worst, 5 = Best].

figures are based on two-way analysis of variance (ANOVA). Further analysis of differences between the two armrests using the Newman-Keuls range test showed that during both keying and mouse use, fatigue in the right trapezius muscle was significantly greater with the use of height adjustable armrests. The changes in perceived discomfort, however, were not significantly different.

DISCUSSION

The results suggest that for computer-intensive tasks, use of chair armrests reduced forearm and shoulder fatigue as well as discomfort. Except for the greater fatigue experienced by the right trapezius muscle, the height adjustable armrests provided comparable forearm support during keyboard use. But during mouse use, the frequency shift with the height and rotation adjustable armrests was significantly lower for the right trapezius and left and right flexor muscles. The differences were also more pronounced. This suggests that for mouse activities, height and rotation adjustable armrests performed superior than only height adjustable armrests. Subjects also rated the height and rotation adjustable armrests superior for "functionality" and "overall rating."

The difference in height adjustment range built into the two evaluated armrest designs -- two inches in height adjustable design compared to five inches in the other design -- likely influenced the results. For keyboard tasks, subjects used the height adjustable armrests positioned at the best possible height. For the shortest subjects, the arm support height appeared to be higher than the optimum height and for the tallest subjects, it appeared to be lower than the optimum height. However, none of them complained about it. During the mouse use, however, subjects could not and did not use the height adjustable armrests and opted for unsupported arm positions. Conversely, all subjects used the height and rotation adjustable armrests. Overall, the addition of rotation as a feature appeared to be useful in the following ways: 1) While keying, subjects positioned the armrests under the wrists or anterior forearms near the elbow, depending on their workstyle. 2) During mouse use, subjects positioned the armrest to support the extended arms in their working positions. 3) Subjects from lower percentiles, based on stature, rotated the armrests inwards whereas those from higher percentiles rotated the armrests outwards to achieve the needed width of forearm support.

This study on the role of armrests for modern office tasks involving both keyboard and mouse highlights the need to evaluate the effect of new input devices on the design of armrests and other aspects of workplace design. Further research should focus on other commonly used input devices like joysticks, voice synthesizers, digitizing pads and graphics tablets. For employees working several hours a day on computers, proper support for the unsupported limbs is important. Strategically, such research should be proactive, not reactive, and conducted before or soon after new technologies are introduced in the market. The exploding and ever-changing technology market makes this a challenging task for human factors researchers.

REFERENCES

Corlett, E. N. and Bishop, R.P. (1976). A technique for assessing postural discomfort, *Ergonomics*, 19, pp. 175-182.

Grandjean, E.(1987). *Ergonomics in computerized offices*. London: Taylor & Francis.

Human Factors Society (1988). *The American National Standard for Human Factors Engineering Requirements for Video Display Terminal Workstations, ANSI/HFES 100-1988*. Santa Monica, California: Human Factors Society.

Occhipinti, E., Colombini, D., Frigo, C., Pedotti, A. and Grieco, A. (1985). Sitting posture: analysis of lumbar stresses with upper limbs supported, *Ergonomics*, 28, 1333-1346.

Paul, R. D. and Menon, K. (1994). Ergonomic evaluation of keyboard wrist pads. *Proceedings of the 12th Triennial Congress of the International Ergonomics Association, vol. 2*, (pp. 204-207). Ontario, Canada: Human Factors Society of Canada.

Weber, A., Sancin, E. and Grandjean, E. (1984). The effects of various keyboard heights on EMG and physical discomfort, In. E. Grandjean (Ed.) *Ergonomics and Health in Modern Offices*. London: Taylor & Francis.

Influence of Food Label Quantifier Terms on Connoted Amount and Purchase Intention

Michael S. Wogalter
North Carolina State University
Raleigh, NC 27695-7801

Michael J. Kalsher
Rensselaer Polytechnic Institute
Troy, NY 12180-3590

Diane M. Litynski
Marist College
Poughkeepsie, NY 12601-1387

ABSTRACT

This research examined the influence of food label quantifier terms (e.g., Low, Reduced, and Free) on people's interpretation of implied quantity. Experiment 1 assessed people's perceptions (connotation) of eight terms to determine whether the terms convey distinct quantities. Results showed significant differences among the lowest quantity-connoting terms. Experiment 2 investigated the influence of three specific quantifier terms placed in the context of nutrient claims on product labels and measured their effect on purchase intentions before and after dietary health concerns were made explicit. Results showed that the quantifier terms influenced people's purchase intentions when health and diet concerns were made salient, but not when general purchase preferences were requested. Implications for consumer comprehension and interpretation of food label are discussed, and suggestions for future research are offered.

INTRODUCTION

Changes in lifestyles in the U.S. over the past few decades, including increased use of packaged foods, has led to heightened public interest in food labeling practices. Consumers are demanding that more and better nutrition information be made available on labels. Food manufacturers have capitalized on this demand by using common words such as Light, Low, or Enriched to promote their products and may have, in doing so, added to consumers' confusion (Kessler, 1991). Until recently, no precise definitions existed to control the use or meaning of quantifier terms.

The U.S. Congress mandated truthful and uniform food nutritional labeling on foods by passing the Nutritional Label and Education Act (NLEA) of 1990. This NLEA went into effect in May 1994. The act gives the FDA the legal foundation to expedite rulings on label policies (Podolsky, Roberts, Silver, and Mukenge, 1991). The NLEA also standardizes food labeling practices including the way in which quantifier terms are used to convey the amount of substances present in foods. Specifically, quantitative definitions have been assigned to certain terms associated with nutrients (U.S. Department of Health and Human Services, 1989, 1991). This means that terms such as Low and Reduced can now be used on food labels only when they conform to FDA mandated definitions (Reece, Sheffet, and Rifon, 1996). Some of the approved terms, however, do not have a consistent meaning across nutrients. For example, when applied to sodium, the term Free means less than 5 milligrams of sodium per serving. In contrast, this same term (Free), when applied to cholesterol (Cholesterol Free), means less than 2 milligrams per serving. In addition, the FDA has

allowed for the substitution of certain terms assumed to be equivalent (e.g., the term No can be substituted for the term Free).

Although quantifiers need objective definitions to fix a standard and enable regulation by government and compliance by the food industry, it is unlikely that consumers know the assigned numerical definitions. Rather, consumers are likely to use a subjective interpretation of the verbal claims to guide their initial decisions of which product has more or less of a particular substance—before possibly looking at the more specific numbers in the nutrition label on the side or back panel (Szykman, Bloom, and Levy, 1996). So despite the implementation of objective definitions of food label terms, consumers may not fully understand their intended meaning. However, they might understand their relative ordering in terms of amount.

Other issues have been raised about the use of quantifier terms to convey nutrient quantities (U.S. DHHS, 1989; Mermelstein, 1990). Opponents of their use in nutrition claims argue that quantifier terms can cause consumer confusion. They contend that quantifier terms are unnecessary as more reliable, quantitative nutrition labels are now available. In contrast, proponents for their use argue that the terms can attract consumers' attention and could serve as a short cut for the information in the quantitative nutrition label. Moreover, the claims may motivate consumers to investigate the nutrition label to clarify their meaning (Szykman et al., 1996). Standardizing the terms can also force consistent use by manufacturers.

A survey sponsored by the Food Processors Association (Opinion Research Corporation, 1990) indicates the importance of understandable quantifier claims. In that study, 20% of consumers reported that they rarely or never read nutrition labels or ingredients for the foods they purchase, and approximately 40% reported reading this information only occasionally. Thirty-two percent of the respondents believed that Low Cholesterol meant low or less calories (which it does not). The survey also indicated that identical claims can produce different interpretations by different persons.

The present research examines the connoted meaning of a set of quantifier terms used to describe nutrient quantity on food labels. The research also investigates whether people's relative interpretation of the terms corresponds with the definitions put forth in the NLEA. Experiment 1 examines people's perceptions (connotation) of eight terms to determine whether the terms convey distinct quantities. Experiment 2 investigates the influence of three specific quantifier terms placed in the context of nutrient claims on product labels and measures their effect on purchase intentions before and after dietary health concerns are made explicit. It was expected that consideration of health using a scenario describing a diet-related disease that is controllable by reduced intake of nutrient would motivate participants to make greater use of nutrient claim. It was further expected that their product choices would be controlled by the perceived meaning of the quantifier terms. However, without explicit consideration of a

diet-related disease, participants' use of the claims in their choice of food products would be much more limited.

EXPERIMENT 1

Method

Participants. Twenty Rensselaer Polytechnic Institute (RPI) students and staff voluntarily participated. Their ages ranged from 18 to 50 ($M = 24.5$, $SD = 6.9$).

Materials and procedure. Participants were given a sheet containing eight randomly-ordered quantifier terms and asked to order them according to implied amount, from lowest to highest. The terms were Free, Less, Low, Lower, Lowered, No, Reduced, and Very Low. Participants were allowed to list together words perceived to have the same meaning.

Results

Ranks of 1 (lowest) to 8 (highest) were assigned. In cases of ties, the scores were based on the average of the displaced ranks.

Mean ranks, standard deviations, and medians are shown in Table 1. An overall analysis using the Friedman Test was significant, χ^2 (7, $N = 20$) = 110.68, $p < .0001$. Paired comparisons indicated that No, Free, Very Low, and Low were significantly different from one another and from the other four terms ($ps < .05$). Lower, Lowered, and Less did not differ from each other ($ps > .05$). The term Lower was significantly higher than Reduced ($p < .05$) but the terms Lowered and Less did not differ from Reduced ($ps > .05$).

Discussion

These results suggest that some of the quantifier terms investigated in this study effectively convey distinct levels of quantity; others do not. Significant differences were noted mainly for terms signifying the lowest quantities. The term Free was significantly lower than the term Low which is consistent with the FDA's objective definitions. The term Reduced is a relative/comparative term, and thus, has a more complex meaning than terms such as Free and Low (which signify absolute magnitudes). However, when placed on the same scale, as in the present study, the results indicate that Reduced connotes higher levels than all but two of the other terms. The term No, which the FDA allows as a substitute for the term Free, is apparently interpreted as indicating a quantity significantly lower than Free.

EXPERIMENT 2

Because the terms in Experiment 1 were tested in the absence of context, the findings may not generalize to their use in actual nutrient claims. In Experiment 2, three

Table 1

Mean Ranks, Standard Deviations, and Medians of the Eight Quantifier Terms Tested in Experiment 1

Quantifier Terms	Mean Rank	Standard Deviation	Median
No[a]	1.15	.37	1.35
Free[b]	1.85	.37	1.85
Very Low[c]	3.45	1.40	3.40
Low[d]	4.35	.93	4.17
Lower[e]	5.60	1.19	5.38
Lowered[e,f]	6.25	1.33	6.33
Less[e,f]	6.45	1.15	6.33
Reduced[f]	6.90	1.29	7.21

Note. Quantifier terms with different subscript letters are statistically significant from one another ($p < .05$).

quantifier terms (Free, Low, and Reduced) with FDA quantitative definitions that were shown to be significantly different from one another in the first experiment were reexamined in a more externally-valid context (i.e., on food package labels) to determine their effect on purchase intentions. In addition, the effect of emphasizing diet and health is examined.

Method

Participants. Sixty-four individuals participated. Half (19 males and 13 females) were RPI undergraduates, ranging in age from 18 to 24 ($M = 21.5$, $SD = 1.2$). The other half (8 males and 24 females) were permanent residents of Troy, NY, including homemakers, maintenance personnel, secretarial workers, and white collar professionals who volunteered when approached at local shopping areas. Their ages ranged from 20 to 66 ($M = 38.3$, $SD = 11.1$). Participants were assigned randomly to conditions.

Materials. An optical scanner was used to digitize the front labels of eight nationally-sold dry breakfast cereals. The images were manipulated by using graphics software. Any preexisting dietary claims were deleted from the original label images. For every product, 13 label images were produced. Twelve contained quantifier-nutrient claims that were produced by factorially pairing the three quantifier terms (Free, Low and Reduced) with four food nutrients (Sugar, Sodium, Fat and Cholesterol), e.g., Sodium Free or No Cholesterol. One image for each product lacked a quantifier/nutrient claim which served as a control. The terms were located in the same place as the labels' original nutrition claim (if there was one). If no nutrition claim was originally present, the quantifier-nutrient claim was located in the least cluttered area of the label. All quantifier-nutrient terms were in 24-point type in a font that resembled the other print on the label. A laser printer and photocopier were used to produce the final experimental materials.

Thirty-two booklets were formed each containing labels of all eight products. Every booklet presented each quantifier term twice in connection with one nutrient term. Except for the control labels which had no nutrient claim, all labels of a booklet referred to a single nutrient term (i.e., only cholesterol or only sodium). Across the 32 booklets, all quantifiers were paired an equal number of times with all products and nutrients. A Latin Square was used to systematically rotate the labels and quantifiers through the booklet orders.

Procedure. Participants were given one of the booklets that contained the following request: "Please rate each product according to the likelihood that you would purchase it if you were to see the product on a grocery store shelf." Participants responded using a six-point Likert-type scale from (0) definitely would not purchase to (5) definitely would purchase.

After completing their ratings to the first question, participants were given one of four medical-dietary scenarios in which they were told to assume the following: "Your doctor has told you that you have a medical condition which requires that you minimize your intake of _____." The blank was replaced with the specific nutrients that corresponded with their booklet assignment. Participants then evaluated each of the products using the same purchase-intention question and scale given earlier.

Results

The two questions were analyzed separately. Table 2 shows the means and standard deviations for the general and dietary concern ratings as a function of conditions.

General purchase intentions. A one-way repeated-measures ANOVA for the general purchase intention ratings failed to show a significant effect, $F(3, 189) < 1.0$. Exploratory analyses were conducted using several additional independent variables: gender, participant group (student vs. nonstudent), age group (two groups formed from a median split), and nutrient. These variables were added individually (or in pairs where the cell sizes were not grossly unequal) to ANOVAs that also included quantifier as an independent variable. Only two significant effects were noted. One was a main effect of nutrient, $F(3, 56) = 3.19$, $p < .05$. Comparisons showed that products with sugar claims ($M = 1.63$) produced significantly lower purchase intentions than

Table 2

Mean Purchase Intention Ratings for General Preference and Dietary Concern Conditions as a Function of Quantifier Term Conditions in Experiment 2

	Quantifier Term Conditions			
Purchase Intention	Free	Low	Reduced	No Quantifier (Control)
General:				
Mean	2.23[a]	2.29[b]	2.19[c]	2.42[d]
SD	1.33	1.30	1.16	1.19
Dietary Concern:				
Mean	3.23[a]	2.86[b]	2.55[b]	2.09[c]
SD	1.40	1.37	1.19	1.20

Note. Quantifier terms with different subscript letters are statistically significant from one another ($p < .05$).

the other three nutrients (Ms = 2.37, 2.45, and 2.50, for sodium, fat, and cholesterol, respectively).

Another analysis showed a significant interaction between participant group and nutrient, $F(3, 56) = 4.29, p < .01$. Simple effects analysis indicated that nonstudents had higher purchase intentions for cholesterol reduction claims ($M = 3.11$) than students ($M = 1.86$).

Purchase intention with dietary concern. For the second set of ratings, a one-way repeated-measures ANOVA showed a significant effect, $F(3, 189) = 15.62, p < .0001$. Comparisons showed that all differences among the quantifiers were significant ($ps < .05$) except between the terms Low and Reduced ($p > .05$).

Additional analyses showed a significant main effect of nutrient, $F(3, 56) = 2.83, p < .05$. Fat claims produced significantly higher purchase intentions ($M = 3.03$) than sugar claims ($M = 2.10$), $p < .05$.

Discussion

Although the first general purchase intention question did not show an effect of quantifier, when a dietary-health concern was subsequently made salient, differences among the terms were evident. This suggests that diet and health are relevant contributing factors to whether quantifiers affect people's purchase intentions.

Participants had lower purchase intentions for claims of lowered sugar, suggesting that it is a nutrient that consumers consider to be less important as a dietary risk relative to the other nutrients. The general purchase intention results indicated that nonstudents preferred cholesterol reduction claims more than students. This result may be due to recently publicized health concerns and information in food advertisements (Reese et al., 1996) about cholesterol in older individuals (represented here by the nonstudents). However, age cannot be the only reason for the difference because there was no evidence of an age by nutrient interaction. Thus the difference found between the groups is probably due to some other variable that cannot be determined by the demographic data collected.

The failure to find an interaction of quantifier and nutrient or of quantifier and participant group suggests that quantifier perceptions are consistent across nutrients and that the results may generalize to other consumer populations.

GENERAL DISCUSSION

The present research examined the influence of quantifier terms used on food labels to indicate various degrees of nutrient quantity, such as No, Free, Low, etc. Experiment 1 showed most of the eight tested terms connote different meanings. Experiment 2 tested three specific quantifiers in a more ecologically valid context and produced effects that were generally in accord with Experiment 1 (except the comparison between Low and Reduced). However, in Experiment 2 differences were shown only when individuals were made aware of a dietary health condition that could be controlled by nutrient intake. The terms had no effect when health problems were not explicitly stated. Thus, while quantifiers terms appear to be potentially useful in conveying relative amounts of nutritional substances, these terms, by themselves, may not influence purchase intentions without the nutrient claim being considered *relevant* to individuals. This concurs with results showing that persons with diet-related diseases are more likely to to develop positive attitudes towards good nutrition than healthy consumers (Russo, Staelin, and Nolan, 1986). Nevertheless, it is interesting to note that several studies have found little or no evidence for a relation between disease status and search for nutrition information (Moorman and Matulich, 1993; Szykman et al., 1996). Feick, Hermann, and Warland (1986) found that poor health increased the search for nutrition information from books, pamphlets, and health care providers, but not food labels.

Although this research shows that the relative ordering of some of the terms (e.g., Free and Low) correspond with the ordering in FDA's quantitative definitions, other findings are less consistent. First, while the FDA allows the substitution of No with the term Free, Experiment 1 showed that these terms are perceived somewhat differently: No is interpreted as a lower quantity than Free. One explanation is that Free is an ambiguous term that has many meanings—only one of which is the absence of something. Another meaning is "gratis"(free of charge). It is therefore possible that individuals who do not have a broad knowledge of English could misinterpret statements such as Sugar Free or Caffeine Free by believing that the product possesses additional sugar or caffeine as a bonus! The term No is clear, as it lacks alternative definitions.

Second, an inconsistency was shown between experiments. In Experiment 1, the terms Reduced and Low differed, but in Experiment 2 they did not. As mentioned earlier, Reduced is a relative term whereas Low refers to a direct numerical (absolute) quantity. Using the FDA's criteria, it is possible for a nutrient claim to be matched with both terms but actually be in the same or opposite direction with respect to actual quantity. Thus, the amount conveyed between the terms is ambiguous and this might partly account for the reason a difference was found between experiments.

Consumers will only benefit from quantifier terms if they understand them. Given the potential ambiguities mentioned above, a relevant issue is how to remove consumer misperceptions. One method is to educate the public. Through the mandate of the NLEA, the FDA and others have developed large-scale educational programs to reduce confusion and increase comprehension of nutrition information (Foulke, 1992; Kessler, 1991). Experimental research (e.g., Jessen and Wogalter, 1992) has shown that such programs can be successful in helping people to make better decisions between similar foods differing in nutrient quantities. However, a better, less formidable, way to facilitate correct interpretation is to use terms that people already know and understand. Large-scale measurement studies on the connoted meanings of alternative terms (Herbert, Kalsher, and Wogalter, 1993; Kalsher, Wogalter, and Gilbert, 1992) document a broad range of well-known terms not currently included on FDA lists. These scaling studies are much less costly than training substantial numbers of consumers on what they are supposed to mean. In fact, if terms are evaluated and then selected for understandability beforehand, education of the population could be less costly in terms of expense, time and effort, and where education is still needed, it could be allocated to more restricted venues.

Three limitations of this research should be noted. First, the current study concerned products that contained only one nutrient claim. Consumers are frequently confronted with labels that simultaneously make more than one claim, and thus, future studies should examine multi-claim judgments. Second, this study focused on purchase intentions and not behavior. Thus, well-designed quasi-experimental field studies are needed to determine the impact of food claims on actual purchase behavior and how they might differentially affect various consumer groups. Third, the manipulation of presence versus absence of dietary concern was confounded with order. However, counterbalancing the order of the two purchase intention conditions would have produced another kind of carryover effect that we wished to avoid. If participants were cued to consider a dietary concern first, it would be difficult to remove this mental orientation for the general (or absent dietary concern) purchase intention condition. Therefore it made sense to have participants make judgments without mention of dietary concern first, and then afterwards, elicit an interest in a concern for nutrition and diet. Future studies incorporating scenarios should consider using a between-subjects design instead.

Finally, this research adds to the rather minimal empirical work on quantifier terms (Lowrey, Gallay, and Shrum, 1996), and provides a basis from which future standards for food labeling can be established The goal is to ensure that nutrition information is clearly conveyed to facilitate proper purchase decisions by consumers. This is particularly important for certain populations who need to limit intake of certain dietary substances—where misinterpretations could produce health risk consequences (Earl et al., 1990).

ACKNOWLEDGEMENTS

Funding for this research was provided to the first author by a grant from the Paul M. Beer Trust, Rensselaer Polytechnic Institute. The authors would like to thank Barbara J. Seipp for her assistance in this research.

REFERENCES

Earl, R., Porter, D. V., and Wellman, N. S. (1990). Nutrition labeling – Issues and directions for the 1990s. *Journal of the American Dietetic Association*, 90, 1599–1601.

Feick, L. F., Herrmann, R. O., and Warland, R. H. (1986). Search for nutrition information: A probit analysis of the use of different information sources. *Journal of Consumer Affairs*, 20, 173-192.

Foulke, J. E. (1992). Wide-sweeping FDA proposals to improve food labeling. *FDA Consumer*, 26(1), 9-13.

Herbert, L. B., Kalsher, M. J., and Wogalter, M. S. (October, 1993). Connoted meaning of descriptive terms across food nutrients. Poster presented at the 37th Annual Meeting of the Human Factors Society, Seattle, WA.

Jessen, D. M., and Wogalter, M. S. (1992). The influence of audio-visual instruction on consumers' selection of nutritious food products. *Proceedings of the Human Factors Society*, 36, 533-537.

Kalsher, M. J., Wogalter, M. S., and Gilbert, C. M. (1992). Connoted quantity of food-label modifier terms. *Proceedings of the Human Factors Society*, 36, 528-532.

Kessler, D. A. (1991). Building a better food label. *FDA Consumer*, 25 (7), 11-13.

Lowrey, T. M., Gallay, R., and Shrum, L. J. (1996). Effects of nutrition labels and advertising claims on product perceptions. *Marketing and Public Policy Conference Proceedings* (Chicago: American Marketing Association), 28-29.

Mermelstein, N. H. (1990). FDA and Congress seek changes in nutrition labeling. *Food Technology*, 44, 54–60.

Moorman, C., and Matulich, E. (1993). A model of consumers' preventative health behaviors: The role of health motivation and health ability. *Journal of Consumer Research*, 20, 208-228.

Opinion Research Corporation (1990). *Food labeling and nutrition: What Americans want - Summary of findings*. Washington, DC: National Food Processors Association.

Podolsky, D., Roberts, M., Silver, M. S., and Mukenge, M. (1991). Hype–free food labels. *U.S. News and World Report*, 110 (Issue 24, June 3), 67–70.

Reece, B. B., Sheffet, M. J., and Rifon, N. J. (1996). The impact of the NLEA on the use of "Light," "Natural, " and other key words in food advertisements. *Marketing and Public Policy Conference Proceedings* (Chicago: American Marketing Association), 30-35.

Russo, J. E., Staelin, R., Nolan, C. A., Russell, G. J., Metcalf, B. L. (1986). Nutrition information in the supermarket. *Journal of Consumer Research*, 13, 48-70.

Szykman, L. R., Bloom, P. N., and Levy, A. S. (1996). A proposed model of the use of package claims and nutrition labels. *Marketing and Public Policy Conference Proceedings* (Chicago: American Marketing Association), 26-27.

U.S. Department of Health and Human Services (1989). Food labeling: Advance notice of proposed rulemaking. *Federal Register* (DHHS Publication No. 21 CFR Ch. 1). Washington, DC: U.S. Government Printing Office.

U.S. Department of Health and Human Services (1991). Food Labeling; General Provisions; Nutrition labeling; Nutrient content Claims; Health Claims; Ingredient Labeling; State and Local Requirements; and Exemptions; Proposed Rules. *Federal Register* (DHHS Publication No. 21 CFR Part 101, et al.). Washington, DC: U.S. Government Printing Office.

SHARPNESS OF LID EDGES

Stephan Konz, Don Aye and Yanci Fu
Dept. of Industrial and Mfg. Systems Engineering
Kansas State University
Manhattan, KS 66506 USA

In Experiment 1, 15 plexiglass lids were constructed, using all combinations of 3 diameters (32, 70, 82.5 mm) and 5 edges (having a radius of curvature of 0, .8, 1.6, 2.4 and 3.2 mm).   The lids were inserted into a torquemeter and held vs 2.3 kg-m of torque. The 44 adult subjects (18 M; 26 F) indicated their discomfort. Gender significantly affected discomfort, with female discomfort values about 70% greater than male values. For the same radius of curvature, discomfort was highest at 32 mm dia, intermediate at 82.5 mm and least at 70 mm.

In Experiment 2, 82.5 mm dia lids with edges of .8, 1.6 and 2.4 mm radii were used by 10 subjects (5 M, 5 F). The torques were 1.1, 1.7, 2.3, 2.8 and 3.4 kg-m. Female discomfort vote increased 1.9/ kg-m while male vote increased 1.2/kg-m.

INTRODUCTION

It generally is accepted that an edge can be too sharp. But what affects perceived sharpness? And what are acceptable numerical values of edge sharpness?

Konz and his students (Kazmi and Konz, 1991; Chen and Konz, 1993) studied various container lids. Among the variables studied were lid shape and knurls. A persistent problem with the lids was "sharp edges".

Brough and Konz (1992) studied pressure sensitivity at 48 locations on the head, using a constant pressure of .95 kg/cm^2. Females (mean vote 66) were more sensitive to the pressure than males (mean vote 62). The temple area was most sensitive.

Fransson-Hall and Kilbom (1993) reported that women, when exposed to sustained externally applied surface pressure on the hand, experienced pain at a lower threshold than males. The pain pressure threshold was measured at 64 different points on the hand; the female threshold was about 70% of the male threshold. Goonetilleke and Eng (1994) studied pressure sensitivity on the top of the foot. The maximum pressure tolerance was 831 kPa for the 5 mm dia probe and 249 kPa for the 13 mm dia probe. That is, contact area affected discomfort. Gender was not significant but they only studied 4 males and 4 females.

METHOD

Experiment 1

Task/Procedure

A numerically-controlled lathe was used to cut 15 plexiglass circular lids. There were 3 diameters (32, 70 and 82.5 mm) and 5 edges of 90° (having radii of curvature of 0, .8, 1.6, 2.4 and 3.2 mm). Thickness was held constant at 12.5 mm.

Each lid had two holes; one end of an adaptor was inserted into the holes and the other end placed in a torquemeter. The torquemeter was mounted on a table so the lid was horizontal. The seated subject turned the lid with the right hand until the torquemeter dial read 20 in-lbs (2.3 kg-m). When the experimenter said "let go", the subject released the lid and voted hand discomfort, using a Borg scale of 0 = comfortable, .5 = extremely weak discomfort, 1 = very weak discomfort, 2 = weak discomfort, 3 = moderate discomfort, 5 = strong discomfort, 7 = very strong discomfort and 10 =

extremely strong discomfort. Each subject did each of the 15 lids in a random order; the experiment took about 15 minutes.

Subjects

The 44 adult subjects (18 male, 26 female) were non-students from Wichita, KS. They volunteered and received no pay. For age, 18 were between 18 to 35, 10 were 35 to 50, and 16 were older than 51.

Results/Discussion

Figure 1 shows the results; male and female data are combined. The first point to make is that sharp edges increase discomfort. But there is an interaction of radius of curvature and lid diameter. For the 32 mm dia lid, all radii caused discomfort. For the 70 and 82.5 mm dia lids, curvature greater than 2 mm gives relatively little discomfort. The larger diameters have larger contact areas; this may give a different threshold pressure. Another possibility is that a different grip may be used for the different lid diameters.

Gender significantly affected discomfort. At a radius of curvature of 0, female discomfort vote was 4.7 vs 2.7 for males; a ratio of 1.7. At a radius of curvature of .8 mm, female discomfort was 3.0 vs 1.6 for males (1.9). At a radius of 1.6 mm, female discomfort was 2.7 vs 1.6 for males (1.7). At a radius of 2.4 mm, female discomfort was 2.3 vs 1.6 for males (1.4). At a radius of 3.2 mm, female discomfort was 2.3 vs 1.6 for males (1.6). The average female discomfort was 1.7 that of a male. This is compatible with the Fransson and Kilbohm results that female pain threshold is 70% of the male pain threshold.

Experiment 2

Task/Procedure

Three of the lids from Experiment 1 were used. The lids were 82.5 mm in dia and had curvatures of .8, 1.6 and 2.4 mm. The seated subjects held the lid at 50 mm below elbow height vs torques of 1.14, 1.70, 2.27, 2.84 and 3.40 kg-m. Each subject had a randomized sequence of the 15 conditions on day 1; on day 2 they repeated the

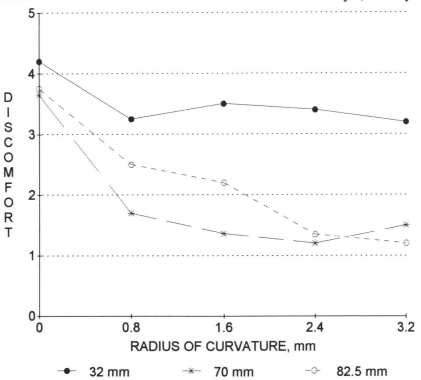

Figure 1. Discomfort is affected by both lid diameter and edge sharpness. For a 32 mm dia, all edges are sharp. For the 70-82.5 lids, discomfort is acceptable if the radius of curvature is over 2 mm.

sequence in a reverse order. During the second trial at a condition, they were not shown their vote on the first trial. A stopwatch was used to ensure that there was 60 s between conditions.

Subjects

One female's data was an outlier; this subject was replaced by another female; the result was 5 males and 5 female subjects with an age range of 20 to 42.

Results/Discussion

In the Analysis of Variance, gender, torque and the gender x torque interaction were highly significant ($p < .0001$).

Fig. 2 shows the mean discomfort vote for .8 mm radius of curvature was 2.08, for 1.6 mm it was 1.91, and for 2.4 mm it was 1.87. Using ANOVA, these means were not significantly different.

Fig. 3 shows the male and female discomfort votes vs torque. The mean discomfort vote for females was 2.7 and for males was 1.2 (2.7/1.2 = 2.2). A linear approximation has male discomfort vote increase 1.2/kg-m of torque. A linear approximation has female discomfort vote increase 1.9/kg-m of torque. The torque of 3.4 kg-m (30 in-lbs) was quite difficult for many females.

Note that our investigations concerned sharpness of lid edges not sharpness of a point.

The gender effect of sharpness could be either physiological or psychological. That is, there may be a greater sensitivity in female hands; perhaps more likely is that males are less willing to report discomfort.

Conclusion

Confirming the literature, there is a gender effect of sharpness; for the same edge, females report more discomfort than males. The sharper the edge, the greater the relative discomfort of females. There also is a gender effect of torque (i.e pressure of the hand on the edge). At low torques the relative discomfort of females is lower than at high torques.

That is, when pressure is higher (sharper edge or higher torque), females report more discomfort than males than when pressure is lower (rounded edge, lower torque).

From experiment 2, the sharpness of edges (within the range of curvatures from .8 to 2.4 mm) is unimportant relative to the variable of torque.

From experiment 1, the effect of edge sharpness varies with lid diameter; larger dia (70-82.5 mm dia) lids are relatively unaffected by edge sharpness. But for 32 mm dia lids, all edges were sharp. Perhaps the grip varies somewhat with lid diameter. Perhaps the pressure on the fingers is less for the larger dia.

To reduce discomfort when opening lids:
1. Keep torque to 2.3 kg-m (20 inch-lbs) or less.
2. Make the radius of curvature of the edge greater than 2 mm.
3. The torque and radius are more critical for small diameter lids and when females turn the lids.

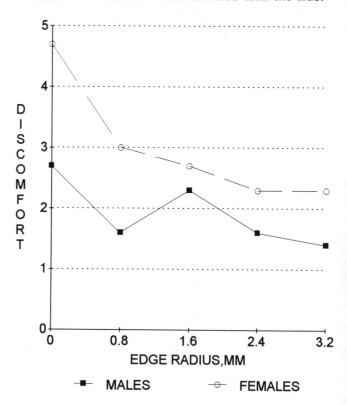

Figure 2. Females had more discomfort than males for all edges. In absolute terms, the difference varies with sharpness. For the sharpest radius (0), females had 2.0 discomfort units higher than males; at 3.2 mm radius, females were 0.7 higher than males.

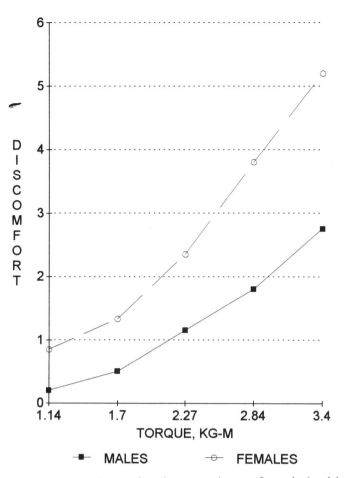

Figure 3. Discomfort in Experiment 2 varied with torque. Using a linear fit, male discomfort increased 1.2/kg-m and female discomfort increased 1.9/kg-m.

REFERENCES

Brough, P. and Konz, S. (1992). Pressure sensitivity of the head. Advances in Ind. Ergonomics and Safety, S. Kumar, Ed., London: Taylor and Francis.

Chen, C. and Konz, S. (1993). Shape of push and turn lids. Proceedings of Human Factors Society 1993, Santa Monica, CA: Human Factors Society, 406-410.

Goonetilleke, R. and Eng, T. (1994). Contact area effects on discomfort. Proceedings of Human Factors and Ergonomic Society 1994, Santa Monica, CA: Human Factors and Ergonomic Society, 688-690.

Fransson-Hall, C. and Kilbom, A. (1993). Sensitivity of the hand to surface pressure. Applied Ergonomics, 24, 181-189.

Kazmi, S. and Konz, S. (1991). Torque on push and turn lids. Designing for Everyone, Y. Queinnec and F. Daniellou, Eds., London: Taylor and Francis, 513-515.

THE RELATIVE SIZES OF PERCENTILES

J.G. Kreifeldt, Ph.D.
Department of Mechanical Engineering
Keoun Nah, Ph.D.
Department of Engineering Design

Tufts University
Medford, MA 02155

In anthropometric design, one usually tries to accommodate the widest range of sizes from small to large. The accelerating ratio of physical size to percentile as one approaches the upper extreme end of the Gaussian distribution, with its generally concomitant cost and/or spatial impacts on the physical system, can often limit the maximum percentile or minimum percentile which can be realistically accommodated. However, what is not generally realized is the relative sizes of the physical dimensions involved because so much work is done with percentiles which are dimensionless quantities. This paper presents several figures which provide some idea of the relative physical sizes of percentiles. The rather small coefficient of variance for anthropometric measurements means that the difference between say the 50th and the 95th percentiles while being a difference of "45 percentile points" is only a matter of a few percent in physical size. Such considerations can help in providing a better understanding of the tradeoff in percentiles and percent and perhaps adds some physical interpretations of percentiles.

INTRODUCTION

In many anthropometric "fitting" design problems, there is a natural conflict between the physical dimensions of the design and the percentage of the population which will be accommodated by the design. At some point, the cost of increasing the physical size outweighs the added benefit from the additional percentage of people "fit" by the design (Kreifeldt, Nah, 1993). Typically, there is advice to use the "95th or 97.5th percentile" as an upper cutoff on the physical dimension of the design. However, this limit is, or should be set by the "diminishing returns" philosophy.

There are three important "fitting" questions relating the percentile accommodated and the physical variables underlying it as for example, the percentage of the population "fit" by the upper range of some variable quantity such as seat height adjustment.

Q1. How rapidly does the percentage of the population accommodated increase per unit increase of the physical dimension?

Q2. What is the percentage increase in the physical dimension per unit increase in percentile value?

Q3. What is the percentage increase in the physical dimension for some specified percentile increase (e.g., from the 75th to the 95th percentile)?

Answers to these questions are obtained (laboriously for Q2 and Q3) from the data itself. However, it is possible to provide several charts for data modeled as Gaussian as adopted for this paper.

RESULTS

Q1. How rapidly does the percentage of the population increase per unit increase of the physical dimension?

The answer to this question is simply the "bell shaped" Gaussian density distribution itself which is the slope or rate of change of percentile per unit of the physical variate of the cumulative distribution. Expressed in "z" form with the mean and coefficient of variation, this is:

$$f(z) = \frac{1}{\sqrt{2\pi}} \frac{1}{COV} \frac{1}{MEAN} e^{-\frac{z^2}{2}} \qquad (1)$$

By definition, the standard deviation is equal to the coefficient of variation times the mean. This form will be useful because the ratio of the standard deviation to the mean (the Coefficient of Variation - COV) of the variate is generally greater than 2.5% and less than about 9-10% (Roebuck, Kroemer, Thomson, 1975). The COV of the "bony" dimensions, such as stature, in particular is on the order of 3.5% - 5%. Taken together, this implies that a large majority (+/- 2 sigma) of the data for height (COV = 3.48%) for example is really about within (+/-) 8% (7.96%) of the average value.

For example, if the mean of some variate is 60 inches and the COV is (say) 4%, then the rate of change of the percentile at the mean value is 16.623 percentile points per inch of change in the variate while at the 95th percentile, the rate of change per inch of the variate is only 4.233 percentile points and at the 97.5th percentile, the rate of change is only 2.435 percentile points per inch change.

Q2. What is the percentage increase in the physical dimension per unit increase in percentile value?

The answer to this can also be derived analytically from the Gaussian cumulative curve with the percentile values on the abscissa. The analytical form for the rate of increase of the physical dimension per unit increase in a percentile value is simply the reciprocal of equation (1).

$$g(z) = \sqrt{2\pi}\,(COV)(MEAN)\,e^{z^2/2} \qquad (2)$$

For example, if the mean is 60 inches and the COV is 4%, then the physical dimension increases at the rate of 6.02 inches per unit increase of percentile change at the mean and at the rate of 23.46 inches per unit increase in percentile at the 95th percentile and at the rate of 41.07 inches per unit percentile increase at the 97.5th percentile. This is the typical problem faced when designing near the "tails" of the theoretical distribution.

Q3. What is the percentage increase in the physical dimension for some specified percentile increase (e.g., from the 75th to the 95th percentile)?

There are two useful forms of this question. (a) the percentage increase in the physical dimension for a one point increase in the percentile value, and (b) the percentage difference in two physical dimension for the corresponding difference in two percentile values. In other words, if percentile value P1 corresponds to the variate value V1 and percentile value P2 corresponds to the variate value V2, the percentage difference between V2 and V1 knowing the percentile difference (P2-P1) is a matter of interest. Expressed in a ratio form, the answer is a dimensionless quantity in which only the COV and the two values {P1, P2}

are of importance. The equation for this percentage difference of V2 to V1 is:

$$\frac{V2 - V1}{V1} * 100 = [\frac{z(P2)(COV)+1}{z(P1)(COV)+1} - 1] * 100 \qquad (3)$$

where (e.g.) Z(P2) is the "z" value corresponding to the percentile value P2. The equation is expressed as the percentage increase of V2 compared to V1.

Figure 1 plots this percentage increase of the variate per unit increase in percentile values beginning at the 51st percentile and increasing to the 99th. The range of COV values common for anthropometric values provides the parametric value. The ordinate (plotted logarithmically) shows that a stepwise unit increase of percentile values causes the variate to increase only by a fraction of a percent until the percentile value is near 95% - 99% depending on the COV. This is different than the rate of increase in the variate per unit increase in the percentile value. The abscissa represents a difference of one percentile point at the value shown - for example, "75" means a change from the 74th to the 75th percentile.

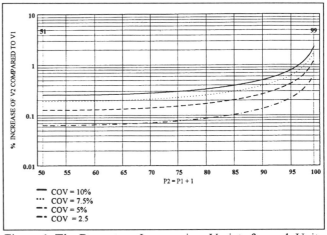

Figure 1 The Percentage Increase in a Variate for each Unit of Percentile Increase Starting from the Mean.

A second computation of interest is the percentage increase of a variate compared to some fixed base value such as the mean. Figure 2 provides a graph of this comparison for increasing percentile departures from the mean.

Note that for most "length" variates which have COV values generally less than 5%, even the 99th percentile is generally less than 10% of the mean value.

Figure 3 shows the percentage increase of a variate compared to the 75th percentile value because this may serve as a reference for the "lower limit" of the upper percentiles.

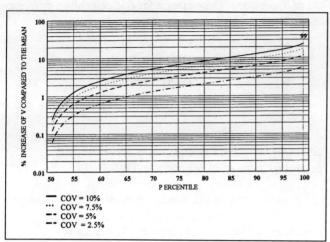

Figure 2 Percentage Increase of a Variate Compared to the Mean Value for Values of the Percentile.

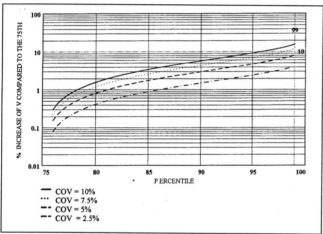

Figure 3 Percentage Increase of a Variate Compared to the 75th Percentile Value for Percentile Values Greater Than the 75th.

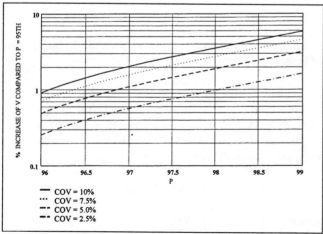

Figure 4 Percentage Increase of a Variate Compared to the 95th Percentile Value for Percentiles Greater Than the 95th.

Figure 4 shows the percentage increase compared to the 95th percentile value because one is often interested in increasing the range as much as possible for accommodation purposes.

BOUNDS ON DISTRIBUTIONS

There is often a need to make some estimate of the percent of the population within some distance centered on the mean value. When the distribution is known, this is a straight forward matter. However, when the distribution is unknown except for its mean and standard deviation, then Tchebychev's inequality provides a guaranteed, if conservative, lower bound. According to Tchebychev's inequality, the percentage of the population within plus or minus "k" standard deviations of the mean is guaranteed to be greater than P(K) where:

$$P(k) > [1 - \frac{1}{k^2}] * 100 \qquad (4)$$

This inequality is only valid for values of k equal to or greater than unity. For k equal to unity (one standard deviation), the formula guarantees that the percentage of the population within one standard deviation of the mean is greater than zero.

Figure 5 graphs this boundary together with the exact value for the Gaussian distribution for comparison. For example, equation (4) guarantees that more than 75% of the data will lie within 2 standard deviations of the mean regardless of the distribution.

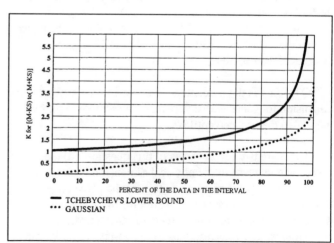

Figure 5 Guaranteed Lower Bound for the Percentage of Data Between Plus and Minus "K" Standard Deviations of the Mean. Gaussian Shown for Comparison.

The Gaussian distribution has 95.45% of its values within this range. Other distributions may have more of less than the Gaussian amount but none can have less than (or even equal to) 75%. In the event that the unknown distribution is known to be symmetrical about its mean (m) and with standard deviation (s), the guaranteed percentage above and below m +/- ks is one half of the value from equation (4). Note that no distribution can lie above the Tchebychev bound in this Figure 5.

CONCLUSIONS

The figures presented can reduce considerably the computational work needed in anthropometric design when trying to place limits on the upper end of a Gaussian anthropometric variate. For example, it may be persuasive for a designer to know that some physical dimension needs to be increased less than 2% in order to increase the population height percentile accommodated from the 95th to the 98th percentile (COV for height is less than 5%).

REFERENCES

J. Kreifeldt and K. Nah, "Optimizing Engineering and Human Factors Costs in Product Design," *Proceedings of INTERFACE '93, 8th Symposium on Human Factors and Industrial Design in Consumer Products*, Raleigh, N.C., 1993

Roebuck, J. A., Kroemer, K. H. E., Thomson, W. G., *Engineering Anthropometry Methods*, 1975, John Wiley & Sons, Inc.

HUMAN FACTORS IN THE PRODUCT DEVELOPMENT PROCESS

Marty Gage
Director
Fitch Inc.

Edie Adams
Ergonomics Manager
Consumer Division
Microsoft Corporation

Robert Logan
Manager, User Interface Design
Thomson multimedia:

Jim Wilson, Ph.D.
Manager
Human Factors & Ergonomics
Motorola Inc.
Land Mobile Products Sector

While the key process to the human factors researcher is the scientific method, the key process for the human factors practitioner is the product development process. For each step in the process there is a variety of methods and deliverables. This panel will begin with an outline of the product development process and discuss some of the relevant human factors issues for each step in the process. Each panel participant will then present his/her perspective on the issues, or methodologies, at each step as it relates to the process at their particular company. Finally, the panel will leave about one half hour for questions from the audience. The goals of the panel are to make the audience aware of:

- the product development process;
- the key issues relevant to the human factors practitioner at each step in the process; and
- some of the typical methodologies used in the process.

PRODUCT DEVELOPMENT PROCESS
Marty Gage
Director
Fitch Inc.

The following is a generic overview of the product development process broken down into three steps:
- *Discovery*
- *Definition*
- *Development*

Each step in the process is subsequently described in terms of the role of the human factors practitioner and the larger goals of each phase as they relate to the overall process.

Discovery

The purpose of the discovery phase is to:
- learn who the user is
- understand the capabilities, limitations, or skill level of the user
- understand how a product is used
- understand the user's task
- understand the user's capabilities and limitations
- determine the user's needs as they relate to the product
- define unmet user needs

in order to:
- generate hypotheses to be analyzed
- determine what to design
- determine the specific attributes of a product that are necessary to ensure ease of use and market success
- define the attributes of a unique and differentiated product
- determine where gaps exist in the way the market or manufacturers address the user or communicate with him/her
- define specific product functionality.

Definition

In the Definition phase, the human

factors practitioner works with the design team to generate solutions that address the needs of the user within the constraints and issues of other disciplines, such as manufacturing and marketing. This work can be to:

- embody the Discovery phase findings in product concepts
- explore solutions that are based upon human factors principles (a key need here is more research based upon issues/hypotheses relevant to product design)
- validate a variety of ideas to determine which one or combination best addresses the user's needs

in order to ensure user needs are addressed early in the process when they can be easily addressed.

While other models of human factors assume a passive role for the practitioner where data or checklists are handed off, this approach involves an active role for the human factors practitioner working with the team to generate user-driven ideas.

Development

In this phase, the human factors practitioner works with the design team to develop and refine a specific idea into a product. Activities in this phase include:

- usability testing
- working with the team to address specific usability issues in detail such as:
  - comfortable forms that fit the body
  - specific interface details
  - actuation forces
  - correct anthropometric dimensions

in order to make sure all details of a product have addressed the needs of the user.

PANELIST 1
Edie Adams
Ergonomics Manager
Consumer Division
Microsoft Corporation

When developing consumer hardware products for computers at Microsoft, we follow the design process stages of *Discovery, Definition,* and *Development.* However, our process accommodates some significant variations to this three-phase model. Three of the factors that contribute to this variation include:

- the short/variable development schedules typical in our industry
- the multidisciplinary team structure we employ to develop products
- our role as in-house, rather than consultant, human factors practitioners.

Discovery Phase

The key issue at this early phase of product development that is common across all our different types of products centers on uncovering sufficiently detailed information to define the intended user, the context of use, and the problem the new product is targeted to solve. Our methods allow us to gather information about the user and the product as a system, with interactions between the user and the product that are physical, cognitive, and emotional/social. These methods borrow from the traditions of the social sciences, marketing, and ergonomics. In-home observation of product use, contextual inquiry, and competitive product analysis are a few of the human factors methods we use to gather the information that will lead to product definition.

Definition Phase

In the Definition Phase, the issues and methods firstly concern the creative development of design concepts and then the evaluation of those concepts using both qualitative and quantitative means. During this phase, the product becomes more and more clearly defined, and the human factors methods used also become more precise in identifying aspects of the product that perform well for users and those that require improvement. Lab-based biomechanical analysis and participatory design sessions are examples of objective and subjective methods used in the Definition Phase.

Development Phase

The Development Phase requires close interaction with other product development team members to ensure that user needs are incorporated into the project decision matrix. The Development Phase often requires that a balance be struck between conflicting goals and the methods of the human factors practitioner at this phase to clarify the outcomes of such trade-offs and reveal acceptable alternatives.

PANELIST 2
Robert J. Logan
Manager, User Interface Design
Thomson multimedia:

The User Interface Design department at Thomson was created four years ago and was the first organization in the company devoted to ease of use. This was a unique opportunity to create a usability program according to a "textbook" process and then improve from that point. In short, we own the user interface for all Thomson America's products including TVs, VCRs, Digital Satellite Systems, Digital Video Disk, and HDTV as well as advanced concepts such as interactive TV. We have adopted a systems approach to product research, design, and development, maintained early and continual involvement by users (through methods such as ethnographic research, market research, participatory design, and usability testing), utilized a relentlessly iterative design process, and worked with our after-sales organization to track consumer

problems.

As part of the panel discussion, we will discuss our research techniques used at various stages of the design process and the types of results we find. For example, early in our processes we have used diverse requirements to gather techniques such as video ethnography, cultural inventories, and pagers to randomly query consumers about their behaviors during a day. We will also discuss the various forms of market research and usability testing we have conducted. To conclude our portion of the panel, we will discuss the pros and cons of the various methods we have used, discuss the areas of our process and tools that need improvement, and discuss the unique aspects of design for entertainment-based products. We believe that for consumer-based products there are two forms of usability— behavioral and emotional—which are mutually compatible but require different research and design techniques.

PANELIST 3
Jim Wilson, Ph.D.
Manager, Human Factors & Ergonomics
Motorola Inc., Land Mobile Products Sector

The presentation will describe a canonical product realization process, and will discuss prototypical human factors and ergonomics activities and deliverables by program phase.

The product realization process, however, is not a rational process. The human factors professional can bring order to this chaos by making explicit the process that he or she uses. The contribution of "process" to a project may be as valuable as the contribution of human factors data.

The human factors professional in industry may be uniquely positioned to understand the need for an explicit product commercialization process, and further, may be able to serve as a catalyst for defining such a process in organizations that do not have one.

High Touch : Human Factors in New Product Design

Myun W. Lee*, Myung Hwan Yun**, Jong Soo Lee*

*: Department of Industrial Engineering, Seoul National University
Seoul, Korea 151-742
** : Department of Industrial Engineering, Pohang University of Science and Technology
Pohang, Korea 790-784

Abstract

In designing a new consumer product, integration of human factors principles is greatly emphasized. However, relatively few attempts have been made to systematically include ergonomic design in the conceptual design phase. High Touch is a product design strategy that uses existing technology systematically to design a new consumer product emphasizing user-friendliness and customer satisfaction. To realize this objective, High Touch design uses several analysis scheme such as identification of implicit needs, realization of potential demand and systematic application of ergonomic considerations into product design. Hierarchical analysis of human variables, product functions and technology attributes is the basic tool of the High Touch design. In this paper, High Touch design process is introduced and a series of new products developed using High Touch design process is demonstrated.

Introduction to High Touch

The tempo of newly emerging state-of-the-art technologies and consequent 'High Tech' products is ever-increasing. Another important trend we are observing is the growing need for user-friendly 'High Touch' products. The term 'High Touch' is first used by Naisbitt(1988) expressing a recent trend of human needs towards more personalized products and services. In this paper, 'High Touch' is operationally defined as the 'transformation of the consumer's implicit needs into product design (Lee et. al. 1991)" To realize this objective, a stepwise design procedure was established. The procedure consists of a series of ergonomic evaluations of a product. Based on the evaluation, High Touch ideas on new product, new functions, and improvements can be conceptualized.

In Table 1, comparison between High Touch and High Tech product development are listed (Lee et. al., 1991). As summarized in Table 1, while High Tech product seeks technology-push, High Touch emphasizes demand-pull by identifying implicit needs of customer. Unlike High Tech product development which is based on large-scale, long term R&D carried out by the central research institutes of a company, High Touch is geared toward small-size, short-term, objective-oriented product strategy that is explicitly focused and directly applicable to the existing product development cycle.

Table 1. The characteristics of High Tech and High Touch product development (Lee et. al., 1991)

	High Tech	High Touch
Strategy	Technology Push	Demand Pull
R &D Period	Long term	Short term (1-5 years)
Investment	Large scale	Small scale
Protectionism	Highly protective	Relatively low
Revenue	Low probability High volume	High probability Intermediate volume

Following the comparison between High Tech and High Touch, an analysis has been made to compare relative distributions between the two trends in consumer electronics. Figure 1 shows the historical development of video cassette recorders (VCRs) since 1973.

New product functions were classified as High Tech and High Touch depending on the nature of new features. The new features were depicted by the order of the market introduction. The horizontal direction shows progressive development of VCR functions by High Tech and vertical direction shows new functions that emphasized High Touch. . The U-matic VCR (1973) and example VCP (Video Cassette Player, 1975) were developed in the U. S. for broadcasting purposes.

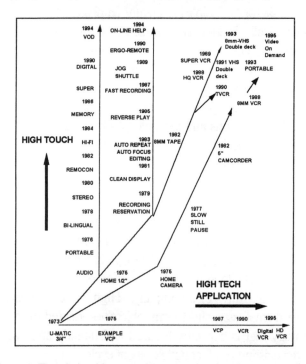

Figure1. Trends in video cassette recorder development - High tech vs. High Touch (Lee et. al., 1991)

However, Japanese industry adopted the concept and developed the 'home video tape recorder' using the half inch tapes (1975). The VCR market grew tremendously with new functions such as program reservation (1979) and stereo sound (1980).

It can be stated that applications of new functions in the VCR were increasingly geared to demand-pull, i.e. realization of implicit needs of customer (e.g. desire to record TV programs) and user-friendly functions (e.g. desire to reserve TV program recording). It seems that High Touch is a distinct trend not only in the conventional consumer products but also in the product

and services that will appear in the future (Lee, 1994). The effort to develop a pioneering product based on more business-driven product development strategy such as High Touch will become the key to product success.

High Touch Design Process

Although consumer needs for better products force manufactures to put emphasis on design, often development of a product has been done without the formal phase to consider human needs.

High Touch was established as an ergonomic design process that enables product development team to consider consumer needs at the initial stage of product planning.

In order to identify the implicit needs of customers and the areas of potential demand on a product, several analysis scheme has been developed. The analysis scheme includes several means to apply human factors disciplines to improve the shortcomings of a product. Variables that tax user convenience most become the primal target of High Touch design.

Several articles have been published to explain the concept of High Touch design in the past [Lee et. al., 1991, Kim et. al., 1990, Lee, 1994). Details of these ergonomic analysis scheme is explained in the following.

Trend Analysis

By observing the evolution of a renowned product systematically, one can identify the future evolution of a newly emerging product. Thus, a proposition can be made:

Proposition: By analyzing the development pattern of an existing product, identification of potential improvement for another of product can be made.

Table 2 shows an example of trend analysis. As shown in the Table, by analyzing the trend of a familiar product (e.g., automobile), the probable trend of a newly emerging product (e.g., computer workstation) can be projected.

As speculated in Table 2, an analogous comparison between automobile seat and computer workstation can provide a direction for future product development for both products.

Table 2. Example of the Trend Analysis
- automobile vs. computer

Category			Automobile	Computer
Performance Criteria	**(Past)**		Power, Speed ↓	Speed(CPU) Memory ↓
	(Present)		Safety Comfort	Software Usability
User Population	**(Past)**		Driver ↓	Programmer, Operator ↓
	(Present)		Whole Population	Whole Population
High Touch	**(Past)** ↓		(Driver seat) Fixed Seat ↓	(Workstation) Fixed Workstation ↓
	(Present) ↓		Adjustable Seat ↓	Interchangeable Workstation, ↓
	Trend ↓		Power Seat ↓	Multi-Media Multitasking ↓
	Future		Computerized Memory Seat Intelligent Seat	Customized Workstation, Information Center)

(dotted box represents projected trend)

Functional Transitivity

Another method to identify new product functions is to adopt a new technology of an advanced product into another product. Application of this principle is defined as Functional Transitivity.

Proposition: By transferring new functions of an advanced product to the product of concern, desirable functions of the product can be visualized.

In Figure 2, the concept of functional transitivity is explained using the functions of TV, VCR, and Audio as an example.

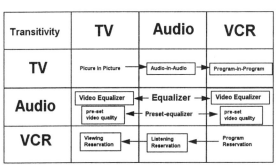

Figure 2. Concept of Transitivity Matrix
(Audio-VCR-TV)

Considering the 'sound equalizer' function of an audio, functional transitivity shows that the concept of equalizer that adjusts the amplitude of sound spectrum can be adapted to video products for a new product function that adjust video spectrum (Lee, 1994). For example, the concept of a pre-programmed audio equalizer which recommends the best sound spectrum based on user's choice of music can be transferred to video functions producing new functions that displays pre-set video effects based on the nature of the video program.

Potential Demand

Potential Demand is operationally defined as 'the expanded demand of a product', i.e. sum of the existing demand and the additional demand generated by adding new functions to the product. Thus, potential demand of a product can be estimated by identifying implicit needs of customer.

Proposition: Potential demand of a product can be estimated by analyzing implicit needs of customer.

Table 3. Example of the Analysis of Potential Demand - Telephone

Needs	Description	Potential New Function
Calling	Time Check Directory Search Select Caller Auditory Monitor Dialing and Monitoring Lifting Multiple Handset Memo Schedule Record Phone Number Record	World time Alarm check Computerized directory Auto call setting Reservation call Sequence call Auto-hook Schedule Record Phone Number Record Alarm setting Auto schedule setting Record and play
Answering	Handset Up Courtesy Answering Person Searching Memo Exchange Answer Auditory monitor Memo Schedule Record Phone Number Record	Auto response Auto call setting Schedule Record Phone Number Record Auto schedule setting Record and Play

Table 3 shows an example of finding the areas of implicit human needs related to a telephone use. Since the functions of a product reflect the ever-increasing nature of consumer needs, many new functions can be conceptualized following the areas of implicit needs.

Hierarchical Analysis

Various human aspects such as capability, capacity and functional limits can be represented by a hierarchical structure. Similarly, functional aspects of a product can be represented by a hierarchy. From these two hierarchies, a matrix of [Human × Product] can be formulated. Given a product, the matrix locates probable areas of implicit needs. Once the area of implicit needs is identified, further analysis follows to implement shortcomings, weaknesses, and necessary functions of the product.

Proposition: By analyzing the relationship between human variables and product functions, targets for High Touch design can be systematically identified.

Table 4. Example of the Hierarchical Analysis - ergonomic remote controller (O:ergonomic design area)

ERGONOMIC VARIABLE		GRIP				CONTROL BUTTON PRIMARY CONTROL					SECONDARY CONTROL		DISPLAY			
		Grp Length	Grp Breadth	Grp Thickness	Grp Shape	Power	Mute	Mode	Volume	Device Select	Channel	Display Button	Display Type	Display Size	Display Angle	Display Location
VISIBILITY	Visual Field					O	O		O	O	O	O	O	O	O	O
	Visual Angle						O		O	O	O			O	O	
	Glare Effect								O	O	O	O		O		O
	Eye Mov't					O		O	O	O	O					O
	Color Percept					O		O	O							
ANTHROPOMETRY	Finger Size	O				O	O	O	O	O	O	O				
	Finger Angle		O		O				O		O	O				
	Hand Grip	O	O	O	O	O	O	O	O	O						
SPATIAL COMPATIBILITY	Motion Economy				O	O	O	O	O		O					
	S-R Compaty						O	O	O	O						
	Control Priority					O	O		O		O	O				
COGNITIVE COMPATIBILITY	Infor'n Load									O			O	O	O	
	S-R Compaty							O	O		O					
	Kinesthetic Feedback			O	O				O							
	Signal Redundancy							O			O					O

(left vertical label: ERGONOMIC FUNCTION FOR REMOTE CONTROLLER)

Table 4 shows an example of hierarchical analysis applied to the design of a new remote controller for a VCR. Even thses days, product design is heavily inclined to an engineer's viewpoint rather than user's viewpoint. For example, a remote controller of a leading VCR model had 125 control buttons. Imagine a typical user who has to struggle with 100 page manuals (Lee, 1994). Table 4

shows the approach used in the design of a new remote controller using an on-board LCD screen.

High Touch Products

High Touch products were began to be developed by the first author since 1984. From 1988 to 1992, High Touch research for consumer electronic products was carried out with the joint efforts between Seoul National University and Daewoo Electronics Co. of Korea.

Shown in Figure 3, the High Touch television (walking-talking television), three kinds of High Touch telephones (quoted in Table 3), and the High Touch remote controller (quoted in Table 4) were all received favorable response from the market. In addition to these, a voice-activated microwave oven, a remote-controlled vacuum cleaner, a voice-instructed VCR, and a computer for pre-school children are also successfully developed.

At present, Seoul National University is carrying out research projects with Samsung Electronics Co. and Lucky-Goldstar Electronics Co. for High Touch products in home appliances, sports/leisure products, medical equipment and multi- media terminals.

High Touch research has received wide attention. "More Future Stuff : Over 250 Inventions That Will Change Your Life by 2001" (Abrams and Bernstein, 1991) featured three High Touch products, the voice-activated microwave oven, the remote-controlled vacuum cleaner and the walking-talking television. The New York Times (New York Times, 1991) called the High Touch idea a new concept for the information era.

In a U. S. consumer survey (Lee et. al., 1991) consumers were willing to pay High Touch products thirty to fifty percent more than the leading high-end models.

Conclusion

We live in a period of transition--a transition to the information era and global economy. The traditional role of product design has to be reinforced to accomplish this transition successfully. Social demand for user-oriented design and a visionary role of conceptual design engineers will be increased dramatically.

Knowledge on human capabilities, physical limits, personal habits, cultural characteristics and individual preferences will be an essential part of new product design. At the same time, mass customization,

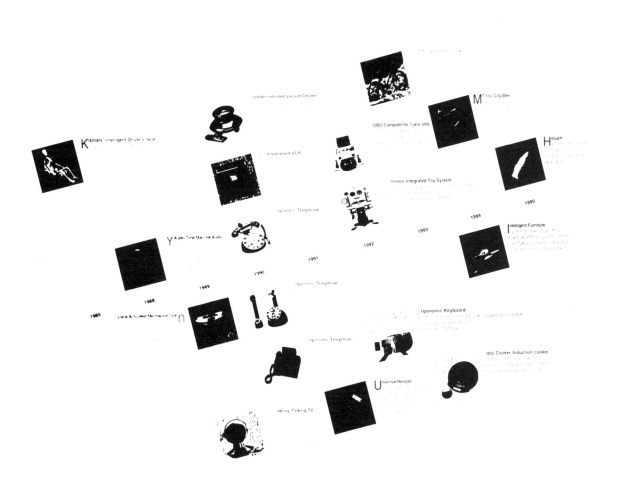

Figure 3. The profile of High Touch products (1986-1994)

personalized products, consumer taste and tailored services will become dominant in most markets. High Touch may soon become a basic requirement for consumer product design.

REFERENCE

Abrams, M., and Bernstein, H., *More Future Stuff: Over 250 Inventions That Will Change Your Life by 200.*, Penguin Press, New York. NY, 1991.

Kim, Y.W., Lee, M.W., and Freivalds, A., EYES- An expert system for the development of High-Touch consumer electronic products, In *Advances in Industrial Ergonomics and Safety II* , B. Das (eds.), Taylor & Francis, pp. 339-346, 1990.

Lee, M.W., Think Theory W, *Keynote Speech for Pan-pacific Forum*, Bali, Indonesia, 1993.

Lee, M.W., Theory W- A New Vision for Human Factors, *Proceedings of the 3rd Pan-Pacific Conference on Occupational Ergonomics*, Seoul, Korea, pp.16-21, 1994.

Lee, M. W., *High Touch for Consumer Electronic Products*, unpublished research report, Daewoo Electronics Co., Seoul, Korea, 1989. (in Korean)

Lee, M.W, Yun, M.H., Park, D.H., Chun, Y.H., Jung, E. S., and Freivalds, A., EYES- ergonomics in a conceptual design process for consumer electronic products, *Proceedings of 35th Annual Meeting of the Human Factors Society* , San Francisco, CA, pp. 1101-1105, 1991.

Lee, M.W., Freivalds, A. Park, D.H., and Yun, M.H., *An Identification and Evaluation of the High Touch Consumer Electronic Products*, unpublished research report, Department of Industrial and Manufacturing Engineering, Penn State University, University Park, PA, 1991

Naisbitt, J. *Megatrends : ten directions transferring our lives*, Warner Books, New York, 1984.

The New York Times, 1991. *Home Section* , P.L. Brown (eds.), Sep.26th., pp. 9., 1991.

IDENTIFYING CUSTOMER EXPECTATION OF
POWER WINDOW SOUND QUALITY

Lijian Zhang, Sharon Huang, Jiaming Du, and Alicia Vertiz
DELPHI Interior & Lighting Systems
6600 East Twelve Mile Rd. W-3
Warren, MI 48092

Sound quality has become an increasingly demanding request in product design. A study was conducted to identify customer requests of power window sound quality and the primary factors that contribute to customer perception and preference of power window sound. Sound samples were recorded from current production vehicles using a binaural recording system in an acoustic chamber. The sounds were then played back to the evaluation panel in a quiet room to rate perceived sound quality using a semantic differential scale. 48 subjects participated in this study. Factor Analysis results of the ratings found three factors: intensity, pitch variation and sharpness. Variables in the intensity factor correlated well to overall preference of the sound, while pitch variation and sharpness were secondary in contributing to the overall perception of power window sound quality.

INTRODUCTION

As product quality improves, customer requests on sound quality become increasingly demanding. To improve customer satisfaction on product sound quality, we need to understand what important features of power window sound are, and how these features affect customer's perceptions of the product.

Research on vehicle interior sound quality, such as vehicle power window sound, is very limited. Previous studies by Takao, *et al.* (1993) and Murata, *et al.* (1993) identified three main factors of sound quality: powerfulness, pleasantness, and sharpness (or, boomingness). However, these sound features are mainly caused by the powertrain system, which are very different from window regulator sound. Earlier studies on Power window sound quality suggested that perceived sound quality is correlated to measured Zwicker loudness (Fridrich, 1989). More recent studies (Cerrato et al., 1995a & b) indicated that the annoyingness of the sound generated by small motors, such as the one used for automobile seat adjusters, is also related to the frequency modulation and impulsiveness of the sound. This study identifies power window sound features based on customer responses, and investigates how these features affect overall perception of sound quality.

METHODS

Sample selection

The power windows of 10 vehicles currently on the market (95) were chosen for the study. Two additional power windows were selected from rejected products due to excessive noise level. Therefore, a total of 12 windows were

used for sound recording. Considering that the sounds may be rather different for power windows moving upward and downward, two window operation conditions were recorded for each window: window up and window down, resulting in a total of 24 sound recordings. All sound samples were recorded using a Head Acoustics' binaural recording system inside the vehicles. The system was placed on the front passenger seat, and the vehicles were parked in an acoustic chamber. The acoustic features of these samples in time and frequency domain were analyzed using the Binaural Analysis System™.

Subjects

48 people were selected for this study, 33 male and 15 female. Their ages range from 20 to 60. All subjects selected have normal hearing ability.

Set-up

The recorded samples were edited into a 10 minutes tape for evaluation. The playback system used in the evaluation includes a *digital audio tape drive, Reproduction Systems*, and *Headphones*.

Procedure

To obtain subjective ratings of the sound characteristics of the power window, a questionnaire was designed, which included a semantic differential scale (see Figure 1) and a open-comments section. Adjective pairs describing sound features were collected from sound quality study literature (Takao, *et al.* 1993; Murata, *et al.* 1993; Bismarck, 1974) and by consulting sound quality experts.

Eight pairs of sound quality descriptors were selected from this list through two pilot trails of 9 subjects. Two pairs of descriptors representing customer overall preference of the sound (*pleasant* and *like*) were then added to the list.

One of the critical concerns of subjective ratings is the reliability of the data. To reduce rating variation due to subjective bias, a pair of power window sound recordings were selected as control sounds: control window up and control window down. Subjects rated these two sounds first. In rating the rest of the samples, subjects were instructed to use the control sound ratings as reference. Among the remaining 11 pairs of power window sound, one pair of recordings was played twice (samples #6 and #11) to check the reliability of the ratings. Therefore, there were total of 24 sound recordings for each subject to rate, not including the control sample.

At the beginning of the data collection, instructions were given to the participants about the meanings of the descriptor pairs in the semantic differential scale and how to use the scale to rate sound quality. Then the two control sounds were played to the participants, and they were instructed to rate the quality of the two sounds separately using the scales on the questionnaire. In rating each sound sample, the control sound was played first, then the sample sound. Subjects used the control sound as a reference when they rated the sample sound. The control sound and sample sound were played repeatedly until all subjects completed the rating of that sample sound. This procedure continued for all 24 sounds, lasting about 2 hours. Considering that a two-hour testing period may result in fatigue, which will affect rating consistency, a 15 minutes break was given to the subjects in the middle of the two-hour session.

ANALYSIS

Reliability

Sample #6 and #11 are the same sound which was played twice to the subjects to check the rating reliability. Figures 2 shows the plots of mean ratings for each sample. In Figure 2, samples #6 and #11 are located very closely to each other, see solid lines in the plots.

The difference of the rating between the two repeated samples for each variable is less than 10% of the total variation among all samples. This indicates that although there is a variation in the subjective ratings, using the control sample, we were able to maintain repeatability to an acceptable level.

Factor Analysis

The subjective sound quality ratings were subjected to Factor Analysis to identify the key factors of the perceived sound quality of the power window sounds. The factor analysis was conducted using SAS Factor procedure with *varimax* rotation. Three factors emerged. The factor loading is listed in Table 1, and the loadings are plotted in Figure 4.

The first factor includes *quiet_loud, gentle_violent, soft_hard* and *smooth_rough*. These variables are located closely together (see Figure 3). These variables are related to the intensity, or, the power of the sound. Although variable *week_strong* also has a relatively higher loading on factor 1, it is a rather isolated item in the plot. The second factor contains two variables: *variable pitch* and *oscillating*, representing pitch variation. The last factor has only one variable, denoting sharpness - the spectral contents of the sound. The first factor explains about 55% of the total communality, indicating that intensity of the sound is a critical factor. Factors 2 and 3 explain about 25% and 20% of the communality respectively.

Table 1. Factor pattern (after *varimax* rotation)

	Factor 1	Factor 2	Factor 3
gentle - violent	**0.887**	-0.113	0.177
soft - hard	**0.874**	-0.023	0.234
quiet - loud	**0.862**	-0.049	0.172
smooth - rough	**0.782**	-0.314	0.076
weak - strong	**0.566**	0.352	0.448
variable pitch - steady	0.013	**0.874**	-0.201
oscillating - constant	-0.256	**0.747**	0.241
dull - sharp	0.292	-0.052	**0.890**
variance by factor	3.376	1.563	1.214

Correlation

Table 2 shows the correlation between sound feature ratings and preference ratings.

Table 2. Correlation between sound feature and preference (n>570)

	window up		window down	
	pleasant	like	pleasant	like
quiet_loud	0.6029	0.5879	0.6920	0.6795
gentle_violent	0.7102	0.6961	0.7325	0.7143
variable pitch_steady	-0.2167	-0.2252	-0.3422	-0.3780
soft_hard	0.6963	0.6898	0.6761	0.6373
weak_strong	0.2735	0.2361	0.3289	0.2902
smooth_rough	0.7299	0.7328	0.7425	0.7626
dull_sharp	0.3304	0.3372	0.4416	0.4371
oscillating_constant	-0.2720	-0.2613	-0.3745	-0.3913

Correlation of sound quality ratings and preference ratings for two operating conditions (window up and window down) showed the same trend, indicating that customer expectations for sound characteristics are the same for the two window operating conditions.

Although the correlation of sound characteristics ratings with preference is not high, the correlation for both conditions indicate that the variables *quiet_loud, gentle_violent, rough_smooth,* and *soft_hard* are better correlated to the overall preference of sound (*pleasant* and *like*). These variables are found in the first factor. Variables representing pitch variation and sharpness are rather poorly correlated to overall preference ratings. This seems to indicate that Factor 1 is the primary factor contributing to customer preference of power window sound, and that pitch variation and sharpness of sound are secondary.

Psychoacoustic Measures

The objective sound features of the 24 samples were analyzed using the Binaural Analysis System[TM]. *Sound intensity, loudness, roughness, fluctuation strength,* and *sharpness* were computed. The computation of these measures are well documented in Zwicker's work (1990).

The results of the study show a consistent trend with subjective ratings. In general, the poorly rated sounds have higher levels of *sound intensity, loudness, roughness* and *fluctuation strength.* The samples that are perceived as *loud* and *violent* had intensity level higher than 60 dB(A) and loudness level greater than 15 soneGF. On the other hand, the samples that are perceived as gentle and quiet had lower intensity < 60 dB(A) and loudness level < 10 soneGF. Similarly, the samples that are perceived as *rough* had roughness level >1.5 asper, and the roughness level of the samples that are rated as *smooth* are below 0.7 asper. Figure 4 and 5 illustrate the differences between the preferred and the poorly rated samples.

CONCLUSION

1. Power window sound features may be classified as three factors: *intensity, pitch variation,* and *sharpness.*
2. Loudness, roughness, and impression of hardness and violence of the sound are primary variables in sound perception; pitch variation and sharpness are secondary.
3. Power window sounds with lower level intensity, lower roughness, and lower impulsiveness in starting and stopping are preferred.
4. Customer expectations of power window sound are essentially the same for window up and window down operation.

REFERENCES

Bismarck, G. Von (1974) Timbre of steady sounds: A factorial investigation of its verbal attributes. *Acoustica,* 30, 146-159.

Cerrato, G., Crewe, A. & Terech, J. (1995a) Sound Quality Assessment of Powered Seat Adjusters. *Proceedings of the 1995 Noise and Vibration Conference,* Traverse City, MI, SAE 951288.

Cerrato, G., Crewe, A. & Terech, J. (1995b) Sound Quality Modeling: A comprehensive Stepped Approach to Target Development. *Proceedings of 3rd International Conference of Vehicle Comfort and Ergonomics,* Bologua, Italy, 95A1006.

Fridrich, R. (1989*) Loudness Calculation (ISO 532) for Evaluating Components and Vehicles,* SAE Noise and Vibration Conference poster, Traverse City, Michigan

Murata, H., Tanaka, H., Takada, H., & Ohsasa, Y. (1993) Sound Quality Evaluation of Passenger Vehicle Interior Noise, *Proceedings of the 1993 Noise and Vibration Conference,* SAE 931932.

Takao, H., Hashimoto T. & Htano, S (1993) Quantification of Subjective Unpleasantness Using Roughness Level, *Proceedings of the 1993 Noise and Vibration Conference,* SAE 931332.

Zwicker, E. & Fastl, H. (1990) *Psychoacoustics, Facts and Models,* Springer-Verlag, Berlin, Germany.

Figure 1. Rating Scale for Subjective Evaluation

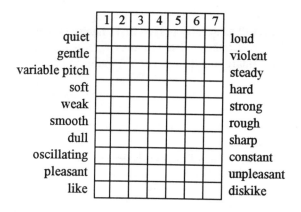

Figure 2. Plots of Mean Subjective Ratings

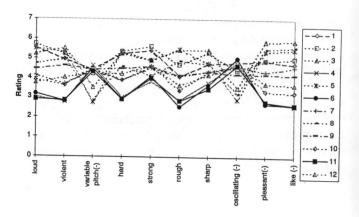

Figure 3. Plots of Factor Patterns
(factor 1 *vs.* factor 2)

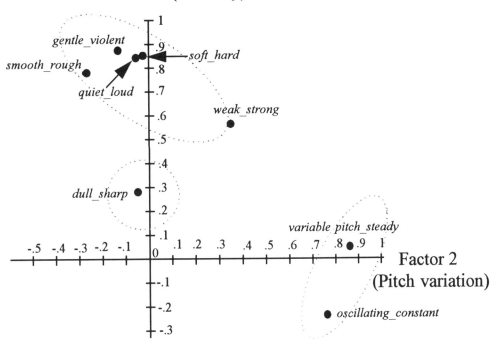

Figure 4. Measured loudness in soneGF

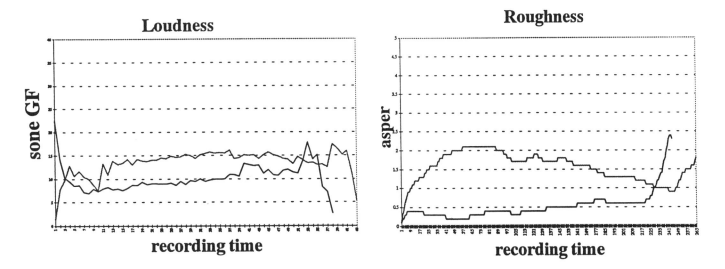

Figure 5. Measured Roughness

INFORMATIONAL CONSTRAINTS OF REMOTE TECHNICAL SUPPORT ENVIRONMENTS: IMPLICATIONS FOR THE DESIGN OF SUPPORT SYSTEMS

Karen S. Wilson
NCR Corporation
Duluth, Georgia

ABSTRACT

Effective remote technical support is a competitive factor in consumer satisfaction. This paper reports the analysis of the remote support domain with respect to the agents, their information requirements, the issues relevant to the transfer of information from one agent to another, and the technology and tools that are currently used. The goal of this work was to understand the current environment with respect to the future direction of such environments and to understand the issues that afflict that environment so that the design of any support system would incorporate the appropriate design requirements. The factors contributing to performance behavior in the remote support task include the problem solving nature of the task, the restricted perceptual context, the distributed knowledge of the remote support team, and the role of communicating to remotely located agents. Research in these issues leads to the conclusion that tools designed to support the collaboration of remote technical support teams must maintain the interactivity of the team member dialogues that are used to define, test, and respond to the problem. But the tools must also be developed to facilitate behaviors exhibited in effective teamwork.

INTRODUCTION

As consumers purchase and use more technology, the availability and quality of technical support become a competitive factor in consumer satisfaction and increased and repeated sales. Likewise, the cost of that technical support becomes a major factor in the business case for the product. To remain competitive, companies are investigating means of providing world class consumer support while trying to minimize the cost of such support. One means of achieving these goals is to design products that need minimal operational support for the consumer and are functionally reliable. The second means is to provide technical support that meets the needs and expectations of the consumer.

There are two general levels of technical support:

- generic support in the form of user documentation and interactive voice messaging systems that are designed to anticipate user needs, and

- specialized technical support in the form of on-site technicians and manned help desks that provide personalized support for specific problems.

A wide range of cost benefit tradeoffs occurs with either implementation. The criterion for implementation would be to determine the effectiveness of the support level and the consumer's perception of the quality of the benefit in comparison to the cost of implementation by the manufacturer. Effective generic support can reduce the volume of customer reported calls by providing information to solve frequent, typical, and simple to resolve problems, but it cannot fully eliminate the need for personalized technical support. Likewise, specialized expert support can potentially be supplemented with extensive knowledge-based support systems that off-load the requirement for individual technical expertise on a large and changing product base. In order to develop and implement such tools, it is essential to understand the tasks inherent in remote support, the informational requirements of workers in such an environment, how that information is required to be redistributed, and what technology is required to implement such a design. This paper discusses an analysis of the current environment with respect to its future direction and the issues that afflict that environment to ensure that the design of any support system would include the appropriate design requirements.

METHODS

To understand the future directions of technical support, it was important to analyze the current domain with respect to the agents, the goals of the agents, their roles in the task, their information requirements, the issues relevant to the transfer of information from one agent to another, and the technology and tools that are currently used. The methods used in this study included:

- analysis of tracked problem reports, which was used to establish problem categories and frequencies
- field observations, which involved "jacking into" consumer problem reports, operational support calls, and field technician support calls to technical product support agents, and "riding" with field technicians to customer sites, and
- informal interviews, which were used to further elaborate observations, to gather agent background information, to access collateral information that was not necessarily obvious from the observations, and to gain information about personnel backgrounds.

The information from the observations and interviews was analyzed to model the existing remote support system. That analysis included:

- cataloging information requirements,
- cataloging information sources,
- mapping information transfers,
- task analysis of process related tasks, and
- a semantic mapping of the agents and the information sources and flow within the domain (Figure 1).

RESULTS

The task domain of remote technical support is one of collaborative problem solving distributed across multiple agents at various locations. The overall goal of the remote technical support system is to determine the nature of the problem, to distribute the appropriate information to the appropriate agent, and to arrive at a solution at the earliest possible moment. Typically, there are five distinct agents: the consumer who places the problem call, the call processor who reports call data from the consumer and distributes the call, the help desk analyst who provides operational support directly to the consumer, the field technician who goes to the consumer's site to provide direct intervention to the problem, and the product expert who provides advanced technical support to the field technician. Each agent has a variety of available information, ranging from the symptom state of the equipment to technical specifications detailing the engineering design. Each agent has a variety of tools that are used to clarify the problem and derive a solution. These agents operate from four different locations: the caller and the field technician are located at the product site, the call processor in a call distribution center, the help desk analyst from a pool of operational specialists; and the product expert from within product engineering. All of the agents rely on the telephone to communicate with one another. Figure 1 illustrates the semantic mapping of the domain with respect to each agent, the agents' tool set, the available information, and the information flow.

There are three basic phases to the task of remote support: the product identification phase, the problem identification phase, and the diagnostic phase. In the product definition phase, information about the product, the owner, the caller, and the equipment location is gathered. This information is used to verify product support entitlement, to initiate a record of the problem report, and to distribute the problem report to the appropriate support personnel. During problem definition, information is gathered about the state of the system so that the scope of the problem may be assessed. On-site diagnosticians use visual information and interaction with the equipment as well as questioning the user about the system state, what the user did prior to the onset of the problem, and what the user has done since the onset of the problem in order to define the problem. Remote support personnel are forced to question the caller to gather all of this information. During diagnosis and resolution, the diagnostician will formulate

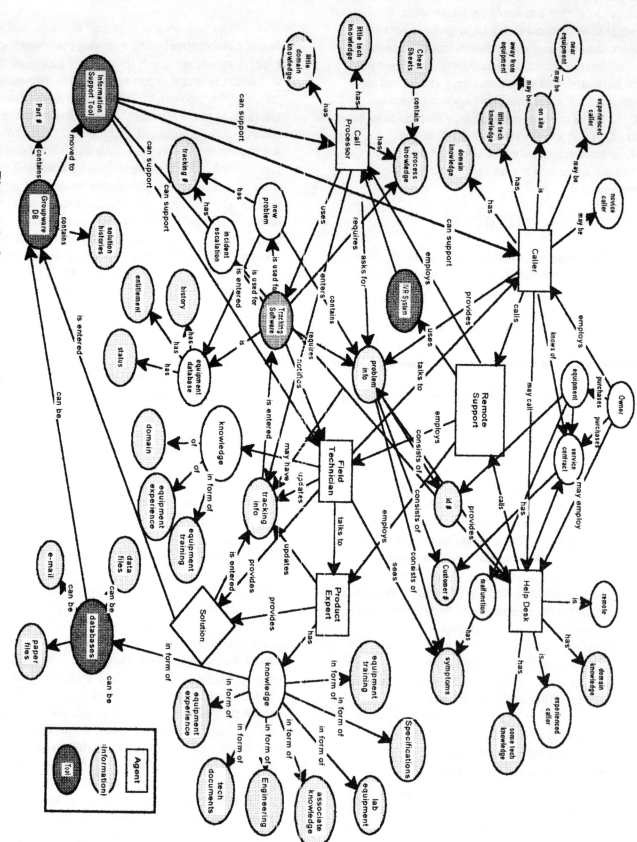

Figure 1. Semantic Map of Remote Support Domain

hypotheses to explain the system state. They may read indicators, reset controls, perform dynamic tests, observe the operations of components, and remove and replace components. Remote support personnel perform these tasks virtually, by instructing on-site personnel to perform these tasks and report the results.

DISCUSSION

The model of the current domain revealed specific factors that contribute to performance behavior in the remote support task, including the problem solving nature of the task, the restricted perceptual context, the information distributed across the domain, the distributed knowledge of the remote support team, and the role of communicating to remotely located agents. Research in these issues provides insight into the complexity of designing tools that can be used effectively to improve the quality of support and to reduce the load imposed on the support teams.

The resolution of the problem is a team endeavor, whether it occurs between the consumer and the help desk or between the field technician and the product expert. The team, interacting dynamically, interdependently, and adaptively toward a common goal, relies on effective communication, mutual performance monitoring, giving and receiving feedback, and coordinating task activities (Salas et. al. 1992). These behaviors are at risk in the remote technical support environment because of the distributed locations of the team members, the variations in team member expertise, the lack of direct feedback from the tasks as they are being performed, and the transitory composition of the team.

Information emerges from the system itself and from observing the actions of others. On-site, the troubleshooting tasks rely on the availability of visual access to the system and the actions of collocated team members, providing the diagnostician with valuable information about the situation and the expected state of the system. Visibility is an essential part of the flow of information for creating and sustaining teamwork and lack of visibility requires team members to adopt different communication strategies, including extracting detailed explanations (Johannesen, Cook, & Woods, 1994).

Knowledge and actions are distributed across various agents within the system. Heath and Luff (1991) defined distributed cognition as a process in which various individuals share an interrelated orientation towards a set of tasks and activities. People use physical cues and one another as sources of information to form

extensions of their own knowledge. Effective collaboration requires a common availability of the same sources of information. Because the team members are not collocated, one team member may be acting solely on the instructions of another, relying solely on the modality of communication and with a lack of mutually shared vocabularies.

The role of communication within problem solving situations is to reduce situational uncertainty through information gathering and exchange (Wellens, 1989). The levels of expertise within the remote support team range from novice to expert in several areas: the product itself, the support calling process, and the domain in which the product is being used. The need to transfer the required information from one agent to another is hampered by the mismatch between the novice's and the expert's vocabulary and conceptual content. This mismatch is compounded by the distribution of agents across locations. Suchman (1987) indicates that one agent in the interaction may not provide the required information because of a lack of understanding of the purposes of the interaction. That agent may not know the plan of the inquiring agent and may not readily make available the required information about the state of the system. The agent on the other end of the voice line must make explicit requests for the required information and be able to reformulate the received information in terms of his/her own needs.

Johannssen et. al. (1994) reported that collocated teamwork contains cooperative exchanges within a common frame of reference that provides common accessibility to the problem state and problem solving technique. Because the distributed team members do not have this accessibility, they must explicitly define the problem, articulate their plans, and obtain relevant feedback. Breakdowns occur when communication does not provide relevant information concerning the state of the problem. Transmission errors, or the loss of information, occur when the intervening agent provides incomplete or inaccurate information. To avoid these miscommunications, the information seeker must precisely formulate questions and be sensitive to the informativeness and relevancy of what is reported. The questions and responses must be reformulated in terms that are understandable, answerable and interpretable. Differences have been reported in the strategies between problem solving activities carried out in face-to-face situations and those mediated by the telephone. Grosz (1981) found that successful communication between an expert and apprentice with limited visual access to each other requires a common focus on the task at hand and that the effectiveness of verbal instruction is more dependent on the interactivity of speech than on its modality.

Mental models are an agent's representation of a system and/or task, are based on previous experience and current understanding, and are used to guide subsequent understanding of the situation. It is generally held that mental models depend on the experience level of the agent, thus each team member has a model of his/her own individual unique expertise, but a naïve understanding of other aspects of the problem. Through interaction, team members build a more comprehensive model of the problem at hand, as well as an understanding of team members' knowledge and skills, producing what is termed a shared mental model (Orasanu & Salas, 1993). Since the remote support team is reformulated as each new problem is reported and again when the call is passed from one support agent to another, the benefits of shared situation and mental models are not available to the team. The shared understanding of the problem, goals, information cues, strategies and member roles must be rebuilt with each incoming problem.

CONCLUSIONS

Tools designed to support the collaboration of remote technical support teams must maintain the interactivity of the team member dialogues that are used to define, test, and respond to the problem. But the tools must also be developed to facilitate behaviors exhibited in effective teamwork. The distributed collaborative team would benefit from structuring dialogs consistently across the tool, from revealing the strategy of the dialog within the tools, from formulating dialog vocabularies that do not require unnecessary reformulations, and from depicting situational features required to select a course of action. The design must represent the expert's reference points in such a way that they will be correctly described, interpreted and identified by the distributed team members. It must also represent the expected outcomes of actions and provide alternative courses of actions. The information gathered in one phase of the problem report should be available to all subsequent interactions, and should be structured in such a way that the gathered information is informative and relevant.

REFERENCES

Grosz, B. Focusing and description in natural language dialogues. In A. Joshi, B. Webber & I. Sag (Eds.), *Elements of discourse understanding*. Cambridge, UK: Cambridge University Press.

Heath, C. & Luff, P. (1991). Collaborative activity and technological design: Task coordination in London underground control rooms. In L. Bannon, M Robinson & K. Schmidt (Eds.), *Proceedings of the Second European Conference on Computer-Supported Cooperative Work* (pp 65-80). Amsterdam

Hutchins, E. (1990). The technology of team navigation. In J. Galegher, R. Kraut, & C. Egido (Eds.), *Intellectual Teamwork: Social and Technical Bases of Cooperative Work*. Hillsdale, NJ: Erlbaum.

Johannesen, L.J., Cook, R.I, & Woods, D.D. (1994) Cooperative communications in dynamic fault management. In *Proceedings of the Human Factors and Ergonomics Society 38th Annual Meeting*, 225 - 229. Santa Monica, CA: Human Factors and Ergonomics Society.

Ochsman, R.B. & Chapanis, A. (1974). The effects of 10 communication modes on the behavior of teams during co-operative problem-solving. *International Journal of Man-Machine Studies, 6*, 579-619.

Orasanu, J. & Salas, E. (1993). Team decision making in complex environments In G.A. Klein, J. Orasanu, R. Calderwood, & C.E. Zsambok (Eds.), *Decision making in action: Models and methods* (pp 327-345). Norwood, NJ: Ablex Publishing.

Salas, E., Dickinson, T.L., Converse, S.A. & Tannenbaum, S.I. (1992). Toward an understanding of team performance and training. In R.W. Swezey & E. Salas (Eds.), *Teams: Their training and performance*. Norwood, NJ: Ablex Publishing.

Steinberg, L.S. & Gitomer, D.H. (1993). Cognitive task analysis, interface design, and technical troubleshooting. *Proceedings of the 1993 International Workshop on Intelligent User Interfaces*. (pp 185-191). New York: Association of Computer Machinery.

Suchman, L. (1987). *Plans and situated actions: The problem of human-machine communication*. Cambridge, England: Cambridge University Press

Wellens, A.R. (1989, September). Effects of telecommunication media upon information sharing and team performance: Some theoretical and empirical observations. *IEEE Aerospace and Electronic Systems Magazine 4*, pp 13-19.

EMOTIONS MATTER: USER EMPATHY IN THE PRODUCT DEVELOPMENT PROCESS

Uday Dandavate, Elizabeth B.-N. Sanders, and Susan Stuart
Fitch Inc.
10350 Olentangy River Road
Worthington, Ohio 43085

ABSTRACT

The reason why some products get intimately linked with people's lives, while others do not, remains a mystery to most consumer products researchers and designers. A shift in thinking from a focus on the rational to the more emotional domains will help us to understand those uniquely human traits that are responsible for people's liking, using, and wanting to live with the products we design.

The gap in the methods and tools available to product development researchers and practitioners is centered upon the emotional domain. This gap exists throughout most of the product development process, both for generative as well as evaluative research. We propose that researchers and practitioners working on product development teams attend to improving their ability to recognize and address the feelings of product users—in particular, the feelings that users have about owning and using products. The success of such products in the future will depend upon the degree to which we learn how to empathize with the product users very early in the product development process.

"What people seek is not the meaning of life but the experience of being alive." — Joseph Campbell

INTRODUCTION

The discipline of human factors has been based on the scientific study of human behavior and thought, with the assumption that rationality and logic dominate our thinking. The concept of emotionality has traditionally been considered "out of bounds." Emotion is more likely to be seen as a nuisance variable—an annoyance to be ignored or controlled across conditions. Human factors textbooks do not typically include "emotion" or "effect" in their indices (*e.g.*, McCormick, 1976; Eastman Kodak Company, 1983; O'Brien and Charlton, 1996).

Yet, often products which were designed based on rational cognitive models, even those which passed usability tests with flying colors do not, in the end, become intimately linked with people's lives on an emotional level. It is this intimate, emotional link that creates awareness, comfort, and satisfaction with the product and the brand, and which propels users to choose to live with that brand again. Perhaps it is time to pay more attention to what provides this intimate link, and motivates users from an emotional level.

RATIONAL AND EMOTIONAL: RECENT STUDIES

There now exists a growing body of evidence that supports the idea of two very different kinds of intelligence: rational intelligence and emotional intelligence (Goleman, 1995; Cytowix, 1993; LeDoux, 1993, 1994; Lemonick, 1995). These studies suggest, in contrast to more traditional thinking, the primacy of emotion over reason. One's emotional state can, for example, either enhance the learning process or actually prevent any new learning from taking place altogether (*e.g.*, Goleman, 1995).

While emotion is often the central focus of clinical and social psychology, researchers and many designers concerned with the relationship between people and their physical environments largely ignore the role of emotion. C.C. Marcus, a professor of architecture states, "Psychologists whose domain is the study of emotional development view the physical environment as a relatively unimportant backdrop to the human dramas of life. Those who are interested in people-environment relations—geographers, anthropologists, architects, and the newly emerging field of environmental psychology—have for most part ignored issues dealing with emotional attachment" (Marcus, 1995). Only a few designers and researchers have addressed emotions and rationality in a balanced manner.

The theme of a product's "intimate link" with its user is found in several studies. Often this link is associated with personal identity, either one of self or others. Mihaly Csiszentmihalyi and Eugene Rochberg-Halton (1981) discussed the intimate relationship between people and products with which they surround themselves, in stating, "Home contains the most special objects: those that were selected by the persons to attend to regularly or to have close at hand, that create permanence in the intimate life of a person and, therefore, are most involved in making up his or her identity. The objects of the household represent at least potentially the endogenous being of the owner. Although one has very little control over the things encountered outside of the home, household objects are chosen and could be freely discarded if they produced too much conflict within the self."

Dandavate (1995) pursued the literature connected with the emotional elements in people's interactions with products, and noted that there seemed to be relationships between people and products that resemble interpersonal relationships. He proposed that some of the methods currently used to investigate interpersonal relationships would be useful to product developers trying to understand the nature of the emotional link between people and products.

THE CURRENT SITUATION IN PRODUCT DEVELOPMENT

Human factors researchers have traditionally focused their efforts on the user's understanding and use of products. Many of their techniques focus on what people *do* when confronted with new products, and look at time and error as the primary measures of performance. Measures of preference, *i.e.*, what people *say* they like, are becoming more common. Usability testing that combines measures of performance as well as the user's (more subjective) verbal protocol is also becoming more acceptable. Hence, the emotional components (as might be expressed in discussions of preference) of the product-person relationship are just now beginning to influence the evaluative phases of the product development process.

One recent example of evaluative research directly addressing the emotional side is the study by Logan, Augatis and Renk (1994) on the design evaluation of television remote controls. This study introduces the notion of "emotional usability," which includes factors such as fun, excitement, and appeal. The authors suggest that "for certain categories such as consumer electronics, an expanded definition of usability is required." Another recent study by Adams and Sanders (1995) focused on the "fun" value of an input device designed specifically for small children.

Studies of usability, including those dealing with emotional usability, evaluate whether or not products meet the needs of users. They do not, however, provide generative, *i.e.*, pre-design, input that can be used to drive the design and development process.

The construction of cognitive models and their subsequent role in the design development of products and interfaces is now an accepted way of ensuring that the rational side is being addressed in the generative end of the product development process. But there are few generative research tools designed explicitly for the purpose of addressing the emotional side of the equation in the early stages of the process. One notable exception is Paul Losee of M+, who employs regression techniques to elicit users' "imprinting experiences," *i.e.*, users' first experiences with a certain type of product or brand. Losee believes these initial experiences carve out lasting impressions with which all reactions to similar products will coincide in the future (Losee, 1996). More information on the "imprinting" notion can be found in Zuckerman and Hatala (1992).

Market researchers are addressing emotional needs, but primarily emotions surrounding the shopping and purchasing of products, in contrast to their use. They focus almost exclusively on the evaluative rather than generative part of the product development process, and on what people *say* about products (*e.g.*, traditional market researchers conduct focus groups, customer surveys, and opinion polls). Market researchers also investigate what people actually *do* when faced with the product in the marketplace by studying real purchase behavior. They are well aware of the discrepancy between what people *say* they will buy and what they actually do end up buying, but still lack methodologies for tapping consumers' true emotions in the generative stage of product development research.

More innovative market researchers, such as Hanan Polansky, an assistant professor of marketing at the University of Rochester's William E. Simon Graduate School of Business Administration, are attempting to uncover new ways to understand this discrepancy. For example, Polansky has developed a computer software system, based on artificial intelligence, called CustoWare, that he claims "distinguishes between what people *say* they will do and what they *actually* will do." The software works on data sources such as surveys, customer panels, hotlines, letters, and sales force feedback.

A FRAMEWORK FOR
PRODUCT DEVELOPMENT RESEARCH

The chart below shows the domain of product development research methods defined by two dimensions: stage in the product development process (*i.e.*, generative to evaluative) and "psychological focus" (*i.e.*, emotional to rational). The chart shows that research methods for the evaluation of the rational content of products are well established. Generative methods for an understanding of cognitive/rational content also exist. Evaluation of emotional content is, on the other hand, not so well established. Finally, the area needing the most effort is in generative research of emotional content.

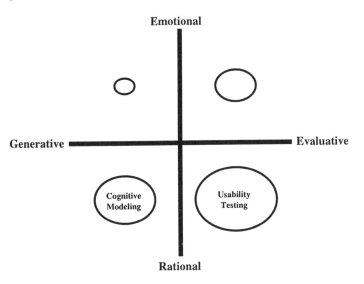

For product development researchers and practitioners, more methods and tools are needed for tapping users' emotions, in both generative and evaluative research. The large gap in generative methods is, perhaps, most critical since it is in this pre-design portion of the product development process that the product concept is identified and defined. We propose that people working on product development teams should attend to improving their ability to recognize and to elicit the emotions and feelings of product users—in particular, the feelings that the users have about products and their uses.

GENERATIVE RESEARCH METHODS
FOR USER EMPATHY

Empathy involves *"understanding or vicariously experiencing the feelings, thoughts, and experience of another without having those feelings, thoughts, and experience fully communicated in an objectively explicit manner."* (*Webster's Dictionary*, 1988).

It is far too early in the evolution of generative research methods to describe precisely which methods to use, how to use them, and when best to use them. Few firms have even embraced the concept of user-empathic generative research methods. Notable exceptions include Interval Research

with Cheskin and Masten (Ireland and Johnson, 1995) and Sandvik (as reported in Leonard-Barton, 1995).

At Fitch, we have been experimenting with user-empathic generative research methods for over ten years, although much of the earlier work in this area was done with children, particularly pre-schoolers. Our focus in the last five years has been on developing these tools and methods for adults. Experimentation and practice in this area are allowing our teams of practitioners to develop unique understandings of, and empathy with, the users of products (as well as interfaces and environments). The tools and methods are being put to use by researchers trained originally in human factors as well as by those trained in market research. The tools and methods are now being used collaboratively by researchers as well as designers. The results of their use is just now beginning to be rewarded in the marketplace.

Some observations based on our experience so far:

Product development researchers and practitioners should use **multiple methodologies** in concert with one another. This is particularly true for generative research methods. The convergence of the knowledge derived from multiple methods can provide much greater empathy with users than can any single methodology alone (*i.e.*, Sanders, 1992).

Projective techniques which offer ambiguous settings, stimuli, and situations to which users are to respond, are particularly useful in eliciting the expression of emotional content. We have found that carefully selected visual stimuli are particularly useful as projective devices when used in appropriate exercises.

Multimodal expression, *i.e.*, encouraging users to express themselves not only verbally, but also visually, and through action, allows them to respond in modes in which they are most comfortable and in which they may most accurately express their emotions. It also encourages people to express themselves in ways they may have enjoyed as children but have not had the opportunity to engage in as adults.

Most of the expressive activities we create for adults make use of **participatory design tools and techniques** in which users make or create objects, collages, composites, spaces, rooms, models, diagrams, etc., with "toolkits" we have designed for them. The toolkits afford users the creation of rational expressions or the creation of emotional expressions, depending upon the components of the kit and the instructions for use. Other types of toolkits, based on games and role-playing activities, can afford users the expression of experiential meaning. Design of these "toolkits" has been a learning experience for our researchers and designers together.

Analysis and the abstraction of meaning from data collected using generative tools and methods have not been adequately addressed. We are continually

experimenting in this domain. This is a wide-open area that can be potentially quite interesting, challenging, as well as useful (in terms of being hired by progressive product development firms) for graduate students in psychology, design, business, human factors, and other relevant areas.

CONCLUSION

A shift in thinking from a focus on the rational to a balance between the rational and emotional domains will help us understand those uniquely human traits that are responsible for people's liking, using, and wanting to live with the products we design. The success of these products in the future will depend upon the degree to which we learn how to empathize with the product *users* very early in the product development process.

BIBLIOGRAPHY

Adams, E. and Sanders, E. B.-N. "An evaluation of the fun factor for the Microsoft EasyBall Mouse." Proceedings of the Human Factors and Ergonomics Society, 39th Annual Meeting, 1995, pp. 311–315.

Csikszentmihaly, M. and Rochberg-Halton, E. *The meaning of things: Domestic symbols and the self.* Cambridge University Press, 1981.

Cytowix, R.E. *The man who tasted shapes: A bizarre medical mystery offers revolutionary insights into emotions, reasoning, and consciousness.* G.P. Putnam's Sons, New York, 1993.

Dandavate, U. M. User response to video images of products: "A prelude to a multimedia-based methodology to record users' emotional responses to new product concepts." Master's Thesis, The Ohio State University, 1995.

Eastman Kodak Company, Human Factors Section, Health, Safety, and Human Factors Laboratory, *"Ergonomic Design for People at Work."* Van Nostrand Reinhold Company, New York, 1983.

Goleman, D. *Emotional Intelligence.* Bantam Books, New York, 1995.

Gordon, W. and Corr, D. "The...space...between...words: The application of a new model of communication to quantitative brand image measurement." *Journal of the Market Research Society*, Vol. 32, No. 3, 1990.

Ireland C., and Johnson, B. "Exploring the Future in the Present." *Design Management Journal.* Vol. 6., No. 2, Spring 1995.

Krippendorff, K. and Butter. "Where Meaning Escapes Functions." *Design Management Journal*, Vol. 4, No. 2, Spring 1993, pp. 30–37.

Krippendorff, K. and Butter. "The Semantics of Form," a special issue of *Innovation*, Vol. 3, No. 2, 1984.

LeDoux, J. "Emotional Memory Systems in the Brain." *Behavioral and Brain Research*, p. 58, 1993.

LeDoux, J. "Emotion, Memory, and the Brain." *Scientific American*, June 1994.

Lemonick, M.D. "Glimpses of the Mind." *Time*, July 17, 1995.

Leonard-Barton, D. *Wellsprings of Knowledge: Building and Sustaining the Sources of Innovation*, Harvard Business School Press, Boston, MA, 1995.

Logan, R., Augatis, S. and Renk, T. "Design of simplified television remote controls: A case for behavioral and emotional usability." Proceedings of the Human Factors and Ergonomics Society 38th Annual Meeting, 1994, pp. 365–369.

Losee, Paul. Personal communication, 1996.

Marcus, C.C. *House as a mirror of self: Exploring the deeper meaning of home.* Conari Press, Berkeley, California, 1995.

McCormick, E.J. *Human Factors in Engineering and Design.* McGraw-Hill Book Company, 1976.

Ortony, A., Clore, G.L. and Collins, A. *The Cognitive Structure of Emotions.* Cambridge University Press, Cambridge, 1988.

O'Brien, T.G., and Charlton, S.G. *Handbook of Human Factors Testing and Evaluation.* Lawrence Erlbaum Associates, Publishers. Mahwah, New Jersey, 1996.

Sanders, E. B.-N. "Converging Perspectives: Product development research for the 1990s." *Design Management Journal*, Fall 1992, pp. 49–54.

Zuckerman, M. R., Hatala, L. J., *Incredibly American Releasing the Heart of Quality.* ASQC Quality Press, Milwaukee, Wisconsin, 1992.

FACTORS AFFECTING PRODUCTIVITY AND ERGONOMICS OF SUPERMARKET CHECKERS

Edwin F. Madigan Jr. and Katherine R. Lehman
NCR Corporation
Atlanta, Georgia

A field study was conducted to examine scanning productivity and ergonomic issues as a function of scanner technology and product type. Cashiers performed both normal and staged activities using bi-optic and flat-bed scanners. In Phase One of the study, package handling and throughput were examined. Overall, bi-optic scanning was faster and some package types were handled more efficiently. Some package types, however, were handled less optimally (i.e., flipping and twisting scans). In phase two of the study, wrist acceleration measures were examined as a function of scanner technology and package type. Lower wrist accelerations in both hands for both flexion/extension planes of motion were found for the bi-optic scanner. In addition, bi-optic scanning also produced flexion/extension accelerations (both hands) and radial/ulnar accelerations (left hand only) that were below industry validated low risk benchmarks. The effects of scanner technology and package type on wrist acceleration and scanning behavior are discussed.

INTRODUCTION

The introduction of optical barcode scanners in the 1970s into retail operations dramatically increased the productivity of Point-of-Sale (POS) operations over traditional key input methods of item processing. Scanners have been responsible for improving inventory management, decreasing pricing errors and transactions times while greatly benefiting customer service.

In recent years, however, increased attention has been directed to musculoskeletal disorders (MSDs) and scanning (Baron, Milliron, Habes, and Fidler, 1991). Some researchers have examined whether causal relationships exist between retail cashiering jobs and the incidence of MSDs (Margolis and Kraus, 1987; Ryan, 1989; Morgenstren, Kelsh, Kraus, and Margolis, 1991; Harber, Pena, Bland, and Beck, 1992). Although a causal relationship between scanning and MSDs has not been demonstrated, research is underway to examine commonly presumed risk factors.

Since the late 1970s, advances have been made to improve read accuracy, productivity, ease of use, and the ergonomics of scanning. Today, several retail system manufacturers offer bi-optic scanners that allow laser beam patterns to emanate from two windows. The advantages of the new optics are: 1) It allows for multidirectional decoding, 2) it requires less barcode orientation, 3) it reads poor quality barcodes, 4) it requires fewer scanning movements, 5) is less fatiguing because it does not require lifting, and 6) it affords continuous item movement through the three-dimensional scan zone.

Several studies have shown productivity advantages with bi-optic scanners (Lehman, 1995; Hoffman, 1992; Hoffman, 1996; Madigan 1995). In addition, research examining the ergonomic impact of different scanners, scanner orientations, and scanner-checkstand combinations has been performed (Marras, Lehman, Greenspan, Nelson, Lee, and Fattalah 1993; Lehman and Marras, 1994). These researchers found that front-facing checkstands with bi-optic scanners allow cashiers to scan with lower wrist accelerations compared with other scanner-checkstand combinations. There has been, however, no controlled, comprehensive, quantitative field research to examine the effects of product type and scanner design on scanning accuracy, throughput times, and wrist biomechanics.

This paper describes two phases of a field experiment performed to quantify scanning activity on two different scanner types (flat-bed and bi-optic). Wrist acceleration, scanning speed, and scanning accuracy were measured to assess productivity and ergonomic risk as a function of scanner technology and product type. The goal of this research is to further quantify critical factors that affect scanning behavior in order to increase productivity while providing a more comfortable work environment. It was expected that productivity advantages found previously with bi-optic scanning would be replicated. However, it was predicted that performance differences would be found across package types. It was also predicted that bi-optic scanners would facilitate scanning by reducing wrist accelerations for some package types.

PHASE 1

Methods

Subjects. The scanning behaviors of fifteen different cashiers ranging in age from 21 to 49 years of age were examined. Eight cashiers (five males and three females) were observed in the flat-bed scanner condition and seven cashiers (one male and six females) were observed in the bi-optic scanner condition. All cashiers had at least one year of experience at their present location and volunteered to participate in this study.

Apparatus. Closed circuit video cameras and video cassette recorders were used to record normal cashier activities. A running time stamp to the nearest 1/100th of a second was encoded on the video tapes. Wireless microphones were used to record auditory feedback from the scanner.

Procedure. Cashier-customer transactions from four different supermarkets in the Midwest were recorded using the NCR 7870 and Spectra Physics Magellan scanners (bi-optic condition) and the NCR 7824 scanner (flat-bed condition). Retail sites with only front-facing checkstands were examined. Three, twelve hour days of activity were recorded.

Analysis. Data were extracted from video tape using mnemonic codes designed to capture the following information: 1) Scanner type, 2) item manipulation, 3) package type, 4) scanning accuracy, and 5) scanning speed. Item manipulation was coded according to six possibilities: continuous sweep passes, discrete sweep passes with a stop at the scanner window, spin passes where the item was not held, twist passes where the item was held, flip passes where the item was flipped in either the horizontal or vertical plane, and unknown for ambiguous motions. Package types were coded according to twenty distinct classes including: cans and cylinders, bottles, boxes, bags, tubs, jugs, small items, barcoded produce, and in-store barcoded items. Scanning accuracy was coded according to the number of passes required for a successful scan (1 pass, 2 passes, 3 or more passes, no-scan, and not-on-file), or keyboard entries (no scan attempted).

Throughput times were calculated for each package type. Throughput is the average scan time weighted according to scanning accuracy percentages for each package type. Frequencies and percentages of occurrence were calculated for the remaining variables.

Results. A total of 3750 items were examined; 1750 items for the bi-optic condition and 2000 items for the flat-bed condition.

Scanning Accuracy. Cashier performance using the bi-optic scanner was superior to the flat-bed scanner. First pass scans occurred more often (83.37% vs. 78.85%) and were faster ($\underline{M}$ = 1.32 sec. vs. $\underline{M}$ = 1.49 sec.) in the bi-optic condition. In addition, there were fewer second pass (7.89% vs. 12.75%) and multiple pass scans (2.46% vs. 3.9%) in the bi-optic scanning condition.

Throughput. Throughput times were examined according to a criterion of 1.7 seconds. Times greater the criterion were considered to be non-optimal because experienced cashiers have been shown to scan a majority of items in 1.7 seconds or less (M.S. Hoffman, personal communication, September, 1995).

Figures 1a and 1b show item throughput for the ten most frequent package types in each scanner condition.

Throughput was greater than 1.7 seconds for 8 of 10 package types in the flat-bed scanning condition (see Figure 1a) and for 4 of 10 package types in the bi-optic condition (see Figure 1b). Comparing Figures 1a and 1b shows that bi-optic throughput was superior to flat-bed throughput for five of the ten most frequent package types (cans and cylinders, rectangular and square boxes, bottles, and tubs). Also note in Figure 1b that 38.8% of in-store barcoded items were entered using the keyboard without attempting to first scan the items.

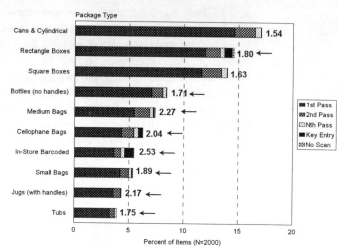

Figure 1a. Item throughput for the flat-bed scanner. (Note. Number indicates throughput in seconds, "←" indicates throughput greater than 1.7 seconds.)

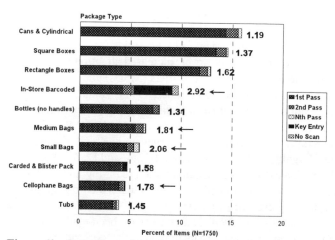

Figure 1b. Item throughput for the bi-optic scanner. (Note. Number indicates throughput in seconds, "←" indicates throughput greater than 1.7 seconds.)

Item Manipulation. The positive effects of the bi-optic scanner design were fewer flipping motions (28.97% vs. 36.3%) and more continuous sweep motions (33.3% vs. 31.3%) compared to the flat-bed scanning. However, more twisting (23.71% vs. 19.95%) and spinning motions (2.34% vs. 0.75) were observed in the bi-optic condition suggesting poor product handling strategies.

When item manipulation is considered according to package type, the data show that three of six scanning strategies were more frequently adopted (see Figure 2). Boxes were scanned with a higher overall percentage of continuous sweep motions (46%) compared with all other package types. High percentages of flipping motions were found for bags (44.3%), in-store barcoded (47.1%), and carded items (50.6%). High percentages of non-optimal twist motions were also found for cans/bottles (52.2%) and tubs (49.3%).

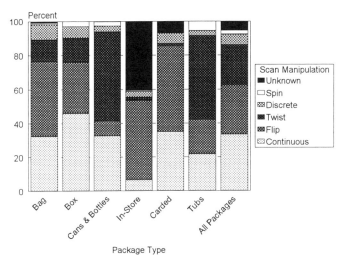

Figure 2. Manipulation by package type (bi-optic scanning).

In summary, bi-optic scanning throughput was superior to flat-bed throughput for five of the ten most frequent package types. Although productivity advantages were found with bi-optic scanning, cans and bottles, bags, in-store barcoded, and carded items were handled less efficiently than other products. Productivity advances are not the only feature of the bi-optic scanner. The bi-optic scanner should also provide greater ease of use, little product orienting, and fewer scanning motions. The objective of Phase Two was to examine the ergonomic impact of the NCR 7870 bi-optic scanner compared to a flat-bed scanner, by collecting wrist acceleration data from checkers performing their normal activities in a field environment.

PHASE 2

Method

Subjects. Nine cashiers from a prominent supermarket in the Southeast volunteered to participate in an 8 week biomechanical study. The group was comprised of eight females and one male ranging in experience levels (6 months to 7 years) and age (21 to 51 years).

Apparatus. Subjects were tested on the NCR 7824 flat-bed scanner and the NCR 7870 bi-optic scanner with a front-facing checkstand performing their normal checkout operations. Wrist monitors developed at the Ohio State University Biodynamics Laboratory were applied to both of the cashiers' hands and wrists to measure radial, ulnar, flexion, and extension deviations. The raw voltage signals

captured from the wrist monitors were sampled at 300 Hz and were ported to an IBM compatible PC for later analysis. The raw voltage signals were converted using an analog-to-digital board and were then converted to angular position, velocity, and acceleration using La Place transformations.

Procedure. Wrist monitor data were collected on cashiers performing normal and controlled activities except bagging. Controlled activities included scanning only bags, bottles/cans, or boxes. Cashiers were initially tested on the NCR 7824 flat-bed scanner. The NCR 7870 bi-optic scanner was then installed and cashiers were instructed on scanning techniques. Wrist monitor data were collected 6 weeks after training to allow for learning and practice.

Wrist acceleration was chosen as the dependent measure because it is a dynamic biomechanical measure of wrist movement and is thought to contain components of commonly proposed risk factors such as: force, posture, and acceleration. Although it does not itself identify risk or injury, wrist acceleration is considered by some researchers to be a predictor of risk for other industries (e.g., factory work).

Analysis. The wrist monitor data were analyzed to calculate position, velocity, and acceleration of the wrist for the radial, ulnar, flexion, and extension planes of motion. The mean and peak acceleration values were calculated according to Schoenmarklin et al., (1994) and compared to MSD risk benchmarks validated in other industries (Marras and Schoenmarklin, 1993; Marras et al., 1993).

Results. The levels of mean wrist accelerations calculated for each subject during normal cashiering were compared to MSD risk benchmarks (Marras and Schoenmarklin, 1993) for each of the two scanners types (bi-optic and flat-bed). As shown in Figure 3, the mean and standard deviations for all planes of motion averaged over all cashiers, were lower for the 7870 bi-optic scanner than for the flat-bed scanner. In addition, the mean values for the bi-optic scanner were below the low risk level for all planes except right radial/ulnar whereas the means were clearly above the low risk level for all planes for the flat-bed scanner.

Controlled trials were also run on the 7870 bi-optic scanner to examine the effect of different product types on wrist acceleration. The most frequent product types were chosen for analysis: boxes, bottles/cans, and bags. These product types comprise 21%, 22%, and 13%, respectively, of all products sold. Figure 4 shows the left and right hand mean flexion/extension wrist accelerations for all subjects according to the three controlled conditions and the live (normal) transaction condition.

It is evident from this figure that the boxes and bags conditions produced acceleration values well below the low risk level and were slightly lower than the live transaction (normal transaction) values. The bottles and cans condition, however, produced higher accelerations that were above the low risk line.

The majority of subjects were observed to simply slide boxes and bags over the scanner with little manipulation. However, some subjects tended to spin bottles and cans while

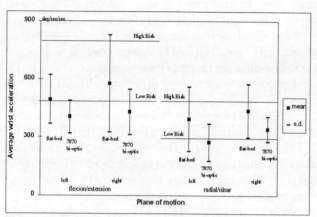

Figure 3. Average wrist acceleration for all planes and both hands comparing flat-bed to bi-optic scanning.

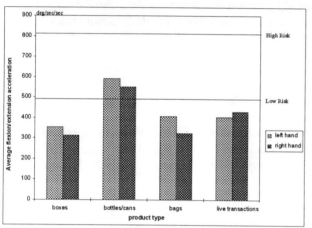

Figure 4. Average flexion/extension acceleration by product type and live transaction data for all subjects combined.

passing them over the scanner. This may have impacted the higher accelerations reported for the bottles/cans condition.

DISCUSSION

Increasing productivity while maximizing cashier comfort during check-out is a multi-faceted challenge. The scanning strategy chosen by an individual cashier is based on many different factors including: the type of checkstand employed, the product to be scanned, barcode location, barcode quality, and scanner style.

The scanning strategy affects scan time, accuracy, and the amount of wrist acceleration. The strategy itself can be qualitatively described by determining whether the product was lifted, slid, spun, twisted, flipped, or scanned in a continuous, or discrete manner. However, a more complete understanding of the effects of item handling on productivity and risk is gained when quantitative performance measures are utilized. This, study was designed to employ both video and biomechanical analysis techniques to quantify scan times, read accuracy, and wrist acceleration.

Bi-optic scanning productivity and wrist acceleration measures were found to be superior to flat-bed scanning.

Cashiers achieved more first pass scans and were faster under bi-optic scanning conditions and bi-optic item throughput was superior to flat-bed throughput. Lower flexion and extension accelerations were found under bi-optic scanning conditions for both hands. It is important to note that left and right hand flexion/extension accelerations and left hand radial/ulnar accelerations were below low risk benchmarks observed in other industries.

While bi-optic scanning productivity was shown to be superior to flat-bed scanning, clear effects of package type were found. Bi-optic scanners showed higher percentages of continuous sweep motions and fewer flipping motions overall. However, more twisting and flipping occurred for some package types under bi-optic conditions. Products that have no natural bottom (i.e., bags and carded items) were handled more frequently with non-optimal flipping motions. In addition, cans, bottles, and tubs were more likely to be twisted while scanning although the vertical scanning window does not require such a motion. It is likely that these items afforded twisting because of their shape and the side location of the barcode. Boxes were the most efficiently handled items and contain barcodes on the sides and/or bottom but do not afford spinning. Boxes also have a natural bottom and therefore allow for more continuous sweep scans. In addition, it was found that a large percentage of in-store barcoded items were entered using the keyboard without attempting a scan. This suggests that either the flexible wrapper degraded the ability to achieve a scan or the barcode quality may have been poor.

Wrist acceleration measures showed that boxes, bags, and all products combined were below the low risk benchmark level. Bottles and cans (side coded, cylindrical shaped products) promoted less desirable scanning motions than all other package types and wrist accelerations were above the low risk benchmark level. The tendency to spin and twist bottles and cans may have contributed to the recorded higher accelerations. The less optimal spinning motions used for bottles and cans are not required to obtain a successful read nor are they efficient scanning motions.

These data suggest that when evaluating cashier performance, usability strategies as a function of package type should be considered. Although the bi-optic scanner design affords more optimal, continuous motions, cashiers did not take advantage of this design feature. It is possible that cashier training could promote more optimal scanning motions. Lehman (1996) for example, found bi-optic scanner training to facilitate the use of smooth, continuous motions for all package types. Lehman also found that promoting smooth, continuous motions is likely to reduce wrist accelerations.

Future research should consider the effects of barcode quality on scanning behavior. Both perceived and true barcode quality problems have been shown to negatively affect scanning performance (Madigan, 1995a; 1995b). Madigan (1995b), for example, measured the barcode quality characteristics of items that were found to be difficult to enter (i.e., 2nd pass, multiple pass, no-scan, or key entry). It was found that only 44% of the items that were difficult to enter actually failed the UCC barcode quality test specifications (UCC, 1994). It is also interesting to note that the remaining 56% of the items that were difficult to scan passed the UCC barcode quality test specifications - further suggesting that package type and package handling contributes to the lack of optimal scanning behavior.

In summary, the bi-optic scanner has been shown to facilitate both productivity and ergonomic comfort associated with cashier scanning behaviors. However, retailers should consider the effects of package type and include training programs to ensure that cashiers are taking advantage of design features.

ACKNOWLEDGMENTS

The authors would like to thank Thomas V. Brady, Nicole I. Miller, and Antonio Saltó of the Uniform Code Council for their technical assistance with this study. Partial funding was made available for this research by the Uniform Code Council.

REFERENCES

Baron, S., Milliron, M., Habes,D., and Fidler, A. (1991). Health Hazard Evaluation Report. New Jersey, Shoprite Supermarkets.

Harber, R. , Pena, L., Bland, G., Beck, J. (1992). Upper extremity symptoms in supermarket workers. American Journal of Industrial Medicine, 22, 1-12.

Hoffman, Mark. S. (1996). Impact of technology and package design on the ergonomics of scanning checkout. Manuscript submitted for publication In Proceedings of the Human Factors and Ergonomics Society 40th Annual Meeting, Santa Monica, CA: Human Factors.

Hoffman, M.S. (1992). Considerations in the design of future retail Point Of Service (POS) terminal systems. In Proceeding of the Human Factors Society 36th Annual Meeting, (pp. 863-866). Santa Monica, CA: Human Factors Society.

Lehman, K. R. (1996). The ergonomic benefits of bi-optic scanning: A comparative study of flat-bed versus bi-optic scanners by evaluation of wrist biomechanics during scanning. Duluth, GA: NCR internal report.

Lehman, K. R. (1995). Bi-optic versus flat-bed scanner productivity: A human factors engineering analysis. Duluth, GA: NCR internal report.

Lehman, K.R. and Marras, W.S. (1994). The effects of human interface design on wrist biomechanics during scanning. In Proceeding of the Human Factors and Ergonomics Society 38th Annual Meeting, (pp. 616-620). Santa Monica, CA: Human Factors and Ergonomics Society.

Madigan, Edwin F. Jr.. (1995a). An examination of UPC location and scanning behavior. Duluth, GA: NCR internal report.

Madigan, Edwin F. Jr. (1995b). An examination of UPC quality. Duluth, GA: NCR internal report.

Margolis, W. and Kraus, J.F. (1987). The prevalence of carpal tunnel syndrome in female supermarket checkers. Journal of Occupational Medicine, 29 (12), 953-956.

Marras, W.S. and Schoenmarklin, R.S. (1993). Wrist motions in industry. Ergonomics, 36 (4), 341-351.

Marras, W.S., Lehman, K.R., Greenspan, J.G., Nelson, J.E., Lee, C., and Fattalah, F.A. (1993). A field biomechanical analysis of three grocery checkstand/scanning operations: Phase III report. Report for the Food Marketing Institute.

Morgenstern, H. Kelsh, M., Kraus, J. and Margolis, W. (1991). A cross-sectional study of hand/wrist symptoms in female grocery checkers. American Journal of Industrial Medicine, 20, 209-218.

Ryan, G.A. (1989). The prevalence of musculo-skeletal symptoms in supermarket workers. Journal of Ergonomics, 32, 4, 359-371.

Uniform Code Council (1992). U.P.C. * symbol location guidelines manual. Dayton, OH: Uniform Code Council.

Uniform Code Council (1994). Quality specification for the U.P.C.* printed symbol. Dayton, OH: Uniform Code Council.

MAY I TAKE YOUR ORDER?
HUMAN FACTORS FIELD STUDIES IN THE FAST FOOD ENVIRONMENT

Michael D. Gravelle and Sally M. Cohen
NCR Corporation
Duluth, GA

Ann M. Bisantz
Georgia Institute of Technology
Atlanta, GA

Karen S. Wilson
NCR Corporation
Duluth, GA

While applied human factors research is frequently conducted in technology-rich advanced environments (e.g., aircraft, nuclear power plants, and marine vessels), this paper describes a successful applied human factors field study in a complex, dynamic, and highly demanding commercial environment—your local fast food restaurant. The primary study objective was to gain a comprehensive understanding of fast food store operations and employee activities in an effort to conceptualize and to develop design guidelines for next-generation technology in this industry. The findings suggested that the current implementation of technology in the fast food industry does not always meet the needs of today's users. Several opportunities exist for human factors professionals to re-engineer store processes and operations and to design more appropriate technology solutions.

PROBLEM STATEMENT

Traditionally, human factors specialists have been involved with assessments of environments replete with complex technology, such as military systems (Briggs and Goldberg, 1995), airplane cockpits (Kelly, 1988), and process control plants (Moray, Lee, Vicente, Rasmussen, Jones, Brock, and Djemil, 1993). These have been the most familiar environments because they offer opportunities to study and impact dynamic and life-threatening systems and to evaluate new and automated equipment with the luxury of well-funded budgets. There is another complex environment familiar to us all that has had little or no attention paid to it by the human factors community. This environment is the quick service or fast food restaurant—an important and major segment of the retail industry. The methods and findings described in this paper represent a successful case study of applied human factors research in a commercial environment.

The fast food restaurant offers an excellent environment for human factors practitioners to study employee activities and interactions with technology. First, this industry is expanding its use of technology and automation because of labor shortages, cost control measures, and advancements in technology. Second, these environments can be considered complex as they possess uncertainty (with respect to customer and product demand),

are dynamic and event driven, have interdependent processes and competing goals, and maintain a business level of risk that motivates store management on a daily basis (Roth and Woods, 1988). Third, many of the organizational constraints that affect the successful implementation of technology, such as cost control and labor concerns, operate at the local restaurant level. From an analysis perspective, this is beneficial because these issues are more accessible than in a large distributed organization. Interdependent functions, such as inventory management, food production, and training, are managed at the local restaurant level and organizational goals play an integral role in the daily operations.

THESIS

A series of field studies were performed in the fast food industry to gain a better understanding of how well technology has been applied to meet employee needs in serving customers. Several business objectives were established at the initiation of the study. The primary objective in conducting the study was to gain a more comprehensive understanding of quick service store operations and employee activities in an effort to conceptualize and develop design guidelines for next-generation technology in this industry. By observing user problems with current technology and by identifying user needs for new technology, we plan to translate these requirements into tangible ideas, concepts, prototypes, specifications, and ultimately, products for the quick service market. Our goal in designing these products is to ensure that they are easy to use and maintain, increase user productivity, decrease user errors, and lead to overall improved customer satisfaction and increased store revenues.

METHOD

Fast Food Environment

In this study, a total of seven individual restaurants representing three different quick service franchises were investigated. The field sites covered the spectrum of restaurant profiles, from those in rural to urban locations; those with high to low customer volumes; and those with spacious and effective to confined and illogical store layouts. Their individual differences, notwithstanding, each fast food restaurant can be characterized as a system of production and service equipment staffed by a team of individuals with varying skill levels. The pace of hourly activity within the restaurants can vary from slow to very busy, and is affected by both expected and unexpected system stressors. The restaurant can be compared to an individual manufacturing facility that converts raw materials into food products. However, quick service restaurants are unlike other types of manufacturing facilities, because they must sell and distribute the products to their customers on demand.

Site Analysis Methods

The methods used in this study were selected to obtain the information necessary to answer the business objectives and to build a baseline store performance model. In doing so, the data collected were intended to help us better understand the store employees, their daily activities, and the environment given the resources (people, information, and time) available. The following user-centered design methods were used: field observations, interviews, focus group discussions, and video analysis.

Field observations. The first phase of the study consisted of field observations in the restaurants. The purpose of these observations was to gain a hands-on understanding of employee activities and each store's unique attributes. The authors were viewed as new employees in training, and in doing so, were educated on store operations and allowed to participate in some team activities, such as taking and packing customer orders and assisting with management tasks. The observer role provided ample opportunities for note-taking, task participation, question asking, and collecting sample materials. Extensive note-taking and informal discussions with employees resulted in detailed descriptions of the store processes being observed.

The initial observations also helped to diagnose potential problem areas for further investigation in the interviews, focus group discussions, and video analyses.

Interviews and focus group discussions. The purpose of conducting personal interviews and focus group discussions was to solicit a range of opinions from actual users and an understanding of various issues related to their jobs, personal and work experiences, and technology. The one-on-one interviews were conducted primarily with individuals representing District and Regional Management, Field Operations, Corporate Operations, and Corporate Information Technology. The focus group sessions were held with groups of six to eight General Store Managers, Shift Supervisors, and Assistant Store Managers. Both the interviews and focus groups centered on six topical areas: day-to-day operations, product and inventory management, team member management, store performance measures, customer service, and employee wishes and needs. The audio tapes from the interviews and focus group discussions were transcribed and submitted to content analyses using MacSHAPA (Sanderson, 1994; Cohen, Gravelle, Bisantz, and Wilson, 1996). Hundreds of statements resulted from the analyses and were used to substantiate problems observed in the field and evidenced in the video analyses. The content analyses also helped reveal previously unknown problem areas, which were investigated by further analyses.

Video analysis. The purpose of the video analysis was to understand and to quantify the tasks users perform in this environment. From the video footage captured during normal working hours, it was possible to identify each task that employees perform, the individual elements or steps composing each activity, and the amount of time users spent performing each task and task element. Between four and eight cameras were located throughout the employee work areas to capture all activities to the level of detail appropriate to address the business objectives. The cameras were focused on each of the major stores processes, such as product preparation and assembly; order taking, packing, and delivery; and facility maintenance and cleaning. By capturing video footage simultaneously and throughout the entire store, we were able to conduct parallel activity analyses to facilitate the study of product and information flow through the system. MacSHAPA was also used to extract data from the videotape footage and to perform various qualitative and quantitative analyses. Summary statistics were used to characterize each restaurant's baseline performance in terms of transactions times, average task times, and transaction component analyses.

FINDINGS

The combined findings from the analyses supported the development of a comprehensive restaurant process description. Briefly, this process description includes activities such as maintaining adequate inventory levels of uncooked food, food staples (e.g., flour, milk, butter), and non-food stock (e.g., napkins, bags, straws); managing cash; hiring, training, and scheduling employees; cleaning and maintaining the equipment and facility; taking, assembling, and finalizing customer orders; and preparing food. Depending on the particular restaurant's menu offering, the food preparation process can vary from simple to complex. It can involve preparing and packing side items with long hold times (several hours to several days) ahead of customer orders, preparing and assembling items with short hold times (less than one hour to several hours) ahead of customer orders, and preparing special orders for customers. The emphasis in preparing food is to maintain the necessary quantity of fresh products while minimizing waste and customer wait time.

Labor was found to be a major concern in the fast food industry. First, the labor pool willing to work for low wages in these restaurants tends to be low skilled. Second, many employees are high school and college students working part-time jobs in the evenings and weekends. The rate of employee turnover for this group of employees is very high. Third, the size of the traditional labor pool for fast food restaurants is shrinking due to a decrease in the number of teenagers, leading to increased competition for employees and thus the need to offer competitive wages. The combination of these factors has sparked the increased use of automation in an effort to maintain customer service levels with declining staff sizes. An understanding of these factors and others should contribute to the effective design of technology for these environments; unfortunately, it often does not as illustrated by this study. For example, equipment is needed for employees with limited skills and often low motivation to work, and that the equipment does not require an extensive training investment.

The activities are fast paced in this environment, and can be considered event driven because products must be delivered in response to customer demands at the time of the request. While these demands are fairly predictable based on seasons of the year, day of the week or month, and time of the day, at any particular time the demand can vary dramatically. Depending on store location, a rush can occur because of community and school events, busloads of travelers pulling off the highway, or sales at a nearby mall merchant. Unfortunately, restaurants (in particular, the backline or cook and assembly area) have a relatively slow response time to changing demands, since the time to prepare and cook food is long relative to acceptable service times.

DISCUSSION

The results of this field study yielded an in-depth characterization of fast food restaurant operations and employee activities. A number of fundamental human factors problems were observed across the restaurants represented in this study. First, a number of employee activities were

identified as time-consuming, highly variable, and frequent—such as taking special orders, toasting sandwich buns, and drawing drinks, to name a few. Focused efforts now should be undertaken to further explore these tasks and task sequences, and to solve these process bottlenecks using new processes, systems, or training programs. Second, the findings indicate that different day periods influence the speed of customer service and employee productivity. Customers are typically served more quickly if they visit the restaurant during the store's busiest day periods (e.g., lunch and dinner) than if they visit during the store's slowest day periods (e.g., mid-afternoon and evening). This happens because cooks were constantly preparing food during busy periods in an effort to keep up with the unrelenting customer demand; therefore, food was typically available at the moment an order was placed. During slow periods, only a small buffer of food was available; therefore, customers were frequently waiting for food to be prepared. Third, store status levels (e.g., customer volume, cooked product quantities) are not always apparent to the employees—especially those preparing and assembling food on the backline. Unfortunately, such a lack of *situation awareness* can lead to under-production of fresh food resulting in increased customer wait times, or over-production of food leading to product waste. Both of these resulting effects ultimately affect the restaurant's primary goal of maintaining or increasing store profits. Fourth, equipment ease of use is a problem and much of the existing technology does not meet the needs of the current users. This is evident in many usability, training, and proficiency problems that were discovered in the site analysis findings. The following examples briefly illustrate how some technology has been poorly applied and offer opportunities where human factors professionals can conceptualize new or improved product designs.

On-Line Cooking Decision Aid

To assist employees with determining the correct amount of food to cook, restaurant management installed a computerized decision aid to provide minute-by-minute cooking instructions based on current and historic sales data and cooking parameters (e.g., cooking and hold times). The decision aid was intended to reduce workload demands placed on cooks and managers because they no longer had to synthesize any factors into the cooking decision but could simply react to the on-line display. However, despite pressure from field and corporate management to follow the cooking instructions, the system went unused. The aid was not used because decision making theories embodied in the aid about potential users conflicted with the actual work practices in the environment.

Point-of-Sale (POS) Devices

The POS devices, located at the front counter and drive-thru window (DTW), are used to input customer food orders, calculate order totals, facilitate the payment process, and in some cases support management activities (e.g., displaying sales volumes, balancing cash drawers). These devices, representing a variety of manufacturers, covered the technology spectrum from electronic cash registers to computerized terminals linked to the back office management system. Both hardware and software problems were observed. In most of the observed stores, the devices were found to be complex, ineffective in supporting user needs, and difficult for training new employees. Some of the specific human factors problems included:

- Key caps were heavily worn and hard to read
- Key labels used extremely small fonts and inconsistent color coding schemes
- Key labels used inconsistent abbreviations (e.g., COUP, CUPON, and CPN for coupon)
- Glare from overhead lighting and exterior sunlight caused readability problems
- Outdated keys cluttered the keypad for menu items that no longer existed on the menu
- Cashiers memorized menu codes or referred to 'cheat sheets' when entering menu items
- Terminals alternated between many different input modes (e.g., breakfast vs. lunch menu)
- Very little visual feedback was provided and error correction was difficult and sometimes impossible without voiding the complete transaction and starting over
- Value meals (containing multiple food items) required lengthy entry of each food item, quantity, and qualifier (e.g., 1 + Sandwich + 1 + Side Item + 1 + Large + Drink) and forced user memorization of complex menu items

Communication Systems

There are a number of problems associated with the communication technology found throughout this environment. Examples can be found with the headset and microphone technology at the DTW and the automated product ordering system that used the telephone as the interface. Rarely can a consumer make it through a drive-thru dialog with a cashier without having to repeat the order or ask the employee to repeat the menu choices, and ultimately receive the wrong product. Miscommunication in the DTW ordering process occurred because the headset and microphone technology is poor and radio interference and ambient interior and exterior noise degrades speech intelligibility. The absence of face-to-face interaction resulted in increased order taking and assembly errors and the need

for dialog repair. The headset device required the cashier to press the ON button when he or she spoke, thus reducing the cashier's manual resources needed for entering in-coming customer orders and packing bags. Likewise, store managers often report great difficulty placing product orders with distributors using automated tele-ordering systems. These systems provide no feedback or confirmation of products and quantities ordered; the automated menuing system is complex; and little or no opportunities are provided to correct errors or backup in the menuing system. The design of the tele-ordering system did not take into account the nature of the manager's job. Store managers rarely have time to sit down for extended periods of time to place an order; they are frequently pulled away from administrative duties to support frontline or backline emergencies. It does not allow managers to pause and save their order, later return to the order, resume it, and complete it.

SUMMARY

In summary, it is clear that the current implementation of technology in the fast food industry does not always meet the needs of today's users. Many areas of opportunity exist for human factors professionals to re-engineer store processes and operations and to design more usable technological solutions. These field studies provided a comprehensive understanding of quick service store operations and employee activities. Future efforts will be directed at interpreting the data and findings further and to begin conceptualizing, prototyping, and developing design guidelines for next-generation technology in this industry.

REFERENCES

Briggs, R., and Goldberg, J. (1995). Battlefield recognition of armored vehicles. *Human Factors, 37* (3), 596-610.

Cohen, S., Gravelle, M., Bisantz, A., and Wilson, K. (1996). Analysis of interview and focus group data for characterizing environments. In *Proceedings of the Human Factors and Ergonomics Society 40th Annual Meeting.* Human Factors and Ergonomics Society: Santa Monica, CA.

Kelly, M. (1988). Performance measurement during simulated air-to-air combat. *Human Factors, 30* (4), 495-506.

Moray, N., Lee, J., Vicente, K., Rasmussen, J., Jones, B., Brock, R., and Djemil, T. (1991). Evaluation of an integrated display for PWRS based on the Rankine Cycle. In *Proceedings of the Topical Meeting on Nuclear Plant Instrumentation, Control, and Man-Machine Interface Technologies* (pp. 573-580). American Nuclear Society: La Grange Park, IL.

Roth, E.M., and Woods, D.D. (1988). Aiding Human Performance I: Cognitive Analysis. *Le Travail Humain, 51,* 39-64.

Sanderson, P.M. (1994). *Exploratory Sequential Data Analysis: Software*. Engineering Psychology Research Laboratory Technical Report (EPRL-94-01). Department of Mechanical and Industrial Engineering, University of Illinois: Urbana-Champaign, IL.

IMPACT OF SCANNING TECHNOLOLGY AND PACKAGE DESIGN ON THE ERGONOMICS OF SCANNING AT THE CHECKOUT

Mark S. Hoffman
Retail Systems Group
NCR Corporation
Duluth, Georgia

The design of bar code scanners has steadily improved scanning performances over the past twenty years. Changes in scan patterns, improvements in decode algorithms, and design features of the hardware user interface have resulted in scanners that are ergonomically superior both in biomechanics and overall performance. This study reviewed scanner performance and technological developments, and compare these to changes in packaging designs. The results showed that future innovations in scanner design will have minimal impact on improving throughput and the accuracy of data captured because of package design and bar code placement.

INTRODUCTION

Bar code scanners have been used in the retail checkout process for the past twenty years. During this time, retailers and suppliers have engineered inventory management systems based on tracking product sales, periodic counts of items on the shelf, and store inventories. Inventory replenishment systems used by today's retailers are dependent upon the daily exchange of sales information between the manufacturer and retailer. Scanners are designed to improve the accuracy of product information captured at the Point Of Sale (POS).

Efficient Consumer Response (ECR), a formal program initiated by the retail food industry has been implemented by many large retailers and manufacturers. The wide spread acceptance of this program during the past four years is an acknowledgment of the commercial dependencies on sharing product sales information between retailers and suppliers. ECR allowed retailers and suppliers to deploy Just In Time (JIT) inventory replenishment based on consumer pull merchandising programs. Successful ECR programs require accurate real-time

capture, recording, and transmission of product movement data from the checkout to the manufacturer.

In the design of scanner, particularly the multi-dimensional scan patterns is a complex process because the scanners have to read bar codes in many different orientations, speeds in excess of 150 inches per second, and label varying degrees of Universal Product Code (UPC) conditions. The ultimate objective in the design of scanners to provide scanners that read all items on the first pass. Even though scanning from the cashier's viewpoint is a target acquisition task, scanning efficiency is dependent upon the user's comfort with blind scanning, this is scanning with minimal visual orientation of the UPC. When the scanning envelope is insufficient for blind scanning, cashiers can develop package handling behaviors that under long-term exposures have the potential to lead to a Upper Extremity Muscular Skeletal Disorders (UEMSD).

In 1991, the Food Marketing Institute sponsored the FMI Ergonomics Task Force (FMI 1992, FMI 1996). This task force was chartered to explore ergonomic issues related to scanning checkout systems used in supermarkets. Two research programs were initiated to further the food industry's understanding in checkout ergonomics: (1) biodynamic research investigating the relationships between wrist and lumbar motions and associated UEMSDs, and (2) measurement of muscle activity levels associated with different scanning checkout designs (FMI 1996).

These research programs showed several significant findings: (1) dynamic research techniques using eletrogoniometers and the Lumbar Motion Monitor were there most effective methods of discriminating the relative UEMSD risks between different scanning checkout designs; (2) scanner design, particularly the position of the window, (vertical or horizontal) did not promote scanning behaviors that had significantly different UEMSD risk estimates; and (3) the checkstand design and work methods promoted by the designs in combination with different types of scanners did result in dramatic differences in UEMSD risk estimates (Marras 1995). This research lead to the development of bioptic scanners, scanners with both horizontal and vertical windows, and to the increased use of a Front Checkstand; this style of checkstand promoted the lowest UEMSD risk behaviors.

The area of research neglected by previous studies in the impact of package design, placement and quality of the UPC on scanning behavior. Package design was briefly explored by Marras (Marras 1995, FMI 1992) in the Phase II study. Packages were selected based on results from previous studies (Hoffman 1992). The size, location, and optical characteristics of the UPCs were based on a small sample of groceries that were representative of those purchased in US supermarkets.

METHOD

Data files from fifteen years of scanning checkout studies conducted by the Human Factors Engineering department of NCR were the used in this research. These data files were created using a video based task analysis developed by NCR (Hoffman 1991). In each of these studies, dependent variables included: data entry accuracy and speed, scanning and key entry techniques, and overall throughput per transaction. Independent variables included: type of scanner and style of checkout, package type, and manpower assignments. Each study had a minimum of 500 transactions with an excess of 12000 packages per study. These studies included NCR and non-NCR scanners. The analysis for this research paper was limited to package movement, method of data entry, and scanning accuracy.

RESULTS

Package distributions were compared over a fifteen year period. These distributions represent customer purchasing patterns, not the packages on the shelf. Packages were grouped into four general types: Cylindrical, Boxes, Flexibles, and Bags.

Scan patterns contain three sets of beams, verticals, horizontals, and diagonals. The arrangement of these has evolved to provide more coverage in the scan zones. Likewise, the standard decode logic, originally developed by NCR in the '70s, has evolved to enhanced decode logic (Pacesetter) that functions similar to a parallel processor. These enhancements to the scan patterns and the decode electronics have significantly improved the ease of achieving First Pass Scans(FPS) on the majority of items scanned.

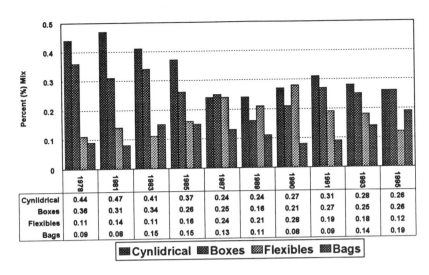

Figure 1. Packages Distributions of Consumer Purchases in US Supermarkets

The chart below show the evolution of scanner technology and its impact on FPS. FPS have been used by the retail industry's standards to judge the quality of scanners. FPS are calculated as a percent of only the packages that were scanned or attempted to have been scanned.

First Pass Scans (FPS) in the late '70's on scanners with Picket Fence scan patterns, were 75.9 percent. In the early 80's the F Model scanners with the X Bar scan patterns were introduced into the market; FPS increased within the range of 82-85%. Enhanced decode electronics (Pacesetter)

were then introduced in the late '80's. FPS rated did not significantly increase, but the overall percent of items scanned did increase. However, UPCs that previously were unscannable because of failing the UPC algorithm, were auto-corrected by the enhanced decode electronics, and thereby scanned.

The use of Vertical scanners also did not improve FPS significantly, but the vertical designs did improve the ease of moving packages through the scan zone. FPS finally did increase to an industry average of 88% with the introduction of Bioptic scanners.

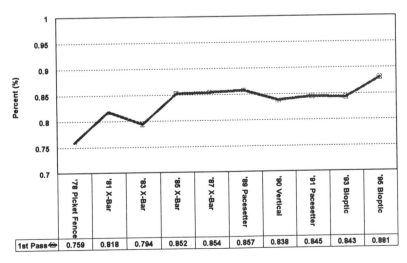

Figure 2. Changes in First Pass Scan Percentages with Evolution of Scanner Technology

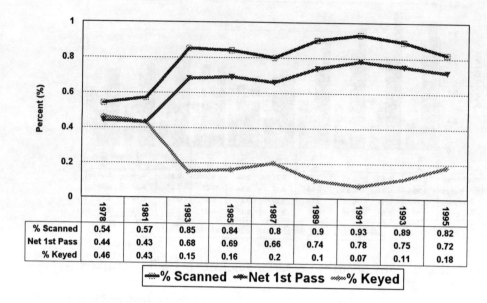

	1978	1981	1983	1985	1987	1989	1991	1993	1995
% Scanned	0.54	0.57	0.85	0.84	0.8	0.9	0.93	0.89	0.82
Net 1st Pass	0.44	0.43	0.68	0.69	0.66	0.74	0.78	0.75	0.72
% Keyed	0.46	0.43	0.15	0.16	0.2	0.1	0.07	0.11	0.18

—% Scanned ⚬⚬Net 1st Pass ⚬⚬% Keyed

Figure 3. Impact of 1st Pass Scans on the Overall Percent of Items Scanned

Examination of the actual percentage of packages scanned, net FPS, and keyed items showed the even though 76% of the items were FPS in the late 70's; this scan percentage only accounted for 44% of the items purchased. The first significant change in the total number of scanned items was in the '83-'85 time frame; the percentage of keyed items dropped to 15-17%. This shows the effect of the increased number of source and in-house marked items, and performance improvements gained through advances in scanning technology.

The introduction of the enhanced decode scanning technology showed a dramatic increase in the number of overall items scanned. However, with the introduction of Bioptic scanners, the FPS increase as previously mentioned; but the number of keyed items also increased This anomaly shows the effect of increases in the amount of produce items, compared to previous years; this finding is also reinforced with the data presented in Figure 1, with the significant increases in Bags and Flexible wrappers.

DISCUSSION AND CONCLUSIONS

Madigan (1995) showed that package type (box, flexible wrapper, and bags) and the quality of the UPC have significant impacts on scanning behaviors. Lehman (1995) found that even with the freedom Bioptic scanners offer, cashiers tended to position the UPC parallel with one of the windows. This behavior was attributed to their mental model of how scanners read the UPC, and that some of these behaviors were encouraging undesirable wrist motions.

These results indicate that future advancements in scanner technology will not probably improve the overall level of scanning performances. Technological improvements in scanners will have minimal impact on scanning behaviors. The rate of keyed product data is increasing, and this contributes to the erosion of the benefits of scanning. UPC quality and code location appear to be a large source of the problem. UPC marking guidelines need to be revisited based on the scanner technologies, particular Bioptic scanners, that are in use today.

Designing scanning systems to promote UEMSD risk free scanning behaviors while improving productivity is a complex process. The results from this research provided a better understanding of the impact of UPC quality on scanning behavior. UPC quality can be either real or perceived, and the decision of scanning or keying is based on the cashier perceptions. Alternative solutions to controlling the quality of UPC printing and placement on packages are needed to improve the ergonomics of scanning, increase the accuracy of scanned data, and provide the error free data for ECR.

REFERENCES

U.P.C. Symbol Specification Manual, *Uniform Code Council*, January 1986.

Suggestions for the Ergonomic Improvement of Scanning Checkstands, *Food Marketing Institute Report*, September 1992.

Final Report of the FMI Ergonomics Task Force, *Food Marketing Institute*, 1996 (in-press).

Hoffman, Mark S. Challenges of applying ergonomic solutions to the retail scanning environment, *Advances in Industrial Ergonomics and Safety III*, Taylor & Francis, 1991.

Lehman, Katherine R. Balancing perception and biodynamics: designing the cashier interface on checkout scanners, *AT&T Human Factors & Behavioral Sciences Symposium '95*, Homedale N.J., 1995.

Madigan, Edwin An Examination of UPC Location and Scanning Behavior: Uniform Code Council, *AT&T Human Factors Engineering Report*, October 1995.

Marras, W.S.; Marklin, R.W.; Greenspan, G.J.; and Lehman, K.R. Quantification of Wrist Motions During Scanning, *Human Factors*, 1995, 37(2), 412-423.

Sluchak, T.J.; Cochran, D.; Grant, K. & Habes, D.; and Hoffman, M.; and Marras, W.S. Ergonomics research efforts in the supermarket industry, *Proceedings of the Human Factors Society*, Santa Monica, Ca., 1992.

Acknowledgments

Partial funding for this research was received from the Uniform Code Council.

PLACEMENT OPPORTUNITIES AND PREPARING FOR THE
HUMAN FACTORS AND ERGONOMICS MARKETPLACE

Symposium Chair: W. F. Moroney, University of Dayton

The first two presentations of this symposium present the results of an analysis of 159 placement announcements listed with the HFES Placement Service during 1994-1995. The first presentation describes the characteristics of 129 industry, government and consulting positions; while the second presentation describes the characteristics of 10 academic and 20 internship positions. The ability to perform as a team member was cited often and is the focus of the third and fourth presentations.

The third presentation discusses the importance of the team experience in today's work place. It also describes how three educators incorporated a team experience into their courses.

The last presentation of this symposium highlights various benefits and potential problems of groups and teams and focuses on strategies for successfully implementing student teams in the classroom.

Placement Opportunities for
Human Factors Engineering and Ergonomics
Professionals: Part I: Industry, Government/Military
and Consulting Positions

W. F. Moroney & C. M. Adams
University of Dayton

This portion of the symposium presents the results of an analysis of 159 placement announcements listed with the HFES Placement Service during 1994-1995. Characteristics of 129 industry, government and consulting positions are described. The features of the position descriptions examined include: degree requirements, major field of study, areas of expertise, required work experience, salary, geographic location, job description and skills required. Most industry, government and consulting positions describe the masters degree as the minimum requirement. The major fields of study most frequently specified were human factors, psychology and engineering. The most frequently cited area of expertise was human computer interaction, while the most frequently cited primary responsibility was interface design.

Placement Opportunities for Human Factors
Engineering and Ergonomics Professionals:
Part II: Academic and Internship Positions

W.F Moroney, A. Sottile, & B. Blinn
University of Dayton

This portion of the symposium presents the results of an analysis of the ten academic and twenty internship positions listed with the HFES Placement Service during 1994-1995. In the academic area, most positions were available in industrial engineering, and human factors/ergonomics. Expertise in the area of industrial ergonomics and consumer products was most often required. The most frequently cited courses-to-be-taught were human factors/ergonomics and statistics. Among internships, human-computer interaction skills were the most frequently expected. Professional skills and expectations of interns are also discussed.

Providing the Team Experience
To Human Factors and Ergonomics Students

W.F Moroney, University of Dayton
P.A. Green, University of Michigan
S. Konz, Kansas State University

The ability to work as a team member appeared repeatedly as a requirement, at all levels and across industry, government/military and consulting positions, in the placement positions analyzed by Moroney and Adams (1996), and among the internship positions analyzed by Moroney, Sottile, and Blinn, (1996). This section describes teamwork experiences provided by three educators, who represent the areas of industrial engineering, psychology and industrial and operations engineering.

Applying in the Classroom What We Know About Groups and Teams

Nancy J. Stone
Creighton University

Human factors and ergonomics specialists need to work in teams. This article highlights various benefits and potential problems of groups and teams. The main focus describes strategies for successfully implementing student teams in the classroom. Key elements include: centralizing the organization structure (with the faculty member in charge), determining group composition, defining group goals and roles, establishing an appropriate reinforcement structure (i.e. the grading system), and dealing with conflict. References for more in-depth team building guidelines are also provided.

References

Moroney, W.F., & Adams, C. (1996). Placement opportunities for human factors engineering and ergonomics professionals: Part II: Academic and internship positions. Proceedings of the Human Factors and Ergonomics Society 40th Annual Meeting. In press. Santa Monica, CA: Human Factors and Ergonomics Society.

Moroney, W.F., Sottile, A., & Blinn, B. (1996). Placement opportunities for human factors engineering and ergonomics professionals: Part II: Academic and internship positions. Proceedings of the Human Factors and Ergonomics Society 40th Annual Meeting. In press. Santa Monica, CA: Human Factors and Ergonomics Society.

Placement Opportunities for Human Factors Engineering and Ergonomics Professionals:
Part I: Industry, Government/Military and Consulting Positions

William F. Moroney & Catherine M. Adams
University of Dayton
Dayton, OH

During the period from November 1994 through October 1995, the Placement Service of the Human Factors and Ergonomics Society distributed announcements describing 159 positions available for human factors engineers and ergonomics professionals. These announcements were divided into two groups according to employment sector and position type. This paper describes industry, government and consulting positions (N=129), while its sequel describes academic and internship positions (N=30). The features of the position descriptions examined include: degree requirements, major field of study, areas of expertise, required work experience, salary, geographic location, job description and skills required. The masters degree was specified as the minimum requirement for most positions. The most frequently specified fields of study were human factors, psychology and engineering. The most frequently cited area of expertise was human computer interaction, while the most frequently cited primary responsibility was interface design.

During the period from November 1994 through October 1995, the Placement Service of the Human Factors and Ergonomics Society (HFES) received announcements describing 159 positions available for human factors and ergonomics (HF&E) professionals. This paper describes industry, government and consulting positions, while its sequel (Moroney, Sottile & Blinn, 1996) describes academic and internship positions.

In order to announce a position, employers completed a "Job Listing" form, provided by the HFES Placement Service. The employer provided information on a variety of factors including: degree requirements, major field of study, areas of expertise, required work experience, geographic location, job description, employment sector and skills required. The analysis of these data is the basis of this article. Please note that this analysis is not a complete listing of all the positions available to HF&E professionals. Related positions are also listed with other professional placement services.

Overview of the Positions

The analysis revealed that most of the positions were available in industry, consulting, and government/military (see Figure 1). Positions in industry, consulting, and

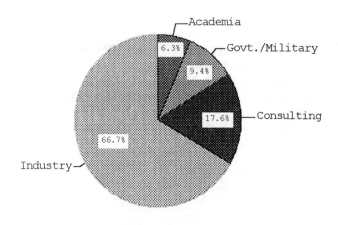

Figure 1: 159 Placement Opportunities Described by Employment Sector.

government/military comprised 93.7% of the 159 positions announced by the HFES Placement Service. Of these positions, 87.4% were full time and 12.6% were internships. The remaining positions were full-time academic positions. No requests for expert witnesses were placed through the Placement Service.

An analysis of the degree requirements revealed that most of the 159 positions specified a masters degree as the minimum requirement, with a bachelor degree running a close second (see Figure 2). However, eleven of these bachelor positions were internships. Fifty-four of the 159 full-time positions were available for individuals with a bachelors degree.

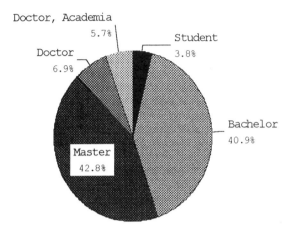

Figure 2: Minimum Degree Requirement for 159 Positions and Internships.

Additionally, it should be noted that several of the positions indicating a bachelors as a minimum degree requirement actually described an individual with a higher degree. As shown in Figure 2, twenty doctoral positions were announced; 9 in academia and 11 in industry, consulting, and government/military. Twenty internships

were also announced. The category "students" was required to include the six internships for students either currently working towards their bachelor degree or for those who have completed their bachelor degree. The remaining internships were included in Figure 2 according to the appropriate degree level.

Because of the diversity of the data, and the expected interest of the readers, the industry, consulting, and government/military positions (N=129) were examined independently of the twenty internships and ten academic positions (See Moroney, Sottile & Blinn, 1996).

Employment Sector

The 129 positions discussed in the remaining portion of this paper were categorized into three employment sectors; industry (66.6%), government/military (10.8%), or consulting (22.4%). The 86 industry positions were further classified according to type of industries (see Figure 3). Most (28.9%) of the positions were in the software industry. As might be expected, many positions were also available in the telecommunications industry. The descriptions of twelve positions were too ambiguous to classify into a specific industry and thus comprise the category "undetermined". Railroad, defense, oil/gas and ergonomic services each represented less than five percent of the total and were not shown in Figure 3. The nature of work in the consulting and government/military sectors is so amorphous that this type of analysis could not be applied to those data.

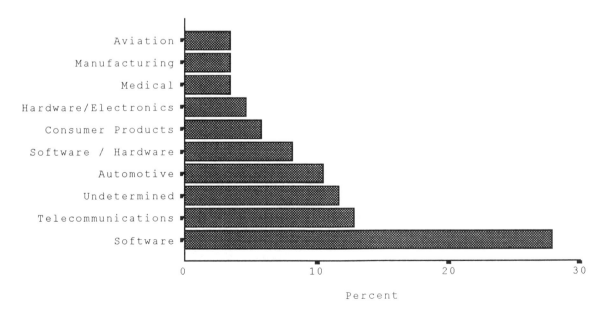

Figure 3: 86 Placement Opportunities in the Industrial Sector By Type of Industry.

Minimum Degree and Minimum Years Experience

With respect to the minimum degree requirements, master and bachelor degrees were requested for 91.5% of these positions, with the master degree being specified for almost half of the 129 positions. Table 1 provides a more detailed description, broken into type of degree by employment sector. The consulting sector, with 29 positions, requires the most experience at the doctoral level (range 3-7 years). Industry had one doctoral position available for an ergonomist, while the four doctoral positions available in the government/military sector were for "fresh-outs" with some experience (range 0-4 years).

Salary & Geographical Location

Ninety percent of the employers described the salary range as negotiable. The salaries specified ranged from $26,000 to $90,000. The median salary range was $43,000 to $55,000. Readers interested in salary are best advised to consult the salary survey conducted by Sanders (1993).

The highest concentration of positions were available in the Northeast (N=21). Thirty-two of the positions were evenly divided between California and the East Central region. Only three positions were located in New England and four in the Midwest. Of the international positions, five were in Canada, one was in Venezuela, and one was in the Netherlands. The remaining positions were evenly distributed throughout other regions in the United States .

Table 1. Minimum years experience needed for employment sector and degree.

SECTOR	DEGREE REQUESTED Number of Positions	MINIMUM YRS. OF EXPERIENCE Median
Industry N=86	Bachelor (38) Master (47) Doctor (1)	2.0 2.0 0.0
Consulting N=29	Bachelor (12) Master (11) Doctor (6)	1.0 1.0 4.0
Govt./Mil. N=14	Bachelor (4) Master (6) Doctor (4)	3.0 3.5 1.0

Areas of Expertise & Responsibility

Employers were allowed to specify up to six areas of expertise needed for the position. Because these areas of expertise were not prioritize, it was impossible to assess the primary needs of the employer. Figure 4 specifies the number of positions requesting a particular expertise. For example "Aerospace" was requested in 13 of the 129 positions. Areas of expertise with less than 6 entries were not reported. These included aging/disabled, individual differences, environmental design and MANPRINT.

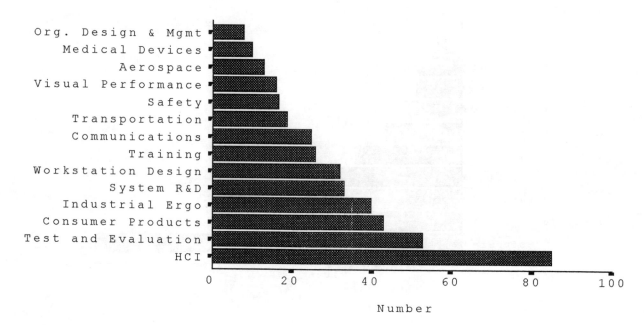

Figure 4: Areas of Expertise Requested for 129 Positions in Industry, Govt./Mil.and Consulting

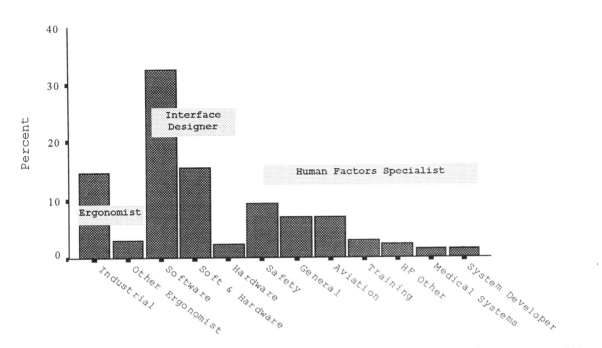

Figure 5: Primary Area of Responsibility for 129 Positions in Industry, Govt./Mil. and Consulting.

In many cases, the position title was not specified, or general terms such as HFE, ergonomist or manager were used. In addition, terms as different as usability designer, usability specialist, user interface designer, and application architect were used to describe positions which required an individual who could develop graphical user interfaces (GUIs). The authors reviewed the announcements, particularly the narratives, and classified each position into a category which they felt reflected the primary area of responsibility for that position (see Figure 5). The greatest demand was for interface designers of which there are three categories: interface designers with GUI, HCI, or graphics expertise (N=42), interface designers for both software and hardware (N=20) and hardware interface designers (N=3). The authors used the term "hardware" to reflect electronics, computer hardware and other consumer products. The classification "Other Ergonomist" includes: two positions in fabrication ergonomics, a position as a rehabilitation ergonomist and a position in the area of accident prevention. The classification " HF Other" includes single positions for human factors specialists in the areas of aging/disabled, communications, and decision support.

Other Skill Requirements

Many basic "tools" are expected to be in the HF&E professional's toolbox. The tools most often cited in these announcements include: a working knowledge of usability techniques for that domain, proficiency with rapid prototyping tools, the ability to conduct user need assessments, basic research skills, and a working knowledge of testing and evaluation techniques. Within the software industry, many employers required computer programming skills. Several skills which transcend particular task domains were frequently cited in the position announcements and include: computer literacy, professional skills such oral and written communication and the ability to work with others in a team setting. Work experience in the employer's domain was also desired. These expectancies can be met, at least in part, through an internship, a cooperative position or some other form of practical experience while in school. Industry's demand for these skills and the themes of effective team work, flexibility, interdisciplinary skills, and practical experience are strongly reinforced in the works of Shapiro, et al. (1995) and Shapiro (1995, 1994).

Additional skill requirements were identified and are discussed in Moroney, Green and Konz (1996). They also describe strategies for incorporating a team experience into the curriculum.

CONCLUSION

The authors hope that they have provided a useful analysis of the placement opportunities available to HF&E professionals seeking positions in industry, government/military and consulting. A similar analysis performed annually should be of use to both HFES members and educators.

References

Moroney, W. F., Green, P. A, & Konz, S. (1996). Providing the team experience to human factors and ergonomics students. Proceedings of the Human Factors and Ergonomics Society 40th Annual Meeting. In press. Santa Monica, CA: Human Factors and Ergonomics Society.

Moroney, W. F., Sottile, A. & Blinn, B. (1996). Placement opportunities for human factors engineering and ergonomics professionals: Part II: Academic and internship positions. Proceedings of the Human Factors and Ergonomics Society 40th Annual Meeting. In press. Santa Monica, CA: Human Factors and Ergonomics Society.

Sanders, M. S. (1993). 1993 Salary survey. Human Factors and Ergonomics Society Bulletin, 32, (11), 1-3.

Shapiro, R. G. (1994). What is it like working in industry? Poster presented at the 38th Annual Meeting of the Human Factors and Ergonomics Society.

Shapiro, R. G. (1995, October). How can human factors education meet industry needs? Ergonomics in Design, p 32.

Shapiro, R. G., Brown, M. L., Fogleman, M., Goldberg, J. H., Granda, R. E., Hale, J. P., Sanders, E. B-N. (1995). Preparing for the human factors/ergonomics job market. Proceedings of the Human Factors and Ergonomics Society 39th Annual Meeting. (pp 379-389). Santa Monica, CA: Human Factors and Ergonomics Society.

Placement Opportunities for Human Factors Engineering and Ergonomics Professionals: Part II: Academic and Internship Positions

W. F Moroney, A. Sottile, & B. Blinn
University of Dayton

This paper describes placement opportunities for HFE and ergonomics professionals in academic and internship positions, which were contained in the position announcements distributed by the Human Factors and Ergonomics Society (HFES) Placement Service during 1994-1995. Ten academic and twenty internship positions were announced. The features of the position announcement examined include: degree requirements, major field of study, areas of expertise, required work experience, salary, geographic location, job description and skills required. Academic positions were most frequently available in industrial engineering, and human factors ergonomics. Expertise in the area of industrial ergonomics and consumer products was most frequently desired. The most frequently cited courses to be taught were human factors/ergonomics and statistics. Among internships, knowledge of human-computer interaction was most frequently cited. Professional skills and expectations of interns are also discussed.

During the period from Nov. 1994 through Oct. 1995, the Placement Service of the Human Factors and Ergonomics Society (HFES) received announcements describing ten academic positions and twenty internships available for human factors and ergonomic (HF&E) professionals. Employers completed a "Job Listing" form, provided by the HFES Placement Service, on which they provided information on a variety of factors including: degree requirements, major field of study, areas of expertise, required work experience, geographic location, job description, employment sector and skills required. This paper describes those academic and internship positions while Part I (Moroney & Adams, 1996) described industry, government and consulting positions.

It should be noted that the positions examined do not represent all of the academic and internship opportunities available to HFE and ergonomic professionals. Related academic and internship positions are listed in the American Psychological Association's Monitor and the American Psychological Society's Observer, among others. While academic positions are usually announced across broad geographic areas (i.e. the entire U.S.), internships are often communicated informally by industry, or by announcements to selected

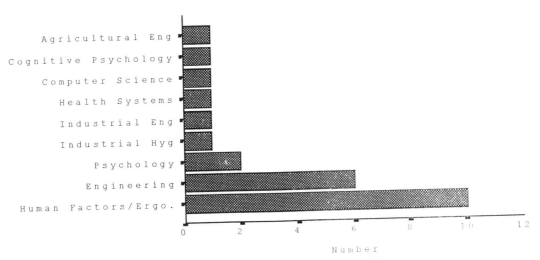

Figure 1: Major Fields of Study Specified for Academic Positions

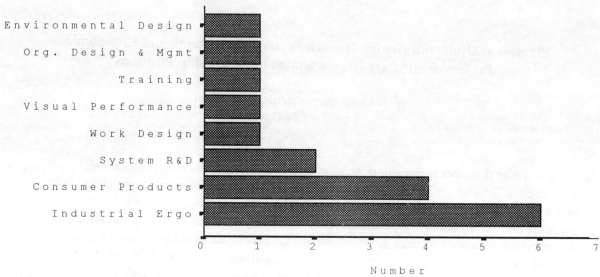

Figure 2: Areas of Expertise Expected Of Academics

universities with which the employer had good relations previously and/or from which employees have graduated. Furthermore, internship opportunities may be announced within limited geographical areas, since the granting agency may not provide travel expenses and may prefer to meet its needs locally.

ACADEMIC POSITIONS

Nine of the ten academic positions required a doctorate, while the one exception required a masters in the area of industrial or agricultural engineering or industrial hygiene. Data describing the major field of study specified by potential employers are contained in Figure 1 (employers could specify up to three major fields of study per position).

All positions were described as full time and eight of the ten positions were described as tenure track. Five positions were at the assistant professor level, one specified that all ranks were available, and two stated that while an assistant professor position was available, higher levels were possible for qualified personnel. The remaining two did not specify a professorial level. Figure 2 contains the number of academic positions requesting a particular expertise. Employers were allowed to specify up to six areas of expertise. Since these areas were not prioritized, it was impossible to assess the primary needs of the employers. Industrial ergonomics was cited most frequently, followed by consumer products. With respect to courses to be taught (see Figure 3), human

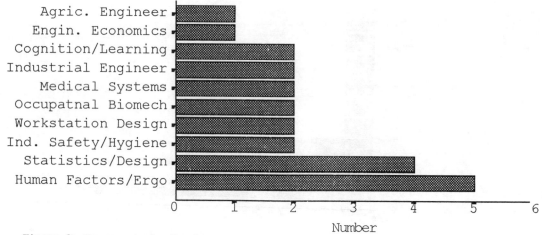

Figure 3: Courses to be Taught.

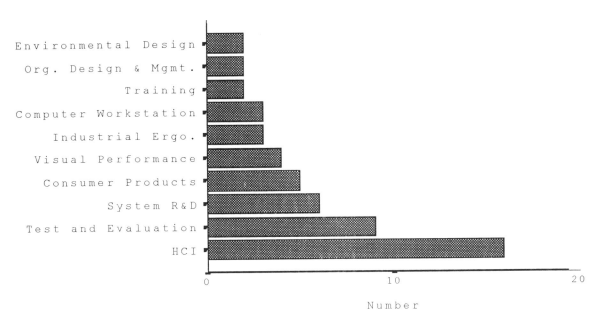

Figure 4: Areas of Expertise Requested of Interns.

factors/ergonomics was most frequently cited as the course to be taught, with statistics/experimental design being the second most frequently cited.

Seven of the nine doctoral positions did not specify an experience requirement, whereas, the remaining two specified: one year of teaching and research experience; and two years for work in the health systems area, respectively. The masters position preferred one year of experience in industrial hygiene or agricultural engineering .

Salary was described as negotiable in all cases, with two listings specifying ranges from $45-61,000 and $45-48,000.

Three positions were available in the both the East-Central (IN,KY,MI,OH) and Mid-Central (AR,KS,MO,NE,OK) areas of the United States. The remainder were distributed at other locations.

INTERNSHIP POSITIONS

Twenty internship positions were announced. The minimum degree requirements for these positions were distributed among three categories: Bachelors (N=9), Masters (N=5) and degree "Not Specified" (N=6). Most positions required a minimum of a bachelors (i.e. these students would be working on their masters or doctoral degrees).

Human Factors/Ergonomics was cited most frequently (18 times) as the major field of study, while Psychology was the second most frequently cited (15 times). Computer Science and Engineering were specified nine and eight times respectively.

Figure 4 contains data describing the desired areas of expertise. Areas of expertise with less than two entries are not reported. As was the case for positions in industry, government/ military and consulting positions (Moroney & Adams, 1996), the area of expertise most frequently requested was Human Computer Interfaces, followed by Test and Evaluation (usually in the area of computer systems). An examination of the narrative portion revealed, not surprisingly, that all positions are related to the computer industry.

Most internships were offered for an academic quarter, a semester, or a summer; although a few were offered for six months to a year. Salary was not specified for 17 of the 20 positions. The three positions for which salary was specified offered: $2475-3070/month for 3 months, $9-14/hr and $12-15/hr. Zero experience was specified for all but two positions which desired 0.5 years experience in computer science, engineering or human factors. Most (N=6) of the positions were located in the Southeast, followed by the Northeast (N=4) and Southwest (N=4). One announcement offered several positions throughout the U.S. to Ph.D students (minimum degree requirement was masters).

The computer skills desired of interns were culled from the placement forms. They included knowledge of operating systems, applications (word processors, statistical packages) and programming (e.g. C++, Supercard, Visual Basic). A listing of professional skills and expectations desired of interns is provided in Table 1. Most of the observations regarding desired skills reported by Moroney & Adams (1996) also apply to the internship positions. Addition details are provided by Shapiro (1995, 1994), and Shapiro, et al.(1995).

The requirement to work as a team member was common to both the internship positions and industry, consulting, and government/military positions, but was not listed as a requirement for academia. This highlights the difference between academia and the workplace: individualism vs. teamwork. Despite this contrast, educators have made efforts to provide their students with opportunities for team work. The sections which follow will describe strategies used by three educators (Moroney, Green & Konz, 1996) and provide reference material (Stone, 1996) for use by interested educators.

CONCLUSION

The authors hope that they have provided a useful analysis of the placement opportunities available to HF&E personnel seeking academic and internship positions. A similar analysis performed annually should be of use to both members and educators.

References

Moroney, W. F. & Adams, C. M. (1996). Placement Opportunities for Human Factors Engineering and Ergonomics Professionals: Part I: Industry, Government/Military and Consulting Positions. Proceedings of the Human Factors and Ergonomics Society 40th Annual Meeting. In press. Santa Monica, CA: Human Factors and Ergonomics Society.

Moroney, W. F., Green, P. A, & Konz, S. (1996). Providing the Team Experience to Human Factors And Ergonomics Students. Proceedings of the Human Factors and Ergonomics Society 40th Annual Meeting. In press. Santa Monica, CA: Human Factors and Ergonomics Society.

Shapiro, R. G. (1994). What is it like working in industry ? Poster presented at the 38th Annual Meeting of the Human Factors and Ergonomics Society.

Shapiro, R. G. (1995, October). How can human factors education meet industry needs. Ergonomics in Design, p 32.

Shapiro, R. G., Brown, M. L., Fogleman, M., Goldberg, J. H., Granda, R. E., Hale, J. P., Sanders,E.B-N. (1995). Preparing for the human factors/ergonomics job market. Proceedings of the Human Factors and Ergonomics Society 39th Annual Meeting. (pp 379-389). Santa Monica, CA: Human Factors and Ergonomics Society.

Stone, N. J. (1996). Applying What We Know About Teams In the Classroom. Proceedings of the Human Factors and Ergonomics Society 40th Annual Meeting. In press. Santa Monica, CA: Human Factors and Ergonomics Society.

Table 1. Professional Skills and Expectations Desired of Interns.

Professional Skills

- Analytical
- Communication
- Problem solving
- Visual presentation
- Writing

Expectations

- Broad background
- Maturity
- Self-motivated
- Show potential for innovation
- Work independently
- Work with a team

Providing the Team Experience to
Human Factors and Ergonomics Students

W.F. Moroney, University of Dayton
P.A. Green, University of Michigan
S. Konz, Kansas State University

The importance of the team experience to human factors and ergonomics students is discussed. Three faculty members describe different approaches, which they used to incorporate the team experience into their courses.

Van Cott and Huey (1992) provide a detailed examination of the education and utilization of human factors specialists in a report to the National Research Council. They reported frequent interaction between both human factors and ergonomics (HF&E) personnel and specialists from other areas including: computer programming, engineering, marketing, systems analysis and system users. Much of this interaction occurs as team members.

This requirement was repeated in *Science* by Kaiser (1995) "But today in industry, Ph.D.s need to be ready to learn new disciplines, work in teams, explain their research to non-specialists, and understand the business impact of their work (p 133)." While Kaiser's comments are focused specifically on the Ph.D., the emphasis extends to HF&E professionals at all academic levels. Borchardt (1996) also emphasizes the importance of a team-orientation in engineering. Shapiro (1994), speaking specifically to the HF&E community, is more direct when he states simply "teamwork is survival". Additional documentation regarding industry's need for HF&E personnel as team players is provided in Shapiro (1995) and Shapiro et al. (1995).

The ability to work as a team member appeared repeatedly as a requirement, at all levels and across industry, government/ military and consulting positions, in the placement positions analyzed by Moroney and Adams (1996), and among the internship positions analyzed by Moroney, Sottile, and Blinn, (1996). The sections which follow describe teamwork experiences provided by three educators, who represent the areas of industrial engineering, psychology and industrial and

operations engineering. The final paper in this symposium (Stone, 1996) reviews the literature on team dynamics and team building and provides references for use by interested educators.

The Teamwork Experience in an IE Program
By Stephan Konz, Kansas State University

The Industrial Engineering Advisory Council at Kansas State University has emphasized the importance of teamwork to the faculty for many years. I have taught a required "senior design project" course (Facility Design) for approximately 25 years. In the "lab" (50% of course grade), students completely designed a factory, starting with a product, selection of machines (including feeds and speeds), material handling, utilities (lighting, climate, noise control), energy, waste, quality policies, labor policies, etc.

There were 4 to 5 students/team. Each report took 10-20 h so there were 15 h x 13 weekly projects = 200 h; in the semester, the lab took about 50 h/student. With each report, the students submitted a peer rating. They did not rate themselves. The procedure was as follows:
* Each team member received a "grade multiplier".
* The average of the multipliers had to equal 1. That is, if one team member received a 0.8, then the others might receive 1.1, 1.05 and 1.05.
* The instructor then averaged the multipliers of the team members.
* No rating was submitted if everyone was graded as "1"; this was to encourage using "1"s and discourage using the "club" (i.e. ratings less than 1).

The instructor graded the report (giving say 45 out of 50), and multiplied that grade by the average rating to get the final grade for that student's assignment.

In a typical year, there would be 13 reports turned in by 6 or 7 teams--80-100/year. Generally only 1-2% would not give a 1 for everyone. However the students used the peer rating with non-ones as a threat "if you don't shape up." When "non-ones" ratings were given, they often severely penalized the "goof off"--cutting the grade to say 50%. Usually this was sufficient warning and it occurred only once in a team. However, twice in the last 10 years, we had to have a group meeting, including the instructor, to "talk it out". The group then resumed giving "ones" but there were hard feelings in the group about the slackers.

Our students knew each other well from previous classes. The groups were formed in two different ways. In some years, the instructor let students pick their own groups at the start of the term. This generally resulted in one or two groups of "leftovers". It was quite a shock for these students during the semester not to have someone else do their work; these groups are those which had the rating conflicts.

The other approach was to have the instructor pick the members of each group. This makes the teams more equal and protects the "slackers" from the "cruel world" as the good people in the group would carry them. My recommendation is to let students pick their own groups.

The Teamwork Experience in a Psychology Program
by W. F. Moroney, University of Dayton

Students participate in a team building exercise as part of the Human Factors in Systems Development graduate level course at the University of Dayton. Usually, eighty percent of the students are psychology majors, while the remainder come from areas such as engineering, computer systems and mathematics. The course is built around redesigning a Terminal Radar Approach Control (TRACON) station in response to a simulated Request for Proposal (Moroney & Cameron, 1995). The course, which makes heavy use of simulation, emphasizes both the process involved in product development and the final products.

Early in the course, the students are introduced to industry's emphasis on teamwork by the use of some of the material presented in the introduction to this article. In addition, they are exposed to "lessons learned" and problems encountered by students who had previously completed the course. Finally, the various roles (coordinator, procedure developer, critic, etc) that team members can play are briefly discussed. Wilson and Hanna (1986), and Stone (1996) provide useful material in this area.

Students are required to: determine their group structure, assign responsibilities, specify deliverables, develop schedules, etc. The only constraint is that they meet the deadlines specified in the simulated RFP. Since, performance evaluation is the center of human factors, students are also required to specify how they want the group performance to be evaluated. This provides an interesting learning experience. In addition, students are required to determine how to allocate bonus points to the team member(s) who contributed most to the final product. This is accomplished by requiring the students to allocate a predetermined number of points among their team members (they cannot allocate any points to themselves). For example, 100 points could be distributed among 10 team members. Thus, a top performer very well might receive 15 points, while a poor performer might receive 5 points. These points are collected by the faculty member after the completion of the design exercise and the predetermined bonus is awarded to the team member(s) with the highest number of points. To facilitate the process of point allocation, students are provided with an evaluation form used to evaluate the performance of Civil Service personnel. The form utilizes a five point scale to evaluate individuals on 14 characteristics including: initiative, dependability, accuracy of work, acceptance of criticism, originality, etc. The group completing the course this year elected to exchange the evaluation forms among the team members midway during product development, so that individuals could be aware of how their performance was perceived by the group and make adjustments as required. This evaluation process is internal to the group and the faculty member does not get involved. Unfortunately, time constraints precluded the mid-term review this year, but the end-of-project evaluation forms were distributed to the team members at the end of the course. Thus the students were informed of the number of points that they received and provided with a diagnostic rating of their performance.

Requiring the students to determine how they will be evaluated and to evaluate the performance of their team members, personalizes the evaluative process. It also allows them to experience, in a relatively benign environment, the difficulties associated with evaluating and being evaluated. While, the team project and evaluation process are well received, they do require considerable explanation and monitoring by the faculty member.

Using Outward Bound-Like Activities to Teach the Teamwork Experience
by Paul Green, University of Michigan

I have taught a one-credit introductory laboratory course (4 sections, enrollment cap of 16) at the University of Michigan (Industrial and Operations Engineering (IOE), Ergonomics Laboratory) for the last 16 years. All sections are typically full. Undergraduate Industrial and Operations Engineering students are required to complete the laboratory and an associated three-credit lecture class.

The laboratory course teaches students how to use equipment available to ergonomists, teaches them how engineers approach problems, enhances the quality of their technical writing, provides applications of the lecture material, and fosters working in groups. The course involves six major experiments covering (1) the measurement of sign lighting and legibility, (2) response time to speedometers, (3) industrial workplace sound measurement and audiometry, (4) work physiology (oxygen consumption and heart rate for a manual materials handling task), (5) anthropometric measurements of classmates, and (6) a survey of symbols. With the exception of the lighting experiment, each group of three to four students submits a report summarizing each exercise. Reports must be written in a format consistent with scientific journals (e.g., Human Factors). Students are provided with a short guide (Green ,1991) that includes a sample report. The most unique aspect of the course is its use of the University of Michigan Challenge Program ("Ropes Course") to foster teamwork.

The Ropes Program, taken by the class as a group, is an "Outward Bound-like" activity conducted at the University's outdoor recreational facility (in the winter, a gym is used). The initial exercise is a game in which students learn the name of every other classmate. Prior to this course, students have only been enrolled in large lectures, so opportunities to meet classmates have been limited. After the name game, a trust exercise is typically scheduled.

Subsequently, students work together to solve a series of engineering problems. Challenges include (a) getting the class across an imaginary chasm using timbers that can be placed on piers, (b) passing classmates through an extended badminton net with large holes, such that no two people go through the same hole, (c) getting the entire class on and off a teeter-totter while it remains level, and so forth. (See Rohnke, 1984, 1989.) To foster bonding with the class, the instructor and the graduate teaching assistant participate, but as silent followers. The final exercise involves getting the entire class over a 14-foot wall without ladders or ropes, a task that seems impossible to complete. The solution usually requires the class to form a human chain. During the exercises, students realize that they all fail if one person fails, and each individual must be included in the planning and execution of every exercise. Each exercise is followed by a briefing given by a trained facilitator in which students are asked questions to provoke discussion of the group problem solving process. (Was this activity a success? Can you point out something someone did that was a positive achievement? What would you do differently next time?)

Initially, students have great misgivings about participating in this exercise. Concerns include the time (Sunday afternoon), potential physical demands, and the unconventional nature of the program. However, when students leave this five-hour program, they know every other member of the class (and trust them). They have a sense of each person's strengths and weaknesses, and they understand the give and take required for successful group work. Student reaction is invariably very positive. In end-of-the-semester evaluations, students have commented that the afternoon spent at the Ropes Program was more beneficial to them than most of their three-credit classes. It is not unusual for lab teams to schedule all of their classes together the following semester because of the camaraderie. In addition, students report that they find the University a more welcome place, as they now know at least four or five other students in any Industrial Engineering class they might elect.

Conclusion

Incorporating a teamwork learning experience into an already crowded syllabus is difficult but rewarding for both students and faculty. Hopefully, some of the approaches described above will encourage other educators to incorporate similar experiences into their syllabi.

References

Borchardt,J.K. (1996). Navigating the new workplace. Graduating Engineer,17 (3), 22-26.

Green, P. (1991). Specifications for Human Factors Reports (Superman Solves a Puzzle), Ann Arbor, MI: The University of Michigan, Department of Industrial & Operations Engineering and the Transportation Research Institute.

Kaiser. J. (1995, October). A business blueprint: How to build a better Ph.D. Science, 270, 133-134.

Moroney, W.F.,& Adams, C. (1996). Placement opportunities for human factors engineering and ergonomics professionals: Part II: Academic and internship positions. Proceedings of the Human Factors and Ergonomics Society 40th Annual Meeting. In press. Santa Monica, CA: Human Factors and Ergonomics Society.

Moroney, W.F., Sottile, A., & Blinn, B. (1996). Placement opportunities for human factors engineering and ergonomics professionals: Part II: Academic and internship positions. Proceedings of the Human Factors and Ergonomics Society 40th Annual Meeting. In press. Santa Monica, CA: Human Factors and Ergonomics Society.

Moroney, W. F., & Cameron, J.A. (1995). Using simulations in teaching human factors: Bridging the gap between the academic world and the world of work. Proceedings of the Human Factors and Ergonomics Society 39th Annual Meeting. (pp 399-403). Santa Monica, CA: Human Factors and Ergonomics Society.

Rohnke, K. (1984). Silver bullets, Dubuque, IA: Kendall/Hunt.

Rohnke, K. (1989). Cowstails and cobras II, Dubuque, IA: Kendall/Hunt.

Shapiro, R.G. (1994). What is it like working in industry ? Poster presented at the 38th Annual Meeting of the Human Factors and Ergonomics Society.

Shapiro, R.G. (1995, October). How can human factors education meet industry needs. Ergonomics in Design, p 32.

Shapiro, R.G., Brown, M.L., Fogleman,M., Goldberg, J.H., Granda, R.E., Hale, J.P., Sanders,E.B-N. (1995). Preparing for the human factors/ergonomics job market. Proceedings of the Human Factors and Ergonomics Society 39th Annual Meeting. (pp 379-389). Santa Monica, CA: Human Factors and Ergonomics Society.

Stone, N. J. (1996). Applying in the classroom what we know about teams. Proceedings of the Human Factors and Ergonomics Society 40th Annual Meeting. In press. Santa Monica, CA: Human Factors and Ergonomics Society.

Van Cott, H.P. & Huey, B.M. (1992). Human factors specialists' education and utilization: Results of a study. National Academy Press, Washington, D.C.

Wilson, G.L., & Hanna, M.S. (1986). Groups in context. New York, N.Y.: Random House.

Applying in the Classroom What We Know About Groups and Teams

Nancy J. Stone
Creighton University
Omaha, NE

Human factors and ergonomics specialists need to work in teams. This article highlights various benefits and potential problems of groups and teams. The main focus is on ways to successfully implement student teams in the classroom. Key elements include: centralizing the organization structure with the faculty member in charge, determining group composition, defining group goals and roles, establishing an appropriate reinforcement structure (i.e, the grading system), and dealing with conflict. References for more in-depth team building guidelines are provided.

The study, development, and use of teams is widespread in industries such as meat packing (May & Schwoerer, 1994a), manufacturing (Reiste & Hubrich, 1995, 1996) as well as in military aircrews (e.g., Beard, Salas, Prince, 1995), health care (Boss, 1991), and education (Rankin & Aksamit, 1994; Thomas, 1991). Unfortunately, the literature on teams or groups in education focuses mostly on teams of teachers, administrators, or support personnel. Nevertheless, the importance of teams is so great that Arizona State University is redesigning its engineering program to reflect more teamwork (Bellamy, Evans, Linder, McNeil, & Raupp (1994). Human factors specialists are not exempt from working in groups and/or teams. These professionals need to be able to work in teams (Kaiser, 1995; Shapiro, Beith, Goldberg, Hale, & Kelly, 1994), specifically when teams are composed of professionals with diverse specialties (Van Cott & Huey, 1992). Because teams are so prevalent, it is important that students be exposed to the group or team process. Hence, the benefits and potential problems of using groups or teams in the classroom and ways to successfully implement teams are discussed.

Benefits of Teams

Benefits of teams may include a division of labor, an ability to offer checks and balances to decisions, process gain with methods such as brainstorming, social support, and greater satisfaction. Work teams will have members who have been selected (Guion, 1983) and possibly trained (Hinrichs, 1983) to meet a certain level of competence. Students, on the other hand, are not members of work teams. Although students begin with a certain level of competence, as class participants they expect to increase their competence. Unfortunately, increasing their level of competence is most commonly described as learning the course content as opposed to learning process skills (in this case how to be a team member). Student teams have the potential of increasing student learning, though, especially that of the lower performing students. According to Laughlin and colleagues (Laughlin & Johnson, 1966; Laughlin, Branch, & Johnson, 1969; Laughlin & Branch, 1972) lower ability students learn better when working with a high ability member. That is, groups perform better when at least one of the members is of high ability relative to the task. Higher ability members introduce novel information to the group/team discussion (Laughlin & Johnson, 1966; Laughlin, Branch, & Johnson, 1969; Laughlin & Branch, 1972). Therefore, the composition of the groups will influence how well they perform. Groups composed of all high ability members performed better than the sum of their abilities, but groups composed of all low ability members performed worse than the sum of their abilities (Tziner & Eden, 1985).

Potential Problems within Teams

Although teams may be beneficial, numerous potential problems may arise in teams. First, individuals have different motivations for group membership. Some individuals may be more motivated to promote themselves, others promote the social aspects of the group, whereas others are concerned with completing the task (Bass, 1962; Carter, 1954). Members may also be defined as competitors, individualists, cooperators, or equalizers (Greenberg & Baron, 1995). The majority of people tend to be competitors, motivated to do better than others. Individualists attempt to better their own performance and are unconcerned with the performance of others. Cooperators tend to be few in number, but attempt to maximize the group outcome. Finally, equalizers wish to minimize any difference between group members' outcomes so everyone receives about the same result. Ideally, members should be task motivated, which does not exclude individuals from being socially or personally motivated, and cooperators. Cooperators are more helpful than individually or competitively oriented individuals (McClintock & Allison, 1989). Similarly, collectively oriented individuals gain more from team interaction than individually oriented members (Driskell & Salas, 1992).

Besides motivation problems, groups and teams often experience process loss or social loafing (Latane, Williams, & Harkins, 1979). Fortunately, social loafing may be reduced by increasing identifiability of members (Williams, Harkins, & Latane, 1981) or task difficulty, whereby members perceive their contributions not to be redundant, but unique (Harkins & Petty, 1982). That is, when individuals believe their input is essential for completion of a task, they are more likely to participate.

Effective teamwork may also be blocked by communication trouble (Standley, 1993). Newly formed groups do not have a developed interaction pattern (Watson, Michaelson, & Sharp, 1991). Kanki and Foushee (1989) found that flight crews which had flown together, though, used more task related communication such as greater information exchange and validation, statements of intent, and acknowledgements, while crews which had not flown together before used more non-task communication in order to get to know each other. Socialization appears to be an important aspect to group development (Bolman & Deal, 1992).

Motivation, social loafing, and communication not only independently affect teams, but indirectly affect teams by possibly creating conflict. Competition among members (Standley, 1993), interpersonal problems, and ambiguity about one's role and responsibility also contribute to conflict (Greenberg & Baron, 1995). If managed effectively, conflict can lead to a better understanding of the problem and individuals' positions, the development of new ideas, improved decision making, and increased cohesion (Greenberg & Baron, 1995).

Implementing Groups and Teams in Educational Settings

Because of these potential problems, some students may prefer to work alone. Yet, the purpose of using teams in courses is to give the students exposure to the group process. Even though learning the course content is important, learning about team building is also important (LaBar, 1995). Therefore, student teams may be used in education in two ways. First, to help students gain a better understanding of the content area. Second, to help students learn about the group/team process. Based on what is known about groups and teams, the following suggestions for successful implementation of student teams in education are provided.

Discuss Process and Provide Guidance. Students placed into groups to discuss certain concepts or issues will often talk about weekend activities. This may be due to inappropriate guidance. These student groups may be considered leaderless discussion groups or self-managed teams. Self-managed teams without initial leadership often flounder (Laiken, 1994). According to Laiken (1994), self-managed teams should be considered an end point to which teams are developed, not a starting point. Initially, consider the class to be an organization of flat structure (faculty member vs. student teams) that is highly centralized (i.e., the faculty member is in charge). This is not the time to advocate the sink-or-swim style of management. In an educational setting students are not yet proficient at the task and are there to learn how to work as a team. Over time, the class should become more decentralized whereby the groups' leaders are in charge, having developed competence to lead the team. That is, the teams should become less dependent on the faculty member and function more independently.

Create Appropriate Team Composition.
Due to the potential problems within teams, some students may not learn as much as possible of the course content nor gain an understanding about group process. Therefore, one of the first tasks is to consider the composition of the teams. Because groups composed of members with homogenous skills are not desirable (Orpen, 1986), if possible, the faculty member should create teams of members with complementary skills. Remember, though, that the selection and training process of these students in not the same as in industry. That is, students are not usually selected for a class because they have the necessary skills to perform the class tasks. So, their competencies may be more variable than work groups.

Define Goals and Roles.
After the teams have been created, these leaderless discussion groups will tend to go through Tuckman's stages of development: forming, storming, norming, performing, and adjourning (Maples, 1988). To help these student teams through the first three stages so they can begin performing, it is necessary to reduce ambiguity by defining the group goal(s) (Scearce, 1992; Standley, 1993) and member roles for each team. Involving the students in this process will increase their commitment to the goals and roles. It may also be necessary to develop team efficacy using verbal encouragement, successful job experiences, and modeling (May & Schwoerer, 1994b). Relatedly, collective efficacy, which is suspected to influence group performance, is hypothesized to be influenced by members' relevant knowledge, skills, and ability (KSAs; Mischel, 1996). Therefore, it is important to provide guidance in the development of these KSAs throughout the term.

Implement Appropriate Reinforcement Structure.
Another means to reduce ambiguity is an appropriate reinforcement structure or reward system. That is, students should be reinforced for learning the course content as well as contributing to the team. In an educational setting, the reinforcement structure tends to revolve around the grading system. Yet, if grades are based on the final project, then those who social loafed or were in constant conflict with other team members may be reinforced for those behaviors by receiving a good grade if the project receives a good grade. It is important to determine how the team members will be evaluated prior to incorporating teams into the syllabus. Informing students at the beginning of the term about how they will be assessed will reduce student anxiety and "protect" the faculty. This discussion should occur on the first day of class as well as each time team assignments are distributed. Johnson and Johnson (1994) briefly address the ethics of experiential learning.

Although having students work in teams is an attempt to prepare students for the real world, the educational setting is not the real world, there are some differences. It is the place to train the students, to allow them to err, without experiencing the severe consequences some encounter in the real world. For example, if a competition is used to motivate students to create the best product, only one can win. If grades are based on placement (first is an A, second is a B, etc.), then there is the assumption that these placements (i.e., rankings) accurately reflect quality. Yet, the last placed product may still represent "A" work, or the first placed team's product may not be of "A" quality. A competition which leads to final rankings of the products can be a great motivator, and placement may award different number of points to the team members, but grades should be based on a more objective criterion or set of criteria by which to judge the quality of the team product.

Additionally, the group product should not be the only measure. There should be some individual measures of performance, which increase identifiability and reduce the likelihood of social loafing. Tests are a good means to assess content knowledge, and peer evaluations are a good means to tap teamwork skills. Understanding that peer reviews have their own flaws, it is also one way to tap the actual process within the group. In order to avoid the mis-use of peer evaluations, it is important to have continuous peer evaluations of the team process. This allows the members to evaluate the feedback and make changes.

Develop Conflict Resolution Skills. Finally, it is important to help students understand how to deal with conflict constructively. First, do not avoid the conflict, because conflict unattended may lead to more problems. Second, members must be objective and present issues in behavioral terms. That is, do not get personal. For example, express one's concern that information is received late not that the member is slow, lazy, or unorganized.

Conclusion

Given that students will likely become a part of a work team, it is essential that students learn the work team process. It is possible to successfully implement student groups or teams in educational settings. To be successful, it is important to give guidance to these student teams, gradually making the teams less dependent on the faculty member as the teams mature. Additionally, potential problems of groups should be discussed. Although some faculty may resist this style of teaching due to the feeling of a loss of control over the class, the faculty member actually establishes control during the initial guidance stage. Therefore, the faculty member does not relinquish complete control, but rather trains the students to be self-managed teams, a skill greatly need for work.

For more in-depth information on team building, see Scearce (1992). For details on specifically preparing students for group work, see Farivar & Webb (1994) or Standley (1993) . Also, a text (Mears & Voehl, 1994a) and instructor's manual (Mears & Voehl, 1994b) provide useful information on team building. McHale (1995) provides a review of various team building videos. Other good books on group process have been written by Johnson and Johnson (1994), Napier and Gershenfel (1993), and McGrath (1984).

References

Bass, B. M. (1962). The orientation inventory. Palo Alto, CA: Consulting Psychologist Press.

Beard, R. L., Salas, E. & Prince, C. (1995). Enhancing transfer of training: Using role-play to foster teamwork in the cockpit. International Journal of Aviation Psychology, 5, 131-143.

Bellamy, L., Evans, D. L., Linder, D. E., McNeil, B. W., & Raupp, G. (1994, March). Teams in Engineering Education. Arizona State University, College of Engineering and Applied Sciences.

Bolman, L. G. & Deal, T. E. (1992). What makes a team work? Organizational Dynamics, 21, 34-44.

Boss, R. W. (1991). Team building in health care. Journal of Management Development, 10, 38-44.

Carter, L. F. (1954). Evaluating the performance of individuals as members of small groups. Personnel Psychology, 7, 477-484.

Driskell, J. E. & Salas, E. (1992). Collective behavior and team performance. Human Factors, 34, 277-288.

Farivar, S. H. & Webb, N. M. (1994). Are your students prepared for group work? Middle School Journal, 25(3), 51-54.

Greenberg, J. & Baron, R. A. (1995). Behavior in organizations: Understanding and managing the human side of work (5th Ed.). Englewood Cliffs, NJ: Prentice Hall.

Guion, R. M. (1983). Recruiting, selection, and job placement. In M. D. Dunnette (Ed.), Handbook of Industrial and Organizational Psychology, New York: John Wiley and Sons (pp.777-828).

Harkins, S. G. & Petty, R. E. (1982). Effects of task difficulty and task uniqueness on social loafing. Journal of Personality and Social Psychology, 43, 1214-1229.

Hinrichs, J. R. (1983). Personnel Training. In M. D. Dunnette (Ed.), Handbook of Industrial and Organizational Psychology, New York: John Wiley and Sons (pp.829-860).

Kaiser, J. (1995, October). A business blueprint: How to build a better Ph.D. Science, 270, 133-134.

Kanki, B. G. & Foushee, H. C. (1989). Communication as group process mediator of aircrew performance. Aviation, Space, and Environmental Medicine, 60, 402-410.

Labar, G. (1995). What makes ergonomic teams work? Occupational Hazards, 57, 83-85.

Laiken, M. E. (1994). The myth of the self-managing team. Organization Development Journal, 12(2), 29-34.

Latane, B., Williams, K., & Harkins, S. (1979). Many hands make light the work: The causes and consequences of social loafing. Journal of Personality and Social Psychology, 37, 822-832.

Maples, M. F. (1988). Group development: Extending Tuckman's theory. Journal for Specialists in Group Work, 13, 17-23.

May, D. R. & Schwoerer, C. E. (1994a). Employee health by design: Using employee involvement teams in ergonomic job redesign. Personnel Psychology, 47, 861-876.

May, D. R. & Schwoerer, C. E. (1994b). Developing effective work teams: Guidelines for fostering work team efficacy. Organization Development Journal, 12(3), 29-39.

McClintock, C. G. & Allison, S. T. (1989). Social value orientation and helping behavior. Journal of Applied Social Psychology, 19, 353-362.

McHale, J. (1995). Unusual videos look a the nitty-gritty of building teams. People Management, 1, 53.

Mears, P. & Voehl, F. (1994a). Team building: A structured learning approach. Delray Beach, FL: St. Lucie Press.

Mears, P. & Voehl, F. (1994b). Team building: A structured learning approach. Instructor's manual. Delray Beach, FL: St. Lucie Press.

Mischel, L. J. (1996, April). Determinants of collective efficacy: The roles of group task and members KSAs. Paper presented at the 11th Annual Conference of the Society for Industrial and Organizational Psychology, San Diego, CA.

O'Neil, M. J. & Jackson, L. (1983). Nominal group technique: A process for initiating curriculum development in higher education. Studies in Higher Education, 8, 129-138.

Orpen, C. (1986). Improving organizations through team development. Management and Labour Studies, 11, 1-12.

Rankin, J. L. & Aksamit, D. L. (1994). Perceptions of elementary, junior high, and high school student assistant team coordinators, team members, and teachers. Journal of Educational and Psychological Consultation, 5, 229-256.

Reiste, K. K. & Hubrich, A. (1995). How to implement successful work teams: Learning from the Frigidaire experience. National Productivity Review, 14, 45-55.

Reiste, K. K. & Hubrich, A. (1996). Work team implementation. Hospital Material Management Quarterly, 17, 47-53.

Scearce, C. (1992). 100 ways to build teams. Palatine, IL: Skylight Publishing.

Shapiro, R. G., Beith, B. Goldberg, J. H., Hale, J., & Kelly, J. F. (1994). I'm graduating, now what? A comparison of work in academics, consulting, government, industrial research, and industrial development. Proceedings of the Human Factors and Ergonomics Society 38th Annual Meeting, 394-398.

Standley, J. (1993). To become a team. Campus Activities Programming, 26(6) 32-36.

Thomas, G. (1991). Defining role in the new classroom teams. Educational Research, 33(3), 186-198.

Van Cott, H. H. & Huey, B. M (1992). Human factors specialists' education and utilization: Results of a study. Washington, DC: National Academy Press.

Watson, W., Michaelson, L. K., & Sharp, W. (1991). Member competence, group interaction, and group decision making: A longitudinal study. Journal of Applied Psychology, 76, 803-809.

Williams, K., Harkins, S., & Latane, B. (1981). Identifiability as a deterrent to social loafing: Two cheering experiments. Journal of Personality and Social Psychology, 40, 303-311.

USING ELECTRONIC DIALOGUE TO AUGMENT TRADITIONAL CLASSROOM INSTRUCTION

Heidi Ann Hahn
Los Alamos National Laboratory
Los Alamos, New Mexico 87545

This paper demonstrates how an electronic dialogue with a panel of human factors experts was used effectively as an augmentation to traditional classroom instruction. Nine students spent a one and one-half hour class session using a variety of commercial electronic mail software packages available on their own desk-tops (not in a university computer lab) to engage in discussion with remotely distributed instructors on topics generated by the students themselves. Ninety eight messages were exchanged, with about 60% having technical content. Interaction content and style were analyzed, and a survey was distributed to participants to evaluate the session. Process observations by this author augmented these data. Strengths and weaknesses of using technology not specifically designed for this function are discussed.

INTRODUCTION

During the Fall semester of academic year 1995, I had the opportunity to teach an introductory graduate level human factors course in the Mechanical Engineering Department at a branch campus of the University of New Mexico. This course was funded as part of an Advanced Research Projects Agency (ARPA) Technology Reinvestment Project (TRP) called "Semiconductor/Electronics Manufacturing Experts in the Classroom." The fundamental premise of the project was that both education at the university level and continuing education of professionals in industry could be enhanced by incorporating practical messages of experts in the subject matter area into more traditional instructional methods. Key activities of the overall project included:
• Developing new courses related to manufacturing engineering, utilizing outside technical experts for a variety of activities such as lectures, tours, practicums, mentoring, or panel discussions; and
• Disseminating such courses via instructional television (ITV) networks not only to the university campus but also to sites at companies, federal laboratories, technology centers, and other universities.

In agreeing to teach the course (which was a new one for the department), I accepted responsibility to bring several (preferably at least five) Ph.D.-level "experts" into the classroom to provide instruction in whatever fashion suited the course objectives. Because the density of human factors practitioners in this geographic area is quite small, my aspirations of bringing local experts into the classroom for live lectures or using instructional television to "pipe" lectures up from the main campus fell short of my commitment.

The desirability of using ITV for project-supported courses led to thoughts about the adequacy of other electronic media, in this case simple electronic mail, as a mechanism for involving students with outside experts. Most of us have probably had experience with using email as a means of communicating with a course instructor or our students, but this tends to be a simple question-and-answer exchange, and is generally invisible to other class participants. On the other extreme, I have also been involved in a project in which the entire course was taught in an electronic classroom, with no live instruction. Although this project was successful, it provided a poor model for the electronic dialogue that I was proposing for the human factors course, as it had the advantages of (1) having a built-in conferencing structure that allowed visibility of dialogue to all participants and (2) allowing dialogue, and its attendant social behaviors, to emerge over time, such that the richness of the dialogue built through the duration of the multi-month course. (For more information, see Hahn, Ashworth, Phelps, and Byers, 1990.)

This paper demonstrates how an electronic dialogue with a panel of human factors experts was used effectively as an augmentation to traditional classroom instruction. Further, the strengths and weaknesses of using technology not specifically designed for this function are addressed.

ABOUT THE DIALOGUE SESSION

On December 5, 1995, three human factors professionals served as members of a panel discussing student-generated topics of interest with nine students in an introductory human factors course. The primary course instructor, also a human factors practitioner (rather than a university professor), served a coordinating function for the dialogue.

All students were adult learners, and all were employed by a federal laboratory operated for the Department of Energy (DOE). Thus, potential guest panelists were invited from within the DOE complex (a total of seven invitations was extended), and a theme for the discussion of "Human Factors in the DOE Complex" was selected. Guest panelists were geographically dispersed, with locations ranging from Washington state to Tennessee.

The dialogue took place during the normally scheduled class time (a one and one-half hour block in the early evening), with students remaining at their home or office computer stations rather than coming to the usual classroom. No special software was distributed, so a variety of commercially available electronic mail software packages were being used simultaneously. Advance information about the backgrounds of the guests was disseminated, but no other pre-work (either computer set-up or framing questions) occurred prior to class time.

At the beginning of the session, a welcome message containing computer set-up instructions was disseminated to all participants, who were asked to check in to make their presence known. When the majority of the class was ready, the coordinator sent out a seed question designed to get the dialogue started. From that point on, all questions were student-generated, until the closing message was sent by the coordinator. This closing message contained instructions for restoring computers for normal use and a survey assessing satisfaction with the dialogue. Both students and instructors were asked to respond to the survey.

SOURCES OF INFORMATION

Data sources include analysis of the content and volume of all message traffic; survey results, including both ratings and comments from the participants; and process observations by the primary instructor.

FINDINGS

Message Traffic Analysis

In all, a total of 98 messages was exchanged. Of these, 16 were generated by the coordinator, 44 by the guest instructors, and 38 by the students. The coordinator intentionally refrained from joining technical content dialogues. As discussed below, this may have been a mistake in terms of the depth of dialogue achieved.

Roughly 40% (41 in number) of the messages had logistics regarding set-up or session closure and/or social exchange as their content, with the remainder addressing technical issues. The messages having technical content can be categorized into three broad subject areas, including education and career development, experiences of the practitioners, and specific technical questions on a variety of subjects prompted by biographical information that revealed areas of expertise of the lecturers. Table 1 shows the distribution of comments in each subject area and further sub-categorizes each area into more specific discussion topics.

Interaction styles were also examined in the technical content areas. For each of the 15 sub-topics, a determination was made regarding the level of dialogue that had occurred. Here, the analysis unit was an "exchange," which is defined as all of the transactions or individual messages that occurred between when a question was posed and when the dialogue was brought to closure.

Exchanges occurred in three categories: (1) seven exchanges (47%) were characterized as dialogue, in which more than one volley of messaging occurred, either through follow-up questioning by a student or addition of comments by an instructor other than the first respondent; (2) six exchanges (40%) were deemed to be question and answer, in which only a single volley of message exchange occurred; this was characterized by a student posing a question and receiving an answer from a single instructor; and (3) two comments (13%) were called "no exchange"; in these cases a question was posed but never received a response. Of these two, one was closely aligned with another sub-topic, and the instructors may have thought it had been adequately answered in that context.

Survey Results

Student responses. Six of the nine students (67%) responded to an eight-item survey which assessed the educational value, pace, and logistics for the session. Survey participation was entirely

Table 1. Frequency of comments by topical area.

Topic	Number of Comments
Education and Career Development	18
-- Engineering vs Psychology Background	8
-- Blending the Disciplines	1
-- Available University Programs	9
Experiences of Practitioners	15
-- Overcoming Resistance to HF	6
-- Identifying HF "Hot Topics"	6
-- Good vs Bad Aspects of Job	3
Specific Technical Questions	24
-- Virtual Reality	7
-- Data Collection Methods	4
-- Scientific Visualization	2
-- Safety Analysis	2
-- Prototyping	2
-- Defining Performance vs Productivity	1
-- Usability Testing	2
-- Sample Size	2
-- Data Applications	2

voluntary, and students were instructed that neither their decision to fill out the survey nor their answers would have any effect on their course grade. (Because the survey was administered via email, anonymity could not be offered.)

As shown in Table 2, median responses were in the neutral to satisfied range for all items. Negative responses (both ratings and written comments) occurred primarily with respect to the number of instructors (students thought the sessions would have benefited by having more than three lecturers) and logistics. Two major concerns arose regarding logistics: (1) that computer set-up should have been done and verified in advance and (2) 80% of the students wrote in comments indicating that they wanted the opportunity to formulate questions in advance.

Instructor responses. Two of the three instructors responded to the same survey that was administered to the students. As was the case with the student ratings, median responses were in the neutral to satisfied range for all items. Contrary to student requests to do pre-work, both instructors

noted that the session had been "painless" in part because of the very limited advance preparation requirements.

Comments. Comparison of the responses of students and instructors shows that the instructors viewed the educational objective aspects of the session more positively than did the students, although the students had a more positive view of the breadth of topics covered -- perhaps because they were generating questions on topics of interest to them.

Students had a slightly more favorable view of the dimensions related to message traffic. Session length and logistics were viewed equally favorably (with a median rating of 4, which means "satisfied") by both groups. The latter result is interesting, as one-third of the members of each group experienced technical difficulties.

Finally, both groups were least satisfied with the number of instructors. All who had a negative view on this dimension thought that having more instructors would have improved the breadth and liveliness of the discussion.

Table 2. Survey responses.

Question (1 to 5 rating scale, unless otherwise noted)	Median Student Response	Median Instructor Response
Educational value of the session objective	4	4.5
Degree to which the session met the session objective	3	4.5
Number of guest lecturers participating	3	3.5
Traffic volume (# of messages)	4	3.5
Traffic pace (how fast it flowed)	4	3.5
Breadth of topics covered	4	3.5
Length of the session	4	4
Logistics (around the use of the technology)	4	4
Would you participate in this kind of a "class" again? (Y/N)	100% Yes	100% Yes

Process Observations

The first observation actually occurred prior to the start of the session, and involves the difficulty in obtaining agreements to participate (without compensation) by panelists. Two rounds of contacting potential panelists yielded only 43% of the possible population. No systematic attempt was made to follow-up on the reasons for non-participation. Most who chose not to participate did not respond to the request at all; one person expressed an interest in what we were doing but had a time conflict that prevented his participation.

Second, a decision was made prior to the session to not do advance computer set-up so as to minimize extracurricular demands for students and faculty. This approach backfired, as at least two of the students and one instructor were unable to broadcast to all participants for some portion of the session.

On a balancing note, however, for those people for whom the computer set-up worked as planned, observation of the exchanges showed that many of the characteristics of electronic bulletin boards were emulated using normal email channels. Given that participants were asked to set their computers to check for new mail every two minutes, responses were rapid -- most checks had at least one new message, and delays in responding were rarely longer than four minutes. Further, an instruction to append new comments to the last relevant message helped ensure that a coherent running dialogue was disseminated with each new transaction.

Fourth, when the session ended, several of the exchanges still appeared to be active. Therefore, the email address was left open for an additional several days so that the dialogues could continue, and participants were made aware of its availability. Interestingly, other than to return surveys, the address was not used.

Finally, as noted above, only about half of the exchanges were real dialogues, and even those were short (usually consisting of only two volleys). Exchanges would have been richer and more satisfying had more and lengthier dialogues occurred.

DISCUSSION

Our experience with this session demonstrated that an electronic dialogue with a panel of human factors experts could be conducted effectively, without the use of specialized equipment, as an augmentation to traditional classroom instruction. Several weaknesses with the approach were identified, including needs to use more instructors and to invest more advance preparation time in setting up and checking out the computer links. One drawback to this advance set-up, however, is that some of the suggested settings (for example, an email-checking frequency of every two minutes) are inappropriate for normal business use. Thus, the computer set-up would have to occur as a dry run that would be reversed rather than allowing the settings to remain until class time.

Differences in time zone created problems for one of the instructors. Here, the balance is between the excitement generated by synchronicity and convenience to the participants. In circumstances similar to what we had in this class, where all of the students and two of the four instructors were in one time zone, I would argue for synchronous exchanges. With greater degrees of geographical dispersion among participants or in cases where participants live in more than just the four time zones of the continental U. S., mandated synchronicity might be inconvenient to the point of preventing some potential participants from engaging in the dialogue at all. There, allowing for asynchronous communication (in which participants could leave email messages at any time they wished during some constrained period, say, of several days) might be appropriate.

In spite of students' opinions that they should have the opportunity to formulate questions and send them to the instructors in advance, I would resist this advice, as it runs a real risk of removing the spontaneity and non-linearity that made the session stimulating. Interestingly, students did have biographical information about one of the instructors several days before the session and received information about the other two teachers a few hours prior to the session, but there was no evidence that

they had used that information to frame questions prior to the beginning of the class period.

At the same time, interventions are needed to emphasize dialogue over simple question-and-answer exchanges. Better instructions to both students and faculty indicating that adding on to someone else's question or answer is appropriate and encouraged would likely help this situation. Also, the coordinator should take a more active role in keeping conversations going, either through joining the conversation him- or herself or by having side conversations (not visible to students) prompting additional comments by instructors.

ACKNOWLEDGMENTS

Thanks are due to Dr. John Wood at the University of New Mexico for supporting this course as part of his TRP grant. The assistance of Drs. John Draper (Oak Ridge National Laboratory), Daniel Donohoo (Pacific Northwest Laboratory), and Susan Hill (Idaho National Laboratory) in conducting the expert dialogue panel was greatly appreciated.

This work was supported in part by the Department of Energy under contract W-7405-ENG-36.

REFERENCES

Hahn, H. A., Ashworth, R. L. Jr., Phelps, R. H., and Byers, J. C. (1990). Performance, throughput, and cost of in-home training for the Army Reserve: Using asynchronous computer conferencing as an alternative to resident training. Proceedings of the Human Factors Society 34th Annual Meeting, Orlando, FL, 1417-1421.

ACQUIRING USER-CENTERED DESIGN SKILLS BY DESIGNING AND EVALUATING WORLD WIDE WEB PAGES

Mark C. Detweiler
Oracle Corporation
Redwood Shores, CA

Stephen M. Hess
Applied Research Laboratory
The Pennsylvania State University
University Park, PA

Andrew C. Peck
Dept. of Psychology
The Pennsylvania State University
University Park, PA

This paper describes the design, development, and outcomes of a project-based human factors course using the World Wide Web (WWW). The primary goal was to give students hands-on opportunities to learn about user-centered design (UCD) and evaluation by using Internet resources and creating and critiquing WWW pages. In addition, students developed sophisticated skills at working together in teams. Although the overhead involved in conducting a course of this type is considerably larger than more traditional lecture courses, students' levels of involvement, motivation, and enjoyment suggest this is an effective and valuable means of teaching UCD principles and techniques.

INTRODUCTION

Over the past decade there has been a growing awareness in the human factors and human-computer interaction (HCI) communities that the standard curricula have not given students sufficient hands-on experience designing and evaluating products and services. There has also been a growing emphasis on the need to help students acquire skills required to work effectively in teams. Further, the past two years have witnessed explosive growth in WWW services, creating demands for general Web literacy, as well as interface and content designers. In view of these developments, an undergraduate course was created to promote learning in the context of doing. The course's explicit learning objectives were to help students gain skills at: 1) assessing human performance issues and their relation to design; 2) applying UCD principles and techniques to HCI; 3) applying empirical research methods and procedures to design and evaluate systems and their interfaces; 4) analyzing tasks and eliciting user knowledge and skills; 5) collecting information, solving problems, and working as effective members of multi-disciplinary teams; 6) developing techniques for representing, organizing, and using information; and 7) understanding and using WWW tools and technologies.

COURSE DESCRIPTION

Engineering Psychology (PSY 432) was offered in the fall of 1995 to twenty-three students drawn from: Psychology, Engineering, and Computer Science. The course met three times a week for 50 minutes each period, and included readings, lectures, class discussions, class exercises, projects, labs, a personal electronic journal, and one exam. Given the diversity among the students and their entry-level knowledge about human factors and computers, the course used a progressive series of projects to develop fluency at using the Internet and acquiring the necessary skills to organize Web content, and to code it into working HTML prototypes.

Orientation

To introduce UCD principles, students read Norman's (1990) book The Design of Everyday Things, and articles by Gould, Boies, Levy, Richards, and Schoonard (1987), and Wurman (1989). After having read the first four chapters of Norman (1990), five teams were formed with the goal of designing a device of the team's choosing. Teams were instructed to carefully define their potential users, to develop paper mockups of the device's look and feel, and to do so with the goal of incorporating Norman's usability principles into their designs. This activity was meant to promote thinking about design, and to help socialize students about the importance of working together as teams.

Course Projects

Following the first design exercise students performed six team-based projects over the semester. The first project, "Exploring the Internet/World Wide Web", had students: Explore various types of information available on the Web, reflect on the types of formats and layouts used to display information; and develop performance and aesthetic criteria for assessing the strengths and weaknesses of Web information sources and services (see Table 1 for examples). Students learned about the Netscape browser, spent a few hours familiarizing themselves with and exploring the Internet, and wrote a short paper describing what they believed made a good Web site--drawing inspiration from Norman (1990) concerning functionality, ease of use, aesthetics, interactivity, etc.

The remaining projects were related to the unified goal of designing and building a WWW site representing the Department of Psychology at Penn State. The second project "Information Needs Assessment and the WWW", consisted of two phases. In the first phase, students browsed approximately 10 academic sites to gather information relevant to making design decisions about the Penn State site. Students then examined 3 or 4 sites in detail, and assessed the types of information available, considered the kinds of things one might do with this information, and reflected on what makes some sites better than others, e.g., in terms of types of information provided, as well as formats and layouts used to display and navigate that information. In the second phase, students wrote a 3-4 page paper in which they identified, rank ordered, and characterized the kinds of information they felt made a good departmental site--justifying their choices based on principles from Norman (1990), Wurman (1989), and criteria developed for the first project.

The third project, "Using Metaphors in Interface Design", was designed to help students develop skills at generating effective interface metaphors, and to critique the strengths and weaknesses of interface metaphors. Students brainstormed 10-15 ideas each that could be used as Web metaphors, and selected their four best metaphors to develop more fully. Next they drew sketches to illustrate how their metaphors might appear, and how users would interact with the interface via the proposed metaphors. All sketches were displayed and discussed in class.

The fourth project, "WWW Design and Prototyping I", involved designing, story boarding, and presenting paper and electronic prototypes of various parts of the Web site. A hierarchical organization was developed for the site, based on the outcomes of assessments in Project 2. And project-sized portions were assigned to teams based on students' interests and skills. Because teams developed and organized the kinds of information they wanted to include in their contributions to the Web site, each part was somewhat different, and there was variability in how information was organized and presented. Teams demonstrated their designs to the class using both paper posters and an LCD projection screen to view their working Web pages.

In the fifth project, "WWW Design and Prototyping II", the teams extended the work they began in Project #4, and implemented their complete designs in HTML. Teams demonstrated their electronic prototypes to the class using an LCD projection screen.

The final project, "Assessing WWW Project Usability", was designed to give students opportunities to practice: 1) developing criteria for evaluating usability, and 2) using these criteria to assess and make recommendations for improving the overall WWW project. In the first part of the assignment, teams developed a rank-ordered list of usability criteria (based on Norman (1990), Molich and Nielsen (1990), Nielsen (1990), and their own design efforts). In the second part of the assignment they used these criteria to evaluate a subset of the overall project, i.e., parts that other teams had developed. Finally, they prepared a short paper of recommendations describing changes they felt would improve the evaluated work. One class period was devoted to reviewing all of the recommendations and discussing their rationale.

Electronic Journal

In addition to course readings, class discussions, exercises, and design projects, all students kept an electronic journal to document their learning activities over the entire semester. Given the importance of being aware of design processes as well as design products, students were encouraged to treat these journals as working field notes in which they regularly gathered and reflected on what they were learning and how they were learning it. Thus, the journals served as personal knowledge bases where a variety of information was recorded, including: Course notes and ideas; questions about human learning and performance; miscellaneous URLs, HTML code they copied, modified, created etc.; interface design ideas; thoughts about hypertext/hypermedia design and evaluation; and insights about ways to improve WWW services, products, etc. from a UCD perspective.

RESULTS

Web Site Design and Evaluation

After Project 2 was completed and discussed in class, the authors pooled all of the teams' suggested organizations, and prioritized information to be included in the Web site. We discovered considerable overlap in the teams' suggestions, and found that we could accommodate the majority of desired content within four general areas: 1) Student Resources, 2) Faculty, Staff, and Student Directories, 3) Areas of Departmental Study/Research, and 4) Affiliated Organizations/Miscellaneous Information. These areas served as the general framework teams used to guide their efforts in Projects 4-6. The Student Resources area included information on undergraduate and graduate courses and degree requirements as well as a brief overview of faculty research interests to help students trying to identify opportunities for independent study and work study. The Directories area provided easy access to information about all departmental faculty, staff, and students, as well as university-wide personnel. The Study/Research area provided an overview of the different programs available in the psychology department. And finally, the Affiliated Organizations/Miscellaneous Information area served as a source of information about both internal and external organizations of potential interest to psychology students, faculty, and staff.

Examples from the projects will be shown and discussed, and the working system will be demonstrated in the accompanying talk. Specific WWW page design considerations will also be demonstrated and discussed (see issues outlined in Table 1).

LESSONS LEARNED

In previous years PSY 432 had consisted of essentially independent projects. For example, although students had worked together in teams and conducted task analyses, created paper prototypes, and performed usability tests, there was little need to coordinate the individual team efforts or to worry about the cascading consequences of working toward a fully functioning prototype. The format of this new course was more labor-intensive for the instructor and teaching assistant than previous versions. In addition, coordinating the HTML and file-management activities would not have been possible without close collaboration with departmental support staff responsible for maintaining the server

software. The talk will outline the theoretical and practical issues one needs to consider when developing a course such as this.

COURSE EVALUATIONS

This course was ranked among the top courses given in the Psychology Department during the fall term. Based on a 7-point rating scale, with 7 being the highest possible score, students rated the overall course 6.47, the overall quality of the instructor 6.76, their freedom to ask questions and express options 6.94, and the amount of information learned compared to other courses taken 6.41. In addition, in open-ended comments, students repeatedly noted that they enjoyed the projects best, and found working on teams exceptionally rewarding.

CONCLUSION

Overall, based on student comments and evaluations, together with our own observations and assessments, we consider this course to have been very successful. Students succeeded at acquiring advanced skills at navigating and using Internet resources; and more importantly, they developed highly effective skills at working in teams, and applying fundamental UCD design and evaluation principles. In contrast to more traditional lecture courses, we feel this type of course is particularly effective at providing students with opportunities to develop the kinds of practical, procedural skills that are so vital to human factors practitioners.

COURSE READINGS

Introduction to User-Centered Design
Norman, D. (1990). The design of everyday things. New York: Doubleday.

Designing Artifacts From a User-Centered Design (UCD) Perspective
Gould, J.D., Boies, S.J., Levy, S., Richards, J.T., and Schoonard, J. (1987). The 1984 olympic message system: A test of behavioral principles of system design. Communications of the ACM, 30, 758-769.
Wurman, R.S. (1989). Hats. Design Quarterly, 145, 1-32.

Representing Information Visually
Barnard, P., and Marcel, T. (1984). Representation and understanding in the use of symbols and pictograms. In R. Easterby and H. Zwaga (Eds.), Information Design (pp. 37-75). New York: John Wiley and Sons, Ltd.

Zwaga, H.J., and Boersema, T. (1983). Evaluation of a set of graphic symbols. Applied Ergonomics, 14, 43-54.

Designing Interfaces

Furnas, G.W., Landauer, T.K., Gomez, L.M., and Dumais, S.T. (1987). The vocabulary problem in human-system communication. Communications of the ACM, 30, 964-971.

Shneiderman, B. (1986). Designing menu selection systems. Journal of the American Society for Information Science, 37, 57-70.

Molich, R., and Nielsen, J. (1990, March). Improving a human-computer dialogue. Communications of the ACM, 33, 338-348.

Nielsen, J. (1990, October). Traditional dialogue design applied to modern user interfaces. Communications of the ACM, 33, 109-118.

Young, D., and Shneiderman, B. (1993). A graphical filter/flow representation of boolean queries: A prototype implementation and evaluation. Journal of the American Society for Information Science, 44, 327-339.

Creating and Evaluating Hypermedia

Marchionini, G., and Crane, G. (1994). Evaluating hypermedia and learning: Methods and results from the Perseus project. ACM Transactions on Information Systems, 12, 5-34.

TABLE 1: SAMPLE EVALUATION CRITERIA

Accessibility
• Does the site support a variety of different browsers, or browsing options (high & low bandwidth)? For example: Can viewers using all-text browsers access and find useful information?

• Can viewers access the site quickly to see or do something without having to wait excessively as images and text are downloaded?

Organization
• Are the site's available contents clearly displayed on the homepage? For example: Are viewers informed about what they might find and what they could do?

• Are the contents well organized and readily understandable? For example: Are topics arranged: 1) alphabetically, 2) temporally, 3) spatially, 4) categorically, or 5) on a continuum or by magnitude?

• Are there multiple ways to access the same content from different perspectives?

Navigation
• Are navigational aids presented at the top and bottom of each page, with easy access to the homepage?

• Are short descriptive titles or headings available to orient viewers about the contents?

• Is scrolling kept to a minimum?

• Are visitors given a preview of what will happen and where they will go when they click on highlighted (underlined, colored, etc.) items?

• Is some type of table of contents, index, or both provided so viewers don't have to return to the homepage to search other contents or topics?

• Is a site map or index used to give viewers a global perspective of available contents and ways to access those contents?

• Where special-purpose icons or images are used, are their functions clearly suggested either in the images, accompanying text, or combination of the two?

• Can visitors explore the site without excessive backtracking or feelings of disorientation?

• Can viewers search within the local site?

• Are useful links provided to other sites of interest on related topics or more in-depth coverage?

• Are enough links provided to enable viewers to access more or related information?

Aesthetics/Readability
• Is the site aesthetically pleasing? For example: Is color used effectively? Is it easy to read the text (size, color, font, etc.) on the background (color, texture, etc.)? Does important information stand out in graphics, bullets, italicized text, characters, etc.?

• Does the site indicate when information was last updated and/or how often it has been updated?

• Is information about the creator and/or maintainer provided, including an electronic address for contacting the person re errors, mistakes, problems, concerns, suggestions, etc.?

• Are attention-getting symbols, images, etc. used to draw viewers' attention to news, special topics, or items of special interest or importance?

MULTI-YEAR, MULTI-UNIVERSITY PROJECTS PROVIDE REAL HUMAN FACTORS EXPERIENCE TO UNDERGRADUATES

Marc L. Resnick, John A. Stuart, Vassilios A. Tsihrintzis, and Elizabeth S. Pittenger
Florida International University
University Park, Miami, FL 33199

Human Factors professionals rarely work in isolation. The success of any product is contingent upon many other groups, including engineers, industrial designers, marketers and salespeople. Formal human factors education rarely includes interaction with members of these groups, and only considers their concerns in passing. Florida International University, as a member of the National Science Foundation Gateway Coalition, has recently created the Multi-Year, Multi-University Projects (MYMUP) group as an experimental program to investigate interdisciplinary training at the undergraduate level. Student project teams from human factors, civil and environmental engineering, and architecture and design collaborated on a project entitled "Smart Streets." The project involved the creation of kiosks in heavily trafficked urban areas to support tourism and way-finding through interactive computer systems and vending. Each student team communicated with teams in other departments and at other universities through telephone, electronic mail, desktop videoconferencing and the internet. The final design incorporated the constraints and ideas of each discipline, simulating the corporate design process. Team interaction followed the model seen in today's virtual corporation.

INTRODUCTION

The skills needed by the modern human factors practitioner range far beyond the application of design tools and methodologies. In order to succeed in today's competitive business environment, practitioners must be able to communicate and collaborate with a wide variety of other groups. Engineers must be part of any design team to determine the technological feasibility of design specifications. Manufacturing engineers must be consulted to insure that any design can be constructed cost effectively. Marketing departments determine what pricing strategies and quality levels are needed to successfully introduce the design into the market. In order for the human factors practitioner to participate effectively in the design process, he or she must be able to work in an interdisciplinary environment. This ability is usually not included in the standard university curriculum.

Zander, Powers and Ackerman (1995) discuss an interdisciplinary design project requirement based in a Civil Engineering department. They assert that these projects broaden the perspective of the students from the technical expertise they obtained in their previous courses to a practical focus that includes considerations of teamwork, economics and the perspectives of the other disciplines. They also gain an appreciation of their role in an engineering team.

However, when this interdisciplinary focus is expanded to include groups from multiple universities, there are additional issues to be addressed. Ferguson (1996)

reports that long distances may adversely affect the interdisciplinary collaboration. However, with the advent of modern communications technologies such as e-mail, desktop videoconferencing and the World Wide Web it is possible for many of these problems to be overcome. This experience may be invaluable to the participating students because of the wide use of these technologies in industry. Virtual offices are becoming common, and may be the predominant office structure of the future (Business Week, 1996).

FIU, in partnership with The Cooper Union and Polytechnic University (both in N.Y.), developed a Multi-Year, Multi-University Projects (MYMUP) program as part of the engineering education Gateway Coalition.

The MYMUP program is intended to provide future engineers with experience in the interdisciplinary environment that permeates the modern corporation. Several teams of students from departments including civil and environmental engineering, industrial engineering, and architecture and design work to develop design solutions which consider all of these areas. Teams from each department collaborate in the development of design specifications, constraints, target end users, and design solutions. Teams can be made up of students from any institution in the National Science Foundation's Gateway Coalition. The final goal is a product or system which satisfies the needs of each discipline and would be a successful product in the market. Students are encouraged to pursue the release of this product or system through independent entrepreneurship.

METHODS

The Overall MYMUP Process

The MYMUP process is streamlined to fit within the academic environment and schedule. There are six steps involved to develop and complete a MYMUP project:

1. development and selection of suitable projects
2. invitation of faculty
3. specification of project scope within each discipline
4. recruitment of student teams
5. project execution
6. project evaluation

Specific design problems are developed by a core faculty team with representatives from several engineering disciplines at Florida International University, Polytechnic University and The Cooper Union. It is envisioned that this core team will be expanded to include members of other Gateway institutions after the completion of the pilot stage in 1996. Each project must require the expertise of at least three disciplines.

Once the project has been outlined, a faculty coordinator from each discipline is recruited. MYMUP guidelines require that faculty from a minimum of three disciplines and two universities are included. The maximum is constrained solely by communication and logistical concerns. The general requirements for the discipline are provided to the faculty coordinator during the invitation stage. Once faculty accept the invitation to participate, they individually determine the specific nature of their student teams' participation, including the number and makeup of the students, the scope of the project, and the procedures to be used. Student teams can participate as part of the requirements for an existing course or as an independent study.

The pilot project was designed to span one semester (about four months). This is envisioned as a typical project length, however projects spanning more than one semester are feasible. Hybrid projects where some disciplines participate for one semester while others participate for longer have also been considered. Though the durations are constrained by the academic calendar, a goal of the MYMUP program is to simulate the corporate environment.

One critical feature of the MYMUP program is communication facilitation. In order for students from different disciplines and in disparate geographical locations to collaborate effectively, communication is essential. This involves two separate issues, transmission media and language barriers. Electronic communication through the use of e-mail, internet and desktop video conferencing is the primary transmission media for MYMUP participants. A Web site was created for use by student groups. Language barriers exist because the terminology and perspective of each discipline can be very diverse. Students may not be able to understand the work of other groups, hindering collaboration. Participating faculty must make a concerted effort to facilitate this aspect of communication.

The specific criteria used to evaluate students' contributions are left to the discretion of the individual faculty participants. The entire project is evaluated by the core MYMUP committee. A continuous improvement model is used to update the MYMUP process each semester. Student entrepreneurship in marketing the products and systems developed within the MYMUP program is encouraged.

The Human Factors Component

Because one of the core MYMUP faculty is a human factors professional, human factors considerations are included in each MYMUP project. The interdisciplinary nature of human factors makes experience in this type of environment especially critical for future human factors practitioners. Students gain experience selling human factors concerns to other disciplines at the design specification stage as well as when promoting the final design.

The human factors team participates as part of a Human Factors Engineering course taken by junior and senior level industrial engineering students. The course provides content in both the design process and in specific human factors theories and practices. Two textbooks are used. *The Design of Everyday Things,* by Donald Norman (1990) is used to provide an overview for the types of human factors issues which are responsible for the usability problems users encounter when using a variety of products and systems. Students read this book during the first two weeks of class while they are formulating the design specifications of their project. For the duration of the semester, *Human Factors in Simple and Complex Systems*, by Robert Proctor and Trisha Van Zandt (1994) is used to provide students with the underlying human performance concepts which explain the limitations of human ability and can suggest solutions. Other materials are included on a just-in-time basis as additional concepts need to be covered. Lectures for the class include detailed study of the design process and many case studies to provide students with experience in human factors methods.

The human factors team of the MYMUP project has three main tasks. In conjunction with other student groups, the team must develop a set of specifications for the design. This is one of the more challenging tasks for students because of the requirement to collaborate with peers who may not be familiar with the importance of human factors. The human factors team must become an advocate in these interdisciplinary discussions to promote the criticality of considering human factors issues. This stage is trial-by-fire practice for the same process in the real world. The second stage requires students to develop a taskflow diagram for each function of the proposed design. This stage involves survey and questionnaire techniques, as well as time and motion studies on existing products, models and prototypes. The final stage is user testing. Students must use valid statistical methods to evaluate the usability of their designs. Changes to the design are made iteratively at this point.

At the completion of each stage, the human factors team must communicate with other groups to insure that each groups' contributions are compatible and follow the original design specifications. A single group member can be selected for this task. An interim report is due at the end of each stage to allow the professor to evaluate the groups' progress and make suggestions for the next stage. The final step of a project in the MYMUP program is for all groups to prepare a presentation to sell its design to the participating faculty. Students are not only evaluated on the scientific merits of the design, but on the realistic prospects for the design's success if it were marketed.

The Smart-Streets Project

During the Spring 1996 term, a pilot MYMUP project was conducted. The project, entitled "Smart Streets," was the design of an information kiosk in a heavily trafficked urban area (Miami and/or Manhattan for the student groups involved) to support tourists and provide way-finding and vending assistance. The components of the kiosk could potentially include an interactive computer system, vending machines with food and newspapers, an information ticker with news and sports headlines, and restrooms. The computer system could include functions such as directions to and information on tourist attractions, an interactive system for making hotel, restaurant and travel reservations, bus and train schedules, and local real-time traffic information from the Florida Department of Transportation Intelligent Highway System. Decisions regarding which components to add were to be made as a result of user surveys. An attempt to include teams from the FIU business school to assist in this aspect was made, however they were not available for the pilot project.

Five groups participated in the project. One human factors group designed the layout of the kiosk interior (including consideration of access to disabled persons, consideration of a drive-through facility, and security). A separate human factors group designed the interface for a group of interactive terminals located within the kiosk. A group of students from architecture and design designed the structure of the kiosk. A group of civil and environmental engineering students considered waste water treatment and structural concerns for weather conditions. A final group of engineering design students considered future developments, such as possible technological innovations which could be incorporated in future versions of the Smart Street Kiosk.

RESULTS

The Smart Streets Design

The first step of the design process was to determine the functions which users would want to see in the kiosk. A survey was conducted by the human factors groups using a list of functions compiled by all groups. The

architecture group performed a site analysis of the location which was mutually agreed upon as a target location by all groups. Based on the results of the survey and the site analysis, the architecture group developed site schematic designs through a thorough analysis of the urban fabric of downtown Miami in the area of the site, a synthesis of this analysis resulting in contextual response to the urban context, and an innovative design solution encompassing context, site and the public space of the kiosk (see Figure 1). Final design drawings for the site and buildings were prepared by the Civil Engineers. The next step was to design the interior of the kiosk.

One human factors group was in charge of the layout of the components. This group used facilities planning as well as human factors principles to organize the flow of the interior. This group was charged with working closely with the architecture group so that all modifications could be integrated into the aesthetic composition of the structure. They also needed to communicate with the civil engineering group to insure that plumbing and electrical needs were supported by the structure and that the traffic flow, pedestrian and automotive, was supported.

The second human factors project was the design of the interface of the interactive reservation system. This group needed less collaboration with other groups, however they needed to keep abreast of all changes in the design to insure overall compatibility of the design. The resulting design needed to include all components determined by the original software, and be user friendly. Other considerations included cost, maintainability and user satisfaction. The design was based on the touch screen interface common in public interfaces of this kind (see Figure 2).

Evaluation

The MYMUP process was evaluated on several levels. Because of the small sample size, and a general dearth of long term success metrics for academic experience, statistical evaluation is not particularly helpful. However, the MYMUP process was evaluated through comparison with other projects in the same human factors class but which did not include interdisciplinary collaboration, comparison of these projects with others completed by the same students in other classes, and through surveys of the students and faculty involved.

In general, the quality of the submitted project reports and presentations was on a par with the others in the class and submitted by these students in other classes. One of the MYMUP projects exhibited significant problems due to the unfortunate occurrence of the team leader dropping the class halfway through the semester. The purpose of the MYMUP program was not to improve the quality of the human factors component of the projects, but rather to prepare students for the inter-disciplinary nature of the real world, so no quality improvements *per se* were expected. The fact that the projects were not impaired by the extra collaborative efforts was viewed as a success.

Figure 1. Architectural Design of the Smart Streets Kiosk

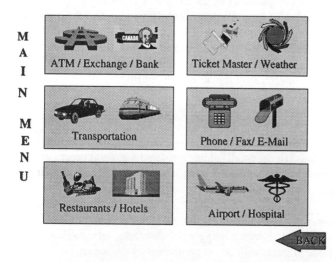

Figure 2. Main Menu of the Smart Street Kiosk

However, the kinds of issues that were dealt with in the projects were expected to be more comprehensive. For instance, the location of the restrooms within the kiosk was determined by a combination of human factors principles and the location of the sewage lines at the proposed site. Without the civil engineering collaboration, the design would not have been realistic. This experience taught the human factors engineers something that they would not have learned in a solely human factors project. The design of the traffic flow around the kiosk also needed the input of both civil and human factors engineering groups. The optimal

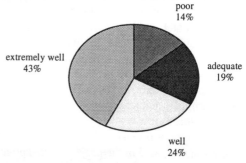

Figure 3. Reports of group effectiveness in working together

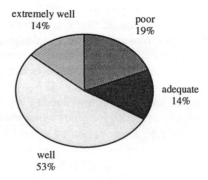

Figure 4. Reports of group effectiveness in communicating with other groups

design could not have been achieved with the input of only one group.

Surveys of the students taken at the end of the pilot project indicated that they viewed the collaboration very favorably. Forty-three percent of the students responded that the team worked together extremely well (see Figure 3). Reports on how well the teams worked with other teams was also high (see Figure 4).

Students provided many examples of the benefits they perceived as a result of the interdisciplinary collaboration. Students reported that the teams were able to see different points of view in looking at the problem. They felt is was beneficial to create a mutual appreciation among human factors engineers and other disciplines.

Faculty responses are taken from a survey of the entire Gateway Coalition team at FIU, so include additional faculty to the ones involved in the MYMUP program. Responses to this survey mirrored the perceptions of the students. Faculty were exposed to aspects of the projects different from their traditional classes. Additionally, they saw an improvement in the education of the students in terms of leadership and communication.

Challenges of interdisciplinary collaboration

The first challenge of the MYMUP process was to get all of the participating students together to establish lines of communication and roles for each group. E-mail addresses and telephone numbers were exchanged among team leaders from each group. One problem encountered was that students felt that their individual responsibilities did not require much communication. Despite being encouraged to communicate with other groups, this was resisted (as seen in many existing commercial organizations). The first 'continuous improvement' for next year is to create more tangible deliverables to coerce more consistent communication. One student suggested a single class with students from each discipline participating. Unfortunately, political and logistical considerations make this highly unlikely for the present, especially for students from different universities. A more practical solution may be an additional class meeting on Fridays (a rare class day at FIU) which will include all groups in one session.

The second challenge was collaboration with other participating universities. Notwithstanding the great potential of electronic communication, there is no substitute for face to face meetings, both for faculty and students. Communication among the FIU faculty was maintained through a combination of twice-monthly meetings and continuous streams of e-mail. With the other universities, communication was much more restrained. Students exhibited the same behavior. Though contact with students in other FIU departments was limited, it was even more so between universities. One reason for this was the late start of the MYMUP program at the other participating universities. They had not assigned student teams to the project until half way through the semester. This problem should not occur now that the pilot project has been completed.

A third challenge was that not all disciplines are at the same level of expertise in comparable areas. For instance, the architecture group requested using double reinforced concrete slabs for the design of the structure. However, the civil engineering department does not cover double-reinforced slabs until the graduate level, so the design had to use single-reinforced slabs.

Despite these challenges, the eventual result of the Smart Streets project was a success. The final design incorporated aspects of each discipline and was received favorably when presented to representatives of local industry. Converting this reception to investment funding may be more difficult; however, several of the students may make the attempt.

DISCUSSION

In order to be successful in flexible, dynamic design organizations, human factors practitioners must be able to work in interdisciplinary teams. This skill is missing from most university curricula and therefore must be learned on the job. The MYMUP program has been established by the Gateway Coalition to provide undergraduate students with some experience in a multi-disciplinary environment.

An additional need for human factors practitioners is to become adept at advocating human factors issues to members of other disciplines. Human factors students in the MYMUP program gain valuable experience in this regard, when during their projects they debate design specifications with other engineering, design, and business students.

One mission of the Gateway Coalition is to improve engineering education so that graduates are more able to become productive engineers without extensive training by their eventual employers. The MYMUP program pursues this goal by providing a realistic design environment to prepare students for the post-academic reality.

REFERENCES

Business Week (1996). The New Workplace. The McGraw Hill Companies. April 29, 106-117.

Ferguson G.L. (1996) Teaming and Long-Distance Learning - A Subjective Look at if They Work. *Proceedings of the ASEE Southeast Section Meeting*. Gatlinburg, TN.

Norman D. (1990). The Design of Everyday Things. United States: Basic Books.

Proctor R., and Van Zandt T.(1994). Human Factors in Simple and Complex Systems. Boston: Allyn and Bacon.

Zanders A.K., Powers S.E., and Ackerman N.L. The Advantages and Organization of Interdisciplinary Design Projects. *Proceedings of the ASEE Annual Conference*. Anaheim, CA.

USING TECHNOLOGY TO TEAM TEACH ACROSS INSTITUTIONS:
THE CIRCLE PROJECT

Deborah A. Boehm-Davis
George Mason University
Fairfax, VA

Kent L. Norman
University of Maryland
College Park, MD

Marc M. Sebrechts
The Catholic University of America
Washington, D.C.

Barry G. Silverman
The George Washington University
Washington, D.C.

The Consortium Interactive Research on Collaborative Learning Environments (or CIRCLE) project, was designed to examine how the distance among remote universities can be bridged electronically and how this bridge can be used to develop truly collaborative learning with shared, distributed student and faculty responsibilities. Although the problem of "distance learning" based on the model of instructional television has been well-studied, the problem of collaborative distance learning, based on a design team model, poses new technological and psychological issues. The CIRCLE project was designed to develop a model of four-way instructional collaboration. A course offering across the four institutions suggests that although the students' learning is enhanced by exposure to multiple experts, teaching through technology does impose constraints on the learning environment that make it more difficult to acquire that expertise.

Learning is often constrained by limited departmental and institutional resources. What students learn depends on the expertise that is available locally and the methods defined by disciplinary boundaries. This has been a particularly problematic issue in the development of advanced technology which requires diverse skills. The need for wide-ranging expertise and cross-disciplinary problem solving in education poses a challenge to curriculum development in general (e.g., see Johnson & Johnson, 1991).

Perhaps nowhere is this challenge more prevalent than in the specific area of computer interface design. In addition, with increased technical and instructional information and decreasing resources for faculty and curriculum development, it is difficult to have adequate expertise at any single location.

However, computer technology offers an opportunity for new forms of educational interaction. It provides a mechanism for establishing a "virtual" faculty, where it is the "access" to expertise rather than its "ownership" that is essential. In addition, it serves as the basis for new types of collaborative learning.

For many institutions, one of the biggest barriers to such inter-university instruction is the administrative overhead. In our case, the four universities with which the authors are affiliated are part of the Consortium of Universities of the Washington Metropolitan Area. Because of this consortial arrangement, registration at any of the universities can be completed through the student's home institution with no direct incremental cost to the student. Despite this arrangement, the need to spend a minimum of forty-five minutes for travel to another campus can serve as a significant impediment to student exchange and to guest lectures by faculty at other institutions in the Consortium. This project is exploring how to eliminate that barrier and to take advantage of alternative models of learning through the use of existing technologies.

With the support of the Consortium, we have been exploring ways to develop new forms of computer-supported collaborative teaching under the title CIRCLE, Consortium Interactive Research on Collaborative Learning Environments. This project addresses two major issues. First, how can we electronically bridge the distance among the institutions? Second, how can we use that bridge to develop truly collaborative learning with shared responsibilities?

In addressing the first question, we considered only "realistic" options that would place relatively small incremental demands on existing resources. Thus, for example, although videoconferencing rooms linked through cable TV provide one potentially compelling solution that has been implemented elsewhere, it was not addressed in our research. Not only is this technology unavailable at our institutions, but it tends to be cost ineffective for most other universities (Jones and Simonson, 1993).

In addressing the second question, we have explored what technologies are available for true collaborative teaching. The problem of "distance learning" on the model of television, where a central facility broadcasts the information beyond the walls of the classroom or institution to individual students, has been well studied (Walsh and Reese, 1995). As a result, many of the technologies are designed to work on this model of dissemination of information from a central point with asynchronous response, usually via electronic mail. Even most desktop videoconferencing tools tend to be point-to-point rather than multi point. In contrast, the model we are interested in pursuing is more that of a design team, with different members having varying expertise and

contributing by collaborating. Such multi point interaction poses new technological and psychological issues (for a discussion of some of these issues, see, e.g., Pea, 1992 or Privateer and MacCracken, 1992).

PREVIOUS WORK

Over the first two years of the project, we conducted a number of individual "trials" of hardware/software technologies available for collaborative work (e.g., microwave broadcast over television, a Multi-User SHared environment (MUSH), FTP, CUSeeMe, and PictureTel) with one to four participants at each of two to four sites. Through a series of such trials, we identified a number of constraints and issues that need to be considered in the development of multi-point collaborative efforts; we then were able to develop guidelines for implementing collaborative efforts based on this information. For example, we discovered that (1) existing technologies are limited in their ability to provide for synchronous communication; their use requires additional time for set-up and introduces the need for a technology manager, who can operate the technology and free the faculty to focus on delivery of the course content; (2) a visual image of participants is important for developing a sense of "group"; we recommend use of a technology that facilitates display of the individual speaking at any given point in time, especially in the early phases of a collaboration; (3) the specific technological tool being used influences the role assumed by the participants in their interactions; thus, the technology should emphasize active engagement; and (4) the task to be accomplished significantly affected the usefulness of particular technologies; specific technologies need to be matched to the task to be accomplished.

CURRENT PROJECT

Our previous work suggested that collaborating to teach a course would be difficult, but possible, given the technologies available across our four institutions. Thus, we decided to offer a semester-long, graduate course for credit. Given our interest in a multi-disciplinary approach to teaching, we chose to offer a seminar that combined expertise in cognitive psychology with expertise in engineering. The seminar was entitled "Human-Computer Interaction (HCI): Designing for Virtual Collaboration and the World Wide Web", and it focused on the way in which cognitive psychology, human factors, and engineering can contribute to improved design of computer systems. In the course, we brought the instructors' expertise (three psychologists and one engineer) to a consideration of alternative theoretical perspectives for the novel field of "cognitive engineering", and the application of these perspectives to the evaluation and design of computer tools relevant to psychology. Specifically, the human-computer issues have been situated within the context of a specific problem -- the design of a world wide web site by student teams distributed across multiple universities.

Forty-five students were enrolled in the course across the four participating universities. Students attended the class at their home institution and were linked to the other campus classrooms through a combination of technologies. Audio signals were carried through speaker phones connected through conference calls. Visual images of the participants were carried over the Internet through CUSeeMe (Anon., 1994). For most of the sessions, the world wide web (www) was used to display the instructor's textual and graphic materials. We also explored a relatively new collaborative technology called LiveBoard, which provides a multimedia computer packaged with a 67" diagonal "whiteboard" screen, and a pen-based input device. One location used this device, while the other three sites used Meeting Desk, a companion software application that allows participants at each remote location to download images which can then be accessed and manipulated by all participants simultaneously. This technology provides for collaborative work by not only sharing pictures, but also sharing workspaces. Although the LiveBoard proved useful within the individual classroom in which it was used and with two-way

communications, it was not possible to establish reliable four-way connections during the course of the semester. Thus, this technology was only used on two occasions during the semester.

Each student was assigned to a "team" that included students from each of the other three universities. The teams were required to identify a client and to develop a web site for that client, following good HCI design principles. Teams also led a discussion one time during the semester on a topic that had been outlined the previous week by the faculty.

Subjective evaluations of the course were solicited in the form of verbal and written feedback. In addition, data were collected throughout the semester on interaction styles, verbal communications during the conduct of the class, and e-mail volume and content. There was no course offered at our universities during this semester which could be used as a "baseline" course for comparison purposes; however, each of the instructors has offered courses of this type before with more traditional delivery techniques.

DISCUSSION

To date, we have analyzed the verbal and written feedback on the course, and we have summarized our observations collected throughout the semester. Quantitative analyses of the verbal and written communications throughout the course and of the e-mail traffic have not yet been completed.

Our findings from the subjective data supported our earlier work and have suggested that many of the constraints identified in the first trials are magnified in a full classroom setting. For example, having over forty students requires even more careful matching of the technology to the task. Team presentations across locations, for example, require a common shared space that can be annotated and updated from each location. In addition, the conduct of a full course provided observations that suggested additional benefits and concerns arising from implementing courses that are

Table 1. Alternative views on computer-based collaborative learning.

BENEFITS	CONCERNS
Audio was the most critical component of the course. Students became quite frustrated and had difficulty with comprehension when the audio was not crisp and clear.	When people disagreed with a speaker, it was easier to ignore them when the audio made it difficult to hear them.
It is best to be honest about our lack of technological superiority since we are all at risk when it comes to technology. Telling the students about our vulnerability pits us both against the technology, not the faculty plus technology against the students.	Students may equate technological problems with incompetence on the part of the faculty.
Technological problems lead to positive experiences when they engaged the students as problem solvers.	Technological problems lead to negative experiences when students sat passively waiting for the system to work.
The use of multiple technologies allowed us to carry different types of information through different media.	The use of multiple technologies lead to confusion about which media to focus on at a given point in time.
It is useful to have all the technology tested, up, and running well before the class starts.	It is expensive to establish and maintain connections prior to class time; in addition, rooms and staff may not be available prior to class time to set up the connections.
Providing each student with an individual monitor for the display of www material increases the clarity of the display.	Providing each student with an individual monitor for www material forced them to split their attention between this monitor and that which displayed CUSeeMe. Students found it difficult to focus on the appropriate source of information.
The number of faculty (four) lead to the students being exposed to more expertise and a greater number of perspectives than possible with traditional courses.	The number of faculty (four) lead to a fair amount of confusion: who's in charge today, who is responsible for what, and who is speaking.
Uncertainty and ambiguity about the course was useful when it was seen as an intellectual challenge to be dealt with in a constructive way.	Uncertainty and ambiguity about the course was a problem when the students considered it as the result of incompetence on the part of the faculty.
It is helpful if www pages are organized, structured, and planned before a semester begins.	Maintaining consistency throughout a semester may compromise or constrain the interactive experience of the course.
All students should learn the technical/hands-on abilities to write code, generate graphics, etc.	The demands of learning the technology may overshadow other learning; further, it may promote individual skills over team skills (where only one student is responsible for the technology).
All students should create a personal home page first so that people can get to know one another.	Early on, students don't know how to produce web pages and they may be particularly sensitive to criticism while learning.

delivered to physically separated locations. These benefits and concerns are summarized in Table 1; they fall into three categories: technology issues (the first six issues), issues arising from multiple experts delivering the course (the next two), and course delivery issues (the remaining three).

The technology issues revolved around how well the technology helped or hindered delivery of the material. As has been found in previous research, audio was the most critical component in delivering the course. Student feedback suggested that improving the quality of the audio would have been one of the single most-valued improvements in the course. Students (and the faculty) were also frustrated with the amount of time required to put the technology connections in place each week, despite the fact that difficulties often led to interesting exercises in problem solving. Also as in our earlier research, we found that the physical arrangement of the technology in the room affected performance. Students in classrooms where they had individual monitors for displaying the world wide web found it quite difficult to split their attention between the image of the instructor (projected on a screen in front of the room) and the instructor's graphics (displayed on the www). However, projecting CUSeeMe adjacent to the www on a single screen (as done in one classroom) or projecting CUSeeMe on monitors in several locations around the room, closer to the individual monitors (as done in another classroom) significantly reduced the problem of divided attention.

Perhaps one of the most positive benefits of our collaborative teaching came from the exposure that students were given to differing viewpoints and interpretations of research studies. Student comments suggested that hearing the four faculty members discuss alternative interpretations of the same study enriched their understanding of the material.

Another interesting observation was the difficulty in developing "balances". For example, feedback suggested that it would have been useful to have a more consistent structure to our web notes from week to week; however, structure can inhibit creativity of expression. Further, it was difficult to determine the extent to which individual students should be involved in the coding of text and graphics in the web site they were building. Although it is important for all students to gain some hands-on experience in every aspect of the task, this can come at the expense of learning to work in a team, where individuals rely on the specific expertise of others in accomplishing a task.

Finally, we encountered administrative problems in scheduling the course. Not only was it difficult to find a time for the class to meet that would satisfy each university's scheduling needs, but scheduling was also complicated by the fact that each of our universities meets on a different calendar. Thus, starting and ending dates for the semester and spring breaks were all quite different across the schools. As a result, the course met "jointly" (with all four schools) only 10 times over the course of the 16-week semester.

Overall, we felt that the course provided an interesting and novel experience for both the faculty and the students. It was obvious that despite our efforts, much of the technology is limited in terms of its actual classroom use. Those limits were seen as substantial problems by a number of students and as challenges to be overcome by others. Overall, however, the students appeared to enjoy the unique opportunities of this novel approach to education.

ACKNOWLEDGMENTS

This project was supported by the Consortium of Universities of the Washington Metropolitan Area, The Catholic University of America, George Mason University, The George Washington University, and the University of Maryland, College Park.

REFERENCES

Anon. (1994). CUSeeMe: Video Conferencing Over the Internet. *CACM, 3 (7)*, 22.

Johnson, D., Johnson, R. (1991). *Learning Together and Alone* (3rd ed.). New York: Prentice-Hall.

Jones, J.I. & Simonson, M. (1993). *Distance education: A cost analysis*. Convention of the Association for Educational Communications and Technology, New Orleans, LA.

Privateer, P.M., MacCracken, C., (1992, October). Odyssey Project: A Search for New Learning Solutions, *T.H.E. Journal, 20 (3)*, 76-80.

Pea, R. (1992, June) Distributed Multimedia Learning Environments: Why and How?, *Interactive Learning Environments, 2 (2)*, 73-109.

Sebrechts, M. M., Silverman, B. G., Boehm-Davis, D.A., and Norman, K. L. (1995). Establishing an electronic collaborative learning environment in a university consortium: The CIRCLE project, *Computers and Education. 25 (3)*.

Walsh, J. And Reese, B. (1995). Distance learning's growing reach. *T.H.E. Journal. 22 (11)*, 58-62.

Development of a temperature control procedure for a room air-conditioner using the concept of just noticeable difference (JND) on thermal sensation

Chang K. Cho*, Hak Min Lee*, Myung Hwan Yun**, Myun W. Lee*

*Dep.t of Industrial Engineering, Seoul National University, Seoul, Korea
**Dep.t of Industrial Engineering, Pohang University of Technology and Science, Pohang, Korea

ABSTRACT

Temperature control for an air-conditioner is an ergonomic design variable. Ergonomic studies on the thermal sensations in room environment are relatively few while many studies are available for the thermal sensations in extremely warm/cold conditions. The objective of this study was to find out factors which affect the cutaneous thermal sensations and so to propose new cooling control procedure using the JND(just noticeable difference) values of cutaneous thermal sensation for the design of an air-conditioner. The JNDs of cutaneous thermal sensation corresponding to the skin, room temperatures and temperature change rates were obtained. Based on the result of the study, a new cooling procedure using the concept of JND on thermal sensation was developed and applied to a new air-conditioner model.

1. Introduction

Though room temperature control is an important ergonomic variable of an air-conditioner, most of the temperature control logic of commercially available air-conditioner was programmed as if it is independent of human thermal sensation.

Advanced techniques such as neural network theory and fuzzy control have been applied to the control of room temperature of air-conditioner. Even with these advanced techniques, the consumer's response to the temperature control was below satisfactory level(Lee et al., 1994).

Therefore, the purpose of this study was to ; (1) investigate the relationship between room temperature changes and human thermal sensation; (2) design and implement a new temperature control logic based on human thermal sensation.

2. Method

Basic Data

The cutaneous thermal sensation is divided into cool thermal sensation and warm thermal sensation. It was assumed that both changes directionally from neutral zone with ambient temperature changes(Figure 1).

Both cool thermal sensation and warm thermal sensation are assumed to be aroused at physiological zero. According to Coren et al.(1978), only warm thermal sensation exists at temperatures above 35 °C and only cool thermal sensation exists at temperatures below 30 °C.

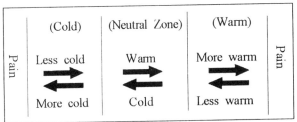

[Figure 1] Change of cutaneous thermal sensation according to temperature change

Some of the major factors which affect the cutaneous thermal sensation are body parts, room temperature, skin temperature, temperature change rate, size of the body exposed area and clothing(Kenshalo et al., 1970; Eissing, 1995).

Room temperature and skin temperature have linear relationship. Figure 2 shows the schematic diagram of JND for both warm and cool thermal sensation. The warm sensation is more sensitive than the cool sensation for the given skin temperature(Tanaka et al, 1986).

JND of cutaneous thermal sensation is also affected by temperature change rate. Figure 3 shows the schematic diagram of JND for warm and cool thermal sensation for the given rate of temperature change. The warm sensation changes more sensitively than the cool sensation(Kenshalo, 1978).

Figure 4 shows the schematic diagram of JND for both warm thermal sensation and cool thermal sensation. Warm thermal sensation changes more sensitively than the cool thermal sensation for the given skin exposure area (Steven et al, 1974 ; Berg, 1975).

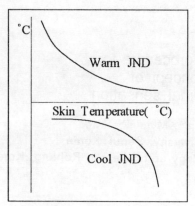

[Figure 2] JND change according to skin temperature

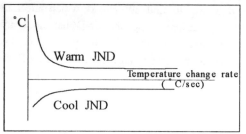

[Figure 3] JND change according to temperature change rate

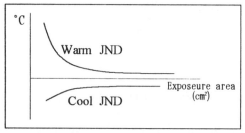

[Figure 4] JND change according to skin exposure area

Increases in the amount of clothing affect insulative and permeability characteristics and can interfere with a person's heat dissipation abilities. Tables have been developed for estimating the equivalent thermal load based on different levels of clothing(Ramsey, 1993).

For the sedentary, normally clothed worker(0.5~0.7 clo) with low air movement(less than 0.3 m/s), the comfort range is approximately 23~27 °C ET(21~26 °C WBGT)(ASHRAE 1977, Fanger 1970) and comfort for a person doing work slightly above the sedentary level would likely be 1~2 °C lower(Ramsey, 1995).

Subject, Apparatus and Procedure

Room temperature and skin temperatures for ten male adults aged 25 to 34 years(mean age 29) were measured by 297 thermocouples. The HR2500E(Yokogawa) was used for thermometer. The experiment was conducted in a thermo-

proof room. As shown in Figure 5, the air-condition was placed on the center side of the thermo-proof room. Subjects were arranged in circular distance from the center of the air-conditioner. Three thermocouples were placed on the subject's skin. Those were at forehead, forearm and foot. Room temperature was measured at locations distributed uniformly throughout the thermo-proof room using the rest 288 thermocouples.

Before the experiment, the thermo-proof room was heated until the room temperature was stabilized at 35 °C and 50 percent relative humidity. Then, the air-conditioner was operated for twenty minutes. Room temperature, skin temperature and relative humidity were measured at every twenty seconds. Subjective ratings and thermal comfort ratings were evaluated at every one minute. Subjective ratings for thermal sensation were divided into 9 scales(1: no difference(hot), 3:cool a little, 5:cool, 7:very cool, 9: cold).

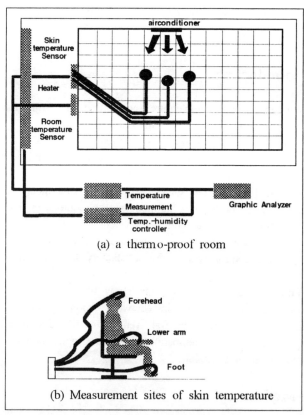

(a) a thermo-proof room

(b) Measurement sites of skin temperature

[Figure 5] Layout of experiment condition

3. Results

Although body temperature has not been reported in most studies, there are means of estimating body temperature function of environmental temperatures, exposure time and level of metabolic activity(Ramsey, 199 5). Thus, if environmental conditions such as room temperatures are known, the body temperature can be estimated. Mean skin temperatures were calculated by equation (1).

$$MST = 0.25 \times T_{Forehead} + 0.5 \times T_{Forearm}$$
$$+ 0.25 \times T_{Foot} \quad \cdots\cdots\cdots\cdots\cdots\cdots (1)$$

The relationships between skin temperatures for all subjects and room temperatures were obtained by regression analysis(Table 1). As expected, skin temperature decreased linearly with room temperature and relationship between skin temperature and room temperature was almost linear.

[Table 1] Regression analysis between room temperature and skin temperature for subjects

subject	regression equation	r^2	ρ
1	Y = 0.57X + 18.17	0.96	0.98
2	Y = 0.25X + 27.96	0.92	0.96
3	Y = 0.29X + 26.91	0.96	0.98
4	Y = 0.34X + 25.25	0.93	0.96
5	Y = 0.31X + 26.32	0.98	0.99
6	Y = 0.35X + 24.42	0.88	0.93
7	Y = 0.49X + 19.88	0.77	0.88
8	Y = 0.19X + 30.02	0.68	0.83
9	Y = 0.33X + 25.71	0.92	0.96
10	Y = 0.28X + 27.05	0.93	0.96

(Y : skin temperature, X : room temperature)

Air-velocity around subject's head was measured 10 times at every five seconds since air-conditioner was operated for 1 minutes. Air-velocities at three different distances from air-conditioner to subject were ;

1m : 1.92~2.24 $^{m}/s$ (average 2.07$^{m}/s$),
2m : 1.50~2.03 $^{m}/s$ (average 1.69$^{m}/s$),
3m : 0.50~1.48 $^{m}/s$ (average 1.03$^{m}/s$).

Figure 6 shows the frequency of thermal sensation change. The changes of thermal sensation occurred more frequently during initial 10 minutes.

Figure 7 shows the result of subjective ratings evaluation on thermal sensation. The JNDs of thermal sensation at the given value of room temperature and temperature changing rate were obtained using data of room temperature, skin temperatures and cooling rate at point of subjective rating scale. The JND values of cool thermal sensation according to initial skin temperature and temperature change rates were shown in Figure 8.

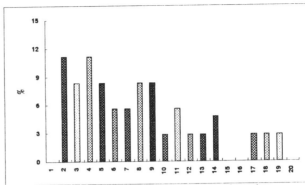

[Figure 6] frequency of thermal sensation change

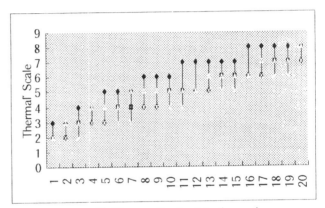

[Figure 7] The result of subjective ratings
for thermal sensation

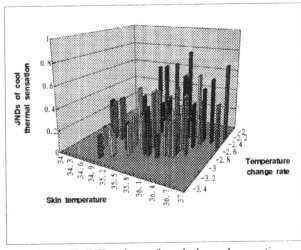

[Figure 8] JND values of cool thermal sensation
corresponding to skin temperature
and temperature change rate

4. Conclusion

In this study, JND model of cutaneous thermal sensation for the decision of JNDs of cooling control procedure was suggested considering factors which were found out from previous studies on thermal sensation.

This model shows that skin temperature and skin temperature change rate can be estimated based on room temperature, room temperature change rate and JND values of warm sensation and cool sensation can be estimated (Figure 9).

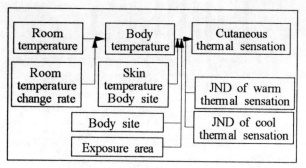

[Figure 9] JND model of cutaneous thermal sensation

Cooling control procedure which lower the temperature rapidly above the JND values of cool thermal sensation and maintains temperature below the JND values of warm thermal sensation is proposed using the relationship among room temperature, skin temperature, cutaneous thermal sensation understood in this study(Figure 10).

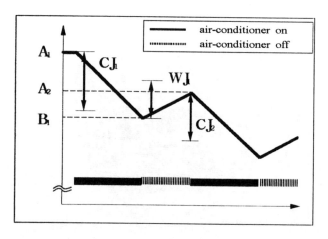

[Figure 10] Conceptual figure of cooling control using JND

Let A_i and B_i be room temperature, CJ_i and WJ_i be JND values of cool and warm thermal sensation at temperature of A_i and B_i respectively .Then, the example procedure of cooling control using this scheme can be as below :

Step 1. Estimate of body temperature from measured room temperature(A_i).

Step 2. Get the JND value of cool thermal sensation(CJ_i) on the basis of estimated skin temperature, room temperature and temperature change rate.

Step 3. Operate cooling control until room temperature reaches below the JND of cool thermal sensation (CJ_i), that is, below A_i - CJ_i.

Step 4. Interrupt cooling control not until room temperature goes up above estimated JND of warm thermal sensation (WJ_i), that is, not above B_i + WJ_i.

Repeat above steps until temperature reaches comfort level.

5. Further study

In this study; (1) factors which affect cutaneous thermal sensation were found out; (2) JNDs of cutaneous thermal sensation for cool thermal sensation were obtained; (3) cooling control method using JND was proposed.

Only skin temperature, room temperature and temperature changing rate were considered in estimation of JND. Revised control scheme which includes body site and exposure area is in progress. A new air-conditioner is being developed, adopting this control scheme and several new convenience functions. Initial consumer response to the prototype air-conditioner was above satisfactory level.

Reference

1. ASHRAE(1993) Physiological principles and thermal comfort. In ASHRAE 1993 fundamentals handbook(SI). Chapter 8.
2. Beshir,M.Y. and J.D.Ramsy (1981) Comparison between male and female subjective estimation of thermal effects and sensations. Applied Ergonomics.Vol.21(1) pp.29~33.
3. Coren,S.,C.Porac and L.M.Ward (1978) Sensation and Perception. Academic Press. New York.
4. Davies,C.T.M. (1980) Influence of air flow and skin temperature on sweating at the onset,during and following exercise. Ergonomics. Vol.23(6).pp559~567.
5. Ebaugh,F. and R.Thauer (1950) Influence of various environmental temperatures on the cold and warmth thesholds. Journal of Applied Physiology.Vol.3 .pp173~182.
6. Johnson,K.O.,I.Darian-Smith and C.LaMotte (1973) Peripheral neural determinants of temperature discrimination in man : A correlative study of response to cooling skin. Journal of Neurophysiology. Vol.36.pp

347~370.

7. Kenshalo,D.R.,C.E.Holmes and P.B.Wood (1968) Warm and cool thresholds as a function of rate stimulus temperature change.Perception & Psychophysics.Vol.3.pp 81~84.

8. Kenshalo,D.R. (1978) Biophysics and Psychophysics of Feeling. In Handbook of Perception. Academic Press. New York.

9. Stevens,J.C. and L.E.Marks (1971) Spatial summation and the dynamics of warmth sensation. Perception & Psychophysics. Vol.9.pp391~398.

10. Tanaka, M., S. Yamazaki, T. Ohnaka, Y. Tochihara and K. Yoshida (1986) Physiological reactions to different vertical (head-foot) air temperature differences.

Ergonomics, Vol.3, No.1, pp131-143.

11. Ramsey, J.D.(1995) Task performance in heat : a review, Ergonomics. Vol. 38(1). pp154~165.

12. Ramsey, J.D.(1993) Heat and clothing effects an stay time, Advanced in Industrial Ergonomics and Safety V, edited by R. Nielsen and K. Jorgensen(Taylor & Francis, London), pp435~441.

13. Eissing, G. (1995) Climate Assessment indices. Ergonomics, Vo.38(1). pp47~57.

14. Cho, C.K., H.M.Lee, M.H.Yun, M.W.Lee (1995) A study on cooling control using JND of cutaneous thermal sensation, Proceedings of the 1995 fall conference of the KIIE (In Korean)

REACTIONS TO USE OF A COMPUTER SCREEN GLARE FILTER

Alan Hedge, Ph.D.
Dept. Design and Environmental Analysis, Cornell Univ., Ithaca, NY 14853-4401.
Daniel McCrobie, Ph.D.
Commercial Flight Systems Group, Honeywell, Inc., Phoenix, AZ 85036-1111.
and
Simone Corbett
Dept. Design and Environmental Analysis, Cornell Univ., Ithaca, NY 14853-4401.

Results from three surveys of almost 200 computer workers are described. An initial survey asked about office lighting and screen glare problems, and results served as a baseline. A second survey was conducted about one-month after two thirds of the workers received a grounded, glass screen glare filter, the rest served as a control group. The final survey was conducted about one month after the remaining control group workers received their screen glare filters. Measures of illuminance and screen luminance contrast were taken at 117 workstations. Results showed that various screen image attributes and visual health measures were substantially improved with the screen glare filter. The filter minimized electrostatic shocks to workers. Most workers liked the screen glare filter and reported improved productivity.

INTRODUCTION

The growth of computer use in offices has been accompanied by increasing complaints of vision and lighting problems that seem to stem from inappropriate office lighting systems (Hedge , 1991; Hedge et al., 1995). Complaints of eyestrain among U.S. office workers are widespread, frequently topping the list of their health complaints among (e.g. Harris, 1989, 1991). Complaints of tired eyes, found to be the most common workplace-related symptom, affect some two-thirds of all workers and these tired eye complaints significantly correlate with hours of computer use (Hedge, 1991). Collins et al., (1990) reported that using VDT screens with poor legibility was significantly associated with ocular symptoms. Levin (1984) reports that as many as 75% of word processor operators may suffer from eyestrain and 55% from headaches.

Many of these vision complaints are associated with specular and veiling glare on the computer screen. Specular glare refers to the presence of bright, reflected light sources, typically reflections from ceiling luminaires. Veiling glare refers to high illuminance that "washes out" screen contrast. Both forms of glare are commonplace in many modern offices. Specular glare on the computer screen increases the time to read relatively easy text passages from the screen (Dugas-Garcia and Wierwille, 1985). Interestingly, this study found that reports of visual discomfort occurred long before any measurable performance decrement was demonstrable. Thus, subjective reports of visual discomfort provide an early warning of potential productivity problems.

Several solutions can be implemented to minimize or eliminate screen glare, such as reorienting the computer screen, covering the top and sides of the screen with a hood, covering the front of the screen with a glare filter, or installing different office lighting (Grandjean, 1987; I.E.S., 1989). Apart from studies of the effects of different office lighting systems (Marans, 1987; Marans and Yan, 1989; Collins et al., 1990; Hedge , 1991; Hedge et al., 1995). Romon et al. (1989) evaluated glass and micromesh filters under normal working conditions. Filter effectiveness was assessed by questionnaire (visual fatigue and visual quality of the characters) and by visual testing in the workplace. Results showed that the display was judged more pleasant to look at with a micromesh filter, but contrast was better with a glass filter.

Another potential problem with conventional computer screens is that they emanate electrical fields that can create electrostatic charging of the user, especially in the vicinity of the face (Wedberg, 1987). Such charging increases the deposition of charged airborne particles on exposed skin, such as the face and hands, and when deposits include particles a skin rash ensues. Placing a grounded conductive filter between the surface of the computer screen and the user should eliminate this electrostatic field effect.

The present study is a field experiment which tests the reactions of computer workers to the installation of a grounded, optically coated glass screen glare filter, designed to reduce screen glare, improve screen contrast, and eliminate electrostatic fields.

METHOD

Survey Sample

One hundred and ninety four computer workers from 8 buildings of the Honeywell Corporation in Phoenix volunteered to participate in this study. Workers were full-time computer users. The survey sample included people working on a broad range of computer tasks. The survey sample also included several different types of computers and computer monitors.

Equipment

The screen glare filters that were installed were sized for each computer screen, but all were made from the same optically coated glass (approx. 33% transmission). The screenside surface was conductive and charge was drained through a grounding lead. Illuminance was measured using a Minolta T-100 illuminance meter. Screen luminance was measured using a Brüel & Kjaer Type 1100 luminance contrast meter.

Procedure

Three self-report questionnaire surveys were administered. For the first survey (pre-installation) a 66 item questionnaire was used that collected data on workers, their computer use, their opinions about environmental office lighting, screen glare, and health. Following this survey, respondents were randomly allocated to 2 groups. About two thirds of respondents were assigned to a treatment group all of whom received a glare screen, based on the severity of their initial ratings of glare problems, and the remainder served as a control group. About one month after installation of the glare screens all workers in both groups were resurveyed (post-installation 1) using a 98 item questionnaire. Upon completion of this second survey, all control group respondents then received a glare screen, and about one month later all workers received a final survey (post-installation 2), in which a 54 item questionnaire was administered. In the final survey, measures of illuminance and screen luminance and contrast, with and without the glare screen, were also taken at each worker's desk.

Luminance contrast was assessed by measuring the luminance of a black and a white area on each computer screen, with and without the glare filter. All measures were taken at the screen brightness level that had been set by each worker. The screen luminance contrast ratio was calculated:

$$\% \text{ Luminance contrast} = \{ (C_L - C_D)/ C_L \} \times 100$$

where C_L = the luminance of the white (light) area
C_D = the luminance of the black (dark) area

RESULTS and DISCUSSION

Survey sample sizes were as follows: survey 1 n = 171, survey 2 n = 130; survey 3 n = 77. Lighting conditions were measured at 117 work stations. Results showed that mean illuminance was 420 lux ± 20.6 lux. Average screen contrast was unaffected by use of the glare screen filter: screen contrast with the glare screen filter was 87.4% ± 1%, and without this it was 88.8% ± 1%.

A summary of the main survey questionnaire results is given below. Results are expressed as valid percent responses to each question item common to the 3 survey questionnaires.

In the pre-installation survey bothersome screen

glare was quite commonplace, affecting some three quarters of respondents. By the final survey, bothersome screen glare problems were reported by less than one-third of workers (Figure 1). Comparable changes were not seen for reactions to reflected glare from work surfaces, or opinions about office lighting, which suggests that the screen glare filter was responsible for this decrease in complaints, rather than any changes in ambient lighting.

Comparison of workers' ratings of office lighting between surveys 1 and 2 showed no differences in affective reactions that could be attributed to the installation of the screen glare filter (Figure 2). Reactions of workers to the effects of the filter on various computer screen attributes are summarized in Figure 3. Results show a majority of workers reported that the filter dramatically improved screen appearance, and reduced veiling and reflected glare. A majority of workers also reported that the filter enhanced screen sharpness, contrast, colors, and brightness. 81.3% of workers said that the filter made it easier to read the computer screen and 73.3% said that screen text was clearer. 33.7% said that their was less screen flicker with the filter.

Figure 4 shows the effects of the filter on workers health reports between the first and last surveys. Results show significant improvements in the prevalence of frequent symptoms of headache, eye symptoms and lethargy/tiredness. Also, 65.9% of the survey 2 control group reported eyestrain, compared with 38.0% of survey 3 respondents. The survey 2 control group reported skin rashes on the face, neck (9.1%) arms and hands (6.8%), but in survey 3 few workers reported skin rashes on the face, neck (1.3%) and none reported rashes on their arms and hands. These results may be associated with the elimination of the electrostatic field at the front of the computer screen by the grounded screen glare filter. 77% of workers said that their filter actually had eliminated static electricity shocks. 85.3% said that there was less dust on their computer screen with the filter.

Most workers (86.6%) rated the appearance of their screen glare filter positively, and 89.2% said it improved the quality of the screen image. 89.3%

said the filter was easy to fit to their screen. Workers were asked whether they agreed or disagreed with a series of statements about their screen glare filter. Only 9.5% thought that this made their computer screen look worse, compared with 73% who said the screen looked better. Indeed, 79.7% of users agreed that the filter made it easier to read what was on their computer screen. 55.2% of workers said that the glare screen filter had helped their productivity, and only 5.5% disagreed with this. Finally, 69.9% of workers said that they liked their screen glare filer 'a lot', 26.0% said 'a little', and only 4.1% said that they didn't like the glare screen filter.

Analysis of written comments from the questionnaires showed that workers generally were enthusiastic about their screen glare filter. One wrote: "The difference is tremendous! Sometimes I lift the glare screen a little as a reminder of what I'm missing out on." Others said: "I'm so glad to have it! Has made a big difference in amount of eyestrain! Thank you!"; "Thanks for the opportunity to participate, never realized how much screen could help - no more headaches!"; "Great for cutting down eyestrain, thanks for including me in your survey." Finally, another worker simply wrote: "I don't have glare problems any more."

CONCLUSIONS

Overall reactions to the screen glare filter were very positive and the use of this type of conductive, coated optical glass screen glare filter, in situations where the ambient lighting creates screen glare, is an inexpensive ergonomic intervention that can help to improve the visual conditions for many computer workers.

ACKNOWLEDGMENTS
This research was conducted with the permission of the Honeywell, Inc. The screen glare filters were supplied by SoftView Computer Products.

REFERENCES
Collins, M., Brown, B., Bowman, K., and Carkeet, A. (1990) Workstation Variables and Visual Discomfort Associated with VDTs. Applied Ergonomics, 21 (2), 157-161

Dugas Garcia, K. and Wierwille, W.W. (1985) Effect of glare on performance of a VDT reading-comprehension task, Human Factors, 27, 163-173.

Grandjean, E. (1987) Ergonomics in Computerized Offices, London: Taylor & Francis.

Harris, L., and Associates (1989) Office Environment Index, Grand Rapids: Steelcase, Inc..

Hedge, A. (1991) Healthy office lighting for computer workers: a comparison of lensed-indirect and direct systems. Healthy Buildings - IAQ '91, ASHRAE, 61-66.

Hedge, A., Sims, W.R. and Becker, F.D. (1995) The effects of lensed-indirect uplighting and parabolic downlighting on the satisfaction and visual health of office workers, Ergonomics, 38, 260 280.

I.E.S. (Illuminating Engineering Society of North America) (1989) VDT Lighting: RP-24-1989, New York: Illuminating Engineering Society of North America .

Levin, C. (1984) VDUs - Glare and eyestrain. Safety Practitioner , 2 (2), 4-5.

Marans, R.W. (1987) Evaluating office lighting environments, Lighting Design & Application, August, 32-51.

Marans, R.W. and Yan, X. (1989) Lighting quality and environmental satisfaction in open and enclosed offices, Journal of Architectural and Planning Research, 6, 118-131.

Romon, R. M., Puppinck, T. P., Hache, J. C., Francois, D. S., Francois, E., Boulengez, C., Duwelz, M., Frimat, P., and Furon, D. (1989) Usefulness of screen filters for use with visual display terminals. Archives des maladies professionnelles , 50 (7), 665-669.

Wedberg, W.C. (1987) Facial particle exposure in the VDU environment: the role of static electricity. In B. Knave and P.G. Widebäck (eds.) Work with Display Unites '86, Elsevier Science Publishers, NY, 151-159.

Figure 1

Reactions to glare on the computer screen for each survey.

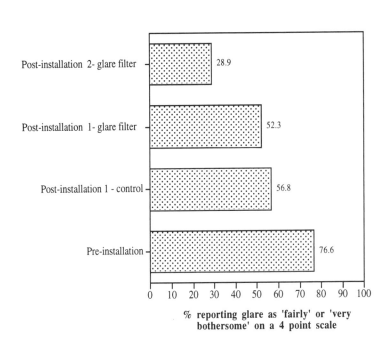

% reporting glare as 'fairly' or 'very bothersome' on a 4 point scale

Figure 2

Reactions to glare on worksurfaces for each survey.

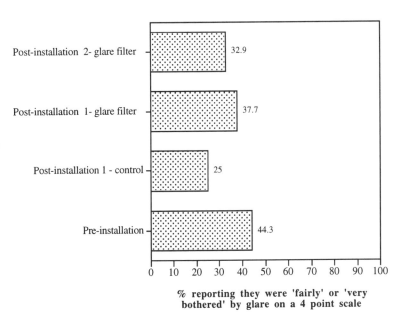

% reporting they were 'fairly' or 'very bothered' by glare on a 4 point scale

Figure 3

Reactions to the office lighting level for each survey.

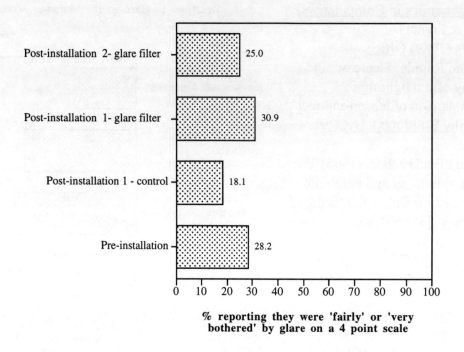

% reporting they were 'fairly' or 'very
bothered' by glare on a 4 point scale

Figure 4

**Post-installation 2 survey opinions on anti-glare
filter effects on screen characteristics**

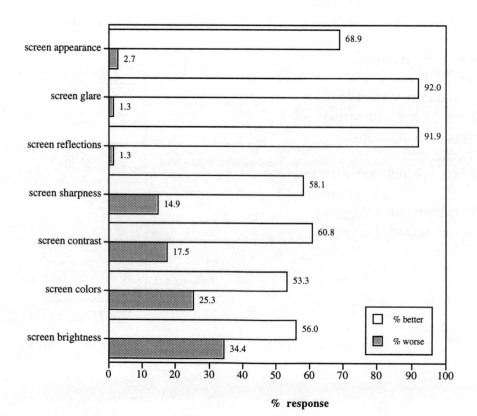

% response

EFFECT OF COMFORT ON MOOD STATES AFFECTING PERFORMANCE

Rajendra Paul   Jane Morrow   Martin Helander
Haworth   Automotive Molding Co.   Linkoping Inst. of Technology
Holland, Michigan   Warren, Michigan   Linkoping, Sweden

This study investigated the hypothesis that excessive physical comfort adversely affects mood states generally associated with superior mental performance. This hypothesis was explored in an experiment on office chairs. Eleven subjects evaluated three chair designs varying in comfort for eight hours on three separate days. The subjects performed a word processing task from 8 a.m. to 5 p.m. Comfort and discomfort were evaluated using the Shackel Scale and the Corlett Scale. Mood states were evaluated using a mood profile survey. The results suggested that the most comfortable chair induced mental mood states such as boredom, tiredness, sluggishness, (lack of) alertness and (less) energetic feeling which are generally associated with lower mental performance. It is suggested that for knowledge workers, comfort and performance may follow an inverted-U relationship. Too little or too much comfort may be detrimental to both physical and mental performance. This moderation on the view of comfort has considerable implications for the design of the modern workplace, particularly for office furniture. Its relevance to the sedentary work-styles prevalent in the United States is also discussed.

Key words: Comfort; Performance; Mood states.

INTRODUCTION

Comfort is an ill-understood construct. It has been associated with body support, anthropometric fit, biomechanical stressors, aesthetics and tactile sensation (Corlett and Bishop, 1976; Lueder, 1983, 1995; Zhang and Helander, 1992; Helander et al. 1987). These studies investigated comfort as a desired goal -- the more comfort, the better the evaluated design or product. In comfort evaluations of most workplaces and products, prevalence of discomfort and the resulting disorders warrant such focus. However, for products such as an office chair or a living room couch, it is perhaps necessary to investigate whether the product is too comfortable. Since our notions about home furnishings focus on coziness and relaxation, the couch is perhaps never too comfortable. But at work, design of the workspace must consider comfort and its effect on performance. Anecdotal evidence suggests that too much comfort induces sleepiness. While this may not be a issue at home, it could be a serious problem at work.

Mood states such as fatigue and boredness are characterized by a reduction in the level of cerebral activation thereby suggesting reduced cognitive function and performance (Grandjean, 1988). This study explored the relation between sitting comfort and mood states generally linked with mental performance. Three chairs, referred to as Chair A, Chair B and Chair C, varying in design and initial perception of comfort, were evaluated. All chairs were highly adjustable and included features such as height and tilt adjustable seatpans and adjustable backrests. Additional details of these chair designs are irrelevant for this discussion.

EXPERIMENTAL PROCEDURE

Subjects

Eleven healthy subjects, five male and six female, average age of 29.5 years and average height of 70.5 inches, participated in this study. They were recruited from a temporary employment agency and had no symptoms or history of musculoskeletal injuries. They had at least one year of experience in office work and 20/20 corrected vision. The demographic characteristics of the subjects are listed in Table 1.

Table 1. Demographic characteristics of subjects

Variable	Mean	Std. Dev.
Age (years)	29.45	6.62
Height (inches)	70.52	2.68
Weight (lbs)	169.60	33.30
Office Experience (years)	7.09	6.07
Computer Experience (years)	7.45	4.06

Procedure

At the beginning of the work day, i.e. 8:00 a.m., the height-adjustable workstation and chair were adjusted to fit the subjects' morphology. Accessories such as a document holder or a footrest were used when necessary. Subjects were then instructed to carry out a self-paced word-processing task. They were also instructed to consistently use the backrest and to support their arms on the armrests. One minute rest breaks were given each hour on the hour during which the subjects walked in the neighboring area. During the lunch hour, from 12:00 p.m. to 1 p.m., subjects were given 40 minutes to eat lunch during which their activities were not restricted. For the remaining 20 minutes, they walked. At 3 p.m. subjects were given a five-minute restroom break, instead of the regular one-minute break. Comfort surveys were administered at 8 a.m. and 5 p.m. Surveys used for subjective evaluation included the Corlett comfort scale (Corlett and Bishop, 1976), the Shackel sitting comfort scale (Shackel et al., 1969) and a mood profile survey. The mood profile survey evaluated four mood states -- boredom, alertness, tiredness and energetic -- which were decided after pilot trials, using the scale [1-Hardly at all 2-A little 3-Some 4-A lot 5-A great deal]. Subjects evaluated the three chairs on three days in a random order.

Previous studies suggest that sitting posture and upper extremity support affect perceived comfort. Variations in the hip joint angle have a significant impact on spinal disc pressure, which in turn affects sitting comfort (Andersson and Ortengren, 1974). Consequently, an effort was made to maintain consistent hip joint angles across the chairs. For each subject, hip joint angle was measured initially using a goniometer and later verified using video recordings of the subjects recorded in the sagital plane. Occhipinti et al. (1985) showed that disc pressures varied with support for the upper limbs, or lack thereof. To control this variable, subjects were instructed to consistently use armrests and backrests.

The hip joint angles collected from analysis of video recordings in the sagital plane are shown in Table 2. The differences were nominal and achieved the objective of maintaining similar hip joint angles across chairs.

Table 2. Average hip joint angle assumed by subjects by chair

Statistic	Chair A	Chair B	Chair C
Average(deg.)	114.41	110.84	112.53
Std. Deviation	3.94	2.64	2.33

Notes:

1. With Chair A, subjects also stretched their torsoes against the chair backrest for 20-30 seconds every hour on the half-hour while sitting in the chair for other design-related reasons. This streching, however, did not reduce discomfort.
2. Spinal shrinkage of the subjects was measured at 8 am, 9 am, 10 am, 12 pm, 1 pm and 5 pm using a stadiometer. The associated results are not discussed in this paper.

RESULTS

The changes in body part discomfort are shown in Figure 1. The changes in mood states are shown in Figure 2. Effect of the chairs were analyzed against the hypothesis $\mu=0$ using two-way analysis of variance(ANOVA) using a subject-condition design. Significance of the contrast differences between the chairs was analyzed using Newman-Keuls range test.

Overall, subjects experienced significant increase in sitting discomfort (ANOVA with $p<0.05$). The increase with chair A was significantly greater than those with chair B. Subjects reported significant ($p<0.05$) but similar overall increase in discomfort in the lower back with all three chairs.

Between the three chairs, Newman-Keuls test showed that the increase in discomfort in the upper back with chair A was significantly greater than that with chairs B and C.

Even though subjects experienced the most sitting discomfort with Chair A, they also reported that they were the least bored (significantly less than that with Chair B, $p<0.05$, but not with Chair C), the least tired (significantly less than that with chairs B and C) and the most energetic (significantly greater than with chairs B and C) after using that chair. On the other hand, subjects experienced the least discomfort with Chair B (not statistically different from Chair C), but they were also the most bored (significant more than with chair A, $p=0.05$, but not chair C), the least alert (significantly less than chair

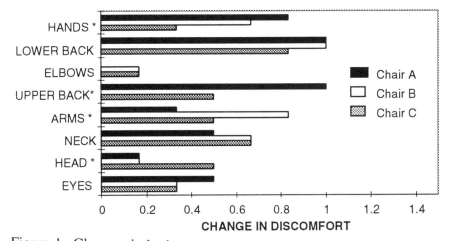

Figure 1. Changes in body part discomfort measured using Corlett Scale.
 * significant at $p=0.05$

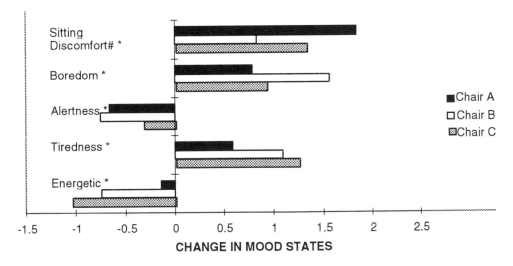

Figure 2. Changes in subjects' mood states. * significant at $p=0.05$
 # measured using Shackel scale.

C, p=0.05) but not A) and significantly more tired (than chair C but not A) after using that chair. Except the alertness, the change in mood states with the least and most comfortable chairs are significantly different.

DISCUSSION

These results hint that high levels of comfort can adversely affect mood states generally linked with superior mental performance. A considerable number of previous studies suggest that discomfort and subsequent pain adversely affect performance (Springer, 1986; Dressel and Francis, 1987; Dainoff, 1990). So discomfort should be minimized to a level less than the recoverable threshold. Zhang and Helander (1992) suggest that comfort and discomfort are orthogonal and exclusive concepts and it may therefore be warranted to consider them separately. They associated discomfort with attributes like biomechanical violations and pain, and comfort with attributes like aesthetics and plushness. In the context of comfort alone, this study provides preliminary indications that comfort above a certain value may deteriorate cognitive performance. Perhaps, comfort and performance -- physical or mental -- follow an inverted-U relationship as shown in Figure 3. More research is needed to investigate this relationship.

The objective of this study was to highlight that an unrestrained drive for comfort alone, under the auspices of ergonomics, might be counter-productive to performance at work and to emphasize the need for further research. The concept of optimum comfort is particularly important in light of the fact that existing notions of comfort promote inactivity. We often consider the comfort of recliner chairs to be an answer to physical stress. However, several studies suggest that activity, not comfort, might be the appropriate answer for our physical woes, including discomfort and disorders at work (Michel and Helander, 1993; Paul, 1995a, 1995b; Paul and Helander, 1995a, 1995b; Winkel, 1986a, 1986b, 1989). This discussion also alludes to the preference versus performance dichotomy suggested Bailey (1993). The dominance of sedentary life-style and work-style in developed countries, and the prevalent focus on comfort at home and work, might in fact be in conflict with the overall objectives of national health, well-being and productivity. The inverted U relationship between comfort and performance hypothesized here has significant implications on the design of workplace furniture in particular. More such research on issues like physical and cognitive states could help us understand the role of physical environment in employee performance.

REFERENCES

Andersson, B.G.J. and Ortengren, R. (1974). Lumbar disc pressure and bioelectric back muscle activity during sitting. . Studies on an Office Chair. *Scandinavian Journal of Rehabilitation Medicine*, 3, 104-114, 122-127 and 128-135.

Bailey, R.W. (1993). Performance versus preference. In *Proceedings of the Human Factors and Ergonomics Society 37th Annual Meeting* (pp. 282-286). Santa Monica, CA: Human Factors and Ergonomics Society.

Corlett, E.N. and Bishop, R.P. (1976). A technique for assessing postural discomfort, *Ergonomics*, 19, 175-182.

Dainoff, M.J. (1990). Ergonomic improvements in VDT workstations: health and performance benefits.

Performance

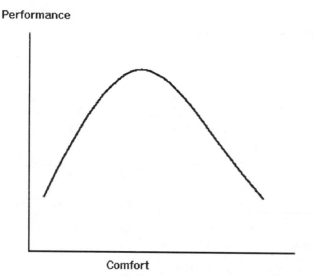

Comfort

Figure 3. Inverted U-shaped relationship between comfort and performance

In S.L. Sauter, M.J. Dainoff and M.J. smith (Eds.), *Promoting health and productivity in the computerized office: Models of successful ergonomic interventions* (pp. 49-67). London: Taylor & Francis.

Dressel, D.L. and Francis, J. (1987). Office productivity: contributions of the workstation. *Behavior and Information Technology*, 6, 279-284.

Francis, J. and Dressel, D.L. (1990). Workspace influence on worker performance and satisfaction: an experimental field study. In S.L. Sauter, M.J. Dainoff and M.J. Smith (Eds.), *Promoting health and productivity in the computerized office: Models of successful ergonomic interventions* (pp. 3-16). London: Taylor & Francis.

Grandjean, E. (1988). *Fitting the task to the man: A textbook of occupational ergonomics* (pp.181-191). London: Taylor & Francis.

Helander, M. G. and Quance, L. A. (1990). Effect of work-rest schedules on spinal shrinkage in the sedentary worker. *Applied Ergonomics*, 21(4), 279-284.

Helander, M.G., Czaja, S.J., Drury, C.G. and Cary, J.M. (1987). An ergonomic evaluation of office chairs. *Office: Technology and People*, 3 (pp. 247-262), Amsterdam: Elsevier Science Publishers B.V.

Lueder, R.K. (1993). Seat comfort: A review of the construct in the office environment, *Human Factors*, 25(6), 701-711.

Lueder, R. (1995). Adjustability in context. In R. Lueder and K. Noro (Eds.) *Hard facts about soft machines* (pp. 25-36). London: Taylor & Francis.

Michel, D. P. and Helander, M.G. (1993). Effects of two types of chairs on stature change and comfort for individuals with healthy and herniated discs, *Ergonomics*, 1175-1188.

Nachemson, A. and Elfstrom, G. (1970). Intravital dynamic pressure measurements in lumbar discs. *Scandinavian Journal of Rehabilitation Medicine, Supplement 1*, Uppsala, Sweden: Almqvist & Wiksells Boktryckeri AB, 3-40.

Occhipinti, E., Colombini, D. Frigo, C., Pedotti, A and Grieco, A (1985). Sitting posture: Analysis of lumbar stress with upper limbs supported, *Ergonomics*, 28, 1333-1346.

Paul, R.D. (1995a). Foot swelling in Sedentary VDT operators. In A. C. Bittner and P. C. Champney

(Eds.), *Advances in Industrial Ergonomics and Safety VII* (pp. 623-630). London: Taylor & Francis.

Paul, R.D. (1995b). Foot swelling in VDT operators with sitting and sit-stand workstations. In *Proceedings of the Human Factors and Ergonomics Society 39th Annual Meeting*, (pp. 568-572). Santa Monica: Human Factors and Ergonomics Society.

Paul, R.D. and Helander, M.G. (1995a). Spinal shrinkage in sedentary and non-sedentary VDT jobs. In A. C. Bittner and P. C. Champney (Eds.) *Advances in Industrial Ergonomics and Safety VII* (pp. 631-638). London: Taylor & Francis.

Paul, R. D. and Helander, M. G. (1995b). Effect of sit-stand schedule on spinal shrinkage in VDT operators. In *Proceedings of the Human Factors and Ergonomics Society 39th Annual Meeting* (pp. 563-567). Santa Monica, CA: Human Factors and Ergonomics Society.

Shackel, B., Chidsey, K.D. and Shipley, P. (1969). The assessment of chair comfort, *Ergonomics*, 12 (2), 269-306.

Springer, T.J. (1986). *Improving productivity in the workplace: Reports from the field*. St. Charles, IL: Springer Associates, Inc.

Winkel, J. and Jorgensen, K. (1986). Evaluation of foot swelling and lower-limb temperatures in relation to leg activity during long-term seated office work, *Ergonomics*, 29, 2,313-328.

Winkel, J. and Jorgensen, K. (1986). Swelling of the foot: Its vascular volume and systemic hemoconcentration during long-term constrained sitting, *European Journal of Applied Physiology*, 55, 2, 162-166.

Winkel, J. and Oxenburgh, M. (1989). Towards optimizing physical activity in VDT/office work. In S.L. Sauter, M.J. Dainoff and M.J. Smith (Eds.) *Promoting health and productivity in the computerized office: Models of successful ergonomic interventions* (pp. 94-117). London: Taylor & Francis.

Zhang, L. and Helander, M.G. (1992). Identifying factors of comfort and discomfort: a multidimensional approach. In S. Kumar (Ed.), *Advances in Industrial Ergonomics and Safety IV*. London: Taylor & Francis. 395-402.

IF I KNEW THEN WHAT I KNOW NOW: THE CONTINUING SAGA OF A CORPORATE ERGONOMIST

Nancy LJ Larson
American Express Corporation
Minneapolis, Minnesota

Convincing Corporate America to allocate funds and resources to an ergonomics program often requires the corporate ergonomist to assume roles of magician, juggler and Madison Avenue marketer. Even though corporations come in a variety of sizes and structures, their ergonomics programs have common concerns, problems and goals. The collective expertise of the Human Factors/Ergonomics specialty area is needed for ergonomics to successfully be incorporated into the business culture of today's corporations. Presented is a case study of the ergonomics efforts of a large financial corporation during the past three years. Included is information about internal partnerships, specific projects and successes (and failures), and a forecast for the future.

INTRODUCTION

To achieve a successful program in a large corporation requires the development of cross-functional alliances within the corporate structure. At American Express Financial Corporation, Risk Management "owns" the workers' compensation/medical management effort, System Support Services and Technology "own" the hardware and software purchases, Corporate Facilities "owns" the furniture standards, Field Real Estate "owns" the field office layout and field furniture standards, Human Resources "owns" ADA accommodations, business departments "own" job organization and design, and upper management "owns" the budget. Implementing an ergonomics program and incorporating ergonomics expertise into everyday work practices when many of these groups hadn't heard of ergonomics five years ago was, and continues to be, the challenge.

BACKGROUND

Ergonomics was introduced to then IDS, now American Express Financial Advisors (AEFA), in 1987 with the hiring of a professional human factors expert with one-third time committed to human factors/ergonomics issues. A full-time ergonomist reporting into the Facilities Department was hired in March of 1993. A support staff position was added in January of 1995. The implementation strategy of ergonomics has been to create temporary project teams comprised of representatives from the appropriate business and expert areas needed to effectively address the specific ergonomic issue. The following is a partial summary of our program

collaborations, major accomplishments and results of our efforts.

INTERNAL PARTNERS

Risk Management.

Ergonomics has partnered with Risk Management to address repetitive strain illness and manual material handing workers' compensation concerns. AEFA is highly proactive in educating and encouraging employees to report concerns and to file a workers' compensation claim if they are experiencing work-related cumulative trauma illness and are seeking medical care.

Procedure.

- Employee identifies possible work-related illness and contacts either Risk Management or Ergonomics.
- Ergonomics performs an evaluation of the employee's workstation and makes recommendations to accommodate.
- Risk Management files a workers' compensation claim if appropriate.
- Facilities responds to recommendations by adjusting workstation.
- Employee submits request for appropriate accessory items from Central Supply.
- Ergonomics and Risk Management continue to work together until the claim is resolved.

As the process has become known, employees now contact ergonomics for assistance prior to filing a workers' compensation claim. We continue to encourage employees who are receiving medical care to file claims. However, we tend to see more employees through the ergonomics process before workers' compensation -- in essence reversing the process.

Results.

- Workers Compensation rates are 2 to 3 times higher than comparable Bureau of Labor Statistics (BLS) rates.
- The number of WC claims has decreased.
- Our developed workers' compensation claim costs in the area of repetitive strain illness has decreased between 10 and 15 fold.
- Employees have access to immediate assistance and appropriate accommodation is provided when needed to address repetitive strain illness.

Corporate Real Estate.

A major AEFA project began in May of 1993 with the decision to restructure our client service departments. Approximately one-fourth of the corporate workforce -- nine hundred employees -- were reorganized into Geographical Service Teams (GSTs), making it necessary to relocate those employees to new workspaces before the end of 1993 to new workspaces.

A collaborative effort of Ergonomics, Design and Construction, Move Management and Property Services resulted in mandatory ergonomics training for all GST employees and customization of employee workstations. There were four goals of the training.
1. Provide information about the new workstations.
2. Allow employees to customize their new workspace to better fit personal and task requirements.
3. Inform about cumulative trauma disorders.
4. Describe computer work posture and equipment guidelines.

Process.
- Employees attend ergonomics training.
- Employees create their personal workstation profile.
- Profile reports are utilized by specialty groups during construction and move to customize employee workstations.

1. Project managers - work surface height installation information.
2. City Desk - accessory requirements.
3. Move management - distribution of accessories.
4. Telecommunications & Workstation Services - installation and placement of phone and computer equipment.

Results.
- This process, with continuing modification, continues to be a basic component of the ergonomics program.
- Training is now offered weekly to any employee. In addition, special sessions associated with relocation or department requests are also available.
- Approximately 80% of all employees have attended ergonomics training.
- Survey results indicate:
  1 80 - 90% of attendees feel ergonomics training was an effective use of their time.
  2 80 - 85% of attendees changed some work habits based upon the information received from ergonomics training.
- Individual employee workstation profile information is used for all employee moves.
- This process was awarded an American Express Silver Quality Award for the process of integrating ergonomics into the real estate process.

Ergonomics also provides assistance in the development of furniture and workstation standards. Usability testing and task analysis are both included in the development of these standards.

Results.

- Work surfaces are maintenance adjustable.
- Work surface colors are within reflectance guidelines.
- Chairs are adjustable and appropriate for task and user size needs.
- Furniture is adjustable and appropriate for task requirements.

Field Real Estate.

There are approximately 6,000 AEFA employees located in hundreds of locations throughout the United States. A task force was formed with members from Risk Management, Field Real Estate, Telecommunications, Graphics, Studio 55 (video studio), Property Services, Ergonomics, Human Resources and an Office Administrator from a Divisional Field Office to develop the program concept and create specific training materials to address field employee's needs.

Process.

- Packets, including the Ergonomics Training Video and copies of the Ergonomics Guidebook for each employee were distributed to each location.
- Information was distributed during staff meetings.
- Each site submitted a purchase order for appropriate accessories.
- Corporate ergonomics is a resource to address specific situations and questions.

Results.

- The training products created were:
  1. A 12 minute training video.
  2. A 20 page VDT Guidebook.
  3. A summary VDT Guidelines wall poster.
- All Field employees were provided ergonomics materials.
- A Field Employee Accessory Program was established.
- Training materials were created for "in-house" employee ergonomics training.
- The Ergonomics Guidebook won a graphics Crystal Clarion award for training materials.

Business Departments.

Business departments have requested ergonomic support when addressing specific workstation or job organization needs. In one instance, job analyses were performed for client service jobs in order to determine the need for larger (22") monitors for multiple displays to be presented while servicing client requests. By evaluating the task demands we were able to help identify which

jobs, based upon task information needs and criticality of errors, required the larger (and more expensive) monitors and which jobs could be performed as and effectively efficiently (at less cost) with average sized monitors.

The business departments also have provided time for employees to attend training.

Property Services.

Property Services proposed and fought to have the full-time ergonomics position created. It continues to provide administrative support and financial funds to operate the program.

OPPORTUNITIES FOR IMPROVEMENT

- Collaborative efforts with Technology have been limited and need to be expanded.
- Increased participation with the business departments, focusing on productivity issues, is potentially the next area of focus.
- Measurement procedures of ergonomics' influence on productivity issues -- not just workers' compensation costs -- need to be created and implemented

CONCLUSIONS

- The collaboration of all specialty areas, business and technical, is critical to the success of an ergonomics program.
- Ergonomics must have a champion to provide leadership and organize efforts.
- Cross-functional efforts and projects are essential
- A program success measurement must be developed. Cost benefits need to be assessed to establish results.
- Priorities need to be established and constantly reassessed to manage available resources.

RECURRING ACCIDENTS IN TRANSPORTATION SYSTEMS

Michael S. Wogalter (Chair)
North Carolina State University
Raleigh, North Carolina

SYMPOSIUM ABSTRACT

The four papers in this symposium all deal with events and accidents in surface transportation systems which display a continuing condition. The recurring aspect of these events allows patterns to emerge, associated environmental conditions to be reviewed and demographics to be considered. Human factors/ergonomics analyses and some suggested remedial actions are included for each of the accident types. All four papers deal with elements related to forensic and litigation contexts and provide insights, information and techniques for human factors practitioners in this field. Some observations and insights into the U.S. legal system are also included.

The specific topics dealt with in this symposium are: 1) Automatic shoulder belt/manual lap belt restraint systems in late model automobiles and their effect on users and accidents, 2) Pedestrian accidents, repeating patterns, and the need for lawyer, prosecutors, and judges to be more sophisticated in the limitation of both drivers and pedestrians in these situations, 3) Individuals falling out of moving passenger trains in both the U.S. and the U.K. and Amtrak's response, and 4) Misconceptions about how the expectancies of drivers affect their detection and perception of objects in night driving conditions.

AUTOMATIC SHOULDER BELT MANUAL LAP BELT RESTRAINT SYSTEMS: HUMAN FACTORS ANALYSES OF CASE STUDIES DATA

Kenneth R. Laughery
Department of Psychology
Rice University
Houston, Texas 77251-1892

Keith A. Laughery
Department of Psychology
Rice University
Houston, Texas 77251-1892

David R. Lovvoll
Department of Psychology
Rice University
Houston, Texas 77251-1892

Automatic shoulder belts combined with manual lap belts satisfy federal requirements for passive vehicle restraint systems. Previous studies show that usage rates of the lap belts in these systems is considerably lower that usage rates for manual three-point belts. Recent years have witnessed a substantial amount of litigation involving the automatic shoulder belt manual lap belt systems. Forty-one legal cases have been reviewed in which an occupant was injured or killed in an accident while wearing the shoulder belt but not the lap belt. Particular attention was given to issues of risk perception and warnings. Analyses of these cases indicate reasons for not wearing the lap belt include: (1) feeling belted or secure when the shoulder belt was in place; (2) forgetting to fasten the belt; and (3) not familiar with the type of seat belt system. The vehicle warning systems designed to encourage lap belt use generally fail to communicate the hazards and consequences of not wearing the lap belt.

INTRODUCTION

Seat belts have been required in passenger vehicles in the United States for approximately three decades. In the 1970's and early 1980's requirements went through various revisions, and in 1984 the Federal Motor Vehicle Safety Standard (FMVSS) was amended to require the use of automatic (passive) restraint systems. A phase-in was permitted with the requirement that by the 1990 model year all vehicles have a passive system. Presumably, one important reason for developing and implementing passive restraints was the low usage rates for seat belts at that time. The objective, of course, was to remove responsibility from the vehicle occupants for activating restraints.

The passive system requirements could be met in two ways, air bags or automatic seat belts. Automatic seat belts come in a variety of configurations. They may be motorized or non motorized, they may or may not include a lap belt, some are detachable and some are not, and some come in vehicles equipped with knee bolsters while others do not. The system of interest in the work reported here is the automatic shoulder belt (motorized and non motorized) and the manual lap belt. In these designs, when the driver or right front seat passenger is in the vehicle with the door closed the shoulder belt closes automatically, while the occupant must manually engage the lap belt. It should be noted that in this design the system is not fully passive. The shoulder belt is passive, the lap belt is active. Thus, in order for the restraint system to be fully functional, the occupant is required to take some action.

Since the introduction of this automatic shoulder belt, manual lap belt system (hereafter referred to as the automatic/manual system), a number of studies have reported usage rates for these systems (Lovvoll, Laughery, Wogalter and Terry, 1994; Reinfurt, St. Cyr and Hunter, 1990; Rosenfeldt, 1988; Streff and Molnar, 1991; Streff, Molnar and Christoff, 1994; Williams, Wells, Lund and Teed, 1989). The data in these studies permit comparisons between usage rates for the manual three-point belt system and the manual lap belt in the automatic/manual system. Generally, the results have shown substantial differences in the rates at which these two types of belts are being used. For example, Reinfurt et al. (1990) reported 74% usage for the manual three point belt but only 29% for the lap belt in the automatic/manual system. Similar results were reported by one of the automobile manufacturers (Rosenfeldt, 1988). Williams et al. (1989)

found usage rates of 58% and 43% for the three-point manual and automatic/manual systems respectively. In another study using self report data and a very limited sample, Lovvoll et al. (1994) reported usage rates of 90% for three-point belts and 67% for the manual lap belt. An exception to this pattern of results is two studies reported by Streff and his coworkers (Streff and Molnar, 1991; Streff et al., 1994). They found usage rates for both the three-point manual belts and the lap belt in the automatic/manual system to be in the 71% - 79% range.

As a result of the introduction of automatic/manual systems, several types of hazards and the injuries associated with them began to be noted (Roh and Fazzalaro, 1993). The hazards and consequences of interest occur when the driver and/or passenger are involved in an accident while wearing the automatic shoulder belt but not wearing the manual lap belt. Obviously, in such circumstances the lower body restraint provided by the lap belt is absent. In certain types of crashes the occupant will be subjected to movement that would be prevented by the lap belt. In frontal crashes the person may "submarine"; that is, the lower body accelerates forward. A reclined seat back can also serve to diminish the effect of the shoulder belt in these situations. Common outcomes of such events are injuries to the lower body (i.e., ruptured spleens or kidneys) and/or broken necks as a result of being "caught" by the shoulder belt. The latter circumstance may result in paraplegia, quadriplegia or death. Another category of hazards and consequences that the lap belt helps to prevent is ejection during side impact or rollover accidents. A long established finding is that occupants who are ejected from a vehicle are 25 times more likely to be killed than those who remain inside.

Partly in response to an early awareness of the low usage rates of the lap belts in the automatic/manual systems and the resulting injuries that occurred from wearing only shoulder belts, vehicle manufacturers using these restraints have employed warnings to alert and inform the driver and passenger to fasten the lap belt. Generally these warnings have appeared in the owner's manual and on the sun visors.

During the past several years there has been a substantial number of law suits against vehicle manufacturers employing the automatic/manual belt sys-

tems. To date, we have been involved in 41 such cases. In all instances, we have been employed by the attorneys representing the plaintiffs, with the first author (Kenneth R. Laughery) serving in the role of expert witness. In the context of this litigation, we have had the opportunity to analyze the 41 events, generally in considerable detail. These case studies have provided a source of data from which a number of interesting human factors issues have been identified. The purpose of this paper is to present the results of our analyses and to discuss some of the findings and issues.

METHOD

The data were gathered in the context of 41 lawsuits filed as a result of injury or death producing vehicle accidents where the front seats were equipped with an automatic shoulder belt/manual lap belt restraint system. The accidents occurred during the period 1988 through 1995 and all involved situations where the driver and/or right front seat passenger (hereafter referred to as the passenger) were injured or killed while wearing the automatic shoulder belt but not wearing a manual lap belt. The data consist of occupant interviews conducted by accident investigators, testimony of vehicle occupants, testimony of fact witnesses who had information about matters related to the accidents, testimony and reports of experts, testimony of manufacturers' representatives including people involved in various aspects of the design and marketing of the vehicles, regulations and guidelines issued by various organizations and agencies regarding the design and use of these restraint systems, and a substantial variety of written materials including correspondence, reports, publications, etc. that address aspects of the restraint systems and/or safety issues associated with them.

Because of the litigation context in which the data was generated, we had limited control or influence on what data was available or its format. Nevertheless, as anyone who has worked in this context can attest, there was generally a great deal of information about the cases. Our approach was typical of case study work: gathering facts, identifying issues, and attempting to draw conclusions. We have made an effort to quantify the data where possible, although as a rule such data do not lend themselves to statistical analyses.

In this type of work it is often the case that facts or information one would like to know are simply not available. An example here concerns what the accident victim knew or did not know about the hazards and potential consequences associated with not wearing the lap belt. Obviously in accidents that resulted in death such information was not directly available.

RESULTS

In this section we will present our analyses and/or understanding regarding several aspects of the cases. As noted, our interests focus on the human factors issues. The results are organized on the basis of demographics, belt use history, accident events, occupant movement and injury, risk perception, and warnings.

Demographics

The following data describe those injured or killed in the accidents:

Gender: 8 Males, 33 Females

Seating Position: 16 Drivers, 25 Front Passengers

Age*:   less than 10 years old: 1

        10 - 20 years old: 9

        21 - 30 years old: 12

        31 - 40 years old: 4

        41 - 50 years old: 5

        51 - 60 years old: 0

        61 - 70 years old: 3

        71 years or older: 1

* known for 35 of the 41 accident victims

Belt Use History

In 18 cases information was available about the belt use history of the person involved in the accident. These people can be divided into two groups: nine who normally drove or were passengers in three-point belt equipped cars, and nine who that were normally drivers or passengers in automatic/manual system vehicles. Seven of the nine three-point belt people normally wore their three-point belt, two did not. Of the nine people who ride in automatic/manual system vehicles, four normally wore the lap belt and five did not.

Accident Event

A description of the accident event was available in 39 of the 41 cases observed. Those 39 accidents were divided into the following three categories:

Front Impact: 23

Rollover: 12

Side Impact: 4

Occupant Injury and Movement

People were fatally injured in 21 of the 41 cases observed. Four people suffered paraplegia and four suffered quadriplegia. Six people sustained liver damage. Thirteen of the 41 were ejected, seven submarined under the shoulder belt, and 15 suffered injuries or a fatality from various types of movement within the vehicle. The occupant injury and movement outcomes can be further analyzed by accident type: front impact, rollover and side impact.

Front Impact Accidents: Descriptions of occupant movement were available for 19 of the 23 front impact accidents. In all 19 cases the person remained in the vehicle. In seven accidents the occupants submarined beneath the shoulder harness. Of those seven, two sustained fatal internal injuries, one sustained a fatal neck injury, three suffered paraplegia, and one suffered severe injuries to both legs. In 12 of the 19 front impact accidents the occupants sustained injuries due to movement within the vehicle while partially restrained by the shoulder harness. Of those 12, eight were fatally injured, two suffered quadriplegia, and one suffered major internal injuries.

Rollover Accidents: Nine of the 12 rollover accidents resulted in ejection from the vehicle. In one case a person was partially ejected from the vehicle. In another case the individual was fatally strangled by the shoulder harness. Seven of the 12 rollover accidents resulted in people being fatally injured, three resulted in paraplegia, and two resulted in major head injuries.

Side Impact Accidents: Three of the four side impact accidents resulted in ejection from the vehicle. Two of the individuals were fatally injured, one by decapitation, and the third was rendered quadriplegic. The fourth side impact accident resulted in severe internal injuries.

Risk Perception

Two types of issues or questions were of interest

to us in this category: What did people know about the hazards and consequences of not wearing the lap belt? and Why did they not fasten the lap belt? In cases where interviews or depositions with survivors were available, they were generally not asked direct questions about the submarining or ejection hazards. However, our clear impression from the available data, such as deposition testimony, is that such considerations never occurred to most of them.

As to why people did not fasten the belt, 13 responses fell into the following categories:

Passengers unfamiliar with this system: 3

Were aware there was a lap belt, but "forgot" to fasten it: 3

Felt secure or belted when the shoulder belt closed: 5

Not a regular belt user: 2

Some of the accident survivor comments help illustrate why they didn't buckle the lap belt. One person provided a response that sounds quite similar to what a non-user of a three-point belt might say: "I just forgot, sometimes I just didn't really think about it". Other responses clearly point to the different problems people encounter with these systems. One person said, "I did not understand why I had two belts", while another commented that "once the automatic shoulder strap engaged, I felt that was good enough."

Warnings

The majority of the vehicles in these accidents had owners manuals and visor labels containing information about the use of the lap belt. Three of the vehicles had no visor labels. None of the manuals or visor labels contained information specific to the hazards of submarining and ejection or to the consequences of broken necks, lower body injuries and increased probability of death with ejection. Regarding the visor labels, only two contained color; that is, most were black and white or shades of gray. Further, most had very poor contrast, some had high density print, and a few had pictorials.

All of the vehicles involved in these accidents had the required auditory and visual reminder warnings that are based on the status of the driver's belts. There was, however, some variation in how they actually worked, at least as described in the manuals. For example, most of the auditory warnings do not come on or

stop if the lap belt is fastened. In at least one vehicle, though, the auditory reminder was based on the status of the automatic shoulder belt.

DISCUSSION

The analyses reported here do not, of course, address questions about the number or percentage of people who do or do not use the lap belt in the automatic/manual system. Rather, the questions of interest here were why they are not being used and the nature and adequacy of the warning systems addressing their use. The information directly available from the injured occupants is consistent with the concepts of forgetting (Lovvoll et al., 1994) and a feeling of security (Reinfurt et al., 1990). The notion that the occupant did not realize a lap belt existed is of interest. In addition to the three occupants who reported this lack of knowledge, at least three of the fatalities who were passengers were not familiar with or regular users of the vehicle in which they were killed. In addition to the possibility that the shoulder belt closing is providing a sense of security, it may also be giving the impression, especially to the unfamiliar occupant, that the belt system is engaged. This may well be one of the hazards associated with a restraint system that is part passive and part active.

Overall, the warning systems employed are judged to be poor. They certainly do not pass a reasonable criterion for getting attention. Also, they contain little information about the hazards and consequences of not wearing the lap belt, and the information provided is not explicit. Generally, they would not satisfy the ANSI Z535.4 standard (ANSI, 1991).

In the context of working on these cases, we developed an on product warning that we believe is appropriate to address the problem of lap belt use in vehicles with automatic/manual belts. The warning is printed on a label that would be displayed on both sides of both visors. The warning label is shown in Figure 1. The three shaded areas represent orange backgrounds in the actual warning label. The label is designed to meet the ANSI Z535.4 guidelines, and we believe it is adequate to gain attention and convey the important hazard and consequence information. This label in conjunction with good warning information in the owners manual and appropriate reminder warnings on the vehicle, would constitute an adequate warning system to address lap belt use in the automatic/manual restraints.

REFERENCES

ANSI. (1991). American national standard for product safety signs and labels: Z535.4. New York: National Electrical Manufacturers Association.

Lovvoll, D.R., Laughery, K.R., Wogalter, M.S. and Terry, S.A. (1994). Risk perception issues in the use of motorized shoulder belt/manual lap belt systems. *Proceedings of the Human Factors and Ergonomics Society 38th Annual Meeting*, Nashville, 456-460.

Reinfurt, D.W., St. Cyr, L.L. and Hunter, W.W. (1990) Usage patterns and misuse rates of automatic seat belts by system type. *34th Annual Proceedings of the Association for the Advancement of Automotive Medicine*, Scottsdale, 163-179.

Roh, L.S. and Fazzalaro, W. (1993). Transection of trachea due to improper application of automatic seat belt (submarine effect). *Journal of Forensic Sciences, 38 (4)*, 972-977.

Rosenfeldt, R. (1988). Passive and active seat belt field survey. Ford Automotive Safety Office.

Streff, F.M. and Molnar, L.J. (1991) Use of automatic safety belts in Michigan. *Journal of Safety Research, 22*, 141-146.

Streff, F.M, Molnar, L.J. and Christoff, C. (1994). Automatic safety belt use in Michigan: A two-year follow-up. *Journal of Safety Research, 25*, 215-219.

Williams, A.F., Wells, J.K., Lund, A.K. and Teed, N. (1989) Observed use of automatic safety belts in 1987 cars. *Accident Analysis and Prevention, 21*, 427-433.

Figure 1. Visor Warning for Vehicles with Automatic/Manual Belts

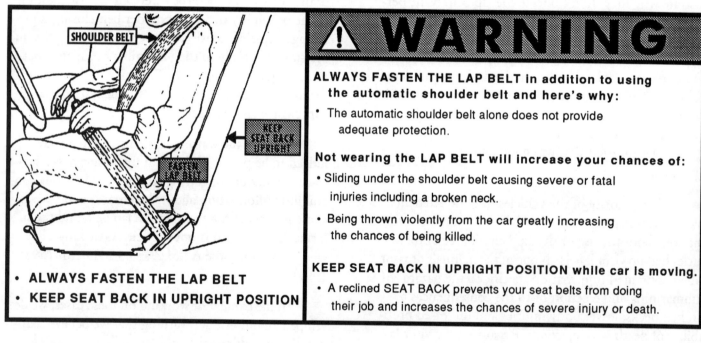

PEDESTRIAN ACCIDENTS: REPEATING PATTERNS AND LITIGATION

Richard A. Olsen
Human Performance: Limited
Santa Clara, California

Human factors analysis can be applied to pedestrian safety, but pedestrian fatalities have continued to be about 15 percent of the total traffic fatalities, many involving litigation. In some cases, pedestrian behavior is the only part of the situation that is likely to be manipulable. Fatalities are more likely at night. It does not seem likely that effective countermeasures will be introduced in the driver or the visibility environment in the near future. Efforts to encourage safe practices in walking and crossing patterns have been made with marginal effects. Litigation should recognize that drivers cannot avoid every pedestrian. The only practical approach seems to be to modify the visibility and behavior of each pedestrian. A concerted program to make every pedestrian visible by devising, distributing, and enforcing use of visibility aids seems promising. Efforts to implement such visibility improvements would recruit every agency that contacts the walking public, especially the young, old, and those using wheelchairs, bicycles, and other mobility aids.

INTRODUCTION

A premise of human factors engineering is that a pattern of repeated loss in a system that involves the human operator reacting to various inputs indicates flaws in the system design or its implementation. Of course, in locating a problem it does not follow that the remedy is necessarily practical or cost effective, given the existing system. The variety of situations presented to pedestrians and to drivers as they are exposed to each other on our streets and highways makes it unlikely that the risks can be summed up and controlled by a small number of approaches. In general—and specifically in litigation—the prevention of (or fault in) a given loss depends on the opportunities for each party to perceive and to avoid the other to prevent the confrontation. Clearly it is not always the pedestrian or always the driver who causes a collision. If opportunity was available for either or both parties, were the parties likely to be able to recognize and utilize that opportunity to prevent or mitigate the loss? If early assessments could be made according to realities of human behavior as seen by HFE specialists and accident reconstructionists, jury trials might be simplified or eliminated, especially in criminal cases where the driver was involved in a pedestrian's death or injury.

While applying human factors analysis to pedestrian safety is not a new idea, losses remain high. US Pedestrian fatalities were 6869 in 1988, out of 42,119 total traffic fatalities, or 16.3%. In 1990, 1992, and 1994 the percentages were, respectively, 14.5%, 14.1%, and 13.5%. In 1994, 90,000 pedestrians were injured (USDOT, 1988, 1990, 1992, 1995). Pedestrians are hit by all kinds of drivers, by all kinds of vehicles, and in all kinds of environmental types and conditions. No radical reduction is expected because of the diversity of causes and conditions, but a few types of accidents involving visibility do recur regularly.

NIGHTTIME VISIBILITY AND SPEED

Pedestrians walking at night along or crossing freeways or rural roads, and some urban roads with overhead lighting, are hit by drivers who claim they never saw the pedestrian at all or until impact. In some cases, witnesses report seeing, or other drivers report avoiding, the same person just before the collision. Often there is no known impairment in the driver, though low levels of alcohol may be reported. While pedestrian accidents on rural roads are 14.3% of the total, they account for 25% of the deaths (Zeeger, 1991, page 18). On freeways, half the drivers were going straight ahead and/or at sustained speed at impact (Zeeger, 1991, page 24), implying that there was no perception of a problem until too late. The serious drinking pedestrian problem (Miles-Doan, 1996) is not emphasized here. The usually ignored problem of poor detectability of pedestrians is.

Questions typically at issue

1.   Was the involved driver necessarily negligent while the witness was alert and other avoiding drivers "good" drivers? Was there a lack of alertness that justifies criminal prosecution?
2 .   Was the pedestrian reasonably visible under the conditions presented?
3.   Did or would overhead lighting help or hinder one observer or the other?
4.   Is *not* using the high-beam headlights by a driver negligence?
5.   When does "too fast for conditions" apply in assigning driver fault?

Topics For Consideration

Lighting. On freeways, the overriding factor is usually speed and the limited time for evasion that results. Commonly the clothing encountered is dark (reflectance: 3% to 15%). A driver in a lone vehicle using low-beam headlights may detect a pedestrian directly ahead at 30 to 100m (100 to 300 ft) or more. While specific headlights differ in beam patterns, in general, for a person off to one side a few degrees—especially to the left—the detection distance will be substantially less.

Expectancy. Competing with the likelihood of detection are the factors of expectancy and other patterns visible in the area. A rural freeway driver does not often experience pedestrians, especially without accompanying signs of trouble nearby. Driving is largely a speed- and steering-monitoring task. If something happens to add new information, the "alert" driver responds. Scanning the forward scene takes both attention and time. If the driver notices something (other than the pedestrian), it must be assessed: if of likely

interest, such as the traffic cues and lights, it may be evaluated while the visual input from the (less likely) pedestrian is deferred or missed.

Perception of pedestrians. If an unexpected pedestrian is to be perceived as a person in space, it usually requires more than a point of light such as a flashlight. Identification requires perception of a pattern, preferably one that does not have to be learned. Even with good overhead lighting, the context may not provide a coherent pattern for identifying a pedestrian in an environment of competing inputs. In common clothing, detection distances may range down to 30 or even 15m (100 or 50 ft). With retroreflective markers, detection can range routinely up to 1000m (3000 ft) or more.

Lay traffic knowledge. The court and jury are unlikely to be familiar with the time-distance relationships of high-speed traffic. For example, at 88 km/h (55 mph), an optimistic detection distance of 36.6m (120 ft) is equal to the distance traveled during the driver's perception-reaction time (1.5 seconds). This leaves no room to stop (41.2m or 135 ft more) or to change lanes (26.2m or 85 ft more).

Lighting geometry. The geometry of the cues and the lighting can be critical to who sees and who does not. Street lighting can have benefits for pedestrian safety, but not all street lighting will. A simple lighting analysis will reveal that the streetlight placed over a crosswalk, for example, is less effective than one set a short distance before or after the crosswalk. With the driver's eyes adapted to the moderate (mesopic) level of light provided by the headlights and overhead on the road surface, lower light levels will not be conspicuous. The driver is driving in his or her own headlight glare, about 23m (75 ft) ahead, not to mention that added by the windshield and eyeglasses or contacts. Almost any retroreflective area bathed by headlighting will be easily detected by the driver.

Headlight beam switching. The protocol of high- low-beam use is set forth in the vehicle code: no high beams with oncoming traffic within 500 ft (150m) or traffic ahead within 300 ft (91.5m). Low beams are "to reveal a person or vehicle" at 100 ft (30m) and high beams at 350 ft (107m), according to sections 24409 and 24407 of the California Vehicle Code (California, 1995). In practice, up to two-thirds of all drivers never use the high beams. Even if one could expect 350-ft (107-m) visibility of pedestrians by use of high beams, a straight-on stop from 108 km/h (67 mph) would require unusual alertness and an immediate panic stop upon perception, with four-wheel locked or ABS braking.

The pedestrian's perception. A person on the roadway is subject to the ambient light and to some headlights. That is usually a lower level than drivers are subjected to from their own headlights, and perception of dim objects will be better for them. The lack of a pitted, scratched, dirty, and tinted windscreen adds to this advantage. For pedestrian observers of a pedestrian collision these advantages are obvious. This, and differences in viewpoint, geometry, backlighting, and similar factors can make the developing situation look obvious and avoidable to some observers. Other drivers presented with the same situation but with these differences may get very different impressions. For persons about to be struck, their better vision gives the false impression that the drivers can see everything the pedestrian can see and that they will avoid any hazards, including the pedestrian. Pedestrians seldom realize how inconspicuous they really are in traffic. Even white clothing has relatively little value at night compared to retroreflective devices. Despite lighted signs, good work zone and equipment markings, cones, flaggers, and reflective vests and helmets, highway work crews continue to have losses.

Assignment of fault. Causation factors are assigned in most accidents, based on the investigating officers' understanding of events and circumstances and on their interpretations of the vehicle code and statutes. Safety researchers probably assume that causation (or "fault" or "blame") is assigned for the ultimate purpose of preventing similar losses in the future. It is not obvious how citing some "violations" could have that result. However they may result in criminal as well as civil litigation.

OBSCURED PEDESTRIAN PATHS

In Traffic

Day or night, pedestrians crossing multiple lanes are presented with multiple sources of threat. One repeating pattern is the pedestrian, often in a marked crosswalk, who has crossed the centerline and additional lanes and is struck by a

driver whose vision was obstructed by another vehicle to the left. This may be related to a "fresh" green traffic light for the striking driver, or to a stop for a crossing pedestrian by the driver on the left which was not anticipated by the striking driver.

Questions typically at issue:
1.   Where no traffic light was present (but there was a marked or unmarked crosswalk) is it reasonable to expect drivers to anticipate pedestrians in positions obscured by other traffic?
2.   When a green light is visible to a driver who is moving toward an intersection, not in the left-most lane, is it reasonable to expect that driver to stop because of the *possibility* that traffic may obscure a pedestrian?
3.   Is a driver who obstructs another driver's vision of a crossing area liable for that action? Does addition of after-market window tinting etc. that obscures a pedestrian make the vehicle owner liable?
4.   Was the crosswalk location and marking such that the driver could reasonably expect a pedestrian to be present? Was it reasonably clear of objects, shadow, and visual clutter?

Dart Outs

In urban and residential areas, a common collision between a vehicle and a pedestrian or a bicycle is the dart-out collision. Typically a young child or infant, but often a grade-school age child and occasionally an adult, the injured party is obscured by a vehicle, building, or other object until immediately before entering the roadway. The person emerging may be visible to a passing driver for only a very short time before being struck. Visibility of the person's clothing may be an added problem. Urban residential streets may have speed limits up to 55 km/h (35 mph) and wide variations in traffic density.

Questions typically at issue:
1.   What could the driver have reasonably done (if anything) to reduce the chance of a collision? Was the posted speed "reasonable?"
2.   To what extent do children understand traffic threats and what options do *they* see? Do drivers know what to expect of children in traffic?
3.   Is a lack of supervision (under about age 10) causative in traffic losses, and how can this

issue be addressed humanely after a child has been injured or killed?

SYSTEM WEAKNESSES

1.   The typical pedestrian in the path is not readily detected when speed is high, contrast is low, expectancy is low, and a "person" pattern is not provided. If detection is not accomplished while steering evasion or stopping are still possible, the collision occurs, except by lucky random deviations by both parties. Active lighting or retroreflective devices can extend detection distances dramatically.

2.   Headlighting alone is not sufficient for detection and avoidance of pedestrians in the current freeway environment or in many arterial or local street environments. Adding fixed lighting may help or hinder detection.

3.   Practical overhead lighting cannot assure visibility of pedestrians and mobility devices (e.g., wheelchairs) along extended roadways or across wide roads without visibility aids.

4.   Pedestrians do not understand their own lack of visibility at night and often fail to provide necessary means for drivers to detect them in traffic.

5.   There is a driver tendency to react to a threat only after it is clear, not at the first sign of something that may be unusual. Because there is liable to be a large cost in routine over-reaction to false alarms, this tendency may be desirable overall; it would be hard to counter in any event.

6.   Assignment of fault can help in loss prevention only if it is logical and reasonably related to training or choices of behavior presented to drivers. Assignment of fault for other reasons may not aid in evaluating evasive actions or strategies. In court it can make the system look frivolous and cause criminal actions that are not warranted.

7.   There is no concerted effort among law enforcement or any other groups to prevent loss by requiring individuals either to be visible or to remain out of vehicular traffic. Visibility aids have not been sufficiently studied and have not been made available to the average pedestrian.

CONCLUSION

Recurring losses imply need for a change in the system. Pedestrian losses are not likely to respond dramatically to any single countermeasure because there are many variables and varieties of collisions. However, for some types of pedestrian losses, it seems repeated jury trials on similar issues could be avoided if lawyers, prosecutors, and judges were more sophisticated in the limitations of drivers and the situations being presented to them. Use of experts in HFE or accident reconstruction can clarify issues both in trial and before charges are filed. This may reduce costs by bringing about changes in civil or criminal charges or complaints, as well as in court where experts are used in the adversary process. While loss of life or function is not to be taken lightly and courts will continue to adjudicate in accident losses, vehicle-pedestrian collisions are at times essentially beyond the reasonable control of the driver. The pedestrian who does not take steps to be visible or to avoid certain traffic situations must accept responsibility, not only for any injury, but also for the trauma and trouble caused in the life of the driver of the striking vehicle.

REFERENCES

California, Department of Transportation, Sacramento, CA (1995). *CA. Vehicle Code.*

Miles-Doan, R. (1996). Alcohol use among pedestrians and the odds of surviving and injury: Evidence from Florida law enforcement data. *Accident analysis & Prevention 28* (1), 23-31.

US Department of Transportation (USDOT) (1988). Washington, DC. *Fatal Accident Reporting System FARS)*

US Department of Transportation (USDOT) (1990). Washington, DC. *Fatal Accident Reporting System (FARS)*

US Department of Transportation (USDOT) (1992). Washington, DC. *Fatal Accident Reporting System FARS)*

US Department of Transportation (USDOT) (1995). Washington, DC. *Traffic Safety Facts.*

Zeeger, C.V. (1991). *Synthesis of Safety Research - Pedestrians.* US Department of Transportation (USDOT), Washington, DC. FHWA-TS-90.

FALLING OUT OF TRAINS

Stuart O. Parsons
Parsons and Associates
Saratoga, California

Approximately ten cases have been documented of people falling out of the side exit doors of moving passenger trains throughout the US. Most of these individuals were elderly passengers and the events usually happened late at night. Amtrak has classified most of these accidents as individuals with mental disorders or suicidal tendencies. Legal actions at the present time have been dismissed by the courts due primarily to the lack of any witnesses to the events. An investigation was conducted of a passenger car allegedly involved in one of these accidents. A number of recommendations were made including: 1) a systems engineering study of the problem, 2) an interlock system which would prevent the side exit door locking handle from being activated while the train is in motion, 3) an "All Green Board" at the engineer's station to indicate that all doors on the train are locked before starting the train, 4) a protective translucent cover installed over the regular door handle to indicate the criticality of opening this door, 5) a new warning sign on the exit doors which follows the ANSI standard, and 6) a training program, related to this dangerous situation, for all Amtrak operational personnel.

BACKGROUND

During the past few years approximately ten cases have been documented of people falling out of the side exit doors of passenger trains throughout the US while the train was moving at about 70 miles per hour (Figure 1). Most of these individuals were elderly passengers and the events usually happened late at night.. In some cases the side door was found to be open or unlatched. Unfortunately, all of these accidents have occurred without any witness observing the event (*Inside Edition*, 1995, Sept. 20).

Amtrak has classified most of these as individuals with suicidal tendencies or persons with mental disorders. One case was described as trespassing on the tracks even though the victim had a ticket and was seen by fellow passengers. Another case involved a 77 year old man who survived the ordeal but today has traumatic amnesia and cannot remember anything about the accident (Eischeid v. Amtrak, 1995).

Figure 1   Amtrak Passenger Car with Side Exit Door

An expert on suicide stated that jumping off a train is not a common type of suicide and that the individuals in this group did not fit the profile of a typical suicidal individual. Legal actions at the present time have been dismissed by the courts due primarily to the lack of any witnesses.   Some cases are being appealed.

In 1995, *Inside Edition*, a TV weekly documentary program, sent representatives out on five cross country trains and found a number of situations in the passenger cars which could be especially dangerous and confusing to older persons. These findings included:

o The latching handle on the side exit door (Figure 2) was frequently in the unlatched position while the train was in motion

o Winding stairs leading to rest rooms which could disorient passengers moving from one level of the car to another

o Sign to rest rooms pointing in the same direction as the side exit door

o Bathrooms adjacent to the exit doors and with similar handles

o Warning signs at a very low position on the exit side door (Figure 2)

Several years ago British Railways had a similar problem with 26 people in one year falling out of trains. The British government directed a study to investigate the problem. Based upon the recommendations of the study group, all doors on British passenger trains now have centralized locking doors and status indicators. The doors can only be opened by a conductor or engineer. The cost for this retrofit was equivalent to thirty-three million dollars and has significantly reduced accidents of this type in the U.K.

HUMAN FACTORS/ERGONOMICS INVESTIGATION

The author was requested by one of the plaintiff's attorneys to review the case (Eischeid v. Amtrak, 1995) and perform a human factors/ ergonomics investigation. An Amtrak equipment plan and specification data for passenger cars was reviewed along with a description of the economy bedroom and layout of the car. The *Inside Edition* video tape, which was shown on national television on September 20, 1995, was viewed, and the Amtrak car (Figure 1), which was alleged to be involved in the accident, was examined. Passenger travel routes in the car were analyzed and numerous measurements and photographs of the layout of the train were taken. Particular attention was directed toward the side exit door, rest rooms, labels and warning signs. Measure-ments were recorded and sketches made of the height and position of the warning signs, the size of the window on the side door, the height and design of the lower handle and the upper locking handle, and maximum distances that a passenger could be from the door when viewing the signs. The side exit door was quite easy to open when the latching handle was in the unlatched position. It would have been interesting to measure the force required to open the door when the train was moving at 70 miles per hour. The car did not have any electrical power in the yard so it was impossible to take illumination measurements. Due to the necessity of reading the instruction and warning signs in the train's aisles and the high percentage of senior passengers whose eyes need more light, it is estimated that approximately 50 fc or 540 lux would be required for an optimal illumination level (Woodson, et al, 1992). The sketches, measurements and photographs were later used to compare against human factors standards/publications and aided in developing the recommendations and conclusions presented below.

DEVELOPMENT OF AN APPROPRIATE WARNING SIGN

The current warning signs on the two side exit doors on the lower level are located below the window at 32-38 inches above the floor (Figure 2). They are red with white lettering and read:

Caution.
For your own safety
do not open door
or window

Emergency
Exit

Figure 2 Side Exit Door From Inside Train

These signs do not meet the American National Standards Institute standard for warning signs (ANSI Z535.4-1991 and FMC Corp., 1990) and are placed too low for optimal viewing. This warning was redesigned by Mrs. Lisa Forsland and the author. After evaluating a number of iterations, the final version is shown in Figure 3. This design conforms to ANSI standards and to experimental studies of warnings related to behavioral effectiveness (ANSI Z535.4-1991, FMC Corp., 1990 and Laughery et al, 1994). The new warning as shown in Figure 3 is 40% of the original foamcore mockup.

In order to be seen from about 10 feet, which is the furthest direct visual distance from the aisleway in the car to the door, and under variable lighting conditions, it is necessary to have the letters .4 inches high. The font should be a bold, non-serif, simple font. The Safety Alert Symbol is specified as a white triangle with a red exclamation point. The signal word, DANGER. is used since this is an imminently hazardous situation which, if not avoided, will result in death or serious injury. The lettering of DANGER should be at least 50 percent greater than the height of the message panel warning, and should use white lettering on a Safety Red background. The Symbol/Pictorial was developed in accordance with the product safety sign and warning system (FMC Corp., 1990) and the message of the hazard using a simple graphic of a person falling out of a door. The Warning Message was composed to meet the three criteria of an effective warning based upon research studies (Laughery, et al, 1994). These are: 1) the hazard classification which in this case is DANGER, since there is an imminently hazardous situation which, if not avoided, will result in serious injury of death, 2) an easy to understand warning associated with the related behavior, and 3) the consequences of not heeding the warning and opening the door.

The normal display or signage range on a door or panel for effective viewing is 41 inches from the floor to 70 inches, with the optimal position being from 50 inches from the floor to 65 inches (Watson, et al, 1992). Since the position under the window of the current warning design (Figure 2) is lower than recommended, our new ANSI conforming warning (Figure 3) is 4.25 inches in width and 15 inches high. This sign would fit on the right side of the window, and would cover the optimal viewing position of 50 inches to 65 inches above the floor.

HUMAN FACTORS/ERGONOMICS RECOMMENDATIONS

Figure 3   New Warning Which Meets ANSI Standard

⚠ DANGER

Do not open this door or window while the train is moving. Such action could result in severe injury or death. Emergency exit only.

The following recommendations are listed in order of importance from a safety standpoint. Good systems analysis should always be the first step, and equipment, designed for eliminating the possibility of human error, is usually more effective than just warnings or training.

1. An in-depth systems study of the problem and potential solutions should be conducted immediately to ensure that the corrective actions are long term, effective solutions. As part of the study some demographically and behaviorally descriptive profiles of individuals involved in each type of accident should be generated so that countermeasures can be tailored for the specific accident classes or types. The study team should include design engineers, operational personnel, a systems safety engineer, and a human factors engineer. The British study should be thoroughly reviewed to incorporate relevant and transferable concepts and technology. During this study period, recommendations 4, 5, and 6 can be immediately implemented at minimal cost. This team should continue after the retrofitting program is approved so that prototypes of recommended hardware and software can be evaluated on representative user populations.

2.. There should be an interlock system (Seminara and Parsons, Oct. 1963, and Seminara and Parsons, Nov. 1964) which would prevent the side exit door locking handle from being activated while the train is in motion.

3. An "All Green Board" (Seminara and Parsons, Nov. 1963) should be installed at the engineer's station to indicate that all doors on the train are locked before starting the train. A red light would come on at the related indicator light for a particular door if the door is opened. An alarm would also activate if any door was opened while the train was moving.

4. A protective, translucent, cover (Seminara and Parsons, Oct. 1963, and Seminara and Parsons, Nov. 1964) could be installed over the regular door handle to indicate the criticality of opening this door and to differentiate it from other similar

door handles on the train. This would also provide two step activation for safety purposes.

5. The new warning sign, which follows the ANSI standard (Figure 3), should be immediately fabricated and installed on the lower level side exit doors on all Amtrak passenger cars. Other features such as the toilet signs which point to the exit door and the similar handles on the toilet and the exit door should be modified and evaluated on a typical user population.

6. A training program should be prepared and conducted for all Amtrak train operational personnel. This would relate to the recent door accidents and emphasize the criticality of locking the safety latch handle while the train is moving and checking on the lock every time a conductor or other personnel walks by a door. This would be an interim corrective action until the new electronic and mechanical devices described in Recommendations 2 & 3 are designed and installed on all Amtrak cars and trains.

CONCLUSION

Currently, this "falling out of trains" situation is enmeshed in litigation, politics and economics. It is not a major cause of casualties in the U.S., but people are dying each year and the problem appears to be preventable. There are well established engineering and human factors methods for correcting the problem which should be initiated immediately by Amtrak.

REFERENCES

American National Standards Institute (1991). *Product Safety Signs and Labels Z535.4.* Washington DC: National Electrical Manufacturers Association.

Eischeid v. Amtrak. (1995). United States District Court Central District of California. Case No. CV 95-0293 RAP.

FMC (1990). *Product Safety Sign and Label System.* Santa Clara, CA: FMC Corporation..

Inside Edition (1995, September 20). An ABC television documentary on Amtrac accidents.

Laughery, K. R. Sr., Wogalter, M.S. and Young, S. (1994). *Human Factors Perspectives on WARNINGS.* Santa Monica, CA: Human Factors and Ergonomics Society.

Seminara, J.L. and Parsons, S.O. (1963, October 10). Accident-proof controls, *Machine Design,* New York: McGraw Hill Inc. 181-182.

Seminara, J.L and Parsons, S.O. (1963, November 7). Display system design. *Machine Design,* New York: McGraw Hill Inc. 167-176.

Seminara, J.L and Parsons, S.O. (1964, November) 17 ways to stop control accidents. *Control Engineering,* New York: McGraw Hill Inc..

Woodson, W.E., Tillman, B. and Tillman, P. (1992). *Human Factors Design Handbook,* New York: McGraw Hill Inc..

THE EFFECT OF EXPECTANCY ON VISIBILITY IN NIGHT DRIVING

Rudolf G. Mortimer
University of Illinois at Urbana-Champaign

The role of expectancy on the perceptions of drivers is an important variable that affects highway safety. But, it is not well understood. An experiment in which the visibility of a pedestrian dummy was measured found that the visibility distance was twice as great when drivers expected the pedestrian than when it was unexpected. The results of that study have been used as though the 2:1 ratio in visibility could be extrapolated to other night driving conditions. Arguments against such generalization are made. They involve analyses of the illumination of headlamp beams, the probability of detection of objects and the effects of speed. Various theoretical approaches are used to evaluate the role of expectancy and other studies are reviewed. No support was found for the 2:1 constant ratio between "expected" and "unexpected" driver visibility in night driving. Further work is needed to study expectancy and to quantify its effect on visibility in a variety of situations.

Expectancy plays a role in affecting human behavior in at least two ways. Firstly, it affects our predictions about the environment and the kinds of responses we may have to make and, secondly, it affects our estimates of the probability that a response will be needed. Thus, expectancy is based on prior experience and affects predictions we may make based on that learning. In design of facilities, such as the road-vehicle-environment in driving, expectancy plays a role in affecting safety.

There seem to be some misconceptions about how the expectancies of drivers affect their detection and perception of objects in night driving conditions, which is the topic of this paper.

The issue was discussed by Hyzer (1993) in relation to the utility of photographs shown to a jury of a scene of an accident which occurred in darkness. Such a photograph, even if it accurately portrayed the actual scene as viewed by an observer, might show the presence of a pedestrian or other hazard at a particular distance, so that the inference might be made by the jury that a pedestrian, or other object, in that same position would have been visible to a driver from that distance. However, the jury has prior knowledge of the events that occurred and has an "expectation" of the presence of the object of interest. There are other variables also, such as the time available to scan the photograph, the lack of competing tasks, etc., that differentiate the

jury's task from that of the driver who was involved. The question that was raised had to do with the quantification of the expectancy effect. Namely, to what extent did the expectation of the driver, that there was an object in the road, affect the distance at which the object would be detected?

The author then referred to a study done by Roper and Howard (1938). These two engineers were working at General Electric Company on the development of headlamps for motor vehicles and had reported on an experiment on the visibility of a "pedestrian" test target. The tests were run on a test track and allowed measurements to be obtained of the distance of the car, which subjects were driving, from the pedestrian dummy when the subjects detected the dummy. At the conclusion of the experiment on the track, each subject was told the test was completed and asked to drive the car back to the reception area. On this return trip the subjects encountered another dummy target. The distance of the car from the dummy when a subject released the accelerator in response to the dummy was obtained. The measures taken during the official experiment on the track were represented as "expected" or "alerted" trials while those obtained after the subject was told that the experiment was concluded were considered to be "unexpected" or "unalerted" trials, since the subject was not expecting a pedestrian object in the road. A finding of the study that relates to the present issue was that in the "expected" cases subjects responded to the dummy at about twice the distance as in the "unexpected" trials. This result has been interpreted by some investigators (e.g. Henderson et al, 1987; Hyzer, 1993) to mean that, at least in night driving conditions, an expected object will be seen at twice the distance of one not expected.

This interpretation means that an object that is perceivable at 100 m by an alerted driver would be perceived at 50 m by an unalerted driver and, similarly, one perceivable by an alerted driver at 300 m would be perceivable at 150 m by an unalerted driver, etc. It is this 2:1 relationship, inferred from the study by Roper & Howard (1938), between the expected and unexpected cases that, I believe, cannot be generalized to all situations in night driving—not even those that use similar headlamp beams as those in the experiments or even to the very same study if it had been done using different car speeds or different reflectances of the dummy.

RATIONALE

It is my hypothesis that the differences between the "expected" and the "unexpected" results of the Roper & Howard (1938) study are largely due to the interaction of a number of variables that have an important effect on the visibility of the pedestrian dummy, and that these relationships do not produce a constant ratio between the "expected" and "unexpected" conditions. Some of these variables are: the characteristics of the headlamps' low beam pattern, the reflectance of the dummy, the size of the dummy and the position of the dummy on the roadway with respect to the headlamps and a number of human factors, such as attention, set, and situation awareness.

ANALYSIS

Headlamp Characteristics.

Let us consider the low beam pattern of the conventional headlamps mounted 24 inches (61 cms) above the road and the illumination they provide at a point 24 inches above the surface of the road, straight ahead of the headlamp on the right side of the car, at a distance of 250' (76 m). The illumination from both headlamps at that point, which can be calculated from the iso-candela plots of the low beam headlamp, is about 0.16 foot-candles (1.7 lux), and is a point on the curve in Figure 1 labeled "B.Low". If we now wish to

find the illumination provided by the same beam at half the distance, i.e., 125 ft. (38 m), we move up the curve to intersect that distance and read the illumination, which is about 0.78 f-c (8.4 lux). If it is assumed that the effect of a lack of "expectancy" of an object is a halving of the detection distance, then the above example shows the change in illumination that would occur if the visibility in the expected condition was 250' and the unexpected was therefore 125'.

In order to estimate the effect of different target positions and headlamp beams we can plot their illumination by distance, as done in Figure 1. Plot "A.Low" is the low beam illumination on a target 6 inches (15 cms) above the pavement in front of the right headlamp. Plot "B.High" is for high beams illuminating a point 24 inches (61 cms) above the pavement in front of the right headlamp. Plot "C.Low" is for one low beam headlamp illuminating a point 6 inches (15 cms) above the pavement located 72 inches (1.8 m) to the right of the lamp, as might apply to a motorcycle headlamp illuminating a point just above the right road edgeline.

It will be noticed that the illumination-distance plots have different slopes, showing that equal changes in illumination will not result in equal ratio changes in distance. For example, the same change in illumination that produced a doubling of the distance in the example described for beam "B.Low" produced ratios of 1.88, 2.25 and 1.74, respectively, for conditions "A.Low", "B.High" and "C.Low".

Thus, to take a simplified case, such as those above, in which it is assumed that the night driving visibility is determined by the illumination on an object, disregarding all other variables, Figure 1 shows that a doubling of visibility due to the "expectancy" factor does not hold up too well even when only target location and beams are varied.

In fact, visibility is determined by many variables and the criteria for "visibility" are also numerous. These factors undermine the concept of a simple effect of expectancy on visibility.

Probability of Detection.

As another example, we have calculated the visibility of the rear of a large trailer at night for a car driver, using typical low beam headlamps, who is approaching from behind the trailer. Given a low reflectance of the trailer it was found that the 50th and 99th percentiles, respectively, of the probability of detecting the trailer were 94' (29 m) and 83' (25 m).

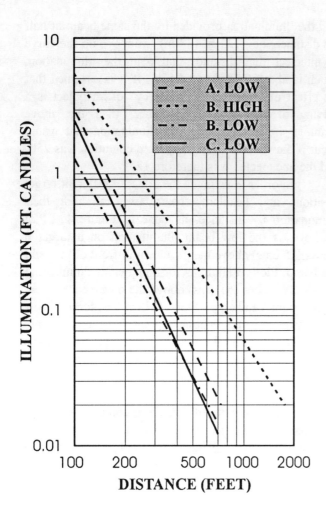

Figure 1. Illumination from headlamps by beam, object location and distance.

What this illustrates is a phenomenon that is clearly noticeable when participating in headlighting visibility tests. Namely, that the object to be detected, such as a pedestrian dummy or a trailer, rapidly looms out of the darkness after it is first faintly detectable. The above example shows that the difference in distance when the trailer was at threshold of detectability and when it was clearly detectable was only 11' (3.4 m), which is a change in distance of about 12%.

Speed.

In the above case, if the observer responds at the 99th percentile detection distance (83') within about 1.0 second to represent the "unexpected" case, in a car traveling 30 mph (48 km/h), the car will have covered another 44' and will be 39' (83'-44') from the trailer.

Using the 50th percentile detection distance as the "expected" visibility distance and a response time of 0.3 seconds in the "expected" case, the car travels about 13' (4 m) and will be 81' (94'-13') from the trailer. In this example, the measured visibility distance will be 39' (94'-55') in the unexpected case and 81' (94'-13') in the expected case. Thus, the expected: unexpected "visibility" distance ratio is about 2:1, as found by the Roper & Howard (1938) data. Obviously, the unexpected case is particularly susceptible to the speed of the car, which has little relevance to visibility in this context. Thus, if the above example is changed such that the speed of the car is 20 mph (32 km/h) and the response times remain the same, the measured visibility distances will be 54' (16 m) in the unexpected case and 85' (26 m) in the expected case. The resultant ratio of the visibilities in the expected: unexpected conditions is 1.6:1, which is a substantial deviation from the 2:1 ratio that has been used.

THEORY

Signal Detection Theory.

Signal detection theory (Green & Swets, 1966) provides another reason why a constant relation is not likely between visibility and expectation. Signal detection theory can help to explain the finding of the Roper & Howard study and the effect of the experimental trials on the subsequent "unexpected" pedestrian condition. The variable d' describes the sensitivity of the observer, such as the contrast threshold in a night driving task or even the effect of the driver's eye fixations, which are not constant. The other variable in the theory, "beta" is a measure of the response criterion of the observer that is analogous to expectation and is a function of the perceived conditions and the payoff or perceived risk. Because beta can have a broad distribution, the effect it has on a complex process like night visibility is not constant.

Expectancy Theory.

The expectancy theory of Baker (1961) can explain some of the reduction in visibility found in the Roper & Howard (1938) experiment's "unexpected" phase. After having been exposed to a series of trials in which pedestrian dummies were frequently encountered, the experiment ended officially for the subjects. They, appropriately, adjusted their beta criterion upward, indicating their prediction that there would be far fewer pedestrians encountered, which resulted in a stronger stimulus being required for detection of the pedestrian and, hence, a lower visibility distance. Certainly, a lack of expectancy will impair perception, but there is no evidence from this theory that it will be a constant effect on visibility in night driving.

Set.

The readiness to respond to a stimulus is also a function of the presence or absence of a warning, or readiness, signal (Karlin & Kestinbaum, 1968). If the readiness signal is presented within a short time of the stimulus it can reduce the response time to the stimulus. That is akin to the situation in the "expected" phase of the Roper & Howard (1938) test in which the subjects knew to look for targets, while no such prior information or warning signal, other than that associated with the driving task was available in the "unexpected" phase. Thus, part of the longer visibility distance in the "expected" phase was attributable to the readiness of the subjects to react, which is not relevant to visibility, but did affect the measure of visibility distance that was used.

Situation Awareness.

Situation awareness is also associated with these processes. It involves an awareness by the driver of the present and future position, speed and direction of the vehicle, other road users and the characteristics of the road environment . This is a phenomenon that is receiving extensive attention in aviation (Endsley, 1988), but is also applicable to driving. Situation awareness will be affected by expectancy and vice-versa.

But, situation awareness varies from moment to moment and, therefore, the level of expectation will also vary. But, the relationship between situation awareness and visibility is not known.

QUANTIFICATION

The quantification of the effects of expectancy, based on the above analyses and theories, will be difficult and cannot readily be generalized. A study (Johansson & Rumar, 1971) of the brake response times of drivers to a horn at the side of the road, which they had been told to anticipate within a few kilometers ("anticipation trials") and to a buzzer in the car of each subject which was activated intermittently with long

intervals ("surprise trials") found median response times of 0.54 and 0.73 seconds, in the "anticipation" and "surprise" trials, respectively. The difference of 0.19 seconds is a measure of the expectation effect for those tasks. Olson et al. (1984) measured the time taken by drivers to release the accelerator on detecting an object, 6 inches high (15 cms) and 36 inches (0.91 m) wide, in their lane after cresting a hill. The median time to release the accelerator ("perception time") for young and older drivers was 0.67 and 0.75 seconds, respectively, and the analogous 90th percentile values were 0.95 and 1.0 seconds. Olson et al. also obtained the times to detect the object with their subjects in an "alerted" state, after they had already had a "surprise" trial. The 50th percentile alerted perception times were about 0.5 seconds for the younger and older drivers. Those data show that the 50th percentile lag times in perception were about 0.17 (0.67-0.5) and 0.25 (0.75-0.5) seconds, respectively, due to lack of expectancy, for the young and older drivers. Clearly, such short lag times will have negligible effects on visibility.

It should be noted that the foregoing studies were done in daytime. It is conceivable that the effect of expectancy on visibility is greater at night and, in more complex driving tasks the effect on perceptual response lags should be greater.

There is one study that was done in darkness, concerning the response of drivers to a weakly reflectorized stop sign, which is pertinent to this discussion (Olson, 1988). The stop signs used were much less reflective than normal so that they would be seen at a relatively short distance and force the drivers to slow the car. The study was done on a rural, two lane road. The subjects drove a car and were told that the purpose of the study was to "take some measurements of driver-vehicle performance." An experimenter in the rear seat told them where to turn, the lane to use and the speed limits over a 10 mile course which finally ended in the test section. Subjects were told the speed there was 50 mph. At the top of a crest in the road the subjects encountered the stop sign. When the subjects released the accelerator pedal a counter was started which measured the distance to the stop sign in the "unexpected" condition. After the subjects had been told the true purpose of the study, the test was repeated on a round trip of about three miles on the test section of the road, during which they encountered the stop signs four times. As soon as a subject could see the stop sign they were told to say "sign" and the distance to the sign was measured in the "expected" condition. There were a few real road signs also present on the test road section that was used. For the ten subjects used in the study, the ratios of the "unexpected" to "expected" response distances varied from a low of 1:1.08 to a high of 1:2.64, with a mean of 1:1.65. The variation in these ratios is of substantial interest and shows that the inter-subject variation is so great that it is not appropriate to take any fixed value to describe the effect of expectation.

Among the four trials in the "expected" condition for each subject there was also substantial variation within subjects (Olson, 1988). As an extreme example, one subject who responded in the "unexpected" condition at 291 feet, responded at 482, ,729, 812 and 1053 feet in the four "expected" trials. What could cause such a large variation in the "expected" trials? It obviously can not be due to expectancy because the subjects were primed to look for the stop signs. It seems reasonable that the variation is due to the search adopted by the subject, attention to other aspects of the driving task and to the variation in the detection criterion that was adopted in the trials. Therefore, the differences between the "unexpected" and "expected" conditions is in part due to the expectation effect and partly to other sources of variation and particularly the detection criteria adopted by the subjects.

We conclude that the effect of expectation alone is much less than the ratio of the "unexpected" to "expected" response distances measured in this test and others of this kind. This is in no way intended as an indictment of the methodology used. On the contrary, the test was ingeniously done. It is merely that the results need to be interpreted in the context of factors, other than expectation alone, that must be operating in such signal detection paradigms.

CONCLUSIONS

Based on the analyses and the theoretical and experimental literature it does not appear that there is any justification for the assumption that the effect of a lack of expectancy reduces visibility distances in night driving by a factor of two, as has been suggested previously. Expectancy, while an important construct, is only one of many variables that affect visibility in night driving and there is no evidence to show that it acts in a constant and linear fashion across the variety of conditions that affect the visibility of drivers at night. There are inadequate data currently available to quantify the effect of expectancy. Each situation will need to be evaluated for its individual characteristics.

Meanwhile, expectancy is a variable that is in need of study. Perhaps the growing interest in and techniques of study of situational awareness, of which expectancy is a part , will increase our knowledge of the role of expectancy in highway safety.

REFERENCES

Baker, C.H. (1961). Maintaining the level of vigilance by means of knowledge of results about a secondary vigilance task. Ergonomics, 4, 311-316.

Endsley, M.R. (1988). Design and evaluation for situation awareness. Proc.,32nd annual meeting of the Human Factors Society, 97-101.

Green, D. & Swets, J.A. (1966). Signal Detection Theory and Psychophysics. New York: Wiley.

Henderson, R.L. (editor), (1987). NHTSA Driver Performance Data Book. US-Department of Transportation, Report DOT-HS-807-121.

Hyzer, W.G. (1993). More on the eye and camera. Photoelectronic Imaging, 35, (2)

Johansson G. & Rumar, K. (1971). Driver's brake reaction time. Human Factors, 13 (1), 23-27.

Karlin, L. & Kestinbaum (1968). Effects of number of alternatives on the psychological refractory period. Quart. J. Exp. Psychol., 20, 160-178.

O'Hare, D. & Roscoe, S. (1990) Flightdeck Performance: The Human Factor. Ames: Iowa State University Press.

Olson, P.L., Cleveland, D.E., Fancher, P.S., Kostyniuk, L.P. & Schneider, L.W. (1984). Parameters affecting stopping sight distance. NCHRP report 270, National research council.

Roper, V.J. & Howard, E.A. (1938). Seeing with motorcar headlights. Trans. Illum. Engr. Soc.,33.

Van Cott, H.P. & Kinkade, R.G. (1972). Human Engineering Guide to Equipment Design. New York: McGraw-Hill.

ADMISSIBILITY OF HUMAN FACTORS EXPERT TESTIMONY IN LIGHT OF *DAUBERT*

Martin I. Kurke, Ph.D., LL.B., CHFP
George Mason University
Fairfax, Virginia

The 1993 Supreme Court decision in *Daubert v Merrell Dow Pharmaceuticals, Inc.* created new, more stringent rules for admissibility of expert witness testimony based upon "scientific" knowledge as opposed to expert testimony based upon "technical or other specialized" knowledge. Because human factors professionals may testify on the basis of either of the two categories, they should be aware of the requirements for admissibility of both categories. Each time one is engaged as a potential expert witness, the human factors-ergonomics professional needs to determine the basis of the testimony he/she is about to provide and to notify the client attorney accordingly. This lecture describes the evolution of requirements for admissibility of both kinds of expert testimony. It also discusses the implications of additional requirements imposed upon all expert witnesses by the 1993 amendment of Rule 26 of the *Federal Rules of Civil Procedure*

Two recent changes in the legal system can be expected to impact upon human factors-ergonomics folks when they prepare to testify in court as experts. One such change is the December 1993 modification of the *Federal Rules for Civil Procedures*, which lists the requirements for reports by expert witnesses that must be complied with before the expert's evidence may be presented (FRCP, 1993). Rule 26 (a), (b) and (c) specify that

- The (written) report must be prepared and signed by the witness.
- The report must contain a complete statement of all opinions to be expressed, including the basis and reasoning therefor. Data or other significant information considered by the expert must be included. Presumably data considered and rejected must be included.
- All exhibits to be used as a summary or as support must be included.
- All publications authored by the witness within the past 10 years, regardless of relevance, must be listed.
- The compensation to be paid for this

testimony and study must be included.
- A list of other cases where this expert gave deposition or trial testimony within the last 4 years must be submitted with the report, without regard to relevance to this subject matter.

Earlier the same year, the U.S. Supreme Court decision in *Daubert v Merrell Dow Pharmaceuticals, Inc.* significantly changed the rules under which experts may testify. This lecturer will consider the implications of both changes to the legal system upon the content and admissibility of expert human factors-ergonomics evidence.

Human factors specialists often are requested by attorneys to apply their expertise to the study of issues surrounding some aspect of their scientific or professional knowledge. The attorney anticipates that the results of those studies may become the basis for expert testimony in a lawsuit. At trial, the attorney proffers to the court that the human factors person is an expert in his/her field and should be qualified as an expert witness in the instant case.

A leading jurisprudential reference book states that "The proper use of the human factors expert is in proving that a product was defectively designed in that the design failed to take into consideration human behavior." (Bliss, 1979, p. 117). The author describes a human factors expert as a specialist in the area of man-machine relationships and interactions. Distinguishing a human factors expert from a clinical psychologist, physician and engineer, the human factors expert is described as specializing in the study of machine design as it affects the behavior of the machine operator. Essentially the human factors is an expert in some aspect of human behavior, anatomy or physiology as those endeavors relate to machine design or machine functioning.

Thus, the human factors person may be proffered to be an expert in the design or operation of a class of devices, products or systems. Depending upon the expert's training and experiential base, he/she may tend to testify in either of two orientations. Testimony may be offered as to the applicability of the (engineering) design to human performance capabilities and limitations. Alternatively, the expert will be prepared to testify as an expert in human cognition and behavior within a system as it is affected by the design of the system or its components. Each type of expert will offer a different type of testimony that may be differentially admissible under rules established by the Supreme Court 1993 decision in *Daubert v Merrell Dow Pharmaceuticals, Inc.* The difference in admissibility may well rely upon whether the testimony is offered as "scientific", "technical" or "other." In the remainder of this lecture I discuss the legal principles concerning who may give expert evidence and the admissibility of such evidence.

In a case in which the qualifications of a psychologist was denied admissibility because she was not a psychiatrist, *(Jenkins v U.S.,*, 1962) the court on appeal ruled that the degree held by

the witness was irrelevant.   The decision noted

"Two elements are required to warrant expert testimony: (1) the subject of expert inference must be so distinctly related to some science, profession, business or occupation as to be beyond the ken of the average layman, and (2) the witness must have such skill, knowledge and experience in that field or calling as to make it appear that his opinion or inference will probably aid the trier in his search for truth."

As to the content of the testimony, it was ruled in *Frye v US* (1923) that the court shall admit

"expert testimony deduced from a well-recognized scientific principle or discovery, The thing from which the deduction is made must be sufficiently established to have gained general acceptance in the particular field in which it belongs"

In 1975, the Federal Rules of Evidence appeared to overturn *Frye's,* "general acceptance" rule, stating:

"If scientific, technical or other specialized knowledge will assist the trier of fact to understand the evidence or to determine a fact at issue, a witness qualified as an expert by knowledge, skill, experience, training, or education may testify thereto in the form of an opinion or otherwise." *Federal Rules of Evidence Rule 702.*

In 1993, *Daubert* established a threshold standard for the admissibility of testimony that is based upon scientific knowledge. It stresses the validity of the underlying scientific theory. It adds four tests on the admissibility of scientific evidence: (1) Falsifiability, i.e. the underlying theory's logical refutability or testability; (2)

Error rate, i.e., a measure of scientific validity, (3) Peer review and publication; and (4) General acceptance. *Frye's* general acceptance test has been replaced by *Daubert's* validity standards and FRE Rule 702. Thus, *Daubert* imposes a roadblock to the "scientific" testimony of experts arising more on the basis of clinical or other professional consensus than on vigorous research.

Because *Daubert* addresses only "scientific knowledge," the admissibility of evidence related to "technical or other specialized knowledge" *(FRE 702)* based upon one's "profession, business or occupation" *(Jenkins v US)* is not affected.

A recent paper (Faigman, 1995) addresses the applicability of *Daubert* to psychology. The article concludes that psychologist's evidence based upon clinical consensus is not "scientific" and need not be admitted as such under *Daubert* rules, while a psychologist who testifies about eyewitness memory relies upon results of rigorous research. Such evidence is subject to those rules.

The question we must now consider is whether or not we consider human factors to be based upon science, technology or "otherwise specialized knowledge" Does human factors testimony based upon an engineering-based orientation differ from psychology-based orientation in admissibility? Must one expert's testimony meet *Daubert* standards while the other expert need only meet the *Jenkins* standard? Federal Rules of Evidence Rule 702 does not require expert witness evidence to be based upon scientific knowledge. The knowledge base may be technical or otherwise specialized so long as it assists the trier of fact to understand or determine a fact at issue.

The preferred way to achieve safety in any system is to design it as free of hazards as possible. Such a goal often may not be possible, and in such cases, the next best thing is to remove the likelihood of injury by a design that will remove the hazard. If the hazard can not be removed by design, then guards or interlocks should be incorporated into the design to minimize the risk. If there is no way to eliminate the hazard by design, then people should be warned by signals, labels, training and instructions how to avoid harm. (Christensen, 1986) Clearly, expert testimony concerning defects in design such as failure to eliminate or guard against hazards may differ in emphasis (along a psychology-engineering continuum) from expert testimony concerning the effectiveness of the same product's warning, labeling or training components.

As forensic human factors professionals we face a dilemma. More often than not, we are engaged by attorneys who may be totally unfamiliar with human factors as a specialty. Many of them do not know how to use us for they are unfamiliar with our capabilities, our limitations, or both. As expert witnesses our job on the witness stand is to educate the trier of fact about these matters. Often, our job also is to similarly educate attorneys who engage us about the same thing *before we are called to the witness stand.* When agreeing to study a case and to offer testimony, the human factors expert would be wise to advise the engaging attorney whether, in the expert's opinion, the opinion to be rendered derives from a scientific, technical or "other professional ..." base. Knowing the expert's base, the attorney may be able to frame the direct examination and to constrain cross examination accordingly,

References:

Bliss, W.D. (1979) *Defective product design -- Role of human factors. American jurisprudence: Proof of facts, 2nd series.* **18:** 117-148.

Christensen, J.M. (1986) Forensic human
    factors. **In** M.I. Kurke & R.G. Meyer
    [eds] *Psychology in product liability and
    personal injury litigation.* N.Y.,
    Hemisphere Publ. Co. pp 33-79

Daubert v Merrell Dow Pharmaceuticals, Inc
    113 S. CT 2786 (1993).

Faigman,, D.L. (1995). The evidentiary status of
    social science under *Daubert:* Is it
    "scientific," "technical," or "other"
    knowledge. *Psychology, Public Policy
    and Law,* **1**: No.4., 960-979.

Federal Rules of Civil Procedure. Rule 26
    (Effective December 1, 1993)

*Federal Rules of Evidence for United States
    Courts and Magistrates* (1975), St. Paul,
    MN, West.. § 702

Frye v United States, 293 F. 1013 (D.C. Cir.
    1923)

Jenkins v United States. 307 F2d 637 (D.C. Cir
    1962)

EXPLAINING TO A JURY HOW AND WHY PEOPLE FALL

David A. Thompson
Industrial Engineering and Engineering Management
Stanford University
Palo Alto, CA 94305

Human factors expert witnesses may be called upon to explain the mechanics of a particular **physical** falling event to a jury using only verbal, **intellectual** tools. Photos or diagrams of the falling event area usually show only the environment, not the human interacting with the environment during the critical falling event. The reconstruction the accident may be done with computer-generated 3-D mannequins to substitute for the humans involved in the falling event. This paper discusses the illustration of the mechanics of the falling process using 3-D mannequins to illustrate the laws of motion and of human behavior as accurately as possible.

Introduction

Falling because of slips, trips, or missteps is a national health problem Estimates of 12,200 deaths a year (Tompkins, 1983) and 187,000 annual injuries (recent CPSC estimate in personal communication) from all falls represents a very large personal and business cost). Much work has been done to reduce these incidents (English, 1996; Gray, 1990; Templer, 1992; and US National Engineering Lab, 1979), but they remain a growing source of litigation (Rosen, 1995).

Whenever a human factors expert witness is called upon to explain how a particular falling event happened, he or she is faced with an interesting "information conversion" problem, explaining a three-dimensional **physical** event to a jury primarily with the one-dimensional **intellectual** tools of logic and words. Photos or diagrams of the area where the failing event are helpful, but they may show only the environment (sidewalk stairs, curb), not the human interacting with the environment.

Explaining a Falling Event

An accurate description of how an individual interacts with his or her environment during the critical few-hundred milliseconds of the failing event is vital to the understanding of the event by one's client and a jury. However, reconstructing the accident with a live human, anthropometrically similar to the injured party, may be difficult or impossible. The original physical environment may not be available or may no longer exist. Finding an "exemplar human" to act out the failing event, and staging the event in a safe manner that also replicates the best evidence about its occurrence, may also be difficult. In addition, explaining Newton's laws of motion in the abstract rather than how they probably acted on a plaintiff is not done easily, even in a classroom. As a last resort, the human factors expert may be left talking his or her way through a rough acting out of the falling event in front of the jury, all the while explaining the ways in which his or her demonstration is not like the real thing (while trying not to be injured during the demonstration).

This paper presents several examples of solutions I have found to solve these problems, using computer-generated 3-D mannequins as a substitute for the humans involved in the falling event. Through a series of frames illustrating the progression of the event's critical stages, the physical realities of the event can be related to the world we personally live in. And they can be presented at a speed or pace that is not too rapid to be digested by someone not familiar with the biomechanics of falling. Like the rest of us, the jury members have had their slips and falls, but generally have no conscious understanding as to why and how they fell, or why and how others fall.

I am most familiar with the mannequins generated by the MANEQUIN[tm] program from Biomechanics, Inc., which has allowed me to recreate the biomechanics of the event. However, this software is only one of a growing body of humanoid models available, and will be used here only for illustration of the principles involved. The MANEQUIN[tm] program is in fact a very difficult one to use, and is no longer technically supported by its source. However, there are fixes around some of the difficulties that make it useable in falling event reconstruction.

The MANEQUIN[tm] program allows the posing of 5th, 50th, and 95th percentile wire frame models (3-D models with hidden line perspectives) of thin, average, and heavy body builds, of the male or female sex. The depiction of people of different ethnic origins is also possible. These models may then be manipulated into any humanly possible posture (e.g., the head will not rotate beyond about 80 degrees, the forearm will not supinate or pronate more than 90 degrees, and knees and elbows bend as expected). The various parts of the body may be manipulated to duplicate a selected pose, or a point in space may be selected for the hand to reach toward in a natural manner. While the program accurately represents nearly the full range of human sizes and postures, a user must have a relatively high frustration tolerance to obtain the desired results. In many cases, however, I have found the results are worth the effort.

Reconstruction of the Falling Event

The reconstructing of the falling event begins with the understanding of the physics and biomechanics of the event, to the extent that the evidence and testimony allow. A major element in this is the initial and subsequent location of the falling party's center of mass. This center of mass, following the precepts of Newtonion mechanics, generally tends to follow a ballistic trajectory to the ground during a slipping event, or a rotational path during a tripping event, unless altered or interrupted by outside forces, facilities such as steps or handrails, or the body's recovery attempts. Following the subject's reaction time delay, motions of the arms, hands and opposite foot (the one not slipping or tripping) during the event should be understood as much as possible, to the extent that they may have or should have aided in the recovery of some degree of balance. The mechanics of the falling process must be as true to the laws of motion and of human behavior as possible, because these are where the biomechanical accident reconstruction is most vulnerable to challenge. Plaintiff's experts must tie this down, because defense experts will certainly question it, and vice versa. [It is not the intention of this paper to discuss the underlying theories and principles of falling.] I have used these models to explain why and how someone fell in the manner in which they testified, or why and how they probably did not fall in the manner they described.

Examples of Falling Event Reconstruction

One slipping event and one tripping event will be used to illustrate the use of 3-D mannequins in reconstructing falling scenarios. In **Figure 1,** the upper left figure shows the primary force vector through the lower leg during stair descent. Also shown are the vertical force component representing the effect of gravity, and the horizontal force component representing the effect of forward momentum. A firm footfall, with adequate friction, allows the leg to rotate about this point while the alternate foot proceeds to its own footfall one step lower. In a classic slipping event, however, inadequate friction does not restrict the sliding of the foot off of the stair tread and the consequent rotation of the body about its center of mass as it falls to the steps.

A tripping event, however, typically traps a toe or foot causing the momentum of the center of mass to rotate the body forward. The injuries are quite different in these two cases; the slipping event normally results in bruising or injury to the pelvis or lower back, while a tripping event may cause knee, arm and shoulder injuries, and other problems depending on the distance fallen and other factors. Analysis of the medical records will indicate whether the injuries are consistent with the claimed manner of falling.

Figure 2 shows the sequence of events involved in a forward trip on an upper balcony in which a person's foot was allegedly trapped in a carpet tear while walking forward carrying a laundry basket. The upper and lower left figures show the position and basic dimensions of the person at the time of the start of their left foot tripping event. The upper right figure shows the beginning of forward rotation; with the left foot trapped, the right foot attempts unsuccessfully to regain balance. Also shown is a slight rotation to the left about a vertical axis through the center of mass caused by the left foot trapping. The final two frames show the subsequent position of the body after the fall. A forensic issue in this case was whether the person's claim of allegedly having fallen over the balcony rail as a result of the left foot tripping was a credible claim, given that the tripping happened 24" from the rail that he was walking parallel to.

Software Modifications

While I have found the MANEQUIN[tm] program to produce very good 3-D wire frame mannequins of men and women of various sizes and ethnicities, it does not allow very elegant drawing capabilities of the surrounding environment. However, manipulation of MANEQUIN[tm] software to achieve good quality graphics is possible. For example, better line definition and greater drawing refinement occurs when the MANEQUIN[tm] files are exported to a PAINT program or to a CAD drawing program. I draw selected human postures in MANEQUIN[tm] and then convert the files to the *.DXF format (or a *.BMP format) and export them to SKETCH (by Autodesk). This allows me to draw elements of the environment (stairs, handrails, entrances, machine elements, etc.) with much greater sophistication in SKETCH than I am able to in the MANEQUIN[tm] program. SKETCH also imports the selected files at the same scale as the environment I have drawn, so that the 3-D postures fit well.

Timing

The models of human postures during a falling event are obviously static, but because they allow the human factors analyst to freeze time in order to understand and explain rapidly occurring events, they misrepresent the effect of time on the behavior of the person falling. During the fall, events occur very quickly. The peak speed of the unobstructed, swinging foot moving past the planted foot may be nine to ten feet per second, or about one foot in 100 milliseconds. The time for an unsupported torso to fall to the ground in a slipping event may be only 350-400 milliseconds. These times are too short for all but the most instinctive reaction responses by the falling person---probably not enough time to reach and grasp a handrail, for instance.

Summary

The human factors specialist in his or her role as a forensic expert has the obligation to explain to the client and to the court, in as sophisticated and professional manner as possible, how an individual involved in an injury event interacted with the environment in a manner that caused the injury. I have found the use of 3-D wire-frame mannequin models to simulate human behavior in the reconstruction of a falling event to be of significant assistance in this regard.

References

English, W. (1996) *Pedestrian Slip Resistance, How to Measure it and How to Improve It*, Wm. English Inc, Alva, FL

Gray, B. E. (1990) *Slips, Stumbles, and Falls; Pedestrian Footwear and Surfaces*, ASTM STP 1103.

Rosen, S.I., (1995) *California Case Law: Slip and Fall, Trip and Fall*. Rugby Hall Books, Solano Beach, CA.

Templer, J. (1992) *The Staircase; Studies of Hazards, Falls, and Safer Design*, MIT Press.

Tompkins, N.C. (1993) Friction and Gravity, *Occupational Safety and Health*, October, p. 51.

US National Engineering Lab, (1979) *Guidelines for Stair Safety*, Report prepared for the Consumer Product Safety Commission, US Department of Commerce, NBS, May.

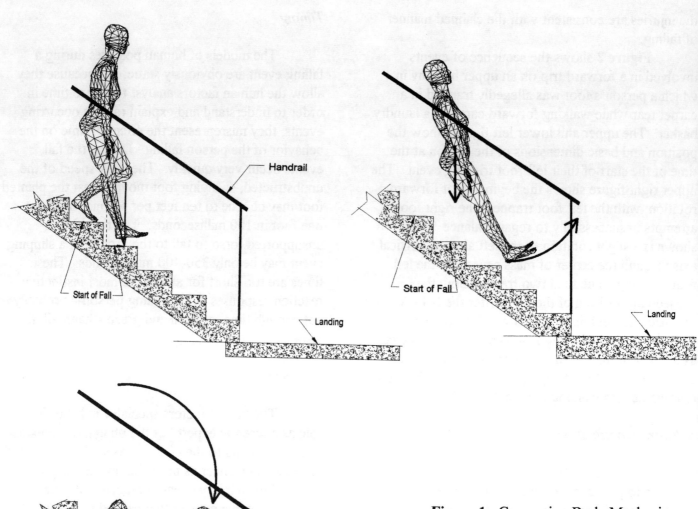

Figure 1. Comparing Body Mechanics during a Slip and Trip while Descending Stairs

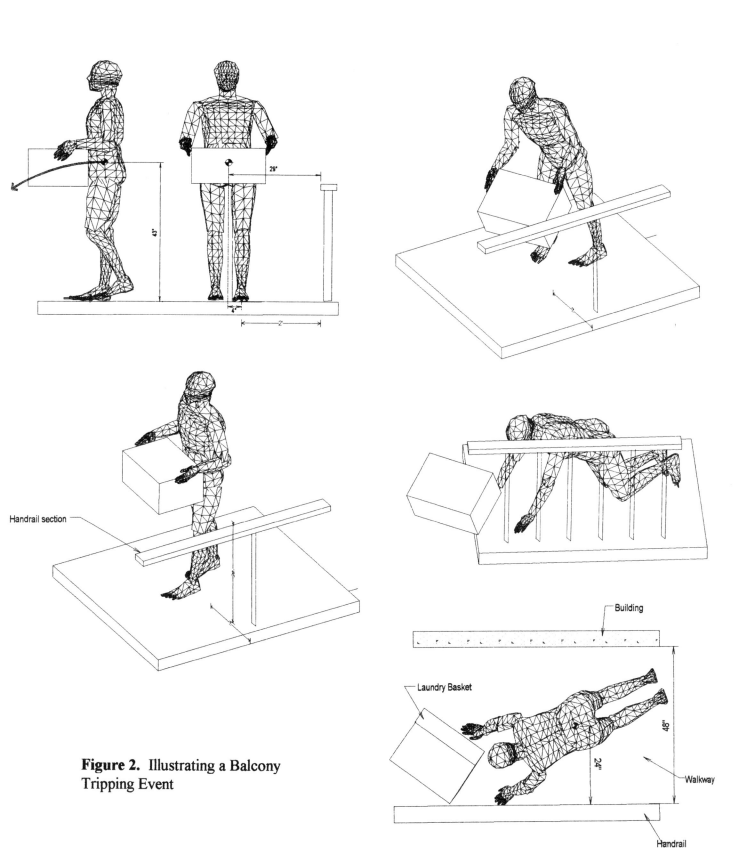

Figure 2. Illustrating a Balcony Tripping Event

COMPARATIVE RISK PERCEPTION OF COMMON ACTIVITIES

Alison G. Vredenburgh, M.S., CPE, Ilene B. Zackowitz, M.S., H. Harvey Cohen, Ph.D., CPE
Error Analysis, Inc.
La Mesa, California

Research on risk perception has been motivated by a variety of practical and theoretical interests, from public policy making and consumer product litigation to psychophysics and cognitive psychology. This paper examines risk perception data for activities that are directly related to public safety. Vredenburgh and Cohen (1993, 1995) have conducted several studies in which subjects responded to risk questionnaires about recreational and work related activities. This paper provides analyses of the risk perception data not included in the previous reports from these studies. The perceived-risk ranks for the recreational and diverse activities are given. Subjects regarded hang gliding as the most dangerous activity and water-skiing as the least dangerous of these sports. Although the full 7-point scale was used by at least some subjects, the means all begin at the middle of the scale and ran towards the high-risk end. Taken at face value, the results of these questionnaires have direct applications to policy and legal issues.

Research on risk perception has been motivated by a variety of practical and theoretical interests, from public policy making and consumer product litigation to psychophysics and cognitive psychology. Consultants in the area of Human Factors and safety often testify in cases concerning the "reasonableness of conduct" of plaintiffs and their "assumption of risk." Some work-related examples of such incidents are: Mexican migrant workers who have been killed as a result of riding in the open bed of a pick-up truck as part of their jobs and workers who fell from ladders while carrying heavy objects. Examples of recreation related incidents include accidents involving ski equipment and All-Terrain Vehicles (ATVs).

The perception of risk may be influenced by the type of warning information presented to product users. Moore (1990) found that while warnings alone increased the perception of risk, risk estimates were highest for participants who received both hazard and system information about the product. Familiarity with an activity is another factor determining whether an individual will read a product warning. The more confident and familiar an individual is with a product or activity, the less likely he or she is to read and comply with warnings (Otsubo, 1988). Godfrey and Laughery (1984) found that consumers may fail to perceive a difference in the level of hazard when changing to a similar but more hazardous product, and that they may also fail to read a stronger warning on the new product.

Products used for recreational activities may bear a greater burden for manufacturers because the danger of the activity itself may be an attractive feature. The consumer may assume responsibility for his or her own actions, but the manufacturer can be held liable in accident cases if no warnings (or insufficient warnings) are supplied with the recreational equipment. Kerr and Svebak (1989) have identified an arousal-seeking personality who prefers to participate in high-risk sports. Both the need for new experiences and attraction to high-risk characterize the sensation-seeker's participation in sporting activities. Manufacturers and facilitators of these activities should ensure that users have access to adequate warnings and instructions, so that participants can then decide whether or not to comply with them.

This paper examines risk perception data for activities that are directly related to public safety. When people choose to engage in a given activity, it is reasonable to assume that they consider that activity in the context of related activities. When considering two recreational activities that are similar (e.g., both offering high speed thrills), perceptions of relative riskiness may be relevant to decision-making behavior and to consideration of right-to-know issues. This argument can also be made for occupational tasks. In order to explore the relationship between actual and perceived risks or the underlying dimensionality of risk-related cognition, the choice of activities or potential threats to be considered, may be somewhat arbitrary, subject only to the demands of good experimental design (e.g., studies by Wogalter, Desaulniers, and Brelsford, 1986; Wogalter, Brelsford, Desaulniers, and Laughery, 1991; and Slovic, Fishoff, and Lichtenstein, 1979). Studies with direct practical goals need to address specific activities of interest.

Vredenburgh and Cohen (1993, 1995) have conducted several studies in which subjects responded to risk questionnaires about recreational and work related activities. One study (1993) obtained perceived risk rankings for six types of recreation activities. Subjects completed surveys during their lunch breaks while participating in either snow-skiing or scuba-diving. The other study (1995) recruited members of four different ethnic groups to complete a questionnaire that included a 7-point rating scale for common activities, most of which were either recreational or work related. Neither of those reports provided results comparing the activities. The first report addressed interrelations with other measures. The second report examined the extent of differences among the groups. This paper provides analyses of the risk perception data from these studies not included in the previous reports.

METHOD

Information about the studies is found in the original reports. Details relevant to the present analyses are described here. For the recreation study, 38 snow skiers (contacted at ski resorts) and 27 scuba divers (contacted at diving facilities and on dive boats) provided rankings (with 1 as *most dangerous*) of a set of six active recreational activities. For the diverse activity study, 316 subjects contacted through church congregations and social clubs provided ratings on a 7-point scale (with 1 and 7 anchored as *no risk* and *high risk*, respectively) for eleven activities; 283 to 300 usable ratings were provided for each activity.

RESULTS

Table 1 provides descriptive statistics for the perceived-risk ranks for the recreational activities. There was high agreement among the subjects concerning the rankings of the activities. The mean rankings of the activities ranged from 1.26 (most dangerous) to 5.02 (least dangerous). At the individual level, each activity was ranked 6 (least dangerous) by at least some subjects. Table 2 provides the percentages of

responses for each activity. The table shows that 95% of the respondents ranked hang gliding and 57% ranked All- Terrain Vehicle riding (ATV) as either the most dangerous or second most dangerous activity. No respondent ranked water-skiing, 6% ranked jet skiing, and 12% ranked snow skiing as the most dangerous or second most dangerous activities. Most respondents (72%) ranked water-skiing in the lowest two perceived danger rankings.

Table 3 provides the descriptive statistics for the perceived-risk ratings of the diverse activities. Although the full 7-point scale was used by at least some subjects, the means all begin at the middle of the scale (3.89) to the high-risk end (6.01). The variability in response was striking: each activity was rated as 1 (*no risk*) by at least some of the participants, and as 7 (*high risk*) by some others. Seven percent (7%) of the subjects gave a rating of 1 (*no risk*) to every activity. Looking at Table 4, the percentage of responses for each rating, it is evident that the numbers jump down from the rating of 1 to 2 and then increase in risk up to a rating of 7. This group lowered the mean ratings for the activities. Drinking and driving was rated as the most dangerous activity, with 80% of respondents perceiving it as high risk (6 or 7).

A significant positive correlation was found between whether subjects saw themselves as risk-takers and whether they enjoyed activities that others may consider high risk (r=.44, p<.001). An overall perceived risk score was computed by averaging the ratings of the 11 scale items. This overall score was used as a dependent variable (DV) to determine if there was a difference in risk perception between subjects who saw themselves as risk takers and those who did not. An analysis of variance (ANOVA) was performed and no significant difference was found between these two groups. A second analysis of variance was performed (using the overall score as a DV) to determine whether there was a difference between those who enjoyed activities that others considered to be dangerous and those who did not; again, no significant difference was found.

DISCUSSION

Taken at face value, the results of these questionnaires have direct applications to policy and legal issues. Sport participant groups are reasonably consistent in their ranking of the dangerousness of the various recreational activities (although the use of ranks rather than ratings prevented discussion of the magnitude of differences in perceived risk); hang gliding and water-skiing occupied the opposing extremes of the scale. Regardless of ethnic background, groups generally perceived the diverse activities to have at least mid-scale risk, with carrying 10 lbs. up a ladder right at mid-scale and drunk driving at the high-risk end.

The variability in response choices is probably a result of individual differences. As past research has demonstrated, a multitude of factors influence risk perceptions. Because familiarity, gender, culture, and knowledge and training affect perceptions at an individual level, care should be used in the application of these data.

It is important to note that the data were not as variable as they first appear if one studies Table 4. Excluding the 7% of subjects that rated all the activities as "no risk," a clear pattern emerges. Subjects were more likely to assign low risk scores to some activities than others. Still, the diversity of responses is of potential interest. Were there really people who considered that riding a motorcycle without a helmet has no risk, or that going down a dark stair case at home belongs at the high-risk end of the scale, above such activities as driving after drinking or using firearms? Without further information, such as might be gained by interviews with the subjects, it is impossible to explain the variability in responses, and the oddness of some of the responses. Perhaps there were differing interpretations of the activities, for example: *working with chemicals* could be interpreted as lifelong occupational exposure to toxic chemicals, or as one afternoon of using paint stripper on a chair.

The variability of the responses warrants caution in the application and interpretation of risk perception data, particularly at the individual level. However, some useful patterns emerged. Many accidents involve people working with heavy machinery, going down dark stairways in their homes or workplaces, and carrying objects up ladders. Yet, the data indicate that subjects saw these activities as comparatively low risk with only 36% or fewer of the respondents assigning a 6 or 7 on the scale. Perhaps, if the public saw these behaviors as higher risk, they would take more heed of their behavior and would have fewer accidents.

ACKNOWLEDGMENTS

Thomas Ayres, Ph.D., provided valuable assistance in data analysis and preparation of this paper.

REFERENCES

Godfrey, S.S., and Laughery, L.R., (1984). The biasing effects of product familiarity on consumers' awareness of hazard. Proceedings of the Human Factors Society 28th Annual Meeting, 483-486.

Kerr, J.H., and Svebak, S. (1989). Motivational aspects of preference for and participation in "risk" and "safe" sports. Personality and Individual Differences, 10, 7, 797-800.

Moore, L.L., (1990). The effect of hazard and system information on estimated risk. Proceedings of the Human Factors Society 34th Annual Meeting, 508-512.

Otsubo, S.M., (1988). A behavioral study of waning labels for consumer products: Perceived danger and use of pictographs. Proceedings of the Human Factors Society 32nd Annual Meeting, 536-540.

Slovic, P., Fishoff, B., and Lichtenstein, S., (1979). Rating the risks, Environment, 21, 14-39.

Vredenburgh, A.G., and Cohen, H.H. (1993). Compliance with warnings in high risk recreational activities: skiing and scuba. Proceedings of the Human Factors and Ergonomics Society 37th Annual Meeting, 945-949.

Vredenburgh, A.G., and Cohen, H.H., (1995). High risk recreational activities: skiing and scuba - what predicts compliance with warnings. International Journal of Industrial Ergonomics, 15, 123-128.

Vredenburgh, A.G., and Cohen, H.H., (1995). Does culture affect risk perception?: Comparisons among Mexicans, African-Americans, Asians, and Caucasians. Human Factors and Ergonomics Society 39th Annual Meeting, 1015-1019.

Wogalter, M.S., Desaulniers, D.R., and Brelsford, Jr., J.W. (1986). Perceptions of consumer products: hazardousness and warning expectations. Proceedings of the Human Factors Society 30th Annual Meeting, 1197-1201.

Wogalter, M.S., Brelsford, J.W., Desaulniers, D.R., and Laughery, K.R., (1991). Consumer product warnings: The role of hazard perception. Journal of Safety Research, 22, 71-82.

Table 1.
Descriptive Statistics of 6 **ranked** recreational activities (*1 = most dangerous to 6 = least dangerous*)

Activity	Mean	Standard Deviation	N
Hang glide	1.26	0.76	65
ATV	2.62	1.21	65
Scuba	3.63	1.56	65
Snow ski	4.15	1.41	65
Jetski	4.28	1.1	65
Water-ski	5.02	0.87	65

Table 2.
Percentages of responses ranking each activity.

Activity	1=most dangerous	2	3	4	5	6= least dangerous
Hang glide	83	12	3	0	0	2
ATV	12	45	25	11	3	4
Scuba	4	25	25	14	14	18
Snow ski	0	12	28	19	15	26
Jetski	0	6	17	34	29	14
Water-ski	0	0	5	23	38	34

Table 3.

Means of **rated** risk of activities (*1 = no risk to 7 = high risk*)

	Means	SD	N
Driving a car after drinking	6.01	1.74	292
Riding a motorcycle without a helmet	5.98	1.75	292
Smoking cigarettes	5.69	2.06	300
Bungee-jumping	5.67	1.89	283
Hitch-hiking	5.58	1.98	286
Using firearms	5.49	2.02	290
Working with chemicals	5.26	1.8	294
Riding in the bed of a pick-up truck	4.94	1.93	292
Working with heavy machinery	4.7	1.85	286
Going down a dark staircase in your home	4.08	1.93	296
Carrying 10 lbs up a ladder	3.89	1.89	300

Table 4.

Percentages of responses rating risk levels of each activity (*1 = no risk to 7 = high risk*)

	1= none	2	3	4	5	6	7= high
Driving a car after drinking	7	0	3	5	5	18	62
Riding a motorcycle without a helmet	8	0	2	3	11	15	61
Smoking cigarettes	13	1	2	3	9	14	58
Bungee-jumping	10	1	2	6	12	18	51
Hitch-hiking	11	2	2	6	10	19	50
Using firearms	9	5	6	6	11	11	52
Working with chemicals	7	3	5	12	22	15	36
Riding in the bed of a pick-up truck	9	5	9	14	18	16	29
Working with heavy machinery	9	5	7	20	23	13	23
Going down a dark staircase in your home	11	15	14	18	15	11	15
Carrying 10 lbs up a ladder	14	11	16	23	15	8	13

A CASE STUDY OF USING HUMAN FACTORS ANALYSES ON WARNINGS

S. David Leonard
University of Georgia
Athens, Georgia

Edward W. Karnes
Metropolitan State College of Denver
Denver, Colorado

On occasion defects appear in products after some period of use. Often, the products have been distributed to many unknown individuals requiring a program to inform them of the hazardous defect. In this study a comparison is made between two warning programs one of which did not utilize good human factors principles (considering them too academic) and a second which concerned itself with the problem of notifying individuals of an unknown and potentially deadly hazard. An analysis of the human factors principles involved in such programs was performed. The human factors principles were discussed in terms of the failures of the one program to reach the target audience with information in an acceptable form. In addition, the procedures for developing an adequate warnings program are described and the techniques for testing the warnings with individuals who are akin to the target audience were presented.

INTRODUCTION

This paper describes the development of a warnings program that resulted from the human factors analyses of inadequate warnings approaches. The basic problem which created the need for warnings was the failure and incipient failure of a type of flexible metal connector used with gas appliances. Due to the failures, several deaths occurred, and ultimately, suits were filed that resulted in court cases which hinged primarily on the issue of adequacy of warnings.

The need for warnings usually involves products whose hazards are known when they are introduced into the workplace or marketplace. However, a variety of products may come to produce unexpected hazards after they have been in use for some time, for example, the automobile. Automobiles are often recalled because a defect producing a safety hazard has been discovered. Typically buyers of new cars have their names and addresses recorded at the time of purchase and are notified by mail of any recall. The present paper involves a similar but more complex problem and describes some efforts that failed and the use of human factors analyses and testing designed to produce a more successful warning program.

Although the term ergonomics is frequently superseding what has generally been called

human factors, a key concept for the activities performed by ergonomics/human factors specialists remains the knowledge of and attention to human capabilities and behaviors. Many strides have been made toward making the human's job easier by automating operations in the workplace and out of it. However, such advances do not come without problems. For example, anti-lock braking systems require behavior contrary to that which many people have learned. Therefore, it is necessary to inform the users of such systems that they need to modify their behaviors. If there were no individuals who had learned to pump their brakes, it would not be necessary, but we know that there are. Knowledge about the behavior of humans is necessary if we are to provide adequate warnings. It is necessary to have knowledge of the general tendencies of humans, the behaviors that may have been learned by particular groups, and also the variability in human abilities and behavior. The present study is concerned with the fact that the characteristics and likely behavior of individuals was not considered in establishing a warning program.

BACKGROUND

A brief history of the problem that produced the human factors effort for the warning program will be useful in order to recognize the relevance of the involvement of forensic human factors experts in the situation. Prior to 1968 flexible metal gas connectors with soldered or brazed end fittings were used in a variety of construction projects, primarily in the Midwestern and Western states of the United States. Although the 1968 ANSI Z21.24 standard prohibited this type of construction, by the time it was disseminated many such connectors had been installed and many of them remained in place. It is also possible some of those in stock at building supply outlets may have been used as replacements by individuals who were not aware of the potential for failure. With the passage of time the connectors began to fail. A gas utility company realized this was a possible cause of fires in homes in its service area. The Consumer Products Safety Commission (CPSC) was notified, and in turn, CPSC notified the American Gas Association which notified all of its public utility company members. One utility estimated that in its service area 40,000 to 45,000 homes were at risk. These were mostly older homes which meant that the inhabitants tended to be low income individuals, and some were located in areas where many inhabitants did not speak English as their native tongue.

The public utility company initiated a campaign to alert the public to the hazard. The campaign occurred over a period of about five years and involved the activities and materials listed in Table 1. Over a period of about seven or eight years including the time of this campaign the utility averaged about 100 incidents a year that could be attributed to the defective connectors with about 30 fires and 70 leaks per year resulting in property damage, injuries, and some deaths. Some of the deaths led to law suits and the entry of human factors forensic experts.

HUMAN FACTORS INVOLVEMENT

During two trials involving the public utility company, human factors testimony was a key factor. Testimony by the human factors expert concerning the initial safety campaign of the utility company dealt with the inadequacy of the campaign's procedures to reach the target audience and the failure of the utility company to evaluate the campaign's effectiveness. Specific warning and safety campaign deficiencies identified included:

- All warnings were written at a very high level with an average grade level of 14.12.

- Customers were told not to inspect their connectors but to call the utility or a plumbing contractor if they suspected a faulty connector; however, they were not told what the indications of a faulty connector were.

- Emphasis was placed on olfactory warnings, that is, smelling the gas. By the time an odor occurred the hazard was imminent.

- The messages were written in terms that did not provide clear indication of the hazard. Rather than terms such as danger, warnings, fires, and explosions, the terms used included defective, potentially hazardous, and deterioration.

- There was no testing of the warning messages.

- There was no monitoring of the public's response to the campaign or its awareness of the risk.

- There was no direct notification of the persons who were unknowingly at risk.

- There was no direct notification of plumbing and heating contractors.

- Despite the fact that there was a large Hispanic population, no messages were presented in the Spanish language media and warnings were not given in Spanish.

To mount an appropriate campaign, the human factors approach requires that a variety of efforts must be made. Given the risk of death or serious injury coupled with the possibility of significant damage to property, extraordinary efforts should have been be employed to see that the message was received by the target audience in a manner that would be understood by them. In the present case it seems reasonable to assume that lower income individuals are also likely to be lower in literacy; therefore, reading material for them should have been written at a low grade level. Further, many of the Hispanic customers of the utility may not have had a reading knowledge of English. Indeed, many whose native tongue is English are functionally illiterate. The use of newspaper ads and billing stuffers (a pamphlet included with the monthly bill) would not be expected to reach such individuals. Further, many individuals in lower

Table 1

Components Involved in Safety Campaigns.

| | First Utility Company | | Utility with |
	Original program	Revised program	Human Factors involvement
News conference	Yes	No	No
Billing stuffer notices	Yes	Yes	No
Direct mailing	No	Yes	Yes
Presentation in Spanish	No	No[a]	Yes
Presentation for illiterates	No	No	Yes
Testing of warning	No	No	Yes
Monitoring of effectiveness	No	No	Yes

[a] A note was included in Spanish to the effect that a Spanish language description was available by calling a designated number.

income housing would be renters. If their heating bill was paid by the landlord, they would not receive the bill from the utility. Thus to reach these persons directly would require a different approach than that taken by the utility company of reliance on billing stuffers.

A salient question raised during the trials was that of the adequacy of the coverage of the target population. One mainstay of the utility company's defense was the contention that the billing stuffers that they sent had included the necessary information. During the campaign they had included the notice about the defective connectors in five of these pamphlets. They contended that this method of communication was very effective because of 90% readership by the target audience. The evidence they presented for this coverage was that they had experience in sending out the billing stuffers and knew their readership.

Unfortunately, the fact that one has made an effort does not necessarily mean that one has solved the problem. It is necessary to monitor the status of the operation. To evaluate the claim of 90% coverage by the billing stuffers the human factors expert surveyed 145 adult college students. Of these individuals 124 indicated that

they received and paid bills for the gas service. As shown in Table 2 only 56% of the bill-paying respondents indicated that they read the billing stuffers at all, and the mean percentage of times they read them was 46%. Further, only 10% indicated awareness of the problem of defective connectors. Inasmuch as the expected percentage of people who were aware of the problem based on the proportion of readers and frequency of reading percentage would be about 26%, we must presume that the readers were very cursory even when they did read the billing stuffers or the information was not presented in a fashion to produce recall. (There is, of course the possibility that the customers overestimated their relative frequencies of reading.) A further question about the adequacy of the billing stuffer message is that it was presented in English only. Because almost all of the college students were fluent in English we might assume that the figures presented above are an overestimate of the reception of the message in the target population.

Despite the amount of deviation from the presumed readership displayed in the survey, a defense objection to the evaluation procedure was that the students were not representative of the

Table 2

Results of Billing Stuffer Evaluation Surveys.

Population	Percent readers	Mean percentage stuffers read	Percent Aware of problem
College Students (N = 124)	56	46	10
Neighborhood canvass (N = 129)	56	51	10
Telephone survey (N = 99)	55	57	7
Total (N= 352)	56	51	10

Note. Sample sizes include only those who were recipients of bills.

population at large. Prior to the second trial a neighborhood canvass of an area of the type likely to have the problem connectors and a telephone survey of individuals with telephone number prefixes located in that area were conducted. As shown in Table 2 the percentages were remarkably similar to the college student sample.

Although the utility company referred to the human factors analysis as soft science and attempted to paint the human factors expert as an ivory tower academic, the lawsuits were won by the plaintiffs. After the second trial (which resulted in substantial punitive damages as well as the compensatory damages) the utility company revised its warning campaign using materials and procedures remarkably similar to those suggested by the human factors expert. These procedures and materials are displayed in Table 1 along with the program developed with the assistance of human factors experts.

Perhaps, it was the knowledge of the outcome of these trials that caused another gas utility to seek the services of human factors experts when they discovered a similar problem in their territory. Their goal was to eliminate all connectors of the type that might become defective. In particular, their interest in using human factors experts was their concern with getting warning information directly to their customers. They devised a three-pronged attack to get direct access to the individuals at risk. Important features of the program were as follows:

- Servicemen on service or inspection calls were directed to remove and replace defective connectors at no cost.

- Warning letters sent to all other customers at both service and billing addresses.

- Mailing of certified letters to customers whose first letters were returned.

- Service department given follow-up list of service addresses of all certified letters returned.

- Mail special letter to all known landlords highlighting free replacement

- Inspect at all turn-ons and each time serviceman is in customer's residence.

In the interests of maximizing the understanding of the letters, human factors experts were engaged to help evaluate the content. Although the letters were to be written in both English and Spanish, there was concern that some of the recipients would be illiterate and that some might not speak either language. Thus, an attempt was made to determine whether or not a pictograph would be comprehended by those individuals. The approach taken by the authors involved three stages. In the first stage different forms of the pictograph were presented to college students and their responses about the meaning were obtained. From this set of items the most promising were presented to other college students, and their responses were obtained. In addition, these groups were used somewhat as focus groups to develop the final version which was then tested in the third stage on individuals enrolled in adult literacy projects and in an English as a second language (ESL) group.

The adult literacy groups include a wide range of individuals, including functional illiterates as well as those who are preparing to take their General Educational Development (GED) tests for high school diploma equivalency. In order to simulate the task of those individuals who do not speak English, the letter was printed in Greek letters. The students were then asked to indicate what they would do if they received such a letter while living in a foreign country. Those individuals in the GED classes who were at the higher levels were asked to fill out forms answering questions about what they might do to obtain additional information. The individuals who were classified as beginning readers were tested individually. In the ESL groups the individuals were tested in small groups, because there were often individuals who had some understanding of English who could translate for the more beginning students.

In the main, the results were encouraging. Although not all the replies stated that they would call the indicated number for the helpline, the other responses indicated that the individuals would call someone to translate the message for them or use their dictionary to translate the message. In the ESL classes only 17% ailed to respond, and the others responded by indicating they would call the number or take some other reasonable step to determine the meaning of the letter. Among the literacy study group there were more failures to response, but this was partly because they had difficulty in framing the response. Members of the very low reading level group were tested individually, and they all responded, 78% correctly. In one or two cases in this group the individuals were not able to comprehend the task. However, these were individuals who would not be able to live outside an assisted living arrangement and would not be the direct recipients of the letter. It was judged that individuals who were the responsible members of the household would be unlikely not to recognize the way in which they could respond to the letter. It is likely that had the other literacy groups been allowed to respond orally, they would have performed better. As it was nearly 70% of these groups responded appropriately in one fashion or another.

CONCLUSIONS

In summary, the authors have presented an example of a program that failed to reach its intended audience, and have described the human factors principles that are relevant to the problem. A description of how those principles were used in a warning program was presented. The basic principles are consistent with human factors principles in general.

A prime consideration is that the target population be served. In our view a warning about a serious hazard must reach an irreducible minimum of the possible target audience. This obviously includes more than the upper 95% that could be acceptable in other circumstances. Just as it would be unthinkable to go into production of a jet aircraft without testing the responsiveness of the controls, it is unacceptable to use a warning in this case without testing its comprehendability. In addition, the program requires monitoring to be sure the procedures that have been developed operate as they should.

Certainly, other situations may involve different approaches from the ones used in this case to achieve these ends, but the principles must be followed.

ORGANIZATIONAL BRIDGES FROM RESEARCH TO PRACTICE: CASES IN MEDICAL PRACTICE

Barrett S. Caldwell, Ph.D., *Panel Chair and Organizer*
Department of Industrial Engineering
University of Wisconsin-Madison
Madison, Wisconsin

Panelists:

John Gosbee, Executive Director, Center for Applied Medical Informatics, Michigan
State University Kalamazoo Center for Medical Studies

Harold S. Kaplan, Professor and Director, Transfusion Medicine, University of Texas
Southwestern Medical Center

Bruce R. Thomadsen, Associate Professor, Human Oncology and Radiophysics,
University of Wisconsin Hospital and Clinics

Human performance and human error in medicine have emerged as critical health care issues affecting the entire practice of medical treatment and health care delivery. Human factors professionals, particularly members of the HFES, can make major contributions to health care systems, but there is a lack of transition between problem domains. Since the human factors profession has been built on the study and improvement of high-risk systems, the current state of medical practice is clearly an opportune environment in which to operate. This panel will address several practical "ironies" of the current challenges to build improved bridges between key aspects of HFES expertise and the needs of the medical community.

INTRODUCTION

A significant aspect of the 1996 Annual Meeting is an exploration of increasing the linkages between human factors research and practice. Several factors are needed in order to improve the quality and effectiveness of these bridges, including increased recognition of research focus areas most needed by practitioners, enhanced cooperation and collaboration between researchers and practitioners to conduct suitable field evaluations, and improved practitioner access to research products. These issues, while strongly dependent on individual initiative, are also at their base critical organizational issues. This panel is intended to examine several of these organizational and practical factors in several areas of medical practice, drawing in part on a combination of HFES and other research initiatives (Bogner, 1994; Schoenfeld, et al., 1993).

Research policy factors are clearly essential at the highest organizational and governmental levels of funding initiatives and broad research calls to address immediate priorities. Although these policy issues are not the primary focus of this panel, significant

parallels can be made with other research policy considerations to address specific human factors issues and needs in areas such as transportation and energy generation. For instance, in addition to ongoing human factors programs in the nuclear power and aviation / aerospace industries, the Transportation Committee of the newly formed National Science and Technology Council has a specific subcommittee charged to address human performance and interface design needs for near-term system development. Work initiated by members of HFES have brought greater attention to the need for policy initiatives to address human factors and human performance in the medical profession. However, the first HFES symposium, held in 1992 does not have published proceedings.

Nonetheless, even effective research policy does not ensure that front line practitioners will be aware of human factors innovations that could significantly affect their performance in medical settings. For example, we often assume that there is broad knowledge of the relevance of human factors research and practice in the general society, even though there are only approximately 5000 HFES members. The Annual Meetings are not largely attended by persons outside of the Society, and therefore only a very limited number of people are aware of the variety of research activities which are presented in the conference proceedings. In fact, the public's true state of knowledge of human factors, especially applied to fields without a long history of human factors involvement, is usually found lacking. For instance, a human factors engineering research group at a major university was asked to provide assistance to a hospital / HMO group contemplating a major redesign of laboratory information systems integration for improved communication and processing of medical records and laboratory tests. The management engineer making the request, however, did not consider any of the information systems integration, such as efforts to involve laboratory technicians in participatory design evaluations, or records annotation and archiving capabilities of the resulting systems, as human factors problems. "Human factors," in his perception, was limited to concerns which would be addressed only after system design and implementation (for example, keystroke errors by technicians).

The current climate of investigation of medical practice has many facets of origin, from economics to ethics to efficient diffusion of innovations. In this climate, the parallels between medical practice and other technologically dynamic, error-critical systems (such as aviation, nuclear and petrochemical facilities, or critical economic and information communications systems) are becoming more obvious. However, the number of human factors professionals (indeed, professionals of any type) who can easily move between domains is small; the multidisciplinary communities which have evolved are often serendipitous rather than systematic.

Since organizations of all types do suffer from "not invented here" and "over the wall" shortcomings, we should not expect very complex medical facilities to recognize the potential benefit of human factors work solely from the human factors journal or proceedings literature. The HFES is exceptionally small compared to, say, the AMA; large numbers of AMA members would therefore not be aware of current HFES promotional materials. In this environment, quantity purchases of $50 research summaries are unlikely. As an analogue, although Cockpit Resource Management (CRM) has been an effective organizational tool in improving aircraft crew performance, its application to other elements of aviation systems (such as aviation maintenance or air traffic control teams) has been much slower in taking hold (Michelle Robertson, personal

correspondence, 29 January 1996). Maintenance operations in other types of industries are even less likely to have applied CRM training techniques to attempt to improve their performance, despite the success seen in aviation maintenance settings. In fact, many of the tools of CRM may not even be recognized as CRM by persons outside aviation settings. Those persons may be more familiar with participatory ergonomics, human relations, TQM, or other terms not universally recognized as within the problem domain of human factors in organizational design and management (ODAM).

TQM is finally beginning to become a recognized contribution to medical practice in mainstream care facilities. The methods and applications of TQM, although clearly within the scope of human factors research and application, are not perceived as "Human Factors" tools. This perception may be intensified by a lack of cross-fertilization between TQM practitioners and other human factors-oriented ODAM professionals.

It is this type of situation that underscores the desperate need for new types of organizational bridges to support the diffusion of human factors knowledge and application across the range of medical practice settings. The panelists in this session represent a variety of levels of experience and professional background in distinct areas of medicine. More importantly, two of the panelists bring a non-HFES perspective to the discussion, with their experiences of learning about human factors research, and specifically the HFES. In addition, the variety of medical practice areas, including radiological oncology and transfusion medicine, will help to demonstrate the span of disciplinary foci (and therefore, areas of human factors application) which coexist in modern medical centers. The differences in domain of medical practice, practitioner jargon, level of time-critical activity, and even catastrophic error potential mediate against assumptions that even immensely successful human factors work in specific areas of medicine (Bogner, 1994) will be carried over to other areas in the medical community. A frequent call is that "our type of practice is different"; only an examination of the common issues across medical practice areas will help to demonstrate the generic potential benefit of human factors applied to medical systems.

The goal of the panel presentations is to suggest new methods of carrying the human factors message and capabilities into the practitioner's realm, as suggested by medical practitioners both familiar and unfamiliar with HFES. After the panelists have presented their remarks, the chair will lead a brainstorming session with the intent of producing action items that can be accomplished by HFES members in collaboration with non-HFES practitioners in a variety of medical system environments.

References

Bogner, M. S. (ed.) (1994). *Human Error in Medicine*. Hillsdale, NJ: Lawrence Erlbaum Associates.

Schoenfeld, I., Morisseau, D., Callan, J. R., Henriksen, K., Jones, E. D., Kaye, R. D., Quinn, M. L. (1993). "Risk Assessment and Approaches to Addressing Human Error in Medical Uses of Radioisotopes." *Proceedings of the 37th Annual Meeting of the Human Factors and Ergonomics Society (Seattle)* (pp 859-862, panel presentation). Santa Monica, CA: Human Factors and Ergonomics Society.

Panelists:

John Gosbee, MD, MS

Executive Director, Center for Applied Medical Informatics, Michigan State University Kalamazoo Center for Medical Studies

Problem areas for the practitioner and the usual methods of continuing education should guide the efforts to bridge human factors research and medical practice. Problem areas include: selecting and implementing usable information systems; avoiding adverse events (errors); and coordination and communication across an integrated delivery system (reengineered health care). Usual routes of continuing education include journals, scientific meetings, and weekly in-house seminars (grand rounds). Internet (WWW), required use of on-line decision support, and interacting with sales people are becoming more influential in changing practice patterns and providing education. Examples of how this panelist has targeted some of these problem areas and applied them in a medical school and hospital setting will be discussed.

Harold S. Kaplan, MD

Professor and Director, Transfusion Medicine, University of Texas Southwestern Medical Center

Although significant work in nuclear power and aerospace industries can be demonstrated to be relevant to safety in medical systems, few medical practitioners are aware of the common issues. It is especially relevant to look at near miss and accident precursor analysis techniques, because of the frequency of medical procedures and relative difficulty of examining specific catastrophic scenarios. HFES members should strongly consider more presentations at medical practitioner meetings, in order to direct information to settings where larger numbers of representatives of clinical practices and medical research facilities are likely to be exposed to it. The panelist's experience in identifying the HFES as a valuable resource is used as a case study, and potentially a counter-example, of HFES links to other organizations.

Bruce Thomadsen, Ph.D.

Associate Professor, Human Oncology and Radiophysics, University of Wisconsin Hospital and Clinics

The increasing interest in both high dose and low dose rate remote afterloading brachytherapy (RAB) has resulted in a stronger emphasis on identifying and correcting misadministration events. Radiological oncology teams working at modern sites have significant experience in calibrating and analyzing anatomical sites and isotope radiation levels. However, these teams have less experience in cross-functional integration, use of computer visualization tools, and coordinating multiple tasks in the time-critical area of RAB. The growing number of RAB sites also indicates a critical need for addressing these human factors and team dynamics factors, especially for medical teams without significant experience in these areas.

Group Discussion

The panel will conclude with an open discussion of current developments and possible opportunities for HFES members in enriching medical practice. Among the possible discussion topics are an examination of the unique roles graduate education and graduate student research projects may have in providing new occasions for conducting human factors research in medical systems. There is a need for not simply introducing human factors, but embedding it in everyday practice. The timing for such occasions seems to be fortuitous, as shown by increasing numbers of popular press articles, research papers, and focused cross-disciplinary conferences. Strategies for continuing this trend will be discussed.

GLOBAL PLANNING FOR ERGONOMICS AND HUMAN FACTORS

Martin G. Helander
Linköping Inst. of Technology
Linköping
Sweden

Hal W. Hendrick
University of Southern California
Los Angeles, CA
USA

Neville Moray
Universite de Valencienne
Valencienne
France

Ian Noy
Transport Canada
Ottawa,
Canada

Christopher Wickens
University of Illinois
Urbana-Champagne, IL
USA

Human factors and ergonomics are quickly evolving and new areas of application are introduced around the world. In industrially developing countries (IDC's) the emphasis was in the past on physical workload. With the introduction of computers there has been a sudden shift in interest, and the problems of HCI and usability of complex systems have become paramount. This panel will discuss the changes in ergonomics, and how they are affected by changing work ethics, and changing needs of the society. It is argued that ergonomics is well suited to adapt to new goals and respond to difficult challenges. As a result we need to prepare ourselves in education and training to meet the new demands in ergonomics. Can certification programs for ergonomists be flexible enough to consider current needs, and can teaching programs incorporate new knowledge? What are some effects on the type of research that will be conducted in the future. How will ergonomics professionals deal with design and applications. These are some of the issues that will be discussed in this panel session.

INTERNATIONAL DEVELOPMENTS IN ERGONOMICS

Martin Helander
Department of Mechanical Engineering
Linköping Institute of Technology
Linköping
Sweden

Due to the diffusion of computer technology and complex machinery a new image of ergonomics is emerging around the world. The trend is clear not only for industrialized countries but also for industrially developing countries. Cognitive ergonomics, usability studies, human reliability, and human-computer interaction seem to be top priorities. Organizational design and the study of industrial change processes and methodology for continuous improvement are also important. Biomechanics and work physiology are less dominating than they were in the past, except there is a renewed interest in biomechanics due to the increasing number of injuries in the workplace due to musculoskeletal disorders.

It behoves our profession to take note of this new emphasis. Our responsibility is the analysis and design of work scenarios, artefacts, communication, training and education programs. The rapid entry into the information age requires that ergonomists complement their professional skills.

In many countries outside the USA, there is a developing crisis concerning professional identification. We need to develop a scheme to firmly identify ergonomics as a profession. This framework must also be flexible to allow ergonomics to evolve with the times.

CERTIFICATION AND ACCREDITATION INTERNATIONALLY AND IEA's ROLE

Hal W. Hendrick
University of Southern California
USA

The developmental history and state of the art of both professional certification of ergonomists and establishing standards for educational programs, including accreditation, around the world are reviewed. The structure and activities of the IEA to both promote and harmonize certification and professional education standards are described. The outcomes of two recent IEA international meetings on professional standards, one in Italy and one during the last IEA Triennial Congress in Toronto, are summarized. Both the commonalities and differences in certification programs are noted; and major issues in harmonizing educational standards are described. It is concluded that IEA efforts have been effective in promoting harmonization among certification programs and clarifying issues regarding educational standards and accreditiation. Strategies for resolving current issues in professional standards are discussed.

THE CULTURE OF ERGONOMICS

Neville Moray
Universite de Valencienne
Valencienne, France

The gulfs between the levels of economic development in different countries is reflected in their needs for and attitudes to ergonomics/human factors. In addition there are wide differences in political, social and educational philosophies between countries which is reflected in the meaning and purpose ascribed to ergonomics. In this paper I will draw on my experience of working in several countries to suggest some of the problems which may face both the development and application of the discipline, and the necessary "missionary" structures for the future.

One can point to several problems at various levels of abstraction. At the level of detailed industrial ergonomics, there is the fact that there are substantial differences in s-R compatibility and cultural stereotypes. At a more abstract level there are problems connected with the degree of sophistication towards high technology, and the availability of technology

- for example in connection with the cost structure of telecommunications. There are certainly large cultural differences in expectations from work, the relation of work to profit and employment etc., and much of the world is increasingly different from the USA in this respect. Finally, there are the new problems, largely associated with the increase in population, including water shortages, medical services, resource depletion, etc.. These in turn are linked to sociological problems, political problems, economic problems and even the impact of religion on behaviour and society.

The human factors/ergonomics is the "executive" branch of psychological research, so to speak, how are our qualifications in terms not just of psychology, but in political philosophy, economics, and languages preparing us for the tasks to come?

THE CHANGING WORLD OF ERGONOMICS SCIENCE AND PRACTICE

Dr. Ian Noy
Transport Canada
Ottawa, Canada

Profound changes are taking place within the field of ergonomics which challenge its continued development and adaptation to new economic, social and political realities. This paper will present a personal view of the factors influencing the reshaping of ergonomics as a discipline and professional practice. Important factors include: changing work ethics, changing business values and strategies, changing needs of society. It will be argued that ergonomics is well suited to adapt to new goals and respond to difficult challenges. The discipline is young and evolving, the professionals have a multi-disciplinary orientation and are accustomed to working in diverse application areas. Moreover, the fundamental principles of user-centred design and systems science are compatible with modern concepts of work re-organization and other socio-technical trends in the workplace. But the change process is difficult and must be managed effectively if ergonomics is to emerge a distinct field of science with an identifiable, accepted and a valued mission by the next century.

FUTURE RESEARCH IN COGNITIVE ERGONOMICS

Dr. Christopher Wickens
University of Illinois
Urbana- Champagne, IL
USA

The future of cognitive ergonomics must focus its efforts on validating models of complex task performance. Two reasons for this effort are as follows. First, it appears that nearly every complex design problem involving the human user will be one that pits one human factors principle against one or two others. For example principles of compatibility may be pitted against those of consistency. For human performance models to have any predictive power, they must have the capability of combining opposing principles. Second, increasingly in certain domains we are seeing computational models that assume that the human operator is a relatively invariant "constant" component. For example, models of traffic flow in the national airspace. It is essential that psychologists be able to offer to those modelers some evidence for what contributes to variance in human response time or information transmission performance, in a way that can effect the performance of the larger system. An example from future predictions of the national airspace capacity will be used to illustrate this requirement.

Describing Faces from Memory: Accuracy and Effects on Subsequent Recognition Performance

Michael S. Wogalter
Department of Psychology
North Carolina State University
Raleigh, NC 27695-7801

ABSTRACT

The present research examines: (a) the accuracy of three face description methods, and (b) the effects of post-exposure description and imaging activities on subsequent face recognition performance. Participants viewed a sequence of six target photographs, and after each, performed one of three description tasks: generated their own set of descriptors, checked-off descriptors from a pre-existing list, or rated the same set of descriptors on bipolar scales. Other participants performed a distractor (control) activity. Additionally, participants were either told or not told to image the targets while they simultaneously performed the description tasks. Results showed that the checklist task lowered subsequent recognition performance compared to the generate task. Imaging with the generate task facilitated recognition, but imaging with the checklist and rating tasks degraded recognition. The generate task produced the highest quality descriptions as determined by other participants' performance in matching the descriptions to face photographs. The checklist decrement is discussed in terms of memorial confusion initiated by the presence of irrelevant face cues. These results indicate that descriptor generation is the preferred method of collecting eyewitness' face descriptions.

INTRODUCTION

Eyewitnesses, having viewed an individual involved in a crime, are frequently asked by police investigators to participate in various memory tests, e.g., examining a lineup or a mugfile. In the intervening period prior to these tests, several things might happen: witnesses may rehearse or hold an image of the face in their mind and/or they might participate in other memory tests such as giving a verbal description or helping to produce a sketch or composite.

The present research examines three types of verbal description instruments with respect to the quality of descriptions that they produce, and whether these verbal description methods (and holding an image of the target face) affects subsequent recognition performance.

Why would different methods of eliciting verbal descriptions be of interest? Witnesses generally do not give many descriptors when describing faces, and the terms that they produce are frequently not very specific (e.g., 'medium nose' or 'thin lips'). These general terms are frequently inadequate to do an effective search for the culprit. One possible way to elicit better descriptions is to provide a list of possible adjective descriptors which they can select from—as opposed to generating the descriptors themselves.

There has not been much empirical work directly comparing different description techniques. One exception is by Goulding (1971) who had police officers make cued or free recall descriptions of target faces. Free recall produced better quality descriptions than cued recall. In the present

study, this effect is re-examined and includes ratings as another kind of cued method.

Another issue is whether the description task influences subsequent recognition. Previous research on this topic is equivocal. Verbal description has been reported to degrade, facilitate, and have no effect on subsequent recognition performance. However, virtually all of the previous studies used different methods of eliciting the descriptions. The exception is Wogalter (1991) who compared two description methods, and found that a cued test which made available specific descriptors (checklist) produced lower subsequent recognition compared to a description test in which participants produced their own terms (generate). One possible reason for this decrement is that the checklist, by its very nature, has numerous descriptors that are irrelevant, or wrong, with respect to any particular target face. By considering these terms, witnesses might get confused on what the target looked like, lowering subsequent recognition. In the present study, this issue is re-examined using a different checklist and generate test, plus adds a third method involving ratings.

Another activity the witness might perform is to visually rehearse or image the target face. Some studies (Graefe and Watkins, 1980; Read, 1979) have shown a *small* facilitative effect on recognition. However, other research has found no effect (Schooler and Engstler-Schooler, 1990) or a negative effect (Hall, 1979) of imaging. The influence of imaging in the present study is examined in a way heretofore not examined: Imaging is concurrent with the assigned post-exposure activity.

Figure 1. *Exposure/Post-exposure sequence for the 6 targets.*

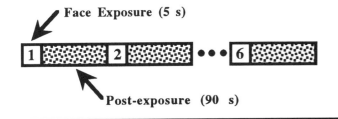

METHOD

Design and Participants

The experiment was a 4 (post-exposure task) X 2 (imaging instructions) between-subjects design A total of 192 undergraduates participated in the main experiment, 24 in each cell. Another 12 and 4 students participated in the in-view description and matching tasks, respectively.

Procedure and Materials

Six white male targets (in frontal poses) were selected at random from a large pool of photographs. Targets were shown in the initial exposure (study) as color slides for 5 s each. Following each target slide, a 90 s period was provided where participants were to perform one of the post-exposure activities described below. Figure 1 displays a representation of the exposure/post-exposure sequence.

Participants were given a booklet containing six pages that differed according to post-exposure condition. Representations of the pages in the three description conditions can be seen in Figure 2.

(1) In the *Rate* condition, the pages of a response booklet contained a list of 10 feature headings. Under each heading was a set of 5-point bipolar scales with adjective descriptor endpoints reflecting various dimensions. The descriptors and dimensions were derived from previous verbal description research, and studies on feature saliency, multidimensional scaling and cluster analyses of faces. The

Figure 2. *Representations of the Post-exposure description forms.*

Rate

OVERALL SHAPE OF FACE
short — long
narrow — broad
bony — fleshy

COMPLEXION
fair — dark
pale — red
unlined — lined
clear — blemished

HAIR
short — long
tidy — untidy
straight — curly
bald — thick/full
black — white

FOREHEAD
low — high
narrow — broad
straight — sloping

EYEBROWS
thin — thick
straight — bent
low — high
meets in middle — set far apart

EYES
small — large
narrowed — opens
close set — wide spaced
deep set — protruding
dark — light

NOSE
small — large
short — long
narrow — broad
concave — hooked
small nostrils — large nostrils
narrow tip — broad tip

MOUTH
small — large
thin upper lip — thick upper lip
thin lower lip — thick lower lip

CHIN
small — large
pointed — square
receding — jutting

Checklist

OVERALL SHAPE OF FACE
bony ___ narrow ___
short ___ long ___
fleshy ___ broad ___

COMPLEXION
pale ___ dark ___
lined ___ clear ___
fair ___ red ___
unlined ___ blemished ___

HAIR
bald ___ white ___
untidy ___ short ___
thick/full ___ black ___
tidy ___ curly ___
straight ___ long ___

FOREHEAD
low ___ high ___
narrow ___ broad ___
straight ___ sloping ___

EYEBROWS
straight ___ low ___
set far apart ___ bent ___
thin ___ high ___
meets in middle ___ thick ___

EYES
open ___ dark ___
mall ___ protruding ___
close set ___ large ___
light ___ narrowed ___
wide spaced ___ deep set ___

NOSE
narrow ___ large ___
hooked ___ short ___
small nostrils ___ broad ___
broad tip ___ concave ___
small ___ narrow tip ___
large nostrils ___ long ___

MOUTH
thin upper lip ___ thick lower lip ___
small ___ thin lower lip ___
large ___ thick upper lip ___

CHIN
jutting ___ small ___
large ___ square ___
pointed ___ receding ___

Generate

OVERALL SHAPE OF FACE

COMPLEXION

HAIR

FOREHEAD

EYEBROWS

EYES

NOSE

MOUTH

CHIN

descriptors/dimensions have been used successfully in the FRAME computer-assisted search system to locate target faces in a mugfile (Shepherd, 1986). Participants were told to complete the form to describe the face just seen. (2) In the *Checklist* condition, the sheets were identical except the adjectives under each heading was randomized. Participants were to check all descriptors that described the previously seen face. (3) In the *Generate* condition, the sheets were identical to the other conditions except the adjective descriptors were deleted. This provided space for participants to write descriptions in their own words. (4) Participants in the *Distractor/Control* condition were told that a second purpose of the study was to measure how fast they could do a visual scanning/perceptual speed task. The pages of the booklet contained a large matrix of random letters. Two different letters were circled on each page. Participants were told that during the post-exposure periods they should mark all other instances of the circled letters on the page. Performance on the control task was not analyzed.

Half of the participants received explicit instructions to generate and hold a mental image of the most recently viewed

---

Figure 3. *Recognition Performance Measures.*

Recognition Responses
N = No, not presented
Y = Yes, presented

Confidence Rating
1 = guessed
2 = probably correct
3 = certain

6-Point Scale that combines the above two measures

"No" responses			"Yes" responses		
"N3"	"N2"	"N1"	"Y1"	"Y2"	"Y3"
I..........	I..........	I..........	I..........	I..........	I
1	2	3	4	5	6

6 Recognition Measures:

- **Hit scores (for 6 Targets):**
  (a) **HM:**   Hit/miss (Mean of targets on 6-point scale)
  (b) **PH:**   Proportion hit (Mean of targets where "Y" = 1, "N" = 0)

- **False alarms scores (for 134 Distractors):**
  (c) **FACR:**   False alarm/correction rejection (Mean of distractors on 6-point scale)
  (d) **PFA:**   Proportion false alarm (Mean of distractors where "Y" = 1, "N" = 0)

- **Discrimination/Sensitivity scores:**
  (e) **H-F:**   HM minus FACR
  (f) **SHM:**   Mean target z-score (after standardizing each participant's responses to all test photographs)

Note: Better recognition performance is indicated by *higher* scores on the *hit* and *discrimination* measures and *lower* scores on the *false-alarm* measures.

---

face while they simultaneously worked on one of the post-exposure tasks. The other half of the participants were not given explicit instructions to image.

After the exposure/post-exposure phase was completed, participants worked on a study strategy questionnaire for 5 min which was immediately followed by the recognition test. The test pictures were comprised of 140 black and white slides: 134 were distractor faces, and the other 6 were the targets. The targets appeared at random positions after the 50th distractor slide. The target photos in the test sequence were similar, but not identical, to those of the shown at study; they were taken by different cameras several minutes apart. Participants were told that the faces they saw at study may or may not be in the test sequence.

In the recognition test, slides were presented at a 7 s rate. Participants indicated "yes" or "no" to each face according to whether they believed the individual was shown earlier, and also gave a 3-point confidence rating. From these scores, 6 recognition performance measures (two hit, two false-alarm, and two discrimination) were derived as shown in Figure 3.

RESULTS

Recognition Performance

Recognition performance was examined using 4 (post-exposure task) X 2 (imaging instructions) between-subjects analyses of variance (ANOVAs). Comparisons among means for significant effects were performed using Fisher's Least Significant Difference test. The comparisons described below are at or below the .05 probability level.

The hit measure means, HM and PH, are shown on the top row of each cell in Table 1. The ANOVAs showed a significant effect of post-exposure task, $F(3, 184) = 13.84$, $MS_e = .672, p < .0001$, and $F(3, 184) = 11.38$, $MS_e = .033, p < .0001$, respectively. The Distractor condition produced significantly lower HM and PH scores compared to each of the three verbal description conditions. Neither hit measure showed a main effect of imaging instructions, $Fs < 1.0$. However, imaging instructions interacted with post-exposure task, $F(3, 184) = 4.25$, $MS_e = .672$, $p < .01$, and $F(3, 184) = 3.58$, $MS_e = .033$, $p < .02$, for HM and PH, respectively. With HM, the Checklist and Rate conditions showed a significant decrement with imaging instructions. With PH, Rating plus imaging produced a decrement, but Generating plus imaging produced facilitation.

The second row of each cell in Table 1 shows the mean false alarm scores. There was a significant main effect of post-exposure task with FACR, $F(3, 184) = 11.38$, $MS_e = .033$, $p < .0001$, but not with PFA. The Checklist produced significantly higher FACR than the Generate or Distractor conditions. Neither false alarm measure yielded a main effect of the imaging instructions or interaction ($Fs < 1.0$).

The two discrimination scores, H-F and SHM, are shown on the bottom row of each cell in Table 1. The

ANOVAs showed significant post-exposure main effects with both measures, $F(3, 184) = 10.25, MS_e = .770, p < .0001$, and $F(3, 184) = 4.94, MS_e = .553, p < .0001$ for H-F and SHM, respectively. The Generate condition produced significantly higher discrimination than the Checklist condition. The Distractor condition produced significantly lower discrimination than the three description conditions (except the Checklist with SHM). Neither discrimination measure showed a main effect of imaging instructions ($Fs < 1.0$), but there was a significant interaction with H-F, $F(3, 184) = 2.80$, $MS_e = .770, p < .05$, that was marginal for SHM, $F(3, 184) = 1.84, MS_e = .553, p < .07$. The Checklist produced significantly lower H-F with imaging instructions than with no imaging instructions. With SHM, the Generate condition tended to be higher with than without imaging ($p = .06$).

Additional analysis examining performance for only the first face that was shown to participants indicated a pattern of results that was virtually identical to those described above. Other analyses showed that target face did not interact with post-exposure task or imaging condition.

Description Quality

To examine description quality, a separate group of participants completed the description forms while the targets were in view (i.e., not from memory). These in-view descriptions and all of the descriptions produced in the main (post-exposure) experiment were randomized, assembled into booklets, and then four participant judges attempted to match the descriptions to the six target photographs mounted on a poster board. From the matching assignments, a measure of quality was derived.

If the descriptions did not provide any useful information, the judges would make their matching assignments at random and performance would be at or near the chance level of 1/6 or .167. Table 2 shows the means. Performance was better than chance for all conditions ($ps < .001$) indicating that the description techniques provide, at least, some useful information.

Post-exposure description quality. A 3 (post-exposure description task) X 2 (image instructions) between-subjects ANOVA using the description quality measure showed a significant effect of description task, $F(2, 138) = 36.25, MS_e = .373, p < .0001$. Paired comparisons indicated that the Generate condition produced higher quality descriptions than the Checklist and Rate conditions. The ANOVA showed no main effect of imaging instructions ($F < 1.0$), but this factor interacted with description condition, $F(2, 138) = 5.48, MS_e = .373, p < .01$. Only one pairwise comparison was significant: Higher quality Checklist descriptions were produced with imaging instructions than without imaging instructions.

In-view description quality. The in-view description quality means are shown on the bottom row of Table 1. A one-way within-subjects ANOVA showed a significant effect of description task, $F(2, 22) = 6.00, MS_e = .246, p < .01$. Pairwise comparisons indicated that the Generate condition produced significantly better descriptions than the Checklist and Rate conditions.

Relation of Description Quality and Recognition

Correlational analyses examined the relation between post-exposure description quality and recognition performance. For the Checklist, none of the recognition measures was significantly related to description quality. However, for the other two description conditions, there were significant positive correlations between quality and recognition discrimination (Rate: H-F, $r = .38, n = 48, p < .01$, SHM, $r = .35, n = 48, p < .05$; and Generate: H-F, $r = .35, n = 48, p < .05$, SHM, $r = .38, n = 48, p < .01$).

Table 1. Mean Recognition Performance as a Function of Post-Exposure Description and Imaging Condition.

		Post-Exposure Tasks							
		Checklist		Rate		Generate		Distractor/Control	
Image Instructions	HM (PH)	4.71	(.81)	4.56	(.76)	5.13	(.89)	4.24	(.69)
	FACR (PFA)	2.45	(.25)	2.09	(.20)	2.08	(.19)	2.14	(.21)
	H-F (SHM)	2.26	(1.50)	2.47	(1.74)	3.05	(2.14)	2.10	(1.45)
No Image Instructions	HM (PH)	5.19	(.88)	5.11	(.88)	4.85	(.81)	3.87	(.63)
	FACR (PFA)	2.45	(.27)	2.35	(.25)	2.12	(.20)	2.12	(.20)
	H-F (SHM)	2.74	(1.78)	2.76	(1.85)	2.73	(1.75)	1.75	(1.32)
Mean	HM (PH)	4.95	(.84)	4.84	(.82)	4.99	(.85)	4.06	(.66)
	FACR (PFA)	2.45	(.26)	2.22	(.22)	2.10	(.20)	2.13	(.20)
	H-F (SHM)	2.50	(1.64)	2.62	(1.80)	2.89	(1.95)	1.93	(1.39)

Table 2. Mean Post-Exposure and In-View Description Quality as a Function of Description and Imaging Conditions.

| | Description Method | | |
	Checklist	Rate	Generate
Imaging Instructions	.46	.43	.62
No Imaging Instructions	.32	.46	.67
Mean	.39	.44	.64
In-view	.65	.60	.77

DISCUSSION

The results show that recognition performance following the Checklist task was lower than the Generate task. This result may seem somewhat surprising given the fact that participants only had to check off appropriate descriptors. Lowered performance in the Checklist condition can be explained in terms of exposure to irrelevant or wrong descriptors. By its nature, the checklist provided extraneous descriptors (in order to describe a range of different faces). Therefore, some adjectives were not descriptive of the particular face they had just viewed. By considering these erroneous terms, participants possibly incorporated some of this information into memory, resulting in confusion about what the target looked like, reducing subsequent recognition performance. The Rate technique provided the same descriptors as the Checklist but produced a less severe decrement. In the rating task the descriptors were ordered along dimensions which might have enabled consideration of a broader range of features, thereby causing less confusion. In contrast, the Generate technique allowed participants to produce verbal descriptions without the confusion of irrelevant descriptors because the terms were not present. Nevertheless, Schooler and Engstler-Schooler (1990) reported recognition interference using a generate-type description task following exposure. However, these researchers provided a much longer period of time to describe the face (5 min) which could promote confabulation of irrelevant face features while composing the description. The theorized confusion of memory by intervening stimuli is similar to the interference reported in other research (e.g., Loftus and Greene, 1980) and supports earlier work (Wogalter, 1991) showing a recognition decrement with a different descriptor checklist.

In this experiment, strong support for an overall benefit of imaging on recognition was not found. However, imaging instruction interacted with the verbal description tasks showing some improvement when it co-occurred with the Generate method, and a decrement when it co-occurred with the Checklist and Rate methods. One explanation for these results is that the request to image in the Checklist and Rate conditions increases the likelihood that participants imaged representations of irrelevant (or wrong) verbal descriptors. But when directed to image in the Generate condition, participants could do so without irrelevant terms to consider.

The description quality results showed that all three description techniques provide some useful descriptive information, but the Generate condition produced the best descriptions, under both post-exposure and in-view conditions. The Generate technique allows greater freedom to use the most effective language to describe the targets. The other two description tasks are more restrictive in the features that could be described.

The results also showed that directing participants to image produces significantly better quality Checklist descriptions than without these instructions. One explanation is that Checklist participants, without explicit instructions to image, might merely check off descriptors with less considered thought than those given image instructions. However, the process of imaging irrelevant items on the Checklist might partially destroy specific target memory — producing degraded recognition in the subsequent test.

Neisser (1987) suggests that free recall tests are more accurate and less likely to produce distorted, constrained, contaminated memorial reports than cued recall and recognition tests. The present results support this notion. When capturing face descriptions, free recall methods (like the generate condition) are preferred over methods that rely on recognition of descriptors (like checklists), because descriptor generation does not degrade subsequent recognition, and it produces the better quality descriptions.

REFERENCES

Goulding, G. J. (1971). Facial description ability. *Police Research Bulletin, 19,* 42-44.

Graefe, T. M., and Watkins, M. J. (1980). Picture Rehearsal: An effect of selectively attending to pictures no longer in view. *Journal of Experimental Psychology: Human Learning and Memory, 6,* 156-162.

Hall, D. F. (1977). *Obtaining eyewitness identifications in criminal identifications: Two experiments and some comments on the Zeitgeist in forensic psychology.* Paper presented at the meeting of the American Psychology-Law Society.

Loftus, E. F. and Greene, E. (1980). Warning: Even memory for faces may be contagious. *Law and Human Behavior, 4,* 323-334.

Mauldin, M. A., and Laughery, K. R. (1981). Composite production effects on subsequent facial recognition. *Journal of Applied Psychology, 66,* 351-357.

Neisser, U. (1987). *The present and the past.* Paper presented at the Second International Conference on Practical Aspects of Memory, Swansea, Wales, U.K.

Read, J. D. (1979). Rehearsal and recognition of human faces. *American Journal of Psychology, 92,* 71-85.

Schooler, J. W., and Engstler-Schooler, T. Y. (1990). Verbal overshadowing of visual memories: Some things are better left unsaid. *Cognitive Psychology, 22,* 36-71

Shepherd, J. W. (1986). A computer-based face image retrieval system. In H. D. Ellis, M. A. Jeeves, F. Newcombe, and A. W. Young (Eds.), *Aspects of face processing.* Dordrect: Martinus Nijhoff.

Wogalter, M. S. (1991). Effects of post-exposure description and imaging on subsequent face recognition performance. *Proceedings of the Human Factors Society, 35,* 575-579.

HISTORY AND CHARACTERISTICS OF HUMAN FACTORS RESEARCH

David Meister
San Diego, California

As a sequel to a study which analyzed the annual meetings of the Human Factors Society (HFS) during the period 1959-1972 (Meister, 1995), the following is a study of the characteristics of Human Factors/Ergonomics (HF/E) research as these have changed over the years 1965-1994.

INTRODUCTION

The rationale for the study was that every discipline must examine its effectiveness in the same manner in which feedback enhances system operation. To paraphrase Plato, the unexamined discipline is a second-rate one. Moreover, as a discipline matures, its history becomes of interest: Moroney (1994) and Meister & O'Brien (1996) have written the history of the discipline, but these have not discussed its research characteristics, which can be considered its "technical history."

The hypothesis underlying this study is that there are certain relatively invariant trends and characteristics of HF/E research, whereas others have changed over the past 30 years. More specifically, the study sought to answer the following questions:

(1) What are the characteristics of past and present HF/E research?

(2) Have there been any significant changes in that research over the years?

(3) What factors influenced the nature of that research?

(4) What motives induced or initiated the research?

(5) What institutions have influenced that research?

(6) What topics have dominated HF/E research, and has technology had a significant effect?

(7) What is the role of theory and modeling in that research?

(8) Is the research useful, and to whom?

METHODOLOGY

The study was conducted by performing a content analysis of 621 papers in the Proceedings (henceforth abbreviated as PROC) of the annual meetings of the HFS (1973, 1984, 1994); and those published in its journal, Human Factors (also abbreviated HF) during 1965, 1968, 1970, 1973, 1984, 1994). Papers from these two sources were analyzed separately, because there are significant differences between the two publications. PROC imposes a 5 page limitation on papers, whereas HF does not; acceptance criteria for HF are more rigorous than are those for PROC.

For both documents, however, one major analytic criterion was imposed by this author: only papers in which subjects were tested were analyzed. Non-empirical papers (e.g., research reviews, comments, speeches) were eliminated from the sample. Non-empirical papers were most common in the early years, but have been progressively reduced. At the present time HF publishes almost no non-empirical papers, although a quarter of PROC papers (as of 1994) were of the non-empirical type. (However, the Editor of HF has recently announced a policy of including research reviews in the journal.)

The papers were analyzed using 12 content analysis categories derived from the initial questions with which this study began. Each category has a number of subcategories, only a few of which can be specified here as examples. These categories are: primary and secondary topics of the paper; source (e.g., university, government agency); venue (e.g., laboratory, simulator, operational environment); subject type (e.g., students, general public);

methodology (e.g., experiment, subjective method, physiological measurement); research initiation (i.e., motive for performing the study); involvement of theory and modeling in the paper; type of research performed (e.g., methodological, theory testing); unit of analysis (e.g., individual, team, system); hypothesis specified (yes/no); results application suggested (yes/no).

Some of the categories were entirely factual (e.g., source); others required judgments of greater or less complexity. As much as possible, judgments were based on the authors' own words. However, because some subjectivity was inevitable, to ensure that the criteria used were stable, and to eliminate bias, all papers were reanalyzed independently after all initial judgments were made, and all judgmental discrepancies then discovered were resolved.

The data analysis on which the findings are based was in terms of frequency and percentage of content categories. The use of more sophisticated statistics like analysis of variance or trend analysis was rejected, because these are pertinent only to highly controlled conditions. (Because of the relatively small numbers of papers and their variability over the years, it is difficult to apply accepted statistical methods. One can therefore determine trends, but not with the absolute certainty of an experiment.) Another caveat is that the frequency of certain research topics was influenced by editorial interventions, such as special issues in HF and the inclusion of new technical groups (and their papers) in annual meeting PROC. Moreover, this study does not pretend that this is a complete picture of all Western HF/E research, because it did not include journals other than those published by HFS.

RESULTS

It is impossible to present all the detailed tables developed in the study. Those who are interested in further detail can write the author for the complete report (Meister, 1996a).

The following are the findings:

(1) The percentage of papers that are empirical (i.e., in which subjects are tested) has increased markedly over the years, from 65% for HF in 1965 to 94% in 1994. The same progressive increase was noted in PROC, but the percentage increase is not as great (45% in 1973 to 74% in 1994). It is possible that the journal was intended to adopt a more "academic" format which excluded non-empirical papers, whereas PROC was supposed to accommodate a greater variety of papers.

(2) The median number of authors for individual papers was two, with as many as 4, 5, or 6 authors sometimes involved. This is a continuing tendency which suggests that American HF/E research is often a team affair.

(3) the range of topics found in the two publications represents a sampling of the entire range of HF/E subject matter (a listing far too lengthy for this paper), with concentration in certain areas. The number of distinctly different primary and secondary topics rose respectively in PROC from 15 and 26 in 1973 to 55 and 125 in 1994. In HF primary and secondary topics rose from 16 and 36 in 1965 to 28 and 36 in 1994, which suggests that HF represents a much smaller palette to depict research. The most important research themes (i.e., those most frequently studied over the years) have been system and function-oriented. The most common system themes are aerospace (far more frequent than others), automotive, and displays; function themes reflect visual performance, information processing, and biomechanics. Although a very great range of topics is described (particularly as secondary topics), very few are studied intensively ("intensively" defined as that in any one year at least two papers on the same topic are published). In consequence, HF/E research is quite broad, but extremely shallow.

One topic category is significant by its absence--maintenance. There are two ways in which one can look at human-machine technology: in terms of the way in which it (and the human) functions routinely; and the way in which it (and the human) performs or fails to perform when the technology breaks down (as it does invariably). Psychology as the study of the normal (abnormal psychology is at best a minor speciality, at least academically) has influenced our discipline to emphasize the normal functioning of technology. The author's viewpoint is that there should have been much greater emphasis on the human role in technological malfunctions.

The other absent topic is the system. True, many research topics involve types of systems, like automobiles or helmet mounted displays, but they are not about the human-machine system in general, which means that the research does not deal with system variables except as these are incidental to the particular type of system research question. If one believes, as many HF/E specialists do, that the human-machine system is the intellectual heart of HF/E, this void is inexcusable.

(4) Initially (1965-1970) more papers were published in HF by people working in industry, government, and as contractors (combined) than by those affiliated with a university. Starting in 1973, however, when PROC was first published by the Society, this trend was reversed in HF (35% university in 1970, 70% in 1973), and now the largest proportion of papers in HF (64% in 1994) is published by university-affiliated people. In PROC. the percentage of papers from the university has always varied between 50% and 60%. Chronologically, there appears to be a correlation between the rise in HF of the university as a research source, the increasing "professionalization" of its papers (defined as a "tone" similar to those published in APA journals), and the introduction of PROC as a publication.

(5) The laboratory has been the preeminent location for data collection in papers published in HF (50% to 70%), somewhat less so in PROC (40% to 50%). It is refreshing that laboratory dominance is being challenged progressively by the operational environment and the simulator, perhaps because of the importance of aerospace and automotive research, which rely heavily on these venues.

(6) Students are most often subjects for research (50% in PROC in 1973, 33% in 1994; in HF, 27% in 1965, 45% in 1994); but those who are nonstudents (e.g., typists, engineers, pilots) and the general public are, when numbers are combined, used just as or even more frequently.

(7) The methodology most frequently employed by researchers is the experiment (in PROC 74% in 1973 and 40% in 1994; in HF, 66% in 1965 and 31% in 1994). Increasingly the experiment is of the within-subjects design type. The drop in experiment percentages is deceptive; it occurs only because other methods are used more than they have been previously. Over the years, subjective methods, principally questionnaires and ratings, have been used increasingly to support or verify experimental results, although few studies are based solely on subjective methods. If the number of techniques employed in any one study is used as a criterion of study complexity, there is a slight tendency for studies to have become increasingly complex over the years.

(8) The impetus or "motive" for the individual research was usually multi-determined. Certain hypotheses can be entertained: (a) that research is initiated to test or illuminate theory; (b) in the case of a discipline so closely tied to technology, one might suppose that much HF/E research would be initiated by new technological developments or design problems; (c) a conceptual or empirical problem might initiate research; (d) finally, a lack of data might be a rationale.

The most frequent motives cited were: an empirical ("real world") problem such as the frequency of carpal tunnel syndrome (in HF, a high of 26% in 1984-1994; in PROC, a high of 45% in 1973 with a low of 15% in 1994). Or a conceptual problem (e.g., what is the most effective way of measuring workload?): In HF 37% in 1968, in 1994, 17%; in PROC a maximum of 19% in 1994). A third major category was the need to add to the database or to resolve data discrepancies (in HF, 18% in 1965, 24% in 1994; in PROC an average of 24%). The exploitation of new technology was a distant runner in the race, and almost at no time was a design problem cited as the reason for performing a study.

(9) Theory and modeling had little connection with HF/E research. Only 1-5% of all studies reviewed tested theories or aspects of a theory; only about 15% of the studies in PROC referenced either a theory or a model; in HF, theory was referenced between 17% in 1965 and 32% in 1994.

(10) The research studies were of several types: methodological studies (approximately 25%); and/or study of variables affecting human performance (approximately 34%); and/or studies determining human performance characteristics (approximately 14%). Many studies were of more than one type. Almost no work was published involving theory or models, product development and testing, or training evaluations.

(11) The unit of analysis was almost always individual performance (90-95%), although data on individuals was usually aggregated for statistical purposes. Almost no performance analysis involved work stations, teams, systems, or organizations. Few papers suggested that hypotheses were developed prior to conducting the studying (before 1994, an average of 15%; in 1994, 44%), nor was there much attempt on the part of researchers to suggest applications for the results they achieved (in HF, an average over the years of only 30%; in PROC 60% in 1973, diminishing to 20% in 1984 and 1994).

CONCLUSIONS

Since 1973, when the Society first published its annual meeting PROC, the papers published in it and in HF have tended to become more "professional," more academic, more "psychological." It is difficult to define operationally the criteria that represent such an orientation, but a visual/verbal comparison of the early (1965) and later (1994) issues of HF shows striking differences between them. This shift to a more "professional" orientation, resembling that of APA journals, has had perhaps the effect of distancing the research somewhat from its former highly empirical foundations. The orientation toward the individual as the research focus (an orientation characteristic of psychology) has also been perpetuated. In short, HF/E research is quite academically professional, but, as in all academic endeavors, somewhat remote from reality.

The recurrent complaint by practitioners that the research has little to offer them seems justified by the relative lack of application noted in the papers (although a recent shift in policy by the Editor of HF may reverse this). At the same time, as the research has become more conceptual over time, the continued absence of theory and theory testing in that research is difficult to understand. Theory is very important (at least superficially) to many people, and these may find this absence disturbing. They may assert that even when theory is not specifically referred to in an individual study, it still exists as context for the study; but this is at best a weak refutation of the data.

RECOMMENDATIONS

If the conclusions resulting from this analysis are considered somewhat negative (although it is entirely possible that a majority of HF/E specialists may not think so), then the first need is for the discipline to determine why its research is as it is, why theory and application are both relatively absent from its research. In this, the Society, as the only single organization of the profession, should play a leading role, by setting up something like a continuing commission of inquiry, supporting its deliberations, and publicizing its findings.

If one looks at the characteristics of HF/E research as a whole, it is individual-centered and deconstructive, in the sense that to perform the research one must break the system down into its components (e.g., in an aircraft system, for example, helmet-mounted displays, cockpit layout, degree of automation). This must be done because of our reliance on the experimental methodology, which demands deconstruction.

There are alternative paradigms, of course. It is possible to develop an HF/E research program based on attributes of the system as a whole; attributes like complexity (the number of informational states the systems presents); autonomy (the freedom given the operator to select modes of system functioning); and visibility (the extent to which internal component functioning is revealed by its design to the operator) (Meister, 1996b). Whether research based on this framework, together with more traditional research, would get us further, is an unknown. It would certainly produce a most interesting research, but whether "interesting" research is what one wants is perhaps questionable.

One has to begin by asking what the researcher wants his/her research to accomplish. Knowledge for its own sake? If research need not be applied or be practically useful, then no more need be said; HF/E research then becomes an art form and should be appreciated primarily for its aesthetic qualities. If it is an art form, then it has minimal relevance to this discipline. At present HF/E research does seem to fit that description.

If, on the other hand, one feels, as some do, that

the research should supply design guidelines for practitioners who cannot do their own research, then the research has failed signally; almost none of the papers stems from or relates to actual or hypothetical design problems. Although we have design guidelines like MIL STD 1472D, these are largely a matter of heuristics. The research leads to no overarching theoretical concepts, or if it does, to none that has practical design significance.

At the very least, it seems, HF/E research should provide regularities that can be used in system development. The data should say: if X technological factor has Y value, then a certain operator performance effect (Z) should be expected. But if such regularities exist in the literature, these have not been exploited by compiling and analyzing that literature-- certainly not in a quantitative probabilistic form, although long ago there was a short lived attempt at this (Munger, et al., 1962).

There are those like Chapanis (1996) who suggest that each design problem is individual, and theory is therefore unimportant. And, if, as Moray (1994) says, HF/E is largely contextual, this means that it is entirely relativistic and no general principles or theory can guide it. If that is the case (and they may be right, although I should regret it), then we must ask what function HF/E research performs. If it is not practically useful, if it produces no general principles or relationships, why does it exist?

It may appear as if the preceding arguments are a return to basics. After all, professionals are supposed to know what their research is for. I think, on the contrary, that hardly anyone knows what HF/E research is for; otherwise, this study would not have revealed what has been found. Whether one agrees with the study conclusions, they demand discussion.

REFERENCES

Chapanis, A. (1996). Personal communication.

Meister, D. (1995). Human Factors-- The early years. Proceedings, Annual Meeting of the HFES, 478-480.

Meister, D. (1996a). Historical and Epistemological Study of Human Factors/Ergonomics Research. Unpublished manuscript.

Meister, D. (1996b). A new theoretical structure for developmental ergonomics. Proceedings, 4th Pan-Pacific Conference on Occupational Ergonomics, Taiwan.

Meister, D. & O'Brien, T.G. (1996). Introduction: the history of human factors testing. In T.G. O'Brien & S.G. Charlton (Eds.)., Handbook of Human Factors Test and Evaluation. Mahwah, NJ: Lawrence Erlbaum Associates, 3-11.

Moray, N. (1994). "De maximis not curat lex" or how context reduces science to art in the practice of human factors. Proceedings, Annual Meeting of the HFES, 526-530.

Moroney, W.F. (1994). The evolution of human engineering: A selected review. In J. Weimer (Ed.), Research Techniques in Human Engineering. Englewood Cliffs, NJ: Prentice-Hall, 1-19.

Munger, S. et al. (1962). An index of electronic equipment operability: Data Store. Report AIR-C43-1/62-RP(1). American Institute for Research, Pittsburgh, PA.

DUAL-TASK PERFORMANCE AND VISUAL ATTENTION SWITCHING

Danny R. Hager
David G. Payne
Binghamton University

Four factors were manipulated to assess the possible performance decrements caused by concurrent performance of a visual and an auditory task. For the visual task, memory load and the degree of attentional switching were manipulated. The nature of the memoric information and the intelligibility of the speech signal in the auditory task were also manipulated. The level of single-task performance was compared with the performance of the same task under dual-task conditions to determine which factor lead to the greatest dual-task decrement. The results demonstrate that memory load and the nature of the memory representation had little effect on the performance of a concurrent task. Visual attention switching had a large effect on the amount of dual-task decrement.

INTRODUCTION

The research presented here evaluates factors that negatively impact performance when visual and auditory tasks are performed concurrently. These experiments extend research by Payne, Peters, Birkmire, Bonto, Anastasi & Wenger (1994).

In that study, subjects performed an auditory task and a visual task both singly and in combination. A significant dual-task decrement occurred when two tasks both tapped central processing resources such as working memory or decision making. We examine in more detail which resources determine the amount of dual-task decrement in auditory-visual environments.

Auditory Tasks

Two auditory tasks were used: a Sternberg memory search task (Sternberg, 1966) and a Category Decision task. The Sternberg task relied on storage and processing working memory whereas the other task relied more on retrieval of information from semantic memory. Difficulty was varied by manipulating the intelligibility of the speech signal.

Visual Tasks

A "same-different" paradigm was used for the visual task. Subjects were shown pairs of displays and decided whether the two were the same or different.

Inter-stimulus interval was manipulated to produce three variations: a simultaneous version (both items present at the time of the decision), a short delay version (the target item disappeared 0.5 s before the onset of the second item) and a long delay version (a 1.5 s inter-stimulus interval). Two processing resources are varied by this manipulation: the reliance on working memory (i.e., longer inter-stimulus intervals increase memory demands) and the amount/type of visual attention switching. By visual attention switching, we refer to the idea that subjects attend to one item and then shift their attention to the spatial location of the second item (perhaps with repeated shifts to facilitate comparison). For the simultaneous task, the attention switching is exogenous (i.e., the subject moves his/her eyes to evaluate the items), whereas both successive tasks involve shifting attention between an external item and the internal representation of an item. The former type should require more effortful processing.

The simultaneous visual condition was performed by subjects in all experiments. The short visual delay condition was used in Experiments I and III whereas the long visual delay condition appeared in Experiments II and IV. The Sternberg search task was performed in Experiments I and II, and the Category Decision task was used in Experiments III and IV. Speech intelligibility of the auditory messages was manipulated identically in all four experiments. The experiments were run identically except as described above. Due to space limitations, discussion of the four experiments will be combined.

EXPERIMENTS I, II, III AND IV

Subjects

Twelve Binghamton University undergraduates participated in each experiment, with experimental sessions lasting one hour.

Apparatus

A 486-50 MHz PC controlled the experiment. Auditory stimuli were recorded using a high quality microphone attached to a Sound Blaster Pro 16-bit sound card. Stimuli were recorded at 16 bits, with a sampling rate of 22,050 Hz. The stimuli were processed by a chopping circuit (described by Peters and Garinther, 1990), amplified using a Radio Shack Model SA-150 amplifier, and presented over Realistic Nova 40 headphones.

The intelligibility of the speech signal was varied using a chopping circuit designed by the U.S. Army Human Engineering Laboratory (described in detail by Peters and Garinther, 1990). Portions of the speech signal are removed by chopping the speech signal for varying durations. The chopping circuit gated the signal at 60 Hz with a duty cycle variable from 0% to 95%. Two setting were used in the study: 25% and 80% (corresponding to 75% and 20% intelligibility, respectively).

Materials

Stimuli for the Modified Rhyme Task (MRT; described below) were taken from the stimulus words developed by House, Williams, Hecker and Kryter (1965). The Category Decision and Sternberg tasks used four categories, with 12 words from each category. Stimuli for the visual tasks were histograms constructed with 6 rectangular bars of varying lengths with a straight line at the base of the rectangles and were presented as white line drawings on a black screen.

Design and Procedure

Each experiment consisted of two phases, with speech intelligibility manipulated across phases.

During one phase, speech intelligibility was chopped at 75% (low intelligibility) and during the other it was chopped at 20% (high intelligibility). During each phase, six tasks were performed: 1) Modified Rhyme task, 2) single auditory, 3) single simultaneous visual, 4) single delayed visual, 5) dual simultaneous visual and 6) dual delayed visual.

Modified Rhyme Test (MRT). Speech intelligibility was measured using the MRT developed by House, Williams, Hecker and Kryter (1965). Subjects were presented with 55 words, each proceeded by a carrier phrase ("The next word is..."). After each target word was presented, six words appeared on the computer. Subjects were to choose the word they had just heard. Speech intelligibility was defined as the percentage of correct responses on the MRT.

Auditory Sternberg task. One word from each of the four categories was randomly selected and those four words served as the *memory set*. Also, one word from each category was selected to be a part of the four-item negative set. The memory set items were presented on the screen at the beginning of the trial, along with instructions to memorize the four words.

Each Sternberg trial consisted of 12 target items and 12 non-target items. The item presented on each trial was chosen randomly from its set, and subjects were instructed to indicate whether the word was part of the memory set. There were 24 five-second intervals defined for the task. The probe item was presented at a randomly determined time up to three seconds into the five-second interval.

Category Decision task. The Category Decision task was identical to the Sternberg except for the nature of the memory set. Subjects were given the name of a category and decided whether probe items belonged to that category. All 12 items from a category served as positive stimuli, and the negative stimuli consisted of four items from each of the remaining three categories.

Spatial Reasoning tasks. The spatial reasoning task consisted of the presentation of 24 pairs of histograms. For the simultaneous task, on each trial a single histogram appeared on the left half of the screen and was followed a moment later by the presentation of another histogram on the right side of the screen. Both items remained on the screen until the end of the trial. For the short and long delay conditions, a delay of 0.5 or 1.5 s, respectively, was interposed between the offset of the first histogram (which occurred 1 s into the interval) and the onset of the second. For a delayed task, only the second histogram was present at the time of the decision. On half of the trials, the two histograms were identical. On the remaining trials, the heights of at least two of the bars differed across the histograms.

Results and Discussion

MRT data. For all experiments, performance (proportion correct) was worse under low intelligibility, $M=0.239$, than under conditions of high intelligibility, $M=0.534$; $p < 0.001$ for all experiments. (Chance performance would be approximately 17%.)

Sternberg search task. Single-task performance was good (M=88%) at high intelligibility but was close to chance (M=59%) at low intelligibility. Figure 1 shows the difference between single-task and the dual-task accuracy (for all analyses of accuracy, dual-task performance was subtracted from single-task performance). Accuracy on the Sternberg was unimpaired by the concurrent performance of either of the delayed visual tasks, but was impaired by the simultaneous task at low intelligibility. Figure 2 shows the difference between dual-task and single-task reaction time (for all analyses of reaction time, single-task performance was subtracted from dual-task performance). Concurrent performance of the simultaneous visual condition slowed responding to the Sternberg task (especially at low intelligibility), whereas the delayed conditions did not.

Category Decision task. The mean single-task accuracy of the Category Decision task was 79% at high intelligibility and 53% at low intelligibility. The delayed conditions had almost no effect on accuracy (Figure 1) or reaction time (Figure 2). The simultaneous visual task impaired both accuracy and reaction time, especially at the low intelligibility level.

Short- and Long-delay tasks. Performed in isolation, accuracy for the short delay condition ($M=85\%$) was higher than for the long delay condition ($M=76\%$). The higher accuracy for the short delay condition was accompanied by a slightly slower reaction time ($M_{short}=1.012s$, $M_{long}=0.957s$). Though the reliance on memory is greater for the long delay task, the decrease in accuracy and increase in reaction time is greater for the short delay condition when combined either with the Sternberg task or the Category Decision task (Figures 3 and 4).

Simultaneous visual task. Performed in isolation, the simultaneous task was performed more accurately ($M=89\%$) than either of the delayed visual task. There was, however, a speed accuracy trade off in that the simultaneous task produced much slower reaction times ($M=1.245$ s) than the delayed visual tasks.

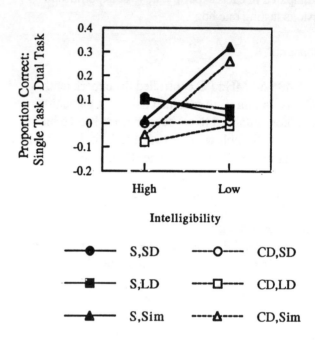

Figure 1. Difference between single-task and dual-task accuracy for the Sternberg task and the Category Decision task (S = Sternberg, CD = Category Decision, SD = Short Delay, LD = Long Delay, Sim = Simultaneous; abbreviations are the same for all figures).

Accuracy for the simultaneous task was impaired when performed with either auditory task (Figure 3). Note that the dual-task decrement is similar to that seen for the delayed visual conditions.

The pattern of reaction times differs from the pattern seen for accuracy: performance on the simultaneous task slowed by approximately one half second when performed with the auditory tasks (see Figure 4).

General Discussion

Attentional switching exerts a major effect on dual-task decrement in an auditory/visual environment. Across all experiments, the simultaneous task generally resulted in greater dual-task decrement than did either the long or short delay conditions. Also, the short delay condition generally resulted in poorer performance than did the long delay condition when it was paired with an auditory task. This pattern can be explained by referring to the type of attentional switching required: for the simultaneous condition, subjects could switch their attention between two spatial locations. For the short delay condition, subjects could switch their attention between the external item and the trace they held in memory. Finally, after 2 s, little distinct visual information may have remained of the target item, prohibiting the subjects from shifting attention from one item to the other. Instead, a more global comparison of the patterns may have been performed. In other words, the effect of attentional switching is mediated by the amount of visual information available and the amount of visual processing performed.

The difference in patterns for the Sternberg and Category Decision tasks were quite small, leading us to believe that that differential use of working memory makes little difference for dual-task decrement. Instead, the simple processing of the input for meaning as well as the generation of a response appears to be the source of dual-task decrement.

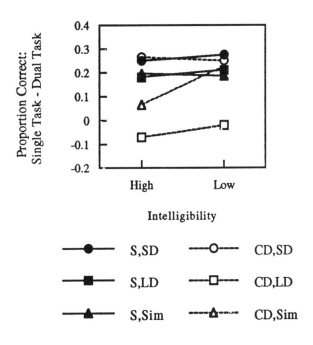

Figure 3. Difference between single-task and dual-task accuracy for the visual tasks when paired with an auditory task.

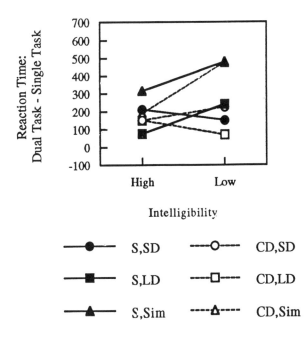

Figure 2. Difference between dual-task and single-task reaction time for the Sternberg and Category Decision tasks.

Unfortunately, the speech intelligibility manipulation resulted in less than interpretable results. A comparison between the high intelligibility and low intelligibility conditions for the visual tasks is not feasible since performance was very near chance under conditions of low intelligibility. If subjects were guessing on a substantial portion of low intelligibility trials, little can be said about the cognitive processes that occurred in that condition.

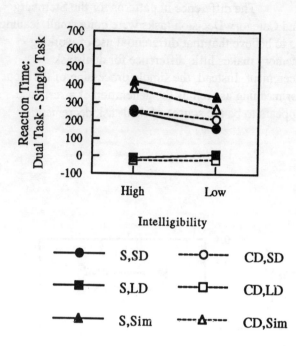

Figure 4. Difference between dual-task and single-task reaction time for the visual tasks when paired with an auditory task.

Implications and Future Directions

The results presented here point to several routes of investigation. First, it should be possible to manipulate the degree of attentional switching within a unitary stimulus so that actual eye movements are eliminated (i.e., the contribution of saccade programming may be separated from the shifting of the attentional "spotlight"). For example, using stimuli that vary on separable attributes (i.e., attributes that can be individually attended) should provide more opportunities for attentional switching than would stimuli that vary on integral attributes (i.e., attributes to which attention may not be individually and separately allocated; Garner, 1978).

Our findings also imply that, in designing an interface to be used in a multi-task environment, it would be preferable to minimize the amount of attentional switching required, even at the cost of increased memory load. This might be achieved, for example, by using more centrally located indicators for which the operator must remember the meaning and characteristics.

REFERENCES

Garner, W. R. (1978). Selective attention to attributes and to stimuli. *Journal of Experimental Psychology: General, 107*, 287-308.

House, A. S., Williams, C. E., Hecker, M. H. L., & Kryter, K. D. (1965). Articulation-testing methods: Consonantal differentiation with a closed-response set. *Journal of the Acoustical Society of America, 37*, 148-166.

Payne, D. G., Peters, L. J., Birkmire, D. P., Bonto, M. A., Anastasi, J. S., & Wenger, M. J. (1994). Effects of speech intelligibility level on concurrent visual task performance. *Human Factors, 36*, 441-475.

Peters, L., & Garinther, L. J. (1990). *The effects of speech intelligibility on crew performance in an M1A1 tank simulator* (HEL Technical Memorandum TMII-90). Aberdeen, MD: U. S. Army, Aberdeen Proving Ground.

Sternberg, S. (1966). High-speed scanning in human memory. *Science, 153*, 652-654.

THE PONZO ILLUSION IN VIRTUAL ENVIRONMENTS: CORRECTLY APPLIED SIZE CONSTANCY?

Robert A. King, Greg E. Fujawa, & Kelly G. Elliott
The Department of Psychology and
The Graphics, Visualization, and Usability Center
Georgia Institute of Technology
Atlanta, Georgia

The perceived size of a stimulus can be greatly influenced by the surrounding depth cues. The effect size of the Ponzo Illusion was tested in a virtual environment with many depth cues (MC) and few depth cues (FC) conditions. The effect of size and the depth cues on the perceived proximal size of the stimulus was measured. **Methods.** A modified Wheatstone stereoscope was used to present the stimuli on a monitor positioned 80 cm from the subjects. The subjects used a 6AFC confidence rating scale to indicate whether the first or second of two sequentially presented stimuli had a greater proximal extent. The 'many cues' (MC) condition included texture, relative height, foreshortening, linear perspective, relative brightness, and relative size. The few cues condition consisted of linear perspective, relative height, and relative size. Stimulus size (proximal extent) was varied independently from all other depth cues. **Results.** Both depth cue context and proximal size were found to have a significant effect on the perceived proximal size for FC conditions. However, for MC conditions only the depth cues, and not proximal size, had an effect on perceived proximal size. The effect size for both depth cues and proximal size had a significant linear trend. But proximal size had a larger trend that was significantly related to the size constancy function.

INTRODUCTION

The perceived size of an object in an environment can be influenced by the surrounding context. Due to the relatively deprived conditions of many virtual environments, the intentional or unintentional effects of the context may exert a greater effect on the perceived distance and size of an object. Factors effecting the strength of the Ponzo Illusion were explored.

A common explanation for the Ponzo Illusion is misapplied size constancy (Gregory, 1963). For example, because of the linear perspective depth cues, one perceives the upper horizontal line in Figure 1 to be at a greater distance than the lower line. Consequently, for it to have the same proximal extent it must be larger, and it is perceived as larger.

Figure 1. The Ponzo Illusion.

Smith (1978) proposed that acute angles tend to be overestimated. Therefore, intersecting lines are perceived to intersect closer to a right angle than the actual angle of intersection. The angle of the inducing lines are said to cause the square in Figure 2 to appear to be a trapezoid with the larger section at top.

Integrated Field Theory (Pressy & Epp, 1992) proposes that the Ponzo Illusion is a result of complex interactions between the inducing lines and the stimulus lines as a result of the limited area of the focus of attention.

Figure 2. Smith's (1978) variant of the Ponzo Illusion.

The last two theories were not directly tested in this experiment. However, if two test lines are to be presented sequentially, direct interactions between the test and comparison lines should be reduced or eliminated. Therefore, without modification, the models of Smith (1978) and Pressy & Epp (1992) do not predict an effect for the Ponzo Illusion for sequentially presented stimuli. The theory of misapplied size constancy is to be tested against the predictions of linear perspective. Stereo vision can potentially provide absolute size information, therefore, the effect of stereo depth will also be tested.

EXPERIMENT 1

Methods and Stimuli

A modified Wheatstone stereoscope was used to present stimuli on a monitor positioned 80 cm from the subjects. Two subjects used a 6AFC confidence rating scale to indicate whether the first or second of two sequentially presented stimuli had a greater proximal extent (1 = least confident the

standard stimulus is larger, 6 = most confident the standard is larger). Two conditions were run with and without binocular disparity: many depth cues and few depth cues. The 'many cues' (MC) condition included texture, foreshortening, linear perspective, relative height, relative brightness, and relative size. The 'few cues' (FC) condition consisted of linear perspective, relative height, and relative size. The depth cues and proximal sizes of the stimuli were consistent with an object located at one of nine distances between 96 and 104 cm. For the MC condition, the stimulus was a 5 cm square. For the FC condition the stimulus was a horizontal line 5 cm long. A standard stimulus had depth cues and a size that indicated a distance of 100 cm. The depth cues and sizes for the 81 comparison stimuli were fully crossed.

Results and Discussion

Stereo disparity had no effect on the perceived size of the standard relative to the comparison stimulus (t = -.232, $p < .8167$). The effect of the size and depth cues were both significant for the FC condition (size: $t = 11.48$, $p < .0001$, $\beta = .344$; cues: $t = -5.61$, $p < .0001$, $\beta = -0.168$). However, for the MC condition only the depth cues had an effect on the perceived proximal extent of the stimuli (size: $t = -1.1$, $p < .273$; cues: $t = -7.12$, $p < .0001$, $\beta = -0.223$). Therefore, the actual size (proximal extent) of the stimulus only had an effect on the perceived proximal extent in the FC condition. This range of

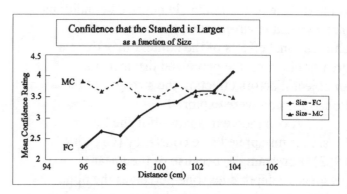

Figure 3. The confidence that the standard stimuli has a greater proximal extent as a function of proximal size than the comparison is shown for both MC and FC conditions.

stimuli is greater than the relative size JNDs found by Elliott, Davis, King, & Fujawa (1996) and Surdick, Davis, King, & Hodges (submitted) for similar stimuli and conditions. The depth cues had a smaller effect than size for the FC condition, but the effect size of the depth cues was greater in the MC condition.

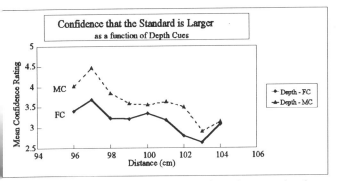

Figure 4. The confidence rating that the standard has a greater proximal extent than the comparison is shown as a function of the depth cues for both MC and FC conditions.

EXPERIMENT 2

The results from Experiment 1 were consistent with the size constancy theory. Furthermore, additional depth cues reduced the effect of proximal size, and increased the effect of depth cues on the perceived proximal extent. To further test the size constancy hypothesis, a greater range of distances was used.

The shape of the size constancy function can be predicted from the principles of linear perspective. Figure 5 shows the rate of reduction in proximal extent as an object increases its distance. The asymptote of this function is a 100% reduction in size at an infinite distance -- the vanishing point. If size constancy were the basis of the Ponzo illusion, the rate of change of the effect size for the Ponzo Illusion should match the slope of this curve.

Methods and Stimuli

The apparatus for Experiment 2 was the same as Experiment 1. Two subjects used a 6AFC procedure to indicate whether the first or second of two sequentially presented stimuli had a greater

proximal extent. The nine stimulus distances were 60, 100, 140, 180, 220, 260, 300, 340, and 380 cm. The MC and FC conditions were run without binocular disparity. Both the MC and FC conditions were replicated three times with a standard stimulus at 100, 220, and 340 cm.

Results

Both proximal size and depth cues had a significant effect on the perceived proximal extent (size: $t = 68.3$, $p < .0001$, $\beta = .662$; cues: $t = -5.86$, $p < .0001$, $\beta = -.0568$). A subset of the data was then used to test the size constancy hypothesis. For the condition in which the standard stimulus had a 100 cm size and depth cues, only data with both size and depth cues in the range of 60 cm to 140 cm were selected. Likewise, for the 220 cm standard, data in the 180 cm to 260 cm range was selected; and for the 340 cm standard, data in the 300 cm to 380 cm range was selected. These subranges were chosen because the ratings showed a high degree of nonlinearity across the entire range due to ceiling effects.

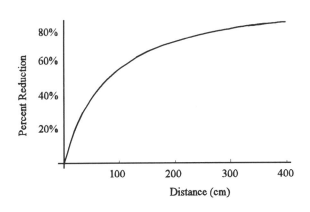

Figure 5. This curve shows the percentage that the proximal extent of an object decreases as it moves to the vanishing point.

This subset of ratings was then regressed onto the size constancy function. The non-linear trend was tested only after any linear trends were removed. The size constancy function was the best single predictor of confidence ratings for proximal size ($t = 6.86$, $p < .0001$, $\beta = 0.817$) although there was an additional linear trend ($t = 2.34$, $p < .0001$, $\beta = 0.420$). The size constancy function was not found to predict the ratings for depth cues ($t = -.289$, $p < .773$) after the linear trend was removed ($t = -6.23$, $p < .0001$, $\beta = -0.846$). The mean confidence ratings and the predicted curve are shown in Figure 6. The predicted size constancy curve has been overlayed on this graph so the relationship between the slope of the data and the slope of the size constancy function can be compared.

There were no significant differences between the FC and MC conditions for the second experiment ($F = 0.4$, $p < .842$).

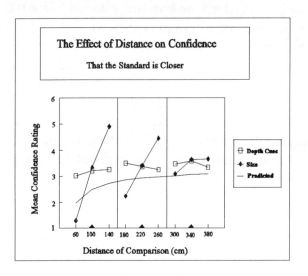

Figure 6. A comparison of the ratings for the size and depth cues are compared with the predicted size constancy function.

CONCLUSIONS

Depth cues have a large effect on the perceived proximal extent of an object. If there are a large number of depth cues, depth cues can overwhelm the actual proximal extent in determining the perceived proximal extent of a stimulus. However, the size constancy function was not related to the effect of depth cues on perceived proximal size after a linear effect was removed. Therefore, the effect of depth cues on perceived proximal extent is not the shape one would predict from a veridical size constancy function. Instead, depth cues have a linear effect on the perceived proximal size.

REFERENCES

Elliott, K. G., Davis, E. T., King R. A., & Fujawa, G. E. (1996).Single and Combined Visual Distance Cues. Proceedings of the Human Factors and Ergonomics Society- 40th Annual Meeting, Philadelphia, PA.

Gregory, R. L. (1963). Distortion of visual space as inappropriate constancy scaling. Nature, **199**, 678-680.

Pressy, A. W., & Epp, D. (1992). Spatial Attention in Ponzo-like patterns. Perception and Psychophysics, 52(**2**), 211-221.

Smith, D. A. (1978). A Descriptive Model for Perception of Optical Illusions. Journal of Mathematical Psychology, **17**, 64-85 (1978).

Surdick, R. T., Davis, E. T., King, R. A., & Hodges, L. F. (submitted) The Perception of Distance in Simulated Visual Displays -- A Comparison of the Effectivenss and Accuracy of Multiple Depth Cues Across Viewing Distances.

Human Factors Consulting: Business and Professional Lessons Learned.
A Panel

K. Michael Dresel,
Chair

Shelley Waters Deppa, Marilyn S. Joyce, Michael E. Maddox, Thomas B. Malone,
Panelists

Purpose

This discussion will assist Human Factors professionals who are contemplating starting to consult, forming a consulting business, or are currently providing consulting services. Consulting has always been an option for the Human Factors or Ergonomics professional. However, it seems that several trends in the corporate and public worlds are bringing more and more of the society members to consider the consulting option. Corporate downsizing, increased expectation of "user-friendly" interfaces, large jury awards for injuries and accidents, all provide an opportunity for the independent consultant. However, consulting is not an easy way to earn a living. There are many ways to do consulting, and many forms of business possible for the Human Factors or Ergonomics professional. Some of these ways are better than others and, it is hoped, we can all learn from those who have boldly gone where no one has gone before. It is the purpose of this panel to provide a forum for exchanging ideas on the topic of the Human Factors consulting business.

Importance of the topic

The annual meeting of the Society is a place where data can be presented, where networks can form, friends can meet, and where people can find jobs or fill positions. However, one of the main functions of the annual meeting, and one that is not carried out very well by other Society elements, is the function of providing a forum for a free-flowing exchange of ideas, among people who get together because of a common interest. This free-flowing exchange of ideas is best served by relatively unstructured time devoted to a single topic that is of common interest to a number of Society members. The sessions in San Diego that related to business matters, ethics, forensics, and the planning focus groups were always well attended, with much valuable discussion and information exchange. These topics are becoming increasingly more important in the light of the trends mentioned above. This panel presentation also clearly meets the desire of the Technical Program committee to provide sessions on issues pertaining to professional development, as highlighted in the Call for Papers.

Format

There is a tremendous amount of information that could be presented and discussed in this area, and there is a tremendous amount of interest on the part of the Society members. The challenge for the moderator will be to provide a forum for information exchange within the time constraints of the session schedule. The panelists feel very strongly that a major benefit of the panel format is the opportunity for interactive discussions with the audience, providing a chance for the panelists to learn as well. The general format for the panel will be for each of the panel members to give a short presentation, with lessons learned and hopefully some good war stories, and then to open the floor for audience statements and questions. The goal will be for active exchange of information and ideas, and so the moderator will not ask questions unless the audience fails to respond. The expectation of the moderator and of the panel members is that there will a great deal of interest and audience participation.

Panel Members

The panel was chosen to provide a range of experience and consulting areas. The panel consists of Tom Malone, of Carlow International Inc., Marilyn Joyce of the Joyce Institute, Mike Maddox of Sisyphus Associates, and Shelley Deppa of Safety Behavior Analysis, Inc. Each of them are principals with their company, and have a number of lessons learned to share, and some questions for the audience as well.

Topics

The range of experience and interests of the panel will allow many subject areas to be considered. Some of the panelists are in the early stages of company development, while others have long histories of growth. Some of the topics that will be discussed are: differences between Human Factors consulting and other businesses, differences between large and small consulting firms, growth of a business, ethical issues, customer relations, the future of consulting, corporate downsizing, education, certification, and hopefully, several topics that we haven't thought of, but that will arise from audience participation.

Abstracts from Panel Members

Michael E. Maddox
Sisyphus Associates

Challenges for the sole proprietor

There are many challenges associated with running any small business. Human Factors consultants, especially sole proprietors, face all of the common small-business-related problems, as well as the uncertainty of the consulting domain, the difficulty of marketing, and the need to maintain extremely high technical work quality. Of course, along with these problems comes the ability to do a variety of interesting work with virtually no political or bureaucratic barriers. At this level of the food chain, people are not interchangeable. In larger consulting organizations, clients might be willing to accept whoever is labeled a "senior scientist". In very small

companies, however, clients expect specific people to perform their work.

Effective marketing is probably the most insidious and widespread challenge for small HF consulting businesses. It is often said that "word of mouth" advertising is the best type. For sole proprietors like myself, word-of-mouth advertising appears to be the *only* way of selling our services. Since it isn't possible to leverage the time of a single individual, I find it very difficult to perform marketing when it should be done, *i.e.*, when I'm busy actually doing billable work. I will discuss these issues, as well as the implications of improved voice and data connectivity for small HF consultants.

Marilyn Joyce
The Joyce Institute

Strategies for Marketing an Ergonomics/Human Factors Consultant

Although most consultants say that they know the importance of marketing, few seriously look at marketing as a challenging, rewarding part of their business. The purpose of this part of the panel is to focus the discussion on the importance of marketing and selling, not only to the success of an individual consultant, but to the success of the profession itself.

The primary discussion points are as follows: Defining the vision and scope of the consulting practice; defining the marketing strategies to support that vision and scope; developing the sales approach; evaluating the success of the approach. Within those general topics, we will consider the following: understanding the differences between marketing and selling; determining geographic and industry focuses; "packaging" and "pricing" of services (do you want to be the cheapest or the best?); deciding the means of delivering services; moving potential clients from general awareness to commitment to purchase; identifying the true "buyers" in an organization; designing collateral pieces and presentations to support the sales effort; and responding to changing market conditions.

Effective marketing truly enables a profession to define itself.

Shelley Waters Deppa
Safety Behavior Analysis, Inc.

Challenges and lessons for the start-up phase.

Four years ago, I left 16 years of government service to establish a very successful Human Factors consulting business focusing on the safety of consumer products and warning label development. I quickly learned that while I was an expert in the Human Factors discipline, I knew little about the business end of establishing and running a business. Hoping to make this learning process easier for others, I will share my insights gained in the process of identifying and working through issues involved in starting and running a successful Human Factors consulting business. I will identify the many business decisions that need to be made, such as type of business entity, need for insurance, when to hire others, and billing. The discussion will then focus on marketing strategies and attributes important to succeeding as a start-up Human Factors consultant, such as self-discipline, risk taking, networking and contacts, and identifying a niche.

Thomas B. Malone
Carlow International Incorporated
Carlow Associates Ltd - Ireland

The Ups and Downs of Owning/Operating a Human Factors Company

The challenges in maintaining a successful company devoted to human factors fall into three areas: finances, management, and marketing. Under finances there are three major issues: cash flow, contracts, and coverage. Of these, the most demanding is cash flow. Given time lags and glitches in procurement actions, billing cycles, and funding, while having to meet payrolls and accounts payable, it is critical that any business have readily available resources in the form of a line of credit or cash. Probably the number 1 killer of new companies is failure to provide for unanticipated cash needs.

In management, the major challenge for a company is to have a vision, and to stick with it. Too often executives tend to drift toward areas that interest them, irrespective of their experience, expertise, or customers in the new area. A second management issue is to take the time and effort to generate a plan for achieving the vision. This plan should be a documentation of the company's goals and objectives, a review of the company's business areas, with discussion of the strengths and weaknesses, and a process for achieving the objectives, capitalizing on strengths and compensating for weaknesses.

The marketing problem is straightforward: establish a customer base, keep them satisfied, and expand the services and products for these customers. Marketing and cash flow represent the life's blood of the company, and deserve a good deal of the manager's attention.

As to what is special about human factors as a business, first of all, the negatives. It is more difficult than with other disciplines to sell to a new customer who has little or no experience with human factors. There is too little information on the benefits of applying human factors. Budgets for human factors support are usually the first to be cut in times of retrenchment.

On the positive side, working in human factors ensures a level of diversity not seen in other disciplines. The range of applications where we can make a significant contribution is broad, which provides a wider span of marketing opportunities and also a continually challenging and always changing working environment. Our involvement can be genuinely rewarding since we are making systems and products safer and easier to use, and more economical to build.

Name	Position	Affiliation and Address
K. Michael Dresel, Ph.D.	Organizer & Panel Chair	Principal Dresel & Mowry, LLC 1420 NW Gilman Blvd., Ste 2209 Issaquah, WA, 98027-7001 (206) 916-4526
Shelley W. Deppa, CHFP	Panelist	President Safety Behavior Analysis, Inc. 21021 New Hampshire Ave. Brookeville, MD 20833 (301) 774-9682
Marilyn S. Joyce, M.E.	Panelist	Director The Joyce Institute 1313 Plaza 600 Bldg. Seattle, WA, 98101 (206) 441-6745
Michael E. Maddox, Ph.D.	Panelist	Principal Scientist Sisyphus Associates PO Box 911 Madison, NC 27025 (910) 427-8124
Thomas B. Malone, Ph.D.	Panelist & Co-chair	President Carlow International, Inc. 3141 Fairview Park Dr. #750 Falls Church, VA 22042 (703) 698-6225

FITNESS-FOR-DUTY TESTING:
ISSUES, CHALLENGES, AND OPPORTUNITIES.

Dary Fiorentino and Marcelline Burns
Southern California Research Institute
Los Angeles, California

Implementation issues of Fitness-For-Duty testing are discussed from the scientific, administrative, and labor viewpoints. Results on the sensitivity of 4 Fitness-For-Duty tests from a 24-subject laboratory experiment using alcohol as the stressor are also reported.

INTRODUCTION

The objectives of this paper are to explore the issues associated with Fitness For Duty (FFD) tests as they pertain to the transit industry. The feasibility of FFD testing in the transit workplace hinges upon three criteria: First, the tests must be scientifically demonstrated to reliably measure impairment. Second, testing programs must be shown to be operationally viable within transit agencies. Third, labor must accept the technology and agree to participate in testing programs. With that in mind, the authors contacted Mr. Steve Andrle, Manager of the Transit Cooperative Research Program of the Transportation Research Board; Mr. Michael Townes, Executive Director of the Peninsula Transportation District Commission; and Mr. Sonny Hall, International President of the Transport Worker Union of America and asked them to provide a short position paper on the matter. The papers were summarized, incorporated into a common format, and are discussed within the context of a recent experiment evaluating the sensitivity of four FFD tests.

FFD Defined

FFD testing compares an individual's test performance upon entering the workplace for a shift to his/her own baseline, for the purpose of determining whether that individual is mentally and physically fit to begin his/her job duties.

Assumptions

FFD testing is based on the assumption that humans receive sensory input, process it, and produce an output (performance). Assuming that performance can be quantified and measured, observed performance decrements can be inferred to occur because of degraded sensory, cognitive, or motor processes. An individual with sufficiently degraded sensory, cognitive, or motor processes, regardless of the cause, may be deemed unfit for duty.

Background

The greatest responsibility of managers at all levels within public transit operations is the safety of passengers. Since the majority of public transit services delivered throughout the world is provided with vehicles requiring some level of human control, the fitness of the human beings providing that control is as critical to the provision of safety as the adequacy of the vehicle's mechanical and structural condition. There are, however, very few reliable tools available to transit managers and supervisors to ensure that workers are indeed fit to provide safe vehicular control.

Current drug screening programs have several limitations. First, the test themselves are expensive and must be administered by trained and certified personnel. There is also a large associated labor cost resulting from the fact that, in most cases, employees must be paid for the time they are being held off from work while awaiting test results, the time associated with resulting disciplinary procedures, and the medical and other costs associated with contractually and regulatory required rehabilitation periods. In addition, the work that the employee would have been doing during those unproductive payroll times must be covered by another employee, typically at premium payroll rates. Also, usually the results from the analyses are received several days after the specimens are collected, thus making detection a deterrent rather than a preventive issue. Second, the detection of drugs may depend on the time between intake and specimen collection and on how common and detectable a particular substance is. Third, drug screening does not detect impairment, but only the potential cause of such impairment. To this day, the precise correlation between specific drug intake and impairment is not very clear. Furthermore, because different drugs affect different people in different ways, it is difficult to generalize across individuals. Fourth, there is no assurance that, after having screened "clean", an employee may not return to work and use a drug there. Finally, drug screening programs have met severe resistance on the part of unions and right-to-privacy advocates.

FFD tests have been developed and applied in the military and in the nuclear industry. They have not, however, been used extensively in the transportation industry and not at all in the public transit industry. Because there is concern about the FFD of safety-sensitive employees in the transit industry, the Transit Cooperative Research Program (TCRP) recently sponsored a research project on the potential application of FFD testing devices to the industry.

The TCRP was authorized by the Intermodal Surface Transportation Efficiency Act of 1991 and was created by memorandum agreement in May 1992 between three participating organizations: The Federal Transit Administration (FTA), the National Academy of Sciences acting through the Transportation Research Board (TRB), and the Transit Development Corporation (TDC), a non-profit educational and research organization established by the American Public Transit Association. The TDC is responsible for forming the independent governing board and designated the TCRP Oversight and Project Selection (TOPS) Committee. The TOPS Committee selects the research projects that will, in their judgment, provide the greatest benefit to the transit industry. The issue of FFD was recognized by the TOPS Committee as a problem confronting the industry that could be meaningfully addressed by research.

TCRP project "Fitness-For-Duty Testing in the Transit Workplace" focused narrowly on the technical feasibility of adding FFD tests to the policies and procedures currently used by transit managers for determining employee fitness for safety-sensitive work. In actuality, however, FFD testing must meet multiple levels of feasibility criteria. It must meet administrative criteria, labor criteria, and scientific criteria.

It must be made clear that the TCRP position is that FFD testing is not proposed as a substitute for drug and alcohol testing procedures required by law, but as an additional safeguard against impairment of safety-sensitive employees.

Scientific Criteria

For FFD testing to be considered viable, the following scientific criteria must be met:

Validity is the ability of FFD tests to measure what they are intended to measure (i.e., impairment).

Sensitivity is the degree to which FFD tests can detect levels of impairment.

Reliability is the ability of FFD tests to consistently measure impairment over time and place.

Specificity is the ability of FFD tests to distinguish between stressors. While the objective of FFD testing is to detect impairment regardless of the cause, a distinction must be made between various stressors. For example, Performance differentials due to drug use are unacceptable, but performance differentials due to circadian rhythms should be allowed. FFD tests must be able to distinguish the two.

Administrative Criteria

Assuming that FFD testing is shown to meet the scientific criteria and is demonstrated to enhance the safety of transit operations, FFD testing must meet four administrative criteria.

Cost. Any level of FFD intervention leads to increased costs. Some of the additional costs that would be incurred in a FFD testing program include:

- Hardware and software set-up time
- Employee baseline time
- Administrator baseline time
- Employee test time
- Administrator test time

- Administrator back-up time
- Administrator archival time
- Maintenance

The cost conflict must be balanced so that attaining enhanced safety does not diminish the transit agency's ability to meet the goal of service delivery. FFD can be successfully implemented only if the cost issues will be addressed by changes in today's standard transit working conditions, by increases in resources allocated to transit service delivery, or by a reduction of services.

Fail Management. The added ability to screen for a wider variety of stressors implied by the emerging FFD technologies causes a further dilemma for cost conscientious transit managers. The potential for higher rates of detection means the need to absorb the payroll cost related to the need to have greater numbers of standby personnel to fill in for the unfit employee. This represents the potential for more unproductive payroll costs. Further, the concept of FFD testing implies that traditional employee discipline techniques cannot be used to mitigate unproductive payroll costs because the focus is on the readiness to perform, not on what is causing the degraded performance. It would be very difficult to issue discipline, for example, if an employee is deemed unfit due to fatigue or as a result of the stress caused by an argument with a spouse.

Availability. This concerns the degree to which FFD tests are functioning correctly and can be used by the employees.

Usefulness. This criterion reflects the benefit to the transport industry derived from the implementation of FFD testing. Usefulness can be estimated in the laboratory, but only determined in field studies.

In summary, to attain the correct balance between the reasonable assurance of safety and efficient service delivery, FFD tests must be administered quickly and produce reliable results. The additional resources which will be needed by the transit agency to use FFD testing to its maximum potential should be minimal. These tests must be delivered in a manner which maintains the civil and contractual rights and dignity of the employee being tested. Preferably, the tests should be non intrusive. The tests should measure performance indicators that relate to the skills needed to do the individual job.

Labor Criteria

Labor criteria relate to the degree to which employees, including unions, would resist introduction of FFD testing programs. Labor criteria can be grouped into safeguards against improper and unfair rules and usability issues. Safeguards against improper and unfair rules include:

Guidelines. The assurance that FFD testing will not be used as a "mixed bag" based on the agenda and goal of the person and/or operation that puts in effect the FFD operation.

Prevention and Rehabilitation. The primary focus of FFD testing should not be disciplinary. It should be prevention, protection of the public, and rehabilitation offered to an employee.

Validity. FFD testing must measure real impaired skills, not occasional performance fluctuations.

Ramification. This concerns the clear understanding of the consequences of a Fail result. This may include additional FFD testing, Standardized Field Sobriety Tests (SFSTs), drug testing, or a combination of these.

Discipline. Employees must be aware of the options available to management in case illegal drug use is detected.

Usability issues include:

Availability. The high probability of successful performance is essential to employees' perception of FFD technology.

Difficulty. Employees should perceive FFD testing within the perceived difficulty range of their job tasks.

Interest. The task to be performed by the employee should be perceived as neither too boring nor too arousing.

Confidence. The degree in which employees have confidence that FFD testing is accurate. A corollary of all the previous measures, confidence ultimately will be determined by the sensitivity, specificity, and reliability of specific FFD tests.

FFD in the Transit Workplace Study

The brief review of FFD testing issues presented above can be further addressed in the context of the recent FFD in the Transit Workplace study. Besides addressing the issue of sensitivity, the study helped to fill some information gaps on the logistics and technical requirements of FFD testing programs.

METHOD

The first decision facing FFD researchers is "what stressor to use?" Because people react to drugs, fatigue, and stress in different ways, it is necessary to create a consensus on a particular stressor, one that is well understood in terms of its effects on performance. Such a "standard" could function as the meter with which researchers could initially begin mapping out this area of human performance. Alcohol is the preferred choice at this time. First, alcohol linearly affects performance (i.e., the higher the dose administered, the higher the level of impairment). Second, alcohol is the most socially accepted drug. There are fewer availability and legal limitations in laboratory testing of alcohol than other drugs. Third, alcohol is the drug most commonly used. Inability of FFD test to detect at least alcohol would certainly limit their usefulness. In this study, alcohol was used as the standard, and impairment was defined as a BAC between 0.04% and 0.08%.

Twenty-four subjects (Ss), 12 males and 12 females, participated as paid volunteers in the study. Each S was given three treatments: A, B, and C. Treatment A was placebo where no measurable alcohol was administered and all measured BACs were 0.000%. In treatments B and C, the target BACs were 0.08%. Ss were tested five times at each treatment: Predose, at the expected peak (EP), EP+1hr, EP+2hrs, and EP+3hrs.

Four commercially available FFD tests were examined in the study, two performance and two physiological. Performance tests measure task performance as the number of correct (or incorrect) responses, reaction times, and mean absolute tracking error, among other measures. Performance tests are of various nature and varying degrees of difficulty. Physiological tests measure involuntary eye reflexes such as horizontal gaze nystagmus, smooth pursuit, pupil's response times, and saccades, among others. Some tests also incorporate a breath sampling instrument to measure an individual's BAC.

RESULTS

Sensitivity. The best performing test failed 79% of Ss at 0.08% BAC, 62.5% of Ss at 0.06% BAC, 38.5% of Ss at 0.04% BAC, and 19.35% of Ss at 0.02% BAC, while failing 4.76% of Ss at 0.000% BAC. The same test, using 0.04% BAC as the impairment criterion, yielded 80% correct decisions and 20% incorrect decisions. Of the correct decisions, 36% were hits and 64% were correct rejections. Of the incorrect decisions, 21% were false positives and 79% were false negatives.

Reliability. The two alcohol treatments in the study allowed for comparison of test results over two sessions. The percentages of same tests results at similar BACs ranged from 85.8 to 72.3.

Specificity. Because only one stressor was used in the study, the issue of specificity was not addressed.

Hardware and software set-up time. Equipment set-up required less than 2 hours. Setting up databases required up to 2 min per S. Set-up time for a mid-sized transportation agency (work force = 300) could, therefore, require in excess of 10 hrs.

Employee baseline time. Establishing Ss' baselines required from 15 min to over 5 hrs. These times, however, were spread out over several days at the vendors' instructions that Ss be allowed to develop a natural learning curve.

Administrator baseline time. Although most of the tests were self-administered, the test administrator needed to instruct the Ss on how to take the tests, monitor the tests, evaluate results, and give feedback. Administrator baseline time required from 1 to 5 hrs per S.

Employee test time. Tests time varied from 1 to 4 mins, depending on the test.

Administrator test time. Some tests required 2 min per S of administrator time.

Administrator back-up time. All vendors suggested that employees data files be copied onto floppy disk. Backing up a single database required from several seconds to several minutes, depending on software. Other tests stored data on individual floppy disks. Copying from disk to disk required up to 3 mins per S.

Administrator archival time. Baseline data, test data, and employee Pass/Fail history must be retrieved, updated, and re-stored for each test. For a mid-sized agency, depending on testing frequency, archival time may require several hours per week.

Maintenance/Availability. During the five-month testing period, two tests required maintenance. In both cases the vendors responded promptly and no significant testing delays were experienced.

Difficulty/Interest. In general, females found physiological tests more difficult, and performance tests less difficult, than their male counterparts. Both males and females rated physiological tests as less interesting than performance tests.

Confidence. Most subjects, when queried, expressed confidence that the FFD tests actually measured impairment.

DISCUSSION

There was great variability among the tests in detecting impairment. Some tests detected impairment with greater sensitivity and reliability than others. The best performing tests failed 79% of the Ss at 0.08% BAC, with 72.3% of the Ss receiving the same result at the two alcohol sessions. Physiological tests yielded better results than performance tests. On the other hand, the most sensitive test was also rated as the most difficult and the second least interesting. The transit industry must now develop a definition of what constitutes "impairment" and, based on such definition, decide if those results are acceptable. In general, using sensitivity as the sole criterion of feasibility, it appears that the answer to the question "is FFD testing in the transit workplace feasible?" may be a qualified "yes".

More significantly, however, there seems to be a commonality in the concerns expressed in the position papers. Allow FFD testing to undergo rigorous scientific evaluation and, if the results warrant it, it will naturally find its place in the transit industry. From the administrative point of view, management would clearly embrace a tool that would more than offset its implementation costs by a diminution of lost revenues due to low productivity, personal injury, and compensatory and punitive damages. Similarly, employees would agree to be tested routinely if it is demonstrated that the technology could prevent accidents, protect the public, and detect addiction.

The FFD area demonstrates the difficulty and importance of sound research when the data almost certainly will influence policy decisions. This is a challenge, as well as an opportunity, for the Human Factors professional.

REFERENCES

Burns, M., Hiller-Sturmhöfel, S. (1995). Fitness-for-Duty Testing. Alcohol Health & Research World, 19, 159-160.

Miller, J.C. (1996). Fit for Duty? Ergonomics in Design, 4(2), 11-17.

SITUATION AWARENESS: A VALIDATION STUDY
AND INVESTIGATION OF INDIVIDUAL DIFFERENCES

Leo J. Gugerty
Galaxy Scientific Corporation
San Antonio, Texas

William C. Tirre
Armstrong Laboratory
Brooks Air Force Base, Texas

The first experiment found that varying the rate of road hazards in a personal-computer-based driving simulator had no effect on subjects' situation awareness, as measured in the simulator. Thus, setting a high rate of hazards does not distort subjects' situation awareness. In the second experiment, the situation awareness test was found to predict driving performance in a realistic simulator. Individual differences in situation awareness were correlated with working memory and psychomotor abilities.

We are studying the cognitive abilities necessary for optimal performance on real-time tasks, such as driving and air-traffic control, that involve tracking a changing situation (situation awareness), allocating attention among multiple subtasks (multitasking), and rapid decision making. In particular, we are focusing on situation awareness abilities, and how these are related to real-time performance. Adams, Tenney, and Pew (1995) describe situation awareness (SA) as a mental model of a dynamic situation that has two elements: (1) explicit focus - active knowledge in working memory, and (2) implicit focus - less active knowledge that is relevant to the current situation, but more accessible than irrelevant long-term-memory knowledge. We have developed both explicit, recall-based, and implicit, performance-based, measures of SA in a driving task (Gugerty, in press). In the experiments reported here, we focused on the implicit measures, where subjects are put in hazardous driving situations in a personal-computer (PC) based simulator and try to avoid the hazards. Their SA is inferred from their avoidance responses.

The research reported here had two goals. In Experiment 1, we investigated a potential problem in assessing SA using implicit, performance-based measures. In order to obtain as much information as possible about subjects' SA in a limited time period, it is helpful to present subjects with a high rate of hazardous situations, or, in signal detection terms, to set a high signal probability. However, this high signal probability may cause subjects to be more alert than they would be in a normal driving situation, where hazards are infrequent. Thus the intervention to assess SA may change subjects' SA. We assessed the extent to which this happens in Experiment 1 by varying the frequency of hazards and measuring subjects' SA and driving performance.

The goal of Experiment 2 was to: (1) validate the SA measures obtained from the low-fidelity, PC simulator by comparing them to measures from a moderate-fidelity driving simulator, and (2) determine whether individual differences in cognitive and psychomotor skills can predict SA abilities.

GENERAL METHODS

The PC driving simulator showed 3-dimensional animated driving scenes in a window on the computer screen. The window was 11.4 by 15.5 cm for Experiment 1 (half screen), and 18.5 by 25.3 cm in Experiment 2 (fullscreen). The subject saw the front view from the driver's perspective and also the rearview, left-sideview, and right-sideview mirrors (see Figure 1). All scenes showed traffic on a three-lane divided highway, with all cars moving in the same direction. Subjects watched scenes lasting from 18 to 35 seconds, and were instructed to imagine that their simulated car was on autopilot. However, subjects could make driving responses while viewing the animated scenes; that is, they could override the autopilot. On some trials, an incident would occur that required a driving response, for example, a car would move into the driver's lane ahead of the driver while moving slowly enough that it would be hit by the driver. Subjects could avoid hazards such as this by accelerating, decelerating, or moving to the lane on the left or right. They indicated these responses with the up, down, left, and right arrow keys, respectively. Subjects could usually make only a single arrow-key press on each trial. When they did this, the moving scene usually ended. After the scene ended, the subjects received textual feedback concerning the correctness of their response.

Implicit (Performance) Measures of Situation Awareness.

We agree with Endsley (1995a) that the concept of SA is best seen as encompassing perceptual and comprehension processes, but not decision-making and response execution processes. The major difficulty in developing performance measures of SA is creating measures that reflect peoples' perceptual and comprehension processes more than they reflect decision and response-execution processes, even though all of these processes must contribute to any measure of real-time performance. Using Endsley's (1995b) terms, we wanted to develop *imbedded task measures*, which reflect particular aspects of SA. In contrast to imbedded task measures, *global performance measures* reflect a more even mix of situation-awareness and decision/action processes.

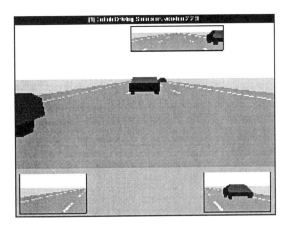

Figure 1. 3-Dimensional scene from driving simulator.

The first imbedded-task measure developed was *hazard detection*, which was calculated using the *A'* nonparametric, signal-detection measure of sensitivity (Grier, 1971). On each signal (hazard) trial, the response interval began when a car entered the driver's lane on a trajectory that would hit the driver and ended when it was too late for the driver to avoid the oncoming car. Following the procedure of Watson and Nichols (1976) for measuring sensitivity and bias with continuous signal-detection tasks, the catch-trial response intervals were set to be equal in duration to those on hazard trials. A hit was defined as any arrow-key response, even an incorrect response, during the response interval of a hazard trial. A false alarm was any arrow-key response during the response interval of a catch trial. For all trials, responses before the response interval, which were infrequent, were ignored in this analysis.

When subjects responded incorrectly to a hazard car, this shows that they were aware of the hazardous situation, but selected and executed an inappropriate avoidance response. Therefore, since the hazard-detection measures counted even incorrect responses to hazards as hits, the measure should reflect subjects' ability to detect hazards (an aspect of situation awareness) more than their decision/action abilities. The hazard detection measure focused on subjects' awareness of vehicles in front of and behind their car, since the hazardous cars always entered the driver's lane from a side lane and then approached the driver.

The second imbedded-task SA measure (*blocking-car detection*) focused on subjects' awareness of cars on their right and left that blocked their escape from the hazard cars. Most of the blocking cars were in the subjects' blindspot. Since the three-dimensional display did not show cars immediately to the right or left of the driver, subjects had a larger blindspot than in real driving. On hazard trials, subjects could usually

only know about blocking cars by remembering that a car had entered the blindspot and had not left it. All cars defined as blocking cars were located such that the subject's car would hit them if the subject tried to avoid the hazard car by moving to the right or left at any time during the response interval.

A small percentage of subjects adopted a strategy of only making accelerate or decelerate responses, probably so as to avoid having to remember cars in the blindspot. Because of this, when subjects made an accelerate or decelerate response (or no response) on a hazard trial, it was difficult to determine whether their response was based on knowledge of blocking cars. Therefore, blocking-car detection was estimated based only on hazard trials where the subject responded right or left during the response interval. For example, on a trial where the hazard car approached from the front and there were blocking cars to the right and left, subjects were considered as detecting one of two blocking cars if they went right or left. Overall, blocking-car detection was estimated by the ratio of the total number of blocking cars avoided over the total number of blocking cars (in each case, these totals were summed over all hazard trials where the subject responded right or left). Blocking-car detection was only estimated for subjects who responded right or left on more than four of the hazard trials.

Scoring high on the blocking-car detection measure did not depend on making a correct response, in terms of global task performance. In the above example, subjects would be credited with 50% blocking-car detection on a trial where they crashed. Thus, blocking-car detection should reflect subjects' awareness of blocking cars more than their decision/action processes.

The main measure of global performance in the PC simulator was *crash avoidance*, the percentage of hazards successfully avoided. A correct response on this measure involved avoiding any hazard cars without hitting blocking cars on hazard trials.

EXPERIMENT 1 - EFFECTS OF HAZARD RATE

Method.

Sixty seven subjects (47 males and 20 females) completed a practice block of 37 trials with a hazard rate of 49% (i.e., 18 hazard and 19 catch trials). Then, each subject completed two blocks of 74 trials. For 36 of the subjects, the hazard rate in the last two blocks was 74%; for 31 subjects, it was 26%. All of the hazards seen by subjects in the low hazard-rate condition were also seen by subjects in the high hazard-rate condition. This allowed performance of the low and high hazard-rate groups to be compared on the same set of hazards. Subjects took about 25 minutes to complete the practice block and 45 to 60 minutes to complete the two test blocks. There was a break of at least 10 minutes between each block.

Results.

Since the high hazard-rate group saw more hazard scenarios than the low hazard-rate group, there were potential differences in the difficulty of the scenarios seen by the two groups. Therefore, the performance of the two groups were compared only based on the common hazard scenarios seen by both groups. All analyses presented below are for these common scenarios. Subjects' performance was evaluated using the global driving performance (crash avoidance) and situation awareness (blocking-car detection) measures described above. In addition, subjects' response times for correctly avoiding hazards were evaluated.

With practice (i.e., across blocks), subjects improved both their global driving performance and their SA, as measured by the blocking-car detection measure ($p < .0005$). In addition, response times for correctly avoiding hazards declined significantly with practice ($p < .0005$).

If the frequency of hazards affected subjects' driving performance and SA, one would expect that subjects with a high hazard rate would improve more than those with a low rate. As shown in Table 1, varying the hazard rate had a slight but nonsignificant effect on global driving performance. The high-hazard group improved their crash avoidance 15% from the first to the last block, almost twice as much as the low-hazard group, which improved 8.5%. However, the block by hazard-rate interaction was not significant ($F(2,130) = 1.82$, $p = .17$).

Table 1. Performance and situation-awareness data from Experiment 1.

Variable	Hazard Rate	Block		
		1	2	3
Crash	25%	56	63	64
avoidance (%)	75%	60	70	75
Hazard	25%	.97	.97	.97
detection	75%	.97	.98	.98
Blocking-car	25%	62	77	82
detection (%)	75%	71	81	86
Response	25%	1.08	0.80	0.75
time (s)	75%	1.06	0.76	0.71

Thus, our data suggest that hazard rate does not affect driving performance, as measured by crash avoidance. Since this conclusion involves accepting the null hypothesis, we calculated the power of the statistical test. Given the variability in the data, we could have detected a significant difference (with $p < .05$) if one group's improvement in crash avoidance from the first to last blocks was at least 7.6% more than that of the other group. Therefore, in practical terms, we can conclude that any greater improvement in driving performance due to the higher hazard rate was less than 8%.

In addition, varying the hazard rate did not affect hazard detection, blocking-car detection or the time subjects took to respond to hazards. As Table 1 shows, performance on these variables improved to the same extent across blocks, regardless of hazard rate.

Discussion.

The main conclusion from this experiment is that almost tripling the hazard (or signal) rate in a driving task had very little effect on subjects' overall driving performance or their SA. One could argue that even a hazard rate of 26% is fairly high, when compared to real driving. However, the signal rates in this experiment are similar to those used in other experiments that sometimes found significant effects on subjects' signal detection performance. Our hazard percentages of 26 and 74% translate into signal rates of 0.6 and 1.8 signals per minute. Using a visual monitoring task, Parasuraman and Davies (1976) found that varying signal rates from 0.53 to 1.0 signals per minute resulted in an increased false alarm rate, a decreased response criterion, and no change in correct detections or detection sensitivity.

Thus, we conclude that setting a high hazard rate will allow researchers to obtain more information about drivers' SA without significantly changing their SA. In the second experiment, we used a hazard rate of 67% in assessing whether SA as measured in the PC simulator predicted performance in a moderate-fidelity driving simulator.

EXPERIMENT 2 - VALIDATION AND COGNITIVE CORRELATES STUDY

Like all real-time tasks, driving involves an interactive cycle of perceiving and comprehending the current situation, planning and deciding upon the next control action, and then executing the action. We measured driving subskills for each of these three stages. For the perceptual stage, we measured drivers' situation-awareness ability using the PC-based driving simulator. For the decision-making stage, we measured drivers' risk-taking tendencies by recording their driving speed while using a moderate fidelity simulator. For the response-execution (road-tracking) stage, we used computer-based tests of psychomotor ability, and also measured drivers' ability to maintain speed and lane position in the moderate-fidelity simulator. Our criterion measures of good driving included a questionnaire that assessed accident and traffic-violation rates, as well as ability to avoid road hazards in the driving simulators. (See Table 2.)

Method.

The 108 subjects were US Air Force basic traineees (64 males and 44 females). They ranged in age from 17 to 30, with an average age of 19.3. Each subject completed a block of 54 trials in the PC simulator. Each subject also drove in a moderate-fidelity simulator which had three 64-cm screens giving a 170 degree field of view, and a steering wheel, brake and accelerator that simulated realistic vehicle dynamics. In the moderate-fidelity simulator, subjects first drove an "obstacle" course, with realistic road hazards, and then a "speed course", with no hazards. In the latter, their goal was to maintain a speed of 60 miles per hour and good lane position. Crash avoidance (the percentage of hazards avoided) was measured, as well as average error in maintaining speed and lane position on the speed course.

Table 2. Tests and measures in Experiment 2.

Test	Measure
Driving questionnaire	accident and ticket rate
Moderate-fidelity simulator: obstacle course	crash avoidance chosen speed
Moderate-fidelity simulator: speed course	chosen speed lane & speed tracking errors
PC simulator	crash avoidance hazard detection blocking-car detection
Multilimb coordination	tracking errors
Rate control	tracking errors
Working memory	percent correct

In contrast to the tests involving the two driving simulators, the working memory and psychomotor tests used more abstract tasks, which were performed on a personal computer. In the spatial working memory test, subjects remembered and integrated visual information from displays. There were two psychomotor tests. In the rate-control task, subjects used the joystick to adjust the speed of an object moving around a circular path so that it matched the speed and position of another, target object. In the multilimb-coordination task, subjects tried to keep a circle centered in the screen despite a random force that tended to move the circle away from the center. To do this, they simultaneously used the joystick to adjust the vertical position of the circle and the foot pedals to adjust its horizontal position.

The tests were administered in a different random order for each subject during a testing session lasting 3.5 hours, including breaks.

Results.

On-road driving errors, in particular ticket rate, correlated with risk taking, as measured by average speed on the speed course of the moderate-fidelity simulator (Pearson's $r = .21$, $p < .03$). (Alpha levels for all correlations are based on one-tailed tests.)

Crash avoidance in the moderate-fidelity simulator correlated with two risk-taking measures, speed in the obstacle

course ($r = -.34$, $p < .001$) and in the speed course ($r = -.25$, $p < .01$) of the moderate-fidelity simulator, and one SA measure, blocking-car detection in the PC-simulator ($r = .20$, $p < .03$).

In addition, subjects' SA and driving performance on the PC correlated with good performance in the speed course on the moderate-fidelity simulator. Blocking-car detection and PC crash avoidance correlated -0.21 ($p < .03$) and -0.29 ($p < .01$), respectively, with errors in maintaining lane position in the speed course. Thus, the PC-based SA measures were predictive of both hazard avoidance and road-tracking ability in the moderate-fidelity simulator

Finally, the PC SA measures were correlated with performance on abstract laboratory tests of spatial working memory ability ($r = .20$, $p < .03$) and psychomotor errors ($r = -.44$, $p < .001$). The psychomotor tests assessed some of the abilities identified by Fleishman (1954), in particular rate control and multilimb coordination.

Discussion.

SA in the PC simulator was significantly correlated with hazard avoidance in the moderate-fidelity simulator. Also, even though the PC driving task was a perceptual task with very little psychomotor involvement, the SA measure correlated with the psychomotor skill of road tracking in the moderate-fidelity simulator. Similarly, the abstract Fleishman tests of psychomotor ability also correlated with SA. Psychomotor skills may be related to this non-motor SA task because the psychomotor tests we used measured psychomotor working-memory ability, and working memory ability is important in performing many cognitive tasks.

Note, however, that the static working memory task correlated less well with SA ability than did the dynamic psychomotor tasks. Perhaps there is something about attending to dynamic stimuli that SA and psychomotor tasks share.

Other measures besides SA were found to be predictive of criterion driving performance. In particular, risk taking tendencies were related to a number of aspects of driving performance. For example, subjects who chose to drive faster in the speed course of the moderate fidelity simulator had a higher rate of on-road tickets and avoided fewer hazards in the obstacle course of the moderate fidelity simulator.

CONCLUSION

Two experiments showed that: (1) a PC-based test of situation awareness ability does not distort subjects' situation awareness, and (2) the test predicts driving performance in a realistic simulator. Individual differences in situation awareness measured by this test were correlated with working memory and psychomotor abilities.

REFERENCES

Adams, M. J., Tenney, Y. & Pew, R. (1995). Situation awareness and the cognitive management of complex systems. *Human Factors, 37*(1), 85-104.

Endsley, M. R. (1995a). Towards a theory of situation awareness in dynamic systems. *Human Factors, 37*(1), 32-64.

Endsley, M. R. (1995b). Measurement of situation awareness in dynamic systems. *Human Factors, 37*(1), 65-84.

Fleishman, E. (1954). *Evaluations of psychomotor tests for pilot selection: The direction control and compensatory balance tests*. Technical Report TR54 131, Skill Components Research Lab, Lackland AFB, TX.

Grier, J. (1971). Nonparametric indexes for sensitivity and bias: Computing formulas. *Psychological Bulletin, 75*(6), 424-429.

Gugerty, L. (in press) Situation awareness during driving: Explicit and implicit knowledge in dynamic spatial memory. *Journal of Experimental Psychology: Applied*

Parasuraman, R. & Davies, D. R. (1976). Decision theory analysis of response latencies in vigilance. *Journal of Experimental Psychology: Human Perception and Performance, 2*(4), 578-590.

Watson, C. & Nichols, T. (1976). Detectability of auditory signals presented without defined observation intervals. *Journal of the Acoustical Society of America, 59*(3), 655-667.

APPLICATION OF MULTIPLE CUT-OFFS FOR
FITNESS FOR DUTY TESTING

Robert S. Kennedy, Julie M. Drexler
Essex Corporation, Orlando, FL
Gene G. Rugotzke
Public Health Laboratory, Cheyenne, WY
Janet J. Turnage
Star Mountain, Inc., Alexandria, VA

"Sensitivity," (percentage correct detections of the number treated) is the most often used index of psychological tests or test battery adequacy, although other factors (stability, reliability, other forms of validity, usability, etc.) are also reported as positive features. This paper reports on a method to improve "specificity" [ratio of persons correctly identified as not treated to total number not treated]. Specificity issues are a great concern, where behavioral testing may be used for managerial or regulatory decisions about workers. In such cases the percentage of workers with appreciable dosages may be < 10% and false positive percentages thereby become important. In two alcohol experiments, we empirically validated different multiple cut-offs seeking good false positive rates. In two additional alcohol studies we cross validated our optimum findings and found false positive rates of 3.6% (96.4% specificity) can be achieved with a combination of 6% decrement on three of four tests while retaining adequate sensitivity.

SUMMARY

Introduction

Microcomputerized test batteries of cognition for behavioral toxicology and occupational safety have a relatively short history (Hanninen, & Lindstrom, 1979)]. Recently, considerable research has been focused on developing further these microcomputer-based neurobehavioral and cognitive test batteries for the assessment of human performance and mental acuity (Kane & Kay, 1992; Russell, Flattau & Pope, 1990, for review). To be useful, a test should be accurate and there must be an understandable way of referring to the loss in performance which occurs on the test. Therefore, the use of a test battery for behavioral toxicology testing also would require normalization techniques to provide a context in order to evaluate the performance loss (impairment) that occurs. Similar requirements for accuracy exist for fitness for duty applications.

In fitness for duty and other behavioral toxicology testing, other than practical issues and economic factors, it is still the chief concern that two kinds of errors (misses and false positives) be avoided. The two measures of accuracy that are customary are "sensitivity" and "specificity". "Sensitivity" refers to the proportion of the treated population that is correctly identified. "Specificity" has to do with the percentages of people correctly identified as not dosed, (or treated), divided by the total number of people in the untreated group. For some concerns, sensitivity can be the more important issue. For other concerns, specificity can be the more important issue.

Our primary focus on test development in the laboratory often emphasizes sensitivity, but from a practical standpoint (i.e., in the workplace), we want to avoid false positives; and in some cases we think false positives can be a more troublesome problem than low sensitivity. Our concern here is with the prospect of being able to tune our false positive rate in order to minimize to the extent possible identifying anybody as treated if indeed they are not.

The reason for this concern relates to the prospect of a low base rate of impairment in the workplace. Several years ago Murphy (1987) pointed out that when the base rate of impairment is low, those who fail the test are more likely to be unimpaired than impaired—even when specificity and sensitivity are both high. Suppose, for example, that the base rate for nonimpairment is 95%—not an unrealistic supposition, if anything perhaps, a little low; that specificity is also 95%; and that sensitivity is 25%. Under these suppositions those who fail the test are almost four (3.8) times as likely to be unimpaired as impaired. If sensitivity were 50%, persons failing the test would still be more often unimpaired than impaired. In fact, sensitivity would have to equal 95% before there would be as many impaired as unimpaired persons among those failing the test.

The most feasible approach to this problem is to look more closely at those who fail the test but are unimpaired. To this point, "impaired" has meant "tests positively on urinalysis." (A positive urine test only means that the person has used drugs in the last few days. It does NOT correlate with improvement at time of collection.) It may be, however, that people who fail the test really are impaired, in the more general sense that their behavioral effectiveness is compromised, but not by drugs or alcohol. Most of them, perhaps, are fatigued. If so, it might be better to redefine the test's purposes to include "impairment for any reason." Such a redefinition would have several good effects. It would definitely increase the base rate of impairment, and it could improve both sensitivity and specificity. Of course, it would also mean a change in how the test was used. The test would now serve multiple purposes. Backed up by urinalysis, it would still serve as a deterrent to substance abuse. But it could also be used as a tool in employee relations and a means of improving productivity. Subjects who turn out to be fatigued, sick, or preoccupied with personal problems could

be identified. Poor performance on the test could signal a correspondingly poor performance overall. Finally, broadening the test's purposes and uses would make a positive result less incriminating, thereby allowing a somewhat more relaxed specificity requirement.

These considerations do not assure that performance tests can be used successfully in fitness for duty assessment. They do, however, suffice in order to state the criteria that performance tests must meet if any such success is to be realized. First, the test must have very high specificity, very few false positives (97% or better). Second, it must have appreciable sensitivity (about 25%). And third, it must be short (not more than 5 minutes or so).

METHODS

General

Four alcohol studies have been performed to determine the relationship between blood alcohol level and performance test scores on a computerized test battery. The first three of these studies are reported in Kennedy, Dunlap, Turnage, and Fowlkes (1993), Kennedy, Turnage, Wilkes, and Dunlap (1993), and Kennedy, Turnage, Rugotzke, and Dunlap (1994). The fourth is in progress. All studies were conducted in conjunction with the Public Health Laboratory of Wyoming.

Although study conditions varied slightly, the following features were common to all four experiments. Subjects were recruited and allowed to participate if they indicated some, but not excessive, experience with alcohol, no past history of chronic dependency of any type, good general health, and low risk for future alcohol-based problems. Selected tests from the DELTA (formerly APTS) test battery, which are completely described in Turnage and Kennedy (1992) were administered.

The tests, which were selected, all possess high predictive validity for holistic measures of intelligence (Kennedy, Dunlap, Turnage, & Wilkes, 1993), and also require only 12 minutes of testing time. We utilized tests that were shown to stabilize rapidly (< 10 minutes), with reliabilities greater than r=.707, for three minutes of testing time.

The nine-test battery was implemented on a portable, battery-operated laptop computer (NEC PC8201A) and consisted of two Tapping Tests (two-hand tapping and nonpreferred hand tapping), Grammatical Reasoning, Mathematical Processing, Code Substitution, Pattern Comparison, Manikin, Short-Term Memory, and Reaction Time (4 choice).

An Intoximeter 3000 breath analyzer was used to estimate alcohol concentrations in the blood. Prior to alcohol testing, subjects would practice the test battery for a minimum of ten trials to insure that they were familiar with testing procedures and had attained peak performance. On the day of testing, subjects received instructions and warm-up-testing and were breath tested. Alcohol was then consumed in a group setting to achieve an ultimate blood alcohol concentration (BAC) of .15 (.08 BAC in experiment four). In all four studies, blood alcohol concentration was confirmed by breath analyzer.

Performance testing was accomplished at predesignated levels of blood alcohol, after which subjects were returned to supervised housing where they were required to stay for the remainder of the evening and abstain from further consumption of alcohol. Experiment 1 used four single dosages (Placebo; .05; .10; .15 BAC) on four different occasions (days); Experiment 2 dosed subjects to .15 BAC and followed the alcohol removal over several hours; Experiment 3 dosed subjects to .15 BAC and followed both the ascending and descending curve; Experiment 4 followed both descending curves of two successive graded dosages of .08 BAC. In all four studies, there were significant correlations between BAC and performance score deficits from baseline performance (an average of the stable practice trials and warm-up trials), indicating the performance was reliably degraded by alcohol dosage.

The subjects then started a cyclic process involving breath test monitoring until the subject reached 0.08%, 0.07%, 0.06%, 0.05%, 0.04%, or 0.03% BAC. Subjects were given the DELTA battery, a breath test, and the field sobriety maneuver battery at each of these levels. A second drink was administered after lunch, or when the

subject finished the 0.03% cycle. The second drink may have been smaller than the first depending on the time it was given. The subjects then started the same cyclic process until the subject had a BAC < 0.03%.

Taking studies 1 & 2 as an empirical validation group, we took each subject's Delta performance scores at each blood alcohol concentration that was available and calculated a percentage decrement score for each test using the average baseline performance prior to alcohol ingestion. (Note that since some learning continues throughout testing, this will underestimate the sensitivity of the cognitive tests.) We then calculated proportion of the decrement as the denominator. We grouped the BAC levels as 0.000; 0.001 -- 0.039; 0.040; -0.059; 0.060 - 0.079; 0.080 - 0.099; ≥ 0.100 and data from all subjects were pooled. We are aware that this procedure mixes within and between subject data, but considered the approach reasonable since it would be used in this fashion in practice and we planned to cross-validate in another sample. Then we evaluated what proportion of tests (out of nine) exceeded disparate levels of proportion of decrement scores for a given BAC level and assessed accuracy by determining the proportion of correct decision. Note that a correct decision for 0.00 would be "no alcohol" and any other decision regardless of dosage would be considered "alcohol - yes." Similar analyses were performed for Experiments 3 and 4.

RESULTS

The data from Experiments 1 and 2 may be seen in Table 1. Note that false positive rates decrease (specificity improves) as smaller proportion of decrements are employed. Moreover, specificity is improved as the proportion of tests on which the proportion decrement is required. As expected, Table 1 also shows a reverse of this relationship when sensitivity is concerned but it appears as though a relatively high specificity can be attained for a 6% cutoff and a proportion of 3 out of 4 tests on which it is required. Specificity and sensitivity in Experiments 3 and 4 for these same parameters are also shown in Table 1.

Table 1

Number of subjects out of the total number of subjects with a percent decrement greater than the criterion on 3 of the 4 APTS tests for each level of BAC

Percent Decrements from Baseline

BAC	Alcohol I			Alcohol II			Alcohol III			Alcohol IV		
	4%	5%	6%	4%	5%	6%	4%	5%	6%	4%	5%	6%
.00 Specificity	0/16 100%	0/16 100%	0/16 100%	1/18 94.4%	0/18 100%	0/18 100%	2/22 90.9%	1/22 95.5%	1/22 95.5%	6/36 83.3%	3/36 91.7%	1/36 97.2%
.001-.059 Sensitivity	2/3 66.7%	2/3 66.7%	2/3 66.7%	3/17 17.6%	2/17 11.8%	2/17 11.8%	26/40 65.0%	21/40 52.5%	16/40 40.0%	119/203 58.6%	111/203 54.7%	103/203 50.7%
.06-.079	2/11 18.2%	1/11 9.1%	1/11 9.1%	11/19 57.9%	11/19 57.9%	8/19 42.1%	21/25 84.0%	19/25 76.0%	15/25 60.0%	62/82 75.6%	59/82 72.0%	57/82 69.5%
.08-.099	2/6 33.3%	2/6 33.3%	2/6 33.3%	6/6 100%	5/6 83.3%	4/6 66.7%	24/25 96.0%	22/25 88.0%	19/25 76.0%	26/33 78.8%	26/33 78.8%	25/33 75.8%
≥.10	23/28 82.1%	22/28 78.6%	22/28 78.6%	44/48 91.7%	44/48 91.7%	40/48 83.3%	64/64 100%	64/64 100%	64/64 100%	4/4 100%	4/4 100%	4/4 100%
TOTAL	45/64 70.3%	43/64 67.2%	43/64 67.2%	81/108 75.0%	80/108 74.1%	72/108 66.7%	155/176 88.1%	147/176 83.5%	135/176 76.7%	241/358 67.3%	233/358 65.1%	224/358 62.6%

DISCUSSION

The multiple cutoff strategy appears to be a technique that minimizes false positives to a greater degree than alternative methods that we have tried. In the Alcohol 3 and Alcohol 4 studies, sensitivity was nearly perfect at the 0.10 and greater levels of BAC, indicating that almost every subject was correctly identified as impaired at this BAC level regardless of whether the percent decrement from baseline was set at 4, 5, 6 or greater percent. Perhaps more important is the ability of the multiple cutoff criterion to guard against false positive identifications. Across the four alcohol studies at levels of percent decrement between 4-6%, the specificity of impairment detection was 96.1% using only four tests from the DELTA battery and a three out of four test decrement criterion.

The presumed reason that the National Research Council (Normand, Lempert, & O'Brien, 1994) and others have disparaged the ability of performance tests to correctly assess impairment (or deception) when low base rates exist may largely be due to the fact that most tests are not given repeatedly, do not possess the requisite reliability to monitor performance changes, and are not multifactorial in makeup. Likewise, the presumed inability to accurately assess impairment at low levels of BAC may well be due to unreliable measures, lack of repeated testing, and variability in scores before performance has had a chance to stabilize (i.e., before initial learning of the test has plateaued and stabilized). We believe the results reported here present a brighter picture for the ultimate ability of performance testing to provide an accurate assessment of fitness-for-duty and afford a step toward answering the National Research Council's recommendation that "research should be conducted on the utility of performance tests prior to starting work as an alternative to alcohol and other drug tests."

ACKNOWLEDGMENT

This work was sponsored, in part, by National Science Foundation, Grant #III-9122907, under Joseph L. Young, Program Official.

References

Hanninen, H., & Lindstrom, K., 1973. *Behavioral test battery for toxico-psychological studies used at the Institute of Occupational Health in Helsinki*, (2nd Ed., Rev. Reviews 1. Helsinki, Finland: Institute of Occupational Health.

Kane, R. L., & Kay, G. G. (1992). Computerized assessment in neuropsychology: A review of tests and test batteries. *Neuropsychology review, 3*(1). Plenum Press, New York and London.

Kennedy, R.S., Dunlap, W.P., Turnage, J.J., & Fowlkes, J.E. (1993). Relating alcohol-induced performance deficits to mental capacity: A suggested methodology. *Aviation, Space, and Environmental Medicine, 64*, 1077-1085.

Kennedy, R.S., Dunlap, W.P., Turnage, J.J., & Wilkes, R.L. (1993). Human intelligence in a repeated-measures setting: Communalities of intelligence with performance. In D.K. Detterman (Ed.), *Current Topics in Human Intelligence* (Vol. 3). Norwood, NJ: Ablex.

Kennedy, R.S., Turnage, J.J., Rugotzke, G.G., & Dunlap, W.P. (1994). Indexing cognitive tests to alcohol dosage and comparison to standardized field sobriety tests. *Journal of Studies on Alcohol, 55*, 615-628.

Kennedy, R.S., Turnage, J.J., Wilkes, R.L., & Dunlap, W.P. (1993). Effects of graded dosages of alcohol on nine computerized repeated-measures tests. *Ergonomics, 36*, 1195-1222.

Murphy, K.R. (1987). Detecting infrequent deception. *Journal of Applied Psychology, 72*, 611-614.

Normand, J., Lempert, R. O., & O'Brien, C. P. (1994). *Under the influence: Drugs and the american workforce*. Washington, DC: National Academy Press.

Russell, R. W., Flattau, P. E., & Pope, A. M. (1990). *Behavioral measures of neurotoxicity*. The National Academy Press, Washington, DC.

Turnage, J.J., & Kennedy, R.S. (1992). The development and use of a computerized human performance test battery for repeated-measures applications. *Human Performance, 5*(4), 265-301.

ANOTHER LOOK AT AIR TRAFFIC CONTROLLER PERFORMANCE EVALUATION

Earl S. Stein Ph.D. and Randy L. Sollenberger Ph.D.
FAA/William J. Hughes Technical Center and Princeton Economic Research, Inc.
Atlantic City International Airport, NJ 08405

This paper describes a study that evaluated the reliability of a recently developed rating form designed to assess air traffic controller performance. Six supervisors from different radar approach control facilities nationwide viewed 20 video tapes of controllers working traffic from a previously recorded simulation study. The observer/raters used a new evaluation form that consisted of 24 different rating scales measuring specific areas of controller performance. An important part of this study was observer training. The training consisted of practice rating sessions followed by group discussions. In discussion, observers established mutual evaluation criteria for each performance area. Inter-rater reliability was assessed using intraclass correlations, and intra-rater reliability was assessed using Pearson product-moment correlations on repeated video tapes. In general, the reliability of the form was quite good, however, a few rating scales were much less reliable than the others. Reasons for the differences in rating scale reliability are discussed.

INTRODUCTION

The evaluation of human performance is a complex process. It is difficult to do systematically and objectively. In a complex command and control system such as air traffic control, the majority of the critical behavior is cognitive and therefore not directly observable. While there are minimum standards established by law and regulation, most of the performance variance is above the minimum and is traditionally evaluated using "my standard". This construct is created as part of the mental model of the evaluator based on tradition, training, and local practice or procedures that are familiar.

In a laboratory environment, researchers can collect objective and absolute measures of performance. However, there is considerable debate about which measures are the correct ones. There is the belief that only those measures which relate in some way to the collective models of my standard could possibly be valid. This creates some confusion when attempting to evaluate new equipment, procedures, or controller selection tests. Measures must meet the minimum standard of face validity or decision makers will not want to use them when evaluating new systems or procedures.

This study was begun as an attempt to develop new performance rating tools that would have face validity, inter-rater reliability, and possibly criterion-related validity using objective simulation-based measures such as those identified by Boone and Steen (1981), Buckley, DeBaryshe, Hitchner, and Kohn, (1983), and Stein and Buckley (1992). The current study also supports a program underway to develop valid measures for selecting air traffic control specialists in the FAA (Nickels, Bobko, Blair, Sands, and Tartak, 1995).

Controllers have always used over-the-shoulder ratings since the early days. They place a strong faith in their ability to observe and evaluate each other. Careers may rise or fall, especially during training, based on the ratings and comments placed on FAA Form 3120-25. This is the ATCT/ARTCC OJT INSTRUCTION/EVALUATION REPORT. It contains 27 scales, of 5 categories: Separation, Control Judgment, Methods and Procedures, Equipment, and Communication-Coordination. Each scale allows for a rating on three points: Satisfactory, Needs Improvement, and Unsatisfactory. There is no room for written observations on the front side of the form; that is reserved for the back side.

Controller culture is such that when they receive a "check ride" by a trainer or supervisor, and if they notice him/her writing, they become worried. Writing is discouraged unless something is wrong. The trainer or supervisor likely feels some subtle pressure to avoid writing and to depend on memory for events related to performance. Depending on memory and using a three-point scale are basic prescriptions for unreliable measurement.

METHOD

The program worked on by Nickels et al. (1995) produced a conceptual model of air traffic controller performance that contributed to the creation of a new observational rating form. The original version of the form was cumbersome and did not provide space or encouragement for written observations. A revised version was created and the scales can be seen in Table 1.

Traditionally, supervisors are not properly trained to evaluate controllers. Writing observations is actively discouraged because long established norms imply that writing means something has been done wrong. This research attempted to change these norms by implementing a detailed training protocol before any observation and rating took place. Six supervisory controllers voluntarily

Table 1. Controller Performance Evaluation Form Rating Scales

I-Maintaining Safe and Efficient Traffic Flow
    1-Maintaining Separation and Resolving Potential Conflicts
    2-Sequencing Arrival and Departure Aircraft Efficiently
    3-Using Control Instructions Effectively
    4-Overall Safe and Efficient Traffic Flow Scale Rating

II-Maintaining Attention and Situation Awareness
    5-Maintaining Awareness of Aircraft Positions
    6-Ensuring Positive Control
    7-Using Control Instructions Effectively
    8-Correcting Own Errors in a Timely Manner
    9-Overall Attention and Situation Awareness Scale Rating

III-Prioritizing
    10-Taking Actions in an Appropriate Order of Importance
    11-Preplanning Control Actions
    12-Handling Control Tasks for Several Aircraft
    13-Marking Flight Strips while Performing Other Tasks
    14-Overall Prioritizing Scale Rating

IV-Providing Control Information
    15-Providing Essential Air Traffic Control Information
    16-Providing Additional Air Traffic Control Information
    17-Overall Providing Control Information Scale Rating

V-Technical Knowledge
    18-Showing Knowledge of LOAs and SOPs
    19-Showing Knowledge of Aircraft Capabilities and Limitations
    20-Overall Technical Knowledge Scale Rating

VI-Communicating
    21-Using Proper Phraseology
    22-Communicating Clearly and Efficiently
    23-Listening to Pilot Readbacks and Requests
    24-Overall Communicating Scale Rating

participated. They were recruited from radar approach control facilities across the country and spent one week (approximately 32 hours) in training after arrival. They were encouraged to write down all the behavioral indicators they could see. Training included a detailed explanation of rating biases. During this training phase, the supervisors observed video tapes of a simulation study recently completed by Guttman, Stein, and Gromelski (1995). They reviewed their observations after each 1-hour video tape and discussed their differences and the behaviors they used to make their judgments.

At the beginning of this process, my standard took precedence, but after one week the supervisors gradually came together. While not totally agreeing on everything, they did find some common ground.

During the second week of the project, the participants observed 20 hours of simulation playback that they had not seen before. They completed ratings at the end of each hour. These ratings were accomplished independently and without post-hoc review. What made this process unique was the use of a multi-screen projection system. One screen provided over-the-shoulder video and the other screen displayed a relatively perfect replay of the controller's radar display. The replay included any changes induced by his/her control instructions.

Participants felt that this was comparable to actually being with the controller during the simulation. The replay technology had the added advantage that the presence of observers did not influence the performance of participating controllers.

RESULTS

The purpose of this study was to determine if reliable observational data could be collected, and if the ratings were related to other performance measures generated in air traffic control simulation. An inter-rater reliability analysis was conducted using intraclass correlation. These were computed and pooled across the 20 1-hour samples of rating behavior that were collected in the simulation. Figure 1

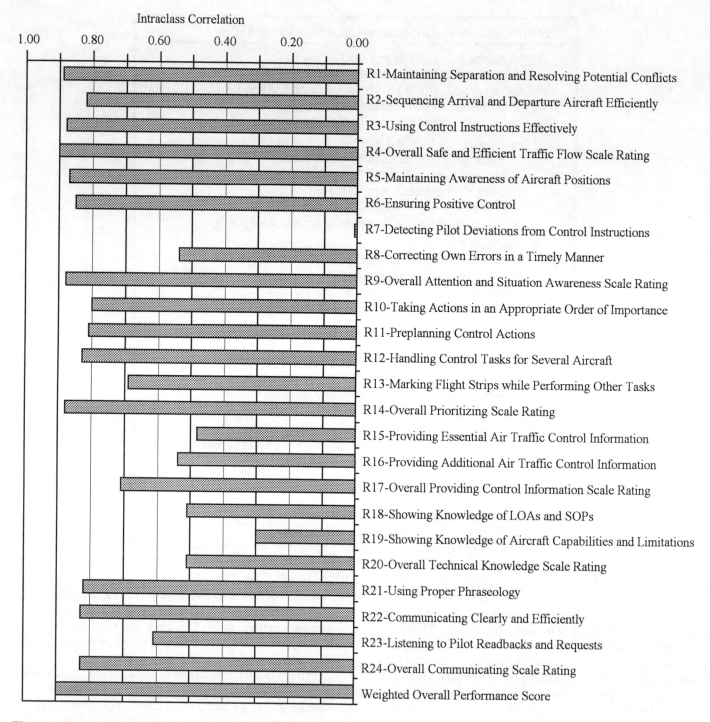

Figure 1. Inter-rater reliability for each of the rating scales using intraclass correlations.

describes the results across the 24 rating scales and a weighted overall performance score. This score was based on using weights provided by the individual observers. The weights were their perceptions of the relative importance of the six major behavioral categories included in the evaluation form.

The obtained reliability coefficients varied across the scales with some scales achieving less than r=0.10 and others approaching r=0.90. This represented a large

improvement over initial training trials in the first week, where observers were in considerable disagreement.

A second analysis was conducted using Pearson product-moment correlations, and the results are presented in Figure 2. These correlations represent intra-rater reliability across four traffic scenarios that were presented twice to observers. There are some notable differences between these coefficients and those in Figure 1. Rating scale R7, Detecting Pilot Deviations from Control Instructions, is

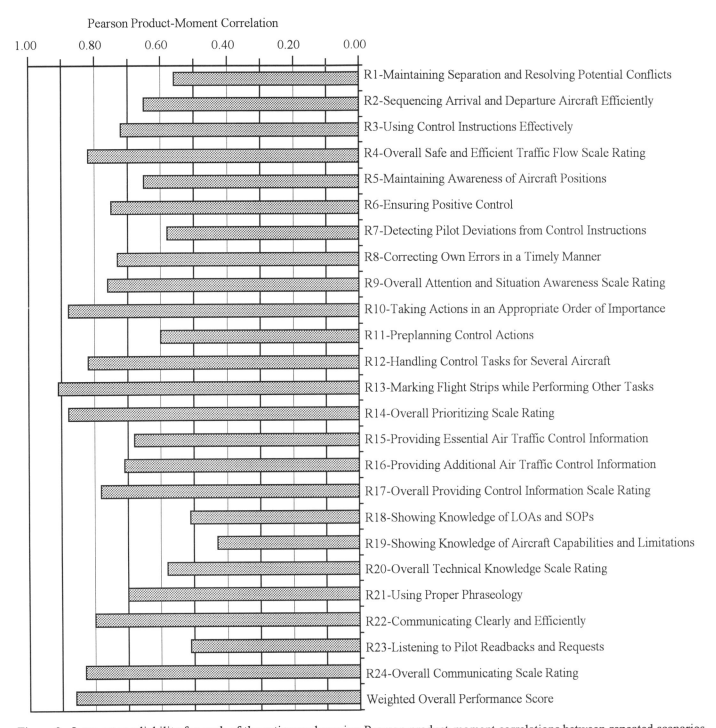

Figure 2. Intra-rater reliability for each of the rating scales using Pearson product-moment correlations between repeated scenarios.

obviously an observational chore that leads controllers to see very different things when evaluating the performance of one of their peers. However, they maintain a moderate level of consistency over time and within their own frameworks or my standard.

The last analysis is represented in Figure 3 and involves using the data collected in the actual simulation as a criterion standard. Selected system performance variables were extracted from the data base and correlated against the weighted overall performance scores. Results indicated inverse relationships that were low to moderate in magnitude. Observers did not have access to the system performance data. They appear to view higher levels of controller activity as negative performance indicators. Further, participants in the original simulation who reported higher levels of workload using the ATWIT system were rated lower in overall performance.

Pearson Product-Moment Correlation

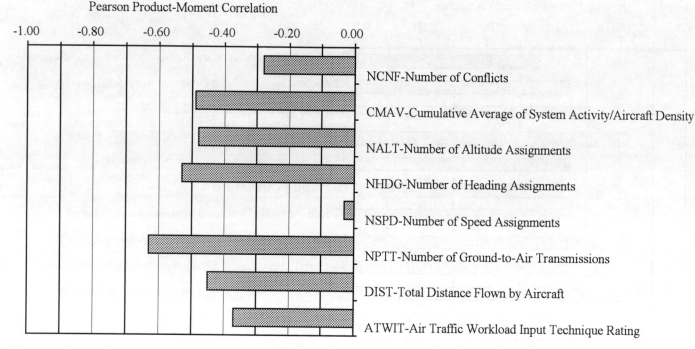

Figure 3. Correlations between the system effectiveness measures and the weighted overall performance scores.

DISCUSSION

Changes in attitude and behavior by the observer/raters were remarkable. They arrived with each individual being certain of my standard and their ability to observe and rate performance. Each had done it before in an operational setting.

They were surprised when they observed the same behavior at the same time, and initially, they produced very different ratings. In discussions, it became clear that they were actually focusing on different behaviors and valued these various behavioral units uniquely. At first, the observers were reluctant to let go of these implicit standards, but gradually, through discussion among themselves and with the encouragement of the research team, they attempted to find some common basis for evaluation.

One factor that facilitated this change was the training and insistence that they write specific examples of behavior that they observed for each rating scale. They were given a separate recording form that was keyed to the rating scales and were asked to complete it while observing. The observers were then instructed to refer to it when making their ratings. The fact that they were viewing the behavior on video tape may have made it easier because they recognized that their writing would not influence the performance of the controller.

Figure 1 demonstrates that the observers were not always successful in obtaining reliability in their ratings for every performance area. One might speculate that one component of this success is related to the observability of the behavior called for in each scale. The least reliable scale, R7, would require an observer to make several cognitive

transformations and inferences prior to coming up with a rating. Observers, no doubt, vary considerably in their ability to accomplish this.

The controllers who served as observers in this study expressed collective satisfaction with what they had accomplished. To a person, they indicated that they had learned something about observing and evaluating and would certainly take the process more seriously in the future.

REFERENCES

Boone, J. O., & Steen, J. A. (1981). A comparison between over the shoulder and computer derived measurement procedures in assessing student performance in radar air traffic control. Aviation, Space and Environmental Medicine, 52(10), 589-593.

Buckley, E. P., DeBaryshe, B. D., Hitchner, N., & Kohn, P. (1983). Methods and measurements in real-time air traffic control system simulation (DOT/FAA/CT-TN83/26). Atlantic City, NJ: DOT/FAA Technical Center.

Guttman, J., Stein, E. S., & Gromelski, S. (1995). The influence of generic airspace on air traffic controller performance (DOT/FAA/CT-TN95/38). Atlantic City, NJ: DOT/FAA Technical Center.

Nickels, B. J., Bobko, P., Blair, M. D., Sands, W. A., & Tartak, E. L. (1995). Separation and control hiring assessment (SACHA): Final job analysis report (DOT/DAA/DTFA0A-91-C-00032). Bethesda, MD: University Research Corporation.

Stein, E. S., & Buckley, E. P. (1992). Simulation variables. Unpublished manuscript.

INDIVIDUAL DIFFERENCES IN DRIVER STRESS AND PERFORMANCE

Gerald Matthews
University of Dundee
Dundee, Scotland

80 subjects high and low in vulnerability to driver stress participated in a study of simulated driving performance. Subjects completed the Driving Behaviour Inventory, which assesses vulnerability through a Dislike of Driving scale. Half the subjects performed in a 'stress' condition, in which they frequently lost control of vehicle steering, the remainder in a non-stressful control condition. Two performance measures were analyzed: response time (RT) on a secondary attentional task, and a measure of lateral tracking. The stress manipulation was more strongly related to longer RTs in high Dislike of Driving subjects than in low Dislike subjects. However, slowing of response was more pronounced on straight than on curved road sections, i.e. when the driving task is relatively undemanding. This finding suggests that stress-related impairment is not simply due to overload of attention. Instead, the stress-vulnerable driver may have difficulties in matching effort to task demands, with under-mobilisation of effort when the task appears relatively easy. Lateral tracking data were also consistent with this hypothesis. Self-report data suggested that the manipulation was generally effective in inducing subjective stress symptoms. However, high Dislike subjects tended to react to the manipulation with particularly high levels of intrusive thoughts and 'cognitive interference'.

INTRODUCTION

This study investigated relationships between stress and individual differences in performance on a driving simulator. The general aim was test whether a self-report measure of stress vulnerability, validated in field studies of driving, predicted objective performance measures obtained in the laboratory. Studies of this kind are important for two reasons. First, they may contribute to the development of information-processing theories of stress and performance. Contemporary stress theory (see Wells and Matthews, 1994) relates stress to a dynamic interplay between the person and the demands imposed upon them by the external environment. The effects on performance of a stressful encounter or transaction may depend on the individual's cognitive appraisal of the encounter, and his or her efforts to cope with its demands. Second, laboratory studies of driver stress may contribute to understanding of the real-world hazards of stress. It is well-established that stress is a risk factor for motor vehicle accidents (Selzer and Vinokur, 1975), but there is little evidence on the processes mediating the relationship between external pressures on the driver and hazardous driving behaviors.

Driver stress may be viewed as the outcome of an interaction between external demands, such as difficult driving conditions, and characteristics of the individual driver. Glendon, Dorn, Matthews, Gulian, Davies and Debney (1993) developed a questionnaire measure of driver stress vulnerability, the Driving Behaviour Inventory (DBI), which comprises several scales, including a broad stress syndrome labelled as Dislike of Driving. Dislike of Driving predicts negative post-drive mood in both field and laboratory studies, especially when

the task requires active interaction with other traffic (Dorn and Matthews, 1995). Matthews (1993) outlines a transactional model of driver stress which relates stress vulnerability to styles of cognition. Vulnerability to distress is attributed to negative self-appraisal, and use of emotion-focused coping strategies such as worry and self-criticism. In empirical studies, high Dislike drivers tend to rate themselves as lacking control over safety and as relatively incompetent drivers (Dorn and Matthews, 1995).

High Dislike drivers show impaired control of lateral position on the road in simulator studies (Matthews, Dorn, Hoyes, Glendon, Davies and Taylor, in press). In general, emotional distress may be associated with performance impairment because negative affect is often accompanied by worries and intrusive thoughts which divert attention away from the task at hand (Wells and Matthews, 1994). Sarason, Sarason, Keefe, Hayes and Shearin (1986) use the term *cognitive interference* to describe the distracting effects of intrusive thoughts. High Dislike drivers may show impairment of vehicle control because they are distracted by their worries about driving. Various specific information-processing mechanisms may contribute to cognitive interference. Matthews, Sparkes and Bygrave (1996) distinguished two alternative explanations for the interference effect in driving. The *overload hypothesis* states that Dislike of Driving is associated with worries about driving which use up attentional resources which might otherwise be allocated to the driving task. The driver may have insufficient attentional resources to process task stimuli effectively, if the task is sufficiently demanding and requires a large quantity of resources. The alternative, *effort-regulation hypothesis*, derives from Hancock and Warm's (1989) proposal that stress impairs the person's ability to adapt to changing task demands. The workload of driving varies considerably, so that the person must regulate actively the mental effort applied. It may be especially difficult to match effort to task demands in conditions of low workload or 'underload' (Hancock and Warm, 1989). Hence, stress may impair the driver's ability to maintain effort and voluntary control when the task is perceived as undemanding. The stressed driver may be most vulnerable to cognitive interference when somewhat 'off-guard'.

Matthews et al. (1996) tested overload and effort-regulation hypotheses using dual-task methods. 80 drivers were periodically required to respond to grammatical reasoning problems by making a spoken response, during simulated driving. Results supported the effort-regulation but not the overload hypothesis. Dislike of Driving was more detrimental to performance when task demands were low, i.e. in single rather than dual-task performance, and also on straights rather than curves. Task-related cognitive interference was measured using a modification of Sarason et al.'s (1986) Cognitive Interference Questionnaire (CIQ). The study confirmed that Dislike is correlated with interference. However, as with Dislike, interference was more robustly related to performance impairment on straight rather than curved road sections.

The present study aimed to conduct a further test of hypotheses for stress-related impairment on the driving simulator, incorporating an experimental manipulation of stress due to periodic loss of control of the vehicle, accompanied by negative feedback. Vehicle control and dual-task performance were assessed following induction of a stress state. A secondary task designed to assess visuo-spatial attention was included. Drivers were required to discriminate single characters presented on road-signs by making a manual response. A spatial cuing procedure was used to test whether Dislike of Driving has any specifically attentional effects, as opposed to inducing a general performance decrement. Overall task demands were manipulated via road curvature. The general prediction was that Dislike should be more strongly related to performance impairment under stress, when high Dislike drivers should be most prone to worry. Specifically, the overload hypothesis predicts that Dislike should have stronger effects on secondary task performance when the road is curved, but the effort-regulation hypothesis predicts stronger Dislike effects when the road is straight, and the task relatively undemanding. Mood and cognitive interference were monitored before and after the drive to test for interactive effects of Dislike and the stress manipulation on subjective distress.

METHOD

40 male and 40 female drivers, aged 18-30, participated in the study. Prior to the simulator testing session subjects completed Glendon et al.'s (1993) DBI in their own time, so they could subsequently be categorized as high or low in Dislike of Driving on the basis of a median-split. Stress was manipulated between-subjects, and task factors within-subjects.

Each subject performed a single drive on the Aston Driving Simulator, a moderate-fidelity, fixed-based computer-controlled driving simulator used in previous studies (see Matthews et al., 1996). The 'driver' views scaled 3-D graphics on a 22" monitor, and controls the 'car' with a steering wheel and brake and gas pedals. The simulator operates with an 80 ms refresh cycle. Stress was manipulated as follows. 40 subjects performed what was described as 'a winter drive', with invisible ice on the road. Initially, a white landscape was displayed, and the driver was required simply to follow a track comprising an unpredictable sequence of straight and curved road sections. In this phase of the task, there were approaching vehicles in the other lane. Periodically, the driver lost all steering control, so that the car skidded frequently, unpredictably and uncontrollably. Obstacles such as pedestrians and stationary vehicles were placed in the driver's lane, so that collisions were likely when control was lost. Each time a skid resulted in a major error, such as a collision or driving off the road, a negative feedback message was displayed. This phase of the task lasted for 5.5 km (typically lasting about 10 minutes).

The driver then encountered three cars waiting at a red traffic light. While the driver waited behind these cars at the light, the landscape reverted to its usual green colour, and normal vehicle control was restored. After about 5 s the light changed to red/amber and then to green. From this point on, the driver was required to follow the cars in front, at a steady 30 m.p.h., so that speed was controlled in the subsequent assessment of performance. The driver followed the lead vehicles for a further 6 km, with sections requiring both driving only (single-task driving) and driving while performing the secondary discrimination task (dual-task driving). Both types of driving were performed on alternating straight and curved road sections. The secondary task required the driver to discriminate single upper-case letter stimuli presented on road-signs while continuing to drive at 30 m.p.h. The letter stimulus might appear on either side of the road, in a pseudo-random sequence. A total of 120 pairs of white road signs were presented, 40 m apart, one on each side of the road. Hence, event rate at 30 m.p.h. was 20/minute. When the driver was 64 m from the pair of signs, an asterisk was presented for 80 ms on one of the signs. The asterisk was a cue, indicating the side of the road on which the letter was likely to appear. On 80% of trials the cue was valid; i.e., it appeared in the same position as the letter which followed it. On the remaining trials the cue was invalid. After a short interval, a single letter was presented on one of the two signs, for 1200 ms. The driver was instructed to press one of two keys set into the steering wheel to indicate whether the letter was a vowel or a consonant, as quickly as possible. Two cue-stimulus SOAs were used, 160 and 560 ms. However, SOA had minor effects only on response, and the present data are averaged across SOA. Separation of left and right positions was approximately 5 degrees at onset of the letter stimulus.

There were two dependent variables. Response time (RT) to the letter stimuli was used to assess attention. (Accuracy data are not reported here as no speed-accuracy trade-offs were found.) The second dependent variable, used to assess vehicle control, was heading error, the mean deviation between the direction of the road and the direction of the road. High heading error indicates poor control of lateral position.

An additional 40 subjects performed a control drive, which differed from the stressful winter drive only in the first phase of the drive. During the initial 5.5 km, the driver had full control throughout, and the objects which served as obstacles in the stressful drive were placed off the road. No feedback was delivered.

Subjective state was assessed before and after driving with the mood and cognitive interference measures used by Matthews et al. (1996), and a state perceived control scale developed at Dundee. These measures were used, first, to check whether the stress manipulation actually influenced subjective stress, and, second, to check whether subjects high and low in Dislike of Driving differed in their subjective reactions to the manipulation.

RESULTS

Speed of Response to Secondary Task Stimuli

RT data were analyzed within a $2 \times 2 \times 2 \times 2$ ANOVA (Dislike $\times$ stress $\times$ curvature $\times$ cue validity), with repeated measures on curvature and cue validity (see Table 1). One subject was dropped from the analysis due to a very high error rate. There was a significant main effect of cue validity ($F(1,75) = 81.6$, P<.01): a valid cue facilitated response speed. There was also a main effect of curvature ($F(1,75) = 9.72$, P<.01), indicating a dual-task interference effect, with response slower on curves than on straights. There were three significant interactive effects on RT involving the stress factors: the Dislike $\times$ stress $\times$ curvature interaction ($F(1,75) = 4.65$, P<.05), the Dislike $\times$ stress $\times$ cue validity interaction ($F(1,75 = 5.54$, P<.05), and the Dislike $\times$ stress $\times$ validity $\times$ curvature interaction ($F(1,75) = 3.95$, P<.05). On straight road sections, the stress manipulation had little effect on the response of low Dislike subjects, for both valid and invalid cues. However, high Dislike subjects tended to respond more slowly in the stress condition than in the control condition, especially on straight road sections. On curves, a similar effect was found when the cue was valid, but the interaction between Dislike and stress tended to reverse for invalid cues. Averaging across cue type, the effect of Dislike was stronger on straights than on curves. On straights, high Dislike subjects were 39 ms slower in the stress condition than in the control condition, on average, whereas the low Dislike subjects were 3 ms faster. On curves, average mean RT was 14 ms slower in the stress condition, compared to the control condition, for both high and low Dislike subjects. Use of the cue was only affected by the stress-related factors on curves. The speed enhancement associated with a valid cue was reduced in two groups during curve driving: low Dislike subjects in the control condition, and high Dislike subjects under stress.

Table 1. RT (ms) on Secondary Attentional Task as a Function of Dislike of Driving, Stress Condition, Curvature and Cue Validity.

Curvature	Condition	Low Dislike			High Dislike		
		Valid	*Invalid*	*Cuing*	*Valid*	*Invalid*	*Cuing*
Straight	Control	605	656	+51	579	637	+58
	Stress	602	653	+51	624	669	+55
Curve	Control	627	645	+18	607	659	+52
	Stress	622	677	+55	641	652	+11

Note. Cuing = Invalid RT - Valid RT

Heading Error

Heading error data were analyzed within a $2 \times 2 \times 2 \times 2$ ANOVA (Dislike $\times$ stress $\times$ curvature $\times$ task combination), with repeated measures on curvature and task combination (single vs. dual task). The main effect of curvature was significant ($F(1,75) = 125.6$, P<.01), with heading error generally greater on curves (see Table 2).

There was also a significant interaction between curvature and task combination ($F(1,75) = 142.9$, P<.01). On straight road sections, heading error was actually lower during dual-task performance than during single-task performance, but dual-task performance tended to be inferior on curves. Three interactions involving stress factors reached significance: Dislike $\times$ task combination ($F(1,75) = 6.90$, P<.01), Dislike $\times$ curvature ($F(1,75) = $

4.83, P<.05), and Dislike × stress × curvature × task combination (F(1,75) = 6.06, P<.05). The two-way interactions reflect Dislike tending to be detrimental to performance during single-task performance and on

straights, but not during dual-task performance and on curves. The four-way interaction derives primarily from the particularly high heading error of high Dislike subjects under stress during single-task straight road driving.

Table 2. Heading Error (Degrees) as a Function of Dislike of Driving, Stress Condition, Curvature and Single- vs. Dual-Task Performance.

Curvature	Condition	Low Dislike		High Dislike	
		Single	Dual	Single	Dual
Straight	Control	1.59	1.24	1.62	1.09
	Stress	1.74	1.22	2.38	1.19
Curve	Control	1.89	2.70	1.95	2.16
	Stress	1.97	2.49	2.01	2.36

Subjective Stress

Subjective effects were analyzed by a series of 2 × 2 × 2 (Dislike × stress × time) ANOVAS, with repeated measures on the time factor (pre- vs. post-task). Effects of stress condition on change in subjective state across time were evidenced by stress × time interactions, significant at P<.01 for several of the subjective state scales. The stress manipulation induced substantial and significant increases in tense and depressed mood: scores on these variables increased by .82 and 1.56 SD respectively. There was also a substantial increase (+1.24 SD) in task-related cognitive interference, and a large decrease in perceived control (-1.55 SD). A moderating effect of Dislike of Driving was indicated by a Dislike × stress × time interaction. This

interaction was significant for the two CIQ subscales. Table 3 presents standardized change scores which illustrate these effects. For task-related interference, the general trend towards increased interference following the winter drive was stronger for high Dislike subjects. For task-irrelevant interference, the general trend, across both conditions, was for task-irrelevant interference to decrease during the drive. This trend may indicate that subjects focus attention on the task. However, the High Dislike subjects failed to show any marked reduction in task-irrelevant interference following the stress manipulation. At the subjective level, High Dislike subjects reacted to the stress manipulation with a high level of intrusive task-related thoughts and a failure to suppress personal worries, a potentially maladaptive pattern of reaction.

Table 3. Cognitive Interference Change Scores (Post-Drive - Pre-Drive), as a Function of Dislike of Driving and Stress Condition.

Type of Interference	Condition	Low Dislike	High Dislike
Task-Relevant	Control	+0.54	+0.22
	Stress	+0.96	+1.52
Task-Irrelevant	Control	-0.37	-0.98
	Stress	-0.73	-0.08

Note. Change scores expressed in standardized units.

DISCUSSION

Performance data were broadly consistent with the transactional model of driver stress outlined previously. Although the loss of control manipulation appeared to be generally stressful, as evidenced by the subjective data, there were no main effects of the manipulation on any performance measure. Instead, the effects of the stress manipulation were moderated by individual differences in stress vulnerability, assessed by the Dislike of Driving

factor. Dislike effects were further moderated by task demands, but, in some task conditions at least, high Dislike subjects under stress showed slower RT, reduced cuing and increased heading error. The data show some degree of correspondence between these effects on objective performance and subjective response to stress. High Dislike of Driving was associated with a seemingly maladaptive pattern of increased cognitive interference under stress. The manipulation was designed to undermine the driver's confidence in ability to control the vehicle, and as the transactional model predicts (Dorn and Matthews, 1995), high Dislike subjects are especially prone to worry under these circumstances.

Moderating effects of task demands on stress effects indicate how cognitive interference may impair performance. The overload hypothesis predicts that detrimental effects of stress should be most evident when task demands are high, but the data are mostly inconsistent with this hypothesis. Averaging across the two cue types, interactive effects of Dislike and stress on secondary task RT appeared to be more robust on the less demanding straight road sections. This finding is consistent with the results obtained by Matthews et al. (1996), in a different paradigm. It supports the effort-regulation hypothesis, that stress may lead to insufficient effort or active control of performance when task demands are low. The data showed also that the cuing effect seemed to become more fragile on curves, perhaps because the greater demands of curved driving make subjects increasingly reluctant to direct their efforts towards processing the cue. In high Dislike subjects, reduced cue processing on curves under stress may be a strategic attempt to cope with a potential overload of attention. Heading error data were also more supportive of the effort-regulation hypothesis than of the overload hypothesis. On straights, performance was actually better during dual-task than single-task performance, indicating that subjects react to increased demands with increased effort. The effects of Dislike of Driving on heading error, which were detrimental mainly in single-task driving on straight roads under stress, suggest a failure to apply sufficient effort to this relatively easy task condition.

It is difficult to see how conventional resource theories can accommodate these results, although some of the findings are reminiscent of Kahneman's (1973) view that total resource allocation increases with task demands. The most promising avenue for future research may be to investigate further how stress-vulnerable individuals choose to exert active control over performance in overload and underload conditions. Desmond and Matthews (1996) have obtained similar results in studies of task-induced fatigue. Fatigue is associated with impaired vehicle control only on straight road sections. Again, a propensity to apply insufficient effort to the task may only be expressed in behavior when the task is appraised as undemanding.

The findings also have implications for road safety. The stressed driver may be most at risk of slow response to a hazard, or committing a vehicle-handling error, when task demands are low, and the driver is likely to appraise driving as relatively safe. Matthews and Desmond (1995) characterize the problem as one of maintaining the stress-vulnerable driver's engagement with the task under such circumstances. In the UK, driver training tends to emphasise traffic contexts likely to involve overload rather than underload, such as negotiating intersections and busy

city streets. Resistance to distraction from worries when task demands are low may require more explicit training. There may also be a role for in-vehicle technology in maintaining safety under stress. Matthews and Desmond (1995) point out that integration of in-vehicle performance monitoring systems with navigation systems may improve diagnosis of impairment, by assessing indices of failing performance under low workload conditions.

REFERENCES

Desmond, P.A., and Matthews, G. (1996) Implications of task-induced fatigue effects for in-vehicle countermeasures to driver fatigue. In *Proceedings of 2nd International Conference on Fatigue and Transportation: Engineering, enforcement and education solutions.* Applecross, Western Australia: Promaco Conventions.

Dorn, L., and Matthews, G. (1995) Prediction of mood and risk appraisals from trait measures: Two studies of simulated driving. *European Journal of Personality, 9,* 25-42.

Glendon, A.I., Dorn, L., Matthews, G., Gulian, E., Davies, D.R., and Debney, L.M. (1993) Reliability of the Driver Behaviour Inventory. *Ergonomics, 36,* 719-726.

Hancock, P.A., and Warm, J.S. (1989) A dynamic model of stress and sustained attention. *Human Factors, 31,* 519-537.

Kahneman, D. (1973) *Attention and effort.* Englewood Cliffs, NJ: Prentice Hall.

Matthews, G. (1993) Cognitive processes in driver stress. *Proceedings of the 1993 International Congress of Health Psychology* (pp. 90-93). Tokyo: ICHP.

Matthews, G., and Desmond, P.A. (1995). Stress as a factor in the design of in-car driving enhancement systems. *Le Travail Humain, 58,* 109-129.

Matthews, G., Dorn, L., Hoyes, T.W., Glendon, A.I., Davies, D.R., and Taylor, R.G. (in press) Driver stress and simulated driving: Studies of risk taking and attention. In G.B. Grayson (ed.), *Behavioural research in road safety III.* Crowthorne: TRRL.

Matthews, G., Sparkes, T.J., and Bygrave, H.M. (1996) Stress, attentional overload and simulated driving performance. *Human Performance, 9,* 77-101.

Sarason, I.G., Sarason, B.R., Keefe, D.E., Hayes, B.E., and Shearin, E.N. (1986) Cognitive interference: Situational determinants and traitlike characteristics. *Journal of Personality and Social Psychology, 51,* 215-226.

Selzer, M.L., and Vinokur, A. (1975) Role of life events in accident causation. *Mental Health and Society, 2,* 36-54.

Wells, A., and Matthews, G. (1994) *Attention and emotion: A clinical perspective.* London: Lawrence Erlbaum.

THE USE OF THE MULTIVARIATE JOHNSON DISTRIBUTIONS
TO MODEL TRUNK MUSCLE COACTIVATION

Gary A. Mirka, Naomi F. Glasscock, Paul M. Stanfield,
Jennie P. Psihogios and Joseph R. Davis

The Ergonomics Laboratory
Department of Industrial Engineering
North Carolina State University
Raleigh, North Carolina

The accurate description of trunk muscle coactivation, and more specifically antagonist muscle activity, has recently been the focus of a great deal of research in the spine biomechanics literature. The research presented in this paper is an empirical approach to the problem. Electromyographic (EMG) data were collected from 28 subjects as they performed simulated lifting tasks. These EMG data were collected from the right and left pairs of the erector spinae, latissimus dorsi, rectus abdominis, external obliques and internal obliques as subjects performed a variety of trunk extension exertions. Nine repetitions of each combination of independent variables were performed by each subject. Included in these exertions were asymmetric postures and dynamic (isokinetic and constant acceleration) exertions. The data collected during these trials were used to develop marginal distributions of trunk muscle activity as well as a 10 x 10 correlation matrix that described how the muscles cooperated in the development of these extension torques. These elements were then combined to generate multivariate distributions describing the coactivation of the trunk musculature.

INTRODUCTION

In an effort to understand the types of stresses placed on tissues of the low back during occupational lifting tasks, researchers have developed biomechanical models of the torso. Typically included in these models are ten primary muscles: erector spinae, latissimus dorsi, rectus abdominis, external obliques and internal obliques. These muscles when activated use the spine as a fulcrum to exert torques to perform useful manual materials handling tasks. One of the questions that presents itself when one is developing these biomechanical models is how to include antagonist muscle activity. Often these forces are just simply assumed to be negligible and therefore are omitted. Another concern regarding many of the existing models is that they typically take a deterministic approach to estimating muscular force, that is for a given set of circumstances (posture, load etc) the muscle activation levels are said to be constant such that equilibrium exists. Given the indeterminate nature of the biomechanical system, we should, at the very least,

consider this variability and the potential impact it may have on stresses in the low back.

An attempt to address both of these issues is found in Mirka and Marras (1993). This method involved collecting empirical muscle activity data from 5 subjects as they performed repetitive simulated lifting exertions under controlled conditions. Histograms of the muscle activity were constructed for each of the muscles under each of the conditions. These histograms were then fit to univariate Johnson distributions as described in DeBrota et al (1989). The distributions that resulted were described by four parameters: γ (a shape parameter), δ (a shape parameter), λ (a scale parameter) and ζ (a location parameter). One of the limitations of this model was that the distributions generated were simple marginal distributions. An attempt was made to describe the coactivity of the muscles by partitioning the data space in such a way that the resulting distributions were conditional on the activity of the right and left erector spinae muscles. From a practical perspective this is limited because it ignores the potential influence that any of the other muscles might exert on any of the other muscles. For example, might not the activity of the right latissimus dorsi affect the activity level of the left latissimus dorsi? Therefore, a much more robust approach would be to generate a 10 dimensional multivariate system that will allow each muscle to influence every other muscle. This multivariate approach is developed in this paper.

METHOD

Subjects

Twenty eight people from the university community served as subjects in this study. There were twenty one men and seven women. None of the subjects had a history of low back disorders (defined as no lost time from work or school due to back pain) and each signed an informed consent form before participating in this study. Experience in manual material handling tasks varied. Basic subject anthropometry is listed in Table 1.

Table 1. Basic Anthropometry of Subject Population

Variable	Mean	Standard Deviation
Age (years)	29.43	8.65
Body Mass (kg)	78.2	14.4
Height (cm)	175.4	8.9

Apparatus

A Kin/Com dynamometer was used in conjunction with a trunk motion reference frame to provide an environment that allowed the researchers to have a great deal of control of the forces, postures and movements of the subjects. (See Figure 1.) An EMG data processing system and a data collection system were used to gather the data describing the signals from the dynamometer (2 load cells, position potentiometer and velocity tachometer) and the muscle activity levels (ten trunk muscles). The EMG signals collected by the electrodes were amplified 1000x by miniature preamplifiers located at the muscle site. The electrode leads to the preamplifiers were kept short so as to reduce the movement noise and the external electrical noise from the surrounding environment. The signal was amplified (total amplification ~ 60,000x) and high and low pass filtered at 80 and 1000 Hz. This filtered signal was rectified and processed using a 20 msec moving average window. These processed EMG data along with torque,

angle, and velocity were collected at 100 Hz by the data collection system.

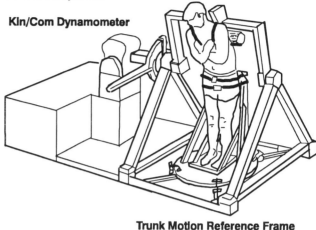

Kin/Com Dynamometer

Trunk Motion Reference Frame

Figure 1. Experimental Apparatus (Trunk Motion Reference Frame and Kin/Com dynamometer)

Experimental Design

Independent Variables. In order to quantify the variability of the muscle forces during lifting, the subjects in this experiment were asked to perform highly controlled bending motions repeatedly. These trials included isometric, isokinetic (10 or 45 deg/sec) and constant acceleration (50 deg/sec/sec) exertions. Torque exerted by the subjects were either 30 Nm or 80 Nm for the experimental trials. (For five of the subjects the 80 Nm condition was beyond their capability and therefore the upper force level was reduced to 60 Nm for those subjects.) Two trunk positions (5 and 40 degrees of forward sagittal bend) and two levels of trunk asymmetry (0 degrees and 30 degrees twisted to the right) were evaluated in this study:. Each combination of independent variables was repeated 9 times per subject. The order of presentation of the combinations of the above independent variables was completely randomized within level of asymmetry and the presentation of asymmetry was counterbalanced across subjects.

Dependent Variables. The dependent variables in this study were the normalized processed EMG values of the ten trunk muscles identified by the transverse cutting plane technique described by Schultz and Andersson (1981). These muscles included the right and left erector spinae (RES, LES), right and left latissimus dorsi (RLAT, LLAT), right and left rectus abdominis (RAB, LAB), right and left external obliques (REX, LEX) and the right and left internal obliques (RIN, LIN) muscles. The inter-electrode distance for each electrode pair was 3.0 cm.

Procedure

Upon arrival the subjects had surface electrodes applied to their skin through standard preparation procedures. The subject was then asked to enter the reference frame so that the adjustable base could be set for the subject's leg length in order to insure that the subject's L5/S1 joint was aligned with the rotating axis of the Kin/Com dynamometer. Once the subject was secured in the reference frame they performed maximum voluntary contractions (MVCs) at four positions (5 and 40 degrees of sagittal bend and 0 and 30 degrees of asymmetry). Both maximum static extensions and flexions were collected as well as the resting values in each of these postures. After these maximal exertions, the experiment began with the subject performing a sequence of randomized trials. Each of these trials dictated that the subjects perform a controlled exertion defined by set levels of torque, posture, angular trunk velocity, and angular trunk acceleration. During these trials the angular position, velocity and acceleration were controlled by the dynamometer. The exerted torque was controlled by the subject within a tolerance of +/- 10% using a graphical video feedback system that displayed their instantaneous torque output as well as the target torque designated for the particular trial. If the subject failed to maintain the designated amount of torque the trial was repeated.

Data Analysis

The EMG data were first normalized with respect to the maximum and resting EMG values that occurred at each particular trunk posture. The main emphasis of this research project was to better understand the effects of the task parameters on the distributions of muscle activity. We were therefore interested in eliminating the inter-subject variability. This was accomplished by standardizing the data across subjects so that the variability between subjects would not influence the results. This was accomplished by calculating a mean and a standard deviation for each subject in each experimental condition. The overall mean and pooled standard deviation were then calculated for each condition. Using these values, the individual EMG values were then standardized using the following formula:

$$SV(j, k, l, m) = MP(j, k) - [STDP(j, k) * (M(j, k, l) - AV(j, k, l, m))] / STD(j, k, l)$$

Where:

SV (j, k, l, m) - standardized EMG value of muscle j, condition k, subject l and repetition m

AV (j, k, l, m) - actual EMG value of muscle j, condition k, subject l and repetition m

STDP (j, k) - pooled standard deviation for muscle j and condition k

STD (j, k, l) - standard deviation for muscle j, condition k and subject l

MP (j, k) - average for muscle j and condition k

M (j, k, l) - average for muscle j, condition k and subject l

Model Development

At this point the data was in the form of 32 - {10 X ROW} matrices containing the normalized, standardized EMG values, where 10 refers to the 10 muscles sampled and ROW refers to the number of trials that met the strict criteria laid out for the acceptability of the data based on the position, velocity, acceleration and torque parameters for the trial. The range across experimental conditions for the number of acceptable trials was 102-180. The 32 different matrices refer to the 32 unique combinations of independent variables that each of the subjects performed.

Each of these 32 data sets were then used to generate a set of multivariate distributions. The procedure used is described in greater detail in Stanfield (1993) and is briefly outlined below.

1) Determine the first four moments of the distribution for each muscle (mean, standard deviation, skewness and kurtosis) and the correlation coefficients between muscles.

2) Develop a lower triangular matrix V such that $V\,V^T = C$, where C is the {10 X 10} correlation matrix.

3) Develop two new standardized {1 X 10} skewness and kurtosis vectors using the following equations:

$s* = (V^{(3)})^{-1} * s$ where s is the original {1 X 10} skewness vector

$k* = (V^{(4)})^{-1} * [k - 6 * \sum_{j=1}^{9} \sum_{l=j}^{10} V_{ij}^2 * V_{il}^2]$ where k is the original {1 X 10} kurtosis vector

4) Using the above standardized skewness and kurtosis vectors, fit a marginal Johnson distribution to each of the muscle distributions (DeBrota et al, 1989).

5) Finally, to generate samples that reflect the true multivariate nature of the data use the following relationship:

$$X = S (V * Y) * \mu$$

Where:

X is a {1 X 10} vector of actual multivariate values

S is a {10 X 10} diagonal matrix containing the original standard deviation for each muscle

V is the {10 X 10} lower triangular matrix as described in 2) above

Y is a {1 X 10} vector of samples from the marginal distributions generated using the Johnson distributions developed in 4) above.

μ is a {1 X 10} vector of the original means

Using the above outlined procedure, multivariate Johnson values are generated for each muscle under each experimental condition. With multiple runs of the simulation the shapes of the best fit distributions can be developed.

RESULTS

The results of this simulation are distributions for each of the trunk muscles in each of the experimental conditions. Displayed in Figures 2 - 5 are a small sample of these fitted distributions. Figures 2 and 3 show the best fit distributions for the right erector spinae and the right rectus abdominis muscles while Figures 4 and 5 show the best fit distributions for the right latissimus dorsi and the right external oblique muscles, respectively. Note how this distribution fitting system models even the very skewed distribution of the right latissimus dorsi.

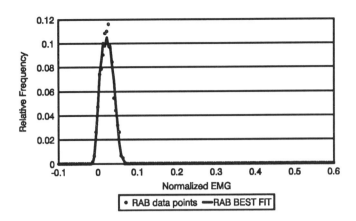

Figure 3. Empirical Data and Best Fit Distribution for the Right Rectus Abdominis (Sagittal Angle = 40, Sagittal Velocity= 10 deg/sec (isokinetic), Torque = 80 Nm)

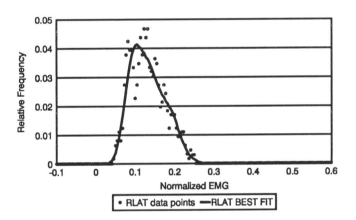

Figure 4. Empirical Data and Best Fit Distribution for the Right Latissimus Dorsi (Sagittal Angle = 40, Sagittal Velocity= 10 deg/sec (isokinetic), Torque = 80 Nm)

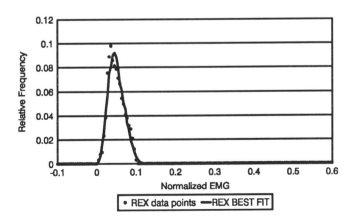

Figure 5. Empirical Data and Best Fit Distribution for the Right External Oblique (Sagittal Angle = 40, Sagittal Velocity= 10 deg/sec (isokinetic), Torque = 80 Nm)

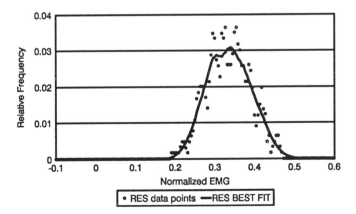

Figure 2. Empirical Data and Best Fit Distribution for the Right Erector Spinae (Sagittal Angle = 40, Sagittal Velocity= 10 deg/sec (isokinetic), Torque = 80 Nm)

The previous model (Mirka and Marras, 1993) was developed using data from sagittally symmetric lifting postures only. Figure 6 shows the response of this model to an asymmetric condition in this study.

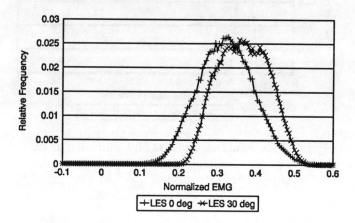

Figure 6. Distributions of the Left Erector Spinae during Symmetric (0°) and Asymmetric (30°) Lifting. (Sagittal Angle = 40, Sagittal Velocity= 10 deg/sec (isokinetic), Torque = 30 Nm)

DISCUSSION

Many of the biomechanical models of the torso that have been developed rely on electromyographic inputs to drive the model (Granata and Marras, 1993; Marras and Sommerich, 1991a,b; McGill and Norman, 1986; McGill, 1992; Reilly and Marras, 1989). However, it is not practical to collect EMG activity in many industrial environments. The model developed in this research has the capability of generating muscle activities during bending and lifting activities and can therefore act as a motor for these EMG driven biomechanical models. Given a set of environmental conditions (weight, moment, trunk posture and trunk dynamics) this model can produce EMG signals that would be generated during these exertions. These signals could then be input into any of the existing EMG driven model to render estimates of the spine reaction forces.

The model developed in this paper marks a significant improvement over it predecessor (Mirka and Marras, 1993) in that it allows for a true multivariate representation of trunk muscle coactivation. It is believed that as this model continues to develop the predictions of the model are going to more closely reflect the actual muscle activities and therefore our understanding of the trunk muscle coactivation patterns will improve

ACKNOWLEDGMENTS

This publication was partially supported by Grant No. KO1 OH00135-03 from The National Institute for Occupational Safety and Health. The contents are solely the responsibility of the authors and do not necessarily reflect the official views of NIOSH.

REFERENCES

Debrota DJ, Dittus RS, Roberts SD, Wilson JR, Swain JJ, Venkatraman S. (1989) "Modeling Input Processes with Johnson Distributions", *Proceedings of The 1989 Winter Simulation Conference*, 308-318.

Granata KP, Marras WS. (1993) "An EMG-Assisted Model of Loads on the Lumbar Spine During Asymmetric Trunk Extensions",. *Journal of Biomechanics*, 26, 1429-1438.

Marras WS, Sommerich CM. (1991) "A Three-Dimensional Motion Model of Loads on the Lumbar Spine: I. Model Structure", *Human Factors*, 33, 123-137.

Marras WS, Sommerich CM. (1991) "A Three-Dimensional Motion Model of Loads on the Lumbar Spine: II. Model Validation", *Human Factors,* 33, 139-149.

McGill SM, Norman RW. (1986) "Partitioning of the L4/L5 Dynamic Moment into Disc, Ligamentous, and Muscular Components During Lifting", *Spine*, 11, 666-677.

McGill SM. (1992) "A Myoelectrically Based Dynamic Three-Dimensional Model to Predict Loads on Lumbar Spine Tissues During Lateral Bending", *Journal of Biomechanics*; 25, 395-414.

Mirka GA, Marras WS. (1993) "A Stochastic Model of Trunk Muscle Coactivation During Trunk Bending", *Spine*, 18, 1396-1409.

Reilly CH, And Marras WS. (1989) "Simulift: A Simulation Model of Human Trunk Motion", *Spine*, 14: 5-11.

Schultz AB, Andersson GBJ. (1981) "Analysis of Loads on the Lumbar Spine", *Spine*, 6, 76-82.

Stanfield PM, (1993) "Stochastic Scheduling in a Remanufacturing Job Shop", Unpublished Master's Thesis, North Carolina State University.

THE EFFECTS OF TRAINING HISTORY AND MUSCLE FATIGUE ON LOAD SHARING IN A MULTIPLE MUSCLE SYSTEM.

Dan Kelaher
Ergonomic Technologies Corporation
Oyster Bay, New York

The effects of exercise training history and localized muscle fatigue on the load sharing patterns in a multiple muscle system were investigated. Six male and six female subjects of various training backgrounds performed fatiguing isokinetic knee extension exercises of quick onset and slow onset fatigue. Electromyographic data collected from the 4 superficial quadriceps muscles as well as the hamstring muscles was fed into an EMG-based model of the knee joint. The relative torque contribution of each muscle as predicted by the model was uniquely different for each of the 5 quadricep muscles modeled. The load sharing patterns were effected by the training group among other factors but they were not significantly effected by the interaction of training group and the onset of fatigue.

INTRODUCTION

Localized muscle fatigue is a very common occurrence in manual laborers. Jobs that require large muscular forces for short durations or intermediate muscular forces for longer durations can both fatigue muscles. Since fatigue has been found to lead to temporarily decreased lifting capacity, movement precision, velocity, range of motion, and more risky movement patterns (Lance and Chaffin, 1971; Parnianpour et al, 1988; Lorentzon et al, 1988; Trafimow et al, 1993), ergonomists often try to design tools and workstations with the specific goal of minimizing localized muscular fatigue. Another common goal of industrial ergonomists is to reduce the internal forces that the joints are subjected to during various physical activities. Training periods are often used to teach manual laborers to develop better lifting techniques and reduce the chance for localized muscle fatigue and acute injury.

The goal of many occupational biomechanical models is to model the compressive and/or shear forces within a joint and to use that as the measurement of musculoskeletal stress (Chaffin and Andersson, 1984). The earlier models utilized kinematic data as the only subject-specific inputs (Anderson et al, 1985) or were based on an assumed physiologically-based optimization scheme (Schultz and Andersson, 1981; Dul et al, 1984). The problem with these earlier models was that they did not account for individual variation. Muscle mechanics, fatiguability, muscle strengths, and neural activation patterns all vary between individuals and can greatly alter the internal joint reaction forces. For this reason, EMG-based biomechanical models have become increasingly popular in the past decade to predict internal joint reaction forces (Norman and McGill, 1985; Marras and Sommerich, 1991). The goal of this project was to employ an EMG-based biomechanical model of the knee to help determine the effects of localized muscular fatigue, exercise training history, and biomechanical factors on the load sharing patterns of the quadricep muscles and the subsequent joint reaction forces.

METHODS

Six men and six women of various training backgrounds from the North Carolina State University community took part in this experiment. The subjects were separated into three training groups: strength trained (at least three strength training workouts per week average), endurance trained (at least three cardiovascular workouts per week average) and untrained (no more than one day of strength or cardiovascular training per week average).

Each subject participated in two fatiguing isokinetic knee extension protocols. The low velocity/low force knee extension protocol ("Low" protocol) was performed at an angular velocity of 10°/sec and at a resistance of 10% of the subject's maximal isometric knee extension torque while the high velocity/high force protocol ("High" protocol) was performed at an angular velocity of 20°/sec a resistance of 20% of the maximal isometric knee extension force. The subjects performed the test on separate days with at least 24 hours of rest between days; the order of protocols was randomized. Before the testing protocols, the subjects were given two days of practice in order to become familiar with the test protocol and equipment. The subjects were instructed to perform continuous concentric/eccentric isokinetic knee extensions from 115° to 175° at the given velocity and force levels until the experimenter stopped the test. The test was stopped when either the target force could not be maintained throughout the range of motion or the subject could not keep the force within a 10% force tolerance throughout the entire range of concentric motion. Immediately before and after the submaximal isokinetic testing, maximal isokinetic knee extensions were performed to provide additional inputs into the biomechanical model. Before the pretest maximal isokinetic exertions and after the post-test maximal isokinetic exertions were performed isometric maximum knee extensions and knee flexions were performed at knee angles of 120°, 130°, 140°, 150°, 160°, and 170° so that the EMG data could be normalized with respect to the knee angle-specific maximal exertion.

Surface EMG of the vastus medialis oblique (VMO), vastus medialis longus (VML), rectus femoris (RF), vastus lateralis (VL), Semitendinosus (ST), and the Biceps Femoris Long Head (BF), were collected from the exercising leg (the right leg). The BF signals contained cross-talk interference from the vastus lateralis in some subjects so the ST normalized IEMG (NIEMG) signal was extrapolated to the BF. Also, the torque contribution of the deep vastus intermedius (VI) was calculated using the NIEMG values from the VL since previous research has shown the activity of the VI and VL to be quite similar in form and function (Salzman et al, 1993). The surface electrodes were placed over the bellies of these muscles at an inter-electrode distance of 1" and in line with the lines of action of each muscle according to the muscle fiber architecture data of Narici et al (1992) and Weinstabl et al (1989). The positions of the electrodes are shown in Figure 1.

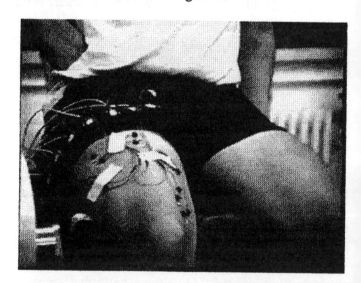

Figure 1. Electrodes and preamplifiers on subject's right quadriceps.

The EMG signals were notch filtered at 60 Hz and low pass filtered at 2000 Hz in hardware to remove the signal power between 60-64 Hz and the signal power above 2000 Hz. These signals were then amplified by a factor of 9000. The raw EMG and integrated EMG (IEMG) signals were then collected along with the force, angle, and velocity data from the Chattecx Kin-Com 125E Isokinetic Dynamometer. These dynamometer and EMG data were collected at 1024 Hz for the isometric trials and at 400 Hz for the dynamic trials by an analog-digital conversion program on a personal computer. This entire setup ensured good synchronization of the dynamometer data with the electromyography data.

MODEL

An EMG-based biomechanical model of the knee was constructed to help determine the load sharing patterns of the quadriceps muscles. This model was based heavily in theory on previous EMG-based models of the low back (McGill and Norman, 1986; Marras and Sommerich, 1991). The model included the five parts of the quadriceps muscles (VMO, VML, VI, VL, and RF) and two hamstring muscles (ST and BF). The cross-sectional area of each muscle was calculated by measuring the thigh circumference and skinfold thickness and multiplying that by the muscle's known percentage of cross-sectional muscle mass and pennation angles (Narici et al, 1992; Friederich and Brand, 1990). Each muscle area was then multiplied by the appropriate length-tension factors calculated by Herzog (1985). The moment arm of each muscle was derived from regression equations by Visser et al (1990) based on the anthropometric dimensions of the leg. The cross-sectional area, length-tension factor, the NIEMG, and moment arm were all multiplied together. These values were then set equal to the force applied to the Kin-Com dynamometer during the fatiguing exercise bout. Regression equations were set up for each subject for each protocol to predict the knee extension torque from the NIEMG and relative torque contribution of each muscle at the angles of 120°, 140°, and 160°. The two coefficients of the regression equations represented the muscular stress of each muscle at the appropriate knee angular velocity. The resulting stress values and relative torque contribution values were then multiplied by the NIEMG at the intermediate angles of 130° and 150° to predict the tensions within each individual muscle. The predicted torque contribution of each muscle was then divided by the total predicted torque about the knee to determine the torque contribution of each muscle; this was the statistic of choice on which the ANOVA procedures were performed. In the analyses, the sum of the contribution of each of the

quadricep muscles is greater than 100% by the amount of hamstring (ST + BF) torque.

Experimental Design

The independent variables in the experiment were the training history group (GROUP in the ANOVA tables; strength trained, endurance trained, and untrained), knee angle (ANGLE; 130° and 150°), fatigue state (PREPOST; pre-fatigue and post-fatigue), protocol (PROTOCOL; High and Low), and the muscle (MUSCLE; VMO, VML, RF, VI, VL, ST, and BF). Separate analyses were performed for the quadriceps and hamstrings. The dependent variable was the relative torque contribution output from the biomechanical model.

RESULTS

The maximal knee extension torque data followed previously published trends as a function of knee angle (Herzog, 1985). The data exhibited a peak torque between 120° and 130° and the trend held consistently for each training group, although the differences between the groups were less dramatic than expected. The amount of work performed during the exercise bouts (i.e., the torque multiplied by the number of repetitions completed before fatigue) showed better discrimination between the three groups.

The results of the ANOVA are shown below in Table 1 for the quadriceps and Table 2 for the hamstrings. The ANOVA showed a strong effect of MUSCLE as an interactive term and a main effect in the quadriceps' analysis. The strong main effect of MUSCLE was confirmed with a Duncan's multiple range test showing that each muscle contributes a percentage of torque significantly different from every other muscle. MUSCLE also showed significant effects in interactive terms with PREPOST, ANGLE, and GROUP.

The finding to highlight from the ANOVA performed on the hamstring torque contribution was that there was a significant main effect of GROUP. The Duncan's multiple range test showed there to be significant differences between the hamstring torque generated by the untrained subjects relative to both the endurance and strength trained subjects.

Table 1. ANOVA results of the relative torque contribution of the quadricep muscles.

Source	DF	F Value	Pr > F
MUSCLE	4	1450	0.0001
PROTOCOL	1	3.78	0.0535
PREPOST	1	3.55	0.1044
MUSCLE*PREPOST	4	4.29	0.0031
ANGLE	1	9.11	0.0025
MUSCLE*ANGLE	4	75.97	0.0001
GROUP	2	6.34	0.0011
MUSCLE*GROUP	8	9.46	0.0001
MUSCLE*PREPOST*GROUP	10	1.61	0.0722

Table 2. ANOVA results of the relative torque contribution of the hamstring muscles.

Source	DF	F Value	Pr > F
MUSCLE	1	248	0.0001
PROTOCOL	1	4.03	0.0142
PREPOST	1	3.79	0.0412
ANGLE	1	9.71	0.0001
MUSCLE*ANGLE	1	7.76	0.0370
GROUP	2	6.75	0.0001
MUSCLE*GROUP	2	4.86	0.0267
PROTOCOL*GROUP	2	4.15	0.0093

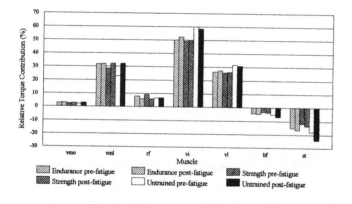

Figure 2: Relative torque contribution of the quadriceps and hamstrings as a function of training group and fatigue state.

Figure 2 shows the increased antagonist torque generated by the untrained subjects being offset by the increased VML torque after fatigue.

DISCUSSION

The data shows that the individual quadricep muscles do contribute different amounts of torque to the total torque about the knee when doing seated knee extensions. This finding could only be found through the use of the biomechanical model as the EMG data (which were not presented here) did not show differences as large as the torque contribution data.

The significance of the MUSCLE* PRE-POST factor suggests that the load sharing pattern is different after fatigue than before fatigue. However, the failure of the MUSCLE*PREPOST* GROUP factor to reach significance at the 0.05 level suggests that the load sharing patterns are not different in the different training groups. Together, these points suggest that if any neural optimization scheme exists such as a shifting of the load to the muscles with higher percentages of slow twitch fibers (as proposed by Dul et al, 1984) then it may not be controlled by regular training. This is not to say, however, that no amount of training could change this, just that "regular" training (i.e., 3 times/week as was the case for these subjects) does not seem to significantly effect the load sharing patterns after fatigue.

The importance of the training history seems to be in the amount of antagonist torque generated by the hamstrings. The higher antagonist torque would require more agonist torque generated by the quadriceps in order to maintain a constant external torque about the knee. Therefore, the high antagonist torque from the hamstrings most likely caused GROUP to be a significant main effect for the quadriceps. The two important ramifications of this higher antagonism is that localized muscle fatigue can occur with lower force requirements and that internal joint reaction forces are going to be higher. As mentioned earlier, muscle fatigue and joint forces are two important criteria that ergonomists use when designing a tool or workstation. This finding highlights the importance of industrial training programs where new workers gradually build up to the more stressful tasks over time. With appropriate training, workers can learn to utilize more efficient muscle activation patterns by utilizing less antagonist activity. It remains to be seen if this pattern holds for muscle systems which are of more direct interest to ergonomists such as the low back musculature.

CONCLUSION

This study tried to integrate the biomechanics and physiology of physical work by investigating the effect of fatigue on the muscle mechanics, specifically the load sharing patterns. Specific training was shown to decrease the amount of antagonist activation with the onset of fatigue, causing more efficient muscle activation profiles.

Training programs for manual material handling jobs typically train the workers until the recommended amount of work can be performed. These results show that even though a person may be able to complete a physically demanding task at a given load level, the long-term training effect of learning more efficient muscle activation patterns may not have taken place. This emphasizes the importance of training programs to provide training past the point of just being able to maintain quota production.

REFERENCES

Anderson, C.K., Chaffin, D.B., Herrin, G.D., and Matthews, L.S., (1985). A biomechanical model of the lumbosacral joint during lifting activities. *Journal of Biomechanics, 18,* 571-584.

Chaffin, D.B. and Andersson, G.B.J. (1984). *Occupational Biomechanics.* New York, NY: John Wiley and Sons, Inc.

Dul, J., Johnson, G.E. Shiavi, R., and Townsend, M.A. (1984). Muscular synergism--2. A minimum-fatigue criterion for load sharing between synergistic muscles. *Journal of Biomechanics, 17,* 675-684.

Friederich, J.A. and Brand, R.A. (1990). Muscle fiber architecture in the human lower limb. *Journal of Biomechanics, 23,* 91-95.

Herzog, W. (1985). *Individual Muscle Force Prediction in Athletic Movements.* Ph.D. Dissertation. Calgary, Canada: University of Calgary Printing.

Lorentzon, R., Johansson, C., Sjostrom, M., Fagerlund, M., and Fugl-Meyer, A.R. (1988). Fatigue during dynamic muscle contractions in male sprinters and marathon runners: Relationship between performance, electromyographic activity, muscle cross-sectional area, and morphology. *Acta Physiologica Scandinavica, 132,* 531-536.

Marras, WS, and Sommerich, CM, (1991). A three-dimensional motion model of loads on the lumbar spine: I. Model structure. *Human Factors, 33,* 123-137.

McGill, S.M., and Norman, R.W. (1986). Partitioning of the L4-L5 dynamic moment into disc, ligamentous, and muscular components during lifting. *Spine, 11,* 666-677.

Narici, M.V., Landoni, L., and Minetti, A.E. (1992). Assessment of human knee extensor muscles stress from *in vivo* physiological cross-sectional area and strength measurements. *European Journal of Applied Physiology and Occupational Physiology, 65,* 438-444.

Parnianpour, M., Nordin, M., Kahanovitz, N., and Frankel, V. (1988). The triaxial coupling of torque generation of trunk muscles during isometric exertions and the effect of fatiguing isoinertial movements on the motor output and movement patterns. *Spine, 13,* 982-992.

Salzman, A., Torburn, L., and Perry, J. (1993). Contribution of rectus femoris and vasti to knee extension: An electromyographic study. *Clinical Orthopaedics and Related Research, 290,* 236-243.

Schultz, AB and Andersson, GBJ. (1981). Analysis of loads on the lumbar spine. *Spine, 6,* 76-82.

Trafimow, J.H., Schipplein, O.D., Novak, G.J., and Andersson, G.B.J. (1993). The effects of quadriceps fatigue on the technique of lifting. *Spine, 18,* 364-367.

Visser, J.J., Hoogkamer, J.E., Bobbert, M.F., and Huijing, P.A. (1990). Length and moment arm of human leg muscles as a function of knee and hip-joint angles. *European Journal of Applied Physiology and Occupational Physiology, 61,* 453-460.

Weinstabl, R., Scharf, W., and Firbas, W. (1988). The extensor apparatus of the knee joint and its peripheral vasti: Anatomic investigation and clinical relevance. *Surgical and Radiologic Anatomy, 11,* 17-22.

TASK EFFECTS ON THREE-DIMENSIONAL DYNAMIC POSTURES DURING SEATED REACHING MOVEMENTS: AN ANALYSIS METHOD AND ILLUSTRATION

Xudong Zhang and Don B. Chaffin
Center for Ergonomics, Dept. of Industrial and Operations Engineering
The University of Michigan, Ann Arbor, MI 48109

This paper presents a new method to empirically investigate the effects of task factors on three-dimensional (3D) dynamic postures during seated reaching movements. The method relies on a statistical model in which the effects of hand location and those of various task factors on dynamic postures are distinguished. Two statistical procedures are incorporated: a regression description of the relationship between the time-varying hand location and postural profiles to compress the movement data, and a series of analyses of variance to test the hypothesized task effects using instantaneous postures with prescribed hand locations as dependent measures. The use of this method is illustrated by an experiment which examines two generic task factors: 1) hand movement direction, and 2) motion completion time. The results suggest that the hand motion direction is a significant task factor and should be included as an important attribute when describing or modeling instantaneous postures. It was also found that the time to complete a motion under a self-paced mode was significantly different from a motivated mode, but the time difference did not significantly affect instantaneous postures. The concept of an instantaneous posture and its usage in dynamic studies of movements are discussed. Some understanding of human postural control as well as the implications for developing a general dynamic posture prediction model also are presented.

INTRODUCTION

The need to model and simulate human physical activities dynamically and three-dimensionally is receiving increasing attention in the field of ergonomics (Chaffin, 1992; Ayoub 1994). An initial but essential step towards fulfilling this need is to establish an empirical basis for the development as well as the validation of models that describe or predict three-dimensional dynamic postures. This step, however, has been hindered by the complexity of empirically studying normal human movements under various operational settings to identify specific task effects. Among previous models that attempt to predict human postures or movements, many are static and/or two-dimensional (Kilpatrick, 1970; Park, 1973; Ayoub and Hsiang, 1992; Jung et al., 1994). The ones developed by computer scientists to drive three-dimensional animation of human motions (Armstrong et al., 1987; Badler et al., 1987; Monheit et al., 1991; Phillips, 1991) are usually not based on population statistical motion data; rather, they employ heuristic algorithms or rough estimates of human postural kinematics. Therefore the application of these models in biomechanical analysis and ergonomic assessment remains limited and questionable under different task requirements.

The complexity associated with the empirical investigation of 3D dynamic postures lies in several aspects of an experimental paradigm. First, the amount of data yielded from multi-segment movements is overwhelmingly voluminous. Without a desirable means of data compression or reduction, the time and effort required to manage, process, and analyze the data can be prohibitive. Secondly, it is difficult to choose the appropriate dependent variables. A straightforward way may be to examine the entire motion profiles under different prescribed conditions, which is rather tedious and cumbersome. On the other extreme, assessing a number of static postures is convenient but rarely provides any information about the dynamics. To date, a measure with both a manageable data scale and dynamic information has not

been presented in the literature. Thirdly, most movements under operational settings are target-directed. The end-effector (usually the hand) location largely affects but not completely determines a posture, due to the kinematic redundancy of the human body. Thus, discerning whether task performance variations are caused by different end-effector locations or various task factors can be problematic.

Two of the most generic task factors inherent to any reaching movement are 1) the time to complete a movement, and 2) the intended or instantaneous direction of the hand motion. Many studies in the motor control domain have addressed the movement direction and speed issues, primarily focusing on isolated limb motions (Morasso, 1981, 1983; Flash and Hogan, 1985; 1990; Karst and Hasen, 1990; Gordon et al., 1994a, 1994b). These studies, however, have yet to be extended to include both torso and upper extremity motions.

The purpose of this study is to establish a methodology which alleviates the complexity associated with the empirical investigation of 3D multi-segment dynamic postures, and allows experimental identification of significant task factors. More specifically, the study examines whether and how hand movement direction and completion time affect dynamic postures. The postures considered are of the torso and right upper extremity during seated reaching movements which are volitional, discrete, and right-handed. The general scheme presented, nevertheless, should be applicable to the study of many other task factors involved in various types of human movements.

METHODS

Model Description

As depicted in Figure 1, the configuration of the torso and right arms is represented by a mechanical linkage system composed of four links: torso, right clavicle, right upper arm, and right forearm. Given a 3D location of the right hand, the configuration of this linkage system is indeterminate, referred to as the problem of kinematic redundancy. It is usually postulated that the determination of preferred postures is accomplished by some unknown inherent strategy used by human beings which also considers various task and environmental factors. This postulation provides the underlying theory of a statistical model that governs the investigative scheme of the current study.

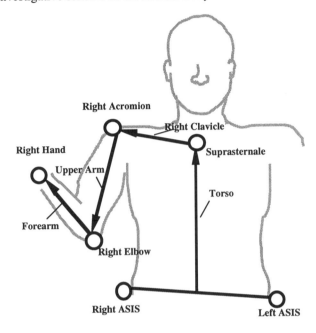

Figure 1. A linkage representation composed of 4 links: torso, right clavicle, right upper arm, and right forearm.

The hand location largely but not completely determines the configuration of a multi-segment system. The extent of this determination depends on the degree of kinematic redundancy of the system. Therefore, three-dimensional hand coordinates are the major, albeit not the sole, determining factors. It is hypothesized that various task factors also play significant roles: these task factors affect the strategic way the kinematic redundancy is resolved. In other words, they influence the form of the relationship between the hand coordinates and the preferred postures.

Let $P_t = \left[\theta_{1,t} \cdots \theta_{i,t} \cdots\right]^T$ be a set of kinematic variables, normally joint or segment angles, that mathematically describes an instantaneous posture at time t. Each variable can be modeled as

$$\theta_{i,t} = f[X_h(t), Y_h(t), Z_h(t)] \mp g[F_1, F_2, \cdots] + \varepsilon_{i,j}$$

The X_h, Y_h, and Z_h are the three time-variant hand coordinates; the F's denote prescribed task factors, which are time-invariant and mainly categorical (e.g., time to complete a motion, motion direction, with or without handling weight and so on). The symbol $\mp$ reflects the recognition that the relationship between f and g may not simply be additive. The function f is a regression description of relationship between the hand coordinates and the joint angles (or coordinates), whereas the function g is synthesized from an analysis of task effects. The latter analysis uses sample instantaneous postures derived from the former by specifying representative hand coordinates. This is based on an

assumption that postural profiles as described by the f are interplotable or extrapolatable across a range of hand locations where the movement condition formed by a specific combination of the factors remains invariant.

Experimental Design and Procedures

An experiment was designed to examine the effects of two specific task factors, 1) hand motion direction and 2) speed mode, on instantaneous postures during discrete seated reaching movement. Five healthy young adults including three males and two females volunteered to serve as subjects. Instructions as well as a training secession were provided to the subjects.

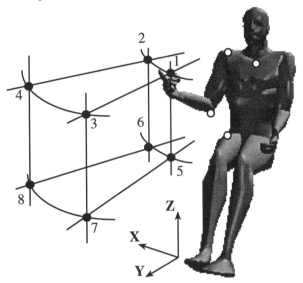

Figure 2. A 3D illustration of hand motion trajectories for the three types of seated reaching tasks.

Three types of seated reaching were performed, distinguished by hand motion directions: anterior-posterior (AP), medial-lateral (ML), and up-down (UD). Each type of the reaching tasks was performed at 4 different positions. Figure 2 presents a 3D illustration of the reaching tasks incorporated in the experiment, where the twelve thin lines delineate the trajectories of hand motions of these tasks. These trajectories followed by simple hand motions were intended to intersect with each other at 8 locations as also identified in Figure 2. In other words, these 8 locations served as the via-points of the hand motions. The 8 locations were characterized by three measures: 1) the height—shoulder or hip height, 2) the radial distance from the right shoulder—60% or 120% of the arm length, and 3) offset angle from the sagittal plane—0-deg or 45-deg, all with respect to a seated person with the torso upright. The former two dimensions were adjusted in accordance with individual anthropometry. Each of the twelve tasks was completed using two speed modes: 1) normal, self-paced, or 2) faster, motivated. Two repetitions were obtained for each movement condition evoked in the experiment. Therefore, a total of 48 trials were performed, in a randomized order, by each subject.

Data Acquisition and Analysis

A MacReflex™ motion analysis system with four cameras was employed to capture the motions of upper torso and the right arms at a sampling frequency of 25 Hz. As illustrated in Figure 1, six reflective markers were placed over palpable body landmarks identifying the right hand, right elbow, right acromion, supersternale, right and left ASIS (anterior superior iliac spine). The two ASIS markers were utilized to locate the bottom of the torso link as their bisection. A four segment linkage was formed as portrayed in Figure 1.

For every instant (time frame), a selected set of five joint angles was derived from the coordinates of the six markers as a collective measure of the instantaneous posture. The definitions (see Table 1) as well as the computations of these angles were facilitated by establishing five coordinate systems as illustrated in Figure 3. Linear polynomial (2nd degree) regression analyses were performed to obtain the f functions. Instantaneous postures were depicted as groups of the five joint angles at the eight common hand locations. Then a 3×2×8 within-subject analysis of variance (ANOVA) on each of the five joint angles was conducted to examine the effects of the two task factors and their possible interplay with each other as well as with the hand locations.

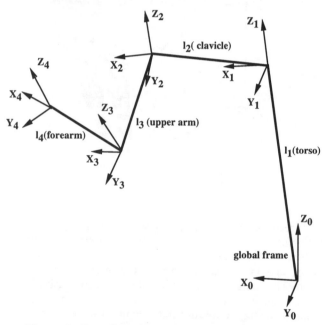

Figure 3. Coordinate systems established to compute the joint angles of the linkage.

Table 1.
Definitions of joint angles selected to characterize a posture

Joint Angle	Definition
1. Torso Flexion	Rotation angle of the torso link with respect to X_0 axis.
2. Torso Lateral Bending	Rotation angle of the torso link with respect to Y_0 axis.
3. Shoulder Extension	Rotation angle of the upper arm link with respect to X_2 axis.
4. Shoulder Abduction	Rotation angle of the upper arm link with respect to Y_2 axis.
5. Elbow Flexion	Rotation angle of the forearm link with respect to Z_3 axis.

* All the angles are 0 when a standard anatomical posture is assumed. Sign convention follows the right hand rule.

RESULTS

Table 2 is a summary of the results from analyses of variance for the five selected joint angles. The most salient and consistent result appears to be a significant two-way interaction between the effects of hand motion direction and the hand location, exhibited by all the five angles. Figures 4-8 graphically illustrate this interaction in a more detailed fashion. Inspection of the graphs resulted in a host of findings including: 1) when the instantaneous hand location is within the arm length (1,2,5,6) substantial torso movements were involved in AP motions but neither ML nor UD motions; 2) consequently at these close-in locations, a more deviated arm posture with more abducted shoulder and more flexed elbow was incurred during AP motions as compared to ML and UD motions; 3) at the far-out hand locations, ML motions were associated with the largest amount torso lateral bending but smallest amount of shoulder abduction in the 0-deg offset plane (3,7), while the reverse was true in the 45-deg offset plane (4,8); 4) at these latter locations (4,8), UD motions were associated with the most substantial amount of torso lateral bending but the least amount of shoulder abduction.

Results from the ANOVA indicate that the speed mode adopted when completing a motion cannot be shown as a significant factor. An additional verification was conducted to show that the time to complete a motion under the self-paced mode was significantly different ($F_{1,4}=209$; $P< 0.0001$) from the motivated mode with the former being approximately 1.5 times the latter, as depicted in Figure 9.

It is also evidenced in the ANOVA results as well as the graphed values of five joint angles that the overall instantaneous postures at the 8 hand locations differed significantly. This was expected since the hand as the end effector largely dictates the linkage configuration.

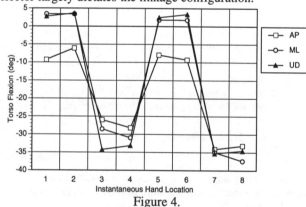

Figure 4.

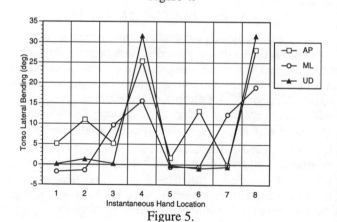

Figure 5.

Table 2. A summary of the analyses of variance results

Effect	Torso Flexion	Torso Lat. Bending	Shoulder Extension	Shoulder Abductn.	Elbow Flexion
1. Motion Direction	$F_{2,8}=1.44$ $P=0.29$	$F_{2,8}=2.28$ $P=0.16$	$F_{2,8}=0.356$ $P=0.71$	$F_{2,8}=9.15$ $P=0.0086$	$F_{2,8}=7.46$ $P=0.0148$
2. Hand Location	$F_{7,28}=68.9$ $P<0.0001$	$F_{7,28}=131.7$ $P<0.0001$	$F_{7,28}=218.89$ $P<0.0001$	$F_{7,28}=12.4$ $P<0.0001$	$F_{7,28}=56.7$ $P<0.0001$
3. Speed Mode	$F_{1,4}=1.115$ $P=0.35$	$F_{1,4}=2.67$ $P=0.177$	$F_{1,4}=0.036$ $P<0.858$	$F_{1,4}=1.92$ $P=0.24$	$F_{1,4}=2.56$ $P=0.185$
Interaction 1*2	$F_{14,56}=4.78$ $P<0.0001$	$F_{14,56}=11.19$ $P<0.0001$	$F_{14,56}=2.29$ $P=0.015$	$F_{14,56}=3.83$ $P=0.0002$	$F_{14,56}=2.15$ $P=0.0224$
Interaction 1*3	$F_{2,8}=0.05$ $P=0.95$	$F_{2,8}=3.74$ $P=0.07$	$F_{2,8}=0.115$ $P=0.89$	$F_{2,8}=2.36$ $P=0.156$	$F_{2,8}=0.97$ $P=0.42$
Interaction 2*3	$F_{7,28}=2.07$ $P=0.08$	$F_{7,28}=0.307$ $P=0.9447$	$F_{7,28}=0.47$ $P=0.85$	$F_{7,28}=1.95$ $P=0.1$	$F_{7,28}=1.04$ $P=0.43$

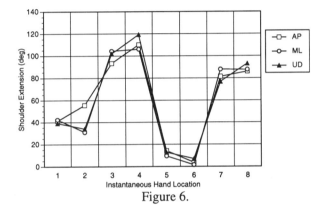

Figure 6.

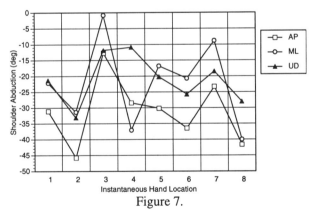

Figure 7.

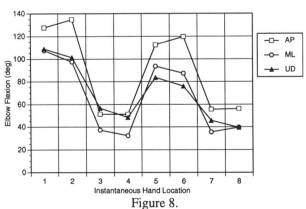

Figure 8.

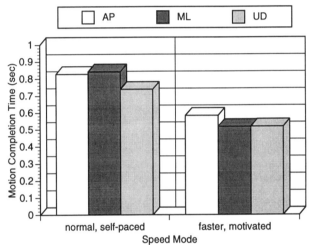

Figure 9. The averaged time to complete three types of motions under two speed modes.

DISCUSSION

Conceptually, there is a distinction between an instantaneous posture and a posture. An instantaneous posture is a configurational entity that carries dynamic information such as acceleration and velocity, whereas a posture is conventionally considered as static, and not associated with any dynamic information. The current study has demonstrated this distinction by the use of a collective postural measure (five joint angles) embedded with the direction and speed information of hand motion. Indeed, as illustrated by this study, the hand motion direction is an important attribute as well as a significant task factor. The use of instantaneous postures also facilitates the separation of the effects of hand location as the kinematic variables from those of the hand motion direction and speed as the task factors. The hand location kinematically dictates but does not completely determine the instantaneous postures, whereas various task factors influence the way the kinematic redundancy is resolved under different conditions evoked during various tasks. It is the latter strategy issue that is of the greatest interest to human movement and posture research.

The proposed analysis method is efficient: the regression analysis serves the purpose of data compression, providing estimates of postural angle profiles during 3D motions;

instantaneous postures are extracted for further analyses of variance. This scheme allows a quick identification of the factors that accompany and affect the instantaneous postures, and avoids arduous comparisons of the entire motion profiles. One limitation of this "sampling of a dynamic process" approach is that some aspects of a movement, such as the timing or coordination between the joints, can not be examined. However, considering the tremendous amount of data generated from a 3D movement study, the ability to compress the data and yet gain useful insights is desirable. The method presented has achieved a good compromise between efficiency and thoroughness.

The findings yielded from this study are helpful to understanding as well as modeling 3D complex, multi-segment human movements. It was evident from the experiment that the hand (end-effector) motion direction significantly affect an instantaneous posture—the difference in individual joint angles can be as large as 20-30 degrees. This necessitates a specification of this attribute in describing an instantaneous posture, which will then reduce the ambiguity in determining how joints or segments are positioned. More importantly, the experiment reveals some aspects of the strategic planning in human movement or postural control. For instance, it was shown that during both the ML and UD movements, regardless of whether the torso was positioned erect or deviated to extend the reach, people largely relied on arm motions to accomplish a reach while minimizing torso participation. On the other hand, once a substantial torso assistive motion was anticipated such as during the AP movements, torso motion was initiated right at the beginning of a reach; this torso motion, however, did not seem to be excessive as compared to the ML or UD motions at far-out locations. These observations pertain to the issue of deducing the objective of an optimal strategy presumably employed by people in postural control. They appear to support the notion that people follow some type of energy-saving strategy: torso motion consumes more of such energy and thus carries a "heavy" penalty function; once a significant torso motion is involved, people execute a multi-segment movement in a synergetic rather than a one-segment-at-a-time way to be "economical". The finding that the movement completion time is not a significant task factor is consistent with the insensitivity of isolated arm motion to speed variation previously reported (Morasso, 1981, 1983; Hogan, 1984; Flash and Hogan, 1985; Flash, 1990). This would appear to exclude the need to study movement speed as a modeling element for normal dynamic postures, as long as the information regarding the sequence of and the coordination between joints is retained.

The findings obtained from the present study can serve as an empirical basis for the development of 3D dynamic posture prediction models. Such models are the driving engine for future dynamic, computerized biomechanical analysis as well as ergonomic design.

ACKNOWLEDGMENT

The authors acknowledge the support provided by Chrysler Corporation, in particular, Dr. Deborah Thompson. Thanks are also extended to James Foulke, Charles Woolley, Rodney Capps at the Center for Ergonomics, and Prof. Julian Faraway at the Statistics Department of the University of Michigan for their assistance.

REFERENCES

Armstrong, W.W., Green, M., and Lake, R. (1987). Near-real-time control of human figure models. IEEE CG&A, 7, 52-61.

Ayoub, M.M. and Hsiang, M.S. (1992). Biomechanical simulation of a lifting task. Advanced in Industrial Ergonomics and Safety IV, Kumar, S. (ed.), Taylor & Francis.

Ayoub, M.M. (1994). Biomechanics of material handling through simulation. Proceedings of The Third Pan-Pacific Conference on Occupational Ergonomics, 376-380.

Badler, N.I., Manoochehri, K.H., and Walters, G. (1987). Articulated figure positioning by multiple constraints. IEEE CG&A, 7, 28-38.

Chaffin, D.B. (1992). A biomechanical model for simulation of 3D static human exertions. Computer Applications in Ergonomics and Safety, Mattila, M. and Karwowski, W. (ed.), North Holland.

Flash, T. and Hogan, N. (1985). The coordination of arm movement: an experimentally confirmed mathematical model. J. Neuroscience, 5, 1688-1703.

Gordon J., Ghilardi, M.F., and Ghez, C. (1994a). Accuracy of planar reaching movements: I. independence of direction and extent variability. Exp. Brain Res., 99, 97-111.

Gordon J., Ghilardi, M.F., Cooper, S.E., and Ghez, C. (1994b). Accuracy of planar reaching movements: II. systematic extent errors resulting from inertial anisotropy. Exp. Brain Res., 99, 112-130.

Hogan, N. (1984). An organizing principle for a class of voluntary movements. J. Neuroscience, 3, 2745-2754.

Jung, E.S., Choe, J., and Kim, S.H. (1994). Psychophysical cost function of joint movement for arm reach posture prediction. Proceedings of HFES 38th Annual Meeting, Nashville, TN, 636-640.

Karst, G.M. and Hasen, Z. (1990). Direction-dependent strategy for control of multi-joint arm movement. Multiple Muscle Systems: Biomechanics and Movement Organization, Winters, J.M. & Woo, S.L-Y. (ed.), Springer-Verlag, New York.

Kilpatrick, K.E. (1970). Model for the design of manual work station. Ph.D. Dissertation, The University of Michigan, Ann Arbor, MI.

Monheit, G. and Badler, N.I. (1991). A kinematic model of the human spine and torso. IEEE CG&A, 11, 29-38.

Morasso, P. (1981). Spatial control of arm movements. Exp. Brain Res., 42, 223-227.

Morasso, P. (1983). Three dimensional arm trajectory. Biol. Cybern., 48, 187-194.

Park, K.S. (1973). Computerized simulation model of posture during manual materials handling. Ph.D. Dissertation, University of Michigan, Ann Arbor, MI.

Phillips, G.B. (1991). Interactive postural control of articulated geometric figures. Ph.D. Dissertation, University of Pennsylvania, PA.

THE EFFECT OF LIFTING VS. LOWERING
ON SPINAL LOADING

Kermit G. Davis
The Ohio State University
Columbus, Ohio

In industry, workers perform tasks requiring both lifting and lowering. During concentric lifting, the muscles are shortening as the force is being generated. Conversely, the muscle lengthens while generating force during eccentric lowering. While research on various lifting tasks is extensive, there has been limited research performed to evaluate the lowering tasks. Most of the research that does exist on lowering has investigated muscle activity and trunk strength. None of these studies have investigated spinal loading. The current study estimated the effects of lifting and lowering on spinal loads and predicted moments imposed on the spine. Ten subjects performed both eccentric and concentric lifts under sagittally symmetric conditions. The tasks were performed under isokinetic trunk velocities of 5, 10, 20, 40, and 80 deg/s while holding a box with weights of 9.1, 18.2, and 27.3 kg.. Spinal loads and predicted moments in three dimensional space were estimated by an EMG-assisted model which has been adjusted to incorporate the artifacts of eccentric lifting. Eccentric strength was found to be 56 percent greater than during concentric lifting. The lowering tasks produced significantly higher compression forces but lower anterior-posterior shear forces than the concentric lifting tasks. The differences in the spinal loads between the two lifting tasks were attributed to the internal muscle forces and unequal moments resulting from differences in the lifting path of the box. Thus, the differences between the lifting tasks resulted from different lifting styles associated with eccentric and concentric movements

INTRODUCTION

Manual material handling (MMH) tasks have been associated with lower back injuries (Snook et al., 1978; Bigos et al., 1986). A typical MMH task requires both lifting and lowering of items that have a variety of shapes, sizes and weights. Lowering involves a different mechanism of muscle operation, in which the muscles elongate as the force is generated while concentric lifting entails muscle shortening. Drury et al, (1982) found that 52% of the MMH tasks in industry were eccentric lowering while concentric lifting consisted of only 32%. While research on various concentric tasks is extensive, there has been limited research performed to evaluate the eccentric task.

Many researchers have found that the muscle activity of the trunk extensor muscles were lower for eccentric than concentric tasks despite the generation of greater external torque (Kumar and Davis, 1983; Marras and Mirka, 1989; Cresswell and Thorstensson, 1994; De Looze et al., 1993). Henriksson et al. (1972) found that the perceived exertion for lowering tasks was less than for lifting tasks. The strength of the various trunk muscles have been evaluated as a function of eccentric and concentric lifting by several authors. In general, individuals have greater strength during eccentric than concentric lifting (Marras and Mirka, 1989; Reid and Costigan, 1987; Smidt et al., 1980).

The objective of the current study was to compare both eccentric and concentric lifting. The evaluation of the two tasks consisted of estimating the moments and loads acting on the spine through the use of an EMG-assisted model that has been adjusted for the eccentric behavior of the muscle.

METHODS

Subjects

The ten subjects who participated in the study were male students with no prior history of low back pain. The ages of the subjects ranged from 22 to 34 years. The averages (sd) for height and weight were 181.0 cm (6.6) and 79.3 kg (12.6), respectively.

Task

Subjects performed the concentric and eccentric exertions while positioned in a pelvic support structure (PSS). The subjects started the eccentric lifts in an upright position (0 degrees of flexion) and lowered the box to a posture of 40 degrees of flexion. On the other hand, the concentric lifts had the subjects start at 40 degrees of flexion and lift until they reached a fully upright position.

Experimental Design

This study was a three-way, within-subject design. The independent variables included the following: box weight (9.1, 18.2, and 27.3 kg.), trunk isokinetic velocity (5, 10, 20, 40, and 80 deg/s), and lifting task. These weights and velocities were chosen to reflect the values commonly found in industry (Marras et al. 1993). In order to account for variability between the subjects, subjects were used as a random effect. The lifting tasks consisted of both eccentric and concentric lifting of the box.

The dependent variables of this study were maximum moments imposed on the spine and spinal loading. All these measurements were computed using the dynamic EMG-assisted model developed at the Ohio State University over the past decade (Marras and Reilly, 1988; Marras and Sommerich, 1991a, b; Granata and Marras, 1993; and Granata and Marras, 1995a, b; Marras and Granata, 1995). The model used the kinematic and electromyographic information about the trunk to estimate the spinal loads as well as the moments imposed on the spine in three dimensional space. The spinal loads estimated were compression, anterior-posterior shear, and lateral shear on the L_5/S_1 joint. The predicted moments were sagittal bending moments about the lumbosacral joint since the lifting tasks were performed sagittally symmetric.

The relationships for length-strength and force-velocity were empirically determined by calculating the error between the predicted to measured torque at the instantaneous lengths and velocities of the muscle during the lifts. A best fit curve was used to estimate both relationships for eccentric and concentric lifting separately.

Apparatus:

The Lumbar Motion Monitor (LMM) was used to collect the trunk motion variables. The LMM is essentially an exoskeleton of the spine in the form of a triaxial electro-goniometer that measured the instantaneous three-dimensional position, velocity, and acceleration of the trunk. For more information on the design, accuracy, and application of the LMM, refer to Marras et al. (1992).

Integrated electromyographic (EMG) activity was monitored through the use of bi-polar electrodes spaced approximately 3 cm apart at the ten major trunk muscle sites. The ten muscles of interest were: right and left erector spinae; right and left latissimus dorsi; right and left internal obliques; right and left external obliques; and right and left rectus abdominis. For the standard locations of the electrode placement for these muscles, refer to Mirka and Marras (1993).

A force plate (Bertec 4060A) was used to measure the kinetic variables of the lifts. The subject was positioned into a pelvic support structure (PSS) that was attached to the force plate. The PSS restrained the subject's pelvis and hips in a fixed position. Also, the relative position of L_5/S_1 to the center of the force plate remained constant for the entire experiment. By knowing the position of L_5/S_1, the forces and moments measured at the center of the force plate were translated and rotated to L_5/S_1.

All signals from the above equipment were collected simultaneously through a customized Windows™-based software developed in the Biodynamics Laboratory. The signals were collected at 100 Hz and recorded on a 486 portable computer via an analog-to-digital board. The data were saved by the computer for subsequent analysis.

An additional computer was used to display the instantaneous sagittal angular velocity recorded by LMM in real time. A target region was provided by displaying two lines at a given slope that corresponded to the velocity of interest. The lines allowed the subject to have a tolerance of 3 percent deviation for the target velocity. The signal was transferred from the LMM to the computer through an analog-to-digital board and converted into velocity by customized software. The computer monitor was positioned directly in front of the PSS in direct view of the subject.

Procedure:

Upon arriving at the Biodynamics Laboratory, subjects completed a consent form and anthropometric measurements were taken. After proper application of the electrodes, a set of maximum exertions was performed for normalization procedures. Subjects were then positioned into the PSS and the LMM was attached to the back. Velocities were controlled by the subject by following a trace through a given region displayed on a computer screen while lifting the box. All subjects were allowed to practice the different velocities until they were able to remain within the tolerances. If the subject's trace fell outside the tolerance levels, the lift was repeated.

RESULTS

The length-strength relationship for both eccentric and concentric exertions was found to be the same. Under the concentric conditions, the force-velocity modulation was found to be an exponential function, as seen by other researchers (Hill, 1938; Wilkie, 1950; Granata, 1993). Conversely, the eccentric force-velocity modulation factor was determined to be constant. The value of the constant was set to be equal to the ratio between the eccentric and concentric gains, thus, resulting in a constant inter-subject gain. This ratio was found to be 1.56.

The EMG-assisted model was then used to predict the external moments and spinal loads with the adjustment for eccentric lifting. As a result, there was no difference between the gains for the two types of lifting. Additionally, the model performance was slightly better for eccentric lifting. The average r^2 between the predicted and measured external moments for eccentric and concentric exertions were 0.95 and 0.88, respectively.

A summary of the analysis of variance (ANOVA) for the maximum predicted moments and spinal loads are in Table 1. Notice, tasks refers to the type of lifting being performed. Under eccentric lifting conditions, the maximum sagittal predicted moment (hereafter referred to as moment) was larger than when lifting concentrically, as seen in Figure 1. The values of moment for eccentric and concentric lifting were 140.5 Nm and 113.1 Nm, respectively (a difference of 27.4 Nm). As the weight of the box increased, the moment also increased. The maximum sagittal moments for the 9.1, 18.2 and 27.3 kg. weights were 105.1, 126.5, and 148.8 Nm, respectively. Therefore, the maximum moment increased about 25 Nm for every 9.1 kg. of increased box weight. For sagittal predicted moment, the velocity of 5 degrees per second (deg/s) had a significantly smaller moment than the other velocities (10, 20, 40, and 80 deg/s). The Velocity*Task interaction indicated that for the 5 deg/s velocity, the task type had no effect on the moment, however, for the other velocities, eccentric lifting produced larger moments than the concentric lifts.

Table 1: Summary of Significant Effects for the EMG-assisted Model Outputs.

Effect	Maximum Sagittal Predicted Moment	Maximum Lateral Shear Force	Maximum A/P Shear Force	Maximum Compression Force
Task (T)	*		*	*
Weight(W)	*	*	*	*
Velocity(V)	*		*	*
W*T		*	*	
V*T	*		*	*
W*V				
W*V*T				

* indicates significant at p < 0.05.

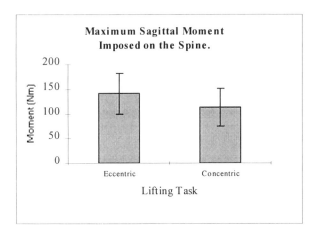

Figure 1: Maximum Sagittal Supported Moment at L5/S1 as a Function of Lifting Task.

The results for the maximum lateral shear forces indicated that the Weight main effect was the only significant effect where an increase in weight corresponded to an increase in lateral shear forces. A 9.1 kg. increase in box weight resulted in approximately 22 N of additional lateral shear force on the spine.

The maximum anterior-posterior (A-P) shear forces placed on the spine during concentric lifting were larger than when lifting eccentrically (Figure 2). The difference between the two types of lifts was approximately 135 N. An increase box weight of 9.1 kg. also produced an increase in A-P shear of about 130 N. The 5 deg/s velocity resulted in lower A-P shear forces than the 20 and 80 deg/s velocities, but was not different than the 10 and 40 deg/s lifts. Similarly, the A/P Shear forces for the 80 deg/s lifts were significantly larger than the 10 deg/s lifts but not the 20 and 40 deg/s velocities.

The A-P shear forces for the 5, 10, 20, 40, and 80 deg/s velocities were 707, 748, 765, 752, and 768 N, respectively. Additionally, the Weight*Task and Velocity*Task interaction indicated that the effects of type of lifting on A-P shear depended upon the weight lifted or the velocity of the lift. There was a larger increase in A-P shear forces with increases in weight for the concentric lifts. The concentric lifts had higher A-P shear forces at all velocities, but a larger difference was seen for the 80 deg/s lifts than the other lifts.

The maximum compression forces were greater during eccentric lifting than when lifting concentrically (Figure 2). The difference in loading between eccentric and concentric lifting was close to 600 N. As with the lateral and A-P shear forces, an increase in loading was experienced when the box weight increased. The type of task produced a larger difference in the maximum compression forces than when the box weight was increased by 9.1 kg. The 5 deg/s lifts produced less compression force than the other velocities. There was no difference between the 10, 20, and 40 deg/s lifts, but a difference was found between the 80 deg/s lifts and 10 deg/s lifts. For velocities at 10, 20, 40 and 80 deg/s, the maximum compression forces were lower for concentric lifting compared to eccentric lifting while no difference was found for the 5 deg/s velocity.

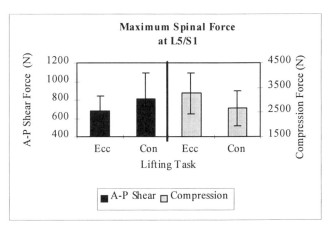

Figure 2: Maximum Spinal Loading at L5/S1 as a Function of Lifting Task.

Many of the differences between the two types of tasks were confounded by the differences in the sagittal predicted moment. Thus, the results for A-P shear force, and compression force were normalized to predicted sagittal moment. This would eliminate any variance attributed to the differences in moment. The results of the ANOVA indicate that there was still a difference between the two types of lifting (Table 2).

The spinal loading for the two types of lifting are shown in Figure 3. The A-P shear force per unit of moment during eccentric lifting was significantly lower than during concentric lifting. This would indicate that the difference between A-P shear loading for the two types of tasks resulted from not only increased moment but also from an increase in muscle loading of the spine (via coactivity). The difference between the two types of lifting was actually reversed for compression when normalized to moment. Initially, eccentric lifting had larger compression forces than concentric lifting, but when normalized to the moment, eccentric lifting was lower. This would indicate that the sagittal moment was the main reason for the increase in compression forces placed on the spine during eccentric lifting.

Table 2: Summary of Significant Effects for the Estimated Spinal Loads Per Unit Moment.

Effect	Maximum A-P Shear Force Per Unit Moment	Maximum Compression Force Per Unit Moment
Task (ecc vs. con)	*	*
Weight		*
Velocity		*
W*T		*
V*T	*	
W*V		
W*V*T		

* indicates significant at p < 0.05.

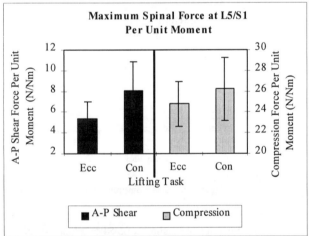

Figure 3: Maximum Spinal Loading Per Unit Moment as a Function of Lifting Task.

Additionally, the interaction between the type of lifting and velocity remained significant for A-P shear forces when normalized to external moment. The amount of box weight significantly influenced the compression forces even after normalization for moment. The internal forces resulting from the muscles contributed more at the lower weight (18.2 kg.) than the heavy weight (27.3 kg.). The compression per unit moment was only significant for the 80 deg/sec condition. Thus, the difference in compression forces between the velocities was attributable to the external moment, except for the 80 deg/sec velocity. For the 80 deg/sec velocity, the internal forces from the muscles contributed additional compression force. The interaction between task and weight was significant for compression per unit moment. The

concentric lifts had a larger increase across the different weights than eccentric lifting. Again, it appears that the difference between the two tasks at the various levels of weight resulted from both external moment and internal force differences.

DISCUSSION

The results were not affected by the fidelity of the EMG-assisted model. The gains for each of the subjects were constant between the two types of lifting. The values of r^2 between the predicted and measured moments (validity) did not differ significantly between the eccentric and concentric lifts. The ratio eccentric and concentric gains was 1.56. Reid and Costigan (1987) found that the ratio of eccentric to concentric strength was 1.2. The difference between the two ratios was that the present study was based on internal estimation of the external torque which reflects both agonistic and antagonistic muscle inputs. Additionally, their ratio was determined by using a KIN/COM to control the exertion, possibly resulting in different motions from free dynamic lifting. The importance of the difference is that the relationship between concentric and eccentric strength is not necessarily universal.

The present results show that the maximum sagittal moment was substantially higher for the eccentric lifts. De Looze et al., (1993) and Gagnon and Smyth (1991) found that concentric lifting produced larger peak moments than eccentric tasks. The difference in the studies can be explained by the different techniques used to control the subject's posture. For the present study, the subjects performed the exertions in a structure that fixed the pelvis and hips in one position. By allowing the hips and legs to move, as in the other studies, the subjects could have adopted a different lifting style for eccentric and concentric exertions. The individuals might have moved into the load during the eccentric lifts.

In the present study, the maximum sagittal moments were equivalent for all velocities except the 5 deg/s velocity. The subjects appeared to move the box to a maximum distance away from the body independently of the velocity. For the 5 deg/s, subjects seemed to keep the box closer to the body throughout the exertion. Granata and Marras (1993) found that the average sagittal moment reduced with increased velocity. These results indicate that the subjects might move the box to the same maximum horizontal distance but bring it closer to the body during the rest of the exertion for the faster velocities.

In the present study, eccentric lifting produced maximum A-P shear forces that were 135 N lower than the concentric tasks. On the other hand, the eccentric tasks resulted in higher compression values (almost 600 N) than concentric lifts. There appears to be a trade-off in the nature of the loading between eccentric and concentric exertions. However, it was hypothesized that the resulting loads occurred from the differences in sagittal moment. When the A-P shear and compression forces were normalized with respect to the sagittal moment, there was still a difference between the two lifting tasks. When normalized to moment, the difference in A-P shear force between the two types of lifting remained after normalization indicating that there was a difference in the mechanism that produced the forces to counter-balance the external forces. The A-P shear per unit moment was attributed to the difference in coactivity between the different lifting motions resulting in higher loads during concentric lifting. For compression forces per unit moment, the eccentric lifts actually produced lower compression forces per unit moment than the concentric lifts indicating that the difference resulted from the subjects using different lifting styles along with a change in the internal force mechanism.

The peak A-P shear and compression forces were the smallest for the 5 deg/sec velocity. There was no difference between A-P shear forces for the faster velocities. The differences between the velocities for A-P shear was explained by the external moment lifted under the various conditions. Thus, the A-P shear forces resulted from changes in the path of movement for the various velocities. Also, there was only a difference between the 10 deg/s and 80 deg/s lifts for compression. Granata and Marras (1993) found that average spinal forces changed as a function of velocity. These researchers found that velocity effects both the path of movement and amount of internal loading. The present study indicated that compression forces per unit moment increased only for the 80 deg/sec conditions. Once again, it appears that the subjects adopted a lifting pattern that resulted in the same peak forces for the lower velocities but the internal loading increased during the fast velocity. Hence, the dynamics of the exertion also influences the lifting style and ultimately the loading on the spine.

CONCLUSION

The present study has shown that eccentric strength was 56 percent greater than during concentric lifting. The type of lifting performed by an individual seemed to effect the nature of the lift and loading on the spine. It was also determined that the dynamics of the lift played an import role in the lifting style adopted under both eccentric and concentric lifts.

REFERENCES

Bigos, S.J., D.M Spengler, N.A. Martin, J. Zeh, L. Fisher, A. Nachemson, and M.H. Wang, (1986), Back Injuries in Industry: A Retrospective Study. II. Injury Factors. Spine, vol. 11(3), 246-251.

Cresswell, A.G. and A. Thorstensson, (1994), Changes in Intra-Abdominal Pressure, Trunk Muscle Activation and Force during Isokinetic Lifting and Lowering. European Journal of Applied Physiology, vol. 68, 315-321.

De Looze, M. P., H.M. Toussaint, J.H. Van Dieen and H.C.G. Kemper, (1993), Joint Moments and Muscle Activity in the Lower Extremities and Lower Back in Lifting and Lowering Tasks., Journal of Biomechanics, vol. 26(9), 1067-1076.

Drury, C.G., C. Law, and C.S. Pawenski, (1982), A Survey of Industrial Box Handling. Human Factors, vol. 24(5), 553-565.

Gagnon, D. and M. Gagnon, (1992), The Influence of Dynamic Factors on Triaxial Net Muscular Moments at the L5/S1 Joint During Asymmetric Lifting and Lowering. Journal of Biomechanics, vol. 25(4), 891-901.

Gagnon, M. and G. Smyth, (1991), Muscular Mechanical Energy Expenditure as a Process for Detecting Potential Risks in Manual Material Handling. Journal of Biomechanics, vol. 24(3/4), 191-203.

Granata, K.P., (1993), An EMG-Assisted Model of Trunk Loading During Free-Dynamic Lifting. Unpublished Doctoral Dissertation, the Ohio State University, Columbus, Ohio.

Granata, K.P. and W.S. Marras, (1993), An EMG-Assisted Model of Loads on the Lumbar Spine During Asymmetric Trunk Extensions. Journal of Biomechanics, vol. 26(12), 1429-1438.

Granata, K.P. and W.S. Marras, (1995a), An EMG-Assisted Model of Trunk Loading During Free-Dynamic Lifting. Journal of Biomechanics, vol. 28(11), 1309-1317.

Granata, G. P., and W. S. Marras, (1995b), The Influence of Trunk Muscle Coactivity Upon Dynamic Spinal Loads, Spine, 20(8), 913-919.

Henriksson, J., H.G. Knuttgen, and F. Bonde-Peterson, (1972), Perceived Exertion During Exercise with Concentric and Eccentric Muscle Contractions. Ergonomics, vol. 15(5), 537-544.

Hill, A.V., (1938), The Heat of Shortening and the Dynamic Constants of Muscle, Proceedings of the Royal Society of Biology, vol. 126, 136-195.

Kumar, S. and P.R. Davis, (1983), Spinal Loading in Static and Dynamic Postures: EMG and Intra-abdominal Pressure Study. Ergonomics, vol. 26(9), 913-922.

Marras, W.S., F. Fathallah, R.J. Miller, S.W. Davis, and G.A. Mirka, (1992), Accuracy of a Three Dimensional Lumbar Motion Monitor for Recording Dynamic Trunk Motion Characteristics. International Journal of Industrial Ergonomics, vol. 9, 75-87.

Marras, W. S. and K. P. Granata, (1995), A Biomechanical Assessment and Model of Axial Twisting in the Thoraco-Lumbar Spine, Spine, 20(13), 1440-1451.

Marras, W.S., S.A. Lavender, S.E. Leurgans, S.L. Rajulu, W.G. Allread, F.A. Fathallah, and S.A. Ferguson, (1993), The Role of Dynamic Three-Dimensional Motion in Occupationally-Related Low Back Disorders. Spine, vol. 18(5), 617-628.

Marras, W.S, and G.A. Mirka, (1989), Trunk Strength During Asymmetric Trunk Motion. Human Factors, vol. 31(6), 667-677.

Marras, W. S., and C. H. Reilly, (1988), Networks of Internal Trunk Loading Activities Under Controlled Trunk Motion Conditions, Spine, 13(6), 661-667.

Marras, W.S. and C.M. Sommerich, (1991a), A Three Dimensional Motion Model of Loads on the Lumbar Spine: I. Model Structure. Human Factors, vol. 33(2), 129-137.

Marras, W.S. and C.M. Sommerich, (1991b.), A Three Dimensional Motion Model of Loads on the Lumbar Spine: I. Model Validation. Human Factors, vol. 33(2), 139-149

Reid, J.G. and P.A. Costigan, (1987), Trunk Muscle Balance and Muscular Force. Spine, vol. 12(8), 783-786.

Schmidt, G.L., L.R. Amundsen, and W.F. Dostal, (1980), Muscle Strength at the Trunk. The Journal of Orthopaedic and Sports Physical Therapy, vol. 1, 165-170.

Snook, S.H., R.A. Campanelli, and J.W. Hart, (1978), A study of Three Preventive Approaches to Low Back Injury. Journal of Orthopeadic Medicine, vol.20(7), 478-481.

Wilkie, D.R., (1950), The Relation Between Force and Velocity in Human Muscle. Journal of Physiology, vol. 110, 249-280.

PULLING ON SLIPPERY VERSUS NON-SLIPPERY SURFACES: CAN A LIFTING BELT HELP?

Steven A. Lavender[1], Sang-Hsiung Chen[2], Yi-Chun Li[3], Gunnar B.J. Andersson[1]
[1]Rush-Presbyterian-St. Luke's Medical Center, Chicago, Illinois
[2]Chang Gung Memorial Hospital, KaoHsiung, Taiwan
[3]University of Illinois at Chicago

Biomechanical research has shown that pull forces are largely generated by leg extension and by leaning (using one's body mass). However, this pulling technique requires sufficient friction in the foot floor interface. When pulling tasks are performed on slippery surfaces the center of gravity must be kept over the feet, thereby increasing the trunk's role in force generation. The purpose of this study was to investigate using electromyography (EMG) the muscle recruitments in the torso when pulling tasks are performed on slippery versus non-slippery surfaces. It was hypothesized that if a lifting belt acts to stiffen the torso, this would reduce the muscle activities detected with the EMG. Twelve subjects pulled at 40% of their maximal exertion values in four postures, with a lifting belt either tight or very loose, and under good footing conditions or extremely slippery footing conditions. When averaged across subjects all EMG activities increased under the slippery conditions. These increases ranged between 25 and 131 percent, although for some muscles the response to the slippery conditions was mediated by the pulling posture. The lifting belt had no effect on any of the muscle activities and created no differential muscle response due to the footing conditions or the pulling postures employed.

Epidemiologic studies frequently cite occupational tasks that require lifting, pushing and or pulling as contributing to low-back disorders (LBD's) (Andersson, 1991). Most of the biomechanics research to date has focussed on the lifting aspect of the manual material handling problem. Perhaps this is because, for most workers, pulling tasks are performed less frequently than lifting tasks. Yet pulling tasks can still create considerable bending moments and compressive forces on the spine (Chaffin et al., 1983; Dempster, 1958).

In order to develop a pulling force from a standing posture the body must be capable of transferring the reaction force from the floor all the way to the hands. This requires the trunk act as a rigid link in the kinetic chain. It is has been demonstrated that lifting belts may act to increase the stiffness of the trunk, particularly in the frontal and transverse planes (Lavender et al. 1995; McGill et al., 1994). Thus, it is possible that a lifting belt will facilitate the force transfer, thereby requiring less internal force from the trunk musculature during whole body pulling tasks.

Much of the work done when pulling is accomplished through the moments created by gravitational forces acting on the leaning body mass and by the leg muscles. Therefore, standing pulling tasks often result in postures that are extremely dependent upon the quality of the foot-floor interface. Hence, when pull forces are generated under slippery

conditions the trunk extensor musculature must play a bigger role as the body's center of gravity is now maintained over the feet. Further, the uncertainty of pulling objects under slippery footing conditions may lead to increased co-contraction of the antagonistic muscles, thereby, further accentuating spinal loading.

The objective of the current study was to test the following hypotheses with regard to isometric pulling tasks on non-slippery and slippery flooring conditions:
1. A lifting belt reduces the trunk extensor muscle forces required to generate a specified pull force.
2. Slippery conditions result in greater trunk muscle forces as determined via electromyography.
3. The lifting belt reduces the co-contraction of the antagonistic trunk muscles when pulling under slippery conditions.

METHODS

Subjects

Twelve subjects, ten males and two females, without a history of LBD volunteered for the study. Their mean age was 29.6 years (range 22 - 46 years), height was 1.77 m (range 1.56 - 1.88 m), and weight was 82.7 kg (range 55 - 132 kg).

Experimental Design

A repeated measures experiment was designed to evaluate three independent variables: the pulling posture, the lifting belt and the footing condition, during controlled submaximal exertions. The dependent measures were the electromyographic data (EMG) obtained from the following bilateral muscle groups: Erector Spinae (ERSR and ERSL), Latissimus Dorsi (LATR and LATL), External Oblique (EXOR and EXOL), and Rectus Abdominus (RABR and RABL).

Apparatus

Subjects pulled on a 30.5 cm long, 4.5 cm diameter bar that was connected to a uni-axial dynamometer via cable attached in the middle. The dynamometer was positioned 22 cm off the floor. This system allowed to handle height to float as the subjects achieved postures that allowed them to maximize their force output. The dynamometer was angled such that the axis was in-line with the anticipated cable orientations based on pilot testing and was adjusted for each defined pulling posture tested.

A reference frame apparatus was used to obtain EMG data during maximal isometric exertions in a standing posture. The apparatus included a floor-mounted stand to affix each subject's pelvis. The subjects were fitted with a chest harness that had attachment points in the center and 15 cm each side of center on the front and the rear.

The lifting belts were supplied by Chattanooga Group, Inc (Chattanooga, TN). The belts had a double layer construction with the inner mesh layer containing sewn in vertical plastic stiffeners across the lumbar section. The inner layer attached anteriorly with velcro. The outer layer was comprised of a two 10 cm wide elastic bands. The bands were offset such that the elastic region was 15 cm wide in the center of the back where they were anchored to the inner mesh layer and narrowed to 10 cm at the velcro attachment points.

Surface EMG data were collected using disposable Niko bipolar electrodes (model 4533, Distributed by Salberg Medical, Minneapolis) with a 9 mm diameter sensor. The inter-electrode distance was 3 cm. The electrodes were connected to small preamplifiers with a gain of 1000 attached with velcro to the lifting belt. These were connected to amplifiers the had a gain of 57. The amplified signals were rectified and integrated with a time constant of approximately 100 ms. The band pass frequencies of the system were 30 to 1000 hz. A Gateway P5-60 computer sampled the EMG data entering through the A/D card at 120 Hz.

Procedure

Electrodes were applied to the each of the four

bilateral muscle pairs along the line of action for a given muscle. The specific electrode placements were as follows: (a) ERSR and ERSL: centered halfway between the L3 and L4 spinous processes and approximately 3 to 4 cm lateral from the midline over the belly of the muscle; (b) LATR and LATL: over the muscle belly at the level of T7 and approximately 13-15 cm lateral from the midline; (c) EXOR and EXOL: at the level of the umbilicus and approximately halfway between the iliac crest and the anterior superior iliac spine; (d) RABR and RABL: level of the umbilicus 2 cm lateral from the midline.

Once instrumented with electrodes the chest harness was applied and the subject was strapped to the pelvic support in the reference frame. The harness was connected to dynamometers via turnbuckles in the front or rear for maximal trunk extension and flexion exertions, respectively. Two dynamometers were connected at the 15 cm lateral connection point, one front and one rear, to provide resistance and measure the peak force during maximal trunk twisting efforts. The subjects were instructed to provide maximal exertions over a five second period. Exertions were repeated as necessary after a two minute rest period to insure maximal values were obtained in each of the four directions: Extension, flexion, twist left, and twist right.

Following the maximal exertions the electrodes were surrounded with a 1 cm dense foam material taped to the skin. These pads prevented the lifting belt from applying direct pressure to the electrodes. Next the subject was fitted with a lifting belt. The subject was instructed that during the "belt" trials the belt should be tightened, first, by tensioning the under layer, and second by stretching the elastic binders as far anteriorly as possible. During the "No Belt" trials the subjects were instructed to remove the tension from the elastic binders and loosen the under layer so that their hand could easily be slid between their belly and the lifting belt.

The subjects were instructed in how to attain the four pulling postures shown in figure 1. The instructions also stressed that foot positions could not be changed during the exertions. All pulling exertions were performed barefoot so as remove potential confounding due to variation in shoe tread material and design. In each posture the subjects were encouraged to lean away from the dynamometer during the pulling exertion if they felt it would increase their force output. The postures were defined in terms of the number of hands used and placement of the feet as follows:

1. "Sagittally Symmetric, Bar over Ankles" (SSBOA), Front pull with both hands on the bar which was positioned over the ankles. Both feet were placed so that tips of the big toes were along a designated line. The cable length was adjusted so that the handle was approximately over the ankles.

2. "Sagittally Symmetric, One Leg Back" (SSOLB), Front pull with both hands on the bar which was positioned over the anterior ankle. One foot (of the subjects choosing) was placed with the big toe along a designated line. The other foot was shifted posteriorly to a position where the subject felt comfortable.

3. "Asymmetric, Two Handed" (ATH), The subject was turned such that the dynamometer was in the mid-frontal plane through the ankles. The left foot was placed such that the long axis was in a sagittal plane perpendicular to the cable connecting the handle to the dynamometer. The right foot was positioned approximately shoulder width apart from, and generally parallel to, the left foot. Subjects were permitted to vary the right foot's orientation if they believed greater pull force could be achieved, although, they were instructed to repeat this placement through all ATH trials. The subject was instructed to twist to the left so that both hands could be placed on the bar.

4. "Asymmetric, One Handed" (AOH), same foot stance as above in the ATH posture, however, the bar was only grasped with the left hand in the center. The subjects were instructed to hold their right hand across their chest.

In the first phase subjects pulled as hard as they could against the dynamometer in each of the four described postures. A two minute rest period was given between each of the pulls. Peak dynamometer readings were recorded by the investigators but these values were not shared with the subjects. Pulls were repeated only if the subject felt that they had not provided a maximal effort and could do better.

In the second phase the subjects were required to generate a pull force that was 40 percent of their lowest maximal exertion value measured in the first phase. For example, if a subject's maximal pull values ranged between 430 and 580 N, the trials during the second phase of the experiment required pull forces of 172 N. The pull forces were controlled using a video display connected to the computer sampling the pull force data. The analogue display provided the subject with feedback regarding the current level of force applied relative to a target level. The subject performed the eight exertions comprising the combinations of the belt and pull posture condition while maintaining designated 40 percent pull force value in a randomized sequence. EMG data were sampled for three seconds once the subject achieved the proper pull force.

The eight trials just described were then repeated in a new randomized sequence but with the subject standing on a slippery floor surface. Two pieces of vinyl flooring material were mounted on the

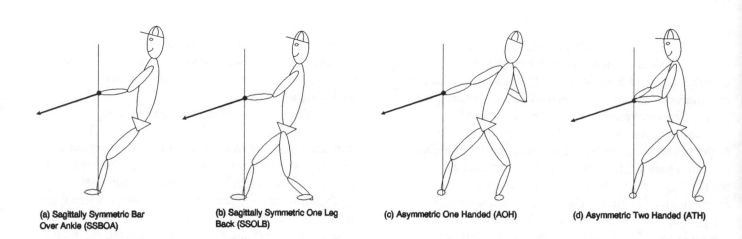

(a) Sagittally Symmetric Bar Over Ankle (SSBOA)

(b) Sagittally Symmetric One Leg Back (SSOLB)

(c) Asymmetric One Handed (AOH)

(d) Asymmetric Two Handed (ATH)

Figure 1. The four pulling postures used in this study.

floor and covered with a generous amount of baby oil. Subjects were assisted onto the platforms so they would not fall. The data were collected as described above.

Data Treatment

The integrated EMG were normalized with respect to maximal and minimal values as follows:

$$NEMG(i) = \frac{IEMG(i) - Min\ EMG(I)}{Max\ EMG(i) - Min\ EMG(i)} * 100 \quad (1)$$

where:

i =	muscle 1 through 8
IEMG=	integrated EMG signal for muscle i
Min EMG =	minimum signal observed for muscle i
Max EMG=	maximum signal observed for muscle i

The NEMG signal is expressed in terms of a percentage of the muscles maximal voluntary contractile capacity.

RESULTS

The peak pull force measured by the dynamometer varied significantly ($p < .05$) between the four postures used. The greatest force, 681 N, was generated in the SSBOA posture followed by the SSOLB (621 N), ATH (564 N), and the AOH (504 N) postures. For most subjects the AOH posture resulted in smallest maximal pull force. Therefore, on the average the subjects exerted 184 N (sd = 26 N) during the submaximal (40 percent) trials.

Significant differences were observed in the analyses of peak EMG data. Table 1 summarizes the five muscles which changed as a function of the pulling posture employed. All of the eight muscles showed a significant increase in their recruitment when the floor surface was oily (Table 2). On the average the oily surface lead to a 73 percent increase in the normalized EMG of the four posterior muscles and a 51 percent increase in the EMG measured from the anterior muscles. The ERSR and the LATL showed a differential response in the four postures depending on whether the oil was present (Figure 2). More specifically, the LATL increased by a factor of 2.8 for pulls made from the AOH and SSBOA postures on the oily surface versus the non-oily surface. The SSOLB posture on the oily surface only increased the LATL by a factor of 1.6. The ERSR, on the other hand, showed no increase electromyographic activity with the ATH posture when oil was present. Although, this

same muscle's activity was 1.4 times greater on the average during pulls in the other three postures when oil was present.

Table 1. Relative levels of muscle activity across the four postures tested in Experiment 1 (A>B>C). Postures with the same letter are not significantly different with regard to peak EMG levels for the tested muscle.

	POSTURE			
MUSCLE	SSBOA	SSOLB	AOH	ATH
LATL	B	B	B	A
ERSR	C	C	B	A
ERSL	B	A	D	C
EXOR	C	C	A	B
EXOL	B	B	B	A

Table 2: The mean and standard errors of the mean for the eight muscles as a function of the presence of the oil on the flooring surface.

Muscle	No Oil	No Oil SE	Oil	Oil SE	%Change
LATR	10.5	0.9	21.1	1.8	101.0
LATL	13.5	1.0	31.2	2.2	131.1
ERSR	37.6	2.0	47.0	1.9	25.0
ERSL	25.5	2.1	34.6	2.1	35.7
EXOR	9.1	1.0	11.5	1.1	26.4
EXOL	5.8	0.3	9.8	0.8	69.0
RABR	3.1	0.2	4.9	0.7	58.1
RABL	4.9	0.4	7.4	0.8	51.0

There were no positive or negative changes in the EMG data from the eight trunk muscles due to the lifting belt. Likewise, the belt had no differential effect on the muscle recruitments, positive or negative, due to the four postures employed, or the presence of oil on the floor.

DISCUSSION

This study has shown that pulling strength is greatest in sagittally symmetric postures. In our paradigm the symmetric postures allow subjects to use their leg strength and their body weight to assist in force generation. The EMG results indicate that there are significant changes in the way pulling forces are generated when working on a slippery surface. The need for stability shifts the role of the trunk muscles from just maintaining the rigid link between the upper and lower extremities to that of force generation.

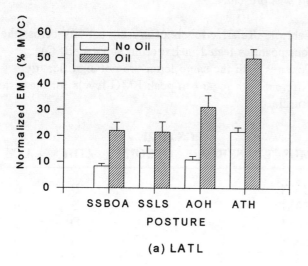

(a) LATL

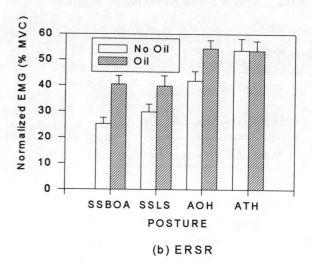

(b) ERSR

Figure 2. The mean response levels of the LATL (a) and the ERSR (b) as a function of the four postures tested and the presence of oil on the floor surface.

Associated with the pull force generation in the torso under slippery conditions is the uncertainty regarding the overall stability of the posture. When such uncertainty exist, the co-contraction of antagonistic muscles has been shown to occur (De Luca and Mambrito, 1987). It is interesting to note that the increase in the EMG sampled from the anterior muscles due to the oil was independent of the posture condition. Thus, even though the external oblique muscles were more heavily recruited in the asymmetric postures, this co-contraction did not preclude the need for additional spine stability from these muscles under slippery conditions.

The consequences of this co-contraction are most evident when the internal forces contributing to spine loading are considered. Chaffin et al. (1983) estimated that the spinal compression for their sagittally symmetric pulls with low handles were approximately 3600 N. However, this model may severely underestimate the compression due to the co-contraction. Take for example the simple SSBOA posture. Under optimal pulling conditions the back muscles just stiffen the torso to maintain the rigidity of the link (Dempster, 1958). Recognizing that the internal muscle forces are the primary contributor to spine compression, the poor footing and the associated need for postural stability increases the spine loading for three reasons. First, the posterior trunk muscles become responsible for generating the pull force as pulling techniques that rely on the body's mass and leg strength cannot be utilized to the same degree. Second, the high slip potential leads to the recruitment of the antagonistic anterior trunk muscles, again increasing the spine compression. And third, this

co-contraction then requires an even further increase in the posterior muscle recruitment as the net internal trunk moment must be sufficient to generate the desired external force. It was hypothesized that the lifting belt would assist in stiffening the trunk, possibly alleviating the need for the co-contraction response. If this had happened the net result would be less compression on the spine as the second and third reasons mentioned above would be reduced or eliminated. This, however, did not occur when our entire group of subjects was evaluated as a whole. But given the variability in human muscle recruitment strategies, further analysis is needed to identify whether our sample can be subdivided into those who utilized the belt for this purpose and those who did not.

REFERENCES

Andersson, G.B.J. (1991). The epidemiology of Spinal Disorders. In J.W. Frymoyer(ed.) The Adult Spine: Principles and Practice. (pp. 107-146). New York: Raven Press.

Chaffin, D.B., Andres, R.O., Garg, A. (1983). Volitional postures during maximal push/pull exertions in the sagittal plane. Human Factors, 25, 541-550.

De Luca, C.J. and Mambrito, B. (1987). Voluntary control of motor units in human antagonistic muscles: coactivation and reciprocal activation. J. Neurophysiology, 58, 525-542.

Dempster, W.T. (1958). Analysis of two-handed pulls using free body diagrams. J. Applied Physiology, 13, 469-480.

Lavender, S.A., Thomas, J.S., Chang, D., Andersson, G.B.J. (1995). Effect of lifting belts, foot movement, and lift asymmetry on trunk motions. Human Factors, 37, 844-853

McGill, S., Seguin, J., and Bennett, G. (1994). Passive stiffness of the lumbar torso in flexion, extension, lateral bending, and axial rotation: Effect of belt wearing and breath holding. Spine, 19, 696-704.

THE ROLE OF POWER IN PREDICTING LIFTING CAPACITY

Patrick G. Dempsey
Liberty Mutual Research Center for Safety and Health
71 Frankland Road
Hopkinton, Massachusetts 01748 USA

M.M. Ayoub
Department of Industrial Engineering
Texas Tech University
Lubbock, Texas 79409-3061 USA

Preplacement strength testing is one of the most viable supplements to ergonomic job design. Previous studies have examined the roles of isometric, isokinetic, and some isoinertial measures in predicting lifting capacity. The goal of the current study was to examine the role of maximal lifting power in predicting maximum acceptable weight of lift (MAWL) relative to previously used isometric, isokinetic, and isoinertial tests. Twenty five male subjects participated in an experiment involving two isometric tests, peak isokinetic strength at velocities between 0.1 and 0.8 m. $\cdot$ sec.$^{-1}$, and isoinertial lifting capacity and peak power measured on an incremental lifting machine. Peak isoinertial power was the measure most strongly correlated with MAWL, followed by isokinetic strength measured at 0.1 m. $\cdot$ sec.$^{-1}$. Overall, the results support previous studies which have shown that dynamic strength measures are superior to static strength measures for the purposes of predicting maximum acceptable weights. Results of regression analyses and measures of prediction accuracy reported further support the use of dynamic measures.

INTRODUCTION

Despite the tremendous amount of research that has been conducted in the area of preventing injuries associated with manual materials handling (MMH), these injuries continue to represent a significant burden to employees, employers, and society. While job design is the most powerful and effective control tool available, it has not alleviated the problem of MMH-related injuries. Given the ineffectiveness of preplacement radiography (e.g., Gibson, 1987) and the confusion surrounding the benefits of training and education (Kroemer, 1992), preplacement strength tests appear to be the most attractive supplement to job design techniques. Additionally, strength tests can be used to prioritize engineering controls, i.e., if a mismatch is found between job demands and worker capacity then the workplace can be altered to reduce job demands to a level consistent with the capacity of the workforce. This latter technique is more desirable than job selection, but is not always an option.

A conclusion of a panel of experts convened by the National Institute for Occupational Safety and Health (NIOSH) was that there is a need for research to "concentrate on improving those worker evaluation methods that are directly related to specific job requirements, as opposed to generic physical performance tests of strength, flexibility and endurance" (NIOSH, 1992). Thus, in order to develop a test for predicting safe lifting capacity, the test should be as close to the lifting task as possible with respect to postural constraints, kinematic constraints, kinetic properties and task geometry. Past research, as well as the results reported here, supports such congruency between the criterion task and predictor test.

Previous studies have indicated that dynamic strength is superior to static strength for the purposes of predicting maximum acceptable weight of lift (MAWL) (e.g., Aghazadeh and Ayoub, 1985; Ayoub et al., 1982, 1987; Duggan and Legg, 1993; Mital, 1985; Mital and Karwowski, 1985). There has been considerable comparison of the predictive abilities of dynamic tests to those of static tests, but there have been few comparisons amongst dynamic measures. Given that fact that lifting is an isoinertial activity, it follows that the closer a predictor test resembles the kinematic and kinetic nature of the lifting task, the more accurate the predictions will be.

Having a preplacement test that closely resembles the task the test was designed for is also important from the perspective of The Americans with Disabilities Act [Public Law 101-336]. Preplacement tests must be job-related, which means they must be a legitimate measure for a specific job. The test must also possess either criterion-related, content, or construct validity properties [See 29 Code of Federal Regulations Part 1607 - Uniform Guidelines on Employee Selection Procedures].

During the past several decades, power capabilities have been studied with considerable rigor by exercise scientists seeking to determine the sources of performance variation for a variety of athletic events (e.g., Coyle et al., 1979; Dowling and Vamos, 1993; Gregor et al., 1979; Margaria et al., 1966; McCartney et al., 1983; Santa Maria et al., 1985; Sewall and Lander, 1992). The rationale for the interest has primarily been driven by the fact that many athletic contests require short, but very powerful, bursts of work to be performed. Although industrial lifting is not a competitive event, there is a great deal of inter-individual variability in MMH-capabilities and many benefits can be achieved by gaining a better understanding of which attributes are responsible for this variation. Power has a great deal of intuitive appeal as a source of furthering this information. Power tests can be developed that are free from the limiting "iso" constraints that have been used previously. MMH activities, particularly lifting, require force exertions under conditions which are by no means isokinetic or isometric. Thus, the goals of this study were to investigate the role of maximal power in predicting lifting capacity and to compare the predictive ability of power to previously used isometric, isokinetic, and isoinertial measures.

METHODS

Subjects

Twenty five healthy male subjects volunteered to participate in the experiment, and were paid for their participation. Table 1 summarizes the characteristics of the subjects. Most subjects were university students, as indicated

Table 1. Anthropometric characteristics of subject sample.

Variable	Mean	SD	Minimum	Maximum
Age	21.64	2.46	18	26
Height (cm.)	179.63	5.81	170.2	195.6
Weight (kg.)	78.45	13.52	61.2	112.5

SD = standard deviation

by the fairly narrow range of ages.

Variable Selection

While the primary goal of the study was to investigate the role of power in predicting lifting capacity, the secondary goal was to compare power to previously used measures. For this reason, a fairly wide array of predictor tests was used.

Two isometric measures were chosen: maximum lifting strength at 15 cm. and 75 cm. Fifteen cm. represents strength at the origin of the lift for the psychophysical task involved, and this strength measure has been used to define capacity in previous research (e.g., Chaffin and Park, 1973). The reason for selecting 75 cm. is that the NIOSH equations (NIOSH, 1981; Waters et al., 1993) use a height of 75 cm. as the vertical dimension for the "standard lifting location."

The velocities chosen for the isokinetic strength tests (0.1, 0.2, 0.4, 0.6, and 0.8 m. • sec.$^{-1}$) represent a fairly wide range of velocities. A velocity of 1.0 m. • sec.$^{-1}$ was going to be used, but it was found that this velocity was too fast for the limited vertical range (0 - 95 cm.) of the isokinetic apparatus. The X-factor incremental lifting test was used due to its popularity in previous research (e.g., Ayoub et al., 1982, 1987; Kroemer 1983, 1985). Finally, a test of maximal power was devised. The X-factor machine was also used for this test because of its safety and ease of attaching transducers. The authors felt that the only way to measure maximum power was by having subjects lift a load as quickly as possible. The 25 kg. load used for the power test was the lightest load that could be used. Heavier loads were not selected for safety reasons.

The dependent measure chosen was MAWL at a frequency of one lift per eight hours. The low frequency was chosen because the predictor measures investigated are appropriate for low frequency tasks. Low frequency tasks are more taxing to the musculoskeletal system than are high frequency tasks, which are taxing to the cardiovascular

system. Thus, all measures were associated with strength and not cardiovascular capacity.

Apparatus

Isometric strength and isokinetic strength were measured using a Cybex® II isokinetic dynamometer, cables, and handles guided by linear bearings. The midpoints of the handles were 35.5 cm. apart, which is equal to the separation of the handles used for the psychophysical procedure.

An aluminum wheel was attached to the isokinetic dynamometer. By attaching a cable to the wheel, guiding the cable with pulleys, and attaching the cable to the handles, the isokinetic dynamometer was converted from rotary to linear (vertical) motion. A load cell was attached between the cable and the handles to measure forces.

A velocity transducer was affixed to the apparatus. The transducer was used so that a digital readout could be used to set the various velocities, rather than using the less accurate analog readout on the control box of the Cybex® II. The transducer was affixed to the apparatus and the signal was fed to the analog-to-digital (A/D) board, as were the signals from the load cell.

This apparatus was also used for collecting the isometric strength data. This was accomplished by placing the handles at the proper vertical height (15 or 75 cm.) and setting the velocity of the dynamometer to zero.

The isoinertial apparatus consisted of the X-factor incremental lifting machine fitted with a load cell to measure vertical forces applied to the handles and a velocity transducer. This apparatus was used for the incremental lifting test and the isoinertial power measurements. For the power measurements, velocity and force were sampled from the respective transducers at 100 Hz. The LabView® software program, which was used to control all data acquisition, converted all digital outputs to appropriate units (N [force], m. • sec.$^{-1}$ [velocity], or m. [displacement]). Linear regression parameter estimates from calibration procedures were used in the conversions. LabView® also was used to calculate power values (dot product of force and velocity). Since the X-factor machine only permits vertical motion, the power measurements represent vertical power.

The apparatus used for the psychophysical assessments consisted of a 30.5 cm. × 30.5 cm. × 30.5 cm. wooden box with handles, lead weights of various shapes and sizes, and a stationary 76 cm. high shelf.

Procedure

Subjects were given a basic questionnaire about various health issues and filled out a consent form. All subjects were examined by a physician prior to participation. The purpose of the examination was to ensure that subjects were free from any musculoskeletal disorders that would predispose them to injury.

The experimentation was divided into 4 sessions with at least 24 hours between sessions. Session 1 involved determination of MAWL for a frequency of one lift per eight hours. The starting load of the box was alternately light (2-18 kg.) or heavy (32-45 kg.) (Ciriello et al., 1990). The subjects were read instructions specific to a frequency of 1 lift per 8 hours (Ciriello and Snook, 1983), and performed two repetitions of the psychophysical procedure. If the two values were not within ±15% of each other, then the procedure was repeated on another day as was done by Snook and Ciriello (1991).

The remainder of the first session involved practice with the X-factor isoinertial lifting machine. Subjects lifted the 25 kg. carriage to a height of at least 183 cm. once per minute for 15 minutes. Subjects then practiced lifting the carriage as quickly as possible once per two minutes for 20 minutes. After a 10-minute rest period, subjects practiced the incremental lifting test. Subjects lifted the 25 kg. carriage to a height of 183 cm., then a 4.5 kg. weight was added and the subject lifted the weight to 183 cm. The experimenter continued to add 4.5 kg. increments until subjects could not lift the load to 183 cm.

During the second session, the isoinertial power measurements (ISOPOW) were taken. Subjects were instructed to lift the 25 kg. carriage as quickly as they safely could, and the score on the test was the peak power value during the exertion. Subjects performed three repetitions of this test, separated by five-minute rest periods. In some cases, a fourth trial was run. Some subjects appeared to need a "warm-up" trial, indicated by a relatively low value on the first trial. For these subjects, the fourth trial replaced the first. If one of the three trials appeared to be significantly lower than the other trials, another trial was run. Following a 10 minute rest period, subjects performed the X-factor incremental lifting test (XFAC). The score on the test was the largest load the subject could lift to at least 183 cm.

The third session involved practicing the isometric and isokinetic strength measurements. Subjects practiced the isometric measurements at 15 cm. (IMET15) and 75 cm. (IMET75) for three trials each separated by three-minute rest periods. Subjects also practiced the isokinetic measurements at speeds of 0.1 m. • sec.$^{-1}$ (IKIN0.1), 0.2 m. • sec.$^{-1}$ (IKIN0.2), 0.4 m. • sec.$^{-1}$ (IKIN0.4), 0.6 m. • sec.$^{-1}$ (IKIN0.6) and 0.8 m. • sec.$^{-1}$ (IKIN0.8) for three trials each separated by three-minute rest periods.

The fourth session involved collecting the isometric and isokinetic strength data. The isometric data were collected first. The order of collecting IMET15 and IMET75 was alternated. The experimenter had subjects build up to their maximum voluntary contraction (MVC) over a period of two to three seconds. Once subjects achieved their MVC, based on the analog torque meter on the Cybex®, the experimenter initiated the collection of 75 samples over three seconds. The strength datum used was the mean value of the 75 samples, as suggested by Caldwell et al. (1974). If any sample was outside ±10% of the mean value, the test was rerun (Caldwell et al., 1974). Two consecutive exertions within ±15% of each other were considered acceptable, and the mean of these two values was used as the datum for a given subject. Subjects were given five minutes rest between all exertions.

Following the isometric testing, subjects performed the isokinetic testing. The order of presentation of the five velocities was random. Due to a lack of instructions and protocols for isokinetic testing in the literature, subjects completed three repetitions at each velocity. Subjects were instructed to exert maximum force on the handles until the handles reached a height of 95 cm., at which point the apparatus restricted the handles from moving any further. The

peak force value during the trial was used as the datum for that trial. If one datum appeared to be considerably lower than the other two, the trial was repeated. There was a five-minute rest between all exertions.

Statistical Analyses

For the correlation and regression analyses, a single value for each predictor variable and the dependent measure was needed. Two or three trials were run for each type of test, so the multiple samples for each subject were averaged and used as the datum.

Pearson product-moment correlations were calculated between MAWL and each of the predictor variables. In order to assess the predictive ability of each independent measure, simple linear regression models were estimated for each predictor variable. The root mean square error (RMSE) and mean absolute residual (MAR) statistics were calculated for each simple regression model. These statistics are indicative of the accuracy associated with predicting MAWL with the various predictor variables.

RESULTS

The means and standard deviations of the various strength measures are presented in Table 2. The results show a clear effect of velocity on force capabilities, but also postural effects. The mean IMET15 value is lower than IKIN0.1 and IKIN0.2, whereas IMET75 is higher than all isokinetic measures. The 75 cm. height provides a distinct strength advantage over 15 cm., as expected from the literature.

Table 2. Summary statistics for dependent and predictor measures.

Variable	Mean	SD
IMET15 (N.)	802.87	190.45
IMET75 (N.)	990.29	198.42
IKIN0.1 (N.)	885.92	194.41
IKIN0.2 (N.)	843.66	201.30
IKIN0.4 (N.)	776.06	183.61
IKIN0.6 (N.)	723.24	183.68
IKIN0.8 (N.)	686.93	199.49
XFAC (kg.)	55.79	12.14
ISOPOW (W.)	1210.89	256.18
MAWL (kg.)	56.06	13.71

The correlations between MAWL and each of the predictor values are presented in Table 3. All of the correlations were significant. Table 4 provides performance measures for each of the models estimated, where MAWL was the dependent measure.

DISCUSSION

The goals of this study were to investigate if power is related to the ability to lift from the floor to knuckle height and to compare the predictive ability of power to previously investigated strength measures. The results indicate that power

Table 3. Correlations between MAWL and predictor variables. Significance levels are below the correlation coefficients.

Variable	MAWL
IMET15	0.62622
	0.0008
IMET75	0.61574
	0.0011
IKIN0.1	0.79042
	0.0001
IKIN0.2	0.62677
	0.0008
IKIN0.4	0.65531
	0.0004
IKIN0.6	0.67281
	0.0002
IKIN0.8	0.73549
	0.0001
XFAC	0.49835
	0.0112
ISOPOW	0.83849
	0.0001

Table 4. Performance measures for simple OLS regression models. MAWL is the response variable for each model.

Predictor Variable	RMSE (kg.)	MAR (kg.)
IMET15	10.92	8.24
IMET75	11.03	8.38
IKIN0.1	8.58	6.61
IKIN0.2	10.91	8.04
IKIN0.4	10.58	7.97
IKIN0.6	10.36	7.25
IKIN0.8	9.49	7.28
XFAC	12.14	8.58
ISOPOW	7.63	5.46

is significantly related to lifting capacity and that power, as investigated here, results in more accurate predictions of MAWL than the isometric, isokinetic, and isoinertial predictor variables investigated.

The performance measures in Table 4 provide an indication of the accuracy of low-frequency lifting capacity prediction using single strength tests. With the exception of XFAC, the trend is clearly that dynamic measures of strength outperform static measures. The reason for the poor performance of the XFAC test may be because the precision of the test is limited to 4.5 kg. A second reason may be that the vertical range of the test is 0 - 183 cm. versus 0 - 76 cm. for the MAWL determination. The isokinetic tests had a range of 0 - 95 cm. and the isometric tests were within the 0 - 76 cm. range. The disparity in vertical range between XFAC and MAWL, coupled with the fact that failure on the XFAC test usually occurs at waist height or higher, indicate that the 183 cm. XFAC test may be more suited for a floor-to-shoulder prediction.

IKIN0.1 performed almost as well as ISOPOW in terms of the correlations, but predictions of MAWL using ISOPOW were approximately one kg. more accurate in terms of the performance measures in Table 4. Many researchers have only reported r^2 values for similar studies. While this measure is important, prediction accuracy is also important. Prediction accuracy and the proportion of variation a model accounts for (r^2) are correlated, but the relationship is not perfect. Thus, both types of measures should be used to assess models of this type.

Earlier, issues of validity were mentioned. The results reported here would be useful for predicting an individual's capacity for low-frequency floor-to-knuckle lifting. Since the predictor tests are significantly correlated with the lifting task, these tests have criterion-related validity.

The research reported here further supports literature indicating that dynamic tests are better predictors of lifting capacity than are static measures. Combined, the results of the present study and previous research clearly indicate that isometric testing for predicting lifting capacity should be replaced with dynamic testing when practical and feasible. The XFAC test was the exception to the trend, and the reason for the poor correlation between XFAC and MAWL were discussed above. Also, the low correlation may simply be a sampling artifact, i.e. another sample of 25 subjects might produce a higher, or even lower, correlation.

One limitation of the results is that only one type of lifting task was investigated. Thus, the validity of the results is limited to these conditions. The results should not be extrapolated to lifting tasks with different ranges or frequencies. Future research will be needed to investigate the relationships for a wider array of task conditions. Additionally, the subjects were college-aged males and future research will be needed to investigate the relationships studied here among a more heterogeneous subject pool. Preferably, the study should be replicated using an industrial subject pool composed of males and females from a wider age range.

Many researchers have developed tests to predict the handling capacity of workers. Unfortunately, few of the tests have been validated epidemiologically. Examples of field validations of such techniques include the investigations by Chaffin and Park (1973) and Liles et al. (1984). Unfortunately, these techniques are based upon capacity definitions that use isometric strength. Certainly there is a need for more accurate techniques based upon dynamic strength measures.

Determining the value of preplacement strength testing with respect to reducing MMH-related injuries can be very beneficial. Given the ineffectiveness of education and training, and the enormous costs that continue to be associated with MMH, it is apparent that job design techniques are not sufficient.

Based on studies such as Liles et al. (1984), it is fair to assume that preplacement strength testing techniques can have a net positive economic benefit. However, determining the benefits associated with these techniques will require several steps. First, further research will be needed to define appropriate predictor variables and equations for numerous handling activities. These equations should utilize dynamic predictors. Second, the predictions will have to be tested with epidemiological studies. Typically, such studies have been performed by looking at the ratio of task demands to operator capacity. Statistically modeling the relationships between such ratios and the incidence and severity of associated injuries can be very beneficial for loss prevention purposes. These models would permit the prediction of the expected cost and incidence rate for a particular task demands to operator capacity ratio.

Although the strength tests described here have been primarily discussed within the context of preplacement screening, it should be noted that another effective use of such techniques can be prioritizing design changes and allocating resources, such as ergonomics personnel, etc. This aspect of strength testing is often overlooked. The advantage of such an approach is that no selection of personnel is required, which completely alleviates the potential for litigation associated with some personnel selection procedures. Additionally, the engineering changes will be permanent and will benefit all workers performing the job.

ACKNOWLEDGMENTS

This project was supported by grant R03 OH03335 from the National Institute for Occupational Safety and Health of the Centers for Disease Control and Prevention and was completed while the first author was at Texas Tech University.

REFERENCES

Aghazadeh, F., and Ayoub, M.M. (1985). A comparison of dynamic- and static-strength models for prediction of lifting capacity. *Ergonomics*, 28(10), 1409-1417.

Ayoub, M.M., Denardo, J.D., Smith, J.L., Bethea, N.J., Lambert, B.K., Alley, L.R., and Duran, B.S. (1982). *Establishing Physical Criteria for Assigning Personnel to Air Force Jobs*. Final Report, Contract No. F49620-79-006, Air Force Office of Scientific Research.

Ayoub, M.M., Jiang, B.C., Smith, J.L., Selan, J.L., and McDaniel, J.W. (1987). Establishing a Physical Criteria for Assigning Personnel to U.S. Air Force Jobs. *Am Ind Hyg Assoc J*, 48(5), 464-470.

Caldwell, L.S., Chaffin, D.B., Dukes-Dobos, F.N., Kroemer, K.H.E., Laubach, L.L., Snook, S.H., and Wasserman, D.E. (1974). A Proposed Standard Procedure for Static Muscle Strength Testing. *American Industrial Hygiene Association Journal*, 34(4), 201-206.

Chaffin, D.B., and Park, K.S. (1973). A longitudinal study of low-back pain as associated with occupational weight lifting factors. *American Industrial Hygiene Association Journal*, 34(12), 513-525.

Ciriello, V.M., and Snook, S.H. (1983). A Study of Size, Distance, Height, and Frequency Effects on Manual Handling Tasks. *Human Factors*, 25(5), 473-483.

Ciriello, V.M., Snook, S.H., Blick, A.C., and Wilkinson, P.L. (1990). The effects of task duration on psychophysically-determined maximum acceptable weights and forces. *Ergonomics*, 33(2), 187-200.

Coyle, E.F., Costill, D.L., and Lesmes, G.R. (1979). Leg extension power and muscle fiber composition. *Medicine and Science in Sports and Exercise*, 11, 12-15.

Dowling, J.J., and Vamos, L. (1993). Identification of Kinetic and Temporal Factors Related to Vertical Jump Performance. *Journal of Applied Biomechanics*, 9, 95-110.

Duggan, A., and Legg, S.J. (1993). Prediction of maximal isoinertial lift capacity in army recruits. In R. Nielsen and K. Jorgensen (eds.) *Advances in Industrial Ergonomics and Safety V*. London: Taylor and Francis.

Gibson, E.S. (1987). The Value of Preplacement Screening Radiography of the Low Back. *SPINE: State of the Art Reviews*, 2(1), 91-107.

Gregor, R.J., Edgerton, V.R., Perrine, J.J. and DeBus, C. (1979). Torque-velocity relationships and muscle fiber composition in elite female athletes. *Journal of Applied Physiology: Respiratory, Environmental and Exercise Physiology*, 47(2), 388-392.

Kroemer, K.H.E. (1983). An Isoinertial Technique to Assess Individual Lifting Capability. *Human Factors*, 25(5), 493-506.

Kroemer, K.H.E. (1985). Testing Individual Capability to Lift Material: Repeatability of a Dynamic Test Compared with Static Testing. *Journal of Safety Research*, 16, 1-7.

Kroemer, K.H.E. (1992). Personnel training for safer material handling. *Ergonomics*, 35(9), 1119-1134.

Liles, D.H., Deivanayagam, S., Ayoub, M.M., and Mahajan, P. (1984). A Job Severity Index for the Evaluation and Control of Lifting Injury. *Human Factors*, 26(6), 683-693.

Margaria, R., Aghemo, P., and Rovelli, E. (1966). Measurement of muscular (anaerobic) power in man. *Journal of Applied Physiology*, 21, 1661-1669.

McCartney, N., Heigenhauser, J.F., and Jones, N.L. (1983). Power output and fatigue of human muscle in maximal cycling exercise. *Journal of Applied Physiology: Respiratory, Environmental and Exercise Physiology*, 55(1), 218-224.

Mital, A. (1985). Use of anthropometry and dynamic strength in developing screening and placement procedures for workers. In H.J. Bullinger and H.J. Warnecke (eds.) *Toward the Factory of the Future*. Berlin: Springer-Verlag.

Mital, A., and Karwowski, W. (1985). Use of Simulated Job Dynamic Strengths (SJDS) in Screening Workers for Manual Lifting Tasks. In *Proceedings of the Human Factors Society 29th Annual Meeting*. Santa Monica, CA: Human Factors Society.

NIOSH (1981). *Work Practices Guide for Manual Lifting*. HEW(NIOSH) Report No. 81-122. Cincinnati, OH: Author.

NIOSH (1992). *A National Strategy for Occupational Musculoskeletal Injuries: Implementation Issues and Research Needs* DHHS(NIOSH) Pub. No. 93-101. Cincinnati, OH: Author.

Santa Maria, D.L., Grzybinski, P., Hatfield, B. (1985). Power as a function of load for a supine bench press exercise. *National Strength and Conditioning Association Journal*. 6(6), 58.

Sewall, L.P., and Lander, J.E. (1992). Biomechanical components of the vertical jump and an analogous task involving the upper body. *Journal of Human Movement Studies*, 23, 77-93.

Snook, S.H., and Ciriello, V.M. (1991). The design of manual handling tasks: revised tables of maximum acceptable weights and forces. *Ergonomics*, 34(9), 1197-1213.

Waters, T.R., Putz-Anderson, V., Garg, A., and Fine, L.J. (1993). Revised NIOSH equation for the design and evaluation of manual lifting tasks. *Ergonomics*, 36(7), 749-776.

MAXIMUM SAFE WEIGHT OF LIFT: A NEW PARADIGM
FOR SETTING DESIGN LIMITS IN MANUAL LIFTING TASKS
BASED ON THE PSYCHOPHYSICAL APPROACH

Waldemar Karwowski
Center for Industrial Ergonomics
Department of Industrial Engineering
University of Louisville
Louisville, Kentucky 40292, USA
email: w0karw03@ulkyvm.louisville.edu

The main objective of this study was to introduce a new concept for setting design limits in manual lifting tasks, i.e., the notion of the maximum safe weight of lift (MSWL), and to examine the effects of different sets of subject instructions for load selection on the amount of weight determined by male subjects in the psychophysical experiment. The results revealed that the MSWL values were significantly lower than the maximum acceptable weight of lift (MAWL) values. It was proposed that the classical approach to setting design limits in manual lifting tasks, based on the psychophysical method introduced by S. Snook, be modified to emphasize the subjects' safety and to reduce the risk of musculoskeletal injury.

BACKGROUND

The psychophysical approach to determining the maximum acceptable weight of lift (MAWL) has been extensively used for almost thirty years (Snook and Irvine, 1967; Ayoub et al, 1978; Snook 1985). This approach aims to quantify human lifting capacity based on subjective perception of exertion, under the assumption that *workers are able to determine with some accuracy the highest acceptable workload* (Gamberale, 1985). The experimental procedures first proposed by Snook and Irvine (1967) and again reported by Snook (1978), require the subjects *to imagine working on the incentive basis, as hard as they can (lifting as much as they can), without straining themselves or becoming unusually tired, weakened, overheated, or out of breath.* Typically, the subjects are given control of the weight (or force) handled, and are asked to adjust the weight of the box (force) over a period of thirty/forty minutes, up to the maximum level they are willing to accept for an eight hour shift at a given frequency of task repetition.

According to Gamberale et al (1987), the assumptions of the psychophysical approach to setting limits in manual lifting tasks are as follows: 1) the individual is able to rate the perceived effort in a lifting task, 2) he/she is able to produce an individually acceptable level of performance on this task, and 3) this level of performance will be safe from manual handling injuries. Unfortunately, validity of these assumptions has never been fully examined with the required scientific scrutiny that

would warrant their wide acceptance in the research community.

SUBJECT INSTRUCTIONS IN DETERMINATION OF MAWL

The original procedure proposed by Snook and Irvine (1967) for determining what is assumed to be the "maximum acceptable weight of lift," focuses on work productivity, based on proportionate pay incentive scheme. Furthermore, the instructions do not explicitly refer to the worker's safety or injury avoidance. With only a few exceptions discussed below, all studies in the field of manual materials handing conducted, since the original study by Snook and Irvine (1967), followed the same assumed premise of self-selected load acceptability. It should be noted, however, that the subjects themselves are never asked to determine the "maximum acceptable weight of lift" or to evaluate the degree of acceptability of the selected weights with respect to lifting the entire working day. The subjects are only asked to adjust the weight of the box to the level they believe they would be willing to lift over an eight hour work day. The resulting weights selected by the subjects were then assumed by Snook and Irvine (1967) to represent the "maximum acceptable weights of lift."

Gamberale et al (1987) pointed out that the individual choice of workload is governed not only by sensory inputs (from tendons, muscles and cardiovascular system), but also the subject's assessment with regard to the risk of musculoskeletal injuries or fatigue leading to accidents. Therefore, it was concluded that assessment of acceptable workload performed using the psychophysical method would rely more on the subjects' cognitive judgement rather than on their immediate perception of exertion. These important issues, however, have not been explored in the subject literature.

OBJECTIVES

The classical psychophysical approach to setting limits in manual lifting tasks (Snook, 1978) implies that the workload or force established using this method represents a balance between health, satisfaction, and performance (Straker, 1994). According to Gamberale and Kilböm (1988), this approach also assumes that a worker is able to estimate with sufficient accuracy the maximum tolerable workload, and that the acceptable workload selected in a simulated task is "safe." However, the validity of such assumptions is to some extent unknown (Karwowski, 1989; Chaffin and Page, 1994; Straker, 1994). To the author's knowledge, no published study has examined validity of the assumption that the weights selected under Snook and Irvine's (1967) original instructions, and called the MAWL (Snook, 1978), would also be perceived by the same subjects as 'safe.' Except for a rather vague warning against overstraining, there are no references as to worker safety and avoidance of back pain and musculoskeletal injury. Therefore, in this study, a new set of instructions to determine the maximum safe weights of lift (MSWL) proposed by Karwowski (1995) was used.

The main objective of this study was to examine the effects of different sets of subject instructions for load selection (called here the MAWL instructions and the MSWL instructions) on the amount of weight determined by the subjects in the psychophysical experiment. The secondary objective of the study was to investigate human perception of load heaviness in terms of linguistic variables (natural expressions), for manual lifting tasks performed under the same experimental conditions and subjects' instructions for load selection.

METHODS AND PROCEDURES

Subjects

Ten male volunteer subjects were selected from the student population at the University of Louisville. The students were compensated for their participation in the study. Each subject filled out an Informed Consent Form, and only those subjects with no history of low back pain or musculoskeletal

disorders were allowed to participate. The subjects were asked to wear comfortable clothes, such as athletic wear, to stretch before the experiment, and refrain from any strenuous activities prior to the experimental sessions.

Experimental Design

A randomized complete block design, with subjects as blocks, was used in both experiments. Subjects were asked to determine the maximum acceptable weight of lift (MAWL) and the maximum safe weight of lift (MSWL) based on the instructions provided in the Appendix. The order of experimental sessions was chosen randomly for each of the subjects. In order to assure uniformity in the presentation of the instructions, the instructions were audiotaped and played to each subject before each experimental session. Only one session per subject was conducted on any given day.

Snook's (1978) proposed instructions for determining the MAWL were utilized. In summary, the subjects were asked to imagine working under an incentive pay scale, getting paid for the work completed, and working as hard as they could without straining themselves or becoming tired, weakened, overheated or out of breath.

The instruction set for the MSWL (for exact wording see Appendix) asked the subjects to imagine working safely over an eight hour shift, and to determine the weight they could safely lift without increasing the risk of low back pain or muscular overexertion, as proposed by Karwowski (1995). Both sets of instructions used the same wording with respect to encouraging subjects to make as many adjustments as necessary, and to caution them against perception of being in any kind of competition with others while performing their own lifting tasks.

Subjects Physical Characteristics

The anthropometric measurements were taken according to NASA (1978) procedures. For each subject, the body weight, height, shoulder height, hip height, knee height, and arm length, were measured. Four static strengths were determined for each subject. These measurements are: static arm strength, static back strength, static composite strength, and static shoulder strength, as recommended by Chaffin (1975). Two dynamic strengths, i.e. dynamic lift strength and dynamic back extension strength, were also measured for each subject. The procedures proposed by Pytel and Kamon (1981) were used for this purpose.

Procedures

Each subject was asked to repeatedly lift identical boxes with handles on the sides, from the floor and place it on the table (69 cm high) during a period of 30 minutes. The box dimensions were 33 x 33 x 26 cm, and each box had a false bottom to minimize visual cues. An electronic timer controlled the lifting frequency of 4 lifts per min (lpm). At the start and end of each experimental session, the box was confidentially weighed by the experimenters. Standard procedures for the weight adjustment process as proposed by Snook (1978) were adopted.

After selecting the preferred load (either MAWL or MSWL), the subject was asked to rank his perception(s) of both the weight just selected and the task performed as follows. The subject selected a rating of perceived exertion (RPE) during the session using the Borg scale (1982). The subject was also asked to choose one value which, in his/her opinion, would best describe heaviness of the selected weight, using one of the eight linguistic values, i.e. *very light, light, more-or-less-medium, medium, heavy, very heavy,* and *extremely heavy*. A 10 cm bipolar scale (marked with *uncertain*, on one end, and *positive* on the other end), was used to determine the subject's confidence in his choice of the linguistic category associated with the selected weight concept (MAWL or MSWL).

In addition, using a series of similar bipolar, 10 cm long scales, the subject was asked to provide the following estimates: 1) the degree to which the selected weight was perceived as acceptable for an 8-hr work-day, 2) the degree to which the selected task was perceived as acceptable for an 8-hr work-day (using the end markers named *absolutely unacceptable* and *fully acceptable*), 3) the degree to which the selected weight was perceived as safe for an 8-hr work-day, and 4) the degree to which the selected task was perceived as safe for an 8-hr work-day (using the end markers named *absolutely unsafe* and *fully safe*).

RESULTS AND DISCUSSION

The Statistical Analysis Systems (SAS) software was used to analyze the experimental data. A summary of results is given in Table 1.

Main Effects in Determination of the Preferred Weight

The analysis of variance (ANOVA) showed significant effects of subject instructions on the preferred weights. The subjects selected less weight when lifting according to the MSWL instruction versus the MAWL instruction (see Table 1). The average MSWL value was 38.3 pounds (17.36 kg), while the average value of MAWL was 46.02 pounds (20.87 kg). This 16.82% average difference (7.74 pounds or 3.51 kg) was found statistically significant (Tukey test) at p < 0.0001 level.

RPE analysis

The effect of instructions for selection of preferred weight of lift (MAWL or MSWL) on the rate of perceived exertion (RPE) was significant at p<0.001 level. Subjects reported higher average RPE values associated with the MAWL values (mean of 15.7), than those associated with the MSWL values (14.7), indicating that they elected to work at lower levels of perceived physical exertion when instructed to pay attention to their own safety.

It is important to note that the RPE values weakly correlated with the MAWL values (Pearson coefficient r = 0.51, p < 0.02), but did not correlate with the MSWL values (r = 0.22, p > 0.1), indicating that when selecting the maximum safe weight of lift, the subjects took into consideration more than just perception of their physical exertion, as they likely did when selecting the maximum acceptable weights of lift. Similar observations with respect to the lack of strong correlation between the RPE values and the preferred weights selected by the subjects during the psychophysical lifting experiments were made by Ljungberg et al (1982) and Gamberale et al (1988).

Table 1. Main effects for dependent variables.

Variable	Mean	S.D.	Range	N
Preferred Weight				
MAWL	46.02	12.12	23-74	20
MSWL	38.28	8.39	18.5-55.7	20
RPE				
MAWL	15.8	0.71	15-17	20
MSWL	14.7	0.71	14-16	20

Estimation of Load Lifted

The effect of instructions on the subjects' estimation of how much weight (ESW) they selected at the end of each trial was statistically significant at p < 0.02 level. On average, the subjects estimated to have selected 51.01 lbs while asked to determine the MAWL values, and only 40.1 lbs while asked to determine the MSWL values.

The Effect of Instructions on Perception of Load Heaviness

Of the forty trials performed in this study, 67% of the preferred loads were independently judged by the subjects as heavy. However, under the MAWL instruction set, 75% of the time the psychophysically determined loads were classified as heavy. Under the MSWL instructions, only 65% of the selected loads were judged to be heavy.

The Safety/Acceptability Scores

Upon completion of each experimental session, the subjects were asked to indicate their degree of belief using a 10 cm bi-polar scale (marked not safe & not acceptable, and absolutely safe & acceptable on each end, respectively), regarding how safe, and, therefore, how acceptable the selected MAWL and MSWL values were. On average, the preferred MAWL values were judged by the subjects to be much less safe and acceptable (p < 0.01) for continuous lifting over an 8-hr period (mean score of 5.6), than the MSWL values (mean score of 6.3).

CONCLUSIONS

Despite relative simplicity of the psychophysical method to determine 'acceptable' limits for manual lifting, which makes this approach popular, caution should be exercised with respect to interpretation and usability of the currently available design limits which were developed based on the psychophysical approach. It seems plausible that with the MAWL instructions, the subjects are mainly focusing on task performance and efficiency. On the other hand, the main focus of attention while lifting under the MSWL instructions is subject's safety. The results of this study revealed that the MSWL values were significantly lower than the MAWL values. Therefore, the classical approach to setting design limits in manual lifting tasks based on the psychophysical method should be modified, as proposed by Karwowski (1995), in order to emphasize the subjects' safety and to reduce the risk of musculoskeletal injury.

REFERENCES

Ayoub, M., Bethea, N. J., Deivanayagam, S., Asfour, S., Bakken, G. M., Lilies, D., Mital, A., and Serif, M., 1978, *Determination and Modeling of Lifting Capacity*. NIOSH Report, Grant No. 5-r01-OH-000545-02, Cincinnati, Ohio.

Borg, G. A. V., 1982. Psychophysical bases of perceived exertion. *Medicine and Science in Sports and Exercise*, 4(5), 377-381.

Gamberale, F., 1985. The perception of exertion, *Ergonomics*, 28, 299-308.

Gamberale, F. and Kilböm, A., 1988. An experimental evaluation of psychophysically determined maximum acceptable workload for repetitive lifting. In *Proceedings of the 10th Congress of the International Ergonomics Association*, A. S. Adams, R. R. Hall, B. J. McPhee, and M. S. Oxenburgh, Eds., Taylor and Francis, London, pp. 233-235.

Gamberale, F., Ljungberg, A. S., Annwall, A., and Kilbom, A., 1987. An experimental evaluation of psychophysical criteria for repetitive lifting work. *Applied Ergonomics*, 18.4, 311-321.

Karwowski, W., 1995. *The Maximum Safe Weight of Lift: A New Design Paradigm for Setting Limits in Manual Lifting Tasks*. Unpublished Technical Report, Center for Industrial Ergonomics, University of Louisville, Louisville, Kentucky, 40292, USA.

NASA/Webb (Eds), 1978. *Anthropometric Sourcebook* (3 volumes). (NASA Reference Publication 1024). Houston, TX, LBJ Space Center.

Pytel, J. L., and Kamon, E., 1981. Dynamic Strength Test as a Predictor for Maximal and Acceptable lifting. *Ergonomics*, 24, 663-672.

Snook, S. H., 1978. The Design of Manual Handling Tasks. *Ergonomics*, 21, 963-985.

Snook, S. H., 1985. Psychophysical acceptability as a constraint in manual working capacity. *Ergonomics*, 28, 331-335.

Snook, S. H., and Irvine, C. H., 1967. Maximum Acceptable Weight of Lift. *American Industrial Hygiene Association Journal*, 28, 322-329.

APPENDIX

Instructions for determination of the Maximum Safe Weight of lift (MSWL) as proposed by Karwowski (1995):

We want you to imagine that you are working over a normal 8 hour shift, and we ask you to determine the maximum level of load that you feel you could SAFELY lift every day without hurting yourself, i.e. without the risk of experiencing any low back pain, or possibly overexerting yourself.

YOU WILL ADJUST YOUR OWN WORK LOAD. You will work only when the timer beeps. Your job will be to adjust the load; that is, to adjust the weight of the box that you are lifting WITH RESPECT TO YOUR OWN PERCEPTION OF HOW SAFE IT IS FOR YOU.

Adjusting your own work load is not an easy task, only you know how safely you feel.

IF YOU FEEL THE LOAD IS TOO HEAVY, AND THEREFORE NOT SAFE, or that lifting of this load could result in low back pain and/or muscular overexertion, you should reduce the load by removing weight from the box.

HOWEVER, IF YOU FEEL THAT YOU CAN WORK HARDER, WITHOUT INCREASING THE RISK OF LOW BACK PAIN OR OVEREXERTION, you should put in more weight into the box.

DON'T BE AFRAID TO MAKE ADJUSTMENTS. You have to make enough adjustments so that you get a good feeling for what is SAFE and what is NOT SAFE for you. You can never make too many adjustments -- but you can make too few.

REMEMBER, THIS IS NOT A CONTEST. EVERYONE IS NOT EXPECTED TO DO THE SAME AMOUNT OF WORK.

WE WANT YOUR JUDGMENT ON HOW HARD YOU CAN WORK SAFELY, I.E., WITHOUT EXPOSING YOURSELF TO THE RISK OF LOW BACK PAIN AND/OR MUSCULAR OVEREXERTION WHEN PERFORMING THIS TASK OVER AN 8-HOUR PERIOD.

THE EFFECT OF LIFT HEIGHT ON MAXIMAL LIFTING CAPABILITIES OF MEN AND WOMEN

Valerie J. Rice, Michelle Murphy, Marilyn A. Sharp,
Randall K. Bills, Robert P. Mello
Occupational Health & Performance Directorate
U.S. Army Research Institute of Environmental Medicine
Natick, MA 01760-5007
phone: 508-233-4850 fax: 508-233-4195

The purpose of this study was to determine the effect of lift height on the one repetition maximum box lifting strength (1 RM) of men and women. Ten men and eleven women lifted to heights of 1, 2, 3, 4, 5, and 6 ft using a 46.5 cm long x 31 cm wide x 23 cm high aluminum box with handles. All lifts were accomplished using correct lifting technique while facing forward. An ANOVA, focused on six heights and blocked for gender was used for analysis. A Newman-Kuels post-hoc analysis was used to examine the significant differences between means. Results revealed that men lifted more than women overall (F = 128.9, p < 0.01), and at each individual height (p < 0.01). When the genders were combined, the two highest lifts (5 and 6 ft) were not different from each other, nor were the two lowest (1 and 2 ft). All other heights differed from one another (p < 0.01). For both genders, the greatest percentage decrease from one height to the next occurred between 3 and 4 feet, when more upper body strength and torso involvement were required.

INTRODUCTION

Knowledge of maximal lifting capacities is important when a reduction in object mass is not possible, mechanical aids can not be used, or in an emergency situation which requires immediate individual action. Such information is also necessary to estimate the percentage of soldiers entering the Army who may potentially qualify for a given military occupational specialty requiring heavy lifting and to estimate manpower needs during conflict. In addition, these data guide contractors in the design and packaging of equipment.

Military and civilian manual material handling tasks require men and women to lift loads to various heights. Prior lifting research has primarily focused on identifying acceptable lifting limits to reduce risks to the musculoskeletal system, as opposed to identifying maximal lifting capabilities. In addition, evaluations of the effect of lift heights on lifting capabilities have used relative heights based on the individual volunteer's anthropometric dimensions, rather than absolute heights (Snook and Ciriello, 1991).

Although Emanuel, Chaffee, and Wing (1956) identified maximal lifting capabilities from the floor to a five foot level (in one foot increments) for men, these capabilities have not been identified for women. Emanuel and his co-investigators had subjects stand beside a custom built staircase, face the rear of the staircase, lift an F-86H Ammunition case, place the end of the case on the designated step and slide it sideways onto the platform. As a result, it appears the lifts required some twisting motion. The objective of the present study was to quantify maximum lifting capabilities of men and women to various heights using correct lifting technique and facing forward.

METHODS

Participants included 10 male and 11 female active duty soldiers. All subjects were medically screened and signed an informed consent form

Table 1. Physical characteristics of subjects.

	Men (mean ± SD)	Women (mean ± SD)	Percent Difference
n	10	11	
age (yr)	21.4 ± 3.5	22.0 ± 4.7	
height (cm)	178.1 ± 4.7	162.1 ± 4.0*	
weight (kg)	81.6 ± 7.1	59.3 ± 7.5*	
body fat (%)	17.8 ± 5.4	27.6 ± 5.8*	
fat-free mass (kg)	66.3 ± 4.3	42.2 ± 3.8*	
dead lift (kg)	132.5 ± 20.5	64.4 ± 17.2*	49%
38cm uprt pull (kg)	153.4 ± 28.6	84.4 ± 8.0*	55%
bench press (kg)	92.3 ± 19.9	33.4 ± 5.0*	36%
hand grip (kg)	56.8 ± 8.1	29.2 ± 3.8*	51%
IDL (kg)	75.6 ± 8.5	42.5 ± 12.3*	56%

* Significantly different from men ($p < 0.01$).

following a detailed briefing. The body composition of volunteers was determined by dual energy x-ray absorptiometry (Mazess, Barden, Bisek, and Hanson, 1990). The maximal strength measures included a dead lift, 38 cm upright pull (uprt pull), incremental dynamic lift (IDL) (Sharp, Rice, Nindl, and Williamson, 1993), dynamic bench press, and isometric hand grip. Table 1 contains the physical characteristics and strength measures of the subjects. The women were shorter, lighter, had a higher percent body fat, and lower quantity of fat free mass than men ($p < 0.01$). Women's strength capabilities were lower than men's on all measures ($p < 0.01$).

All box lifts were performed using an adjustable shelf, which allowed subjects to face forward when lifting (Teves, McGrath, Knapik, and Legg, 1986). A 6.1 kg aluminum box with handles was used. The box was 46.5 cm long x 31 cm wide x 23 cm high. The warm-up consisted of one set of three lifts at 30% or less of the volunteer's predicted 1 repetition maximum (1 RM), followed by a second set of three lifts at less than 50% of their predicted 1 RM (Semenick, 1994). Weight was added

according to each volunteer's subjective assessment of his or her ability and was generally in 1-10 kg increments. After a failed attempt, weight was removed to yield an intermediate load to assess 1 RM as accurately as possible (to the nearest 1.0 kg). Maximum load was reached when the subject judged the weight as too heavy, could not physically complete the lift, or could not maintain a safe lifting technique. A minimum of three minutes rest was given after each attempt. The investigator monitored the weight added so that proper procedures and appropriate increments were used.

Volunteers lifted the box from the floor to heights of 1 to 6 feet, in one foot increments. No more than two lifts were performed in one day. If two lifts were performed during the same day, a minimum of 3 hours rest was given between lifts.

An ANOVA, focused on six heights and blocked for gender was performed. A Newman-Kuels post-hoc analysis examined the significant differences between means. An independent t-test was used to determine differences in strength and anthropometric measures between men and women.

RESULTS

A significant height effect was observed (F = 93.2, p < 0.01). Figure 1 shows the loads lifted to each height by men and women. Loads lifted to the lowest two heights (1 and 2 ft) and loads lifted to the two highest heights (5 and 6 ft) were not significantly different from each other (p > 0.05). All other heights were significantly different from each other (p < 0.01).

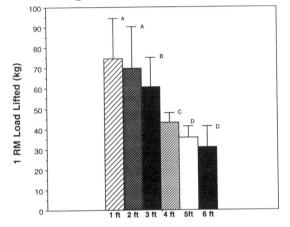

Bars with different heights are different from each other (p < 0.01).

Figure 1. Main Effect for Height

Overall, men lifted more than women (men = 72.5 kg, women = 34.2 kg, F = 128.9, p < 0.01), and they lifted more than women at each height (p < 0.01). Figure 2 shows a significant gender x height interaction (F = 17.2, p < 0.01).

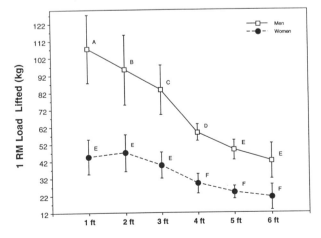

Different letters indicate significant differences (p < 0.01)

Figure 2. Height by Gender Interaction.

Men showed a steady decline in load lifted with an increase in lifting height with the exception of lift heights 5 and 6 feet. Women did not demonstrate this same pattern. Instead, they lifted similar loads for lift heights 1 to 3 feet and a lower load to lift heights 4 to 6 feet, with no steady decline across lift heights. The load lifted by women at heights up to 3 feet was not significantly different from that lifted by men to 5 or 6 feet.

Table 1 shows the female-to-male percent differences (women/men*100) for descriptive strength measures. The female-to-male percent differences for each 1 - 6 foot lift were 42%, 49%, 48%, 49%, 49%, and 50% respectively. That is, at a height of 1 foot, women lifted 42% of the load that men lifted. Table 2 shows the percent decrease ((higher lift-lower lift/lower lift)*100) in weight lifted with each increment in height for men and women. Except for the 1 foot height, the percent decrease at each height was similar for men and women. The greatest percent decrease occurred between the 3 and 4 foot lifts for both genders (men = 29.7%, women = 27.1%).

Table 2. Percent change in lift with each increase in height .

Height	Combined Gender	Men	Women
1-2 ft	- 6.2	-11.4	+ 5.0
2-3 ft	-13.4	-12.4	-15.3
3-4 ft	-28.8	-29.7	-27.1
4-5 ft	-17.3	-17.3	-17.3
5-6 ft	-12.9	-13.4	-12.1

Another perspective for evaluating the effect of lift height is to examine the percent difference in load lifted to the lowest height with the load lifted at each of the higher heights. When the genders were combined, the greatest load was lifted to the height of 1 foot. At 2 feet, volunteers lifted 6.2% less than they lifted to 1 foot. When lifting to 3 feet, volunteers lifted 18.8% less than they lifted to 1 foot. At 4 feet, they lifted 42.2% less, at 5 feet -

52.2% less, and at 6 feet - 58.4% less than the load they lifted to 1 foot.

DISCUSSION

This study demonstrated the maximal capabilities of men and women lifting to heights of 1, 2, 3, 4, 5, and 6 feet. When both genders were combined, no differences were seen between loads lifted to the two lowest heights (1 and 2 feet) or to the two highest heights (5 and 6 feet). This indicates that when designing tasks, weight limits for these heights could be combined. It is suggested that the lower limit capabilities be used, ie. the 2 foot load for the lower height and the 6 foot load for the higher height.

Men lifted more than women at each height. Previous studies that involved maximal box lifts from the floor to shoulder height demonstrated that women were able to lift 60% that of men (Beckett and Hodgdon, 1987; Myers, Gebhardt, Crump, et al., 1984; Teves, Wright, and Vogel, 1985). This is considerably more than the strength differences for box lifts (42-50%) seen in this study. The isometric strength of female soldiers has been reported to be more comparable to male soldiers for lower-body strength (67%) than for upper-body strength (60%)(Knapik, Wright, Kowal, et al., 1980; Sharp, 1994). These female-to-male percentages are greater than those found in this study (lower body 49-55% and upper body 36-51%). It appears the upper-body strength was less for women and slightly greater for men in this study than seen in previous studies. Because of these issues, we might expect the gender difference in weight lifted to be exaggerated at higher lift heights. However the percent differences were approximately the same (42-50%) throughout the range of heights.

The greatest difference in both absolute weight lifted and percentage of weight change between each height occurred between the 3 and 4 foot heights for both men and women. Men lifted progressively less as the height increased, except for the final two heights of 5 and 6 feet. Women on the other hand, lifted similar weights to 1, 2, and 3 foot

heights, and to 4, 5, and 6 foot heights. However, the percent change for each height was similar for men and women. The greatest change in upper body strength requirements, torso involvement, and upper extremity biomechanics occurs between 3 and 4 foot lifting heights.

Table 3 shows the maximum weight lifted to each height for male subjects from this study and for the research conducted by Emanuel, et al. (1956). The results were similar at the two lower heights; however, our subjects lifted more at the 3 - 5 foot heights. This may be the result of the method of lifting, placing, and sliding the box onto the shelf in the study by Emanuel, et al. (1956). Although these authors state that their subjects held their backs straight and faced straight ahead, it would be necessary to turn one's head slightly to see the stair on which they were to place the box. Even this slight movement could involve some twisting of their torso and have influenced their maximum loads. In addition, the box size and configuration differed from ours. They used an F-86H aluminum ammunition case 64.8 cm long x 32.8 cm wide (box height was not noted), with handles. The longer box may have required the volunteers to adduct their shoulders and extend their elbows, thus putting their upper extremities at a greater biomechanical disadvantage during higher lifts.

Table 3. Comparison of maximum lifting capabilities (kg) of men from this study compared with Emanuel, Chaffee, and Wing, 1956.

Height	Present study (mean ± SD)	Emanuel, et al. (mean ± SD)
1 ft	107.4 ± 19.9	104.9 ± 21.3
2 ft	95.2 ± 20.3	87.6 ± 18.2
3 ft	83.4 ± 14.5	54.0 ± 14.1
4 ft	58.6 ± 4.8	36.8 ± 8.6
5 ft	48.5 ± 5.7	26.3 ± 7.3
6 ft	42.0 ± 10.2	not tested

CONCLUSION

Maximal lifting capabilities for men and women at heights of 1, 2, 3, 4, 5, and 6 feet were identified in this study. The maximal lifting capabilities for men exceeded those reported previously (Emanuel, et al., 1956). Since there were no significant differences in the load lifted at the two lowest heights and the two highest heights when genders were combined, it is possible that the weight limits for these heights could be combined (based on the lower value for each). The results further demonstrate that the greatest changes in individual lifting ability occur as the lift approaches and exceeds waist height.

The views, opinions and/or findings contained in this article are those of the author and should not be construed as an official Department of the Army position, policy, or decision. The investigators adhered to the policies regarding the protection of human subjects as prescribed by 45 CFR 46 and 32 CFR 219 (Protection of Human Subjects).

REFERENCES

Beckett, M.B., and Hodgdon, J.A. (1987). Lifting and carrying capacities relative to physical fitness measures (Technical Report No. 87-26). San Diego, CA: Naval Health Research Center.

Emanuel, I., Chaffee, J.W., and Wing, J. (1956). A study of human weight lifting capabilities for loading ammunition into the F-86H Aircraft. (Technical Report No. 56-367). Dayton, Ohio: Aero Medical Laboratory, Wright-Patterson Air Force Base.

Knapik, J.J., Wright, J.E., Kowal, D.M., Vogel, J.A. (1980). The influence of U.S. Army basic initial entry training on the muscular strength of men and women. Aviation, Space, and Environmental Medicine, 51, 1086-1090.

Mazess, R.B., Barden, H.S., Bisek, J.P., and Hanson, J. (1990). Dual-Energy X-Ray Absorptiometry for total-body and regional bone-mineral and soft-tissue composition. The American Journal of Clinical Nutrition, 51, 1106.

Myers, D.C., Gebhardt, D.L., Crump, C.E., and Fleishman, E.A. (1984). Validation of the military entrance physical strength capacity test. (Technical Report No. 610). Alexandria, VA: US Army Research Institute for the Behavioral and Social Sciences.

Semenick, D.M. (1994). Testing protocols and procedures. In T. Baechle (Ed.), Essentials of strength training and conditioning (pp 258-273). Champaign, IL: Human Kinetics.

Sharp, M.A. 1994. Physical fitness and occupational performance of women in the U.S. Army. Work 4(2):80-92.

Sharp, M.A., Rice, V.J., Nindl, B.C., and Williamson, TL. (1993). Maximum team lifting capacity as a function of team size. (Technical Report No. T94-3). Natick, MA: US Army Research Institute of Environmental Medicine.

Snook, S.H., and Ciriello, V.M. (1991). The design of manual handling tasks: Revised tables of maximum acceptable weights and forces. Ergonomics, 34, 1197-1213.

Teves, M.A., McGrath, J.M., Knapik, J.J., and Legg, S.J. (1986). An ergometer for maximal effort lifting. Proceedings of the Eighth Annual Conference of the IEEE/Engineering in Medicine and Biology Society.

Teves, M.A., Wright, J.E., and Vogel, J.A. (1985). Performance on selected candidate screening test procedures before and after Army basic and advanced individual training (Technical Report No. T13/85). Natick, MA: US Army Research Institute of Environmental Medicine.

EFFECTS OF A CHAIR-MOUNTED SPLIT KEYBOARD ON PERFORMANCE, POSTURE AND COMFORT.

Alan Hedge, Ph.D. and Gregory Shaw

Dept. Design & Environmental Analysis

Cornell University, Ithaca, NY 14853-4401

In a repeated measures experiment, twelve female typists performed typing and cursor positioning tasks using a chair-mounted split keyboard (FAK) and a conventional keyboard (CK) on an articulated tray. Results showed that the FAK significantly reduced ulnar deviation, but did not reduce wrist extension compared with the CK arrangement. Typing speed was slower for the FAK than the CK, but accuracy was unaffected. Videomotion analysis showed that more subjects sat fully back in their chair with their shoulders relaxed and their hands in less ulnar deviation when using the FAK to type. Other postural changes are described. Subjective comfort and strain ratings are described. Results are discussed within the limitations of a short-term laboratory study.

INTRODUCTION

Split keyboard designs supposedly improve the user's upper body posture and reduce ulnar deviation during typing (Kroemer, 1972; Nakaseko *et al.*, 1985). Studies testing either fixed angle or adjustable angle split keyboards with conventional office furniture have not confirmed substantial postural benefits (Chen *et al.*, 1994; Hedge and Ng, 1995; Honan *et al.*, 1995). The Kinesis keyboard, which laterally separates the alphanumeric keys into left and right hand keypads does decrease ulnar deviation, although short-term typing performance also decreases (Chen *et al.*, 1994). Wrist rests may not improve keyboarding posture and may create painful pressure on the underside of the forearm and wrist (Parsons, 1991; Paul and Menon, 1994). They do not decrease forearm and shoulder muscle strain (Fernström *et al.*, 1994). A flat keyboard tray, with or without a wrist rest, does not necessarily improve wrist posture (Hedge *et al.*, 1995). Tilting a keyboard tray away from the user reduces wrist extension (Hedge and Powers, 1995), and improves upper body posture (Rudakewych *et al.*, 1994; Hedge *et al.*, 1995).

The chair mounted split keyboard system ('Floating Arms' keyboard - FAK) consists of a conventional keyboard layout that is split into two separate sections, mounted to the arms of an office chair (Figure 1). Each section is fully articulated and adjusts laterally, vertically, and twists and/or tilts into almost any position. A trackball or trackpad is also part of the right hand section of this keyboard. The FAK system provides a unique opportunity for investigation. Mounting the keyboard onto a chair is a novel design that supposedly improves upper body posture by encouraging the user to sit back in the chair with their arms supported by arm rests. If postural improvements are confirmed, the FAK will provide another alternative to conventional keyboards for those at high risk for musculoskeletal injury. This study was performed to compare postural differences between using the FAK and an IBM compatible keyboard mounted on a conventional articulating tray.

METHODS

Subjects

Twelve female subjects, who were proficient typists, were studied. All did jobs requiring substantial daily typing. All were free from any symptoms or disorders relating to keyboard use, and were free from any injury which would affect keyboard use. They were paid $50 for participating in the study. Seven subjects were 35 or over, and 5 were under 35 years old.

Equipment

The following equipment was used to measure the experimental variables: wrist movement and posture were dynamically measured using the Exos Gripmaster system; body posture was evaluated using Peak 2D Videomotion analysis system; typing speed and accuracy were recorded by software.

Procedure

Subjects worked with either the FAK or a conventional keyboard on an articulated tray fitted to an office desk. Subjects attended separate sessions, each lasting about 90 minutes, on different days for each of the keyboards. For the typing tasks input text was presented on the upper half of the screen and the output text appeared on the lower half of the screen. Errors were highlighted in the output text to provide the subjects with feedback on their performance. Subjects were told not to correct any typing errors. Eight separate 10 minute typing tasks were presented in counterbalanced order. Subjects typed all of the 8 tasks once. Mouse and trackball cursor positioning tasks, where subjects clicked on a moving object on the screen, were also performed. Performance was determined by the number of misses per 3 minute session. Only one subject was present at each experimental session. All sessions were completed in a two week period. The sessions were identical for all 24 experimental trials except for keyboard adjustment and setup. Six subjects were given written instructions on keyboard and chair adjustment including the manufacturer's recommendations. The six remaining subjects had the keyboards and chair adjusted for them by the experimenter. All subjects were allowed to make changes to the keyboard adjustments until after the second ten minute typing trial. For the final typing and cursor positioning trials in each session the wrist exoskeleton apparatus was placed on the subject's right hand/wrist. Subjects completed questionnaires on to the CK and the FAK immediately at the end of the corresponding session. At the end of the study they also completed a questionnaire which directly compared the two keyboards.

Video Analysis of Posture

For each session body posture was assessed using video motion analysis. Images of the upper body during typing and cursor positioning were captured using two video cameras, one mounted overhead and one to the right side of the subject. Peak 2D video digitizing software was used to analyze postural data for the CK and FAK arrangements. Three angles were measured from the side videotape: wrist extension, elbow angle, and shoulder/neck angle. Ten frames, randomly selected at approximately equal one minute intervals were analyzed for each typing trial, and five frames for each cursor positioning trial.

Posture Ratings

Five trained raters rated upper body posture from the video tapes of subjects, using a battery of postural checklists that were specifically developed for the study. Each checklist combined a predefined rating scale with posture diagrams. Each rater assessed posture for each of the 96 individual video clips (2 keyboards x 12 subjects x 2 task types - typing and mousing, x 2 views - overhead and side views). Three checklists were used to rate posture from the side for all trials. Separate checklists were used to rate the overhead videotaped clips for the typing and cursor positioning trials. The side view checklist rated subject's relative upright torso angle (upright; forward; backward); degree of resting of torso against backrest (not at all; partially; completely); shoulder position (upward; forward; relaxed); upper arm position (extended back; vertical; extended forward); forearm position (raised; level; lowered); degree of forearm resting (not at all; partly; fully); and neck position (extended back; vertical; flexed forward). The overhead view typing checklist rated forearm position; right wrist position; the portion of the forearm resting on the armrest; upper arm abduction; and the degree to which the hands moved while typing. The overhead view cursor positioning checklist was similar, but rated how a subject used the mouse or trackball rather than how they typed.

Data Analysis

Comparisons of the FAK with the CK speed and accuracy data were made using repeated measures ANOVA. Effects of between subjects variables were test using independent t-tests. Pearson correlations and multiple regression analysis were used to test for associations between variables. Kruskal Wallis 1-way analysis of variance was used to test for differences between observers in their ratings of upper body posture.

RESULTS AND DISCUSSION

Typing Trials

Typing speed was slightly faster for the CK (67 wpm) than the FAK (59 wpm) ($F_{1,11} = 11.89$, p =0.005). Typing accuracy was comparable for the CK (96%) and FAK (92%). Given the short-term nature of this study these result may be of little practical consequence.

Average ulnar deviation was reduced by the FAK (1.7°) compared to the CK (8.5°) (t = -5.8247, df = 11, p < 0.0001). The % time spent with the hands deviated while typing was calculated. Results showed a significant difference between the CK and FAK for ulnar deviation ($\chi^2 = 55.40$, df = 2, p <0.0001), but no difference between the CK and FAK for wrist extension. The percentage of movements either greater than 15% wrist extension or greater than 15% ulnar deviation were computed for each keyboard. Subjects made 27.5% of such movements with the FAK and 38.9% of such movements with the CK. No significant associations were found between armrest height, keyboard height, keyboard slope, and any of the wrist angle data. None of the comparisons between self-adjusted and experimenter adjusted keyboard arrangements was statistically significant for any of the wrist angle data. There was no difference in average wrist extension between the FAK (9.7°) and CK (10.8°) arrangements.

Subjects over 35 years old showed less wrist extension with the FAK (5.7°) than did those under 35 years (15.3°) (t = 3.08, df = 10, p = 0.0116). No other age comparisons were statistically significant for any of the wrist angle data.   Video motion analysis showed that when typing at the FAK subjects' neck and shoulder (t = -3.781, df = 11, p <0.003) and elbow angle (t = -9.083, df = 11, p <0.0001) were significantly improved. There was, however, no effect of the FAK on wrist extension, which corroborates the dynamic measurement data.

There were no significant differences between the mouse (M) versus trackball (T) use for either

dynamically measured wrist extension (M = 19.3°; T = 21.9°), or ulnar deviation (M = -7.1°; T = -5.1°). Video motion analysis showed that when using the FAK trackball subjects' neck and shoulder (t = -7.826, df = 11, p <0.0001) elbow angle (t = -11.488, df = 11, p <0.0001) were significantly improved, but wrist extension was significantly greater for the trackball (t = 2.537, df = 11, p <0.0276).

Observer ratings

Upper body posture (side view).

For typing, there was no difference in ratings of forearm, neck/head, or seated posture for the CK and FAK conditions. With the FAK more subjects rested completely against the back of the chair ($\chi^2 = 6.71$, df = 4, p <0.0349), with relaxed or slightly raised shoulders ($\chi^2 = 11.44$, df = 4, p <0.0220), with their forearms resting on the armrests ($\chi^2 = 84.84$, df = 4, p <0.0001), and with their upper arms straight down or extended backwards, whereas with the CK most subjects sat with their upper arms extended forward of their body ($\chi^2 = 65.13$, df = 4, p <0.0001).

For cursor positioning tasks with the FAK more subjects rested completely against the back of the chair ($\chi^2 = 18.82$, df = 4, p <0.0001), sat with relaxed shoulders or shoulders slightly raised ($\chi^2 = 32.66$, df = 4, p <0.0001), with their upper arms straight down or pulled backwards ($\chi^2 = 120.00$, df = 4, p <0.0001), and their forearms at or below elbow level ($\chi^2 = 15.33$, df = 4, p <0.0041). With the mouse more subjects leaned forwards to use this on the desktop ($\chi^2 = 21.53$, df = 4, p <0.0025), and sat with their neck bent forwards ($\chi^2 = 26.39$, df = 4, p <0.0001).

Upper body posture (overhead view).

With the FAK all subjects worked with their lower arms in a less deviated posture ($\chi^2 = 120.00$, df = 4, p <0.0001), their right hand in less ulnar deviation ($\chi^2 = 80.33$, df = 4, p <0.0001), and their forearms or elbows resting on the armrests ($\chi^2 = 112.57$, df = 4, p <0.0001). However, ratings of shoulder abduction were higher while typing with the FAK ($\chi^2 = 68.38$, df = 4, p <0.0001), although given that the forearms were resting on the chair arms this may not reflect any increase in shoulder load.

For the cursor positioning task with the FAK trackball most subjects either worked with their lower arms straight ahead or bent slightly at the elbow ($\chi^2 = 91.11$, df = 4, p <0.0001), and their right hand in less ulnar deviation ($\chi^2 = 11.48$, df = 4, p <0.0217). More subjects showed shoulder abduction with the mouse than the FAK trackball ($\chi^2 = 40.73$, df = 4, p <0.0001). Only half of the subjects using the FAK moved their arms while typing, whereas all of those using the CK moved their arms while typing ($\chi^2 = 38.35$, df = 4, p <0.0001).

Individual keyboard questionnaires

There were no statistically significant differences in ratings of postural strain, how comfortable each hand and arm felt, how the rest of their body felt, and overall comfort between the keyboards. Also, there were no statistically significant differences in ratings of typing accuracy or how easy it was to press all of the keys. However, there was a significant difference favoring the CK in how easy it was to reach all of the keys for both left hand reach ($\chi^2 = 11.57$, df = 5, p <0.0412) and right hand reach ($\chi^2 = 13.71$, df = 5, p <0.0330).

Finally, there was a statistically significant difference in subjects' opinions about the appearance of each keyboard ($\chi^2 = 15.11$, df = 5, p <0.0099). Four subjects said that they 'slightly liked' the FAK and 4 that they 'liked' the FAK, but 3 'slightly disliked' and 1 'disliked' this design. Six subjects were 'neutral' about the CK, but 5 said they 'liked' this and 1 said they 'strongly liked' this.

Comparative questionnaire

Comfort ratings and postural strain ratings for body regions were comparable for the FAK and CK arrangements, but more subjects reported feeling strain in their shoulders, upper arms, and forearms with the FAK (Figure 2).

DISCUSSION

Results from this short-term laboratory test of the FAK, compared with a CK arrangement, show several potential benefits for trained typists. Compared with the CK, the FAK significantly reduced average ulnar deviation and the amount of time spent in extreme ulnar deviation. The FAK did not, however, reduce average wrist extension (although in a subsequent pilot study we have found that

sloping the FAK downward at a 15° angle does encourage a vertically neutral typing posture). There was an age effect, with less wrist extension for subjects over 35 years. There was no difference in the effects of the FAK on wrist posture between subjects who self-adjusted the FAK and those for whom the FAK was adjusted by the experimenter. The FAK slightly slowed typing speed but did not affect typing accuracy, and this may reflect a familiarity effect.

Videomotion analysis showed that the FAK significantly reduced average elbow angle, because the upper arms were extended further back, posterior of the coronal (frontal) plane, with the forearms level. The FAK significantly increased average shoulder/neck angle, which indicates that the upper arms and head were in a more neutral posture. However, there were no differences in rated seated posture when using the FAK, although more subjects sat completely back in their chair when using the FAK and had their shoulders in a relaxed or slightly raised posture. Other postural ratings show that the FAK encouraged a more neutral seated posture when typing. On balance, it was easier for some subjects to work in less deviated and supported posture with the FAK than with the CK arrangement.

There were some differences in comfort and postural strain ratings for body regions between the FAK and CK arrangements. Questionnaire responses showed some differences, with more subjects reporting some shoulders, upper arm, and forearm strain with the FAK. This may be because these subjects tried to hold their arms above the armrests of the chair to allow for easier access to the keyboard. We did note a tendency for the arm rests of the chair to interfere with subject's ability to effectively reach all of the keyboard keys, and several commented that their arms felt 'stuck to the chair arms'. Future research should record EMG measures of shoulder and upper arm activity to assess the effects of chair arm design.

ACKOWLEDGEMENTS
The FAK and the chair used in this study were supplied by Workplace Designs Inc.

REFERENCES
Chen, C., Burastero, S., Tittiranonda, P., Hollerbach, K., Shih, M. and Denhoy, R. (1994) Quantitative

evaluation of 4 computer keyboards: wrist posture and typing performance, Proceedings of the Human Factors and Ergonomics Society 38th Annual Meeting, Vol. 2, 1094-1098.

Fernström, E., Ericosn, M.O. and Malker, H. (1994) Electromyographic activity during typewriter and keyboard use, Ergonomics, 37, 477-484.

Hedge, A., Ng, L. (1995). Effects of a fixed-angle split keyboard with center trackball on performance, posture and comfort compared with a conventional keyboard and mouse. In press.

Hedge, A. McCrobie, D., Land, B., Morimoto, S. and Rodriguez, S. (1995) Healthy keyboarding: effects of wrist rests, keyboard trays and a preset tiltdown leyboard on wrist posture, seated posture, and musculoskeletal discomfort, Proceedings of the Human Factors and Ergonomics Society 39th Annual Meeting, Vol. 1, 630-634.

Hedge, A. and Powers, J. (1995). Wrist posture while keyboarding: effects of a negative slope keyboard system and full motion forearm supports. Ergonomics, 38, 508-517.

Honan, M., Serina, E., Tal, R. and Rempel, D. (1995) Wrist postures while typing on a standard and split keyboard, Proceedings of the Human Factors and Ergonomics Society 39th Annual Meeting, Vol. 1, 366-368.

Kroemer, K.H.E. (1972) Human engineering the keyboard, Human Factors, 14, 51-63.

Nakaseko, N., Grandjean, E., Hunting, W. and Gierer, R. (1985) Studies on ergonomically de-signed alphanumeric keyboards, Human Factors, 27, 175-187.

Parsons, C.A. (1991) Use of wrist rests by data input VDU operators, In Lovesey, E.J. (ed.) Contemporary Ergonomics 1991: Proceedings of the Ergonomics Society's 1991 Annual Conference, London, Taylor & Francis, 319-321.

Paul, R. and Menon, K.K. (1994) Ergonomic evaluation of keyboard wrist pads, proceedings of the 12th Triennial Congress of the International Ergonomics Association, Vol. 2, 204-207.

Rudakewych, M., Valent, L. and Hedge, A. (1994) Field evaluation of a negative slope keyboard system designed to minimize postural risks to computer workers, Work With Display Units'94, Milan, Italy, October 2-5, C17-C19.

Figure 1
The chair-mounted split keyboard

Figure 2
Questionnaire responses by subjects

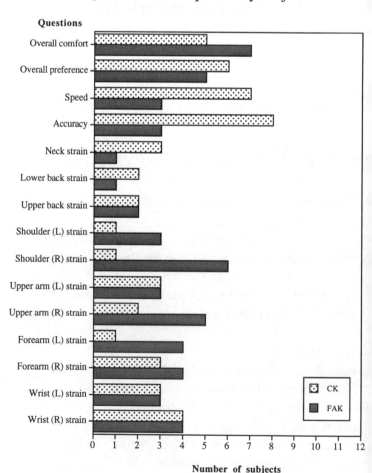

Changes in wrist postures during a prolonged typing task

M Honan, M Jacobson, R Tal, D Rempel

Ergonomics Laboratory,
University of California, San Francisco and Berkeley

Twenty experienced typists participated in a laboratory based study to determine whether wrist and forearm postures changed over a 4 hour period of intensive keyboard use. Subjects were randomly assigned to use a conventional keyboard or a fixed split keyboard. Posture data was acquired using electrogoniometers after a 10 warm-up period and at the end of each hour. Wrist and forearm postures did not change significantly over the four hour period among subjects using the split geometry keyboard. On the conventional keyboard, all joint postures were stable except right wrist extension and left forearm pronation. The right wrist extension increased by 5° over the four hour period (p=.002) and left pronation decreased by approximately 9° (p=.001). Wrist postures among typists exposed for the first time to a split keyboard remained constant throughout a four hour period of intensive typing. On the conventional keyboard, some postures drifted over the four hour period.

INTRODUCTION

The risk of developing symptoms of upper extremity musculoskeletal disorders has been associated with the number of hours of computer keyboard usage (Bernard 92) and with wrist postures deviating from neutral. In previous studies quantifying wrist posture for both standard and alternative keyboards, the typing time evaluated ranged from ten minutes up to one hour (Sauter 91, Hunting 80, Nakaseko 85, Rempel 94). However, it is unknown whether arm postures adapt over a longer period, when typing on an unfamiliar or alternative keyboard. It has also been suggested, but not measured, that arm postures change during the work day due to fatigue. The changes in posture over a prolonged period of typing have not been measured.

The purpose of this study was to measure wrist and forearm postures while subjects used a standard keyboard or an alternative keyboard over a four hour period of intensive typing. The final aim was to compare the mean joint angles of wrist extension, ulnar deviation and pronation for changes over time. The null hypothesis is that there is no significant change in posture during four hours of intensive typing.

METHODS

Twenty experienced typists participated in a laboratory based repeated measures study with one factor: time. Subjects were recruited from a temporary employment agency. Two keyboard configurations were tested: the conventional keyboard (equivalent to Apple Extended™), or an alternative split keyboard (the Microsoft Natural Keyboard with Leveler™ extended). Subjects were randomly assigned into a keyboard group. None of the subjects had experience typing on alternative keyboards. The conventional keyboard and alternative keyboard were tested at worksurface height of 26.5 inch (67.3 cm) and 25 inch (63.5 cm) respectively, so that the D/K key height on the home row were the same. The split keyboard had a built-in wrist rest, while the conventional keyboard did not. Subjects were instructed not to use the wrist rest during the study and were observed to use a "floating method" when typing.

Anthropometric body measurements were taken and the workstation chair was adjusted to match the subject's popliteal height. Subjects were subsequently permitted to make minor adjustment to the chair for comfort. Four electrogoniometers were used to gather data on wrist and forearm postures. Methods of calibration, mounting and accuracy testing have been previously described (Smutz,1994).

Prior to beginning the prolonged typing task and data acquisition, there was a 10 minute typing warm-up period on the assigned keyboard configuration at the specified height. In total, there were five periods of data acquisition. The first was during the first ten minutes of the hour, after the

warm-up period. Data was also acquired during the last ten minutes of each subsequent hour of typing. The subject was given a ten minute break after each hour and a 30 minute lunch break after two hours of typing.

A one factor repeated measures analysis of variance with no grouping factors and one "within subject" factor (time) was performed. Statistically significant findings (p < .05) were followed-up using Tukey's studentized range test at a procedure-wise error rate of .05.

RESULTS

Twenty subjects were tested: 3 male and 17 female. The mean age was 45 years: age range 19 to 65 . The racial distribution was 5% Asian, 25% African American, 10% Hispanic, and 60% White.

Mean postures among those using the conventional keyboard (Figure 1) showed a significant increase in right wrist extension over time (p=.002). Left wrist extension showed no similar trend. Ulnar deviation did not vary significantly over time. Left pronation decreased significantly from the beginning to the end of the four hour typing period (p=.001). Right pronation showed a similar trend but the change was not statistically significant.

Time	Wrist Extension		Ulnar Deviation		Pronation	
	Left	Right	Left	Right	Left	Right
Post warm-up	33.3°(12.2)	**21.6°^{a,b}(7.1)**	15.1° (3.7)	11.6° (2.9)	**76.6° ^{c}(8.4)**	68.9° (9.7)
1 hour	28.9° (14.1)	**20.5° ^{a}(6.3)**	12.6° (4.6)	12.1° (4.4)	**72.8° ^{b,c}(9.6)**	66.6° (11.3)
2 hours	27.5° (13.6)	**20.3° ^{a}(7.3)**	11.9° (5.8)	12.1° (4.8)	**70.0° ^{a,b}(9.8)**	65.4° (10.6)
3 hours	28.2° (12.7)	**25.1° ^{b}(9.3)**	13.0° (5.2)	11.6° (3.6)	**66.6° ^{a,b}(9.4)**	65.6° (9.9)
4 hours	30.9° (14.2)	**25.2° ^{b}(9.1)**	13.1° (3.2)	10.5° (2.7)	**67.6° ^{a,}(8.0)**	65.2° (11.5)
p-value	.20	**.002**	.67	.37	**.001**	.16

Figure 1. Wrist and forearm postures in degrees (standard deviation) during 4 hours of typing on a conventional keyboard (n=10). Significant results in bold. Values in a column sharing a common superscript are <u>not</u> significantly different.

Among the subjects assigned to the alternative keyboard there were no significant postural changes over time (Figure 2). Both extension and ulnar deviation varied over a

three degree range. Again a trend toward decreasing pronation with time was noted bilaterally, though the changes were not significant.

Time	Wrist Extension		Ulnar Deviation		Pronation	
	Left	Right	Left	Right	Left	Right
Post warm-up	27.2°(10.2)	25.2° (12.7)	.6° (5.0)	4.2° (4.9)	72.2° (6.7)	69.9° (8.5)
1 hour	25.9°(10.0)	23.9° (11.2)	.6° (7.6)	4.4° (4.7)	68.4° (7.3)	69.1° (8.5)
2 hours	27.0° (11.8)	24.2° (11.8)	1.0° (7.9)	5.6° (4.8)	66.7° (7.8)	71.4° (5.8)
3 hours	26.9° (11.5)	25.5° (6.8)	3.0° (5.8)	2.7° (3.3)	68.1° (10.7)	68.6° (6.4)
4 hours	29.3° (12.7)	26.8° (7.5)	2.0° (7.4)	2.0° (5.8)	64.4° (11.3)	66.3° (6.9)
p-value	.99	.53	.10	.33	.17	.39

Figure 2. Wrist and forearm posture in degrees (standard deviation) during 4 hours of typing on an alternative keyboard (n=10).

DISCUSSION

Wrist and forearm postures did not change significantly over the four hour period among subjects using the split geometry keyboard. On the conventional keyboard all joint postures were stable, except right wrist extension and left forearm pronation. Right wrist extension increased by 5° over the four hour period and left pronation decreased by approximately 9°. Ulnar deviation remained stable for the conventional and split keyboard.

Moffet et. al. (1995) had observed that no substantial change in posture occurred when typing on a portable computer after 15 minutes or one hour of typing. Our results were similar for the first hour. However, postural changes for right wrist extension and left pronation were seen with the conventional keyboard during the last two hours of the four hour typing period.

The urge to compare between the wrist posture results of the conventional and alternative keyboards should be resisted due to keyboard design differences. The alternative keyboard had a built-in wrist rest that may function as a guide or block against increasing extension. The conventional keyboard tested had no built-in wrist rest. The presence of the wrist rest may explain the lack of change in wrist posture on the split keyboard.

These results provide some confidence that postures measured during the first 20 minutes of exposure to a new keyboard will be the same as postures measured over a day. However, whether postures changes over a longer time are unknown. The finding that some postures drift over a four hour period on the conventional keyboard deserves further exploration and may coincide with psychophysical measures of peripheral fatigue.

REFERENCES

Bernard, B., Sauter, S., Peterson, M., Fine, L., Hale, T.,(1992), Los Angeles Times, NIOSH Health Hazard Evaluation, HETA 90-013

Hunting, W., Grandjean, E., Maeda, K., (1980), Constrained postures in accounting machine operators. Applied Ergonomics, 11.3, 145-149.

Moffet, H., Hagberg, M., (1995), Variation in physical exposure measures during keyboard work on a portable computer. A methodological study to determine optimal test duration. Premus 95: Book of abstracts, 447-449

Nakaseko, M., Grandjean, E., Hünting, W., Gierer, R. (1985), Studies of ergonomically designed alphanumeric keyboards. Human Factors, 27, 175-187.

Rempel, D., Honan, M., Serina, E., Tal, R., (1994), Wrist postures while typing on a standard and split keyboard. HFES 95 Proceedings, Vol. 1, 366-368

Sauter, S., Schleifer, L., Knutson, S., (1991), Work Posture, Workstation Design, And Musculoskeletal Discomfort In A VDT Data Entry Task. Human Factors, 33(2),151-167

Smutz, P., Serina, E., Bloom, T., Rempel, D. (1994), A system for evaluating the effect of keyboard design on force, posture, comfort and productivity. Ergonomics, 1649-1660.

ERGONOMIC EVALUATION OF KEYBOARD AND MOUSE TRAY DESIGNS

Rajendra Paul
Haworth
Holland, Michigan

Chandra Nair
Ergonomic Technologies Corporation
Oyster Bay, New York

Surface electromyography and electrogoniometry were used to evaluate four designs of keyboard and mouse trays. The four designs were: winged design, sliding design, two-tier design and combination of a keyboard tray and a mouse tray. For comparison, data were also collected with the keyboard and mouse placed on a plain work surface. Eight subjects, four male and four female, evaluated the designs in random order while performing a word processing task. Subjects' working postures were recorded using a high-resolution video camera. Body part discomfort was evaluated using the Corlett scale. The subjective and objective results suggested that the sliding design and the two-tier design performed better than the other two designs. The work surface performed inferior to all four keyboard and mouse tray designs. Since mouse input devices are widespread in modern offices, this study could provide some useful information to practicing ergonomics, health and safety professionals.

INTRODUCTION

Adjustable keyboard trays are used to retrofit non-adjustable workstations so that employees can use keyboards in comfortable upper extremity positions. Some recent studies (Hedge, 1995a, 1995b) investigated the effect of keyboard trays on wrist posture and visual distances, reporting favorable results. In recent years, with the introduction of a mouse -- a computer input device -- in most office workstations, several keyboard and mouse trays (KMTs) were introduced in the marketplace, claiming ergonomic benefits without any supporting studies. Abundance of such accessories in the marketplace has confused the users and made the selection process difficult.

This study focused on the design evaluation of KMTs. The primary objective was to identify the design features that reduced musculoskeletal stressors experienced by users. Four designs of keyboard and mouse trays shown in Figure 1 were evaluated. The KMT # 1, referred from now on as the 'winged design,' could be set up for the left-handed or right-handed users. An eight inch area at the each end could be either aligned with the centerpiece or individually swiveled inward at an angle of 45°. The KMT # 2, referred hereafter as the 'sliding design,' also could be set up for the right-handed or left-handed users. A platform placed underneath the keyboard support surface slid out either on the right or left side of the centerpiece and served as the mouse pad. In these two designs, the mouse was at about the same level as the keyboard. The KMT # 3 was a combination of two separate units -- a height and tilt-adjustable keyboard tray and a height-adjustable mouse tray. The mouse tray could be adjusted to the same level or at a different level relative to the keyboard. In design # 4, referred hereafter as the 'two-tier design,' the mouse pad was located two inches above the keyboard. It could be swiveled so that the numeric 10-key pad on the keyboard is either covered or accessible. For users who do not use the numeric keypad, the mouse pad could be located above those keys to reduce shoulder abduction. Along with these four designs, a reference condition of working on a plain work surface, mounted 29.5" above the floor, was also studied.

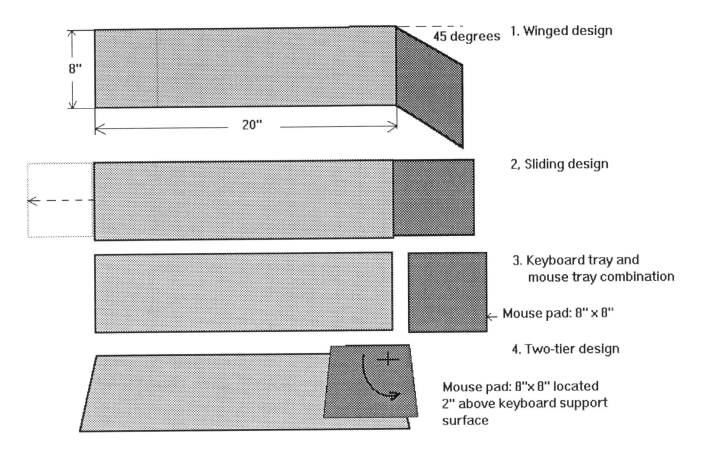

45 degrees   1. Winged design

8"

20"

2. Sliding design

3. Keyboard tray and
   mouse tray combination

Mouse pad: 8" x 8"

4. Two-tier design

Mouse pad: 8"x 8" located
2" above keyboard support
surface

Figure 1. The keyboard and mouse tray designs evaluated in this study (not to scale).

EXPERIMENTAL PROCEDURE

Subjects

Eight subjects, four male and four female, free from symptoms or history of cumulative trauma disorders (CTDs) voluntarily participated in this study. They were full-time office employees at a mid-size manufacturing company in the mid-western United States. Their average age and height were 36.0 (standard deviation, SD = 8.6) years and 68.3 (SD = 4.7) inches. They had 10.6 (SD = 7.1) years of experience on a computer. To carry out their regular job duties, they used computers, on an average, for 4.2 hours per day.

Equipment

Surface electromyography (EMG) was used to measure muscle activity of two forearm muscles - flexor carpi radialis and extensor carpi radialis. The sensors were placed on the dominant hand, viz. the right hand for all subjects. The EMG system was pre-calibrated and the collected data was curve-fitted to the muscular contraction level. The Penny and Giles biaxial electrogoniometer, accurate within ±5 percent, was placed on the dominant hand to measure wrist deviations. It was calibrated to zero with a straight wrist. A high resolution video camera was used to record elbow and shoulder postures assumed by the subjects. In addition, Corlett comfort evaluation scale (Bishop and Corlett, 1976) and product rating surveys were employed to collect subjective data.

Procedure

Each subject was familiarized with the study, equipment used for data collection and the test procedure. Subjects were informed about the freedom to withdraw and confidentiality of their identity. Before evaluating a KMT design, subjects were trained on the adjustments and given the necessary time to adjust it to fit it to individual morphology and preferences. The workstation, chair, KMT, and accessories, e.g. foot rest, were adjusted so that the subjects could consistently assume a slightly reclined and comfortable posture.

Subsequently, each subject performed a word-processing task requiring use of a keyboard and a mouse for 30 minutes, during which EMG and postural data were collected. At the beginning and end of each trial, the subject reported any body part discomfort. They also completed feature rating questionnaires at the end of the trial. Subsequently, the procedure was repeated for other KMT designs. The order in which a subject evaluated the KMT designs was randomized. When working on the work surface, a height-adjustable footrest was employed to accommodate the elbow rest height variations between subjects.

RESULTS

The results of surface EMG and goniometry are shown in Figures 2 and 3. Body joint angles derived from video records of working positions are shown in Figure 4. The subjective results are shown in Tables 1 and 2. To understand the differences between the designs, the results were analyzed using two-way analysis of variance (ANOVA). The two independent variables were subjects and KMT designs. Differences between the conditions were analyzed using Newman-Keuls range test.

The participants ranked the work surface significantly below all four KMT designs ($p < 0.05$). The shoulder flexion, shoulder abduction and elbow flexion were also the greatest for the work surface. The forearm muscle activity and wrist deviations with the work surface, however, were not significantly greater than those with the KMTs.

Of the four KMT designs evaluated in this study, the sliding design resulted in significantly less flexion and extension at the wrist joint ($p < 0.05$). Analysis using the Newman-Keuls range test showed that the extensor muscle activity and ulnar deviation were not significantly different from those when

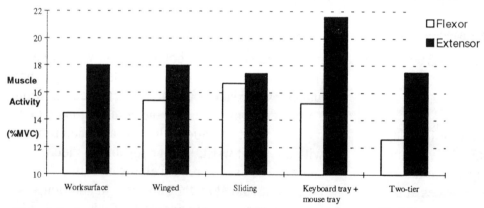

Figure 2. Average muscle activity.

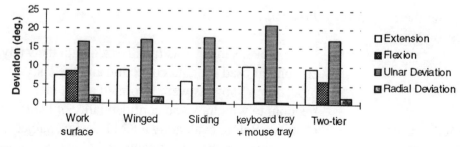

Figure 3. Average wrist deviations.

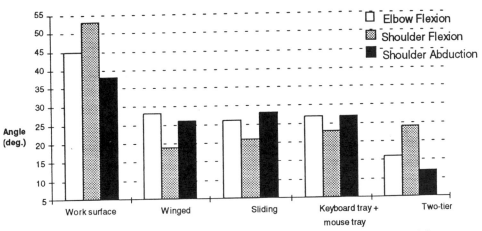

Figure 4. Average elbow and shoulder joint angles assumed by subjects.

Table 1.   Subjects' ratings for the features and overall utility of the designs.
Format: Average (standard deviation). [Scale: 1-Very poor ... 4-Average ... 7-very good]

Variable	Winged		Sliding		Keyboard tray + Mouse Tray		Two-tier	
Keyboard and mouse tray:								
Stability	4.4	(1.8)	5.4	(1.1)	4.2	(1.7)	4.7	(1.2)
Aesthetic appeal	3.7	(1.6)	4.4	(1.8)	4.7	(1.2)	4.2	(1.4)
Functional rating	4.3	(1.6)	4.8	(1.2)	4.2	(1.1)	4.9	(1.0)
Overall rating	4.1	(1.7)	4.5	(1.3)	4.4	(1.0)	4.8	(1.0)
Mouse pad:								
Location	4.8	(1.4)	4.9	(1.0)	3.8	(1.5)	5.0	(1.9)
Ease of height adjustment	3.7	(1.2)	4.0	(1.1)	4.3	(2.7)	3.9	(1.4)
Ease of adjusting location	4.4	(1.5)	3.7	(1.4)	4.7	(1.0)	5.2	(1.1)
Aesthetic rating	3.9	(1.7)	4.4	(1.4)	4.8	(1.0)	3.9	(1.6)
Functional rating	4.3	(1.6)	4.7	(1.4)	4.1	(1.4)	5.2	(1.3)
Overall rating	4.4	(1.6)	4.7	(1.3)	4.3	(1.2)	4.8	(1.3)

Table 2. Subjects' rankings of the five design
conditions [Scale: 1-Best .. 5-Worst]

Test Condition	Rank	
Work surface	4.0	(1.3)
Winged	2.6	(1.4)
Sliding	2.2	(1.3)
Keyboard Tray + Mouse Tray	3.2	(1.7)
Two-tier	2.9	(1.1)

using the winged and two-tier designs, but were significantly less than the combination design. The flexor muscle activity, with this and other three designs, however, was significantly greater than that with the two-tier design. Subjectively, this design was ranked significantly better than the other designs.

The two-tier design caused the least amount of shoulder abduction and elbow flexion. The flexor muscle activity was also the lowest for this design (all significant at p=0.05). This and the sliding designs were rated significantly superior to the other two designs for functional design and overall utility. Based on these results, the biomechanical performance of two-tier design could be judged to be similar to that of the sliding design.

Compared to the previous two designs, the winged design was rated significantly lower on functionality, aesthetic appeal and overall design. The majority of the ratings for its mouse pad were also lower. This design caused significantly more wrist extension than the sliding design. Compared to the two-tier design, it caused significantly more flexor muscle activity, elbow flexion and shoulder abduction. The keyboard tray and mouse tray combination also performed poorly. Subjects recorded significantly greater ulnar deviation and extensor muscle activity compared to those with other designs. Subjectively, the design was ranked significantly inferior to the other designs for overall design and functionality.

DISCUSSION

These results suggest that placing the keyboard and mouse on a work surface posed similar postural and muscular stresses on the forearm and hands, but significantly greater postural stresses on the elbow and shoulder. Also subjects rated is a significantly inferior solution. Among the KMTs, none of the designs resulted in the least value of all the postural and muscular stresses evaluated in this study. On wrist extension, wrist flexion and overall rank, the sliding design performed better than the other designs. On flexor muscle activity, elbow flexion and shoulder abduction, the two-tier design performed better than the other designs. Majority, not all, of the user ratings for these two designs were similar. The winged design, except for the wrist flexion, and the combination performed inferior to the sliding and two-tier designs. Considering this performance pattern, the sliding design and two-tier designs were judged superior to the winged design and the keyboard tray and mouse tray combination.

The design objective behind the winged design was to reduce lateral movement of the arm necessary for accessing the mouse by bringing the mouse pad closer to user's body. However, it appeared to require greater accuracy in arm movement from keyboard to mouse and the imposed constraints outweighed its benefits. The objective behind the two-tier design was to locate both keyboard and mouse within or close to the 10"x 10" optimal work area proposed by Ayoub(1973). To that extent, this significantly reduced shoulder abduction and elbow flexion and was an improvement. The sliding design was hypothesized as a simple and intuitive, though not necessarily the best, solution. Its simplicity and habituated pattern of use promoted postural and biomechanical stresses comparable to those with other designs. The separate keyboard and mouse trays used in combination promoted greater freedom in individual fit, but performed more poorly than the sliding design in all categories. This suggests that integrated KMT designs promoted user behavior and response that facilitate concurrent use of keyboard and mouse. Use of a work surface imposed significantly greater postural deviations at elbow and shoulder joints, likely due to the chair armrests. The armrests were height adjustable within a two inches but precluded subjects from locating themselves closest to the input devices.

The ANSI/HFES 100-1988 standard (Human Factors Society, 1988) does not provide recommendations for KMTs. This study provides some guidance to the practicing ergonomics, health and safety professionals in the selection of these accessories. Additional research in this area should include muscle fatigue in the neck and shoulder area, and other input devices like trackballs, digitizing tablets and newer keyboard designs.

REFERENCES

Ayoub, M.M. (1973). Workplace design and posture, *Human Factors*, 15(3), 265-268.

Corlett, E. N. and Bishop, R.P. (1976). A technique for assessing postural discomfort, *Ergonomics*, 19, 175-182.

Hedge, A. and Powers, J.R. (1995a). Wrist postures while keyboarding: effects of negative slope keyboard system and full motion forearm supports, *Ergonomics*, 38 (3), 508-517.

Hedge, A., McCrobie, D., Land, B., Morimoto, S. and Rodriguez, S. (1995b). Healthy keyboarding: Effects of wrist rests, keyboard trays, and a preset tilt-down system on wrist posture, seated posture, and musculoskeletal discomfort. *Proceedings of the Human Factors and Ergonomics Society* (pp. 630-634), Santa Monica: HFES.

Human Factors Society (1988). *American National Standard for Human Factors Engineering Requirements for VDT Workstations ANSI/HFS 100-1988*. Santa Monica, CA: HFS.

Paul, R. D. and Menon, K. (1994). Ergonomic evaluation of keyboard wrist pads. *Proceedings of the 12th Triennial Congress of the International Ergonomics Association*, vol. 2, (pp. 204-207). Ontario, Canada: Human Factors Society of Canada.

RISK FACTORS AND THEIR INTERACTIONS IN VDT WORKSTATION SYSTEMS

[1]Hongzheng Lu and [2]Fereydoun Aghazadeh
[1]BCAM International, Inc., Melville, New York
[2]Louisiana State University, Baton Rouge, Louisiana

This study examined important risk factors and their interactions associated with physical symptoms reported by VDT users. A research model was developed. A survey was designed and conducted among 88 computer users. The results show that risk factors associated with various physical symptoms are different. Screen glare, awkward working posture, and fatigue are important factors related to physical symptoms. Psychosocial factors significantly interact with other variables, such as demographics variables, and contribute to awkward work posture and psychological stress. Workstation design variables significantly affect working postures.

1. INTRODUCTION

A large amount of research has been carried out to investigate the risk factors of cumulative trauma disorder injuries, physical symptoms, and psychological stress experienced by VDT workers. Past studies can generally be divided into three categories: (1) Investigation of the tasks, workstation design and physical work environment, and their association with the injuries and symptoms experienced by VDT workers (Laubli et al., 1981; Lu et al., 1993; Schleifer et al., 1990); (2) Examination of the psychosocial factors and their relationship with the psychological stress experienced by VDT operators (Järvenpää et al., 1993; Miezio, et al., 1987; Rogers et al., 1990); and (3) Examination of the relationship of psychosocial factors and psychological stress and ergonomics risk factors, and their combined effect on the physical discomfort (Lim and Carayon, 1993). Little research has been conducted to examine the physical work environment and psychosocial factors together and their combined effect.

The objectives of this study were: (1) to examine important risk factors of the physical discomfort experienced by VDT operators in the physical and psychosocial environment; and (2) to examine the interrelationship among these risk factors.

2. METHOD

A research model was developed for the hypothetical relationship among the risk factors and their relationship to the physical symptoms. A survey was then designed using questionnaire, posture analysis, measurements and a checklist. The survey was conducted among VDT users in a local hospital and an university. The data was then analyzed using factors analysis and regression analysis.

2.1 Research Model

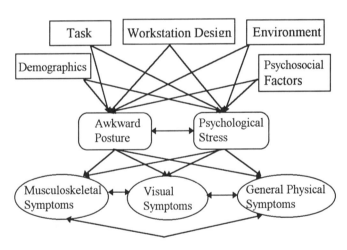

Figure 1   Research model

A work system consists of the following elements: (1) the worker, (2) the equipment, (3) the work environment, (4) tasks, and (5) the work organization. These various elements interact when work is being done. Figure 1 illustrates a research model which incorporates the variables of worker, task, and physical and social environment. It was hypothesized that the interaction of the system components have effects on the physical symptoms via their effect on the awkward work posture and psychological stress. Awkward work posture and psychological stress are directly related to physical symptoms.

2.2 Survey Design

A survey was designed to investigate the physical symptoms experienced by VDT workers and related work and environment factors in the workplace. The survey consists of three parts: (1) a questionnaire; (2) posture recording; and (3) anthropometry and workstation measurements.

3. RESULTS

3.1 Background Information

Eighty eight (88) VDT users, 77 females and 11 males, age 21-63 (average 38), participated in the study; of which 80 participated in anthropometry and workstation measurements, and 74 participated in video recording of working postures in the workplace. They were all full-time employees with experience at current job from 3 months to 36 years (average 55 months). Seventy-two subjects (82%) were from the Business Office, Accounting & Payroll, and other office of a local hospital. Twelve subjects (18%) were from various department and laboratories of an university.

3.2 Test of The Research Model

Factor analysis was applied to the following categories of variables to identify common factors: physical symptoms, work postures, workstation design variables, and psychosocial variables.

Four factors, which explain 62% variances of physical symptom variables, were identified among the physical symptoms. They are: (1) Ocular symptom factor (M1), which included the symptoms of burning eyes, tired eyes, tearing/itching eyes, and dry eyes; (2) General musculoskeletal stress factor (M2), which included the symptoms of lower back, neck, shoulders, and headache, (3) Upper extremity factor (M3), which included the symptoms of wrists, upper back, shoulders, and neck, and (4) Other physical symptom factor (M4), which included blurred vision, ringing ears and stomach ache.

Two factors, which explain 61% of the variance of the posture variables, were identified among work posture variables. They are (1) upper body posture factor (P1), which included the postures of head/neck, trunk, and upper arm, and (2) extremity posture factor (P2), which included the variables of lower arm, wrist and foot posture.

Two factors, which explain 68% of the variances among workstation design variables, were identified among workstation design variables. They are (1) screen glare factor, which included subjective and objective measurement of screen glare, and (2) workstation layout factor, which included the layout of keyboard and screen, subjective rating of comfort with screen position, and subjective rating of comfort with keyboard position.

Two factors, which explain 51% of the variances of psychosocial variables, were identified among psychosocial variables: (1) job satisfaction factor, which reflected various aspects of satisfaction with the job including satisfaction with job challenge, job responsibility, supervisor support, supervisor feedback, and interaction with other people, and (2) workload pressure factor, which reflected subjective ratings of work load including the variables of time pressure and surges of work load.

The following relationships were tested using multiple regression method:

Regression models 1, 2, 3, 4 :

Physical symptoms (M1, M2, M3, M4)

$= f$ (demographics, task, workstation design, work environment, psychosocial factors, work posture, psychological stress)

Regression models 5, 6:

Awkward work posture (P1, P2)

$= f$ (demographics, task, workstation design, work environment, psychosocial factors, psychological stress)

Regression models 7, 8, 9:

Psychological stress (Depression, Anxiety, Extreme fatigue)

$= f$ (demographics, task, workstation design, work environment, psychosocial factors, awkward work posture)

Various regression methods were used to determine the predictors for each regression model: forward, backward, stepwise, and adjusted-R^2. The results from the different methods were compared and the final regression model was determined based on the following criteria: (1) high adjusted R^2, which is an alternative to R^2 representing the proportion of variance that can be explained by the model that has been adjusted for the model degrees of freedom; (2) reasonable interpretation; (3) partial R^2, which is the portion of variance that can be explained by the selected parameter; and (4) Cp, which is a measure of total squared error. When the right model is chosen, the parameter estimates are unbiased, and this is reflected in Cp near the number of parameters p in the model. Tables 1 lists the regression results.

Table 1 Regression Analysis Results

Reg. Model	Dependent Variable	Significant Independent Variables	Parameter Estimate	Partial R²	Prob. > F
1	Ocular discomfort factor (M1)			R²=.39; Adj.R²=.34	.0001
		Screen glare factor	.149	.101	.0068
		Length of time using computer continuously x Layout of screen and keyboard	.012	.087	.0080
		Extremity posture factor	-.355	.081	.0081
		Discomfort with illumination	.274	.045	.0394
		Job satisfaction factor	-.171	.039	.0505
		Average luminance around VDT	-.016	.038	.0610
2	General musculoskeletal stress (M2)			R²=.42; Adj.R²=.38	.0001
		Extreme fatigue	.357	.186	.0002
		Upper body posture factor	.203	.125	.0021
		Age	-.025	.061	.0179
		Length of time using computer continuously	-.244	.043	.0490
3	Upper body discomfort factor (M3)			R²=.28; Adj.R²=.26	.0016
		Extremity posture factor	.362	.141	.0094
		Depression	.159	.068	.0242
		VDT work history	.004	.038	.0816
		Upper body posture x Layout of screen and keyboard	.031	.032	.104
4	Other physical symptom factor (M4)			R²=.29; Adj.R²=.27	.0001
		Extreme fatigue	.243	.156	.0014
		Computer usage per day	.157	.077	.0256
		Age x type of eye wear	-.003	.059	.0510
5	Upper body posture factor (P1)			R²=.43; Adj.R²=.41	.0001
		Keyboard and screen position x Screen glare	.031	.233	.0001
		Sex	.934	.076	.0083
		Avg. illumination level at workstation	.020	.074	.0050
		Sex x Work pressure	.115	.043	.0286
6	Extremity posture factor (P2)			R²=.31; Adj.R²=.29	.0001
		Sex x Work pressure factor	-.238	.067	.0003
		Work space rating	-.473	.054	.0029
		Working hours/day	.360	.051	.0085
		Length of time using computer continuously	.160	.038	.0009
		Average illumination at workstation	-.016	.025	.0156
7	Depression			R²=.39; Adj.R²=.36	.0001
		Job satisfaction factor	-.491	.164	.0004
		Upper body posture factor	.212	.073	.0124
		Length of time using computer continuously x Position of screen and keyboard factor	.013	.045	.0440
		Average luminance around VDT	-.020	.047	.0312
		Age x Work pressure factor	.008	.033	.0788
		Work space rating	.325	.026	.0991
8	Anxiety			R²=.31; Adj.R²=.29	.0008
		Sex x Work pressure factor	.333	.148	.0003
		Job satisfaction factor	-.508	.091	.0029
		Average luminance around workstation	-.023	.040	.0085
		Type of VDT tasks	.114	.026	.0009
9	Extreme fatigue			R²=.24; Adj.R²=.21	.0027
		Sex x Work pressure	.158	.068	.0274
		Eye wear type x Average luminance around workstation	-.005	.064	.0280
		Job satisfaction factor	-.258	.038	.0818
		VDT task type x Length of time at present job	.001	.036	.0869
		Age	-.025	.032	.1016

5. DISCUSSIONS

5.1 Ocular Discomfort

Ocular discomfort includes the symptoms of tired eyes, burning eyes, tearing/itching eyes, and dry eyes. Screen glare is an important factor related to the ocular discomfort. The interaction of the length of time using computer continuously and the position of screen and keyboard factor is another important factor accounting for the variance of ocular discomfort. The result, parameter estimate = 0.149, $R^2 = 0.10$, p=0.068, shows that as the time of using computer increases, the ocular discomfort increases. Luminance and illuminance around VDT workstations are also important risk factors of ocular discomfort. Discomfort with illumination level and low luminance level are associated with more ocular discomfort. Job satisfaction and extremity posture also contribute to the symptoms.

Interestingly, the symptom of 'blurred/double vision' which was defined as a visual symptom does not belong to this factor. However, it was not surprising because this symptom was also found to be apart from 'ocular symptom' and was called 'perceptual symptom' by other researchers (Schleifer et al., 1990).

5.2 General Musculoskeletal Symptoms

General musculoskeletal symptoms include the symptoms of low back, headache, neck and shoulders. Extreme fatigue is the most important factor in this category of discomfort. Another important risk factor is the upper body posture factor (P1). The positive coefficient (0.203) indicates that the poorer the upper body posture (i.e., increased head/neck tilt, increased trunk angle and upper arm angle), the more musculoskeletal symptoms were found in the areas of lower back, neck, shoulder and headache. Age is also a risk factor of general musculoskeletal symptoms.

5.3 Upper Body Musculoskeletal Symptoms

This category of variables includes all symptoms of the body part above low back, i.e. neck, shoulders, wrists, and upper back. Awkward extremity posture accounts for a large amount of variance of the upper body musculoskeletal symptoms. This result suggests that deviation from neutral position of low arms affects the upper body musculoskeletal symptoms. Another risk factor is depression, a psychological stress factor. This finding agrees with the result of another study by Lim and Carayon (1993) which found that psychological stress affects musculoskeletal symptoms. VDT work history is also an important risk factor of upper body symptoms. This result suggests that poor upper body posture at VDT workstation may result from long-time computer use. Upper body posture interacting with the layout of screen and keyboard also affects the upper body symptoms.

5.4 Other Physical Symptoms

This category include the symptoms of blurred vision, ringing ears and stomach ache. The regression model M4 indicates that fatigue is the most important factor for this category of variables. This result suggests that these symptoms are stress related. Another risk factor is the total time of computer usage per day. The longer time using a computer is associated with higher scores of "other physical symptoms." This result suggests that long time computer use is related to stress. The interaction of age and the type of eye wear also affects the "other physical symptoms." Examination of the interaction shows that operators wearing contact lenses have high complaints of these symptoms.

5.5 Risk Factors of Awkward Work Posture

The most important risk factors of awkward upper body postures, i.e., head/neck, trunk and upper arm posture, are the interaction of the layout of screen and keyboard and screen glare. Among the poorly designed workstations, screen glare significantly affects awkward upper body posture. Other risk factors are sex, average illumination level at workstation and the interaction between sex and work pressure factor (S2).

The most important risk factors of awkward extremity postures are the interaction of sex and work pressure factor (S2), work space comfort, working hours per day, time of using computer continuously and the illumination level at VDT workstation.

The above risk factor variables related to awkward work posture come from the categories of workstation design, demographics, tasks, work environment and psychosocial factors. The results indicate that working posture is determined by the interaction of many factors in the work place. Among these factors that affect work posture, the variables of workstation layout (e.g. the height, orientation and location of VDT, keyboard and supporting surface) are most important as these factors determine how a worker must position his/her body when performing a task.

5.6 Risk Factors of Psychological Stress

The following factors are related to psychological stress: psychosocial factors, work environment, workstation design, tasks and demographics. Among these factors, psychosocial factors, i.e. job satisfaction and work pressure factors, are most significant to all the variables used to measure psychological stress, i.e. depression, anxiety and extreme fatigue. This result agrees with past studies in which psychosocial factors are found to be significant predictors of psychological stress (Järvenpää et al., 1993; Miezio, et al., 1987; Rogers et al., 1990).

5.7 Interactions among Risk Factors

Many factors may create harmful loads on an individual in a VDT workstation system. These factors interact when work is being done. Among the interactions that this study examined, the layout of screen and keyboard is found to interact with other factors, such as screen glare, time of using computer continuously and extremity posture. The interactions affect awkward upper body posture, general musculoskeletal symptoms, ocular discomfort and psychological stress. Psychosocial factors interact with other factors and affecting the psychological stress and awkward working posture. Among the variables of demographics, the variables

of sex and age interact with other variables affecting the psychological stress.

6. CONCLUSIONS

1. Screen glare is the most important risk factor contributing to ocular symptoms; fatigue and awkward posture are the most important risk factors to affect general musculoskeletal symptoms. Awkward posture is the most important risk factor of upper body symptoms. Fatigue is the most important factor related to other physical symptoms.
2. Psychosocial factors interact with other variables and contribute to work posture psychological stress. The effect is more significant among female workers than among male workers.
3. Workstation design significantly affects working posture which in turn contributes to physical symptoms.

7. REFERENCES

Järvenpää, Eila, Carayon, Pascale, Hajnal, Catherine, Lim, Soo-Yee, and Yang, Chien-Lin, 1993, A cross-cultural framework for the study of stress among computer users: comparison of the USA and Finland. In Human-Computer Interaction: Applications and Case Studies. (Edited by Michael J. Smith and Gavriel Salvendy), Elsevier Science Publishers B.V., 874-879.

Läubli, Th., Hünting, W. and Gradjean, E., 1981, Postural and visual loads at VDT workplaces II. Lighting conditions and visual impairments. Ergonomics, 24, 933-944.

Lim, Soo-Yee, and Carayon, Pascale, 1993. An integrated approach to cumulative trauma disorders in computerized offices: the role of psychosocial work factors, psychological stress and ergonomic risk factors. In Michael J. Smith and Gavriel Salvendy (Eds.): Human-Computer Interaction: Applications and Case Studies. Elsevier Science Publishers B.V., 880-885.

Lu, H., Caylor, J., and Aghazadeh, F., 1993, "VDTs in offices: A Field Study". In: Michael J. Smith and Gavriel Salvendy (Eds.), Advances in Human Factors/Ergonomics, 19A, Human-Computer Interaction: Applications and Case Studies. Elsevier Science Publishers B.V., 786-790.

Miezio, K., Smith, M. J. and Carayon, P., 1987, Electronic performance monitoring: behavioral and motivational issues. Trends in Ergonomics/Human Factors IV. Editor S.S. Asfour. Elsevier Science Publishers B.V. (North-Holland). 253-259.

Rogers, Katherine, J.S., Smith, Michael J., Sainfort, Pascale C., 1990, Electronic performance monitoring, job design and psychological stress. Proceedings of Human Factors Society 34th Annual Meeting. 854-858.

Schleifer, Lawrence M., Sauter, Steven L., and Smith, Randall J., 1990, Ergonomic predictors of visual system complaints in VDT data entry work. Behavior & Information Technology, Vol.9, No.4, 273-282.

THE ROLE OF PSYCHOSOCIAL FACTORS IN OCCUPATIONAL MUSCULOSKELETAL DISORDERS

Fadi A. Fathallah (**Chair**)
Liberty Mutual Research Center for Safety & Health
Hopkinton, Massachusetts

Sheila Krawczyk (**Co-Chair**)
Aluminum Company of America (ALCOA)
Newburgh, Indiana

George E. Brogmus (**Panelist**)
Liberty Mutual Research Center for Safety & Health
Hopkinton, Massachusetts

Stover H. Snook (**Panelist**)
Liberty Mutual Research Center for Safety & Health
Hopkinton, Massachusetts

Soo-Yee Lim (**Panelist**)
National Institute of Occupational Safety and Health
Cincinnati, Ohio

Naomi G. Swanson (**Panelist**)
National Institute of Occupational Safety and Health
Cincinnati, Ohio

William S. Marras (**Panelist**)
The Ohio State University
Columbus, Ohio

Ernest Volinn (**Panelist**)
Liberty Mutual Research Center
Hopkinton, Massachusetts

INTRODUCTION

Occupational musculoskeletal disorders (MSDs) still constitute a major concern in terms of both cost and human suffering. Despite the numerous attempts to reduce these disorders, their prevalence is still alarming. In recent years, there has been an increased indication that the role of physical factors may not be sufficient to explain the etiology of musculoskeletal disorders. Other factors may play an important role. "Psychosocial" factors have received attention in the clinical and industrial research communities due to their potential to provide some answers in cases where addressing the physical factors alone was not a satisfactory approach. It is a scientific challenge to ascertain the relative importance between these different types of factors. Several researchers have emphasized that a multi-disciplinary approach to the problem may be the right vehicle to understand the relative importance of these factors. This in turn may help prevent MSDs in occupational settings.

This panel is an attempt to inform researchers and practitioners in the field of industrial ergonomics about the importance of psychosocial factors and highlight research needs. Some of the specific questions that the panel will address include:

1. What is the state of knowledge about various classes of factors that affect occupational MSDs, and the mechanism by which these factors interact and affect each other?

2. How might psychosocial, legal, and cultural factors influence the reporting of musculoskeletal disorders?

3. What is the role of psychosocial and organizational factors in the development of office-related MSDs?

4. What is the distinction between low back/ MSD pain and low back/MSD disability? What is the role of psychosocial factors in the extent and length of disability?

5. Are there some prevention guidelines related to psychosocial factors?

6. Where should we focus our priorities for future research in the area?

PANELISTS

Following is a synopsis of each panelist's primary focus and views on the topic of psychosocial factors in occupational MSDs:

Fashion Trends in Workers Compensation Claims Reporting

George E. Brogmus
Liberty Mutual Research Center for Safety & Health
Hopkinton, Massachusetts

Psychosocial factors are increasingly being recognized as having a significant impact on the reporting and disability associated with workers compensation claims. Workers compensation data on occupational mental stress, hearing loss, and cumulative trauma disorders in the United States and in Australia provide an indicator of the impact of unemployment, legislation, standards enforcement, litigation, reporting definitions, public awareness, and the awareness and attitude of medical practitioners on reporting and disability associated with these types of claims. Occupational mental stress claims seem inseparably tied to unemployment, litigation trends, and especially legislation. Hearing loss claims reporting shows a

pattern of correlation with changes in legislation. Reporting of cumulative trauma disorders seems strongly influenced by awareness, medical practitioner attitudes, reporting definitions, expanding workers compensation laws, and litigation precedence. Although individual practitioners can have little influence over these factors, they can promote mitigating interventions. Organizations with a healthy corporate culture and benevolent management/labor relations are most likely to be successful at avoiding the fashion trends in workers compensation claim reporting.

The Role of Psychosocial Work Factors, Ergonomic Risk Factors, and Stress on Musculoskeletal Disorders

Soo-Yee Lim
National Institute of Occupational Safety and Health
Cincinnati, Ohio

This abstract proposes a conceptual framework depicting the role of psychosocial work factors on musculoskeletal disorders. This framework or model suggests that psychosocial work factors have an indirect relationship on musculoskeletal disorders via ergonomic risk factors and psychological stress. Specifically, it means that psychosocial work factors can simultaneously have an impact on ergonomic risk factors and stress. Furthermore, this model also suggests that psychological, psycho-physiological, and behavioral mechanisms, are used to support the above proposed relationships. Very limited research studies have been conducted in this area. However, there is some research evidence indicating that psychosocial work factors, such as work pressure and task control, are related to repetition and awkward work postures, and to psychological stress such as anxiety and fatigue (see for example, Lim, 1994). In a similar study, Lim (1994) also found that psychological stress, such as anxiety, was related to upper extremity musculoskeletal discomfort, and to repetition and awkward work postures, which in turn were related to upper extremity musculoskeletal discomfort. There are many challenges ahead in studying the role of psychosocial work factors on musculoskeletal disorders. Currently, the most important need is to improve and establish assessments methods that would certainly distinguish psychosocial and ergonomic (physical) exposures. Finally, the relationships can be better understood and can benefit by longitudinal study design.

Psychosocial-Biomechanical Link in Occupational Low Back Disorders

William S. Marras
The Ohio State University
Columbus, Ohio

Low back disorders (LBDs) in occupational setting have long been recognized to constitute a major musculoskeletal problem in the US. Trying to fully explain the etiology of these disorders is a formidable task. However, given the state of the art, it has been well established that physical risk factors (weight, frequency, duration, etc.) play a major and rudimentary role in association with LBDs. From a biomechanical standpoint, these factors generate complex loading patterns on the spinal structure which in turn could compromise the integrity of that structure and hence lead to injury. Other factors may interact with these basic mechanical factors and could alter their magnitudes and patterns. Psychosocial factors are an example of a category of such "altering" factors. The exact nature of interaction between the two classes of factors is not well understood. The field is in dire need to assess the relative importance of these factors and the interactive mechanism among them.

The Role of Psychosocial Factors in Low Back Pain and Low Back Disability

Stover H. Snook
Liberty Mutual Research Center for Safety & Health
Hopkinton, Massachusetts

When discussing musculoskeletal disorders, it is important to distinguish between pain and disability. Many people experience musculoskeletal pain, but continue to work. The role of psychosocial factors appears to be more important

in reducing musculoskeletal disability than in reducing musculoskeletal pain. The decision to work or not to work may be related to pain, but is also heavily influenced by the characteristics of the individual worker, the nature of the job, and the policies and practices of management.

It is difficult and sometimes impossible to control the characteristics of the individual worker (e.g., age, gender, attitude). However, changes in job design and management factors can result in reduced disability. Many of these changes are psychosocial in nature. Recent findings from the IASP Task Force on Pain in the Workplace, the Upjohn study on disability prevention in Michigan, and the Canadian study of disability due to occupational low back pain will be reviewed.

The Relationship Between Workplace Psychosocial Factors and Musculoskeletal Disorders

Naomi G. Swanson
National Institute of Occupational Safety and Health
Cincinnati, Ohio

Evidence continues to accumulate indicating that stress and psychosocial factors may play an important role in the etiology of health disorders reported by VDT workers. This appears to be particularly true with regard to musculoskeletal disorders. Sauter and Swanson (1995) have developed a model of musculoskeletal disorders among office workers which delineates a number of pathways by which psychosocial and work organization factors may directly or indirectly influence the experience of musculoskeletal symptoms and disorders. This talk will focus on the two pathways for which the strongest evidence currently exists: 1) The introduction of the VDT into the workplace can change important aspects of the psychosocial work environment (e.g., workpace, work demands, work organization, etc.). These new psychosocial stresses may then change the actual physical demands of the job, which may result in musculoskeletal symptoms. 2) The psychosocial work environment may create cognitive or emotional strain, which may then result in musculoskeletal symptoms via neuromuscular mechanisms, such as increased muscle tension.

Psychosocial Factors Implicated in Back Pain Disability

Ernest Volinn
Liberty Mutual Research Center for Safety & Health
Hopkinton, Massachusetts

What are the psycho-social factors implicated in back pain disability? This is a different issue than psycho-social factors implicated in self-reported perceptions of back pain itself. In attempting to understand back pain disability, the scope of inquiry has expanded beyond bio-mechanical and physiological factors to psycho-social factors. But, as the Boeing study (Bigos et al., 1992) illustrates, data on psycho-social factors collected from workers at the workplace are not sufficient to understand back pain disability. It is necessary to expand the scope of inquiry still further and consider the larger cultural context that in turn affects individual physical and psycho-social factors.

SESSION FORMAT

After introducing the panelists, the chair will give a brief (five minute) introduction of the problem and the need for this panel discussion. The following fifty minutes will be dedicated to addressing the specific questions listed above. Five-to-ten minutes will be allocated for each question. Any panelist could respond to each question; however, the chair will first address the question to the most appropriate panelist(s). The last thirty minutes will be allocated to questions from the audience.

SUMMARY

In recent years, there has been a lot of attention paid to the role of psychosocial factors in relation to occupational musculoskeletal disorders. We believe that this panel would be an excellent platform for both practitioners and researchers alike in gaining a better understanding of the role of these factors in the etiology of MSDs. We also foresee that the panel would underline the multi-disciplinary nature of the problem. This is an important step towards attaining a complete understanding of the mechanism leading to MSDs and finding ways to reduce both their prevalence and the disability associated with them.

REFERENCES

Bigos, S., Battie, M.C., et al. (1992). A longitudinal, prospective study of industrial back injury reporting. *Clinical Orth Rel Res*, 279, 21-34.

Lim, S.Y. (1994). *An integrated approach to upper extremity musculoskeletal discomfort in the office environment: The role of psychosocial work factors, psychological stress and ergonomic risk factors.* Unpublished doctoral dissertation, University of Wisconsin, Madison, WI.

Sauter S.L., and Swanson, N.G. (1995). An ecological model of musculoskeletal disorders in office work. In Moon S., and Sauter S. (Eds.), *Psychosocial factors and musculoskeletal disorders in office work.* New York: Taylor and Francis.

THE EFFECTS OF BOX WEIGHT, SIZE, AND HANDLE COUPLING ON SPINE LOADING DURING DEPALLETIZING OPERATIONS

William S. Marras, Kevin P. Granata, Kermit G. Davis, W. Gary Allread, and Mike J. Jorgensen
Biodynamics Laboratory
The Ohio State University
Columbus, Ohio

It is widely known that the order selection task in food distribution centers places the worker at risk of occupationally-related low back disorders (LBDs). One approach to controlling this risk consists of manipulating the characteristics of the object or box to be handled in the food distribution center. However, it is currently unknown what effect these changes to the box characteristics would have on the loading of the spine and the subsequent risk of low back disorder. Hence, the *objectives* of this study were to determine the change in spine loading at L5/S1 associated with selecting boxes that varied as a function of: 1) weight (40, 50, and 60 lbs), 2) size (2681 or 1584 cu. in.), and 3) the existence of handles or hand holds. In addition, these variables were explored as a function of where the box was on a pallet. Ten experienced order selectors were recruited from a local food distribution center and were evaluated as they selected boxes of different characteristics from a slot (bin) on to a pallet jack. Workers were monitored for their trunk motion characteristics as well as the EMG activity of 10 trunk muscles as they performed the task. The kinematic and EMG information was used as inputs to an EMG-assisted model that was used to predict the three-dimensional spine loadings that occurred during the task. The results indicated that conditions where a worker must reach to a low level of the pallet significantly increases spinal load. Thus, spinal loads were significantly great and only of a magnitude that would be expected to lead to low back disorders when workers lifted form the lowest layer of the pallet. Handles had the affect of reducing the spinal loading by an amount that was equivalent to a reduction in box weight of about 10 pounds. This effort has also facilitated our basic understanding as to why spine loading increases under the various conditions studied in this experiment. Nearly all differences in spinal loading can be explained by a corresponding difference in coactivation of the trunk musculature. This in turn significantly increase the synergistic forces supplied by each muscle to the spine and resulted in an increase in spinal loading.

INTRODUCTION

It is widely known that the order selection task in food distribution centers places the worker at risk of occupationally-related low back disorders (LBDs). This job is associated with one of the greatest incidence rates of LBD in the United States. The National Association of Wholesale Grocers of America (NAWGA) and the International Foodservice Distribution Association (IFDA) disclosed that 30% of the injuries reported by food distribution warehouse workers were attributable to back sprains/strains (Waters, 1993). In addition, over a five year period, it was found that back injuries could account for nearly 60% of lost work days (NIOSH Interim Report, HETA 91-405, March 1992). Hence, grocery item selectors have an incidence of low back pain that is at least as severe as other manual materials handling jobs.

One approach to controlling this risk consists of manipulating the characterisitcs of the object or box to be handled in the food distribution center. A committee organized by the Food Marketing Institute is currently considering the various options available to them in order to help mediate the risk of work related LBD in these food distribution centers. Among the options considered are: 1) reducing the weight of the boxes, 2) reducing the size of the boxes, or 3) incorporating handles into the boxes. However, it is currently unknown what effect these changes to the box characteristics would have on the loading of the spine and the subsequent risk of low back disorder.

Hence, the *objectives* of this study were to determine the change in spine loading at L5/S1 associated with selecting boxes that varied as a function of: 1) weight (40, 50, and 60 lbs), 2) size (2681 or 1584 cu. in.), and 3) the existence of handles or hand holds. In addition, these variables were explored as a function of where the box was on a pallet.

METHODS

Subjects

Ten experienced order selectors ages 19 to 49 years of age were recruited from a local food distribution center and were evaluated as they selected boxes from a slot (bin) on to a pallet jack. The average (SD) weight and stature of theworkers was 176.3 lbs (18.5 lbs) and 71 in. (2.8) respectively. Their work experience in warehouse settings ranged from 0.25 to 23 years.

Experimental Task

During the different experimental trials the box weight, box size, and box coupling (handles) conditions were varied. Workers were instructed to pick the entire compliment of boxes from the pallet so that they could be observed picking from all locations on a pallet. While the workers were lifting boxes they were being continuously monitored so that trunk loading could be assessed.

In order to simulate a "realistic" warehousing depalletizing task, subjects transferred boxes from one pallet to another. The depalletizing task started when the subject grasped the box and ended when he crossed an imaginary line that coincided with the point at which the subject was upright and facing the "palletizing" pallet. Data were collected for only this interval of time, although subjects completed the task. The lifting rate for all subjects was set at 166 boxes handled per hour and was determined from the minimum loading rate required at the local warehouse where the subjects were employed. The actual lifting cycle was one box lifted every 10 seconds (360 per hour) which was signaled by a computer tone; however, the actual lifting rate was adjusted to 166 lifts/hour by including any down time (e.g., moving pallets, filling out body part discomfort surveys by the subjects, lunch, and additional rest breaks).

Experimental Design

The experimental design consisted of a four-way, within subject design. The *independent variables* included box size, box weight, box weight and pallet region. Subjects served as a random effect. Two sizes of boxes were evaluated to represent "small" boxes and "large" boxes in typical distribution center environments. The "small" box dimensions were 8 in. by 16 in. by 12. in (H x W x D), whereas, the "large" box dimensions were 11 in. by 19.5 in. by 12 in. (H x W x D) which corresponded to volumes of 1584 in^3 and 2681 in^3, respectively. Handle conditions consisted of boxes with cut-out handles and boxes without handles. The cut-out handles were 3.5 in (8.9 cm) wide and 1 in (2.5 cm) high, positioned at the center of the sides of the boxes, 2 in (5.1 cm) below the top of the box. The position and size of the handles were similar to those commonly found on boxes in warehouse environments. The weights of the boxes in this study were 40, 50, and 60 lbs. (18.2, 22.7 and 27.3 kg.). These weights were at the upper percentiles of typical box weights in a common warehouse setting. Therefore, these weights were chosen to evaluate the effects of heavier loads on the low back and the subsequent risk of LBD.

Each of the pallets were divided into six regions corresponding to front-top, back-top, front-middle, back-middle, front-bottom, and back-bottom areas. Figure 1 shows a schematic view of these six regions on a standard pallet. The handles of the boxes in each of the regions remained at a set level corresponding approximately to: regions A and B at a height of 52.7 in (133.8 cm) from the floor, regions C and D at a height of 37.5 in (95.3 cm) from the floor, and E and F at a height of 18.75 in (47.6 cm) from the floor. The number of boxes in each region depended on the size of the box. A pallet of large boxes had four boxes in region A, three boxes in region B, and seven boxes in regions C, D, E, and F, while the pallets of small boxes had eight boxes in all six regions. In order to simplify the analysis, the six regions were combined into three layers. The top layer contained regions A and B, the middle layer contained regions C and D, and the bottom layer contained regions E and F.

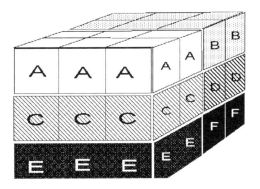

Figure 1. Six Regions of a Pallet.

Dependent variables consisted of the spine loading at the lumbosacral junction (L5/S1). Spine loading variables of interest included compression, lateral shear, and anterior-posterior (A-P) shear forces. These forces were determined via the Biodynamic EMG-assisted model developed at the Ohio State University over the past decade (Marras and Reilly, 1988; Reilly and Marras, 1989; Marras and Sommerich, 1991a; 1991b; Granata and Marras, 1993; Mirka and Marras, 1993; Granata and Marras, 1995a; 1995b; Marras and Granata, 1995). The model uses kinematic information about the trunk along with electromyographic information about the trunk musculature to estimate spinal loads as well as predict the moments imposed on the spine in three dimensional space. The spinal loads estimated in this study were the maximum values of compression force, anterior-posterior shear and lateral shear forces on the lower back at the lumbosacral joint. The trunk moments supported during the lifts were also included as dependent measures. The maximum values of sagittal bending, lateral bending, and axial twisting moments were considered in this study.

Apparatus

The Lumbar Motion Monitor (LMM) was used to collect kinematic information about the trunk. The LMM is essentially an exoskeleton of the spine in the form of a triaxial electro-goniometer that measured instantaneous position, velocity, and acceleration of the trunk in three dimensional space. The light-weight design of the LMM allowed the data be collected with minimal obstruction to the subject's movements. For more information on the design, accuracy, and application of the LMM, refer to Marras et al. (1993).

Electromyographic (EMG) activity was monitored via bi-polar surface electrodes spaced approximately 3 cm apart at the ten major trunk muscle sites. The ten muscles of interest were: right and left erector spinae; right and left latissimus dorsi; right and left internal obliques; right and left external obliques; and right and left rectus abdominis. For the standard locations of electrode placement for these muscles, refer to Mirka and Marras (1993).

A force plate (Bertec 4060A) and a set of electro-goniometers measured the external loads and moments placed on L_5/S_1 during the various calibration exertions that permitted one to "tune" the model for the individual subject. The purpose of the calibration exertions was to determine the individual gain to be used in the "open-loop" exertions. The term "open-loop" referred to exertions that use a predetermined gain to calculate internal moments and forces, rather than calculating a specific gain for each exertion. The electro-goniometers measured the relative position of L_5/S_1 with respect to the center of the force plate, along with the subject's pelvic angle. The forces and moments were translated and rotated from the center of the force plate to L_5/S_1 in this manner (Fathallah, 1995). The internal moments were adjusted to equal the external moments through the use of a gain factor. The value of the gain represented the force output of the muscles per unit area for the particular subject.

All signals from the above equipment were collected simultaneously through customized Windows™-based software developed in the Biodynamics Laboratory. The signals were collected at 100 Hz and recorded on a 486 portable computer via an analog-to-digital board.

The boxes were stacked on a standard pallet generally found in a warehouse. The pallet was constructed of wood with a width of 40 in. (101.5 cm) and a depth of 44 in. (112 cm). The small box conditions used a double-stacked pallet to allow the handles of the boxes in various regions to correspond to the heights of the large boxes which were stacked on a single pallet. The small boxes contained 5 lb boxes of salt while the large boxes contained plastic bottles of water. In order for the boxes to have the desired weight, specific amounts of material were removed from the

Table 1. Significance Summary of Biomechanical Variables.

	SPINE LOAD			MOMENT				EMG
	Shear (Lat)	Shear (A/P)	Compression	Sagittal	Lateral	Twisting	Resultant	Coactivity
Size (S)		*						
Handle (H)		*	*	*		*	*	*
Weight (W)	*	*	*	*	*	*	*	*
Region (R)	*	*	*	*	*	*	*	*
S x H								
S x W								
S x R	*	*	*	*				*
H x W								
H x R	*	*	*	*			*	*
W x R		*	*	*	*	*	*	
S x H x W								

* Significant at $\alpha \leq 0.05$

inside containers (i.e., water was drained from the center bottle). The pallets of small boxes contained six rows of ten boxes, while pallets of the large boxes comprised five rows of seven boxes.

RESULTS

The results of the analysis of variance (ANOVA) which evaluates the influence of the box characteristics and box location upon spinal loading are shown in Table 1. For the most part, compression, lateral shear, and A-P shear responded significantly to the same box characteristics. This analysis indicates that spinal loading responded primarily to changes in handle conditions, weight, and the position of the box on the pallet. It is was particularly interesting to note that all significant interaction terms involved the pallet region variable. Table 1 also shows that these same variables significantly influenced the imposed moment about the trunk. The analysis also indicated that these same three variables increased trunk muscle coactivity which, in turn, increased spinal loading.

A summary of maximum spine compression is shown as a function of the significant independent variables in Figure 2. Spine shear forces responded in a similar manner to spine compression but with a lower magnitude. Several insights can be gained from examining Figure 2. First, position on the pallet has a dramatic effect on maximum spine compression. When lifting from the bottom layers of the pallet, regions E and F in Figure 1, spinal compression is greatest. In fact, few compression forces over 3400 N (typically less than 30% of the observations regardless of box weight) were observed at the top or middle

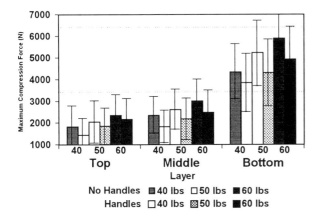

Figure 2. Maximum Spinal Compression Forces as a Function of Handle Condition, Box Weight and Pallet Layer.

layer lifts. However, depending upon the weight and handle conditions of the box between 60% and 97% of the lifts from the lower layer of the pallet resulted in spine compression values above 3400 N and would be expected to increase the risk of an occupationally related low back disorder. This figure also shows that the increases in spine compression for

every 10 lb increase in box weight was more dramatic at the bottom level of the pallet (W x R interaction).

Second, as expected as box weight increases, spine loading increases on average by about 470 N for every 10 lb increase in box weight. However, as shown in Figure 2, the increase in loading is far more dramatic when one considers the point on the pallet from where one is lifting (weight x region interaction). For example the average increase in compression for each 10 lb increase in box weight in region F (far bottom layer) was greater than 655 N or a 130 percent increase in spine compression per 10 lb increase in box weight over what was seen in region A (top layer). Thus, the position from which one lifts is far more important to spine loading than the mere weight of the box.

Third, handles had a significant load relieving effect on the spine. In general, when handles were included in the box the spine compression was equivalent to a box weight which was 10 pounds less than that actually lifted. However, as with box weight, the region on the pallet significantly influenced the effects that handles had on spine compression loading (H x R interaction). In region F, over all box weights, handles reduced spine compression by an average of 770 N, whereas, in region A the reduction was only of the order of 25 N.

Table 1 also indicated a significant influence of box size on A-P shear. Even though this was statistically significant this finding was of little practical value. Post-hoc analyses indicated that the significance of this variable was dictated by changes in A-P shear at the top level of the pallet. However, none of the shear loads at this level were large enough to exceed the spine tolerance to shear. Thus, there is little impact of this finding.

DISCUSSION

This work has facilitated our understating of which specific conditions in a depalletizing operation contribute to excessive spinal loading. In general, conditions where a worker must reach to a low level increases spinal load. This fact in itself is not new. However, we have now been able to document that within these box weight ranges the lowest level of the pallet is the only level that presents a substantial risk. In addition, we have now been able to quantitatively assess the degree or magnitude of spine loading that occurs once one bends down to the lowest level of a pallet. This work has also been able to delineate the added benefit that one could receive from the existence of hand holds or handles in a box. On average, they have the effect of reducing spine loading by an amount equivalent to about a 10 pound reduction in box weight. We have also been able to determine the collective influence of pallet location and box handles on spinal loading. Thus, this work can serve as a guide for workplace design involving pallet unloading such as is done in distribution centers and warehouses.

This effort has also facilitated our basic understanding as to why spine loading increases under the various conditions studied in this experiment. Nearly all differences in spinal loading can be explained by a corresponding difference in coactivation of the trunk musculature. When workers bend to the lower levels of the pallet they must co-contract their muscles in order to increase trunk stability. This increase in cocontraction causes the muscles within the trunk to contract simultaneously, thereby opposing each other and increasing their inefficiency. This process, in turn increases the loading of the spine. This logic adds an additional aspect to the traditional beliefs that forward bending was hazardous because of the reduction in muscle strength that occurs when flexed due to the length-strength relationship of the muscle. A combination of these two effects may explain why risk of low back disorder increases at these low levels of lifting. This same coactivation principle can explain why handles reduce the apparent loading of the spine. The effects of including hand holds and in a box is to effectively require the worker to bend to a lesser flexion angle. This, in turn, reduces the amount of coactivation as discussed previously which reduces the spine loading.

CONCLUSIONS

This study has been successful at pinpointing which box parameter variables are worthy of consideration for inclusion in the food distribution environment for the purposes of reducing the risk of work-related LBD. In general, the following conclusions can be drawn:

- Risk of LBD increases linearly as box weight increases.
- The greatest risk and loading of the spine occur during lifts from the bottom layers of the pallet. The other layers of the pallet pose acceptable lifts regardless of the box weight.
- Box size has a significant effect on risk but the difference has no practical meaning. There is no reason to control box size based upon the range of sizes explored in this study.
- Handles has a significant effect upon spine loading. The effect is particularly significant when lifting from the lowest levels of the pallet. The 40 pound box when combined with handles represents the condition with the lowest level of risk even when lifting from the lowest level of the pallet. Handles have the effect upon spine loading of lowering the box weight by 10 pounds. Thus, a 50 pound box with handles can reduce spine loading to the level of a 40 pound box without handles.

These findings provide some practical solutions to the design of the distribution center. It is clear that given the success of this study, further studies have the potential to further reduce risk through the investigation of other box features.

REFERENCES

Granata, K.P. and Marras, W.S.. (1993), An EMG-Assisted Model of Loads on the Lumbar Spine During Asymmetric Trunk Extensions. *Journal of Biomechanics*, 26(12), 1429-1438.

Granata, G. P., and Marras, W.S., (1995a), W. S., An EMG-Assisted Model of Trunk Loading During Free-Dynamic Lifting, *J. Biomechanics*, 28(11), 1309-1317.

Granata, G. P., and Marras, W. S, (1995b), The Influence of Trunk Muscle Coactivity Upon Dynamic Spinal Loads, *Spine*, 20(8), 913-919.

Marras, W. S. and Granata, K. P., (1995b), A Biomechanical Assesment and Model of Axial Twisting in the Thoraco-Lumbar Spine, *Spine*, 20(13), 1440-1451.

Marras, W.S., Lavender, S.A, Leurgans, S., Rajulu, S., Allread, W.G., Fathallah F. and Ferguson, S.A., (1993), The Role of Dynamic Three Dimensional Trunk Motion in Occupationally-Related Low Back Disorders: The Effects of Workplace Factors, Trunk Position and Trunk Motion Characteristics on Injury, *Spine*, 18(5), 617-628.

Marras, W. S., and Reilly, C. H., (1988), Networks of Internal Trunk Loading Activities Under Controlled Trunk Motion Conditions, *Spine*, 13(6), 661-667.

Marras, W.S. and Sommerich, C.M., (1991a), A Three Dimensional Motion Model of Loads on the Lumbar Spine: I. Model Structure. *Human Factors*, , 33(2), 129-137.

Marras, W.S. and Sommerich, C.M.,(1991b), A Three Dimensional Motion Model of Loads on the Lumbar Spine: I. Model Validation. *Human Factors*, 33(2), 139-149

Mirka, G.A., and Marras, W.S., (1993), A Stochastic Model of Trunk Muscle Coactivation During Trunk Bending. *Spine*, 18(11), 1396-1409.

Reilly, C. H.,and Marras, W. S., (1989), SIMULIFT: A Simulation Model of Human Trunk Motion During Lifting, *Spine*, 14(1), 5-11.

Waters, T.R., Putz-Anderson, V., Garg, A., and L.J. Fine (1993) Revised NIOSH equation for the design and evaluation of manual lifting tasks. *Ergonomics*, 36(7): 749-776.

THE EFFECTS OF BOX DIFFERENCES AND EMPLOYEE JOB EXPERIENCE
ON TRUNK KINEMATICS & LOW BACK INJURY RISK DURING DEPALLETIZING OPERATIONS

W. Gary Allread, William S. Marras, Kevin P. Granata, Kermit G. Davis, Michael J. Jorgensen
The Ohio State University
Columbus, Ohio

Workers from a local food distribution center were studied depalletizing boxes from a pallet. The objectives of this study were to determine the change in trunk kinematics associated with selecting boxes having different characteristics and to observe if there was a relationship between trunk kinematics and employee job experience. The boxes varied in terms of: size; presence/absence of handles; weight; and location on a pallet. Worker job experience also was recorded. Kinematic trunk motions and subsequent risk of low back disorder (LBD), assessed using a risk model, were studied as dependent measures. Results indicated that the weight and layer conditions influenced most of the kinematic variables. The size and handle conditions influenced fewer dependent measures. All main effects but the handle condition had an influence on LBD risk. Most of the significant interaction effects were related to layer, illustrating the tremendous influence that box location on a pallet had on trunk kinematics and LBD risk. In fact, at the bottom pallet layers, LBD risk was the same regardless of the weight lifted or the size of the box. In studying job experience, inexperienced workers were found to have LBD risk values that were, on average, 5% higher than the experienced group. This study has been successful at pinpointing which box parameters are worthy of consideration to include in a food distribution environment for the purposes of reducing the risk of work-related LBD.

INTRODUCTION

It is widely known that the order selection task in food distribution centers places the worker at risk of occupationally related low back disorders (LBDs). This job is associated with one of the greatest incidence rates of LBD in the United States. The National Association of Wholesale Grocers of America (NAWGA) and the International Foodservice Distribution Association (IFDA) disclosed that 30% of the injuries reported by food distribution warehouse workers were attributable to back sprains/ strains (Waters, 1993). In addition, over a five-year period, it was found that back injuries accounted for nearly 60% of lost work days (NIOSH Interim Report, HETA 91-405, March 1992). Hence, grocery item selectors have an incidence of low back pain that is at least as severe as other manual materials handling (MMH) jobs.

One approach to controlling this risk consists of manipulating the characteristics of the object or box to be handled in the food distribution center. A committee organized by the Food Marketing Institute is currently considering the various options available to them in order to help mediate the risk of work-related LBD in these food distribution centers. Among the options considered: 1) reducing the weight of the boxes; 2) reducing their size; or 3) incorporating handles into the boxes. However, it is currently unknown what effect these changes to the box characteristics would have on trunk kinematics and the subsequent LBD risk.

As with many MMH jobs, grocery warehouse work requires a significant amount of knowledge and expertise. It has been the experience of the authors that there are different ways in which warehouse workers perform the same job. These observations led to the hypothesis that personal characteristics, namely, job experience, may be related to motions patterns as well.

Hence, the objectives of this study were: 1) To determine the change in both trunk kinematics and LBD risk associated with selecting boxes that varied as a function of *size*, existence of *handles*, box *weight*, and the *layer* at which the box was located on the pallet; and 2) To observe if there was a relationship between trunk kinematics and employee job experience.

METHODS

Subjects

Ten experienced, male order selectors, ranging from 19 to 49 years of age, were recruited from a local food distribution center. The average (SD) weight and stature of the workers were 80.0 (8.4) kg and 180.3 (7.1) cm, respectively. Their work experience in warehouse settings ranged from 0.25 to 23 years.

Experimental Task

During the different experimental trials, the box size, coupling (handles), and box weight conditions were varied. Subjects were instructed to depalletize the entire compliment of boxes so that they could be observed picking from all locations on the pallet. They were instructed to completely pick from one layer of the pallet before unloading a new layer. While the workers were lifting boxes they were being continuously monitored so that trunk loading could be assessed.

Boxes were stacked on a standard wooden pallet as they generally would be found in a warehouse. Small boxes were placed on a double-stacked pallet to allow their handles to correspond to the heights of the large boxes, which were stacked on a single pallet.

The depalletizing task started when the subject grasped the box and ended when he crossed an imaginary line that coincided with the point where the subject was upright and facing the "palletizing" pallet. Data were collected for only this interval of time, although subjects completed the task. The lifting rate for all subjects was set at 166 boxes handled hourly. This frequency was selected for the study, as it was the minimum acceptable work level at which subjects have to perform on their jobs to "make rate."

Experimental Design

The experimental design consisted of a four-way, within-subject design. The *independent variables* included box size, handle usage, box weight, and pallet layer. Subjects served as a random effect. Two sizes of boxes were evaluated, representing "small" boxes and "large" boxes found in typical warehouse environments. Box dimensions were 20.3×40.6×30.5 cm (H×W×D) for small boxes and 27.9×49.5×30.5 cm (H×W×D) for large boxes. The position and size of the handles were similar to those commonly found on boxes in warehouse environments. Box weights studied were 18.1, 22.5, and 27.2 kg. These were at the upper percentiles of typical box weights common in warehouse settings, and they were chosen to evaluate the effects of heavier loads on trunk motions.

Each of the pallets was divided into three layers. Figure 1 shows a schematic view of these layers on a standard pallet. The bottom of each box remained at a constant level from the floor, corresponding to a height of 123.0 cm for the top layer, 89.0 cm for the middle layer, and 41.0 cm for the bottom layer. The handle cut-outs were 5.1 cm below the top of each box.

A between-subjects analysis also was performed to assess the effects of subject job experience in combination with the aforementioned box characteristics. Subjects who had one year or less experience working in the distribution center were grouped as "inexperienced" while all others were categorized as "experienced." However, one subject reported working in the warehouse for 13 months, so he was placed in the inexperienced group. This resulted in five subjects in the inexperienced group (ranging from 0.25 to 1.08 years of experience) and five subjects in the experienced group (2.58 to 23.00 years of experience).

The *dependent variables* consisted of trunk kinematics (positions, velocities, and accelerations) of the low back in

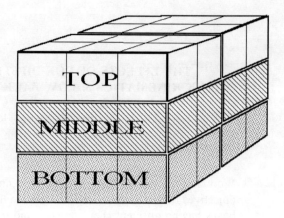

Figure 1. The three layers of a standard pallet.

the three cardinal planes of the body (lateral, sagittal, and transverse). The maximum external trunk moment during lifting was included as a dependent measure and determined using a tape measure. An assessment of LBD risk (probability of high risk group membership), as determined by Marras et al (1993), also was determined. Three kinematic parameters (maximum sagittal flexion, maximum lateral velocity, and average twisting velocity), along with lifting frequency and the maximum moment generated, comprised this risk assessment.

Apparatus

A Lumbar Motion Monitor (LMM) was used to collect kinematic information about the trunk. The LMM is essentially an exoskeleton of the spine in the form of a triaxial electrogoniometer that assessed instantaneous position, velocity, and acceleration of the trunk in three-dimensional space. The light-weight design of the LMM allowed the data be collected with minimal obstruction to the subject's movements. For more information on the design, accuracy, and application of the LMM, refer to Marras et al (1992).

RESULTS

The results of the analysis of variance (ANOVA), which evaluated the influence of the box characteristics and location upon trunk kinematics, are shown in Table 1. It lists the main effects and interactions along the first column; subsequent columns list the kinematic parameters studied. Also included are effects on the measured moment generated during box depalletizing and the LBD risk.

In general, Table 1 shows that, of the four main effects (size, handle condition, weight, and pallet layer), the weight and layer conditions influenced nearly every dependent measure. Box size mostly affected transverse plane motions but none of the lateral plane variables and only average sagittal velocity. Handle presence/absence influenced most most of the sagittal plane variables but none of the transverse plane kinematics and only average lateral velocity.

Table 1.   Summary of the ANOVA results. Main effects and interactions are listed in the first column; dependent measures are listed in subsequent columns. An asterisk (*) indicates that the effects were significant at α=0.05.

	Lateral Plane				Sagittal Plane					Transverse Plane				Moment	LBD Risk
	ROM	Avg Vel	Max Vel	Max Acc	Flex Pos	ROM	Avg Vel	Max Vel	Max Acc	ROM	Avg Vel	Max Vel	Max Acc	Moment	LBD Risk
Size (S)							*			*	*	*		*	*
Handle (H)		*			*	*		*	*					*	
Weight (W)	*	*	*	*	*	*		*	*	*	*	*	*	*	*
Layer (L)	*	*	*	*	*	*	*	*	*	*	*	*	*	*	*
S×H					*										
S×W															
S×L	*	*	*		*	*	*	*		*	*	*	*	*	*
H×W	*														
H×L	*	*	*		*	*	*			*	*	*	*	*	*
W×L	*					*	*			*	*	*		*	*
S×H×W	*									*	*	*	*		*

Note: ROM = Range of Motion

Table 2.   Summary of the ANOVA results, including worker job experience. An asterisk (*) indicates which effects were significant at α=0.05.

	LBD Risk
Size (S)	*
Handle (H)	
Weight (W)	*
Layer (L)	*
Experience (E)	*
S×H	
S×W	
S×L	*
S×E	
H×W	
H×L	*
H×E	
W×L	*
W×E	
L×E	*

Both size and handle conditions significantly affected moment, but only the size condition affected LBD risk; the handle presence/absence main effect had no influence on LBD risk. Table 1 shows that several interaction effects were present, and most were related to interactions involving pallet layer. This finding illustrates the tremendous influence that box location on a pallet had on trunk kinematics and the subsequent risk of LBD. For example,

though the handle main effect was not significant, the handle×layer interaction was significant for most kinematic variables and LBD risk. Post-hoc comparisons (not shown here) found LBD risk to be significantly higher (5%) when lifting without handles from the middle pallet layer. Handle presence/absence had no effect at the top or bottom layers.

LBD risk results from Table 1 are shown graphically in Figure 2. Several features of this chart are of note. First, LBD risk increased as depalletizing occurred at lower layers. On average, this risk was 10% higher for depalletizing from the middle layer as compared with the top layer, and another 10% higher when lifting from the bottom layer as opposed to the middle layer. Second, LBD risk increased as more weight was lifted. This amounted to about a 4% increase in risk with each 10 lb added to the load. Third, lifting small boxes produced significantly higher LBD risk values than large boxes, but only at the top layer. This accounted for the significant size×layer interactions found across several of the dependent variables. Finally, at the top and middle layers, increasing the weight handled also increased the LBD risk of the task; but at the bottom pallet layer, risk was essentially the same (71%-75%). This suggests that load weight is less of a factor than location in determining risk when the loads were placed at more extreme (i.e., lower) positions.

Table 2 shows the added effects of worker job experience on LBD risk. This table indicates that the two experience groups produced a significantly different risk value (α=0.05). In addition, the only significant two-factor interactions involved the layer condition. No higher-order interactions were found to be significant.

The effects of job work experience on LBD risk are presented in Figure 3. Results are shown for the main

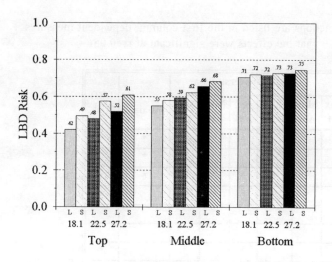

Figure 2.   Effects of box size, weight, and location on a pallet on LBD Risk.

effects. Figures 2(a) and (b) show that inexperienced subjects produced significantly higher risk (about 5%) across both box size and handle conditions. In Figure 2(c), only during lifts of 22.5 kg loads did inexperienced subjects produce higher risk. For the pallet layer condition, Figure 2(d) shows that only at the bottom layer was risk higher for the inexperienced group. This difference was over 10%. These results show that methods used by experienced workers produced less trunk motion and lower LBD risk. Also, as shown in Figure 2(d), the effects of work experience was most evident at pallet locations (bottom layer) traditionally presenting the highest risk of injury.

DISCUSSION

This work has contributed to our understating of the specific situations in a depalletizing operation that contribute to differences in trunk kinematics and the resulting risk

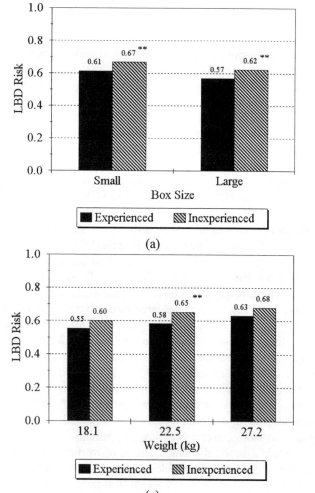

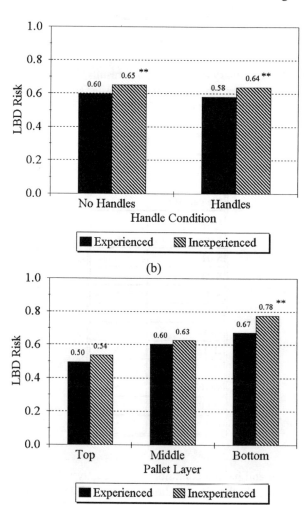

Figure 3.   Effects of job experience on LBD risk, in combination with box size (a), handle condition (b), weight (c), and pallet layer (d). The asterisks (**) indicate where inexperienced subjects produced a significantly higher LBD risk value than experienced subjects (post-hoc tests with a family-wise error rate of α=0.05).

of experiencing a LBD. In general, conditions where a worker reached to near floor increased LBD risk. This finding is not new; however, with this study we have been able to document and quantify more specific conditions that contributed to LBD risk. The study also has advanced our knowledge of how work experience may affect trunk motions and the subsequent risk of low back injury.

In terms of specific lifting conditions, this study found that the weights of the loads lifted and their pallet locations most affected trunk kinematics. Box size and the handle type main effects were less significant factors, though handle presence did result in a lower LBD risk value at the middle pallet layer. The interaction of box size and pallet layer on LBD risk indicates the difficulty of handling small boxes at the top pallet layers. This result questions the use of the layer-by-layer approach to depalletizing, which is standard policy in many warehouse environments. Further study is needed to determine if alternative approaches (i.e., a pyramiding scheme) may affect injury risk when handling boxes. The large and consistent LBD risk values found at the bottom pallet layers (above 70%) indicate that neither box weight nor size had as much influence as box location in determining risk. This finding more specifically suggests where redesign efforts should first be focused to minimize LBD risk, namely, at raising load levels.

This effort also has facilitated our understanding of the effects of job experience on trunk motions and injury risk. Across most conditions studied here, experienced workers produced lower LBD risk than those having about a year's experience or less. This difference was most dramatic at the bottom pallet layer, in which inexperienced subjects produced an LBD risk value of over 10% more than those experienced doing the work. This suggests that more experienced workers have learned ways of doing the job that reduce their risk of injury. This result also could be used to design better and more specific training for new workers on the conditions of their jobs that pose the greatest risk.

CONCLUSIONS

This study has been successful at pinpointing which box parameter variables are worthy of consideration for inclusion in the food distribution environment for the purposes of reducing the risk of work-related LBD. In general, the following conclusions were drawn.

- LBD risk increased linearly as box weight increased.
- The greatest risk occurred during lifts from the bottom layers of the pallet; however, risk was 10% lower as loads were depalletized from the middle layers of the pallet, and another 10% lower when lifted from the top layers. LBD risk values were consistent at the bottom pallet layers regardless of the box's size or weight.
- The presence of box handles produced a lower LBD risk value, but only at the middle pallet layer. Considering that raised box heights have been found to reduce

trunk motions and injury risk, and that these findings may result in MMH being done at higher levels, use of handles may indeed play a role in reducing LBD risk.
- Experienced subjects produced lower LBD risk values (about 5%) for most of the study conditions. The difference between experience levels was most dramatic at the bottom pallet layers (inexperienced subjects had a risk value over 10% greater than the experienced group), which is commonly a high risk factor regardless of work experience.

In conjunction with the companion paper to this article, these findings provide some practical solutions to the design of the distribution center. Estimates of risk were assessed both from a biomechanical and an epidemiological perspective, and similar results were achieved. It is clear that, given the success of this study, further studies have the potential to further reduce risk through the investigation of other box features.

LIMITATIONS

This study used a relatively small sample size (ten subjects). A larger sample of workers may have produced other results. Also, even though the laboratory was designed to simulate a warehouse environment, kinematics of the trunk could be different if subjects were tested at their place of work. However, experimental constraints made this impractical.

ACKNOWLEDGEMENTS

The authors would like to thank the Food Marketing Institute for funding this study.

REFERENCES

Marras, W.S., Fathallah, F.A., Miller, R.J., Davis, S.W., and Mirka, G.A., (1992). "Accuracy of a Three-Dimensional Lumbar Motion Monitor for Recording Dynamic Trunk Motion Characteristics," Int J Ind Ergo, 9, 75-87.

Marras, W. S., Lavender, S. A., Leurgans, S., Rajulu, S., Allread, W. G., Fathallah F. and Ferguson, S .A., (1993). The Role of Dynamic Three Dimensional Trunk Motion in Occupationally-Related Low Back Disorders: The Effects of Workplace Factors, Trunk Position and Trunk Motion Characteristics on Injury, Spine, 18(5), 617-628.

Waters, T.R., Putz-Anderson, V., Garg, A., and L.J. Fine (1993). Revised NIOSH equation for the design and evaluation of manual lifting tasks. Ergonomics, 36(7): 749-77.

RELATION BETWEEN BIOMECHANICAL SPINAL LOAD FACTORS AND RISK OF OCCUPATIONAL LOW-BACK DISORDERS

Kevin P. Granata, William S. Marras, Susan A. Ferguson
Biodynamics Laboratory
The Ohio State University
Columbus, Ohio

ABSTRACT

The objective of this study was to identify individual and combinations of biomechanical parameters which are associated with the probability of risk for occupationally related low-back disorders. Ten subjects performed lifting tasks simulating warehouse order selection. During the lifting exertions dynamic trunk motion, EMG, and workplace data were collected. Risk of low-back disorder was assessed from an epidemiologic model incorporating workplace factors and trunk motion data. Comparison with biomechanical results indicated that static estimates of spinal compression poorly predicts the probability of risk. Spinal compression computed from dynamic models improve the correlation, and regression models including multi-dimensional dynamic spinal loads and load rates best predict the probability of risk. The results agree with biomechanical research demonstrating vertebral failure modes are influenced by complex biomechanical interactions. This study highlights the limitations of ergonomic assessment via static compression, demonstrates the influence of biomechanical interactions upon risk of LBD, and advocate greater efforts toward understanding complex dynamic interactions in human factors research.

INTRODUCTION

Spinal compression is traditionally presumed to be the principle biomechanical mechanism associated with occupationally related low-back disorders (LBD). The NIOSH lifting guide discriminates between safe and hazardous tasks based primarily upon the static compressive loads on the spine (NIOSH 1981). Research examining the causative nature of ergonomic risk of low-back pain most often focuses on axial compressive loads associated with occupational tasks (Schultz and Andersson 1981; Freivalds et al 1984). However, epidemiologic studies indicate that repetitive twisting or lateral bending and lifting, even with relatively light loads, are significant risk factor for LBD (Kelsey et al 1984; Punnet et al 1991). These findings suggest that spinal shear and torsional loads associated with asymmetric lifting postures may be under-appreciated. Similarly, based upon the high correlation between task dynamics and the risk of LBD (Marras et al 1993), spinal and biomechanical load dynamics may be associated with the mechanism of injury.

Biomechanical literature indicates spinal injury solely due to spinal compression is unlikely. Brinkman (1986) demonstrated that compressive loads applied to in vitro lumbar vertebra failed to produce clinically relevant injuries unless pre-existing endplate damage was present. Adams et al (1987) stated that compression in the lumbar spine, in the absence of forward bending moments, cannot "injure the soft tissue without first causing gross damage to the vertebrae." Thus, tissue failure and associated LBD is more likely generated by combinations of multi-dimensional spinal loads. In a theoretical analysis, Shirazi-Adl (1989) found that bending moments and shear forces combined with axial compression significantly increased the risk of injury to the lumbar disc. Yingling et al (1995) demonstrated load rate influences the ultimate strength, stiffness, displacement, and failure mechanisms. Clearly, occupationally related low-back pain associated with vertebral tissue injury is unlikely caused by static compression alone.

The objective of this study was to identify the multi-dimensional, dynamic biomechanical factors associated with LBD risk. Specifically, answers to two questions were sought. First, does axial compression in the lumbar spine correlate well with the risk of LBD associated with a prescribed task? Second, does the inclusion of three-dimensional, dynamic spinal load and load rate data more accurately describe the probability of risk than static compression alone? Identifying possible biomechanical parameters capable of discriminating between safe and hazardous tasks may contribute to the development of more accurate and robust ergonomic analyses and reduced incidence of LBD in the workplace.

METHODS

Ten experienced warehouse order selectors 19 to 49 years of age were recruited from a local food distribution center. The weight and stature of the subjects was 80.1 ±8.4 kg. (176 lb.) and 180.3 ±7.1 cm. (71 in.) respectively. The subjects' experience as warehouse selectors ranged from 0.25 to 23 years.

In order to simulate realistic warehouse working conditions, subjects were required to lift boxes ranging from 18.2 kg. (40 lb.) to 27.3 kg. (60 lb.) from one pallet to another until the entire pallet load, an average of 35 boxes, was transferred. Twelve pallets of boxes were moved at a frequency of 166 lifts per hour, simulating a "slow" five hour work day. Dynamic, three-dimensional trunk motion data were collected from the Lumbar Motion Monitor (LMM) (Marras et al 1993) and integrated myoelectric (EMG) activity of ten trunk muscles were collected from bi-polar surface electrodes (as per Mirka and Marras 1993) during the depalletizing tasks. Prior to beginning each pallet, a set of "test" exertions were performed while standing on a force plate (Bertec 4060A) and with added electrogoniometers to measure the location and orientation of the lumbo-sacral spine relative to the center of a force plate. The test exertions were designed to permit data quality assurance and supply calibration data for biomechanical analyses.

Each lifting task was assigned a probability of being at high risk for occupationally related LBD. The assessment was achieved from a multiple logistic regression model of dynamic trunk motion parameters and workplace factors (Marras et al 1993). This epidemiologic model was developed from a database developed from on site measurements over 400 industrial workers, and incorporated factors including the lifting moment, lift rate, multi-dimensional trunk range of motion and velocities. As subjects' lifted each box, a measure of risk was assigned to that task by the epidemiologic model, and saved for comparison with a

variety of biomechanical parameters including spinal loads and load rates.

An EMG-assisted biomechanical model was employed to determine the dynamic spinal loads associated with the lifting exertions (Granata and Marras, 1995). The analysis incorporates normalized EMG data, anthropometrically scaled muscle cross-sectional areas and vector directions as well as force-length and force-velocity relations to determine the force supplied by ten dynamically co-contracting muscles. Three-dimensional spinal loads were determined from the vector sum of the muscle forces, and trunk moments from the sum of vector products of muscle forces and moment arms. Direct comparison of dynamic trunk moments determined from the force plate data collected during the test exertions, with predicted trunk moments determined from the biomechanical model during those exertions provided subject dependent calibration values and model validation parameters. Model output included peak spinal loads as well as the peak value of the rate of change of spinal load, i.e. load rate.

For comparison with the dynamic biomechanical data, static compressive forces were computed for each task using the method outlined in Chaffin and Andersson (1981). External moments were determined from the product of box weight and moment arm distance from the trunk measured during each lifting exertion. Upper body mass and center of mass were determined from subject anthropometry and multiplicative coefficients cited in Chaffin and Andersson (1981). The restorative force generated by a single extensor muscle and spinal compression was determined from the muscle moment arm and total trunk moment. Thus, static estimates of spinal compression were determined for comparison with probability of LBD risk and dynamic biomechanical values.

Linear regression analyses were used to examine the association between the probability of a high risk classification and spine biomechanics factors. Correlation values were achieved for all possible models of combined biomechanical factors, including regression models of individual factors. The association between biomechanical parameters and occupationally related LBD were evaluated based upon the correlation coefficient (R^2) between the regression model of biomechanical values

RESULTS

The correlation between the probability of risk represented in the simulated warehouse tasks and independent biomechanical factors are presented in table 1. Results indicate that dynamic spinal compression represented the strongest individual correlation with probability of risk at $R^2 = 0.443$. Static compression represented the poorest individual correlation with risk, $R^2 = 0.137$. Lateral and AP shear forces were also poorly correlated with the probability of risk as individual factors. R^2 values for load rate in the compressive direction were slightly lower than spinal compression at $R^2 = 0.429$. However, shear load rate in the lateral and AP directions were better correlated with predicted risk than the shear forces.

Table 1. Correlation (R^2) between individual biomechanical factors and probability of high risk for LBD. Results indicate dynamic compression and spinal load rates are better correlated to LBD risk than static estimates of compression or dynamic estimates of shear load.

	X	Y	Z
Static Load	--	--	0.137
Dynamic Load	0.193	0.197	0.443
Load Rate	0.344	0.354	0.429

X = Lateral Shear Force,   Lateral Shear Load Rate
Y = AP Shear Force,   AP Shear Load Rate
Z = Compressive Force,   Compressive Load Rate

Incorporating multi-dimensional factors and dynamic parameters dramatically improves the predictive power of the regression model. Table 2 identifies the three best regression models containing two, three and four biomechanical factors each. Models with greater than four biomechanical factors demonstrated reduced predictive power. The strongest two parameter model includes dynamic compression force and AP shear force, which improved the correlation with probability of risk to 0.473. Combined load rate data also demonstrates improved correlation with probability of risk. Predicted probability of risk is best represented by a combination of dynamic spinal compression, AP shear force, lateral shear load rate and AP shear load rate, $R^2 = 0.517$. Combinations of biomechanical factors with static compression failed to generate the levels of correlation demonstrated by equivalent models with dynamic compression.

DISCUSSION

Analyses of LBD risk has traditionally focused on static compressive loads in the spine. However, epidemiologic and biomechanical data suggest the risk associated with shear and torsional loads on the spine significantly enhance the risk of occupationally related LBD. Similarly, dynamic loading parameters may be related to LBD risk. Thus, examination of dynamic, and multi-dimensional spine biomechanics factors may improve the ability to identify hazardous occupational tasks.

Axial compressive loads on the spine determined from a fundamental static biomechanical model were significantly ($p<.001$) lower and poorer predictors of risk than dynamic spinal compression. The average static compression, 2700 N., was 20% lower than the average dynamic value of 3400 N. Freivalds et al (1984) and McGill and Norman (1985) similarly indicated static analyses underpredict dynamic spinal compression by 20 to 40%. As a single variable, dynamic compression correlated with the results from the epidemiologic risk model at $R^2 = 0.443$, much better than the statically determined compression, $R^2 = 0.137$. This demonstrates the static and dynamic determination of spinal load are largely unrelated ($R^2 = 0.31$), as illustrated by the scatter of points when plotting the dynamic compression versus the static values (Figure 1). Since the static determination of compression fails to account for the variability associated with lifting dynamics, it becomes difficult for this variable, or combinations including it, to discriminate between safe and hazardous tasks in a dynamic work environment.

Table 2. Correlation (R^2) between combinations of biomechanical factors and task probability of high risk for LBD. Only the three best regression models in each category are provided.

Regression Models	R^2
Two Factor Models	
Fy + Fz	0.473
Fx + LRz	0.466
LRy + LRz	0.433
Three Factor Models	
Fy + Fz + LRx	0.501
Fx + Fy + LRz	0.475
Fy + LRx + LRz	0.474
Four Factor Models	
Fy + Fz + LRx + LRy	0.517
Fx + Fy + LRx + LRz	0.494
Fy + LRx + LRy + LRz	0.490

F_X = Lateral Shear Force LR_X = Lateral Load Rate
F_Y = AP Shear Force LR_Y = AP Load Rate
F_Z = Compressive Force LR_Z = Compr. Load Rate

Results indicated that a regression model including multi-dimensional spinal loads improves the predictive power for LBD risk when compared to a model including only axial compression. This agrees with the theoretical assessment of Shirazi-Adl (1989) who concluded that shear forces combined with compression increases the risk of vertebral disc failure. The influence of bending and torsional moments supported by the spine may also enhance the risk of injury. In this study, the relation between passive spinal moment and risk of LBD could not be quantified. However, considering the trend demonstrating greater fidelity when shear loads are included, and the biomechanical evidence relating spinal bending and torsional moment with vertebral failure, one might be tempted to hypothesize that inclusion of those parameters would further enhance the predictive power of the biomechanical regression model.

Lifting dynamics may directly influence the load and tolerance of the spine to injury. Analyses of dynamic exertions have demonstrated spinal load increases with velocity and acceleration (Freivalds 1984; Granata and Marras, 1995). Furthermore, those increased loads must be supported by a spinal column wherein the tolerance may be influenced by the load rate (Yingling et al 1995). There are few studies examining the influence of load rate upon injury mechanisms in the lumbar spine, but our results indicate that load rate may adversely affect the safety of a lifting exertion in terms of the probability of LBD risk. Passive spine bending moment rate was not examined in this study. Considering the known response of biologic tissue to viscous loading, future research might demonstrate moment rate offers significant insight into the injury mechanism associated with occupational tasks.

The interpretation of these results is limited by the use of an epidemiologic model as a baseline. This study examined the correlation between dynamic biomechanical parameters and predicted LBD risk, which in turn represented the actual risk values measured in industry (Marras et al 1993). This method to estimate LBD risk undoubtedly introduced variability into the analyses. However, in order to achieve the appropriate biomechanical data it was necessary to simulate the warehouse selection environment, requiring an epidemiologic model to estimate probability of risk. The predictive ability of the epidemiologic model has been determined to be quite high, more than three times better than the NIOSH lifting model. Thus, the data represents an association between biomechanical factors and

probability of risk, but cannot be interpreted as causative.

The research methods allow several potential interpretations of the results. The most direct conclusion is that LBD risk is better predicted from dynamic, multi-dimensional biomechanics than static assessment of compression alone. Other possibilities include potential limitations or interactions in the biomechanical and epidemiologic models. However, both these models have been exhaustively validated in separate research efforts. Thus, the enhanced ability to predict LBD risk from multi-dimensional dynamic spinal load and load rate factors is believed to be valid. Furthermore, the implication agrees with the biomechanical literature wherein spinal tolerance and material failure has been associated with multi-dimensional loads on the functional units of the spine.

The improved predictive ability associated with multi-dimensional, dynamic analyses of coactive load and load rate is generated at the high cost of biomechanical complexity. This technical cost-benefit may appear prohibitive for simple ergonomic assessments of the workplace. However, identifying the one-dimensional static parameters may ignore the injury mechanism associated with the greatest number of workplace injuries. Thus, these results advocate greater efforts toward understanding complex dynamic interactions in human factors research.

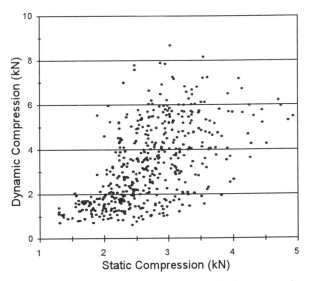

Figure 1. Plot of dynamic estimates of spinal compression versus static estimates of compression. Results indicate the static and dynamic compressive values are poorly correlated ($R^2 = 0.36$). Consequently, LBD risk factors associated with workplace dynamics may be overlooked when using static biomechanical analyses.

REFERENCES

Adams MA, Dolan P, Hutton WC. Diurnal variations in the stresses on the lumbar spine. *Spine* 1987; 12 (2), 130-137

Brinkman P. Injury of the anulus fibrosis and disc protrusions. An in vitro investigation on human lumbar discs. *Spine* 1986; 11 (2), 149-153

Chaffin D.B. and G.B.J. Anderson (1984) Occupational Biomechanics, John Wiley and Sons, N.Y.

Freivalds A, Chaffin DB, Garg A, Lee KS, A dynamic biomechanical evaluation of lifting maximum acceptable loads. *J Biomechanics* 1984; 17 (4): 251-262

Granata KP, Marras WS. An EMG assisted model of biomechanical trunk loading during free-dynamic lifting. *J. Biomechanics* 1995; 28 (11): 1309-1317,

Kelsey KL, Githens PB, White AA III, Holford TR, Walter SD, O'Conner T, Ostfeld AM, Weil U, Southwick WO, Calogero JA. (1984) An epidemiologic study of lifting and twisting on the job and risk for acute prolapsed lumbar intervertebral disc. *J Ortho Res*, 2(1): 61-66

Marras WS, Lavender SA, Leurgans SE, Rajulu SL, Allread WG, Fathallah FA, Ferguson SA. (1993) The role of dynamic three-dimensional trunk motion in occupationally-related low back disorders: The effects of workplace factors, trunk position and trunk motion characteristics on risk of injury. *Spine*, 18 (5), 617-628

McGill S.M., Norman R.W. Dynamically and statically determined low back moments during lifting. *J. Biomechanics* 1985; 8 (12): 877-885.

Mirka GA, Marras WS. A stochastic model of trunk muscle coactivation during trunk bending. *Spine* 1993; 18 (11): 1396-1409

NIOSH (1981) National Institute for Occupational Safety and Health: A Work Practices Guide for Manual Lifting Tech. Report No.81-122, U.S. Dept. of Health and Human Services (NIOSH), Cincinnati, Oh.

Punnet L., Fine L.J., Keyserling W.M., Herrin G.D. and D.B. Chaffin (1991) A case-referent study of back disorders in automobile assembly workers : The health effects of non-neural trunk postures. *Am. J. Epidemiology*,

Schultz A, Anderson G. Analysis of loads on the lumbar spine. *Spine* 1981; 6 (1): 76-82

Shirazi-Adl A. Stress in fibers of a lumbar disc, analysis of the role of lifting in producing disc prolapse. *Spine*, 1989; 14 (1): 96-103.

Yingling VR, Callaghan JP, McGill SM. The effect of load rate on the mechanical properties of porcine spinal motion segments. *Proceeding of the 19th Annual Meeting of the American Society of Biomechanics* 1995; 119-120

THREE-DIMENSIONAL SPINAL LOADING DURING COMPLEX LIFTING TASKS

Fadi A. Fathallah[*], William S. Marras, and Mohamad Parnianpour

Biodynamics Laboratory
The Ohio State Univerity
Columbus, Ohio

[*]Currently at the Liberty Mutual Research Center for Safety & Health
Hopkinton, Massachusetts

Knowledge of the complex *three-dimensional* loads imposed on the spine during typical manual materials handling (MMH) tasks could provide more insights about the mechanical etiology of low back injuries in occupational settings. Comprehensive treatment of such information has been lacking. Most previous studies quantified spinal loading in terms of compressive forces alone. However, there is enough empirical and epidemiological evidence to indicate that the shear forces imposed on the spine may be more important than mere compression. Hence, the purpose of this study was to assess, *in-vivo*, the three-dimensional complex spinal loading associated with lifting tasks. Subjects performed simulated lifting tasks with varying workplace characteristic. An EMG-assisted model provided the continuous three-dimensional spinal loads. Asymmetric (complex) lifting tasks showed distinctive loading patterns from those observed under symmetric conditions. Simultaneous occurrences of spinal loads in all three directions (compression and shear forces) were patterns unique under the "risky" asymmetric lifting conditions. These situations could be identified and abated through proper workplace design. In conclusion, this approach allow the determination of the magnitudes and temporal occurrence(s) of complex spinal loading, and assess the sensitivity of these loading patterns to workplace characteristics.

INTRODUCTION

Pope et al. (1991) best described the state of low back pain and its effect on industry as: "Low back pain (LBP) has been called the nemesis of medicine and the albatross of industry." Back pain has been established to be one of the most common and significant musculoskeletal problem in the US, leading to substantial levels of disability, morbidity, and economic loss (Webster and Snook, 1994; Praemer et al., 1992; Andersson, 1991; Hollbrook et al., 1984). Finding means to abate or at least reduce the risk of low back disorders in occupational settings would benefit workers and employers alike.

There is significant epidemiological and biomechanical evidence that implicates complex motions and loading as important risk factors for LBD. Several epidemiological studies have shown that risk increases under these *combined* risk factors conditions. Magora (1973) has found that twisting and lateral bending were significant risk factors only when occurring simultaneously with sudden movements (dynamic activities). Kelsey et al. (1984) have also indicated that occupational LBD risk increases in jobs involving lifting activities when the lift is combined with twisting action. Other epidemiological studies have also indicated combined (asymmetrical) motions as potential risk factors for developing low back disorders (e.g., Marras et al., 1993; Drury et al., 1982; Damkot et al., 1980; Brown, 1975). The effect of

complex motions on the spinal structure *in-vivo* is not well understood. These complex spinal motions generate complex loading patterns on the spinal elements (e.g., combined lateral shear and compressive loading). Pope et al. (1991) have expressed the importance of quantifying such complex combined loads; however, the authors indicated that such a task is difficult to investigate under both field and laboratory situations. Biomechanical changes in spine tolerance are also expected to occur when these risk factors occur simultaneously. Shirazi-Adl (1994, 1991, 1989) has demonstrated how strains in spinal disc annulus fibers are dramatically increased under combined lateral and twisting conditions, reaching levels that may exceed the tissue tolerance limits. These latter studies as well as other *in-vitro* studies have implicated combined loading of the spinal structure as a mechanism for back injury. Therefore, for prevention of LBD, it is essential to quantify the types and magnitudes of the three-dimensional mechanical loading experienced by the spinal structure when subjected to dynamic combined or complex motions. Knowledge of this type of information allows us to identify, in detail, the situations that compromise the integrity of the structure, and thereby help reduce injuries resulting from these conditions. Hence, the main objective of this study was to investigate the magnitudes and patterns of the three-dimensional spinal loading under complex lifting conditions and assess their potential risk.

METHODS

Subjects

Eleven healthy male subjects volunteered to participate in this experiment. Mean (standard deviation) age was 28.4 (4.4), mean stature and weight were 180.7 cm (3.7) and 78.6 kg (10.8), respectively. A questionnaire was administered to each subject to ensure that there was no recent history of back disorders.

Apparatus

An EMG system collected signals from ten pairs of bipolar silver-silver/chloride surface electrodes affixed over the specific locations of the ten muscles of interest. Five pairs of muscles were studied: 1) Right/left latissimus dorsi, 2) right/left erector spinae, 3) right/left rectus abdominus, 4) right/left external obliques, and 5) right/left internal obliques.

The EMG signals were first amplified 1000x by preamplifiers placed at short distances from the muscle sites (less than 25 cm). The signals were further amplified in the main amplifier between 30x to 55x, depending on the muscle and the subject under consideration. Also, to eliminate undesired signal artifacts, the signals were low pass filtered at 1000 Hz. The filtered signals were rectified and processed via a 20 ms moving average window (integration constant). The window was moved at 2 ms increments. An asymmetric reference frame (Marras and Mirka, 1992) was used to solicit static maximum voluntary contractions (MVCs) in six directions (flexion, extension, left and right twist, left and right lateral bending).

Three-dimensional continuous position, velocity, and acceleration of the trunk were determined using the Lumbar Motion Monitor (LMM) developed at the Ohio State University Biodynamics Laboratory. Three-dimensional external forces and moments about L5/S1 were monitored by the combination of a Bertec 4060A force plate (Bertec, Worthington, Ohio) and two electrogoniometers used to determine the continuous location and orientation of the L5/S1 joint in three-dimensional (3-D) space (Fathallah, 1995). This system provided a mechanism to monitor torque or moment about L5/S1 during a lift. The EMG-assisted model provided estimates of the internal moments required to achieve the balanced equilibrium condition(s) (e.g., Granata and Marras, 1995).

During the experiment the subject lifted a wooden box that was filled with the proper weight of the prescribed condition. For each condition, the subject was provided with an auditory signal (loud tone) indicating the lift pace by identifying the start and end of each lift phase. This tone was necessary to control the lift speed (duration). All the analog signals gathered from the devices described above were collected at 100 Hz via a 12-bit 32-channel Analog-to-Digital (A/D) converter connected to a 386-based microcomputer. Figure 1 shows a subject performing a symmetric lift.

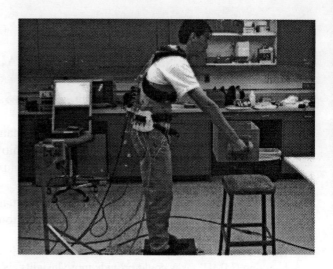

Figure 1. *A subject shown performing a typical symmetric lift.*

Design

The experiment consisted of a three-way within-subject design. The independent variables included: 1) Lift type, 2) speed of lift (duration of movement), and 3) weight handled. These variables were chosen to solicit complex motion conditions under varying workplace parameters similar to those observed in industrial settings (Marras et al., 1993). The lift types investigated were symmetric lifts and asymmetric (complex) lifts. Lift speed was set at three levels: low (2 sec per lift), medium (1.5 sec per lift) and high (1 s per lift). Three weight levels were considered: low (22 N), medium (67 N), and high (156 N).

The dependent variables consisted of either "average maximum" (peak) or "continuous" three-dimensional spinal loading magnitudes at the L5/S1 level in terms of the compressive, anterior/posterior shear, and medial/lateral shear forces.

Procedure

The subject was provided with written instructions detailing various parts of the experiment. The subject first performed the six *static* MVCs in a randomized order. There was a two-minute rest period given between two exertions. After all six MVCs were collected, the subject was ready to perform the dynamic lifting tasks. The subject was first fitted with the LMM along with the proper attachments of other equipment. The subject was given time to familiarize himself with the tasks, and to ask any pertinent questions. Each task consisted of two lifts and one lowering with tones indicating the start and end of each part.

In the symmetric condition, the box (weight) was placed on a platform in front of the subject just above knee height, and at a horizontal distance equal to his arm length. At the onset of the tone, the subject was asked to lift the box from its location to a position as close as possible to the body. For the complex (asymmetric) conditions, the box was placed in front of the subject in the same manner as the symmetric condition; however, in this case the subject was asked to set the box down at another platform placed to his right at an angle perpendicular to the mid-sagittal plane. The platform height was set at the level of the subject's iliac crest and was placed at about arm length distance horizontally. Within a given type of lift (symmetric or asymmetric), the three weights and speeds were presented in random order to control for carryover effects.

Data Analysis

In order to assess the three-dimensional loading on the spine, the current study utilized an extension of the EMG-assisted modeling approach developed in Biodynamics Laboratory at the Ohio State University (e.g., Marras and Reilly, 1988; Granata and Marras, 1995). In order to investigate the effects of weight, speed and lift type on the internal loading of the spine, multivariate analysis of variance (MANOVA) was conducted. The internal spinal loading was characterized by average maximum compression, anterior shear, and lateral shear. The *simultaneous* occurrence of combined loading was quantified by bivariate distributions of lateral and anterior shear forces. Note that, in this paper, only the lifting portions of the tasks were considered in all analyses. The bivariate distributions were statistically compared among conditions using the two-dimensional Kolmogorov-Smirnov test. (2-D K-S) (Fasano and Farnceschini, 1987).

RESULTS

The results of the MANOVA for the *magnitudes* of maximum three-dimensional loading revealed a significant interaction effect between weight and lift type (p < 0.0001) (Table 1). This interaction was significant for both compression and lateral shear (p < 0.0001), but not for anterior shear. Figure 2 depicts such interactive pattern. With the exception of lateral shear under symmetric lifts, maximum spinal loading significantly increased with increases of the box weight (p < 0.001). For the low weight condition, only lateral shear showed a significant difference between the two lifting types. With increase in the lift weight, the anterior shear was significantly different between the lifting types; and compression was significantly different only for the medium weight condition (Newman-Keuls; p < 0.05).

Figures 3a and 3b depict the continuous three-dimensional loading profiles of a typical "high weight" symmetric and asymmetric lifts, respectively. Both types of lifts had increased compressive and anterior shear at the beginning of the lift with low levels of lateral shear. This corresponded to the phase where the subject lifted the load off the first platform. However, as the subject progressed in executing the task, loading magnitudes in the symmetric condition continued to systematically decrease until reaching considerably low loading levels (see Figure 3a). On the other hand, the asymmetric lift exhibited rather complex loading patterns

TABLE 1

Multivariate analysis of variance (MANOVA) and analysis of variance (ANOVA) for maximum spinal loading : Compression (COMP), lateral shear (LATSHR), and anterior shear (ANTSHR). Type I error probabilities are shown only for significant effects.

Factor	MANOVA	ANOVA COMP	LATSHR	ANTSHR
L	0.00001	0.00001	0.00001	0.00002
W	0.00001	0.00031	0.00001	----
S	----	----	----	----
L X W	0.00001	0.00001	0.00001	----
L X S	----	----	----	----
W X S	----	----	----	----
L X W X S	----	----	----	----

L = Lift Type; W = Weight; S = Speed

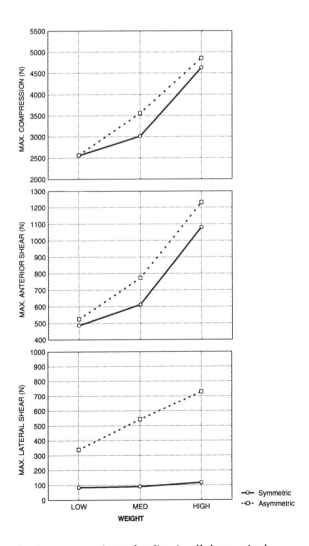

Figure 2. *Average maximum loading in all three spinal loading directions under each weight*

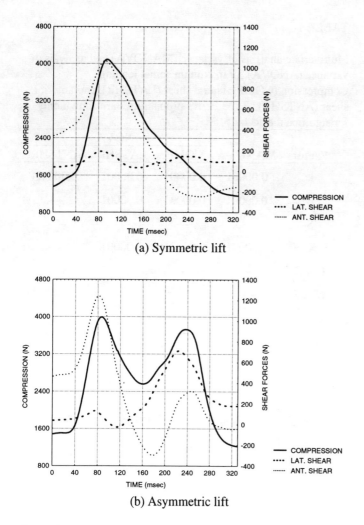

Figure 3. *Example of three-dimensional continuous profile for a "high weight" symmetric lift (a), and asymmetric lift (b).*

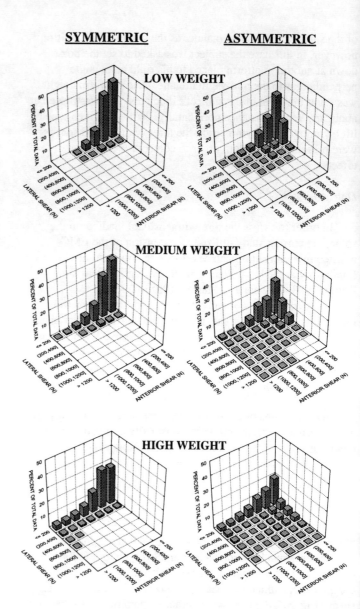

Figure 4. *Bivariate lateral and anterior shear distributions for each weight and lift type (symmetric/asymmetric). Note that under each weight condition, the distributions were statistically different between the two lifting types (2-D K-S, p < 0.001)*

especially in the last phase of lift (most asymmetric region). In that phase, there was a notable period of **simultaneous** occurrence of compressive and shear forces (see Figure 3b). To fully explore the combined (complex) loading conditions, the simultaneous continuous combined anterior and lateral shear forces were investigated. Figure 4 shows the bivariate distributions of anterior and lateral shear under each of the three weight levels for all subjects. This figure provided a means to investigate the nature of complex loading on the spine. As shown earlier, the magnitude of spinal loading significantly increased with increase in the weight; however, for each weight level, the patterns of combined loading differ between the symmetric and asymmetric lifts. Under symmetric lifts, lateral shear forces were maintained at rather low levels. For all three weights, over 90% of the time the lateral shear force was maintained under 200 N during symmetric lifts. In contrast, during asymmetric lifts, lateral shear was observed at increased levels for a considerable proportion of the time. For instance, for both the medium and high weight asymmetric conditions, about 40% of the total data was observed at lateral

shear levels exceeding 200 N with some instances reaching higher than 1200 N **combined** with substantial magnitudes of anterior shear loading.

DISCUSSION AND CONCLUSIONS

This study provided us with estimates of the combined three-dimensional loads experienced at the L5/S1 spinal unit during various lifting conditions. Although the compressive forces reported in this study were within the range reported by several studies that shared similar experimental conditions (e.g., Adams and Dolan, 1995; Cholewicki et al., 1995), very few dynamic lifting studies have quantified the three-dimensional loading experienced on the spinal structures. Most studies report only the compressive loads without quantifying the shear forces. Also, the study is unique due to its capability to account not only for the effects of dynamic motion variables and inertial forces on the net external moment, but also the effects on the internal loading and muscular recruitment parameters.

The complex asymmetric lifting conditions in general and those with increased weight in particular exhibited elevated levels of complex spinal loading patterns. More specifically, elevated magnitudes of compressive loading **combined** with lateral and anterior shear forces seemed to be a key element in distinguishing such "risky" situations. This approach allows the identification of loading patterns that expose the spinal structures to loading levels which may exceed their tolerance limits and hence, could lead to injury. These situations could be identified and abated through proper workplace design. In sum, the level of detail provided by the present approach allows us to determine the **magnitudes and temporal occurrence(s)** of complex or combined loading, and assess the sensitivity of these loading patterns to workplace characteristics.

ACNOWLEDGEMENTS

Partial funding for this study has been provided by the Ohio Bureau of Workers' Compensation, Division of Safety and Hygiene. The authors would like to thank Dr. Kevin Granata for his invaluable assistance in the experimental phase of this study.

REFERENCES

Adams, M.A., and Dolan, P. (1995). Recent advances in lumbar spinal mechanics and their clinical significance. *Clinical Biomechanics,* 10,3-19.

Andersson, G.B. (1991). The epidemiology of spinal disorders. In J.W. Frymoyer (Ed.), *The adult Spine* (pp. 107-146). New York: Raven Press.

Brown, J.R. (1975). Factors contributing to the development of low back pain in industrial workers. *Amer Ind Hyg Assoc J,* 36, 26-31.

Cholewicki, J., McGill, S.M., and Norman, R.W. (1995). Comparison of muscle forces and joint load from an optimization and EMG assisted lumbar spine model: Towards development of a hybrid approach. *J. Biomechanics,* 28,321-331.

Damkot, D.K., Pope, M.H., Lord, J., and Frymoyer, J.W. (1984). The relationship between work history, work environment and low back pain in men. *Spine,* 9(4), 395-399.

Drury, C.G., Law, C.H., and Pawenski, C.S. (1982). A survey of industrial box handling. *Human Factors,* 24, 553-65.

Fasano, G., and Franceschini, A. (1987). A multidimensional version of the Kolmogorov-Smirnov test. *Mon Not R Astr Soc,* 255, 155-170.

Fathallah, F.A. (1995). *Coupled spine motions, spine loading and risk of occupationally-related low back disorders.*

Unpublished doctoral dissertation, The Ohio State University, Columbus, OH.

Granata, K.P., and Marras, W.S. (1995). An EMG-assisted model of trunk loading during free-dynamic lifting. *Journal of Biomechanics* 28, 1309-1317.

Hollbrook, T.L., Grazier, K., Kelsey, J.L., and Stauffer, R.N. (1984). The frequency of occurrence, impact and cost of selected musculoskeletal conditions in the United States. *American Academy of Orthopeadic Surgeons.* Chicago, IL, pp. 24-25.

Kelsey, J.L., Githens, P.B., White, A.A., et al. (1984). An epidemiological study of lifting and twisting on the job and risk for acute prolapsed lumbar intervertebral disc. *J. Orth. Res.,* 2, 61-66.

Magora, A. (1973). Investigation of the relation between low back pain and occupation: 4. Physical requirements: Bending, rotation, reaching and sudden maximal effort. *Scand. J. Rehabil. Med,* 5, 186-190.

Marras, W.S., and Reilly, C.H. (1988). Networks of internal trunk-loading activities under controlled trunk-motion conditions. *Spine* 13,661-667.

Marras, W.S., and Mirka, G.A. (1992). A comprehensive evaluation of trunk response to asymmetric trunk motion. *Spine,* 17,318-326.

Marras, W.S., Lavender, S., Leurgans, S., Rajulu, S.L., Allread, W.G., Fathallah, F.A., and Ferguson, S.A. (1993). The role of dynamic three-dimensional trunk motion in occupationally-related low back disorders: The effects of workplace factors, trunk position and trunk motion characteristics on risk of injury. *Spine,* 18(5), 617-628.

Pope, M.H., Andersson, G.B.J, Frymoyer, J.W., and Chaffin, D.B. (1991). *Occupational Low back pain: Assessment, treatment, and prevention.* Chicago: Mosby Year Book.

Praemer, A., Furner, S., and Rice, D.P. (1992). Musculoskeletal conditions in the United States. *American Academy of Orthopeadic Surgeons.* Park Ridge, IL, pp. 23-33.

Shirazi-Adl, A. (1989). Strain in fibers of a lumbar disc: analysis of the role of lifting in producing disc prolapse. *Spine,* 14, 96-103.

Shirazi-Adl, A. (1991). Finite-element evaluation of contact loads on facets of an L2-L3 lumbar segment in complex loads. *Spine* 16,533-513.

Shirazi-Adl, A. (1994). Biomechanics of the lumbar spine in sagittal/lateral moments. *Spine,* 19(21), 914-927.

Webster, B.S., and Snook, S.H. (1994). The cost of 1989 workers' compensation low back pain claims. *Spine* 19,1111-1116.

PERSPECTIVES ON U.S. ARMY PHYSICAL REQUIREMENTS: PROBLEMS AND CURRENT APPROACHES TO SOLUTIONS

Donald B. Headley (Chair)
U.S. Army Research Laboratory-
Human Research & Engineering Directorate
Aberdeen Proving Ground, MD

Valerie J. Rice (Co-chair)
Military Performance & Neuroscience Division
U.S. Army Research Institute of Environmental Medicine
Natick, MA

SYMPOSIUM ABSTRACT

Overview

Of the various types of occupational standards (e.g., cognitive, formal education, manual dexterity) two of the more important considerations for military operations are physical demands and anthropometric characteristics. Fifty-seven percent of Army entry level jobs have a physical demands rating of heavy or very heavy (i.e., occasional lifting of 100 or more pounds, & frequent or constant lifting of 50 pounds). Also, the issue of fit is important given that current assignment opportunities for women have greatly increased, and given that typically the design envelope on a given dimension for most equipment has been the 5th to 95th percentile male. Inadequate matching of personnel capabilities and task assignments results in increased costs in the form of supplemental training, inadequate performance, and work-related injuries.

Summary of the Four Presentations

In order to address these issues, the mismatches in physical demands and anthropometric design parameters must first be clearly defined. The first two presentations in this symposium address these issues. The initial presentation demonstrates an occupation specific data base management system which can be used to assist research efforts, identify jobs most in need of redesign, and set training and performance standards. The second presentation describes a research effort that identifies anthropometric dimensions of military clothing, equipment, and workstations and the level of compatibility (height-reach characteristics) of these fielded systems with female soldier anthropometrics.

Once the potential problems are clearly defined, remedial methods must be explored. Task/equipment redesign, personnel selection, and training are methods typically suggested for addressing ergonomic issues and enhancing job performance. Two of these key methods are addressed in this symposium. The third presentation discusses physical training issues, including general versus task-specific training regimens and their potential influence in improving military performance, increasing soldier safety, and decreasing attrition. The final presentation describes a procedure for identifying, evaluating, and redesigning physically demanding tasks and equipment. This solution is the recommended procedure for ensuring soldier safety by designing a task to eliminate the hazard itself.

Project Goals

This symposium follows an ergonomic systems approach of analyzing the problem, designing and implementing appropriate interventions and establishing methods for evaluating the effectiveness of those interventions (e.g., analyzing cost-benefit tradeoffs of potential implementations). By closely matching job demands with human capabilities and limitations, the ability of all soldiers to perform their duties is enhanced. The goals of these projects and similar military research efforts represent a unified approach to enhance personnel utilization, sustain performance, and ensure the health and safety of our service members.

PHYSICAL TRAINING AND MANUAL-MATERIAL HANDLING:
LITERATURE AND MILITARY APPLICATIONS

Joseph J. Knapik
U.S. Army Research Laboratory
Human Research and Engineering Directorate
Aberdeen Proving Ground, MD 21005

Manual material handling (MMH) is performed in 83% of all U.S. Army enlisted military occupations. Studies that have examined the influence of physical training on MMH can be separated into two types: those that use the same task for testing and training (task-specific training studies) and those that do not (general training studies). Reported relative improvements in maximal symmetric lifting and repetitive lifting are 26% to 99% in task-specific training studies and 16% to 23% in general training studies. Psychomotor learning probably accounts for a large proportion of performance gains in task-specific training while both psychomotor learning and muscle hypertrophy account for gains in general training studies. While both types of training are effective and currently practiced in the military, general training may be useful for improving a wide range of MMH tasks while task-specific training results in larger gains in targeted MMH tasks.

INTRODUCTION

Traditional ergonomic approaches to reducing worker job stress during manual material handling (MMH) have largely focused on redesigning the working environment through changes in equipment or task requirements (NIOSH, 1981). Until relatively recently, there has been little work examining how improving the physical capacity of the worker might influence MMH capability (Asfour, Ayoub, & Mital, 1984; Genaidy, Bafna, Sarmidy, & Sana, 1990a). Worker physical capacity can be improved by specific physical training designed to improve certain components of physical fitness, especially muscular strength and muscular endurance (Asfour, et al., 1984).

In the U.S. Army, 83% of all military occupational specialities (MOS) have manual material-handling (MMH) requirements. More than 175 MOS require occasional lifting of 44 kg or more and frequent lifting of 23 kg or more. Examples include 1) the field artilleryman who may be required to lift 44-kg artillery rounds as often as 275 times per day, 2) the cargo specialist who is required to lift

and carry 240 kg in four-soldier teams, and 3) the chemical operations specialist who must lift oil drums weighing 108 kg from the ground onto a truck in two-soldier teams (Army Regulation 611-201). Beside these occupational requirements other, infrequent physical demands are placed on all soldiers during operational exercises or deployments. These include such tasks as loading equipment onto trucks, setting up large canvas tents, covering tents and vehicles with cumbersome camouflage netting, and evacuation of casualties with litters or by body carriage.

The purpose of this paper is to review and analyze studies that have examined influence of physical training on MMH capability. Mechanisms for training-induced changes are examined and military applications of these investigations explored. Before proceeding with the literature review, it is useful to define and clarify what is meant by physical training.

Physical Training

Physical training can be defined as muscular activity targeted at enhancing the physical capacity of the individual by improving one or more of the components of physical fitness. The components of physical fitness include muscular strength, muscular endurance and cardiorespiratory endurance (aerobic capacity). Muscle strength is the ability of a muscle group to exert a maximal force (e.g., lifting as much weight as possible). Muscular endurance is the ability of a muscle group to perform short-term, high intensity physical activity (e.g., repetitively lift 44-kg artillery shells as fast as possible). Cardiorespiratory endurance is the ability of the circulatory and respiratory systems to supply fuel to sustain long term physical activity (e.g., road marching, long distance running, bicycling). (Caspersen, Powell, & Christenson, 1985). In order for physical training to effectively increase or maintain physical fitness, it must be performed regularly and be of sufficient frequency, intensity, and duration to induce changes in specific fitness components (Wenger & Bell, 1986).

Limitations in performance imposed by inadequate muscle strength and muscular endurance are probably the most common problem in military and industrial setting (Hogan, 1991). These fitness components can be developed simultaneously or individually using progressive resistance training (PRT). PRT involves exercising with resistances or loads that just fatigue particular muscle groups. The resistance is continually increased as the individual improves his fitness (hence the term "progressive"). PRT is based on exercising with "repetition maximums". A one-repetition maximum (1RM) is the maximum load that can be moved through a range of joint motion just one time. A 10-repetition maximum (10RM) is the maximum load that can be moved 10 times (the individual is not able to perform an eleventh repetition because of fatigue). There is a continuum of repetition maximums that have different effects on strength and endurance. A 1RM builds primarily strength, a 10RM a combination of strength and endurance, and a 20RM primarily muscular endurance (Fleck & Kraemer, 1987).

REVIEW OF THE LITERATURE

Studies that have examined physical training and MMH capability can be divided into two categories. The first category involve studies that use the same MMH task for testing and training (task-specific training studies). The second category of studies are those that use more generalized and traditional training programs that do not include the MMH tasks in the training program (general training studies).

MMH and Task-Specific Training Studies

In the earliest investigation, Asfour et al. (1984) had 10 male college students train for a total of 30 sessions (5 days/week, 6 weeks) doing a symmetrical lifting task. For strength training they performed three sets of a 6-repetition maximum (6RM), lifting a box to three different heights (nine exercise sets total). For muscular endurance training they performed 10 minutes continuously lifting 14 to 20 kg at rates of six to nine lifts/min. For cardiovascular endurance training, they performed cycle ergometer exercise for 30 minutes. At the end of the program, improvements in a 1RM box lift was 41% for the floor to 76-cm lift (78 to 110 kg, $p<0.01$), 99% for the 76- to 127-cm lift (44 to 88 kg, $p<0.01$) and 55% for the floor to 127 cm lift (51 to 79 kg, $p<0.01$). Cardiorespiratory endurance (VO_2max estimated from heart rate) also improved 23%.

Sharp and Legg (1988) used a psychophysical approach. Eight male soldiers selected the maximal mass they thought they could lift to a distance of 132 cm for 1 hour at a rate of six lifts/min. Subjects were trained with the self-selected loads (continuously subject-adjusted) in 20 sessions (5 days/week, 4 weeks), lifting in two 15-minute periods each session. At the end of training, the self-selected box mass had increased 26% (25 to 31 kg, $p<0.05$), 1RM box lift increased 7% (64 to 68 kg, $p<0.05$), and there was no change in perceived exertion on the psychophysical task.

A number of studies have been performed by Genaidy and coworkers (Genaidy, Davis, Delgado, Garcia, & Al-Herzalla, 1994; Genaidy, 1991; Genaidy, et al., 1990a; Genaidy, Gupta, & Alshedi,

1990b; Genaidy, Mital, & Bafna, 1989; Guo, Genaidy, Warm, Karwowski, & Hidalgo, 1992). All of these investigations used tasks involving a complex series of lifting, carrying, pushing and pulling tasks. Subjects trained for periods of 2.5 to 6 weeks (8 to 24 sessions) in the same task for which they were tested. In general, training resulted in a) progressive improvements in endurance time (time to volitional exhaustion) ranging from 46% to 1350%, b) little or no change in the rating of perceived exertion and c) a decrease in activity heart rate suggesting an improvement in cardiovascular endurance.

MMH and General Physical Training Studies

Sharp et al. (1993) trained 18 male soldiers for 36 sessions (3 days per week, 12 weeks), using 10 traditional weight training exercises. For each exercise, the men performed three to five sets of a 10RM. MMH tasks consisted of 1) 10 minutes of lifting a 41-kg box as many times as possible from floor to chest level and 2) a 1RM for the same distance. After the training program, there was a 17% improvement in the 10-minute task (79 to 92 lifts per 10 minutes) and a 23% improvement on the 1RM task (73 to 89 kg). This study was the first to demonstrate that a well-designed general training program fashioned to improve muscle strength and endurance could augment the performance of men on MMH tasks.

Knapik and Gerber (1996) trained 13 female soldiers for 36 sessions (12 weeks). The women trained with certified instructors, performing resistance training 3 days per week, and running with interval training 2 days per week. Resistance training involved nine exercises (each three sets of a 10RM) using exclusively free weights. Compared to values obtained before training, subjects increased their maximal ability to lift a box from floor to knuckle height by 19% (68 to 81 kg, $p < 0.001$) and from floor to chest height by 16% (49 to 57 kg, $p < 0.001$). They improved by 17% their ability to lift a 15-kg box from floor to chest height as many times as possible in 10 min (167 to 195 lifts, $p < 0.001$). Cardiorespiratory endurance was also enhanced since maximum effort 2-mile run time improved 9% (20.3 to 18.4 minutes, $p < 0.001$).

ANALYSIS OF SPECIFIC AND GENERAL TRAINING STUDIES

Specific physical training programs appear to effect larger changes in MMH than generalized training programs. Specific training programs resulted in improvements of 26% to 99% in maximal symmetric lifting and repetitive lifting (Asfour, et al., 1984; Genaidy, et al., 1990a; Sharp & Legg, 1988). This contrasts with relative improvements of 16 to 23% reported in generalized physical training studies (Knapik & Gerber, 1996; Sharp, Harman, Boutilier, Bovee, & Kraemer, 1993). Investigations employing endurance time as a dependent measure (and involving complex motor tasks) report improvements of 34% to 1350% (Genaidy, et al., 1994; Genaidy, 1991; Genaidy, et al., 1990a; Genaidy, et al., 1990b; Genaidy, et al., 1989; Guo, et al., 1992).

The improvements seen in specific training studies may have been largely attributed to enhanced psychomotor learning. In fact, several authors (Asfour, et al., 1984; Genaidy, 1991; Genaidy, et al., 1990a; Genaidy, et al., 1989) noted that at least some of the gains in endurance and lifting capacity were attributable to improved MMH "technique". This would be especially true of the studies of Genaidy and coworkers because the complexity of the tasks may have allowed for many adjustments to achieve longer endurance time (e.g., alterations in body position, pacing of the task, pushing force, etc.). Further, the specific training studies cited were conducted for no longer than 6 weeks, and most for 4 weeks or fewer. It has been demonstrated that neural adaptations account for the majority of strength gains in the first few weeks of resistance training, with hypertrophy becoming a more dominant factor later in training (Moritani & deVries, 1979). Early neural adaptations include fuller activation of muscle prime movers, reduced co-contraction of antagonistic muscles, improved coordination of muscle involved in the intended movement, and removal of inhibitory influences (Sales, 1988).

Improvements in MMH capability seen in the general training studies may involve both a combination of neural adaptations and improvements in muscle hypertrophy. The two general training studies performed were both 12 weeks long, allowing sufficient time for hypertrophy to become a dominant factor in training (Moritani & deVries, 1979). Muscle hypertrophy is important because absolute muscle strength and muscular endurance are proportional to the cross-sectional area of muscle tissue (Maughan, 1984).

MILITARY APPLICATIONS OF STUDIES IN MMH AND PHYSICAL TRAINING

Both general and specific physical training are currently practiced to varying degrees in the U.S. Army. General physical training is an integral part of the daily routine. Army Regulation 350-41 prescribes vigorous exercise three to five times per week during the normal duty day. There is a strong institutional pressure to adhere to this requirement. The importance of physical training is further emphasized to the individual soldier by the Army Physical Fitness Test. This three-event test (pushup, situps, and 2-mile run) must be completed and passed twice a year; promotion and retention in service are tied to the results. Generalized training programs can improve performance on many tasks provided that a wide variety of muscle groups are included in the training program. This could be important in the military (as well as occupations such as police and fire fighting) in which individuals are often called upon to perform non-routine tasks.

Task-specific training is also used in the U.S. Army but is not always of sufficient frequency, duration, or intensity to improve physical capability. Creative methods of introducing task-specific physical training could be extremely useful. A unique application of this type of training was conducted with field artillerymen (Sharp, Knapik, & Schopper, 1994). The soldiers participated in a 45-hour continuous operations exercise involving repeated manual lifting of 44-kg artillery rounds. During the course of the exercise, time required to complete fire missions was reduced 14% (28 to 24 minutes, p<0.01) and energy cost was reduced 23%

(8.0 to 6.2 kcals/min, p<0.01), despite increases in self-reported fatigue and perceived exertion. Thus, a short, intense exercise improved soldier performance and energy efficiency, probably reflecting improvements in psychomotor learning.

CONCLUSIONS

The advantages of regular physical activity include increased worker health and productivity with reduced absenteeism and medical costs (Shephard, 1992). Regular physical training may also help reduce injuries since strength training (Lehnhard, Lehnhard, Young, & Butterfield, 1996) and higher levels of strength and endurance (Barnes, Reynolds, Dettori, Westphal, & Sharp, 1995) are associated with fewer injuries. Besides these benefits, the present review demonstrates that both task-specific and general fitness training programs can improve MMH capability, a common occupational task in the military.

REFERENCES

Asfour, S.S., Ayoub, M.M., & Mital, A. (1984). Effect of an endurance and strength training programme on lifting capability of males. Ergonomics, 27, 435-442.

Barnes, J., Reynolds, K., Dettori, J., Westphal, K., & Sharp, M. (1995). Association of strength, muscular endurance and aerobic endurance with musculoskeletal injuries in US Army female trainees. Medicine and Science in Sports and Exercise, 27, S77.

Caspersen, C.J., Powell, K.E., & Christenson, G.M. (1985). Physical activity, exercise and physical fitness: definitions, and distinctions for health related research. Public Health Reports, 100, 126-131.

Fleck, S.J., & Kraemer, W.J. (1987). Designing Resistance Training Programs. Champaign IL: Human Kinetic Publishers.

Genaidy, A., Davis, N., Delgado, E., Garcia, S., & Al-Herzalla, E. (1994). Effects of a job-simulated exercise programme on employees performing manual handling operations. Ergonomics, 37, 95-106.

Genaidy, A.M. (1991). A training program to improve human physical capability for manual handling jobs. Ergonomics, 34, 1-11.

Genaidy, A.M., Bafna, K.M., Sarmidy, R., & Sana, P. (1990a). A muscular endurance program for symmetrical and asymmetrical manual lifting tasks. Journal of Occupational Medicine, 32, 226-233.

Genaidy, A.M., Gupta, T., & Alshedi, A. (1990b). Improving human capabilities for combined manual handling tasks through a short and intensive physical training program. American Industrial Hygiene Association Journal, 51, 610-614.

Genaidy, A.M., Mital, A., & Bafna, K.M. (1989). An endurance training programme for frequent manual carrying tasks. Ergonomics, 32, 149-155.

Guo, L., Genaidy, A., Warm, J., Karwowski, W., & Hidalgo, J. (1992). Effects of job-simulated flexibility and strength-flexibility training protocols on maintenance employees engaged in manual handling operations. Ergonomics, 35, 1103-1117.

Hogan, J. (1991). The structure of physical performance in occupational tasks. Journal of Applied Psychology, 76, 495-507.

Knapik, J.J., & Gerber, J. (1996). Influence of physical fitness training on the manual material handling capability and road marching performance of female soldiers (Technical Report No. No. ARL-TR-1064). Aberdeen Proving Ground, MD: Human Research and Engineering Directorate, U.S. Army Research Laboratory.

Lehnhard, R.A., Lehnhard, H.R., Young, R., & Butterfield, S.A. (1996). Monitoring injuries on a college soccer team: the effect of strength training. Journal of Strength and Conditioning Research, 10, 115-119.

Maughan, R.J. (1984). Relationship between muscle strength and cross-sectional area. Sports Medicine, 1, 263-269.

Moritani, T., & deVries, H.A. (1979). Neural factors versus hypertrophy in the time course of muscle strength gain. American Journal of Physical Medicine, 58, 115-130.

National Institute of Occupational Safety and Health (1981). Work practice guide for manual lifting (Report No. No. PB82-178948). Cincinnati, OH: U.S. Department of Health and Human Services, National Institute for Occupational Health and Safety.

Sales, D.G. (1988). Neural adaptation to resistance training. Medicine and Science in Sports and Exercise, 20, S135-S145.

Sharp, M.A., Harman, E.A., Boutilier, B.E., Bovee, M.W., & Kraemer, W.J. (1993). Progressive resistance training program for improving manual materials handling performance. Work, 3, 62-68.

Sharp, M.A., Knapik, J.J., & Schopper, A.W. (1994). Energy cost and efficiency of a demanding combined manual materials-handling task. Work, 4, 162-170.

Sharp, M.A., & Legg, S.J. (1988). Effect of psychophysical lifting training on maximal repetitive lifting capacity. American Industrial Hygiene Association Journal, 49, 639-644.

Sharp, M.A., Rice, V., Nindl, B., & Williamson, T. (1993). Effects of gender and team size on floor to knuckle height one repetition maximum lift. Medicine and Science in Sports and Exercise, 25, S137.

Shephard, R.J. (1992). A critical analysis of work-site fitness programs and their postulated economic benefit. Medicine and Science in Sports and Exercise, 24, 354-370.

Wenger, H.A., & Bell, G.J. (1986). The interaction of intensity, frequency and duration of exercise training in altering cardiorespiratory fitness. Sports Medicine, 3, 346-356.

A DATA BASE OF PHYSICALLY DEMANDING TASKS PERFORMED BY U.S. ARMY SOLDIERS

Marilyn A. Sharp, John F. Patton and James A. Vogel
Occupational Physiology Division
U.S. Army Research Institute of Environmental Medicine
Natick, MA

The Department of Defense spends approximately one million dollars annually on research to enhance soldier physical performance (LTC K.E. Friedl, personal communication, Jan 1996). To most effectively direct this research effort, an accurate understanding of the physical demands of Army jobs is needed. The physical demands are available in printed form, however, there is no computerized means to quickly access and compile this information. The purpose of this paper is to describe the creation of a series of data bases containing the physically demanding tasks of Army occupations and to provide a preliminary summary of a selected data base.

INTRODUCTION

The U.S. Army is comprised of approximately 200 entry level military occupational specialties (MOSs). A description of the physical tasks for each MOS is published in Army Regulation 611-201 (AR 611-201) "Enlisted Career Management Fields and Military Occupational Specialty" (Department of the Army, 1995). The physical tasks for each MOS are updated on an as-needed basis, and a new regulation published. All MOSs are classified into one of five physical demand categories based on the lifting requirements of the MOS. These physical demand categories are listed in Table 1 and represent a modification of the Department of Labor Classification scheme (Office of the Deputy Chief of Staff for Personnel, 1982). While other physically demanding tasks such as carrying, digging, pushing or pulling are listed in AR 611-201, the physical demand categories are based solely upon lifting demands, with no reference to lifting height.

AR 611-201 is a valuable tool to describe a single military job; however, it's use is cumbersome in making generalizations across physical demand categories or types of physically demanding tasks. This paper describes a series of

data bases created to compile the information available in AR 611-201 into a useable tool to

TABLE 1 Army Modified Department of Labor Physical Demand Classification System

	Occasional[1]	Frequent[2]
Light	20 lbs	10 lbs
Medium	50 lbs	25 lbs
Mod heavy	80 lbs	40 lbs
Heavy	100 lbs	50 lbs
Very heavy	>100 lbs	>50 lbs

[1] <20% if time, [2] 20<80% of time.

make generalizations regarding the physical demands of MOS. There are several potential uses for this information: 1) to design appropriate models for research purposes; 2) to identify the jobs most in need of redesign; 3) to set physical performance standards for groups of MOS with similar physical demands and 4) to develop job related physical training programs to improve soldier performance.

METHODS

The first step in data base development was to ensure that the MOS descriptions found in AR 611-201 were current and accurate. To do this, the Army Training and Doctrine Command (TRADOC, Ft Monroe, VA) organized a teleconference with subject matter experts from each of the twenty-three training schools (i.e., Infantry School, Engineer School, Armor School, etc). Each training school is responsible for the training curriculum and manuals which detail job tasks and performance standards for each MOS in that school. Training school representatives were instructed to review the physical tasks listed in AR 611-201 for accuracy, make them as specific as possible, and revise them to reflect recent changes in equipment or task performance. Form DA 5643-R, published in AR 611-201, is routinely used by the schools to update MOS physical requirements and was used for the current update request. Examples were provided during the teleconference to illustrate the level of detail desired. The revisions received for each MOS were annotated in AR 611-201.

In step two, separate data bases were constructed for each of the following task categories: lifting, lifting and carrying, pushing/pulling, climbing, digging, and walking/marching/running. All data bases included the following information: MOS number, MOS nomenclature, career management field and physical demand category. The following fields were included as appropriate to the data base: total load (lifted, carried, shoveled, packed, etc) in lbs, team size (one or more), load handled per person in lbs, distance moved in feet or miles, rate of movement, torque applied, volume dug and frequency of task performance. Vertical lifting distances were categorized as follows: few inches to two feet, two feet to waist height, above waist to shoulder height and above shoulder to full reach. Rate of movement was categorized as crawl, walk/march, run/sprint. Task frequency was rated as occasional (less than 20% of the time), frequent (more than 20%, but

less than 80% of the time) or constant (more than 80% of the time). It is recognized that these frequencies are vague, and that this is a limitation of the data base.

In constructing the data bases, the most demanding task for each task category was selected for each MOS. If a MOS had multiple tasks from a given category the most demanding task at each frequency was entered into the data base. In addition, if a MOS had both team and individual tasks within a task category, the most physically demanding team and individual tasks at each frequency were included.

The lifting and carrying data base will be used to illustrate some of the capabilities of these data bases. Descriptive statistics for the key fields will be presented and will be broken down by task frequency, team size and physical demand category.

RESULTS

The lifting and carrying data base contained 232 tasks from 172 different MOS. Forty-seven MOS had two lift and carry tasks, three MOS had three tasks, one MOS had four tasks and one had five tasks entered in the data base.

Figure 1 is a scatterplot of the load carried per person and the distance carried. There was no apparent relationship between the load carried and the distance it is carried. The majority of lift and carry tasks (52%) involved carries of 30 feet or less. In eighty-four percent of the tasks, loads were carried 50 feet or less, while only 6.6% of the tasks involved loads carried more than 100 feet. The shortest carry reported was two feet and the longest 300 feet. It should be noted that loads carried more than 400 ft were placed in the walking/marching/running data base.

The loads per person ranged from 10 to 187.5 lbs. The 25th percentile load was 50 lbs, the 50th percentile 66 lbs and the 75th percentile 89.5 lbs. Eleven percent of the loads (26 tasks) were in excess of 100 lbs, thus falling into the very heavy physical demand category.

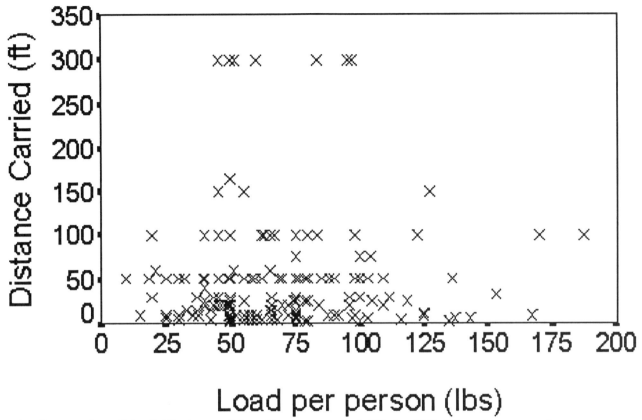

Figure 1. Scatterplot of load (lbs/person) by distance carried (ft) for 232 lift and carry tasks.

Of the 232 lift and carry tasks, 51% were individual tasks, 26% two-person tasks, 7% three-person tasks and 13% four-person tasks. The median load and distance lifted and carried by one-, two-, three- and four-person teams was 67 lbs and 50 ft, 63 lbs and 20 ft, 90 lbs and 15 ft, and 88 lbs and 30 ft, respectively.

Eighty-nine percent of tasks involved lifting to heights at or below waist level. Nine percent of the lifting heights were between waist and shoulder height, and 2% were above shoulder height.

There was little or no difference in the load or distance carried due to the frequency of task performance. The mean load and distance carried by frequency of task performance was 70 lbs and 45 ft for occasional; 72 lbs and 52 ft for frequent; and 55 lbs and 34 ft for constant tasks. Sixty percent of the lift and carry tasks were performed on an occasional basis, 37% frequently

and 2% constantly.

Table 2 lists the mean and range of loads and distances carried for each physical demand category. By definition, the load carried increases with the physical demand category. There are several instances when the physical demands exceed or are less than the physical demand category weight range. The mean and median values for the light category are both greater than the category occasional limit of 20 lbs, while those of the heavy and very heavy categories fall below the category occasional weight limits.

DISCUSSION

The distance soldiers carry loads is long compared to industrial standards. Drury, Law and Pawenski (1982) reported the median distance for an industrial carry was 5 feet compared to 30 feet for all military lift and carry tasks. It should be

TABLE 2. Load and distance carried by MOS physical demand category.

Category (n)	Load (lbs/person)			Distance carried (ft)		
	mean ± SD	range	median	mean ± SD	range	median
Light (7)	31.7 ± 15.7	10-50	32	111.4 ± 129.8	10-300	50
Medium (30)	46.8 ± 13.9	18-66	50	34.4 ± 56.5	3-300	10
Mod Heavy (43)	56.0 ± 17.2	21-138	55	46.9 ± 53.0	2-300	25
Heavy (36)	65.0 ± 23.2	25-119	67	52.7 ± 68.9	2-300	50
Very Heavy (116)	85.7 ± 29.8	30-188	80	45.2 ± 51.5	3-300	50

noted that military tasks are often performed in a field setting where environmental conditions or physical terrain may preclude the use of vehicles or other materials handling aids.

The median load lifted in an industrial setting was reported to be 20 lbs (Drury, Law and Pawenski, 1982) as compared to our finding of 66 lbs for all military lift and carry tasks. By definition, the most physically demanding tasks were selected for inclusion in the data base, therefore, it is difficult to make a valid comparison of loads with the data of Drury, Law and Pawenski (1982).

In a previous of study by Sharp, et al. (1995) the average maximum acceptable load selected for lifting and carrying boxes a distance of 24 ft was 67 lbs for one person, and 131 lbs for two person teams (at a rate of 1 x/min). This compares favorably with the average loads for one and two-person lift and carry tasks (66 lbs and 122 lbs, respectively in this data base). The job requirements appear, therefore, to be a reasonable match to the loads soldiers are willing to lift and carry in two person teams.

There were a number of MOS with lift and carry tasks that either exceeded or fell short of the weight range of the assigned physical demand category. There are several reasons for this mismatch. It will be recalled that the physical demand category was taken directly from AR 611-201, but that many of the tasks were based on updated information received from the training schools. The loads lifted and carried may have increased or decreased since the last publication of

AR 611-201. Therefore, the updated load information and the published physical demand category no longer match. A second explanation is that the physical demand category assignment was based on a different task. While multiple lifting and carrying tasks were entered for many MOS, they may not all have been of the same level of difficulty. In addition, the qualifying task may have been in the lifting data base, rather than the lifting and carrying data base. A third reason is that the MOS was not properly classified. Seventy-six percent of the MOS were correctly classified. Of the misclassified MOS, 19 were misclassified as they appeared in AR 611-201, while an additional 25 MOS were misclassified due to updated information.

Confirmation of the information contained in the data bases is a continuing process. The data bases will need to be updated as tasks change due to equipment changes, changes in standard operating procedures for task performance and as the MOS themselves are merged or become obsolete. The data bases represent a convenient means to examine the range of MOS physical demands in terms of task variables and provides an important tool to confirm the MOS physical demand category assignments.

REFERENCES

Department of the Army. (June 26, 1995). Enlisted Career Management Fields and Military Occupational Specialty. Army Regulation 611-201. Washington, DC:

U.S. Government Printing Office.

Drury, C.G., Law, C., and Pawenski, C.S. (1982). A survey of industrial box handling. Human Factors, 24, 553-565.

Office of the Deputy Chief of Staff for Personnel. (1982). Women in the Army Policy Review Group Final Report. Washington, DC: U.S. Government Printing Office.

Sharp, M.A., Rice, V.J., Nindl, B.C., and Mello, R.P. (1995). Maximum acceptable load for lifting and carrying in two-person teams. In Proceedings of the Human Factors and Ergonomics Society 39th Annual Meeting, (pp. 640-644). Santa Monica, CA: Human Factors and Ergonomics Society.

FEASIBILITY OF MOS JOB ANALYSIS AND REDESIGN TO REDUCE PHYSICAL DEMANDS IN THE U.S. ARMY

Rene J. de Pontbriand, Ph.D., and Joseph J. Knapik, Sc.D.
U.S. Army Research Laboratory, Human Research and Engineering

Heavy physical requirements characterize a large number of military occupational specialties (MOSs). Current efforts at reducing some of these physical requirements stem from concerns with health and safety, the need to conserve soldier strength and endurance for other battlefield tasks, and the need to optimize personnel utilization. This paper describes an ongoing feasibility study aimed at identifying and attempting to reduce physical demands in five MOSs. Three data-collection phases involve (a) review of publications describing occupational tasks, (b) structured interviews with subject matter experts (SMEs), and (c) filming the most physically demanding tasks. The fourth phase involves use of SMEs and ergonomists to identify specific redesign solutions. An illustrative example is provided. This technique coupled with others (job selection and physical training) can enhance military operational capability by reducing physical requirements.

INTRODUCTION

The U.S. Army currently has 277 military occupational specialties (MOSs) (see Army Regulation 611-201). Each MOS has its own particular set of mental, physical, and skill oriented requirements. As in industry, some specialties require high mathematical aptitudes, while others demand more manual dexterity or long-term training in selected skills. However, it is the extreme physical demands which often distinguish Army occupations from those in the civilian sector. There are many reasons for this, some having to do with the exigencies of the battlefield and the operating environment. A missile round may weigh 100 or 130 pounds because it needs sufficient mass and explosive power to destroy its tactical objective; the weight is based on mission demand, not on human capability. For infrequently performed tasks or sudden emergencies, adequate personnel may not be available to distribute the work.

In other cases, tasks that are easily performed in garrison on hard surfaces are made difficult in the field environment because the surface may be uneven, rocky, loose, muddy, snow-covered, and generally unpredictable. Tasks such as tank track repair, portable bridge emplacement, casualty extraction and mobile kitchen setup become difficult.

This paper describes the rationale for and approach to redesigning specific military tasks. The goal is to reduce the physical requirements on the soldier. The preliminary methods used to identify "high driver" tasks, analyze those tasks, and recommend modifications are presented. An example of a solution strategy is provided.

JOB REDESIGN PERSPECTIVE

Job redesign options complement personnel placement and physical training approaches. Personnel placement may not provide enough personnel to handle all necessary tasks, unless supported by extensive and costly recruiting efforts.

Physical conditioning increases individual capacity to meet physical requirements and provides benefits even away from the job (e.g., increased health, longevity, and productivity). Indeed, the major military response to the physical demands of military life has been to maintain highly conditioned troops through regular physical fitness training. Well designed physical training programs have been shown to improve the physical capability of soldiers (Knapik & Gerber, 1995; Sharp, Harman, Boutilier, Bovee, & Kraemer, 1993), and are related to the concept of "work hardening" as practiced in occupational therapy. Major limitations of physical training, however, are training time constraints and inherent biological limits to improvements.

Task redesign has a unique advantage, namely, reducing physical requirements. Lowering demands allows soldiers to conserve strength and endurance, extending performance and providing energy for emergency situations. Also, once the task has been made less difficult, everyone benefits without further involvement.

RATIONALE FOR LOWERING DEMANDS

The battlefield requires great effort and even greater sacrifice not found in most other human endeavors; it is expected to be difficult. However, there are strong reasons to reconsider its physical requirements.

A principal reason is safety and health: physically demanding tasks are dangerous. While these are among the most difficult type of data to quantify from a cost-benefit perspective, they provide vivid pictures of the impacts of high physical demands. It is known that injuries are indirectly related to the amount of exposure to heavy physical demands (e.g., Jones, Cowan & Knapik, 1994). Heavy loads are prone to being dropped, to unexpected shifting, and to being improperly lifted, carried, and emplaced. Resulting injuries can incapacitate individuals and jeopardize readiness and mission success. Lowering requirements would reduce these potential risks and enhance battlefield effectiveness.

Equally critical is the need for performance sustainment. Some Department of Defense (DoD) restructuring changes call for reduced crew sizes. An example might be going from a six- to a four-person field artillery crew. The reduced squad is expected to support the same rate of fire, maintenance, or resupply activities as in its previous. One way to sustain a given workload with fewer people is to reduce "heavy lift" (anaerobic, quickly fatiguing) tasks to "light lift" (aerobic, low, or insignificant fatiguing). A related benefit would be performance maintenance: The influx of new technologies (e.g., the 60-mph tank, command and control on the move) has the effect of increasing the operational tempo for everyone--more must be done in a given period of time.

Reduced demands are also key to more complete personnel utilization. Of the 277 current MOSs, 175 (63%) require occasional lifting of 100 pounds or more and frequent lifting of 50 pounds or more. An indication of the severity of these lift requirements is given by the fact that some 20% of the military-age male population are not capable of such lifting, and only a small portion (approximately 10%) of the military-age female population are capable (Fischl, 1993). Lowering lift requirements will have the result of making a higher proportion of those in any given MOS eligible to perform all necessary lifting tasks. This provides more backup and distributes the load for performing those tasks among a larger group of individuals. An associated benefit relates to NATO or UN interoperability. To the extent that DoD downsizing and strategic change requires greater interaction with and reliance upon joint coalition missions, the potential for success of those missions is enhanced if the load can be shared with international counterparts. This can be accomplished if U.S. Army tasks, materiel, and equipment are designed to the lower boundaries of physical stature and strength.

APPROACH TO JOB REDESIGN

For the current feasibility study, five MOSs were selected. Selection was based on the types of physical demands and work conditions represented, as well as the availability of subject matter experts (SMEs) with which to interact. The five MOS were Chemical Operations Specialist, Track Vehicle Mechanic, Motor Transport Operator, Medical Specialist , and Food Service Specialist. Collection of data involved three steps:

1. Locate principal source documents and derive physical requirement information to determine high driver tasks. Documents included the Military Occupational Classification and Structure regulation (Army Regulation 611-201), Soldier Training Publications (STPs), Army Training and Evaluation Programs (ARTEPs), Programs of Instruction (POIs), Army Occupational Survey Program results, and accident reports from the Army Safety Center and other available documents.

2. Develop and administer to MOS SMEs a structured questionnaire focused on clarifying high driver tasks selected from publications. Questions were designed to solicit the most physically demanding tasks from the SMEs, based on their knowledge and experience (Adams, 1989).. The critical incident technique was used (Flanagan, 1954; Meister, 1985).. Further ideas about task modifications were sought. If demanding tasks had been identified in publications and not mentioned by SMEs, these were also discussed.

3. Go to field, garrison, or other operating sites as appropriate and videotape and observe high driver tasks being performed. Note the presence of workarounds or other job aids used to lower the physical demands.

After the data were collected, analysis proceeded in three overlapping phases with both SMEs and ergonomists:

1. Each high driver task was reviewed on the basis of observed or reported:

    a. Task segment of the heavy lift requirement
    b. Restriction of movement, hearing, vision
    c. Load or equipment bulkiness, awkwardness
    d. Personal balance and stability factors
    e. Hand-hold suitability or feasibility
    f. Load-shifting potential
    g. Load adjustability
    h. Observed degree of required (a) asymmetric lift; (b) movement above shoulders or below knees; and (c) movement of the load away from the body (reach requirement)
    i. Possibility that the task may be performed in protective gear

2. The feasibility of redesign options was developed and reviewed. Consideration was given to solutions involving:

    a. Engineering (e.g., mechanical: bearings, winches, pulleys, ramps, liquid-transfer pumps; optimal handles, grip shape and texture, surface grip and texture; combined functions; workplace layout)
    b. Packaging (modularizing, , single-use packs that do not need to sustain repeated usage)
    c. Physical (biomechanical: hand-hold placement, body posture guidance; lift-height requirement changes; passive job aid, e.g., to shift weight from weaker upper body to stronger lower body, such as a hip-mounted harness upon which to rest the load)

3. Solutions were rank ordered via utility analysis on the basis of::

    a. Physical requirements (how changed)
    b. Redesign costs
    c. Staffing changes involved
    d. Time to perform tasks, missions
    e. Proportion of newly eligible personnel
    f. Effects on performance sustainment

TASK EXAMPLE

An illustrative task involves the Chemical Operations Specialist. The physical requirements of the job are described in Army Regulation 611-201: "Many tasks routinely performed in mission-oriented protective posture levels I through IV (MOPP I-IV) [nuclear, biological and chemical protective clothing]; frequently moves 474 lb. (2-person lift); frequently lifts

237 lb. [1-person lift]; occasionally lifts and carries 86 lb., 50 ft."

One high driver task was identified as loading smoke-oil barrels onto a general purpose truck bed. Each barrel has a 474-lb mass. The truck bed is 4-1/2 feet from the ground. Eighteen barrels are loaded per truck using two people to lift each barrel. The task takes about 30 minutes to perform. A 12-foot ramp is recommended (but not required) so the barrels can be manually rolled onto the truck bed. Concerns include the mass of the barrel (which increases the potential for hand and low back injury), load imbalances caused by individuals of different heights, and fatigue induced by repetitive manual handling.

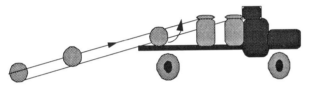

A possible redesign solution is to mount a winch on a truck to pull the barrel over a "pivot" point, righting the barrel when more than half its weight crosses the pivot. Electric or mechanical (with adequate leverage) winches are most affordable; others result in major equipment maintenance concept changes and cost. Health, safety and energy savings include the fact that ramp-roll, load-shift, and barrel-righting hazards are minimized. Operations are moved to the truck bed, away from potentially unstable ground. The remaining requirement is to "walk" the barrel into its location on the bed. No lifting is required so soldier energy expenditure is considerably reduced. The effects of body stature differences are minimized. Performance time becomes a function of winch power and mechanical leverage.

An alternate solution is to eliminate the requirement for barrels altogether by changing the delivery system. A tanker truck with a large capacity to hold smoke oil could be loaded by a hose from a main storage, and go onsite to refill barrels located near the smoke oil generators. Only one person would be needed for the operation. A potential disadvantage of this solution is the high cost of maintaining a separate vehicle that has less flexibility than the general purpose truck. A cost trade-off analysis would be in order before proceeding.

DISCUSSION

Redesign of physical tasks is achievable and would be effective in enhancing safety, readiness, and battlefield effectiveness. The current approach is a hybrid based on epidemiological, physiological, psychophysical, and biomechanical considerations (Ayoub & Mital, 1989; Chaffin & Andersson, 1991). This technique, coupled with findings related to improved fitness methods and job selection, may allow soldier capabilities to be better matched with the given tasks.

It is clear that some tasks are more amenable to redesign than others. Those least amenable tend to be afflicted with tight workspace layout requirements (e.g., working in vehicles); require asymmetric lifting and out-of-position lifting (e.g., changing tires); tend to be part of a series of rapid, interdependent activities; or are critically affected by the wearing of protective gear (e.g., chemical operations). Improvements of these latter tasks may warrant additional personnel or require adopting a new technology or otherwise eliminating the hazardous tasks.

Given the large number of affected MOSs and the different tasks and operational conditions involved, a streamlined approach to examining additional MOSs would be useful. Such an approach could include a checklist based task categorization scheme, such as that under development at the National Institute for Occupational Safety and Health

(NIOSH). This would allow for quick diagnosis of which tasks can be redesigned resource effectively and the type of remedy required (e.g., equipment or workspace redesign), and which types of tasks would benefit from other approaches. Existing references (e.g., Mital, Nicholson, & Ayoub, 1993) provide good examples of lift and carry aids and methods used in the private sector, and should be evaluated for use in military settings. Another approach might be survey based. This would entail a structured orientation and questionnaire package to elicit perceived high driver task information from individuals in the target MOS. Such surveys require a high level of command support; however, proper followup on survey responses with appropriate SMEs would provide coverage of a wide variety of tasks.

The greatest difficulty is in establishing a valid and useful utility assessment methodology. The difficulty stems from the complexity of tasks, operating environments, and high heterogeneity among the target population. Relative weightings of such factors as energy savings, injury hazard reduction savings, fielding time, engineering or training costs or savings differ across SME subgroups. Further, in many cases it is extremely difficult and time-consuming to obtain adequate, balanced data; the process may cost more than the product returns. The DoD Ergonomics Working Group (1996) is exploring a software program to help determine the "return on investment" of redesign or other risk-lowering efforts. Cost-benefit analysis is one topic that we will explore further in subsequent efforts. It is critical in gaining needed command support and to help guide efficient use of the project's resources.

The next phase of this study uses a broader, more quantitative approach. Dynamometers, energy expenditure, and other types of instrumentation, as well as computer simulation and modeling, will be used to help analyze certain tasks and categories of tasks. Working with the broader community and tapping into other databases (e.g., the Army Physical Disabilities Database) will provide increased, mutual leverage. In this way, the scope of the Army's physical demands requirements can be better understood, and the most negative of its impacts can be controlled to better reduce physical demands.

REFERENCES

Adams, J.A. (1989). Human factors engineering. New York: Macmillan Publishing Co.

Ayoub, M.M., & Mital, A. (1989). Manual materials handling. Philadelphia: Taylor and Francis.

Chaffin, D.B., & Andersson, G.B.J. (1991). Occupational biomechanics. New York: John Wiley & Sons, Inc.

Fischl, M. (1993). Human Resources Department Briefing, Office of the Deputy Chief of Staff of Personnel, Department of the Army, Washington, DC

Flanagan, J.C. (1954). The critical incident technique. Psychological Bulletin, 51, 327-358.

Jones, B.H., Cowan, D.N., & Knapik, J.K. (1994). Exercise, training, and injuries. Sports Medicine, 18: 202-214.

Knapik, J.J., & Gerber, J. (1996). Influence of physical fitness training on the manual material handling capability and road marching performance of female soldiers (Technical Report No. ARL-TR-1064), Aberdeen Proving Ground, MD: Human Research and Engineering Directorate of the U.S. Army Research Laboratory.

Meister, D. (1985). Behavioral analysis and measurement method. New York: John Wiley and Sons.

Mital, A., Nicholson, A.S., & Ayoub, M.M. (1993). A guide to manual materials handling. London: Taylor and Francis.

Sharp, M.A., Harman, E.A., Boutilier, B.E., Bovee, M.W., & Kraemer, W.J. (1993). Progressive resistance training program for improving manual materials handling performance. Work, 3, 62-68.

STUDY OF COMPATIBILITY OF ARMY SYSTEMS WITH
ANTHROPOMETRIC CHARACTERISTICS OF FEMALE SOLDIERS

Wendy L. Todd
GEO-CENTERS, INC.
Natick, Massachusetts

Steven P. Paquette, M.A.   Carolyn K. Bensel, PhD
U.S. Army Natick Research, Development and Engineering Center
Natick, Massachusetts

Many Army clothing and equipment systems in use today were designed to accommodate male soldiers within the 5th - 95th percentile range for Stature and other critical body dimensions. Thus, female soldiers whose body dimensions are outside this design envelope may be compromised with regard to clothing and workspace fit. The objective of this study was to determine the compatibility of Army female height/reach dimensions with currently fielded, representative clothing, individual equipment, and workstations. These items were selected using the results of subjective questionnaires, theoretical disaccommodation rates, and interviews with the developers. Data were collected on female soldiers 5'5" and shorter (workstations, n=202; clothing and equipment, n=202). Compatibility was determined by the difficulty experienced by the soldiers as they performed specific reach tasks in workstations, and by the acceptability of static and functional fit of the clothing and equipment. Problematic items were identified and solutions were developed, along with associated implementation costs.

INTRODUCTION

Many U.S. Army systems currently in the field were designed some years ago when the primary users of the systems were male soldiers. The typical design standard was to accommodate the 5th through the 95th percentile values for critical body dimensions. As the number of women in the Army increased, the disparity between male and female dimensions became increasingly apparent. For example, the 5th percentile value for Stature (height) of male soldiers is 5'5"(165 cm). This value, which exceeds by 5" the 5th percentile value for Stature of female soldiers, corresponds to the 65th percentile value for the females (Gordon et al., 1989). The disparity indicates that approximately 65% of the Army female population is likely to be outside the typical design envelope for many Army systems. Given these anthropometric comparisons, concern was expressed by the Department of the Army as to whether women who must use Army systems will be able to perform their jobs without impediment.

The objective of the study was to determine the compatibility of female anthropometry with currently fielded, representative U.S. Army systems; specifically, workstations, clothing, and individual equipment. Workstations included vehicles, consoles, and equipment with fixed user interfaces; clothing and individual equipment (CIE) included protective items worn by soldiers. Compatibility assessments focused on the relationships between the item and female height/reach characteristics. In the case of items that were found to disaccommodate females, possible solutions were developed and estimates of costs involved in implementing the solutions were generated.

METHODS

Survey Items

To identify candidate workstations, clothing, and equipment to be studied, surveys regarding ease of use, functionality, and compatibility were administered to active duty military personnel at three Army posts: Ft. Drum, NY, Ft. Devens, MA, and Ft. Bragg, NC. In addition, literature reviews were conducted, and discussions were held with item developers. As a result, workstations representing each of five occupational classes were identified for evaluation: M577 Light Tracked Command Post Carrier (Artillery/Defense); M10A Rough Terrain Forklift (Engineering/ Construction); M1070 Heavy Equipment Transporter (Transportation); M978 HEMTT Fuel Tanker (Industrial Support); and MKT75 Mobile Kitchen Trailer (Supply/Support).

Eleven CIE items were identified for the study. They represented a broad range of protective functions, as well as linear and circumferential fit issues for various segments of the body. The CIE items studied were: Cold Weather (CW) Trigger Finger Mitten, Combat Vehicle Crewmen's (CVC) Coverall, Mechanics' Coverall, Field Pack (ALICE) with External Frame, Ballistic (PASGT) Vest, Enhanced Tactical

Load Bearing (ETLB) Vest, Extended Cold Weather (ECWCS) Parka, Wet Weather Trousers, Light Duty Work Glove, Ballistic (PASGT) Helmet, and MC1-1 Parachute Harness.

The workstation and CIE data were collected separately at different Army posts. The workstation evaluation included a sample of 202 female soldiers from Ft. Hood, TX, all of whom were 5'5" and shorter. The CIE items were evaluated on a separate sample of 202 female soldiers from Ft. Bragg, NC, all of whom were 5'5" and shorter. Data collected were anthropometric measurements and qualitative assessments of accommodation. Assessment criteria were established using technical manuals, military specifications, interviews with item developers and users, and human factors design principles. Data collection protocols were tested and standardized before data collection began.

Workstation evaluation. The body dimensions measured in the workstation evaluation were Stature, Weight, and length/reach variables that characterize the major linear segments of the body (Eye Height-Sitting, Functional Leg Length, Crotch Height, Hand Length, Overhead Fingertip Reach-Extended, Popliteal Height, and Thumbtip Reach). Typical work tasks for each workstation were identified using technical manuals and the suggestions of maintainers and operators. Each task was evaluated for level of difficulty to accomplish the task. The evaluator determined level of difficulty using biomechanical cues, as well as subjective input from the subject. Task performance was determined to be unacceptable if the task was extremely difficult or could not be completed.

CIE evaluation. The body dimensions measured were Stature, Weight, and those used to issue items (Chest Circumference, Waist Circumference, Hand Length, and Hand Circumference). The assessments focused on evaluation of fit, with the subject assuming a static standing posture, as well as performing simple movements, such as raising the arms, bending at the waist, and squatting. An item was declared to be unacceptable if it did not satisfy a predetermined number of fit factors. Judgments regarding each factor were made by the evaluators, based upon their observations and the subject's opinions.

Experimental Design

Because this study was limited to collecting data on females 5'5" and shorter, an experimental/control group design was not possible. Furthermore, random sampling was not possible because unit commanders selected the participants, who were usually enlisted personnel and had a tendency to be the shortest females in the unit. Stature represented the independent variable; level of difficulty performing workstation tasks and acceptability of CIE fit were dependent variables. Thus, this study employs an *ex post facto* case study design wherein the independent variable, Stature, is presumed to drive differences in

accommodation, and is not directly manipulated to test the hypothesis. No conclusions are made about males of a similar Stature, or about females above 5'5" in Stature.

Treatment of the Data

Prior to analysis, the data were checked for accuracy and edited. Statistical tests were also carried out to determine whether the study samples were representative of the Army population of females 5'5" and shorter with regard to race, age, and body dimensions. The source of the data for the Army population was the 1988 Anthropometric Survey of U.S. Army Personnel (ANSUR) conducted by Gordon et al. (1989).

It was found that the Race and Age proportions of the samples were not representative of those in the Army female population 5'5" in Stature and shorter. This is an important consideration because Race and Age can greatly influence body size and shape (Finch & Hayflick, 1977; Gill and Rhine, 1990). Each sample was analyzed to identify any differences in anthropometric values within the sample that were attributable to Age or to Race. In the workstation sample, Race, but not Age, explained significant anthropometric differences (using a Bonferroni correction of $p<.05/9$ anthropometric variables $= p<.0055$). Therefore, it was necessary to weight the workstation sample on population Race proportions. After weighting, comparisons of the body size variables revealed that the sample differed significantly (Bonferroni correction, $p<.0055$) from the ANSUR population only on Popliteal Height; the mean Popliteal Height of the ANSUR population exceeded that of the workstation sample by 2 cm.

For the CIE sample, Age, but not Race, was associated with significant differences (Bonferroni correction of $p<.05/4$ anthropometric variables $= p<.0125$) in Weight, Chest Circumference, and Waist Circumference. Therefore, the sample was weighted on population Age proportions. After weighting, comparisons of the body size variables indicated that Chest and Waist Circumferences of the CIE sample were significantly larger (Bonferroni correction, $p<.0125$) than those of the ANSUR population. Also, compared with the ANSUR population, a higher proportion of the sample exceeded the Army's weight-for-height standards (AR600-9, 1986), and by a greater amount.

Analysis of data. The weighted data were used in all analyses and the data for each workstation task and each CIE item were analyzed separately. For analysis of a workstation task, subjects were divided into two groups, those whose performance on the task was found to be acceptable and those whose performance was unacceptable. Similarly, for analysis of a CIE item, subjects were divided into two groups as a function of whether item fit was determined to be acceptable or unacceptable. If the unacceptable group included more than 15% of the sample, that task or CIE item was defined as a compatibility problem likely to impact a substantial proportion of the Army female population. Analyses were carried out to determine if acceptable and unacceptable task performance and CIE fit were related to subjects' body dimensions. If the F-test for homogeneity of variance revealed that variances were not equal, the Mann-Whitney U-test was used; otherwise, an analysis of variance (ANOVA) was performed. Because the CIE sample was, on average, comprised of larger females than are in the Army population, as represented by the ANSUR data, the Fisher Exact Test ($\alpha = .05$) was performed to determine if acceptability of the fit of CIE items was related to whether subjects met or exceeded the Army's weight-for-height standards. The workstation data were also analyzed for conformance with the standards.

RESULTS

Workstation Evaluation

Every workstation disaccommodated the subjects in this study to some extent. The workstation with the most problems was the Mobile Kitchen Trailer (MKT). The task with the highest percentage of unacceptable performances was reaching the V7 valve in the M978 HEMTT Fuel Tanker. Table 1 summarizes the tasks which posed unacceptable difficulty for more than 15% of the sample.

Table 1. Problematic Workstation Tasks
(Weighted $n = 205.0$)

Task	Accept. (%)	Unaccept. (%)	Missing (%)
Install MKT Utensil Holder	20.4	58.0	21.6
Replace MKT Fire Ext.	47.1	50.3	2.6
Remove MKT Fire Ext.	54.4	43.5	2.2
Release MKT Range Prop.	68.6	29.3	2.2
Lower MKT Range Cover	69.2	28.7	2.2
Raise MKT Range Cover	78.2	19.6	2.2
Reach M978 V7 Fuel Valve	5.6	63.3	31.1
Reach M978 V8 Fuel Valve	23.7	45.2	31.1
Close M978 Rear Hatch	77.8	17.4	4.8
Close M1070 Hood	24.5	63.1	12.5
Sight M10A Forkends	82.4	15.9	1.7
Sight 15ft Rear of M10A	70.8	26.5	2.7
View Out of M577 Hatch	77.0	15.1	7.9

The first 10 tasks in Table 1 involved reaching an object located above the head and/or forward of the subject. Typical postures of subjects who experienced unacceptable difficulty were characterized by fully extended arms and legs, standing on toes, reaching with fingertips, and hyperextension of the back and neck. For example, the fire

extinguisher in the MKT was mounted on a hook 221 cm above the standing surface. Subjects who experienced extreme difficulty removing the extinguisher from the hook or who failed to complete the task were able, at best, to grasp the bottom of the extinguisher even when standing on the tips of their toes and extending their arms fully.

The last three tasks listed in Table 1 entailed sighting objects outside a vehicle from the driver's compartment. Seats were adjusted so that the subject could reach foot pedals and other controls. In some instances, views were obstructed; subjects had to rise up in the seat, removing their feet from the pedals, in order to sight the objects.

The results of the Fisher Exact Test carried out on each workstation task revealed that the acceptable and the unacceptable performance groups did not differ significantly ($p>.05$) in the proportions of subjects in each group who met or exceeded the Army's weight-for-height standards. However, on some tasks, body dimensions of the two groups did differ significantly (Bonferroni correction, $p<.0055$). The significant findings are summarized in Table 2. In those instances in which differences were obtained, subjects with acceptable task performance had larger body measurements than subjects with unacceptable performance. Based upon the differences in Stature between the acceptable and the unacceptable performance groups, it appears that subjects who were 5'3" (160 cm) or more, accomplished the workstation task successfully and subjects who were 5'2" (157 cm) or less, did not.

Table 2. Workstation Tasks with Significant Differences in Anthropometric Dimensions

Tasks	Stature	Eye Ht., Sitting	Func. Leg Lgth.	Crotch Ht.	Hand Lgth.	Ovhd. Ftip. Rch.	Popliteal Ht.	Thumb-tip Rch.	Weight
MKT Utensil Hldr.	X		X	X	X	X	X	X	
MKT Fire Ext.	X	X	X	X	X	X	X	X	
MKT Range Prop.	X		X	X	X	X			
M978 V7 Valve	X		X	X	X	X	X	X	
M978 Hatch	X	X	X	X	X	X	X	X	X
M1070 Hood	X		X	X		X	X	X	
M577 Vision	X	X			X	X	X		

Suggested retrofits to address the problems in task execution identified during the workstation evaluations are summarized in Table 3. The modifications proposed are based on observations made during the evaluations, as well as on suggestions from biomechanists, physical anthropologists, developers and manufacturers of the workstations. The estimated cost of implementing a retrofit of a workstation includes the cost associated with making the modifications on all equipment of that type currently in the Army inventory.

Table 3. Workstation Retrofits and Estimated Costs

Equipment Item	Suggested Retrofit	Est. Cost
MKT	Relocate fire ext., install battery-powered lighting, install new range cover prop	$3,300,000
M978 HEMTT	Install strap pull on hatch	$125,500
M1070 HET	Unknown	Unknown
M10A Forklift	Install convex mirror	$126,300
M577 Carrier	Reposition seat post	$152,400

Table 4. Results of Fit Assessment of CIE Items
(Weighted $n = 201.6$)

Item	Accept. (%)	Unaccept. (%)
*CW Trigger Finger Mtn.	3.9	96.1
*CVC Coverall	11.5	88.5
*Mechanics' Coverall	28.3	71.7
*Field Pack w/Frame	38.5	61.5
*PASGT Vest	56.9	43.1
*ETLB Vest	71.0	29.0
*ECWCS Parka	73.5	26.5
*Wet Weather Trouser	84.5	15.5
Light Duty Work Glove	89.6	10.4
PASGT Helmet	100.0	0.0
Parachute Harness	93.1	0.3

*Problematic Items.

Clothing and Individual Equipment Evaluation

The fit characteristics of 8 of the 11 CIE items evaluated were unacceptable on more than 15% of the sample (Table 4). Only the Light Duty Work Glove, the PASGT Helmet, and the Parachute Harness disaccommodated less than 15% of the subjects. The item posing the most severe fit problems was the CW Trigger Finger Mitten, which provided an acceptable fit for only 4% of the sample.

The mitten, which is made in only two sizes, Medium and Large, was too large and too long. In general, the other problematic items were found to be too long and wide in the torso, and the leg lengths of trouser garments were also too long. The looseness of the garments, in some cases, posed the risk of entanglement. In addition, movements such as climbing, raising the arms, and marching in place were often hindered by the poor fit of these items. The PASGT Vest extended well below the waist on many subjects, thus hindering bending from the waist and squatting. The belts on the Field Pack with Frame and the ETLB Vest were designed to be worn below the lower edge of the PASGT Vest. Because of the extreme length of the PASGT Vest on the subjects, these belts were at or below the level of the hips. These observations illustrate the incompatibility of items designed for male-body proportions (longer broader torso, narrower hips, higher crotch height) with female size and shape (shorter narrower torsos, wider hips, shorter crotch height).

The Fisher Exact Test performed on each CIE item revealed that the acceptable and the unacceptable fit groups differed significantly ($p<.05$) in the proportions of subjects who met or who exceeded the Army's weight-for-height standards on only one item, the PASGT Vest. Analyses of body dimensions yielded some significant differences (Bonferroni correction, $p<.0125$) between the acceptable and the unacceptable fit groups. The significant findings are summarized in Table 5. The ETLB Vest was the only item for which a difference was obtained in a circumferential dimension or in weight. Subjects in the unacceptable fit group had higher values for both measurements. With regard to the differences in Stature, the acceptable fit group had a higher mean Stature than did the unacceptable fit group. As was the case with the workstation data, it appears that subjects who were 5'3" (160 cm) or more, were fit successfully in the CIE and subjects who were 5'2" (157 cm) or less, were not.

Unlike the workstation items, most of the clothing items cannot be field-modified to address the fit problems observed in this study. The development of new sizes is recommended for all problematic CIE items. The estimated costs for statistical determination of the number of sizes; establishment of key sizing dimensions and pattern design dimensions; and anthropometric fit testing of prototypes, are summarized in Table 6. It should be noted that Table 6 does not include the actual cost of procuring additional garment sizes or the subsequent increase in logistics management costs associated with additional items of inventory.

Table 5. CIE Items with Significant Differences in Anthropometric Dimensions

CIE Item	Stature	Chest Cir.	Waist Cir.	Hand Lgth.	Hand Cir.	Weight
CW Mitten	X					
Mechanics' Coverall	X					
Field Pack w/Frame	X					
ETLB Vest	X	X				X

Table 6. Summary of CIE Suggested Actions and Estimated Costs

CIE Item	Suggested Action	Est. Cost
CW Trigger Finger Mitten	Develop smaller sizes	$50,000
CVC Coverall	Investigate female/integrated sizes	$150,000
Mechanics' Coverall	Investigate female specific sizes	$80,000
Field Pack, ETLB Vest, PASGT Vest	Ballistic Protection Program; Modular Body Armor/Load Bearing Program	$178,000
ECWCS Parka	Shorten sleeve pattern length in XSmall sizes; authorize field modification of snow skirt	$50,000
Wet Weather Trouser	Develop smaller sizes; issue suspenders	$80,000

DISCUSSION

The results presented here support the hypothesis that females outside a design envelope of 5th-95th percentile U.S. Army male Stature are disaccommodated in a representative sampling of workstations and CIE items. Based on the criteria applied in this study, subjects were disaccommodated in some way by every workstation tested, and by 8 of 11 CIE items. Disaccommodation was significantly related to Stature. Overall, females about 5'3" and taller were likely to be accommodated, while those 5'2" and shorter were not. However, females taller than 5'5"

may also be disaccommodated due to proportional differences between males and females. This segment of the population requires further study.

Actions taken to increase the accommodation of shorter females in both workstations and CIE will increase the safety and efficiency of task performance, as well as the numbers of soldiers who can be assigned to a particular task. It is possible that males of relatively short Stature also experience the reach and fit incompatibilities observed among the females in this study. Joining the U.S. Army in increasing numbers are women and minority group members characterized by shorter Stature and reach dimensions (e.g., Hispanic and Asian/Pacific Islanders). As a consequence, the mean Stature of the Army population may tend to decrease over time and disaccommodation problems may be exacerbated. To address the accommodation issues identified in this study, it is estimated that a minimum of $4.3 million would be required. This cost applies only to the representative items in this study. A logistical analysis to extrapolate costs across the entire spectrum of equipment and clothing in the Army inventory would be needed to obtain a more complete estimate of the cost of accommodating U.S. Army personnel at the lower end of the frequency distribution for Stature.

ACKNOWLEDGMENTS

This study was funded by Headquarters, Department of the U.S. Army, Office of the Deputy Chief of Staff for Personnel, in conjunction with the MANPRINT Office. Dr. Mike Fischl and Dr. Don Headley served as study sponsors. The work of GEO-CENTERS, INC. was carried out under a contract with the U.S. Army Natick Research, Development and Engineering Center.

REFERENCES

AR600-9. (1986). Army Regulation 600-9 "The Army Weight Control Program", 1 Sept 1986. Washington, DC: The Department of the Army.

Finch, C.E. & Hayflick, L. (Eds.). (1977). The Handbook of the Biology of Aging. New York, NY: Van Nostrand Reinhold Co.

Gill, G.W. & Rhine, S. (Eds.). (1990). Skeletal Attribution of Race. Albuquerque, NM: Maxwell Museum of Anthropology.

Gordon, C.C., Churchill, T., Clauser, C.E., Bradtmiller, B., McConville, J.T., Tebbetts, I., Walker, R.R. (1989). 1988 Anthropometric Survey of U.S. Army Personnel: Method and Summary Statistics (TR-89/044). Natick, MA: U.S. Army Natick Research, Development and Engineering Center.

REPORTS OF REGIONAL BODY DISCOMFORT DURING CARPENTER APPRENTICE TRAINING

Peregrin Spielholz and Steven F. Wiker
University of Washington Ergonomics Laboratory, Box 357234, Seattle, WA 98195-7234

Regional discomfort questionnaires were administered to apprentice carpenters at three month intervals for a duration of six months following an ergonomics awareness training as part of apprenticeship school. Reports of frequent musculoskeletal discomfort were reported by between 20 and 29 percent of carpenters for each of the nine body regions with the exception of higher levels for the lower back. Severity ratings and frequency of discomfort were highest for the lower back and hands/wrists. There was no significant difference in reports of musculoskeletal discomfort among the baseline and follow-up questionniares (p > 0.05). The lower back was the only body region showing a decrease in the ratings of discomfort severity during follow-up. Further study of training effects on work methods and discomfort are recommended.

INTRODUCTION

Carpentry is a demanding and dynamic occupation which deserves attention from the ergonomics community. The variable nature of the work and workforce makes it a challenging environment in which to perform assessments and implement solutions. However, despite variable work conditions most workers report significant levels of body discomfort across construction trades. Regional body discomfort, in construction workers, is most frequently reported in the low back, knees and right wrist (Eastern Iowa Construction Alliance, 1991; Cook and Zimmerman, 1992). Construction workers have also reported significant discomfort, and occupational disorders, in each of the following body parts ranked in order of severity with the first being the most severe: low back, knees, hand/wrist, shoulder, neck and upper back (University of Iowa, 1994). More than 50% of workers questioned reported symptoms of disorders in most of these body locations. This data points to the high physical stress involved in most carpentry work and the need to ergonomically assess the high risk areas and identify possible solutions to the problem.

Training has been argued as a means to reduce injuries in the workplace for the last 50 years. Many have questioned the effectiveness of training programs in changing unsafe behaviors and decreasing workplace-related disorders. Very little information exists in the literature evaluating the efficacy of ergonomics training, though it is often recommended as part of an ergonomics management program (OSHA, 1990; Shahnavaz, Abeysekera and Johansson, 1993) and is implemented on some level in many major companies in the United States. There are anecdotal reports which show a positive reception to programs (Wick and Weige, 1995), but little information is given which evaluates measurable outcomes.

Training programs must be implemented which address the primary issues specific to the work environment in order to maximize the probability of a favorable outcome. Assessment of musculoskeletal discomfort is a frequently used correlate for identification and ranking of ergonomic problems in a given work environment. This is a necessary component of training and intervention programs to target critical areas with the most efficient allocation of resources.

BACKGROUND

The ergonomics training program and discomfort evaluation is part of a project directed by the United Brotherhood of Carpenters Health and Safety Fund funded through a NIOSH grant. The project has included ergonomic hazard assessment and evaluation of intervention strategies as part of the development and implementation of training programs for carpenters. The work has been performed with and under the oversight of a joint labor-management focus group consisting of union carpenters, contractor safety and health representatives, union safety and health representatives, apprenticeship school instructors, and state government representatives which was formed in early 1993.

An apprenticeship ergonomics awareness curriculum and training manual were developed by the UBC Health and Safety Fund. The training consists of a four-hour class given to apprentice carpenters by school instructors when they enroll in the apprenticeship program. Information on ergonomics in carpentry is related in a four-hour lecture and contained in a manual kept by the apprentices. The manual and training evaluation discussed shortly were developed by the UBC Health and Safety Fund with review and administration by members of the focus group.

METHOD

Subjects

Two-hundred and five apprentice carpenters participated in the initial ergonomics training and evaluation as part of apprenticeship school requirements. Approximately 172 of the carpenters were male and 34 were female, ranging in age from 23 to 52 years of age (Mean age of 36). Those participating in the training had spent an average of 4.6±4.1 years working in construction and 1.7±1.4 years as an apprentice prior to the class.

Apparatus

A questionnaire was developed containing questions relating to symptomatic discomfort by body region. Respondents were asked to rate discomfort in each of nine specified body regions. The carpenters were first asked whether they experience any frequent pain or discomfort in each body region that they believed to be related to work. Subsequent questions for each body region asked which side was the worst, how many times this had occurred in the last year and how long it usually lasts. Further queries asked whether the pain had occurred in the previous seven days and if it was caused by a sudden accident. The apprentices were also asked to rate how painful the discomfort is on a scale of zero to five and to report how many days of work have been lost or restricted due to the problem in the last year.

Procedures

The discomfort questionnaire was administered by the instructors to the same group of apprentices the day of the training, and at three months and six months following the ergonomics awareness training during the regularly scheduled apprenticeship school.

RESULTS

Only 104 subjects who completed both the baseline and the six month questionnaire were included in the analyses due to a loss of almost half of the respondents during follow-up. The data for reports of chronic discomfort was aggregated by test administration and side of discomfort for each body region as ratios for comparison. An ANOVA was performed on the arc sine transformed data with the interaction as the error term. Figure 1 shows the percentage of apprentices reporting chronic work-related discomfort. Reports of discomfort were relatively consistent across body regions, except for the lower back which was higher. No significant differences were found between questionnaires for any body region ($p > 0.05$). Figure 2 summarizes the carpenters' reports of discomfort in the last seven days by body region. The lower back and the hands were the most frequently reported body regions for each questionnaire.

The apprentices rated their discomfort by body region on a scale of 0 to 5 (no pain to severe pain) only for regions that experienced chronic discomfort. Table 1 details the responses by questionnaire administration and body region. A one-way ANOVA was performed for each body region. The lower back was the only location showing a significant change in ratings of discomfort between questionnaires $F(2,287) = 6.86$. A Tukey multiple comparison test showed a significant decrease in rating reports among the baseline questionnaire and the three and six month follow-ups at the 0.05 level.

The percentage of respondents reporting that the regional discomfort had been caused by a sudden accident on the baseline questionnaire ranged between 2.1 and 4.3 percent for all body regions except the lower back, for which 9.7 percent attributed the cause to an accident. These results are summarized in figure 3.

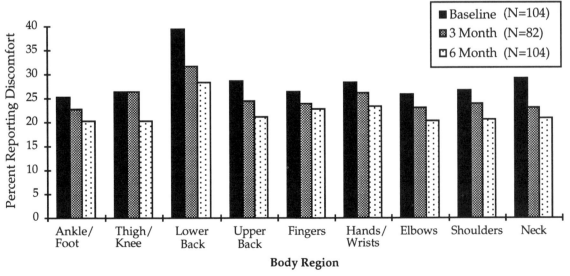

Figure 1.   Percentage of Carpenters Reporting Frequent Discomfort by Body Region for Baseline and Follow-up Questionnaires.

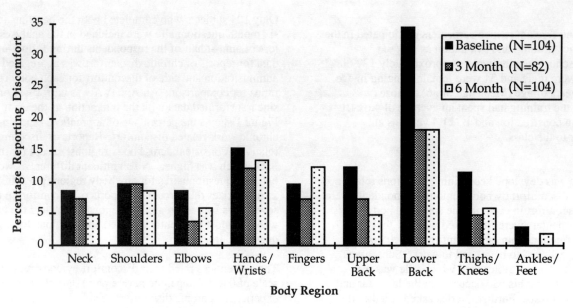

Figure 2.   Percentage of Carpenters Reporting Discomfort in the Last 7 Days by Body Region for Baseline and Follow-up Questionnaires.

Table 1.   Mean Ratings of Discomfort on a Scale of 0-5 (no pain - severe pain) by Body Region for Baseline and Follow-up Questionnaires

Body Region	Baseline (N=104)	3 Month (N=82)	6 Month (N=104)
Neck	0.43±0.95	0.21±0.66	0.18±0.68
Shoulders	0.51±1.10	0.39±0.94	0.24±0.81
Elbows	0.35±0.82	0.30±0.80	0.20±0.67
Hands/Wrists	0.63±1.13	0.68±1.29	0.46±1.06
Fingers	0.29±0.75	0.37±0.95	0.32±0.92
Upper Back	0.39±1.04	0.26±0.93	0.18±0.75
Lower Back	1.40±1.68	0.79±1.38	0.69±1.35
Thighs/Knees	0.43±1.13	0.23±0.85	0.19±0.76
Ankles/Feet	0.06±0.34	0.00±0.00	0.04±0.31

The mean number of workdays reported lost and restricted in the last year were evaluated for the baseline questionnaire by body region. The lower back is higher than any other region, at a little more than three lost workdays attributed to lower back pain in the previous year. Reports of lost or restricted workdays were less than one for all other regions. Mean lost workdays and injuries due to accidents did not show any significant differences among the three reports (p > 0.05).

Reports of frequency and duration of discomfort also did not show significant differences among the baseline and three and six month follow-up questionnaires (p > 0.05). Responses showed that 12.7±6.9 percent of the carpenters had experienced the reported episodes of discomfort at least once a week across body regions. The two most frequently reported locations of discomfort at least once a week were the lower back with 25.5 percent and the hands/wrists with 20.7 percent of the total responses. Discomfort lasting longer than one day was experienced by 7.3±3.2 percent of the carpenters across all body regions. Lower back discomfort was the highest with 14.5 percent of all respondent reporting durations lasting greater than one day.

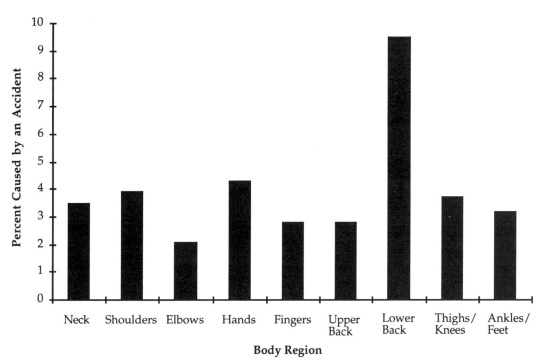

Figure 3.   Percentage of Carpenters Reporting That Current Discomfort was Caused by a Sudden Accident for Each Body Region.

DISCUSSION

The lower back is described as the most frequent and severe problem by every measure in the questionnaire. These results show that carpenters are experiencing very high levels of back pain, consistent with reports of the construction industry as a whole. The hands/wrists were consistently reported as being the second most frequent and severe location of chronic discomfort across questionnaires. These are important findings for the development of training curriculums and intervention strategies.

Weighting intervention and training resources towards the lower back and the hands/wrists could provide the best opportunity for reducing discomfort and injuries in the greatest proportion of carpenters. However, the consistently high reports of musculoskeletal discomfort across body regions indicates that all aspects of the work require attention in the development and implementation of ergonomics strategies.

Less than one-quarter of the reports of discomfort could be attributed to a specific accident or traumatic event. In addition, the ratings of discomfort were of a relatively low grade with the exception of the low back. This points to the existence of general job-related stressors contributing to the development of chronic discomfort. These patterns should also be taken into account in the development of countermeasures.

The power of the evaluation could not discern any differences in the frequency and severity of reports of musculoskeletal discomfort beyond a reduction in ratings for the lower back.

There was a non-significant decreasing trend in frequency of discomfort, however no additional measures were taken which might explain this pattern. Reporting of musculoskeletal discomfort can be sensitive to several job-related factors. It is possible that differences in job demands over projects and seasons could affect reporting of discomfort. In the future, a description of current work conditions should be included in the measurement tool.

CONCLUSIONS

Reports of musculoskeletal discomfort are widely distributed across body regions with the greatest frequency and severity occurring in the lower back. Increased knowledge of discomfort and injury patterns relating to the work can help most effectively target training and intervention resources. Results of this study imply that a comprehensive approach is needed to alleviate discomfort and the development of injuries in carpenters. Study of changes in work methods and job design as well as knowledge and discomfort following ergonomics training could help evaluate the efficacy of our current efforts.

ACKNOWLEDGMENTS

This study was completed under a grant from the National Institute for Occupational Safety and Health (NIOSH). In addition, the authors would like to thank SHARP in the Washington State Department of Labor and Industries and Western Washington general contractors for their participation and contributions to the research.

REFERENCES

Cohen H, and Jensen R, (1984), "Measuring the Effectiveness of an Industrial Lift Truck Safety Training Program", Journal of Safety Research, 15, pp.125-135.

Cook, T.M., and Zimmerman, C.L., (1992), "A symptom and job factor survey of unionized construction workers", Kumar, S., ed., Advances in Industrial Ergonomics and Safety IV, Taylor and Francis.

Eastern Iowa Construction Alliance/University of Iowa Joint Project on Reduction of Work-Related Injuries and Illnesses Through Ergonomic Intervention, (1991) Final Report, Phase I, (unpublished).

Ergonomics for Carpenters, United Brotherhood of Carpenters Health and Safety Fund.

Fellner D, Azaroff B, (1984), "Increasing Industrial Safety Practices and Conditions Through Posted Feedback", Journal of Safety Research, 15, pp. 7-21.

Fiedler F, (1987), "Structured Management Training in Underground Mining- Five Years Later", Proceedings: Bureau of Mines Technology Transfer Seminar, Pittsburgh, PA, Bureau of Mines Information Circular 9145, pp. 82-94.

Gotsch A, and Weidner C, (1994), "Strategies for Evaluating the Effectiveness of Training Programs", Occupational Medicine, 9(2), pp. 171-188.

Hellman C, Budd M, Borysenko J, McClelland D, and Benson H, (1990), "A Study of the Effectiveness of Two Group Behavioral Medicine Interventions for Patients With Psychosomatic Complaints", Behavioral Medicine, Winter, pp. 165-173.

Holmstrom E, Lindell J, and Moritz U, (1992), "Low Back and Neck/Shoulder Pain in Construction Workers: Occupational Workload and Psychosocial Risk Factors, Part 1", Spine, 17(6), pp 663-671.

Holmstrom E, Lindell J, and Moritz U, (1992), "Low Back and Neck/Shoulder Pain in Construction Workers: Occupational Workload and Psychosocial Risk Factors, Part 2", Spine, 17(6), pp 672-677.

Mattila M, and Hyodynmaa M, (1988), "Promoting Job Safety in Building: An Experiment on the Behavior Analysis Approach", Journal of Occupational Accidents, 9, pp. 255-267.

Occupational Safety and Health Administration, (1990), Ergonomics Program Management Guidelines for Meatpacking Plants, US Department of Labor (OSHA) Document Number 3123.

Robins T, Hugentobler M, Kaminski M, and Klitzman S, (1990), "Implementation of the Federal Hazard Communication Standard: Does Training Work?", Journal of Occupational Medicine, 32, pp. 1133-1140.

Saarela K, Saari J, and Aaltonen M, (1989), "The Effects of an Informational Safety Campaign in the Shipbuilding Industry", Journal of Occupational Accidents, 10, pp. 255-266.

Schurman S, Silverstein B, and Richards, S, (1994), "Designing a Curriculum for Health Work: Reflections of the United Automobile, Aerospace and Agricultural Implement Workers-General Motors Ergonomics Pilot Project", Occupational Medicine, 9(2), pp. 283-304.

Shanavaz H, Abeysekera J, and Johansson A, (1993), "Solving Multi-Factorial Work Environment Problems Through Participation", The Ergonomics of Manual Work, W Marras, W Karwowski, J Smith, and L Pacholski Ed., London: Taylor and Francis, pp 499-502.

Sedgwick A, Davies M, and Smith D, (1994), "Changes over Four Years in Musculoskeletal Impairment in Men and Women", Medical Journal of Australia, 161, pp. 482-486.

Tan K, Fishwick N, Dickson W, and Sykes P, (1991), "Does Training Reduce the Incidence of Industrial Hand Injuries?", Journal of Hand Surgery, 16B(3), pp. 323-326.

University of Iowa (1994), Symptom and Job Factor Survey of Unionized Construction Workers, Preliminary Findings, November, 1994.

Vojtecky M, and Schmitz M, (1986), "Program Evaluation and Health and Safety Training", Journal of Safety Research, 17, pp. 57-63.

Wick J and Weige L, (1995), "A Strategy for Providing Ergonomics Awareness Training and Identifying Individuals in Need of Ergonomics Intervention in Office Environments", Advances in Industrial Ergonomics and Safety VII, A Bittner and P Champney Ed., London: Taylor and Francis, pp. 369-373.

AN ERGONOMIC EVALUATION OF A DRYWALL BOARD TRANSPORT HANDLE: SINGLE-PERSON TRANSPORT TASK

Mark A. Stuart and Kerith K. Zellers
State of Washington Department of Labor and Industries
Safety and Health Assessment and Research for Prevention (SHARP) Program
Olympia, Washington 98504-4330

The objective of this study was to determine how the use of a drywall board transport-handle affects physical stress on the human body, subjective reports of pain, discomfort, and physical exhaustion, and task performance compared to a no transport-handle condition. Eight construction workers lifted and transported 20 drywall boards in a simulated single-person construction task. Results indicated that there were no significant differences between the two board transport conditions on any of the subjective and physical data collected. Average task performance times were slower for the handle-use condition (232 seconds compared to 207 seconds), but this difference was not statistically significant. A serious shortcoming of the transport handle evaluated in this study is that it does not eliminate the major risk factors associated with drywall installation: frequent lifting and carrying of very heavy objects, and awkward postures of numerous types. Drywall installers' work-related musculoskeletal disorders will likely continue as long as commercial construction continues to require single-person board lifting and transporting and as long as the boards stay as large and heavy as they are currently.

INTRODUCTION

Lifting and transporting drywall boards exposes construction workers to a number of biomechanical stressors, such as the frequent handling of excessively heavy loads (typically between 90 and 150 pounds) and numerous types of awkward postures (Schneider and Susi, 1994; Spielholz, Narayan, and Wiker, 1995). The majority of injuries in drywall installation occur in the lower back, neck, and shoulder regions of the body (Cook and Zimmerman, 1992). Drywall installers represent 1.41% of all construction workers, yet their compensable injury rate is nearly three times the injury rate for all other construction occupations combined (Hsiao, Stanevich, and Fosbroke, 1996). Sprains and strains, which comprised 43.3% of all compensable drywall injuries in the United States in 1994, are one of the major types of injuries reported (Pan and Fosbroke, 1996). Review of drywall installers' workers' compensation claims in the State of Washington also indicate that the injuries are typically more severe than non-drywall construction injuries. It is reasonable to assume that this trend is consistent across other states as well. An ergonomics intervention in the drywall installation industry may be extremely helpful if proven engineering controls can be implemented in construction sites which reduce or lower the frequency and severity of injuries.

An activity of the State of Washington's Department of Labor and Industries Safety and Health Assessment and Research for Prevention (SHARP) program is to perform research on new work tools and methods to determine those which increase worker safety and productivity. When considering the problem of work-related musculoskeletal disorders (WMSDs) in the drywall industry, there are numerous new tools on the market which purportedly reduce WMSD risk factors. Unfortunately, there is often little or no empirical evidence to support some of these advertised claims. A goal of SHARP is to perform laboratory and field research on a number of drywall installation tools which, at least initially, show indications of possibly being able to reduce physical stress on the body.

The objectives of this study were to determine how the use of a commercially available drywall board transport-handle affects physical stress on the human body, subjective reports of discomfort and handle characteristics, and task performance while lifting and transporting drywall boards in a simulated task. A drywall transport handle was selected for this specific evaluation due to strong interest in their usage among construction labor and management representatives in the State of Washington. This particular tool evaluation is only one of many construction tool evaluations which SHARP plans to conduct. Other drywall installation tool evaluations will follow. These as well as other ergonomic tool evaluations are important because it is necessary to first determine the tool's effectiveness before broader ergonomics interventions can be implemented at actual construction sites.

METHOD

Subjects

Test subjects were eight construction workers enrolled in a drywall installation apprenticeship program at a technical college in the state of Washington. All subjects were males, in good health, and ranged in age from 20 to 33 years.

Experimental Design

The experimental design utilized was a 2 x 2 full factorial repeated measures design. The two independent variables were carrying technique (no use of a handle and the use of a drywall board transport handle) and trial (first and second). Order effects were controlled by having four test subjects begin the testing with the no-handle condition and four test subjects begin with the handle condition.

Apparatus

The drywall board handle depicted in Figure 1 was evaluated. This handle weighs 7.5 pounds and measures 30 inches long by 7 inches wide. This specific handle was selected for evaluation due to interest expressed in its usage by labor and management representatives in the State of Washington.

Twenty drywall boards measuring four-feet wide by eight-feet long and 5/8 inches in thickness were used. Each board weighed 90 pounds. This is a typical drywall board size used in commercial operations, therefore, test subjects were not exposed to risks during this experiment which exceeded what they normally experience everyday on their jobs.

Body area discomfort and handle characteristics questionnaires were used to gather subjective information. Body area discomfort questionnaires were used to collect discomfort information using a ten point scale on various body areas, plus overall discomfort and physical exhaustion. Subjective information on the handle's characteristics was also collected.

Metrosonics pm-385 heart rate monitors were used to collect heart rate data during the performance of the board transport tasks and during a post-transport recovery period. The Metrosonics systems provided a continuous measurement of heart rate data averaged every minute for each test subject throughout the test sessions.

Task and Procedure

All testing took place at a technical college. Test subjects were briefed on how the board handle is used and were given practice trials on attaching and lifting with the use of the handle. Subjects were asked to perform the board transports at a pace similar to a normal work pace.

For each experimental condition, the basic task consisted of moving 20 drywall boards one-at-a-time from a stack (vertically oriented) to a location 30 feet away while using the predetermined carrying technique (handle or no-handle). The transport course made a ninety-degree turn before the endpoint. The transport task was divided into two trials: Trial One -- the first 10 boards and Trial Two -- the second 10 boards. A one-minute break separated the two trials. Test subjects vertically stacked the drywall board at the endpoint, then walked back to the stack and continued the task until each trial was completed. The carrying technique (handle or no-handle) remained the same across Trials One and Two. This experimental task simulated occupational drywall installation tasks such as loading a scissor-lift scaffold and transporting drywall board from pallets to another location within a construction site.

Test subjects participated in only one experimental condition per day to control fatigue buildup from one experimental condition to another. All testing took place in the evening to accommodate the drywall installers' work schedules. This also helped control time-of-day testing effects.

Data Collection

Subjective, performance, and physiological data were collected at various points throughout each experimental session. Questionnaires were administered before transporting the boards and immediately after transporting all of the boards. Heart rate data were collected throughout the performance of each experimental condition and for a five-minute recovery period after transporting all the boards. The five-minute recovery period followed a protocol similar to one used by Jensen and Dukes-Dobos (1976). Finally, time to complete Trial One and Trial Two board transport tasks were recorded. All trials were video taped.

RESULTS

Body Area Discomfort

All pre-test and post-test body area discomfort scores were statistically analyzed with 2-factor repeated measures analyses of variance (ANOVAs). The only body area discomfort questionnaire items which yielded statistical significance were the two lower back questions. There was a significance increase in subjective discomfort when comparing pre-trial and post-trial right lower back and left lower back discomfort responses, $F(1,7) = 6.3, p < 0.05$, and $F(1,7) = 6.22, p < 0.05$, respectively. No significant differences were found due to the use of the transport handle for the right and left lower back questionnaire items, $F(1,7) = 0.32, p = 0.58$, and $F(1,7) = 0.1, p = 0.75$, respectively.

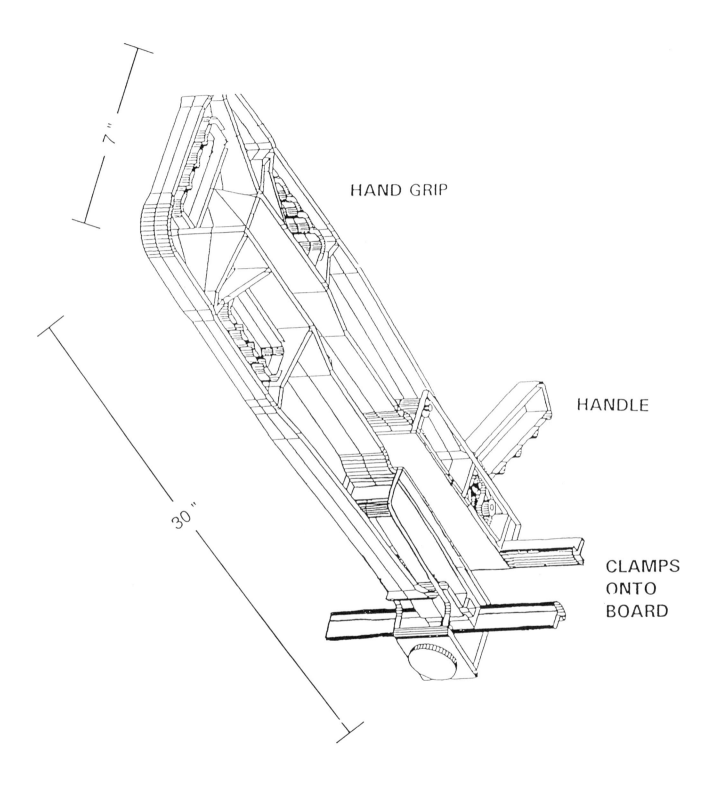

Figure 1. *Illustration of drywall board transport handle used in this experiment.*

Analysis of the physical exhaustion subjective responses data exhibited significance when comparing pre-test to post-test responses, $F(1,7) = 44.55$, $p < 0.05$. As to be expected because of the nature of the task, post-trial responses were significantly worse than pre-trial responses. There were no differences due to the handle-use conditions.

Performance Time

Trial One and Trial Two drywall board transfer performance times are listed in Table 1. After analyzing these data with a repeated measures ANOVA, it was revealed that there was significance due to the effect of the two trials, $F(1,7) = 12.32$, $p < 0.05$, where Task Two performance times were faster than Task One times. There was no significant difference due to the effect of the handle or no-handle condition, $F(1,7) = 2.23$, $p = 0.17$, although the handle condition's average task time (232 seconds) was higher than the no-handle condition's average time (207 seconds).

TABLE 1

Task performance times (in seconds) per test subject within Trial 1 and Trial 2 per handle condition

Ss	NO HANDLE		HANDLE	
	TRIAL 1	TRIAL 2	TRIAL 1	TRIAL 2
1	223	231	295	261
2	200	201	185	167
3	191	171	185	180
4	200	208	228	235
5	264	265	300	250
6	165	158	189	166
7	251	222	307	318
8	195	182	237	221
TRIAL AVG.	211	204	240	224
COND. AVG.	207		232	

Handle Characteristics

This section contains a summary of test subject subjective impressions concerning the transport handle. No statistical analyses were performed on these data due to only one handle being evaluated in this experiment. Test subject opinions about the drywall handle are summarized by the following comments:
- Handle is too big and bulky
- Test subjects felt worse after using handle compared to   not using the handle
- Handle won't speed or slow production, but it should reduce back pain in the long run
- Weighs too much
- Handle needs better grips
- Handle should have padding or cushioning to avoid contact stress with hip and thigh.

In summary, subjective information collected from the construction workers strongly indicated that they think there are numerous problems with the drywall transport handle. It was also revealed that the construction workers generally disliked the handles to the extent that they would rather lift and transport the boards manually instead of using this specific transport handle.

Heart Rate Data

Heart rate data were analyzed two different ways: (1) average heart rate per trial, and (2) heart rate recovery after a five-minute rest.

(1) *Average heart rate per trial*. The average heart rates of test subjects per trial per handle condition were computed and are listed in Table 2. These data were then statistically analyzed with a repeated measures ANOVA. Data analysis revealed a significant heart rate increase from Trial One to Trial Two, $F(1,7) = 8.14$, $p < 0.05$. No significant difference was found due to the use of the handle.

TABLE 2

Average heart rates (beats per minute) per test subject within Trial 1 and Trial 2 per handle use condition

Ss	NO HANDLE		HANDLE	
	TRIAL 1	TRIAL 2	TRIAL 1	TRIAL 2
1	155	158	154	165
2	172	181	179	191
3	161	176	153	169
4	150	146	162	166
5	116	116	113	120
6	170	166	149	158
7	163	162	145	143
8	158	169	143	157
TRIAL AVG.	157	159	149	158
COND. AVG.	158		153	

(2) *Heart rate recovery*. Heart rate data continued to be collected from seated test subjects for five minutes after the drywall board transport trials while they completed the post-test questionnaire. Table 3 contains each test subjects average heart rate collected during Trial 2 and their heart rate after five minutes of rest. Statistical analysis of these data revealed a significant reduction in average heart rates due to the recovery period, $F(1,7) = 158.27$, $p < 0.05$, although heart rates did not decrease to the original baseline levels after the five-minute rest. There were no significant differences due to the handle-use conditions, $F(1,7) = 0.81$, $p = 0.39$.

TABLE 3

Heart rate data (beats per minute) for Trial 2, after a five-minute rest, and amount of heart rate change (Δ)

	NO HANDLE			HANDLE		
Ss	TRIAL 2 HR	AFTER REST HR	HR Δ	TRIAL 2 HR	AFTER REST HR	HR Δ
1	158	103	53	165	109	56
2	181	120	61	191	102	89
3	176	120	56	169	117	52
4	146	90	56	166	91	75
5	116	82	34	120	79	41
6	166	112	54	158	98	60
7	162	111	51	143	113	30
8	169	99	70	157	96	61
AVG	159	105	54	158	100	58

DISCUSSION

For an engineering control or change in work methods to be truly ergonomically effective in occupational settings, a number of issues must be considered. First, the new tool or work method must result in a reduction in ergonomic and physical stresses on the human body, assessed through objective and subjective measurements. Second, it must be determined how the new tool or work practice affects job performance or productivity. Third, it must be determined how the new tool or work practice affects psychosocial factors.

This ergonomic analysis of a tool designed to improve the common industrial task of lifting and transporting drywall boards did not find any supporting evidence for the use of this specific device. Additionally, it was determined that the use of the tool resulted in more time (although not statistically significant) to perform the board transport task. It must be noted that familiarity with this new board transport method should be considered when assessing this result. Given more time to practice with this specific tool may result in improved task performance times, but it is unlikely that task times will ever be as good as manual transport times simply due to the mechanics of using this handle: it takes time to put the transport handles on and off the boards and this time must then be added onto the normal board transport time.

In the highly competitive commercial construction industry, it is hard to conceive of using a tool or procedure which has no ergonomic benefits in addition to its use taking more time compared to currently used work practices. Therefore, the investigators in this evaluation can not recommend the use of this particular drywall board transport handle for occupational tasks similar to the one performed in this study.

This study did find significant increases in heart rates and subjective discomfort for several body regions after the board transports which was to be expected considering the nature of this task. For example, there was a significant increase in physical exhaustion complaints when comparing pre-trial and post-trial data. There was also a significant increase in lower-back discomfort due to transporting the boards. It should be emphasized that an increase in fatigue may not necessarily result in work-related musculoskeletal disorders, but it may increase the likelihood of acute injuries.

It appears that drywall transport handles with this particular configuration may not make the job of moving drywall boards less stressful on the human body. This is primarily because this type of handle does not address the fundamental problem of moving drywall -- it requires repetitive, awkward postures to lift and move very heavy loads. Drywall installers' work-related musculoskeletal disorders will likely continue as long as commercial construction continues to require single-person board lifting and transporting and as long as the boards stay as large and heavy as they are currently.

As mentioned earlier, the State of Washington's SHARP program plans to perform ergonomic evaluations of other new drywall tools at actual construction sites as a follow-up to this research project. SHARP will also ergonomically study new tools and work practices in other construction jobs as well.

ACKNOWLEDGMENTS

This study was supported in part by the United Brotherhood of Carpenters Health and Safety Fund of North America. We would like to thank Jim Gay of the Western Washington Lathing, Acoustical and Drywall Systems and Thermal Insulation Installers Joint Apprenticeship and Training Program for providing the test facility for this evaluation. Thanks also goes to the construction workers who participated in this study. The efforts of SHARP colleagues Marty Cohen and Mike Cotey in helping with data collection during this study are also highly appreciated.

REFERENCES

Cook, T., and Zimmerman, C. (1992). A symptom and job factor survey of unionized construction workers. In S. Kumar (Ed.), *Advances in Industrial Ergonomics and Safety IV*, Taylor and Francis.

Hsiao, H., Stanevich, R., and Fosbroke, D. (1996). Determining focus of research for reducing overexertion injuries in the construction industry. Unpublished manuscript.

Jensen, R., and Dukes-Dobos, F. (1976). Validation of proposed limits for exposure to industrial heat. In S. Horvath and R. Jensen (Eds.), *Standards of occupational exposure to hot environments: Proceedings of symposium*, DHEW (NIOSH) publication no. 76-100.

Pan, C. S. and Fosbroke, D. (1996). Personal communication about the injury and illness rate for drywall installers based on Bureau of Labor Statistics data, May 15, 1996; National Institute for Occupational Safety and Health, Morgantown, West Virginia.

Schneider, S., and Susi, P. (1994). Ergonomics and construction: A review of potential hazards in new construction. *American Industrial Hygiene Association Journal, 55*(July), 635-649.

Spielholz, P., Narayan, M., and Wiker, S. (1995). Assessing ergonomic hazards in unstructured work using work sampling techniques: An application in construction of concrete formwork. In A. Bittner and P. Champney (Ed.), *Advances in Industrial Ergonomics and Safety VII*, Taylor and Francis.

ERGONOMICS COST BENEFITS CASE STUDY IN A PAPER MANUFACTURING COMPANY

Dan MacLeod
Director of Ergonomics
Clayton Environmental Consultants
Edison, New Jersey

Anita Morris
Ergonomics Coordinator
Crane & Co.
Dalton, Massachusetts

This company initiated a comprehensive workplace ergonomics program in 1991. After five years experience, total investment is estimated at $2.5 million and total benefits at $3.5 million, based on reduced workers compensation cost savings plus improvements in productivity. Lost-time CTD cases were reduced over 80%. CTD recordable cases initially increased because of heightened employee awareness, then decreased. Employee discomfort surveys showed a 40% drop in reported pain; greater results were noted in those areas where improvements have been made, and no change in those areas where tasks have not yet been addressed. Examples of specific workstation improvements and results are presented, including instances where CTD risk factors were reduced and productivity more than doubled. The implication is that employers should institute practical programs in ergonomics whether or not OSHA promulgates a regulation on CTDs.

BACKGROUND

This company is a 200-year-old paper manufacturer located in New England with approximately 1200 employees. Several tasks in the company were labor intensive and involved high repetitions of the hands and arms, resulting in Cumulative Trauma Disorders (CTDs). The company began to address these injuries, but a subsequent OSHA inspection resulted in a citation and fine. The company responded positively by developing a more aggressive and comprehensive ergonomics effort. The company initiated this comprehensive program in 1991 and the formal settlement agreement with OSHA was signed in December, 1991. The results reported in this paper are thus based on five years of experience.

PROGRAM DESCRIPTION

Key aspects of the company's ergonomics program included the following. Management formed various committees, involved employees in all committees, retained a professional ergonomist, provided training at all levels of the organization, instituted a communication network, started evaluating all tasks, instituted annual discomfort surveys, upgraded their medical management program, and made workstation, process and procedure improvements.

The orientation of the process was low tech and common sense, using the term "Yankee Ingenuity" to make the field of ergonomics more accessible and within the traditions of this long-established New England firm. This phrase also tapped into the workforce's heritage of innovation and creativity. Initial task evaluations were non-quantitative in nature and focused on simple identification of ergonomic issues and brainstorming for possible improvements. As the program evolved, quantitative approaches to evaluating risk factors were introduced in several cases to verify that risk factors were being reduced.

Medical management of CTDs is technically not part of the field of ergonomics, but certainly an integral component of CTD reduction. The company upgraded its existing system to the equivalent of that recommended in OSHA's *Ergonomics Program Management Guidelines for Meatpacking*. The on-site nurse was involved on the corporate ergonomics committee, reviewed jobs with CTDs and made recommendations for improvements. Employee education focused on early recognition and reporting of symptoms; the approach to treatment was conservative. A myotherapist was retained to provide early, hands-on treatment and to teach stretching exercises. The Medical Department was given total control of work activities and restrictions for injured employees.

RESULTS

Total Program Costs and Benefits

Total investment in ergonomics over a five-year period is estimated at about $2.5 million, including cost of new machinery and equipment. Total benefits over this same five year period are estimated at $3.5 million, based primarily on worker's compensation cost savings plus improvements in productivity. Thus, the return on investment for this ergonomics program is approximately 40%.

Injury/Illness Data

The following graphs (Figures 1-5) show injury and illness trends based on OSHA recordkeeping data:

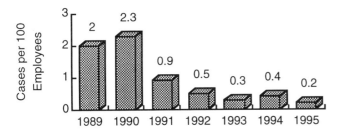

Figure 1. CTD Lost-Time Rates (upper extremity repetitive motion disorders) were reduced approximately 80% .

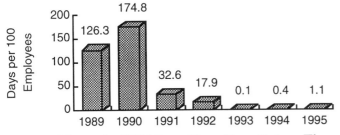

Figure 2. CTD Lost-Time Days Rates. The rates for lost time days were reduced dramatically.

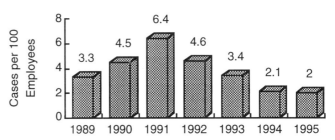

Figure 3. CTD Recordable Rates initially increased because of heightened employee awareness and a more responsive medical program, then decreased to nearly half of the number of initial cases. Current cases remain "elevated" because of the active medical program.

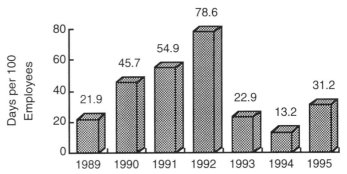

Figure 4. CTD Restricted Workday Rates also rose initially and then decreased, again due to the effect of an active medical program.

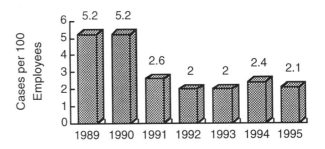

Figure 5. Strains and Sprains Rates (back injuries and upper extremity overexertion injuries) were cut approximately in half.

Workers Compensation Costs

Workers' compensation costs savings totaled an estimated $2.8 million over a five-year period. These changes enabled the company to switch to a self-insurance system, which entails further financial benefits.

Employee Surveys

Discomfort surveys have been administered to all employees annually for four years. In general, results show positive changes in those areas where improvements have been made, and no change in those areas where tasks have not yet been addressed. Figure 6 shows an overall decrease of 40% in discomfort scores for the company as a whole. Figure 7 shows one site of the corporation that has not been as active in the ergonomics program as the rest of the company; corresponding discomfort scores are stagnant. Figure 8 shows another site, with decreasing discomfort rates in initial years, then a rise which correlates to the lack of focus on ergonomics and CTDs in that year.

A conclusion is that the employee discomfort survey can be a valuable tool as part of an overall ergonomics tool kit. This type of survey appears to have face validity, that is, it appears to measure the real trends in the workplace. Survey results are useful for tracking accomplishments and for highlighting areas where additional work needs to be done. As a final comment, in this company the experience has been that the survey has helped management keep a focus on employee concerns, since no one could deny that problems existed in their own work areas.

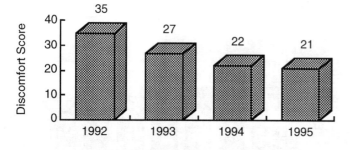

Figure 6. Company-wide employee discomfort survey score averages.

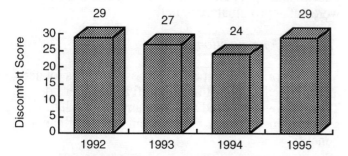

Figure 7. One inactive site discomfort scores.

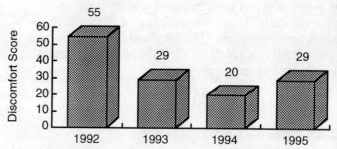

Figure 8. One site with initial progress reducing discomfort scores, then lapse in focus.

Productivity

Productivity improvements ranged from 0% to over 200% for specific projects. The average productivity improvement is estimated at about 25%, which is roughly equivalent to $700,000.

Other Savings

Other cost savings were also achieved, including reduced errors (and potential loss of customers), less down time because of medical absences of trained employees, and related administrative and legal costs. However, these costs are difficult to quantify and estimates are not included in this review.

Workstation Improvements

Examples of specific workstation improvements and results are presented below.

a. A machine feeder reduced arm repetitions from 5000 per day to zero, and output increased from 5000 pieces per day to about 15,000 (300% increase).

b. Improvements in a paper counting task reduced finger repetition from 45,000 per day to near zero, and productivity doubled.

c. A unique device to tie ribbons eliminated much fastidious hand repetitions and sustained pinch grips, plus doubled productivity. (This device, invented by a female production employee, lies in the finest tradition of ingenious 19th Century tools like apple peelers and cherry pitters.)

d. Unconventional tables for a precise, hand-intensive task enabled employees to alternative between sitting and standing, plus eliminated some reaches and motions. Modifications in hand tools have reduced grasping force.

e. Mechanical changes and automation in a packing operations reduced hand motions from approximately 32,000 per day to 3200.

f. A variety of other, more standard ergonomic devices were procured: vacuum-assisted hoists, various lift tables, anti-fatigue mats, improved chairs, numerous changes in heights and reaches, and automation.

DISCUSSION

Particular organizational factors which led to the success of this program included:

- The company provided training for all levels of the organization, aimed at each group's required involvement. The focus on training lasted one year, and is considered to be key for the overall success of this program.

- Top management was committed to integrating the tools and perspectives of ergonomics into a way of doing business. Special attention was given to each manager's training and accountability and to including formal ergonomics factors and injury reduction into individual goal-setting and performance evaluations.

- Employees were involved at all levels, including training sessions, suggestions, serving on committees, videotaping, job analysis, administering the discomfort survey, auditing, recommending improvements and monitoring progress. This was the first formal effort of the company in employee involvement and has led to other ways in which employees are involved in activities traditionally reserved for managers.

- Professional consultants were involved to provide training, guidance and analytic tools, but were not relied upon to provide the actual solutions to problem tasks. These outside consultants proved particularly valuable in being able to see changes that were occurring in the organization and the physical plant that company personnel were too close to recognize.

- The importance of good communications was highlighted throughout the program development, including monthly features in the company newsletter. These features included information on specific improvements, reports on status of program development, and results of injury/illness and discomfort survey findings.

- Focus was given to the recognition of individuals and groups that contributed to the effort. Celebrating milestones also contributed to promoting a sense of momentum, especially necessary to change mindsets in an old, established firm.

- Specific aspects of the program were changed as the program evolved, such as structure of committees, approaches to risk factor evaluation, how often meetings were held, and the relative focus between engineering and administrative controls.

CONCLUSION

The overall conclusion is that this workplace ergonomics program proved effective on a variety of levels, including financial. The implication is that employers should institute practical programs in ergonomics whether or not OSHA promulgates a regulation on CTDs. The most important factors of an ergonomics program are the core elements: corporate commitment to injury/illness reduction; good organization, involvement and accountability; training at all levels of the organization; effective communications and feedback; a systematic effort to evaluate all tasks in the company; making practical improvements focusing on low-tech workstation modifications; and monitoring progress. Specific approaches to implementing these program elements cannot be required, but left to each employer to make given its own circumstances.

THE EFFECTS OF GLOVE FRICTIONAL CHARACTERISTICS AND LOAD ON GRASP FORCE AND GRASP CONTROL

Lisa Bronkema-Orr[1] and Ram R. Bishu[2]
[1]Applied Risk Management, Oakland, CA
[2]University of Nebraska, Lincoln, NE

The objective of this experiment was to determine the effect of varying levels of glove surface friction, glove type, and various loads lifted on submaximal holding performance. The independent variables were glove type, friction level, load lifted, trial, and gender. Ten males and ten females performed two trials of lifting a device similar to a standard hand dynamometer under each of these conditions. All the main effects were significant at the 0.0001 level for the dependent variables of stable force and peak force. The results indicate that the surface friction of a glove affects the amount of force with which the subject feels he needs to grasp an object. In addition, the amount of force exerted per pound lifted decreased with increasing weights, perhaps indicating over exertion at the lighter levels of the weight lifted, or under exertion at the higher levels of load.

INTRODUCTION

Through the dexterity and versatility of the hand, humans are able to perform a variety of tasks. The hand allows a person to perceive information through its sensory mechanisms, manipulate objects through the mobility and flexibility of its fingers, and lift or squeeze objects through its tendon and muscle mechanisms. However, partly because of its versatility, the hand is sometimes subjected to harsh working environments. Industries which are concerned about employee safety may provide workers with various types of gloves, and while many recognize the protective merits of gloves, some may not realize that performance can be hindered by the use of gloves. The effects of gloves on multiple hand functions have been well documented in the literature. However, when looking at strength capabilities, it appears as though most research has focused on maximum voluntary contraction capabilities rather than on submaximal or "just holding" types of capabilities.

Only a few studies have looked at submaximal types of forces. Lyman and Groth (1958) found that wearing gloves caused subjects to exert more force on a peg when inserting it into a pegboard. For a similar task, Groth and Lyman (1958) found that the force exerted was directly related to the surface friction of the glove worn.

An experiment was performed by O'Hara, et al., (1988) in which subjects grasped and held objects under three glove conditions: barehanded, an unpressurized extra-vehicular activity (EVA) glove used by shuttle astronauts, and a pressurized EVA glove. The authors found that the unpressurized

glove condition increased the safety margin (defined as the difference between the stable grip force and the slippage grip force) by 86% over the barehanded condition, while the pressurized glove condition increased it by 111%.

A similar study was reported by Bishu, Bronkema, Garcia, Klute, and Rajulu (1994). The effects of glove type, glove pressure, and load on both stable holding exertions and maximal force exertions were determined through a series of experiments. The first experiment was performed under the hypothesis that holding force could be a measure of tactile feedback. In other words, as a person picks up an object, he or she would adjust the amount of force used according to how well he or she could feel the object being grasped. The independent variables included in this experiment were glove type (two types of EVA gloves and the barehand), glove pressure, and load lifted (3.5, 8.5, and 13.5 lbs.). Subjects were asked to lift and hold a device similar to a hand dynamometer for 20 seconds, while the peak grasp force and stable grasp force were measured. Unlike the O'Hara et al., (1988) experiment, the only effect that was found to be significant was the load lifted. Because it was expected that tactility would be function of both glove and pressure (increasing pressure causes the gloves to be inflated and very stiff), the hypothesis seemed to be incorrect.

The results of this experiment led to some changes in a second experiment reported by Bishu, et al. (1994). Perhaps a glove or pressure effect was not present because the subjects could see the load that they would be lifting prior to each trial. Therefore, the visual cue was eliminated.

Additionally, a thinner and more slippery meatpacking glove was included. Subjects performed a similar task, and for this experiment glove type was found to be significant. The two EVA gloves (both very thick and rubbery) were not significantly different from each other or the barehanded condition. Only the meatpacking glove was grouped differently from the other conditions. So, if thickness and pressure do not affect stable holding force, then it appears that holding force and tactility may not be related. These results seemed to indicate that glove surface friction may affect holding force more than glove type.

In addition to the friction effect, Bishu, et al., (1994) noted that as the load lifted increased, the amount of force exerted per pound lifted decreased. The tendency to exert more force per pound lifted was greater for the more slippery glove conditions. A similar pattern of grasp was reported by Grant (1994), who asked subjects to grasp a cylindrical aluminum handle, barehanded, and move it against various levels of weight. Subjects were found to apply more force than theoretically necessary, especially at the lighter loads.

Objective

From these experiments, it appears that the surface friction of the glove affects "just holding" types of exertions, while glove thickness or stiffness does not. However, the effect of glove friction, apart from glove type, has not been studied. Objects having different frictional characteristics were lifted in O'Hara, et al., but only two levels of friction were used, and no significant effects of friction were found.

Research has also shown that different patterns of grasp seem to exist at different levels of loads lifted (for relatively small magnitudes of load). Therefore, the objectives of this research are to provide some insight into the relation between glove friction, load lifted, and the grasp force applied by:

- Determining the effect of varying levels of glove surface friction on submaximal holding performance.
- Determining the effect of glove type on submaximal holding performance.
- Determining the effect of various magnitudes of load on submaximal holding performance, specifically magnitudes of 0.5, 5.5, 10.5, 15.5, and 20.5 lbs.

METHOD

Subjects

The subject population consisted of 10 healthy male and 10 healthy female volunteers, between the ages of 20 and 32.

Apparatus

To accomplish the objective of this experiment, a device was needed that would accommodate the loads to be lifted by the subject and at the same time measure the force with which the subject was grasping it. Such a device, identical to the larger of the two used in Bishu et al. (1994), was fabricated at the University of Nebraska-Lincoln College of Engineering and Technology Machine Shop. It consisted of two aluminum halves to which a small plate was attached so the appropriate weights could be added as needed. Between the two halves, at the top and bottom of the device, load cells (Sensotec® non-amplified transducers) were placed to measure the forces applied by the hand. The output of the load cells was then channeled through a real-time data recording system.

Two types of gloves were used in this study: cotton canvas and smooth leather. To achieve different levels of friction for both glove types, rubbery textured silicon was applied to the gripping surface of the gloves. This was done to obtain gloves with two different levels of friction, with a larger surface area of silicon offering a higher coefficient of friction on the grasping surface. A third glove with no added silicon was also included. Hereafter the gloves will be denoted as "cotton 1" (cotton glove with no silicon), "cotton 2" (cotton glove with a surface area of 4.8 square inches of silicon), "cotton 3" (cotton glove with a surface area of 7.2 square inches of silicon), "leather 1" (leather glove with no silicon), "leather 2" (leather glove with a surface area of 4.8 square inches of silicon), and "leather 3" (leather glove with a surface area of 7.2 square inches of silicon).

Experimental Design

The experiment was a factorial design in which the following were independent variables:

- Glove type: Cotton Canvas, Leather, & Barehanded
- Friction: A plain glove, a glove with a surface area of 4.8 square inches of silicon on the grasping surface, and a glove with a surface area of 7.2 square inches of silicon on the grasping surface
- Load: 0.5, 5.5, 10.5, 15.5, and 20.5 lbs
- Gender: Male and Female
- Trial: Trial 1 and Trial 2

Each subject performed one trial of each of these 35 conditions in a random order on one day (trial 1) and one trial of each of these 35 conditions in a random order on a second day (trial 2). Therefore, a total of 70 trials were performed by each subject over two different days.

For each condition, the subject grasped the handle which had been loaded with the appropriate weight, lifted it, and held it for 10 seconds. While holding the device, the subjects were instructed to keep their elbows close to their sides and their forearms at a 90° angle to their upper arm. After 10 seconds, the subjects were advised to gradually release the handle. Subjects rested three minutes between trials.

The dependent variables measured were the peak grasp force, stable grasp force, ratio of peak force to stable force, and the ratio of stable force to the load lifted.

RESULTS

An analysis of variance was performed and a summary of the results follows in Table 1.

Table 1. Summary of ANOVA Results.

Source	Stable Force Pr > F	Peak Force Pr > F	Peak/ Stable Pr > F	Stable/ Load Pr > F
Friction	0.0001	0.0001	0.4914	0.0001
Glove Type	0.0001	0.0001	0.4570	0.0131
Load	0.0001	0.0001	0.4580	0.0001
Trial	0.0001	0.0001	0.2861	0.0001
Gender	0.0001	0.0001	0.3704	0.0001
Friction*Glove	0.0001	0.0001	0.3245	0.6794
Friction*Load	0.1446	0.0229	0.3611	0.0001
Friction*Trial	0.0335	0.2237	0.4201	0.2928
Glove*Load	0.0010	0.0001	0.2786	0.6433
Glove*Trial	0.8863	0.8156	0.3213	0.4341
Load*Trial	0.5275	0.7025	0.4966	0.0001
Friction*Glv*Load	0.0007	0.0001	0.3529	0.9741
Friction*Gen	0.0516	0.0706	0.4018	0.4696
Glove*Gen	0.0136	0.2773	0.4253	0.0731
Load*Gen	0.1609	0.0046	0.3616	0.0001
Trial*Gen	0.0666	0.0371	0.3922	0.9695
Friction*Glv*Gen	0.0322	0.2199	0.4791	0.9754

All the main effects (friction, glove type, load, trial, and gender) were significant at the 0.0001 level for the dependent variables of stable force and peak force. For the ratio of stable force to load lifted, all the main effects except glove type were significant at the 0.0001 level. None of these effects were significant for the dependent variable of peak force/stable force. Plots of all of the significant main effects follow.

Figure 1 shows the effect of the three levels of friction on the dependent variables and Table 2 contains the Tukey groupings of the friction levels. For friction level 1, significantly higher peak grasp force and stable grasp force was required. Friction level 1 also required significantly higher stable force

per pound of load lifted, about 136% that of the average for friction levels 2 and 3.

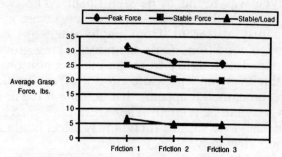

Figure 1. Effect of Friction on Grasp Force.

Table 2. Tukey groupings for friction levels.

Friction	Stable Force Group	Stable Force Mean	Peak Force Group	Peak Force Mean	Stable/Load Group	Stable/Load Mean
Friction 1	A	25.06	A	31.78	A	6.54
Friction 2	B	20.45	B	26.31	B	4.85
Friction 3	B	19.85	B	25.94	B	4.78

The effect of glove type can be seen in Figure 2, and the Tukey groupings are shown in Table 3. For both the peak force and the stable force measures, the leather glove condition was not significantly different from the bare hand condition.

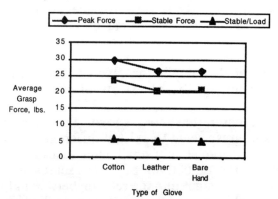

Figure 2. Effect of Glove Type on Grasp Force.

Table 3. Tukey groupings for glove types.

Glove	Stable Force Group	Stable Force Mean	Peak Force Group	Peak Force Mean	Stable/Load Group	Stable/Load Mean
Cotton	A	23.40	A	29.69	A	5.64
Leather	B	20.17	B	26.33	A	5.13
Bare hand	B	20.46	B	26.44	A	5.00

Figure 3 and Table 4 illustrate the effect of the load lifted on grasp force. Figure 3 shows a steadily increasing stable grasp force and peak grasp force as the load increases, each load being significantly different from the other. The ratio of stable force to load lifted, however, decreases with increasing loads. The most dramatic decrease in the ratio is from the 0.5 lb. load to the next level of load (5.5 lbs.).

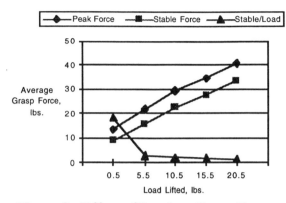

Figure 3. Effect of Load on Grasp Force.

Table 4. Tukey groupings for the load lifted.

Load	Stable Force		Peak Force		Stable/Load	
	Group	Mean	Group	Mean	Group	Mean
0.5	A	9.19	A	13.76	A	18.38
5.5	B	15.85	B	21.98	B	2.88
10.5	C	22.48	C	29.26	B C	2.14
15.5	D	27.76	D	34.27	C	1.79
20.5	E	33.77	E	40.88	C	1.65

Trial had a significant effect on grasp force, and this effect can be seen in Figure 4. Trial 1, which was performed first, had a significantly higher peak force, stable force, and ratio of stable force/load lifted than did trial 2.

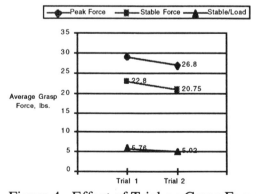

Figure 4. Effect of Trial on Grasp Force.

Figure 5 shows the effect of gender on grasp force. Peak force, stable force, and the ratio of stable force to load lifted were all significantly affected. Females exhibited 78% as much stable force as males.

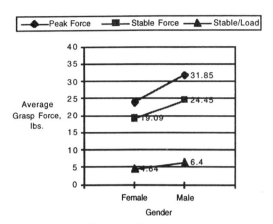

Figure 5. Effect of Gender on Grasp Force.

DISCUSSION

It was hypothesized that glove surface friction, glove type, and the load lifted would all influence the force a person uses to hold an object. Each of these factors were indeed found to have an effect on grasp force, and in addition, trial and gender were also found to be significant variables. These effects as well as how they relate to each other are discussed individually below.

Glove Surface Friction

As indicated in Figure 1, the application of silicon to the surface of the glove significantly affected subjects' peak and stable holding forces. Only 4.2 square inches of silicon reduced the holding forces involved to an average of 82% of the glove condition with no silicon. This is an important finding for glove designers as well as those concerned about reducing the risk of developing a cumulative trauma disorder (CTD), since the application of high forces is thought to be a CTD risk factor (Putz-Anderson, 1991). A word of caution is in order, however. The forces measured in this experiment were the actual forces applied by subjects; they were not the minimum forces necessary to hold the object. In other words, the subjects may not have been grasping with exactly the amount of force necessary to hold the object. They may have, in fact, been over-grasping in order to provide some sort of "safety factor" when holding the object. The minimum amount of force necessary was not measured, and this force may or may not be related to the force the subjects actually grasped with. It can be said, however, that the surface friction of a glove can help reduce the amount of force with which the subject feels he needs to grasp an object.

The friction*glove interaction also illustrates important glove design principles. The cotton glove required much more peak and stable holding force in the friction level 1 condition than the leather glove.

However, with the added 4.8 in^2 of silicon (friction level 2), the difference between the gloves was reduced to almost 1/3 of its previous value. With the application of 7.2 in^2 of silicon, both gloves allowed subjects to exhibit identical forces. These results indicate that for tasks similar to this, glove material may not be as critical as a grasping surface with high friction. A variety of glove materials can be considered according to the type of protection desired as long as a frictional grasping surface is provided (for a task such grasping and holding).

Glove Type

Because the leather glove and barehanded conditions were in a Tukey grouping different from the cotton glove condition, then perhaps thicker gloves with a coefficient of friction similar to that of the barehand will allow performance similar to the barehand (as measured by this task). This is consistent with the findings of Bishu, et al., (1994). This is not to say that when designing a glove for a holding task, the glove can be thick for protective reasons without considerably decrementing performance. The thickness of the glove may still affect fatigue rates or muscular exertions of the flexors and extensors, neither of which were studied in this experiment. The glove*load interaction indicates that glove type becomes increasingly important when heavier loads are lifted (above 0.5 lbs.).

Load Lifted

As expected, the force used to grasp an object increased as its weight increased (see Figure 3). This is also consistent with the results of Bishu, et al., (1994). One of the most interesting findings is that the ratio of grasp force to load lifted decreased with increasing weights. Does this mean subjects are overgrasping the lower weights, undergrasping the higher weights, or both? This is difficult to determine. Perhaps because the force used to support a lower weight represents such a low percentage of the subjects' maximum ability, subjects may tend to exert more force than necessary to hold the object. Likewise, perhaps the force required to support the heavier weights is very close to a maximal effort, so subjects may be exerting as much force as possible.

The load*trial interaction tends to support this argument. For the lightest load (0.5 lbs.), subjects grasped with more force on their first trial than on their second trial, suggesting that subjects were becoming more acquainted with the conditions and had better "optimized" their performance.

Trial

As mentioned previously, the second trial required on the average less force than the first trial (see Figure 4). This suggests that as humans become familiar with a task, they are able to perform it more efficiently. The ratio averages also support this; subjects were able to grasp with significantly less force per pound lifted in the second trial.

Gender

For nearly identical conditions, female subjects used significantly less peak and stable holding forces than the male subjects. In addition, the ratio of stable holding force to load lifted was smaller for females. This may have been due to the fact that females are physically not able to exert forces as high as males. Both genders lifted the same loads, but these loads were a larger percentage of the females maximum capacity. Therefore, the females did not "overexert" as much as the males did. The effect of gender on maximum grip strength is similar. Females have been cited to exert anywhere from 51% (Hunter, et al., 1978) to 76% (McMullin and Hallbeck, 1991) that of males.

CONCLUSIONS

The overall conclusions of this research are:
- The surface friction of a glove affects the amount of force with which the subject feels he needs to grasp an object. The higher the coefficient of friction, the lower the holding force exerted.
- Thicker gloves with a coefficient of friction similar to that of the barehand seem to allow performance similar to the barehand (as measured by this task).
- The amount of force exerted per pound lifted decreased with increasing weights, perhaps indicating over exertion at the lighter levels of the weight lifted, or under exertion at the higher levels of load. This has also been seen to occur with barehanded performance in other experiments (Grant, 1994).
- Males tend to overexert more than females, perhaps because they have a larger maximum capacity than the females.
- The trial effect showed that as humans become familiar with a task, they are able to perform it more efficiently.
- Optimization (or learning) of a task may be more likely to occur with lighter loads (0.5 lbs.) than with heavier loads (20.5 lbs.) since subjects are closer to their maximum ability at the heavier loads and have little room to overgrasp.

REFERENCES

Available upon request.

Gender, Grip Span, Anthropometric Dimensions, and Time Effects on Grip Strength and Discomfort

By M. Peggy Pazderka, Melissa Henderson, and M. Susan Hallbeck

Department of Industrial and Management System Engineering
University of Nebraska
Lincoln, Nebraska, USA

ABSTRACT

The objective of this study is to examine the relationship among anthropometric dimensions, grip span, discomfort, gender, and grip strength. The 24 volunteer subjects (12 males, 12 females) performed five grips squeezing their hardest for 2 minutes at each of the five grip spans on the Jamar grip dynamometer. The grip strength was recorded using the UPC software and then averaged for each of the 30 second intervals. The data was analyzed using ANOVA, post-hoc (Tukey) hypothesis tests, and regression. In the ANOVA analysis gender, grip span, time, and the interactions of gender-grip span, grip span-time, and time-gender were determined to be the significant effects. In all four of the 30 second intervals, average grip strength was significantly higher for males than females. Female average grip strength was found to be 70% of male average grip strength. The post-hoc (Tukey) tests showed that grip spans 3, 4, and 2 were significantly higher than grip spans 5 and 1. The anthropometry of several segments of the hand were found to be important predictors of grip strength and discomfort in the stepwise regressions. Grip span 4 had the highest average severity of discomfort, while grip span 1 had the most areas of the hand experiencing discomfort.

INTRODUCTION

Improperly designed hand tools and devices can cause serious injuries and illnesses to a user. The repetitive use of these poorly designed tools can contribute to cumulative trauma disorders. In order to decide the actual design of a tool and to prevent cumulative trauma disorders several factors need to be measured. These measurements need to be studied and analyzed to find a reasonable force that a user would have to exert on the tool and to find a reasonable size for the grip of the tool. The factors include gender, grip span, discomfort, anthropometric measures of the hand, and grip strength.

Gender: Studies have shown that there is a significant difference in the grip strength between male and female. Some studies have shown that female grip strength is approximately 50-60% of male grip strength (Mathiowetz, Kashman, Volland, Weber, Dowe, and Rogers, 1985). However, other studies have found that the female grip strength is approximately 70% of the males grip strength (Harkonen, Piirtomaa, and Alarnata, 1994; Ramakrishnan, Bronkema, Hallbeck, 1994).

Grip Span: Many studies have been conducted on grip strength using the level 2 on the Jamar, but there are a limited number of studies conducted on the effects of all five Jamar levels. Crosby, Wehbe, and Mawr (1994) found that 62% of men and women have their maximum grip at level 2, but 38% of the men and women have their maximum grip at a level other than 2. When studies are conducted using level 2 only, more than 1/4 of the population is excluded.

Discomfort: Discomfort is an important aspect to consider when using a product. Therefore it is important for designers to understand the experience of discomfort during the exertion of maximal forces by users of these products. One way to measure discomfort is by the use of Corlett and Bishop's (1976) body part discomfort scale. For Corlett and Bishop's (1976) study, operators performed a task and were asked to indicate on the body diagram the areas that were causing pain while performing that task. Instead of rating discomfort on body parts, this study focused on a segmented hand similar to Bishu, Hallbeck, King, and Kennedy (1995). For this study, a 10 point scale based on Corlett and Bishop's body part discomfort scale and Borg's CR-10 scale was used in rating discomfort (Borg, 1990). The 0 on the scale represented no discomfort and 10 represented extreme discomfort.

Anthropometric measure of the hand: Each section of the hand and palm should be measured separately because each section of the hand has its own "force times coefficient of friction" (Kroemer, Kroemer, Kroemer-Elbert, 1994). This means that when applying pressure on an object each area of the hand exerts its own amount of force. Only a few studies have shown the effects that these measures have on grip strength.

The objective of this study is to determine the main effects of grip strength and discomfort in relation to grip span (5 levels), phalange length, hand size, palm size, hand thickness, and gender for a 2 minute maximal exertion.

METHODS

Subjects: The study consisted of 12 male and 12 female student volunteers. The age of the subjects ranged from 21-32 years old. Based on a power calculation (95%) of the data from the pilot study of 8 subjects, 24 subjects were used in the final study.

Apparatus: The materials used in this experiment consisted of the following: digital Jamar hand dynamometer model 95048-554, UPC software and card, PC computer, GPM anthropometric kit, discomfort chart, segmented hand picture, table, and chair.

Experimental Design: In this experiment, the hypothesis was that the most comfortable and strongest grip span were related to the anthropometric dimensions of the phalanges. The independent variables were phalange length (digits 1-5), hand length, palm length, hand thickness, and grip span. The dependent variables were mean grip strength and discomfort in sections of hand.

Procedure: Each subject was asked to perform five grip strength tasks. The subject was in a seated position with his/her upper arm adducted and a 90 degree angle at the elbow. The forearm was positioned half-way between supination and pronation and the wrist was also in the neutral position. The forearm and dynamometer were supported by the table. The subject used his/her dominant hand to perform the tasks. The task consisted of gripping the Jamar as hard as he/she could for 2 minutes. Five different grip spans were used with grip span 1 being the smallest and 5 being the largest. The order of grip spans was randomized for each subject. Every 30 seconds during the task, the subject was asked to rate his/her discomfort using an adapted Corlett and Bishop (1976) body part discomfort scale. After the task was completed, the subject was asked to rate his/her discomfort on the segmented hand. The hand was divided into 22 regions and had been adapted from Bishu et al. (1995). Then the subject was permitted to recover at least 3 minutes before beginning the next task. The anthropometric dimensions were recorded for the 14 phalanges, hand length, hand thickness, and palm length. The collected data was analyzed using an ANOVA with significant results further analyzed using post-hoc

(Tukey) tests. Four ANOVAs were used to examine each of the 30 second intervals over the total 2 minute exertion. Four stepwise regressions were used to examine grip strength over the four time intervals using the anthropometric dimensions of the hand and grip span. Four more stepwise regressions were used to examine discomfort over the four time intervals again using anthropometric dimensions and grip span. The discomfort in sections of the hand was analyzed using descriptive statistics.

RESULTS

An ANOVA was conducted using grip strength as the dependent variable and gender (G), subject (S), grip span (GR), time (T) and their interactions as the independent variables with blocking on subject. The ANOVA results can be found in Table 1. Using an a-level of 0.05, the significant effects were gender, grip span, time, gender-grip span interaction, gender-time interaction, and grip span-time interaction. Post-hoc (Tukey) tests were performed on the main effects of gender (Table 2), grip span (Table 3), and time (Table 4). Mean grip strength for males and females were found to be significantly different. The female mean grip strength was 70% of the male mean grip strength. The Tukey examination of grip span showed that the grouping of grip spans 3, 2, and 4 had significantly higher grip strength than the grouping of grip spans 4 and 5, which were higher than grip span 1. For the Tukey test on time, the first time interval had significantly higher grip strength than the grouping of the second and third time intervals and the grouping of the third and fourth intervals.

Source	DF	Anova SS	F Value	P-value
G	1	6906.825	9.28	0.0059
S(G)	22	16373.86	.	.
GR	4	2426.838	15.75	0.0001
GR*G	4	938.4477	6.09	0.0002
GR*S(G)	88	3390.684	.	.
T	3	17182.81	70.32	0.0001
T*G	3	929.4895	3.8	0.0141
T*S(G)	66	5376.089	.	.
GR*T	12	228.8492	2.16	0.0139
GR*T*G	12	138.4148	1.31	0.2142
GR*T*S(G)	64	2330.144	.	.

Table 1. ANOVA summary table for Jamar grip strength.

Groupings	Gender	Mean Grip Strength (lbs)	% of Male Grip Strength
A	Male	25.02	100%
B	Female	17.43	70%

Table 2. Post-hoc (Tukey) test for mean grip strength (in pounds) by gender.

Groupings		Grip Span	Mean Grip Strength (lbs)
A		3	23.73
A		2	22.64
A	B	4	22.31
	B	5	20.01
	C	1	17.44

Table 3. Post-hoc (Tukey) test for mean grip strength (in pounds) for grip span level.

Groupings		Time	Mean Grip Strength (lbs)
A		0-29 sec.	31.10
	B	30-59 sec.	20.64
	B C	60-89 sec.	17.64
	C	90-120 sec.	15.54

Table 4. Post-hoc (Tukey) test for mean grip strength (in pounds) and time interval.

Four ANOVAs were performed using average grip strength for each of the thirty second intervals as the dependent variable. The independent variables were gender, subject, grip span, and their interactions with blocking on subjects. The ANOVAs hypothesis test results are summarized in Table 5. Using an a-level of 0.05, significant effects were found to be gender, grip span, and the gender-grip span interaction in all four time intervals. Post-hoc (Tukey) tests were performed on the main effects of gender (Table 6) and grip span (Table 7). As shown from Table 6, the male and female average grip strength were significantly different for all of the four time intervals. The post-hoc (Tukey) test performed on the significant effects of grip span showed that for the first time interval grip spans 3, 2, and 4 as a group had a higher grip strength than the grouping of grips spans 4 and 5, and the grouping with the lowest grip strengths was grip spans 5 and 1. For the second interval the grip spans 3, 4, and 2 were significantly higher than grip spans 5 and 1. The third time interval showed that the highest grip strength was grouping grip spans 3,4, and 2, and next highest was the group of grip spans 2 and 5, and the lowest was grouping of grip spans 5 and 1. The fourth time interval showed the grouping of grip spans 3, 4, and 2 as having higher grip strength than the grouping of grip spans 4, 2, and 5. Grip span 1 had significantly lower grip strength than the other groupings. A similar pattern was shown for all four time intervals.

ANOVA SUMMARY TABLE FOR JAMAR GRIP STRENGTH FOR EACH 30-SECOND TIME INTERVAL

1st Time Interval (0-29 seconds)

Source	DF	Anova SS	F Value	Pr > F
G	1	4488.65	7.31	0.0130
S(G)	22	13516.6	- - -	- - -
GR	4	1159.08	7.85	0.0001
GR*G	4	430.675	2.92	0.0256
GR*S(G)	88	3246.00	- - -	- - -

2nd Time Interval (30-59 seconds)

Source	DF	Anova SS	F Value	Pr > F
G	1	1533.32	8.84	0.0070
S(G)	22	3817.52	- - -	- - -
GR	4	551.601	12.65	0.0001
GR*G	4	265.156	6.08	0.0002
GR*S(G)	88	959.597	- - -	- - -

3rd Time Interval (60-89 seconds)

Source	DF	Anova SS	F Value	Pr > F
G	1	1042.09	9.35	0.0058
S(G)	22	2452.35	- - -	- - -
GR	4	539.279	14.23	0.0001
GR*G	4	216.025	5.7	0.0004
GR*S(G)	88	833.857	- - -	- - -

4th Time Interval (90-120 seconds)

Source	DF	Anova SS	F Value	Pr > F
G	1	763.560	8.63	0.0076
S(G)	22	1945.76	- - -	- - -
GR	4	398.543	13.02	0.0001
GR*G	4	162.782	5.32	0.0007
GR*S(G)	88	673.664	- - -	- - -

Table 5. ANOVA summary table of time intervals for Jamar grip strength.

	Time Interval	1st	2nd	3rd	4th
A	male	37.22	24.22	20.56	18.06
B	female	24.99	17.07	14.67	13.01

Table 6. Post hoc (Tukey) tests for average grip strengths (in pounds) by gender and time interval.

Grouping	0-29 sec.	Grip Span	Grouping	30-59 sec.	Grip Span
A	34.39	3	A	23.18	3
A	34.30	2	A	21.87	4
A B	31.26	4	A	21.82	2
B C	29.39	5	B	19.03	5
C	26.17	1	B	17.32	1

Grouping	60-89 sec.	Grip Span	Grouping	90-120 sec.	Grip Span
A	19.90	3	A	17.41	3
A	19.24	4	A B	16.85	4
A B	18.39	2	A B	16.07	2
B C	16.50	5	B	15.12	5
C	14.04	1	C	12.23	1

Table 7. Post hoc (Tukey) tests for average grip strengths (in pounds) at five grip spans by time interval

Four stepwise regressions were conducted for grip strength (GS) for the four 30-second intervals using anthropometric dimensions of the hand and grip span. The following equations were found from the results:

1st Time Interval (0-29 seconds)
R-squared = 0.7118
GS = -207.14 + 28.11(5th distal) - 54.64(5th middle) + 34.52(5th proximal) - 30.98(4th middle) + 15.19(4th proximal) +52.19(3rd middle) + 60.70(3rd proximal) - 38.54(2nd proximal) - 131.34(1st distal) + 8.80(1st proximal) +13.22(hand thickness) + 37.92(palm size) + 9.49(gender)

2nd Time Interval (30-59 seconds)
R-squared = 0.4124
GS = 25.83 + 13.13(3rd middle) + 5.08(2nd proximal) + 2.55(1st proximal) -3.35(hand length) + 10.63(gender)

3rd Time Interval (60-89 seconds)
R-squared = 0.4761
GS = -22.56 - 20.46(4th distal) + 20.05(3rd distal) + 6.65(3rd middle) + 7.51(2nd proximal) - 2.68(hand length) + 4.96(palm size) +3.99(gender) + 0.58(grip span)

4th Time Interval (90-120 seconds)
R-squared = 0.4849
GS = -27.17 - 17.12(4th distal) + 18.14(3rd distal) +5.55(3rd middle) + 4.48(2nd proximal) - 2.45(hand length) + 5.44(palm size) + 2.73(gender) + 0.66(grip span)

Four more stepwise regressions were performed for discomfort (DC) using anthropometric dimensions of the hand and grip span as the predictors. The following equations were found from the analysis:

1st Time Interval (0-29 seconds)
R-squared = 0.3753
DC = 4.73 + 0.02(grip strength) + 0.20(grip span) + 2.78(5th proximal) + 1.75(3rd middle) + 2.56(3rd proximal) -1.61(2nd middle) - 3.46(1st distal) + 1.63(1st proximal) - 0.85(hand length) + 2.64(hand thickness)

2nd Time Interval (30-59 seconds)
R-squared = 0.4366
DC = 15.16 + 0.07(grip strength) + 0.18(grip span) + 2.30(4th middle) + 6.05(3rd distal) - 5.14(2nd distal) + 1.04(1st proximal) - 1.60(hand length) + 2.15(hand thickness)

3rd Time Interval (60-89 seconds)
R-squared = 0.4427
DC = 18.68 + 0.21(grip span) + 2.70(4th middle) + 9.67(3rd distal) - 7.53(2nd distal) + 1.65(2nd proximal) +0.98(1st proximal) - 2.16(hand length) + 2.60(hand thickness)

4th Time Interval (90-120 seconds)
R-squared = 0.4911
DC = 17.50 + 0.24(grip span) + 2.65(5th distal) + 2.38(4th middle) + 12.36(3rd distal) + 1.87(3rd proximal) - 11.73(2nd distal) - 2.61(hand length) + 2.18(hand thickness) + 1.08(palm size)

For average discomfort on the sectioned hand, grip span 4 had the highest average severity of discomfort and grip span 1 had the most areas of the hand that had discomfort. See Table 10 for average severity and frequency discomfort ratings of the hand. The severity was the average discomfort rating over the 22 hand locations and the frequency was the number of locations reported experiencing discomfort for each subject.

AVERAGE SEVERITY AND FREQUENCY OF DISCOMFORT ON SECTIONED HAND ACCORDING TO GRIP SPAN					
average	grip span 1	grip span 2	grip span 3	grip span 4	grip span 5
severity	1.76	1.51	1.48	1.80	1.55
frequency	9.05	8.23	7.95	8.00	7.14

Table 10: Average severity and frequency discomfort ratings of the hand.

DISCUSSION

Gender: The findings of the experiment in relation to gender agreed with the expectations based on previous research done on grip strength. It has been shown that the maximum female grip strength is about 70% of maximum male grip strength (Ramakrishnan et al, 1994; Harkonen et al, 1993). In this experiment the mean grip strength of the females over the two minutes was 70% of the mean grip strength of the males and approximately that for each time interval.

Grip Span: In the results, grip spans 2,3, and 4 were found to yield higher grip strengths than grip spans 1 and 5. This was also seen in all four time intervals. These results support the bell-shaped theory on grip strength and grip span (Crosby et al., 1994). Males had their highest grip strength average for the two minutes at grip level 3 and females had highest average grip strength at level 2. In the current study 38% of males and females had their highest grip strength at level 2. This supports Crosby et al. (1994) argument that more than a 1/4 of the population is excluded from the point of their maximal exertion when only grip span level 2 is used in the study. Although grip spans 3, 4, and 2 were not significantly different, after the 1st time interval, grip spans 3 and 4 had higher grip strengths than 2. This result shows a maximum grip strength at a larger handle than Crosby et al. (1994). This may be due to the length-tension relationship of the muscles.

Task: The task of exerting maximum force on the Jamar hand dynamometer for two minutes seemed to have an impact on the overall maximum force exerted in the first three seconds. These values were lower than the normative maximum grip strengths of 62.8lb and 104.3lb for female and male (Mathiowetz et al., 1985). The values in this experiment can in no way be compared to the normative values. We hypothesize that this difference is due to the subjects' perception of the task. The subject did not exert his/her maximum but a level that he/she thought could be sustained for the two minutes.

Anthropometry of hand: As expected the anthropometric dimensions of the hand played a significant role in the prediction of grip strength. These results support the conclusion that each section of the hand has its own "force times coefficient of friction."(Kroemer et al., 1994). Grip span became a significant factor in predicting grip strength after 60 seconds. This confirms the notion that the grip size of a tool becomes a critical factor with repetitive and extensive usage of the tool.

Discomfort: The results show that the anthropometric dimensions of the phalanges, along with hand length, hand thickness, grip span and grip strength (for the first 60 seconds) had a significant effect on overall discomfort. This could be due to the fact that the phalanges are positioned on the Jamar so as to exert the majority of the force. Overall discomfort could be related to hand thickness because the thicker the hand the greater the padding to protect the hand from mechanical tissue compression. It is understandable that hand length would be a factor in predicting the overall discomfort of the hand. As hypothesized for this experiment, grip span and grip strength had a major role in predicting overall discomfort. In the first 60 seconds, the regression model showed that as grip strength increased discomfort increased. The relationship of higher grip strength to higher discomfort is a logical assumption. As expected areas of discomfort on the hand changed as the grip span level changed. In the proximal phalanges of the second, third, and fourth digits, the

discomfort decreased as grip span increased. In the distal phalanges of the second and third digits, the discomfort increased, in general, as grip span increased. The results of grip span 4 having the highest discomfort severity rating was surprising. The expected result was that either extreme, grip span 1 or grip span 5, would have had the highest average discomfort severity rating. The results of grip span 1 having the highest frequency rating maybe due to more area of the hand being exposed to the Jamar.

CONCLUSION

As shown from this study, gender, grip span, discomfort, anthropometric dimensions of the hand, and grip strength are important factors to consider when designing hand tools. The conclusion drawn above about slightly larger handles than other research found being easier to exert a higher grip strength over time needs to be investigated further. The finding that subjects experience the most severe discomfort at grip span level 4 needs to be examined more thoroughly.

REFERENCES

Bishu, R. R., Hallbeck, S., King, K., Kennedy, J. (1995). Is 100 %MVC truly a 100 % effort? Advances in Industrial Ergonomics and Safety VII, 545-550.

Borg, G. (1990). Psychophysical scaling with applications in physical work and the perception of exertion. Scandinavian Journal of Work Environmental Health, 16, 55-58.

Corlett, E. N., Bishop, R. P. (1976). A technique for assessing postural discomfort. Ergonomics, 19, 175-182.

Crosby, C. A., Wehbe, M. A., Mawr, B. (1994). Hand strength: Normative values. The Journal of Hand Surgery, 19A, 665-670.

Harkonen, R., Piirtomaa, M., Alarnata, H. (1993). Grip strength and hand position of the dynamometer in 204 finnish adults. The Journal of Hand Surgery, 18B, 129-132.

Kroemer, K., Kroemer, H., and Kroemer-Elbert, K. (1994). Ergonomics: How to Design For Ease and Efficiency. New Jersey: Prentice Hall.

Mathiowetz, V., Kashman, N., Volland, G., Weber, K., Dowe, M., and Rogers, S. (1985). Grip and pinch strength: Normative data for adults. Archives of Physical Medicine and Rehabilitation, 66, 16-21.

Ramakrishnan, B., Bronkema, L. A., and Hallbeck, M. S., (1994). Effects of grip span, wrist position, hand and gender on grip strength. Proceeding of the Human Factors and Ergonomics Society 38th Annual Meeting, 554-558.

EFFECTS OF INTERDIGITAL SPACING, WRIST POSITION, FOREARM POSITION, GRIP SPAN, AND GENDER ON STATIC GRIP STRENGTH

Angela Hansen and Susan Hallbeck
University of Nebraska
Lincoln, Nebraska

To analyze the effects of factors which may decrement static grip strength, a study was performed using 20 subjects (10 males and 10 females). Three levels of interdigital spacing were examined: 0, 4, and 8 mm to correspond to bare hand, a thermal/spectra knit layered glove, and a chemical glove. Three wrist positions (neutral, 45 degree extension, and 45 degree flexion) and three forearm positions (neutral, full supination, and full pronation), as well as two levels of grip spacing (6.1 and 7.4 cm on the hand dynamometer) were varied in the study. Thus, each subject performed 54 trials, which were presented in random order.

The results indicate that the following were significant main effects: gender, grip span, forearm position, wrist position, and interdigital spacing. As expected, males exerted 63.7% more strength, on average, than females. Also, full pronation had a significantly lower mean grasp strength than either neutral or full supination. The 8 mm spacing differed significantly from both 0 and 4 mm spacing and both 4 and 8 mm spacing had a lower mean grasp strength than 0 mm. Also, the 7.4 cm position on the hand dynamometer had significantly lower mean grasp strength as compared with the 6.1 cm position. Wrist position affected mean grasp strength as well; with a neutral wrist position, the strength was significantly higher than either flexed or extended, with a greater decrement from flexion than extension.

INTRODUCTION

Cumulative trauma disorders are believed to have several causes, such as repetitive motions, the application of large forces, and awkward positions (Putz-Anderson, 1988). Therefore, it is useful to understand the relations between various factors and grip strength. These factors include deviated digit, wrist, and forearm postures, anthropometric dimensions, grip span, gender, and their interactions.

A literature survey indicates that the relationship between grip strength and individual factors such as wrist position, gender, grip span, forearm position, and interdigital spacing have been studied by several researchers. (Schmidt and Toews, 1970; Lunde, Brewer, and Garcia, 1972; Miller and Freivalds, 1987; Putz-Anderson, 1988; Hallbeck and McMullin, 1993; McMullin and Hallbeck, 1991; 1992). A neutral wrist position is preferable to a flexed or extended position because higher strengths can be exerted and it is less likely to contribute to the development of a CTD (Putz-Anderson, 1988). While in a 45 degree extended position, strengths of approximately 75-82% of the neutral position are possible, while strengths of only 60-75% can be exerted when the hand is in a 45 degree flexed position (Hallbeck and McMullin, 1993; McMullin and Hallbeck, 1991; and Putz-

Anderson, 1988). The reduced strengths with a deviated wrist have been attributed to the fact that the ability of a musculotendinous unit to generate strength is dependent upon its functional length (Hazelton, Smidt, Flatt, and Stephens, 1975).

Gender may also affect grip strength. Female grip strength has been found to be approximately 50-60% of male grip strength (An, Chao, and Askew, 1983; Kellor, Frost, Silberberg, Iverson, and Cummings, 1971; Mathiowetz, Kashman, Volland, Weber, Dowe, and Rogers, 1985). However, in 1991 and 1992, McMullin and Hallbeck found that females could generate 76% and 66%, respectively, as high a power grasp strength as males. Hallbeck et al. (1993) found that over several glove conditions females averaged 74% of male grip strength.

The distance spanned by the dynamometer handle was studied by Wang (1982). The dynamometer used in his study was set at spans of 3.5, 4.7, and 6.0 cms. Paired t-tests that were performed on this data showed the 3.5 and 4.7 cm handles to be in one grouping with lower mean strength than the 4.7 and 6.0 cm handles in the other tukey grouping. Härkönen, Piirtomaa, and Alaranta (1993) also studied grip span, including 5 spans in their study. The third handle position (6 cm) of the Jamar hand dynamometer was found to yield the highest grip strength for both male and female subjects.

Three postures of forearm rotation (full pronation, full supination, and neutral) in combination with wrist flexion/extension and radial/ulnar deviation were examined by Terrell and Purswell (1976). They found that forearm pronation reduced grip strength. Marley and Wehrman (1992) also studied forearm rotation, including seven positions ranging from 90 degrees pronation to 90 degrees supination. They found that forearm posture was significant with 90 degree pronation differing from both the 0 degree (neutral) and 30 degree supination postures.

Several attempts have been made to derive empirical equations for grip strength, associating with it the subjects' anthropometric dimensions. Schmidt et al. (1970) found grip strength to be proportional to stature and body weight, up to a maximum of 75 inches and 215 pounds. Wang (1982) correlated grip strength to anthropometric data of the hand for males and females, separately. Thumb circumference was found to be a correlate ($R^2=0.5$) of grip strength for males, and hand breadth and finger crotch length to be a good correlate ($R^2=0.8$) for females. Härkönen et al. (1993) found hand length did not significantly predict grip strength in their study. Ramakrishnan, Brokema, and Hallbeck (1994) found palm thickness and hand breadth to be a significant correlator of grip strength for wrist positions of neutral, 45 degree extension, and 45 degree flexion. Wrist circumference had a good correlation only with the non-dominant hand and a grip span of 6.0 cm. For the same grip spans as Wang (1982), they also found that as grip span increases the forearm length was a good correlate.

Hallbeck, Muralidhar, and Ramakrishnan (1994) studied the effect of interdigital spacing on grip strength. Five levels of interdigital spacing (0, 3, 5, 7, and 10 mm) were used in order to represent the bare hand and different types of gloves. The interdigital spacing of 0 mm had the highest mean grip strength with each increase in interdigital spacing decreasing grip strength.

One common and inexpensive way to protect workers from vibration stress and/or minor abrasions commonly found in today's workplace is to provide them with gloves. However, the use of gloves often results in a reduction of capabilities. Another factor for this decrement is increased interdigital spacing due to varying thickness of gloves.

This study attempts to identify the combination of factors which may contribute to the loss of grip strength. These factors include: gender, interdigital spacing, wrist position, grip span, forearm position, and all of their interactions.

METHOD

Subjects

Twenty subjects (10 male and 10 female) volunteered for this study. All subjects were in the age group of 21 to 40 and reported no previous wrist injuries.

Apparatus

A Jamar (Model 1) hand dynamometer with a modified straight handle was used for the strength measurement tasks. The straight handle was used in place of the curved handle to facilitate the use of finger spacers. The anthropometric dimensions were taken from Garrett (1971) and were measured using a standard anthropometric measuring kit (GPM - Swiss). A table with an arm restraint and positions marked for neutral, 45 degree flexed, and 45 degree extended positions of the wrist was used. The restraint helped position the arm at a 90 degree included elbow angle with the upper arm adducted, and the marking enabled the participant to position their hands at a repeatable angle while the study was conducted. To represent the bare hand and a variety of glove types, three levels of interdigital spacing were chosen: 0, 4, and 8 mm. The interdigital spacing was achieved by placing the spacers between the digits at both the distal and proximal interphalangeal joints.

Procedure

The subjects placed their forearm in the restraint and were instructed to align their arms in the same fashion for each trial. Then, the subjects were given instruction and guidance to arrange the hand and fingers in the correct position. The trials were performed according to the Caldwell regimen (Caldwell, Chaffin, Dukes-Dobos, Kroemer, Laubach, Snook, and Wasserman, 1974) and were administered in a random order. The Caldwell regimen consists of the subject applying pressure for one second (warm-up period), exerting steady pressure for four seconds, applying pressure for one second (cool-down period), and releasing pressure. Two minutes of rest time was given between each trial, with additionnal rest given at the subject's request.

Experimental Design

This study evaluated the loss of grip strength due to factors associated with gloves. The study included an evaluation of wrist positions, grip span, forearm position, and variation in interdigital spacing. Three levels of interdigital spacing were examined: 0, 4, and 8 mm, to represent bare hand, thermal/spectra knit layered gloves, and chemical gloves. Three wrist positions (neutral, 45 degree extension, and 45 degree flexion) and three forearm positions (neutral, full supination, and full pronation), as well as two levels of grip span (6.1 and 7.4 cm on the hand dynamometer) were also varied in the study. The resultant design was a 3 (interdigital spacing) by 3 (wrist position) by 3 (forearm position) by 2 (grip span) by 2 (gender) mixed factor model, with blocking on subjects.

Analysis of variance (ANOVA) was run to identify the nature of the variance of the independent variables mentioned above with respect to grip strength. Tukey tests were also performed on significant main effects to determine which level or condition of the variables were significantly different from the others.

To find the relationship between anthropometric dimensions and grip strength, a forward stepwise regression analysis was carried out for the neutral wrist position, neutral forearm position, and 0 mm interdigital spacing with a significance level of entry of 0.15. A correlation analysis was performed for each of the 34 upper limb anthropometric dimension and grip strength by gender and grip span. Based on common practice in ergonomics, adequation correlation was required to be a Pearson product-moment coeffecient (r) of at least 0.7 (Kroemer, Kroemer, and Kroemer-Elbert, 1994).

RESULTS

Analysis of variance:

The ANOVA results are summarized in Table 1. It may be noted that all the main effects, namely the forearm position, grip span, wrist position, interdigital spacing, and gender are all significant at the 0.01 level. Significant interaction terms that are significant at the 0.05 level are gender*wrist position, forearm position*grip span, forearm position*wrist position, and grip span*wrist position. Tukey tests were performed on significant main effects.

Table 1. ANOVA Results

Variable	DF	SSE	F-Value	p
Gender = S	1	41218.13	20.52	0.0003
Subject(Gender) = Sub(S)	18	36162.59	---	---
Forearm Position = FP	2	1125.72	14.63	0.0001
FP*S	2	81.24	1.06	0.3586
FP*Sub(S)	36	1385.45	---	---
Grip Span = GS	1	2178.85	124.39	0.0001
GS*S	1	2.31	0.13	0.7204
GS*Sub(S)	18	315.28	---	---
Interdigital Spacing = DS	2	657.76	20.68	0.0001
DS*S	2	83.29	2.62	0.0868
DS*Sub(S)	36	572.59	---	---
Wrist Position = WP	2	10185.12	43.83	0.0001
WP*S	2	922.91	3.97	0.0276
WP*Sub(S)	36	4182.39	---	---
FP*GS	2	85.25	5.84	0.0064
FP*GS*S	2	5.70	0.39	0.6794
FP*GS*Sub(S)	36	262.72	---	---
FP*DS	4	18.21	0.43	0.7901

Tukey analysis:

The Tukey analysis was performed on all of the significant main effects with the following results. First, females averaged 63.7% of male grip strength (see Table 2).

Table 2. Tukey test of strength by gender.

Gender	Mean Strength (kg)	Tukey Grouping
Male	34.033	A
Female	21.678	B

Second, the grip strength of the subjects for the grip span of 7.4 cm was found to average 90.3% that of the 6.1 cm position (see Table 3).

Table 3. Tukey test of strength by grip span.

Grip Span (cm)	Mean Strength (kg)	Tukey Grouping
6.1	29.276	A
7.4	26.435	B

Third, the forearm position in full pronation was found to be significantly different from both full supination and neutral. Full supination was not found to be significantly different with grip strength averaging 97.6% of the neutral position and full pronation was found to be significantly different with grip strength averaging 91.6% of the neutral position (see Table 4).

Table 4. Tukey test of strength by forearm position.

Forearm Position	Mean Strength (kg)	Tukey Grouping
Neutral	28.892	A
Full Supination	28.208	A
Full Pronation	26.467	B

Fourth, the wrist position in 45 degree flexion was found to be significantly different from both neutral and 45 degree extension. Fourty-five degree extension was found to not be significantly different with grip strength averaging 95.0% of the neutral wrist position and 45 degree flexion was found to be significantly different with grip strength averaging 76.8% of the neutral wrist position (see Table 5).

Table 5. Tukey test of strength by wrist position.

Wrist Position	Mean Strength (kg)	Tukey Grouping
Neutral	30.744	A
45 Degree Extension	29.219	A
45 Degree Flexion	23.603	B

Finally, the interdigital spacing at 8 mm was found to be significantly different from both 0 and 4 mm. The 4 mm spacing was found to not be significantly different with grip strength averaging 99.5% of the 0 mm spacing and the 8 mm spacing was found to be significantly different with grip strength averaging 93.9% of the 0 mm spacing (see Table 6).

Table 6. Tukey test of strength by interdigital separation.

Interdigital Separation (mm)	Mean Strength (kg)	Tukey Grouping
0	28.483	A
4	28.328	B
8	26.756	C

Regression Analysis:

For males at a grip span of 6.1 cm:
Strength = -88.7 + 11.6 Wrist Breadth - 6.8 Digit 2 Length + 9.7 Digit 3 Length +14.3 Digit 3 Breadth with $R^2 = 0.4863$ and p = 0.0223.

For males at a grip span of 7.4 cm:
Strength = -110.3 + 8.0 Wrist Breadth - 3.6 Digit 2 Height - 2.5 Crotch 1 Height + 12.2 Crotch 3 Height + 7.5 Digit 1 Circumference with $R^2 = 0.5310$ and p = 0.1275.

For females at a grip span of 6.1 cm:
Strength = -42.8 + 15.6 Hand Breadth - 8.0 Crotch 4 Height - 1.2 Digit 5 Length - 18.4 Digit 1 Breadth + 8.1 Digit 4 Breadth + 1.6 Forearm Length with $R^2 = 0.5725$ and p = 0.0303.

For females at a grip span of 7.4 cm:
Strength = -27.5 + 10.2 Hand Breadth - 5.4 Crotch 4 Height - 3.9 Digit 1 Length - 0.8 Digit 5 Length - 2.0 Digit 1 Circumference + 2.2 Forearm Length with $R^2 = 0.4555$ and p = 0.0619.

Correlation Analysis:

Correlation analyses were conducted between grip strength and each of the 34 upper limb anthropometric dimensions by grip span and gender. No correlations were deemed adequate (r >= 0.7 and p <= 0.05) for any of the anthropometric dimensions and grip strength.

DISCUSSION

As expected, it was found that grip strength was significantly higher when the wrist was in the neutral position as compared to when the wrist was either flexed or extended. Other points noted are that the grip strength in flexion was about 76.8% that of the neutral wrist grip strength, which is about 5% higher than in previous studies (Hallbeck et al., 1993; McMullin et al., 1991; Putz-Anderson, 1988). The grip strength with the wrist extended was about 95.0% of neutral grip strength, which was about 10% higher than previous studies. The flexed and extended positions show a significant difference, as was noted both by Hallbeck et al. (1993) and Putz-Anderson (1988), with flexion causing a greater strength decrement than extension.

A significant gender effect was found with respect to the grip strength. Females were found to have 63.7% of the grip strength of males, which is about 4% higher than most previous studies (An et al., 1983; Kellor, et al., 1971; Mathiowetz, et al., 1985) but is consistent with McMullin et al. (1992). This decrease in grip strength could well be explained by the natural anthropometric differences between males and females. For a grip span of 6.1 cm, the female participants were found to have about 64.9% as much strength as that of the male participants. As the grip span increased to 7.4 cm the female participants were found to have about 62.3% as much strength as that of the male participants. Therefore, as the grip span reaches an optimal point, the females may narrow the strength gap.

A significant grip span effect was found with respect to the grip strength. The 6.1 and 7.4 cm positions on the dynamometer were found to be significantly different on grip strength with the 7.4 cm position averaging 90.3% of the grip strength at the 6.1 cm position. These results agree with Härkönen et al. (1993) that the grip span of approximately 6.0 cm yields the highest grip strength in both female and male subjects.

A significant interdigital spacing effect was found with respect to grip strength. The interdigital spacing of 0 mm had the highest mean grip strength with each increase in interdigital spacing decreasing grip strength, which agrees with Hallbeck et al. (1994). The average grip strengths at 0, 4, and 8 mm were found to be significantly different with each partition being 100%, 99.5%, and 93.9% respectively. This reduction in grip strength as interdigital spacing increases agrees with previous research of the decremental grip strength performance when wearing gloves (Hallbeck et al., 1993; McMullin et al., 1991).

A significant forearm position effect was found with respect to the grip strength. The full pronation position was found to be significantly different from neutral, while the neutral and full supination positions appear not to be significantly different, which agrees with Terrell et al. (1976) and Marley et al. (1992). The grip strength in the supine position averaged 97.6% of the grip strength in the neutral position, while the grip strength in the prone position averaged 91.6% of the grip strength in the neutral position.

For males, the regression models developed here have an R^2 value between 0.45 and 0.55 indicating that 45-55% of the variation in grip strength is being explained by the anthropometric measurements, namely wrist breadth, digit 2 and 3 length, digit 3 breadth, crotch 1 and 3 height, digit 2 height, and digit 1 circumference. For females, the regression models developed here have an R^2 value between 0.45 and 0.60 indicating that 45-60% of the variation in grip strength is being explained by the anthropometric measurements, namely hand breadth, forearm length, dit 1 and 4 breadth, digit 1 and 5 length, crotch 4 height, and digit 1 circumference. For males and females, the low R^2 values around 0.50 suggest that either there are other anthropometric measurements not considered here which will represent some of the variation in grip strength or anthropometric measurements are not a good predictor of grip strength.

For males and females, none of the 34 upper limb anthropometric dimensions tested were correlated with grip strength at an adequate level ($r >= 0.70$ and $p <= 0.05$) by gender and grip span. This differs from Schmidt et al. (1970) and Wang (1982), who found anthopometric dimensions correlated with grip strength.

This study demonstrates that interdigital spacing is a significant factor in grip strength decrement as are other, more established, factors. This implies that to keep grip strength at a higher output level, the glove thickness should be reduced to a minumum. Factors not considered in this study were the variations in grip strength due to age and handedness. Further studies would definitely be beneficial in designing workplace changes.

REFERENCES

An, K.N., Chao, E.Y., and Askew, L.J. (1983). Functional assessment of upper extremity joints. IEEE Frontiers of Engineering and Computing in Health Care, 136-139.

Caldwell, L.S., Chaffin, D.B., Dukes-Dobos, F.N., Kroemer, K.H.E., Laubach, L.L., Snook, S.H., and Wasserman, D.E. (1974). A proposed standard procedure for static muscle strength testing. American Industrial Hygiene Association Journal, 35 (4), 201-206.

Garrett, J.W. (1971). The adult human hand: some anthropometric and biomechanical considerations. Human Factors, 13, 117-131.

Hallbeck, M.S., Muralidhar, A. and Ramakrishnan, B. (1994). Effect of interphalangeal separation on grasp strength. International Ergonomics Association '94, 2, 83-85.

Hallbeck, M.S., and McMullin, D.L. (1993). Maximal power grasp and three-jaw chuck pinch force as a function of wrist position, age, and glove type. International Journal of Industrial Ergonomics, 11, 195-206.

Härkönen, R., Piirtomaa, M., and Alaranta, H. (1993). Grip strength and hand position of the dynamometer in 204 Finnish adults. The Journal of Hand Surgery, 18B(1), 129-132.

Hazelton, H.J., Smidt, G.L., Flatt, A.F., and Stephens, R.I. (1975). The influence of wrist position on the force produced by the finger flexors. Journal of Biomechanics, 8, 301-306.

Kellor, M., Frost, J., Silberberg, N., Iverson, I., and Cummings, R. (1971). Hand strength and dexterity. American Journal of Occupational Therapy, 25, (2), 77-83.

Kroemer, K.H.E., Kroemer, H.J., and Kroemer-Elbert, K.E. (1994). Ergonomics: How to Design For Ease and Efficiency. New Jersey: Prentice Hall.

Lunde, B.K., Brewer, W.D., and Garcia, P.A. (1972). Grip strength of college women. Archives of Physical Medicine and Rehabilitation, 55, 491-493.

Marley, R.J., and Wehrman, R.R. (1992). Grip strength as a function of forearm rotation and elbow posture. Proceedings of the Human Factors Society 36th Annual Meeting, 791-795.

Mathiowetz, V., Kashman, N., Volland, G., Weber, K., Dowe, M., and Rogers, S. (1985). Grip and pinch strenth: normative data for adults. Archives of Physical Medicine and Rehabilitation, 66, 16-21.

McMullin, D.L., and Halbeck, M.S. (1991). Maximal power grasp force as a function of wrist position, age, and glove type: a pilot study. Proceedings of the Human Factors Society 35th Annual Meeting, 733-737.

McMullin, D.L., and Hallbeck, M.S. (1992). Comparison of power grasp and three-jaw chuck pinch static strength and endurance between industrial workers and college students: a pilot study. Proceedings of the Human Factors Society 36th Annual Meeting, 770-774.

Miller, G.D., and Freivalds, A. (1987). Gender and handedness in grip strength - a double whammy for females. Proceedings of the Human Factors Society 31st Annual Meeting, 906-910.

Putz-Anderson, V. (1988). Cumulative Trauma Disorders: A Manual For Musculoskeletal Diseases of the Upper Limbs. London: Taylor & Francis.

Ramakrishnan, B., Bronkema, L.A., and Hallbeck, M.S. (1994). Effects of grip span, wrist position, hand and gender on grip strength. Proceedings of the Human Factors and Ergonomics Society 38th Annual Meeting, 554-558.

Schmidt, R.T., and Toews, J.W. (1970). Grip strength as measured by jamar dynamometer. Archives of Physical Medicine and Rehabilitation, 51, 321-327.

Terrell, R., and Purswell, J.L. (1976). The influence of forearm and wrist orientation on static grip strength as a design criterion for hand tools. Proceedings of the International Ergonomics Association, 28-32.

Wang, M. (1982). A Study of Grip Strength from Static Efforts and Anthropometric Measurement, Unpublished masters thesis, University of Nebraska, Lincoln, NE.

VALIDATION OF A FREQUENCY WEIGHTED FILTER FOR REPETITIVE WRIST FLEXION TASKS AGAINST A LOAD

Mei-Li Lin
Wright State University
Dayton, Ohio

Robert G. Radwin
University of Wisconsin-Madison
Madison, Wisconsin

This experiment validates an instrument that implements a force and frequency weighted filter for biomechanical stress exposure assessment. This filter network was developed based on a previously established discomfort model that associated physical stress of force, posture and repetition with subjective discomfort. A simulated industrial task was used in the current study to test the instrument involving repetitively transferring a peg and inserting it into a hole against a controlled resistance. Ten subject performed the task for six conditions. Continuous wrist angular data were recorded using an electrogoniometer and processed through the filter. Subjective discomfort was reported after performing the task for one hour using a 10 cm visual analog scale. The discomfort model was shown to estimate relative discomfort for the experimental conditions tested ($r^2 = 0.98$, $p < .05$). Linear regression analysis showed that the instrument aptly predicted subjective discomfort ($r^2 = 0.87$, $p < .05$). Applications and limitations of this instrument are explored.

INTRODUCTION

Numerous exposure assessment approaches have emerged for preventing discomfort and musculoskeletal disorders in repetitive hand intensive tasks. These include both biomechanical (Moore, et al., 1991; Gilad, 1995; Moore and Garg, 1995; Wells, et al., 1995;), and psychophysical (Saldaña, et al., 1994; Snook, et al., 1995) approaches. The objective of this study was to develop an exposure assessment instrument that could process and integrate continuous multi-factor biomechanical data, including wrist flexion angle and force into a quantitative index proportional to psychophysical acceptable levels.

This exposure assessment approach is based on previous work. Radwin, et al. (1994) modeled the effects of posture and repetition on subjective discomfort and demonstrated how frequency weighted filters could be applied for reducing large amounts of biomechanical data into a single quantity. A high pass filter weighted repetitive wrist flexion angle, measured using an electrogoniometer, in proportion to subjective discomfort. Lin, et al. (1994) established a discomfort model that associated physical stress of force, wrist flexion and repetition with subjective discomfort. Lin, et al. (1995) further introduced force factors for adjusting the filtered data to account for force. Lin and Radwin (1995) broadened the discomfort model from Lin, et al. (1995) by expanding the range of exposure and increasing the levels of control within each factor. The discomfort model was used to shape a force and frequency weighted filter. The current study utilizes this filter network and psychophysical data published by others (Marley and Fernandez, 1995; Snook, et al., 1995) to develop an exposure assessment instrument that processes and integrates continuous biomechanical measurements into an exposure index proportional to psychophysically derived acceptable levels. A simulated industrial task is used to validate this instrument. Applications and limitations of this instrument are explored.

METHODS

Exposure Assessment

Lin and Radwin (1995) investigated the relative effects of force, posture, and repetition on subjective discomfort in order to form a continuous discomfort model. The resulting discomfort regression model was:

$$D = 10^{(-0.046 + 0.393 \log E + 0.225 \log A + 0.275 \log F + 0.047 B)} - 1,$$

where D was discomfort (0 to 10), E was exertion (N), A was wrist flexion angle (degrees), F was frequency (Hz), and B was block level ($F(4,18) = 60.20$, $p < .01$, $r^2 = 0.93$). This model was used to generate the attenuation slope for a frequency weighted angle filter and to determine force factors for processing continuous wrist angular data.

Frequency weighted filter. A frequency weighted angle filter weighed postural signals by the corresponding frequency in proportion to the equal discomfort function, so that the filter angular data accounted for angle and repetition. The attenuation slope for the frequency weighted angle filter was 24 dB/decade which was determined by the relative relationship between angle and repetition and was obtained by algebraically solving the discomfort model at a given exertion and discomfort level (Lin, et al., 1995). The frequency weighted angle filter was modeled using MATLAB™. The characteristics of the filter are illustrated in Figure 1. The cut-off frequency for this filter was arbitrarily set at 1 Hz because of limitations in the filter design algorithm which was constrained by the desired attenuation slope and band width. The upper bound for wrist flexion was set to 75° to accommodate the range of motion for the wrist. This angle was used as the reference for determining angular attenuation (dB). Discomfort level ten was used to define the 0 dB attenuation at the cut-off frequency.

Force Factor. Since the frequency weighted filter does not account for force, filtered angular outputs must be adjusted by a force factor in order to agree with the discomfort model. Because the discomfort model is linear and the filter attenuation slope is a constant, the necessary adjustment (dB) for a given exertion is constant and independent of frequency. The adjustment (dB) for a given exertion E was determined by:

$$a_E = A_{E,F} - A'_F$$

where a_E is the adjustment (dB) for exertion level E (N), $A_{E,F}$ is the angular attenuation level (dB) for exertion E and frequency F (Hz) based on the discomfort model, and A'_F was the angular attenuation level (dB) for frequency F of the frequency weighted filter. The difference between $A_{E,F}$ and A'_F was be determined by:

$$A_{E,F} - A'_F = 20 \log(\frac{X_{FE}}{X'_F})$$

where X_{FE} was the angle from the discomfort model for frequency F and exertion E, and X'_F was the angle from the filter without accounting for force. The linear equation between a given exertion level and its corresponding force adjustments for discomfort level ten was:

$$a_E = 34.88 \log E - 58.67.$$

The adjustment a_E was derived by solving the linear equation between the points ($\log E_1$, a_{E_1}) and ($\log E_2$, a_{E_2}), where a_{E_1} (dB) was the force adjustment for E_1 (N) and a_{E_2} (dB) was the force adjustment for E_2 (N). The force adjustment for 35 N exertion is demonstrated in Figure 1.

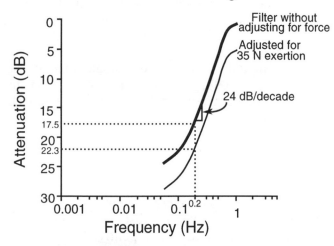

Figure 1. Characteristics of the frequency weighted filter.

The force factor f_E for a given exertion level E was determined by converting the force adjustment (dB) into a ratio:

$$f_E = \frac{X_{FE}}{X'_F},$$

substituting,

$$a_E = 20 \times \log(f_E)$$

and solving,

$$f_E = 10^{\left(\frac{a_E}{20}\right)}.$$

Therefore the force factor f_E for exertion level E was equal to:

$$f_E = 10^{\left(\frac{34.88 \log E - 58.67}{20}\right)}.$$

The root-mean-square (RMS) of the force and frequency weighted postural data ($X_{FE}(nT)$) was used to described the relative exposure level $\overline{X}_{FE}$. The $X_{FE}(nT)$ was equal to:

$$X_{FE}(nT) = X_F(nT) \ f_{E(nT)},$$

where $E(nT)$ was the exertion level corresponding to sample time nT. Force was assumed to be held constant in the current study.

Exposure Index. A discomfort level of 3.5 (Lin and Radwin, 1995) corresponded to posture and force conditions that were acceptable to 90% of the female subjects in Marley and Fernandez (1995) and Snook, et al. (1995). In order to provide an absolute output, the relative exposure level corresponding to discomfort level 3.5 was used to define the exposure limit (X_L) for repetitive wrist flexion and was anchored as an exposure index of 1. The exposure index for a repetitive wrist flexion task was obtained by:

$$\text{Exposure Index} = \frac{\overline{X}_{FE}}{X_L},$$

where $\overline{X}_{FE}$ was the relative exposure level for the task. Exposure indices less than unity should correspond to maximum repetition, postures, and forces that were acceptable for 90% of the subjects in these studies. The exposure assessment instrument is shown schematically in Figure 2.

Validation Experiment

Apparatus. Special peg boards were designed for controlling force during peg insertion (Lin, et al., 1995). Resistance was controlled by ball plungers which were calibrated using a load cell. Wrist flexion angle during peg insertion was controlled by adjusting the height of a horizontal bar in front of the peg board. The peg boards were located on a table which was adjusted so they were at seated elbow height. Subjects received a supply of pegs from a chute next to the seat on the dominant side. Continuous wrist flexion angles were measured using a Penny and Giles model M110 strain gauge twin axis wrist electrogoniometer. A MacAdios 12-bit analog-digital converter, LabView® software (National Instruments Corp.) and a Macintosh II/fx microcomputer were used for sampling posture signals and for implementing the frequency weighted filter. Wrist flexion angular data were sampled at 20 samples/s.

Experimental Design. The experimental task involved repetitively transferring a peg across a horizontal bar and inserting it into a peg hole against a controlled resistance. The experimental conditions (see Table 1) were presented to each subject in a random order. Only one condition was presented to a subject on a given day. A two-minute warm-up period was provided at the beginning of each session. Subjects performed each condition continuously for one hour. Five minutes before the end of the one-hour period, subjects were asked to assess discomfort while performing the task. Discomfort ratings were taken at the conclusion of the one-hour period. Continuous wrist angles were processed through the instrument to obtain relative exposure levels.

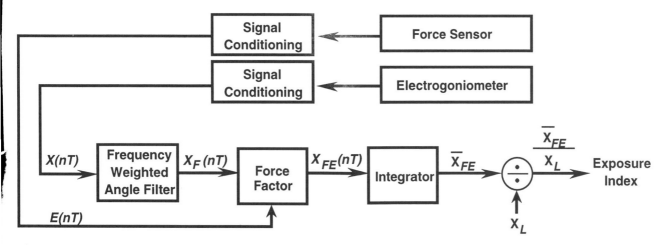

Figure 2. Block diagram of the exposure assessment instrument.

Discomfort was defined to include sensations like fatigue, soreness, stiffness, numbness, or pain. Subjects were required to be free of discomfort at the beginning of every session. Discomfort was measured using a ten centimeter visual analog scale anchored as "none" at 0 cm, and as "very high" at 10 cm. Subjects were advised that "none" meant that none of the sensations were experienced during the experiment and "very high" corresponded a very high level. A thin 0.5 cm vertical line was placed on the scale to indicate the mid-point. Subjects drew a vertical line across the horizontal scale to indicate discomfort. Subjective discomfort was regressed against expected discomfort predicted by the discomfort model based on the experimental conditions in order to test the model. A linear regression equation was fit using subjective discomfort ratings as the dependent variable and the relative exposure level as the independent variable in order to validate the instrument.

Table 1. Experimental conditions

Experimental Condition		
Force (N)	Wrist Flexion (°)	Pace (s/motion)
5	10	10
5	45	10
5	45	4
45	10	4
45	45	7
45	45	4

Subjects. Ten subjects, six males and four females, ranging between 21 to 24 years of age, were recruited by broadcasting electronic announcements and posting signs on the university campus. Subjects were required to have no restriction of hand/arm motion and were paid on an hourly basis.

RESULTS

Results from linear regression analysis showed that the discomfort model predicted relative discomfort for the experimental conditions tested. The resulting regression model was:

$$D = 0.460 + 0.894 \, \hat{D}$$

($r^2 = 0.983$, $F(1,4) = 230.818$, $p < .05$), where D was subjective discomfort and $\hat{D}$ was the expected discomfort levels. Mean and standard deviation of the subjective discomfort and the relative exposure levels are shown in Table 2. A linear regression model was fit using mean subjective discomfort as the dependent variable and mean relative exposure levels as the independent variable. The resulting regression model was:

$$D = 1.411 + 0.527 \, \overline{X}_{FE}$$

($r^2 = 0.873$, $F(1, 4) = 27.617$, $p < .05$), where D was subjective discomfort and $\overline{X}_{FE}$ was the relative exposure level. According to this regression model, the relative exposure level corresponding to discomfort level 3.5 was 3.96. This level represented the exposure limit (X_L) for repetitive wrist flexion in this study and was anchored as exposure index 1. Mean relative exposure levels for all experimental conditions were divided by X_L to obtain the exposure indices (see Table 2).

Table 2. Mean and standard deviation for subjective discomfort ratings and the corresponding relative exposure levels and exposure indices.

Condition	Subjective Discomfort (0-10 scale) Mean (*s.d.*)	Relative Exposure level (0-10 scale) Mean (*s.d.*)	Exposure Index
1	0.90 (0.88)	0.08 (0.02)	0.02
2	1.48 (1.14)	0.10 (0.02)	0.03
3	2.21 (0.95)	0.17 (0.04)	0.04
4	3.26 (1.42)	5.31 (0.84)	1.34
5	4.49 (1.12)	5.40 (1.18)	1.36
6	5.57 (1.19)	6.87 (1.20)	1.73

DISCUSSION

The strong correlation between subjective discomfort and the expected discomfort showed that the discomfort model was apt. The results also showed that relative exposure levels corresponded to subjective discomfort. This finding demonstrates the feasibility of using force and frequency weighted filters for processing and integrating continuous biomechanical data in to a value that accounts for physical stress of force, flexion angle, and repetition in proportion to relative discomfort.

Studies have shown that the risk of injuries increases when psychophysical acceptance is exceeded (Liles, et al., 1984; Snook, 1985; Herrin, et al., 1986). Psychophysical acceptance levels were used in this investigation to define the exposure limit for repetitive wrist flexion. Snook, et al. (1995) estimated the maximum acceptable force for wrist flexion tasks. Marley and Fernandez (1995) estimated maximum acceptable frequency for a drilling task. These psychophysical acceptability data were incorporated into the discomfort model to determine maximum acceptable discomfort levels. Since the exposure assessment instrument described in this study processes continuous biomechanical measurements directly from electrogoniometers and force sensors, it eliminates potential bias from observers and provides an objective assessment for single subject studies. An exposure index is proportional to psychophysical acceptable levels and may be used to evaluate and prioritize repetitive manual tasks for ergonomic intervention.

While results from the current study show that this instrument provides a promising approach for exposure assessment, these results are still preliminary. The discomfort model was established using repetitive wrist flexion tasks with a power grip, therefore it may not be applicable to tasks outside of this scope. The model only considered physical stress of force, flexion angle, and repetition, while there were other task and environmental factors may contribute to discomfort. Future studies should be conducted to refine and broaden the scope of this model. Although the scope of this study is limited, the methodology proposed may be applicable to biomechanical measurements at articulations other than the wrist and may ultimately lead to the establishment of ergonomic guidelines for repetitive upper extremity tasks.

ACKNOWLEDGMENTS

This project was supported by grant K03 0H00170 from the National Institute for Occupational Safety and Health of the Center for Disease Control.

REFERENCES

Graf, M., Guggenbuhl, U., Krueger, H. 1995, An assessment of seated activity and postures at five workplaces, *International journal of Industrial Ergonomics*, **15**, 81-90.

Genaidy, A. M. and Karwowski, W. 1993, The effects of neutral posture deviations on perceived joint discomfort ratings in sitting and standing postures, *Ergonomics*, **36**, 785-792.

Gilad, I. 1995, A Methodology for functional ergonomics in repetitive work, *International Journal of Industrial Ergonomics*, **15**, 91-101.

Herrin, G. D., Jaraiedi, M. and Anderson, C. K. 1986, Prediction of over exertion injuries using biomechanical and psychophysical models, *American Industrial Hygiene Association Journal*, **47**, 322-330.

Liles, D. H., Deivanayagam, S., and Mahajan, P. 1984, A job severity index for the evaluation and control of lifting injury, *Human Factors*, **26**, 683-693.

Lin, M. L., Radwin, R. G., Snook, S. H. 1994, Development of a relative discomfort profile for repetitive wrist motions and exertions, *Proceedings of the 12th Triennial Congress of the International Ergonomics Association*, 219-221.

Lin, M. L., Radwin, R. G., Snook, S. H. 1995, A single metric for quantifying biomechanical stress in repetitive motions and exertions, Submitted to *Ergonomics*.

Lin, M. L. and Radwin, R. G. 1995, A relative discomfort model for biomechanical stress in repetitive wrist flexion against a load, Submitted to *Ergonomics*.

Marley, R. J. and Fernandez, J. E. 1995, Psychophysical frequency and sustained exertion at varying wrist postures for a drilling task, *Ergonomics*, **38**, 303-325.

Moore, A., Wells, R., Ranney, D. 1991, Quantifying exposure in occupational manual tasks with cumulative trauma disorder potential, *Ergonomics*, **34**, 1433-1453.

Moore, J. S. and Garg, A. 1995, The strain index: a proposed method to analyze jobs for risk of distal upper extremity disorders, *American Industrial Hygiene Association Journal*, **56**, 443-458.

Radwin, R. G., Lin, M. L., Yen, T. Y. 1994, Exposure assessment of biomechanical stress in repetitive manual work using frequency-weighted filters, *Ergonomics*, **37**, 1984-1998.

Saldaña, N., Herrin, G. D., Armstrong, T. J., Franzblau, A. 1994, A computerized method for assessment of musculoskeletal discomfort in the workforce: a tool for surveillance, *Ergonomics*, **37**, 1097-1112.

Snook, S. H., Vaillancourt, D. R., Ciriello, V. M., Webster, B. S. 1995, Psychophysical studies of repetitive wrist flexion and exertion, *Ergonomics*, **38**, 1488-1507. Wells, R., Moore, A., Potvin, J., Norman, R. 1994, Assessment of risk Factors for development of work-related musculoskeletal disorders (RSI), *Applied Ergonomics*, **25**, 157-164.

AUTHOR INDEX

Adams, C. M. 436
Adams, E. M. 398*
Aghazadeh, F. 637
Aiello, F. III 1212
Akers, J. W. 1193
Albin, T. J. 772*, 870*
Allen, R. W. 943
Allendoerfer, K. 1252*
Allread, W. G. 646, 651
Alton, J. 20
Amundson, D. 1278*
Andersson, G. B. J. 604
Andre, A. D. **, 1170*
Andrle, S. J. 559*
Ankrum, D. R. 1270*
Archer, R. D. 987
Aretz, A. J. 91
Armony, L. 1060
Armstrong, T. J. 870*
Arndt, S. R. 1141
Asdigha, M. 1260*
Auflick, J. L. 869*
Avans, D. 1259*
Aye, D. 390
Aykin, N. 348
Ayoub, M. M. 609
Ayres, T. J. 947
Azof, H. A. 1267*

Baar, C. 1279*
Baker, C. C. 977
Baker, D. R. 1141
Barickman, F. S. 1017
Barnes, M. J. 1261*
Baron, R. A. 1282*
Bartolome, D. S. **, 1261*
Battiste, V. 997
Baxter, K. K. 1267*
Beaton, R. J. 1188, 1247
Becker, A. B. 1279*
Beith, B. H. 1022
Belyavin, A. J. 39
Benne, M. R. 1267*
Bennett, K. B. 1165
Bensel, C. K. 683
Berbaum, K. S. 1126
Bergman, E. 323*
Beringer, D. B. 86
Besco, R. O. 54*
Bilazarian, P. 1290*
Billings, C. 98
Billingsley, P. A. 323*
Bills, R. K. 619
Birkmire, D. P. 727*
Bisantz, A. M. 424, 957
Bishop, P. J. 1212
Bishu, R. R. 702, 772*, 1102
Bittner, A. C., Jr. 102

Blinn, B. P. 441
Bliss, J. P. 1237
Boehm-Davis, D. A. **, 468
Bogart, E. H. **, 1261*
Bogner, M. S. 727*, 752*
Boies, S. J. 343
Bonto-Kane, M. A. 1264*
Bookman, M. A. 742
Bost, J. R. 977
Boston, B. N. 57
Bowers, C. A. 49, 155*
Braun, C. C. 57, 830, 1122
Braune, R. J. 102
Brickman, B. J. 30
Brink, L. 727*
Broach, D. L. 1092
Brock, J. F. 929, 933, 938
Broen, N. L. 900
Brogmus, G. E. 642*
Bronkema-Orr, L. 702
Brown, S. 1275*
Broyles, J. W. 1293*
Brunetti-Sayer, T. M. **
Buchholz, B. 1274*
Bullemer, P. 283
Bunting, A. J. 39
Burdette, D. W. **, 1261*
Burdick, M. D. 204
Burns, J. J. 1288*, 1290*
Burns, M. M. 559*
Buzzard, J. 1266*

Caldwell, B. S. 530*
Cano, A. R. 786
Carswell, C. M. 1258*
Carter, R. J. 1017
Carter, S. E. 1270*
Carvalhais, A. B. 1285*
Casali, J. G. 840
Catrambone, R. 1041
Cavanaugh, K. 1285*
Ceplenski, P. J. 1222, 1271*
Chadwick, N. J. 830
Chaffin, D. B. 594
Chandler, T. N. 1087
Chappell, A. R. 264
Chen, S.-H. 604
Chervak, S. G. 303
Chiang, D. P. 900
Chin, A. 1102
Cho, C. K. 473
Christ, R. E. 757*
Chrysler, S. T. 923
Clarke, D. L. 1022
Clegg, C. W. 982
Coates, G. D. 1212
Coble, J. R. 1291*
Cohen, H. H. 520

Cohen, M. S. 155*, 158*, 179
Cohen, S. M. 424, 957
Cole, L. 1280*, 1286*
Collins, T. R. 1267*
Collopy, M. T. 1272*
Collura, J. 905
Comer, S. 862*
Conant, R. C. 169
Conlan, R. W. 1260*
Connolly-Gomez, C. 1097
Converse, S. A. **
Conzola, V. C. 1269*, 1281*
Cook, C. A. 982
Corban, J. M. 967
Corbett, S. 478
Corbridge, C. 328, 982
Corso, G. M. 1065, 1267*, **
Cortez, J. P. 1267*
Coury, B. G. 333
Cregger, M. C. **
Cress, J. D. 1131
Crosland, M. B. 1065
Crowell, R. R. 737
Cunningham, J. A. 1131
Cuomo, D. L. 370
Curl, C. 1284*
Curnow, C. K. 1288*
Czarnolewski, M. Y. 1287*

D'Souza, M. E. 781
Dahl, S. G. 962
Dainoff, M. J. 1263*, 1268*
Dandavate, U. 415
Danielson, S. M. 923
Davis, E. T. 1184, 1193
Davis, J. R. 584
Davis, K. G. 646, 599, 651
Day, E. 1289*
de Pontbriand, R. J. 678
Deffner, G. P. 353
Dember, W. 1286*
Dempsey, P. G. 609, **
Denning, R. 98, 357*, 1289*
Deppa, S. W. 555*
Detweiler, M. C. 459, 1232, 1263*
Dietrich, D. A. 732
Dingus, T. A. 860*, 896, 1107
Dischinger, H. C., Jr. 1254*
DiVita, J. 1292*
Dodson, M. L. 1258*
Donchin, Y. 752*
Downer, K. F. 1285*
Downs, M. 997
Drake, K. L. 1281*
Draper, M. H. 1146
Dresel, K. M. 555*

Drexler, J. M. 569
Drury, C. E. 1275*
Drury, C. G. 303, 772*, 796, 882, 1082, 1275*
Du, J. 406
Duncan, P. C. 1288*
Dunlap, W. P. 1126
Dyre, B. P. 1207

Edworthy, J. 845
Eggemeier, F. T. **
Ehrlich, J. A. 1290*, 1292*
Eischeid, T. M. 1271*
Eisenburg, R. 318
Elias, B. 1227
Elliott, K. G. 551, 1184
Ellis, R. D. 1256*
Ellis, S. R. 1197
Emery, C. D. 1255*
Endsley, M. R. 82, 228, 1077, 1170*
Entin, E. E. 233
Entin, E. B. 233
Eustace, J. K. 1155
Evans, A. M. 1254*

Fairclough, S. H. 1283*
Farmer, E. W. 39
Fathallah, F. A. 642*, 661
Ferguson, S. A. 656, 737
Fine, L. J. 870*
Fiorentino, D. D. 559*
Fischer, U. 1258*
Fisher, D. L. 123, 118, 905
Fisk, A. D. **, 128, 184
Flin, R. 158*, 1262*
Foley, J. P. 887
Forsythe, C. 365
Fox, J. M. 77
Fozard, J. L. 138
Franks, J. R. 840
Frederick, L. J. 1276*
Freeman, J. T. 179
Freivalds, A. 1276*, 1286*
Fu, Y. 390
Fujawa, G. E. 551, 1184
Funk, K. H. II 254
Furness, T. A. III 1146

Gage, M. 398*
Galante, G. T. 1252*
Galinsky, T. L. 1279*, 1280*
Galushka, J. 1252*
Gardner-Bonneau, D. **, 323*, 1279*
Garland, D. J. 1170*

Papers numbered 727 and above appear in Volume 2.
*Abstract only
**Manuscript not submitted for publication

Garness, S. A. 1207, 1294*
Garrott, W. R. 896
Garvey, P. M. 1285*
Gawron, V. J. 72, 1146
Geddes, N. D. 357*
Getty, R. L. 772*
Gilbert, D. K. **
Gillan, D. J. 1242
Gilmore, B. J. 1286*
Gilson, R. D. 850
Glasscock, N. F. 584, 1273*
Gleason, G. A. 1294*
Glenn, F. A. III 1012
Glover, B. L. 910
Glusker, S. A. 830
Goenert, P. N. 67, 1287*
Goettl, B. P. 1097, 1289*
Goldberg, J. H. 861*
Gonzales, N. D. 1267*
Gopher, D. 1060, 1151
Gorman, M. F. 118
Gosbee, J. W. 530*, 1279*
Graham, J. M. 259
Gramopadhye, A. K. 772*, 1070*, 1072
Granata, K. P. 646, 651, 656
Gravelle, M. D. 424, 957
Green, G. K. **
Green, P. A. 445
Greene, S. L. 343
Greenshpan, Y. 1060
Greenstein, J. S. 781
Gregory, M. H. 1254*
Grose, E. M. 365
Gross, T. P. 752*
Gubler, K. D. 1278*
Guerlain, S. A. **, 283
Gugerty, L. J. 564
Guilkey, J. 34

Haas, E. C. 845
Haas, M. W. 30, 1131
Hager, D. R. 546
Hahn, H. A. 454, 869*
Hall, S. 559*
Hallbeck, M. S. 707, 712
Hambrick, D. Z. 133
Hancock, P. A. 96*, 106, 111, 1170*, 1202, 1296*
Hankey, J. M. 860*, 896
Hanowski, R. J. 877
Hansen, A. 712
Hardcastle, J. 1275*
Hardiman, T. D. 228
Harmon, C. M. 862*
Harper, B. D. 747
Harpster, J. L. 891
Harris, D. H. 1032
Harris, W. C. 67, 1287*
Harrison, B. L. 375
Hartsock, D. C. 15
Hashemi, L. 952

Headley, D. B. 666*
Hedge, A. 478, 624, 1269*
Heers, S. T. 204
Helander, M. G. 483, 533*, 882
Henderson, M. R. 707
Hendrick, H. W. 1, 533*, **
Hess, S. M. 459, 1232
Hettinger, L. J. 30, 1126, 1131
Hicinbothom, J. H. 863*
Hink, J. K. 1155
Hinson, G. E. 133
Hoag, D. W. 1257*
Hoffman, M. S. 429
Holt, R. W. 44
Honan, M. 629
Howell, W. C. 11
Howie, D. E. 972
Huang, S. 406
Hubbard, D. C. 1253*
Hudak, M. J. 887
Hudlicka, E. 1260*
Huey, R. W., Jr. 891
Hunter, D. R. 34, 54*
Huntley, M. S., Jr. 25
Hursh, S. R. 1260*
Hurts, K. 1160
Hurwitz, J. B. 223
Hutchins, S. G. 199
Hutton, R. J. B. 274

Illgen, C. 1260*
Inman, W. E. 1295*

Jablonka, E. 370
Jackson, T. S. 1273*
Jacobson, M. 629
Jahns, D. W. 102
Jahns, S. K. 928*
James, T. M. 1269*
Jamieson, B. A. 146, 1255*
Jamieson, G. A. 162*, 173
Janovics, J. 1295*
Jarboe, C. 1284*
Jeans, S. M. 1237
Jebaraj, D. 1072
Jensen, R. S. 34
Jentsch, F. G. 49, 820
Jilig, J. M. 214
Johannsen, C. 91
Johnson, R. F. 1217
Johnson, S. L. 1275*
Joines, S. M. B. 1273*
Jones, L. 343
Jones, M. M. 1247
Jones, P. M. 357*, **
Jones, T. N. 752*
Jordan, J. 67
Jordan, C. S. 39
Jorgensen, M. J. 646, 651
Joseph, B. 870*
Joyce, M. S. 555*

Kalsher, M. J. 385, 1282*
Kanis, H. 1265*
Kantowitz, B. H. 877
Kaplan, H. S. 530*
Kaplan, J. D. 962
Kappé, B. 1293*
Karnes, E. W. 525
Karwowski, W. 614
Katsikopoulos, K. V. 123
Kaufman, L. S. 353
Kelaher, D. P. 589
Kelley, J. F. 343
Kelley, M. L. 1212
Kelly, R. T. 199
Kennedy, R. S. 298, 569, 1126
Kidwell, D. L. 259
Kies, J. K. 338
Kim, J.-Y. 1278*
King, R. A. 551, 1184, 1193
Kingsley, L. C. 1285*
Kinney, C. A. 1202
Kira, A. **
Kirby, V. M. 923
Kirlik, A. 184
Kirschenbaum, S. S. 214
Kleid, N. A. 1264*
Klein, G. A. 155*, 158*, 178*, 274
Kleiner, B. M. 786
Kline, P. B. 1112
Knapik, J. J. 668, 678
Knapp, B. G. 1179
Knecht, W. 106
Knott, B. A. 1288*, 1291*
Kobus, D. A. 1278*
Kolasinski, E. M. 1292*
Konz, S. 390, 445
Koonce, J. M. 54*
Korteling, J. E. 1293*
Kotval, X. P. 861*
Kraft, E. M. 835
Krahl, K. 1271*
Kramer, A. F. 77
Kraus, D. C. 1072
Krawczyk, S. 642*
Kreifeldt, J. G. 394
Krieser, A. M. 1267*
Kurke, M. I. 511

Larson, N. L. J. 488
Lassiter, D. L. 133
Latorella, K. A. 249
Lau, E. C. 947
Laughery, K. A. 492
Laughery, K. R., Sr. 492, 801, 810, 814
Laux, L. F. 348
Lavender, S. A. 604
Le Mentec, J.-C. 742, 864*
Lee, H. M. 473
Lee, J. S. 401
Lee, K. M. 1253*

Lee, M. W. 401, 473
Lehman, K. R. 419
Leonard, S. D. 525
Lerner, N. D. 891
Levas, R. 1290*
Levine, O. H. 1136
Lewin, J. E. K. 1296*
Lewis, G. W. 987
Li, Y.-C. 604
Liddle, R. 54*
Liggett, K. K. 15
Lim, S.-Y. 642*
Limaye, J. 380
Lin, G. 1036
Lin, M.-L. 717
Lipshitz, R. 189
Litynski, D. M. 385
Liu, Y. 872
Llaneras, R. E. 929, 933, 938
Lockett, J. 987
Logan, R. J. 398*, 862*
Loughead, T. E. 1254*
Lovvoll, D. R. 492, 801, 810, 814
Lowe, B. D. 1276*, 1286*
Lu, H. 637, 1036
Lueder, R. 380
Lund, A. M. 323*, 348
Lussier, J. W. 1260*

Mackenzie, C. F. 158*, 218, 747
MacLeod, D. 698
Maddox, M. E. 555*, 1266*
Madigan, E. F., Jr. 419
Magurno, A. B. 732, 910
Mahach, K. R. 1283*
Mahan, R. P. 214
Maisano, R. E. 1288*
Major, D. A. 1222
Malone, T. B. 555*, 977
Mandeville, D. E. 1267*
Manser, M. P. 1202
Marino, C. J. 214
Mark, L. S. 1263*
Markert, W. J. 1179
Marras, W. S. 642*, 646, 651, 656, 661, 737, 1278*
Marsh, R. 77
Marshall, R. K. 1283*
Matthews, G. 579
Mazzae, E. N. 896
McCann, R. S. 997
McCloskey, M. J. 194, 308
McConkie, R. 77
McCoy, B. 254
McCoy, E. 98, 158*
McCrobie, D. E. 478
McDaniel, J. W. 1026
McDonald, D. P. 850
McEvoy, J. C. 1268*
McGehee, D. V. 860*, 896
McGrew, J. F. 1259*
McMillan, G. R. 1131

Papers numbered 727 and above appear in Volume 2.
*Abstract only
**Manuscript not submitted for publication

McMullin, D. L. 1276*
McNeese, M. D. 767
McQuilkin, M. L. 801, 810, 814
Mead, S. E. 146
Mears, M. G. 840
Meeker, D. 1285*
Meiman, E. 44
Meister, D. 541
Mello, R. P. 619
Menges, B. M. 1197
Merullo, D. J. 1217
Merwin, D. H. 77
Metalis, S. A. 1252*
Meyer, B. 1041
Meyer, J. 1151
Miller, D. 742
Miller, D. L. 1002, 1007
Miller, L. A. 1027
Miller, M. 133
Millians, J. 151
Mirka, G. A. 584, 1273*
Mitchell, C. M. 209, 264, 357*,
    1264*
Modrick, J. A. 757*, 758
Mogford, R. 1252*
Molloy, R. 1272*
Montgomery, D. A. 1295*
Moore, P. 328
Moore, R. A. 868*
Moray, N. 164, 533*
Moroney, W. F. 436, 434*, 441,
    445
Morris, A. 698
Morrison, J. G. 199, 868*
Morrow, D. G. 117*, 133
Morrow, J. 483
Mortimer, R. G. 506
Morzinski, J. A. 869*
Mosier, J. N. 370
Mosier, K. L. 204, 238*
Mouloua, M. 850, 1272*
Mourant, R. R. 1136
Muldoon, B. 1272*
Munro, I. 1260*
Murphy, M. 619
Murray, L. T. A. 910
Murray, J. 872
Muto, W. H. 752*

Nah, K. 394
Nair, C. M. 632
Najd, N. S. 1267*
Neale, D. C. 338, 1117, 1282*
Neale, V. L. 1107
Nelson, K. 1287*
Nelson, W. T. 30
Nemeth, K. J. 1268*, 1270*
Netsch, S. 1287*
Nevo, I. 752*
New, M. D. **
Newman, P. 39
Nickerson, R. S. 162*
Nittoli, B. 1165

Norman, K. L. 468
North, M. M. 1291*
North, S. M. 1291*
Noy, Y. I. 533*
Nugent, W. A. 1174
Nygren, T. E. 1258*

O'Hara, J. M. 992
Ober, K. 91
Obradovich, J. H. 357*, 1055
Oesterling, B. 1286*
Ohnemus, K. R. 360
Older, M. T. 982
Olsen, R. A. 497
Oman, C. M. 25
Orasanu, J. 98, 155*, 158*
Orr, C. E. 1026
Ottensmeyer, M. P. 1277*
Ouellette, J. P. 303

Pan, C. S. 1280*
Pang, G. 1278*
Paquet, V. L. 1274*
Paquette, S. O. 683
Parasuraman, R. 1272*
Parnianpour, M. 661, 1278*
Parseghian, Z. 943
Parsons, S. O. 491*, 501
Patterson, E. S. 967
Patton, J. F. 673
Paul, R. D. 380, 483, 632, 791
Pauls, J. L. 818
Payne, D. G. 546
Pazderka, M. P. 707
Pearsall, S. 1262*
Peck, A. C. 459
Pence, H. R. 1270*
Penner, R. R. 762
Peters, L. J. 308, 727*
Phipps, D. A. 1257*
Pietrucha, M. T. 1285*
Pineau, J. 825, 1256*
Pines, E. 1256*
Pittenger, E. A. 463
Plocher, T. 757*
Plott, B. M. 962
Pollack, J. G. 398*
Pollack-Nelson, C. 855
Prabhu, G. V. 882
Prabhu, P. V. 1070*
Pracharktam, T. 1046
Priest, J. E. 72
Prince, C. 155*
Prioux, H. J. 1237
Proulx, G. 825, 1256*
Pruitt, J. S. 1288*, 1290*
Psihogios, J. P. 584
Pullen, M. T. 123
Punnett, L. 1274*
Pupons, D. E. 1271*
Purvis, B. D. 1026

Quinn, M. L. 868*

Radwin, R. G. 717, 870*
Rantanen, E. 861*
Rao, P. 1072
Rascona, D. 1278*
Rasmussen, S. A. M. 25
Raspotnik, W. B. 1253*
Ratner, J. A. 365
Redding, R. E. 269, 1092
Redmond, D. P. 1260*
Reed, P. S. 323*
Reid, D. M. 1267*
Reising, D. V. C. 293
Reising, J. M. 15
Rempel, D. M. 629
Renner, G. 20
Resnick, M. L. 463
Reynolds, K. C. 1179
Riccio, G. E. 1131
Rice, V. J. 619, 666*
Rich, C. J. 1122
Riegler, J. T. 1294*
Riley, D. D. 1012
Riley, M. W. 772*
Ritchie, E. 1279*
Roberts, J. 67
Robertson, M. M. 1077
Robertson-Schulé, L. 1097
Rodgers, M. D. 82
Rodvold, M. 98
Roe, M. M. 30
Rogers, W. A. **, 146, 238*, 239,
    1255*, 1271*
Rogers, W. C. 938
Rosa, R. R. 1286*
Rosenfeld, J. P. 1290*
Rothrock, L. 184
Rourke, C. J. 1274*
Rousseau, G. K. 146
Rueb, J. D. 1254*
Rugotzke, E. G. 569
Ryder, J. M. 278

Sabol, M. A. 1288*
Salthouse, T. A. 116*
Sanders, E. B.-N. 415
Sanderson, P. M. 288, 293, 752*
Sauter, S. L. 1279*, 1280*
Savage, P. A. 1262*
Scallen, S. F. 96*, 111
Scerbo, M. W. 1212, 1222, 1271*
Scheel, M. 1287*
Schmidt, R. A. 947, 1051
Schmidt, W. 1242
Schnell, T. 915, 919, 1046
Schreiber, B. T. 20, 1253*
Schremmer, S. D. 864*
Schutte, P. C. 244
Schwartz, A. L. 313
Schwartz, J.-C. 742

Schwerha, D. J. 1276*
Scott, K. L. 732
Seagull, F. J. 752*
Seamster, T. L. 44, 269, 1092
Sebrechts, M. M. 468, 1291*
Seifert, C. M. 313
Selcon, S. J. 39, 228, 1170*
Self, S. 1284*
Selner, A. 380
Semmel, R. D. 333
Serfaty, D. 158*, 178*, 233
Shafto, M. G. 313
Shalin, V. L. 882
Shamo, M. K. 1151
Shanks, C. R. 39
Shapiro, R. G. **
Shapiro, S. J. **
Sharp, M. A. 619, 673
Shaw, G. 624
Shebilske, W. L. 1097, 1289*
Shen, W. 1277*
Sheridan, T. B. 1277*
Silverman, B. G. 468
Sinclair, C. B. 1253*
Sind-Prunier, P. 865
Singer, M. J. 1290*
Sit, R. A. 128, 146
Skitka, L. J. 204
Skriver, J. 1262*
Smith, J. A. 1265*
Smith, K. 96*, 106, 111, 1259*,
    1296*
Smith, M. J. 776
Smith, P. J. 98, 158*, 259, 357*,
    1289*
Smith-Jentsch, K. 1288*
Snook, S. H. 642*, 870*
Sojourner, R. J. 1281*
Sollenberger, R. L. 574
Solz, T. J., Jr. 15
Sottile, A. J. 441
Sparre, E. 1065
Spelt, P. F. 1017
Spencer, C. A. 1263*
Spielholz, P. O. 688
Spilich, G. J. 1284*
Spraragen, S. L. 343
Stahlhut, R. 1279*
Stanfield, P. M. 584
Stanney, K. M. 298, 1027
Steckler, L. S. 1265*
Stein, A. C. 943
Stein, E. S. 574
Steinberg, R. 1179
Stewart, J. 353
Stone, N. J. 434*, 449
Stout, R. J. 155*
Strauss, O. 189
Stuart, J. A. 463
Stuart, M. A. 693
Stuart, S. 415
Stuart-Buttle, C. 870*
Stubler, W. F. 992
Suantak, L. 1261*

Papers numbered 727 and above appear in Volume 2.
*Abstract only
**Manuscript not submitted for publication